ENVIRONMENTAL ISSUES AND MANAGEMENT OF WASTE
IN ENERGY AND MINERAL PRODUCTION

PROCEEDINGS OF THE SIXTH INTERNATIONAL CONFERENCE ON ENVIRONMENTAL ISSUES AND MANAGEMENT OF WASTE IN ENERGY AND MINERAL PRODUCTION
SWEMP 2000/ CALGARY/ALBERTA/CANADA/MAY 30 – JUNE 2, 2000

Environmental Issues and Management of Waste in Energy and Mineral Production

Edited by

Raj K. Singhal
Federal Government of Canada & Université Laval, Québec, Canada
International Journal of Surface Mining, Reclamation and Environment

Anil K. Mehrotra
University of Calgary, Department of Chemical and Petroleum Engineering, Calgary, Alberta, Canada

A.A. BALKEMA / ROTTERDAM / BROOKFIELD / 2000

The texts of the various papers in this volume were set individually by typists under the supervision of each of the authors concerned.

Published by
A.A. Balkema, P.O. Box 1675, 3000 BR Rotterdam, Netherlands
Fax: +31.10.413.5947; E-mail: balkema@balkema.nl; Internet site: www.balkema.nl
A.A. Balkema Publishers, Old Post Road, Brookfield, VT 05036-9704, USA
Fax: 802.276.3837; E-mail: info@ashgate.com

ISBN 90 5809 085 X

Printed in the Netherlands

Environmental Issues and Management of Waste in Energy and Mineral Production, Singhal & Mehrotra (eds)
© 2000 Balkema, Rotterdam, ISBN 90 5809 085 X

Table of contents

Environmental issues in open pit and underground mining

Emerging technologies for environmental protection

Mine site closure – Tailings disposal and rehabilitation

Mine site closure – Acid rock drainage

Environmental impact – National reports

Computer modeling and applications

Waste characterization and modeling

Environmental Issues and Management of Waste in Energy and Mineral Production, Singhal & Mehrotra (eds)
© 2000 Balkema, Rotterdam, ISBN 90 5809 085 X

Foreword

This Symposium on Environmental Issues and Waste Management in Energy and Mineral Production (SWEMP) is the sixth of a series of bi-annual symposia on the subject matter.

The basic aim of this series of symposia is to contribute to the development of methods and technologies for assessing, minimizing and preventing environmental problems connected with mineral and energy production.

This symposium has come to be recognized as a leader in promoting international technology transfer. It was last held in Calgary in 1992 (second of the series). The third and fourth SWEMP were held in Australia and Italy, respectively. The fifth SWEMP was held in Turkey in 1998. In view of the expressed interest of professionals from around the world, this series of symposia has become a biannual event. The next three symposia have been scheduled: Cagliari, Italy in 2002, Ankara, Turkey in 2004, and New Delhi, India in 2006.

A wide range of high quality papers from North and South America, Europe, Australia, Africa and Asia have been attracted. Major topics to be covered are: Environmental impact assessment, permitting and management; Waste management practices; Environmental issues in open pit and underground mining; Emerging technologies for environmental protection; Mine site closure; Tailings disposal and rehabilitation; Acid mine drainage; Environmental impact; National reports; Computer modeling and applications; and Waste characterization and modeling.

Environmental Issues and Management 2000 is supported by a number of organizations. To be noted are:

Dipartimento di Geoingegneria e Technologie Ambientali, University of Cagliari; The International Journal of Surface Mining, Reclamation and Environment; Département de Mines et Metallurgie, Université Laval; Henry Krumb School of Mines, Columbia University; Universidad Politechnica de Madrid, Spain; Virginia Polytechnic Institute and State University; Western Australian School of Mines, Curtin University of Technology; University of Calgary; T.H. Huxley School of Environment, Earth Sciences and Engineering, Imperial College of Science Technology and Medicine, United Kingdom; Center for Environmental Engineering Science and Technology, University of Massachusetts; National Technical University, Greece; Luleå University of Technology-CENTAK; Technical University of Ostrava; Atilim University, Turkey; American Society for Surface Mining and Reclamation: Glukauf Mining Reporter and Mining Environmental Management-Mining Journal Ltd., and National Mining University of Ukraine.

This volume of proceedings contains a listing of papers in alphabetical order (by first author's last name) for each of the eight major themes of the symposium. We hope this will make it easy to locate specific papers during presentation.

The organization and success of such a symposium is due mainly to the tireless efforts of many individuals, authors included. All members of the Organizing Committee and conference chairpersons have contributed greatly. The support of our plenary session and invited speakers and co-chairs is gratefully acknowledged. My greatest appreciation goes to Professor Mehrotra who has worked tirelessly to ensure that proceedings appear on time and who has single-handedly developed the technical program.

In addition, particular recognition is accorded to Dr Meena Singhal, Margaret-Anne Stroh and Amber Spence of the University of Calgary, and our publisher A.T. Balkema.

Raj K. Singhal
Chair, International Organizing Committee

Environmental Issues and Management of Waste in Energy and Mineral Production, Singhal & Mehrotra (eds)
© 2000 Balkema, Rotterdam, ISBN 90 5809 085 X

Organization

Chair: Dr Raj K. Singhal, Canada
Co-Chairs: Prof. R. Ciccu, Italy
Prof. A. G. Pasamehmetoglu, Turkey
Prof. Hilary Inyang, USA
Prof. Joan Osborne, Australia

Technical Program Chair: Prof Anil K. Mehrotra, Canada

Maria de los Angeles Morelli, Argentina
Prof. Yun Qing-Xia, P.R. China
Gert Asmund, Denmark
Dr G. N. Panagiotou, Greece
Prof. S. Sakurai, Japan
Prof. L. Puchkov, Russia
Prof. Ramirez Oyanguren, Spain
Prof. Semyon Shkundin, Russia
Prof. Tuncel M. Yegulalp, USA
Prof. Vladimir Pavlovic, Yugoslavia
Dr E. A. Wright, Zimbabwe
Prof. Martin J. Haigh, United Kingdom
Dr Robin Chowdhury, Australia
Prof. Michael Duchene, France
Olga Zakrzewska, Poland
B. P. Singh, India
Prof. M. Karmis, USA
Dr Didier Hantz, France
Prof. David J. Spottiswood, Australia
Prof. Wildor T. Hennies, Brazil
Dr Newton Amegbey, Ghana
Tsolo Voutov, Bulgaria
Prof. Gennadiy G. Pivnyak, Ukraine
Dr Uta Gerlach-Laxner, Germany
Dr Richard Poulin, Canada
M. Singhal, Canada
Dr Thomson Sinkala, Zambia
Zou Jian, P.R. China
Enid Gamboa Robles, Costa Rica
Prof. Per Nicolai Martens, Germany
Prof. Mircea Georgescu, Romania
Prof. Sevket Durucan, United Kingdom
Magnus Ericsson, Sweden
Dr Lidia Gawlik, Poland
Prof. Michael Zhurakov, Belarus
Sven Erik Osterlund, Sweden
Dr Kostas Fytas, Canada
Dr Mario Sanchez, Chile
Dr Marie Vrbova, Czech Republic
Prof. A. B. Szwilski, USA
Dr Chris Chiwetelu, Canada
Gordon R. McKenna, Canada
Dr Z. Bzowski, Poland
Dr Richard L. McNeary, USA
Prof. Donald H. Graves, USA

Plenary Session and Invited Speakers

Dr Jim Carter, President and C.O.O., Syncrude Canada Ltd., Canada
Paul R. Clark, Director, Fuel supply, TransAlta Energy Corporation, Canada
Michael P. Davies, AGRA Earth & Environmental Ltd, Canada
Prof. Dr Ing. Per Nicolai Martens, Technical University of Aachen, Germany
Jim Popowich, Vice President, Fording Coal, Canada
Patrick Reid, President, Ontario Mining Association, Canada
Prof. Rene Van Berkel, Curtin University of Technology, Australia
Dr Erdal Yildrim, ACR/CO_2 Synergies Research Network & CONRAD, Canada

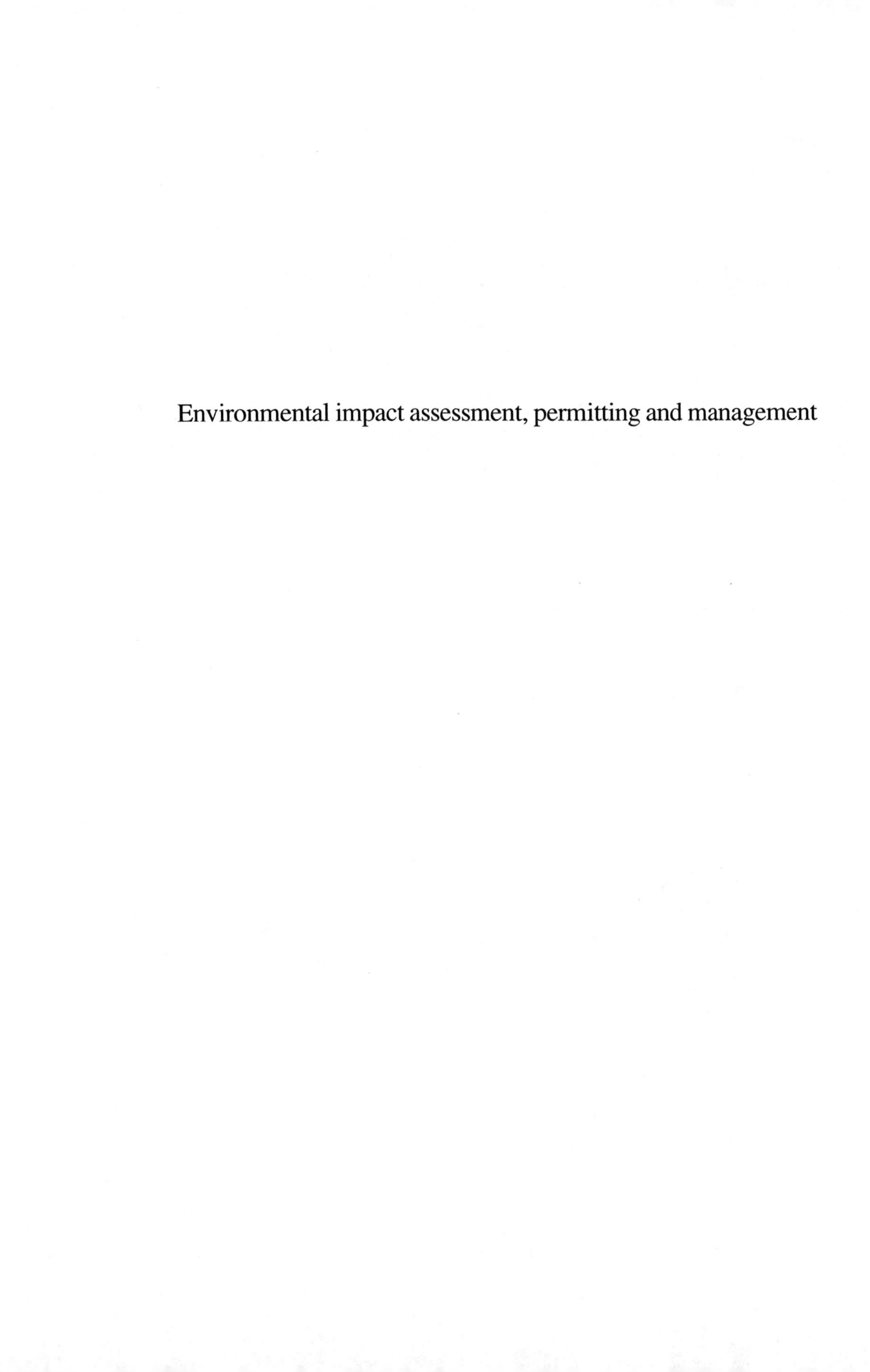

Environmental impact assessment, permitting and management

Environmental Issues and Management of Waste in Energy and Mineral Production, Singhal & Mehrotra (eds)

Reliability and productivity enhanced systems of the P&H 4100XPB electric mining shovel for overburden excavation

T.D.Barnes
P&H Mining Equipment, Milwaukee, Wis., USA

ABSTRACT: Moving large amounts of overburden and waste material in large open pit mines, often on the order of hundreds of millions of tons per year, requires mining companies to make significant investments in loading and hauling equipment. The 4100XPB shovel has been designed with availability and production enhanced control, power transmission, diagnostic, and communication systems that make it economically feasible to move the massive amounts of overburden and waste material in a modern open pit operation

1 INTRODUCTION

Electric mining shovels have gradually grown larger in terms of payload capacity since their introduction in the early 1900's. This growth in size has been paralleled by increases in haul truck capacity. Throughout the 1990's, haul trucks with a capacity in the 218 metric ton range became the standard size from several manufacturers, and electric shovels with nominal payload capacities of 77 metric tons were developed to load these trucks in three even passes.

These electric shovels would typically take 30 to 35 seconds to load one dipper payload into a haul truck depending on the type of digging and operator skill level. While numerous smaller shovels and haul trucks still operate throughout the world, the combination of three pass loading of 218 ton haul trucks represented the lowest cost per ton for a truck shovel operation. Fueled by the worldwide success of this truck shovel size combination, mining operations began in the mid 1990's to exert considerable pressure on equipment builders to produce the next larger combination. Trucks became available with 290 metric ton capacity, and even with four or five pass loading, the cost per ton decreased to a lower level. Now trucks in the 330 metric ton range are becoming available with 360 metric ton on the horizon. Clearly, a larger shovel which can load these large trucks even quicker and get them back on the haul roads and to the dumping sites or mill will pay off by ratcheting down the cost per ton to a new lower standard. The 4100XPB has a nominal payload capacity of over 90 metric tons and cycles nearly four seconds faster than earlier shovels. This, combined with availability and productivity enhanced systems, allow the 4100XPB to achieve higher production levels with the new larger haul trucks.

2 AVAIALBILITY

2.1 *Loading capacity when needed*

As mines begin to convert fleets to the new larger trucks and larger electric shovels, their annual production requirements can be achieved with fewer pieces of equipment in the pit. For example, where once a fleet of, say ten or fifteen shovels loaded a fleet of trucks, now the same production could be made with six or eight shovels. This goes against the age old warning of putting all your eggs in one basket, but the economics of the fewer but larger shovel and truck fleets cannot be disputed. Therefore, mines must be able to rely on each shovel in the smaller fleet of equipment to be available for loading more than before, since the options of shifting production to other shovels has diminished. The reasons that a shovel could be unavailable for loading trucks are many: shift change, scheduled preventative maintenance, breakdown, service downtime for lubrication and other maintenance. In addition to being larger and more productive, the new generation of larger shovels must be more reliable and be available for use more than ever.

2.2 *Maintenance and repair*

In mines, things get broken. The loads on the equipment are large and the operating conditions are severe. Therefore a maintenance program is a vital part of the success of a mine. When a shovel is taken out of production for maintenance, it can be compared to race car in for a pit stop. Getting that car back on the track as fast as possible wins the race. Diagnostic systems and modular assemblies on a shovel contribute to fast and effective maintenance. When maintenance planners have detailed information at hand relating to the specific maintenance needs of the shovel, the work orders are complete, replacement parts are on the site, and the result is a fast and complete repair. As will be described, the P&H 4100XPB has been designed with systems to provide *actionable* information to the operator and maintenance engineer. The mechanical and electrical assemblies are modular to facilitate quicker repairs and change outs.

3 AVAILABILITY ENHANCED SYSTEMS

Reliability of electrical and mechanical systems results in more hours that the shovel is available for production and less time in maintenance. The built in diagnostic capabilities on the 4100XPB make maintenance faster and which serves to maximize availability.

3.1 *DC systems with digital control*

The power to drive the large electric motors of the 4100XPB is provided by an on- board Digital DC power conversion system. Three phase electric power is converted to controlled power for the large DC motors. The computer revolution has impacted DC drive technology and enables better utilization of system and motor performance characteristics.

These advances in control technology maintain the edge DC drive technology has historically held. DC continues to use fewer components, both power and control, with a single power conversion process and performance comparable to that of the AC system. The combination of all the above results in a lower operating cost per ton for the mine owner.

3.2 *Computerized control systems*

The advancement of computer technology has impacted the logic and sequencing of operational functions of the machine. When shovels used relay control, information technology was a book of prints and a series of test equipment such as a multimeter in front of an open cabinet door.

Now that shovels use Programmable Logic Controllers, PLC's, for logic and sequencing, the information technology may vary by manufacturer but usually includes a computer display of the program in real time format. Better tools include displays of a number of parameters all in one place that impact the operation of the shovel which would be difficult to gather without the use of electronic data gathering. A couple of examples of this are machine operating hours, Reactive Power Compensation (RPC) step display and the machine propel parameters.

The use of PLC's for the control of equipment is certainly not new in a relative sense. But it's use as an enabling technology on mining shovels to provide greater assistance to the operator and maintenance personnel is relatively new. The development of this ability is primarily driven by the need for a higher level of reliability and maintainability that results in higher up time and lower operating cost per ton.

3.3 *Alarms and Faults*

As a direct result of using the PLC's capability, extensive diagnostics and trouble shooting ability is now possible as compared to the use of analog drives and relay control. It is possible to program microprocessors to communicate system status and health information on Graphic User Interface (GUI) and Man Machine Interface (MMI) displays on the shovel.

On-line help provides step by step instructions for troubleshooting and in some cases, the nature of the fault allows control algorithms to specify a faulty component for replacement. Another component in maintenance programs is machine operational history. With computerized alarm and fault collection, mine maintenance personnel have a tool to identify problems at an early stage and plan corrective actions as opposed to operating in a reactive mode.

One example of the PLC's capability of being used as a planning tool is the ability to record and trend electric motor and bearing temperatures. By collecting information on a constant basis, temperature trends are able to be displayed for each motion of the shovel. This information is logged by period and is able to be viewed either on the GUI or downloaded. This information can be used to determine if temperature changes are attributable to either a particular shift, particular operator, location in the face, time of day, etc. By utilizing this information, mine operation personnel can determine what action needs to be taken.

Information eases the task of the maintenance troubleshooter. By utilizing the PLC capability, no problems go unreported since the computer automatically keeps the log of events. The sorting and analysis of downloaded fault logs by way of existing software applications such as spreadsheet or database programs provides additional planning tools. The use of this information in conjunction with other maintenance requirements allows a more thorough approach to be taken to ensuring the shovel is reliable and reduce unexpected downtime.

3.4 *Remote Communications*

With the development of the monitoring devices mentioned above, the need to have information readily available to the maintenance and mine planners without having to physically proceed down to the shovel armed with laptop computer, has lead to the requirement for remote transmission of the information from the shovel to the mine office. The primary driver for this requirement is the need for higher availability, closer monitoring of machine performance and planning information - all factors in reducing overall operating cost per ton.

Typically when a shovel has a fault, the maintenance personnel rely on the operator to relay the information leading up to the fault to the best of his ability. Following receipt of a radio call, normally a trip to the shovel is required, in most instances, a trip back to the workshop for parts and then back to the shovel to carry out the repair. What if the technician could integrate the fault history prior to leaving the office to determine what the problem was and what parts maybe required? What if he could reset the fault without actually going down to the shovel?

In the case of production information, typically the data is gathered from a combination of mine survey and truck counts. What if the actual tons moved could be accumulated onboard the shovel and transmitted to the mine office?

With the use of spread spectrum communication technology operating at a frequency of 2.4 Gigahertz, the ability to perform this function is now becoming reality. This high frequency resist interference since most radio communication frequencies are well below this. The spread spectrum technology allows multiple users to share the same frequency band. The ability to interface this system with other existing systems which may be present on the mine site, such as modular mining, is also possible.

All machine control signals and functions are accessed from the PLC via a modem. This means that the personnel in the remote location have access to all aspects of the shovels operating status, fault history, machine parameter trends and troubleshooting information. The ability to reset non-critical faults and alter non-critical machine parameters such as lubrication settings remotely are also possible. Access to these same functions by and OEM support technician are also possible.

4 PRODUCTIVITY

4.1 *Bigger payload and faster cycle time*

The 4100XPB has the highest production capability available. This is achieved by not only by the bigger payload that matches the larger trucks, but by making the shovel faster. The time that is required to move a mass, whether that mass is the payload being dug from the bank, or the entire mining shovel plus payload swinging to dump in the truck, is a function of three parameters. The magnitude of the mass itself, the force (or torque), and the peak speed of the motors. All three variables have been closely considered in the design process of the 4100XPB. The mass of the shovel – primarily the steel structures and counterweight required to balance the shovel – has been optimized through a combination of finite element analysis and painstaking attention to the geometry of the *digging attachment.* The digging attachment has been designed to bring the dipper and payload closer to the shovel which requires less counterweight. The force for digging and the torque for swinging have been increased significantly through bigger DC motors and bigger power electronics. The motions have been made faster through the larger motors, and moreover, through the advanced Digital DC drive.

4.2 *Smarter systems*

To take full advantage of the production capability from the bigger payload and faster cycle time, the 4100XPB has production enhanced "smart" systems. These systems are designed to allow the shovel operator to use the machine more efficiently, easily set operating parameters, and prevent costly damage. In addition, systems to measure the payload in the dipper provide real time productivity feed back, and more importantly, help avoid over or under loading the haul trucks.

5 PRODUCTIVITY ENHANCED SYSTEMS

5.1 *Digital DC drive enables faster digging*

The torque and speed of a DC motor is controled by the electric current and voltage supplied to the armature and the field. The field is controlled separtely from the armature to allow more torque in some conditions and more speed in other conditions. Prior to today's smart digital drives now used on 4100XPB, DC technology used either fixed or two step field excitation. This scheme limited DC motor output levels to those allowed by torque per armature ampere and speed per armature volt available at the selected field excitation. The 4100XPBs high production digital DC drive systems use modulation of the field supply as well as the armature. Shovel engineers can now obtain a much broader range of peak horsepower through motor application and wise manipulation of all DC motor parameters.

Quick response digital regulators limit overshooting to enable extended operating ranges of DC motors possible without commutator distress. In a static power conversion system, this extended range allows better utilization of motor capacity and a broader range of operation in all motions. All of this achievable without increasing the complexity of the power conversion equipment or adding components.

The results of the Digital DC drive technology of the 4100XPB are noteworthy. Despite having a significantly higher payload than its earlier cousins, the 4100XPB will actually move the dipper *faster*. What once was a 30 to 35 second cycle time has been reduced by 4 seconds. For a shovel that operates over six thousand hours per year, four seconds translates to millions of tons. Bigger and faster.

5.2 *OptiDig system – computer assisted digging*

With older, non-computer automated control systems, the productivity of the dipper in the bank was highly dependent upon the operator. Operators developed the skill and recognized changing conditions quickly in order to keep the dipper moving and filling without stalling the motion. When digging conditions changed, maintaining productivity was entirely dependant on the operator to recognize the situation and change accordingly. A computer assisted digging system is available for the 4100XPB that automatically senses the digging conditions and the status of the dipper and drive motors, and then makes slight adjustments to the operators control signal. All this takes place almost instantaneously and almost without the operator even noticing. Besides increased productivity an added benefit is reduced stalling of the shovel's main which in turn reduces heating and power consumption.

5.3 *Impact protection*

Despite the operator's best efforts to smoothly dig and dump the payload, bank conditions and other factors often contribute to sizable impact loads being imparted on the shovel. And the impacts are large. When collisions occur, they involve payloads in excess of 90 metric tons and equally massive structures. While the 4100XPB is designed with such loads anticipated, these type of loads often produce damage to the machine. Smart systems are built in to the 4100XPB that allow the operator to easily adjust the range of motion on a touch screen computer. Once adjusted, the operator can freely dig, swing, and dump within a prescribed envelope, and if he or she is forced to leave this envelope, they are warned or stopped before impact occurs. An even more protective system senses when the operator has started to force the digging attachment out of its normal position. Then, automatically, the computerized system backs off of the appropriate motor before damaging impact occurs. Like the computer assisted digging feature, this is almost undetectable by the operator.

5.4 *LoadWeigh system measures payload*

Most mining operations use material moved as a fundamental measure of their production and haul truck fleet's performance. Precise quantification of the material moved is necessary to assure accuracy in their evaluations and to serve as the basis of numerous other metrics. Various means of measuring material movement are used and vary from engineering surveys to monitoring truck loads via mine dispatch systems.

Many of the mine operators use truck suspension weigh systems to determine the amount of material moved. While these systems are deemed to be accurate to within + or - 5% per truck load, the systems do require periodic scheduled maintenance to maintain their accuracy and reliability. The systems accuracy is also subject to variations in the ground and haulage conditions presented to the truck. These systems rely on either relaying the information from the truck to the base station for reporting or back to the shovel for operator feedback.

With the intent of giving the operator immediate feedback, it is possible to gather this information from the shovel and display it in front of the operator on a interactive screen. This is made possible by the use of a high powered computer installed on the shovel which constantly monitors motor feedback and via extensive algorithms, calculates the weight in the dipper. This system has an accuracy of + or - 2% and allows the shovel operator to monitor his progress during the truck loading cycle.

6 CONCLUSION

The 4100XPB has been designed to raise the bar once again of shovel productivity. Productivity is the product of production capability and availability. High production capability has been gained through increased payload, increased speed, and through extensive use of computer enhanced systems. High availability is assured on the 4100XPB through built in features that make it easier to maintain and less susceptible to damage. As mining companies are faced with the challenge of moving millions of tons of overburden in a safe, environmentally responsible manner while at the same time satisfying their shareholders, they need to seek ways to lower costs and increase production. Loading tools like the 4100XPB are pushing the envelope of productivity and availability.

Environmental Issues and Management of Waste in Energy and Mineral Production, Singhal & Mehrotra (eds)
 ISBN 90 5809 085 X

Geophysics comes of age in oil sands development

Paul Bauman, Richard Kellett, Eric Gilson, Russ Pagulayan & Anil Sharma
Komex International Limited, Calgary, Alb., Canada

ABSTRACT: Most energy forecasters predict that the development of shallow heavy oil reserves in Canada's enormous oil sand deposits will play a vital role in bridging the gap between North America's reliance on conventional oil in this century, and the full integration of alternative energy supplies in the next century. Last year, a number of surface geophysical techniques have been successfully applied to oil sands exploration and development. These applications include exploration and detection of oil sands, calculation of bitumen saturation, exploration for water supplies beneath the oil sands, geological mapping, mapping and imaging of thick clays and shales for geotechnical purposes, and non-intrusive monitoring of leachate plumes. Geophysical techniques successfully applied to these problems include 2-D electrical resistivity imaging, transient EM, ground penetrating radar, and high-resolution seismic reflection. This paper reviews present applications of these techniques in the surface mineable ore reserves of the Athabasca deposit.

1 INTRODUCTION

Conventional oil and gas production is steadily declining in North America. In Canada's Western sedimentary basin, the decline of conventional oil production is particularly rapid. Presently, Alberta's oil sands contribute over 20% of Canada's petroleum requirements. The Alberta Energy and Utilities Board predicts that the Alberta oil sands will be the principal source of Canada's crude oil beginning within the next decade.

The oil sands, often incorrectly referred to as "tar sand," are most accurately described as bitumen saturated sands. The bitumen content in the deposits is classified as lean, intermediate, and rich, according to minimum saturation values by weight of 3, 5, and 10% (Minken 1974). There are an estimated 1.2 trillion barrels of bitumen in place, more oil than found in the reserves of the entire Middle East, approximating one-third of the world's known reserves (Alberta Community Development 1998).

Alberta's oil sands are found in four major deposits: Athabasca, Cold Lake, Peace River and Wabasca. Together, they comprise nearly 31,000 square miles, an area greater than the landmass of Ireland. The Athabasca deposit is the largest, and the only, reserve which is shallow enough to be surface-mined. Presently, there are two oil sand mining operations which, together, account for about 350,000 barrels of oil per day.

The oil sand mines of northern Alberta represent the largest mining operations of any kind on the planet. As a result, there are numerous engineering concerns of enormous proportions including mine planning, tailings containment, water supply, and land reclamation. Geophysical methods play an important role in each of these areas and will continue to play an expanded role as new techniques find appropriate applications.

2 GEOTECHNICAL SITING OF MUSKEG PILES

One of every six dollars spent on oil sands development is allocated to land reclamation. A principal component of land reclamation is stripping the muskeg cover and storing it for future reclamation purposes. Geophysics plays important roles in estimating the volume of muskeg to be stripped and in siting the optimal location for building the muskeg piles.

Ground penetrating radar (GPR) is a rapid applied technique for profiling the muskeg/till, muskeg/oil sand, or muskeg Clearwater shale contact. Depths from the profiles can be compiled together into a muskeg thickness map from which volumes can be calculated and stripping planned.

Muskeg piles are preferably built above remnants of the Clearwater Formation, a marine and shore face sedimentary deposit consisting of glauconitic fine to medium sand, silt, clay and shales. The majority of the Clearwater was eroded during Quarternary glaciations. Remnants of the Clearwater Formation conformably overlie the Cretaceous oil bearing sands, thus decreasing the mineable reserves in these areas. Mapping the location of Clearwater remnants is, therefore, critical in siting muskeg piles and consequently minimizing the area of the mine to be sterilized. Delineating the thickness of these clay remnants may be a critical geotechnical concern as large earth works may lead to instabilities and failure. Application of 2-D electrical resistivity surveying has been used extensively in delineating the Clearwater Formation.

3 WATER EXPLORATION

The extraction process of the bitumen from the sands uses vast amounts of water. Groundwater is the preferred water source in oil sands development. As groundwater's use minimizes pipeline and surface intake costs, impact to surface water bodies important to the native communities is minimized, and siltation is not a significant problem. Geophysics is used extensively in developing water resources. These applications include EM31 and EM34 terrain conductivity mapping of shallow, Pleistocene channels; 2-D electrical resistivity mapping of river-connected, deep, bedrock Pleistocene paleovalleys; and 2-D resistivity mapping of basal aquifers beneath oil sand bodies.

4 TAILINGS CONTAINMENT

Bitumen is separated from the oil sands using vast quantities of hot water. Through this hot-water extraction process, the water becomes contaminated with clays and bitumen. As a portion of this contaminated water cannot be recycled through the extraction process, there is a net accumulation of liquid tailings (Page 1974). These tailings are contained in settling basins by dikes and berms that, together, comprise the largest earth-filled structures in the world.

A number of environmental issues relate to these impoundments including monitoring leachate migration, stability of the geotechnical structures, consolidation of the tailings, and salt migration within the tailings. Electromagnetic (EM) terrain conductivity mapping and 2-D electrical resistivity surveying have been used extensively for leachate mapping. Push probe conductivity surveys are also planned for this application. Geophysical logging and 2-D resistivity surveys are planned for investigating the interior of the dikes and berms. Ground-penetrating radar surveys are planned for monitoring tailings consolidation. EM and 2-D resistivity surveys are planned for monitoring salt migration within the tailings piles.

5 OIL SAND EXPLORATION

More than 650 billion barrels of heavy oil are in place in the basal Cretaceous McMurray sands of the Athabasca deposit. The McMurray Formation can be up to 150 m thick (Jardine 1974). Delineation of the McMurray sands, themselves, is a relatively easy task for seismic reflection, transient EM, and DC resistivity surveys. The McMurray is generally bounded from above by the electrically conductive Clearwater shales and from below by surprisingly conductive Devonian limestone and marls, which also provide strong seismic reflection boundaries.

Despite the relative ease of delineating the top and bottom of the McMurray Formation, the direct detection of heavy oil is not trivial. The McMurray, itself, is divided into an Upper Unit and Lower Unit. The Lower Unit is composed of thick, massive fluvial sands and swampy, lacustrine facies. The Lower Unit, which is usually barren of oil, has pore waters, which may vary in salinity from fresh to brine. The Upper Unit can be divided into three facies including fluvial, estuarine, and marine. The richest oil sands are usually in the coarser-grained deposits; nevertheless, the vertical distribution of the bitumen is very complex. The oil/water contacts vary from normal horizontal to vertical boundaries (Jardine 1974). The bitumen may exist in clean sand interbeds. Thick water sands may exist within bitumen deposits. Because of the complexity of the ore deposit, seismic, geo-electrical, and borehole geophysical techniques all provide important and complementary information. A nuclear magnetic resonance (NMR) pilot program is planned for mapping bitumen saturations from surface.

6 DEVONIAN SURFACE MAPPING

As part of the oil sands mining process, thick sections of the Cretaceous McMurray Formation require de-watering down to the underlying Devonian limestone. The top of the Devonian is usually less than 200 m below ground surface, and often less than 120 m below ground surface. The limestone surface relief can be very irregular due to faulting, karst features, salt dissolution and consequent collapse beneath the limestone. High-resolution seismic reflection surveys have been found to be the most useful technique for mapping the top of the Devonian. The availability of sonic and density logs is invaluable in the creation of synthetic seismograms and subsequent time-to-depth conversion. The 2-D electrical resistivity and transient EM surveys have been a useful and often complementary crosscheck of the seismic data.

7 CONCLUSION

The Athabasca oil sand deposits will likely be the most important petroleum source for North America within the next decade. As 10% of this deposit lies at depths shallow enough (less than 250 m of overburden) to be recovered by open-pit mining techniques, the ore is a good target for a number of different geophysical techniques. The gargantuan scale of these mining projects creates various environmental, geotechnical, exploration and mine planning problems that are amenable to the application of near surface geophysical methods. The shallow nature of these deposits provides an opportunity for the 1% of near surface geophysicists to participate in the oil exploration sector, which has traditionally employed 95% of all geophysicists.

REFERENCES

Alberta Community Development 1998. *Alberta's Oil Sands*. Fort McMurray Oil Sands Interpretive Centre: 32.

Jardine, D. 1974. Cretaceous Oil Sands of Western Canada. *Oil Sands Fuel of the Future.* Canadian Society of Petroleum Geologists, Calgary, Alberta: 50-67.

Minken, Douglas F. 1974. The Cold Lake Oil Sands: Geology and a Reserve Estimate. *Oil Sands Fuel of the Future.* Canadian Society of Petroleum Geologists, Calgary, Alberta: 84-99.

Page, H. W. 1974. The Environmental Impact of Alberta's Tar Sands Industry. *Oil Sands Fuel of the Future.* Canadian Society of Petroleum Geologists, Calgary, Alberta: 222-233.

Environmental Issues and Management of Waste in Energy and Mineral Production, Singhal & Mehrotra (eds)

Regional environmental assessment and management of mines in India

T. Biswas, B. Babu Rao & B. P. Sinha
Mining Research Cell, Indian Bureau of Mines, India

F. Cottard & B. Coste
Environment and Process Division, BRGM, France

C. Jucker
ANTEA International, France

ABSTRACT: In 1997, the Indian Bureau of Mines (IBM) entered into a collaboration with BRGM, France, with a view to developing its technical capabilities for environmental studies in the fields of mining and ore processing. Emphasis was placed on the acquisition of state-of-the-art analytical equipment and on technology transfer to the IBM technical staff, in particular regarding analytical techniques, mine-site oriented environmental impact assessments, environmental management methods, and pollution abatement technologies. Among the mining districts identified for their significant impacts on the local environment and ecosystems, two subject areas were selected for conducting environmental studies: 1) iron ore mines in the State of Goa, and 2) Sukinda Valley chromite mines in the State of Orissa. The main objectives of the project in both areas were to undertake a Regional Environmental Assessment (REA) of existing and potential environmental impacts of mining operations, characterize the risks for human health and the environment based on a source-pathways-target approach, and develop an Environmental Management Plan (EMP) incorporating regional and mine-site-specific solutions in terms of both environmental management and technical improvement. The measures proposed in the context of sustainable development are commensurate with acceptable levels of environmental impact and improved economic operation under the technical, legislative and socio-economic conditions of India.

1 INTRODUCTION

India has an extensive mineral extraction industry exploiting a wide range of ore deposits, and its mineral production grows on average by 5% per year with coal occupying first place in terms of tonnage (production of 260 Mt/y) and value. The extracted metal ores are iron, copper, chromite, zinc, gold, manganese and bauxite, whereas for the industrial minerals sector, 90% of total production is accounted for by limestone, magnesite, dolomite, barite, kaolin and gypsum.

Iron ore is the third most important mineral mined, after coal and limestone, with a production of 55 Mt/y from some 300 operating mines making India the world's 4th largest iron-ore producer. Chromite, with a production of 1.4 Mt in 1997, is also a valuable source of foreign exchange earnings for the country.

As in many places in the world, mining, in particular strip mining, and ore beneficiation processes disturb large areas of land and generate various types of waste material: large volumes of waste rock/overburden, substantial quantities of process tailings, airborne dust, mine water effluents, etc. These wastes are considered as significant sources of environmental pollution, adversely affecting landscapes, down-stream ecosystems and human populations in degrees ranging from nuisance to acutely toxic. Surface-water runoff and seepage from poorly managed mine waste dumps is generally the source both of suspended solids and, in some cases, of acid rock drainage. Process tailings can contain high concentrations of metals and residual ore-processing reagents that may be released to the environment as highly concentrated liquid effluents or spread as wind-blown dust.

These environmental pollution concerns, coupled with the fact that many mineral deposits in India are in environmentally fragile contexts (forests, wetlands, wildlife sanctuaries) or sensitive social environments (human settlements), create complex situations that must be dealt with through a systematic approach and at many levels. This requires the application of environmental and decision-making tools such as Environmental Impact Assessments (EIA), risk assessments, regulatory decisions, risk mitigation measures, planning and environmental auditing conducted at both the regional and mine-site levels. It also implies constructive partnerships with local communities and mine operators.

The Indian Bureau of Mines (IBM) is the principal government agency in charge of regulation, inspection and mineral conservation, responsible for providing access to the latest information in respect of mineral resources, for delivering mineral rights and also for ensuring that *mining development is environmentally sustainable.* Being thus in a key position for developing and undertaking environmental-protection actions related to the mining sector, it entered into a cooperation project with BRGM, France, with the aim of enhancing its technical capabilities in this domain. The project, implemented within the framework of the Indo-French Working Group for Mineral Exploration and Development, required a multi-phased and highly multi-disciplinary approach including three objective-oriented components:

1. Procurement of essential laboratory analytical equipment and field instrumentation to carry out comprehensive monitoring programmes, along with the appropriate training and transfer of methods, protocols and procedures.

2. Best practice demonstration actions based on the implementation of EIAs and other environmental management tools conducted in the field at different study scales.

3. Improvement of existing ore-beneficiation processes and establishment of potential processes for recovering valuable metals from mine waste dumps and process tailings.

The object of this paper is to present the main results obtained, in particular as regards component 2 of the project.

2 METHODOLOGY

Among the mining districts identified in India for their significant impacts on the local environment and ecosystems, two areas were selected for conducting the project, as shown in Figure 1:

- a group of 11 iron ore mines in the State of Goa,
- a cluster of 12 mines located in the chromite belt of the Sukinda Valley (State of Orissa).

In each project area, the mines lie relatively close together and their stresses and impacts were to be assessed at the regional scale through a Regional Environmental Assessment (REA) integrating multimedia investigations and multisource and

Figure 1. Map of India showing the location of the two project areas.

multirisk assessments. This REA was to encompass both environmental and social issues with the goal of producing an Environmental Management Plan (EMP).

Although significant background information on the economic domain (IBM 1997), some environmental aspects (Ralha & Vankataraman 1995; Noronha 1996; MECON 1991) and social issues (TERI 1997) was available for each study area, the first task of the project was to characterize, through sampling and analysis, the different environmental components (air, soil, sediment, water, climate, etc.) and their interrelationships with the study-area ecosystems.

A baseline study was thus conducted in the framework of a watershed approach to provide new data as a benchmark for later remediation. This included a seasonal monitoring programme related to the different environmental media, with the collection of more than 3000 field data between January and December 1998. The results were stored in a database and processed to produce thematic data maps based on regional numerical plans prepared from satellite imagery.

The seasonal monitoring programme was accompanied by socio-economic surveys in order to ascertain the expectations of the local population, and was followed by a risk assessment that included a characterization of the potential pollution sources (mining areas), the main pollution pathways (air, surface water, groundwater) and the effects of contamination on the target receptors (human beings, water resources, ecosystems).

The priorities for the EMP strategy were selected directly from the results of the REA.

3 APPLICATION TO STUDY AREAS

3.1 *The North Goa iron ore mines*

The 11 iron ore mines selected for the project study in the State of Goa are contained within a mining belt that includes more than 40 operating open-pit mines with a combined annual production between 14 and 17 Mt of iron ore, most of which is exported. This belt extends 65 km from southeast to northwest, spanning some 700 km² underlain by Precambrian metavolcanic and metasedimentary rock units, locally intensely weathered and covered by lateritic formations. The iron ore beds consist mainly of powdery hematite, limonite and goethite and are closely associated with clay formations (Gokulam 1972). The climate is tropical and is greatly influenced by the southwest monsoon that occurs from June to October with an annual rainfall varying from 2700 to 4000 mm over more than 100 rainy days.

Although the mining industry has had a positive and significant impact on the economic development of Goa, it has also given rise to several negative environmental impacts, some directly related to the unique features of the mining industry in Goa (small leases, cluster mines, road transportation, etc.) and others to bad mining practice and poor environmental management. Important differences exist in the environmental awareness of the mine operators, but they are all prepared to take environmental pollution very seriously and to give high priority to impact management and improving mining practices.

The extent and severity of the mining-related pollution, which is apparent to any visitor to the area, mainly originates from:

- The large volumes of mining waste generated (16 Mt in 1997) because of the high overburden-to-ore ratio (about 3:1). This induces a major problem concerning disposal and land-use management, especially as regards the optimal use of space available for waste dumps. The dumps are generally located in the upper parts of the valley areas, and consequently represent a major source of erosion/sediment loading in the watercourses, especially during the monsoon season.

- The fact that open-pit mining is carried out to depths of more than 60 m. This is below the water table and requires dewatering of the pits, thus contributing to the degradation of surface water quality (high concentrations of particulate matter).

Figure 2. Ore transportation by trucks in the North Goa iron ore district.

- The use of more than 1000 trucks (10-t tippers) to transport the ore to the beneficiation plants and barge loading points situated along the Mandovi River (one of the main perennial navigable rivers of Goa). This traffic is the main contributor to dust and SO_2/NOx emissions along the public roads, with frequent traffic congestions of large truck fleets using high-sulfur diesel fuel (Fig. 2).

Synthesis of the information collected during the REA identified four main environmental issues to be tackled in the EMP (Table 1).

Mitigation measures aimed at reducing these environmental impacts include the implementation of actions mainly related to:

1. Ore-transportation management; i.e. controlling dust and SO_2/NOx emissions by restriciting truck loading and traffic and also by designing and constructing new access roads.
2. Land-use management; i.e. implementing known state-of-the-art practices in the fields of waste disposal and management (dump planning, benching, terracing, runoff drainage, etc.) and dump reclamation (reshaping and planting).

3.2 *The Sukinda Valley chromite mines*

Located in the State of Orissa near the Bay of Bengal (Fig. 1), the Sukinda Valley generates 90% of India's chromite production (1.3 Mt in 1997). The valley, bordered by two NE-SW-trending ridges with elevations exceeding 800 m in the Daitari hills, has an area of about 180 km² and is drained by the Damsal nala. The climate is tropical to subtropical with a heavy rainfall (about 2000 mm on average) during the monsoon season from mid-June to mid-October.

Nine open-pit chromite mines are located in the

Table 1. Summary of the priority list of issues to be tackled by the EMPs.

Identified risk	*Source of risk*	*Priority list of the issues to be tackled by the EMPs*
Air quality (RPM, SPM) does not meet environmental standards at any of the stations	Waste dumps (wind erosion) Transportation Dry screening	Treat the surface of the waste dumps to reduce wind erosion (covering, revegetation) Develop green belts Cover the trucks Water spraying
Air quality (SO_2, NOx) does not meet environmental standards at several villages.	Truck transportation	Truck maintenance (pollution control)
Fine particles, transported by surface runoff, render paddy fields barren and deteriorate surface-water quality	Discharge of material from tailings and dumps into watercourses d paddy fields	Install sediment control and containment facilities to manage the water that leaves the mining sites
Contamination of groundwater with oil and grease	Washing and maintenance of the trucks	Develop truck-specific washing facilities
Coliforms in groundwater	Open dug wells	Use of drilled wells for drinking water

upstream part of the Damsal nala catchment area, and one underground mine and two other open pits are located in the southwestern plain. The chromite exploited by these mines is contained within highly serpentinized dunite of a Precambrian ultramafic belt (Ziauddin et al. 1979) and forms bands along the limbs and hinge of a pluri-kilometre-size fold (Fig. 3). The intensely weathered geological profile, generally capped by a lateritic formation, includes limonitic and nickeliferous clay, and weathered serpentinite closely associated with powdery chromite ore.

One of the main salient features in this densely vegetated area is the overlap between extensive surface mining, beneficiation plants and waste dumps on the one hand, and protected forests, habitations and farmland on the other. Consequently, the major problems to be resolved are waste rock dumping in protected forest or close to nearby villages, and pollution of surface waters in contact with the paddy fields. Since forests and croplands represent about 80% of the valley area, these aspects are a source of conflict between the miners and the local communities whose farmland has been diminished and degraded to accommodate the increasing amount of mining waste which, in 1997, reached 12 Mt.

The results of the integrated REA for the Sukinda

Figure 3. Open pit mine exploiting a chromitite band in the Sukinda Valley.

Valley show that the key environmental problem raised by mining activities is chemical pollution of the waters, soils and sediments by chromium and, to a lesser extent, other metals (nickel). This contamination results from mine water discharge, the release of beneficiation-plant effluents and slurries into the environment, and the leaching of waste dumps and ore stockpiles. Although the presence of hexavalent chromium (CrVI) in the waters was already known (Kar 1997), it has now been widely confirmed at the watershed-scale.

Table 2 summarizes the monitoring results obtained in 1998. It shows that, as far as water resources are concerned, most wells and watercourses in the central part of the valley are contaminated by CrVI (up to a value of 3.4 mg/l in surface water and 0.6 mg/l in groundwater), and in few places by nickel (up to 0.6 mg/l in groundwater). As far as chromium is concerned, these concentrations generally exceed the Indian guidelines for drinking water, surface water (protection of aquatic life) and soils used for agriculture. Analysis of biota (aquatic grass) and foodstuff (fish, milk, paddy, fruits and grass) also revealed that CrVI also exists in the food chain, exceeding the toxicological standards used in the risk assessment study.

Consequently, the main risk for the population is associated with drinking water, bathing, irrigation, aquatic life and the consumption of local agricultural produce.

In response the EMP must address the most urgent concerns related to the presence of high chromium concentrations in the environment and its bioaccumulation in the food chain. Based on the assumption that both mining and agriculture are operated within the framework of sustainable development, it must consider remedial actions including preservation, maintenance and improvement of the natural environment.

Table 2. Main results of the REA monitoring for chromium

Media	*Main features*
Nala sediment	• Cr_{tot} = 0.3–27%; CrVI = 30–104 mg/kg; Ni = 190–1090 mg/kg
Paddy field sediment	Cr_{tot} = 0.04–2%; CrVI = 6–90 mg/kg; Hg = 0.04–14.4 mg/kg; Ni = 40–580 mg/kg
Wastes and Ore	• Ore: Cr_{tot} = 48%; Tailings: Cr_{tot} = 25%; Dump: Cr_{tot} = 7%, Ni up to 0.6%
Leaching tests	• Leachate of river sediments: Cr_{tot} = CrVI = 2–12 mg/l • Leachate of paddy sediments: Cr_{tot} = Cr VI = 0.7 mg/l
Surface water	• 23 out of the 31 monitoring stations exhibit CrVI above drinking water standards (up to 3.4 mg/l) • No specific trend season wise
Groundwater	• 32 out of the 33 monitored wells exhibit CrVI above drinking water standards (up to 0.6 mg/l). • 6 wells exhibit Ni above drinking water standards (up to 0.6 mg/l)
Pit water and seepage water	• CrVI concentrations = 0.07–2.14 mg/l.
Fauna and flora	• Paddy: Cr_{tot} = 16–740 mg/kg; CrVI = <0.001–142 mg/kg • Fish: Cr_{tot} = 180–5700 mg/kg; CrVI = 14–115 mg/kg • Milk: Cr_{tot} = <0.001–0.69 mg/kg • Fruits: Cr_{tot} = 10–112 mg/kg; CrVI = 5–28 mg/kg • Aquatic grass: Cr_{tot} = 430–22,300 mg/kg; CrVI = <0.001–29 mg/kg
Background levels*	• Soils: in laterite, Cr_{tot} up to 5-7% and Ni up to 0.5%

* Background level: natural concentration of metals in pre-mining conditions

Among the solutions proposed from the REA at the watershed level, investigations have already been made into the management of CrVI-contaminated surface water and the management of paddy field sediments. As far as the levels of groundwater contamination are concerned, investigations have shown that the CrVI content in groundwater may not only result from mining activities in the valley, but could correspond in part to the regional and geochemical background prevailing naturally before mining occurred. Therefore, the EMP should advocate the exploration for new sources of drinking water rather than the treatment of the contaminated shallow aquifer groundwater, a solution that would hardly be feasible at anything but an exorbitant cost.

Surface-water quality, on the other hand, can be improved using a dual approach:

1. At the mine-site scale, by a) managing the impacts at their sources, i.e. managing both the water and runoff on site and the water leaving the site (dams, settling ponds, water treatment plant), and b) introducing new practices in waste disposal and management as well as in dump reclamation;

2. At the watershed scale, by setting up one or more waste-water treatment plant(s) to treat the effluents collected from a number of mines identified as having a deleterious effect on water quality and biota, and so ensuring the discharge of water of acceptable quality.

The available water treatment technologies currently being tested in the Sukinda Valley are of two types:

- an energy demanding technique that has been recently set up at a mine site by the Regional Research Laboratory of Bhubaneswar in the form of an electrolytic pilot plant (capacity of 500 l/min),
- a chemical demanding technique that has been installed in water treatment plants (capacity of 250 l/min) at two mine sites and is based on hexavalent chromium reduction using iron sulphate.

Both technologies are in common use by industries that produce low-volume effluents with a high chromium content (plating, leather industries, etc.), but they do not represent cost-effective solutions for the mining industry because of the high-volumes of water discharge that are involved. In Sukinda, the discharged flows to be treated from each mine is between 1,000 and 12,000 m^3/day and so other solutions need to be investigated.

In this context, the cooperation project has begun to evaluate the capacity of micro-organisms for reducing metals in the environment and several experiments have recently been carried out both in the laboratory and in the field in order to understand the bio-physico-chemical mechanisms that occur in the Sukinda Valley paddy fields. Preliminary results show that CrVI contained in irrigation water is, in the presence of paddy plants, reduced to trivalent chromium by micro-organisms occurring naturally in the soil. The process, shown on Figure 4, is still under investigation but could represent a simple, effective and affordable solution to remove chromium pollution from the surface water (Ignatiadis et al., personal communication). Such a phytoremediation technology based on the cultivation of paddy using special agronomic practices could be easily applied at the mine-site or watershed scale. Processing the harvested plants by drying, ashing or composting would also be easy and safe.

4 CONCLUSIONS

Mining in India is still one of the core sector industries playing a positive role in the country's

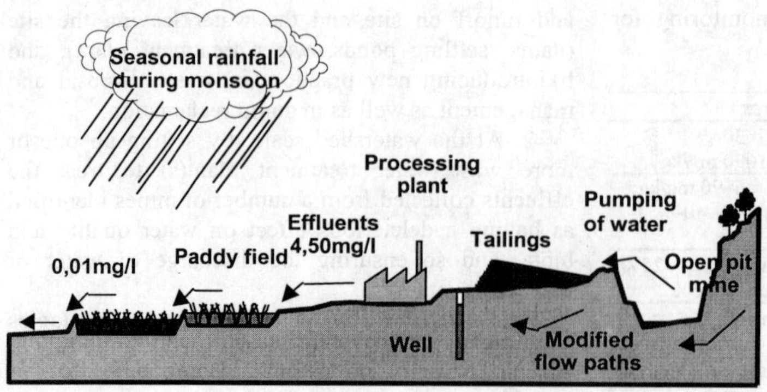

Figure 4. Natural reduction of Cr 6 bearing effluent in Paddy fields of the Sukinda Valley.

development. This paper has outlined the content and the main results of a recent project carried out by the Indo-French cooperation, emphasizing the continued upgrading of government environmental institutions through integrating sustainable development approaches at the regional scale. The use of integrated REAs, followed by the implementation of cost-effective remedial solutions set out in EMPs, will help the main stakeholders to achieve sustainable improvements.

REFERENCES

Gokulam, A.R. 1972. Iron Ore Deposits of Goa. *Bull. of GSI*: *Economic Geology 37*.

Indian Bureau of Mines 1997. Monograph on Iron Ore: Nagpur, India.

Kar, A. 1997. Hydrological conditions in parts of Sukinda Valley, Cuttack district, Orissa. *Central Groud Water Board, Ministry of Water Resources, Bhubaneswar, India.*

Metallurgical and Engineering Consultants Limited 1991. Environmental impact assessment and environmental management plan for Sukinda chromite bearing region. *Technical Report, Directotate of Mining and Geology, Govt. of Orissa, India.*

Noronha, L. 1996. Mining in India: Ecological Management in Goa. *Mining Environmental Management, June 1996:17-20.*

Ralha, D.S. & G. Vantakaraman 1995. Environmental Impacts of Iron Ore Mines of Goa. *Environmental J. Studies, 47: 43-53.*

Tata Energy Research Institute 1997. Areawide Environmental Quality Management. *TERI Report 96/EE/63.*

Ziauddin, M., P. Kar, N.R. Dutta, D.B. Ghosh, S.Roy, K.K.Mishra & B.B. Mallick 1979. Nickel Mineralisation in the Sukinda ultramafic field, Cuttack district, Orissa. *Bull. Geol. Surv. India 43.*

Environmental Issues and Management of Waste in Energy and Mineral Production, Singhal & Mehrotra (eds)

Using ISO 14001 as a structure for integrated environmental training

J.O. Burman
Luleå University of Technology, Sweden

S-E. Österlund
CENTEK, Luleå, Sweden

ABSTRACT: Since the mid-1990s the ISO 14001 standard has won wide recognition as a management tool for companies and organisations aiming for better environmental performance. However, the benefits of the standard are not only restricted to providing a structure for improved efficiency of the environmental work in organisations. It is also a favourable concept for the design of environmental educational programmes. The paper describes how the structure of an environmental management system (EMS) is used to achieve a coherent curriculum embracing various topics related to environmental issues. The methodology has been developed in a short course for professionals active in the mining sector and the experiences gained will be applied in a new 3-year university programme starting in 2001.

1 INTRODUCTION

Increased environmental awareness in the marketplace has compelled companies to voluntarily improve their environmental performance. In the mid-1990s, environmental management systems like ISO 14001, were developed by and for industry, and by the end of 1999, as many as 13,500 companies worldwide were certified according to the ISO 14001standard. The annual increase rate, in number of certified companies, was around 150% during both 1998 and 1999. The standard has won recognition as a management tool helping companies and organisations to improve their environmental work.

2 BACKGROUND

Centek is the industrial liaison office at Luleå University and handles, in that capacity, the University's so-called third mandate, i.e., interaction with society (traditional academic education and research being the other two). In principle, that means organising continuous training programmes for professionals in the private and public sectors and development and commercialisation of innovations and research results. Within the former activity a number of international training programmes have been carried out, notably on mining and environmental engineering. This reflects the strong position of Luleå University of Technology in these subject areas.

Since 1982 Centek has organised an international training programme on mining-technology for mining professionals from developing countries. The programme is financially supported through scholarships provided by the Swedish International Development and Co-operation Agency, Sida. Although mining engineering is still the core topic, the programme has developed to include more managerial subjects, including, among others, environmental issues. However, the introduction of environmental laws in many countries and a growing environmental awareness in general, with the mining industry often in focus, necessitated an even greater emphasis on environmentally related topics.

As a consequence it was concluded that a new programme dealing specifically with the environmental aspects of mining, inviting regulators as well as operators, should be developed to meet the evolving demands from industry and society. The short course Mining and the Environment was thus organised for the first time in 1991, which coincided, incidentally, with the UN round table in Berlin on the same theme (Berlin Guidelines).

The course will be organised for the tenth time in September 2000 and as a result some 250 professionals from countries in which mining is often an important contributor to the national economies have received training that aims to improve the environmental performance of the industry.

Table 1. The framework of ISO 14001 is used as a foundation for the short course Mining and the Environment

Elements in ISO 14 001	Modules in training course
4.1 General requirements	EMS
4.2 Environmental policy	EMS
4.3 Planning	EMS
4.3.1 Environmental aspects	Licensing and permitting
Location	EIA and mine site closure
Soil and ground water	Acid mine drainage
Water	
Energy	Energy and water conservation
Raw material	
Chemical substances	Public health and mining
Emission to air	
Noise	Vibration/dust and noise control
Discharges to water	
Waste	Solid waste handling
Fire, spillage and other uncontrolled situations	
Transport	
Working environment	
Suppliers and subcontractors	
4.3.2 Legal and other requirements	Legislation
4.3.3 Objectives and targets	
4.3.4 Environmental management programme(s)	EMS
4.4 Implementation and operation	EMS
4.4.1 Structure and responsibility	
4.4.2 Training, awareness and competence	Information retrieval
4.4.3 Communication	EMS
4.4.4 EMS documentation	
4.4.5 Document control	
4.4.6 Operational control	
4.4.7 Emergency preparedness and response	Environment risk analysis
4.5 Checking and corrective action	
4.5.1 Monitoring and measurement	Monitoring
4.5.2 Non-conformance, corr. & preventive action	
4.5.3 Records	
4.5.4 Environmental management system audit	EMS
4.6 Management review	EMS

3 COURSE STRUCTURE

From the beginning the course was largely oriented towards technical topics related to the mining industry's environmental impact like, e.g., waste and tailings disposal, mine site rehabilitation, acid mine drainage, monitoring, noise & vibration control, dust suppression, etc. A minor part of the curriculum touched on other disciplines, like environmental law, economics and management.

A highlighted part of the curriculum during the first years was environmental auditing, initially according to the definition made by the International Chamber of Commerce. Gradually the focus shifted from environmental auditing to training in implementation of environmental management systems, reflecting the shift in the industry from a reactive towards a proactive approach. Currently, the concept of environmental management, in particular the ISO 14001 standard, is used to structure the curriculum in order to create a logical framework for the technical and non-technical issues included in the programme.

In principle, many of the training modules have been the same since the start of the training programme. In the right column of Table 1, the present modules are sorted according to the structure of ISO 14001.

The specific training activities associated with the environment management lessons are omitted in the table in order to help identify possible overlaps between the earlier established training modules and the standard.

The specific and practical EMS training for the students will also in future focus on a number of groupwise training exercises covering the following:

- Quality evaluation of an existing environmental review for a mine, including setting up an environmental aspect register and evaluation of significant environmental aspects;
- Writing an environmental policy;
- Defining environmental objectives;
- Design of an environmental management programme;
- Writing an environmental report;

The reporting of the exercises is an important part of the EMS training programme. Each group makes a short written report as well as an oral presentation in the classroom.

The course modules presented in the right column of Table 1 will continue to function as more in-depth studies and amplification of important elements that are more generally represented in the ISO 14001 standard. As an example the module "Monitoring" involves geochemistry, sampling procedures, remote sensing, geophysics and GIS tools.

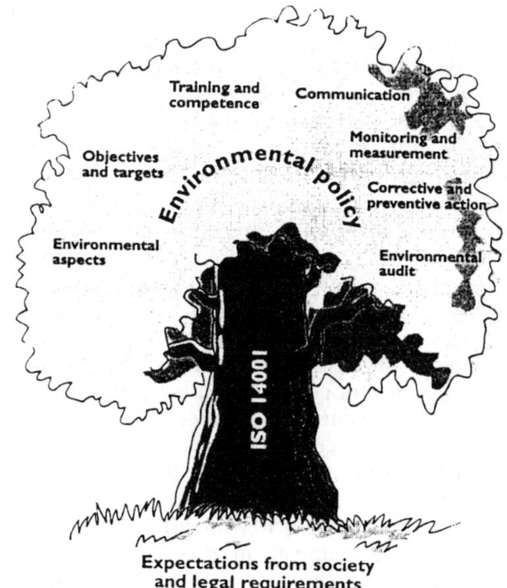

4 THE WAY AHEAD

Today (soft) issues like ethics and environment have taken on increasing strategic importance for companies. This trend will continue and is mainly a result of changing expectations and requirements in the society in which the companies operate.

The managerial methodology used as a framework in the modified course gives the participants a combined technical and management training. This approach has been received very favourably among the participants, many of them with 10-20 years of professional experience, in that it provides a common platform for joint efforts to improve the environmental performance of the mining industry, regardless of whether they represent an operating mine or a supervisory authority. In essence, this reflects the widespread acceptance of the ISO 14001 standard among the various stakeholders.

In an attempt to visualise the refinement of the curriculum it can be viewed as a "knowledge tree" having the standard ISO 14001 as the trunk and the branches representing the in-depth training modules covering a broad spectrum of environmental topics. It is believed that this view fosters a continuous and sustainable development of the course.

5 EDUCATORS' RESPONSE TO DEMANDS

There is presently an increased need within the industry and among supervisory authorities to recruit staff who are generally more well-qualified in terms of both environment and management. However, it is paramount that they should have a basic understanding of science and engineering principles as well technical disciplines relevant to environmental matters. In addition they should also be knowledgeable in environmental law, environmental management systems, environmental communication and marketing, as well being trained to develop social competence.

In order to address this it is evident that a holistic approach should be applied through the integration of environmental training in existing programmes, e.g., education for: economists, lawyers, civil engineers, mining engineers, metallurgists and geologists. This process will probably continue to be rather slow, due to the built-in conflict that follows from the need to exclude existing content from present curricula. The lesson learned is that it is easier to establish new programmes than to change established ones.

Mainstreaming environmental issues into traditional topics often requires a change in attitudes, which is normally a time-consuming process. In this context, the educational institutions themselves can act as catalysts by adopting better environmental practices in their own organisational structures. As a consequence, several departments at Luleå University of Technology will introduce the ISO 14001 standard with an aim to implementing it fully on campus in the future.

6 A NEW ACADEMIC PROGRAMME

The experiences gained during the modification of the short course Mining and the Environment, described above, have been very valuable in the planning of a new 3-year programme at the university; Environmental Co-ordinator (BSc).

The methodology is partly similar to the one described for the short course and the underlying structure of the curriculum is going to be influenced by the structure of ISO 14001. A table resembling Table 1 was constructed with the right column made up of all the appropriate divisions and departments of the university having links to environmental subjects.

This way of making an interdisciplinary inventory of the resources presently available within the university was found to be a good starting point for the scanning of potential curriculum content. The number of possible cross-border connections and overlaps largely exceeds what can be accommodated in a programme restricted to three years.

The programme will have an unmistakable base in natural sciences, but it will differ from existing engineering programmes in that it will have an integrated perspective of social sciences, exemplified by, e.g., management, communication, conflict resolution, jurisprudence, economics and others.

The new programme will also be based on a larger integration of subjects then generally found in existing programmes and the courses will combine, e.g., mathematics, physics and chemistry in order to increase training efficiency.

7 CONCLUSIONS

The EMS concept applied on the structure of the continuous education programme Mining and the Environment has been well received by the professionally active participants. Using ISO 14001 as the framework for the revised curriculum has given a better coherency between technical, environmental and managerial subjects taught in the course.

Luleå University of Technology is planning to use a similar methodology in an upcoming 3-year environmental programme. The university will take advantage of the cross-border connections suggested by the framework of the standard. This platform will probably facilitate a further integration of different sciences and disciplines in the new programme.

Mainstreaming of environmental issues into established programmes is an equally important way to develop the existing curricula at the university, however this will probably require a long-term implementation period.

REFERENCES

Australian Environment Protection Agency Booklet series 1995-1999. *Best Practice Environmental Management in Mining.*

Burman, J-O, Österlund, S-E. & Ulyanov V.P. 1997. EMS training in Russia. *Mining Environmental Management*, June: 14-17.

International standard ISO 14 001, 1996. Environmental management systems - Specification with guidance for use ISO 14001.

UN/DTCD 1992. *Mining and the Environment; The Berlin Guidelines.* Mining Journal Books.

UNEP 1994. *Environmental Management of Mine Sites - A Training Manual.* Technical Report No. 30.

Van Berkel, R. 2000. Minerals education. *Mining Environmental Management.* January: 10-13.

Environmental Issues and Management of Waste in Energy and Mineral Production, Singhal & Mehrotra (eds)
© 2000 Balkema, Rotterdam, ISBN 90 5809 085 X

Environmental impact assessment: Evaluating the follow-up phase

E.G.C.S. Dias & L.E. Sánchez
Escola Politécnica, Universidade de São Paulo, Brazil

ABSTRACT: This paper reports a study about the effectiveness of the environmental impact assessment (EIA) procedures applied to mining projects in the State of São Paulo, Brazil. Since procedures of EIA were regulated in Brazil, in 1986, and an EIS has to be filed as a requisite for permitting several activities, 77 mining projects have been approved in the State of São Paulo. This study examines six cases in full detail, with the purpose of evaluating the follow-up phase, especially the implementation of mitigation and compensatory measures.

The study widely confirmed its initial hypotheses that the implementation of mining projects is actually faulty, severely harming the whole process. Many evidences were collected and it was possible to identify the main causes for the inefficacy of the follow-up actions: (i) lack of articulation among government organisms and (ii) poor quality of the agreement.

1 INTRODUCTION

Since environmental impact assessment (EIA) was established in Brazil, mining was included among the activities whose permission depended on the previous submission of an environmental impact study (EIS). This fact had a strong repercussion on the different players on the mineral sector, bringing out environmental issues on top of the concerns both on the government side and on the entrepreneurial side.

In São Paulo State, where a compulsory environmental license for extracting and processing minerals has been in force for over twenty years, the government agencies' delay in regulating the EIA procedures, particularly the initial phases of the project, resulted in the accumulation of a number of documents to be analyzed and, consequently, process terms incompatible with the dynamics of the activity.

There has been a lot of discussion concerning the various phases of the EIA process and its effectiveness. The post-decision environmental management is of special concern here. This phase encompasses the implementation of mitigatory or compensatory measures, which are legally binding conditions of project approval.

In other words, the environmental impact assessment process is known to be complex and slow until EIS is approved. What comes next, though? There lies one of the key issues to assess the effectiveness of the process, once the large amount of human resources and materials employed in the several activities prior to the final decision may not be turning into effective mitigation of the negative environmental impacts of mining.

This article describes a study on the effectivity of the EIA procedures applied to São Paulo State mineral sector, focusing on the implementation of the terms and conditions of project approval.

2 SÃO PAULO MINERAL SECTOR

São Paulo mineral sector is characterized by a number of small and medium-sized enterprises producing industrial minerals or construction materials. According to recent official data (DNPM, 1996), in 1995, State mineral production placed São Paulo in the fourth place in the Brazilian mining scenario. Approximately 99.5% of this production came from non-metallic minerals. There were 22 groups of substances with strong focus on eight: crushed stones, limestone, sand and gravel, common and plastic clays, mineral water, industrial sand, kaolin and phosphatic rock.

Though the official data are highly imprecise, their analysis shows other outstanding characteristics of the sector in the State. The bulk amount produced in 1995 reached 143 million metric tons of material. Having in mind that for every bulk ton produced a certain amount of waste material was generated, one may realize the high potential for environmental impacts caused by the activity.

Another aspect, particular of the São Paulo mineral sector, is the high number of individual enterprises, frequently grouped in regional production poles due to geological characteristics and land use constraints.

Although there are no precise data on the number of active mines, because of the high level of informality, some studies give rough figures. In 1979, São Paulo had 1,387 active mines, with an absolute predominance of construction materials mines (DIAS, 1980). Nearly a decade later, in 1988, the number of mines was estimated to be 1,363 by RUIZ; ATEM (1990).

More recently, MELLO et al. (1997), analyzing the construction materials produced in São Paulo State, estimated the production of sand, common clay, crushed stone and limestone to be about 150 million tons a year. The estimated number of mines was 500 for sand and 150 for crushed stone.

Despite the virtual absence in the State of big scale mines, of radioactive mineral mines, of mineral deposits with a high potential for acid generation, that is, projects whose environmental impacts are consensually considered critical, São Paulo mineral sector represents a not negligible environmental adverse impact potential, due to its cumulative potential and the location of its mines, usually close to urban areas.

3 MINING ENVIRONMENTAL LICENSING

Environmental licensing first appeared in Brazil in the State of São Paulo, when Law # 997/96 was enacted, establishing the environmental pollution prevention and control system and the licenses for setting up and operating sources of pollution. Regulations set standards for emission and environment quality.

The extraction and processing of minerals was considered to be a source of pollution, whose environmental licensing was made compulsory. However, as a consequence of the characteristics of the State economy, predominantly industrial, the pollution control system consolidated procedures concerning only the control of industrial pollution. Thus a number of mining enterprises settled in the State after this legislation without the necessary environmental licenses, hardly ever undergoing intervention from the environmental control agency.

After the Environment National Policy Law was enacted in 1981 (Law # 6938), environmental licensing was made compulsory all over the country, with the adoption of a three tier licensing scheme - previous license (LP), in the designing phase; installation license (LI), which allows the setting up of a new project, and operation license (LO), which allows for the enterprise operation or activity. This law, considered a milestone in the Brazilian environmental legislation, included environmental impact assessment among its instruments. In 1986, the National Council for the Environment (CONAMA) passed Resolution # 1/86, which established the requirement of a previous environmental impact study (EIS) to be filed as a requisite for permitting several activities, including mining.

From then on, in São Paulo State, several government agencies began to act in environmental licensing for mining projects, each performing a different role. Despite their all being subordinate to State Secretariat for the Environment (SMA), these agencies enjoy a certain independence of action, which many times implies in conflicts and discontinued actions and policies. The main agencies taking part in this licensing are:

1) DAIA - Environmental Impact Assessment Department, responsible for reviewing EIS, provides a technical statement with recommendations to CONSEMA.
2) CONSEMA - State Council for the Environment, which approves or disapproves EIS, based on the DAIA statement.
3) CETESB - Environmental Sanitation Technology Company, which provides the installation (LI) and operation (LF) licenses and controls the implementation of both the project and the environmental control measures: CETESB may apply penalties when irregular procedures are detected.
4) DEPRN - State Department for the Protection of Natural Resources, participates in the review of the project in matters concerning the vegetation; provides statements and authorizes the removal of vegetation, when necessary. It is responsible for controlling the fulfillment of the conditions for the project approval within its competence.
5) DUSM - Metropolitan Land Use Department, whose action is restricted to controlling the Water Sources Protection Law in the Metropolitan Area of São Paulo. When the project area lies within an area protected by this law, DUSM takes part in the project review, provides a statement, licenses and controls the project implementation and operation.

4 EIA IN SÃO PAULO STATE

The EIA process in São Paulo State is made up of different phases : screening; scoping; EIS preparation; RIMA preparation; EIS review; consultation and public participation; final decision; and follow-up monitoring and management.

The screening procedures must be able to distinguish the projects that may cause significant environmental degradation (which must be submitted to a thorough environmental impact assessment process), from those whose environmental control can be carried out under a simple licensing, avoiding the burden of a time-consuming process and the hindering of the agencies in charge of conducting the process.

The scoping step aim to select the issues to be considered by an EIS and, together with the previous phase, can make its preparation process more efficient, allow for a multidisciplinary team fit for developing the study in its different fields of knowledge, considering the depth of analysis each theme will require and provide opportunity for public participation at the beginning of the assessment.

The EIS preparation phase must be capable of identifying the adverse impacts to the environment, propose mitigation, recovery and compensation measures, and evaluate the importance of the impacts, even with the implementation of the preconized measures. This phase has the highest technical and scientific content in the whole process; however, specially when the importance of the impacts is assessed, it is strongly dependent on the opinion of the professionals involved.

RIMA is a document for the general public and as such must be prepared in fairly easy language, using visual communication techniques, so that everybody can understand "the advantages and disadvantages of the project, as well as the environmental consequences of its implementation" (CONAMA Resolution # 1/86). In spite of the simple definition, to attain the goal is a complex task, once the public interested is composed of individuals with varying information and education levels, shows interests in different aspects of the problem and, consequently, requires different communication techniques.

The EIS review phase aims to evaluate the adherence of the documents to the laws, norms, general guidelines and terms of reference, the quality of the study and the project environmental viability. In this phase, alterations to the original project are common as a consequence of negotiations between the applicant and the environment agency, which may get subsidies from the public and from other government agencies. The technical statement prepared by the Environmental Impact Assessment Department at the end of this phase usually contains a summary of the information and documents that underwent the process, and is therefore of great importance to the final decision phase.

Public consultation is one of the most important phases in the process and must not only inform the public, but also collect subsidies for EIS review, and for the final decision-making. Although public participation may occur in several forms and come from different initiatives, in Brazil public hearing is the only formal mechanism established and aims to "debate, learn and inform the public opinion on the implementation of a project or activity likely to cause significant environmental impact" (CONSEMA Resolution # 50/92).

In São Paulo, the final decision over the project environmental viability belongs to CONSEMA, a council that gathers representatives from different government organisms, universities, professional associations and non-governmental organizations. In case the project is approved, a previous license is conditioned by the fulfillment of the mitigating measures proposed in the EIS and on any conditions formulated by CONSEMA.

After being given the green light by the government, the permittee must implement the mitigation and compensation measures. From a technical and scientific point of view, it is possible, at this point, to rate the righteousness of both the impact predictions and the efficacy of the recommended measures.

This is also the phase when the fulfillment of the conditions for project approval are to be checked. Every process, even if very well designed and carried out, is ineffective in case the enterprise is settled and operates contrary to the approved conditions or if the recommended measures have not been implemented. In the Brazilian institutional scene, especially in the São Paulo mining sector, there are evidences of a great gap between the theoretical project - proposed, discussed, negotiated and approved within the EIA process - and the actual settled and operating enterprise. This, precisely, is the main focus of the study.

4.1 EIA implementation in São Paulo

Although EIA was effectively implemented in Brazil in January 1986, the first environmental impact studies were submitted to the approval of São Paulo State Secretariat for the Environment in January 1987, a year later.

Data consolidated until December 1997 by RONZA (1998), show that 470 EIS were submitted in the period, at the end of which just over 40% had been approved and about 15% were still waiting to be reviewed. The remaining 45% had been denied, returned for modifications and never filed again, or taken back by the proponent.

The predominance of mining projects is outstanding and 259 EIS (55,1%) have been accounted for during the eleven years analyzed by RONZA (op. cit.), who grouped them in: energy, sanitation, solid wastes, transportation systems, drainage systems, urbanization and leisure, industrial, agriculture and mining. Table 1 summarizes these data.

The number of EIS submitted to the São Paulo State Secretariat for the Environment reached its peak in 1990 when it began to decline. Though the decrease may also be related to the general behavior of the economy, within São Paulo State environmental system it can be explained by the establishment of screening criteria for the exemption of presenting an EIS. These criteria had been ruled in 1993 by SMA Resolution # 26/93.

Table 1: EIS presented according to project type (from Jan/87 to Dec/97).

TYPE OF PROJECT	# OF EIS	(%)
Mining	259	55,1
Solid wastes	46	9,8
Transportation systems	43	9,1
Urbanization and leisure	39	8,3
Industrial	22	4,7
Sanitation	19	4,0
Energy	18	3,8
Draining systems	15	3,2
Agriculture	9	1,9
TOTAL	470	

Source: RONZA (1998)

Before that, filing an EIS was compulsory for mining any substance, by any mining method, scale or location. Even enterprises already settled that failed to have all the necessary licenses or altered the project after getting its permission could be made to carry out a detailed EIS to clear their situation with the environment organisms. The gradual development of EIS presented is shown in Table 2.

Table 2: Gradual development of EIA presented (from Jan/87 to Dec/97)

YEAR	MINING	TOTAL	(%)
1987	19	34	56
1988	41	68	60
1989	23	47	49
1990	59	100	59
1991	70	98	71
1992	31	60	52
1993	6	17	35
1994	5	19	26
1995	2	5	40
1996	0	13	0
1997	3	9	33
TOTAL	259	470	55

Source: RONZA (1998)

5 MINING PROJECTS APPROVED

Since 1987, 77 mining projects have been granted permission. Table 3 shows the year-by-year development of the number of EIS approved and their respective average time of review.

These figures reveal the increase in review time which, as from 1993, exceeded 2 years and went on increasing until it exceeded 5 years in 1997. Figure 1 illustrates the behavior of these terms.

Table 3: EIS approved year-by-year

YEAR OF APPROVAL	# OF EIS APPROVED	AVERAGE REVIEW TIME (months)
1987	2	7,7
1988	7	4,8
1989	9	11,9
1990	11	15,7
1991	12	18,0
1992	5	14,4
1993	18	25,7
1994	2	29,8
1995	7	59,3
1996	3	37,6
1997	1	42,6
TOTAL	77	
AVERAGE		22,2

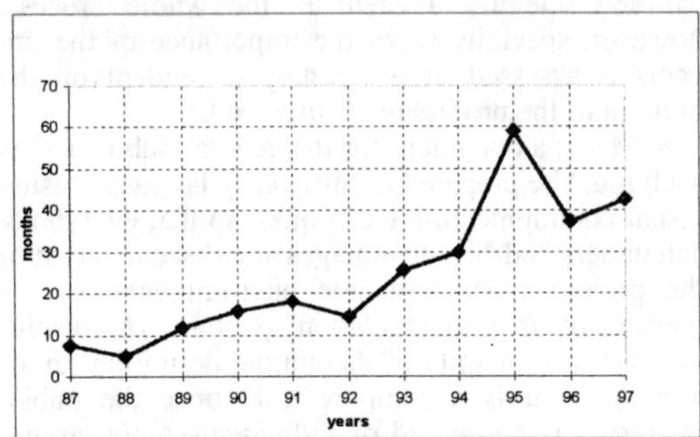

Figure 1: Development of review time according to date of approval.

Most of these studies concerned one enterprise only. Three important exceptions to the rule, however, deserve comments. They are the so called multiple EIS which, in spite of not provided in the regulation, were accepted or even demanded by SMA for situations in which there was a large number of similar enterprises in a certain geographic region. The three cases dealt with enterprises that had been operating for several years.

In Table 4 the distribution of approved EIS and the number of enterprises per mineral substance exploited is highlighted. The predominance of sand and crushed stone is remarkable. Industrial sand stands in third place.

This distribution does not provide a clear profile of São Paulo State mining. Important substances in terms of amounts produced and value of production, such as carbonate rocks, appear marginally, while others, such as phosphatic rocks, do not appear at all. Such fact may be explained by the linking of EIS to the permitting process, that is, enterprises already licensed by the environment organism need to sub-

Table 4: Mining EIS approved according to mineral substance

SUBSTANCE EXPLOITED	# OF EIS	%	# OF MINES	%
Sand	36	46,7	116	73,9
Crushed stone	20	26,0	20	12,7
Industrial sand	8	10,4	8	5,1
Industrial clay	5	6,5	5	3,2
Common clay	2	2,6	2	1,3
Dolomite	2	2,6	2	1,3
Dimension stones	2	2,6	2	1,3
Peat	1	1,3	1	0,6
Bauxite	1	1,3	1	0,6
TOTAL	77		157	

mit an EIS only in case of expanding business or major project modification.

But several EIS were in fact related to mines which had been operating for years, without the environmental licenses due. EIS approval was therefore necessary to clear the enterprise situation, which is called corrective licensing.

A more thorough analysis of the projects reveals the mining methods adopted as well the monthly production rates, which vary widely. Table 5 illustrated this aspect for the projects concerning sand, crushed stone and industrial sand, representing more than 80% of the EIS approved.

Table 5: Mining methods adopted and monthly production rates

SUBSTANCE: Sand MINING METHODS: riverbed dredging; riverside dredging; hydraulicking and mechanical excavation PRODUCTION RATES: 600 to 35,000 m^3/month
SUBSTANCE: Crushed stone MINING METHODS: Open pit mining by drilling and blasting PRODUCTION RATES: 5,500 to 110,000 m^3/month
SUBSTANCE: Industrial sand MINING METHODS: Mechanized excavation associated, in one case, to hydraulicking PRODUCTION RATES: 2,500 to 100,000 t/month

6 CASE STUDIES

Among the 77 mining EIS approved since the EIA implementation in the State, six cases were selected for a thorough study. The selection method initially consisted of three elimination criteria, for later identifying relatively homogeneous groups concerning relevant parameters to the theme and that could be characterized with reasonable reliability in the data collection phase.

The elimination criteria were: (i) multiple EIS, for not being provided in the legislation and its regulation; (ii) EIS presented before 1987, for being the first studies presented and reviewed, characterized by the precariousness of the study and inexpertness of the environment organism; and (iii) projects which would today be exempted of an EIS due to their features and according to the regulations now in force.

Three sand pits were chosen, each representing one mining method - hydraulicking, riverbank dredging and riverbed dredging-; two quarries - one in urban area and the other in industrial zone - and an industrial sand mine.

Within the relatively homogeneous groups chosen, the study case was selected at random. Projects that had already been implemented when EIS was prepared or that had not been implemented when the research was started were rejected. Table 6 shows the six cases chosen and their main features.

Table 6: Cases selected for study

Case 1	Substance: sand Mining method: hydraulicking Installed capacity: 7,500 m^3/month	Area: 50 ha
Case 2	Substance: sand Mining method: riverbed dredging Installed capacity: 5,000 m^3/month	Area: 24 ha
Case 3	Substance: crushed stone Mining method: open pit with blasting Installed capacity: 80,000 m^3/month	Area: 5 ha
Case 4	Substance: sand Mining method: riverbank dredging Installed capacity: 24,000 m^3/month	Area: 42 ha
Case 5	Substance: industrial sand Mining method: open pit, mechanized excavation Installed capacity: 30,000 t/month	Area: 249 ha
Case 6	Substance: crushed stone Mining method: open pit with blasting Installed capacity: 50,000 m^3/month	Area: 3 ha

Each case then underwent an analysis previously defined in general lines and simultaneously detailed to the study of the first case. The first case studied thus besides being a pilot case provided elements for the detailed program of data collection and treatment.

In a broad sense, the study consisted in examining all documents that are part of the administrative dossiers, at the different government organisms that take part in the EIA process. Documents review aimed mainly to know the enterprises features and the measures proposed to mitigate the negative environmental impacts identified, as well as the control actions performed after the EIA process. When the analysis was completed, visits to the enterprises were carried out, aiming to check the implementation of the mitigation measures defined by the EIA process as a condition to approve the project.

7 CHRONOLOGY OF THE PROCESS

One of the points receiving severe criticism by the opponents of the application of EIA procedures to mining projects in São Paulo is the slow pace of the process. Actually, in the cases studied it is observed that the necessary time span for obtaining an operation license (LF) for the enterprises varied between 17 to 122 months, the time for EIS approval included. Table 7 summarizes the chronology of the processes in the six cases studied.

Table 7: Chronology of the processes

CASE	LENGTH (months)			
	EIS approval	LI issuing	LF issuing	Total
1	19	18	5	42
2	16	2	88	106
3	14	2	36	52
4	14	15	56	85
5	7	2	18	27
6	13	9	100*	122*

* In these cases, LF had not yet been issued when the research was carried out, the length registered corresponds to the issuing of a favorable statement to the LF granting.

The first part of the process - EIS approval - was carried out by DAIA. The length for its completion varied little and included, in five out of the six cases, the preparation of complementary reports by the applicant. Most of the time, however, was taken by the environment organism, in reviews or consultations to other departments.

The granting of licenses for setting up (LI) and operating (LF) is a CETESB duty, as well as the main checking and control actions. The long lengths between the approval of EIS and the granting of licenses were due, in these cases, to several factors:

1) increasing strictness in the formulation of terms and conditions for obtaining licenses, partly due to the EIA process;
2) need for adjustment and reformulation of the demands for the EIA process, so as to make them more accurate and feasible;
3) complacency or omission by the government organism, allowing the operation of enterprises without the compulsory licenses;
4) reactive attitude by the enterpreneur, who only implemented the mitigation measures stated in the EIA process when notified or threatened of being fined or having operations halted;
5) disarticulation among the government departments taking part in the licensing process.

Table 8: Development of terms and conditions of project approval

DOCUMENT	Case studied					
	1	2	3	4	5	6
EIS	27	43	35	17	63	22
Complementary report	35	38	9	19	na	na
DAIA statement	28	18	42	21	36	16
CONSEMA decision	10	2	3	3	10	4

na: non-available document, either for not being demanded, or for being lost and therefore not found in the environment organism archives.

8 TERMS AND CONDITIONS OF PROJECT APPROVAL

Concerning the terms and conditions for approving the project, it is worth mentioning that the previous license, granted after EIS approval, depends on the fulfilling of the mitigation measures proposed in the environmental impact study and on the demands stated by CONSEMA. Moreover, it is licit to understand that the project approved is that which is described in EIS and other documents comprehended in the process.

This study tried to pinpoint this type of information in the documents examined and to analyze aspects likely to influence the efficacy of the phases after the EIS approval. Table 8 presents the development of the number of mitigation measures observed in each of the main documents in the process: EIS, complementary report, DAIA statement and CONSEMA decision. RIMA was not included in this analysis because in all cases studied it was just a concise transcription of EIS, presenting the same conditions with identical or summarized content.

The formulation of a number of environmental measures dispersed in different items of the documents was generally observed in EIS and complementary report. The main problems observed in the formulation of these measures are:

1) vague and imprecise language;
2) general guidelines or intention statements;
3) routine operational measures;
4) measures concerning health protection and workers' safety (which are not within the legal scope of EIA);
5) environmental control measures not rated in enough detail for further control actions;
6) excessively detailed measures, with minute definition of equipment, materials, techniques and procedures for implementation (which in principle precludes later use of innovative or more efficient technologies);
7) measures involving other actors without their previous manifestation;
8) high cost or technically unfeasible measures.

The DAIA statement, anyway, tried to recover the conditions formulated by the enterpreneur, presenting them in an orderly but concise fashion, besides formulating new propositions and demands. This effort actually contributed very little to the solution of the formulation problems observed in the applicant's documents and introduced new ones, such as general impoverishment of the text, already little detailed, addition of new elements to the project without checking its technical or economic feasibility.

The CONSEMA decision is a concise document, refers to EIS and RIMA or to the DAIA statement; in some cases it reproduces the demands formulated by the latter and occasionally adds new ones.

Both in the applicant's documents, and in the environment organisms, a broad conceptual muddle is observed. Statements of the environmental policy of the enterprise, environmental impact mitigation measures, degraded areas recovery measures and monitoring program are presented under the same label.

This generalized imprecision *a priori* impairs the next phase, the one that follows up the implementation of the mitigation and recovery measures, as well as the monitoring project.

9 FOLLOW-UP PHASE

In São Paulo, overseeing the implementation of mitigation measures is the attribution of a pollution control agency, CETESB, set up in the 70s to conduct the environmental licensing and the State pollution control actions. Other units of the State Secretariat for the Environment, in theory, should also act at this stage; however, in the six cases studied this action showed to be incipient.

After the EIS approval, CETESB is due to grant the installation and operation licenses, following the fulfillment of the conditions to the project stated in the EIA process. The first problem faced for the completion of this task is the identification of these conditions, scattered in the different documents forming the EIA dossier. Formally, these documents are not substitute in character, that is, the conditions presented in the first documents are not substituted by new formulations, presented in later documents. This way, in principle, every condition is valid from EIS filing to the final decision. For control purposes, its identification therefore requires examination of the whole dossier.

Although CETESB gets copies of EIS, RIMA and the complementary report, it was observed that this material is not filed together with the licensing process and the technical team in charge of following up the project generally do not know the content of these documents or even its location. The information concerning the conditions to the project come exclusively from the DAIA statement and the CONSEMA decision. In some cases, the technical team ignored even the DAIA statement, and based their judgement on the CONSEMA decision only. Thus the detailed descriptions of the project and environmental control measures, as well as maps, plans, drawings and sketches attached to the applicant's documents, are not examined by the CETESB technical team.

Therefore, starting from the information available in the DAIA statement and the CONSEMA decision, summarized and containing several imperfections, the control and fulfillment of the conditions are difficult and many times depend on subjective evaluations, subject to controversies and dispute. The way the conditions are formulated is often inadequate in terms of control, fail for inaccuracy, excessive details, minute description of equipment, material, techniques and procedures. This kind of formulation does not allow for technological improvements to be incorporated to the solutions made at the project making phase and its control, in some cases, would demand the constant presence of an inspector at the enterprise.

Adding to that, CETESB technical team are trained for applying the pollution control laws, referring to pollutant emission or environmental quality standards. In their view, they cannot count on juridical or administrative instruments to ensure proper implementation of terms and conditions of the project approval as defined in EIA process except when these conditions adhere to the pollution control law.

The result is that CETESB bureaucratically treats the conditions for project implementation that do not fit the pollution control law and focuses its inspection actions exclusively on the law own regulatory requirements. In the pollution control aspects, CETESB many times modifies the conditions previously formulated, exempting the enterpreneur from fulfilling those it considers unnecessary and being stricter when it deems convenient.

In some cases, points that received the most attention during the EIA process, resulting in conditions technically difficult to fulfill or economically costly, were forgotten after the EIS approval, with express consent from CETESB.

EIA process in Brazil is public (documents are available in public places, there is public hearing when demanded) and its final decision is collegiate, whereas the environmental licensing process, though public according to São Paulo State legislation, presents several hindrances even for a simple consultation. Even so, the alterations introduced by CETESB on the projects conditions are not informed to either DAIA or CONSEMA.

10 THE IMPLEMENTED PROJECT

Out of the six cases studied, only five mining sites

were visited, as one of them denied permission to the visit. The aim was to observe the enterprise features, confronting them with the licensed project, as well as the environmental measures implemented, comparing them to the conditions defined in the DAIA statement and the CONSEMA decision. Great differences were observed between the project approved and reality, in items such as production, equipment, effluent treatment method etc.

Concerning the conditions defined in the EIA process, several measures were simply not implemented or were implemented differently to what had been established. In some cases, only a *pro forma* fulfillment of the demand is carried out. That is the case of vegetation barriers or measures that involve the planting of trees in general. They are carried out, but in an insufficient manner for the original purposes (type and number of trees, spatial distribution).

The entrepreneurs, in a way, incorporate the environmental discourse, recognizing the importance of the environment organism action to curb abuses and environmental degradation; nevertheless they consider the process confusing, morose and several measures demanded are deemed unbecoming or impossible to implement. In practice they try to work as they have always done, doing the least possible concessions to the environment.

The entrepreneurs believe that the fact of being submitted to an EIA process made them a greater focus of attention by the environment organism than similar enterprises in the region. They thus feel they are in disadvantage before their competitors.

11 CONCLUSIONS

The study widely confirmed its initial hypotheses that the follow-up phase is actually faulty, severely harming the whole process. Many evidences were collected and it was possible to identify the main causes for the inefficacy of the follow-up actions.

One of the great problems in the process is the transition among the phases that precede the final decision and the follow-up phase. Here a major break takes place: the guidance of the process is transferred to another government organism. The pollution control agency, despite being in charge of following up the implementation of the project in accordance with the directions established in the EIA process, has an incipient participation in the previous phases and is not competent, from the technical and administrative point of view, to deal with all the issues comprehended by the environmental impact study. Its competence is restricted to pollution control. The government organisms that have this complementary competence, besides participating very little in the previous phases, do not practically act in overseeing project implementation.

The problems in this transition could be lessened if the conditions established in the EIA process were adequately organized and clearly defined, so as to prevent contradictory interpretations or value judgement. Nevertheless, both project description and the environmental measures proposed and negotiated are scattered in the different documents composing the process. Moreover, they are formulated in vague and imprecise language. To make matters worse, the pollution control technical team bases their follow-up action solely on the final documents of the EIA process - technical statement and final decision - in which the project and its conditions are only summarized.

The EIA process follow-up phase not only presents faults concerning the mining projects; many of the problems stressed in the study also occur in other types of project. Its solution necessarily includes the redefinition of the duties of the different government organisms acting on the process, and on the management of the transition among the previous phases and the enterprise implementation and operation.

12 ACKNOWLEDGEMENTS

The authors are indebted to FAPESP for providing financial support for this research.

REFERENCES

DNPM 1996 *Brazilian Mineral Yearbook.* Brasília, v. 25. [In Portuguese]

Dias, E. G. C. S. 1980 São Paulo supply mineral market assemblage and analysis. *Proceedings of the Brazilian Congress of Geology*: 1964-73. São Paulo: SBG. [In Portuguese]

Mello, I. S. C. et al. 1997 Producer centers of construction materials suppliers in the State of São Paulo. *Proceedings of the Symposium of Geology of Sud East*: 445-6. Penedo: SBG. [In Portuguese]

Ronza, C. 1998. *The environmental policy and the government contradictions:* the environmental impact assessment in São Paulo. Campinas: Geosciences Institute of UNICAMP. (Master tesis) [In Portuguese]

Ruiz, M. S. & S. M. Atem 1990. Supply of mineral commodities characterization in the State of São Paulo. In: M. S. Ruiz & M. R. Neves (coord.), *Supply of Mineral Commodities in the São Paulo State.* São Paulo: IPT. [In Portuguese]

Environmental Issues and Management of Waste in Energy and Mineral Production, Singhal & Mehrotra (eds)
© 2000 Balkema, Rotterdam, ISBN 90 5809 085 X

Sustainable mineral development and environmental conservation: A framework for decision-making

R. Dimitrakopoulos
WH Bryan Mining Geology Research Centre, The University of Queensland, Brisbane, Qld, Australia

K.S. Richardson
Gamma-Sim Pty Limited, St. Lucia, Qld, Australia

ABSTRACT: Implementing sustainable mining strategies is a sought after goal by the mining industry. As both developed and developing countries strive towards sustainable development, a framework to assist the mining industry to examine the trade-offs of location, size and impact of a mine in light of certain environmental issues is required. Such a framework is proposed based on a valuation of the landscape in a given area that would include both mining operations and, for example, natural areas to be put aside for conservation. Prime mining sites, potential mineral deposits, exploration leases and environmental variables are assessed using advanced heuristic algorithms, which assign values of "irreplaceability" across a region of interest. This framework is implemented using a transparent, GIS-based methodology that generates alternative maps displaying and quantifying various options based on the valuation of irreplaceability. An example from Guyana, South America, demonstrates that the framework allows the mining industry to participate in sustainable practices, while maximizing both mining options and addressing conservation needs. Hence, the proposed framework maximizes economic profitability and long-term sustainability.

KEYWORDS: Sustainable mining, irreplaceability, environmental conservation

Introduction The development of strategies for sustainable and socially responsible mining is an issue well appreciated in the mining industry, particularly when both developed and developing countries strive towards sustainable development (e.g. Wiber, 2000; Raczynski, 2000; McAllister et al. 1999; Placer Dome Inc. 1998). Sustainable development is the concept widely popularized as the "development that meets the needs of the present without compromising the ability of future generations to meet their own needs" (World Commission on Environment 1987). It has been seen as a multi-facet problem to be addressed and has been approached from various viewpoints ranging from pure economics (e.g. Simon 1995) to concerns on growth limits and the environment (e.g. Meadows et al. 1992). An important aspect of sustainable mining is the development of a robust decision-making framework that addresses issues of sustainable development, specifically in terms of environmental principles related to land use and reconciliation of mining with the conservation of species and natural habitats. This framework can lead to a socially responsible, defendable and objective decision-making process for land use, that simultaneously integrates mining and conservation needs. However, in order to satisfy mining needs while considering conservation concerns, key issues need to be identified and suitable technologies developed. Key issues include how to "value" land for mining and conservation or how "irreplaceable" land is in terms of both mineral deposits and diversity of biological species; how to adequately trade mining needs off with conservation needs; and how to integrate mining and conservation needs in decision-making while engaging all stakeholders in the process. The practical implementation of a framework needs to be based on the use of advanced computer technologies that are capable of examining various land uses, mineral resources and distribution of biological diversity. Furthermore, the technologies need to be data-driven, flexible and transparent, such as modern optimisation algorithms, to assess mining sites, potential mineral deposits, exploration leases and environmental conservation variables in terms of values of "irreplaceability" across a region of interest (Pressey et al. 1993). These technologies also have to facilitate the participation of all stakeholders in the discussion of trade-offs that maximise economic benefits and conservation value per land area.

The following sections present a framework for decision-making in the context of sustainable mining and environmental conservation, stressing the quantification and valuation of irreplaceability. This paper first overviews pertinent concepts relating to biological diversity and related algorithms for decision-making. Then, the steps of the proposed decision-making process are outlined. Finally, an example is presented from Guyana, South America, which demonstrates practical aspects of the methodology.

1 KEY ELEMENTS IN DEVELOPING A FRAMEWORK

The basic concepts and elements required in developing a decision-making framework are summarized in this section.

1.1 *Some Basic Concepts for Understanding Biological Diversity*

Biological diversity, or biodiversity, is defined as the variety and variability of living organisms and the ecological complexes in which they occur. Biodiversity is most commonly measured as the number of species in a given area. As with most land use issues, conservation of biodiversity is largely a matter of real estate and location is everything. With the estimates of the number of species in the world ranging from 10-100,000 million (Primmack 1993), determining which areas are most species-rich or contain the most 'important' species to conserve is proving to be a difficult task, as the distribution of much of the world's species is poorly known. Assigning a "value" to biodiversity has also proven to be difficult. The value of biodiversity has been defined as:

Biodiversity Value= Direct Value + Indirect Value+ Option Value+ Existence Value

where, direct value is measured by sustainable forestry, non-timber forest products, recreation, tourism, medicine, plant genetics (crops), etc; indirect value includes the nutrient cycle, watershed protection, air pollution reduction, carbon sink, etc; option value refers to future uses as per the direct and indirect value; and existence value is the intrinsic value of forest/biological diversity (McNeeley 1996). Quantifying most of these variables is extremely difficult and has yet to be done for large regions containing diverse biological entities.

The simplest approach that has been used repeatedly in the literature is to use the spatial distribution of species to determine which areas contain either large concentrations of species (very species-rich) or species with very restricted distributions (endemic species) (Williams et al. 1996). If species data are limited due to sampling effort, species distributions can be modelled based on a point-to-point similarity metric that predicts the distribution of a species across a given area. When no species data are available, unique ecosystems or unique environmental domains, which means clusters of similar climatic, geological and topographical characteristics, are used (Belbin 1993).

1.2 *Technological Elements in Decision-Making*

Recent developments in land management and environmental conservation decision-making have been based on using GIS technologies and algorithms that use species distributions or environmental classifications to determine areas that are irreplaceable in terms of their uniqueness in a given area. The irreplaceability value of an area in this case refers to the importance of the area for achieving an explicit conservation target. This target is defined for each situation and can include both a conservation component and a resource component (mining, forestry, agriculture, etc.).

The algorithms currently available to map the uniqueness or irreplaceability of biodiversity can be grouped into five categories. (i). Richness-based "greedy" algorithms. Here the first site chosen for conservation is the one that has maximum species richness. The subsequently chosen sites add the greatest additional species. Iterations stop when all known species are captured in the sites. (ii) Rarity-based algorithms. Sites for conservation are selected on relatively rarity of the species in each site. Subsequently, sites are selected upon highest rarity value and the algorithm stops when all rare species are captured in the sites. (iii) Irreplaceability-based algorithms where the sum of the irreplaceability values are calculated using the formula of Ferrier et al. (2000) and used to rank sites, the algorithm then selected the most irreplaceable sites first; (iv) Simulated annealing. Annealing starts from an initial sets of "seed sites" chosen arbitrarily. Annealing is then performed on the seed sets by setting a radius about each site in the seed set and randomly searching the space within the radius. The search radius decreases in an analogous way as to the decrease in temperature in classical annealing. Annealing stops when the goal is met, for example, areas are selected so that all species are captured. Last, (v) branch and bound algorithms can be used, such as the linear programming-based branch-and-bound algorithm, which looks for an optimal solution for a used defined objective function. From recent studies comparing different algorithms (Underhill 1994), it appears that no one algorithm is best for all situations and that the best algorithm is situation-dependent.

2 PROPOSED STEPS FOR DECISION-MAKING

The proposed decision-making process can be rationalised in 10 steps, outlined in this section. Note that the scale of analysis is critical. Most mines operate at a local scale and unless they are unusually large, occupy only a small percent of land. Land planning and conservation can be done at the local, regional, national or continental scale. To examine both types of land uses simultaneously, the scale must be the same. In addition, the concept of irreplaceability is key to the decision-making. In a straightforward case where data exit on species distribution, a map of irreplaceability values can be derived indicating those areas that are irreplaceable in terms of their biodiversity conservation value, those areas that are irreplaceable in terms of their mining potential (present and future) and those areas that are less than 100% irreplaceable for either biodiversity conservation or mining and can be exchanged to meet realistic conservation and mining targets. Common conservation targets include representing each species several times in a network of conserved areas (e.g. 3 times) or conserving a percent of land (e.g. 15 %). Mining targets may include protecting the interests of all existing mines and the top 10 % of sites identified new areas (e.g. making these areas "irreplaceable"). There may be several solutions that meet a specific target, however it is critical that for the system to work, areas that are 100 % irreplaceable for both biodiversity conservation and mining are protected.

The proposed steps are:

(i) Determine the scale of and select grid size to correspond with scale of analysis.
(ii) Gather biological data at selected scale and map. Interpolate species data if necessary.
(iii) Map existing land uses including, mines, exploration tenements, agricultural lands, forestry concessions, towns, infrastructure, and native title areas.
(iv) Overlay, using a GIS, land uses on species distribution maps to look at possible areas of conflict.
(v) Establish targets for selecting biodiversity conservation areas and mining areas.
(vi) Establish a "stopping rule" which satisfies the conservation target for the algorithm (e.g. the rule to satisfy the above stated target).
(vii) Using the concept of irreplaceability, map the irreplaceability value of all grids in a given area.
(viii) Establish rules for how other land uses outside of biodiversity values and mining are to be incorporated.
(ix) Use an iterative heuristic minimum set algorithm to nominate a notional network of areas that satisfy the established rules and targets.
(x) Examine the location of selected grids and "trade" (exchange similarly-valued areas based on expert knowledge) areas where necessary to satisfy both targets.

3 AN EXAMPLE

To elucidate on the proposed framework, an example from Guyana, South America demonstrates how the framework could be applied. Mining in Guyana includes gold, diamonds and bauxite. Gold mining alone represents over 30% of GDP and provides substantial revenue to the Government of Guyana. Recent studies suggest mineral production could be several times the present level (Sizer 1996). For a developing country, Guyana has a well documented database on plant and animal species, collected over several decades with the assistance of the Biological Diversity of the Guyana's Program managed by the Smithsonian Institution, Washington, DC. Guyana has yet to establish a network of conservation areas and is still in the planning stages, thus a future network can accommodate mining interests (Richardson & Funk 1999). The interest of the mining sector in establishing areas to conserve biodiversity in the country is obvious. It includes, for example, (i) avoiding potential land use conflicts, especially for undetected mineral resources; (ii) extending mining licences and eliminating pending and existing short term ones; (iii) dividing royalties if a conservation area should cover a mining site; (iv) collaborating on conservation issues where appropriate; and (v) assisting the Government in its effort towards sustainable development.

3.1 *Options, Sustainable Development and Decision-Making*

In reconciling sustainable development, biodiversity protection and economic needs in Guyana, one may consider three options. The first is an "all or nothing scenario" which can lead to banning all mining in the areas to be assigned to conservation or, at the other extreme, discard conservation. This option is not desirable by anyone, as it does not provide a

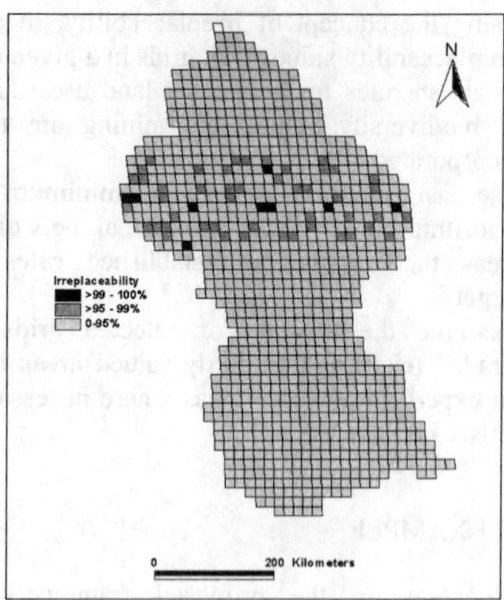

Figure 1. Irreplaceability values using species distributions.

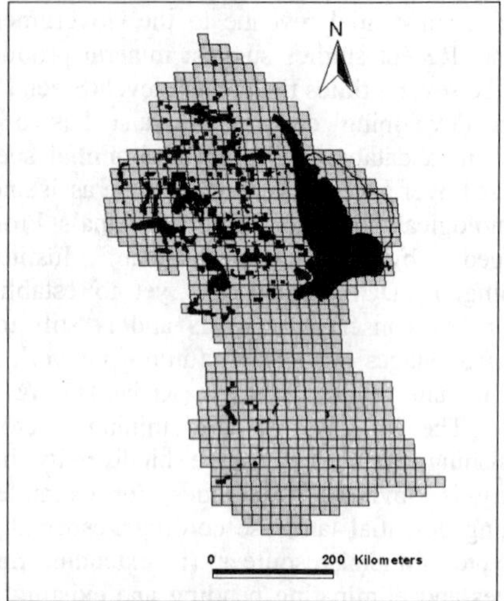

Figure 2. Irreplaceability values for biodiversity when all mines and exploration leases are deemed irreplaceable.

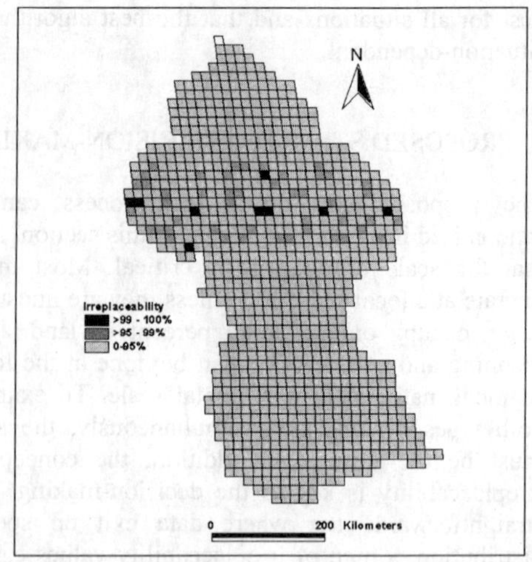

Figure 3. Recalculated irreplaceability values when 25 mine sites are deemed irreplaceable.

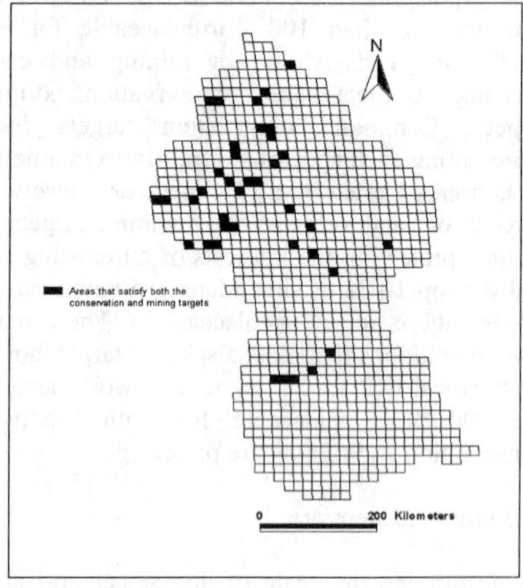

Figure 4. Minimum set of areas required satisfying both the conservation target and mining target.

long-term solution and it could lead the country to hardship. The second option is to assign land to whatever is considered "most economically profitable". It is, however, nearly impossible to provide meaningful long-term calculations on the economic profitability of an area with regards to the price of the potential future uses of a conservation area. In addition, one can usually show that the "biodiversity value" defined in Section 2.1 is higher than the monetary value of short-term mining for possibly the wrong reasons. This suggests that a 'direct valuation' is highly questionable, if not flawed. A third option is a compromise and more realistic approach. This approach selects areas that are irreplaceable in terms of conservation and irreplaceable in terms of mining using set targets. Then, look at remaining areas in order to trade off areas valuable to both type of use and possibly create multiple use areas. The compromise option provides both flexibility and efficiency. It allows the accommodation of the conservation need in as small an area as possible and with the most degree of flexibility in location. This is the same for mining needs, for example which mines can be incorporated into multiple-use areas.

Before elaborating on the compromise option, it is interesting to show some results of the process outlined. Figure 1 shows Guyana to be represented by 942 grid cells of 250km^2 each. The cells grey scale colour (white is zero, dark grey is 100%) shows the value of irreplaceability for the distribution of 320 known species using a conservation target of capturing each species at least three times in a network of areas. The figure as such excludes any mining activities in the northern, eastern and southern part of the country where there are already existing mines and exploration leases. Figure 2 shows that if the all existing and future (measured in this case by exploration leases) mine locations are included as irreplaceable mining areas, all of the remaining part of the country becomes irreplaceable for biodiversity. These figures demonstrate the further need for the flexible compromise approach.

3.2 *Alternatives, Irreplaceability and Land Allocation*

Figure 3 shows the map of irreplaceability values for each grid cell with a target of protecting all present mining sites and 10 % of future sites (as defined by exploration leases), which cover a grid by 25 % (25/942 grids). When this map is combined with the map of irreplaceability for the distribution of species (Figure 1) and a minimum set algorithm is run that selects sites based on their highest summed irreplaceability values, the results is a map of the grid cells required to meet both targets (Figure 4). This map represents the minimum set of areas required. If, however, for some reason one area becomes unavailable due to some other constraint (e.g. land use), then that area can be excluded and the analysis can be run again, resulting in another minimum set (although all the targets may not be satisfied). All stakeholders can examine the results of the minimum set and refinements to the set can be made in a participatory fashion. Since all stakeholders using the methodology have the same data, the results are transparent and repeatable by any user.

4 SUMMARY AND CONCLUSIONS

Strategies for sustainable mining and, in particular, a decision-making process for land use integrating mining interests and conservation targets is presented in this study. The key in this process is the understanding and expression of mining and conservation concepts as well as the availability of suitable computer technologies and algorithms. The concept of irreplaceability of land for both mining and biodiversity is a key attribute in the decision-making process together with the targets and goals sought. The example in Guyana shows that the process proposed provides an effective and efficient tool for land allocation. In addition, the proposed framework engages all stakeholders (Grimble & Chan 1995). Thus it enhances compromise while it provides objective tools for decisions and understanding of "what-if" type scenarios. For these reasons the proposed framework assists sustainable mining practices. It should be noted that irreplaceability framework is suitable for any land use valuation, it is transparent and can be used by all decision-makers.

REFERENCES

Belbin, L. 1993. Environmental representativeness: Regional partitioning and Reserve selection. *Biological Conservation* 66: 223-230.

Ferrier, S., Pressey, R.L & Barrett, T.W. 2000. A new predictor of the irreplaceability of areas for achieving a conservation goal, its application to real-world planning, and a research agenda for further refinement. *Biological Conservation* in press.

Grimble, R. & Chan, M. K. 1995. Stakeholder analysis for natural resource management in development countries. *Natural Resources Forum* 2, 113-124.

Meadows, D.H. et al 1992. *Beyond the Limits*. Post Mils: Chelsea Green Publishing.

Simon, J.L. (ed.) 1995. *The State of Humanity*. Cambridge: Blackwell.

McAllister, M.L., Scoble, M. & Veiga, M. 1999. Sustainability and the Canadian mining industry at home and abroad. CIM Bulletin 92: 85-92.

McNeely, J. A. 1996. Investing in biological diversity: assessing methods for setting conservation priorities. *OECD International Conference on Incentive Measures for the Conservation and the Sustainable Use of Biological Diversity.*

Placer Dome Inc. 1998. Sustainability Policy.

Pressey, R. L., Humphries, C.J, Margules, C.R, Vane-Wright, R. I. & Williams, P.H. 1993. Beyond opportunism: Key principles for systematic reserve selection. *TREE* 8:124-128

Primmack, R. B. 1993. *Essentials of conservation biology*. Sunderland: Sinauer Associates Inc.

Raczynski, A.M. 2000. Mining, sustainable development and the role of international financial institutions. *CIM Bulletin* 93:111.

Richardson, K.S. & Funk, V.A. 1999. An approach to designing a protected area system in Guyana. *Parks* 9:7-16.

Sizer, N. 1996. *Profit Without Plunder: Reaping*

Revenue from Guyana's Tropical Forests Without Destroying Them. Washington, D. C., World Resources Institute.

Underhill, L. G. 1994. Optimal and suboptimal reserve selection algorithms. *Biological Conservation* 70: 85-87.

Wiber, M. 2000. Environmental Management – A getaway to social acceptability. *CIM Bulletin* 93:110-111.

Williams, P., Gibbons, D., Margules, C., Rebelo, A., Humphries, C. & Pressey, R. 1996, A comparison of richness hotspots, rarity hotspots, and complementary areas for conserving diversity of British birds. *Biological Conservation* 10: 155-174.

World Commission on Environment 1987. *Our Common Future*. Oxford: Oxford University Press.

Environmental Issues and Management of Waste in Energy and Mineral Production, Singhal & Mehrotra (eds)
 ISBN 90 5809 085 X

Sustainable development for mineral and energy industries (from a legal point of view)

W. Frenz
Unit for Mining and Environmental Law, Aachen University of Technology, Germany

ABSTRACT: The principle of sustainable development has become a central idea of environmental law. The idea has been around in legal discussion and political declarations for some time. Now, the principle has been legally fixated. This leads to serious consequences for the legal framework in which mineral and energy industries operate. The concept of sustainable development emerged towards the end of the 70s. It has been incorporated into political declarations, first of all, into the Brundtland report from 1987, later into the Rio Declaration on Environment and Development and into Agenda 21 from 1992. According to these documents the needs of future generations must be safeguarded. Furthermore, ecological, economic and social interests must be reconciled. Significantly, the principle could demand from mineral and energy industries to limit the extraction of non-renewable resources. This could imply new restrictions for the mining and energy industrial sector. The following presents ideas which have been developed in connection with Collaborative Research Center 525 "A Resource-Orientated Analysis of the Material Flow of Metallic Raw Materials".

1 POLITICAL DEVELOPMENT

The concept of sustainable development has firmly shaped the national and international environmental discussion of the last years, especially on a political level. Starting point , in the first place, was the report titled "Our Common Future“, which was written by the World Commission on Environment and Development in 1987 (WCED 1987). A commission that had been established before by the United Nations. In this report the English term „sustainable development“ was shaped, which had impact on a lot of nations, like Germany, where the term "Nachhaltige Entwicklung“ represents what "sustainable development“ stands for in the Anglo-American countries. "Sustainable development“ must be interpreted in the light of the understanding of the aforementioned commission. The commission is usually called the Brundtland commission, named after its chairman, the former Prime Minister of Norway Gro Harlem Brundtland.

2 THE CONCEPT OF "SUSTAINABLE DEVELOPMENT"

The commission defines "sustainable development" as a "development that meets the needs of the present without compromising the ability of future generations to meet their own needs" (WCED 1987). This statement contains an ecological, an economic and a social dimension. This does not only mean that the needs of future generations have to be safeguarded. Basically, the principle of "sustainable development" is a triangle of objectives which have to be pursued as well as reconciled. In the mining sector the respect for the needs of future generations has a special importance, because this could mean a limitation of the extraction of non-renewable resources. Non-renewable resources are those resources which are not renewable in a certain foreseeable period of time.

Using the aforementioned three elements, the principle of "sustainable development" was brought into first concrete terms by the German government which developed the socalled "three rules of management" (see below).

A significant element of "sustainable development" as understood by the Brundtland commission lies in the fact that the concept is not only an ecological one. The concept is, moreover, a global concept which is a means for reconciling all aspects, the ecological, the economic and the social one. This means: Every economic decision has to take into account aspects of ecological and social wholesomeness. On the other hand economic effects play an important role when it comes to the protection of environment.

3 THE RIO-DECLARATION

The concept of sustainable development was picked up by the United Nations Conference on Environment and Development that took place in Rio de Janeiro in June 1992. The Conference decided several international agreements, one of them being the Rio Declaration on Environment and Development (so called Rio Declaration).

The importance of this Declaration lies in the fact that the concept of "sustainable development" has been introduced into public international law. The Declaration is "soft law", merely a declaration of intent. Although the concept is mentioned in several passages the Declaration gives no independent definition of the concept of "sustainable development". The Declaration does not expressively fall back upon the Brundtland commission´s understanding of the concept of "sustainable development". But the 27 principles of the Declaration correspond conceptually with the Brundtland commission´s understanding. Principle 1 states that human beings are the centre of concerns for sustainable development. This clarifies the anthropocentric approach of the concept. In the very sense of the Brundtland commission´s definition of "sustainable development" Principle 3 which states that the right to development must be fulfilled so as to equitably meet developmental and environmental needs of present and future generations, contains the idea of justice and equity between the different generations. Principle 4 contains the constitutive requirement that "environmental protection shall constitute an integral part of the development process and cannot be considered in isolation from it."

According to Principle 2 States continue to have the sovereign right to exploit their own resources pursuant to their own environmental and developmental policies, and the responsibility to ensure that activities within their jurisdiction or control do not cause damage to the environment of other States or of areas beyond the limits of national jurisdiction.

For "sustainable development" measures which are conceived for a long period of time are necessary because this is the only way to safeguard the needs of future generations. This requires foresight which is not limited to the environmental hazards of today but also takes into account future developments which may endanger the natural resources in 30 to 40 years time. Those developments are rarely foreseeable if actions are taken. This is why the precautionary approach in order to protect the environment shall be widely applied by States according to their capacities. Where there are threats of serious or irreversible damage, lack of full scientific certainty shall according to Principle 15 not be used as a reason for postponing cost-effective measures to prevent environmental degradation.

4 AGENDA 21

A further important document establishing the concept of "sustainable development" is Agenda 21, which was also decided on at the Rio Conference. The Agenda is an extensive political but not legally binding programme (see Preamble 1.3 and 1.6) which "adresses the pressing problems of today" and also aims at preparing the world for the challenges of the 21st century. The most important issue of Agenda 21 is the global consensus and political commitment at the highest level on development and environment cooperation. The concept of "sustainable development" stands also in the focal point of the Agenda. The concept runs through the entire programme. Chapter 2, for instance, deals with international cooperation to accelerate sustainable development in developing countries. Chapter 7 deals with the promotion of sustainable human settlement developments. Section II of the four sections of the Agenda consists of fourteen chapters. The whole section deals with conservation and management of resources for development. The exploitation of fossil fuels, however, is not dealt with in this section which concentrates on the careful handling of the resource soil and several ecosystems.

5 MANAGEMENT RULES

The principle of "sustainable development" has been interpreted especially in Germany in a way that mineral and energy industries are obliged to limit on the long run the exploitation of non-renewable resources to a certain extent. This has been developed from an interpretation of "sustainable development" that natural capital has to be conserved and that the exploitation of non-renewable resources has to be economical. This has been discussed widely and on an international scale. According to a wider interpretation it is allowed to exploit non-renewable resources, but only if artificial energy sources are provided to the same extent (Daly 1995). The German government developed the so called "three rules of management" on the basis of the Brundtland Commission in a report about the New York-Conference on implementing Agenda 21 (Bundesministerium für Umwelt, Naturschutz und Reaktorsicherheit 1997). This development was later on picked up by a parliamentary commission in Germany. The rules could represent a model after which the industrial states try to shape their future energy policy. The "three rules of management" demand the following:

"First rule: The use of renewable resources (e.g. forests or marine fish stocks) must on the long run be only used at rates within their viable rates of regeneration. Otherwise future generations would lose those resources.

Second rule: The use of non-renewable resources (e.g. fossil fuel, energy sources or agriculturally productive land) must be limited on the long run to an extent which must not be greater than the extent of their substitution by secondary resources (e.g. substitution of fossil fuels may be possible by solar electrolysis).

Third rule: Pollution by substances and energy must be limited to the extent which the environment can cope with (e.g. green house gases in the atmosphere or acidic substances in the forest soil)".

On the first glance the second rule is of the greatest significance for the energy and mineral industries in Germany. But also the other two rules could have an impact on the energy and mineral industry: If mining leads to the destruction of renewable resources on the surface like forests, there could be an impact on the rate of regeneration of this resources. If forests are not able to recreate themselves, then the mining of non-renewable resources also has an impact on renewable resources. But at least there are possibilities of a later redevelopment. The third rule could also have an impact on the mining industries as it could lead to a restriction for (or a sinking demand in) those energy sources which cause a lot of pollution.

The fact that the rules are derived from an international public legal framework makes the rules also significant for the mineral and energy industries throughout the world. The rule is the consequence of the fact that a long time responsibility has been taken serious: If you cannot replace non-renewable resources physically, you have to replace their functions in order to ensure there are energy resources left for future generations. A question remains: Does "sustainable development" imply new restrictions for the mining and energy industrial sector?

The first management rule can be traced back to the Programme for the further Implementation of Agenda 21, chapter 9, where an improved management is demanded in order to promote the exploitation of renewable resources within their viable rates of regeneration.

The third rule can be traced back to the Programme´s chapters 10 and 42 which deal with "increasing levels of pollution" and the need for "sustainable patterns of production, distribution and use of energy".

Chapter 42 of the Programme, however, also states that "fossil fuels (coal, oil and natural gas) will continue to dominate the energy supply situation. Thus, the crucial second rule as regards fossil fuels has a weaker programmatic international foundation than the other two rules. The Brundtland commission recommends to use primary energy in a more effective manner and to develop renewable sources of energy. The report states that "renewable energy sources require a much higher priority in national energy programmes" (WCED 1987) and encourages States to take every effort to develop the potential for renewable energy, "which should form the foundation of the global energy structure during the 21st century."(WCED 1987) It, however, does not recommend a strict restriction of the exploitation of non-renewable resources.

A strict limitation of the exploitation of mineral resources could also be very problematic under the general aspects of "sustainable development".

The strict wording of the second management rule indicates a strict restriction of non-renewable resources. But if you take a closer look concessions are made. The fact that an exploitation of non-renewable resources is only allowed to an extent which must not be greater than the extent of their substitution by secondary resources is confined to a development "on the long run". This means there is a temporal concession to the strict management rule. Does it also mean that a greater exploitation of non-renewable resources is allowed for a certain period of time as long as it is not greater than the extent of their substitution on a wider time scale? Is an exploitation of non-renewable resources allowed as long as their substitution can be expected? How possible must be such an expected substitution? One must bear in mind that a replacing of primary resources by secondary resources is not reasonably foreseeable at this moment. Firstly, the rule is derived from an international point of view. On a global level it is impossible to know whether there are new beds of non-renewable resources and where they can be found. Secondly, the worldwide development in substitution technologies must be also taken into account. The management rule cannot ignore a global development in the discovery of new resources, renewable as well as non-renewable. But a flexible development cannot be governed by a strict management rule. This may be the reason why the second management rule has such an unprecise outline.

Moreover, the second management rule provokes also doubt under the aspect of the international notion of "sustainable development". According to what has been said about the Brundtland commission´s interpretation of "sustainable development" the commission´s interpretation sees the concept of "sustainable development" as a reconciliation of ecological, economic and social needs of the different generations. A strict limitation of the extraction of mineral resources would have an impact on the international market situation of the mining industry. In addition, political decisions could affect the output of the mining industry. If the output is seen as to high States could interfere by imposing new taxes on the companies instead of subsidising them. States could think of restricting the mining sector by prolonging the procedure of licensing. States could also use their entire research money for renewable resources only and stop investing into the mining sector. From this, however, non-renewable resources could benefit. Every new technology that improves means of substitution allows the further exploitation of non-renewable resources. Apart from that, the generations of today could not benefit from the "second rule of management".

In summary, one can say the management rules must be considered as political strategies which have as independent rules no legal effect. However, the entire conception of those rules shows a direction which decisively shapes a future environmental and economic policy. It may be possible, therefore, that the management rules may lead to legally binding provisions in the future.

6 CONCLUSION

Taking into account, what "sustainable development" stands for, the strict restriction of the exploitation of non-renewable resources cannot be the answer to the problems future generations may face.

The claim that only a rigid restriction of mining activities can be the key to future problems is a similar exaggeration to the claim that the needs of today´s generations can only be looked after by not restricting the extraction of primary resources. The answer to today´s and tomorrow´s problems lies in the careful reduction by considering also the needs of today´s generations.

REFERENCES

Bundesanstalt für Geowissenschaften und Rohstoffe 1986. *Untersuchungen über Angebot und Nachfrage mineralischer Rohstoffe*, Band XX.

Bundesministerium für Umwelt, Naturschutz und Reaktorsicherheit, Government report on the occasion of the UN Special Session of the General Assembly on nature and development, New York, 23-27 June1997. *Auf dem Weg zu einer nachhaltigen Entwicklung in Deutschland – Promoting sustainable development in Germany* –. BT-Drucks. 13/7054.

Daly, H. E. 1995. On Wilfred Beckermann´s Critique of Sustainable Development. *Environmental Values* 4: 49.

Frenz, Walter & Unnerstall, Herwig 1999. *Nachhaltige Entwicklung im Europarecht.* Baden-Baden: Nomos.

General Assembly of the United Nations 1997. Paper presented at the UN Special Session of the General Assembly on Nature and Development, New York, June 1997.

United Nations Conference on Environment and Development (UNCED) 1992. *Rio Declaration on Environment and Development (Rio Declaration).* Paper presented at the United Nations Conference on Environment and Development, Rio de Janeiro, 3-14 June 1992.

United Nations Conference on Environment and Development (UNCED) 1992. *Agenda 21.* Paper presented at the United Nations Conference on Environment and Development, Rio de Janeiro, 3-14 June 1992.

World Commission on Environment and Development (WCED) 1987. *Our Common Future.* Oxford: Oxford University Press.

Environmental Issues and Management of Waste in Energy and Mineral Production, Singhal & Mehrotra (eds)
© 2000 Balkema, Rotterdam, ISBN 90 5809 085 X

Environmental impacts of high voltage power lines and stations

N.Gillich & G.R.Gillich
'Eftimie Murgu' University, Reşiţa, Romania

ABSTRACT: Environment pollution due to high voltage power lines and stations (over 400 kV) shows up in several ways: high frequency (radio and TV range) radio waves; sound pollution (noises); various direct and indirect effects on living beings; aestethic pollution.
The indirect effects of electromagnetic field may result in inducing high electric potential to earth insulated objects as cars, shelters and farming equipment, fencing, etc. which on human touch lead to discharge currents which only disappear by interrupting the contact. At high currents, due to muscle contraction, the man often cannot release the touched object, hence serious or even lethal accidents may happen. In depth analysis of such phenomena is possible by separating the electric and magnetic field effects. We shall concentrate on the electric field since the magnetic field effects are much less significant

1 INTRODUCTION

The continuos development of high voltage power facilities has lead their impact over environment to become an increasingly important issue. The importance given to this is uneven from country to country or even from region to region. In many cases, the influence of high voltage facilities over the natural environment is deemed equally important as their design.

The disturbing effects of electrical facilities are produced by the electromagnetic field, radio and TV broadcast disturbance, acoustic noise of Corona discharge, voltage induction in vicinity bodies, electric field on ground surface, effects of biological nature. Also within the issue fall physical space taken by the facilities and safety distances, phase layout and esthetical impact over the landscape. Analysing the impact of electrical facilities over environment one may distinguish between qualitative aspects (unmeasurable) and quantitative or measurable elements, which form the ground setting the norms in estimating the said impact.

2 QUALITATIVE ELEMENTS.

Environment disturbance due to aerial power lines comes from their very perceptible presence, the space taken by them being called „aesthetically disturbed space". This space can be confined by visual obstacles and depends on the size of the facility over the landscape, being an approximate qualitative mean to determine the magnitude of visual disturbances on the environment.

Aerial power lines follow routes dictated on a engineering and economical basis, on which the visual and ecological impact be minimal. Though a power line may disturb some activities, such as agricultural work, it won't be an obstacle for other constructions as motorways or railroads which create a environment discontinuity.

There are however certain restrictions as concerns the insulation distances between the lines and bodies in the vicinity and living beings. One would consider three vicinity probabilities defining three distances depending on voltage (Dragan1987):

l_1=0,0025 U when the vicinity probability is low;

l_2=0,005 U when the vicinity probability is medium;

l_3=0.0075 U when the vicinity probability is high;

For example, let us consider a 400 kV line, and l_2=2m distance; the probability for a commutation overvoltage wave be triggered within this distance is 0,005, that is once every 200 years.

As concerns the impact of aerial lines over landscape, the main blame is put on the pole with its odd shape. The design of a pole, observing the fuzziness of the handsome or ugly notions when applied to electrical facilities must so consider the technical and commercial criteria which dictate the characteristics of the pole. Anyway, the issue stays open and may be the object of a multiple discipline study.

3 QUANTITATIVE ELEMENTS.

Within this category there are the measurable elements, analytically or experimentally determinable elements such as radio and TV disturbances, acoustic noise, electric field on ground level.

3.1 *Radio and TV disturbances*

Analysing the 400 kV lines exploitation data, one may see that only in rare cases have been reported significant radio disturbances; as concerns the 750 kV (or more) lines however, there were stated norms for radio frequency disturbance at 100 m from line measured sidewise.

The magnitude of electric field, E_{adm} on to the surface of phase wires with an r_0 [cm] radius which meet the radio disturbance limiting results from expression (Alexandrov 1989):

$E_{adm}=32-17{,}4\lg r_0$ [kV/cm] (1)

At an altitude H_0 [m] above 500 meters, we must subtract $H_0>690$ from the value given by (1). If using a wire comprising more thinner wires per phase (strands), the radio frequency disturbance level stays the same, provided that the electric field magnitude on the wire surface stays also at the same level. It was experimentally proved (Alexandrov 1989) that radio disturbance level is independent to the number of strands, due to the electromagnetic screening effect, for a number of 1 to 18 strands per phase.

The disturbances produced by very high voltage facilities in the TV frequency range have been less studied since the practical impact is lower than that of radio frequencies. Experiments have revealed that the TV disturbances (PTV) and radio disturbances (PR) are linked by the expression:

PTV-PR=const (2)

or, more precisely:

$$PTV = PR - 20\lg\left(f\sqrt{\frac{1+(D/h)^2}{1+(15/h)^2}} \right) + 3{,}2 \qquad (3)$$

where: h is line height from ground;

D is the distance, sidewise, from the closest phase wire;

F is the frequency in question [MHz].

Such disturbance are significant in medium voltage lines, since was found an electromagnetic noise of 60 dB under the line at a 1 MHz frequency and are highly influenced by the weather conditions.

3.2 *Acoustic noise*

The acoustic noise of power facilities is given by the Corona discharge and by operating electrical devices in power stations.

The method to compute the acoustic pressure generated by a high voltage line involves two factors: noise magnitude and noise propagation process. Considering those, the acoustic noise level Nz [dB] in respect to the reference pressure of 20 μPa is (Kircham 1983):

In case of a single wire or two wires per phase:

$$N_{zg} = 20\lg n + 44\lg d_0 - 665/E + k_n + 75{,}2$$
$$-10\lg D - 0{,}02D \qquad (4)$$

in case of three or more wires per phase ($n \geq 3$):

$$N_{zg} = 20\lg n + 44\lg d_0 - 665/E +$$
$$+22{,}9(n-1)d_0/D + 67{,}9 - 10\lg D - 0{,}02D \qquad (5)$$

where: d_0 is the wire diameter [cm];

R is the radius [cm];

E is the intensity of the electric field on to the wire surface [kV/cm];

D is the distance from line to the given point;

n is the number of wires per phase.

$k_n=7.5$ for $n=1$; $k_n=2.6$ for $n=2$; $k_n=0$ for $n \geq 3$;

The expressions (4) and (5) apply to single phase and heavy rain conditions. In case of more phases, the total acoustic pressure is given by:

$$P_t = 10\lg\sum_{i=1}^{m} 10^{P_i/10} \qquad (6)$$

since:

$$N_{zg} = 20\lg\frac{P_i}{20}$$

For usual values of electric field magnitude we may employ the simplified expression:

$$N_{zg} = 16 + 1{,}14E_M \qquad (7)$$

where E_M is the minimum magnitude of electric field in the side phase [kV/cm].

The data resulted from experience in exploitation of 400 and 750 kV lines have revealed the noise pressure is within the allowed limits and its higher magnitude under rain conditions doesn't matter much since the rain itself makes noise.

3.3 *Electromagnetic field at the ground level*

The importance of this issue results from the fact

that electromagnetic field on the ground produces biological effects which affects the health, therefore is important to know those effects in order to set limitations.

Knowing that the maximum magnitude Em of electric field occurs when lying horizontally under a side phase of the line, it is given by the expression (Nekrasov 1974):

$$E_m = 1{,}05\frac{21{,}8U_m C_1 10^9}{H}\cdot \left[1-\frac{0{,}5}{1+(D/H)^2}-\frac{0{,}5}{1+(2D/H)^2}\right] \quad (8)$$

where: U_m is the maximal line voltage;
H is height over ground;
D is the distance between line phase wires;
C_1 is the line capacitance;
$C_1 = 1{,}02 \cdot 5{,}55 \cdot 10^{-8} \lg (D_e/r_e)$;
$D_e = 1{,}26$ D;
r_e is the equivalent radius of strands;

Computations in case of a 750 kV line have revealed the maximal values of the electric field for various heights, as shown in table 1.

Table 1. Values of the electric field for 750kV line.

H[m]	E_m [kV/m]
9	12,8
15	7,7
17	8,18
21	6,31

When considering the distance x from the line center, one may use simplified formulas to compute the electric and magnetic field within a 10...20% approximation:

$$E_x = 0{,}4\frac{U}{\sqrt{3}}\left[A-\frac{1}{2}(B+C)+\frac{\sqrt{3}}{2}j(B+C)\right] \quad [kV/m](9)$$

$$E_x = 0{,}2\cdot 10^{-6} I\left[A-\frac{1}{2}(B+C)+\frac{\sqrt{3}}{2}j(B+C)\right] \quad [T]\ (10)$$

where: $A=\frac{h_1}{(x-d)^2+h_1^2}$; $B=\frac{h_2}{(x-d)^2+h_2^2}$

$C=\frac{h_3}{(x-d)^2+h_3^2}$

where: h_1, h_2, h_3 are the heights over ground of the three phase wires [m];

d_1, d_2, d_3 are the distances between the individual wires and the pole axis.

Based onto the computed or measured values of electric and magnetic field under continuous regime at 50 Hz, the limits of human exposure stated presently by international standards are listed in tables 2 and 3.

Table 2. Values of electric field for human exposure

Standard	Short time[kV/m]	8 or 24h [kV/m]
IRPA		
- professional	30	10
- public	10	5
CENELEC		
- professional	30	10
- public		10
NRPB		12
BFE (professional)	30	21,3
ACGIH(professional)		25
ICNIRP		
- professional		10
- public		5

Table 3. Values of magnetic field for human exposure.

Standard	Short time[mT]	8 or 24h [mT]
IRPA		
- professional	5	0,5
- public	1	0,1
CENELEC		
- professional		1,6
- public		0,64
NRPB		1,6
BFE (professional)	4,24	1,36
ACGIH(professional)		1
ICNIRP		
- professional		0,5
- public		0,1

IRPA-International Radiation Protection Association, ICNIRP- International Commission on Non-Ionizing Radiation Protection, CENELEC-European Communities, NRPB-Anglia, BFE-Germany, ACGIH-USA.

4 CONCLUSIONS

1 The influence of electromagnetic field over the environment imposes along with quantitative measurement of main parameters a multi discipline approach involving biophysics, electrochemistry and pathology.
2 The lack of evidence (resulted from rigorous mathematical demonstration) mustn't lead to abandoning this field since the analysis made should be subjected to statistical data processing over long periods.
3 Presently, nowhere in the world was evidenced any harm due in particular to electromagnetic field; there is however a supposed acceleration of existing

illnesses in the presence of electric and even more in the magnetic field, experienced mainly by old persons.

5 REFERENCES

Alexandrov, G.N., 1989. *Proiectarea liniilor electrice aeriene de foarte inalta tensiune*, Bucharest: Editura Tehnica.

Cortina, R., Cristina, S. 1983. A general method for the analysis of radiointerference measurements on short multiconductor lines, *Trans IEEE Pas 102.no.1*

Dragan, G. 1987. *Cercetari privind instalatiile eelectrice de 750 kV*, Bucharest: Editura Tehnica.

Kircham,H.&Gajda, W.1983.A mathematical model of transmission line audible noise, *Trans IEEE Pas 102.no.3.*

Maddock, B.J. 1998. A summary of standards for human exposure to electric and magnetic fields at power frequency, *Electra, no.179, August*:51-65.

Nekrasov, A.&Rokotean, S.S. 1974. *Dalnie electroperedaci 750kV.Sbornik statei*, Moscow, Energia.

Environmental Issues and Management of Waste in Energy and Mineral Production, Singhal & Mehrotra (eds)
© 2000 Balkema, Rotterdam, ISBN 90 5809 085 X

Assessing air emission impacts on groundwater quality

Lucien S. Lyness
Komex International Limited, Calgary, Alb., Canada

ABSTRACT: This paper identifies the inter-relationships between air emissions and groundwater quality that may need to be evaluated during Environmental Impact Assessment (EIA) preparation. Emissions-to-atmosphere can affect groundwater quality. Precedents documented in the groundwater literature include chlorofluorocarbons, sulphate from SO_2 emissions and tritium from thermonuclear testing. Despite such demonstrated linkages, this mechanism of groundwater contamination has not been extensively studied.

Air emissions have potential to impact several Valued Environmental Components (VECs). Such VECs typically include soils, vegetation, surface waters and aquatic resources. Thus, air emissions that will accompany a proposed industrial development will usually be evaluated in the context of these VECs during EIA preparation. Groundwater is a VEC that is not commonly required to have an evaluation from an air emissions perspective. This paper gives guidelines on how to assess possible air emission impacts to groundwater quality if such an evaluation is a requirement.

1 INTRODUCTION

In many jurisdictions, prior to regulatory approval, a proposed industrial development of any significant magnitude must be preceded by an Environmental Impact Assessment (EIA). Air emissions commonly emanate from industrial operations. That being the case, EIA preparation will typically include detailed evaluations of air emission aspects. Detailed evaluations are necessary because air emissions have genuine potential to impact several Valued Environmental Components (VECs). VECs that are most immediately susceptible to potential air emission impacts are soils, vegetation, surface water bodies and aquatic resources. However, in the context of air emissions, groundwater is a VEC that does not often warrant documentation.

Emissions-to-atmosphere can affect groundwater quality. Precedents reported in the groundwater literature confirm that such influences do indeed take place. The classic example is tritium. Tritium levels in rainfall have been enhanced since 1953 with the advent of thermonuclear testing. Pre- and post-1953 groundwaters can thus often be distinguished simply by comparing their tritium content. Despite such demonstrated linkages between air emissions and groundwater quality, this mechanism of potential groundwater contamination has tended to receive little attention by way of formal study.

Circumstances may dictate that possible linkages between air emissions and groundwater quality be examined during EIA preparation. This paper identifies the inter-relations that are applicable for consideration. Guidelines are given on how to quantitatively assess possible impacts. In addition, to provide a practical example, a case history is described.

2 GROUNDWATER CONTAMINATION SOURCES

There are many possible sources of groundwater contamination. However, some sources have much more potential than others to cause measurable groundwater contamination. Fetter (1993) ranks the potential of a relatively wide range of groundwater contamination sources. The three leading sources in the USA are underground storage tanks, septic tanks and agricultural activities.

Although atmospheric pollutants do not feature in Fetter's ranking exercise, he discusses how they can contaminate groundwater. In short, atmospheric pollutants can reach the ground surface as dry deposition or as dissolved or particulate matter contained in rainfall. Infiltrating rainwater may then convey such substances into the soil and groundwater.

Typical, well-known sources of atmospheric pollutants include automobile emissions, smokestacks and incinerators (Fetter 1993). Pollutants include hydrocarbons, synthetic organic chemicals, natural organic chemicals, heavy metals, and sulphur and nitrogen compounds.

Substances that have the potential to reach groundwater following atmospheric washout, or from ground level leaching, tend to have a relatively high solubility and a low susceptibility to sorption. An example would be methyl tertiary butyl ether (MTBE). MTBE is added to gasoline as an octane enhancer and to help reduce atmospheric contamination. The potential for MTBE to reach groundwater via wet precipitation from atmosphere is discussed by Davidson (1995). For comparison, polychlorinated biphenyls (PCBs) are much less soluble than MTBE (e.g. 0.0027 mg/L (PCB-1260) versus 43,000 mg/L (MTBE)) and much less mobile (e.g. K_{oc} - 349,462 mL/g (PCB-1260) versus 10.23 mL/g (MTBE)). Wet precipitation is not considered to be a significant contributor to elevated MTBE concentrations in groundwater (Davidson 1995).

Leaching of chemicals spilled, deposited or applied to the land surface is a viable mechanism for groundwater contamination to occur. The classic areally-extensive example would perhaps be the application of fertilizers containing nitrogen, phosphorous and potassium (potash). Phosphorous is not very mobile in soil, hence it does not represent much of a threat to groundwater quality. In contrast, nitrogen in the form of nitrate can be a major cause of groundwater contamination (Fetter 1993).

Overall, the concept of chemicals applied to the land surface, whether incidentally spilled, purposely added or indirectly arriving from atmosphere, being leached by infiltrating water to reach groundwater, is therefore valid.

3 MECHANISM

For air emissions to impact quality groundwater, a mechanism or pathway is needed, whereby the fugitive chemical(s) can access the groundwater regime. Examples of groundwater contamination that are most obvious are usually associated with point sources at or below ground level (e.g. leaking underground gasoline storage tanks). However, wider-scale sources are also common, but are comparatively insidious (e.g. leaching of nitrate from fertilized fields). For air-borne contaminants to reach the groundwater environment, they must firstly access the groundwater recharge mechanism.

Groundwater is recharged when meteoric water (i.e. rainfall and/or snowmelt) infiltrates into the subsurface. A contaminant can be assimilated by this water in two ways: atmospheric washout (by wet precipitation); and/or deposition to a ground level substrate followed by leaching. Indeed, that chemicals emitted to atmosphere can have an effect on groundwater quality is widely accepted by hydrogeologists (Brown & Sharp 1992).

Although it is recognized that anthropogenic emissions-to-atmosphere can affect groundwater quality, this topic has actually received little attention by way of study. Nevertheless, there are sufficient precedents in the groundwater literature to confirm that such influences can, and do, take place. Examples include acid rain (e.g. Edmunds & Kinniburgh 1986), sulphate from SO_2 emissions (Robertson et al. 1989), radioactive fallout from thermonuclear testing (e.g. Fetter 1994) and chlorofluorocarbons (e.g. Kazemi et al. 1998).

4 GROUNDWATER RECHARGE

For atmospheric-emission contaminants to enter the groundwater regime, they must somehow firstly gain access to the local recharge mechanism. Identification of possible pathways therefore requires a consideration of "groundwater recharge" as a topic. An authoritative manual on groundwater recharge is provided by IAH (1990). Groundwater recharge may be defined as the downward flow of water into the subsurface, thus eventually reaching the water table and thereby adding to the groundwater reservoir.

Recharge to groundwater may take place naturally. Examples of natural processes include infiltration of precipitation (i.e. rainfall/snowmelt) and losses to subsurface from those rivers and lakes that are in an influent condition. Anthropogenic influences also apply. Examples include irrigation, most notably related to agricultural activities, and infiltration increases and decreases related to industrial and/or urban activity. Infiltration increases can occur as a result of vegetation removal and topsoil disturbance, while decreases can occur as a result of interception of rainfall and runoff by paved areas.

There are two principal types of recharge:

1. Direct Recharge: When a snowmelt or rainfall event occurs, a certain amount of the available water will typically be lost to evaporation and surface runoff. The remaining portion will enter the soil. This portion will be subject to further depletion due to vegetation demands. As well, evapotranspiration processes generate a soil moisture deficit. If sufficient infiltrating water is still available to temporarily overcome the soil moisture deficit, direct vertical percolation of the excess water through the unsaturated zone may occur. If the unsaturated zone has access to an aquifer, direct recharge of the aquifer occurs.

2. Indirect Recharge: Indirect recharge may occur following runoff. Runoff gathers in low-lying areas and lakes, or contributes to stream flow. Seepage to subsurface from these features can provide recharge to groundwater.

Recharge is controlled by meteorological, geological and terrain factors. Estimating a recharge rate requires assessment of each of these factors separately and in tandem. For this reason, recharge rate is probably the most challenging variable to determine in the field of groundwater resources. However, as outlined below, such detailed levels of assessment may not actually be necessary for EIA purposes.

The initial step is to assess recharge potential. Firstly, if it is determined that the study area is in fact a groundwater discharge zone, the possibility of impacts from air emissions can immediately be assessed as an invalid scenario. The presence of flowing artesian wells, for example, would indicate a groundwater discharge zone.

Next, geologic conditions may, for all practical purposes, preclude the possibility of recharge taking place. For example, the groundwater-bearing strata in the study area may simply be overlain by a relatively thick blanket or sequence of low permeability strata. In this scenario, natural recharge and discharge would occur at distant zones well beyond potential influence of the proposed industrial development. Local geology should therefore be assessed to confirm that recharge can in fact occur.

Having determined that conditions are conducive to recharge, the recharge rate can be quantitatively assessed. Guidelines on how to estimate groundwater recharge are given in the groundwater literature (e.g. IAH 1990). For simplicity, annual recharge will typically be a small percentage of annual precipitation. This water will have assimilated a certain amount of the contaminant, and a concentration would be estimated. However, if the study area is the sole recharge zone for the aquifer, then recharge volume may not have much relevance, and a concentration estimate would be sufficient.

5 GROUNDWATER VULNERABILITY

Intuitively, some land areas will be more vulnerable to groundwater contamination than others. For example, on the basis of spatial variation in geologic conditions alone, it follows that there must be some corresponding variation in groundwater vulnerability. An authoritative discussion on the topic of groundwater vulnerability is given by IAH (1994).

There are two key types of groundwater vulnerability:

1. Intrinsic (or natural) Vulnerability: Natural vulnerability is purely a function of an aquifer's inherent characteristics and the characteristics of the overlying natural protective materials. Protective units may include a soil interval, a vadose zone and an interval of low permeability strata.

2. Specific Vulnerability: Specific vulnerability addresses the potential for impact to an aquifer in the context of the particular characteristics of an individual contaminant. In other words, the natural vulnerability of an aquifer will vary depending on its varying attenuation capacity for different chemicals. Thus, a contaminant-specific vulnerability could apply.

The four main natural factors, or attributes, used in assessment of intrinsic groundwater vulnerability are recharge, soil properties, unsaturated zone characteristics and saturated zone characteristics. To assess groundwater vulnerability at a selected location, these attributes or their component parameters can be assigned different weights according to their relative influence on vulnerability potential.

The two main attributes that have a primary direct bearing on assessment of specific vulnerability are style of release and the attenuation susceptibilities of the contaminant. The physical environment typically affords at least some protection against contaminants entering the subsurface and thereby reaching groundwater. Contaminated water entering the subsurface will be purified naturally to some degree during its percolation through the soil and subsoil materials. The degree of attenuation that occurs between source and aquifer decides the relative potential for groundwater contamination. The attenuation or purification capacity of a subsurface interval comprises the interactions of numerous physical, chemical and biological processes in a soil-rock-groundwater system.

Edmunds & Kinniburgh (1986) provide an excellent paper on groundwater vulnerability in the context of acid deposition from atmosphere. Their paper is especially relevant because it adopts geologic and attenuation perspectives. The phenomenon of acid rain is relatively well known and offers some close parallels as to the possible fate of various chemicals following emissions to atmosphere.

Sulphur dioxide (SO_2) and nitrogen oxides (NO_x) are common gases that contribute to acid deposition. For resulting acidic infiltration to persist in the subsurface, the acid neutralizing capacity (ANC) of the soil and subsoil must be low. Ultimately, ANC is largely a function of the surficial and/or bedrock geology. The groundwater itself may have some buffering capacity (i.e. above-neutral pH); however, even that is typically a function of geologic conditions. Thus, the areal variation in groundwater vulnerability to acid deposition can be broadly assessed by simply examining the geology of a given region.

The presence or absence of calcium carbonate is the single most important factor in determining how vulnerable an aquifer is to acidification (Edmunds & Kinniburgh 1986). Likewise, PCB mobility would be similarly related to a single attenuation feature of the geologic matrix. In the case of PCBs, the potential attenuation mechanism is sorption to the geologic matrix rather than the chemical reaction with carbonate associated with acidic infiltration. Thus, for PCBs, the organic carbon content of the geologic matrix is analogous to the relationship between acid and carbonate content.

6 CASE HISTORY

6.1 *Background*

The Swan Hills treatment centre disposes of special wastes, including PCBs. The facility was opened in 1987, and initially dealt with special wastes from within Alberta. Authorization to accept wastes from all Canadian jurisdictions was granted in 1994. PCBs, being a notoriously persistent group of chemicals, are destroyed using high temperature incineration. In 1997, to address incidental-emission concerns, the operators were required under Enforcement Order 97-01 to assess the potential impacts to various VECs. Groundwater was included in this requirement.

There are 209 PCB congeners (Erickson 1997). The key inherent mobility parameters controlling subsurface PCB movement are solubility, K_{oc} (organic carbon/water partition coefficient) and K_{ow} (octanol/water partition coefficient). However, the literature reviewed did not contain a complete listing of values for all 209 compounds. Chou & Griffin (1986) appear to provide a representative selection of solubility, K_{oc} and K_{ow} values for this group of compounds. This information was supplemented with data from Cortes et al. (1991).

6.2 *Conceptualization*

Potential impacts were assessed using a specific-vulnerability approach. In other words, the physico-chemical properties of the contaminant were used to assess its mobility. Natural-vulnerability was assessed, and was conservatively determined to be a valid impact pathway.

Groundwater flow velocity can be estimated from the following equation:

$$v = \frac{Ki}{\eta_e} \qquad (1)$$

where v = groundwater flow velocity; i = hydraulic gradient; η_e = effective porosity; K = hydraulic conductivity.

However, a chemical that is susceptible to sorption will be delayed relative to the groundwater flow velocity. Thus, the velocity of the dissolved chemical (solute) will be governed by:

$$v_c = \frac{v_x}{R} \qquad (2)$$

where v_c = velocity of the solute (simplified); v_x = groundwater flow velocity; R = the delay, or retardation factor.

The retardation factor is determined from the following equation:

$$R = 1 + (\rho_b/\phi)(K_d) \qquad (3)$$

where ρ_b = dry bulk mass density of the matrix; ϕ = volumetric moisture content of the matrix; K_d = distribution coefficient for the solute in that matrix.

The solute's distribution coefficient for the given geologic host medium or matrix can be calculated from:

$$K_d = f_{oc}(K_{oc}) \qquad (4)$$

where f_{oc} = fraction of organic carbon; K_{oc} = soil partition coefficient.

6.3 *Calculation of Travel Time*

The most commonly occurring and most mobile PCB observed from routine monitoring of the various VECs at Swan Hills appears to be PCB 77. A K_{oc} value for PCB 77 was obtained from the literature cited earlier (K_{oc} = 25,633 mL/g). Laboratory analyses of field samples determined a site-specific representative value for f_{oc} (0.009 mg/mg). Using equation (4) above yields a K_d value of 231 mL/g for PCB 77.

Values for ρ_b and ϕ were determined by laboratory analyses of field samples to be 2.0 g/mL and 0.36 (i.e. 36%), respectively. Thus, R was calculated using equation (3) above.

The groundwater flow velocity was calculated using equation (1). A value for vertical hydraulic conductivity was determined in the laboratory using falling-head permeability tests on core samples (K_v = 2.2×10^{-5} m/day). The hydraulic conductivity value carries a conservative bias because unsaturated conditions actually prevail in the field. Vertical hydraulic gradient was selected as i = 1. Effective porosity was assumed to be a conservative value of 0.1 (i.e. 10%). Hence, the vertical groundwater flow velocity was estimated to be 0.08 m/year.

Applying the estimated groundwater flow velocity and calculated retardation factor in equation (2), the solute velocity is $v_{PCB\ 77} = 6.3 \times 10^{-5}$ m/year. Thus, the transit time for PCB 77 to move vertically through one metre of sediment is t_{1m} = 16,000 years. For all practical purposes, the solute is therefore considered to be immobile.

7 CONCLUSIONS

Processes governing atmospheric contamination of groundwater resources comprise a relatively complex blend of inter-related factors. Assessing groundwater recharge, in itself, is typically a challenging task. When subtleties associated with the physico-chemical attenuation susceptibilities of a given chemical are added, the challenge is compounded.

However, a screening level evaluation may be sufficient. For example, hydrogeologic conditions may simply preclude the possibility of recharge occurring. Hence, a screening level review may find that an air–groundwater contamination linkage is invalid. A screening level approach would, in effect, be assessing attributes of natural vulnerability.

If screening of natural attributes were to identify that a possible linkage could apply, then a second level of screening can be implemented. Second-level screening would examine contaminant-specific attributes. For example, if the contaminant itself simply cannot survive in a subsurface environment, then, in terms of specific vulnerability, the linkage would be deemed invalid.

If natural conditions and contaminant properties both support a linkage, then more detailed consideration may be warranted. However, there appears to be sparse guidance in the literature as to how this can actually be accomplished. Therefore, it seems that the linkage between air emissions and groundwater quality would benefit from further research.

ACKNOWLEDGEMENT

The author would like to thank Mr. Graham Latonas, Bovar Waste Management, for permission to publish material obtained during studies relating to the Swan Hills Treatment Centre.

REFERENCES

Brown, T.J. and J.M. Sharp 1992. A Model for the Effects of Point-Source Emission of Aerosols on Groundwater Systems. *Applied Hydrogeology*. 1: 33-46.

Chou, S.F.J. & R.A. Griffin 1986. Solubility and Soil Mobility of Polychlorinated Biphenyls. In J.S. Waid (Ed.). *PCBs and the Environment*. 1: 101-120. Boca Raton, Florida, CRC Press.

Cortes, A., J. Riego, A.B. Paya-Perez & B. Larsen 1991. Soil Sorption of Co-Planar and Non-Planar PCBs. *Toxicological and Environmental Chemistry*. 31-32: 79-86.

Davidson, M. 1995. Fate and Transport of MTBE – the Latest Data. In *Proceedings of the Petroleum Hydrocarbons and Organic Chemicals in Ground Water: Prevention, Detection and Remediation*: 285-301. National Ground Water Association, Dublin, Ohio.

Edmunds, W.M., & D.G. Kinniburgh 1986. The Susceptibility of UK Groundwaters to Acidic Deposition. *Journal of the Geological Society, London*. 143: 707-720.

Erickson, M.D. 1997. *Analytical Chemistry of PCBs* (2nd Edition). CRC, Lewis, Boca Raton.

Fetter, C.W. 1993. *Contaminant Hydrogeology*. MacMillan.

Fetter, C.W. 1994. *Applied Hydrogeology* (3rd Edition). MacMillan..

International Association of Hydrogeologists–IAH (Lerner, D., A. Issar & I. Simmers, Eds.) 1990. Groundwater Recharge: A Guide to Understanding and Estimating Natural Recharge. *International Contributions to Hydrogeology*: 8. Verlag Heinze Heise.

International Association of Hydrogeologists–IAH(Vrba, J., and A. Zaporozec, (Eds.)) 1994. Guidebook on Mapping Groundwater Vulnerability. *International Contributions to Hydrogeology*: 16. Verlag Heinz Heise.

Kazemi, G.A., W.A. Milne-Home & K. Keshwan 1998. Groundwater Age-Dating with Chlorofluorocarbons in the MacQuarie River Basin, New South Wales, Australia: Implications for Dryland Salinity Management. *Proceedings of the Joint IAH/AIH 18th Conference, Las Vegas, Nevada, USA, Sept 28-Oct 2, 1998*: 549-554.

Robertson, W.D., J.A. Cherry & S.L. Schiff 1989. Atmospheric Sulfur Deposition 1959-1985 Inferred from Sulfate in Groundwater. *Water Resources Research:* 25: 1111-1123

Environmental Issues and Management of Waste in Energy and Mineral Production, Singhal & Mehrotra (eds)
 ISBN 90 5809 085 X

Bauxite mining and its effects

P.N. Martens, M. Röhrlich, M. Ruhrberg & M. Mistry
Institute of Mining Engineering I, Aachen University of Technology, Germany

K. Schetelig, C. Bauer & P. Sliwka
Department of Engineering Geology and Hydrogeology, Aachen University of Technology, Germany

ABSTRACT: The sustainable development challenges facing the bauxite mining industry have become a critical concern for mining companies, governments and NGOs. Considering the scale of bauxite mining activities, it is not surprising that such activities have a wide range of social, economic and environmental impacts. Bauxite is mined at more than 70 sites in 25 countries. Most of the deposits are shallow and the mining methods – open pit mining using scrapers, shovels front-end loaders, trucks, etc. – are all very similar. Differences in the thickness of the topsoil, overburden and bauxite layer lead to varying mass flows and land use. A selection of social, economic and environmental effects associated with open pit mining are presented and discussed. The paper focuses on the quantification of social issues related to employment opportunities and effects on the indigenous population. Further emphasis is placed on the classification of bauxite mining economies in terms of income structure. Finally, descriptive indicators to analyse the environmental effects in the context of land use and diversity of ecosystems are presented and put into concrete terms.

1 INTRODUCTION

The identification of options for the resource-sensitive supply and processing of metallic raw materials is the long-term goal of the Collaborative Research Center 525 (CRC 525). The research programme was established in 1997 at the Aachen University of Technology and is funded by the Deutsche Forschungsgemeinschaft (DFG). An integrated resource management system for important metallic raw materials is to be developed and tested by the CRC 525 with respect to the applicability of this system as a framework for providing useful and efficient tools for decision makers by considering technical developments and economic and ecological aims. Ten institutions at the Aachen University of Technology and one programme group from the Forschungszentrum Jülich collaborate in nine sub-programmes (SP) within this project.

The initial three-year phase of the programme (1997-1999) focused on the aluminium flow analysis, extending from bauxite deposits through mining to alumina and aluminium processing and also including the process chains of secondary production and waste disposal.

The Institute of Mining Engineering I (SP 2) is analysing, modelling and balancing the demand for resources at bauxite mining. Besides bauxite, further resources such as energy and environmental media as well as human and financial resources are used or stressed in the mining process. The Department of Engineering Geology and Hydrogeology (SP 5) deals with local, regional and global investigations of the environment. The aluminium anylsis will be completed during the second three-year phase of the programme, beginning in 2000.

2 BAUXITE PRODUCTION

Bauxite is mined throughout the world at more than 70 sites in 25 countries. Major mining production areas are shown in figure 1.

Figure 1. Bauxite-producing countries and sites. The size of the bullets corresponds to the annual production of the region.

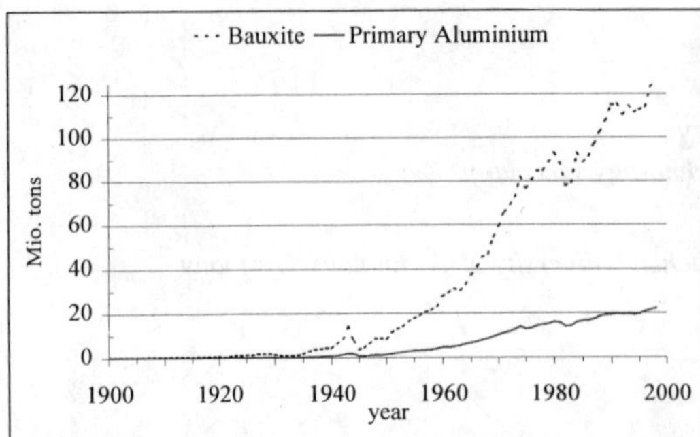

Figure 2. World bauxite and aluminium production since 1900 (Metallgesellschaft 1997; U.S. Geological Survey 2000).

Mining sites in equatorial regions and in the southern hemisphere account for an increasingly large share of production. Major producers are Australia, Guinea, Brazil and Jamaica.

The main purpose of bauxite mining is the production of aluminium. Apart from the aluminium industry, bauxite is also required by the chemical and refractory industry and for the production of abrasives. The demand for aluminium has grown over the last century. Considering the total quantities produced, the increase was quite moderate in the first 40 years. During the following decades, production increased rapidly from less than 1 Mio. tons p.a. to more than 22 Mio. tons p.a.. Since bauxite production is closely linked to the development of the aluminium market, these figures also rose from around 4 Mio. tons in 1940 to 125 Mio. tons in 1998 (Fig. 2).

If one relates the overall mining production of bauxite to the estimated world population of 5.8 billion in 1998 (Worldbank 2000), the consumption of bauxite per capita amounts to 21.6 kg per year. Projecting this figure over an expected 60–year lifetime, every human being will roughly consume 1300 kg of bauxite.

3 BAUXITE EXTRACTION

Most of the bauxite deposits are very shallow and are mined in an open pit. There are also a few underground mines which are not taken into account here because of their small share of world bauxite production. Deposits of lateritic bauxite commonly belong to the blanket type, whereas karst bauxites often appear as bauxite pockets. The overlying strata usually consist of loose soil and uncemented or minor consolidated rock. Thus, most operations do not need any explosives to break up the overlying strata. Some operations have to use rippers to loosen overburden or bauxite before extraction.

Topsoil and overlying strata are removed by scrapers, front-end loaders, hydraulic excavators or draglines. Front-end loaders and hydraulic excavators are also used for bauxite extraction. The in-pit-transport is characterised by trucks. Dozers prepare areas for extraction and reclamation. The biggest share of the equipment is driven by diesel engines. Only a small number of mines also use electric equipment. There is a trend towards an increase in the equipment capacities, leading to lower specific fuel consumption.

In order to analyse bauxite mining, data and information are collated, compiled, and generated in a database. Data from numerous bauxite mines - operative and closed down - has been collected during field-trips and extracted from literature. This data forms the basis for estimations of site-specific values which are inadequately documented in literature. Each data-set comprises information on the identification of the site, geological information (e.g. thickness of overburden and bauxite layer, bauxite density, etc.), data on bauxite quality, production, equipment fleet, fuel consumption, and number of employees. Furthermore, the database contains site-specific information on the stages of the bauxite extraction process.

This database is used to calculate values for land use, energy consumption, labour-input as well as associated material flows of overburden and emissions. Table 1 shows selected figures as a result generated by a database-report.

Table 1: World bauxite production in 1998 and selected cumulative parameters for the production of metallurgical bauxite

total mine production	125	Mio. t
metallurgical bauxite	106	Mio. t
land use	13.7	Mio. m²
overburden mass flow	69	Mio. m³
fuel consumption (diesel)	120	Mio. l
labour force (extraction only)	15,000	workers

All of the aforementioned parameters are part of the stress field of economy, ecology and social aspects and have several effects.

4 EFFECTS OF BAUXITE MINING

Considering the scale of mining activities, it is not surprising that they have a wide range of social, environmental and economic impacts. A selection of potential effects associated with the open pit mining of bauxite is shown in Figure 3.

Impacts of open pit mining have been widely discussed in literature. For further information see (Eggert 1994), (Sengupta 1993), (UNEP 1997), and (Martens 1998). We distinguish between four mining life cycle stages with respect to the origin and nature of effects: exploration, mine development, mining, and mine closure.

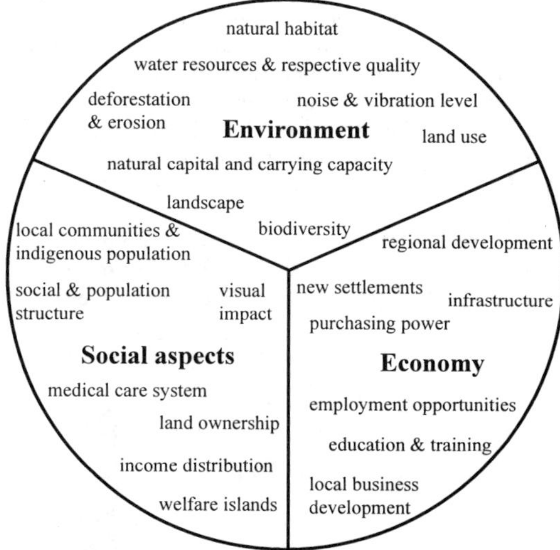

Figure 3: Selection of potential economic, social and environmental aspects affected by bauxite mining.

Before a bauxite deposit can be mined, it has to be identified and its economic and technical viability demonstrated. The exploration phase causes almost negligible environmental damage due to drilling, grab and bulk sampling and pilot plant operation.

Once a workable bauxite deposit has been identified, mining and processing facilities as well as essential supply systems and an infrastructure have to be set up. The mine development phase is characterised by construction activities which usually cause a significant change in the landscape, ranging from a mere visual impact through to severe deforestation and the destruction of natural habitats. The environmental impacts of mine development are more relevant than those arising from the previous life cycle stage but much less than those related to bauxite extraction.

Strictly speaking, activities related to bauxite mining following mine development can affect all three environmental media - land, water and air. In general, "open pit operations produce far more waste per tonne of ore than underground operations, where there is no overburden and where some of the removed material can be used to backfill excavations as work progresses" (UNEP 1997). This does not necessarily hold true for open pit bauxite mining due to the shallowness of ore layers, resulting in less overburden and enabling successive excavation and reclamation sequences. Since neither acid rock drainage or potentially hazardous tailings nor waste material occur, the severest environmental problems related to bauxite mining usually arise from land degradation. Excavation, extraction and waste disposal can lead to substantial soil degradation, deforestation and destruction of wildlife habitats. Fortunately, most ground water levels are below the basic level of superficial bauxite deposits. Consequently, a lowering of the water table is not a major impact question for open pit operations. Water effluents from mine operation include drainage, wastewater from bauxite washing and surface run-off, carrying suspended solids. Significant dust emissions arise from road transport, ore processing, and wind erosion from uncovered top-soils unless appropriate measures are implemented. Gaseous air emissions are mainly caused by fuel combustion.

Due to the non-renewable nature of mineral deposits, mines have a finite life span. The mine closure and decommissioning stage directly follows the cessation of mining. The objective of the rehabilitation and reclamation phase consists in creating a productive and sustainable post-mining land use for the site. Simultaneously, a number of goals related to the protection of public health and safety and to the minimisation of the environmental impact associated with seepage and drainage must be pursued. Pits and waste piles have their slopes stabilised and may be revegetated. In addition, plant growth and water quality have to be monitored. The future land use potential is usually dependent upon topography, drainage, vegetative species mixes and surface texture. Furthermore, the success of the reclamation efforts depends on precipitation patterns and physical and chemical properties of surface and near-surface layers.

Apart from environmental impacts, mining activities also have economic and social effects.

After successful exploration, changes in land use and land ownership may restrict former access to natural resources for local communities. In this context, a participation of all interested parties is desirable during the process of social and environmental compliance prior to mining.

During mine development, the construction of facilities, roads, settlements, and energy and freshwater supply systems may lead to a noticeable regional development in terms of improved living conditions, including the establishment of medical care and education centres. In some cases, a relocation of indigenous communities will be inevitable. Further economic and social consequences of mining arise from direct employment opportunities and the following stimulation of local business development due to the increased purchasing power of the local population. The distribution of the economic benefits of mining changes the income distribution and social structure of the surrounding settlements. Education and training as well as improved living conditions may contribute to a profound alteration of cultural values and former life styles. Furthermore, indigenous people are very sensitive to any disturbance of their local environment. The migration of workers and others taking advantage employment opportunities directly or indirectly related to mining may significantly affect the former local communities.

When mining ceases, a serious economic decline in the area is inevitable unless alternative employment opportunities are available for the redundant mine workers and the related support services. Otherwise, the site and its surroundings dispose of facilities, buildings and infrastructure that may be useful to other industries or business forms. Alternative investment and re-employment programmes stimulated by the mining companies and government can help create new employment opportunities and thus alleviate the adverse socio-economic impacts of mine closure.
In the following, selected social, economic and environmental impacts will be presented and discussed from a global point of view.

5 DISCUSSION OF SELECTED IMPACTS

Social, economic and environmental effects are strongly interdependent and should not be seen as strictly separate problems.

An environmental information system has been developed by the SP 5 of the CRC 525 to quantify the global effects and impacts of bauxite mining in a global material flow study (Bauer & Sliwka, 1998). Spatial data on environmental properties has been collected from across the globe based on a geographic information system. In combination with the bauxite mining database, site specific indicators are quantified which can be used to discuss selected effects of bauxite mining. Unlike former panel-based studies (Atkins, 1993), which gathered a high information content for a sub-set of sites, this approach enables the consideration of almost all sites. This system had already been used to characterise bauxite deposits on a global scale (Hausberg et al. 1999). Core elements of the spatial database are small scale survey data on soils, land cover characteristics, hydrology, morphology and population densities. These are important key indicators not only for the consideration of social and environmental impacts but also to discuss endpoints for a variety of impact pathways.

5.1 *Social Impacts*

5.1.1 *Employment Opportunities*

World-wide bauxite mining operations provide approximately 15,000 direct employment opportunities. In terms of productivity, the average employee produces roughly 4.7 t/h of bauxite. The productivity varies significantly from site to site. Local productivity depends on various site-specific factors, e.g. operating performance, bauxite quality and transport distance between pit and processing plant. Thus, no evaluation of local or regional productivity can be carried out without a further analysis of site-related economic and technological characteristics.

A macroeconomic analysis of the employment structure may be carried out based upon the definition of small and medium-sized enterprises recommended by the European Commission (European Commission 1996):

- small enterprises: <50 employees
- medium enterprises: 50 – 249 employees
- large enterprises: >249 employees

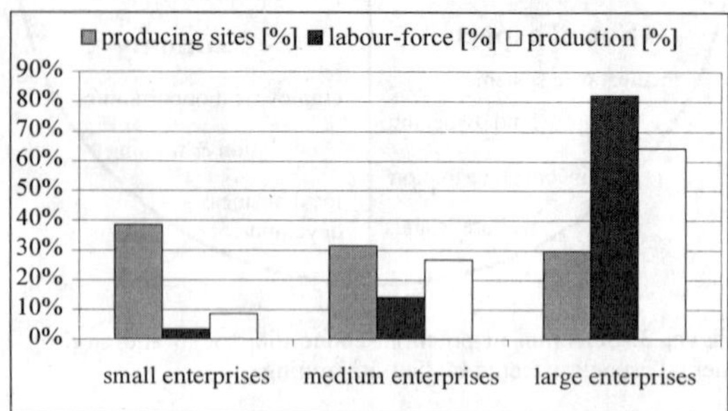

Figure 4: Bauxite mines, production and labour-force classified according to size of enterprise.

More than 80% of employees work in large enterprises which represent two thirds of the total production. 70% of mining sites claim to be small or medium-sized firms employing less than 20% of the total labour-force (Fig. 4).

5.1.2 *Population and Indigenous People*

Besides the provision of jobs, mining has further social effects on the local population, e.g. through resettlements or changes of land cover. At peripheral sites with low population densities, company towns are built whose inhabitants are unfamiliar with the local environment, living conditions, and cultural structures of the original social community. These sites often attract itinerant workers and others who take advantage of the developed infrastructure and considerably enhance the percentage of foreigners in the vicinity of the site. Indigenous people in particular, e.g. Indians in South America or Aborigines in Australia, whose culture is closely related to the stability of their local environment, are very sensitive to any disturbance. Indigenous people are protected by several international and national conventions which have to be considered in all stages of mining projects.

Sites in more densely populated areas are not likely to interfere with undisturbed ecosystems. In these cases, important effects are resettlements and concurrent utilisation of local resources (roads, land, water).

Available data on population density and distribution on a global scale provide no information on the presence or absence of indigenous people in the

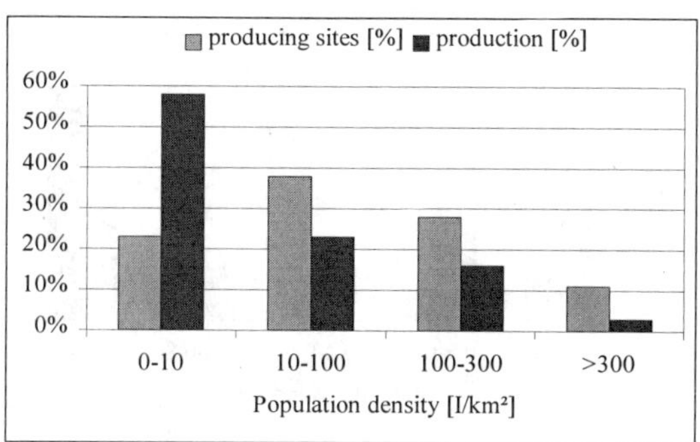

Figure 5: Bauxite mines and production according to population density.

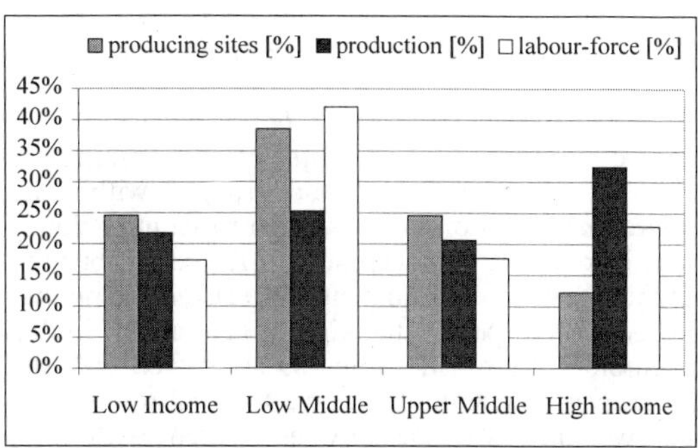

Figure 6: Bauxite mines, production and labour-force according to income structure of economies (GNP per capita).

vicinity of a particular site, but it is almost certain that the probability of their presence increases with a decrease in population density. This relation was proved in field studies conducted by the CRC 525 at bauxite mines in Australia, Brazil and Venezuela.

Population density is hereby used as a key indicator to differentiate between social effects of bauxite mining. In Figure 5, production sites and the proportion of the worlds annual production are according to population density, quoted in inhabitants per km² [I/km²].

Population densities of 0 to 10 inhabitants per km² were chosen to indicate social effects on indigenous people. Population densities of more than 300 inhabitants per km² indicate social effects by concurrent utilisation of local resources. Although 58% of the world's bauxite production takes place at peripheral sites, at only 23% of the current mining enterprises is an interference with indigenous people probable. Higher population densities (>100 I/km²) indicate the presence of established rural or urban structures at the majority of producing sites. Their contribution to the global bauxite market amounts to around 20%.

5.2 *Economic Impacts*

A significant, country-related economic impact of mining may result from the export of bauxite to other economies. The value added by alumina and aluminium production in almost every bauxite mining country normally has to be considered. However, in order to estimate the potential contribution of bauxite, we have related total bauxite production and bauxite price to the total export of commodities. Significantly different shares result for the major producers of bauxite in 1997, e.g. Guinea >47%, Guyana >18%, Jamaica >17%, Australia and Brazil <1%. Note that the price of bauxite varies greatly from region to region, due not least to bauxite quality, ore grade, productivity, transport, and mining technology. The low income economies of Guinea and Guyana in particular are very dependent on bauxite exports and resulting revenues, whereas more developed and diversified economies such as Australia and Brazil are expected to be less vulnerable to changes in market structure.

For operational and analytical purposes, the World Bank's main criterion for classifying economies is gross national product (GNP) per capita. "GNP is the sum of gross value added by all resident producers plus any taxes (less subsidies) that are not included in the valuation of output plus net receipts of income from abroad" (Worldbank 2000). GNP per capita is gross national product divided by midyear population and is converted using the World Bank Atlas Method. Every economy is classified as low income (GNP $760 or less), lower middle income (GNP $761- $3,030), upper middle income (GNP $3,031- $9,360), and high income (GNP $9,361 or more) (Worldbank 2000).

A classification of producing sites, production and workforce is shown in Figure 6. One third of bauxite production takes place at 12% of mining sites in high income economies, providing employment opportunities for almost a quarter of the total labourforce.

Two thirds of sites operate in low and low middle income economies and account for less than 50% of total bauxite production. Less than 20% of the total bauxite-related labour-force is employed at mining sites in low income economies. In terms of production bauxite mining claims equal importance in all kinds of economies.

5.3 *Environmental Impacts*

The main environmental effects of bauxite open pit mining relate to the use of land. Not only the cleaning of the surface in the pit but also the construction of roads and stockpiles constitute disruptions to the local ecosystem. The restructuring of drainage systems or fragmentation of ecosystems by transportation networks may cause negative effects far beyond the mining site. The extent of these impacts depends on the land cover characteristics on the site and its vicinity which are altered by the mining process.

5.3.1 *Dominant Land Cover Characteristics*
Land cover types in the vicinity of each mining activity have been collected and grouped within the environmental information system to identify dominant land cover characteristics affected by bauxite mining. Several environmental effects are being discussed with respect to the kind of land cover. In this discussion, the world's forest is associated with the global warming and air quality. Tropical rain forests and wetlands are specially associated with high biodiversity. Agricultural land was classified because of a potential concurrent use of the surrounding population for a limited period of time. The group "others" is used to summarise remaining land cover types such as tundra, deserts, grassland, etc (figure 7). The specific percentages of each group at all sites were summed up. Figure 7 shows the overall percentage of each group scaled by site and land use. The relations between site percentages and land use percentages within each group can be explained by large differences in the operation size. Considering that a single site can be surrounded by different land cover types, at 50% of the sites mining takes place on agricultural affected land. In terms of the total annual land use of 13.7 km² (calculated by geometric deposit properties and annual production), agriculture accounts for only 25%. These are mostly small mines in arable areas (e.g. China, India).
If one now considers forest cover, and especially tropical rain forest, it can be seen that only a few sites with a greater share of annual land use are dominated by this ecosystem. Wetlands are, as expected, marginal in this discussion.

5.3.2 *Naturalness Indicator*
In order to analyse the diversity of ecosystems affected by mining operations, an indicator was developed to measure the naturalness in the vicinity of each site.

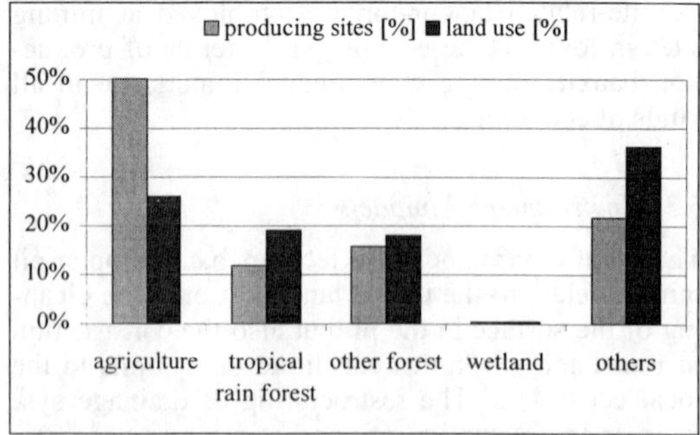

Figure 7: Bauxite mines and production according to affected ecosystem.

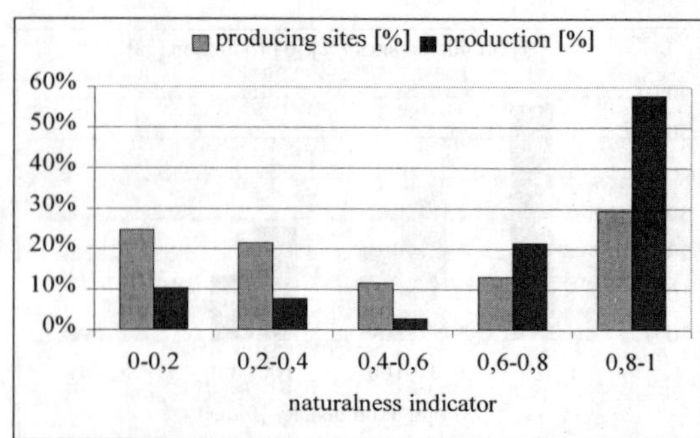

Figure 8: Bauxite mines and production according to naturalness indicator.

The naturalness indicator [N_g = 0-1] is based on the ratio between the area of land cover characteristics unaffected by humans and total area in a unit circle with a 50 km radius of the location. A naturalness indicator level of 1 for a location means that the vicinity has not been affected by human activities so far. A value of 0 indicates a strong manmade alteration.

Figure 8 shows the percentage of current mining sites and their shares of world production according to the naturalness indicator. Approximately 30% of the sites are situated in peripheral regions with very low previous human impact (N_g = 0.8-1). These sites produce 58% of the total annual bauxite production. The application of this indicator is manifold for any discussion of the effects related to bauxite mining. On the one hand, it can be seen that small mines are generally situated in regions where human activities have already altered the environment. These areas are not supposed to be specially sensitive towards additional impacts atttributed to the mining activity. On the other hand, nearly 80% of bauxite mined in natural and as yet undisturbed ecosystems (N_g > 0.6) by 43% of the mines. These predominantly major producers are expected to have sufficient human and financial resources as well as organisational structures and technical equipment at their disposal to cope with the environmental, economic, and social challenges related to bauxite mining.

6 CONCLUSION

The effects of bauxite mining have become a critical concern for mining companies, governments and other stakeholders. Environmental permits and regulations as well as social commitment are important to bauxite mining companies because they increase time, costs, and risks associated with bringing a mine into production. There is growing evidence of the further need to study and disseminate the positive and negative effects of bauxite mining. Apart from a mere compilation of more relevant data, existing concepts and indicators will have to be improved and new approaches and strategies developed in order to promote technical change and foster economic and environmental efficiency.

There is still considerable research to be undertaken to quantify the nature and extent of external effects of bauxite mining. However, as a general trend changes in industrial and governmental policy can already be observed towards more environmentally and socially sound mining operations.

As shown above, bauxite mining activities in socially and environmentally sensitive areas are mostly carried out by large companies due to the complex and challenging character of mining operations in remote and less developed areas. Many of the potentially negative impacts can be prevented or diminished by adopting appropriate measures and management systems. Others, such as habitat destruction or land degradation, can only be dealt with after exploitation by means of rehabilitation and reclamation.

Most of the global players in bauxite and aluminium are committed to meeting the challenges of continual environmental improvement and have already introduced environmental management systems. Moreover, there is a shift towards the adoption and compliance of uniform international technical and environmental standards for production sites all over the world. It is, however, inevitable that further policy and science directions be specified to address the concerns through partnerships with companies, governments and stakeholders.

"One of the most powerful ways mining companies and their associations can contribute to environmental performance is to assist in the development of more effective technologies" (Miller 1997). Technological and organisational innovation offers a potential escape from the oft-presumed trade-off between economic growth and environmental quality.

The sustainable development challenges facing the minerals and metals industry require a comprehensive, integrated and multidisciplinary approach based on shared decision-making and a reliable information base, as well as close co-ordination and co-operration of all interested parties. The integrated approach of the CRC 525 may offer an opportunity to address and cope with some of these challenges by supporting sustainable development-based decision-making.

REFERENCES

Atkins, P.R.: Bauxite mining–a world wide environmental study. *Light Metals*, 20, p. 5-9, 1994.

Bauer, C.; Sliwka P: Estimation of site-specific parameters to describe the potential environmental impacts of material flows of metallic raw materials. In *Risk Analysis. WIT press,* ed. by Rubio J.L., Brebbia C. A., Uso J.-L., p. 307-315, Southampton 1998.

Commission of the European Communities: Commission Recommendation of 3rd April 1996 concerning the definition of small and medium-sized enterprises. C(96) 261 final, Brussels, 1996.

Eggert, R. (ed.): Mining and the Environment – International Perspectives on Public Policy. Resources for the Future, Washington, 1994.

Hausberg, J.; Bauer, C.; Sliwka, P.; Hahn, C.C.: Environmental aspects of bauxite supply in global material flow studies. Proceedings of the MPES & MEEI 99 - International Symposium on Mine Planning and Equipment Selection & International Symposium on Mine Environmental and Economical Issues, Dnipropetrovsk, Ukraine, ISBN 966-7476-12-X, S. 161-167, 1999.

Martens, P.N.; Koch, H.; Röhrlich, M.; Willmen, T.; Schetelig, K.; Bauer, C.; Sliwka, P.: Resource orientated view on bauxite mining by means of a process chain analysis. MPES'98: Seventh International Symposium on Mine Planning & Equipment Selection Calgary, Canada, 1998.

Metallgesellschaft (ed.) 1997. Metallstatistik - Metal Statistics 1986-1996

Miller, G.: Mining and sustainable development: environmental policies and programmes of mining industry associations. In: UNEP publication: Industry and Environment, Volume 20, N°4, October – December 1997, p.14-17

Sengupta, M.: Environmental impacts of mining – monitoring, restauration, and control. CRC Press LLC, Boca Raton, Florida 1993.

UNEP: Mining facts and figures. In: Industry and Environment, Volume 20, N°4, October – December 1997, p.4-9.

U.S. Geological Survey Minerals Information: Statistical Compendium - Bauxite and Alumina [http://minerals.usgs.gov/minerals/pubs/commodity/bauxite/index.html]. January 2000

Worldbank Group: Development data by topic. [http://www.worldbank.org/data/databytopic], January 2000.

Environmental Issues and Management of Waste in Energy and Mineral Production, Singhal & Mehrotra (eds)
© 2000 Balkema, Rotterdam, ISBN 90 5809 085 X

The influence of fluvial geomorphology and particle size on health risk

R. Perona, M. Tardiff, D. Michael & T. McFarland
Neptune and Company, Los Alamos, N.Mex., USA

ABSTRACT: Risk assessments involving contaminated sediments sometimes neglect the influence that geomorphological processes may have on the spatial distribution of contamination. In the arid southwest of the United States, sediment particles in ephemeral stream systems are sorted by water flow into a mosaic of distinct geomorphological features, often with different characteristic particle size fractions. Contaminant concentrations, as well as intake via a variety of exposure pathways, may be correlated with particle size. In a recent investigation of a stream system impacted by mining operations, a sampling design based on fluvial geomorphology was used to collect data for risk assessment purposes. Chemical concentration data for different particle size fractions were obtained within geomorphological features over several miles of an ephemeral stream. Several exposure pathways were employed to evaluate the potential significance of particle size and geomorphological sorting on risk estimates.

1 INTRODUCTION

In order to perform a credible risk assessment for exposure in a dynamic physical environment, one must first demonstrate that the sample data used in the risk calculations capture the spatial heterogeneity of contamination in the environment. However, because one data use is calculation of health risk, the data must also conform to an appropriate model of contaminant exposure and intake. Too often, sampling designs emphasize support of either a physical site model or exposure model and neglect the other. In situations where models of the physical redistribution of contamination and exposure are each potentially quite complex, the sampling design should be based on both models.

In addition to current contaminant spatial heterogeneity and exposure conditions, one may also need to consider changes over time. Information related to exposure conditions and contaminant concentrations support calculation of chronic health risk that pertain to periods of several years or longer. Predicting changes in contaminant concentrations with time in a dynamic physical system is also critical for devising effective and lasting remedial strategies.

The attributes of the site inferred above are a dynamic physical model (*i.e.* contamination is episodically redistributed in space and time by site processes) and a complex exposure model (*i.e.* the types and intensity of land use activities and associated exposure pathways are potentially numerous and uncertain). An optimum sampling design for a risk assessment at such a site cannot be generalized; data needs must be systematically defined for a specific site based on the needs of the risk manager and the physical/exposure characteristics of that site.

The site that is the subject of this paper is a portion of ephemeral stream in the arid southwestern United States that has been impacted by releases of metals associated with a variety of mining-related operations and events. Metals have entered the stream system in both dissolved and particulate form at several distinct locations and at various times along the affected portion of the basin. Location, intensity, and duration of rain events in the basin influence the redistribution of contaminated sediments in unpredictable ways. Land use conditions across the site include rural residential, recreational, rangeland for cattle, and restricted access areas.

A preliminary physical site model and exposure model were developed prior to consideration of the sediment sampling plan. The preliminary physical site model identified three common geomorphic features within the stream system; overbanks, bars, and active channel. It was recognized that these features were likely to contain different proportions of relatively fine- and coarse-grained sediments. Studies of contaminant concentration with grain size in a similar environment (Nyhan *et al* 1976) have shown that contaminant concentrations may be significantly higher on fine-grained particles. In the preliminary exposure model, different particle size fractions were associated with specific routes of intake. Exposure scenarios may also differ among geomorphic feature. For example, only one of the geomorphic features (overbanks) hosts residences. A sampling design accounting for geomorphic features and particle size was developed for supporting evaluation of both the physical and exposure models associated with the site.

2 RISK ASSESSMENT FRAMEWORK

One aspect of the sampling design, collecting contaminant concentration data for multiple particle size fractions, is not common practice in the United States for evaluating risk from soil and sediment. However, EPA's Risk Assessment Guidance for Superfund (EPA 1989) lists particle size among those soil and sediment parameters for which information may be required. The two primary drivers for environmental remediation of industrial sites in the U.S. have different definitions of what particle size fraction constitutes a soil sample. The Superfund program defines soil as having a particle diameter less than 2 mm, while the Resource Conservation and Recovery Act (RCRA) specifies 9 mm as an upper bound on particle size (EPA 1996). Practically, though, soil samples are generally sieved in the analytical laboratory through a 2 mm screen. The value of 2 mm originates from USDA definitions of soil particle size; 2 mm is the size where coarse sand is differentiated from gravel. However, the size fraction of less than 2 mm has no basis in exposure assessment.

An exception to usual focus on 2 mm size fractions in environmental investigations supporting human health risk assessments of soils and sediments is in the investigation of soil lead levels. Wixson & Davies (1993) have summarized several studies describing the correlation of particle size to lead bioavailability. However, these studies relate specifically to the bioavailability of lead in the gastrointestinal tract rather than intake of particulates into the body via ingestion. Although the relationship of particle size to uptake is important, it is an issue that arises subsequent to identifying the size fraction associated with exposure.

The first question that might be asked regarding differentiating particle size fractions for specific exposure pathways is, "Does it matter"? The significant impact of increasing contaminant concentrations on finer particles for human exposure via soil adhering on the skin has been described previously by Sheppard & Evenden (1994). A second concern might be whether the introduction of yet another variable in a risk assessment will further undermine the semi-quantitative risk estimation process and limit it's utility to support remedial decisions. In this case, the affected portion of the stream is large and supports many possible exposure pathways and a variety of ways in which one might combine the environmental data to calculate exposure concentrations.

The defensibility of a risk assessment for a complex site can benefit from a more sophisticated approach to sediment sampling and exposure assessment. Rather than attempt to define exposure areas prior to evaluating field data, for example, we ask the question, "Does the assumed size and location of an exposure area, within reasonable bounds, affect risk-based decisions"? This approach embraces the fact that an exposure assessment supporting the calculation of prospective risk is inherently subjective and there is therefore no "correct" or "true" risk estimate awaiting discovery.

3 TECHNICAL APPROACH

The length of the potentially affected stream system is approximately 17 miles. Stratification of the entire stream system for collecting sediment samples involved several variables including the location of historic sources of releases and changes in topography along the stream length. Furthermore, the particle affinities of the contaminants and the depositional dynamics of the basin were included in the sampling design. Metals concentrations were collected relative to particle sizes and geomorphic features of the system were classified into channel, bars and overbanks. For the purpose of describing the influence of geomorphology and particle size on risk, the overall complexity of the problem can be reduced by considering sediment data from relatively localized areas defined by physical and exposure model characteristics.

3.1 *Hypotheses*

EPA's Data Quality Objectives process (EPA 1994) was employed as a framework for identifying key gaps in existing site knowledge and the inputs necessary to support interim and final site decisions. The relative lack of *a priori* information regarding contaminant distribution in stream sediments resulted in a planned two-phase approach to data collection. In the first phase, data were collected to evaluate a series of hypotheses relating to the spatial distribution of historic contamination in sediments.

Hypotheses specifically related to the topic of this paper include:

1. Different geomorphological features will contain different proportions of each particle size fraction, and;
2. Contaminant concentrations will increase with decreasing particle size fraction within each geomorphological feature.

Our initial assumption is that metal concentration is related to particle size because the surface area per unit volume of sediment increases with a decrease in (spherical) particle size. We also expect that overbanks will contain proportionally more fine-grained particles than either bars or the active channel.

Sample support during Phase I was intended to allow primarily subjective evaluation of these hypotheses. Data were collected within three particle size fractions; 0-63 μm, 63-250 μm, and 250-2000 μm. The data evaluation began with identifying the likely risk drivers within the analytical suite using screening-level soil criteria, followed by exploratory data analysis involving plotting and statistical tests for correlations and trends.

The risk assessment hypothesis relating to the two hypotheses on particle size, geomorphic feature, and concentrations is simply that these factors will have a significant impact on calculated risk.

3.2 *Exposure Models for Particle Intake*

The exposure model for this site accommodates a variety of exposure pathways within residential, recreational, and grazing land use scenarios. In this paper, we focus on exposure via three pathways; soil ingestion, dust inhalation, and ingestion of garden produce.

There are two obvious mechanisms by which soil particles may be ingested; direct hand-to-mouth contact and swallowing of particles trapped in the nose, mouth, larynx, and pharynx during inhalation. For hand-to-mouth contact, research suggests that the upper limit of particle size adhering to the hands is on the order of a few hundred microns (Kissel *et al* 1996; Duggan and Inskip 1985). The trap-and-swallow mechanism is limited primarily to particles with diameters below approximately 50 μm; larger particles are quickly settled due to gravity and so are available for inhalation only under very windy conditions. We adopted the conclusion of Duggan and Inskip (1985) that hand-to-mouth contact is likely to dominate in soil exposure and selected 250 μm as the cut-off particle size relating to soil ingestion.

Because of it's importance in occupational exposure, considerably more information is available on the respirable (*i.e.*, penetration past the terminal bronchioles of the lung) particle size fraction. Although the respirable size fraction differs somewhat depending on inhalation rate and whether inhalation is predominantly via the nose or mouth, a cutoff value of 10 μm for particle diameter (PM_{10}) has been traditionally applied for ambient air sampling and is the basis of EPA standards for airborne particulates. During Phase I sampling, data were obtained for the smallest size fraction that can be easily sorted by sieve (63 μm). Concentration data associated with this size fraction are used for calculation of risk via dust inhalation.

The appropriate particle size fraction associated with plant uptake is uncertain. It is reasonable to assume that as soils dry, water is lost preferentially from the larger soil micropores. Nutrient levels may also be higher in the smaller micopores where soil particle surface area is higher. The influence of these factors is uncertain, however, and variables such as soil moisture content, soil type, and particle size analyzed may also be expected to vary among the studies that were reviewed for recommendation of soil-to-plant uptake factors (Baes *et al* 1984).

A summary of potentially complete exposure pathways for each of the particle size fractions collected in the investigation is provided in Table 1.

Table 1. Summary of exposure pathway by particle size.

Particle Size Fraction (μm)	Soil Ing.	Inhalation	Plant Ing.
0 – 63 μm	yes	yes	yes
63 – 250 μm	yes	no	maybe
250 – 2000 μm	no	no	maybe

3.3 *Intake Equations and Parameter Values*

The intake equations for soil ingestion, dust inhalation, and ingestion of produce are provided as Equations 1, 2, and 3, respectively. Equations and parameter values pertain to an adult receptor and residential land use conditions.

Soil Ingestion:

$$\text{Intake} = \frac{C \times IR \times EF \times ED}{BW \times AT} \quad (1)$$

where, intake = chronic daily chemical intake (mg / kg body weight / day), C = chemical concentration (mg / kg soil), IR = soil ingestion rate, EF = exposure frequency, ED = exposure duration, BW = body weight, and AT= time over which exposure is averaged for experiencing adverse effect.

Dust Inhalation:

$$\text{Intake} = \frac{C \times InhR \times DL \times ET \times EF \times ED}{BW \times AT} \quad (2)$$

where, InhR = inhalation rate, DL = atmospheric dust loading, and ET = exposure time.

Ingestion of Garden Produce:

$$\text{Intake} = \frac{C \times K_{p\text{-}s} \times IR_p \times EF \times ED \times CF}{AT} \quad (3)$$

where, $K_{p\text{-}s}$ = plant uptake factor, IR_p = produce ingestion rate, and CF is a conversion factor of 0.001 kg / g.

To obtain a hazard quotient for noncancer effects, intake is divided by a reference dose (mg/kg–d). To calculate cancer risk, intake is multiplied by a slope factor $(mg/kg–d)^{-1}$. Toxicity values were obtained from EPA's Integrated Risk Information System (1999), with the exception of a site-specific inhalation reference dose of 0.0044 mg/kg–d for copper. No inhalation toxicity value is available for zinc. The parameter values used for the calculations described in Section 4 of this paper are provided in Table 2.

Table 2. Parameter values for risk calculations.

Parameter	Value
IR	50 mg / day
EF	350 day / year
ED	30 year
BW	71.8 kg
AT (carcinogen)	75 year x 365 day
AT (noncarcinogen)	= ED x 365 day
InhR	0.63 m^3 / hour
DL	1E-07 kg / m^3
ET	20 hour / day
K_{p-s}	Chemical specific (mg / kg plant per mg / kg soil)
IR_p	0.9 g / kg BW – day

Chemical specific values for plant uptake factors were obtained from Figure 2.2 of Baes *et al* (1984).

4 RESULTS

The Phase I sampling results support the conclusion that both hypotheses described in Section 3.1 are correct. Figure 1 shows the relative proportion of particles in three size fractions in each of the three geomorphic units. In this figure it can be seen that the percent mass of the 250-2000 μm size fraction remains relatively constant among the geomorphic features, while the mass of fine particles is proportionally larger in overbanks than in the stream channel.

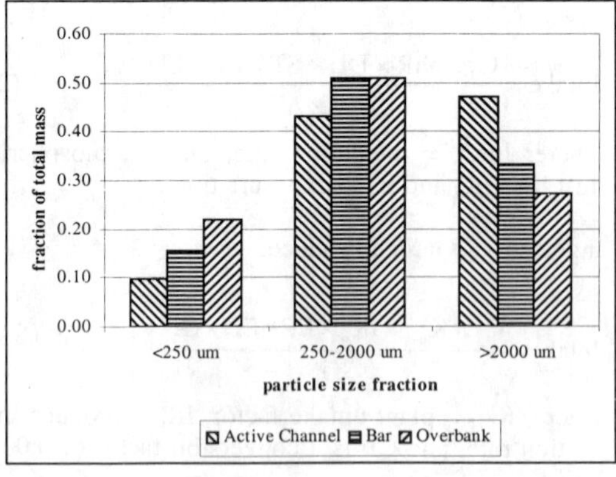

Figure 1. Particle size fraction distribution among geomorphic features.

From Figure 1 we may infer that exposure is likely to be of greatest concern for overbank sediments. Although not included in the figure, the average percent mass for the 0-63 μm particle size fraction (collected only in overbanks) was 4.6 percent of the total sample mass, or about 15% of the 0-250 μm overbank fraction.

The effect of particle size on contaminant concentration is shown in Table 3 for three metals of concern at this site; arsenic, copper, and zinc. The effect of increased concentration with smaller particle size fraction is consistent across all three metals and was observed in virtually all other metals that were analyzed. Although Table 3 was created using only overbank samples, the same relationship was observed in the other geomorphic features using 0-250 μm and 250-2000 μm size fractions.

Table 3. Change in average contaminant concentration (mg/kg) with particle size fraction (μm).

Chemical	0-63	63-250	250-2000
arsenic	5.4	2.9	2.4
copper	1470	930	870
zinc	950	630	540

As described in Section 3.1, based solely on surface area one expects that a halving of particle diameter produces a doubling in the concentration of adsorbed metals. The Phase I data for arsenic, copper, and zinc were analyzed to determine the actual relationships among particle size fractions and metal concentrations using regression analysis. The metals concentration data for smallest particle size fraction (0-63 μm) were regressed against the metals concentration data for each of the other fractions, (63-250 μm, 250-2000 μm, and composite 0-2000 μm). The linear model fit, as measured by r^2, ranged from 0.78 to 0.95.

Figure 2 presents the concentration ratios by metal and size fraction. The figure also shows the median percent of the total sample for each of the size fractions. For example, the median contributions of the 0-63 μm, 63-250 μm and 250-2000 μm fractions to the total sample mass are 4.6%, 30.5%, and 64.2%, respectively. For arsenic, the concentration in the 63-250 μm fraction is 0.43 of the 0-63 μm fraction. The 250-2000 μm fraction is 0.24 of the 0-63 μm fraction. The composite 0-2000 μm fraction is 0.33 of the 0-63μm fraction.

4.1 *Effect of Particle Size on Risk*

Because risk is linearly related to concentration, the relative risk associated with the use of size-fraction-specific concentration data, for soil ingestion, dust inhalation, and plant ingestion pathways, is simply based on the appropriate size fraction for each pathway (Table 1) and the relative concentrations for each metal by size fraction (Table 3). Thus, the risk from copper via soil ingestion is 1.6 times greater for particles of 0-63 μm size than for those 63-250 μm in size.

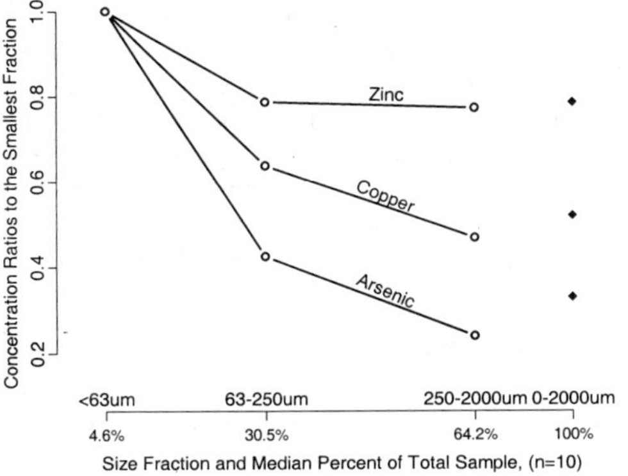

Figure 2. Metals concentrations relative to particle size and percent of total sample.

An example of the potential impact that use of particle size specific data can have within a multi-pathway assessment is also provided. A soil concentration of 10 mg/kg was assigned to the 0-63 μm size fraction of each metal, and the concentrations associated with the other size fractions were scaled according to the information in Figure 2. These concentrations, shown in Table 4, were used as input to the risk equations shown in Section 3.2.

Table 4. Relative concentrations of metals among particle size fractions.

Chemical	0-63	63-250	0-2000
arsenic	10	4.3	3.3
copper	10	6.4	5.2
zinc	10	7.9	7.9

The results of the risk assessment showing the collective impact of using pathway-specific particle size fraction data for copper, arsenic, and zinc are shown in Table 5. In this example, exposure to only the 0-63 μm size fraction was assumed for dust inhalation. Exposure to both the 0-63 and 63-250 μm size fraction was assumed for soil ingestion. Similarly, it was assumed that plants absorbed dissolved contaminants from pore spaces associated with the 0-63 and 63-250 μm size fractions. The incremental cancer risks (ICRs) and hazard quotients (HQs) associated with the use of the traditional 0-2 mm size fraction concentrations for all exposure pathways are also shown for comparison.

Because soil adherence on the skin is higher for finer particulates within the 0-250 μm size fraction, exposure to the 0-63 μm size fraction was weighted twice that of the 63-250 μm size fraction. This type of weighting is consistent with the findings of Sheppard & Evenden (1994) who observed that particles greater than approximately 50 μm in diameter did not adhere well to dry skin. We also weighted these size fractions in this same manner for the plant uptake pathway, for reasons discussed in Section 3.2.

Table 5. Comparison of risk assessment results: pathway-specific particle size fractions vs 0-2 mm size fraction.

Pathway	0 – 2 mm		Pathway-Specific	
	ICR	HQ	ICR	HQ
soil ing.	1.1E-06	0.0074	2.6E-06	0.018
dust inh.	2.6E-07	0.000019	7.8E-07	0.000037
plant ing.	1.6E-06	0.016	3.9E-06	0.033
TOTAL	3E-06	0.02	7E-06	0.05

5 DISCUSSION

A simplistic model of sediment particle sizes using spherical particles and equal bulk densities results in a doubling of surface area with a halving of particle diameter. If an experimental system were constructed with equal-sized particles and a single exchangeable cation, then one might expect sorbed cation concentrations to be related to surface area per unit mass in this manner.

Concentration ratios are used to compare particle fractions instead of absolute values because the metals concentrations show spatial trends. The primary interest, for this analysis, is the influence of particle size on concentration, not concentration trends along the axis of the stream. Ratios were used to remove those trends.

The estimated surface area ratios, using average particle diameters of 30 μm, 150 μm, and 1100 μm for our three size fractions, are 1.0 : 0.20 : 0.027. These ratios are substantial departures from the empirical ratios shown in Figure 2. Explanations for this departure include multiple cations competing for exchange sites on the particles, unknown distributions of particle sizes within the sieve fractions, and unsaturated particle surfaces.

5.1 *Association of Particle Size and Risk*

Differences in cancer risk and hazard quotient for each pathway and for total risk and hazard are approximately a factor of two to three. These differences are specific to these chemicals, pathways, and exposure assumptions, however, and may not pertain to other conditions. For example, although inhalation risk for arsenic was clearly less important than soil or plant ingestion, this would not necessarily be the case for chromium or if higher concentrations of suspended dust were assumed. One might also expect that if data were available for the actual inhalation exposure fraction (less than 10 μm rather than less than 63 μm), the difference in inhalation pathway risk would be still larger.

Another important caution for interpreting these results is that they are based on the relationships of particle size to concentration observed in this particular system. As discussed above, there are several explanations for why changes in concentration were not nearly as dramatic as would be predicted based solely on surface area effects. At a site with lower soil concentrations of cations, or with different types of contaminants (*i.e.* organic chemicals), the observed ratios of concentration and particle diameter might be quite different.

5.2 *Effect of Particle Mass Fraction on Risk Estimates*

The differences in the relative mass of each particle size fraction in the different geomorphic features (Figure 1) is likely to influence receptor exposure to contaminated sediments. As an example, the arsenic concentration for the 0-63 μm fraction is four times the concentration of the 250-2000 μm fraction, but the first fraction represents 4.6% of the total sample and the latter fraction represents 64% of the total sample. These differences may be particularly important for evaluating the dust inhalation pathway, since the presence of large objects that are resistant to erosion can impact the resuspension of fine particles.

At a site where the inhalation pathway is a potentially significant contributor to risk, the actual mass percent of fine particulates can be very important in predicting future dust loading. The threshold friction velocity at which wind generates suspended dust tends to be higher at sites where there is an appreciable quantity of nonerodible (large) particles. Such sites may also represent "limited reservoirs" for resuspendable particles rather than "unlimited reservoirs", where soils consist of deep deposits of fine particles (Cowherd *et al* 1984). These distinctions can result in appreciable differences in predictions of dust loading under future conditions.

The soil ingestion exposure pathway, by contrast, is comparatively less influenced by the mass percent of fine particulates in soils. So long as there is sufficient quantity of 0-250 μm material available, it is likely that particle adherence on the skin will occur. What specifically constitutes a sufficient quantity is debatable as information relating soil adherence to the percentage of fines in a soil is lacking. Similarly, how the relative abundance of fines may affect plant uptake, and how one might apply such knowledge in a calculation that employs a generic plant uptake factor (the ratio of plant chemical concentration to soil concentration), is uncertain.

There is a simple calculation that can be performed to identify an upper bound on the potential impact on calculated risk of the amount of fine particulates in a sample. As discussed in Section 4.1, the risks calculated using pathway-specific particle size fractions are generally greater than those that are calculated by using data for the traditional 0-2 mm size fraction. However, if one weights the 0-63 and 63-250 μm particle size concentrations (Table 3) by their relative percent mass within the entire 0-2 mm sample portion, the risk values will actually be lower than those calculated from the 0-2 mm sample. There is no basis for selecting 2 mm (or any such specific value) for this weighting, but intuitively the notion that risk increases with the amount of fines is appealing. The effect of the relative amount of fines can be accounted for in a dust resuspension model, but remains the subject of speculation for most other exposure pathways.

6 CONCLUSIONS

The differences in total risk resulting from substitution of pathway-specific particle size concentrations for a traditional 0-2 mm concentration are approximately two- to threefold. Given the magnitude of other sources of uncertainty commonly associated with prediction of future risks (*i.e.* chemical toxicity data, land use characteristics, fate and transport modeling), the variability introduced by particle size may be considered relatively small. There are, however, at least three important points to consider when valuing data on the relative abundance of particle size fractions and their associated concentrations.

The results described in Table 5 are specific to the assumptions and models used in the risk calculations, as described in Section 5.1. Therefore, the applicability of these findings to another site must carefully considered before concluding that particle size effects are likely to be minor contributors to overall uncertainty in calculated risk values.

Secondly, the uncertainty associated with changes in concentration with particle size is reducible with relatively little effort, whereas many uncertainties in a predictive risk assessment are not. For example, uncertainties in future land use activities are often ignored within a risk model.

Finally, information on the relative risk associated with different geomorphic features and particle sizes also provides an excellent basis for cost-benefit analyses of risk reduction. If one can target a specific particle size fraction and/or geomorphic feature for remediation and substantially reduce risk, there is an opportunity for considerable reductions in site disturbance and disposal costs.

The use of physical features such as geomorphic elements in a sampling design also has value in increasing the defensibility of a risk assessment by establishing a physical basis for the sampling strata. Exploring contaminant heterogeneity on the basis of physical features can facilitate refinement of the physical site model. This model may then be applied for predicting future trends in contaminant occurrence, which is critical for identifying remedial alternatives that will reduce site risk not only immediately but over extended periods of time.

Acknowledgments: The authors recognize the valuable contribution of individuals in the Tucson, AZ office of Golder Associates, Incorporated in data analysis and development of a physical site model.

REFERENCES

Baes, C.F., *et al* 1984. *A review and analysis of parameters for assessing transport of environmentally released radionuclides through agriculture*. Prepared for the U.S. Department of Energy. Health and Safety Research Division, Oak Ridge National Laboratory, Tennessee.

Cowherd *et al* 1984. *Rapid assessment of exposure to particulate emissions from surface contamination sites*, Midwest Research Institute, Kansas City, Missouri. EPA Report. 1984.

Duggan, M.J. & M.J. Inskip 1985. Childhood exposure to lead in surface dust and soil: A community health problem. *Public Health Rev*. 1985; 13:1-54.

Kissel, J.C., *et al* 1996. Factors affecting soil adherence to skin in hand-press trials. *Bulletin of Environmental Contamination and Toxicology*, 56:722-728.

Nyhan, J.W., *et al* 1976. Distribution of plutonium in soil particle size fractions of liquid effluent-receiving areas at Los Alamos. *Journal of Environmental Quality*, Vol. 5, no. 1.

Sheppard, S.C. & W.G. Evenden 1994. Contaminant enrichment and properties of soil adhering to skin. *Journal of Environmental Quality*, 23:604-613.

U.S. Environmental Protection Agency (EPA) 1989. *Risk assessment guidance for superfund (RAGS), volume I, human health evaluation manual (Part A), interim final*. OSWER Directive 9285.7-01A. Washington, D.C.: U.S. Environmental Protection Agency, Office of Emergency and Remedial Response.

U.S. Environmental Protection Agency (EPA) 1994. *Guidance for the data quality objectives process*, EPA QA/G-4. Washington, D.C.: U.S. Environmental Protection Agency, Quality Assurance Management Staff.

U.S. Environmental Protection Agency (EPA) 1996. *Soil screening guidance: user's guide*. OSWER Directive 9355.4-23. Washington, D.C.: U.S. Environmental Protection Agency, Office of Solid Waste and Emergency Response.

U.S. Environmental Protection Agency (EPA) 1999. *Integrated risk information system* (IRIS). www.epa.gov/iris/

Wixson, B.G. & B.E. Davies, (eds) 1993. *Lead in soil, recommended guidelines*. Society for Environmental Geochemistry and Health, "Lead in Soil" Task Force. Northwood: Science Reviews.

Environmental Issues and Management of Waste in Energy and Mineral Production, Singhal & Mehrotra (eds)
© 2000 Balkema, Rotterdam, ISBN 90 5809 085 X

The economics of sustainable development for Caras Severin county

N. Potoceanu & I. Ruja
University 'Eftimie Murgu', Reşiţa, Romania

I. Chincea
Environmental Protection Agency, Romania

ABSTRACT: This paper involves the application of the principles of sustainable development and conventional economic analysis to particular environmental problems (foe mineral and energy industries) in particular geographical and cultural setting for Caras Severin county. Having selected a specific case study the authors will review this study to identify its objectives, methods and conclusions. What were the strengths and weaknesses? What problems were in implementing these principles encountered?

1 INTRODUCTION

The sustainable development requires the preservation of the existing legacy and at the same time the ensuring the welfare of the present and future generations.

The Romanian society which has been undergoing a transition for ten years now witnesses phenomena apart from the economically advanced countries. There are badly polluted areas as a result of an excessive industrialization that had taken place before 1989. Due to the scarcity of the funds and an unreasonable running, these regions have not been subjected to a dispersal of polluants. Although Romanian politicians have come to know the concept of sustainable development ,the professional estimations (like that done by Camelia Cămăsoiu in 1994) have shown that a total decrease in polluant emissions needs a 5 – 6 % out of the G.N.P and a quick ecological renewal needs allowing 8-10% from the G.N.P.

As a matter of fact not even instruction or health department have been allocated such high ratio in Romania.

There are chronic irreversible phenomena as the followings:

- regions with job-sickness produced by the main polluant as lead, asbestos, DTT, met in towns like Baia Mare, Copsa Mica, Isalnita and others, and where mutations in the local children's health occur;
- severe soil erosion as a result of the excessive and uncontrolled exploitations; the latest years Romania witnessed the most disastrous landslides, of which most extensive are in Moldavia;
- a reduction of the moist territories (the Danube Delta, the lakes, the water meadows) and the environmental deterioration;
- the collapse of pisciculture, mainly in fresh waters, due to unwise fishing but mostly due pollution (the Danube carries every year about 60,000 tons of phosphorous and 340,000 tons of non organic nitrogen).

A better legislation is needed for ecology. The Law of privatization in Romania is deficient with regard to the environment. There is no mention made in it. This is of extreme gravity since the former owner – the state – should assume its responsibilities towards environment when the transfer of possessions takes place.

Yet, the year 1999 was rich in passing legal instruments connected with ecology: 44 resolutions, 3 bills, 5 governmental ordinances , 10 governmental decisions, 15 departmental orders, 6 bilateral agreements and 5 international conventions.

A law that will certainly stir passionate debates and arise difficult problems in the law called after its initiator "Lupu Law" because according to it the former proprietors are to be granted over two million acres of forest, that is one third of the country's wood-covered surface.

2 PECULIARITIES OF SUSTAINABLE DEVELOPMENT IN CARAS – SEVERIN COUNTY

Caras Severin county lies in the south-west of Romania .It is here that the Danube enters into Romania forming the country's natural border with Serbia (Yugoslavia). It covers an area of 971,976 acres (of which 409,864 acres of woods, 399,620 acres of farming lands, 9,884 acres consisting of waters, etc)

The principal hydrographic ways are represented by the Danube (64 km) and its confluents: the Cerna, the Timis, the Barzava, the Caras, the Nera. There are artificial lakes as well as natural ones, some of which are of karst origin, like the Devil's lake covering 700 square meters. Some underground karst lakes area the result of natural dams in the galleries of the caves as Plopa and Buhui in the Anina mountains. From an economic point of view , the county was intensely industrialized. The main products still produced are internal combustion engines of Diesel type used in railway and naval transport, hydro-units, bogies, complex equipment's, rolled goods. One- branched industry was and continues to be a feature of the county. From a technological point of view the equipment and products are those of lesser developed countries as compared with leading countries in the mentioned branch.

After 1989 the newly founded enterprises are strictly dealing with wood, ready-to -wears, food. If we examine the distribution of the plots of land and the geographical aspect of the county we understand that there are inexhaustible environmental assets that can ensure a necessary touristic affluence (if the necessary infrastructure is developed) forest, fish and underground resources. The ore resources include copper, iron, uranium, cool, marble, etc. For the time being all mining zones are considered disadvantaged areas because mining activity was closed down. The marble work at Ruschita is under litigation because the state gave ownership (without auction) to an enterprise from another part of the country.

3 POLLUTION AROUND THE COUNTY

3.1 *Existing polluting sources*

The case study has similarities with the impact study revealing a state of facts on basis of the existing data obtained while carrying measurements and during statistic processing.

3.1.1 *AIR QUALITY*

The Earth's atmosphere is one of the environmental factors that are difficult to control because polluants once in the atmosphere disperse rapidly without any possibility of retaining them for purification.

Due to old technologies leaking and strongly corroded equipment 30 % of the measured values for dispersed colloids indicators exceed the allowed limits with 40% in the area of Resita and Otelu Rosu

As for the sedimentary colloids, they exceed permanently with 50 % . The values of SO_2, NO_2, NH_3 rarely exceed quotes. On the whole the air pollution in Caras Severin is alike all over the country.

3.1.2 *WATER QUALITY*

Considering that each zone depends on a source of water, we shall analyze the situation of the running waters, the lakes, the impurification of the subterranean water.

Impurification can happen either naturally or artificially, the latter being accepted as pollution caused by human activity. The natural impurification of the aquatic lands is generated by physical and chemical processes that take place naturally by transferring impurities from air, land or surface water. The subterranean waters remaining in aquatic grounds that communicate hydraulically with geologic or volcanic structures will be influence by the latter due to natural flow of the subterranean water .The artificial impurification of the aquatic grounds is manifest through various forms as a result of the human activities , and the polluting sources can be considered:

1.industrial sources – the greatest threat for the subterranean water. In Caras Severin the industries are : siderdurgy (Resita, Otelu Rosu) ,and mechanical engineering (Resita, Caransebes, Bocsa). The underground storing spaces the petrol pipes (the neighboring of Yugoslavia and the growing number of cars tripled them) has a certain degree of unsafety in exploitation which is aggravated in time because of the operating condition, ending in polluant leakage.

These substances contain hydrocarbons with highly water-soluble degree (20 – 80 mg/l of petrol) and they are present even in concentrations smaller than 0.005 mg/l;

2. The mineral sources of pollution for the subterranean water are the waters evacuated from the mining galleries and, last but not least, the seeping – in that take place in the sterile waste dumps (copper – Moldova Noua, cool – Anina, uranium – Ciudanovita, iron, marble, etc.)

3. Agricultural resources have becom the principal polluants of the subterranean water in the last wears because of chemical fertilizer and insect- killers. The irrigation water (which hardly exists in Caras Severin) and rain water facilitates the dilution of the chemicals, and by infiltration, their transfer into the phreatic layers and here from, depending on hydraulic conditions, down into the deep layers.

The big live-stock farms from Bocsa, Berzovia and Bozovici are operating on water using technologies. They are great sources of polluants since their used water contain high concentration of various chemical compound and there are great difficulties in cleaning them ;

4. Domestic sources of pollution are represented by refused waters from crowded centers which pollute subterranean water in different ways : the storage in septic tanks has uncontrolled seeping, leakage from old drainage in towns which for the last ten years were repaired only in extreme urgencies. Resita's purification capacity is limited, the station in Calnic

supplies only 35 % of the necessities and its efficiency is of only 20 % for organic substances and 25 % for colloids. We have to mention that all over Romanian has been used one and the same procedure of purification since 1900 - that is using chlorine. The rubbish dumps of all towns are to be included here too because they are a mere enclosure which allows rain water to seep in and pollute the subterranean waters.

Table 1. Main water polluters.

Sources	Possible pollutors
Domestic refuse storage	Heavy metals, Na^+, $CaZn^+$
Industrial refuse dumps	Organic and anorganic compounds
Urbane drainage	Organic compounds, Detergents, Solvents, Sulfates, Microbiologic compounds
Agriculture	Chemical fertilizers, Pesticides, Herbicides
Waste water irrigation, Furrowing fertilization, Mud	Heavy metals, Organic and anorganic compounds
Rain water infiltration	Organic compounds, Heavy metals, Petrochemical compounds
Radioactive storage	Radioactive waste, radioactivity.

3.2 *Measurements in Caras Severin County*

The results of various measurements made throughout 1999 was an average for three month and it is shown in table 2 and 3

Table 2. Polluters measured in running waters

River	Section	Found substance	
The Timis	Upstream Caransebes	P	0.124
The Bistra	Obreja	Phenols	0.002
The Poganis	Brebu	Phenols	0.003
The Birzava	Moniom	NH_4	2.10
		P	0.227
The Caras	Varadia	Phenols	0.005
		P	0.104
	Carasova	Phenols	0.003
The Nera	Naides	$CCOMn/O_2$	13.9
		Phenols	0.004
		P	0.233
The Dunarea	Bazias	Phenols	0.002
	Moldova Noua	P	0.15
		Zn	0.036

All measurements are given in mg/l

4 ECONOMIC INSTRUMENTS

Social practice shows that the social level of the economic activity does not coincide with personal benefit, if high costs are involved. On a global scale the following economical instruments of the environment policy are classified in conformity with OECD methodology in : taxes (of pollution, of running, on product and administration), subventions (allowances, low interest loans), saving and refunding systems (repayable credits when recycling containers), trading licenses to pollute and financial penalties. These instruments affects the costs and benefits of the alternative actions of the polluters influencing their behavior.

Table 3. Lake polluters around Resita

Artificial Lake	Substances	
Secu	$CCOMn/O_2$	1.8 ... 3.4
	NH_4	0.27... 0.66
	Chloride	2.8 ... 6.8
	Sediments	42 ... 59
Grebla	$CCOMn/O_2$	1.9 ... 3.2
	NH_4	0.21... 0.68
	Chloride	2.8 ... 6.4
	Sediments	37 ... 60

All measurements are given in mg/l

The county's Environment Protection Agency uses as instruments the environment agreements, licenses, guiding manuals, environment surveys, supervisions. The air, land and water are monitored without any implications in the economical development. Considering the protected areas in the county, when thinking out the possibilities for an economic increase, we have to take note of the principles of the sustainable development. When answering the two fundamental questions of the environmental economy – if economic growth is possible and if desirable when possible – and when relating to our case, Caras Severin, we find that for the time being there are no supplemental reasons of fear, except forest and fishery. We don't have to worry as the desired and possible economic growth is low level due to high fiscality and the lack of founds needed for zonal investments.

Asking ourselves again to what extend market and technology intervene it the process of development regarding resource exploitation, we underline that at the moment there are incertitude in foreseeing future and we find ourselves in the situation of letting market forces impose the resource employ model and rate. Of course market can be prevented from a correct functioning due to information and knowledge gap or uncertainty regarding future development. Inadequate macroeconomic policies can lead to a short lived development, but at present as far as our county is concerned this problem needs no discussing.

5 CONCLUSIONS

The economic system through its motivating force can offer a part of the real and viable solutions to obtain the economically ecologic balance. Taking

into consideration the concept of sustainable development similar to other countries it is necessary:
-a better awareness of the need to study the medium and long term impact on using the resources management decisions<
-a greater attention paid to generation problems and the transnational impacts when we consider the Danube and everything connected with it;
- Avoiding noxious developments and projects because too strong trust in market can bring under optimal results on a social level,
- Regional co-operation;.
The implementing costs of the sustainable development are very high. Among them are the lost of short term potential big profits if resources run out and the adjacent behavior or some private costs can be implied: home high prices, electricity, food and other goods or services which directly depend on the natural resource quantity.

For example, the preservation of the fields, woods and other public lands brings along an extension of housing lands in unprotected zones, similar effects being visible in agricultural sphere. A clear example of the effect is the rising prices for lands around Dognecea(where flat water bottling is being taken in view),or along the Danube Straits where tourism and fishing is booming.

As far as international cooperation is concerned the conditions and requests like acquiring and spread9ing knowledge and information needed or surveying the biosphere prove its complex character. The sustainable development concept requires moderation and impartiality and its functioning implies self decoding of the needs and ways to be covered, not forgetting the specific, local elements of development .Romania`s long hesitating economic transition and some rules imposed by I.MF has as a result the fact that inspite of a correct real evaluation of the environment condition the starting of a coherent protection strategy is delayed.

The Government ought to imply more in the ways the resources are exploited. We know they are limited and the state as the main proprietor of them is directly interested in obtaining long term benefits. As for the marketing instruments, not only in Caras-Severin, but whole Romania does not possess efficient operational market system that could force pollutors to pay externalities and encourage them to reduce the polluting quantities .Modern technology plays an important role for better health conditions. Inside the EMU was founded a center for technology transfer which has in view not only economically efficient technologies, but also non polluting ones.

REFERENCES

Camasoiu, C. 1994, *Economia si sfidarea naturii*, Bucuresti, Romania, Editura Economica, pages 11-53.

Potoceanu, N., 1999 *Elemente de protecția mediului îm județul Caras Severin*, in Volumul dedicat Zilei economistului, Herculane , volume I, pages 307-311

Environmental Issues and Management of Waste in Energy and Mineral Production, Singhal & Mehrotra (eds)
© 2000 Balkema, Rotterdam, ISBN 90 5809 085 X

Environmental management in hard coal mine group in the Upper Silesian Coal Basin, Poland

Marek Pozzi
Institute of Applied Geology, Silesian University of Technology, Gliwice, Poland

Jacek Węglarczyk
Institute of Mineral Processing and Waste Utilisation, Silesian University of Technology, Gliwice, Poland

ABSTRACT: Mining activity and the other branches of heavy industry existing in the USCB for over 2 centuries have made large unfavourable changes of environment. Prevention of its further degradation needs the solution for the following main problems: utilisation of high saline mine drainage water (a problem unique in the world scale), treatment of solid wastes, land reclamation (mainly treatment of areas of ground subsiding). Market economy introduced 10 years ago and the necessity that all fields of life conform to the requirements of the European Union force the process of deep restructurisation of mining industry. One of the conditions for success of restructuring is the solution of ecological problems. The possibility of environmental management system implementation according to the ISO 14000 standard in the coal mine group condition was discussed. The chances and presumed results of these activities were presented in this paper.

1 INTRODUCTION

The Upper Silesian Coal Basin (USCB) is Poland's greatest Variscan coal basin. Coal exploitation began over 2 centuries ago and reached its apogee (200 million Mg) in the 80-ies. Long-term mining in around 60 mines leading the exploitation mainly longwall mining with roof caving, concentrated on the area of 10,5 thousand hectares (Dulewski & Wtorek 1996) resulted in environmental degradation. Additional factor affecting the degradation was concentration of heavy industry: coking plants, metallurgy, power industry and chemical industry on a small area (corresponding to the area of Katowice province). In the opinion of many institutions and research institutes it was an area of ecological disaster. Market economy being introduced for over 10 years and the necessity of conformance of the state of environment with the requirements of the European Union force the process of deep restructuring of mining industry and substantial improvement of the environment.

Every next government tries with determination to create the basics of law and economy for adapting the mining industry (that lies within planned economy and brings financial losses estimated at 3,0 billion PLN) to the requirements of market economy. The basic assumptions of this project are:

- decrease of employment from 415 thousand people in 1989 to 128 thousand employees in 2002,
- decrease of coal output from 135 million of Mg in 1995 to 100 million of Mg in 2002,
- complete liquidation of 15 mines and partial liquidation of 9 mines.

It should result in increase of efficiency of this branch and bring in a profit in 2002.

Environmental protection system in Poland uses the legal and economic instruments based not only on the system of charges for waste transfer into the environment and penalties for violating the requirements of environmental protection but also on subsidies and preferential credits given thanks to receipts from penalties and charges. Expenditures on proecological investments have increased a few times in the recent years reaching about 1,7% of GDP, that is the level in Western European countries. Coal mining industry spent about 157 millions PLN in 1997 on proecological activities. At the same time the World Bank's prognosis predict, that the cost of conforming to the EU laws in conversion to 1 resident should be 500-1000 ECU during the following 10-20 years and reach 3-4,6 % of GDP.

A thesis was accepted, that a proper Environmental Monitoring System usage could substantially decrease environmental costs of coal mines liquidation.

2 CURRENT ISSUES AND CONDITION OF ECOLOGY IN COAL MINING INDUSTRY

The main ecological problems that mining industry in Poland contends with, are:

- saline mine waters coming from mining drainage system development,
- mining wastes management,
- land reclamation,
- mining damages.

2.1 *Saline waters development*

Saline waters, that include mine waters characterised by mineralisation higher than 3 g/dm^3 and at the same time having over 1,8 g/dm^3 of $Cl^- + SO_4^{2-}$, are undoubtedly mining's greatest ecological problem. Coal mines transferred in 1997 over 550 thousand cubic metres of unused mine waters to drainage systems of two greatest Polish rivers, Vistula and Odra, from which about 350 thousand m^3 were saline waters. Load of salt transferred with saline waters to surface reservoirs was about 4200 t/24h (Chaber 1998). Piping off the saline waters from the USCB coal mines causes water salinity over acceptable borders (500 mg/dm^3 Cl^-) and reaches Warsaw when level of water in the river is low. Maximum concentration of Cl^- ions in Vistula water in Cracow reaches even the values over 2000 mg/dm^3.

Saline waters transfer done directly to the rivers causes in consequence harmful influence on the river biocenosis, increased machines corrosion and less possibilities of water usage in forestry, agriculture, industry and municipality. Mines transferring saline waters to ground waters are obliged to pay charges for environment usage and penalties proportional to the quantity of Cl^- + SO_4^{2-} ions in water. The total of charges and penalties in 1998 was about 108 million USD and it was about 2,1 % of coal production value.

The experience in solving the problems of saline waters is so far connected with liquidation or limitation of these water's influence or with rational development.

The firstly used method has been storing saline mining waters in retention reservoirs and then controlled transfer to the rivers when the level of water is high. Another method of hydro technical protection is transfer of waters by a collector and then dropping them to rivers where the flow is big enough to dilute the salt. Increasing depth of mining and building coal mines in new regions caused such a raise of salt quantity in mining waters that hydro technical protection is no longer sufficient – the quantity of transferred salt exceeds the absorption of rivers specified by acceptable saline of river waters.

Along with the method of pyrotechnical protection, starting from the 60-ies, there was developed a method of saline waters utilisation. At the beginning, it was a technology of selective crystallisation of sodium chloride and calcium sulphate with the recycle of J, Br, K, Mg compounds, and in the 80-ies it was a method of reverse osmosis (RO) in the only one Desalination Water Company at the coal mine "Dębieńsko" (4600 m^3/24h). At the moment a new Desalination Water Company in Oświęcim is being designed with economical capacity of 32,7 thousand of m^3/24h. Having in mind high costs of saline waters utilisation and predicted financial condition of coal mines it seems that they won't be able to bear the costs of water desalination (Chaber 1998).

In connection with the presented limitations of possibilities of hydro technical and utilisation methods usage others methods are being introduced, mainly mining and hydro geological, to resolve the problem of saline waters.

Mining methods consist in such a modification of mining activities that allows for decreasing the content of salt transferred to the mine with incoming waters. It can be reached by:

- change of location of exploitation work – practically moving the exploitation from deeper level to more shallow,
- isolation of excavations with substantial saline waters supply,
- keeping the saline waters in underground excavations, mainly by using them for ash-water mixtures and using them for backfilling and gob sealing (Pozzi, in press).

Hydrogeological methods consist in forcing saline waters back to the rock massif, understood as forcing into the isolated aquifer located beyond the draining reach of underground excavations (used in the southern part of the USCB in one bore-hole on the depth 1005-1463 m in Devonian bed) or as a recirculation (the design put into practice in "Czeczot" colliery).

Saline waters flowing to some of the mines contain raised (reaching 390 kBq/m^3) radium ^{226}Ra and ^{227}Ra isotopes concentration. Two types of radium waters are distinguished: A type – containing bar compounds and B type – not containing bar compounds but containing sulphate ions. From A type waters after mixing with sulphate waters a residue of bar and radium (up to 400 kBq/kg) sulphate is precipitating, that is used for limitation of radium concentration in drainage waters piped off the mines. In the result of various activities taken by mines the load of saline waters to rivers is smaller than their flow to the mines. Quantity of waters transferred to the Vistula river drainage system is decreased by

30% to natural supply to the mine, and in the Odra river drainage system by 75%.

2.2 *Mining wastes management*

In the geological conditions of USCB the quantity of produced mining wastes is about 40% of coal mining. In 1997, the mining industry having total production of 137 million Mg of coal, generated 55 million Mg of wastes, 50 million Mg which included the processing wastes given off the coal in processing facilities. From the quantity of 55 million Mg of generated wastes, 19,9 million Mg were stored on the surface and only 4,8 million Mg were stored underground. Remaining wastes (30,3 millions Mg) were used for engineering works. Moreover in 1997, the mines accepted for storage in underground excavations about 4,4 million Mg of foreign wastes.

Problem of mining wastes management due to their quantity and limited area for safe storage has essential meaning for proper functioning of mines. We can distinguish three basic trends in wastes management:

- industrial development,
- non-industrial development,
- storage in environment.

The following wastes can be rated among industrially developed wastes:

- used in mining (storage in underground excavations),
- used in building and cement industry (production of building elements, light crusher, ceramic elements and raw material for cement production).

It is supposed that the most important and perspective trend in wastes management is their underground storage. Cubature for possible usage is estimated on 11 million m^3 per annum. Wastes underground storage is a technology of environmental protection and also a mining technology – as an efficient way of decreasing the subsidence and surface deformation and the way of mass wastes storage under conditions of fulfilling the requirements of environmental protection and technological conditions.

Similar to European Union strategy, the state regulations expect at first minimisation of wastes production and economic usage of wastes and then their disposal (including storage). Having in mind the restrictions in gaining new areas for storage, it should be presumed that underground storage would be a method resolving this problem. The second trend of industrial waste development – building materials production, promoted in the 70-ies, has not awaited expected growth, and presently it's rather disappearing. Cement industry takes annually about 200-300 thousands Mg of wastes using them as thinning resource.

Non-industrial development consists in:

- engineering development of wastes (ground work including work on the areas deteriorated by mining activity, roads, railways, hydro technical work). It should be supposed that in connection with the beginning of motorways building in Poland, wastes usage would gain greater importance.
- wastes reclamation by other receivers.

The third main trend of wastes management consists in their storage on the heaps that are recultivated and developed.

2.3 *Land reclamation*

Management of recultivation and post-exploitation areas and their handing over to local government or different users result from present legal regulations (Act of Protection of the Environment, Geological and Mining Law).

It is estimated that the total area taken by over 110 heaps of mining wastes in the USCB is more than 3200 hectares (Makowski et al. 1998). In 1997, the coal mines did recultivation work and development on the surface of 605 hectares, of which on 167 the works were finished. The main goal of mining industry in this range is intensification of recultivation work and wider introduction of new technologies of storage and recultivation (modern transport, heap material build up, current recultivation).

2.4 *Reclamation treatment of subsidence and repair of mining damages*

A rise of mining damages and their range has tight connection with geological conditions (big number of coal seams) and with method and intensity of exploitation (almost 90% of coal output is led by longwall mining with roof caving). Mining damages are shown in ground surface lowering (in some regions up to 30m), creation of seasonal and everlasting accumulated water areas (especially in southern and western part of the USCB, where soil waters are well isolated from deeper aquifers, area of impounded water in the subsidence troughs is estimated at 650 hectares) and in consequence damages in building infrastructure. To restrict these negative results mining and building prevention is undertaken.

Repair of mining damages is financed from the resources of the mines. Annual costs of damages removal are about 300 million PLN (about 100 million USD), whereas the arrears are estimated at about 80 million PLN. Donations from the state budget are used for lowering these arrears.

Mining industry is not able to realise current charges for using the environment (360 millions PLN in 1997), that's why it is expected that the cancellation of debts should take place as an important element of the reform of this sector. The goal of introduction of EMS in mining industry is the improvement of efficiency of activities led by individual collieries and coal companies for the environment protection.

3 ENVIRONMENTAL MANAGEMENT SYSTEM (EMS)

EMS as one of the formalised systems of sector management was designed in the 90-ies using the experience coming from functioning of quality assurance systems, commonly known as ISO 9000 standard. Its goals and application are defined by the following statement:

Environmental Management System is:

"... an organisational structure, competences division, practice, processes and resources serving introduction and application of environmental matters of a company".

However the ISO 14001 standard says:

"... it is a process of aiding EMS aiming at reaching global improvement of efficiency – not necessarily in all fields of environmental activity, but accordingly to ecological policy of a company".

In the interpretation of both definitions we should consider the following elements:

- EMS is mostly an organisational structure in which every element has precisely ascribed competences.
- EMS goal is to coherently control functioning of industry and to allow proper management in the environmental range, also in emergencies.
- EMS is supposed to be efficient. Immediate and expensive activity for environmental protection is not required, however it is required to design and apply environmental policy and strategy, designed for actual possibilities of industry.

The last element is very favourable for companies that need complex and sometimes expensive activities for the environment protection – it allows the creation of environmental programme suited for the conditions, and accordingly to third party's requirements. Further consequence of execution of this programme is required – it makes the planning the most important element in the company.

Number of systems put into practice can only be estimated. The state for the end of 1999 was about 8000 of environmental systems all over the world. Main "locations" of these systems are United States, European Union countries, Japan. In Poland over 60 systems are running and certified – in very different branches, and over 100 projects are being introduced. Among the countries that we can compare to, Czech Republic is ahead of Poland, but Poland is ahead of Hungary and Slovakia (Węglarczyk 1999).

4 EMS AS A METHOD OF STRATEGIC MANAGEMENT

Definition defines strategic management as "knowledge and art of controlling for joining strategies and activities to reach strategic goals in tight cooperation with present and future environment". As a process of management it consists of two co-depending stages: formulating the strategy and introducing and realising the strategy".

EMS designed on the base of Deming circle (known also as PDCA process) has identical structure – besides formulating and realising the strategy (policy) it requires constant actualisation (shown on the diagram).

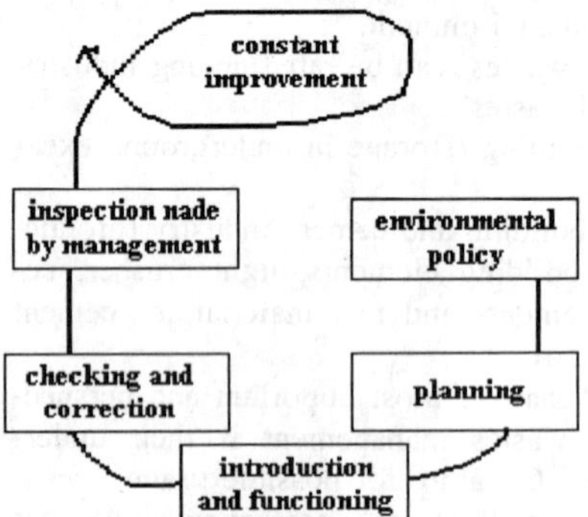

Figure 1. Deming circle

It can be assumed that in the company activity range described usually as "environment" or "environmental protection" and managed by functioning system of environmental management, the rules proper to strategic management apply. They ensure its operational efficiency and decrease the margin of uncertainty of final results.

The rules can be described in a few points.

1. The company needs to have an appropriate database in the range making an efficient management possible.
2. The company needs to define its future position and statute in the environmental range.
3. The company has to identify and if necessary create internal conditions of strategy realisation.

4. The company needs to choose an optimal variant of strategy.
5. The company needs to design strategic plan on the basis of the chosen variant.
6. The company management should gain acceptance of third parties or staff for the strategic plan.
7. The company management should provide operational control of plan realisation.
8. The company management should have the possibility of verification of this plan in a long-term period.

These rules expressed in the same or slightly changed words are presented in the ISO 14001 standard as a philosophy of EMS functioning in the company (Tab. 1).

Table 1. Expression of the rules of strategic management in the ISO 14000 standard

Rules item n°	Element of ISO 14001 standard
1-5	planning
including 4	policy
6-7	introduction and functioning
8	checking, correction and inspection made by management

Application of presented rules should be complex. Skipping any element causes upset of entire structure – and that's the standard tries to avoid, by means of making the requirements of external acceptance – content-related and formal (EMS certificate).

During the company restructurisation the legitimacy of application of described rules increases, because of the need of optimisation of means allocated for changes and – the most often – because of the need of creation and controlled realisation of restructurisation plan.

In our judgement, the more serious range and greater degree of uncertainty in case of restructurisation, the greater is the need of strategic methods of management application – with reference to environmental matters; methods recommended by EMS.

Such situation is currently in Polish mining industry, especially on the area of Upper Silesian coal mines.

5 FUNCTIONING OF EMS IN CONDITIONS OF RESTRUCTURISATION OF A COAL MINE

Coal mines effect on the environment was described in the chapter 2. It comes out of this description, that restructurisation of coal mines needs to have proecological character. This statement is important not only in the context of new facility operation (for example joint venture of mining and energy), mine liquidation (for example necessity of waters drainage), but also in case of environment condition recovery to possible normality (that includes for example recultivation of waste storages).

Let's try to consider the usefulness of precedent rules on the example of liquidation of a middle-size coal mine, mining power coal within the confines of company. This example will surely neither be complete nor ready for applying, but it should point out obvious trends, advantages and disadvantages.

Planning of ecological activities should be made for the period of at least few years and take into consideration the following elements:

- design of coherent and complete database, identifying ecological standing of the coal mine as a company and a part of greater economy system,
- design of the vision of existence of restructured facility or existence of another economic entity that takes over ecological commitments of a mine (company),
- design of methods of proceeding from actual to planned condition, in dependence of internal possibilities and external aid (also in the confines of consortium),
- choice of optimal variant of strategy for realisation – it is suggested to choose on the level of company because of other coal mines needs,
- design of a short-term and long-term executive plan,
- final description of realisation conditions.

The basic circumstance determining the success of preceding activities is keeping of proper works order, their cohesion and complementing one another and completion of individual stages (elements, phases). In other words, there should be used and kept proper procedures of execution – such procedures are proposed by EMS.

Classic example of results of lack of proper procedure is the base of environmental data.

In most cases the elements of such base are existing and they can constitute the basis for assessment and planning the activity concerning environmental protection. The problem is, that the integrity is not available when it is required, mostly because of strong dispersion of databases. The reports concerning environment are in various places (mine departments, by-coal mine companies, consortium, etc.) and practically nobody knows where to look for adequate study in case of emergency. Very often instead of searching for the existing report a new one is made. It surely has its financial measurement.

The practice of work of people responsible for environmental protection in coal mines shows relatively high awareness of practical aspects of impact on environment made by their company. This awareness unfortunately does not translate into elements of management or direct activity that could take effect in environmental condition improvement – mostly because of lack of a systematised database. Functioning of formalised system is foremost:

- definition, on the basis of a strategic plan, execution of schedule with identification of content-related, organisational and financial responsibilities,
- creation of controlling (monitoring) system, allowing fast reaction in case of disagreement between the plan and execution,
- creation of nearly automatic mechanisms working within the confines of mentioned controlling,
- rise of staff awareness by training system that allows gaining their support or at least neutrality.

Checking and correction as well as inspection are executed on the basis of basic data and data gained by monitoring (in this case, of system and environment). Only closing the circle with this element allows in the perspective of a few years to get an optimal shape of both environmental strategy and results, obtained by its realisation.

The fact, that the described activities allow the mining industry to change the presently unfavourable image, should be appreciated. Showing the sector of mining industry managed in the modern way and according to EU regulations can initiate the change of this industry image.

There is no doubt that in the present organisational structure the introduction of modern, formalised system of management will be difficult for many reasons. Organisational structures of standard coal mine are very distant from the system described in ISO 14001 standard, and are, in our opinion, very resistant to changes.

6 CONCLUSIONS

1. In modern management, especially when the company is during restructuring or deep changes, application of strategic management methods is required. It comes out mostly of the necessity of determining the needs and rules of resources allocation, needed for effective carrying out the restructurisation.
2. Foregoing statement concerns fully the case of coal mines restructuring and its implementation is essential.
3. In conditions of mine restructurisation one of the most serious sector problems is environmental activity.
4. The most useful instrument of management within the confines of "environment" sector during the restructurisation is the system of environmental management, constructed on the basis of the ISO 14000 standard.
5. Application of this instrument should be consequent, it means that the system should be introduced and should function in accordance to the requirements of the standard; it should be also certified.
6. Theory and experience of existing systems indicate, that running in the EMS and obtaining the effects lasts for a few years, what means that in Polish, actual conditions running in such systems in coal mines should start immediately.

REFERENCES

Chaber, M. 1998. *Present Issues and Ecology Condition in Coal Mining Industry*. Science Works of Central Mining Institute, series: Conferences, n° 24, p. 7-16, Katowice

Dulewski, J., Wtorek, L. 1996. *Recultivation of Areas Deteriorated by Mining Activity.* Works of Main Inspectorate of Mining, n° 7, p. 4-9

Makowski, A., Cempiel, E., Pozzi, M. 1998. *La reconquete des sites – exemple polonais.* Northern geological Society, Special Publication

Pozzi, M. May 16-18 2000 (in press). *Environmental Aspects of Ash-Fly Underground Location in Katowice District, Poland.* Beijing International Symposium on Land Reclamation

Pozzi, M. May 16-18 2000 (in press). *Protection of Environment in the Light of Polish Law.* Beijing International symposium on Land Reclamation

Węglarczyk, J. 1999. *Integrated Management in Coal Mines – Environmental Management System.* Science Journals of Silesian University of Technology, series: Mining, special edition: Mining 2000, Gliwice

Environmental Issues and Management of Waste in Energy and Mineral Production, Singhal & Mehrotra (eds)
 ISBN 90 5809 085 X

A probabilistic approach for modeling environmental liability

S.C.S.Rimbey, W.A.Funk, P.Weeks, J.Wells & J.C.M.Patterson
Komex International Limited, Calgary, Alb., Canada

ABSTRACT: This paper discusses the use of a modelling and statistical approach that has been developed to estimate environmental liabilities. This technique has been used successfully to evaluate upstream oil and gas assets ranging in size from a single wellsite to multiple major production fields of varying age, geology and ecology. The application and limitations of this method are discussed based on experience gained from environmental liability evaluations in the oil and gas sector in western Canada.

1 INTRODUCTION

An environmental liability (EL) is defined as an obligation to make a future expenditure to abandon and reclaim affected property as per applicable standards set out by the regulatory bodies (Funk 1999).

Consequently, the estimation of EL exposures is an important consideration for property valuation in support of divestment/acquisition negotiations, due diligence and environmental liability management and reporting (e.g. corporate balance sheet liability provisions).

An evaluation to estimate ELs is based upon provincial and federal legislation, guidelines and standards for environmental management at industrial or commercial properties. These liabilities include all future expenses necessary to achieve a reclamation certificate or equivalent (e.g. downhole abandonment, lease reclamation, access reclamation and site decontamination). The environmental costs most commonly estimated are land reclamation and remediation of contaminated soil and groundwater. ASTM (1999) is currently in the process of standardising a practice for the estimation of EL.

Environmental costs are heavily dependent on impacted soil volumes, probability and extent of groundwater contamination, effectiveness of equipment and required remediation techniques and timeframes. These, and many other variables, exhibit a large amount of uncertainty. Consequently, EL exposures are difficult to quantify and are subject to great variability when generated. Methods used to estimate EL include expected value, most likely value, range of values or known minimum values (Konar 1999).

With the exception of the expected value approach, these methods neglect the financial risk inherent in the uncertainty and variability of the estimate. For example, two assets may have identical predicted EL values. However, a different degree of certainty may be associated with the values. Asset A might have considerable exposures to environmental risk with little certainty of achieving the predicted value, while Asset B may have a high certainty associated with the predicted value. The increased certainty translates directly to reduced financial risk.

To better address these difficulties with EL evaluations, Komex International Ltd. has developed and packaged a methodology offering impartiality, defensibility, quantification of uncertainty and decision analysis. This method is identified as the ***LIABILITY ENGINE***™ (trademark pending). This tool uses Monte Carlo Simulation (MCS) methods to seed (or input) values into spreadsheet-based cost estimation routines. These routines model the activities and costs required to resolve environmental issues. The key to this method is that MCS can model complex situations using simulation when the value of essential variables is uncertain or fluctuating. The tool allows the user to define the uncertainty and variability of input variables; thus, enabling scrutiny of uncertainty and variability in the result. Consequently, the risks inherent to environmental issues can be translated into the EL evaluation.

2 OVERVIEW

The evaluation of environmental liabilities using the *LIABILITY ENGINE*™ follows several steps. First, oil and gas assets (e.g. wellsites, compressor stations) are identified and aggregated by common features such as asset type, product type, location, ecoregion, forest area and age to form asset classes with similar characteristics.

Second, a representative sampling of each class (or stratified random sampling) is conducted to gather data on environmental conditions. This sampling can include the collection of Phase I Environmental Site Assessment (ESA) data including ecology, land use, reported incidents, site visits and operational history (CSA 1994 & ASTM 1997a). Additionally, Phase II ESA data collected with intrusive techniques (e.g. soil and groundwater sampling) can help to limit uncertainty as potential contamination is characterised and delineated (CSA 1998 & ASTM 1997b). Notably, the intensity of the investigation (i.e. population sample size, asset classes and level of site assessment) is determined by the requirements, purpose and accuracy demands of the project.

Third, the assets' environmental condition data are summarised and used as inputs to the model. The model has layers of linked spreadsheets and subroutines, with hundreds of formulae that estimate the costs of land reclamation and contaminated soil and groundwater remediation. These costs include such items as contaminated soil volumes and hauling distances and are represented in the formulae as variables. The model simulates the chronological completion of all stages of reclamation and remediation for each asset class.

Fourth, an associated MCS program inputs ranges of values into the formulae variables by selection from specified frequency distributions. The ranges of values and the shape of the frequency distributions are set by remediation and reclamation experience, available data and professional judgement. They are then calibrated with survey and in-house data.

Fifth, using the MCS program, the model is run with several thousand iterations (i.e. random combinations of values from the frequency distributions assigned to the formulae variables are run through the spreadsheet routines) to produce a data set of remediation and reclamation costs for statistical analysis. The cost data are then aggregated and reported using statistical measures. These measures describe the resultant frequency distributions of remediation and reclamation costs in terms of a probability weighted average. Sensitivity analyses are used to summarise the important factors that affect the estimate.

3 APPLICATION

The model allows for the evaluation of large numbers of assets and/or environmental issues in a short time period using pre-constructed costing routines. The result is a carefully defined, concise and bounded estimate of environmental liabilities represented by a probability distribution of EL values. This type of output allows for in-depth analyses of exposure, benefit and uncertainty.

The strengths of the probabilistic modelling approach are that the method:

1. Incorporates uncertainty and variability (i.e. risk) into estimates;
2. Allows user-defined input to increase specificity;
3. Models relationships, which may be difficult to express mathematically;
4. Standardises and defines ELs based on a chronological work break-down structure;
5. Models and systematically accounts for anticipated factors;
6. Provides sensitivity analyses that prompt the user to review variables that are important contributors to variance;
7. Flags extreme outcomes resulting from combinations of rare events and thereby calls attention to significant exposures that may not have been recognised;
8. Reports the results in a statistical manner; and,
9. Produces a database of output cases for further review, monitoring, investigating and reporting.

4 LIMITATIONS

Some limitations of the probabilistic approach for modelling EL include:

1. The costs associated with diminished or lost production, downhole abandonment, land or legal costs and statutory penalties or litigation may not be included in the evaluation;
2. The model is an expert system using multiple software titles and programmed subroutines, rather than a dedicated program. The result is, this model cannot readily be manipulated by the users of its estimates; and,
3. To avoid misunderstandings during negotiations, the interpretation and application of results requires a basic understanding of statistics and probabilistic analyses.

5 APPLICATION EXAMPLES

Komex has used probabilistic modelling to estimate the EL of more than 10,000 oil and gas assets in North America. In general, these projects fall into three categories: environmental liability management and reporting, due diligence and property negotiations. Example applications of the process are presented below.

5.1 *Environmental liability management and reporting*

The disclosure of ELs as a balance sheet liability provision is required by the Canadian Institute of Chartered Accountants (CICA 1996). This annual reporting requirement reflects the accrued liability for site abandonment and restoration (A&R) obligations. The site A&R provision can be affected by a number of issues, including changes to net asset holdings and environmental conditions at corporate sites. Further to the CICA requirements, the Securities Commissions of Ontario, Quebec and Saskatchewan require public disclosure of environmental protection activities in their Annual Information Form and the Management's Discussion and Analysis (Tohana 1995).

5.1.1 *Corporate balance sheet - 1998*

Following the merger of two major Canadian oil and gas corporations, the new business entity required a reliable basis for estimating the site A&R obligations of its combined pool of assets. This would be used to ensure that the reclamation and abandonment provisions for both operations were reported consistently.

The project consisted of the evaluation of environmental liabilities for ~5200 assets comprised of ~4100 wellsites, ~220 facilities and ~890 single well batteries across western Canada and the northern USA.

Model results were compared to data collected through a survey of environmental costs. The survey was conducted through an in-depth analysis of environmental (i.e. reclamation and remediation projects) and accounting records in 13 intermediate and senior oil and gas companies.

Many difficulties were found in interpreting the survey data. These included:

1. A lack of activity-based cost accounting (i.e. accounting data for reclamation and remediation activities was too generalised to be useful);
2. Cost spreading over a limited number of general ledger (GL) accounts (i.e. reclamation and remediation costs were not consistently assigned to their GL accounts); and,
3. Salvage values were not consistently applied to project budgets or their reconciliations.

However, where reliable data could be extracted, the agreement between the survey data and model estimates established the predictive utility of the probabilistic modelling approach.

5.2 *Due diligence*

EL evaluations are also driven by the concept of *caveat emptor* ("buyer beware"). In Canada, environmental statute law (legislation) and the common law have established that buyers and sellers be aware of a site's potential ELs. In the case of buyers, it is important to know the nature and value of these liabilities for both environmental risk management and property negotiations. Furthermore, the onus is on the buyer to be aware of the site's patent defects, i.e. a defect discoverable through reasonable inspection (Davis 1992).

In terms of due diligence, the application of EL estimates to the purchase price can also benefit the vendor. In this case, the vendor divests a site's EL via a reduction in the sale price (this may act to protect the vendor from assuming ELs for the site in the future). Importantly, an EL evaluation may include both patent and latent defects. A latent defect is a defect the vendor has no knowledge of, and could not be expected to have acquired such knowledge (Davis 1992). Accounting for "potential" latent defects provides the vendor with further protection from future site EL.

5.2.1 *Asset divestment - 1999*

To provide negotiations data for the vendor and to comply with internal due diligence requirements, an EL evaluation was conducted prior to the divestment of ~1200 oil assets located in Alberta and Saskatchewan.

The project focused on existing and potential ELs associated with known and suspected soil and groundwater impacts resulting from operations. Included in the Phase I ESA method was a field inspection for a stratified random sample. The Phase II ESA program detailed assets known or suspected of having significant ELs through soil, groundwater and geophysical investigations.

The collected data were compiled and used to model ELs of the asset base.

5.3 *Property negotiations*

As seen earlier, the future cost of environmental damage at a site may be used by the purchaser to negotiate a reduced purchase price. An impartial EL evaluation provides a reference for all parties to begin negotiations and safeguards both the purchaser and vendor.

5.3.1 *Single site liability: third party review - 1999*

In negotiations for the acquisition of a battery, an EL estimate was completed by an outside consultant contracted by the purchaser. Due to the significant projected liability ($1,200,000) the vendor requested that the estimate be re-calculated.

Using the *LIABILITY ENGINE*™ [1] it was concluded that 97.5 times out of 100, the EL would fall below the previously estimated amount. It was recommended that a value ~$450,000 (or ~38%) lower be used in negotiations.

Using the results of this modelling process, the vendor was able to negotiate a substantial increase in market value of the asset. This illustrates the improvement in negotiating power that a Monte Carlo approach can offer in the face of economic uncertainty.

5.3.2 *Take-over: no access to field or headquarters data - 1999*

Using previous modelling experience, the EL of ~1000 assets in western Canada was estimated to support a take-over. This project was completed based on available data from the public domain without visiting a single site or contacting the target. The methodology used to access data sources in concert with the predictive capacity of the model makes this approach feasible in hostile take-over scenarios.

5.3.3 *Asset acquisition - 1999*

In support of the purchase of ~520 assets, including ~480 wellsites, ~30 small facilities and 4 gas plants, an EL evaluation was conducted using data collected through Phase I ESA investigations. The evaluation methods included site inspections of a stratified random sample of the assets.

The resulting data were summarised to present ELs for each operating field and were used in the final calculation of the bid price.

6 CONCLUSIONS

Due to the complex and multivariate nature of EL evaluation, it has been our experience that a probabilistic modelling approach is better suited to estimate the range of potential outcomes than traditional approaches. Specifically, the probabilistic approach for modelling environmental liability:

1. Has proven itself as a defensible and acceptable means of estimating ELs for negotiations;
2. Can be modified to address the unique requirements of the assets under evaluation;
3. Is continually evolving towards providing concise, bounded estimates; and,
4. Provides many advantages over traditional approaches, including reduced evaluation costs and a more rapid turn-around.

REFERENCES

American Society for Testing and Materials Testing (ASTM) 1999. *Committee E50 on Environmental Assessment. Standards Tracking Document.* http:// www.astm.org/ COMMIT/ E50TRAC.doc. January 10, 2000.

American Society for Testing and Materials Testing (ASTM) 1997a. *E1527-97 Standard Practice for Environmental Site Assessment: Phase 1 Environmental Site Assessment Process.* West Conshohocken, PA.

American Society for Testing and Materials Testing (ASTM) 1997b. *E1903-97 Standard Guide for Environmental Site Assessments: Phase II Environmental Site Assessment Process.* West Conshohocken, PA.

Canadian Institute of Chartered Accountants (CICA) (1996). *CICA Handbook.* CICA. Toronto, Ontario.

Canadian Standards Association (CSA) 1994. *Standard Z768-94: Phase 1 Environmental Site Assessment.* CSA. Toronto, Ontario.

Canadian Standards Association (CSA) 1998. *Standard Z769-98: Phase 2 Environmental Site Assessment.* CSA. Toronto, Ontario.

Davis, T. 1992. "Successor Liability for Toxic Real Estate" in *Environmental Law and Intensive Short Course for Practitioners: June 8 to June 12, 1992.* Calgary: Canadian Institute of Resources Law, University of Calgary.

Funk, W.A. 1999. *Estimating and Managing Environmental Liability in the Upstream Oil and Gas Industry.* A Master's Degree Project submitted to the Faculty of Environmental Design, University of Calgary. Calgary, Alberta.

Konar, S. & R.L. White 1999. Proposed new ASTM Standards for Estimating Environmental Liabilities Signal a Preference for the use of Decision Analysis and Expected Cost Analysis. *Strategic Environmental Management.* 1(3): 185-206.

Tohana, B. 1995. *Impact of Environmental Issues on the Existing Financial Reporting Framework in the Environmental Manual for Business and Professionals.* D. Wainman and G. Fords eds. CICA, Toronto, Ontario.

United States Environmental Protection Agency (USEPA) 1996. *Valuing Potential Environmental Liabilities for Managerial Decision-Making: A Review of Available Techniques.* EPA 742-R-96-003. Office of Pollution Prevention and Toxics (MC7409), Washington, D.C.

Environmental Issues and Management of Waste in Energy and Mineral Production, Singhal & Mehrotra (eds)
© 2000 Balkema, Rotterdam, ISBN 90 5809 085 X

Economic model for environment protection in lignite mines

J.N.Singh, S.Nakra, R.Arora & N.Parmar
Gujarat Industries Power Company Limited, India

ABSTRACT: In recent years, environment concerns have attracted a great deal of attention worldwide. It has been acknowledged that the danger of environmental degradation is threatening the very existence of mining industry. Simultaneously ecological awareness and sensitivity have spawned a new commitment, on part of the industry, towards the environment. Pertinent to note in this paper is that an economic model has also been developed, which describes the exploitation of Lignite and its interaction with rest of the economy, including the environmental consequences, to ensure that the damage caused to the environment and society is contained within acceptable limits. This model has been implemented as a workable and viable alternative for environment protection in a lignite mine in India.

1. INTRODUCTION

In recent years environment concerns have seriously threatened the very existence of mining industry. Environmental concerns have assumed a great deal of significance at the same time ecological awareness and sensitivity has thrown up a series of questions that has spawned a new commitment.

The key questions raised in environment analysis are:

1. What is the likely damage to the environment from a particular project?
2. How can the damage to the environment be contained within acceptable limits?
3. How can damage to the environment be compensated for?
4. What is the cost of restoration measures?

This paper outlines the influence of Mining activities on the environment, including the socio-economic aspects of the region as well as to find out a reasonable cost of mitigation to ensure that the damage caused to the environment and society is contained within acceptable limits. In this regard an economic model has been developed, which describes the exploitation of Lignite and its interaction with rest of the economy including the environmental consequences.

The first question mentioned above is addressed in the *Impact assessment,* and the next three questions which are related to *Technical Aspects, Socio-economic Aspects and Financial Aspects* are addressed in the course of development of the *economic model for Environment Protection.*

2. IMPACT ASSESMENT

Mining is divided into three phases' viz. Pre-mining phase, Active mining phase and Post mining phase. The sources for different component activities of mining environment discharge are listed in Figure 1.0. There are gaseous discharges like CO_2, CO, NO_X, SO_X, Pb and air borne dusts due to plying of heavy equipment fueled by diesel. The liquid effluent (water) discharged from the mines pollutes surface water bodies and can effect the soil. Solid wastes (overburden dumps) are a long-term source of pollution to surface/ ground water, air and aesthetics. The environmental pollution in the post-mining phase continues for a long time. For the above three phases of mining, effects of environmental pollution need to be quantified. At the same time resources deployed should also be quantified and then converted to monetary values. Such a study, conducted at the planning stage, will help in choosing the most suitable mining scheme, environmental protection scheme, resource conversion, recycling strategy, environmental management, reclamation scheme and scheme for post-mining land use.

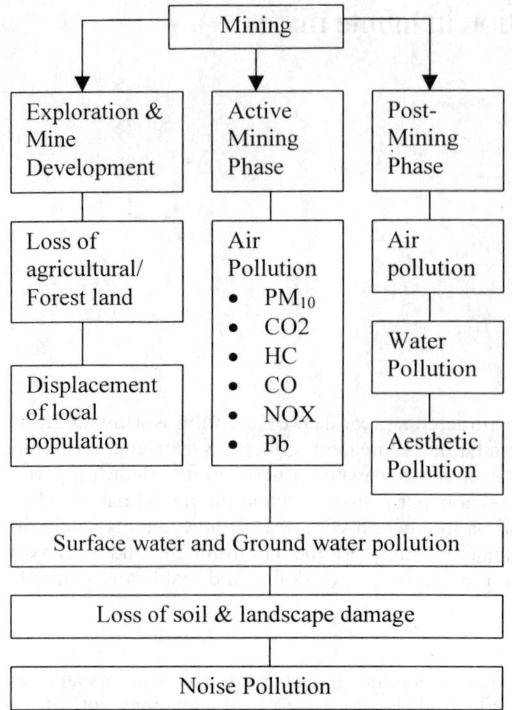

Figure 1.0: Environmental discharge model

3. ECONOMIC MODEL FOR ENVIRONMENT PROTECTION

While gigantic measures may be undertaken to ensure that the impact of industrial activity is mitigated, one can never claim that no environment degradation has resulted as a consequence of the activity. In view of this undisputed hypothesis an effort was made by GIPCL to meet the situation via a three-pronged strategy, viz., Technical, Socio-economic and Financial. Economic model for Environment Protection is thus, built on 1. *Technical Aspects,* 2. *Socio-economic Aspects,* and 3. *Financial Aspects.*

In the light of what has been acknowledged in the foregoing an attempt has been made to build an economic model to address the issue rationally. The model states in detail, and in quantitative terms, the way in which the various aspects of the economy are interrelated. The economic model provides the forecaster with a record of the prediction and a clear statement of the assumptions concerning exogenous variables and the model solution.

3.1 Technical Aspects

It is reasonable and relevant to examine the critical effects due to mining activity, its cause, and design remedial action accordingly. Environmental implication from open cast lignite mining could involve issues like, Land degradation, affects on hydrological conditions and existing water courses, water quality and drainage, air pollution, noise pollution, flora and fauna, health and safety, re-settlement of the ousted population, if any, etc. Remedial measures to restore environment are listed below. The technical aspects are schematically shown in Figure 2.0.

- Environmental monitoring for SPM, NOx, SOx, CO emission, water quality, Dust fall measurement, noise quality and quality of the water pumped out during monsoon help the management in exercising control and applying corrective measures in time.
- Control measures such as water sprinkling on haul roads, lignite stockpile, around infrastructure and good preventive maintenance schedules help to reduce air pollution.
- Provision of garland drains around the mine and waste dumps,. proper drainage arrangement for mine water, installation of sewage treatment systems for domestic effluents from colony and industrial area, provision of sedimentation tank for water discharged from mine, etc., are suggested to ensure that water quality remains within prescribed standards.
- Control measures for noise level abatement includes good preventive maintenance practices, greenbelt development, provision of earmuffs and earplugs, etc.
- Development of green belt and plant nursery ensures maintenance of good ecological and environmental quality after the commencement of full-scale mining operations in the area.
- Plantation activity on surface of dump yard goes a long way in ensuring dump stability and also arrest the run-off of loose soil. This plantation can be carried out by physical plantation of saplings and also by broadcasting mixed species of seeds on flat and sloping surfaces of dump yards.

Remedial measures as suggested in the Environment Management Plan (EMP) are expected to improve the environmental conditions by suitably reducing the negative effects and giving positive impact on some of the environmental attributes. In this connection environmental impact matrix, due to mining and allied activities, was drawn and analysed. The Matrix, which was prepared giving comparative weightage without adoption of mitigatory measures, shows an index value of (–) 1225. However, the Environmental Impact Matrix

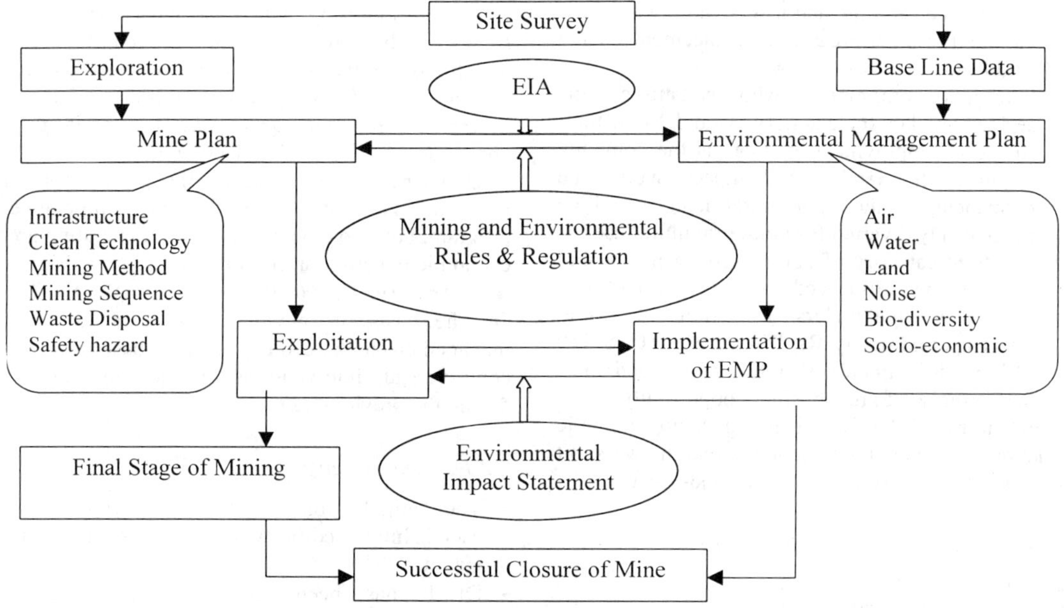

Figure 2.0: An approach for Sustainable Closure of Mine

with adoption of control measures scales up the index value to (+) 667 from (–) 1225, thereby showing that the project will have a net positive effect on an overall basis, mainly on account of positive impact on socio-economic factors and greenbelt development.

3.2 Socio-economic Aspects

Besides direct employment opportunities in the project, indirectly a large number of people benefit by way of opportunities in transport sector, ancillary industries, etc. Gujarat and Central Governments also derive financial benefits through receipt of royalties, cessess, excise, etc. As such impact on socio-economic factors, due to the project, will be positive. Accordingly, overall impact on the environment will be minimised due to project implementation.

However, setting up of a 250 MW lignite based power plant with its captive Lignite mines was certainly expected to directly or indirectly impinge upon the socio-economic structure of the surrounding villages. The project being set up exclusively in an agricultural area, threw up some basic issues ranging from, sustenance of livelihood, business prospects for local population, availability of surplus funds with the community to inflow of huge sums of money in the local market.

Thus, GIPCL, as a responsible corporate citizen, decided to share social responsibility of this area with the state.

3.2.1 Development Efforts for Rural Economy & People (DEEP) – an success story

Realising its responsibility, at the inception of the project, the Company had promoted an organisation named 'DEEP' to exclusively undertake Infrastructure & Community Development Programs in its area of operation.

DEEP is a secular, Not-for-Profit, Non Government Organization (NGO). It closely works at grassroots level and formulates appropriate projects, through participation of local community. Its focus is to facilitate empowerment of the local community and to create, socio-economically, a better society.

3.2.1.1 Situation analysis

(a) Mining of lignite required acquisition of land in the surrounding villages to the extent of 1536 hectares. The consequences of this activity were bound to permanently change the course of life of the community.
(b) Land owning community would loose its land (and with it, its earning capacity) but get in return more than reasonable monetary compensation. It was anticipated that, with the

money received, the land losers would be able to buy land in other areas as a replacement for the land lost.

(c) Land-less community, who constitutes the agricultural laborer category, would loose their source of employment as soon as the company acquired the land. The impact would be permanent, as the agricultural land would be permanently diverted for non-agricultural use.

(d) The third category of community, which would be positively influenced by the arrival of this project, was the traders. The company was to inject approximately Rs 1500 crores (US$ 350 Million) in a matter of a few years, and thus, held promise of great business opportunity.

In the midst of the fast changing socio-economic equation it was noted with concern, that the worst hit section of the society would be the land-less class.

3.2.1.2 DEEP's Journey - Initial Years

- *Year 1997-98*

The basic goal of DEEP, in the first year of its inception, was to build rapport, develop goodwill, and establish credibility amongst the surrounding community. Nevertheless, income generation activity for land-less class of project affected people remained the top priority and included projects like:

(a) Livestock Development: This scheme infused enthusiasm amongst other members of the community too.

(b) Green Belt Development: The project was perceived as a drive hard income generating scheme as substantial returns were to come at the end of the project period of 2 years.

(c) Community Farming: The project had a conceptual deficiency but was carried out albeit with marginal returns.

Similarly, infrastructure development projects to address the basic needs of the community included

(a) Scheme for supply of Drinking Water, and

(b) Improvement of School Infrastructure.

This intervention created a positive impact and revealed the humane face of the organisation.

Though this was the first year of operation for DEEP, efforts were focused towards streamlining its direction. Response from people especially poor & underprivileged was slow but encouraging.

- *Year 1998-99*

Aforesaid programs were repeated with a modified design. A few new programs were included:

(a) Sanitation

(b) Village Infrastructure Development

(c) Water Resources Development

(d) Natural Resources Management

Community farming, which was lauded as a success story of the previous year, proved to be an anticlimax in the second year as it led to class conflicts. Benefits derived by land-less community during the previous year were too good to be appreciated by the original landowners.

Green belt development (GBD) project not only was a successful environment improvement program, but also emerged as model generating great economic benefits for the community.

- *Year 1999-2000 (Present)*

On the basis of previous two years experience management took recourse to need based and least controversial, interventions on the same lines as during the previous years.

3.2.1.3 Achievements

- The initial goal of building rapport and establishing credibility was achieved to a great extent.
- DEEP has been recognised as a reliable implementing agency for rural development programs of Government of India.
- A new intervention in the form of "services to promoters" has emerged as the most promising source of funds for achieving the goal of sustainable development.

3.2.1.4 Lessons learnt

In ones preoccupation to carry out selfless services towards the community disharmony between various power centers in the locality can arise.

3.3 Financial Aspects

To determine the financial justification of environment protection, economic model is developed. The model is based on input output flow diagram. Initially a schematic line diagram (Figure 3.0) is made on which a suitable mathematical model is built.

3.3.1 Cost – the main issue in environment protection.

Estimates of the costs, as a proportion of total production costs, vary from negligible to crippling. Those who are trying to demonstrate that the costs are low, point to examples of mines where investments attributable for environmental reasons have led to higher productivity and lower costs. Whereas, those who aim to show that the costs threaten the industry's existence quote the full costs of investments, including elements that have little, if

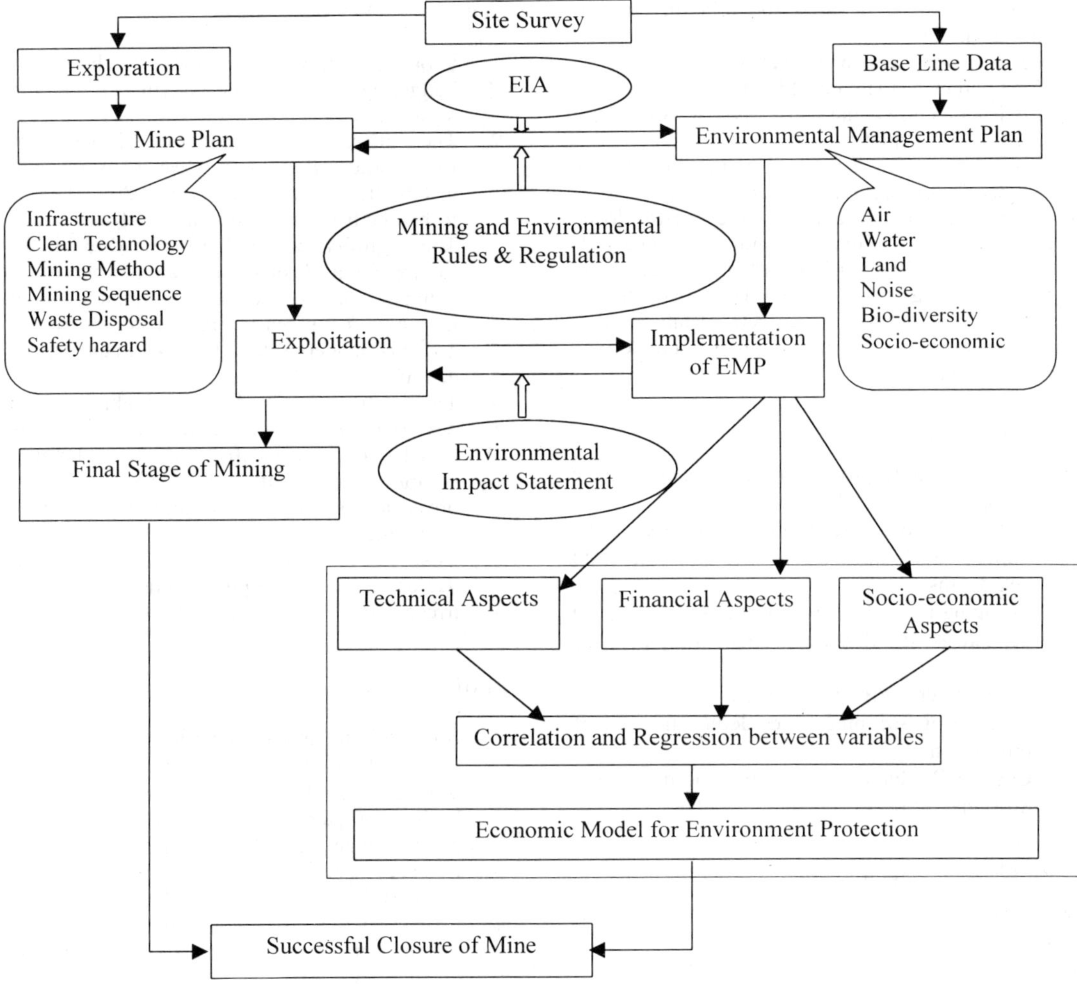

Figure 3.0: Economic Model for Environment Protection

anything, to do with environmental requirements. It appears reasonable to assume that the costs of environmental protection are indeed positive and significant, but that they do not constitute an important constraint on the expansion of production.

The industry has demonstrated an impressive capacity to adjust to regulatory changes (usually after first having complained bitterly about the costs) and has achieved results, when it comes to reducing environmental impacts, which would hardly have been believed two decades ago. Clearly, the industry has realized that the costs of project delays and possibly closures, due to the need to take mitigating measures that were not foreseen but which impose themselves as a result, for instance, of changed regulations, are likely to be higher than the cost of doing it right from the outset.

3.3.2 Mathematical Model for Environment Protection in Lignite Mines

The Mathematical model states in detail and in quantitative terms the way in which various aspects of the Mining scheme are interrelated. The model provides the forecaster with a record of the prediction and a clear statement of the assumptions concerning exogenous variables and the model solution.

The key issues involved in development of an Model for Environment Protection in Lignite Mines are as follows:

- Is there any relation between stripping ratio, depth, quantum of excavation, cost of production and cost of environment protection measures?

- If relation exist, what is the relation between these variables?

The design parameters considered are:

- Cost parameters (Fixed and Variable)
- Environmental parameters (Noise level, Air quality, water quality, etc)
- Socio-Economic factors (Health, Population-Migration, Employment, Literacy etc.)

Based on these factors and the policy of the local, state and central governments, a model is conceived.

Correlation and regression analysis is carried out from the mining scheme and Technical, Socio-economics and Financial aspects. Mathematical Model for Environment protection is then derived from these analyses.

3.3.2.1 Assumptions

1. Annual Lignite production is same for all the stripping ratio and depths.
2. Lease area or mine property size and shape remains unchanged for all the stripping ratio and depths.
3. Stripping ratio changes with the size of lignite deposit, seams thickness, strike, dip, geology etc.
4. Environment Protection Measure Cost is a function of Stripping ratio, depth and cost of production.
5. Cost of Production is function of depth and Stripping ratio.

3.3.2.2 Correlation and Regression

The following relationship is derived from correlation and regression analysis:

1. Relationship between Cost of Environment Protection Measures v/s Stripping ratio
2. Relationship between Cost of Environment Protection Measure v/s Cost of Production at 40, 80 and 110 meter depths.

Out of these relationships Mathematical Model for Environment Protection Measure is derived which is shown below:

$Y_{40} = 0.08517\ X_1 + 0.00212\ X_{40} + 3.575$ ------ I

$Y_{80} = 0.08517\ X_1 + 0.00187\ X_{80} + 3.571$ ----- II

$Y_{110} = 0.08517\ X_1 + 0.00158\ X_{110} + 3.569$ ---- III

Where,

$Y_{40,80,110}$ = Cost of Environment Protection Measure in Rs. per Tonne of Lignite for 40, 80 and 110 m depth

X_1 = Stripping ratio

X_{40} = Cost of production of lignite upto 40 meter depth

X_{80} = Cost of production of lignite upto 80 meter depth

X_{110} = Cost of production of lignite upto 110 meter depth

4. CONCLUSIONS

1. Economic model provides a basis of budgeting for Environment Protection Measures.
2. The efficacy of Economic Model for Environment Protection Measure is realized in GIPCL's Lignite Mines in India, where it forms the basis of action plan for future.
3. The significance of Economic Model for Environment Protection cannot be over-emphasized. Irrespective of uncertainties that may exist or arises the model is worthwhile and advantageous to all future management actions.
4. The principles of sustainable mine development in Vastan Lignite Mines, is based on scientific mine planning complimented with due weightage to protection of environment and also to socio-economical development.
5. Cost of Environment protection measure decreases as stripping ratio decreases, irrespective of change in depth.

5. LIMITATIONS

1. Since mining is site specific, it may not be possible to use a known model at all places.
2. This exercise has been carried out on the basis of projections for 10 years, to be more precise longer periods are advisable.
3. It is necessary to obtain authentic data so that precise relationship is established.
4. In an attempt to accommodate as much information as can be in this short space the quintessential have been dispensed with.

REFERENCES

Dhar, B.B. (1999) *Mining Environment Scenario Beyond 2001(1999)*; Proceeding of International Conference on Mining Challenges.

Dhar, B.B. (1996): *Environmental Management System for Closure and best practice in Indian Mining Industry*: Sixth Workshop of the Mining and Environmental Research Network 4-9 August, Harare.

Dube, A.K. (1999): *Life Cycle Assessment Analysis for Indian Mining Industry _ A Clean Environment Approach*: International Conference on Mining Challenges of the 21st Century during 25-27 November 1999 at Hotel Taj Palace, Delhi.

Ghosh, R. & A. Rani (1999) *A step toward Eco-friendly coal mining in India*, proceedings of the International Symposium on Clean Coal Initiatives, N. Delhi, pp 587-592

Hendnickson, C., Horvath, A. Joshi, S & Lave, L. (1998). *Economic Input-Output models for Environmental Life-Cycle Assessment*, Environmental Science & technology, April 1, p p 184A.

Environmental Issues and Management of Waste in Energy and Mineral Production, Singhal & Mehrotra (eds)
© 2000 Balkema, Rotterdam, ISBN 90 5809 085 X

Using environmental management systems to systematically improve operational performance and environmental protection

Tony B. Szwilski
College of Information Technology and Engineering, Marshall University, W.Virg., USA

ABSTRACT: Environmental management systems (EMSs) provide industry an opportunity to manage their environmental affairs more effectively. An EMS is an effective management tool to measure and manage organizational operational performance and hone industry 'best management practices'. In the United States, government regulatory authorities are working with industry to establish innovative permit processing using ISO 14001 EMS specifications a minimum required. The objective is to improve environmental performance by providing industry the flexibility to be innovative and use the most appropriate and economical options. EMSs are expected to be a vehicle for an environmental regulatory enforcement paradigm shift.

1 ENVIRONMENTAL MANAGEMENT SYSTEMS (EMS)

1.1 *Introduction*

Industry and governments worldwide are increasingly adopting "credible" environmental management systems (EMSs), especially since the development of the eco-management and audit scheme (EMAS) in Europe and ISO 14000 standards. Anecdotal evidence indicates that EMSs can meet the needs of industry and environmental regulatory authorities. Since developing an EMS and attaining certification to ISO 14001, Vauxhall Motors have reduced energy and water-usage costs by 5% in the first year, and Hewlett Packard has reduced energy costs 25% and water usage 50% over eight years (Business Standards). Over a period of 18 months since its ISO 14000 certification, Beers Construction Co. saved $230,000 in waste removal costs through reducing, reusing, and recycling; 50,000 cubic yards of waste was diverted from landfills (IESU December 1999).

EMSs can provide state and federal regulatory authorities the capability to achieve greater environmental protection utilizing scarce resources. Current regulatory initiatives in the United States are provoking a new regulatory policy paradigm change, away from the three decades old 'command and control' to a more industry self-manage system.

1.2 *What is an EMS?*

An environmental management system (EMS) is an integral part of an organization's management structure that addresses the immediate and long-term beneficial and adverse impacts of a company's business activities (products, services) on the environment. An EMS is "the organizational structure, responsibility, practices, procedures, processes, and resources for developing, implementing, achieving, reviewing an maintaining the environmental policy."[1] The total quality management (TQM) based 'plan-do-check-act' EMS provides a credible mechanism to systematically improved the operational efficiency of a firm and environmental protection.

1.3 *International Standards for EMS Development*

The International Organizational for Standardization (ISO) develops voluntary technical standards that add value to business operations and are required by the market place. It is made up of national standards institutes from large and small countries (ISO Publication 1998). A principal objective of the ISO standards is to contribute to the operational efficiency, safety and environmental friendliness of business activities.

ISO/TC 207 is the technical committee responsible for developing and maintaining the 'family' of standards:

- environmental management systems
- environmental auditing
- environmental performance evaluation
- environmental labeling
- life cycle assessment

[1] ISO 14001 EMS: 3.Definitions

- environmental aspects in product standards

1.4 *Relationship between ISO 9000 and ISO 14000*

The first ISO EMS standard ISO 14001 was published on September 1, 1996, and follows the success of ISO 9000 quality management system standards which were published in 1987. Predictably, the manufacturing industry has demonstrated considerable interest in the EMS initiative, whereas the mineral industry has shown just a luke-warm interest (IESU August 1999). See Table 1 (ISO Survey 1998). The Basic Metal and Optical industrial sector (EAC 39) has the highest number of ISO 9000 and ISO 14000 certifications: 36,563 and 2,147 respectively (December 1998).

Table 1. World ISO 9000 and ISO 14000 Certifications. 1998

Certifications	Mining & Quarrying*	Non-metallic Mineral Products**	Total+
ISO 9000	1,052	2,325	229,846
ISO 14000	88	88	7,112

* Industrial sector (EAC 22). ** Industrial sector (EAC 35).
\+ Total world certifications

Table 2 lists the countries with the highest number of ISO 14000 EMS certifications awarded, as of October 1999.

Table 2. ISO 14000 (EMS) Certifications.

Country	Certifications	Country	Certifications
Japan	2,531	United States	520
Germany	1,460	Taiwan	506
United Kingdom	1,009	Canada	115

2 EMS: A TOOL TO IMPROVE OPERATIONAL PERFORMANCE

As previously noted, anecdotal evidence accrued from 3 years of experience with ISO 14000 and EMAS guided EMSs indicates that there are numerous benefits for companies that fully commit to a credible EMS.

An EMS is a management tool for a company to systematically identify the key processes, procedures and activities associated with *the significant environmental aspects*.[2] Appropriate metrics are selected and monitored. This continuous feedback process provides information to improve environmental performance and an opportunity to codify valuable technical knowledge. Although for industries, such as the mining industry, knowledge and know-how required to in performing tasks and operations is largely tacit in nature and not easily codified.

[2] ISO 14001: 4.2:Planning.

For optimum operational performance the principal processes should be effective and controlled. This can be done by

- Monitoring
- Auditing and corrective action
- Continually identifying potential problems and search for new and innovative solutions
- Collecting, analyzing and communicating opportunities for lessons learned and best practices (Husley 1999)

3 ESSENTIAL TOOL FOR IMPROVED ENVIRONMENTAL PERFORMANCE

3.1 *Environmental Performance Metrics*

With the development of the ISO14000 EMS standards and the Environmental Performance Evaluation (EPE) standards (ISO 14031) in particular, more attention is being giving to the use of metrics to measure environmental performance. Interest in industrial environmental performance metrics is increasing as companies find new internal uses for information in marketing and product development applications. More companies are beginning to view environmental performance as an area of potential competitive advantage. However, the true potential and advantages of EMSs remain invisible to most companies and industries such as the mineral extraction industry. A recent study (NAE 1999) identified several challenges to the development of a broad, robust and effective system of environmental metrics: improving standardization of metrics both within and across industry sectors; disseminating best practices more widely; applying metrics across the entire supply chain and product life-cycle; develop new analytical tools; and ecosystem health and sustainability.

The most important aspect of environmental performance metrics is to measure what management needs to enhance decision-making. Information generated should be directly linked to key activities of the business: operation processes, logistics, natural resource utilization. The EMS should perform the dual function of providing continuous feedback of information from the business activities that indicates progress toward minimizing environmental impacts or maximizing environmental protection, and improving best practices in key processes. For example, ISO 14031 environmental performance evaluation standards includes types of metrics: environmental performance indicators (EPIs), management performance indicators (MPIs), operational performance indicators (OPIs), and environmental condition indicators (ECIs) (Wilson 1999). MPIs should provide information on a company's implementation of the EMS, such as, progress towards objectives and targets; progress toward training of employees; allocation of resources for programs within the EMS. The OPIs should provide management with information on the environmental performance of its operations, such as, quantity of hazardous wastes per year per unit product. The ECIs should provide information on the condition of the environment, such as, properties of major water bodies and regional air quality.

4 BEST MANAGEMENT PRACTICES AND ENVIRONMENTAL MANAGEMENT

Best management practices (BMPs) are a discrete set of processes and procedures involved in the production of a product and/or service. An effective EMS will be able to provide continuous feedback by which BMPs can be reviewed, modified and enhanced

4.1 *Australian initiative: BMPs for mining*

At the International Association of Impact Assessment (IAIA) meeting in Ottawa 1994, attended by representatives from the World Bank and the United Nations Environmental Program (UNEP), Australia was asked to take the lead in show-casing examples of best environmental practice in all aspects of the mining industry.

The Australian Environmental Protection Agency partnered with the mining industry to publish a series of very useful modules and videos which demonstrate the best practice of the leading environmental managers in mining and energy production in Australia. As described in the publications (Overview 1995), '*The best practice environmental management in mining*' focuses on the principle of environmental impact assessment and environmental management. Utilizing case studies, it demonstrates how these principles can be integrated through all phases of resource development from pre-resource planning, through construction, operation, closure and post-mining monitoring and maintenance.' This was the first attempt by the Australian mining industry in benchmarking environmental performance across a broad range of mining life-cycle issues with a view to encourage further improvements in environmental performance.

One of the modules examines how mine planning for environmental protection can help mining projects meet community expectations for minimal environmental impacts (Mine Planning 1995). Factors taken into account are air, water quality, noise, transport, biological resources, socio-economic issues, and land use. In the module, *Best practice in mine planning for environmental protection* is defined as "the application of a process of continuous testing and evaluation of different mine design options to satisfy community expectations, government requirements, engineering and costs considerations, and the condition of minimal environmental impacts." Mine plans are evaluated through a public environmental assessment process.

4.2 *Forestry Best Management Practices (BMPs)*

The state of Wisconsin Department of Natural Resources has published forestry BMPs (1995) for water quality. The booklet is intended to help loggers, land owners and land managers be good stewards by protecting water quality during forest management activities. In the past couple of years sustainable forestry management criteria and indicators have been developed (IESU December 1999). This is being done together with the Forest Steward Council standard and American Forest and Paper Association Sustainable Forest initiative, using the ISO 14000 EMS standard as a framework.

5 FEDERAL GOVERNMENT AND STATE INITIATIVES

5.1 *Greening of Government*

In November 1999, President Clinton gave notification to all federal agencies to begin environmental management systems pilot projects by March 31, 2002 and establish an effective EMS at all the federal agencies by the year 2005. The federal mandate will effect all 16 departments in the executive branch which includes more that 15,000 federal facilities (IESU November 1999).

5.2 *Multi-state Working Group*

In 1996, a voluntary group of participants from several states including California, Oregon, Pennsylvania and federal environmental protection agency (USEPA), nongovernmental, business and higher education organizations established an informal voluntary "multi-state working group" (MSWG). The purpose of MSWG is to conduct research on the ability of EMSs to improve the state of the environment and the economy and to evaluate utility in public and private policy innovation (Andrews 1999).

5.3 *Green Permits*

In 1997 the State of Oregon legislature created a Green Permits system to encourage regulated facilities to achieve environmental results that are significantly better than would otherwise be expected by law. The legislation encourages the Environmental Quality Commission (EQC) to use innovative environmental approaches or strategies to accomplish better environmental protection. The approach is based on the use of EMSs such as ISO 14000, and a 'tiered' system in which greater demonstrated environmental performance is acknowledged with increasing regulatory flexibility or 'benefits'(Oregon DEQ 1999).

The principal reason for the regulatory innovation in Oregon is the realization that:

- The existing regulatory system cannot adequately address many environmental issues and does not encourage or reward environmental stewardship.
- Voluntary, market-driven, outcome-based approaches can be effective in accomplishing the desired environmental results.
- Many companies have the knowledge, skills and resources to significantly reduce environmental impacts of their business activities.

The Green Permits program was developed after Oregon studied several regulatory reform projects in the United States and other countries. The key principles of the program are:

- Environmental performance should exceed minimum compliance requirements.
- Significant and measurable environmental performance goals should be established.
- Meaningful stakeholder involvement is expected.

Two types of Green Permits are being considered. The *Custom Waiver Permit* allows limited waivers of environmental laws if the waiver is necessary to achieve superior environmental results. The *Green Environmental Management System Permit*, or GEMS Permit also allows waivers of environmental requirements, but requires the use of a formal environmental management system to achieve results.

Similarly, the state of Wisconsin has established a *Green-Tier* system for the state's businesses using contracts, flexibility, incentives and EMSs to achieve grater environmental performance (Meyer 1999). Entry into the Green Tier by regulated businesses is earned by accomplishments recorded in the Control Tier. Performance contracts must contain an EMS and involve business, government and include public-interest groups. Performance incentives include regulatory relief, direct financial tools such as tax credits and indirect financial tools such as procurement, leasing and construction policies.

5.4 *The Second Generation of Environmental Improvement Act of 1999 (HR 3448).*

This new bipartisan bill is being considered by the U.S. Congress. The objective of the bill is to reform the present command and control environmental regulatory enforcement paradigm that has existed for more than three decades to a credible new environmental management model for the next century. The new bill emphasizes promoting greater creativity in the private sector to enhance environmental improvement. The basic premise being that the most appropriate means to optimize environmental performance are affected by market-based incentives and regulatory flexibility and effective methods of collecting, managing and disseminating environmental information. Presently, reporting required under air, water and waste regulations produce a vast amount of information and is a significant economic burden to business. The bill is expected to become legislation in the year 2001.

5.5 *ISO 14000 and Global Climate Change*

In October 1999, Bonn, Germany, the ISO and the United Nations Framework Convention on Climate Change (UNFCCC) engaged in preliminary discussions on how ISO 14000 might be applied to the issue of global climate change. That is ISO 14000 applied to gather greenhouse gas emissions (GHG) data and establish emissions reduction programs that include monitoring, reporting and certification (WTO Report 1999). The FCCC (Thomas 1999) commits industrialized countries to:

- Adopt policies to mitigate climate change by reducing GHG emissions and enhancing sinks.
- Communicate their policies and measures with respect to the aim of returning GHG emissions to 1990 levels.

An ISO 14001 EMS may be used to identify energy use contributing to GHG emissions and natural resources that impact carbon sinks.

6 INDUSTRY AND BUSINESS INITIATIVES

6.1 *Energy sector*

Energy sector entities such as electric utilities, oil and gas companies are being faced with a high level of uncertainty regarding possible binding GHG emission constraints. Among the energy sector companies committed to ISO 14000 (Thomas 1999) are:

- BP Amoco: To certify all European sites by 2000.
- Con Edison: Transportation & stores certified.
- Epcor: First fossil-fuel power plant in Canada.
- Mobil: Facilities in Germany, Scandina via, Singapore, Egypt.

6.2 *Motor industry*

Ford Motor Corp. and General Motors Corp. announced in October 1999 a mandate requiring their supply chains to certify to ISO 14001. As a result, companies such as Bethlehem Steel are preparing for ISO 14001 certification. Suppliers to large companies and government agencies that are certified to ISO 14001 specifications will be expected to follow suite.

6.3 *London Stock Exchange*

Due to an initiative by the Institute of Charted Accounting (ICA) the London Stock Exchange (LSE) has issued a mandate requiring the 2,900-members to include in their annual financial reports, the identity, scope of the financial and environmental (and other) risks.

The annual report to the LSE must:

- Disclose the internal process for identifying, evaluating and managing significant risks;
- Identify the board of directors as responsible for risk management;
- Describe how the board dealt with risk problems during the period of the report;
- Disclose if the board failed to conduct an internal review of the risk management system.

6.4 *World Trade Organization (WTO)*

In a recent speech (Ruggiero, WTO 1999) the concept of a World Environmental Organization (WEO) was suggested. The WEO would be an institutional and legal counterpart to the WTO. That is, global environmental management should be an integral part of trade globalization.

7 CONCLUSIONS AND RECOMMENDATIONS

The worldwide use of environmental management systems (EMSs) is providing industry an opportunity to better manage their environmental affairs and imprive environmental performance. Incorporating EMSs as the basis of innovative permit processes they give organizations the flexibility to use the most appropriate and economical options to achieve beyond regulatory compliance and continuously improve its environmental performance. For federal and state regulatory authorities, EMSs promise to be a valuable mechanism to improve environmental protection with minimal resource use.

Significantly, EMSs provide a mechanism to employ knowledge gathered from 'lessons learned' from business activities leading to improving operational performance. An EMS is an information system that provides continuous feedback on the effectiveness of key procedures and processes from monitoring activities. Correction action incrementally adjusts and improves the appropriate business activities (procedure or process) and EMS.

Stakeholder involvement and full commitment from corporate management are essential elements of an EMS. Keeping the community involved and informed is an action that all industries should regard as a common business practice. Environmental impacts of mining activities, such as Mountaintop mining in West Virginia, is a growing issue (Szwilski). Following the initiatives of the Australian mining industry and green permit developments in numerous state regulatory agencies, a 'Green Mining Permit' (GMP) is recommended as tool for state and federal regulatory authroities and the mining industry to commence working toward long-term shared goals. The GMP would require a mining company to developing a credible EMS as a minimum requirement. The desired outcome is satisfactory environmental protection through continuously improvements in mining environmental performance and operational effectiveness.

ISO 14000 EMS standards are becoming an integral part of world trade and an accepted factor of business by small and large companies, and government. An effective EMS adds value to an organization.

REFERENCES

Business Standards published by BSi, November/December 1999.

IESU *(International Environmental Systems Update)*, Vol.6, No.12, December 1999.

Publication: ISO14000-Meet the whole family, 1998.

International Environmental Systems Update (IESU), Vol. 6, No. 8, August 1999.

The ISO Survey of ISO 9000 and ISO 14000 certificates. Eighth Cycle. 1998.

IESU: Vol. 6, No, 11, November 1999.

MSWG: http://www.mswg.org

Andrews, R. et al. The Effects of ISO 14000 Environmental Management Systems on the Environmental and Economic Performance of Organizations: Project Summary 1. March 1999.

June 1999 Update, *Oregon Department of Environmental Quality.*

Meyer, G. E . A Green Tier for Greater Environmental Protection, *Wisconsin Department of Natural Resources*, June 1999.

The World Trade Organization Report, October 1999.

Wisconsin Forestry Best Management Practices for Water Quality, 1995.

Overview of Best Practice Environmental Management in Mining, *Australian EPA & Mining Industry Publication (AMEEF),* June 1995.

Thomas, William, L., The Kyoto Protocol and Energy Sector Competitiveness: A Strategy for Flexible Response, *1999 Energy Conference*, San Antonio, Texas, October 1999.

Industrial Environmental Performance Metrics: Challenges and Opportunities, *National Academy*

of Engineering, National Research Council, National Academy Press, 1999.
Wilson, E. G., Environmental Performance Evaluation as an EMS Management Tool, *Multi-State Working Group Workshop*, Milwaukee, July 1999.
Mine Planning for Environmental Protection, *Australian EPA*, June 1995.
Husley, G., Implementing Lessons Learned Systems (CPI) @ Halliburton, *1999 Energy Conference*, San Antonio, Texas, October 1999.
Ruggiero, R., Director-General, World Trade Organization, *High-Level Symposium on Trade and the Environment*, 15 March 1999.
Szwilski, A.B. et al, Mountaintop Coal Mining in West Virginia: Environmental Issues, *Sixth International Symposium on Environmental Issues and Waste Management in Energy and Mineral Production*, Calgary, Canada, May 2000.

Environmental Issues and Management of Waste in Energy and Mineral Production, Singhal & Mehrotra (eds)
© 2000 Balkema, Rotterdam, ISBN 90 5809 085 X

Mountaintop coal mining in West Virginia: Environmental issues

Tony B. Szwilski, Betsy E. Dulin & James W. Hooper
College of Information Technology and Engineering, Marshall University, W.Virg., USA

ABSTRACT: Mountaintop coal mining in West Virginia is facing increasing scrutiny concerning its environmental impacts. The principal environmental issues regarding mountaintop mining, repercussions for future coal mining in the United States, and the potential for a application regulatory innovative 'green permit' concept based on a company developing a credible environmental management system are discussed.

1 INTRODUCTION

The environmental impacts of mountaintop coal mining in the Appalachian region is a crucial issue that could have serious repercussions for coal mining in the region. In October 1999, in response to a U.S. District Court's ruling on the legality of valley fills, the state of West Virginia Division of Environmental Protection (WVDEP) ordered that no new fill permits would be approved and no existing or permitted fills could be advanced in intermittent or perennial streams. On November 1, 1999, the WVDEP suspended the Order after the same Judge granted a Stay of his prior ruling pending appellate action in the U.S. Court of Appeals. These events are causing a great deal of consternation in the mining industry and the state administration as they potentially affect the future of coal mining in West Virginia and the Appalachian mountain region.

At the time the Surface Mining Control and Reclamation Act of 1977 (SMCRA)[1] was passed, there were very few mountaintop mining operations in the Appalachian mountains. However, the 1990's have witnessed an increase in the number and size of these operations, especially in West Virginia and Kentucky (see Figure 1). This is largely due to market forces and the availability of capital to larger mining companies to be able to utilize larger and more efficient earth excavation and moving machines such as draglines, shovels, dozers and trucks.

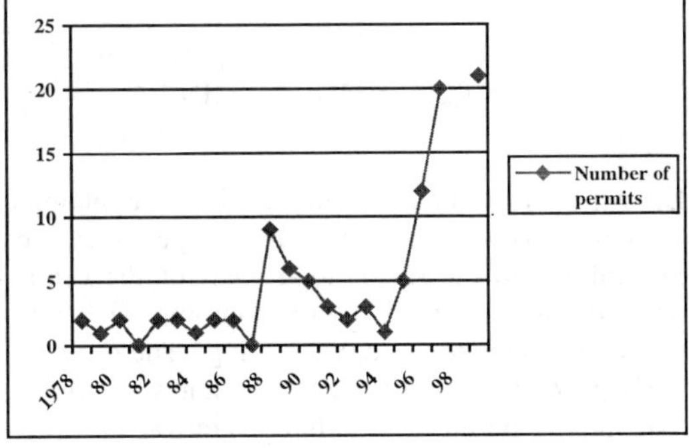

Figure 1: Number of permits issued to currently active mountaintop coal mining operations in West Virginia from 1978 to 1999. (After WVDEP)

1.1 *Value of Coal Mining to the Economy of West Virginia.*

The economy of West Virginia is more dependent on the mostly high volatile bituminous coal than any of the other twenty six coal producing states, contributing significantly to the state economy. In 1997, 631 mines in West Virginia produced 181 million tons: 125 million tons from 400 underground mines and 56 million from 231 surface mines, employing 18,279 persons. The U.S. coal production was 1.09 x 10^9 tons. West Virginia is the largest underground coal producing state in the United States, producing about seventy percent of the total and currently ranks second to Wyoming in terms of total annual coal production. Figure 2 shows the principal coal fields of the United Sates.

The total estimated coal reserves in the state amount to 54 x 10^9 tons, and the state has produced

[1] The Surface Mining Control and Reclamation Act of 1977 created the Office of Surface Mining (OSM), a federal agency that regulates surface mining in the United States.

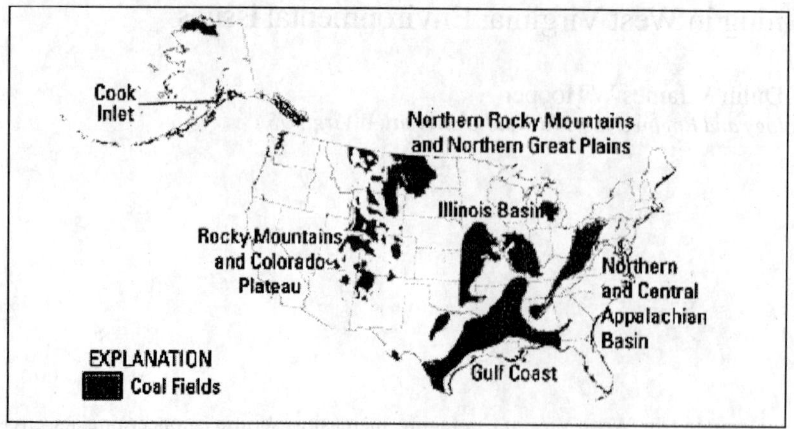

Figure 2: Principal Coalfields of the United States (After USGS)

12×10^9 tons since 1833. Ninety-nine percent of the state's (at \$0.08 per kwh) and fifty-six percent of the nation's (\$0.13 per kwhr) electricity is generated by coal.

2 MOUNTAINTOP REMOVAL MINING METHOD

The term "mountaintop mining" or "mountaintop operations" generally refers to three types of surface coal mining in the steep slope areas of the central Appalachian mountains: contour mining, area mining and mountaintop removal mining. The Office of Surface Mining (OSM) and the States regulating coal mining define mountaintop removal as a particular mining method where the top of the mountain is removed to totally extract underlying coal seams and the reclaimed land is left in a flat or gently rolling configuration that is capable of supporting certain post-mining land uses such as industrial, commercial, residential, agricultural, or public facilities. Since the reclaimed mine site normally consists of flat and gently rolling land, a variance from the regulatory requirement that mined lands be put back to their "approximate original contour' (AOC) is usually necessary. Steep slope area and contour surface mines normally can meet the AOC requirement more easily and, consequently, do not require variances as often.

Of the total acreage of active surface mines in West Virginia[2], 44 percent is mined by mountaintop mining; 24 percent area mining; and 33 percent contour/auger mining. Of the mountaintop mining units 19 percent had a mountaintop removal approximate original contour variance, 8 percent had a steep slope AOC variance, and 68 percent mined without an AOC variance.

2.1 *Environmental permits required to construct a valley (excess spoil) fill.*

The permitting process is an increasing burden and cost to mining companies. For example, currently four environmental permits (approvals) are required for a mining operation proposing to place mining material (excess spoil in the '*waters of the United States*', after the CWA) to construct a valley fill. These include:

- a surface coal mining permit issued under the authority of the Surface Mining Control and Reclamation Act (SMCRA) – OSM has granted the WVDEP primary authority to administer the surface mining regulatory program in the state.
- three permits (approvals) issued under the authority of the Clean Water Act (CWA): Section 402 National Pollutant Discharge Elimination System (NPDES) permit for discharges from sediment control ponds and other drainage structures; a section 401 water quality certification from the state, assuring that the federal permits will not cause a violation of the state's water quality standards; and Section 404 permit for discharge of fill material in streams.

Four federal agencies and one state agency have primary responsibility for mountaintop mining operations: the Fish and Wildlife Service (FWS), Environmental Protection Agency (EPA), Office of Surface Mining (OSM), U.S. Army Corps of Engineers, and the West Virginia Division of Environmental Protection (WVDEP).

[2] Total acres surface mined in West Virginia 216,896, as of September 1998.

3 ENVIRONMENTAL ISSUES AND IMPACT ON MINE PERMIT ISSUANCE.

Until recently, the regulatory issues raised in the permitting processes by governmental agencies, environmental groups, or by other citizens had been minimal. However, a recent citizen suit has stalled the permitting process for numerous mining permits.

3.1 *Aesthetics issue.*

Although mountaintop mining operations accounts for a small percentage of the overall mining in these states, the increase in size of mining operations has created concern among many of the nearby communities (Report to Congress 1999). A contributing factor is the scale of the loss of vertical height or elevation limit to which the final contour of the reclaimed mined out area must be restored after mining. This noticeable change in aesthetics has provoked some in the community to declare that the mining companies are taking away their mountains.

Up until the early 1990's, a 50 feet guideline was typically used in West Virginia, whereby mine operators were required to return the final reclaimed surface contour to within 50 feet. In recent years this guideline has generally not been followed resulting in larger vertical height differences. Actually, the regulations (SMCRA) do not specify a vertical limit to which the final contour must be restored after the entire coal seam is mined. The final surface profile may be a level or gently rolling contour adequate to support the proposed land use and provide for the drainage area.

3.2 *Citizen suit and settlement agreement*

The rising public concern over mountaintop mining came to a head in April 1998, when the Highlands Conservancy, an environmentalist coalition, gave notice of intent to sue the Director of the West Virginia Division of Environmental Protection[3]. As a result, a court hearing and settlement agreement has led to four United States federal agencies[4], in cooperation with the WVDEP, undertaking the development of an Environmental Impact Statement (EIS). The principal environmental concern is the loss and degradation of West Virginia's waters resulting from valley fills[5] associated with mountain top removal, steep slope surface mining and other multiple seam surface mining activities. The Highlands Conservancy, and plaintiffs in the citizens suit, claim that the size and number of streams being filled by excess spoil, in the form of valley fills, is unprecedented. Also, they allege that the procedures to obtain the appropriate permits were not in accordance with the Clean Water Act and associated regulations. Principal among thirteen counts cited in the complaint are:

- Valley fills are subject to section 402 of the Clean Water Act (rather than section 404). Specifically, the placement of earth and rock into a stream must be regulated as a discharge of a pollutant under the NPDES[6] program administered by the EPA rather than as the placement of dredged and fill material in the navigable waters, which is regulated by the Corps of Engineers.
- Valley fills that cover streams violate state water quality standards and the anti-degradation policy[7].
- West Virginia is approving permit applications without variances from approximate original contour (AOC) that fail to ensure that after mining the mine site, including the valley fill areas, will closely resemble the original surface configuration.

The settlement agreement committed the five agencies (EPA, OSM, COE, FWS, WVDEP) to two principal activities: prepare programmatic environmental impact statement (EIS)[8]; develop an interim permitting process that minimizes adverse environmental impacts of valley fills.

The original agreement provided that the U.S Army Corps of Engineers take the lead in permitting valley fills: fills impacting an upstream watershed of 250 acres or more will generally require an individual permit; less than 250 acres, a Nationwide (simpler) permit. However, this portion of the agreement was arguably invalidated by the opinion issued in the citizen suit, as discussed below.

3.3 *Governor's Citizen Task Force*

On June 5, 1998, the Governor of West Virginia appointed twenty citizens of the State to a task force to study mountaintop removal and related mining practices. The task force, chaired by the president of Marshall University, presented its recommendations on December 4, 1998. The task force made a total of 22 recommendations. Principal among them were:

[3] Under SMCRA): Section 520(a)(2), provides for citizen suits against OSM or state regulatory authority for failure to perform nondiscretionary duties. 30 U.S.C. δ 1270(a)(2)-

[4] Office of Surface Mining (OSM), Environmental Protection Agency (EPA), Army Corps of Engineers, Fish and Wildlife Service.

[5] Fill areas created from the disposal of excess spoil.

[6] National Pollution Discharge Elimination System.

[7] State policy required under 40 CFR 131.12 and the Clean Water Act.

[8] Under the National Environmental Policy Act, environmental impact statements are required for any federal actions that will significantly impact the environment. 42 U.S.C § 4332

carry out further study and data compilation on the specific environmental impacts of mountaintop removal mining; establish a new office within the Division of Environmental Protection to regulate the community impact of mountaintop removal mining; rescind recent State legislature (Senate Bill 145) regarding guidelines, standards, and procedures for stream mitigation; discontinue fish and wildlife habitat as a post-mining land-use. As a result of the task force's recommendations the state legislature passed laws to create two new offices: the office of blasting and explosives established within the Division of Environmental Protection that will regulate blasting at surface mining sites; and the office of community development to develop community impact statements that will be included in the mining permit application. The legislature also rescinded the Senate Bill 145.

3.4 *Environmental Impact Statement*[9]

In December 1998, a settlement agreement was reached between the plaintiffs in the citizen suit and the U.S. government, apparently without mining industry input, with the idea of creating a new permitting system to valley fills associated with mountaintop mining. The agreement committed the U.S. government to completion of an Environmental Impact Statement (EIS) over the next two years that will evaluate state agency policies and coordinated decision-making to minimize the adverse effects of mountaintop mining on the waters of the U.S., and fish and wildlife resources.

The EIS studies will assist in determining action to be taken pursuant to SMCRA and the Clean Water Act (CWA). Data collected will include: an inventory of valley fills; impact assessment of fills on terrestrial and aquatic habitats; flood potential determination; stream mitigation regulations identification and extent of their implementation; long-term stability assessment of fills. Until the EIS is completed an interim program will be administered.

3.5 *U.S. District Court Ruling*

On October 28, 1999, the United States District Court for the Southern District of West Virginia issued an opinion in the case of *Bragg v. Robertson* (mountaintop mining citizen suit). The court held that the 'buffer zone'[10] requirement in regulations promulgated under SMCRA protects the entire length of intermittent and perennial streams, and that the placement of valley fills in these streams violate federal and state water quality standards by eliminating the buried stream segments for the primary purpose of waste assimilation. Importantly, the court held also, that excess spoil is a pollutant as defined by the Clean Water Act and also *waste material*, and not 'fill material' subject to the Corps of Engineers (COE) jurisdiction under section 404. Consequently, EPA, and not the Corps of Engineers, has primary regulatory authority over the placement of valley fills into the nation's waters. Upon motion of the parties, the court's opinion has been stayed pending appeal to the United States Circuit Court of Appeals for the Fourth Circuit[11].

4 APPROXIMATE ORIGINAL CONTOUR

Although the recent controversy over mountaintop mining addresses the effectiveness of the implementation of SMCRA and the CWA, and other laws, there are concerns regarding the application of approximate original contour requirements, and how to interpret the postmining land use limitations. The AOC is the configuration of the surface of the final reclaimed area achieved by backfilling and grading the mined area so that the reclaimed area, including any terracing or access roads, closely resembles the general surface configuration of the land prior to mining. SMCRA also requires that the surface blends into and complements the drainage pattern of the surrounding terrain, with all highwalls and spoil piles eliminated.

4.1 *AOC Variance*

All mined areas are to be returned to AOC, unless they receive a variance from it. A mountaintop removal variance can be granted by the regulatory authority only if the entire coal seam or coal seams running through the upper fraction of the hill, ridge, or mountain is removed, and a level plateau or gently rolling contour is created with no highwalls. The site granted such a variance must be capable of supporting certain post-mining land uses. In May 1999, the OSM disapproved West Virginia's proposed amendment for "Fish and Wildlife Habitat and Recreation Lands" as a post-mining land use for mountaintop removal mines, which was commonly used by the mining industry (News Release May 1999).

4.2 *Guidance Document*

In March 1999, a guidance document was developed by an AOC/Excess Spoil Guidance Team, comprising representatives from WVDEP and OSM.

---

[9] National Environmental Policy Act (NEPA)

[10] Buffer zone: No disturbance within 100 feet of a perennial or intermittent stream. 30 CFR. Section 816.57; 817.57.

[11] See U.S.C §1291(Z)

(Mountaintop Reclamation, 1999) The document provides an objective and systematic approach to achieve AOC on steep-slope surface mine operations and determine excess spoil quantities

The document outlines an AOC model that is based on placing the maximum excess spoil material from the mining operation onto the mined out area (mining footprint) and the remaining excess spoil into a valley fill area. The maximum excess spoil to be placed on the remaining mined out area is defined approximately by the spoil pile with slopes 1 unit vertical to 2 horizontal with appropriately placed terraces; or a slope that will result in a long-term safety factor of at least 1.3. The total amount of spoil material (TSM) is determined by the product of the in-situ and interburden volume and a 'bulking factor' (BF). The bulking factors account for the 'swell' of the broken material and the 'shrinkage' during compaction following placement into the valley fills.

5 DISCUSSION

After consideration of all of the recent events associated with the mountaintop mining in West Virginia, one may well determine that most of the controversy originated over aesthetic concerns. However, once the mining practice was brought to the attention of the public, many other issues came under scrutiny. Specifically, both environmental groups and regulatory agencies have become increasingly concerned about various environmental impacts associated with mountaintop mining and valley fills. Likewise, the industry and its employees became increasingly concerned over the continuing economic viability of the mining practice in the state in view of the enhanced public opposition and regulatory oversight.

The ensuing litigation and public controversy underlined the importance of communication between the coal industry and its primary stakeholders: employees and the community. Many of the comments at the public hearings conducted in connection with the Governor's task force and federal agency EIS indicated that the public did not feel included in the development of public policy regarding mountaintop mining. Industry comments indicated that the public may have misperceived many of the engineering and economic constraints faced by the mining industry.

Even after many of the legal issues are finally resolved and the environmental issues are identified and addressed by the agencies, the public policy issues surrounding the mining practice will remain. In fact, both the Governor's task force and the District Court recognized that, ultimately, mountaintop mining and valley fills are policy issues that need to be addressed by Congress and the state legislature.

The EIS is a valuable first step in assessing the current state of the mountaintop mine environment. However, a longer time-frame than two years is needed for a definitive evaluation of the environmental effects of mountaintop mining.

In the interim, however, the mining industry could enhance its public image and make the regulatory path a little smoother by employing environmental management systems (EMSs) that are being embraced by government and industry alike worldwide. Credible EMSs are being acknowledged by state, federal regulatory agencies and industry as a proven mechanism to achieve continual improvements in both environmental protection and operational efficiency.

In 1997, the state of Oregon (and other states) legislature created a Green Permits system, based on an organization developing a credible EMS as a minimum requirement. The objective of the Green Permits system is to encourage regulated facilities to achieve environmental results that are significantly better than would otherwise be expected by law. One of the drivers for this change is the acknowledgment that the existing 'command-and-control' regulatory system cannot adequately address many environmental issues effectively, and does not encourage or reward environmental stewardship.

Recent developments with EMSs and ISO 14000 standards provide a mechanism and opportunity for the mining industry to utilize their unique knowledge, skills and resources to reduce environmental impacts of their operational activities (Szwilski).

6 CONCLUSIONS

The ensuing litigation and public controversy involving mountaintop mining in West Virginia has highlighted the importance of communication between the coal industry and its primary stakeholders: employees and the community.

Environmental management systems (EMSs) and ISO 14000 standards are credible management tools that can improve environmental protection and mining operational effectiveness over the long-term Stakeholder involvement and full commitment from corporate management are essential elements of a credible EMS. Keeping the community involved and informed is an action that all industries should regard as a common business practice.

Ultimately, mountaintop mining and valley fills are policy issues that need to be addressed by Congress and the state legislature. The challenges facing the mountaintop mining in West Virginia requires critical thinking, innovative environmental policy and an effective government and industry partnership.

REFERENCES

Report to Congress on Mountaintop Mining, *Department of Interior*, February 23, 1999.

Mountaintop Mining-Work Plan Approach ww.epa.gov/reg3esd1/mtntop/index.htm.

News Release: *Department of the Interior,* May 14, 1999.

Mountain Reclamation: AOC and Excess Spoil Determinations. Appendix D of the Final Report: Evaluation of Approximate Original Contour and Post Mining Land Use In West Virginia, May 1999, *U.S. Department of Interior*, OSM, Charleston Office, WV

Szwilski, A.B., Using Environmental Management Systems to Systematically Improve Operational Performance and Environmental Protection. *Sixth International Symposium on Environmental Issues and Waste Management in Energy and Mineral Production*, Calgary, Canada, May 2000.

Environmental Issues and Management of Waste in Energy and Mineral Production, Singhal & Mehrotra (eds)
 ISBN 90 5809 085 X

The South Alligator valley, northern Australia, then and now: Rehabilitating 60's uranium mines to 2000 standards

Peter W.Waggitt
Office of the Supervising Scientist, Darwin, N.T., Australia

ABSTRACT: From 1956 to 1964 the upper South Alligator valley, an area in northern Australia about 200 km south east of Darwin, was the location for 13 operating uranium mines and a number of prospects. In each case the exploration involved drilling, costeaning and development of adits and shafts. The sites were abandoned in 1964 along with a small mill and solvent extraction plant and a gravity battery plant. There were no rehabilitation requirements under the regulations in force at that time.

In 1986 a survey of abandoned mines was undertaken by the Commonwealth Government to establish the size and scope of a possible rehabilitation project. The area lay within the boundaries of the World Heritage Kakadu National Park and visitor numbers were increasing annually. The funds available from the Commonwealth rehabilitation fund were insufficient to permit a major rehabilitation program. In 1988, after discussions between the various agencies involved, it was agreed that a hazard reduction program would be undertaken. This was to include reductions in physical, as well as radiological hazards for visitors to the area. The program required that old mine workings be made physically safe and that potentially hazardous radioactive material be buried.

The works took place from 1990 to 1992. The methodology and outcomes of the program are described. Monitoring was undertaken during and after the program and results of these studies for erosion, radiation and revegetation aspects of the work are described and discussed.

In 1996 the land was returned to the Aboriginal Traditional Owners and leased back to the Commonwealth Government to remain in Kakadu National Park. The lease requires all evidence of former mining to be rehabilitated by 2015. The paper concludes by describing the planning and consultation processes for this rehabilitation program that have been taking place since 1997 and outlines the proposed schedule of works due to commence in 2001.

1 INTRODUCTION AND BACKGROUND

In the post-war rush to develop atomic power and weapons much uranium exploration was carried out in the Northern Territory of Australia. Uranium deposits were found at Rum Jungle in 1949, Sleisbeck in 1952 and in the Upper South Alligator valley, at Coronation Hill, in 1953 (Annabel, 1971). The valley is located about 200 km south east of Darwin and lies in the geological province of the Pine Creek geosyncline (Figure1). The remoteness of the area and poor access had limited development to cattle ranching. Mineral exploration had taken place since the 1940's and small tin and silver/lead/zinc deposits were worked in the region in the late 40's and early 50's. Uranium mining at Coronation Hill began in 1956, but was soon followed by mining operations at up to thirteen other locations in the valley as well as numerous other exploration sites (Fisher, 1968 & 1988) (Figure 2).

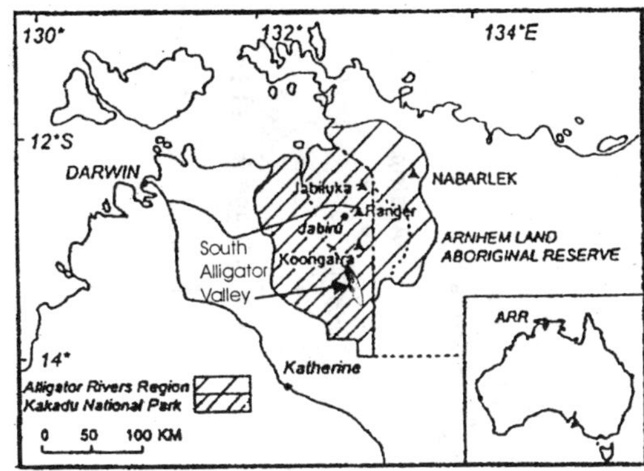

Figure 1. Location Map

These sites were all abandoned at the completion of works with little or no environmental restoration,

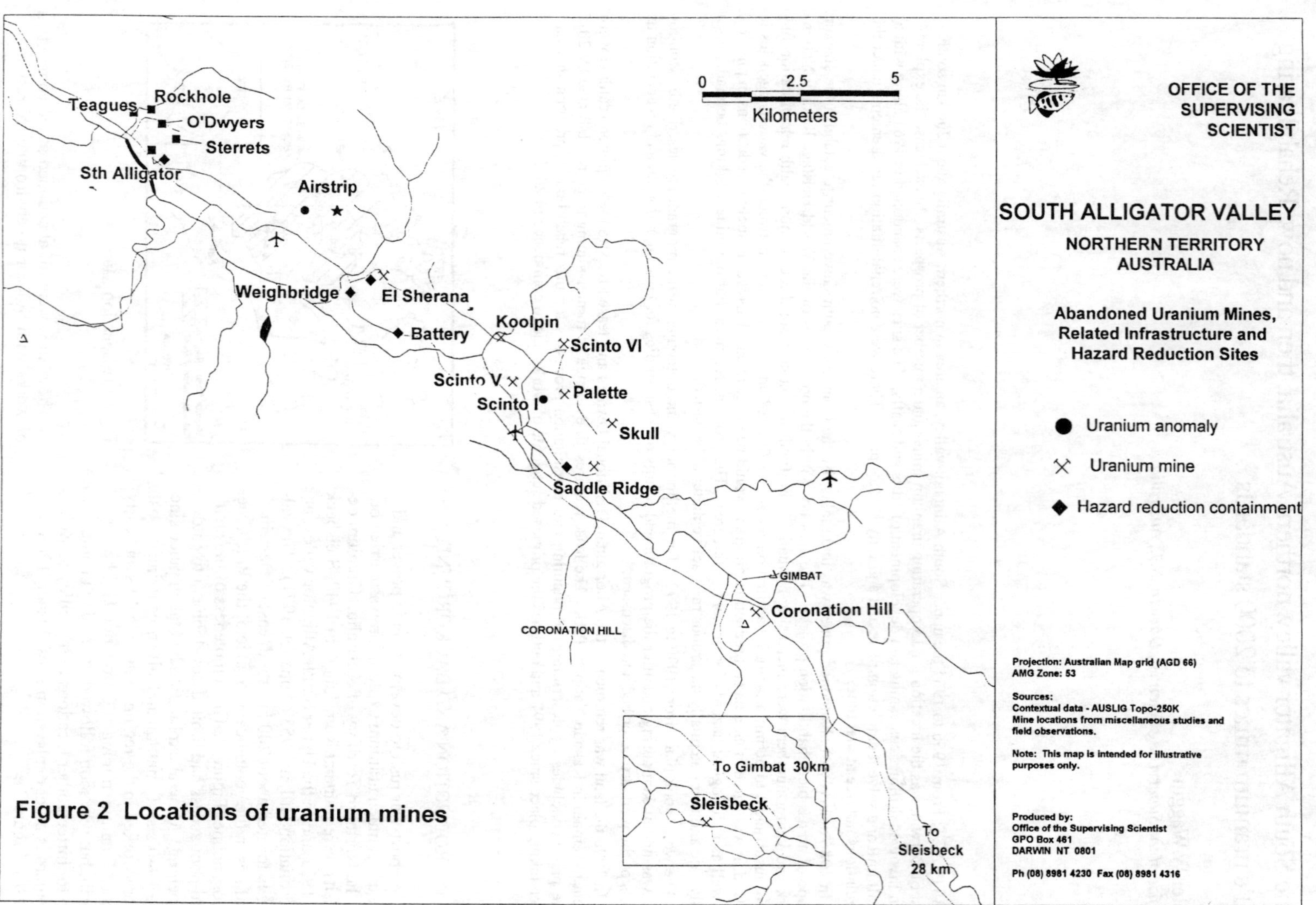

Figure 2 Locations of uranium mines

a regrettable situation but in accord with the regulations in force at that time. Most sites were relatively benign from a chemical point of view and presented only minor hazards from both the physical and radiological standpoints. However, the locations included settlements and processing facilities. These latter sites did present extra hazards in the form of chemical and processing wastes as well as the decaying buildings, some of which contained asbestos.

In 1978 the Northern Territory (NT) was granted self government and the Federal Government agreed to undertake some rehabilitation of former uranium mining facilities within the NT. Consequently all the former mining sites were later surveyed by the Commonwealth Government Department of Housing and Construction (1986).

At the same time the area in which the South Alligator Valley is located was acquired by the Commonwealth Government who resumed the pastoral leases. In 1987 65% of the lease areas was incorporated into the existing and adjacent Kakadu National Park (declared in 1979) as Stage III and the remainder, including the area of the former uranium mining, was set aside as a mineral exploration conservation zone for 5 years (OSS, 1987). Some exploration for gold, platinum and palladium was carried out in the zone over the next few years. The whole area was finally incorporated into the Park in 1991.

The valley lies wholly within the Alligator Rivers Region and the Office of the Supervising Scientist (OSS) is the Commonwealth Government Agency charged with ensuring that the Commonwealth's Environmental Requirements are met i.e. that there should be no adverse environmental impacts as a consequence of uranium mining in the region.

After evaluation of the various reports the South Alligator Valley sites were considered to require only hazard reduction works rather than complete rehabilitation. A follow up radiological assessment of the South Alligator Mill was carried out by the NT Department of Mines and Energy to establish what were the levels of radioactive contamination present, and the consequent risks (Bromwich, 1987). A further study of the area was undertaken by the Commonwealth Department of Primary Industries and Energy (1988) which provided the basis for a specific hazard reduction program. A prioritised schedule of tasks was drawn up which formed the basis of a call for tenders to carry out the work. The contracts were let in 1990 in two phases starting with the mill site and the former associated mining village; the second phase included all the remaining mining and exploration sites. Work was carried out in the dry seasons of 1991 and 1992, i.e. between April and November when access to the sites was freely available.

In 1996 the land was returned to the Traditional Aboriginal Owners following a successful outcome of a claim under the Aboriginal Land Rights Act. The land was immediately leased back to the Commonwealth to remain as a part of Kakadu National Park, now a World Heritage area. A specific clause in the lease required that all evidence of former mining activity be rehabilitated. In particular the plan of rehabilitation was to be agreed by the end of 2000 and all works were to be completed by 2015.

2 REGIONAL ENVIRONMENT

The South Alligator Valley lies in the wet/dry tropics with an annual average rainfall of about 1200 mm, more than 90% of which falls between November and April. In the wet season roads in the area are frequently impassable and there is a risk of heavy plant and machinery causing severe damage to the environment just by moving around. Soils are fragile and highly susceptible to erosion when disturbed. The natural vegetation of the area is a dry schlerophyll savanna forest dominated by trees of various *Acacia* and *Eucalyptus* species in dry areas with *Pandanus* and *Melaleuca* in lower lying and poorly drained areas. The landform is a relatively narrow valley through the Kambolgie sandstone of the Arnhemland escarpment with steep sides rising up to 250 metres above the river. Temperatures average about 21°C throughout the year with minima rarely below 14°C and maxima often around 40°C.

3 HAZARD REDUCTION PROGRAM

The program of hazard reduction works concentrated on both physical and radiological safety to achieve a final condition which would be acceptable for the ongoing land use as a National Park with frequent short term transit by visitors and occasional traditional use for camping and hunting and gathering activities by the Aboriginal owners. The program has been reported in detail elsewhere (Waggitt, 1998) and is summarised here.

3.1 *Physical safety*

Wherever practicable, the intention had been to collapse all adits and tunnels and to block off all shafts. However, surveys carried out by staff of the Commonwealth Australian Nature Conservation Agency (now known as Parks Australia North (PAN)) had revealed the presence of endangered bat species in some locations. Where bats were present in breeding colonies it was decided to place suitable sized grilles across openings, both shafts and adits, which would prevent easy human access whilst still allowing safe passage for the bats. Grilles were finally placed at three sites. Collapsing of openings such as adits, and some shafts, was undertaken using earth-moving machinery.

Open cuts presented a significant safety hazard but it was impractical to consider back filling in all cases. This was due to a combination of such factors as a lack of suitable and easily obtained fill materials, lack of access for plant and machinery and a prohibition on the use of explosives in the area for cultural reasons. The final solution was for fences to be erected around open cuts to prevent casual access by both humans and livestock. In some locations existing tracks were re-aligned or rehabilitated to either deter access or to make it safer. For example, at the former Koolpin open cut there was originally a track passing within a metre of the edge of the 12 metre deep open cut, with no safety barrier of any kind. The program moved the track back some 5 metres, placed a 1m high bank at the edge of the track where it passed the site and a fence around the open cut. Finally warning signs were erected at all work sites in the program.

At all sites the ground surface was finally prepared for revegetation with particular attention being paid to the issue of soil erosion. Precautions were taken to ensure that the sites were left in a condition which would not favour the development of erosion features. Revegetation was carried out by direct seeding with native species and wherever possible planting material was collected from the surrounding countryside.

3.2 *Radiological safety*

The major radiological concern was the former South Alligator mill site. Not only were there process residues and even small amounts of product in pipes and reaction vessels but there were also tailings at the site. During production the tailings had been deposited on the flat area between the mill and the South Alligator River. In times of flood and high runoff tailings were carried away in the water. In 1986 almost all the tailings were removed to the former Moline mine, some 35 km to the west, and re-processed to extract gold. However, some small quantities of tailings remained at the site and these represented a hazard to park visitors. Although the mill was a significant item in the mining heritage of the region the problems of making it safe and decontamination were considered too great to be reasonably overcome. Consequently the facility was dismantled and buried at a site adjacent to the former South Alligator township. The containment was essentially a deep trench dug by bulldozer. The final cover was a minimum of 1.5 m of soil. The area was left over filled to allow for later subsidence. Finally the area was seeded in accordance with the standard practice adopted throughout the program.

At other locations there were varying levels of radioactive contamination. These varied from scattered ore fragments on the ground surface in former truck dumping grounds and ore-sorting stations to process residues at an earlier battery and processing plant, to very low levels of contamination in open cuts and camp areas. All these areas were surveyed by OSS staff prior to the works commencing. This enabled the necessary work areas to be marked out as well as establishing the basis for later comparisons to determine the effectiveness of the work.

3.3 *Worker safety*

Throughout the program of works the contractor was required to monitor conditions for the workforce. In addition OSS carried out some spot checks on dust exposure and radon levels. Dust sampling was carried out using personal air-pump/samplers on workers and attached to plant. Radon levels were measured using an RGM2 and a number of Rolle grab samples. All the results were satisfactory with no worker receiving a dose in excess of the appropriate limit. All workers were required to wear facemasks unless working in air-conditioned cabs and all staff wore thermoluminescent dosimeter badges which were checked regularly.

4 HAZARD REDUCTION WORKS

The works program was carried out by a small contractor using light plant. Vehicle access to many of the sites was restricted and in some cases all construction and fencing materials had to be carried in by hand. As a result the program was built around simple solutions.

4.1 *Approach to works*

In the first instance each site was studied by the contractor to estimate the likely quantities of waste to be generated, this information then enabled a suitable containment site to be located in conjunction with officers from the OSS and PAN. The intention was to limit the number of containment sites to the minimum. Each site was chosen to be away from a likely camping site, clear of surface drainage lines, above the 1 in 100 year flood line and to enable all waste to be buried with a minimum of 1.5 metres of cover.

The radioactively contaminated areas were identified by surveying with a hand held doserate meter. The contaminated material was then scraped up and removed to the containment. Also scrap metal and abandoned mining equipment were collected and placed in the containments. There was an issue of preservation of mining heritage. However, the presence of contamination and hazardous materials such as asbestos meant that no major buildings could be safely left on site.

A number of items of mining plant were considered for salvage on heritage grounds but as there was no safe or effective way of decontaminating them on site they too were consigned to the containments. Before each containment was closed out, it was checked, together with the associated work sites, by OSS staff in the company of the contractor to ensure that the work had been completed satisfactorily.

4.2 *Completion standards*

The works program was carried out with the intention of making the area physically safe, as previously described, and to reduce the dose to a critical group to less than 1 mSv per year. The critical group was considered to be an Aboriginal family camping in the area for up to six weeks to carry out traditional hunting and gathering activities. The target gamma dose rate to be achieved at each site was set at $1 \mu Gh^{-1}$.

5 MONITORING PROGRAM

After the hazard reduction program had been completed the OSS undertook to monitor the works on an annual basis. The intention was that all sites should be regularly inspected for integrity of works as well as being checked for success of revegetation, signs of erosion and ongoing security of those areas where radiological contaminated wastes had been contained. The erosion and integrity inspections were carried out every year in the dry season. The program called for inspections to be early in the season in order that there would be sufficient time to carry out any necessary remedial work before the onset of the next wet season. The radiological inspections, primarily surveys of gamma radiation 1m above ground surface, were carried out in the first dry season after works had been completed and at intervals of three years thereafter. However, the annual inspection crew always carried suitable equipment to be able to check for radiological contamination as a precaution or if erosion was found to be threatening the integrity of a containment.

Over the years few problems have been encountered with the sites. In one instance a contractor was found about the excavate a site as a source of fill for road works and was advised to source material from another location. At another site some additional drainage works were required after two seasons to reduce the risk of an erosion gully developing and threatening the integrity of a containment.

Over the years the revegetation has been generally successful with original pioneer species, such as *Acacia holosericera* being replaced by more 'permanent' species such as *Eucalyptus tetradonta*. At several locations these "permanent" species are already flowering and seeding with evidence of self-seeding being successful. Some re-seeding was carried out at one site after two years following damage by a bush fire and a subsequent flood. Again seed was obtained from local provenances.

6 REHABILITATION PLANNING

Following the handover of the land to the Jawoyn people and the signing of the new lease for the National Park it was determined that planning of the complete rehabilitation would need to commence quickly. The production of the plan was seen as an activity that would require extensive and thorough consultation with all the stakeholders. The overall administrative responsibility for management of the planning process rests with Parks Australia North (PAN), a part of the Commonwealth Environment Department, who manage the National Park and signed the lease on behalf of the Commonwealth Government. OSS is the technical adviser to PAN as well as having a statutory role under the *Environment Protection (Alligator Rivers Region) Act.*

6.1 *Early Consultation*

In October 1997 a meeting was arranged so that OSS could brief as many stakeholders as possible on the issues to be considered in the development of the agreed rehabilitation plan for the South Alligator valley. The meeting was held under the auspices of PAN and the Northern Land Council, an umbrella organisation providing support to all Aboriginal Traditional Owner groups, and the Jawoyn Association. Representatives of the Northern Territory Government were also present. From the outset it became apparent that whilst the stakeholders had a fair appreciation of the general area their site specific knowledge in respect of the former mines was very variable. Also there was not a good appreciation of the scale of the issue and the wide range of options open to the group for carrying out the works program.

At the same time the government representatives obtained a greater appreciation of the range of social and cultural issues which would have to be considered when drawing up the plan of works. Again there was a prohibition on the use of explosives to collapse faces and fill open cuts or to develop quarries as sources of suitable rockfill. There was also a request to limit the size of earth moving machinery that would be used in the works program. This was to take into account cultural considerations of reducing noise and ground disturbance as much as possible.

The meeting ended with visits to two of the more easily accessible sites and discussions on the general expectations of the Traditional Owners for the final land use in the valley. It was also agreed that there should be a Steering Committee set up to manage the day-to-day preparation of the rehabilitation plan and to prepare materials for presentation to the assembly of stakeholders which would make the decisions and final agreement on the plan.

6.2 *The Steering Committee*

The first meeting to create the steering committee was held in early 1999. The final group composition included a majority of Traditional Owners as well as representatives from departments of both Commonwealth and NT Governments and the Northern Land Council. The meeting agreed a schedule for meetings and report production to meet the requirements of the lease. In particular it was agreed that a combined meeting and site inspection would be arranged for the late dry season. This would enable many of the Traditional Owners to become familiar with the characteristics of the area and the sites as well as obtaining a better understanding of the logistical difficulties of working in the area.

6.3 *Supplementary Studies*

The first item on the agreed program was to have the former contractor from the hazard reduction works visit as many of the sites as possible and estimate the amount of work that would be required for various rehabilitation options. These inspections included video footage for use in presenting the report to the steering committee. The initial estimates were based on three main options, do nothing, minimal rehabilitation and complete filling in of open cuts and landscaping.

The annual monitoring program showed that, with one exception, all the sites were in a satisfactory state from an erosion as well as a radiological point of view. The exception was in the vicinity of the former South Alligator mill site. A visual inspection showed that an erosion gully had opened up across an embankment at the side of the road and mill tailings were exposed at the surface. A more detailed inspection showed that there was up to 1 metre thickness of tailings under the embankment over an area of approximately 200 sq. m. Also during recent remedial works on the road, the grader had cut into this deposit and spread tailings in the windrows for several hundred metres along the road.

OSS staff undertook a specific survey of the nature and extent of this contamination and produced a draft schedule of works to be incorporated into the overall rehabilitation program. A suitable site for the necessary containment was located and agreed with the Traditional Owners. This site was some 1300 metres away and adjacent to a containment used in the hazard reduction works.

7 PRESENT PROGRAM STATUS

The major outcomes from the September 1999 meting gave some firm indications of the sort of results the Traditional Owners were expecting from the program.

It was agreed that the rehabilitation program for each site would be drawn up separately. In particular the original request for every open cut to be filled in had been modified. The lack of suitable material would have meant the opening of new quarries to provide rockfill. These in turn would have required further excavations in the countryside and the use of explosives, neither of which was an acceptable activity. Also the eventual rehabilitation of any such sites would extend the period of works.

It was agreed that the small open cut at Sleisbeck would be filled to the maximum extent possible using material from some small waste dumps nearby. There would be no new excavations. Again the preference is for small plant to be used. This is for several reasons including better employment and training opportunities, easier to work around existing trees and needs smaller scale road works to bring machinery and materials to the site, thus reducing environmental impacts as much as possible.

The remainder of the sites are to be remediated using the same principles. This will mean that probably no other open cut will be filled, no new access roads will be constructed and the emphasis will be on revegetation, using native tree and shrub species. The stock for the re-planting program would be grown in a nursery to be developed by the Jawoyn Association as a new business enterprise. This would also create employment and training opportunities for the young people in the community. The Jawoyn Association is already involved in resource development on its lands, usually as a joint venture partner. The potential market for a minesite rehabilitation contracting business is therefore considerable.

The plan is being finalised at the time of writing. Works programs for the tailings removal are complete and work should be commissioned in the early dry season of 2000. The radiological standards for completion of the rehabilitation that have been adopted assume that the mining was a practice. Thus the final radiological objective is to ensure an annual dose rate to the critical group of less than 1mSv per year, in accordance with the provisions of ICRP Publication 60 (ICRP, 1990).

8 CONCLUSIONS

In the early fifties resource development proceeded at a tremendous pace and little attention was paid to issues of environmental impact assessment or management. There was no pollution control legislation, little community consultation and the environment was not an issue for the mining industry. The uranium mines of the upper South Alligator valley were no different. When mining ended in the mid-sixties the sites were simply abandoned.

By the mid-eighties the community was expecting that minesite rehabilitation would be the normal state of affairs and that older, abandoned sites would be included in rehabilitation planning. The lack of funds from government and the "orphan" nature of the sites concerned, i.e. no current operator or owner to accept responsibility, meant that complete rehabilitation was not an option. A program of hazard reduction works was put in place to deal with safety and radiological issues. Throughout this period there was little or no consultation with the Traditional Owners of the land.

Following changes to legislation about Aboriginal Land Rights and the granting of the land to the Jawoyn people, the land was leased back to be used as a National Park. The new lease included an agreement that the Commonwealth Government would rehabilitate the former uranium mine sites within 15 years. However, the Traditional Owners are being consulted at every stage of the process. A system of committees and communication networks has been set up to facilitate the planning and implementation process. The options being developed not only address the rehabilitation of the mines but also aim to create lasting business opportunities for the Jawoyn people well into the future.

REFERENCES

Annabel R I 1977. The uranium hunters. Rigby, Adelaide.

Bromwich D 1987 Radiological Hazards at the abandoned South Alligator Uranium Mill. Northern Territory Department of Mines and Energy Report OH 87/4.

Construction Group 1988 Rehabilitation proposals for abandoned uranium mines in the Northern Territory Report 88/2, Department of Administrative Services, unpublished report.

Department of Housing and Construction SA/NT Region 1986 Report on survey of abandoned mines on Gimbat pastoral lease Northern Territory. Department of Administrative Services, unpublished report.

Fisher WJ 1968 Mining practice in the South Alligator Valley *in Proceedings of a symposium "Uranium in Australia'* AusIMM Rum Jungle Branch 16-21 June 1968. AusIMM , Melbourne.

Fisher WJ 1988 A brief history of mining, exploration and development in the South Alligator river valley. WJ & EE Fisher Pty Ltd Darwin, unpublished report.

ICRP 1990. 1990 Recommendations of the International Commission on Radiological Protection, Pergamon Press, Oxford.

OSS 1987 Supervising Scientist for the Alligator Rivers Region-Annual Report 1986-1987. Austraaian Government Printing Service, Canberra.

Waggitt PW 1998 Hazard reduction works at abandoned uranium mines in the upper South Alligator valley, Northern Territory in Radiological aspects of the rrehabilitation of contaminated sites, eds, Akber RA & Martin P, (1998) pub. South Pacific Environmental Radioactivity Association (SPERA), Christchurch, New Zealand.

Waste management practices

Environmental Issues and Management of Waste in Energy and Mineral Production, Singhal & Mehrotra (eds)
© 2000 Balkema, Rotterdam, ISBN 90 5809 085 X

The tailings pond failure at the Aznalcóllar mine, Spain

N. Eriksson
Boliden Environmental Staff, Aznalcóllar, Spain

P. Adamek
Sevilla, Spain

ABSTRACT: The tailings pond failure at the Boliden Apirsa mining operations in southern Spain in April 1998 captured the attention from the media, industry and the public. Both the immediate effects and the potential long-term effects of the accident were severe. This paper summarises almost 2 years of the work done by Boliden Apirsa after the accident including the clean up, the environmental impact assessment, the closure of the failed tailings pond, the re-start of the operations, as well as the human aspects related to this work.

1. INTRODUCTION

Boliden Apirsa is located 35 km west of Seville in the region of Andalucía in the south of Spain, Figure 1. The mineralisation forms part of the Iberian Pyrite Belt (Leistel et al. 1998) with more than 80 known mineral deposits.

Mining has been active in Aznalcóllar for thousands of years. Intensive mining was carried out already by the Romans. Large-scale mining started in 1976 when the Aznalcóllar open pit was developed. Boliden Apirsa purchased the property in 1987. The Aznalcóllar open pit was mined out in 1996 at which time the production began in the closely located Los Frailes open pit. The mine and the concentrator are designed for an annual production of 4.1 Mton of ore, containing zinc, lead and copper. The operations employ 500 persons. Until the accident in April 1998 the tailings were deposited in the 160 ha tailings pond, located on the riverbank of the Rio Agrio. The pond was designed and built in 1977-78 with a design capacity of 33 Mm^3. The pond contained 15 Mm^3 of tailings in April 1998. In 1996 the tailings pond was subject to a full-scale stability study conducted by independent experts and the Spanish authorities, whereby no signs of instability were detected. In addition, it was subject to regular third-party inspections, the last one on April 14, less than two weeks before the accident, without detecting any signs of instability.

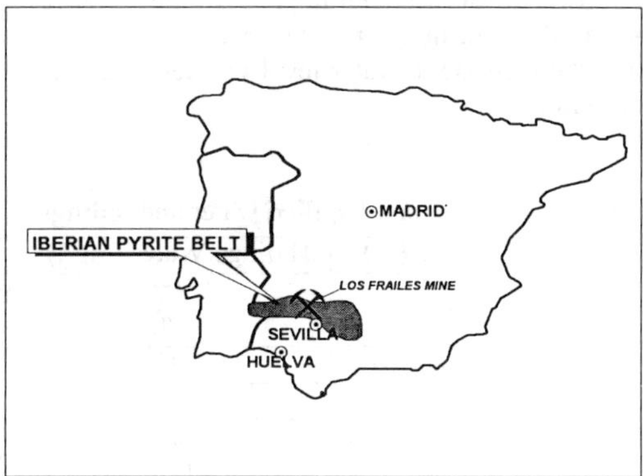

Figure 1. Map of Spain, location of the site and the Iberian Pyrite Belt.

The general geological sequence at the site is composed of Quaternary deposits on top of Miocene and Palaeozoic deposits. The Quaternary deposits consist of alluvial material of varying thickness, typically in the area of the tailings pond between 5 and 10 m. The alluvial is overlying a layer of marls (mud, lime and clay). The thickness of the marls increases towards the south and is around 50 m in the area of the tailings pond. Below the marls there is a thin layer, around 10 m, of bioclastic limestone, which forms the regional aquifer Niebla-Posadas. Further below this Miocene sequence there are

Paleozoic shales, Volcanic sedimentary complexes and Devonian or Carboniferous slates.

The climate in the region is subtropical Mediterranean with warm and dry summers and mild winters, when most of the annual rainfall occurs. The annual average temperature is 18 °C, the average rainfall is 650 mm/year.

2. THE ACCIDENT

On the night between April 24 and 25 1998 a 600 m section of the downstream dike of the tailings pond suddenly slid up to 60 m. The slide created a breach in the dam through which water and tailings were flushed out. In a few hours, 5.5 Mm^3 of acid and metal rich water flowed out of the dam, Table 1. The amount of tailings that was spilled has been estimated to be between 1.3 and 1.9 Mton. Due to the fine particle size of the tailings ($k_{80} < 45\ \mu m$) they were easily transported in suspension with the flood wave. At the flow gauging station El Guijo, 7 km downstream of the tailings pond, the water level increased 3.6 m in 30 minutes. Moreover, 12 hours later, the increase in water level was less than 0.2 m at El Guigo.

Table 1. Composition of spilled water and tailings.

	Tailings (%)	Tailings Water (mg/l)
Zn	1	450
Fe	45	80
Cu	0.2	17
Pb	1	3.5
As	0.6	0.2
S	45	1200
pH	n.a.	2.9

2.1 The cause of the accident

The causes of the accident were assessed by three independent investigations. One commissioned by Boliden Apirsa and conducted by the consulting company EPTISA. This investigation was continuously reviewed and supervised by an independent international expert panel. The regional authorities initiated an investigation that was conducted by the governmental research organisation CEDEX. The third investigation was initiated by the judge leading the legal procedures around the accident and carried out by professors from the University of Barcelona (not completed at the time of writing this paper).

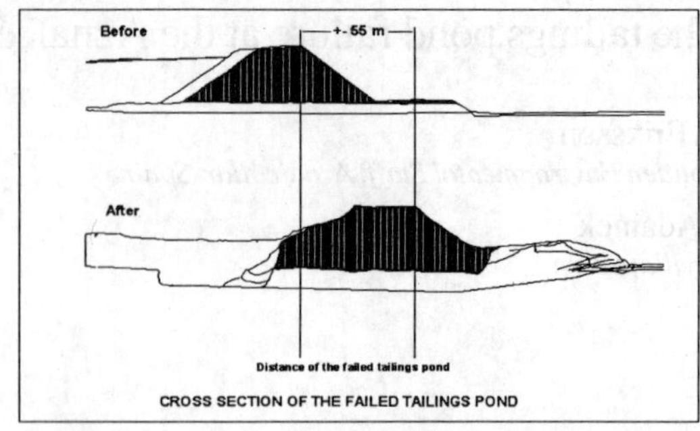

Figure 2. Cross section of the tailings pond.

EPTISA and CEDEX arrived at similar conclusions regarding the cause of the accident. The direct cause of the accident was a fault in the marls 14 m below the ground surface, Figure 2. This fault was the result of surplus pressure in the interstitial water of the clay due to the weight of the dam and the tailings deposit (EPTISA 1998).

2.2 Immediate effects

The spill flooded the riverbanks along the rivers Los Frailes, Agrio and Guadiamar, from a point located 300 m upstream the tailings pond in the river Los Frailes, down to Entremuros, 40 km south of the mine, Figure 3. In total 4634 ha of land were affected, of which 2600 ha were covered by tailings. When the water had withdrawn the thickness of the deposited tailings ranged between 4 m close to the tailings pond to a few millimetres near Entremuros. The flood wave was contained in the 2000 ha Entremuros area by an emergency containment wall built between the banks. The contaminated water did not reach the Doñana National Park.

The aquatic life in the affected section of the river was depleted. No deaths or injuries were caused by the accident, nor was any livestock reported missing. Damages to constructions was very limited and no major bridges open for public access were damaged. Acid water and tailings flooded more than 50 irrigation wells, built in the alluvium. The affected land was riverbanks, grazing fields, agricultural land and fruit plantations. Entremuros, which was affected by the acid and metal rich water, is part of the pre-park to Doñana and was before the accident an important resting site for migrating birds.

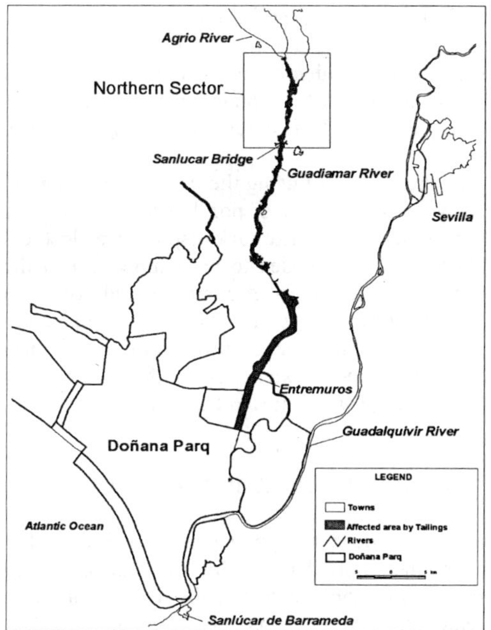

Figure 3. Map of the affected area with reference points and rivers.

2.3 Immediate actions

The plume of contaminated water was stopped in the Entremuros area by the wall built by the Spanish authorities, as described above. The authorities banned all form of land-use and the use of all wells for irrigation in the affected area.

Boliden Apirsa managed to seal the breach within 36 hours. This was required because there was still intensive rainfall expected for the end of the rainy season, which could mobilise more tailings and acid water. Mining and milling operations were stopped immediately and a large part of the working force was laid off (Lindvall et al. 1999).

Boliden Apirsa organised a number of working groups to address various issues such as: investigating the causes of the dam failure, environmental impact, clean-up of the spilled tailings, insurance and legal issues, information issues, restart of the mining operations and decommissioning of the failed dam (Lindahl 1999). The organisation at the mine was not dimensioned to handle this working load. Competent people were rought in from within the Boliden group and external help was seeked.

Being fully aware of the seriousness of the situation, Boliden Apirsa undertook the fastest possible actions to address the existing and potential effects of the accident. The immediate damage was a fact. The priority was then to avoid any secondary damage in the medium and long-term. Three days after the accident a plan for the clean-up work was presented to the authorities. Boliden Apirsa offered to make funding available, even though stating that the financial and legal responsibility for the accident was a separate issue and had to be determined in court at a later stage. Boliden Apirsa bought the entire harvest of fruit for 1998 from the affected area, to minimise the effects for the affected farmers and to calm down fears that crops from the affected area would reach the market (Lindahl 1999).

3. THE CLEAN-UP

Boliden Apirsa's plan for the clean-up led to intensive discussions and collaboration with the authorities. On May 2 the permit for the clean-up was issued. It was decided to divide the responsibility for the clean-up, Boliden Apirsa assumed the responsibility for cleaning the northern sector (from the mine and 14 km downstream to the Sanlúcar bridge) and the authorities became in charge of the southern sector. Although the northern sector was smaller, 780 ha, it contained approximately 80% of the spilled tailings.

Boliden Apirsa's objective with the clean-up was to return the land to a state in which the previous land-use could be resumed. The clean-up had to be finished before the onset of the autumn rains, which normally occurs in October. This meant that there were only 5-6 months available to clean the entire affected area.

The tailings were excavated together with as little underlying soil as possible and trucked to the Aznalcóllar open pit for final sub-aqueous disposal. If the method of removing the tailings was simple, the logistics around the operation were proportionally more complex. A total of more than 450 conventional highway trucks were used to transport the tailings together with 26 mining trucks. More than 250 excavators, motor graders and other machinery were used to collect and load the tailings on the trucks. Even though provisional roads were constructed on both sides of the river, some public roads had to be used for the transport of the tailings back to the Aznalcóllar open pit. All trucks had to cover their buckets, nonetheless there was increased generation of dust, mainly brought to the roads by mud sticking to the tires. There were five lethal accidents during the clean-up operation, all of them were traffic accidents.

The clean-up was relatively easy in the agricultural fields and other flat areas where the thickness of the tailings was comparatively limited.

There, motor graders could be used to scrape the tailings into rows. On the riverbanks and other uneven areas the vegetation was removed to facilitate the clean-up and large areas had to be drained before the tailings removal. Especially complicated was the cleaning of gravel pits, which could not be drained completely.

The clean-up left a completely barren landscape without ground vegetation, except for some large trees that could be saved. All vegetation in the river was also removed together with the tailings as the river was drained and cleaned in sections. By October 15, the original deadline for the clean-up, 90% of the area had been cleaned. The clean-up was finally completed on December 1. By that date 10 Mton of tailings and soil had been recovered in approximately 500,000 truckloads.

The clean-up was conducted without any criteria for the residual metal concentrations. The Guadiamar intervention criteria (Table 2) were set on December 18, 1998. In order to direct the clean-up in the northern sector Boliden Apirsa implemented a soil sampling program, described in section 6.2, which provided fast feedback on the success of the clean-up. It resulted in the immediate re-cleaning of approximately 65 ha.

The clean-up also included 45 irrigation wells in the northern sector. This was carried out through liming the wells in-situ and pumping out water and tailings with large pumps.

During the summer of 1999 a second clean-up was undertaken targeting areas where the residual metal concentration exceeded the intervention criteria, mainly in the river channel. In this second clean-up, approximately 200 ha were re-cleaned and 1 Mm^3 of material taken to the Aznalcóllar open pit for deposition.

4. THE RE-START OF THE MINE

Re-starting the mining and milling operations at the site was the second highest priority for Boliden Apirsa, besides from cleaning up the spill. The operations were intact, with the important exception of the tailings pond. The possibility to resume tailings disposal in the failed tailings pond was soon ruled out. Instead, the efforts were concentrated on finding a new tailings disposal facility. The preferred alternative showed to be the mined out Aznalcóllar open pit.

Investigations were conducted addressing mainly the permeability and the water balance of the pit. In November 1998 an application was submitted to the authorities demonstrating the suitability of the Aznalcóllar open pit as tailings disposal facility. After an extended round of questions and additional investigations, Boliden Apirsa received a provisional permit in March 1999. Production in the mine was resumed by the end of March and milling started in June.

The drawbacks of using the Aznalcóllar open pit exclusively for tailings disposal were that a large volume allocated to waste rock disposal was lost and that process water needed to be pumped out of the mine to assure that the water level would not reach above the permitted level, i.e. 0 masl. An application for raising the water level in the open pit to +34 masl and the subsequent extension of the dumps was submitted to the authorities in February 2000.

5. CLOSURE OF THE TAILINGS POND

After the accident, the pond had been drained but still contained more than 13 Mm^3 of tailings, causing concerns about the stability of the dam. A program for emergency works was therefore implemented, including building a stabilising berm around the perimeter of the dike, filling the erosion channels created by the outflow of tailings, covering the tailings surface closest to the breach in the dike and installing a system for pumping the surface run-off from the pond. Practical problems arose, incurring significant costs due to the fact that the work had to be carried out on water-saturated tailings.

The emergency program evolved into a complete decommissioning of the failed dam that, in addition to the above mentioned procedures, has the following main components: diversion of the Rio Agrio; building an impermeable seepage cut-off wall around the north and east sides of the dam; installation of a hydraulic barrier including a back-pumping system on the inside of the cut-off wall; cutting and re-sloping the dike to 3:1 and covering it; remodelling of the tailings surface to minimise the infiltration and to control the surface run-off; and constructing a vegetated composite cover over the remodelled tailings surface. Starting from the tailings, the cover consists of a geo-textile, 0.5 m waste rock, 0.1 m blinding layer, 0.5 m compacted clay, 0.5 m protective soil layer and vegetation.

The final closure plan for the tailings pond was submitted to the authorities in December 1999, for their approval. The decommissioning is scheduled to be finished in September 2000. A comprehensive control system is being established including inclinometers, piezometers and fix-points to control the stability of the dike and the cover, as well as monitoring wells to control the performance of the cut-off wall and the hydraulic barrier.

6. ENVIRONMENTAL IMPACT

Initially the work was focused on determining and quantifying the immediate impact of the accident. However, the focus shifted quickly towards predicting and preventing medium and long-term effects and monitoring the recovery. The fieldwork program was also designed to provide the best possible input data for a comprehensive risk assessment. Using initial monitoring data and a series of assumptions, a screening assessment of human health and ecological risks was conducted (O´Connors 1999). A list of 8 COCs was identified in this work (Ag, As, Cd, Cu, Hg, Pb, Tl and Zn). In the human health screening, 3 of these COCs (Pb, Tl and Zn) were identified as possibly exceeding the benchmark dose for a farmer in a worst case scenario. In addition, a possible increased risk for skin cancer was indicated for As. The ecological risk assessment identified As, Ag and Tl as metals for which exposure could exceed the benchmark dose for rabbits, which form an important part in the food chain in the area for lynx and imperial eagle.

Based on the results from the screening assessment, it was evident that further site characterisation was required in order to accurately quantify the potential exposure arising from the affected soils and sediments. In particular, more site-specific data on metal uptake in vegetation in the affected area, groundwater, along with data on food consumption habits, will improve the accuracy of the estimated exposures. The follow-up also includes ground and surface water monitoring, soil characterisation and monitoring of the recovery of the ecosystem after the clean-up. The results to-date are summarised below.

6.1 Surface and groundwater

The surface water monitoring program initially included 14 monitoring stations in the Rio Agrio, Rio Guadiamar, Rio Guadalquivir, with reference sampling in the upper Guadiamar, Guadalquivir and Doñana Natural Park. The monitoring frequency was, following the accident, daily and was reduced gradually.

Results showed a fast recovery of the water quality in the Guadiamar river system. At the Aznalcázar official flow gauging station located 20 km south of the mine, the zinc concentration was in the order of 400 mg/l immediately after the accident. It decreased rapidly and was two weeks after the accident below 10 mg/l. Since the end of August 1998, i.e. three months after the accident, the Zn concentration has been below 1 mg/l. The fast recovery was due to the carbonate-rich sediments in the Guadiamar riverbed that buffered the pH and caused precipitation and adsorption of metals.

Figure 4 shows the measured Zn concentration at El Guijo from 1980 to 1999, an official flow gauging station 7 km downstream of the tailings pond. In individual samples it was as high as 90 mg/l before the accident, and average quarterly Zn concentrations ranged up to 50 mg/l before 1993. The likely reason for the enhanced concentrations was probably the diffuse drainage of acid waters from the mining area, documented already for the period 1983- 1986 by González et al (1990). The annual Zn load in the river before the accident ranges between 300 and 1200 tonnes/year between 1980 and 1997. Two trends can be noted. Whilst the increasing trend can most likely be attributed to the accumulating amounts of acid-generating sulphuric waste coincident with the advance of the large-scale mining, the decreasing trend is likely to be the result of remedial measures taken by Boliden Apirsa at the site.

Following the clean-up, the Zn concentrations are not higher than before the accident. No concentration peaks have been detected in the river after rainfall events as was feared. In this perspective, the accident has had no significant lasting impact on the river water quality. However, it is obvious that the water quality in the river is objectionable. Therefore Boliden Apirsa is implementing a comprehensive program addressing this situation.

A series of monitoring wells were drilled in 4 profiles in the alluvial aquifer in the northern sector. The profiles stretched from the river, through the affected area and into the unaffected land. The

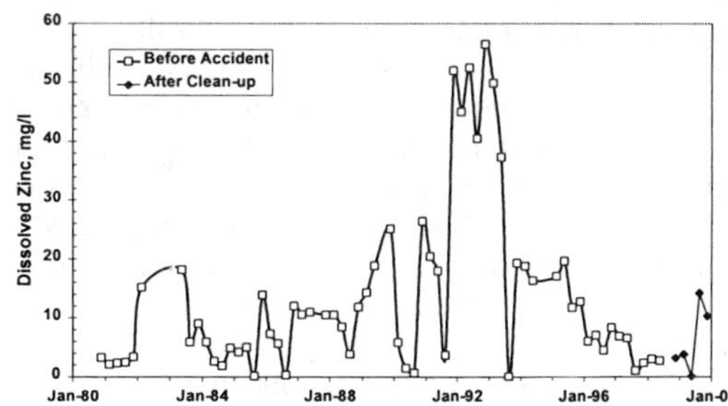

Figure 4. Quaterly average zinc concentrations in the Guadiamar river at El Guijo, 7 km downstream the mine, before and after the accident.

results show no signs of the accident having affected the alluvial aquifer. All these wells have background concentrations of metals and sulphate except the profile located closest to the mine. The reason for the elevated metal concentrations in this profile is, most likely, historical contamination of the aquifer. High Zn concentrations, up to 35 mg/l, in that area had been detected already in 1995 by the Spanish Geological Survey (ITGE).

6.2 Soils

In order to control the clean-up and to form a basis for the risk assessment, a comprehensive soil sampling program was implemented. The northern sector (780 ha) was divided into 159 sampling units (SU), approximately 5 ha each. In every SU 15 discrete soil cores were obtained to a depth of 0.3 m. In total, 3225 discrete samples were taken, 215 composites prepared and analysed. The whole sampling program and sample preparation was supervised and documented by an independent consultant. Sample preparation was done at the Boliden Apirsa assay lab, where a composite sample was produced for each SU and split into 3 parts. One was analysed at the Boliden Apirsa assay lab to provide immediate feedback to the clean-up operation, one was sent to a certified laboratory for analysis and the third was stored as a reference. In 32 SU´s all the discrete samples were analysed to provide the basis for a statistical evaluation. 27 SU´s were re-sampled after re-cleaning or for quality control purposes. The statistical analysis of the data shows that the results obtained are within ± 25-30% in 90% of the time. To obtain a meaningful improvement of the results, more than 50 samples in each SU would have been needed.

The results of the clean-up of the northern sector, completed in December 1998, are given in Table 2 in relation to the Guadiamar intervention criteria. According to these results, metal concentrations generally meet the established criteria, except for arsenic. The success was greater in former agricultural land than in the riverbed and in former waste land and gravel pits. The results imply that 97-98 % of the tailings were recovered in the first clean-up. For example, in the northern sector the total residual load of Zn was around 700 ton.

It has been difficult to obtain pre-accidental baseline concentrations for the affected area, (Ramos et al. 1994). In the study that was completed to set the official intervention criteria (González-Aurioles 1999) the baseline concentrations were concluded to be relatively high in the area, Table 2. It is worth mentioning that for arsenic, the criteria for sensitive land use are identical to the baseline values. This implies that large areas did not pass these criteria before the accident occurred.

6.3 Biota

The scope of the ecosystem monitoring program carried out by Boliden Apirsa includes the main trophic levels, individual organisms and specific population components of the aquatic and terrestrial ecosystems. The program includes monitoring of metal uptake in selected indicator species, and effects/succession monitoring of benthic invertebrates and periphyton communities, as well as vegetation and birds (Boliden Apirsa & Ekologigruppen AB, 2000).

The results of community monitoring of benthic macro-invertebrates in the Guadiamar river have shown that a relatively diverse community of macrofauna has been established already half a year after the accident. This macrofauna is affected by various anthropogenic impacts from discharges of municipal sewage and waste from olive oil processing which overshadow the residual impacts from the tailings spill. In fact, neither on the basis of calculated diversity indexes nor on the relative abundance of macro-benthic communities could the various impacts be clearly distinguished. A better indicator is provided by composition and structure of periphyton communities, as well as by the derived diatom indexes which demonstrate that water quality in the affected area is less favourable than at the reference sites. However, this sampling media also indicates a strong influence of anthropogenic effects other than mining.

Fish (mugils, gambusia and carp) have returned amazingly fast to the river and an otter family established permanently in the river during the winter 1998/1999.

Also the presence of kingfisher birds indicates that a food source is available.

In spite of very sparse rainfall during the autumn-winter 1998/1999, in the spring of 1999, a vegetation cover has been spontaneously re-established on about 20% of the northern sector (compared to reference areas where the mean vegetation cover is about 60%). Regarding species biodiversity, the mean value for the affected area was found to be 13 species per 100 m^2, compared to 37 species per 100 m^2 in the reference areas.

The selected wild plants sampled during autumn 1998 and spring 1999 have shown generally elevated metal content, 2 to 30 times, compared to values obtained at two reference stations outside the

Table 2. Guadiamar Baseline Concentrations, Intervention Criteria and Clean up Results.

		Criteria		**Clean up results**	
	Baseline*	**Sensitive land use***	**Less Sensitive land use***	**Sensitive land use**	**Less Sensitive land use**
	(mg/kg)	(mg/kg)	(mg/kg)	(% Passing criteria)	(% Passing criteria)
As	52	52	100	3	59
Cd	-	5	10	100	100
Cu	120	250	500	92	100
Pb	86	350	500	76	99
Zn	366	700	1200	86	100

*González-Aurioles 1999.

affected area.

Controlled field vegetation tests have been performed during the winter vegetation season of 1998/1999 as a supplement to ecosystems monitoring and in order to supply site-specific data for human health risk assessment (Departamento Agronomía, UCO, 1999). The tests utilised three edible crops, barley (*Hordeum distichum*), triticale (*Triticum aestivum* X *Secale cereale*) and rape (*Brassica napus*), and one supposedly bio-accumulating plant - Ethiopian mustard (*Brassica carinata*).

The results of the tests indicate that metal extraction from soils with representative residual metal concentrations, by physiologically mature plants, is approximately 2 to 8 times higher than metal up-take in a control area. An exception is the extraction of thallium by the rape plants which was found to be as much as 200 times higher. The concentration of metals in the edible parts of the plants, i.e. grains and seeds, is negligibly higher than the concentrations in the control area, with the exception of thallium concentration in seeds of rape and mustard which is approximately 30 times higher than in the control area. The mustard seeds also show 20 times higher concentration of arsenic. Nevertheless, metal concentrations in seeds and grains do not exceed MACs of foodstuff (note, criteria for thallium are not known).

7. HUMAN ASPECTS

The accidental dam failure posed a big challenge for the entire Boliden company. The working load was large and decisions had to be made under time pressure. It was unavoidable that many persons involved were often under heavy stress. In order to learn from the experience, Boliden decided to follow up on the personal experiences of those involved. 65 persons received detailed questionnaires, which were returned anonymously. Some of the conclusions were that (Englyst et al 1999): the needs for information, for both Boliden employees and hired consultants, were not fully met; the majority of the involved personnel noted crisis reactions in themselves; the supporting structure was strong both from the company and from families; one year after the accident, 9 persons still thought that their health was adversely affected by their involvement in the project. Naturally, the feeling that the information was not sufficient also made people feel that the responsibilities and the objectives of the work were not clearly defined. Fortunately, the prevailing opinion amongst the involved persons is that the involvement in the project has been personally and professionally developing and that the results of the clean-up are to be considered as a success.

8. COSTS

The cost of the clean-up of the northern sector finishing in December 1999 was 25 MUSD. According to the newspaper information, the authorities have spent 165 MUSD cleaning the southern sector but that also includes the cost for expropriating the area for the Green Corridor project (of which the affected area forms one part) in which the Doñana Park will be connected to the Sierra Morenã Park by a 40 km long green corridor. The objective is to reduce the isolation of the animal populations in Doñana. The estimated cost for the tailings pond closure is 37 MUSD, project management and quality control included. The cost for buying the fruit harvest for 1998 was 9.9 MUSD. Even though there was nothing wrong with the fruit, it was deposited in a hazardous-waste deposit.

9. DISCUSSION AND CONCLUSIONS

The need for internal and external information cannot be overestimated in this type of situations. Significant resources have to be allocated to dealing with the mass media.

It is unlikely that the accident will have any significant long-term effects on the river water quality. The affected soils contain elevated concentrations of metals and analyses of the vegetation from the affected area show increased accumulation of metals compared to the control areas. However, the metal concentrations in grains from cereals grown in the affected soils do not exceed MACs. Whether these concentrations imply any human health or ecological risk still needs to be evaluated. A good baseline study, conducted before the accident, would have significantly facilitated the evaluation of the effects of the accident.

Considering the starting point for the clean up operation and the limited time available before the rains, the clean up was a major challenge. However, although the situation was very much facilitated by the dry weather conditions, the job was done by the people. These people, who during the hard work went through many emotional stages, rightfully feel proud of their job and consider it a success.

ACKNOWLEDGEMENT

Concha Ortiz, Ismael Barbero, Manfred Lindvall and Lars-Åke Lindahl are gratefully acknowledged for their input to this paper.

REFERENCES

Boliden Apirsa & Ekologigruppen AB, 2000. Ecosystem monitoring program: Results from 1998 and 1999. *Final report.*

Departamento Agronomía, UCO, 1999. Experimentación para el diagnóstico y recuperación de los suelos afectadaos en la cuenca del rio Guadiamar: *Informe de resultados.*

Englyst V., N. Renström, M. Alcatraz, L. Lindahl, & Sicilia J.M. 1999. The Apirsa follow up project. *Boliden internal company report.*

EPTISA 1998. Investigation of the failure of the Aznalcóllar tailings dam. *EPTISA report.*

Feasby G., D. Chambers, R. Fernandez Rubio, J. Gascó Montes, T. P. Hynes & Stuczynski T. 1999. Environmental impact and reclamation planning following the April 25, 1998 accidental tailings release at Boliden Apirsa mine at Aznalcóllar, Spain. *Proceedings to Mine, Water & Environment*, Sevilla, Spain, 13-17 September 1999, 279-290.

González-Aurioles J.M. 1999. Definition of the intervention levels (law of wastes 10/1998, of April 1998) for contaminated soil in the Guadiamar river basin (Spain) and the situation of the soil after the removal of the mine tailings. *Proceedings to Mine, Water & Environment*, Sevilla, Spain, 13-17 September 1999, 247-254.

González M.J., M. Fernández & L.M. Hernández 1990. Influence of acid mine water in the distribution of heavy metal in soil of Doñana National Park. Application of multivariate analysis. *Environ. Technol.* 11:1027-1038.

Leistel J.M., E. Marcoux, D. Thiéblemont, C. Quesada, A Sánches, G.R. Almódovar, E. Pascal & Sáez R. 1998. The volcanic-hostad massive sulphide deposits of the Iberian Pyrite Belt. *Mineralium Deposita* 33: 2-30.

Lindvall M. 1999.The Aznalcóllar tailings pond accident. *Proceedings to CIM*, Calgary, 1999.

Lindvall M., J. Ljungberg, N. Eriksson & Hamrén U. 1999. Händelseförlopp, åtgärder och miljö-konsekvenser. *Proceedings to Mineral-teknik*, Luleå, February 1999.

Lindahl L. 1999. The dam accident in Spain: A case study. *Proceedings to Tailings – Corporate Risk and Responsibility*, Perth, Australia, March 1999, section 8.

O`Conner Associates Environmental Inc. 1999. Screening assessment of human health and ecological risks posed by residual contamination remaining following excavation of soil and sediment contaminated with metals from the Aznalcóllar mines tailings spill: *Final Report.*

Ramos L., M. Hernández & M.J. González 1994. Sequential fractionation of copper, lead, cadmium and zinc in soils from near Doñana national park. *J. Environ. Qual.* 23:50-57.

Environmental Issues and Management of Waste in Energy and Mineral Production, Singhal & Mehrotra (eds)
© 2000 Balkema, Rotterdam, ISBN 90 5809 085 X

Recovering of fine coal particles from tailing ponds of Ankara – Alpagut – Dodurga coal washing plant

H. Koca
Bozuyuk MYO, Mining Department, Anadolu University, Eskisehir, Turkey

S. Koca & M. Karaoglu
Engineering Faculty, Mining Engineering Department, Osmangazi University, Eskisehir, Turkey

ABSTRACT: In this work, samples from tailing ponds of TKI Alpagut-Dodurga Lignites Company Coal Washery were subjected to extensive research to recover clean fine coal particles, by means of gravity methods. The optimum conditions of separation were determined under the laboratory conditions by using shaking tables and Multi-Gravity Separator. Several parameters, thought to be effective on the separation were tested. After the shaking table experiments clean coal products, containing 13.52 % ash with 50.16 % combustible recovery and 16.78 % ash with 53.12 % combustible recovery were obtained at –3.0+0.6 and –0.6+0.3 mm fractions, respectively. After the investigations carried out by Multi-Gravity Separator, clean coal fractions, containing 20.53 % ash with 83.94 % combustible recovery and 25.58 % ash with 88.39 % combustible recovery were obtained at –0.3+0.15 and –0.15+0.05 mm fractions, respectively. Cleaning stage experiments were conducted on the –0.15+0.05 mm fraction and ash percentage of the clean coal was reduced down to 17.39 % with 63.08 % combustible recovery according to head sample.

1 INTRODUCTION

A vast number of coal refuse ponds, which are considered to be environmentally harmful, include significant amount of fine coal particles. These cleanable fine coals generally exist in the refuse ponds due do the inability of conventional technologies to effectively separate the fine coal from the associated gangue particles (Kirnarsky 1998.

It was claimed that over 50 million tons of fine coal particles are disposed to the refuse ponds, annually only in the USA which represent a significant economical resource base (Honaker et al. 1998).

The situation is similar in Turkey, and large tailing ponds were formed where coal washery exists. Alpagut-Dodurga Coal Lignites Company, which is located in the central part of Turkey, produces 400.000 tons of lignite annually and these lignites are cleaned in a washing plant with a capacity of 100 ton/hour. Lignite is washed in coarse fraction, -150+18 mm, and fine fraction, -18+0.5 mm, with a drewboy bath and heavy medium cyclone, respectively. Coarse and fine fraction cleaning circuits are preceded by 2 mm and 0.5 mm washing screens, respectively.

Undersizes of the washing screens are directly disposed to tailing ponds after thickening cyclones. The disposal of the minus 2 mm and 0.5 mm size fractions without any treatment cause 21 % of the valuable losses (Unlu et al. 1998). In the fine and coarse lignite washing circuits, approximately 15 % of the combustible recovery losses are added to the overall loss, which accounts for over 35 %. Apart from losses, tailing ponds cause an environmental problem as they cover a large area, which was used to be for agricultural purposes.

It is the main purpose of this study to gain fine coal particles from tailing ponds and to reduce the negative environmental effects of tailing ponds.

2 EXPERIMENTAL

2.1 *Sample*

Sample, used in this work, was taken from tailing ponds and wet sieved into five size fractions. The sieve analysis and their ash, sulphur and lower calorific value (LCV) contents are given in Table 1. -3.0+0.6 and –0.6+0.3 mm fractions were cleaned with shaking tables, and –0.3+0.15 and –0.15+0.05 mm fraction were cleaned with Multi-Gravity Separator (MGS). Preliminary experiments revealed that –0.05 mm fraction was not effectively cleaned by physical methods and therefore was not used in the experimental work and discarded.

Table 1. The sieve analysis of the sample on dry basis.

Size mm	Amount %	Ash %	Total Sulp.,%	Comb. Rec. %	LCV Kcal/kg
-3.0+0.6	13.53	25.96	1.77	19.72	4383
-0.6+0.3	15.97	32.05	1.68	21.36	4009
-0.3+0.15	12.25	36.31	1.60	15.35	3662
-0.15+0.05	8.65	44.43	1.54	9.46	3246
-0.05	49.60	65.05	1.07	34.11	1386
Total	100.00	49.19	1.37	100.00	2650

2.2 *Equipment and method*

Shaking Table, used in the experimental work was a laboratory type standard Wilfley rectangular deck with 620 mm feed side, 490 mm mechanism end, 370 mm heavy mineral discharge end and 600 mm light mineral discharge end. Feed to shaking table was prepared by mixing 1000 grams of dry sample with 3 liters of water which gives 25 % solid concentration by w/w feed density. The solids were kept in the suspension throughout the experiment by manual stirring. The prepared suspension was fed to the shaking table for 1.5 minutes while washwater was allowed to run until the separation was completed on the table, which takes about for 6 minutes.

The laboratory C 900 Multi-Gravity Separator (MGS) consists of a slightly tapered open-ended drum. Its structure and operating conditions were given elsewhere (MGS Application Guide 1991; Chan et al. 1991). The feed to MGS was prepared by mixing 500 grams of dry sample with one litter of water, giving 33 % solids concentration by w/w. The mixture was agitated manually during the experiment. Frequency, amplitude, washwater flowrate and rotational speed of drum was adjusted and MGS was started. The feed was then fed to the MGS feed vessel at on steady rate for 45 seconds while the separator was kept running until the material flow was finished, which took 5 minutes, and MGS was stopped. Upper cover of the separator was removed and remaining material in the drum was taken into equivalent fractions. Heavy product, which was collected through front launder, referred to as tailings, and light product, which was collected through back launder, referred to as concentrate.

3 RESULTS AN DISCUSSION

3.1 *Shaking table experiments*

Length of stroke, frequency of stroke and side tilt or cross slope was varied to determine the optimum conditions of shaking table. Three products were obtained after each experiment, namely concentrate, middlings and tailings.

It was given in the literature that the amplitude and frequency of the stroke are at major significance both in the mechanism of bed dilation and stratification, and for particle transportation (Burt 1994). The effect of the amplitude of the stroke on separation can be seen in Tables 2 and 3 for -3.0+0.6 and -0.6+0.3 mm fractions, respectively. The amplitude of the shaking table was varied from 10 to 25 mm while frequency and side tilt were set as 300 rpm and 3 degrees, respectively.

The effect of the frequency of the stroke on separation is illustrated in Tables 4 and 5 for -3.0+0.6 and -0.6+0.3 mm fractions, respectively. The frequency of the shaking table was varied from 240 to 300 rpm while amplitude and side tilt were set as 10 mm and 3 degrees, respectively. As it can be seen from Tables 2-5, the best combination was obtained as 10 mm stroke amplitude with 300 rpm stroke frequency for both fractions.

Table 2. The effects of stroke amplitude on separation for –3+0.6 mm fraction.

Amplitude mm	Products	Amount %	Ash %	Combus. Rec., %
10	Concentrate	42.94	13.52	50.16
	Middlings	25.68	24.18	26.30
	Tailings	31.38	44.44	23.54
	Feed	100.00	25.96	100.00
15	Concentrate	43.97	14.06	51.04
	Middlings	26.08	24.25	26.68
	Tailings	29.95	44.92	22.28
	Feed	100.00	25.96	100.00
20	Concentrate	45.07	15.32	51.55
	Middlings	26.37	24.65	26.84
	Tailings	28.56	43.96	21.61
	Feed	100.00	25.96	100.00

Table 3. The effects of stroke amplitude on separation for –0.6+0.3 mm fraction.

Amplitude mm.	Products	Amount %	Ash %	Combus. Rec., %
10	Concentrate	43.37	16.78	53.12
	Middlings	26.74	25.62	29.27
	Tailings	29.89	59.96	17.61
	Feed	100.00	32.05	100.00
15	Concentrate	45.07	17.73	54.57
	Middlings	25.36	26.05	27.60
	Tailings	29.57	59.02	17.83
	Feed	100.00	32.05	100.00
20	Concentrate	46.32	19.17	55.10
	Middlings	26.15	25.95	28.50
	Tailings	27.53	59.52	16.40
	Feed	100.00	100.00	100.00

Table 4. The effects of stroke frequency on separation for –3+0.6 mm fraction.

Frequency rpm	Products	Amount %	Ash %	Combus. Rec., %
240	Concentrate	30.43	17.63	33.85
	Middlings	35.35	22.47	37.02
	Tailings	34.22	36.97	29.13
	Feed	100.00	25.94	100.00
270	Concentrate	34.37	16.53	38.75
	Middlings	33.48	20.45	35.97
	Tailings	32.15	41.78	25.28
	Feed	100.00	25.96	100.00
300	Concentrate	42.94	13.52	50.16
	Middlings	25.68	24.18	26.30
	Tailings	31.38	44.44	23.54
	Feed	100.00	25.96	100.00

Table 5. The effects of stroke frequency on separation for –0.6+0.3 mm fraction.

Frequency rpm	Products	Amount %	Ash %	Combus. Rec., %
240	Concentrate	31.75	21.15	36.84
	Middlings	31.62	28.44	33.30
	Tailings	36.33	44.61	29.86
	Feed	100.00	32.05	100.00
270	Concentrate	40.46	19.66	47.84
	Middlings	29.91	30.17	30.74
	Tailings	29.63	50.87	21.42
	Feed	100.00	32.05	100.00
300	Concentrate	43.37	16.78	53.12
	Middlings	26.74	25.62	29.27
	Tailings	29.89	59.96	17.61
	Feed	100.00	32.05	100.00

The effect of tilt or cross slope of the shaking table on the separation is illustrated in Tables 6 and 7. The amount of tilt given to a table has only limited effects on the separation mechanism, especially within the riffles. This parameter essentially effects the transport of the particles across the deck (Burt 1994). Ideally the side tilt is set at the minimum at which a good distribution of material can be obtained on the deck. High tilts cause more material to discharge along the lights discharging end. A good agreement was obtained with the above general remark at the experiments. At low side tilts, ash content of the concentrates were obtained as 13.52 % and 16.78 % for coarse and fine fractions, respectively. At high side tilts, the amount of concentrates, lights fraction were high, 45.87 % and 48.56 % for coarse and fine fractions, respectively. The ash content of the concentrates was also increased with increasing cross slope.

Table 6. The effects cross slope on separation for –3+0.6 mm fraction .

Cross Slope degrees	Products	Amount %	Ash %	Combus. Rec., %
3	Concentrate	42.94	13.52	50.16
	Middlings	25.68	24.18	26.30
	Tailings	31.38	44.44	23.54
	Feed	100.00	25.96	100.00
4	Concentrate	44.15	15.24	50.54
	Middlings	28.34	23.31	29.35
	Tailings	27.51	45.89	20.11
	Feed	100.00	25.96	100.00
5	Concentrate	45.87	16.09	51.99
	Middlings	27.43	23.12	28.48
	Tailings	26.70	45.83	19.53
	Feed	100.00	25.96	100.00

Table 7. The effects of cross slope on separation for –0.6+0.3 mm fraction.

Cross Slope degrees	Products	Amount %	Ash %	Combus. Rec., %
3	Concentrate	43.37	16.78	53.12
	Middlings	26.74	25.62	29.27
	Tailings	29.89	59.96	17.61
	Feed	100.00	32.05	100.00
4	Concentrate	45.28	18.23	54.49
	Middlings	28.29	25.36	31.08
	Tailings	26.43	62.89	14.43
	Feed	100.00	32.05	100.00
5	Concentrate	48.56	20.15	57.06
	Middlings	27.65	26.47	29.92
	Tailings	23.79	62.83	13.02
	Feed	100.00	32.05	100.00

3.2 *Multi-gravity separator experiments*

Washwater flowrate, drum speed, shake frequency and amplitude were tested to determine optimum conditions of separation. The tilt of the MGS was set at 0° that the separator was kept at horizontal position. The effect of washwater flow rate was tested under following conditions:

Washwater flowrate	: 1.5-3.5 l/min
Drum speed	: 250 rpm
Shake frequency	: 4 cps
Shake amplitude	: 15 mm
Feed density	: 33 %
Feeding duration	: 45 sec
Washing duration	: 5 min

Results for both fractions were given in Figure 1 and shows that combustible recoveries were increased with increasing washwater flowrate. The best results were obtained at 3 l/min and 2.5 l/min for –0.3+0.15 mm and –0.15+0.05 mm fractions, respectively.

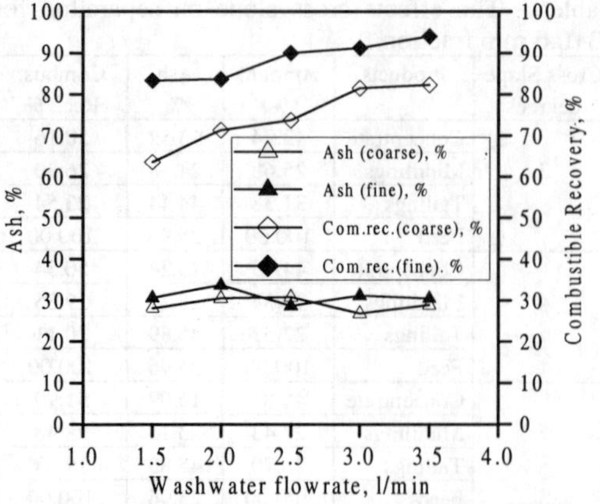

Figure 1. The effect of washwater flowrate on separation.

The optimum combination of shake amplitude and shake frequency was determined at the same conditions given above and washwater flowrate was set as 3 l/min and 2.5 l/min for –0.3+0.15 and –0.15+0.05 mm fractions, respectively. Shake amplitude was set at 10 mm, 15 mm and 20 mm and shake frequency was varied from 4 cps to 5.7 cps at each settings.

Results, illustrated in Figures 2, 3 and 4 show that combustible recoveries were significantly increased with increasing shake frequency for all combinations. The lowest ash content was obtained at 4 cps shake frequency with 10 mm shake amplitude and 4.8 cps shake frequency with 10 mm shake amplitude combinations, as 20.53 % with 83.94 % combustible recovery and 25.58 % with 88.39 % combustible recovery for –0.3+0.15 and –0.15+0.05 mm fractions, respectively.

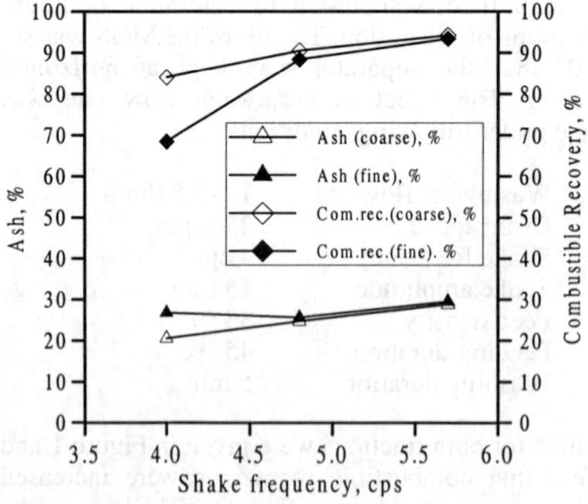

Figure 2. The effect of shake frequency at 10 mm shake amplitude.

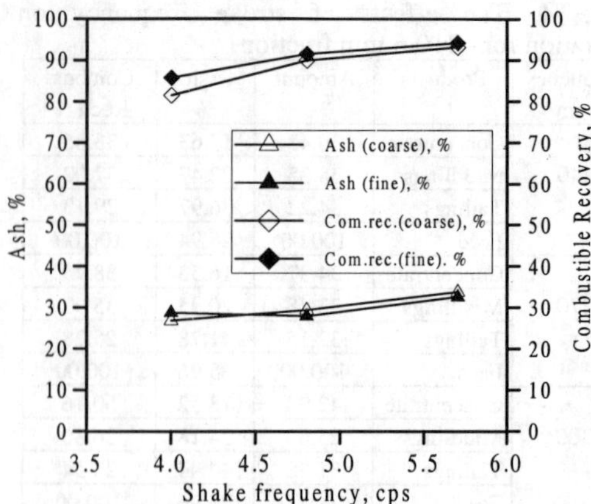

Figure 3. The effect of shake frequency at 15 mm shake amplitude.

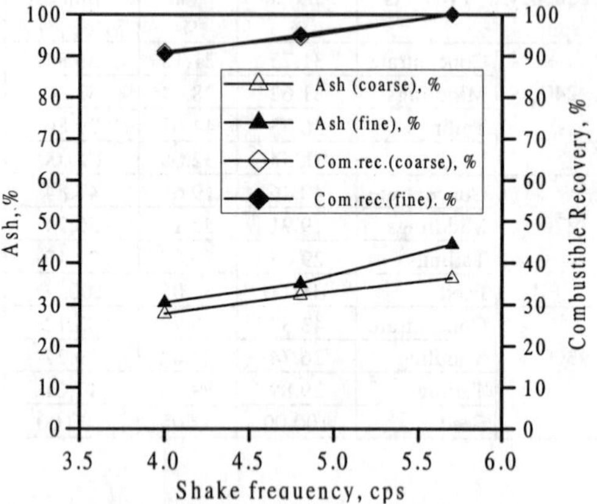

Figure 4. The effect of shake frequency at 20 mm shake amplitude.

Finally the effect of rotational speed of drum was evaluated under the same conditions given above and shake frequency and amplitude combinations were adjusted as 4.0 cps and 10 mm, and 4.8 cps and 10 mm for –0.3+0.15 and -0.15+0.05 mm fractions respectively. Both combustible recovery and ash content of the concentrates were decreased with increasing drum speed (Fig. 5). It is noteworthy to point out that low drum speed; 220 rpm, no separation was obtained and all material was reported to lights fraction.

As it can be seen from the Figures 1-5, the concentrate, obtained at optimum conditions for –0.15+0.05 mm fraction, contain 25.58 % ash with 88.39 % combustible recovery. Both combustible recovery and ash content are high in comparison

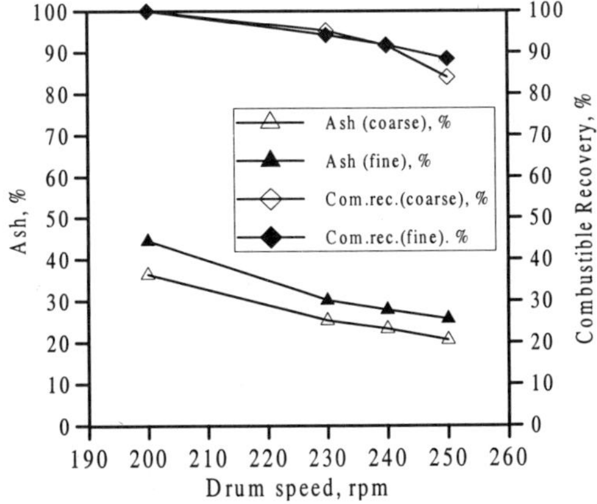

Figure 5. The effect of drum speed on separation.

with the other concentrates obtained from the other fractions. It is therefore decided to conduct a cleaning stage with the fine fraction at the same conditions applied in the first stage experiment and the results are given in Table 8. As can be seen from Table 8 a concentrate, including 17.39 % ash with 63.08 % combustible recovery according to head sample was obtained after cleaning stage.

4 CONCLUSIONS

Recovering of fine coal particles from Alpagut-Dodurga Washing Plants tailing ponds was successfully achieved by utilising gravity methods. Sample was divided into 5 size fractions and –3+0.6 mm and 0.6+0.3 mm fractions were cleaned with shaking tables while –0.3+0.15 and –0.15+0.05 mm fractions were cleaned with multi-gravity separator. –0.05 mm fraction was discarded as the fraction included very high ash content, mainly ultrafine clay particles.

Table 8. The results of cleaning stage experiments for -0.15+0.05 mm fraction

Product	Amount, %		Ash	Comb. Rec., %	
	ATE[1]	ATHS[2]	%	ATE[1]	ATHS[2]
Concen.	64.29	42.43	17.39	71.37	63.08
Tailing	35.71	23.57	40.33	28.63	25.31
Total	100.00	66.00	25.58	100.00	88.39

[1]According to experiment

[2]According to head sample

The best results obtained at optimum conditions and their lower calorific value contents were given below.

Size Fraction mm	Ash %	Combus. Rec %	LCV, Kcal/kg
-3+0.6	13.52	50.16	5580
-0.6+0.3	16.78	53.12	5254
-0.3+0.15	20.53	83.94	4836
-0.15+0.05	25.58	88.39	4408
-0.15+0.05*	17.38*	63.08**	5082*

*After cleaning

**According to head sample after cleaning

5 REFERENCES

Burt, O.R. 1984. *Gravity concentration technology:* 288-316. Amsterdam. The Netherlands: Elsevier science publishers B.V.

Chan, B.S.K., Mozley, R:H. & Childs, G.J.C. 1991. *The MGS-A mine scale machine* UK: Richard Mozley ltd

Honaker, R.Q., Mohanty, M.K. & Patwardhan, A. 1998. High quality coal extraction and environmental remediation of fine coal refuse ponds using advanced cleaning technologies. *Proceedings of the fifth international symposium on environmental issues and waste management in energy and mineral production.* Ankara. Turkey: 573-579. Rotterdam: Balkema.

Kirnarsky, A. & Sbitnev, M. 1998. Development of spiral separation technology for coal slimes treatment. *Proceedings of the fifth international symposium on environmental issues and waste management in energy and mineral production.* Ankara. Turkey: 659-662. Rotterdam: Balkema.

MGS Application Guide. 1991. *How to get the best from your C900 MGS.* UK: Richard Mozley ltd.

Unlu, M., Unal, S., Dogan, H., Tetik, T. & Kur, M. 1998. *Determination of amount and quality of coal fines discarded from Alpagut-Dodurga coal washing plant and dewatering studies.* Ankara. Turkey: MTA publication.

Environmental Issues and Management of Waste in Energy and Mineral Production, Singhal & Mehrotra (eds)

Induced seismicity: Tool for monitoring of nuclear waste storage stability

Vladimir A. Mansurov
Siberian Aerospace Academy, Krasnoyarsk, Russia

ABSTRACT: Underground depository (storage) of nuclear waste presents an excavation in the rock massif which disturb a natural stress state of ambient rock. The acoustic emission/microseismic activity (AE/MA) observation method is proposed for the remote monitoring of stressed state. Two kinetic approaches for the simultaneous interpretation of AE/MA observation data were proposed. As was shown the use of these approaches permits one to evaluate accumulation of stress in the ambient rock and bearing elements of underground NW depository as well as forecast their long-term behavior.

INTRODUCTION

Selection of sites for the deep disposal of nuclear waste (NW) presents a complex scientific, technological, economical and political issue. The main question to be answered may be formulated in the following way: what set of obtained geological and geophysical data will be sufficient for the proper choice of the NW depository site?

Acting international and national recommendations related to a long-term risk assessment are written in a very general form as follows:

- minimal individual dose/risk should be assessed for a long-term period;
- no sudden and dramatic increase of doses for times $>10^4$ years ;
- isolation potential of the depository should be assessed for times $> 10^1$ years;
- assessment of very limiting release of radionuclides into biosphere for a long period.

It is clear the more data related to site selection we try to obtain, the higher cost we shall pay for that. There is a strong bond between expenditures and obtained information. Apparently, one of the ways of the cost reduction is the use of relatively inexpensive but reliable methods of characterization of deep rock at depth in locations which are unavailable for direct measurements. Nondestructive AE/MA testing looks very attractive for that because it allows one to reduce deep drilling and transportation costs with the use of mobile geophysical observatories.

The intent of this paper is to present an application of the measurement of acoustic emission (or microseismic activity) (AE/MA) of rocks to the nondestructive control of rock stressed state. The method is based on the detection of pulses of elastic energy released during dynamic changes in rock structure under load. For rocks the brittle fracture and, consequently, AE/MA is mainly connected with a process of crack formation. According to Kuksenko et al. (1983) a generation of cracks in rock under load leads to the irradiation of impulses of elastic energy followed by quick unloading in micro zones around the cracks. So, using this method one can estimate the energy released in separate locations of the rock massif with the purpose to correlate this energy with threshold crack concentration. If the crack concentration exceeds the threshold value, then loss of stability of the given part of the massif may occur.

The general purpose of this paper is an application of methods of nondestructive characterization of rock (Mansurov & Anikolenko (1996), Mansurov (1999)) by means of acoustic emission/ microseismic activity (AE/MA) observation to the selection of sites which could be potentially used for NW disposal and for evaluation of current state of rock massif including NW storage.

THEORETICAL AND EXPERIMENTAL APPROACHES TO THE OBSERVATION OF FAILURE PROCESS

An experimental study of the rock fracture process on the microscopic level showed that this process may be subdivided in two main stages, see Regel et al. (1974) and Yanagidani et al. (1985). The first one corresponds to the generation and accumulation of stable (noninteracting) microcracks. The second stage is connected with processes of their enlarging and pooling to form a fracture nuclei. However, in natural conditions, the microheterogeneity of rocks prevents the generation and tangible accumulation of new cracks near the boundaries of internal structures of rock mass. Nevertheless, the fracture nuclei may occur just in case a certain specific threshold concentration of cracks is reached.

For the sake of convenience let us introduce the dimensionless concentration factor $K=C^{-1/3}/l_o$, where C is an average concentration of cracks; $C^{-1/3}$ is an average distance between cracks; and l_o is a provisional linear crack size. So, the value of K presents an average distance between cracks measured in units of crack sizes. According to Zhurkov et al. (1981), the interaction of cracks as well as their growth starts when K reaches or exceeds 3. After that the fracture process repeats at the higher size scale level following the same law. The experiments have demonstrated a self-reproducing nature of the rock fracture process for a wide range of materials of different composition, structure and size of initial microcracks, see Sobolev & Zavialov (1980), Ruzhich et al. (1985) and Mansurov (1993).

It is clear that the methods which could be used for the monitoring of the fracture process in rock massif have to be (i) nondestructive, and (ii) remote, because of the conditions of observation. There are a lot of direct and indirect methods of observation of the fracture process. However, most of them would be applicable only in a narrow range of crack dimensions. Some of the methods are not nondestructive ones and may be used only for the post-fracture studies. Comparative analysis shows that the acoustic emission (AE) method is best suited to the monitoring of rock massif fracture. It is based on elastic wave emission, coming as a result of cracking and frictional sliding along faults. It is applicable for any crack size beginning from microscopic levels and finishing at the macroscopic ones. So, nondestructive monitoring of the rock fracture process demands information about the size and concentration of cracks generated under loading. The studies carried out by Frolov et al. (1980), as well as by Stanchitz & Tomilin (1983), have established a correlation between the frequency of the elastic impulse generated when the crack appears or propagates and its size. Here "by the frequency" means the frequency corresponding to the first half wave of the elastic impulse after its first arrival.

There is also strong evidence of the correlation between AE energy (E) and the size of a crack (L) generated under loading, see Kuksenko et al. (1983). According to the laboratory experiments an energy of AE event detected as a combined signal which is an envelope of AE impulses of higher frequency may be estimated as follows:

$$E = \alpha A^2 T \qquad (1)$$

where α is a constant; A and T, correspondingly, amplitude and duration of the combined AE signal, Kuksenko et al. (1990).

So, the parameters of an acoustic signal, generated as a result of cracking could give information about cracks locations and its dimensions and elastic energy emitted from the fracture nuclei as well.

REMOTE EVALUATION OF DILATANCY IN ROCK : RESULTS OF EXPERIMENTS

A detailed description of the acquisition and processing system of AE data is given in the paper of Mansurov (1993).

According to Stanchits & Tomilin (1983) the released elastic wave energy E is approximately proportional to the volume of the generated crack V:

$$E \cong \beta V \qquad (2)$$

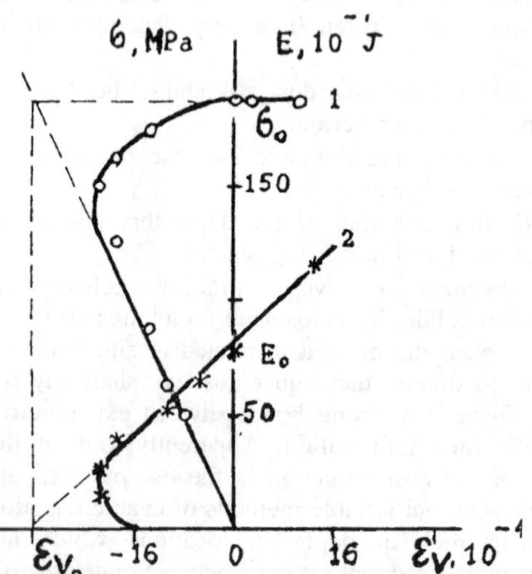

Figure 1. Stress (1) and cumulative energy of AE signals (2) plotted as a function of volumetric strain at uniaxial loading of granite sample with testing machine.

It is natural to suppose that this relation and, consequently, the way of rock fracture should also depend on the conditions of rock loading.

Let us consider the two experimental curves shown in Fig.1. Curve 1 shows the dependence of volumetric strain ε_v on stress σ for the rock sample under load. Correspondingly, curve 2 demonstrates the dependence of volumetric strain versus the cumulative elastic energy of AE signals. As is easy to notice, curve 1 remains in the region of negative values ε_v until the load reaches level of about 0.7 of the strength σ^*. According to Beniavski (1967) this effect is connected with a compaction of the sample material. Then the curve goes into the region of positive values of ε_v. It is established, that the reason is cracking of the rock, which begins to prevail over the compaction processes at higher loads.

Based on the above considerations let us try to explain the behavior of curve 2. As it was mentioned, the left parts of the plots correspond to the situation when the sample compaction prevails over the relaxation processes caused by the cracks' formation. It is easy to understand that in the vicinity of the ordinate axis the processes of compaction and relaxation are playing equal roles. This means that here the volume decrease caused by the compaction of rock is compensated by its increase due to the cracking.

To evaluate the value of ε_{vo} one can draw a line extending the straight portion of curve 1 to the intersection with the ordinate axis. The right part of curve 2 is practically straight due to the dilatancy caused by the formation of new cracks. If we extend this part of the straight line up to the intersection with the horizontal axis, we shall get the value of ε_{vo} too.

As was established values of ε_{vo} obtained by means of the two methods are of the same magnitude. Mathematically this straight line may be written in the following manner:

$$E = E_0 + \beta \varepsilon_{Vo} V_S \tag{3}$$

where V_S is the initial volume of the rock sample, and β is a constant:

$$\beta = E_0 /(\varepsilon_{V0} V) \tag{4}$$

Experiments have shown that values of E vary in the range from 1 J/m^3 to 100 J/m^3. It should be noted that usually energy E was measured in arbitrary units. Nevertheless the absolute values of energy E were obtained with the use of a specially calibrated testing machine and AE recording system, see Kuksenko et al. (1986).

Now let us consider Fig.2 showing dependencies of the cumulative energy release in rock on the applied stress.

For the convenience of comparison these curves were plotted using normalized values. The horizontal axis shows current stress normalized to the failure strength limit and the vertical axis shows values of released energy normalized to the energy corresponding to the strength limit. As can be seen from the plot, the energy release during testing of the marble sample starts at relatively low stress. Correspondingly for the granite sample a noticeable release of elastic energy starts from 0.7-0.8 of σ^*. Then the cumulative energy release quickly increases and the absolute value of energy released in this case appears to be three times higher than that for marble.

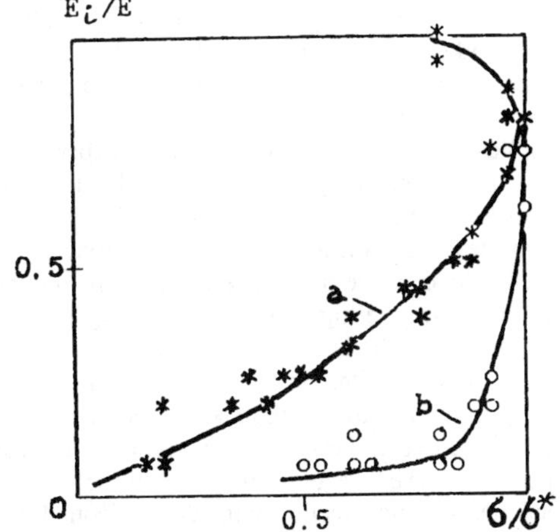

Figure 2. AE energy release under loading for marble (a) and granite (b) samples.

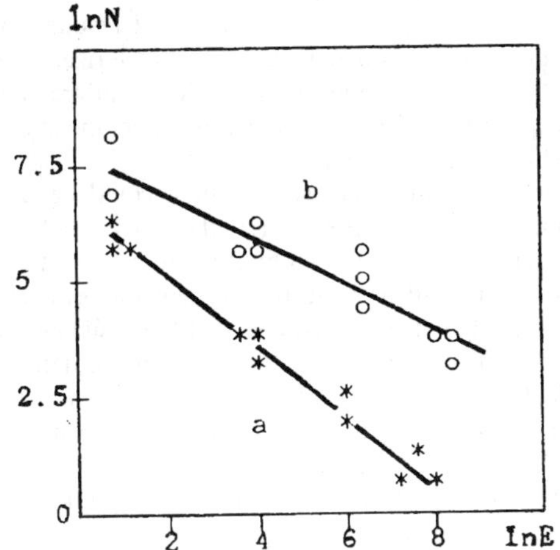

Figure 3. Recurrent intervals for marble (a) and granite (b) samples.

AE emission in the case of marble takes place during the whole period of loading, whereas for granite it occurs only at large stress. This is an evidence of the fact that granite can accumulate elastic energy without noticeable cracking. In contrast, deformation of marble is followed by the fracture process even at low stresses.

Application of the results of laboratory experiments to the induced seismicity monitoring can be discussed by considering the energy distribution of AE events of different level occurring in rock under load. Fig.3 shows the dependence of the logarithm of the number of AE events on the logarithm of the energy released during the event. This distribution presents a linear plot which is similar to the distribution of the recurring intervals for earthquakes. The higher the energy of the event, the lower is the probability of its occurrence. The extension of these linear dependencies to intersection with the horizontal axis gives an evaluation of maximum energy which might be released during a single event. As was established, the values of maximum energy correlate well with the characteristics of brittleness of rock.

Let us consider the above experimental results from the standpoint of nondestructive characterization of rock and their tendency to brittle failure and, consequently, to dangerous dynamic phenomena such as induced seismic events.

The two main conclusions which could be done instantly are as follows: (i) the rock that can accumulate more energy without microfracture during deformation, will be have the big dynamic effects at the final stage of macrofracture; (ii) the realization of big value of energy during the formation of unit size crack shows the more burst – proneness of the rock.

As was established macrocracks and macrofracture arise when the concentration of microcracks reaches the threshold level, then formation of the higher level defects of rock structure will occur. The threshold concentration could be defined as $N^*=1/(KL_o)^3$, where $K\sim3$, and L_o is a microcrack size. If we assume that the number of AE signals corresponds to the number of cracks N having sizes lower than L_o, it is easy to estimate L_o as

$$L_0 = K(N^*)^{1/3} \tag{5}$$

The minimal crack sizes, which were recorded with the data acquisition system differ considerably for different kinds of rock although they corresponded to the same values of released energy $E = 10^{-12}$ J. For granite $L_0 \sim 0{,}1$ mm, and for marble $L_0 \sim 0{,}4$ mm. Of course small size cracks also occur in marble under load. However, in this case the level of released elastic energy is below the sensitivity threshold of the data acquisition system.

Using the equation (2) written as $E=\beta{\cdot}L^3$, we may evaluate the constant β for a given kind of rock. Since β corresponds to the energy released in a unit volume during crack formation, we may understand why it is lower for marble than for granite. The amplitude of the AE signal can be calculated as $A\approx\sigma^2Lc^{-1}$, where σ is the stress in the crack's location; L is the length of the crack and, c is the velocity of propagation of an acoustic wave in the given material. It means that at a given level of applied stresses local stresses depend on the rock structure. So, in marble the level of local stresses is not so high, because marble consists of calcite grains, which begin to slide along one another under load. This prevents concentrations of stresses in rock and they can easily relax due to microshears. For granite, which consists of quartz minerals, the high level of local stresses can be easily achieved at a relatively low level of applied stress. That is why the fracture process the taking place in overstressed areas is followed by the emission of AE signals of high amplitude and energy.

Rock testing experiments carried out with the use of samples of 1m^3 volume and a loading force up to 50,000 ton have not shown any noticeable difference in the behavior and parameters of rock failure.

The linear graphs of recurrent intervals were obtained for different kinds of rocks and different size levels. As was established a qualitative similarity of processes of fault or crack accumulation exists at any size level. Quantitative differences are caused by physical and mechanical properties of rocks and by different stress states. Recurrent intervals for dynamic phenomena of different scale and energy levels were obtained for the following processes where the energy release ranged over 15 orders of magnitude.

To evaluate the maximal elastic energy that could be released during the fracture process one can extend the graphs plotted in the coordinates of Fig.3 to the intersections with the horizontal axis E. The points of the intersections give values of the maximum energy E_{max} which could be emitted instantly in a single event. The relation between E_{max} and the volume of the rock mass under observation can be derived from the linear dependence of the recurrent intervals graph:

$$E_{max} \sim V^{1/2} \tag{6}$$

So, the parameters E_{max} and β can be used for finding of zones in rock masses which accumulated sensible amounts of elastic energy. For this purpose all microseismic events occurring in the observed volume must be plotted in the double logarithmic coordinates as in Fig.3 to obtain values β and E_{max}. Then the current values could be compared with the threshold parameters in every point of the rock massif to map potentially dangerous locations, see Mansurov & Anikolenko (1996).

THE KINETIC MODEL OF LONG-TERM ACCUMULATION OF CRACKS IN ROCK MASS

It is established that the rock fracture process is to a great extent controlled by microstresses which exist in rocks due to their heterogeneity and anisotropy. In a stressed state, microcracks are generated in rocks which then grow in size as deformation develops. During this process pairs of cracks are joining to create cracks of a higher level by overcoming potential barriers caused by interatomic interactions. The physical model and kinetic equations showing the cumulative number of crack generation events versus time were described in details by Anikolenko (1992) . In a general form the equation can be written as:

$$N(t) = \int \{1 - \exp[-k(r)t]\} dr \,, \qquad (7)$$

where $k(r) = k_0 \exp(-2r/a)$, r = distance between two joining cracks; k_0 and a are constant in time.

In the simplest case of uniform distribution $f(r) = 1/\Delta r$ we obtain the following equation for $t >> t_0$:

$$\frac{N(t)}{N(t_0)} = \frac{a}{2\Delta r} \ln(\frac{t}{t_0}) \qquad (8)$$

Here Δr = width of the distribution, and t_0 is the start time of observations. This approximate isotropic solution corresponds to a frequently occurring case when the distance between two interacting cracks is considerably shorter than the distance to another pair of interacting cracks. Isotropic 3-D and 2-D solutions of the equation (7) have been also obtained by Anikolenko (1992). However they are not given here for ease of presentation.

The above equations were obtained on the assumption of the Markov process of defect formation, i.e. defect size and location depend exclusively on the parent defects' parameters and are independent of the history of parent defect generation. Solution of an inverse problem for equations (7) and (8) gives us the parameters of distribution *f(r)*, as well as the value of parameter a which is an average cluster size currently involved in the fracture process. So, a is a scaling parameter which value characterize typical defect' size involved in the fracture process at the moment. Useful information about rock fracture process can be also obtained by analyzing formal kinetic dependencies.

COMPARISON WITH THE EXPERIMENTAL RESULTS

As was mentioned above, monitoring of seismic events of any energy level (which may lead to the loss of stability) requires information about the preparatory process preceding a strong event. Basically in every separate case it is important to determine the energy level above which an event will be considered as strong. Once such a level has been defined, all lower level events occurring prior to the strong event may be considered as a part of the preparatory process. Based on this, let us consider the results of rock testing experiments.

Cylindrical rock samples of 6 centimeters in length and 3 cm in diameter were uniaxially loaded by a testing machine. Cracking was registered by the acoustic emission method. The details of the experimental technique are given in the work of Mansurov (1994) .

Figure la shows a typical differential acoustic emission spectrum of failure of a sandstone sample. Fig. 4b demonstrates integral spectrum plotted in semilog coordinates of the equation (2) for the same experiment. The two plots shown in Fig. 4 relate to the sandstone testing. As was established cumulative kinetic curves were practically independent of the sample's size as well as of rate of loading. All the curves when plotted in the semilog coordinates of the equation (8) look very similar. Similar cumulative curves were obtained also using catalogues of rockbursts and earthquakes, Anikolenko (1992). They give one a possibility of forecasting trends in development of processes of energy accumulation and relaxation in rock masses. As was shown the critical failure corresponds to the transformation of experimental kinetic curves from 3-D to 2-D solution of the kinetic equation (7) due to the generation of the major fault.

EVALUATION OF THE CURRENT STRESS STATE OF ROCK MASSIF AND FORECASTING OF STRONG SEISMIC EVENTS

Two stage model of solid fracture and experimental results of the cracking investigation have allowed to formulate technique both for evaluation current stress state of the rock and for forecasting of strong seismic events that can occur in rock massif caused by creation of underground NW storage (Kuksenko et al.1986). The model suppose similarity of fracture process on all existing scale levels of heterogeneity. Cracks arising during loading may be conceived as flow of discrete events, each of them characterize by coordinate on time axis and in space as well, and by size of corresponding crack. Such flow can correlate a sequence of registered AE/MA where characteristics of crack sizes are parameters of signals elastic energies. The fracture process can be

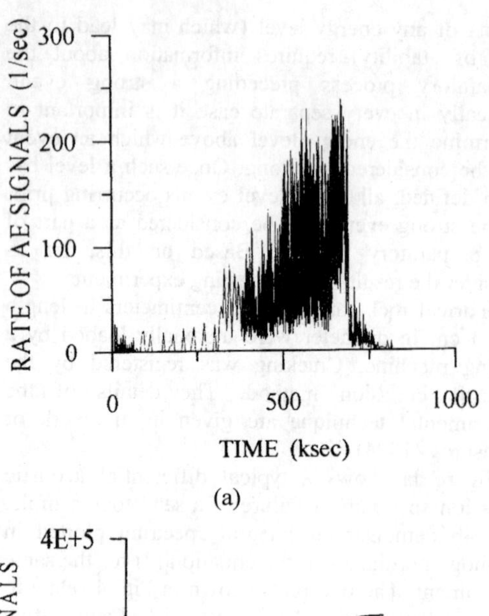

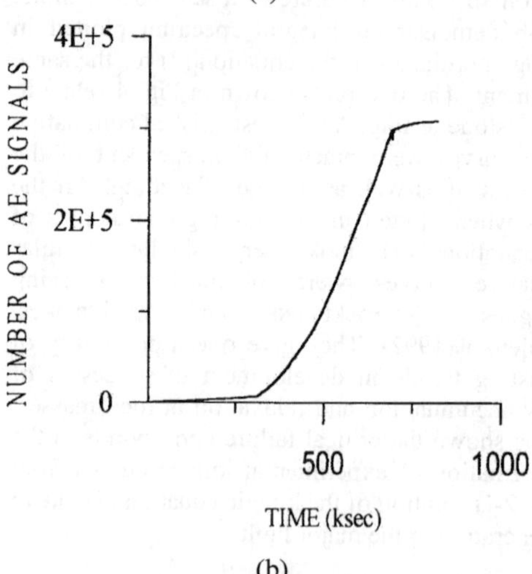

Figure 4. (a) Rate of AE Signals vs Elapsed Time (failure of sandstone sample with a testing machine); (b) Dependence of Figure la Plotted as a Function of the Logarithm of Time

presented like statistical flow of discrete events. According to this assumption, the first stage is satisfying to quasi- stationary Poisson process and disturbance of these conditions is a criterion of formation of the fracture nucleus. The average magnitudes of time intervals (Δt) between chronologically subsequent events and their variation ratio ($V_{\Delta t}$). On Fig.5 showed that the first stage (appearance of cracks) is Poisson's flow with rate Δt. This is corresponding with time interval $t<T1$. When cracks reach the threshold concentration in observing volume of material ($t=T1$), then arise conditions for their interaction and cracking stimulated by it. This stage reflects the simultaneous increasing of the parameter $V_{\Delta t}$ and decreasing of the parameter Δt. The crack of next level of heterogeneity forms due to retardation of fracture nucleus stability ($t=T2$).

Time moment $T2$ is not obligatory for finish of the second non-stationary stage. The result of the nucleus relaxation is inverse tendency ($t=T3$) that is increasing of Δt and decreasing of $V_{\Delta t}$. At a later time we shall call the considered characteristic change of parameters $V_{\Delta t}$ and Δt as the fracture nucleus image.

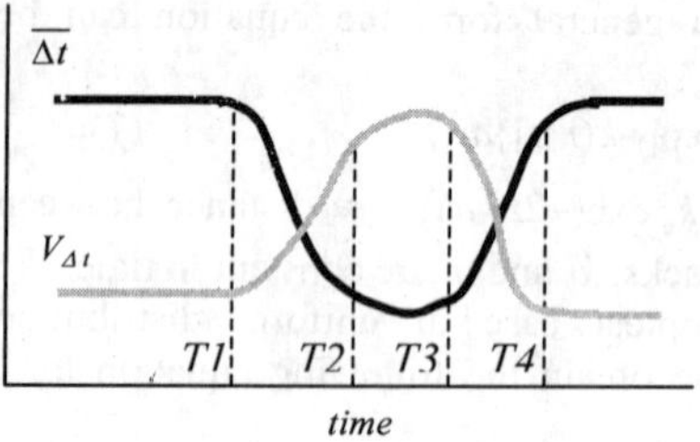

Figure 5. Image of fracture nucleus

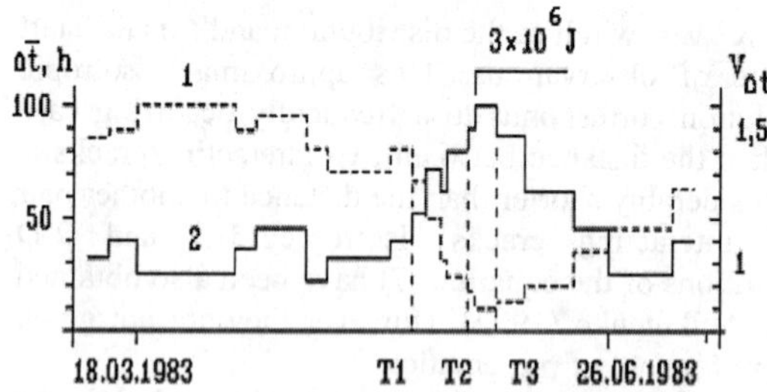

Figure 6. Real image of the rock burst nucleus (1- Δt, 2 - $V_{\Delta t}$)

In common case, the fracture realizes simultaneously on different scale levels of heterogeneity (Kuksenko et al.1990) and on each of them the some stage following after other one. Therefore detection of considered trends is possible at performance of two conditions. Its consist that analyzed sample should include only events that comes to the area of fracture nucleus preparation. Energies of these events are corresponding to the certain rank of the fracture process. Practically, performance of these conditions consists in double (spatial - energetic) selection of events flow, where criterion is revealing and optimization of parameters trends that described above. The existence of correlation dependencies (9, 10) between both the time of nucleus development and the size of nucleus preparation area and value of radiated elastic energy at the moment of stability loss by fracture nucleus (E):

$$\lg T = a \lg E + b \quad (9)$$

$$\lg R = c \lg E + d \quad (10)$$

These criteria are not included physical-mechanical properties of materials (rock) because of many troubles for its determination and for remote control also, and only universal statistical parameters are applicable. The differences in physical-mechanical properties of rock influence certainly upon features of fracture but according to both an our opinion and experimental data its are not disturbing of universality of fracture process statistical relationships.

CONCLUSIONS

- The results of laboratory and field experiments have demonstrated good performance characteristics of acoustic emission/microseismic activity method for the monitoring of stress state in rock;
- Physical models based on the kinetic concept of rock fracture process gives one a possibility of monitoring and long-term forecasting of rock mass behavior;
- Proposed software and system design concept demonstrate their applicability to the monitoring of stress state in ambient rock and structural elements of NW depository;
- The approach could be also used for the monitoring of fracture process in the near-field, eg. in blocks of vitrified and solidified NW.

REFERENCES

Anikolenko V.A. 1992. Kinetic study of rock fracture in laboratory press experiments. Acta Montana Ser.A 3(89): 63-72.

Bieniavski Z.T. 1967. The influence of specimen size on compressive strength of coal. Intern. J. Rock Mechanics and Mining Science 4: 395-430.

Frolov D.N., Kilkeev R.S., Kuksenko V.S. & Novikov S.V. 1980. Correlation between parameters of acoustic emission signals and rupture size in the course of fracture of inhomogeneous materials. J. Mechanics of Composite Materials 5: 907-911 (in Russian).

Kuksenko V.S., Stanchits S.A. & Tomilin N.G. 1983. Evaluation of size of growing cracks using parameters of acoustic signals. J. Mechanics of Composite Materials 3: 114-120 (in Russian).

Kuksenko V.S., Manzhikov B.T., Mansurov V.A. & Stanchits S.A. 1986. Burst-prone characterization of rock using energy release measurement. Physico-Technical Problems of Natural Resourses Mining, Novosibirsk 5: 26-32 (in Russian).

Kuksenko V.S., Mansurov V.A., Manzhikov B.T. & Tomilin N.G. 1990. Similarity of elastic energy radiation in rock on different scale levels. Izvestia Academii Nauk SSSR, Fizika Zemli 6: 66-70 (in Russian).

Mansurov V.A. 1993. The study of rock deformation in post failure conditions. Proc. intern. symp. Istanbul. Rotterdam: Balkema.

Mansurov V.A. & Anikolenko V.A. 1996. Evaluation of dilatancy in rock for forecasting burst prone zones in mines. Mater. science forum 210-213: 527-534.

Mansurov V.A. & Manzhikov B.T. 1999. A scheme for the preparation of strong induced seismic events. Annali di Geofisica. 42 (5): 801-808.

Petukhov I.M. 1983. Rock burst in mines; Proc. 5th intern. congress on Rock Mechanics, Melbourne. Rotterdam: Balkema.

Regel V.R, Slutsker A.I. & Tomashevsky A.E. 1974. Kinetic nature of the solid strength. Uspekhi Fizicheskih Nauk 106: 193-228 (in Russian).

Ruzhich V.V., Mansurov V.A. & Babichev A.A 1985. Seismotectonic criterion of Earth crust destruction in Baykal region. Doklady Akademii Nauk SSSR 281: 566-569 (in Russian).

Sobolev G.A. & Zavialov A.D. 1980. Concentration criterion of seismic. Doklady Akademii Nauk SSSR 252: 69-71 (in Russian).

Stanchits S.A. & Tomilin N.G. 1983. Correlation between the released elastic energy and size of the crack in rock fracture process. Earthquake Prediction, Dushanbe-Moscow 4: 31-45 (in Russian).

Yanagidani I.T., Ehara S., Nishizava O. & Kusunose K. 1985. On rock failure process. J. Geophys. Res. 90 (138): 6840 – 6855.

Zhurkov S.N., Kuksenko V.S. & Petrov V.A. 1981. Physical principles of mechanical failure prediction. Doclady Akademii Nauk SSSR 259(6), 43-50 (in Russian).

Environmental Issues and Management of Waste in Energy and Mineral Production, Singhal & Mehrotra (eds)
© 2000 Balkema, Rotterdam, ISBN 90 5809 085 X

Use of 'A guide to the management of tailings facilities' at Syncrude

J.G. Matthews, S.Gaudet & B. List
Syncrude Canada Limited, Fort McMurray, Alb., Canada

ABSTRACT: Efficient and effective management of mine tailings is an essential ingredient for the continued development and prosperity of the Canadian mining industry. Recognizing this, the Mining Association of Canada (MAC) recently worked with mining industry representatives to develop and publish "A Guide to the Management of Tailings Facilities" in September 1998. The MAC tailings management guide provides direction and assistance with the development and assessment of effective tailings management systems. As a member of MAC, Syncrude Canada Ltd. actively supported the development of the guide. This paper describes how the guide was used at Syncrude to assess the effectiveness and integrity of in-house tailings management systems. The MAC guide was found helpful in providing focus on, and commitment to, tailings management excellence.

1 INTRODUCTION

Syncrude Canada Ltd.'s oil sand operations are located in northeastern Alberta, Canada, approximately 500 km north of Edmonton and 45km north of Fort McMurray (Fig. 1). Syncrude is the world's largest producer of light, sweet crude oil from oil sand and the largest single source of oil in Canada.

In 1999, Syncrude Canada Ltd. produced 81.4 million barrels of light sweet crude oil (Syncrude Sweet Blend) from the Athabasca Oil Sands deposit bringing cumulative production from the Mildred Lake operation to approximately 1.1B barrels since start-up in 1978. As a result of this activity, Syncrude currently manages an inventory of approximately 360M m^3 of fluid fine tails. In meeting steadily increasing annual throughput targets since the start of operations at the Mildred Lake site, Syncrude has done a consistently excellent job managing increasing volumes of the sand, process affected waters, and fluid fine tails components from the extraction process.

As a member of the Mining Association of Canada (MAC), Syncrude was given an opportunity to work with other MAC members to produce a broadly applicable tailings management guide that would be readily accessible to Canadian and foreign mining organizations. The compilation of over two years of collaborative work is "A Guide to the Management of Tailings Facilities" (Mining Association of Canada 1998). The guide was written to provide guidance and insight into the components that should be considered in developing effective tailings management systems.

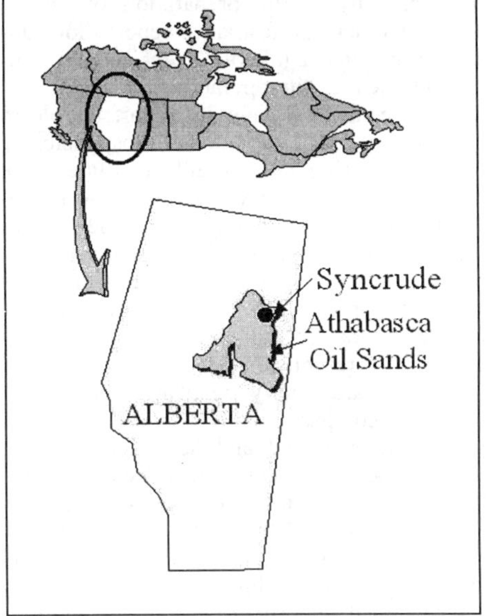

Figure 1 – Syncrude Canada Mine Site Location

The guide is not a prescriptive document nor does it delve into technical detail to any significant degree (Martin et al. 2000). Further, the MAC guide is not prescriptive on the matter of management norms or standards. The objective of the MAC guide is to assist tailings facility management in the development of effective tailings management systems for their respective operations. Syncrude regards the MAC guide as a useful and complementary resource to established tailings management systems and processes.

2 TAILINGS MANAGEMENT AT SYNCRUDE

Syncrude Canada Ltd. operates several large, integrated tailings facilities and regards the operation, maintenance, and monitoring of these facilities in a safe and environmentally acceptable manner as a matter of great importance.

The effective management of the tailings facilities is coordinated within an integrated group of technical support, field operations, and management teams. Each of the major tailings storage facilities has a dedicated geotechnical engineer working within a technical network to provide support to tailings operations. A key responsibility of these engineers is the surveillance and monitoring of tailings impoundment performance to ensure the tailings are stored according to regulatory license & permit conditions and accepted engineering practice. An additional responsibility is the preparation of annual tailings-impoundment performance reports for submission to provincial regulatory agencies as a condition of continued facility operations.

Operations teams also have an important role in tailings management at Syncrude with responsibility for specific monitoring and surveillance duties. The technical support and operations teams work together with management to ensure that the tailings are managed in a safe and environmentally acceptable manner. In this way, the facilities are well monitored and managed within an integrated organizational framework. Coordination is critical to ensure that specific responsibilities for reporting and action responses are clearly delineated and understood within the company.

In addition to a skilled and dedicated in-house workforce, semi-annual Geotechnical Review Board (GRB) sessions are held at the mine site on a bi-annual basis (McKenna 1998). This group of recognized geotechnical consultants provides review and guidance on significant technical aspects pertaining to the geotechnical integrity of the impoundments. Responsible facility engineers, tailings facility operator representatives, and senior management participate in these sessions wherein presentations are given respecting the performance of the tailings impoundments. Through these meetings, the in-house culture of "due care" respecting geotechnical and environmental integrity of the tailings facilities is maintained in support of continued excellent environmental performance.

Complementing these tailings management processes is a dedicated research and development team focussed on improving tailings management through development and implementation of innovative techniques in support of continual improvement in tailings management performance.

Thus Syncrude is proud of three cornerstones of effective tailings management:

1. A professional, dedicated, and knowledgeable staff indoctrinated with a desire and commitment to manage tailings in a safe and environmentally acceptable manner,
2. An expert review board to provide management assurance that Syncrude tailings facilities are being managed with due regard for geotechnical integrity in support of safety and environmental stewardship, and,
3. A dedicated research and development group with a commitment to pursue and develop promising tailings technology opportunities.

These three components work in a complementary manner under the corporate loss management system to provide a systematic, flexible, and effective tailings management program.

2.1 *Syncrude Loss Management System*

Syncrude is a large, integrated organization. To ensure appropriate safety, environment, and fiscal management control, Syncrude manages all operations under a company-wide loss management system. This loss management program was adopted in the early 1980's through the Det Norske Veritas/International Loss Control Institute and is called the International Safety Rating System (ISRS). This program identifies program elements that, in combination, comprise the foundation of the Syncrude loss management program. Aspects addressed in the program elements include inspections, investigations, team meetings, emergency preparedness, change management, and risk analysis (Cascio 1998).

As stated in the corporate environment, health, and safety policy (Fig. 2) Syncrude is committed to safe and responsible operations, including tailings facilities. Stewardship of the corporate environmental policy is maintained through the application of the corporate loss management standards in all corporate undertakings. Thus, Syncrude has implemented a comprehensive and effective loss management/ loss control system into which specific environmental management objectives have been incorporated (Cascio 1998).

Given the effectiveness of the existing loss management program, and to reduce administrative redundancy, Syncrude would be reluctant to adopt additional, parallel standards beyond these established corporate standards (Cascio 1998). Rather, the corporate focus has been on identifying specific deficiencies or weaknesses in the application of the corporate minimum specification standards and then addressing these on an as needed basis within the framework of established loss management norms. This approach was determined to be most appropriate for implementation of the MAC tailings management guide at Syncrude.

Environment, Health and Safety Policy

All employees of Syncrude Canada Ltd. are fully committed to excellence in environment, health and safety performance in the conduct of business and in support of a safe, reliable and profitable operation.

We believe our success in meeting our business objectives is highly dependent upon our ability to avoid loss and we will strive to ensure all our stakeholders share this understanding and commitment.

We will comply with all governing legislation and regulations.

We will provide safe and healthy working conditions and demonstrate continuous improvement toward eliminating incidents and reducing the risk to people, environment, production and our facilities.

We will achieve performance excellence through the systematic application of loss management practices and the active participation of all employees and contractors.

We will comply fully with all environmental obligations as contained in applicable acts, regulations and the terms and conditions of Syncrude's various government licenses and approvals.

We will endeavour to perform as a leader of the resource industry in Alberta with respect to environmental protection including being proactive with the appropriate governmental agencies and area communities.

Figure 2: Syncrude Loss Management Policy

2.2 Environmental policy commitments

Syncrude stewards to a number of environmental management commitments including the corporate environmental policy, the MAC environmental policy and the Canadian Association of Petroleum Producers (CAPP) environmental policy. Within this environment, Syncrude has demonstrated a clear obligation to responsible, legal, and environmentally responsible management across all operating areas, including process tailings (Syncrude Canada Ltd. 2000; Mining Association of Canada 2000; Canadian Association of Petroleum Producers 2000).

These policy statements do not, however, provide guidance on how to meet the stated environmental objectives for tailings. In this regard, the MAC tailings management guide provides a framework for a process to assess the effectiveness of established in-house tailings management systems or to develop new tailings management systems.

3 MAC GUIDE IMPLEMENTATION STRATEGY

To facilitate the use and implementation of the MAC guide, a core team was assembled comprised of tailings facility engineers and leaders representing tailings management interests. The general task identified for the core team was to review the MAC document in parallel with established tailings management practices at Syncrude to determine if significant changes to established tailings management processes were warranted.

The implementation team started with an in-house tailings management assessment. This included team review of established processes and procedures supporting "continuous tailings management improvement". This review identified a dedicated, well skilled, and committed group comprised of operations, technical support, and management representatives. Specific management initiatives identified to be supporting tailings management excellence included semi-annual geotechnical review board (GRB) sessions, extensive performance monitoring programs, and annual regulatory performance report submissions. The team also identified a number of recently completed tailings management excellence initiatives including:

- utilization of the Canadian Dam Association (CDA) Dam Safety Guidelines in support of design and monitoring of impoundment facilities, and,
- completion of a site-wide dam inventory assessment (including consideration of failure consequence ratings).

Additional tailings management continuous improvement initiatives were identified including:

- development of operation, maintenance, and surveillance (OMS) manuals for critical impoundment facilities on site, and,
- development of emergency response (ERP) and emergency preparedness (EPP) plans for impoundment facilities.

Following the in-house review of the MAC guide with senior management, it was decided that "implementation" would be comprised of a comparison of established tailings management systems at Syncrude with the MAC guide recommendations to identify significant tailings management improvement opportunities. Major restructuring of established corporate loss management procedures and standards (inclusive of tailings management) was not considered however management did commit to a review of the gap analysis findings and to strengthen specific tailings management elements where warranted. Thus, it was decided to use the MAC guide as a tool to conduct a gap analysis review of tailings management systems. The primary implementation task was to complete gap analyses for key Syncrude tailings facilities. The objective of the gap analyses was review of the completeness of established tailings management systems relative to the MAC guide recommendations and to identify value-added continuous improvement opportunities. The MAC tailings management guide checklists were used as a template for this exercise.

Using this approach, Individual gap analyses were completed for each of Syncrude's major tailings impoundment facilities. The major tailings facilities were reviewed on a case-by-case basis to accommodate different jurisdictional aspects and to recognize unique environmental aspects associated with each facility.

The task of completing the gap analysis was completed by a team comprised of senior technical leaders (senior technical and supervisory staff) with experience in tailings management. The results of the gap analysis were reviewed with senior management to determine what changes or improvements to the established tailings management systems would be warranted.

3.1 *Gap Analysis Findings*

The gap analysis review process was determined to be a valuable exercise. A number of improvement opportunities were identified based on comparison of the existing tailings management systems with the MAC tailings management guide recommendations.

Specifically, the team found the MAC tailings management guide implementation exercise to be useful in a number of ways. Completion of the checklists provided:

- clarification of primary authority and responsibility for tailings and impoundment design-related decisions,
- a detailed listing of the most important management aspects associated with each of the facilities evaluated, and
- impetus to develop and manage operating manuals for all major tailings facilities.

Another significant improvement made as a result of the implementation exercise was the establishment of annual tailings management-review sessions. These will be above and beyond semi-annual geotechnical review board sessions to provide ongoing focus on, and support to, continual improvement of tailings management at Syncrude.

4 CONCLUSIONS

Though Syncrude has an effective tailings management system, within the context of an established corporate loss management system, use of the MAC tailings management guide was found to provide incremental improvements in the robustness of established tailings management systems.

The use of the guide is recommended for all organizations with an interest in safe and environmentally responsible management of tailings storage facilities.

REFERENCES

Canadian Association of Petroleum Producers 2000. [Online]. 2000, February 1. Available at http://www.capp.ca.

Cascio, J. (ed) 1998. *The ISO 14000 Handbook*: 1998 ASQ Quality Press. pp326 - 327.

Martin, T.E. & M.P. Davies 2000. Trends in the stewardship of tailings dams. In: *Tailings & Mine Waste '00.* Rotterdam: 2000 Balkema. *Fort Collins, January 2000*. pp 393-407.

McKenna, G. 1998. Celebrating 25 years: Syncrude's geotechnical review board. Geotechnical News, Vol 16(3), September, pp. 34-41.

Mining Association of Canada 1998. *A Guide to the Management of Tailings Facilities.*

Mining Association of Canada 2000. [Online]. 2000, February 1. Available at http://www.mining.ca.

Syncrude Canada Ltd. 2000 [Online]. 2000, February 1. Available at http://www.syncrude.com.

Environmental Issues and Management of Waste in Energy and Mineral Production, Singhal & Mehrotra (eds)

Tailings disposal and management practice at Bogoso Gold Limited, Ghana

Samuel Ndur, William Buah & Newton Amegbey
U.S.T. School of Mines, Tarkwa, Ghana

Thomas Morris & Christian Debrah
Bogoso Gold Limited, Ghana

ABSTRACT: Tailings disposal and management are major operational concerns affecting all open pit mines in Ghana today. Bogoso Gold Limited (BGL), one of several new surface mining operations in Ghana commenced operations in 1990 as Canadian Bogoso Resources. The company has completed four lifts of the tailings storage facility since the commencement of operations and has demonstrated a commitment to sound tailings disposal and management practices in spite of special difficulties with open-pit mine activities in the tropical rain belt. This paper discusses the tailings management practices at Bogoso Gold Limited highlighting aspects of dam stability and seepage control monitoring. Minesite water management practices, cyanide decay mechanisms within the dam, the use of various plant species to stabilize dam walls and erosion prevention practices are discussed. The benchmark standards of tailings disposal and management practices at this site have resulted in the adoption of similar practices throughout the Ghanaian mining industry.

1 INTRODUCTION

The introduction of new processing techniques like the Carbin-in-Leach (CIL), Carbon-in-Pulp (CIP), Heap Leaching and Bacteria Oxidation (BIOX) have augmented high gold production and have allowed the treatment of ores of considerable lower grades, which hitherto were regarded as wastes, to be mined at a profit. However, considerably more tailings are produced per ounce of gold with these methods and their proper disposal is very important.

Recent structural failures of tailings dams within Europe and SE Asia highlight the risk factors associated with tailings retention practices. Tailings disposal and management are therefore major operational concerns at all surface mining operations today.

In recent past not much attention was given to environmental management in Ghana. Thus exploration, prospecting, mining and processing techniques were at the discretion of the Mining Companies concerned. Since launching the country's Economic Recovery Programme in 1983 with its attendant increase in investment in the mining sector, new environmental guidelines and policies were enacted in 1988 to ensure sound environmental management so as to change the course of the hitherto damages caused by Mining activities.

Bogoso Gold Limited (BGL), a mine situated around the town of Bogoso 35 km north of Tarkwa in the Western Region of Ghana, has demonstrated a sound commitment to environmental issues. The benchmark standards of tailings disposal and management practices at this site have resulted in the adoption of similar practices throughout the Ghanaian mining industry.

2 BACKGROUND OF BOGOSO GOLD LTD.

Mining in the Bogoso area has a relatively long history, although there is no official record of any workings prior to 1882. Small scale mining operations started near Bogoso in 1906 on a commercial basis and continued to the early 1930s when high grade oxide deposits were developed by the then British Group Marlu Gold Mining Area Limited. Denison Mining Limited of Canada acquired the concession in 1986 and started operations as Canadian Bogoso Resources (CBR) in 1991. The company changed hands to Bogoso Billiton Gold Limited, Ghana with equity shareholding at Billiton Internal Metals B.V. (BIM) 81%, International Finance Corporation (IFC) 9%, and Government of Ghana, 10%. The shareholdings at present are as follows;

Gencor Bogoso Holding (BVI Limited) 82.0%

The Republic of Ghana 10.0%
The International Finance Corporation 8.0%

Financing was arranged through shareholdings' equity, shareholders advances and loans provided by the International Finance Corporation, leading a group of banks, and the German Investment and Development Corporation (DEG).

2.1 *Climate*

The climate of the area is hot and humid, with a temperature range of 20^{0}C to 35^{0}C and humidity of 85.5%. The highest humidity occurs in August/September, while the lowest occurs in January/February. The fifty year average rainfall is 1,800 mm, with a monthly range of plus zero to 400 mm. A drier period occurs in December/January /February. The evaporation rate is 1,500 mm per year. The wind direction is predominantly from the south-west to north-east.

2.2 *Topography and Geology*

Relief in the Bogoso area ranges from 30-170 m above sea-level. The main feature is a series of north-east trending hills, located roughly along the central axis of the concession.

This tectonic province is dominated by sedimentary and volcanic rocks subdivided into three sequences: the Lower and Upper Birimian series and the Tarkwaian system. The Bogoso concession lies along the major structural feature which can be traced for at least 200 km. Lode mineralisation has been related to this north-east trending tectonic crushed zone, which longitudinally bisects the concession. The surface expression of mineralisation is within hills trending along the crushed zone.

Gold mineralisation is of hydrothermal origin and is usually limited to the sulphide minerals and their weathered products, either as crystalline inclusions within sulphide grains or in complete solid solution within the sulphides.

Near surface oxidation of the deposit extends to a maximum of 50 m below the surface, depending upon the topography and distance to the water table. A transition zone ranging from a few metres to 20 m lies between the oxide and the sulphide zones.

2.3 *Mining*

The open pit mining operations consist of drilling and blasting, removing the broken material by a hydraulic excavator and loading it into rear dump haul trucks. The haul trucks carry either the gold ore to the processing plant or the waste material to the waste dumps.

Precautions are taken to ensure that mining of the oxide ores does not penetrate the transitional ore zones, in an effort to avoid acid water generation. A minimum layer of oxide ore is left above the interface. The ore is hauled either to the primary crusher at the treatment plant or to holding stockpile close to the crusher. The grade of the ore varies considerably within the ore zone, in vertical and horizontal directions and also from one deposit to another deposit. Mining activities are carried out currently at several locations on the lease. The ore from the different locations are blended at the primary crusher area to ensure that a stipulated average ore grade is delivered to the treatment plant. This assists in the stable operations of the plant, which allows maximum recovery of the gold from the ore. The waste material from the mining locations is hauled to waste stockpiles.

Total material movement since the commencement of operations to the end of December 1997, has been 52,055,302 tonnes consisting of 9,491,414 tonnes of ore 42,563,888 tonnes of waste to produce 665,748 ounces of gold.

Current oxide ore mining rates are 2 mtpa of ore and 6.5 mtpa of waste, for a total movement of 8.5 mtpa. Where practical, progressive rehabilitation is undertaken in all mining related areas.

2.4 *Treatment of ore*

The processing plant at BGL was originally designed to process sulphide ore by means of a sulphide concentration circuit and a two-stage roaster. Unfortunately, technical and ore type problems resulted in the permanent closure of the roasting operations in early 1994 resulting in an only oxide ore capability.

The run-of-mine (ROM) ore is fed into the primary jaw crusher and the crushed ore is carried to the milling circuit, via a crushed ore stockpile, by conveyors. Ore is continuously withdrawn from under the stockpile onto the mill feed conveyor belt, which feeds directly into the semi-autogenuous grinding (SAG) mill. The milling circuit comprises two-stage milling with a primary SAG mill and a secondary ball mill. The combined discharge from both mills is classified in a bank of hydrocyclones. The underflow (+75 microns) gravitates to the ball mill for further grinding, while the overflow (-75 microns) passes over a trash removal screen before discharging into the five-stage carbon-in-leach (CIL) agitated tanks. A batch of loaded carbon is removed on a daily basis from the first CIL stage (tank one) for the recovery of gold.

After treatment with hydrochloric acid to remove inorganic foulants the loaded carbon is subjected to a pressure stripping procedure, utilising caustic cyanide solution at an elevated temperature (135^{0}C) and pressure (400 kpa). The gold is desorbed from the carbon by the solution. The desorbed gold in the strip solution is electrically plated out onto steel wool cathodes in a pressurised electrowinning cell during the stripping cycle. At the end of the stripping process, usually lasting 12 hours, the carbon is regenerated in a vertical kiln at the temperature of 700^{0}C, to remove organic foulants and then returned to the CIL circuit.

The gold loaded cathodes are removed from the electrowinning cell on a weekly basis, calcined in an oven at 750^{0}C, to remove moisture and to convert base metals to oxide form, and smelted with fluxes to produce gold dore bars with a gold content of between 90-95%. The gold production rate is in the order of 100,000-110,000 ounces per year. Tailings exit the final CIL stage (tank five), pass over a carbon safety screen and are pumped to the tailings dam via two 355 mm PVC pipes.

3 TAILINGS IMPOUNDMENT

The BGL Tailings Dam is located approximately 8km south of Bogoso town and 1km east of the main Bogoso/Prestea road. Geotechnical investigations for the site were completed during 1988 and 1989 by Soil and Rock Engineering Pty Ltd and the initial four embankments were constructed the following year. The tailings impoundment currently consists of a series of embankments and saddle dams around the periphery of the storage basin. Approximately 10 million tonnes of tailings had been deposited by the end of 1997 at a deposited tailings density of 1.45 t/m^3.

Embankment walls were originally designed to have an overall slope of 1V:3H. However, historical construction techniques have formed embankment sections to steeper profiles ranging from 1V:2.2H to 1V:2.8H. Successive upstream embankment lifts between 1996 and 1998 have generally used selected lateritic fill, compacted in 150mm thick layers to 95% standard compaction dry density. A longitudinal drain constructed at the foundation level and connected to a series of transverse drains that daylight at the downstream toe has been incorporated within the design as a means to reduce pore pressures through the embankment and to provide a path for seepage to migrate to a downstream collection sump.

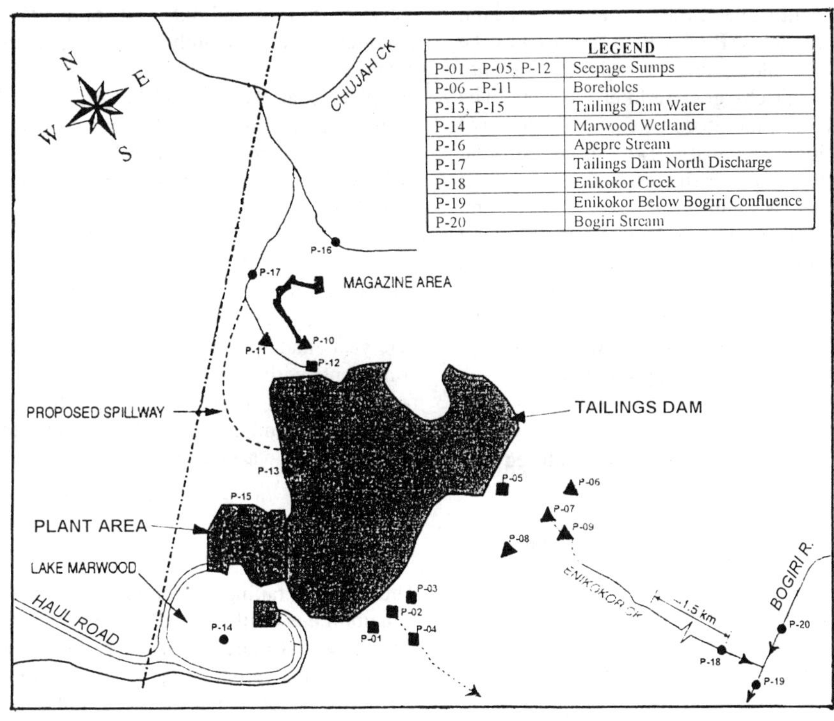

Fig. 1. Tailings dam area: Water quality sampling sites

3.1 *Tailings Impoundment Management*

Tailings deposition within the impoundment is currently running at a rate approximating 180,000 tonnes per month. Tailings are spigoted on a rotational basis from five operational embankments with areas being rotated as tailings deposition depth approaches 200 mm. Rotational spigoting practices aim to maintain the supernatant pond within a small area and prevent encroachment within 100 metres of dry embankment walls. Monsoonal rainfall patterns prevail throughout Ghana and exert a strong influence on water management practices. Excess water accumulation within the tailings impoundment is common during the wet season and operational practices are in place to discharge excess supernatant water to a constructed wetland area (Marwood wetland). Piezometric standing water levels and supernatant pond elevation are recorded at weekly intervals. Seepage collection sumps have been established below the southern and south eastern walls of the impoundment and waters collected in these sumps are pumped to the impoundment on a daily basis.

Figure 1 is a plan view of the tailings dam area showing the various seepage collection sumps (P-01 – P-05) to the south of the tailings dam, monitoring boreholes (P-06 – P-11), as well as the monitoring points along creeks and streams into which seepage or discharges are likely to flow. The phreatic level is monitored via eleven (11) piezometric holes located at various points around the dam walls. Results over the last one year showed standing water levels ranging between 73.58 m and 76.37 m at the southern part of the wall, 65.31 m and 72.51 m to the northern part of the walls and values between 67.53 m and 82.56 m to the south eastern impoundment of the tailings dam. In Table 1, results of the seepage collection sump water quality, monitored over the last year, are presented.

Table 1. Typical ranges of parameters monitored at the seepage sumps.

Parameters	Max. value	Min. value
pH	7.86	3.35
TDS (ppm)	1347	700
Cond. (mS)	2.85	0.21
CN, (ppm)	0.16	0.00
Fe, (ppm)	65.59	0.01
As, (ppm)	3.7	0.005

3.2 *Cyanide Decay within Tailings Impoundment*

Since 1994, ore feed to the Bogoso milling circuit has been confined to oxide ore types which generally comprise zones of leached lateritic material containing residual gold. This ore type is depauperate in transition metals such as copper and zinc which form stable weak acid dissociable (WAD) complexes with cyanide. Given the absence of WAD cyanide complexes at Bogoso, free cyanide measurements are used to quantify cyanide decay rates within the impoundment. The warm climate and long flow path over dry tailings beaches facilitates rapid photo catalytic decomposition of cyanide within the impoundment and domestic sewerage is also fed to the storage to promote bacteriological breakdown of residual free cyanide. Mortalities of avian fauna are known to occur in some tailings impoundments particularly those with high concentrations of copper WAD cyanide complexes (Castrantas, 1988). However, at Bogoso mortalities are rare and just two have been observed during the previous 18 months.

3.3 *Progressive Revegetation*

Revegetation of tailings embankments at Bogoso commences immediately upon the completion of each embankment lift. Dense rows of vetiver grass are planted at the top and bottom of each wall lift with infill rows of citronella grass every two metres added to stabilise the outer wall and prevent erosion. Following colonisation of grasses, tree species including *leucaena leucocephala*, *cassia siamea*, *gliricidia sepium* and *gmelina arborea* are planted to provide additional erosion protection and to mitigate seepage through transpiration. Growth rates of trees on the tailings embankment walls exceed 2.5 meters per year and provide a dense vegetative cover consistent with secondary forest and subsistence farming areas which surround the concession. Creeping ground covers and additional tree species are used for varying site specific conditions as required.

3.4 *Closure Planning*

The current tailings impoundment is expected to remain in use until the end of the year 2001 at which time a new storage facility may be constructed for sulphide ore processing. Food crop trials are currently ongoing using oxide tails to determine likely metal uptake in root crops planted on the decommissioned tailings surface and results of the study will be used to determine the need, if any, for final capping of the tailings dam. Concurrent with this work, the viability of citronella grass plantations and a local oil extraction facility are being evaluated in line with corporate sustainable development objectives.

Design of a final spillway and passive treatment options for seepage water chemistry are currently under review in line with long term closure planning.

4 CONCLUSION

On-going monitoring of phreatic levels along the tailings dam walls indicate that the walls are well drained inspite of the heavy rains in the region. Revegetation techniques for the embankments are effective and dam failure has not been experienced over the years and there are no signs of it happening in the near future. Water quality in the collection sumps show low levels of contaminants. Cyanide concentrations range between 0.00 and 0.16ppm, and this is a fair indication of effective cyanide decay within the impoundment. This was also shown in results obtained earlier in a paper by Amegbey et al. (1994).

It may be too early to comment on the closure planning trials, however, results obtained so far are encouraging.

REFERENCES

Amegbey, N., Ndur, S. A., & Badu F., (1994). Some Aspects of Environmental Management at Billiton Bogoso Gold Limited, Ghana". *Proceedings of the Third International Conference on Environmental Issues and Waste Management in Energy and Mineral Production, Perth, Western Australia*, pp. 159-168.

Castrantas, H., (1988). Cyanide Detoxification of Gold Mining Tailings Pond with Hydrogen Peroxide. Paper presented at the Randol Gold Forum 88, January 23, 1988.

Environmental Issues and Management of Waste in Energy and Mineral Production, Singhal & Mehrotra (eds)
© 2000 Balkema, Rotterdam, ISBN 90 5809 085 X

Design of safety channels in mining for small drainage basins

E. Orche
Department of Natural Resource Engineering, University of Vigo, Spain

ABSTRACT: A combined application of both meteorological and civil engineering techniques is presented in order to design mining safety channels which could take surface water out of open pit or underground mines. The application is specially developed for small drainage basins on which many mines have been implanted, which adhere to Spanish technical standards and safety regulations. This method saves quite a lot of calculation time compared with the long traditional one because of its simplicity.

1 INTRODUCTION

One of the most effective methods to prevent runoff waters from entering mine pits or from getting into contact with materials whose stability could be reduced is by constructing perimeter guard channels which effectively collect such surface waters and channel them to other areas where their effect is benign. This system is used to prevent water entry into underground and open pit mines as well as surface installations and services, dump sites, mineral ore storage sites, etc.

Mining operations in Spain are very often carried out within little hydrographic basins. In such cases, a maximum drainage flow corresponding to a certain downpour can be calculated by specifically designed simplified methods for smaller basins (<75 km^2). These methods have the advantage that their use is practically universal but the disadvantage is that their application is restricted exclusively to the type of basins cited.The maximum drainage flow is calculated for a set return period, and from its value, one can calculate the dimensions of the channel or ditch required to dispose off such waters. In such cases, the calculation sequence is based on the selective and linked application of the many well contrasted methods used in meteorology and civil engineering, which can be summed up as follows:

1. Selection of return period.
2. Calculation of maximum rainfall in 24 hours for the selected return period and by applying Gumbel's method.
3. Calculation of the maximum flow produced by the rainfall using the rational method.
4. Dimensioning the channel using Manning's method.

2 SELECTION OF RETURN PERIOD

Mines are subject to multiple harsh weather types during the many years of their life expectancy. In as far as rainfall is concerned, it is a well known fact that maximum precipitation values occur cyclically at well defined time intervals. The rains produce greater floods when rainfall is more intense. Therefore, when there arises a need to protect the mine from such maximum drainage flows that could occur during its life expectancy, a peripheral channel should be dimensioned such that it would be able to contain at least the flow corresponding to the time the mine has been operational. In other words, the reference flow which must be projected on to a surface drainage is related to the frequency of its appearance, and can be defined by its return period: that is, the higher the return period the greater the flow.

In this context, the said return period T is defined for a flow of value x, as the average interval expressed in years, in which the extreme flow value reaches or surpasses x just the one time. The return period thus defined must not be confused with the average interval between occurrences of maximum values equal or greater than x, because in such a case, the year of such maximum value occurrences is not taken into account, since there could be some years with two or more maxima and others without any maxima. The probability P that the flow x be exceeded at any time during a certain time interval C, for example, that of a mine lifetime, also depends on the duration of the interval and can be expressed as (Llopis 1999):

$$P = 1 - [1 - \frac{1}{T}]^{C}$$

Therefore, for a flow Q with a return period of 50 years (T = 50), there exists a 2 % probability that in any year (C = 1) of the said period, there could be at least one flood with a flow value equal or greater than Q. The probability increases to 86% for a period of a hundred years (C = 100).Thus, in order to guarantee that such an event does not occur, return periods far greater than the mine life expectancy are considered by the mining industry. So it is not unusual to see return periods of up to 100 years for mines whose operational life expectancy is just 20 years. In such a case, the probability of a greater flood flow than that designed would be at 18% (T = 100, C = 20). The risk could be lowered to 4% by increasing the return period to 500 years (T = 500, C = 20), but this would mean constructing a channel of enormous dimensions whose huge costs would not justify its feasibility.

Thus, the selection of corresponding return periods does not obey fixed rules since, they definitely have economic connotations which can be important due to the sizeable investment to be made in the guard channels. Consequently, the criteria normally used is that of maintaining an equilibrium between assumed risk of the presence of maximum calculated flood flow and the investment in a channel able to contain such a flow.

3 TWENTY-FOUR HOUR MAXIMUM PRECIPITATION CALCULATION

The 24 hour maximum precipitation is a data that is required in order to apply the rational method for calculating maximum flows. This data is obtained from a series of historical data from one or many meteorological stations around the site.

The problem is based on the fact that the safety channel must be constructed to receive the maximum rainfall flood corresponding to that expected in 50, 100 or more years, that is, for a return period which usually is of a higher duration than the series of historical data available. In order to resolve this problem such that extrapolation over bigger time periods becomes possible, one has to resort to statistics. Statistical methods consider that maximum values of random variables (such as daily precipitation P_d), adapt well to a Gumbel's distribution. This distribution has been successfully used for independent extreme values of meteorological variables and seems to adjust well with maximum precipitation values at different time intervals. Their utility in solving practical engineering problems of drainage network dimensioning seems to have been confirmed after years of use (Castillo & Ruiz 1979).

The probability F(x) [where F(x) values vary from 0-1], such that a maximum precipitation value be less than a set value x is given by the expression (Gumbel's distribution):

$$F(x) = \text{Prob}\,\{P_d \text{ máx} \leq x\} = \exp\,[-\exp[-\alpha\,(x - u)]] \quad (1)$$

Consequently, the probability that this value be exceeded is 1 - F(x) (MMA 1998).

The values of α and u, when the number of data is sufficiently high are:

$$\alpha = \frac{\pi}{\sigma\sqrt{6}} = \frac{1.2826}{\sigma}$$

$$u = \overline{X} - \frac{0.5772}{\alpha} = \overline{X} - 0.45\,\sigma$$

where σ = typical deviation of maximum daily precipitation values P_d in each year of the available historical series, $\overline{X}$ = arithmetic average of said values.

A precipitation F(x) has a return period T, when on average it is equalled or exceeded once every T years. Therefore it can be written as:

$$T = \frac{1}{1 - F(x)}$$

From where one can deduce that:

$$F(x) = 1 - \frac{1}{T} \quad (2)$$

On the other hand, x can be derived from equation (1):

$$\text{Ln}\,F(x) = -\exp\,[-\alpha\,(x - u)]$$

$$-\text{Ln}\,F(x) = \exp\,[-\alpha\,(x - u)]$$

$$\text{Ln}\,[-\text{Ln}\,F(x)] = -\alpha\,(x - u)$$

$$-\frac{\text{Ln}\,[-\text{Ln}\,F(x)]}{\alpha} = x - u$$

and finally:

$$x = u - \frac{1}{\alpha}\,[\text{Ln}\,[-\text{Ln}\,F(x)]]$$

By substituting F(x) in (2), and by substituting values of α and u, the expression becomes:

$$x = \overline{X} - 0.45\,\sigma - 0.78\,\sigma\,[\,\mathrm{Ln}\,[-\mathrm{Ln}\,(\,1 - \frac{1}{T})]] \quad (3)$$

In this way, by starting from a series of statistical values P_d and by knowing the average value and typical deviation of each of them, one is able to estimate the individual maximum probable values for precipitation in 24 hours, for a return period T.

4 CALCULATION OF THE MAXIMUM FLOOD FLOW

When designing a safety channel, it is not only important to have data on 24 hour maximum precipitation but also the maximum average intensity with which it rains during those 24 hours, that is, the value of the most intense shower within the considered return period, which gives rise to the runoff flow that has to be evacuated by the channel. In order to determine the flow corresponding to such a maximum shower, the rational method i.e., the following expression is used:

$$Q = \frac{C\,I\,A}{3}$$

where Q = peak flow corresponding to a given return period measured in m^3/s, A = the catchment drainage area in km^2, I = average precipitation intensity for time interval of duration T_c (concentration time), for the considered return period in mm/h, C = average runoff dimensionless coefficient for the drainage area. The said formula includes a 20% increase in Q in order to account for the peak precipitation effect.

From here on an important concept; concentration time Tc comes into play, which can be defined as the time taken by the last rainfall runoff drop to exit the catchment area drainage, at a given moment of time.

The average intensity of a strong downpour (I) must be calculated as if this downpour would last for a time T_c, an unknown value that must be calculated to posteriorly estimate the flow rate Q. To this effect, the duration of concentration time T_c must be determined.

There are many formulas that can be used to calculate the same but the most commonly used one in Spain, for small area basins, is the one that is equivalent to that of the U.S. Corps of Engineers:

$$T_c = 0.3\left(\frac{L}{J^{1/4}}\right)^{0.76}$$

where Tc = concentration time in hours, L = length of the principal basin watercourse in km., J = average slope of the principal watercourse, scaling from 0-1.

Knowing Tc, the factors that are useful to determine the flood flow are calculated as under:

4.1 *Calculation of I*

The average precipitation rainfall intensity I can be calculated or estimated in many ways, but the method commonly used in Spain is as under (Témez 1978; MOPU 1990).

The first step is to define the values of I_1/I_d for the study area, which is the relationship between the intensities of hourly rainfall and the daily average for the same return period. This value is specific to each place and characterises the curve for the rainfall intensity/rainfall duration at that place. To calculate the said coefficient, we need the historical data series corres-ponding to the return period. Since this data is not always available in Spain, the different iso-value curve distributions of I_1/I_d are determined from many meteorological stations data.

The relationship bet-ween I/I_d can be estimated as a function of the said quotient, where I is the average rainfall intensity at time T_c, by using the expression:

$$\frac{I}{I_d} = \left(\frac{I_1}{I_d}\right)^{\frac{0,1^{\,}}{}\frac{1.395 - Tc}{0.395}}$$

Upon knowing the I/I_d quotient, the value of I, which is needed to determine the flow rate by the rational method, is obtained by multiplying the said quotient by I_d. The value of I_d can be calculated by dividing the maximum daily precipitation P_d corresponding to the return period by 24 hours, i.e., $I_d = P_d/24$. The value of P_d is obtained from the series of rainfall data or from interpolated values from the Gumbel method.

4.2 *Calculation of C*

The runoff coefficient C is the quotient between the flow that gushes on the land surface and the total rainfall flow. C varies with time and depends on land characteristics (vegetation, permeability, slope, etc.) and that of the zone (temperature, relative humidity, wind velocity, sunlight hours, etc.).

Of the three components from the rational method formula, the runoff coefficient is the one that has been determined with the least precision, resulting in loss of credibility of the rational type formulas. Spain has developed a method based on the Soil Conservation Service of the U.S.A..The original formula has been adapted to national conditions and therefore is valid exclusively for Spain (Témez 1978; MOPU 1990). For purposes of preliminary calculations or for areas located outside the Iberian Peninsula, when information is scant, C can be estimated with reasonable precision from Tables1& 2 (MOP 1965; Hernández et al. 1996). In the first case, depending on the terrain through which water flows, the value of C can be directly estimated within the considered interval.When using the second Table, the different aspects of the terrain (slope, permeability, vegetation and water retention capacity) are taken into consi-deration thus permitting us to determine the coefficient K which is related to C.

4.3 *Calculation of A*

The basin area can be calculated directly from the map using a planimeter, and by ensuring that the units coincide with that in the rational method formula.

4.4 *Calculation of Q*

Once the values of the three parameters that intervene in the rational formula are known, we can then calculate flow rate Q, taking care to ensure that the units are the same in each of the parameters.

5 CHANNEL DIMENSION MEASUREMENT

The numerous formulae known so far for channel measurement can be grouped together as:

$$v = C R^m J^n$$

Table 1. Runoff Coefficient C

Surface type	C
Concrete and Bituminous pavements	0.70 - 0.95
Macadamised pavement	0.25 - 0.60
Cobbled stone surface	0.50 - 0.70
Gravel surface	0.15 - 0.30
Areas with trees and woods	0.10 - 0.20
Areas with dense vegetation:	
Granular terrain	0.05 - 0.35
Clay terrain	0.15 - 0.50
Areas with average vegetation:	
Granular terrain	0.10 - 0.50
Clay terrain	0.30 - 0.75
Areas without vegetation	0.20 - 0.80
Cultivated areas	0.20 - 0.40

where v = average water velocity, C = the coefficient that varies according to the formula considered, R = hydraulic radius, J = slope of principal watercourse m, n = indexes which vary in some formulas.

The widely used formula and the one most recommended by the drainage norms in Spain is that of Manning, which is as follows:

$$Q = S\,v = S\,\frac{1}{n}\,J^{1/2}\,R^{2/3}$$

where Q = flow rate in m^3/sec, S = the current cross-section (or that of the channel) in m^2, n = Manning number or watercourse Coefficient of rugosity in s/$m^{1/3}$, J = principal watercourse slope or the line joining the centres of gravity of the current sections (for purposes of calculating safety channels, it can be roughly considered to be the same as the slope at the far end of the channel. It is expressed as the tangent of the inclination angle), R = hydraulic radius or the quotient between the damp section (S) and the damp perimeter (p), expressed in m.

If we consider a generic channel with a trapezoidal section, where the lower smaller base is a, the surface width of the water film is b, the height of the water film with respect to a is h and the inclination of the lateral channel walls with respect to the horizontal is an angle φ (Figure 1), the values that must be carried over to Manning's formula for the described generic channel are:

1. Current cross-section:

$$S = \frac{(a+b)\,h}{2}$$

2. Hydraulic radius:

$$R = \frac{S}{p} \quad \text{where}$$

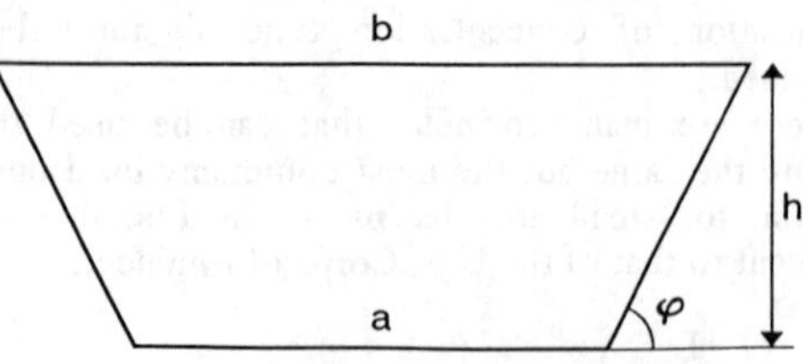

Figure 1. Generic channel with a trapezoidal section

Table 2. Runoff coefficient C as a function of index K

Concept	K – Value			
Relief	40 Very uneven Slope >30 %	30 Uneven Slope 10-30 %	20 Wavy Slope 5-10 %	10 Flat Slope < 5 %
Permeability	20 Very permeable Rock	15 Impermeable Clay	10 Very permeable Normal	5 Very permeable Sand
Vegetation	20 None	15 Scant < 10 % surface	10 Much < 50 % surface	5 A lot < 90 % surface
Water retention	20 None	15 Scant	10 Much	5 A lot
K-Value	75 - 100	50 - 75	30 - 50	25 - 30
C-Value	0,65 - 0,80	0,50 - 0,65	0,35 - 0,50	0,20 - 0,35

$$p = a + \frac{2h}{\sin \varphi}$$

One can deduce from the channel configuration that:

$$b = a + \frac{2h}{\tan \varphi} \qquad (4)$$

Substituting this value of b in the formulae for S and R, we get:

$S = [a + (h / \tan \varphi)]\, h$ and

$$R = \frac{[a + (h / \tan \varphi)]\, h}{a + (2h / \sin \varphi)}$$

The rugosity coefficient n can be determined from Table 3.

By substituting S & R in Manning's general formula, we get the following flow rate:

$$Q = \frac{[[a + (h / \tan \varphi)]\, h\,]^{5/3}\; J^{1/2}}{n\, [a + (2h / \sin \varphi)]^{2/3}} \qquad (5)$$

In channels with a triangular section a = 0 and the above expression gets simplified to:

$$Q = \frac{h^2\, J^{1/2} \cos^{5/3}\varphi}{1.587\, n \sin \varphi}$$

In the general equation (5) the unknowns are a and h, although one of them can be fixed at the beginning (normally a). In channels with a trapezoidal section, there exists an optimum hydraulic section, viz.; that where the hydraulic radius is a maximum. In such a case, the crown width of the damp section (b) is twice the length of the corresponding damp lateral wall, i.e.:

Table 3. Manning's coefficient n

Channel material	n ($s/m^{1/3}$)
Channel without lining:	
On normal bare terrain	
Flat uniform surface	0.020 - 0.025
Irregular surface	0.025 - 0.035
On terrain with vegetation	
Scant	0.035 - 0.045
Dense	0.040 - 0.050
On rocks	
Uniform flat surface	0.030 - 0.035
Irregular surface	0.035 - 0.045
Lined channels:	
Gravel bottom	
Concrete walls	0.017 - 0.020
Stone block walls	0.023 - 0.033
Concrete	0.013 - 0.017
Projected concrete	0.016 - 0.022
Stone blocks	0.020 - 0.030
Bituminous lining	0.013 - 0.016

Table 4. Maximum admissible velocity for water circulation in channels

Channel surface type	V.máx (m/s)
Hard clay	1.2
Clays with gravel	1.2
Fine sand or silt (little or no clay)	0.2 - 0.6
Hard clay sand, Hard marl	0.6 - 0.9
Gravel	1.2
Conglomerate of hard slate and soft	1.4 - 2.4
rock	3.0 - 4.5
Hard rock, stonework	4.5 - 6.0
Concrete	0.6 - 1.2
Terrain partially covered by vegetation	1.2 - 1.8
Grass	

$$b = \frac{2\,h}{\sin \varphi}$$

This equation relates b with h. Because according to (4):

$$a = \frac{2\,h\ (1 - \cos \varphi)}{\sin \varphi} \qquad (6)$$

So, in the case of hydraulic optimisation, there exists just the one unknown, namely h.After substituting (6) in (5), and solving it further, we get:

$$Q = \frac{(2 - \cos \varphi)\, h^{8/3}\, J^{1/2}}{1.587\, n \sin \varphi}$$

The minimum recommended slopes for ditches or channels are:

1. Lined ditches: 0.2%
2. Unlined ditches: 0.5%

Nevertheless, it is recommended that the slope not be less than 1%, whenever possible, to prevent terrigenous sedimentation at the bottom of the channel. For slopes J greater than 0.5%, those channel sections that have a greater depth are in the downhill direction. When slopes are less than 0.5%, the water film height at the drainage point is in an uphill direction.In such conditions, the greatest depths are found within the initial section (uphill) and its estimation is done by adding the depth at the final section (downhill), which is calculated using Manning's formula where J = 0.005, an increase equal to:

$$\Delta = (0.005 - J)\, \beta\, L$$

where L = channel length between sections, β = coefficient whose value is 0.5 in ordinary cases where the water flow rate is progressively incorporated along the channel, and its value is 1.0 if all of the flow enters through the top end of the channel.

In as far as type of materials is concerned, the channels and ditches can be made of either: a) earth or mine wastes, b) either dug in rock or soil, and c) can be lined with vegetation, stones, concrete, etc. Depending on the type of lining or construction material used for the channel, its protection from erosive currents can be achieved by keeping current velocities lower than those indicated in Table 4.If the calculations show a higher current velocity, then the general channel slope would have to be reduced or the rugosity of its walls will have to be increased. In any case, and with rare exceptions, the minimum recommended water velocity is 0.25 m/s.

The determination of the channel dimensions and that of the resulting exact flow rate is done by iterative estimates performed using a computer. To sum up, we use Manning's formula to calculate the cross-section, slope and the construction material for a channel that would be able to manage a very similar flow rate to that of the maximum predictable flood flow using the rational formula.

This is done in the following manner:

1. By the introduction of the desired flow rate, J, n, φ and a or the hydraulic optimization condition.
2. The water film height is taken to an initial value of h.
3. The h increase interval is selected for successive iterative estimates. The estimates are carried out from an initial value of h.
4. The cross-section, the hydraulic radius and velocity are calculated.
5. The flow rate is determined and is then compared with the initial value; if lower, newer incresed predictable h heights are estimated.
6. When the flow rate calculated by using Mnning's formula is equal or greater than the maximum flow rate, the section value is approximated by gradually reducing depth values until optimum h value is achieved.
7. The velocity is checked to see if lies within the admissible limits.
8. If so, the section is adequate. If not, a new estimate has to be carried out by reducing the channel slope, by increasing the rugosity of the walls and bottom, or by trying to construct the channel using newer materials which would permit higher water flow velocities.

9. Once the optimum section has been achieved, the crown width of water b is then calculated.

10. Finally, one calculates the free upper end of the channel.

6 CONCLUSIONS

1. The method is founded on a trustworthy and contrasted technical base.

2. The usage of the proposed method permits us to resolve channel design problems in a fast and easy manner.

3. Using this method, we could eliminate the normal practice of constructing mining channels using the excavator bucket as the sole dimensional reference (the channel is made to the size of the excavator bucket).

REFERENCES

Castillo, F.E. & L. Ruiz 1979. Precipitaciones máximas en España. Estimaciones basadas en métodos estadísticos *Monografía 21*: Madrid: Servicio de Publicaciones Agrarias Ministerio de Agricultura.

Hernández Muñoz, A., A. Hernández Lehmann & P. Galán 1996. *Manual de depuración URALITA*: 19-34. Madrid: Paraninfo.

Llopis, G. 1999. Control de la erosión y obras de desagüe. In C. López (ed.), *Manual de estabilización y revegetación de taludes*: 287-388. Madrid: C. López.

Ministerio de Medio Ambiente (MMA) 1998. *Las precipitaciones máximas en 24 horas y sus periodos de retorno en España.*

Ministerio de Obras Públicas (MOP) 1965. *Instrucción 5.1-IC. Drenaje.*

Ministerio de Obras Públicas y Urbanismo (MOPU) 1990. *Instrucción 5.2-IC. Drenaje superficial.*

Témez, J.R. 1978. *Cálculo hidrometeorológico de caudales máximos en pequeñas cuencas naturales.* Madrid: Servicio de Publicaciones Ministerio de Obras Públicas y Urbanismo.

Environmental Issues and Management of Waste in Energy and Mineral Production, Singhal & Mehrotra (eds)
© 2000 Balkema, Rotterdam, ISBN 90 5809 085 X

Secondary resources of solid fuel in Ukraine

P.I. Pilov, A.N. Shashenko, A.B. Ivanov & A.S. Kirnarsky
National Mining University of Ukraine, Dnepropetrovsk, Ukraine

ABSTRACT: This paper presents the results of field investigations of coal wastes in Ukraine. The most technologically effective, commercially profitable and ecologically unharmful for treatment of coal waste with a size up to 15 mm is wet spiral separation.

1. INTRODUCTION

Coal in Ukraine – single powercarrier of local genesis (reserves – 47-50*10^9 tons) which can supply the national industry for 100-200 years under modern consuming fuel rate.

Ukrainian energetics which is based on solid fuel utilization experiences the sharp deficit of coal due to coal output fall and significant decreasing of coal quality. For example, the ash of coal delivering to power stations for the last 15 years raises from 26 to 36 % that decreases the calorific power till 17 – 19*10^6 jouls/kg and requires the additional consumption of natural gas and mazut. Energetical units of bituminous coal with ash 20-25 % that supply maximum technical-economical effects.

Such coal can be delivered to power station without ins mining by means of wet spiral separation (WSS) of coal wastes of coal mines and coal preparation plants.

2. WET SPIRAL SEPARATION THEORY

Separating medium under WSS-process has autogenous character i.e. self-separation which depends upon three main factors: hydrodynamics of flow reological properties of pump and substance composition of washing materials. The distinctive features of WSS-process hydrodynamics are turbulent character of working flow, acting of transverse circulation and influence of chute from. Some researches (Kizevalter, 1979; Ivanov, Trebukhovsky, 1961) adhere to the opinion that hydrodynamical regime of wet spiral separation includes two flows: turbulent and laminar. We disagree with such viewpoint. Let us prove it. All water and pulp flows under coal and other minerals separation have only turbulent character. The exception to our approach is dewatering process in finegrained medium where water motion occurs due to laminar regime. Laminar flow is unable to transport and separate particles. Any motoin of grain during wet separation may be demonstrated as advance and rotary motion simultaneously. Rotation of particles is a feature of vortex motion. Typical case of turbulent regime is screw motion which takes place in wet spiral separation.

The second distinctive feature of WSS-process is transverse circulation, which can be analyzed on the base of following equation (Rozovsky, 1975):

$$\frac{V_\theta}{r}\cdot\frac{\partial V_r}{\partial\theta}-\frac{V^2{}_\theta}{r}=gI_r+\frac{\partial}{\partial z}\left(\nu_s\frac{\partial V_r}{\partial z}\right) \quad (1)$$

where g – gravity acceleration; I_r – transverse slope; r, θ, z – cylindrical coordinates of point; V_r, V_θ - velocity projection on the coordinate line directions; ν_s – kinematic coefficient of turbulent viscosity.

By means of expression (1) it is possible to obtain the magnitade of circular velocity

$$V_r=V_{r_o}\left[1-e^{-\frac{\sqrt{g}}{c}\cdot\frac{r}{h}\theta}\right] \quad (2)$$

where h – depth; C – Chery friction coefficient; V_{r0} – radial component of velocity at initial part of circulation damping.

Due to transverse circulation the WSS – separation has countercurrent character. Essential distinctive feature of countercurrent systems is presence of specific natural bed which significantly differs from traditional gravity analogy. At working zone of such separator heavy grains and concretions are accumulating and their content is greater then in initial product. Natural bed in countercurrent separator is heavy suspension which is supported by additional pressure gradient. The last factor increases effective medium density, i.e. the density of fraction which supply the most effective separation of materials under autogenious washing medium.

3. THE ACTUAL STATE OF COAL WASTE

Coal quality deterioration in Ukraine was caused by increasing the share of coal mined from thin (less 1,0 m) seams (till 70-80 %) that leads under existing coal extraction technology to mixing of coal with high-ash waste. The second negative of consequence of such mining technology was rise of fine coal and slimes content in run-of-mine coal.

At present time in Ukraine coal quality improvements by means of high waste extraction from run-of-mine coal are carried out of sixty coal preparation plants with a total capacity of 150 Mt/g. The quantity of waste in mine waste heaps and coal slimes dumps is presented in table 1.

The mine waste heaps due to burning and erosion processes unsuitable for combustible mass recovery but burned rocks can be used for needs of construction. For example, retreatment of one coal mine waste heap during 7-8 years allows to obtain $3,5*10^6$ bricks with a size 250x120x88 mm. Such quantity of bricks allows to build 10 thousands of flats i.e. the whole mining town.

The unburned waste and coal slimes of current coal preparation and stocked in heaps and ponds must be treated by means of wet spiral separation. The most preferable variant is coal slimes waste. The field sampling (figure 1) of Avdeeyvka coking – chemical plant coal preparation plant flotation dumps proved the opportunity of obtaining of WSS coal concentrate with such characteristics:

ash – 8,9 %;
sulfur content – 1,0 %;
volatile matter yield – 26,4 %;
calorific power – 7940 kcal/kg;
thickness of plastic layer, y – 14 mm
x - 33 mm;

Therefor, WSS – process may supply effective washing of coal slimes with obtaining coking and energetical concentrates.

4. WET SPIRAL SEPARATION OF COAL WASTE

Since in nearest years it is impossible to wait significant increasing of coal output on Ukrainian mines and reconstruction of mines require 10-15 years, thus enhancement of coal delivery for power station is possible only by utilization of waste. The preparation of waste allows to obtain annually 1,820 Mt of concentrate with ash 25-45%, including 0,4 Mt with ash till 30 % - from mine waste heaps, 0,32 Mt with ash till 35 % from waste heaps of coal preparation plants, 0,5 Mt with ash till 45 % - from slimes dumping, 0,37 Mt with ash till 25 % - from gravity preparation waste and 0,23 Mt with ash till 30 % – from flotation tailings of current coal preparation.

The retreatment of coal slimes include preliminary hydroclassification in low diameter hydrocyclones, spiral classifier, hydrosizers and subsequent wet spiral separation of thickened

Table 1. Actual state of coal waste in Ukraine

Object	Waste properties			Fuel resources		
	Quantity	Mass, Mt	Ash, %	Mass, Mt	Annual volume, Mt	Ash, %
Coal mines						
Mine waste heaps	940	2200	75	50	0,4	30
Coal preparation plants						
Waste heaps	80	830	72	16	0,32	35
Slimes dumps	35	120	40-70	18	0,5	till 45
Current coal preparation						
Gravity separation waste	64*	23	74	-	0,37	till 25
Flotation tailings	30**	3	65	-	0,23	till 30

* quantity coal preparation plants
** quantity of flotation departments of coal preparation plants

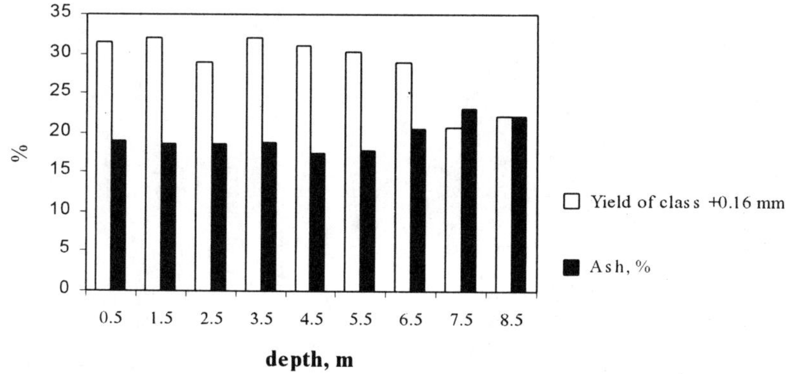

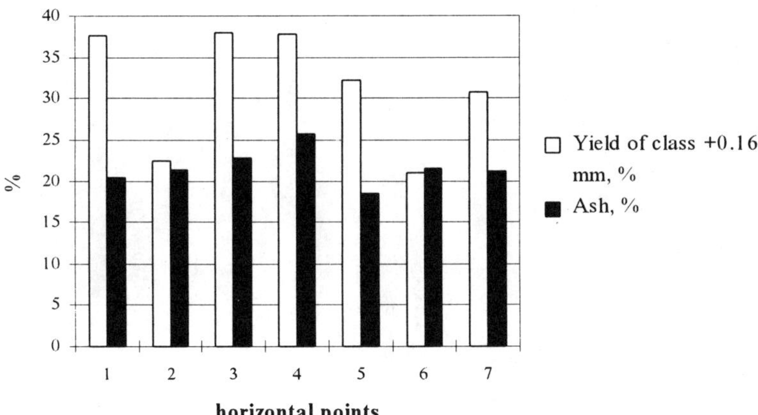

Figure. 1. The results of field investigation of coking chemical plant coal preparation plant flotation tailings stocked in dump

product with a solid content 250 – 400 kg/m^3.

The redressing of coarse material with a size greater 15 mm includes preliminary crushing to size 10 – 7 mm with subsequent wet spiral separation of crushed material that allow to obtain concentrate with ash 25 – 30 %

5. CONCLUSIONS

1. Preliminary preparation and delivery of concentrate to power station may supply for 15 – 20 years additional solid field in quantity 6 –8 MWt annually that supply operation of 10 –12 blocks of power station with 200 MWt power.
2. Wet spiral separation of coal with size up to 15 mm allows to obtain concentrate with ash 20 – 40 %.
3. Burned rocks of mine waste heaps may used for obtaining of construction materials.

REFERENCES

1. Kizevalter B.V. 1979. *Theoretical basses of gravitational benefication processes*. Noscow: Nedra.
2. Ivanov V.D. & Trebukhovsky G.I. 1961. Development of design parameters and process regime for spiral sluice. *Kolyma*. # 3: 32-36.
3. Rozovsky I.A. 1957. Movement of water flow at the turn of open channel. Kiev: Edition of A.Sc. of Ukr. SSR.

Environmental Issues and Management of Waste in Energy and Mineral Production, Singhal & Mehrotra (eds)
 ISBN 90 5809 085 X

Tailings disposal at Syncrude Canada Ltd's Aurora Mine

B.C. Purcell – *Klohn-Crippen Consultants Limited, Canada*
J.P. Barlow – *Agra Earth and Environmental Limited, Canada*
D. Nordin – *Norsand Consulting Limited, Canada*
S. Fleming – *Agra Monenco Incorporated, Canada*

ABSTRACT: Syncrude Canada Ltd's (Syncrude) new Aurora Mine will begin operating in May 2000. The new mine includes a 10 km^2 external tailings pond and recycle water facility. The tailings facility has undergone several significant planning and design enhancements since the initial Design Basis Memorandum was completed in 1996. Some of the more important design features include:

- Founding a 55 m high dyke on muskeg that varies from 0 to 4 m in depth.
- An innovative debris management system.
- A 5 km long, 7 million cubic metre zoned overburden starter dyke.

This paper provides a brief overview of Syncrude's vision for Aurora, background to the Alliance contracting strategy and an introduction to breakthrough performance technology used at Aurora to meet some very challenging targets. In addition, we will provide details of important planning/design issues we faced at Aurora to produce a high quality "fit for purpose" tailings design.

1 INTRODUCTION

The Aurora Mine project organization is divided into five areas:
1. Materials Handling
2. Extraction and Hydrotransport
3. Tailings
4. Utilities
5. Infrastructure

Each area has its own budget, is managed by a different area manager and shares technical resources across the entire project.

The Tailings Area is the main topic of discussion in this paper. Before discussing Tailings planning and design issues we will first provide a brief background on Syncrude's decision to develop Aurora – what were the drivers and why Syncrude decided on an alliance contracting strategy to execute the project.

We will then describe the two key contributing factors to the Aurora Tailings Area success – alliancing and breakthrough performance.

And finally, we will provide some detailed examples from the tailings area of how the alliance approach has resulted in high quality, "fit-for-purpose", innovative design elements, capital and operating cost savings, an exemplary safety and environmental record and all within a fast-track schedule.

Future papers will provide more detailed design performance information as operating experience and data is gathered on the performance of the facility and its various components.

2 BACKGROUND

2.1 *Aurora Mine*

New bitumen extraction process technologies, larger mining equipment and a revamped fiscal environment have set the stage for new growth in the oilsands industry (Syncrude, 1996). This growth will involve a move away from expensive, integrated mine-extraction-upgrading mega-project facilities to:

- Regional mining and bitumen production sites,
- Upgrading sites with economic lives extended indefinitely through continued bitumen supply and reinvestment of new technology, and
- A variety of production streams.

Syncrude continued to explain in its Aurora Mine Application to the Alberta Energy and Utilities

Board that Aurora is the first step in realizing this new vision and it is a vital step in Syncrude's oil sands development for two reasons:

1. It will fulfill immediate bitumen requirements to replace existing supplies from the Mildred Lake Mine that are nearing depletion as well as provide for near term production increases; and
2. It introduces economic, energy-efficient surface mining and extraction technology on a large, attractive resource base to facilitate growth of a more diverse business in the longer term.

3 THE AURORA ALLIANCE APPROACH

3.1 *The Aurora Alliance*

Although Syncrude has always had effective cooperative working relationships with their Contractors and, from such relationships been able to produce high-quality results, Aurora provided an opportunity to push the benefits of cooperation even further and in so doing look for extraordinary results. Syncrude recognized that neither themselves nor any of their contractors were able to achieve this level of performance on their own, and only through the combined effort of all project participants would the breakthroughs necessary to achieve such results be possible. To this end, the Aurora Alliance concept was conceived.

The Aurora Alliance is defined as "A group of companies working together, aligned on minimizing costs, increasing profitability and contributing to each other's long term future." In essence, it meant people from different companies aligned on a common goal that would produce a win-win result.

The Aurora Alliance companies consist of Syncrude (Client and Operator), AGRA-Monenco/Simons Joint Venture (Engineering and Procurement), North American Construction (Earthworks), PCL Industrial (Mechanical and Electrical Construction), Honeywell Ltd. (Advanced Control and Systems).

Based on research done on other alliance projects, Syncrude determined early on in the project development phase the key elements for a successful Alliance organization to be:

- A fully integrated alliance contract between the owner and the alliance members
- Alignment of business, project and operational goals
- Risk and reward sharing
- Alliance members share responsibility for success
- Open and non-adversarial relationships

Once this organization framework was established the Aurora Alliance Management Team set about creating an environment that would challenge all aspects of project performance.

The first step was to develop project goals or "stretch targets" that, if achieved, would constitute "Making History". Table 1 describes the stretch targets that were established for the Tailings Area in January 1998.

Table 1. Tailings stretch targets

Category	Stretch Target
Capex	20% reduction from $28.8 million to $23 million
Opex	20% reduction in Tailings disposal costs
Schedule	Tailings plant on line May 2000
Safety	Beat all WCB standards for this type of work
Environment	Zero non-compliance incidents
Quality	Zero field re-work

The second step was to develop alliance operating principles that would guide our actions and help develop the Aurora culture each company and Aurora staff member could buy into. We have listed some of the more key principles here:

- Early involvement of all key stakeholders including engineering, construction and operations
- No duplication of roles
- Minimum of organization levels and interfaces
- Short lines of communication
- Best-in-Class resources
- Team approach with training and facilitation provided to all staff
- Implement incentives, risk and reward compensation
- A streamlined decision making process employing appropriate limits of authority

Having developed and bought into these principles the bottom line commitment from everyone on the project was to do what is right for all stakeholders.

3.2 *Breakthrough performance*

As the project goal was to build a facility that exceeded "business as usual" expectations, we understood from the start that following business as usual practices would not deliver the results we were committed to. We needed to change how we approached our work, our behaviours and our attitudes.

To help us in this area we enlisted the services of JMW Consultants Inc., a consulting and education company that worked with us to establish an environment of extraordinary performance and to produce breakthrough results.

Setting stretch targets (for both the project and for each area) was one way to access the desired change. Stretch targets provided a target to strive for as well as a measure of how well we are doing. Although seemingly impossible at the outset to achieve, it was important not to fixate on the target itself, rather we focused more on trying new and different ways to work, investigate, explore, challenge and create that would lead to different results. If these actions did not result in the desired outcome we changed our actions again until we got the result we wanted or we ran out of time.

It is beyond the scope of this paper to describe the details of this process and the science behind it, however, we have listed below in point format a brief outline of what we believe were the most effective tactics in producing breakthrough results in the tailings area:

- Set targets that challenge how people think
- Give up having to be right. Look for the opportunities in what people are saying
- Take action, even when afraid
- Talk straight and be true to your word
- Deal in facts, not interpretations
- Be a player, not a spectator
- Focus on results
- Take ownership
- Have it all
- Have fun

As effective as these tactics were in creating the breakthrough environment we needed to produce our stretch targets, there was one key aspect of breakthrough performance that probably contributed 80% of the results and as such deserves some special attention.

3.3 *Breakdowns and breakthroughs*

More often than not when things do not go according to plan, i.e. when there is a breakdown, people's first reaction is to see this as a problem and immediately devise ways to get rid of the problem (solve it, hide it or ignore it). In breakthrough projects it has been found that the source of every breakthrough is a breakdown (Scherr, 1989). This has most certainly been the case in the Tailings area on Aurora.

A breakdown is nothing more than a demand for action. Whether the action taken leads to a breakthrough or not depends on the attitude of the people involved and their objective. For example, a change in a dyke design caused by a change in regulatory requirements or a scope change could lead to higher costs if accepted as a done deal. On the other hand, if the people involved challenge the proposed design change for fit-for-purpose requirements, are willing to explore other possibilities for achieving the same goal with all stakeholders and are true to their commitment, anything is possible.

Breakthroughs need not be truly momentous inventions or discoveries. More often than not they are simply rediscoveries of previously known techniques, slight variations to existing practices or an accumulation of "small" suggestions gathered during the challenge and review process. The important thing is they would not be uncovered if it were not for the chance to look for them that is afforded by breakdowns.

4 OVERVIEW OF TAILINGS FACILITY

Planned startup of the Aurora Plant is May 2000. The first production train will process 58 M tonnes oil sand per year. A second train is planned to come on stream in mid 2003, bringing total ore production to 116 M tonnes. The Aurora external tailings disposal area is designed to store all processed oil sands until sufficient room is available in-pit (Figure 1).

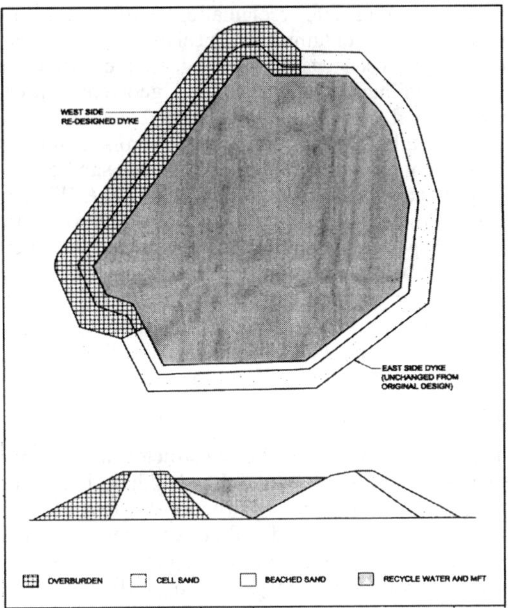

Figure 1. Aurora Tailings pond footprint and section

4.1 *Site selection*

The external tailings disposal facility, including an abutting overburden waste dump, is located immediately east and adjacent to the first Aurora open pit - the East Mine. Minimum setbacks from the toe of the dykes were established at 300 m from the Muskeg River to the east, 100 m to the Shell Lease boundary to the south and 50 m to crest of the East

Mine. The footprint of the tailings disposal area and the waste dump is approximately 12.4 km^2. The site topography slopes downward at approximately ½% from the northwest to southeast and varies from elevation 300m to 283 m.

4.2 *Tailings disposal technology*

Syncrude senior management recognized during the EDS phase that Aurora represented a potential opportunity to change from the existing conventional tailings disposal methods used at Mildred Lake. They also recognized that changes to the tailings disposal process would not just affect the Tailings area, but also the Extraction, Mining, and possibly Utilities areas.

In 1997 a joint Syncrude/Aurora technical group undertook an extensive and comprehensive investigation into alternative tailings disposal technologies, as applied to the Aurora site, followed by an economic feasibility study of implementing the technologies at Aurora. The goal of the study was to develop the best overall tailings disposal plan that achieved rapid site reclamation while requiring minimal fluid containment. Specialized teams studied all proposed cases, focusing on the extraction process, tailings planning, costing, geotechnical, environmental and reclamation issues.

The study group recommended Aurora Train 1 to implement conventional tailings (coarse sand deposits plus accumulating Mature Fine Tails (MFT)) in the external pond and Composite Tailings (CT) for in-pit disposal. The group also recognized that there may still be an opportunity to implement an alternative tailings disposal technology for Train 2 in 2003 and recommended Syncrude to continue to look for and study alternative tailings disposal techniques.

4.3 *Operating plan*

The external pond will be constructed in a similar manner to Syncrude's existing Mildred Lake Settling Basin: centerline construction for two lifts followed by upstream sand cell construction to contain accumulating volumes of MFT, beached sand and process water inventory. When sufficient in-pit space is available in 2007, tailings streams will be converted to CT and placed in-pit.

Beyond 2007, the external pond will remain active for approximately 6 to 7 years, continuing as the water clarification pond. The inventory of MFT in the external pond will be slowly depleted and replaced by an equivalent volume of cycloned overflow beach and/or CT. By 2013 all of the MFT will be depleted from the pond and will be either removed in-pit or used in a dilute CT stream.

The above planning parameters result in the external pond dykes being constructed to final elevation 337m by 2007. Reclamation of the pond surface will begin in 2009 and will be completed in 2013. Dyke side-slopes will be reclaimed as they are completed.

5 TAILINGS BREAKDOWNS AND BREAKTHROUGHS

As is often the case, some of the major design modifications in tailings during the Aurora EDS phase were driven by project budget restrictions. Although cost savings initiatives were always a focus, it was the monetary "breakdowns" (unbudgeted items) which concluded in major design changes. The significant changes were not so much a "Eureka!" as they were an accumulation of ideas from all Alliance members.

Alliancing also allowed more direct communications between engineering and the field without the need for detailed drawings. There are countless examples where design ideas/modifications to the starter dyke, road construction, drainage plans etc., were accomplished by faxed hand drawn sketches followed by a phone call. This open-for-discussion atmosphere encouraged all stakeholders to bring forward suggestions and ideas for improvement in design and construction.

In this regard the experience of the Alliance earthworks constructors, North American Construction, was indispensable. Over the 2 year construction program of Winter 1997 to Fall 1999 North American Construction moved 7 million cubic metres of overburden at a savings of $2.5 million with no lost time to weather and 4 months ahead of schedule.

A key to this success was NAC's early and ongoing involvement in the engineering design process, where their willingness to listen and work with engineering on complex design and construction issues was invaluable.

5.1 *Muskeg pre-load*

The construction of the initial layer of dyke fill on top of the muskeg, termed a "pre-load", was a good example of the benefits of the synergetic relationship between design and construction personnel. Over 90% of the starter dyke was constructed on muskeg that was left in place. The muskeg over the area of the starter dyke was generally saturated and up to 3 m in depth. In order for the placement of fill on top of the muskeg surface to be successful, it was crucial that the integrity of the fibrous muskeg layer be maintained (no ripping or shearing of muskeg layer) and that all muskeg below the pre-load thaw completely in the following summer. The integrity of the fibrous mat needed to be preserved to have acceptable shear strength in the muskeg and allow further dyke construction. The muskeg needed to be in an unfrozen state for the construction of subsequent lifts so that pore pressures generated by dyke construction could be dissipated up into the sand preload.

Historically, the challenge with building on top of muskeg is that the muskeg has very low shear strength in its in situ state and generally cannot support construction equipment unless it's frozen. However, once its frozen it has essentially zero permeability and will not allow any of the water to drain away under the compression from the dyke fill. One approach to constructing on muskeg is to place only in non-freezing months to avoid the problem of building in an ice layer that will inhibit consolidation. This can only be done in a carefully controlled operation with very small equipment. It involves placing a geotextile fabric on the muskeg surface and floating a relatively thin lift of sand over it. Extreme care is required to avoid punching through the muskeg during the advancement of the first lift.

In order to avoid the expense and timing restrictions that such an operation would create, an alternative approach was adopted that involved placement in the wintertime. The earthworks team developed a clearing and grubbing procedure that was done in a way that carefully controlled the frost penetration into the muskeg. If done in an uncontrolled manner, frost thickness in excess of 1.0 m can easily be incurred. However, the construction personnel were able to keep the frost penetration down to 200 to 300 mm, which was found to be just enough frost to allow bulldozer traffic and support the initial 0.5 to 1 m lift of sand. In the following Spring and Summer, the surface of the sand was ripped with a bulldozer to speed the thaw and the frost layer was largely gone by July/August. This allowed timely construction of the next lift and the fast dissipation of pore pressures from the water laden muskeg.

The time that it takes muskeg to thaw when buried under sand is extremely sensitive to the depth of the frost and the thickness of sand fill placed on top of it. By minimizing the frost layer to provide just enough support for construction traffic, and minimizing the sand pre-load thickness to the minimum required to push the second lift in place, it was possible to achieve a timely thaw and proceed with a cost effective earthworks operation.

5.2 *Starter dyke*

The largest component of the capital cost of the settling basin was the earthworks associated with starter dyke construction. The site for the settling basin has little topographic relief and a majority of the area is covered by saturated, highly compressible muskeg up to 4 m thick. The flat topography necessitated the placement of 7 million cubic metres of compacted fill to form a 5.5 km long starter dyke on the muskeg foundation. This created storage for the 10 million cubic metres of water required for plant startup. The combination of substantial fill volumes required before startup and wet, compressible foundation that could only be traversed when frozen presented a major challenge to the project.

Given the challenging nature of the site, and the need to control capital costs, the integrated team of design engineers and earthworks contractor led to design that was both able to safely accommodate large settlements and be efficient and cost effective to build. This was accomplished in the following ways:

- The sand deposit that occurs at the surface of the mine opening cut was selectively stripped. This was used to construct a central sand section in the starter dyke that would act as a chimney drain and crack stopping element so that any cracks formed by settlement of the structure would not provide a continuous pathway across the dyke.
- The requirement to zone the structure and place a selectively mined sand section actually led to a cost reduction rather than an increase. One of the significant cost hits in dyke construction is the time lost to weather. Both snow and rain soaked material typically need to be removed from dyke surfaces before placement can occur. However, incorporation of moisture into the sand section is permissible, as it does not negatively impact the properties of the sand. Therefore, the areas of the dyke consisting of lean oil sand and clay were constructed during dry weather, and the sand elements were reserved for wetter weather. This led to a very productive consistent operation with essentially no loss of production for weather.
- The positioning of the sand element was also adjusted so that it was a convenient location for the haul road (Figure 2).

Both design and construction personnel were aligned in the objective of placing a high quality core in an efficient manner because it was in everyone's interest to construct a safe dyke for the lowest possible cost, and each party had a thorough understanding of both design and construction issues.

5.3 *Debris control*

The tailings pond is the primary source of clean water required by extraction to process the oil sand. As the tailings pond is developed loose debris from forest clearing operations, and displaced muskeg from tailings overboarding, float to the surface, become coated with small amounts of bitumen and are drawn toward the pump suction such that they effectively shut the pumps down. Experience at the Syncrude's South West Sand Storage facility indicates that only a minor amount of debris and bitumen is required to

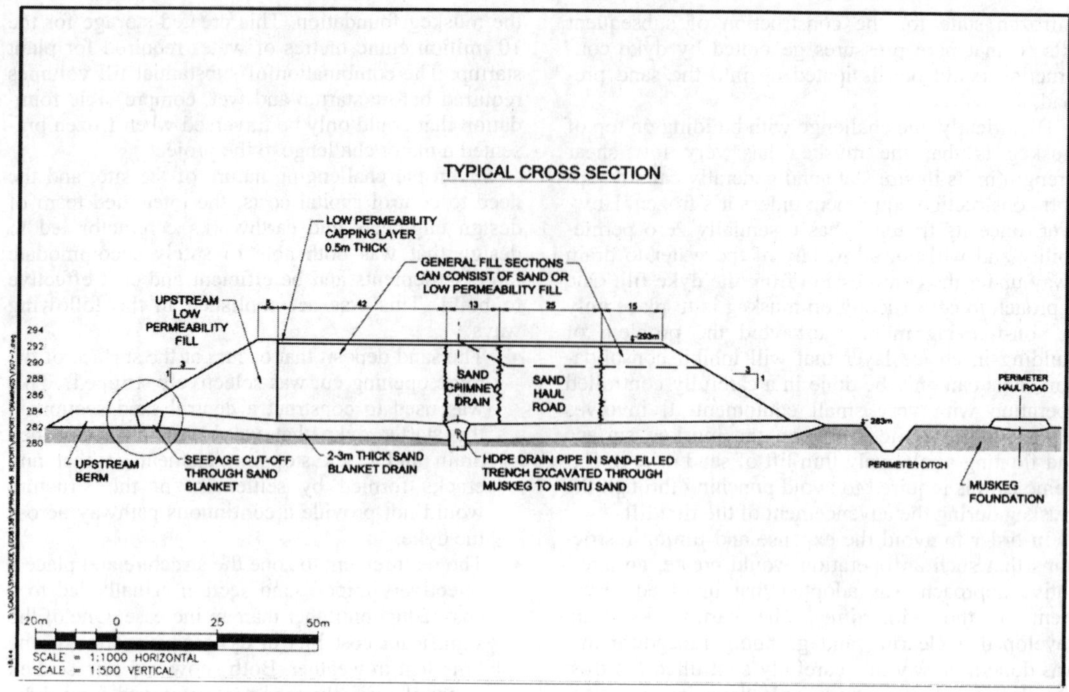

Figure 2. Starter dyke cross section

cause severe pumping problems.

These pumping problems will directly and adversely affect extraction operation even to the point where the process may need to be shut down for lack of water. In light of the high impact debris can have on the bitumen production process, the Alliance undertook a rigorous review of all possible debris management alternatives and in the end selected an observational approach that provides a high probability of success without spending a fortune to achieve it.

A team of Aurora engineering and construction and Syncrude operations and maintenance personnel came together in early 1998 to develop a debris management strategy that provides a reliable source of recycle water within the approved budget. The team used several professionally facilitated processes (situation, decision and risk analysis) to arrive at the final recommendation.

The only option which was guaranteed to work was to remove all vegetation within the entire 10 km^2 pond area above mineral soil for a total cost of >$100 million. Another high cost option, with a lower probability of success, was to remove the trees only and then cover the pond bottom with 0.5 m of overburden for a total cost of $18 million. In the end the team settled on a plan that cost less than $1 million to construct, provided a reasonable probability of success and offered operations an opportunity to fine tune the approach as they learned more about the nature of the debris.

The approved plan involves a staged construction strategy to develop a series of dyke-enclosed clean water ponds that allows water to pass through from the main pond while leaving bitumen and debris behind (Figure 3).

Water is collected from the tailings pond in a buried perforated collection header running along the dividing dyke and transferred into the clean water pond through lateral pipes running across the dividing dyke. The relative position of the clean water layer in the tailings pond and the elevation of the collection header is key to the success of the process. The Aurora tailings plan includes provision for a 3 m clean water layer. The collection header is designed to follow the rising clean water layer thus always ensuring it is located within the clean water. Isolation valves are installed in the cross dyke laterals that allow these pipe to be shut off when the header is submerged in the lower fine tailings layer – thus preventing the transfer of solids in to the clean water pond.

Due to the large contributing area of the collection header system (the dykes are 500 m to 700 m long), the transfer of water from the tailings pond into the clean water pond is a low energy process, resulting in small flow gradients towards the header. This low energy transfer system reduces the poten-

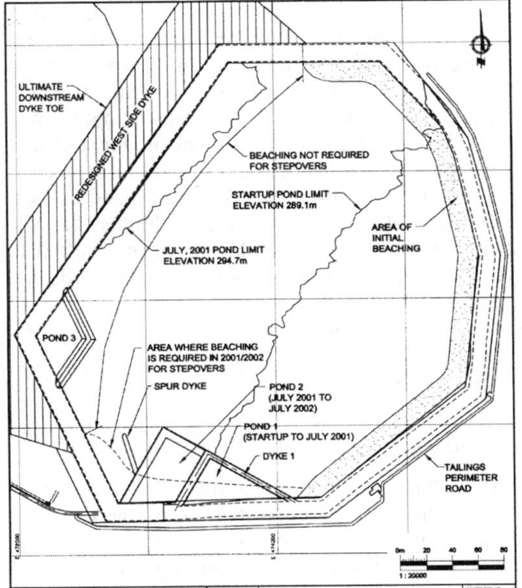

Figure 3. Clean water pond arrangement.

tial for floating debris and bitumen to migrate towards the dykes and cause clogging of the flow system.

This strategy had the affect of not only dramatically reducing the capital cost of the debris control system, but created a "learn-as-we-go" environment to adjust and improve the operation over time. Rather than commit to a costly, and possibly unnecessary solution (before verifying what the actual impact of the debris would be) this strategy gives operations a chance to test other methods (i.e. screens, bubbler systems) if the dyke and drainage piping proves to be inadequate to the task.

5.4 *West side redesign*

A design change was made to the dyke system on the western side of the settling basin to capture several important opportunities that could be realized by integrating tailings and overburden operations. Over this section, the dyke was changed from being a strictly tailings sand structure to being a combined overburden/tailings sand dyke.

This involved a shift from upstream construction with placement of tailings sand cells stepping onto upstream beaches, to centerline dyke construction with the compacted tailings sand cells placed centrally in the dyke, flanked on either side by overburden fill. With the original design, the overburden dump along this section would have been placed against a largely self-supporting tailings sand dyke slope. The new layout involved essentially the same footprint as the original settling basin and adjoining overburden dump had before. The difference is that the overburden has been incorporated into the design to create a centerline dyke on the west side that does not depend on an upstream beach.

The most significant benefit of the new design is the elimination of the upstream beach requirement along the west side. This change has significantly increased the flexibility of the tailings plan and reduced risk in the following key areas:

- It will be much easier to keep the east side beaches above water with a more relaxed beaching schedule. The above water beaches are required for upstream stepover construction and dyke integrity on the east side.
- With the new design, the available fluid storage volume is larger with the same overall tailings footprint, and the fluid storage volume does not diminish with increasing dyke height as it did when upstream construction was planned around the entire perimeter. This means that there is flexibility to raise the height of the structure to accommodate potential increases in tailings volume, rather than create more out-of-pit disturbance.
- By eliminating the requirement for beaching along the west side, the debris management scheme discussed earlier involving the construction of separate clean water ponds became much easier to execute and can be extended to the ultimate life of the settling basin if required. In the old plan, the clean water pond could only last for a few years due to the need to beach around the entire perimeter, eventually burying the pond.
- By locating the tailings sand centrally within the overburden section, higher sand capture rates can be achieved, which will make it easier to meet the required cell sand construction volumes each year.

In order to realize the above benefits, some flexibility was sacrificed with the overburden stripping operation. However, the volumes and timing of the overburden required are well within the ability of the stripping operation to deliver.

5.5 *Starter dyke redesign*

Another enhancement to the plan resulted in no capital cost savings to the Alliance but added to the flexibility and operability of the tailings disposal plan and did reduce Syncrude's post start-up operating costs.

Rather than construct a flat-topped tailings starter dyke to final design elevation (to contain the start-up water inventory and initial three months tailings deposits), the design was modified to incorporate cell-enclosing dry dykes. Two dry dykes were constructed on the upstream and downstream edges enclosing a reconfigured dyke surface which was a few metres below that of the original design.

During the summer of 2000, as cell construction begins advancing along the strike of the dyke, the pond level continues to rise to within approximately one metre of the cell bottom (top of the starter dyke). However, the dry dykes will maintain the 3 m freeboard requirement (regulatory condition for dam construction) but only to the elevation of the cell bottom. To be successful, the scheme is dependant on Operations achieving a relatively high sand capture in the cells during the first few months (i.e. completing the first cell lift before the pond reaches the level of the spillbox discharge pipes). The dry dykes (constructed with lean oil sand) will require considerably less dozer maintenance than dozed tailings sand thereby increasing the placement time in the cells.

The redesign to construct overburden lean oil sand dry dykes will allow tailings operations to asses how they are performing and potentially further reduce overburden placement after startup. The redesign was accomplished because of the willingness of Syncrude Operations to evaluate their ongoing performance rather than construct a "foolproof" design at the outset, and North American Construction's commitment to the Alliance goals and ingenuity in constructing the redesign to result in no increase to the capital budget.

6 RECYCLE WATER SYSTEM

6.1 *Tailings recycle water systems*

There are three main methods of reclaiming water from tailings ponds:

1. Pumps installed on a floating barge and taking water from the surface
2. If elevation permits, decanting the water through the tailings dam via an underground pipeline
3. If elevation permits, using a siphon pipeline over the top the tailings dam to some point lower in elevation at the other side.

Operating constraints at Aurora include:

- The initial pond elevation of 289.1 m will be 7 m below the plant site, which is the destination of the reclaimed water.
- The pond and impoundment dykes will rise approximately 50m over the 7-year life of the facility.
- The location of the clean pond within the tailings pond will change during the life of the project.

The pump and barge method tends to be the most versatile as the pumping power can cope with all kinds of geometric configuration. However, this method also tends to be the most expensive in terms of capital and operating expenditure. Pontoon and bridge fabrication as well as pump maintenance are examples of high cost items.

Decanting water from a tailings pond is a very cost effective method, however since the initial decant location would lie under 40 m of sand within a few years, as well as the relatively low initial elevation of the tailings pond, meant that this option was not technically or economically feasible for start-up.

A siphon system is also considered to be a cost-effective way to transfer water from a pond as long as the pond elevation is sufficiently higher than the destination. This elevation difference is required to overcome pipeline flow requirements. Other design issues include:

- Negative pressure,
- Degassing and
- Winterization

6.2 *Aurora recycle water system*

In 1997 during basic engineering the team budgeted for the pump and barge system as it was the most versatile and could cope with the initial low pond elevation. An allowance of $1 million was identified for purchase and installation of a suitable barge.

Two pumps were then purchased for the task. At this time in the project, the expected water quality was not well defined and it was believed that muskeg and other debris might be present in the water. Syncrude experience with return water systems at their Southwest Sand Storage facility led the team to select slurry pumps to handle large pieces of debris, bitumen and even rock.

As engineering progressed from basic to detailed the quality of water that the barge would receive became a concern and it was decided to split the single barge into two individual barges each with one pump and separated by some distance, thereby imparting a measure of redundancy for RCW supply.

The choice of horizontally mounted slurry pumps meant that the pumps had to be set within a dry well below the water surface. This made for a more expansive pontoon structure than would have been the case with the more common (for water usage) vertical axis pumps. This, however, was the concept carried through detailed design.

The Aurora environment of constant challenge led to the question "can we eliminate the high cost barge and associated pump maintenance costs by locating the pumps on land and feeding them through a siphon system"? After a short investigation confirmed the possibility of significant cost and maintenance savings, the barge idea was put on hold and a more detailed design of the siphon was commenced.

6.3 *Siphon system design*

The main design issues around the siphon/pump design were as follows:

- Pump NPSH with a long suction line.

- Degassing at the pipeline low pressure point.
- Atmospheric pressure as a driving force.
- Pond icing.
- Vortex formation entraining air.
- Aggressive dyke raising schedule.

Using siphon technology, developed with AGRA E&E (project consultant) and utilized at Suncor Inc., the alliance engineering team together with AGRA E&E personnel developed a system that was significantly less expensive than the barge option and more operator friendly.

The two slurry pumps that were originally purchased for the barges were placed in a single relocatable building at the toe of the impoundment dyke to enable relocation as the tailings pond evolves and the clean water pond location moves. A siphon suction line extends over the dyke to a floating intake vessel situated in the clean water pond (Figure 4).

The siphon intake vessel (Figure 5) can operate while being completely frozen into an ice cover of more than 1 m. A long radius elbow helps ensure that water is drawn at a sufficient submergence. The connection of a radiused entry in the form of a bell mouth extends submergence and reduces separation losses. A mud plate below the bell mouth sets up radial flow which ensures that water is drawn from the surface and not from the pond bottom where sediments may be entrained. Added to the bell mouth was a ring plate that extends the diameter of radial flow and thus reduces the intake velocity low enough to avoid vortex formation. Two anchor cables, each at 135° to the pipeline, keep the vessel in the correct location during periods of wind and ice movement. Tension control for the anchor cables are winches, located on the debris control structure for ease of access.

The upstream portion of the pipeline is made of High Density Polyethylene (HDPE) which enables the pipe to flex and therefore the system to function as the water (or ice) elevation rises.

The pipeline crest is the system low pressure point and a valve station is located here. This allows for the attachment of a vacuum pump to establish a siphon prime and a compressor to force water from the pipeline as part of a winterization procedure. The vacuum system is not continuously operating like many other siphon systems of this type. Instead, a higher fluid velocity is used to entrain air into the flow that may be drawn into the pipeline, degass out of solution or enter through leaking seals and valves. This portable unit concept yielded savings in the order of $100K, mostly from building, MCC, PLC, instrumentation and controls costs. The pipeline splits into two lines just upstream of the pumps with a 36"X30" true-wye connection. As it is the intention to operate only one pump for the majority of the time, the wye is heat traced to prevent the stagnant leg form freezing.

The moveable pump-house building has a total moving load of approximately 160 Tonnes and presents no significant problems to Syncrude.

Summing up the costs of the Siphon Intake Vessel, suction pipeline, miscellaneous equipment and pump-house building, the project realized a savings of approximately $400k over the barge system.

7 OTHER NOTEWORTHY ACHIEVEMENTS

Although this paper focuses on the earthworks side of the Aurora Tailings scope, there were many significant breakthroughs in other disciplines also. Some of the more noteworthy achievements in Electrical, Structural and Piping are described in the following sections.

7.1 *Electrical*

Through their on-going discussions and search for cost saving opportunities, Tailings Electrical uncovered a huge potential in used or recycled materials from Syncrude's base mine operation. Because Aurora Tailings was amenable to the benefit of using these materials in a new installation, we were able to make use of an existing E-house building including distribution panel and transformer, lights, and unit heater. We put into service used trailing cable from the mine, used metal sawhorses to support the cable, and used tailings pipe to act as mechanical protection for cables under roads. All recycled or used material saved money and resulted in less environmental impact than if we had used new materials.

7.2 *Structural*

Aurora Structural was faced with designing above ground horizontal in-line anchors for the main tailings slurry pipeline. The pipeline design required anchors spaced at regular intervals between the expansion barrels and at the pipeline bends. The 30 inch diameter Aurora tailings pipeline is a larger diameter and is operating at a much higher pressure than past Syncrude installations. This resulted in very high theoretical anchor loads making for a very challenging anchor design. Traditionally very large and expensive pile and concrete block anchors with elaborate bolt-down brackets are used for this type of application. This approach is very costly and was found to perform poorly operationally. By working cooperatively with Aurora Piping, Syncrude Operations and Syncrude Geotechnical, a risk assessment was used to develop an anchor design basis, which was both practical and economical. Economical straight shaft driven steel and cast-in-place concrete pile anchors were then designed incorporating the results of the risk analysis. The final result was a

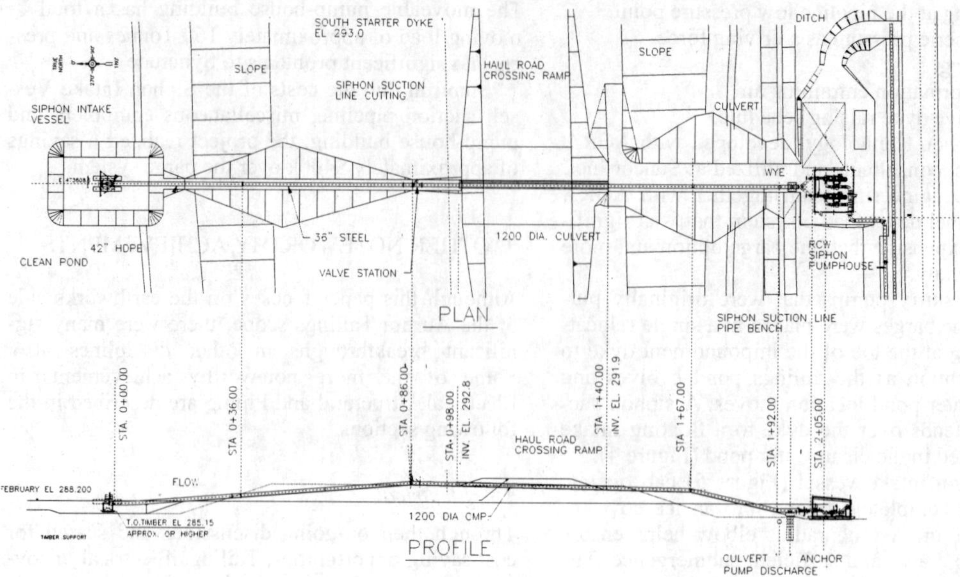

Figure 4. Siphon system plan and profile

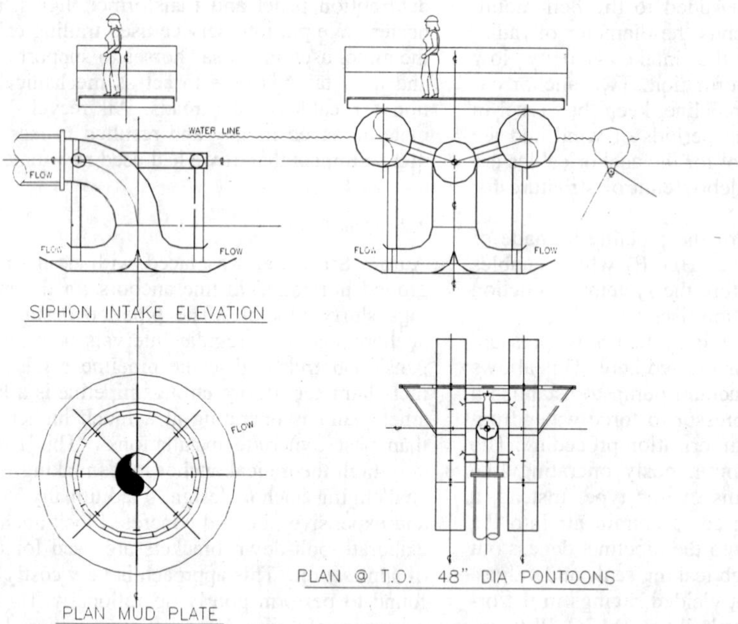

Figure 5. Siphon intake vessel

pipeline anchoring system, which was much less expensive and more constructable than previous installations.

7.3 *Piping*

During the EDS phase Aurora conducted an extensive review of all applicable Syncrude standards and specifications to ensure the requirements were consistent with the Aurora fit-for-purpose design approach. Aurora Piping (in consultation with Syncrude operations and piping engineers) elected to follow ASME B31.11, Code for Slurry Piping Transportation System instead of ASME B31.3, Code for Process Piping. This allowed Aurora to use a higher allowable stress value for the mechanical design of the piping system that resulted in a lower requirement for pressure containment thickness and hence a saving in material cost.

Aurora Piping also developed a project specific piping specification by selecting a pipe with a higher yield stress value (A 139 Gr. C) and at the same time applying some of the API 5L quality requirements. The end product was a higher strength material with higher quality but for a lower cost.

8 CONCLUSION

Table 2 provides a summary of the combined results achieved in all disciplines in the Tailings Area versus our stretch targets.

Table 2. Tailings performance against stretch targets

Category	Stretch Target	Result
Capex	$5.76 M savings	$6.5 M savings
Opex	20% reduction	16.5%
Schedule	May 2000	April 2000
Safety	Beat WCB	Zero lost time
Environment	Zero incidents	1 incident
Quality	Zero re-work	?

The one incident reported under the environmental target was due to failure to take a water sample from the sedimentation pond in December 1998. The sample was not taken because we were not discharging from the sedimentation pond at the time. The unknown under the quality target will be answered after start-up. Feedback from a 1999 customer survey indicated "proof will come post start up but I feel we have a good "fit for purpose" design".

This remarkable achievement was made possible due to the dedicated group of people from operations, engineering and construction working together from conception through construction toward a common objective. It required hard work, getting out of our comfort zones, creativity and persistence to continue to refine our designs until we achieved a solution that was supported by all stakeholders.

REFERENCES

Scherr, A.L. 1989. Managing for breakthroughs in productivity. In Human Resource Management, Fall 1989, Vol. 28, Number 3, Pp. 403-424.

Syncrude Canada Limited, June 1996. Aurora Mine Application to AEP/AEUB

Environmental Issues and Management of Waste in Energy and Mineral Production, Singhal & Mehrotra (eds)
© 2000 Balkema, Rotterdam, ISBN 90 5809 085 X

Design, construction and rehabilitation of spoil dumps – A case study

Y.V. Rao, M. Aruna & H. Vardhan
Department of Mining Engineering, K.R.E.C., India

ABSTRACT: For developed as well as developing countries, mining of minerals to sustain national development is of paramount importance. As per Kautilya, "minerals are treasure of the nation". India has unique blend of big and small, manual and mechanized, opencast and underground mines. Goa is one of the major iron ore producing states in the country. All the deposits in Goa are worked by opencast mining methods. In process of winning these huge quantity of iron ore from the mines, about double of this quantity of waste are generated and all these wastes are piling up in and around the mining pits covering a vast stretches of land. Engineered slopes of mine spoils may be stable at the end of construction, but they can deteriorate over time. An efficient engineering design of spoil dumps aims to ensure the long term stability of the dumps at the end of the construction. This paper presents the analysis of the industrial survey carried out on Goan spoil dump management practices and their design, construction and rehabilitation.

1 INTRODUCTION

Mining all over the world a definitely a much required process for advancement of technology and it is one of the most pollution causing operation. Many countries have stringent regulation but unfortunately in India and in particular in Goa the mining operation is going on without any proper check, till recently. The Goa is one of the major steel producing state in the country. The iron exports from India dates back to 1951, initially from Goa. The Goa exports 15Mt of iron ore which is nearly 50% of the total ore exports. At present almost 82 mines were operating through out the area, in which 26 are manganese mines(Poduval,1997).

Of all the industrial activities in Goa, mining is the most environmentally destructive activity. Whether it is strip or opencast system the damage to the environment and ecosystem is the same in the long run. One critical point in Goan mining industry is over burdening, requiring large volumes of materials to be handled, large number of various earth moving machinery and transportation facilities and extensive requirement for dumping. In process of winning, it generates rock wastes, which is about double the quantity of ore produced, annually(Senguptha,1987). These wastes are piling up as huge dumps due to want of any utility.

Mining operations are being carried out by forming systematic benches on the hill top and along hill slopes and the pits are laterally extended in stages in all directions with increasing depths. Engineered slopes of mine spoils may be stable at the end of the construction, but they can deteriorate over time. There is thus the need to increase the base knowledge on the existing practices of spoil dump design and rehabilitation. In this regard an industrial survey has been carried out to collect the information by sending the questionnaires to all the mining companies of Goa region. The response from the mining industry amounted to 70% and 24 questionnaires were answered completely and used in this analysis. The information derived from this spoil dump design, management and rehabilitation industrial survey is discussed in this paper.

2 STUDY AREA

Goa is one of the few places in India where we have wet and moist evergreen forests. Mining which is regarded as the backbone of Goa's economy has done utmost harm to the ecology of this area.. Geographically, the State of Goa is located along the mid west coast of India and is bound between the co-ordinates 14° 53′ 57″ to 15° 47′ 59″ N and 73° 40′ 54″ to 74° 20′ 11″. Goa covers an area of 3700 square kilometer. A total area of 3350 ha of land is under mining and associated activities of which 1300ha is under dumps. Forest ecosystem in Goa, according to the latest land use classification (Govt. of Goa, 1995), occupy an area of 125,473ha constituting over one-third (34.8%) of the total geographical area of the state.

The general climate of the area is mainly tropical and is influenced to a large extent by the conditions in the Arabian Sea. The climate is characterized by high humidity and less extremes of temperature. The temperature varies between 20 to 33 degrees. The annual rainfall varies from 2700 mm to 3500 mm and the approximate number of rainy days are 110. The distribution of the general topography from the area surrounding the surveyed mine sites indicates that the majority of the sites surveyed have an gentle to moderate slope topographic profile (Table 1).

Table 1.Topography of the surveyed mine sites.

Undulating	Gentle to moderate slopes	Concave depression and valleys
25%	55%	20%

3 OVERBURDEN SPOIL DUMP CHARECTERSTICS

The state is endowed with a wide range of physiography, landforms, geology and vegetation, which have influence on the genesis of the soils. Nearly 2/3rd of Goa is covered by laterite, ranging in thickness from a couple of meters to over 25 meters. The state is rich in minerals such as iron ore, manganese, bauxite and silica sand. At present iron and manganese are the major extractive industries. The overburden material consists of laterite, lateritic clay, manganiferous clay and phyllites. The results arrived at after studying these 24 mines, can be fairly representative of whole Goa (Figure.1).

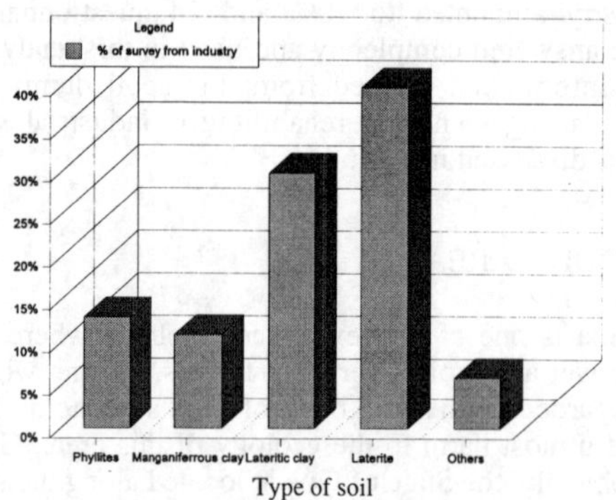

Figure 1. Spoil dump soil type at surveyed sites

4 MINING IN GOA

All the deposits in Goa are worked by opencast mining methods. To extract a tonne of powdery ore, an average of 2.5 cu.m of overburden material is required to be removed. This overburden material is dumped generally within the lease hold along the hill slopes, roadsides and valleys. In some cases, overburden material is being transported to dumping sites outside the lease hold area. It is estimated over 850 million tonnes of laterite overburden/waste dumps are already accumulated and their management is posing a major problem in Goan mining industry. In addition every year about 30 million tonnes of overburden is generated. Normally overburden dumps are up to 30m height but because of non availability of land, mine operations have raised the height of dumps beyond 30 m. Field study reveals that majority of dump heights are varying between 40 to 45 meters(figure 2). Most of the dumps are typically steep with slopes greater than 30 degrees. Many waste dumps are situated in the upper part of the valley regions and during monsoon, runoff from the dumps is common which blankets agricultural fields and settles in water courses.

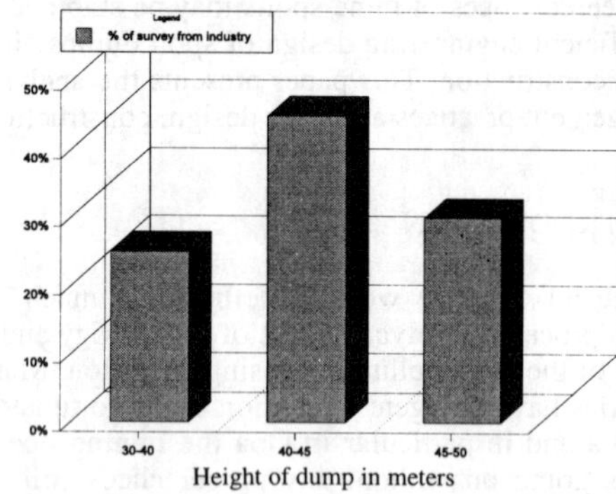

Figure 2. Industrial responses for spoil dump height.

Slope is a major factor controlling the rate of erosion during rainy period in any mining area(Dubey and Rath, 1997). The waste dumps are more susceptible to erosion due to their steep side slopes. In one of the study, it was found that the rate of sediment loss from steep waste dump slopes was 10 to 12 times more that in non-mining/agricultural/forest lands in the same drainage area during rainy period. During rainy season, rate of erosion/sediment loss varies from place to place depending on various factors. The extent of rain water erosion, is largely a function of the slope, but the length of the slope is also equally important(Hussain, 1990). Change in slope of the non-cohesive loose waste is observed during mining process. As slope increases, soil and water loss also increases(Tegwani and Bharadwaj, 1982). Intensity and duration of rain fall are important factors of soil erosion(Singh and Verma, 1975).

4.1 *The problem*

The iron ore industry operates under certain difficult conditions specific to Goan iron ore mines. The major technical difficulties encountered are -

- restricted lateral development due to smaller areal extension since the lease area of individual mines is about 100 hectares or less.
- high overburden to ore ratio of an average of about 2.5:1 implies that a large amount of overburden is generated when ore is extracted. Since the mining leases are less than 100 hectares, there is very limited space or none at all available within the lease area to dump the waste material. Land being in short supply, dumps are typically steep and are of greater heights.

5 SPOIL DUMP DESIGN AND MANAGEMENT

The spoil dumps are generally categorized as dead dumps and active dumps. Dead dumps are old dumps where no further dumping is being done. These dumps may consist of overburden material, reject ore, sub-grade ore or a mix of all these and tailings from beneficiation plants. Active dumps are running dumps where dumping of all these materials is continuing. An efficient engineering design of spoil dumps aims to ensure the long term stability of the dumps at the end of construction. It also seeks to reduce the erosion potential through correct selection of an initial stable landform scheme(Goh, Aspinall & Kuszmaul, 1998).

The industrial survey shows that the principal method used for overburden removal is truck-shovel system. The dumps do not provide suitable substrate for vegetation and are usually susceptible to slides and erosion during rains especially in heavy monsoon. To make the dumps hydrologically compatible, terraces are made at intervals of 10-12 meters of about 8-10 meters width. The terrace are generally sloping inward to allow rain water to concentrate at the center and infilter instead of running down the slope. In addition, efforts are made to establish an adequate cover of vegetation to stabilize the dump slopes and prevent or control erosion. The sloping sides of the dumps are covered with laterite boulders of assorted sizes thereby splashing of particles by rainfall and their dislodgement are avoided. In addition, a laterite wall of sufficient thickness(about 3 meters) and height(2.5-3 meters) forms a girdle around the base of dump to arrest water flowing down the slope. Often a trench, 1.5 -2 meters deep and about the same in width, runs parallel to the girdle, a few meters away and a second retention wall of laterite blocks is built several meters away. In some of the mine areas a bed of laterite pebbles of few meters precede the trench and the second wall. These are meant to filter the suspended material to the extent possible and then let the water flow out of the area into a channel or river in a regulated manner.

6 SPOIL DUMP REHABILITION

The rehabilitation of dead dumps is mostly being done through some bio-engineering measures. Control of erosion is given importance in the rehabilitation programme. The runoff and erosion control measures, as indicated earlier, are being adopted in most of the mines. In addition, efforts to cover the dumps adequately with vegetation have been taken up in the recent years. Figure 3, shows the type of machinery utilized for reshaping of spoil dumps. The spoil dumps are reshaped to suit the surrounding landscape prior to revegetation for improved erosional and mass stability. The findings of the survey indicate that the track type dozer is the most suitable equipment for the reshaping of spoil dumps.

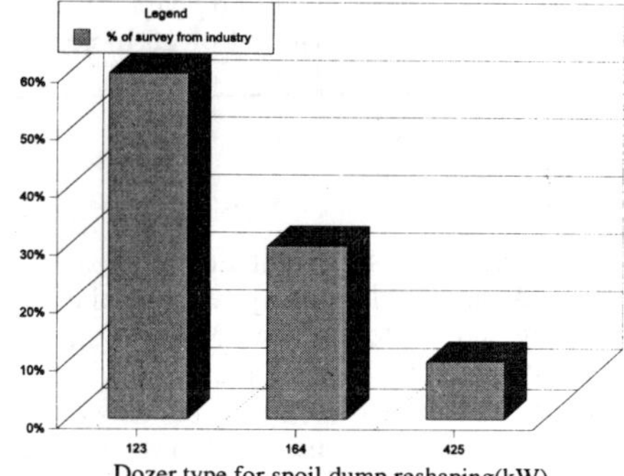

Figure 3. Power of dozers used for spoil dump reshaping.

Rehabilitation of waste dumps have gained considerable importance over the years and by and large, most of the companies are enlarging their afforestation activities by opening an environment cell. Presently many of the mine owners have resorted to long term remedial measures such as physical obstruction to the movement of silt by making laterite tie walls, and planting of dead dumps with suitable plants.

6.1 *Vegetation*

Environmental regulations require that mined out areas and inactive or dead dumps be rehabilitated. In Goa as in the rest of India, the favoured form of rehabilitation has been revegetation. On reject dumps where no further dumping is possible the plantation of various species have been developed by the mining firms. The forest department had undertaken afforestation on dead dumps and according to them acacia auriculiformis is the most suitable species for these dead dumps. Not only it establishes itself in this inhospitable conditions, it also invades such similar areas in the neighborhood besides regenerating in the area where it is planted. Moreover, the rate of growth is satisfactory, and it acts as a catalyst in the process of succession. Cashew dominated the choice of species planted in the first phase of revegetation of dumps from 1983 -87, acacia auriculiformis came to dominate this choice over the period since then, with casuarina remaining the favoured second best. Very clearly, there has been a shift towards faster growing trees. Table 2 shows the growth rate of some of the species planted in the Goa region.

Table 2. Growth rate of species in the region

Type of species	Growth rate
Acacia auriculiformis	Height of 10 meters in 5-6 years
Casuarina	height of 11 meters in 10-12 years
Cashew	Crown width of 5 meters in 8 years
Acacia mangium	Height of 15 meters in 8 years

7 DISCUSSION

No availability of sufficient dumping space at a stretch is posing problem in systematic planning of the dumping and waste disposal. Most of the mining activities are on the slope of the hills wherein the reclamation will be problem due to heavy rainfall. The fast running water along the slopes, removes the fine newly dispersed soil particles, leaving behind the hard pebbles of laterite, which makes a poor seed bed and provide little or no anchorage for the plant roots. The height,slope and form of these heaps affected their ability to support plant life in most of the old dumps. The stabilization of a slope is the function of a many characterisation like: type of debris material, its angle of repose (wet and dry), particle size distribution, slope, age of dump, length, water retention capacity etc. Small heaps composed of large fine particulate material, mild slope and smaller lengths are susceptible to become stable much sooner than large heaps of steeper angles. Top soil is often the most important factor in successful rehabilitation, particularly where the objective is to restore a native ecosystem. However, no where in the mining belt of Goa, top soil was found to have been stripped separately and preserved for future rehabilitation program. So far more than 1120 ha of area is brought under various plantation, including fruits yielding trees.

8 CONCLUSIONS

Information presented in this paper could be viewed as a quantitative guide to the spoil dump design, construction and rehabilitation management practices in the Goan mining industry. Efficient spoil dump design and rehabilitation practices are acknowledged as important phases in mine design and planning. More than double the quantities of iron ore, which Goa has exported since 50's are the waste rocks. They are not used, and hence are piling up. As none of these waste fall into any specification categories for industrial use, a different approach is needed to be adopted, that is usage's have to be found, for the material available. With limited studies conducted so far , it appears that some of the wastes can find their usage's to manufacture roof tiles, bricks, soil conditioners, building blocks, abrasives, paints, glass and refractories (Senguptha,1996).

However this is a gigantic task and needs a effort from all concerned, to find out commercially viable processes to use these wastes, and this will go a long way in removing these waste dumps from the beautiful landscape of Goa. Geological Survey of India in recent years has been playing a very useful role in geoenvironmental planning. It is embarking upon a plan of samples from these mine dumps and analyze them for any possible noble metal/base metal and REE content which if found would indeed turn the heaps of waste into hills of wealth.

ACKNOWLEDGEMENT

Thanks are due to all the mining industries who contributed necessary information towards this waste dump research project and particularly to the Director of Mines and Industries, Goa Government. The authors would also like to extend their sincere thanks to all those who helped directly or indirectly in carrying out the project successfully.

REFERENCES

Ashutosh Dubey & Rath, R. 1997. *Effects of slope and rainfall in waste dump/overburden erosion in opencast mining area - A case study.* The Indian Mining and Engineering Journal. (10): 19-34.

Goh, E.K.H., Aspinall, T.O. & Kuszmaul, J.S. 1998. *Spoil dump design and rehabilitation management practices(Australia).* International Journal of Surface Mining, Reclamation and Environment. (12):57-60.

Hussain, A. 1990. *Biological plant colonization on lateritic waste dumps of bauxite mines - A case study.* Journal of Mines, Metals & Fuels.(12): 391-395.

Hussain, A. 1990. *Results of land reclamation schemes adopted in the worked out areas of BALCO's mines.* The Indian Mining and Engineering Journal. (3): 15-21.

Poduval, A.M.K. 1997. *Pollution in mining with special reference to Goa.* AICTE-ISTE winter school at K.R.E.C. Surathkal.(12).

Sengupta, B.S. 1986. *Blockwise forecasting of grades in an iron ore deposit of Goa.* The Indian Mining and Engineering Journal. (3): 15-20.

Sengupta, B.S. 1987. *Characteristics of waste rocks associated with Goan iron ores and their possible usages.* The Indian Mining and Engineering Journal. (1): 9-18.

Singh, N.T. & Verma, K.S. 1975. *Effect of rainfall intensity and surface condition on water erosion in foothill soils of the Punjab.* Journal of Indian Society of Soil Sciences. Volume 23(1): 27-30.

Tejwani, K.G. & Bhardwaj, S.P. 1982. *Soil and water conservation on research.* In Review of Soil Research in India Part-II. 12th International Congress of Soil Sciences. India. 608-621.

Environmental Issues and Management of Waste in Energy and Mineral Production, Singhal & Mehrotra (eds)
© 2000 Balkema, Rotterdam, ISBN 90 5809 085 X

Environmental issue and waste management in sago industry for energy production

R. Saravanane, D.V.S. Murthy & K. Krishnaiah
Environmental Engineering Laboratory, Department of Chemical Engineering, Indian Institute of Technology, Madras, India

ABSTRACT: The starch manufacturing industrial units, such as, sago mills, both in medium and large scale, suffer from inadequate treatment and disposal problems due to high concentration of suspended solid content present in the effluent. in order to investigate the viability of treatment of sago effluent, a laboratory scale study was conducted. the treatment of sago effluent was studied in a continuous flow anaerobic fluidized bed reactor. the start-up of the reactor was carried out using a mixture of digested supernatant sewage sludge and cow dung slurry in different proportions. the effect of operating variables such as cod of the effluent, bed expansion, minimum fluidization velocity on efficiency of treatment and recovery of biogas was investigated. the treated wastewater was analysed for recycling and reuse to ensure an alternative for sustainable water resourse management. The maximum efficiency of treatment was found to be 82% and the nitrogen enriched digested sludge was recommended for agricultural use.

INTRODUCTION: Anaerobic technology for the treatment of wastes and wastewater was known in India from the beginning of this century. The upflow anaerobic sludge blanket (UASB) and anaerobic contact filter reactors were successfully implemented for the treatment of high strength industrial effluents. The fluidized bed systems were not thoroughly studied at laboratory and pilot scale level. Hence there is ample scope to adapt the anaerobic fluidized bed systems for industrial effluent treatment.

Sago, the edible starch in the form of globules, is manufactured from tapioca tubers (*Mannihot Ulillisema*). In the southern region of India alone there are about 800 sago mills which were medium and large scale units producing 15 to 30 tonne of sago per unit per day. These units discharge about 40000 to 50000 litres of sago per tonne of sago processed. Many investigators observed that the wastewater from sago mills is of serious concern from stream pollution point of view (Murthy, 1961, Sastry, 1963, Saroja, 1972). Most of the low cost and conventional treatment methods reported (Sastry, 1964, Saroja, 1972, Pescod, 1977) to have low treatment efficiency due to high concentration of suspended solids and insoluble fibres present in the effluent. Therefore it has become necessary for these units to treat the wastewater for safe discharge. Hence there is ample scope for an efficient and complete treatment system which will ensure a safe effluent standard limit with potential energy recovery in the form of biogas and recycling of wastewater for sustainable water resource management. The start-up of anaerobic fluidized bed process is initiated by the development of biomass and subsequent attachment to carriers. A review by Hickey (1991) on the start-up of high rate anaerobic treatment systems reported the development of biofilms due to the influencing parameters such as liquid flux rate, scale-up of the reactor, gas flux and organic loading rate. Heppner (1992), Hsu (1993) and Shieh (1996) have compared the start-up performance with propionate and acetic acid as substrates. The performance of anaerobic fluidized bed treatment of cornstarch wastewater was studied (Chen 1988) to demonstrate the effect of operating variables on biomass activity. have also studied The effect of operating variables on the performance of fluidized bed reactor was studied (Sreekrishnan, 1991) to reveal biofilm formation and their attachment to the carries. The experimental studies conducted by Liang (1993) and Morgan (1991) on brewery and ice cream wastewater,

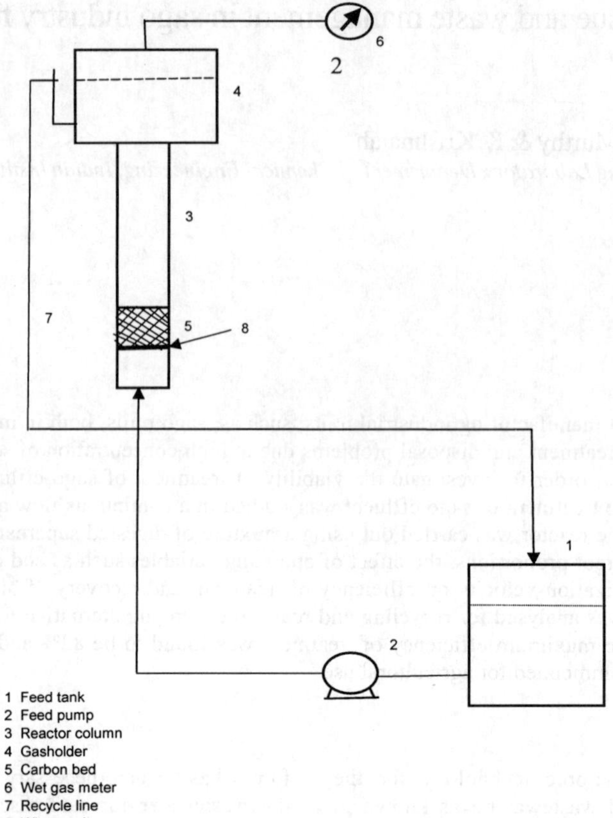

Figure 1. Schematic diagram of anaerobic fluidized bed reactor

respectively evaluated the efficiency of treatment and changing microbial activities, leading to methanogenic biofilm formation during start-up of reactor.

The substrate composition and organic loading rate on the process performance during start-up and steady state was studied by Matsumoto (1991) and Converti (1993) using fluidized bed reactor. The optimum values of operating variables for treating hog wastewater were reported (Chen, 1997) for the anaerobic fluidized bed treatment of hog wastewater. The feasibility of treatment of monosodium glutamate fermentation wastewater was evaluated (Tseng, 1990) in terms of removal efficiency and methane content in the biogas. A BOD removal efficiency of 90% was attained with a methane content of 80.8% and OLR of 10.1 - 31.1 kg COD m^{-3} day^{-1}. Matsumoto (1991) and Converti (1993) have demonstrated the start-up and steady state performance of a fluidized bed process interms of substrate composition and organic loading rate by varying the influent flow rate to the reactor. particles into the effluent and also to serve as a gasholder.

2. MATERIALS AND METHODS

2.1 *Experimental Set-up*

A Schematic diagram of the experimental set up is shown in Figure 1. The bioreactor consisted of a this, an upper section of 300 mm diameter was mounted to prevent carry over of suspended bed porosity was determined to be 0.4. Bed material of 2 kg of carbon was loaded in the reactor.

2.2 *Carrier Material.*

Activated carbon particles of size 700 μm and density 1500 kg/m^3 was used as carrier material. The porosity was determined to be 0.4. The bed material of 2 kg of carbon was loaded in the reactor.

2.3 *Sago wastewater.*

The wastewater was analysed as per Standard Methods and the characteristics are presented in Table 1. The seeding material required was collected from the municipal sewage treatment plant at Kodangaiyur, Madras and stored at 20°C of 24 hours before using for test.

Table 1. Characteristics of synthetic sago wastwater

Parameter	Value
pH	7.3
COD (mg/l)	2500 - 4000
BOD (mg/l)	1500 - 3200
Alkalinity (as $CaCO_3$, mg/l)	1550
Total Solids (mg/l)	4500
Total dissolved solids (mg/l)	2110
Total suspended solids(mg/l)	1410
Volatile suspended solids(mg/l)	1350
Kjeldahl Nitrogen (mg/l)	80
Total phosphorus (mg/l)	15

2.4 *Start-up of the Reactor*

The reactor was loaded with activated carbon and supernatant of digested sludge with diluted feed at 250 mg/L. The reactor was operated under total recycling conditions for one week before being switched to continuous operation. The bed expansion was maintained at 25% during the entire start-up period by adjusting the upflow velocity from 20 to 25 m/hr.

3 RESULTS AND DISCUSSION

3.1 *Effect of Influent COD*

Initially the influent COD was kept within the range 165 to 287 mg/L so as to shock loads and to facilitate gradual build-up of COD value. The influent pH was kept within 7.1 to 8.2. It was observed from Table 2 that the COD of the reactor contents decreased with time of operation for all the start-up runs. Hsu et al. (1993) also reported that COD values were found to decrease with increase in time of operation during the start-up of the reactor. The start-up was completed at the end of 98th day and the various values of parameters monitored were presented in Table 2.

The COD values were varied between 250 to 4000mg/L in continuous flow experiments. The pH and Alkalinity values were plotted in Figures 2 and respectively. Alkalinity was found to decrease with increase in time of operation. The COD removal (%) and COD loading rate were plotted in Figure 4. It was observed from this figure that the COD removal percentage was increased with time of operation and was varying from 44 to 93%. This was in general agreement with investigations of Chen et al. (1988) and Sreekrishnan et al. (1991) on continuous fluidized bed reactor fed with corn starch wastewater and synthetic glucose as substrates respectively. An efficiency of 82% was obtained at a maximum COD loading of 60.5 kg/m^3/day. The treated effluent was recycled for industrial use after mechanical filtration.

3.2 *Biogas Recovery*

The biogas yield during start-up and steady state conditions are discussed below.

The variation of biogas with time of operation for start-up and steady state conditions was plotted in Figure 5. It was observed that the biogas yield was increased with increase in organic loading rate and attained steady state at the end of 98th day. The maximum biogas yield was observed to be 59 to 66.3 L/day. The composition of the biogas was analysed using a high performance gas chromatograph and the consistent methane percentage was found to be 55 - 65%. The biomass activity was assessed by volatile solids variation observed during start-up and steady state conditions. The volatile solids were found to increase with increase in time of operation and organic loading rate as represented in Figure 4. The nitrogen content of the digested sludge was determined to be 2.2 to 2.45% and hence recommended as manure for agriculture use.

4. CONCLUSION

The start-up of the reactor was completed at the end of 98th day and the maximum COD removal (treatment efficiency) was found to be 82% at a concentration of 4000 mg/L and it could be recommended that higher strength wastewater could be diluted and fed to the reactor. The biogas yield was found to be 0.2 to 0.25 m^3/kg COD/day and the maximum rate of generation to be 59 to 66.3 L/day. The alkalinity requirement was found to decrease with increase in OLR. The biomass activity was found to decrease with increase in OLR. The nitrogen enriched digested sludge was recommended as manure for agriculture use. The treated wastewater was analysed and recommended for recycling and reuse within the plant.

5. ACKNOWLEDGEMENT

The authors wish to acknowledge the Kodangaiyur sewage plant, Madras, India for the supply of anaerobic sludge required for the experimental work.

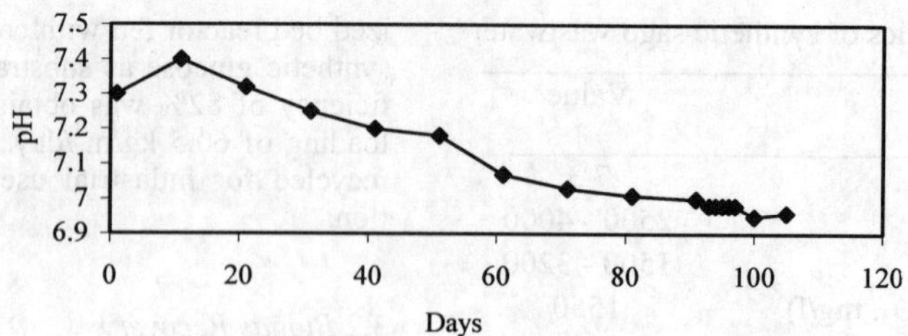

Figure 2 Variation of pH during start-up

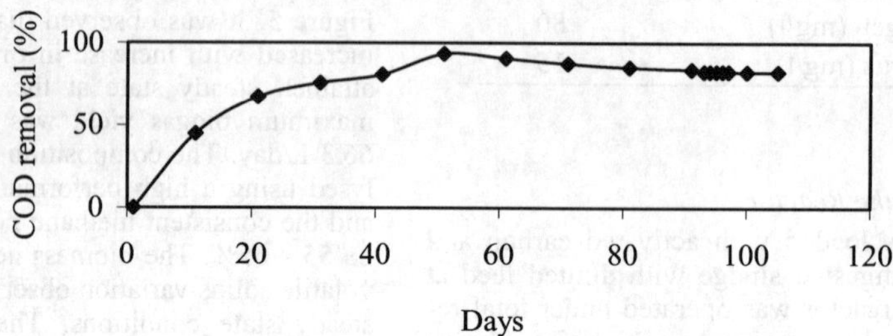

Figure 3 Variation of COD removal (%) during start-up

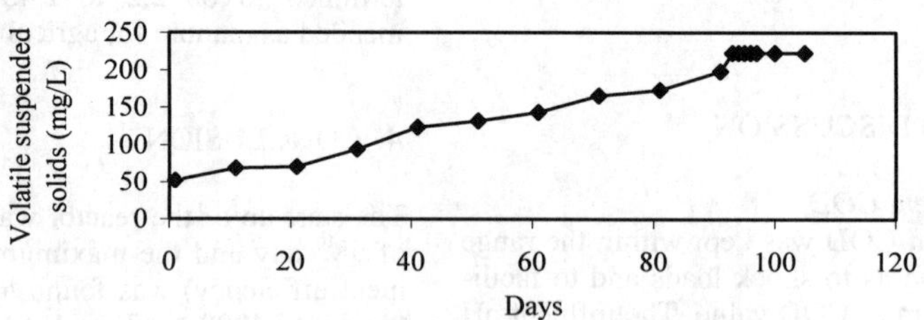

Figure 4 Variation of volatile suspended solids during start-up

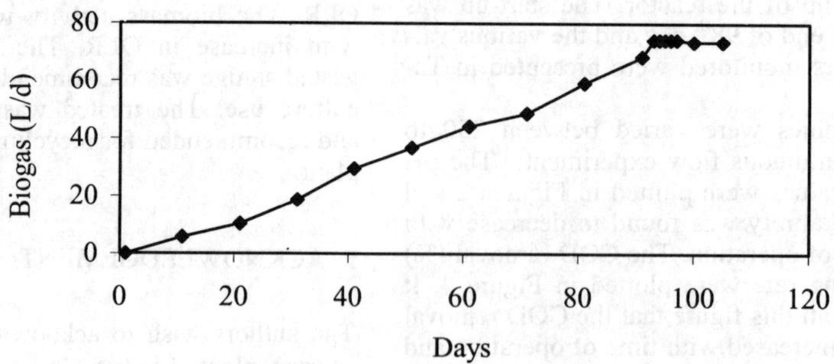

Figure 5 Variation of biogas during start-up

Table 2. Characteristics of synthetic effluent during start-up

Days	OLR ($kg/m^3/day$)	pH	Alkalinity (mg/L)	VA (mg/L)	VSS (mg/L)	Bio-gas (L/d)	COD removal (%)
1	3.2	7.3	1550	188	52.3	0	0
11	6.6	7.4	1602	294.7	68.7	5.7	44.8
21	13.2	7.32	1592	328.5	70.5	10.3	67.2
31	19.8	7.25	1598	373.2	94.2	18.6	75.6
41	26.4	7.2	1722	481.4	123.6	29.4	80.7
51	33	7.18	1695	496.3	132	36.7	93.4
61	39.6	7.07	1632	512.4	143.3	44.1	90.5
71	46.2	7.03	1628	573.2	165.8	48.6	87.3
81	52.8	7.01	1642	697.5	173.5	59.2	85.1
91	59.4	7	1757	784	198.2	68.3	83.7
93	66.3	6.98	1723	826	223	74.2	82.5
94	66.3	6.98	1723	826	223	74.2	82.5
95	66.3	6.98	1724	826	223	74.2	82.5
96	66.3	6.98	1723	826	223	74.2	82.5
97	66.3	6.98	1725	826	223	74.2	82.5
100	66.3	6.95	1721	824	222.6	73.5	82.2
105	66.3	6.96	1722	823	222.6	73.5	82.2
107	66.3	6.95	1721	824	222.6	73.5	82.2

6. REFERENCES

Chen, C.Y., Li, C.T. and Shieh, W.K. 1997. Anaerobic fluidized bed pretreatment of hog wastewater. J. Env. Eng. 23(4): 389-394 .

Chen, S.J., Li, C.T. and Shieh, W.K. 1988. Anaerobic fluidized bed treatment of an industrial wastewater, J. Water Poll. Contl. Fed. 60: 1826-1832 .

Converti, A., Borghi, D.M. and Ferraiolo, G. 1993. Influence of organic loading rate on the anaerobic treatment of high strength semisynthetic wastewater in a biological fluidized bed. Chem. Eng. J. 52: B21-B28.

Heppner, B., Zellner, G. Diekmann, H. 1992. Start-up operation of a propionate-degrading fluidized-bed reactor. Appl. Microbiol.Biotechnol. 36: 810-816.

Hickey, R.F., Wu, W.M., Veiga, M.C. and Jones, R. 1991. Start-up, operation, monitoring and control of high-rate anaerobic treatment systems. Water Sci. Technol. 24(8): 207-255.

Hsu, Y. and Shieh, W.K. 1993. Start-up of anaerobic fluidized bed reactors with acetic acid as the substrate. Biotech. Bioeng. 4: 1347-353.

Liang, Y., Yi, Qian, Jicui, and Hu 1993. Research on the characteristics of start-up and operation of treating brewery wastewater with an AFB reactor at ambient temperatures. Water Sci. Technol. 28(7): 187-195.

Matsumoto, A. and Noike, T. 1991. Effects of substrate composition and loading rate on methanogenic process in anaerobic fluidized bed systems. Water Sci. Technol. 23: 1311-1317.

Morgan, J.W., Evison, L.M. and Forster, C.F. 199.1 Changes to the microbial ecology in anaerobic digesters treating ice cream wastewater during start-up. Water Res. 25(6): 639-653.

Murthy, Y.S. and Patel, M.D. 1961. Treatment and disposal of sago wastes. Central Public Health Engineering Research Institute, India, 1-50.

Pescod, M.B. and Thanh, N.C. 1977. Treatment alternatives for wastewaters from the tapioca starch industry. "Progress in Water Technology, 9(3): 563-574.

Shieh, W.K. and Hsu, 1996. Biomass loss from an anaerobic fluidized bed reactor. Water Res. 30(5): 1253-1257

Sreekrishnan, T.R., Ramachandran, K.B. and Ghosh, P. 1991. Effect of operating variables on biofilm formation and performance of an anaerobic fluidized bed reactor. Biotech. Bioeng. 37: 557-566

Saroja, K. and Sastry, C.A. 1972. Report on treatment of sago wastes. National Environmental Engineering Research Institute, Nagpur, India.

Sastry, C.A. and Mohan Rao, G.J. 1963. Anaerobic digestion of industrial wastes. Environmental Health, 5(3): 20-25.

Sastry, C.A., Murahari Rao and Saroja, K. 1964. Studies on the treatment of sago mill wastewater, Seminar on Environmental Pollution, Kerala, India.

Tseng, S.K., and Lin, M.R. 1990. Treatment of monosodium glutamate fermentation wastewater with anaerobic biological fluidized bed process. Water Sci. and Technol. 22(9): 149-155.

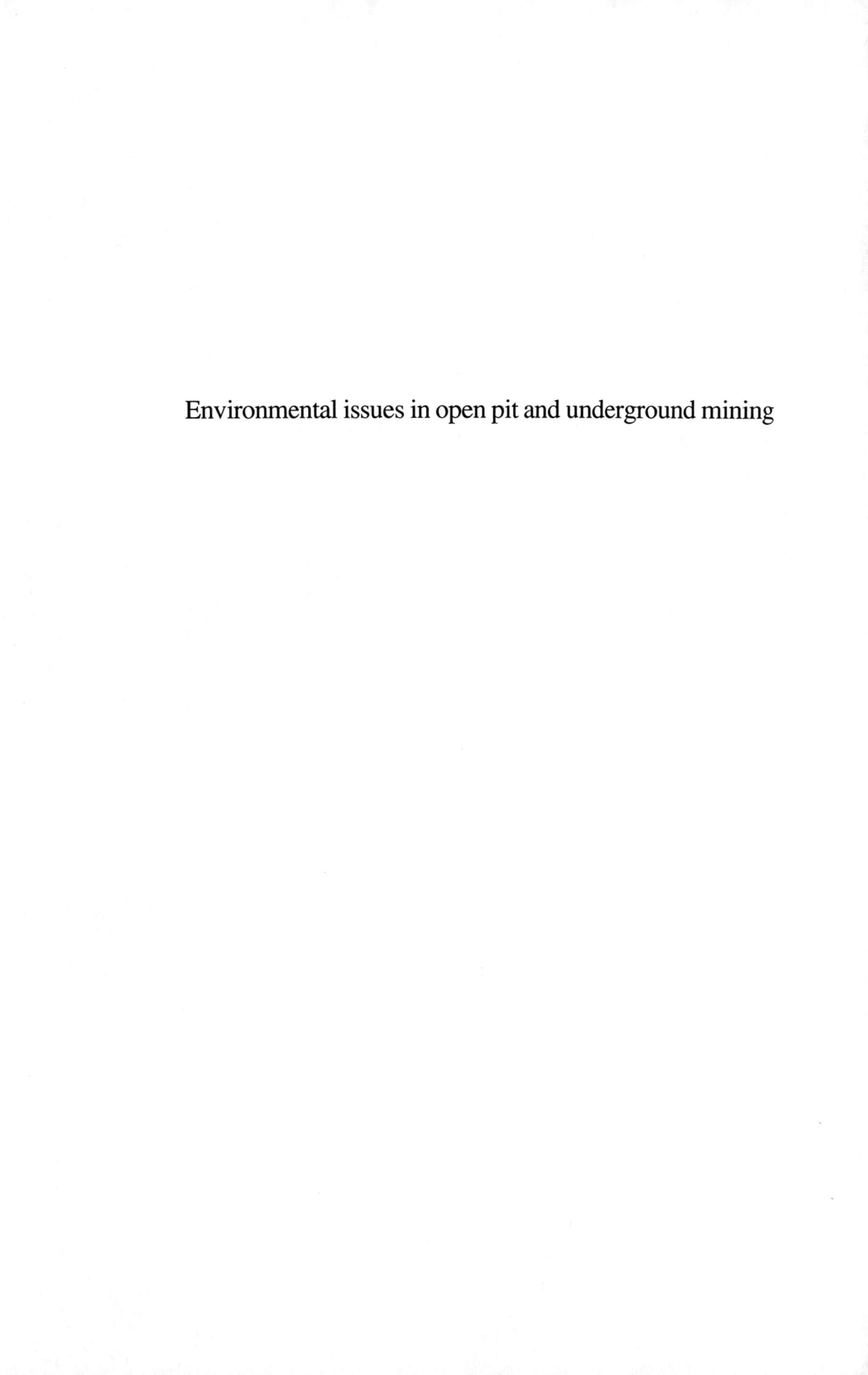

Environmental issues in open pit and underground mining

Environmental Issues and Management of Waste in Energy and Mineral Production, Singhal & Mehrotra (eds)

Air particulate matter monitoring in a major Ghanaian mining town – The case study of Tarkwa

N. Amegbey & S. Ndur
School of Mines, University of Science and Technology, Tarkwa, Ghana

ABSTRACT: Tarkwa has been a major mining town in Ghana for over a century now. It is currently surrounded by a number of mining activities - Manganese mining some 7 km to the south, two gold mines within 8 km to the Southwest and another to the North some 1.5 km away. The surface mining activities around the town have contributed to increased airborne particulate matter within the township. The paper reviews particulate matter monitoring within the township. Dust fallout measurements as well as monitoring of Particulate Matter below 10 μm (PM 10) were considered. The results, though interesting, did not establish any clear trends.

1 INTRODUCTION

Pollution from airborne particulate matter remains one of the major concerns of communities located in mining areas. Surface mining activities, such as removal of overburden, drilling, blasting, loading, hauling and crushing of ore do contribute quite significantly to increase airborne particulate matter by generating fugitive dust. Increased vehicular movement as a result of the presence of the mines would also increase soot and smoke in the ambient air.

The effect on the neighbouring community of persistent dust from open cast mining is its potential to cause annoyance and nuisance. Due to dust deposits, people invariably feel a loss of environmental amenity. Airborne particulate matter when inhaled, does also have the potential for producing respiratory diseases or exacerbating respiratory disorders such as asthma and irritation of the lungs and bronchial passages.

Dust monitoring is necessary to ascertain the level of dust concentrations. Fine air particulate, responsible for causing pneumoconiosis or dust related respiratory disorders, cannot be seen by the naked eye; various dust sampling instruments capable of providing results for assessing the extent of ambient air particulate concentration are therefore used to detect their presence.

Tarkwa, a typical West African mining town in Ghana, has been associated with dusty environments for sometime. This is partly due to untarred roads within the township and also due to activities of mining companies surrounding it. Though dust monitoring was carried out at work places within the mine, using conventional instruments such as konimeters and currently various gravimetric instruments, no attempt was made at carrying out any dust monitoring in the township. Currently, all major roads are tarred and the problem does not appear to be as severe as it used to be. Nevertheless, when ambient dust monitoring equipment became available this case study was undertaken in order to provide some information upon which future studies could be based. The study focuses on dust fallout measurements as well as measurement of concentrations of particulate matter below 10 μm (PM 10).

2 TARKWA AND MINING

2.1 *Location, Climate and Geological setting*

Tarkwa, a town located in the tropical rain forest of south-western Ghana is one area where mining activities have been concentrated for over a century now. It is about 62 km north of Takoradi and can be accessed by both rail and road. The estimated

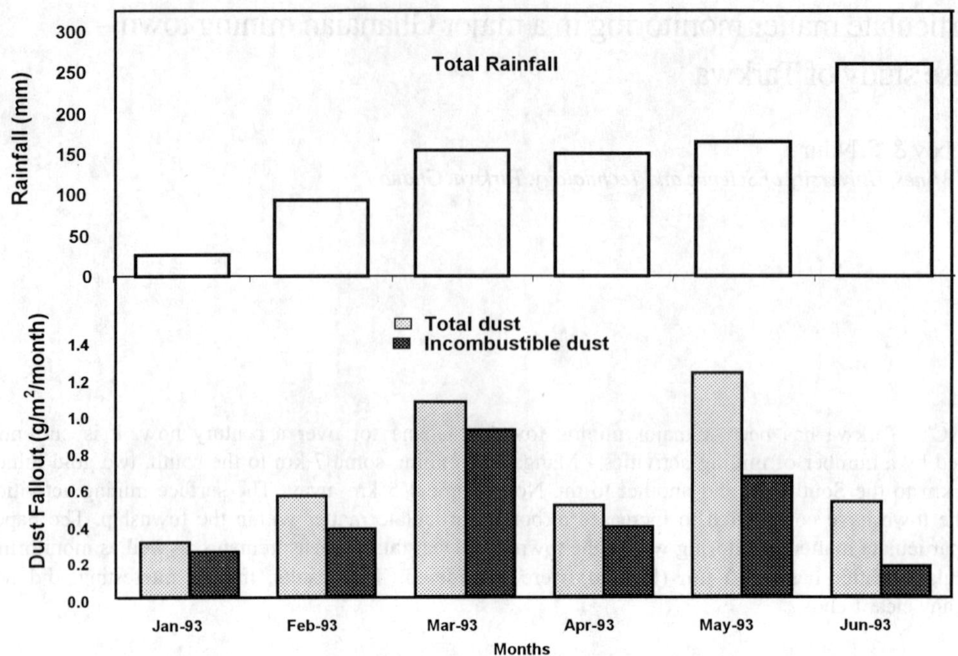

Figure 1: Dust Fallout and Total Rainfall

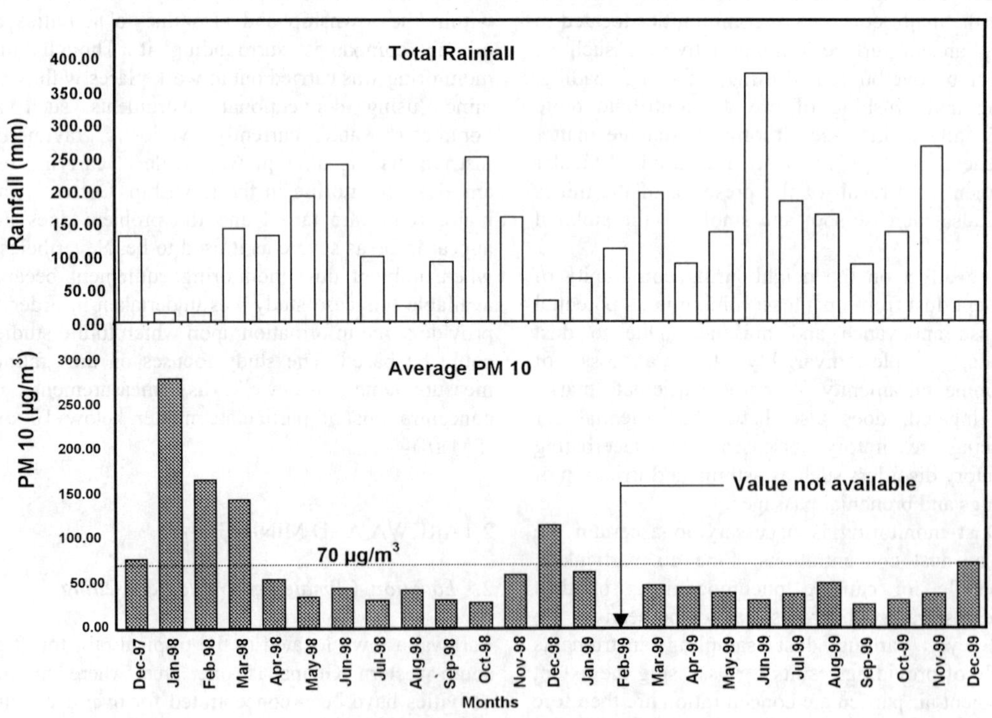

Figure 2: Average PM 10 and Total Rainfall

population is about 110,000.

Rainfall is almost throughout the year with maximum occurring between the months of June and July. The annual rainfall averages about 160 cm. There is no complete dry season in real terms since the period supposed to be the dry season, December to February, has intermittent rains.

Temperature varies between 24 °C during cold nights and 34 °C on hot days. Average monthly Relative Humidity are between 60% and 90%. The once evergreen forest has now been reduced to more or less farmlands. The vegetation consists of perennial shrubs, trees and grasses. The most common plant seen in this area is Raffia palm. Apart from mining activities, the local Wassa people use the land for small maize cultivation, farming of settlement crops such as cassava and a type of cocoyam locally called *Brobe,* etc.

The rock system concentrated in the area, known as the Tarkwaian System, includes the Banket series where Gold deposits occur as Reefs or Conglomerate beds. Favourable Gold bearing areas are found around the contacts of the Tarkwaian System with the Birimiam System. In addition to Gold, alluvial Diamonds are also found in the area (Kesse, 1985).

2.2 *Operating Mines*

Currently, there are four large scale operating mines in the Tarkwa area. Ghana Manganese Company (GMC) operates some 7 km to the South, Gold Fields Ghana Limited (GFGL) is 1.5 km to the North, Teberebie Goldfields Ltd. (TGL) is 8 km to the Southwest and Ghana Australian Goldfields (GAG) is 10 km, also to the Southwest of Tarkwa. All four mines employ conventional open pit mining methods in their operations. GFGL, until 1999, operated an underground mine as well. After the striping of overburden, rock breaking is done using drilling and blasting techniques. Ore recovery is preceded by crushing of the mined material, and haulage of the ore is done with dump trucks covering various distances, up to 4 km on mine haulroads.

3 DUST SAMPLING

Two types of dust sampling methods were used in evaluating the dust level concentration within the Tarkwa township. Dust fallout were estimated using the British standard dust deposit gauge and PM 10 concentrations were determined using a High Volume PM 10 dust sampler with a volumetric flow controller (1200/VFC HV PM 10).

Daily rainfall amounts were obtained from the UST School of Mines synoptic weather station.

3.1 *Dust Fallout Monitoring*

The British standard dust deposit gauge was mounted at a synoptic weather station on the University campus and monitoring was carried out for six months - from January to June 1993. The collecting bottle of the sampler, into which the collected dust is washed, was analysed for total suspended matter (total dust) as well as for ash weight (incombustible dust), and the results were expressed in g/m^2 per month.

3.2 *PM 10 Monitoring*

A high volume dust sampler PM 10 was installed at a selected point within the premises of the Tarkwa Government Hospital located in the Tarkwa township and monitoring has been going on since December 1997. This paper considered PM 10 monitoring results over a period of two(2) years, i.e. December 1997 to December 1999.

The sampler was equipped with a Type E Automatic 8 Port Air Valve Unit and a model G302 Timer/Programmer. Sampling was at six(6) days interval and over a 24-hr period. During the monitoring period under review, 119 measurements were taken.

The net mass of particulate collected after each day of sampling was determined using a precision balance (Sartorius BP211D). The PM 10 concentrations were then calculated on hand, with the volume flow rates in each case from a volume flow recorder on the equipment.

4 RESULTS

The results for dust fallout together with total rainfall during the respective month are shown in Figure 1.

Monthly PM 10 results presented in Figure 2 represent average values obtained from 5 or 6 measurements, as the case may be, taken during the respective month. Also shown are the monthly total rainfall amounts during the monitoring period. The measured values ranged between 7.16 μg/m^3 (on 16th July 1998) and 515.34 μg/m^3 (on 17th January 1998). Thirty-one(31) out of one hundred and nine(119) measurement days had PM 10 concentrations exceeding 70 μg/m^3 (EPA-Ghana recommended limit). In December 1997 there were three(3) days that recorded values higher than recommended, in 1998 there were in all twenty-one(21) days with higher than recommended values and in 1999 seven(7) days had values exceeding the recommended limit.

5 DISCUSSIONS

Dust deposition gauges are used to measure dust fallout, and this gives an indication of the extent of loss of environmental amenity. Noticeable decrease in amenity will occur if the combination of background dust and contribution from dust pollution activities does not exceed an average of 4 g/m^2/month (EPA-Australia, 1995). Dust fallout levels during the period of monitoring did not exceed 1.3 g/m^2/month (Figure 1). Though not much data is available for an elaborate statistical analysis, it appears rainfall amounts do not influence significantly dust fallout concentrations. Water spraying or rainfall is not effective in "knocking down" airborne particulate, but could reduce the amount of settled dust that get airborne.

Particulate matter less than 10 micrometers in size (PM 10) are respirable and large amounts or concentrations can be a health hazard. It may be observed from Figure 2 that PM 10 concentrations are generally high during the months of December, January and February. These months represent the Harmattan months, during which period dry North-easterly winds from the Sahara region of North Africa blow over the country, thereby increasing background dust levels. The Environmental Protection Agency (EPA), Ghana, in its air quality guidelines recommends 70 μg/m^3 as the threshold for a 24-hr sampling period. This value has been indicated in Figure 2 and it may be seen that average monthly values for some Harmattan months exceed this limit. From individual measurement records 26 % of the measurement days recorded values exceeding the threshold. The 24-hr concentration on 17th January 1998 exceeded seven times the recommended level. In 1999, with the exception of the month of December, PM 10 levels were generally below recommended limits, and monthly averages showed reduced levels compared to the previous year.

6 CONCLUSIONS

Dust levels within the Tarkwa township appear to have reduced compared with previous years, and most inhabitants would agree with this general observation. Dust from untarred roads within the township is very minimal since most roads are tarred. Dust from mining activities surrounding the town are believed to be the main contributing factor. This being so, the declining airborne dust levels would then appear to be a result of effective dust control mechanisms being instituted in the mines. Indeed, current legislative requirements and environmental consciousness of the general population must have contributed to this development.

Further studies would be required, to include dispersion modelling and particulate analysis to enable the determination of the sources contributing to current dust levels. Efforts could then be targeted at such sources to further reduce ambient particulate levels within the Tarkwa township.

ACKNOWLEDGEMENTS

The authors gratefully acknowledge: EPA-Ghana, for providing the equipment; Messrs Kweku Zinatunor, Kobina Gyan Budu and Nathaniel Akuayi for their assistance in data collection.

REFERENCES

EPA-Australia (1995), Mine Planning for Environmental Protection, One module in a series on Best Practice Environmental Management in Mining, Australian Federal Environment Department.

Kese, G. O. (1985), The Mineral and Rock Resources of Ghana, Balkema Rotterdam, Boston.

Environmental Issues and Management of Waste in Energy and Mineral Production, Singhal & Mehrotra (eds)
© 2000 Balkema, Rotterdam, ISBN 90 5809 085 X

A multicriterial geographical approach for the environmental impact assessment of open-pit quarries

P. Berry & A. Pistocchi
Department of Chemical, Mining and Environmental Engineering, University of Bologna, Italy

ABSTRACT: The paper shows an application of distribute modeling techniques using a GIS package, together with multicriteria analysis (MCA) for decision support, in the environmental impact assessment of asurface quarry in Tuscany. A comparison is drawn with more traditional techniques now widely in use, such as matricial lumped judgements, thus highlighting the advantages of coupling a geographical approach with MCA in order to minimize the bias due to subjective evaluation.

1 INTRODUCTION

In quarrying activities, strategic impact assessment (SEA) tools are not very commonly used at present. In addition, most environmental impact assessments[1], [2] use simple, synthetic matrix methods[3], strongly influenced by the subjective judgement of the analyst. According to a broad class of these methods, the impact score is assigned through a single score representing the judgement about the effects of an action on the environment. As a consequence, the full range of intensities of the effects, and their spatial distribution in the study area are lost[4].

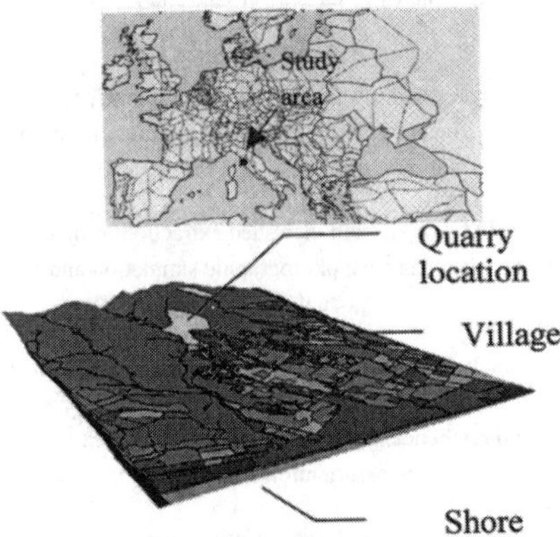

Figure 1

It seems a major issue to characterize the spatial features of the effects due to a quarry, both for a better choice of its location, and for the mitigation of its impacts in the areas where they're more intense, given a geographical location of the activity.
The advances in information technologies during the last decades have made widely available GIS tools that allow to cope with such spatial analysis issues, by forming up a spatial decision support system (SDSS). The construction of a SDSS asks for integration of [5]:

a) A database management system (DBMS);
b) A digital cartographical representation of data;
c) A mathematical model to describe all relevant physical phenomena deriving from choices assumed in the planning process.

Traditional distributed models try to describe reality through a set of complex differential equations to be integrated numerically (using i.e. finite element or finite difference techniques); on its side, GIS-based modeling uses close-form, simple mathematical relations, capitalizing on elementary spatial features of the variables, such as distance, contiguity, gradients and variations, within the conceptual and formal computing environment of map algebra[6].

2 DESCRIPTION OF PHYSICAL PHENOMENA, AND IMPACT ASSESSMENT

It seems pretty useful to point out that it is common practice to draw impact assessments directly based on the planned actions, while a more appropriate approach to impact assessments would require evaluation of the effects of such actions before defining their impacts on the environment. It is in fact quite clear that the evaluation of effects is very often easier to be achieved in objective, or at least shareable terms, while a more subjective judgement, also more difficult to be shared, lies in the assignment of impacts.

Table 1 shows the main effects of open pit quarrying activities on the environment, and helps to highlight how available methods for the description of these effects are quite limited in number, and in cases where no single 'objective' method exists also – as in economical evaluations-. It is relatively easy to achieve a shareable perspective of the actions on the effects themselves.

The judgement on impacts, on its side, always introduces subjectivity due to the ethical choices of the evaluators, the social context and political addresses of the decision makers, and the force equilibria between opposite parts. At present, it must be said

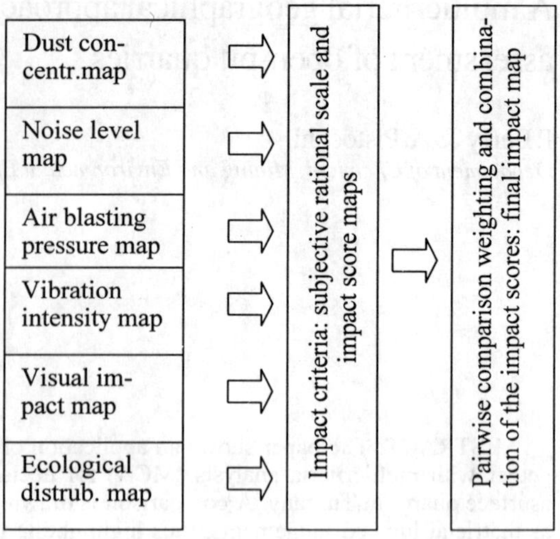

Dust concentr.map
Noise level map
Air blasting pressure map
Vibration intensity map
Visual impact map
Ecological distrub. map
Impact criteria: subjective rational scale and impact score maps
Pairwise comparison weighting and combination of the impact scores: final impact map

Figure 2

Table 1 – Physical effects of quarrying activities on the environment

Issue	Physical effects	Available methods for the description
Noise	Noise production by excavation and transport; air blasting	Empirical distance-based algorithms, analytical models or numerical integration of the wave equation (for non-impulsive sources); acoustical equivalence modeling for impulsive sources
Air pollution	Production and dispersion of particulate	Box models, gaussian models, lagrangian or eulerian integration of the dispersion-advection parabolic equation
Surface- and groundwater	Hydrological and hydromorphological effects; increase in sediment load to streams; potential pollution due to wastewater from mineral treatment plants	Empirical models(e.g. Universal Soil Loss Equation);mass balances; distributed physical models in hydrology; quality models in surface and groundwater (e.g. QUAL2E, MT3D)
Lithosphere	Slope instability in waste disposal and excavation sites, induced slope instability due to morphological modifications; vibrations	Geomechanical models for slope stability (probabilistic, deterministic); empirical vibration propagation models (e.g. IRSM)
Landscape perception	Visual impacts due to excavation fronts and plants	DTM analysis and viewshed extraction; projective geometrical analysis; photographic simulation and virtual reality
Landscape ecology	Ecological disturbance due to modifications in landscape pattern	Analysis of landscape ecological patterns and their modifications based on synthetic indicators;disturbance and resilience predictive modeling; multicriterial transformation evaluation
Socio-economics	Employment and socio-economical induced effects	Ad hoc analyses, descriptive statistics.
Stock of natural resources	Reduction of local stocks (soil, geomass, biomass…)	Mass balances, ad hoc evaluations

that no single method can be indicated for optimal solution of multicriterial evaluation problems[7].

The environmental impact assessment case study hereby presented has been faced using:

- A set of descriptive-predictive models based on GIS analysis tools in order to define the effects of quarrying activities;
- A set of 'impact laws', in order to assign each pixel of the study area an impact score based on (subjective) judgement for each effect separately;
- A priority scale, so to assign weights of relative importance to the various elementary impacts. The weights are drawn from a pairwise comparison technique (Saaty [4]).

Figure 2 describes the overall flowchart of the study.

3 CASE STUDY DESCRIPTION

The above considerations have been kept in account during the construction of a decision support system for an open pit quarry in Tuscany, Italy (figure 1). Such open pit produces limestone.

The quarrying is performed through benching with a progressive rotation of the advancement direction, which allows mitigation of the morphological impacts on the area (a hilly region with elevation ranging 150-500 m a.s.l.), and gradual rehabilitation of the landscape during the quarrying itself.

Physical effects of the excavation can be listed as follow: noise, dust, ground vibrations and air blast, visual impacts, landscape ecological disturbance. Figure 2 shows the general structure of the study.

3.1. Noise

The most relevant noise sources are the mineral treatment plant, the different material transport phases, both within the quarry and around the cable line used for transfering limestone to the railroad downslope, for further transport.

A noise level map has been computed from the LeqA (equivalent noise level according to the standard weighting curve A [8]) of each elementary source, taking in account the attenuations due to different environmental factors [8]: geometrical divergence, attenuation due to air, ground, and diffraction. Equivalent noise level (dBA) is given by:

$$\mathrm{LeqA\ [dBA]} = 10\log\left\{\sum_{i=1}^{n}\left[\sum_{j=1}^{8} 10^{0,1\left[L_{p(ij)} + A_{f(j)}\right]}\right]\right\} \quad (1)$$

Where:

n = n° of sources;

j = index for each of the 8 standard mean values of frequency (63 Hz - 8 kHz) [8];

A_f = weight according to curve A for frequency f [8];

Lp(i,j)= elementary pressure level.

Attenuation (dBA) is given by the sum:

$$\mathrm{Att} = \mathrm{Att}_1 + \mathrm{Att}_2 + \mathrm{Att}_3 + \mathrm{Att}_4 \quad (2)$$

where Att_1 represents geometrical divergence, Att_2 accounts for the effect of air, Att_3 gives the absorption of ground, Att_4 is used to take in account diffraction due to acoustical barriers. The following relations show:

$$\mathrm{Att}_1 = 20\log(r) + 11 \quad (3)$$

$$\mathrm{Att}_2 = a\, r/1000 \quad (4)$$

$$\mathrm{Att}_3 = \mathrm{Att}\,\sigma + \mathrm{Att}\,\mu + \mathrm{Att}\,\rho \quad (5)$$

$$Att_4 = 5 + 20\log\frac{\sqrt{2\pi|N|}}{\tanh\sqrt{2\pi|N|}} \quad (6)$$

Where r= geometrical straight line distance from the source, α = air absorption coefficient (dB/m), λ=wavelength (m); the terms in (5) depend on the distance from the source and can be found in all details in [8]; N is an index given by eq. (7) (where symbols refer to figure 3, representing a barrier between source S and receptor R, and A, B, r are segment lengths):

3.2. Dust production and dispersion

Dust production can be regarded basically as due to the movement of material in the quarry site and along the cable line from the quarry to the railway.

Dust distribution can be estimated using geostatistical techniques provided an appropriate number of sampling points and measures of dust concentration in air. For exploratory purposes, this case study chose to use literature values for dust production in cases very similar to the one in question. The estimates used for the case study are as follows: Highwall Removal- 80 kg/h; Material Loading and Movement- 25 kg/h; Stockpiling Material- 0.0275 kg/h.

In a first approximation dust dispersion has been studied using a box model, in order to obtain screening level indication of the atmospheric dilu-

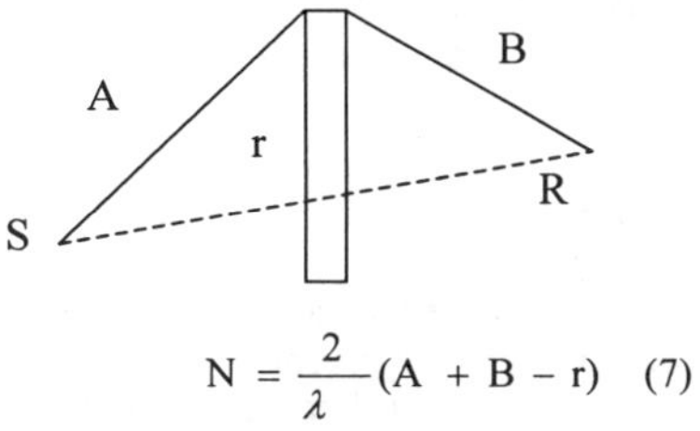

$$N = \frac{2}{\lambda}(A + B - r) \quad (7)$$

Figure 3

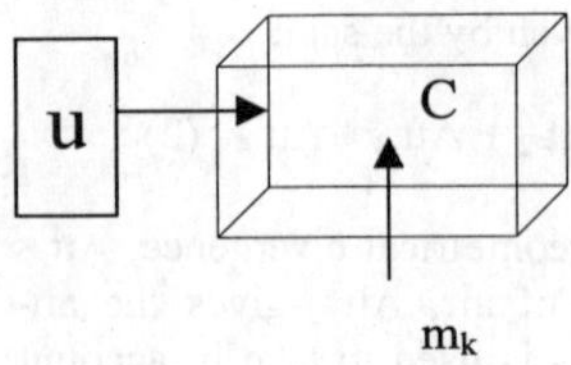

Figure 4

tion processes, using the following hypotheses:

- The system is assumed in simplified form as a regular volume of atmosphere where a mass input rate m_k of pollutant A_k is prescribed; the volume expands isotropically in all directions;
- Air in the volume is considered completely mixed (air density is constant, as well as dust concentration);
- Wind velocity u is constant, and directed perpendicular to one side which is long B (a first estimate of u = 3 m/s has been assumed on the basis of general meteoclimatological considerations over the area, while experimental surveys are being carried at present);
- Stationarity is assumed about overall fluxes of dust across the volume boundary (mass input rate equals gravitational deposition);
- There's no chemical reaction;
- Base level of dust concentration is zero.

Computing for the volume in figure 4 the mass balance, and the concentration (in µg/m^3) results:

$$\rho_k = \frac{m_k}{u * B * H} \quad (8)$$

where B and H are the base side and the height of the volume, or mixing height (in m).

If we assign B as the distance from the source, and we assume the mixing height as H=100 m above ground surface (considering that meteorological surveys on atmospheric stability were not available for the case study), equation (8) allows us to evaluate, at a first screening level, the atmospheric dilution effect which dust is subject after production to.

It is apparent how the simple model discussed here leads to an isotropic prediction of dust concentration. This model needs to be replaced with a more detailed one, once data become available for a gaussian or higher approach.

3.3 Ground vibrations

According to most theoretical and experimental research, the best parameter representing the impact of vibrations is the maximum soil particle velocity at the structures' location. The maximum velocity criterion is assumed in many regulatory acts, as discussed in literature [10], and in the ISRM recommendations [11].

The experimental laws linking the distance from the explosion, and the mass of explosive in use, can be represented in exponential form as follows:

$$V_{max} = K \cdot R^{-n} \cdot Q^{b} \quad (9)$$

Where:

- K, n, b are parameters depending on the kind of explosive, explosion geometry and rock type;
- R is the distance from the explosion (m);
- Q is the explosive mass (kg).

Many values have been proposed for b and n[10]. Using the "preset distance" parameter, eq. 9 can be rewritten as:

$$V_{max} = K \cdot \left(\frac{R}{Q^c}\right)^{-b} \quad (10)$$

where R/Q^c is the preset distance, that will be indicated with D_s .

According to literature, c ranges from 0.33 to 1.

Existing regulations give the minimum safety distance R as a function of the preset distance D_s, and the explosive mass, Q:

$$\frac{R}{Q^c} > D_s \quad (11)$$

Coefficients K and b can be expreimentally valuated or taken from literature. Table 2 gives the values of K and b assumed for each velocity vector component, in the case of hard limestone.

Land can be classified according to land use, as in [13], so that a specific impact can be assigned to vibrations in dependence on the building type vulnerability.

3.4 Air blast

Explosions also generate air blast, which can be linked with high acoustic disturbance. Process analysis is very complex, and theoretical difficulties emerge in describing the sound pressure peak, due to the impulsive nature of the acoustic wave source, that makes the phenomenon clearly different from previously treated noise aspects. The following empirical equation has been proposed [14, 15, 16, 17]:

$$P = K \cdot \left(\frac{D}{\sqrt[3]{Q}}\right)^{-n} \quad (12)$$

Where:
P = instant noise pressure (Pa);
D = distance from explosion (m);
Q = instant explosive mass (kg);
K, n = experimental coefficients.

In this case study, it has been assumed that atmospheric conditions are isotropic. In reality, many different anisotropic atmospheric features can affect propagation. The topic is undergoing further research, but the needs of environmental impact assessment can justify the use of this simplified model.

Table 2 – Coefficients for the vibration velocity equation [12]

Velocity component	B	K
V_v	1.553	37.042
V_l	1.632	68.905
V_t	1.654	56.517

3.5 Landscape ecological disturbance

All previous phenomena have impacts over both the human and natural environment, but this section describes the specific effects given by the quarry on the spatial pattern of ecosystems in their landscape organization.

It has been chosen to evaluate an index (the Btc index, [18]) representing a measure of resistance stability of the landscape, prior and posterior to the quarry excavation. The Btc index, expressed in $Mcal/m^2/year$, can be computed from site specific biomass measures, respiration and primary gross productivity. For exploratory analysis the Btc index has been computed from literature values suggested for mean conditions in the most common ecological settings of southern Europe [18]. It must be reminded that site specific surveys at an initial qualitative analysis level might allow for a better index characterization according to a simplified methodology [18-bis].

Another point was to cope with gradual variation of the index across the ecological patch boundaries in the landscape. This has been managed with by dividing the patches into a core and a buffer, and prescribing a linear variation of the index between two adjacent patch values. In this way, a Btc map prior and posterior to the quarry has been computed, and the Btc value variation has been assumed as an indicator of the impact.

3.6 Visual Impacts

Visual impacts have been evaluated from the ratio between the total area of the quarry visible from each point, and the human visual field in the direction of the quarry. The total visible area of the quarry has been computed through a viewshed analysis [19], after dividing the quarry front into constant height vertical stripes, and summing up for each pixel the indicator variable (0/1) derived from the viewshed analysis and referred to all stripes, weighted with the area of the stripe. The human visual field has been computed by assigning standard visual angles and using the distance of each point from the quarry.

4 EVALUATION AND COMBINATION OF THE IMPACTS

There is no unique rule to move from the indicators of impact (the intensity of the phenomena) to the impacts themselves. In this study, it has been chosen to assign conventional impact values to specified thresholds of the phenomenon intensity value, and to link these points in the (intensity, impact) plane with an 'impact law' [20]. Impact values have been assigned individually for each phenomenon in a scale ranging between 0 and 10.

As a general rule, the score 10 is assigned to regulatory values, since regulations for the phenomena in study are drawn keeping the safe side, thus assuming that being close to the limits gives rise to impacts but not yet to risk. The score 5 is assigned to guideline values and the score 0 is given to null intensity of the phenomenon. In cases where no reference regulation is available, or thresholds are senseless, the overall values have been normalized into the range 0-10.

As an example, the impact assessment of the dust production is reported. Italian regulation [21] gives the quality thresholds referred to dust concentration in atmosphere, that amounts for dust to 50 (guideline value) and 150 (imperative limit value) ug/mc, respectively.

Impact scores are assigned considering that dust concentration values should not overpass the guide value (50 $\mu g/m^3$), assumed already present in the atmosphere.

The impact scores assigned to relevant points are:

- Concentration =0 ug/mc: impact score= 0
- Concentration =50 ug/mc: impact score= 5
- Concentration ≥ 90 ug/mc: impact score= 10

As noticed here, impact is maximum when concentration approaches the imperative limit value.

Table 3 – Impact indexes for dust concentrations

Impact index (y)	Dust concentration due to quarrying works(μg/m³)	Presumed total concentration (μg/m³)
0	0	50
5	50	100
10	≥ 90	≥ 140

The scores between two points have been drawn using the best-fit linear regression curve over the points. In the specific case, the impact law assigning the score y to concentration x is:

$$y_p = 0.0003 \cdot x^2 + 0.0861 \cdot x \quad (13)$$

All elementary impact scores have been computed in an analogous way.

The overall impact has been extracted via weighted sum of the scores, using the Analytical Hierarchy Process [23] multicriteria decision analysis technique, performed as well known via pairwise comparison of alternatives.

According to this approach, a reciprocal matrix is constructed, assigning to position (i, j) an integer value between 1 and 9 (Table 3) according to relative importance of impact i over impact j. When impact j is more important than impact i, position (i, j) receives the reciprocal of position (j, i). For the case study, the matrix is shown in table 4.

It can be shown [22] that the vector of weights is the normalized eigenvector associated to the only non-zero eigenvalue of the matrix, that was in this case $\lambda_{max} = 6.1378$

The obtained weights are:

- Noise: 0.1630;
- Dust: 0.0816;
- Air blast: 0.0500;
- Vibrations: 0.0326;
- Visual Impacts: 0.4271;
- Ecological Disturbance: 0.2457.

In order to check for the consistency of the judgement, the consistency index can be computed [22]: this must be less than 1 and as close to 0 as consistency increases. In case this index is greater than 1, inconsistencies in the judgement cannot be disregarded and the matrix must be re-examined. The consistency index for the matrix shown in table 6 is $\mu = 0.0276$, that is a very low value. The pairwise comparison technique has been chosen among the many existing multicriterial methods because of its simplicity and the possibility to get an easier and wider agreement of the stakeholders in this elementary stepwise judgement, with respect for other alternative ranking methods.

Table 4 – Pairwise comparison impact scores

Level of importance	Definition
1	Same importance
3	Moderately higher importance
5	Higher importance
7	Very higher importance
9	Absolute predominance
2,4,6,8	Intermediate values

5 THE USE OF IMPACT INDEXES IN DECISION MAKING

The aim of an environmental impact assessment is to detect critical areas of the region where the effects of the project are more intense. A major issue is the use of impact indicators in driving the design of mitigative actions and their location.

This allows to chose, on a participative and negotiative basis, the more sustainable alternative and its implementation modalities. In this attitude, the simulation of the material phenomena deriving from different actions allows us to draw the indicators for many design alternatives.

Figure 5 shows the mitigation of impacts obtained with the construction of a noise and visual barrier, and is the result of the previously described computations, including in the input data the presence of the barrier itself.

6 CONCLUSIONS

The use of GIS based modeling techniques for impact assessment supports optimal exploitation of:

a) Knowledge of distributed parameters;
b) Predictive modeling of the different case-specific phenomena.

GISs form the ideal computing environment due to their capability in the management of distributed variables through the simple conceptualizations of map algebra, rather than sophisticated and detailed data- exigent numerical models [24].

Further improvements in the case study here presented will concern the integration in the procedure of a more complex algorithm for dust dispersion.

Table 5 – Pairwise comparison reciprocal matrix

	Noise	Dust	Air blast	Vibrations	Visual Impacts	Ecological disturbance
Noise	1	3	4	5	1/4	½
Dust	1/3	1	2	3	1/5	1/3
Air blast	¼	1/2	1	2	1/7	1/5
Vibrations	1/5	1/3	½	1	1/9	1/7
Visual Impacts	4	5	7	9	1	2
Ecological disturbance	2	3	5	7	1/2	1

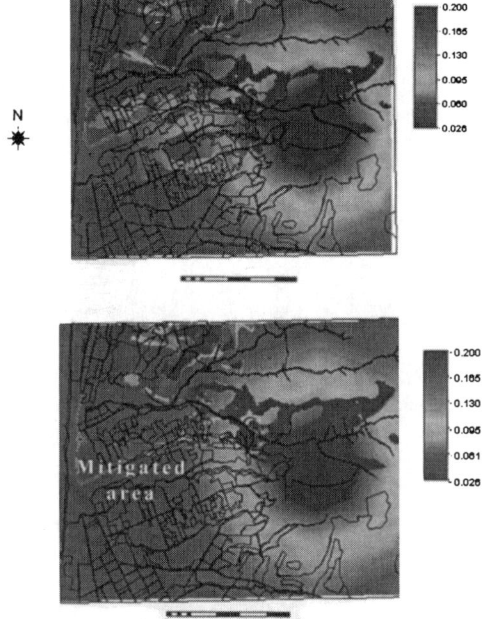

Figure 5

At present, the procedure allows a complete description and characterization of the impacts of a quarry over its context, and it constitutes an expert-system-like environment in which further improvements can be easily integrated in an open and easily adaptable structure which is conceived as a support for social negotiation and rational discussion of the alternatives both at a technical and a political level.

7 ACKNOWLEDGMENTS

The authors wish to acknowledge Eng. Dante Neri, during whose graduation research the case study was developed under their supervision. Dr. Giovanni Gambini, a geologist at Solvay Italia Inc. and the technical correspondant from the studied quarry, is also greatly acknowledged for his help in data acquisition.

8 REFERENCES

[1] Morris, P., Therivel, R., Methods of Environmental Impact Assessment, UCL Press, London, 1995

[2] Vanclay, F., Bronstein, D., Environmental and social impact assessment, J.Wiley & Sons, Chichester, 1995

[3] Prezioso, M., La base geoeconomica della Valutazione di Impatto Ambientale, Pacini, Pisa, 1995

[4] Malczewski, J., GIS & Multicriteria decision analysis, J.Wiley & Sons, New York, 1999

[5] Ye, Z., Maidment, D., McKinney, D. , Map-based surface and subsurface flow simulation models: an object-oriented and GIS approach; CRWR Online Report 96-5, Austin, Texas, 1996

[6] Burrough, P.A., Opportunities and limitations of GIS-based modeling of solute transport at the regional scale; in Applications of GIS to the Modeling of Non-point source pollutants in the vadose zone, SSSA special publication No. 48, 1996.

[7] Munda, G., Multicriteria Evaluation in a Fuzzy Environment, Physica-Verlag, Heidelberg, 1995.

[8] Norm ISO 9613 – 2: Acoustic attenuation of sound during propagation outdoors – general methods of calculation, International Standard 1996

[9] Italian Parliament - Legge 26 ottobre 1995, n. 447, "Legge quadro sull'inquinamento acustico"

[10] Cancedda, A.; Dantini, E.M.; Carastro, M.; Scavo subacqueo nel porto di Olbia. Problemi di sicurezza connessi con le volate. Quaderni dell'IAM, ottobre, 1980 ed. Istituto di Arte Mineraria, Facoltà di Ingegneria, Università di Roma "La Sapienza"

[11] ISRM, Raccomandazioni per il monitoraggio delle vibrazioni indotte dal brillamento di cariche esplosive, Ed. sc. Italiane, Rivista Italiana di Geotecnica, anno XXVIII, n°4, ott. – dic. 1994

[12] Dantini, E.M., Carastro, M., Onde sismiche dovute a volate in cava: ricerca sperimentale di una distanza di sicurezza, L'ingegnere, novembre 1976.

[13] Norm DIN standard 4150, "Threshold values for the vector of maximum vibration velocity components according to different structural types"

[14] Kinney, F. K.; Graham, K. J., Explosives Shocks in Air, Springer – Verlag, N.Y., 1985

[15] Baker, W. E., Explosions in Air, University of Texas Press, Austin, Texas, 1973

[16] Granstrom, S. A., Loading Characteristics of Air Blasts from Detonating Charges, Handlingar n° 100, The Royal Institute of Technology, Stockholm, Sweden, 1956

[17] Berry, P., Dantini, E.M., Stima delle sovrappressioni in aria generate da volate in galleria, Quarry and Construction, XXXII, n° 12, pp. 79 – 87, 1994

[18] Ingegnoli, V., Fondamenti di ecologia del paesaggio. Studio dei sistemi di ecosistemi, CittàStudi, Milano, 1994

[18-bis] Ingegnoli, V. (Ed.), Esercizi di ecologia del paesaggio, CittàStudi, Milano, 1996

[19] Patrono, A., and Saldaña, A., Modeling with neighbourhood operators, ILWIS 2.1 application guide, ILWIS Department-ITC, Enschede, 1997.

[20] Vismara, R., Ecologia applicata, Hoepli, Milano, 1992

[21] Italian Parliament - DPR 24 maggio 1988, n. 208, “Attuazione delle direttive CEE concernenti norme in materia di qualità dell’aria, relativamente a specifici agenti inquinanti, e di inquinamento prodotto dagli impianti industriali, ai sensi della legge 16 aprile 1987, n.183”

[22] Saaty, T. The Analytical Hierarchy Process for decision in a complex world, RWS Publication, Pittsburgh, 1980

[23] Saaty, T.L., A scaling method for priorities in hierarchical structures, Journal of mathematical psychology 15, 234-281, 1977.

[24] Pistocchi, A., Sistemi informativi geografici e pianificazione delle attività estrattive, Quarry and Construction, XXXVII, n° 6, 1999

[25] Ballestrazzi, P. e E. Imolesi, *Criteri di stima dei rischi ecologici dell'attività estrattiva: analisi costi benefici e valutazione d'impatto ambientale*. Quarry & Construction, XXII, n°5, 1996

Environmental Issues and Management of Waste in Energy and Mineral Production, Singhal & Mehrotra (eds)

Stability analysis of an abandoned open pit mine for landfill design

R.Ciccu, M.Costa, P.P.Manca & A.Muntoni
Dipartimento di Geoingegneria e Tecnologie Ambientali, Università di Cagliari, Italy

ABSTRACT: The paper discusses the results of a stability analysis carried out in a mining area that has been considered for the disposal of the slag produced in a nearby smelter. Although the site is suitable for this purpose owing to the presence of an abandoned pit having a potential storage volume of about 1.5 million tonnes, some risks of potential instability have been identified, related to the steepness of the pit slope, often consisting of poor and weathered rocks and to the presence underground openings below. The vertical displacement of the fill layer to be placed inside the pit for the sake of bottom and walls shaping represents a further problem. The stability has been assessed through the limit equilibrium and stress-strain analyses. The possibility of increasing the safety conditions and safeguarding the integrity of the landfill barrier system has been proved through an iterative check of landfill design in face of the results of stability analysis. The implementation of numerical computer models showed that the critical aspects are the shear and the bulk modulus of the backfill material. A special attention has also been devoted to the subsidence history of underground openings and to the predictable displacement of the loose materials involved (backfill, shaping and barrier layers, disposed of slag).

1. INTRODUCTION

The landfill regulations and design criteria are defined in compliance with all the environmental issues affected during the operative and aftercare phases. From this point of view, a surface cavity offers the best opportunity concerning the embankment and waste stability, the impact on landscape and the water circulation control. Nevertheless, in case of an artificial depression produced by mining, either underground or at surface, further attention is required due to the specific peculiarity of the excavation activity. The presence of large underground openings, the dip and the weathering conditions of the rock slopes and the constrains regarding the shape of the landfill bottom could require special solutions technically demanding and even economically burdensome.

The present paper discusses the problems encountered and the solutions adopted for the construction of a smelter slag disposal system into an abandoned open pit mine in the South West of Sardina (Italy), 40 m high, 450 m long and 150 m wide at the centre of the crest, having a total storage capacity of 1.5 Mt. The open pit stretches along a North-South direction according to the elongation of the exploited orebody.

The ore mined consisted of interlocked sulphide minerals (chiefly sphalerite and pyrite with minor accessory components) and the mining activity lasted over 15 years starting from 1964 with the discovery of the deposit. In 1977 an underground development was tried, but it was soon abandoned due to support and safety problems. The exploitation was completed at surface until reaching the limiting waste-to-ore ratio in 1980.

The open pit before undertaking the landfill construction is shown in figure 1.

Today the area is characterised by a considerable deterioration of the land: the presence of waste dumps and pyrite stockpiles, a local weathering of the open pit walls, the release of large amounts of mine drainage waters and the roof collapse of underground openings are the most important consequences of the past mining activity on the environment. The quality of the waters is characterised by an extremely low pH (roughly 2) and by a high content of heavy metals (mainly Pb and Zn).

Therefore the operations required by the construction of the landfill will improve the environmental quality of the site since the project also envisages the reclamation of the rock dumps, the removal of the polluting stockpiles and a strict control of the water flow to and from the landfill area.

The bottom of the landfill will be shaped according to an average slope downstream varying between 1 and 3%, in order to allow the final collection of leachate at the Southern end of the area. The slope angle of the cross sections after walls profiling will be around 25 °.

A stability analysis has been carried out aiming at elucidating the following aspects:

- the conditions of the open pit walls regarding the final stability of the slope and the safety conditions during the construction of the landfill system and the process of slag disposal;
- the subsidence of the landfill bottom in order to prevent damages to the lining and draining system and to avoid the risk of a sudden collapse of the stored mass.

The investigation plan consisted of a geomechanical characterisation of the site through a geological, structural and geophysical survey and the application of a stability analysis aimed at suggesting the suitable measures in order to overcome the above mentioned risks.

Concerning the procedures adopted for this study, the limit equilibrium model has been used for assessing the stability conditions of the cavity walls, while the stress-strain method has been adopted in the case of the roof underground and of the filling material to be placed at the bottom of the pit.

The development of the stability analysis required the preliminary gathering of a broad information through site inspections and survey: geological, geophysical, topographic, hydrogeologic and structural data, as well as mine maps and documents.

2. GEOLOGICAL FEATURES

The geology of the site is characterised by silty sandstone with thin bedding and dolomite partially oxidised. The lead and zinc mineralisation is generally appears near the bottom of the pit, between the hosting formations. Owing to different phases of tectonic stressing, the structural features of the rocks are represented by faults with main N-S strike, which can be associated with the cataclastic formation derived from sandstone and siltstone.

3. DESIGN CONCEPT

The solutions to the potential instabilities must be found in order to control the risk of collapse and to avoid unacceptable displacements, in compliance with the basic requirements of landfill design. In particular, storage capacity, leachate drainage, stability and integrity of the lining system and economic aspects must be taken into account in order to make the best decision.

The strict correlation between the stability analysis and the design requirements is shown in the flowsheet reported in Figure 1.

Figure 1. The Genna Luas open pit. The rock dumps and the pyrite stockpile in the background.

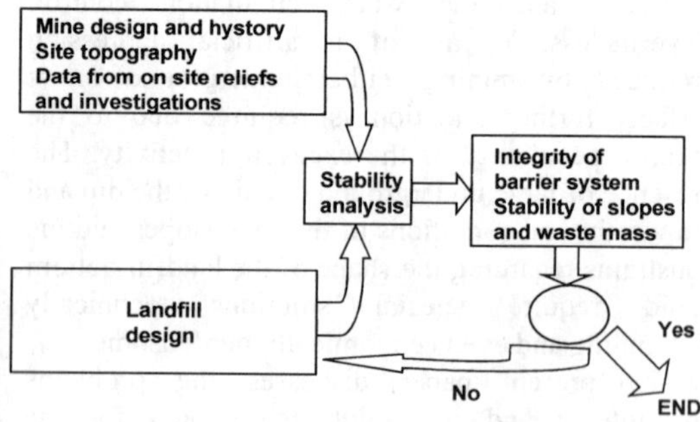

Figure 2. Flowsheet of the links between instability and design solutions

The landfill design envisages a multiple-layer barrier system, lining and drainage, including clay and synthetic materials (HDPE geomembranes, geotexiles, geocomposites).

The integrity of the leachate drainage system must be preserved against the possible vertical displacement resulting from the deformation of the system as a whole, including the subsidence of underground openings and the compaction of the loose materials.

It has been found that the hydraulic backfill, even partial, of the mine openings underground is capable of preventing the roof rock from caving-in, while a final slope angle of 35° could resolve the problem of wedge instability at the open pit walls.

In fact the final slope of the cavity before placing the waterproof linings will result from:

- the removal of the weathered and poor rocks by blasting, ripping and mechanical excavation in the protruding portions of the present pit;
- the filling of the hollowed sides by means of appropriate crushed material placed by horizontal layers and compacted every 15 cm.

However the connections between drainage elements could be interrupted as the consequence of an angular distortion at the landfill bottom, in addition to the fact that the lining elements in the sloping sides could be torn by a possible landslide of the rock walls.

4. SLOPE STABILITY

According to the limit equilibrium analysis suggested by rock mechanics (E. Hoek & J. W. Bray, 1977), an interpretation of the structural data has been made aiming at pointing out the kinematics of rock blocks and the safety factor related.

The structural features of the rock massif have been identified at 17 distinct zones or wall portions, equally distributed along the 450 m of total axial length of the open pit and for each of them the stability analysis has been carried out.

Structural data together with the wall orientation of the open pit are shown in Table 1. Four distinct discontinuity families have been put into evidence. Their dip (D) and dip direction (Dd) are reported in the same table.

The geological survey enabled to single out the presence of dolomite, locally weathered, sandstone and pyrite, this last often deeply oxidised. All these lithotypes are more or less widely fractured. Clay is also found near the Northern end of the pit.

The kinematics of potential instability was predicted on the basis of the stereographic projection of the discontinuities, by evaluating the possible occurrence of wedge sliding or toppling.

Taking into account these main types of slope failure, the risks of instability shown in Table 2 have been put into light.

Wedge stability analysis was performed using the limit equilibrium method at 7 vertical cross sections in critical zones along the open pit slope.

Table 1 - Structural data with dip and dip direction (D/Dd) of the pit slope and of the identified joint families

Stat	Slope	F1	F2	F3	F4
	D/Dd	D/Dd	D/Dd	D/Dd	D/Dd
01	80/100	53/041	84/153	-	90/338
02	42/316	65/055	87/148	-	45/290
03	34/121	51/066	87/148	74/217	45/290
04	40/125	70/064	72/122	74/217	68/292
05	26/086	62/058	64/122	89/346	86/086
06	50/125	60/067	65/103	60/200	78/356
07	38/090	59/062	65/117	-	85/350
08	40/090	58/055	61/113	-	-
11	50/106	78/053	61/131	-	-
12	37/126	62/080	38/096	60/186	-
13	26/351	52/071	48/131	-	-
14	61/275	60/010	71/109	85/180	80/276
16	31/272	65/041	78/105	75/238	75/350
18	34/290	-	-	80/240	82/294
19	50/273	75/071	-	90/260	59/304
20	45/226	25/043	-	-	85/291
21	41/230	-	70/155	85/291	60/315

Table 2 - Stereographic interpretation of structural data (Kinematic analysis).

Station	Toppling	Wedge	
	Tension cracks	Intersecting Systems	Sliding D/Dd
01		F1/F2	50/070
02		F1/F4	32/342
03	F4	F1/F3	25/135
04	F4	F1/F3	36/139
06		F1/F3	34/103
12		F1/F2	19/159
14	F2		
16	F2		
19	F1		
21		F2/F4	20/237

The study was carried out in both dry and wet conditions to verify the detrimental effect of water on the safety factor. The friction angle was assumed taking into account the different characteristics of the joints.

For all the cases examined the safety factor resulted to be equal or higher than 1.33 in dry conditions, except for station 01, as shown in table 3. In this last case the wedge instability is chiefly affected by the steepness of the sliding plane. The value of cohesion was considered zero.

Table 3 Results of the stability analysis by limit equilibrium method

Station	Friction	SF dry cond.	SF wet cond.
01	30°	0.70	0.37
02	30°	1.33	0.91
03	25°	1.97	1.03
04	25°	1.71	1.34
06	30°	1.38	0.97
12	25°	4.16	1.64
14			
16			
19			
21	25°	2.79	0.87

5. STRESS STRAIN ANALYSIS

As the first step of the study, a geomechanical characterisation of the site has been made on the basis of the results of the structural and geophysical survey. The results obtained in terms of RMR (Rock Mass Rating) index (E. Hoek et al., 1995) are presented in Table 4. Then the mechanical properties of the rock have been evaluated as shown in Table 5.

As the second step, the behaviour of the rock was simulated using the FLAC 2D code (Fast Lagrangian Analysis of Continua, 2 Dimensions, based on finite difference method) for two parallel cross sections 160 m apart, representative of the worst situations according to the indication of the structural and geophysical investigation. To this purpose the entire 30 years history of the mining activity and the evolution of the site shaping works in progress have been considered.

Two numerical models have been applied to evaluate the stress-strain conditions of the rock mass surrounding the underground openings (Ciccu and Manca, 1997) and of the fill material to be placed at the bottom of the cavity. The stability was checked taking into account the presence of the sub-level drifts underground and considering the layer of material (crushed rock) placed for smoothing the pit bottom, as a function of the load resulting from the slag disposal during he operation of the landfill.

Accordingly, the stability analysis was carried out in order to assess the vertical movements caused by each variation of load conditions due to:

- the development of past underground mining activity;
- the progress with time of the open pit geometry;
- the back-filling of the underground space and the shaping of the pit bottom;
- the process of slag disposal (up to 40 m height).

The hypotheses adopted for computer simulation (FLAC, 1998) are the following:

- continuum model of the material;
- planar strain and three-axial stress;
- elastic model for the pre-mining stress calculation;
- elasto - plastic model for the post-mining stress conditions and evolution (underground and open pit operations);
- elastic model for the materials used for filling the underground openings and the landfill bottom;
- boundary constraints of null velocity in y-direction at the bottom and in x-direction at both sides;
- original stress-strain state produced by gravity, the ratio between horizontal and vertical stress components being defined by the relationship: $\sigma_{xx} = \sigma_{zz} = 1/3\ \sigma_{yy}$;
- cross area of the study volume in the model equal to 350 x 200 m^2
- number of grid elements equal to 30.000;
- absence of underground waters due to the planned control of the water flow to and from the landfill area.

The mechanical properties of the rocks have been obtained from direct mechanical tests as well as from the geological, structural and geophysical characterisation. Data are reported in Table 5 together with those of the materials used for pit shaping.

Table 4. RMR index obtained in the main section of the landfill

Rock	RMR
Sandstone	55
Dolomite	50
Ore	30
Oxidised rock	20
Weathered rock	40

Table 5 – Mechanical properties of the rock mass and of the materials used for roof support and for landfill shaping.

Lithotype	Bulk density (kg/m^3)	Young Modulus, (GPa)	Poisson coeff.	Internal Friction (°)	Cohesion (MPa)	Tensile str. (MPa)
Sandstone	2650	15.0	0.25	35	5.0	1.0
Dolomite	2600	10.0	0.25	35	4.0	0.90
Ore	4000	1.0	0.30	30	1.0	0.10
Oxidised rock	4000	1.0	0.30	30	1.0	0.10
Weathered rock	2500	3.0	0.30	30	1.0	0.10
Backfill (fine sand)	1800	0.005	0.35			
Bottom fill	2800	0.050	0.30			
Waste slag	1800	0.050	0.35			

The volume at the bottom of the pit will be filled using crushed dolomite and sandstone coming from the slope profiling operation.

The pit shaping operations in progress are shown in Figure 3.

In order to evaluate the effects of placing the filling material at the bottom of the pit, a sensitivity analysis for the maximum shift was performed as a function of six different assumptions for the modulus of elasticity: 20, 40, 60, 80, 100, 120 MPa.

Results are reported in Figure 4.

Angular distortion for every 10 m along the bottom profile for each of the above mentioned values of the Young Modulus was estimated and the results are graphically reported in Figure 5.

6. RESULTS

The following considerations can be drawn:

- considering a slag bulk height of 40 m, the pressure on the landfill bottom will be 0.6 MPa;
- considering a cross section, a deformation asymmetric with respect to the pit slopes can be noticed;
- the wide underground openings close to the bottom of the pit can be considered stable and are not expected to cause any considerable subsidence;
- mining openings more than 6-10 m wide have already collapsed and, when completely filled, they may cause shifts of the order of 1-2 cm;
- the maximum vertical shift deriving from the deformation of the backfill placed underground is equal to 2 cm and does not depend on the characteristics of the material employed;
- the maximum vertical shift deriving from the deformation of the filling material used for shaping the bottom of the landfill is within 6 and 16 cm as a function of the modulus of elasticity ranging between 20 and 120 MPa;

the maximum angular distortion on the landfill bottom is within 1.8/1000-1.2/1000, the modulus of elasticity ranging between 60 and 120 MPa;

Figure 3. Work in progress for the construction of the landfill at the Genna Luas mine.

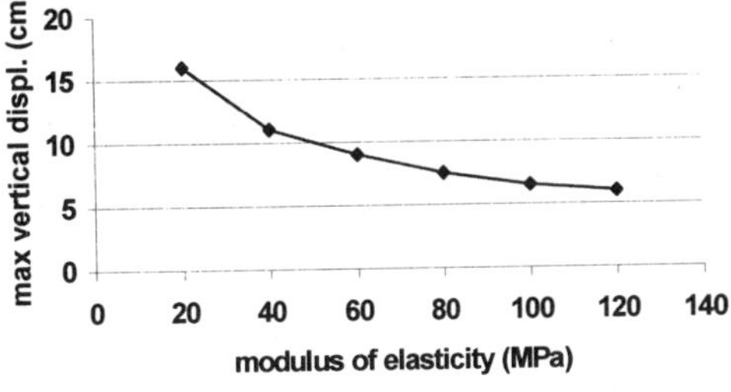

Figure 4. Maximum vertical displacements versus Elasticity Modulus

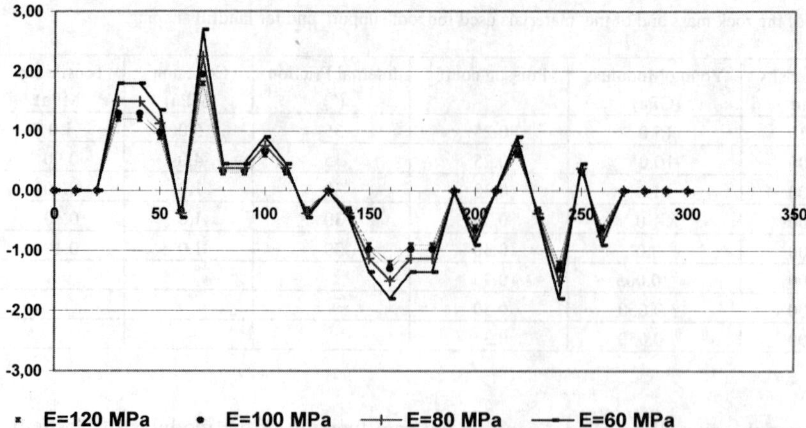

Figure 5. Angular distortion for every 10 m along the bottom profile for each 4 values of the modulus of elasticity.

The pit shaping operations in progress are shown in Figure 5.

CONCLUSIONS

The slope stability analysis and the computer simulation using numerical models confirmed that the slope angles suggested by the structural and functional specifications of the landfill fully meet the stability requirements too. The design foresees slopes with a dip angle always ranging between 25° and 30°, only in a few cases exceeding 35°. The related safety factor is always higher than 1.8.

Concerning the stability of the bottom, the following conclusions can be drawn:

- the first two sub-level drifts near the surface should be filled with a material having a modulus of elasticity equal to at least 5 MPa after placement and drainage; this value can be reached even by low compacted fine sands. The study shows the need for total and effective filling of the openings since their presence should lead to excessive vertical displacement; the need to avoid any residual space left empty underground must be highlighted;
- the material resulting from slope shaping can be used for bottom filling provided that the placement procedure ensures a final stiffness corresponding to a Modulus not lower than 80 MPa since these conditions allow a maximum angular distortion not exceeding 1.5/1000 (the suggested value for iron pipe with lead caulked joints is 4.0 x 10^{-3}, Madan M. Singh, 1992), consistent with the integrity of the bottom barrier system (clay, geomembrane and slotted pipes).

ACKNOWLEDGEMENTS

The authors would like to thank Enirisorse SpA for the permission to publish this paper.

REFERENCES

Ciccu R., Manca P.P. 1997. Disposal of smelter slag into a mined-out pit: A rock stability study. . *In Proceedings of Environmental and Safety Concerns in Underground Construction.* Lee, Yang & Chung (eds), Balkema, Rotterdam, ISBM 9054109106. pp. 41-46.

Madan M. Singh, 1992. Mine Subsidence. In SME Mining Engineering Handbook, 2nd Edition, Howard L. Hartman Senior Editor, SME, Littleton, Colorado, pp. 938-971.

E. Hoek & J. W. Bray, 1977. Rock Slope Engineering. The Institution of Mining and Metallurgy, London.402 p.

Flac Manual 1998. Version 3.4, Itasca Con., Minneapolis, USA.

E. Hoek, P. K. Kaiser, W. F. Bawden, 1995. Support of Underground Excavations in Hard Rocks. A. A. Balkema, Rotterdam. 215 p.

Environmental Issues and Management of Waste in Energy and Mineral Production, Singhal & Mehrotra (eds)
© 2000 Balkema, Rotterdam, ISBN 90 5809 085 X

Environmental problems at the underground limestone Baltar mine, São Paulo, Brazil

S.M.Eston, W.S.Iramina & W.T.Hennies
Escola Politécnica, Universidade de São Paulo, Brazil

R.T.Nakamura
S.A. Industrias Votorantim, São Paulo, Brazil

ABSTRACT: Baltar mine is the largest Brazilian underground limestone producer. The mining method is sublevel longhole stopping, with empty stopes averaging 110 x 40 x 200 meters. The first production panel has main haulage levels up to 150 meters of depth and ventilation problems are minor and easily solved. Development of ramps and access drifts for the second panel, at depth of 300 meters or more, has generated environmental problems related to fresh air quantity, dust and thermal comfort conditions. An environmental data survey was carried out and results indicated that the old ventilation system could not support further deepening of the mine, and thus new fans, main and auxiliary, are be required.

1 INTRODUCTION

Votoran Cement plant is located at Santa Helena, close to Votorantim town about 120 km from São Paulo. Limestone has been mined for decades and present cement nominal production is 7700 metric tons per day and limestone may be consumed up to a maximum of 4 million tons per year.

The three main mines are the open pits of Placa and Pastinho, and Baltar which is an underground mine.

The Baltar limestone deposit has a lenticular shape, northeast direction and dips to the southeast. Mine openings started in 1975 and production began in 1981. The initial mine layout was defined by Outtokumpu Oy company.

2 MINE LAY-OUT

The main access to the stopes is a ramp of 1850 m long, cross-sectional area of 7,5 m by 5 m and grade of 12,4%. This ramp has an entrance at level 616 and connects the surface to level 397, were several important operations such as crushing and equipment maintenance are performed. From this area a new ramp has been developed, with 1100 m long, cross-sectional area of 6,5 m by 5,5 m and inclination of 12,5 %. This secondary ramp is the main access to the second level of mining panels that will be mined in the next years.

Crushed ore haulage to the surface is done by belt conveyor, using a straight ramp that has surface entrance close to the main access ramp, as shown in Figure 1.

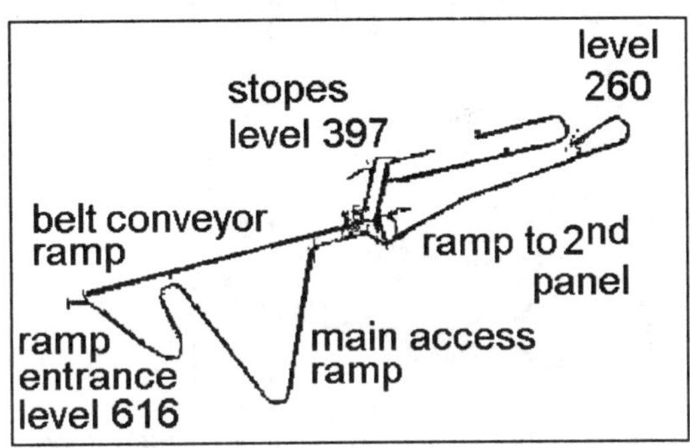

Figure 1 – Top view of main ramps of Baltar Mine.

3 MINING METHOD

Mining method is longhole stopping, with rooms averaging 40 m width, 200 m length and 110 m height.

Production drilling is performed in 3 sublevels drifts and pillars between rooms average 32 to 40 m. Explosive is ANFO and ore haulage is performed by front end loaders and 35 t trucks to feed the underground crusher. Figure 2 presents a sketch of a production stope.

The first panel is at an average depth of less than 200 m and is composed by several developing and production rooms. A second panel below the first one is being developed and production should start in a couple of years. The access ramp to this second panel is already completely opened.

4 UNDERGROUND ENVIRONMENTAL PROBLEMS

No serious environmental problem was encountered during the first panel development and production phases. However the secondary ramp development and the second panel will require a new ventilation plan since depth over 300 m will be attained. The main problems are associated with thermal comfort, gas removal and dust.

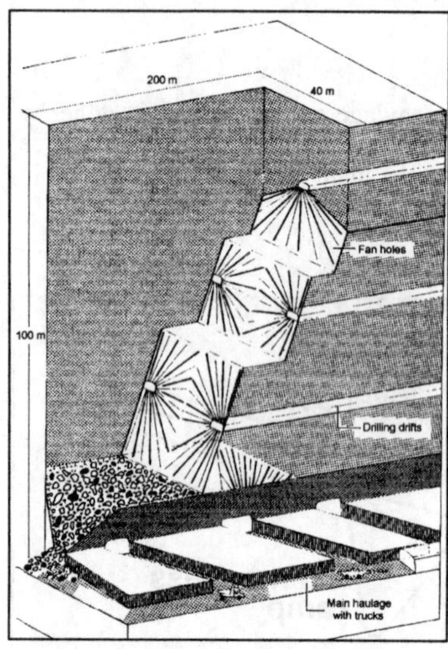

Figure 2 – Drilling and haulage at production stope.

A consulting company was hired to design the new ventilation system and the Mining Engineering Department of the Polytechnic School of the University of São Paulo was authorized to perform independent work to verify and compare the old and new conditions underground.

During the months of January through June 1999 dozens of monitoring visits were done, with several parameters being measured. These parameters included wet and dry bulb temperature, moisture, air velocity, pressure, cross-sectional areas of openings, gas and dust concentrations.

Several difficulties were encountered to analyze the ventilation old conditions, the main one related to the fact that fan operation constantly changed, due to tests that were weekly performed. An average condition was adopted considering 6 fans, 5 operating under exhaust conditions and one under blowing condition.

These fans were located at ventilation shafts and fresh air entered the first panel by the main ramp and one shaft, a condition considered inappropriate as soon as the secondary ramp giving access to the panel below began to be opened.

5 AIRFLOW

Table 1 summarizes the location and characteristics of the old system operating fans, all 6 being axial type fans manufactured by Higrotec-Chicago Blower.

The only fan operating under blowing conditions was installed at the shaft that connected the surface to the south end of level 530.

Four fans operated under exhaust conditions and were installed at 3 shafts connecting the surface to the north portion of levels 420, 500 and 530.

Table 1. Old system fan location and characteristics. Nominal airflow refers to expected values.

Shaft location	number	pressure (Pa)	airflow (m^3/s) nominal
N530 South	1	61	43,2
N420	2 (parallel)	741	27,8
N530 / N500	2 (parallel)	681	28,5
crush exhauster	1	246	30,6

Airflow distribution in the first panel was analyzed by surveying the air velocities at ramps, drifts and shafts, measuring velocities at 9 points in each section. Table 2 summarizes the airflow at the shafts and ramp.

A first set of measurements was taken in order to compare real air distribution with the nominal expected airflow (table 1). Expected total fresh air was 155,8 m^3/s and measured total fresh air is presented in table 2.

For the first panel the total fresh air influx was of 163,8 m^3/s, measured at main ramp entrance and blowing shaft. Total air exiting was measured at exhaust shafts and was 148,1 m^3/s. The difference in measured values was less than 10%, an acceptable result.

Expected flow value of 155,8 m3/s lied between both measured values, indicating that the old ventilation scheme was operating under reasonably good conditions.

Table 2. Measured airflow for old ventilation scheme for the first panel.

Location	airflow (m^3/s)	observation
access ramp entrance – P1	111,6	30% difference due to losses associated with connecting drifts to conveyor ramp
access ramp end – N420 - P2	81,8	
N420 north – exhaust 1	67,8	
N530 north – exhaust 2	55,4	
N500 north – exhaust 3	24,9	
N530 south – blower 1	82,0	

Measurements at main access ramp entrance and bottom indicated a difference in airflow of about 30%, due to leakage associated with drifts connecting the main ramp and the conveyor belt ramp. This is considered too much and leakage should be diminished.

A second set of measurements included psychrometric parameters and gas concentrations. All measurements indicated compliance with legal standards. The only exception was in the truck repair shop were NOx concentration would be above legal standards when all engines were working together.

Fresh air quantity which was adequate for the first panel turn out to be inadequate as soon as the secondary ramp to the second panel began to be developed. There were complaints about thermal comfort, dust and gas. The total fresh air delivered by the old scheme was clearly not enough for the mine expansion.

6 NEW VENTILATION SYSTEM

Due to problems with the old ventilation scheme a new one is being tested, involving 7 fans, all operating by exhaustion at several shafts.

Fresh air will enter the mine through two ramps, the main access ramp and a recently reopened tunnel (TAB). All polluted air will leave the stopes and rooms and will be directed to the closest ventilation shaft.

Table 3 presents the characteristics of the new ventilation system, as proposed by the contractors and which will operate with an estimated total airflow of 228 m^3/s. This is in accordance with the contractors estimate of the required airflow of 224 m^3/s.

Table 3. New ventilation system fans.

location	sublevels connected	shaft length (m)	number of fans (50HP)	airflow in one fan (m^3/s)	total airflow (m^3/s)
R4B north	420-600	210	2	32	64
R5B north	530-600	70	1	30	30
inv 1	530-600	70	2	32	64
inv 2	530-600	70	1	35	35
inv 3	530-600	70	1	35	35
total airflow in the min = 228 m^3/s					

7 FRESH AIR REQUIREMENTS ANALYSIS

The new required fresh airflow was analyzed according to the criteria involving combustion engines operating underground, dilution of the gas generated by detonation and thermal comfort at the deepest drifts and stopes.

7.1 *Combustion engines*

Airflow requirement calculation considered 60 m^3/s of fresh air for each megawatt of power underground. Table 4 presents the values obtained for each machine and the total calculated airflow of 236,4 m^3/s. This value is somewhat greater than the value adopted by the contractors.

Table 4. Air flow requirements due to combustion engines.

Machine	Power (MW) and number of machines	Total power (MW)	Fresh airflow required (m^3/s)
small truck	0,067 (2)	0,134	8,0
truck	0,082 (2) / 0,097	0,261	15,7
truck	0,324 (5)	1,62	97,2
loaders	0,280 (2)		
	0,127 (3)	0,929	55,74
Jumbo drill	0,056 (2)	0,112	6,7
auxiliary equipment	0,112	0,112	6,7
drill	0,043 (2)	0,086	5,2
new trucks	0,216 (2)		
/ drills	0,127 (2)	0,686	41,2
total required airflow = 236,4 m^3/s			

7.2 *Detonation gases*

Detonation generally occurs at lunchtime, between 12 AM and 1 PM or after 5 PM. Normal dilution time is 60 minutes but sometimes delays in blasting preparation makes necessary to dilute gas in 30 minutes.

Adopting the value of 0, 57 m^3/s of fresh air for each kg of explosive, with an average blast with 200 kg of explosive, an airflow of 114 m^3/s will be required. This value is far below the requirement due to combustion engines.

7.3 *Thermal comfort*

Measurements of psychrometric parameters at several locations in the first panel indicated that effective temperatures were all below 27° effective grades, confirming that thermal comfort was not critical at the first panel. Measurements in the second panel should start as soon as the first stope in this second panel is developed. Table 5 presents some psychometric parameters measured in the first panel.

Table 5. Example of thermal comfort parameters in the first panel.

Location	Tbn (°C)	Tbs (°C)	V air m/min	P kPa	moisture (%)	Tef o
N420 access ramp (ambulatory)	16,5	20	210,6	99,2	75,5	14
N420 –ramp split at access for levels N330 e 280	16	19,1	82,2	99,2	74,1	15
N420 – main haulage drift	17,4	21,2	115,8	99,1	74,9	17
access ramp entrance to level N460	18,6	20,7	31,8	98,3	78,3	19
level N500 entrance	18,6	21,4	51,6	98,1	78,4	18
level N530 entrance	18,2	22	27,6	97,8	70	20

8 CONCLUSIONS

The old ventilation scheme worked satisfactorily for the first panel of rooms and stopes. Expected and measured fresh airflow values were reasonably close and above fresh air estimated requirements. Workers complaints were not significant and thermal comfort and gas concentrations were below legal values.

The development of a secondary ramp to a second panel of stopes below the production ones indicated that the old ventilation system was not enough to meet gas, thermal comfort and dust concentration standards. Workers complaints corroborated the need for a new ventilation system.

A consulting firm proposed a new ventilation scheme with estimated fresh air requirement of 224 m^3/s and a fan system operating with 228 m^3/s.

Analysis of airflow requirements indicated that combustion engines would require at least 236,4 m^3/s, not considering thermal comfort in the second panel production stopes.

The authors suggest that the new ventilation system should operate with a larger fresh air influx in order to allow for leakage and future thermal comfort requirements. The new system should consider 250 m^3/s as minimum fresh airflow. Future monitoring will permit to check the second panel working conditions. Leakage should be decreased to below 10% in the main ramp.

REFERENCES

BRASIL. Ministério das Minas e Energia. Departamento Nacional da Produção Mineral. Ventilação em minas subterrâneas. Brasília, DNPM, 1986.

DASYS, A., HARDCASTLE, S.G; Mine ventilation management "expert system". [on line]. In: 6th Mine Ventilation Congress, Pittsburgh, USA, May 17-22, 1997. Disponível: http://icewall.vianet.on.ca/comm/eei/ minesim_doc.html. [Accessed July 21, 1998].

HARDCASTLE, S.G.; LEUNG, E.; DASYS, A., Integrated mine ventilation management systems. [on line]. Ontário, CANMET. Disponível: http://icewall.vianet.on.ca/comm/eei/ ventmgmt_doc.html. [Capturado em July 21, 1998].

HARTMAN, H. L. Introductory mining engineering. New York, John Wiley, 1987.

LACASEMIN TECHNICAL REPPORT MI/587 - 03/93._Análise da rede de ventilação forçada da mina subterrânea de Ipueira III, Lavra de Cromita, Bahia. 3o Relatório. EPUSP

VOTORANTIM: 5 km de túneis para chegar ao calcário. Minérios: Extração & Processamento. N. 18, julho de 1978. p. 24-8.

Environmental Issues and Management of Waste in Energy and Mineral Production, Singhal & Mehrotra (eds)
© 2000 Balkema, Rotterdam, ISBN 90 5809 085 X

Natural ventilation of a tunnel-shaft system

S.M.Eston & W.S.Iramina
Escola Politécnica, Universidade de São Paulo, Brazil

N.Lins
Shaft Consulting, Rio de Janeiro, Brazil

ABSTRACT: A tunnel-shaft system is under construction in the southern state of Santa Catarina, Brazil. Inside the underground openings a gas pipeline will be fitted, with length of 786 m in the tunnel and 321 m in the shaft.

Monthly maintenance crews will work at the bottom of the shaft and tunnel end, requiring adequate ventilation during working days.

Three alternatives are under analysis: natural ventilation, natural eolic ventilation and forced duct ventilation.

Decision on which system will be chosen is based on fresh air requirements, investment costs, maintenance costs, safety and reliability of the system that will not operate continuously. At present forced duct ventilation seems the most appropriate solution..

1 INTRODUCTION

Hydroelectric plants presently generate most of the Brazilian energy and about 92% of the nominal output are being presently used.

In the south and southeast river basins, were the main industrial and population centers are located, 80% of the hydrological potential power has already been used.

In the northern amazon region, over 54% of the hydrological potential still remains without use but their use are associated with a great variety of environmental impacts.

As an example, Balbina hydroelectric plant has methane emission associated with forest underwater degradation that contributes to global warming with the same intensity as a thermoelectric plant of the same size.

The Bolivia-Brazil natural gas pipeline (GASBOL) was designed to transport natural gas (NG) from the gas fields of Bolivia to several Brazilian southern states, as Mato Grosso do Sul (MS), São Paulo (SP), Paraná (PR), Santa Catarina (SC) and Rio Grande do Sul (RS). Figure 1 presents the main states that are crossed by GASBOL.

GASBOL and the Brazilian gas production from Merluza (SP) and Campos (RJ) fields will be distributed to other states such as Minas Gerais. The natural gas will be used in the industrial, residential

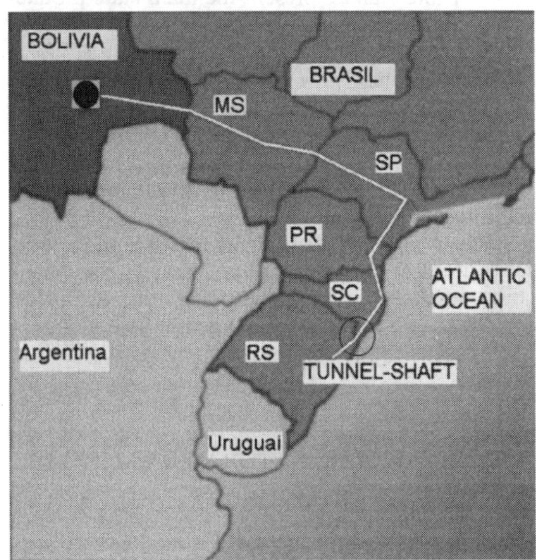

Figure 1. Natural gas pipeline from Bolivia to Brazil

and transportation areas and the GASBOL project is part of the federal government efforts to increase NG participation in the country total energy supply from 2,5% to 11,8%.

2 NATURAL GAS

Natural gas (NG) is a stable mixture in which methane (CH_4) is the main component, with a percentage that varies from 83 to 99%. It is a primary fossil fuel energy source such as coal or oil. It is extracted from underground deposits and often associated with oil fields. It is normally transported to consuming areas by pipeline or by ships.

It is an alternative energy to petroleum, with significant world proven reserves that can last for over 65 years. There is a large geographic dispersion of the deposits, which are located over more than 90 countries and the main ones are in Russia and in the Middle East.

In 1996 NG proven reserves in Bolivia were estimated to be of the order of 93,6 x 10^9 cubic meters, meanwhile possible and probable reserves could be over another 90 x 10^9 cubic meters.

Its main characteristics are low production costs, a clean burning and no toxicity. It has the cleanest burning of all the fossil fuels, with almost no generation of sulfur compounds and particulate ash and smaller emissions of oxides such as CO and NOx family. Table 1 and 2 present the main state producers and the uses of NG in Brazil.

Table 1. Natural gas production in Brazil (1996)

State	Production (10^3 m^3 / day)	Percentage (%)
Rio de Janeiro	9 979	39
Bahia	4 767	19
Rio G. do Norte	2 607	10
Sergipe	1 994	8
São Paulo	1 759	7
Alagoas	1 756	7
Amazonas	1 008	4
others		6

Table 2. Main uses of NG in Brazil.

Use	Quantity (10^3 m^3 / day)	Percentage (%)
Combustible	7 827	83,3
Petrochemical field	682	7,3
Reduction	383	4,1
Residential	378	4,0
Fertilizer	20	0,2
Automotive	108	1,1

Natural gas has however safety problems associated to asphyxiation, fire and explosion hazards. Normally it has no smell and sulfur compounds are introduced in very small quantities in order to facilitate leakage detection during production, transportation and distribution.

Since it is lighter than air, it may accumulate in higher areas of closed environments, and if concentration reaches the explosivity triangle, explosion may occur. If it burns in an oxygen deficient atmosphere, carbon monoxide may be generated which is very toxic.

3 BOLIVIA-BRAZIL NATURAL GAS PIPELINE (GASBOL)

GASBOL was designed to have a total length of 3 150 km consuming 540 x 10^6 t of steel pipes. The pipes were located underground, at an average depth of 1,2 m and welded together. It will cross four rivers, Rio Grande, Paraguai, Paraná and Tietê, and some very ecological sensitive regions as the Pantanal wetlands.

The NG will be compressed in four stations in Bolivia and in 12 stations in Brazil, travelling through 30 pressure reduction and monitoring stations. Two main control stations will be settled, one in Bolivia and one in Rio de Janeiro, with the whole operation being controlled with the help of satellites.

Pipeline diameter in Bolivia is 32 inches and final diameter in Rio Grande do Sul State is 16 inches. Initial operation will transport 8 million of cubic meters of NG per day, and in eight years, this volume may increase to 16 million cubic meters. Further on, this volume may increase up to 30 million.

Its construction generated debates regarding environmental impacts in wild life and indigenous areas, and defending arguments ranged from poverty diminution in Bolivia due to revenues (estimated at 200 to 300 million dollars) to decrease in São Paulo's atmospheric pollution.

The area focused in this paper was a tunnel-shaft system at the top of the Sierra Mountains at the border of the states of Santa Catarina and Rio Grande do Sul. The area is near the end of the GASBOL pipeline in the Rio Grande do Sul state, as presented in figure 1 with a circle.

4 TUNNEL – SHAFT SYSTEM

Figure 2 presents this underground system, were a tunnel of 786 m long is connected to a vertical shaft of 321 m of length. The GASBOL is located under the tunnel floor, at its left side, leaving the right side

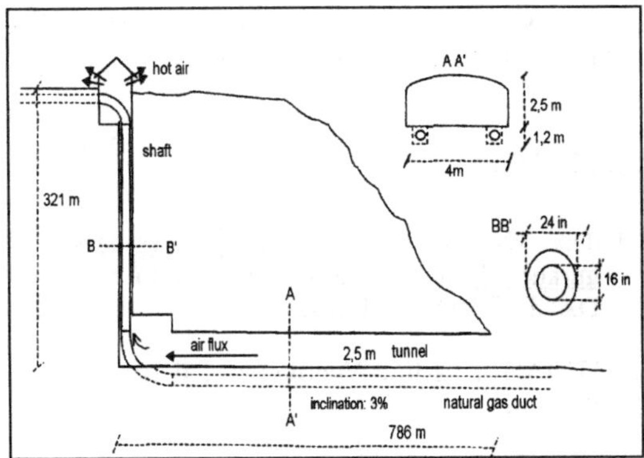

Figure 2. Tunnel-shaft system dimensions.

Figure 3. Tunnel main entrance with blower fan during construction

to a future system expansion. Figure 3 presents a picture of the tunnel entrance.

After the end of the construction, the contractor will remove all material and equipment. Maintenance and monitoring crews will enter the system every couple of months and the main problem was which ventilation solution would best satisfy engineering, economic and safety requirements.

Tunnel entrance could not be totally sealed, since leakage could fill the whole tunnel with methane and create a serious explosion hazard. Steel fences and a locker will protect the entrance.

Maintenance crews entering the tunnel must pay attention to any equipment or instrument that could ignite methane and must carry explosive gas monitoring instruments. Any leakage detection should imply cutting off GASBOL flux prior to any inspection and personal oxygen respirators should be used.

5 VENTILATION ALTERNATIVES

Three ventilation alternatives were analyzed during the lasts months of construction work. They were natural ventilation, eolic ventilation and forced duct ventilation.

Ventilation was first analyzed considering the shaft drilled in the rock and completely empty, having an inner diameter of 76,2 cm, a friction factor of 0,0167 kg/m^3 and a airflow resistance of 135,5 kg/m^7.

A second analysis was made considering the shaft filled with the GASBOL pipeline plus some structural fixing elements and other cables such as those carrying optic fibers. This filling will severely obstruct the shaft, increasing its resistance to over 2 x 106 kg/m^7. Figure 4 presents a top view of a steel pipeline with fixing elements around its perimeter.

The tunnel by itself was considered as having a relatively low resistance of 0,47 kg/m^7.

With these data it was clear that the shaft, after being filled with all planned elements would have such a high resistance that natural ventilation would be ineffective.

Brazilian federal mining laws require a minimum air velocity in all underground openings above 0,2 m/s, thus ensuring turbulent flow and gas mixing.

Measured parameters inside the tunnel were wet and dry bulb temperatures and humidity. They ranged from 18,2°C to 22,8°C, and from 63% to 92 %. Estimated natural ventilation pressure was 141 Pa, which would generate an airflow of 1,0 m^3/s

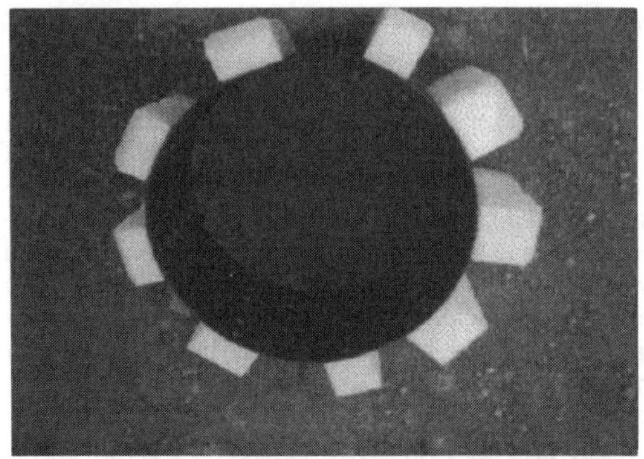

Figure 4. Gas pipeline with different size fixing elements unevenly distributed around its perimeter (top view).

(velocity of 0,12 m/s), for a tunnel-shaft system with the shaft free from obstructing elements.

In this situation air velocity would be close to legal requirements and airflow could be enhanced by the using some type of eolic exhaust fan.

Nevertheless for an air velocity of 0,2 m/s in the tunnel and with the shaft filled with airflow obstructions, the average shaft section would have an estimated area of 0,0046 m^2 imposing an air velocity in the shaft above 360 m/s. In practice this is unacceptable

For this reason forced duct ventilation was the proposed solution, even though it had higher costs than natural ventilation. A blowing system was preferred over an exhaust system for safety and economic reasons.

Being methane lighter than air it would tend to accumulate at the end of the tunnel and a pressure increase at that location would favor the gas flow up the shaft. A little methane trapping could occur inside the irregular shaft geometry and a turbulent flow could help decreasing this trapping.

Moreover a blowing system would require only a flexible plastic duct; cheaper and easier to transport and install in the at the tunnel which is located at a far distant mountain area.

Fan should be operated for a minimum air velocity of 0,2 m/s and if any leakage is indicated by gas detectors carried by maintenance crews, airflow should be increased in order to maintain methane concentration below 1%. Natural gas flow in the pipeline should be also immediately stopped until leakage is corrected.

6 CONCLUSIONS

Transportation of natural gas from Bolivia to Brazilian southern states required the construction of a 3 150 km long gas pipeline (GASBOL). At the top of the sierra at the Rio Grande do Sul's border a tunnel-shaft system would require ventilation for the maintenance and monitoring crews. Natural ventilation was not a feasible solution since the shaft would be almost completely filled with structural elements and other types of obstructions, imposing a very high airflow resistance.

The proposed solution was a forced blowing system with flexible plastic ducts of 0,80 m diameter. Minimum air velocity should be 0,2 m/s according to federal regulations and should be increased if leakage were detected by means of portable gas monitors carried by the working men.

REFERENCES

FUNDACENTRO. Curso de supervisores de segurança do trabalho.3. ED. São Paulo, FUNDACENTRO, 1984. 2 v.

HARTMAN, H. Mine Ventilation and Air Conditioning.

McPHERSON, M.J. Subsurface ventilation and environmental engineering. 1993.

THE MINE VENTILATION SOCIETY OF SOUTH AFRICA. Environmental Engineering in South African Mines. Cape Town, 1982.

LACASEMIN TECHNICAL REPPORT 06/99, EPUSP.

Environmental Issues and Management of Waste in Energy and Mineral Production, Singhal & Mehrotra (eds)
© 2000 Balkema, Rotterdam, ISBN 90 5809 085 X

Characterization of fluidized bed combustion residues for pneumatic backfill in underground mines

N.Ghafoori
Department of Civil Engineering, Southern Illinois University, Carbondale, Ill., USA

B.Paul
Department of Mining and Mineral Resources Engineering, Southern Illinois University, Carbondale, Ill., USA

ABSTRACT: The paper presented herein is intended to describe (1) physico-chemical characteristics, (2) mechanical properties, and (3) leachate compounds and hydraulic conductivity of fluidized bed combustion (FBC) trial mixtures in order to ascertain their suitability for pneumatic placement in underground mines. Various composites of FBC spent bed and fly ash were prepared under two different compaction efforts (25 and 50 psi) and four distinct moisture conditions (25, 30, 35, and 40%). Test specimens were cured in a curing chamber simulating mine conditions for temperatures of 32 ± 2 °FC and relative humidity of 90 ± 10%. The trial matrices were evaluated for workability, consistency, heat of reaction, specific weight, compressive strength, hydraulic conductivity, and leachate concentrations.

1 INTRODUCTION

In 1990 the United States produced nearly 90 million tons of coal combustion by-products mainly from conventional coal combustion units. The 1990 Clean Air Act Amendments required electric utilities and co-generation plants to adopt advanced combustion and flue gas desulfurization technologies, such as fluidized bed combustion (FBC), wet scrubbers, dry sorbent duct or furnace injection, etc. These new technologies significantly increase the amount of flue gas desulfurization residues whereas a marginal increase in the amount of conventional coal combustion by-products is expected.

The management of coal combustion residues presents a major challenge to the coal and electric utility industries. Traditional surface management is both costly and environmentally disadvantageous. Underground disposal of these by-products can overcome most of the disadvantages of surface disposal and, in addition, offers (1) significant potential for reducing the likelihood of future subsidence caused by underground coal mining and (2) potentially controlling acid mine drainage.

In 1994, a 5-year cooperative agreement was made between the United States Department of Energy and Southern Illinois University at Carbondale to evaluate the viability of managing dry residues from coal combustion flue gas desulfurization (FGD) processes in underground mines with a goal to demonstrate reduction of surface subsidence and acid mine drainage often associated with such mines. This paper presents the results of a laboratory investigation (engineering properties and environmental studies) associated with evaluation of fluidized combustion fly ash and spent bed mixtures for high-rate placement in the underground mines by dry pneumatic injection.

2 MATERIALS AND TESTING

The matrix constituents were FBC fly ash and spent bed. Their chemical compositions are shown in Table 1. The FBC spent bed had oven-dry (OD) and saturated surface dry (SSD) specific gravities of 1.92 and 2.19, respectively. The difference between the OD- and SSD-specific gravities is indicative of the ability of the FBC spent bed to absorb a large amount of water. The actual absorption was calculated at 14.6%. Based on ASTM C40, no organic impurities were detected in the FBC spent bed. The size distribution of the FBC spent bed resulted in a fineness modulus of 1.80, resembling that of fine sand. The physical properties of the FBC fly ash are shown in Table 2.

Table 1. Chemical Test Data of FBC Spent Bed and Fly Ash.

Chemical Composition	"FBC" Spent Bed	"FBC" Spent Bed	Fly Ash Specifications
Silicon Oxide (SiO_2)	9.70	22.10	-
Aluminum Oxide (Al_2O_3)	3.69	6.80	-
Iron Oxide (Fe_2O_3)	2.16	6.67	-
Total ($SiO_2 + Al_2O_3 + Fe_2O_3$)	15.55	35.57	>50% Class C >70% Class F
Sulfur Trioxide (SO_3)	24.42	15.67	50.0 max.
Calcium Oxide (CaO)	53.10*	38.70*	<10% Class F <10% Class C
Magnesium Oxide (MgO)	0.88	1.29	-
Loss on Ignition (LOI)	0.80	5.46	6.0 Max.
Free Moisture	0.00	0.11	3.0 Max.
Water of Hydration	2.65	0.71	
Total Na_2O	0.16	0.71	
Total K_2O	0.39	1.12	
Others ($TiO_2 + P_2O_5 + SrO + BaO$)	2.04	0.83	
pH	12.00	12.20	

*Mostly unhydrated lime

The FBC fly ash and spent bed were blended in a counter-current mixer while the predesignated amount of water was added slowly. Upon obtaining a homogenous matrix, the blended materials was placed in 100 x 200 mm cylinders and compacted in two layers using a static compaction machine. Once fabricated, the samples were wrapped in a plastic sheet, while remaining in molds, for a period of three days. They were then extruded and placed in a curing chamber (temperature 32 ± 2 °C and relative humidity of 90 ± 10%) for different curing periods. Compressive strength was measured using ASTM C39. To measure penetrometer index, using a penetrometer, trial specimens were compacted in 203 x 203 x 508 mm beam-shaped molds in two layers, each compacted at 0.175 Mpa. The samples were placed in the curing chamber and the indices were obtained at different curing periods, up to four weeks.

3 RESULTS AND DISCUSSION

The nominal and actual moisture content, average demolded density, and 28-day compressive strength of the test specimens compacted at 0.35 Mpa static pressure are shown in Table 3. Heat of reaction of all matrices were fairly uniform at nearly 74°C. The average demolded unit weight, which varied from 1067.5 to 1647.5 kg/m^3, improved with an increase in the moisture content and a decrease in the fly ash content of the matrix. The optimum 28-day compressive strengths were attained at 25, 30, 35, and 40% nominal moisture content for the mixtures containing 100, 80, 70, and 50% fly ash, respectively. On the whole, an increase in moisture content provided a steady increase of compressive strength. The strength improvement, with an increase in the spent bed content of the matrix, can be attributed to a better matrix gradation and acceleration of the cementitious reactions of the FBC fly ash.

Based on the mixture consistency, it was concluded that 30% moisture content provided a type of matrix ideal for a pneumatic placement of the FBC by-product residues. Moreover, field conditions and available site equipment resulted in adaptation of 0.175 Mpa static compaction effort for the laboratory samples.

Table 2. Physical Properties of FBC Fly Ash.

#325 Sieve Fineness (%)		Specific Gravity	Autoclaved Expansion (%)			Water Requirement %		7-day Compr. Str. (%) of Standard Sample		
Actual	Limit Max.		Unhydrated	Hydrated	Limit		Max.	Actual	ASTM Min.	AASHTO Min.
40.2	34	2.61	0.12	0.0	0.8	96.3	105	61.2	75	60

*Standard Sample has a 7-day and 28-day Compressive Strength of 31.5 Mpa and 39.0 Mpa, respectively

Table 3. Properties of Various FBC Fly Ash-Spent Bed Mixtures Compacted at 0.35 Mpa Static Pressure.

Fly Ash-Spent Bed Wt. Ratio	Nominal Moisture (%)*	Actual Mositure (%)*	Heat of Reactions (°C)	Average Demolded Density (kg/m³)	28-Day Compr. Str. (kPa)
100-0	25	9.00	71.1	1067.4	256.5
	30	10.42	71.1	1117.4	33.1
	35	14.10	72.2	1183.2	783.9
80-20	25	8.55	75.6	1132.1	77.2
	30	10.27	75.0	1143.5	636.4
	35	15.20	76.1	1223.3	885.3
	40	18.38	76.7	1307.4	1651.4
70-30	25	7.45	73.4	1175.4	21.4
	30	12.14	73.4	1204.2	233.7
	35	14.57	73.9	1307.0	745.3
	40	18.03	74.5	1452.5	2082.9
50-50	25	7.45	73.4	1175.4	40.0
	30	11.40	72.8	1267.9	268.2
	35	16.87	73.9	1334.7	513.7
	40	17.71	75.0	1647.5	3052.4

*% by weight of total dry solid

Table 4. FBC Fly-ash-Spent bed mixtures Prepared at 30% Moisture Content and Compacted under 0.175 Mpa Static Pressure

Mixture Proportion	Immediate Unit Wt. (kg/m³)	28-Day Unconfined Compr. Str. (kPa)	60-Day Unconfined Compr. Str. (kPa)	Percentage increase in Compr. Str.
100[1]-0[2]	1057.0	345.2	397.5	15.16
80[1]-20[2]	1172.3	107.6	232.7	116.34
70[1]-30[2]	1183.0	80.1	139.3	74.0
50[1]-50[2]	1205.3	39.5	169.2	328.27

1 FBC fly ash ratio by weight
2 FBS spent bed ratio by weight

The results of the second set of the FBC spent bed-fly ash trial mixtures are presented in Table 4. There was no significant improvement in the immediate unit weight as the compaction effort changed from 0.175 to 0.35 Mpa. With the exception of the mixtures containing 100% fly ash, the remaining matrices prepared at 0.175 Mpa pressure displayed a significantly lower 28-day compressive strength when compared to that of the mixtures produced at 0.35 Mpa static pressure. However, the 60-day strength improved substantially as compared to that obtained after 28 days curing.

The penetrometer index of various FBC spent bed-fly ash mixtures is shown in Table 5. As can be seen, within four days of curing, all mixtures displayed indecies in excess of 430 kpa.

Since FBC mixtures were examined as a substitute for conventional grouts, for subsidence control of underground mines; it was important that the quality of water resources underlying the stabilized ground be protected. The coal burned to produce FBC spent bed and fly ash was high in sulfur and the plant used a significant abundance of limestone above the stoicheometric requirements. As would be anticipated, the FBC fly ash had more reactive surface area and dissolved more than spent bed. Trace element partitioning between FBC fly ash and spent bed fractions were less pronounced than in the ash from typical boilers, most likely due to the lower operating temperature of fluidized bed combustors. The most pronounced constituents to dissolve from the FBC material were calcium sulfate and salt. Calcium rose nearly 1800 ppm above initial concentration in mixes with 80% fly ash to about 1000 ppm in pure spent bed and 50-50 mixtures. Sodium rose about 1200 ppm in mixes with 80% fly ash to as little as 900 ppm in pure spent bed. Potassium rose around 40 ppm but less reactive alkali metals like magnesium did not increase. There were also traces of aluminum and silicon leached (less than 0.5 ppm). None of these elements would be a surprise since they are part of

Table 5. Penetrometer Index (Kpa) for FBC Fly Ash-Spent Bed Mixes Prepared at 30% Moisture Content and Compacted under 0.175 Mpa Static Pressure.

Curing Age	100[1]-0[2]	80[1]-20[2]	70[1]-30[2]	50[1]-50[2]
1	415	430	430	265
4	430+	430+	430+	430+
7	DO	DO	DO	DO
11	DO	DO	DO	DO
14	DO	DO	DO	DO
18	DO	DO	DO	DO
20	DO	DO	DO	DO
25	DO	DO	DO	DO
28	DO	DO	DO	DO

1 FBC Fly Ash ratio by weight
2 FBC Spent Bed ratio by weight
+ Exceeded capacity of the apparatus
DO Same as the previous curing period

the major chemical composition of FBC ash. Trace element leaching included about 0.25 to 0.5 ppm of boron, 0.4 ppm manganese, molybdenum around 0.12 ppm, and zinc < 0.1 ppm. There were also indications of increases below effective quantification levels for cadmium, copper, and selenium.

The mine demonstration site was located in almost dry strata below the brines level and isolated from the above fresh and brine water tables by aquiclude strata with hydraulic donductivity below 10 E-10 cm/sec. The hydraulic isolation of the near surface fresh water table from the underlying brines is attested to by the absence of salt migration into the fresh water table. The hydraulic isolation of the mine from the fossil water brines in the fracture zone about 10 feet above the mine is demonstrated by the almost universal absence of seepage into the mine from either above or below. Water collected from the shaft rings and into a mine level sump provide an indication of the water composition that might over the very long term seep into the mine through fractures and incompletely sealed shafts. The water is a mixture of the seepage from all the overlying water bearing layers and represents a significant dilution of the brines in the fracture zone above the mine. This water was used as a leaching medium in 18 hour shake tests to determine what types of reactions might occur.

The sample water was collected on several different days and analyzed by ICP. Background equivalent level concentrations were reported for silver, aluminum, arsenic, boron, barium, beryllium, cadmium, cobalt, chromium, iron, manganese, lead, antimony, selenium, vanadium, and zinc. The most conspicuous elements were alkali metals from the brines including around 1000 ppm calcium, 35 ppm potassium, 55 ppm magnesium, and 270 ppd of sodium. The water also contained about 5 ppm of silicon, 0.1 ppm of copper, and 0.14 ppm of molybdenum.

The most critical threat to groundwater quality clearly comes from dissolution of the major element constituents, namely the alkali metals and sulfate. Since these concentrations are below the brines in the fracture unit above the mine there is no chance of groundwater contamination of the valuable fresh water resources above the field mine site, but many of the sites where subsidence prevention are needed will be below suburban areas with underground mines 100 feet or less below the surface and well into the fresh water table. A dilution and attenuation factor of about 100 to 1 would be needed to bring these leachate concentrations to an acceptable limit for drinking, and these concentrations are high enough that unlike many trace elements, surface adsorption onto fine soil grains cannot be looked to for rapid attenuation.

The key factor in the acceptability of FBC fly ash and spent bed for stabilization below suburban areas with near surface valuable high quality ground water resources is the hydraulic conductivity. Most commercial grade water extraction wells are screened in intervals with around 10 E-4 cm/sec hydraulic conductivity or better. A variety of column experiments of the trial FBC fly ash-spent bed mixtures were run to determine long term hydraulic conductivity. They all lost all measurable hydraulic conductivity within a few weeks. The long term hydraulic conductivity of the trial mixtures was clearly on the order of 10 E-8 cm/sec, which is 4 orders of magnitude lower than the conductivity of a typical well supporting aquifer. This means, that for a given head, the aquifer will support 4 orders of magnitude higher flow virtually assuring a 100 to 1 dilution except in geometrically controlled locations where mixing has not yet occurred. The low hydraulic conductivity also reduces to near zero the possibility that anyone will ever screen directly into the subsidence control fill without picking up dilution water from above and/or below the fill. The problem associated with mineralogical breakdown of the trial FBC fly ash-spent bed mixtures are also negligible because the underground fills are almost always below any frost lines where thermal cycling could occur.

4 CONCLUDING REMARKS

Test results conclude that an increase in moisture content had little effect on the heat of reaction, whereas it provided a steady increase in compressive strength. The strength improvement, with an increase in the FBC spent bed of the matrix, is attributed to a better matrix gradation and acceleration of the cementitious reactions of the FBC fly ash. The trial matrices produced the strength properties and penetrometer indecies sufficient to reduce surface subsidence of underground mines. Leachate studies indicate relatively few elements of concern with either the components or the mixtures themselves. The most pronounced metals in the leachates were alkali metals such as calcium. Boron leaching was relatively minor and calcium leaching was persistent though diminishing over time. Pneumatic mixes had very limited hydraulic conductivity and, thus, produced little leachate.

ACKNOWLEDGEMENTS

This investigation was funded with a grant made possible by the U.S. Department of Energy Grant Number DE-FC21-93MC-30252.

CONVERSION FACTORS

1 mm = 0.039 in
1 KPa = 0.145 psi
1 kg/m^3 = 0.0624 lb/ft^3
1 N = 0.225 lb (force)
1 MPa = 0.145 ksi
1 kg/m^3 = 1.68 lbf/yd^3

REFERENCES

Electric Power Research Institute (EPRI) 1984. Coal Combustion By-Products Utilization Manual, V.1: Evaluating the Utilization Option, Report No. CS-3122, Prepared by Michael Baker, Jr., Inc.

Ghafoori, N. & Y. Cai 1998. Laboratory-Made RCC Containing Dry Bottom Ash: Part I – Mechanical Properties. *Journal of Materials (ACI)* 95(2):121-130.

Ghafoori, N. & Y. Cai 1998. Laboratory-Made RCC Containing Dry Bottom Ash: Part II – Long Term Durability. *Journal of Materials (ACI)* 95(3):244-251.

Ghafoori, N. & C. Garcia 1998. Compacted Non Cement Concrete Utilizing Fluidized Bed and Pulverized Coal Combustion By-Products. *Journal of Materials (ACI)* 95(5): 582-592.

Ghafoori, N. & L. Wang 1998. Laboratory Investigation of Coal Wastes for Secondary Roads. *Journal of Transportation Research Record (TRB)* 1(1611):19-27.

Ghafoori, N. & S. Kassel 1999. Composite Paving Slabs for Roadways Containing Coal Combustion Dry Bottom Ash. *Proc. of the 5th Materials Engineering Congress, Materials and Construction – Exploring the Connection, Cincinnati, OH, USA May 10-12 1999:* 835-840.

Ghafoori, N. & S. Kassel 1999. PCC Bottom Ash for Structural-Grade Concrete Pavements. *Proc. of the Int. Conf. on Infrastructure Regeneration and Rehabilitation, Sheffield, England, June 28-July 2 1999:* 1211-1224.

Ghafoori, N., Honaker, R. & H. Sevim 1999. Processing, Transportation, and Utilization of Coal Combustion Ash. *Proc. of the 1999 International Ash Utilization Symposium, October 18-20 1999.*

Helmuth, R. 1987. *Fly Ash in Cement and Concrete.* Portland Cement Assoc., SP 040 OIT.

Hznke, K.R., Zobeck, B.S., Moretti, C.J. & M.D. Mann 1987. Leaching Studies of Solid Wastes Generated from Fluidized Bed Combustion of Low-Rank Western Coals. *Proc. of the 9th Int. Conf. on Fluidized Bed Combustion, Boston, 1987:* 954-959.

Environmental Issues and Management of Waste in Energy and Mineral Production, Singhal & Mehrotra (eds)
© 2000 Balkema, Rotterdam, ISBN 90 5809 085 X

Diesel emissions evaluation for the surface mining industry

D. R. Leslie
Fording Coal Limited, Calgary, Alb., Canada

ABSTRACT: An evaluation of diesel emissions in open-pit mines was undertaken by the Surface Mining Association for Research and Technology (SMART), a group formed through the co-operation of most of the large open-pit mining companies in Canada. The impetus for the study was to address some of the issues related to diesel emissions before they become business issues with the potential to increase operating costs. One of the main issues that needed addressing was the commitment by our federal government to limit the production of "greenhouse" gases in support of the theory of "global warming". As an industry, it was important to understand and quantify the levels and trends of emissions created by diesel mining equipment. We needed to show that we use energy wisely to minimize costs and environmental effects, and it is in the mining industry's overall best interest to be as energy efficient as current technology and business economics allow.

This study has shown that, in the absence of government imposed regulations, mining equipment in open-pit mines continues to show increasing efficiency in its use of fuel, and that emissions, on a per unit moved basis, have continued to fall over time. Larger, more productive mining equipment and continued advancements in diesel engine technology will ensure that the trends of increasing energy efficiency and decreasing emission levels will continue into the future, as long as the economics of improvement are not lost due to over-regulation.

1 INTRODUCTION

An evaluation of diesel emissions in large open-pit mines was undertaken by the Surface Mining Association for Research and Technology (SMART), a group formed through the co-operation of most of the large open-pit mining companies in Canada. With the increasing concerns, whether real or perceived, regarding "global warming", and the commitment by our federal government to limit the production of "greenhouse" gases, it is time the mining industry understands and can quantify the levels and trends of emissions created by mining equipment. Comprehensive facts regarding emissions were needed in order to combat any potential moves by government to impose regulatory costs and constraints, which would increase mining costs. We needed to prove that as an industry we use energy wisely, and that it is in our own overall best interest to be as energy efficient as current technology and business economics allows. The goal of this study was to prove with real data that, in the absence of government imposed regulations, the mining industry as a whole continues to show more efficient use of fuel in open-pit mines, and emission levels, on a per unit moved basis, have fallen over time. The mining industry refers to not only the users of the equipment at actual mine sites, but the entire industry including all the manufacturers and suppliers that work together to accomplish the goal of increasing efficiency. The alternative to this pro-active approach is to wait for anti-development interests to convince the government that the mining industry wastes energy and emits dangerous levels of emissions, and that the only way to protect the environment is by imposing strict emission controls and increased energy costs. With larger, more productive mining equipment and continued advancements in diesel engine technology, the trend of increasing energy efficiency while decreasing emission levels will continue into the future.

1.1 *Purpose and Scope of Study*

The first phase of the study into diesel emissions from mining equipment was to gather data from the major manufacturers of diesel engines used in open-pit mining equipment in order to create an historic technical evaluation of the evolution of diesel equipment up to current technology. A comprehensive database of equipment with engine types, horsepower ratings, fuel consumption, and emission levels including Carbon Monoxide (CO), Oxides of Nitrogen (NOx), Oxides of Sulphur (SOx), Hydrocarbons (HC), other Organic Compounds (OOC), and Particulate Matter (PM) was to be created. In addition, new or emerging engine technologies would be evaluated to understand what the future holds. Underground diesel engine technology would be assessed for its application in the surface environment in collaboration with the Diesel Emissions Evaluation Program (DEEP) being carried out by eastern Canadian underground operators.

The second phase was the compilation of all the data to analyze trends in efficiency and emission levels. Technological improvements in equipment design would be summarized and evaluated on their contribution to higher efficiencies on diesel powered equipment, and reduced emissions on a per unit of material moved basis. The health and safety or irritant aspects pertaining to various emission types and levels in relation to equipment operators would be addressed in the study. With the historic and current data available, and some insight into the near future, some realistic future projections of engine efficiency and emission levels would be made. With the assistance of consultants that specialize in the field, an evaluation document would be produced that would be used as a reference for the environmental, health, and safety aspects of the mine permitting process.

2 BACKGROUND

2.1 *Combustion Reaction Components*

The diesel combustion reaction consists of hydrocarbon chains (C_xH_y) being oxidized in an explosive reaction to form carbon dioxide (CO_2) and water (H_2O). However, the reaction is not 100% efficient and the constituents are not pure. The air used to supply the oxygen (O_2) contains about 80% nitrogen (N_2) and the diesel fuel contains a small percentage of sulphur (S). The result is that trace amounts of other chemicals are formed in the reaction. The generalized reaction equation (Equation 1) shows the constituents formed from the combustion of diesel fuel and air. All of the trace constituents are of concern to the environment or can pose a health risk in higher concentrations.

$$C_xH_y(S_z) + (O_2 + N_2) \Rightarrow CO_2 + H_2O + (O_2 + N_2) + \{NOx + HC + OOC + C + CO + SOx\}$$

Combustion Reaction ⇒ Major Constituents + { Trace Constituents }

Of the major constituents of the reaction, CO_2 is a concern as a potential "greenhouse" gas and a theoretical contributor to "global warming". Since it is an integral part of the reaction, the way to reduce output is to increase the efficiency of the engine and reduce fuel consumption. CO_2 is produced at rate of approximately 2.73 kg/litre of fuel (JAQUES, 1992). There are no health risks or direct environmental risks from the production of CO_2. The trace constituents are of more concern to the environment and health. The production of SOx is directly related to the amount of sulphur in the diesel fuel. In certain regions, SOx emissions are significant and can contribute to environmental concerns related to the formation of acidic compounds in the atmosphere. NOx is produced in the high temperature, high-pressure diesel fuel combustion chamber since there is excess oxygen available to combine with nitrogen during the reaction. Oxides of nitrogen can react in the atmosphere to form acidic compounds as well as low level ozone, a major component of smog. Ozone is a lung irritant and can cause serious health effects and breathing difficulty in higher concentrations. CO is not formed in large quantities due to the excess amount of oxygen available during combustion, and is generally not the concern it is in gasoline powered engines. The other trace constituents are the result of incomplete combustion. Unburned hydrocarbons or the products of partial combustion account for the HC, OOC, and free carbon (C) components of the trace constituents. These compounds contribute to the formation of smog after further reactions in the presence of sunlight, and can pose health risks as concentrations rise. Hundreds of separate OOC can be formed when the combustion reaction is not complete. These OOC are reported as the soluble organic fraction (SOF), and the volatile organic fraction (VOF). The trace constituent products also account for the majority of the particulate matter (PM) in diesel exhaust. The visible portion of PM, the black smoke, is larger carbon particles that are formed under acceleration and heavy load due to insufficient air or low combustion temperatures. Electronic engine controls can minimize the formation of black smoke such that it is rarely visible. The majority of the PM is too small to be seen, and is composed of very fine carbon particles, often less than 1 micron in diameter. These carbon particles can have

Table 1: EPA Emission Standards for Non-Road Engines Greater than 560 kW (LEVELTON, 1999)

Tier	Model Years	Emission Standards g/kW-hr (g/bhp-hr)										Emissions Warranty Period**
		NOx		HC		NMHC+NOx		CO		PM		
		g/kW-hr	g/bhp-hr	g/kW-hr	g/bhp-hr	g/kW-hr	g/bhp-hr	g/kW-hr	g/bhp-hr	g/kW-hr	g/bhp-hr	
Actual Unregulated*	Pre-1996	14.3	10.7	0.4	0.3	14.8	11.0	1.7	1.3	0.17	0.13	
Actual Pre-Tier 1*	1996-2000	11.3	8.5	0.3	0.3	11.7	8.7	0.8	0.6	0.15	0.11	
Tier 1	2000+	9.2	6.9	1.3	1.0			11.4	8.5	0.54	0.40	5yrs/3000 hr.
Tier 2	2006+					6.4	4.8	3.5	2.6	0.20	0.15	5yrs/3000 hr.
SOP Long-term goal		2.0	1.5							0.07	0.05	

* Approximate values based on averaged manufacturers (CAT) data

** Tier 1 and Tier 2 engines must comply over "Useful Life" of 10yrs/8000 hr.

many other exhaust compounds, formed from HC, OOC, and SOx, adsorbed on their surfaces (NAUSS, 1997). There are also secondary transformations of NOx and SOx into nitrate and sulphate particulate in the atmosphere. When PM are less than 10 microns in diameter, they are respirable and a lung irritant. As the particles decrease in size to less than a 2.5 microns, they can be deposited deep in the lungs where there is potential to impair function and cause damage, or increase the risk of future serious illness.

2.2 *Diesel Engine Development*

Engine manufacturers have been driven over the years by the need to reduce the operating costs of the engines they produce. This has been accomplished by producing engines that require less maintenance time and burn less fuel. In an unregulated environment, the drive to reduce emissions was not an issue except for the visible components of particulate matter, or black smoke. Even without a focussed effort, most emission component levels were reduced with this strategy. By producing engines that are more efficient, the manufacturers reduced products of incomplete combustion including CO, HC, and OOC, as well as CO_2 since less fuel was consumed. The only component of the emissions that did show any potential for increase was NOx, since its formation occurs as temperatures rise in a combustion chamber.

In the United States, the Environmental Protection Agency (EPA) has used legislation to force diesel engine manufacturers to limit emissions with a primary focus on the reduction of NOx. This has been accomplished through a phased approach that started with smaller horsepower engines for highway applications. The volume of engines produced for those applications was sufficient to bear the research and development costs without huge increases in the cost of the engines. The engines used in off-highway applications, and in particular the large horsepower engines for mining application, have benefited from this strategy of having a phased approach to implementation. Sufficient time is required to implement changes to engine technology, especially where engine volumes are low. The economics of upgrading equipment would be seriously jeopardized if the cost of new engine technology pushed the price of emission regulated engines to unrealistic levels. The benefits of purchasing new larger equipment are that they are more cost effective in mining operations, and the new engines use fuel more efficiently and produce fewer emissions.

Upgrading to larger, more efficient and productive equipment has played a major role in the continued reduction of emissions from mining operations. Increased size and efficiency produce lower fuel consumption and fewer emissions for the same amount of work performed. Even before the EPA regulations were introduced, more efficient diesel engines were being produced for the larger equipment, and overall emissions on a unit moved basis were falling. The EPA regulations for typical large mining equipment sized engines (> 560 kW or 750 hp), which Canada intends to follow under a memorandum of understanding, are shown in Table 1.

Diesel engines greater than 450 kW (600 hp) have a wide range of applications in large open-pit mining operations. Three manufacturers, Caterpillar (46%), Detroit Diesel (27%), and Cummins (26%), supply 99% of the U.S. market, which constitutes only 3% of the total non-road market (EPA, 1998). Large haul trucks, dozers, loaders, and diesel hydraulic shovels typically are in the greater than 560 kW (750 hp) category which constitutes only 0.7% of the total non-road market. During the years from 1991 to 1995, Generator Sets accounted for over 70% of the average annual sales in the greater than 560 kW (750 hp) category, where off-highway trucks accounted for 13%, loaders accounted for 0.7%, and crawlers accounted for 8% (EPA, 1998). The fraction of these total engine sales used in surface mining operations would be considerably smaller. The Canadian breakdown of engine sales is expected to be similar in that there is a very small percentage of the total non-road diesel engines involved in surface mining activities. As well, surface mining operations generally are in sparsely populated areas of the country where the production of NOx and the formation of smog is not the environmental issue that it is in urban centres.

2.3 *Emission Reducing Strategies*

From the time all large diesel engines were naturally aspirated and mechanically injected, numerous improvements have been made to increase horsepower and reduce fuel consumption. At rated power, about 30% of the energy from a diesel engine is lost through the exhaust (EPA, 1998). Turbocharging utilizes some of that lost energy to boost air intake pressure, which increases engine efficiency. Aftercoolers, both air-to-air and air-to-water, were then added to increase the air density going back to the engine, which further increased power ratings. Aftercoolers have the added benefit of reducing NOx production by reducing air intake temperature (EPA, 1998). Constant improvements in combustion optimization, including injection timing, combustion chamber geometry, and increasing air intake turbulence or swirl, have all contributed to reduced output of HC, PM, and lowering fuel consumption. Full authority engine electronic management combined with very high-pressure fuel injection can allow shaped or multiple injections at optimized timing to reduce NOx production without increasing PM. Exhaust gas recirculation is a recent strategy designed to reduce NOx production but some problems still need to be addressed. Introducing exhaust and PM back into the air intake system and combustion chamber can lead to increased turbocharger and engine wear. Deposits on the air intake system can also reduce the efficiency of aftercoolers (EPA, 1998).

At this point, exhaust gas after-treatment technologies to remove HC and PM are not deemed necessary to meet EPA regulations. Oxidation catalysts in conjunction with low sulphur diesel (< 0.05% S) to remove HC and CO, which are already in very low concentrations, can have the unwanted reactions of creating SO_3 from SO_2, and NO_2 from NO (EPA, 1998). Particulate traps definitely show some promise for diesel equipment operating in confined spaces such as underground mines. These technologies are currently not economically viable for the large horsepower engines used in surface mining operations, nor are they necessary for air quality.

3 STUDY RESULTS

3.1 *Data Gathering*

Visits were made to the manufacturing facilities of Detroit Diesel Corp. (DDC), Caterpillar Inc. (CAT), and Cummins Engine Company Inc. (Cummins) in June 1998. CAT is a vertically integrated company; it is the only manufacturer of large non-road diesel engines that also builds the equipment they power. Cummins and DDC only manufacture engines, which they then supply to many different original equipment manufacturers. Information on horsepower growth, fuel consumption and exhaust emissions was requested to detail the evolution of engine product lines used in large surface mining equipment. All of the companies responded with some information, but the amount of information available seemed limited. Aside from some historical fuel efficiency and smoke index numbers, and current technology emission data, there seemed to be little to share. In an unregulated environment, there was no apparent need to perform the complicated and expensive testing required to collect emissions data, and there was no agreement on standard testing procedures. The Tier 1 EPA legislation (59 FR 31306) for non-road diesel engines greater than 37 kW (50 hp) was set in 1994 with compliance required starting in 1996 (NERA, 1997). Similar emission levels were required from heavy-duty highway engines in 1990. With this lead-time before the larger horsepower engines had to comply, it was evident that new technology would have to be employed to meet the standards, and there was no need to conduct rigorous emissions testing. In some cases, new test cells had to be constructed before these large engines could physically be accommodated for testing. DDC and Cummins decided that technology dictated totally new engine designs utilizing new technology proven with their heavy-duty highway engines. DDC has dropped its two-cycle line for equipment to be certified, and introduced the S4000 series. Cummins has introduced the QSK60 series. CAT has made changes to their existing 3500 series engines with advances to the electronic controls and injection. All the manufacturers are confident that these current engine series can be modified to comply with the Tier 1 EPA regulations that come into effect in the year 2000.

The information gathered was reviewed by Levelton Engineering Ltd. to ensure that any conclusions derived from the data were done in an appropriate manner.

3.2 *Engine Manufacturers Data*

Historical data from DDC on fuel consumption for the two-stroke 16V-149 engine, from 1966 through to the introduction of the four-stroke 16V-S4000 engine, is shown in Figure 1. The 16V-149 engine has been a workhorse in the mining industry, especially in powering haulage trucks. The major engine modifications responsible for the drop in fuel consumption are noted. For clarification, DDEC is full authority electronic control and Common Rail is a very high-pressure electronically controlled fuel injection sys-

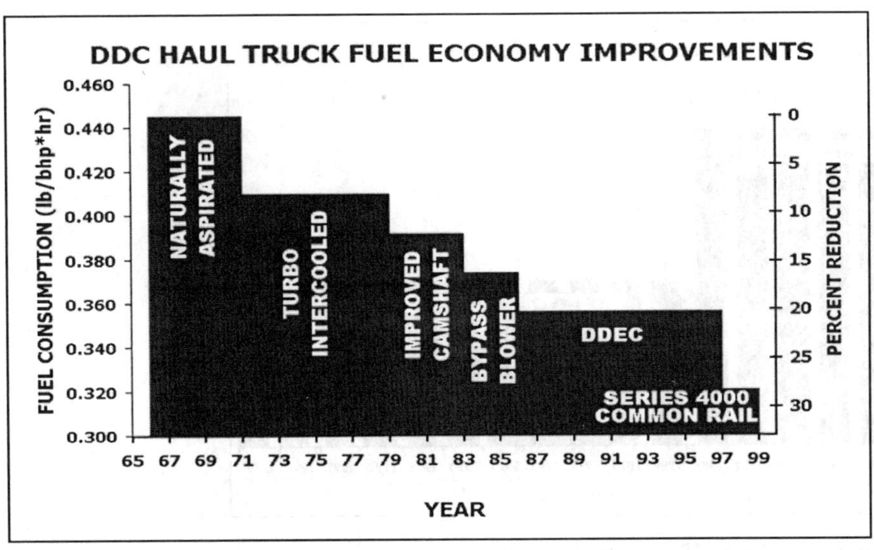

Figure 1: Fuel Consumption Improvements with the DDC 16V-149 Engine Series (DDC, 1998)

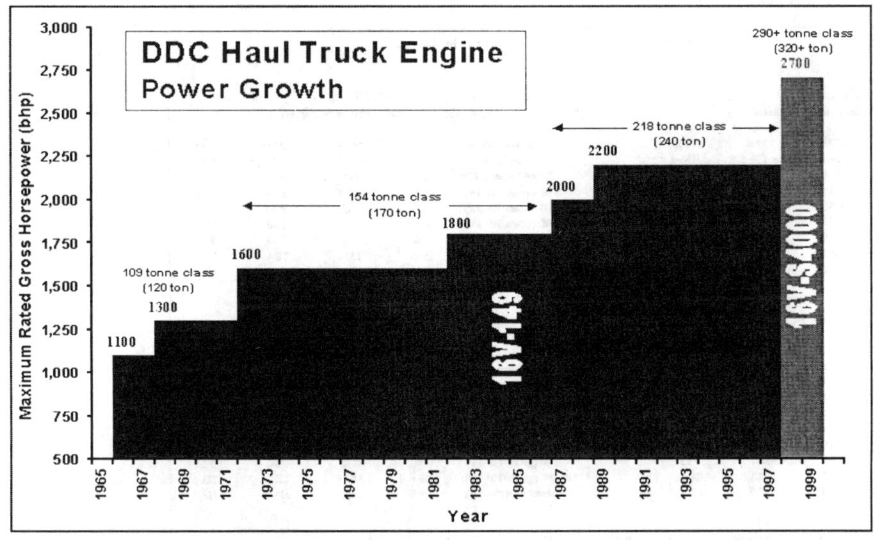

Figure 2: Horsepower Growth with the 16V-149 Engine Series (DDC, 1998)

tem. Fuel consumption on a per horsepower basis dropped over 20% from 1966 to 1986 in this series of engine, and with the introduction of the S4000 series of engines a further drop of more than 9% was achieved. From 1966 to 1998, a drop in fuel consumption per horsepower of over 27% was achieved in unregulated engines. Emissions of CO_2 are directly related to fuel consumption, so there has been a corresponding reduction on a per horsepower basis. The 16V-149 at 2200 hp is one of the engines currently used to power mine haulage trucks with a capacity of up to 218 tonnes (240 ton).

As can be seen in Figure 2, maximum rated horsepower doubled from 1966 through to 1989. The introduction of the S4000 series saw another jump of 22% in maximum rated horsepower in a similar sized engine. The horsepower increases correspond with the fuel consumption improvements due to technological improvements in the engines. The series S4000 with a maximum of 2700 horsepower is now available to power the next generation of haul trucks in the 290 + tonne range (320 + ton).

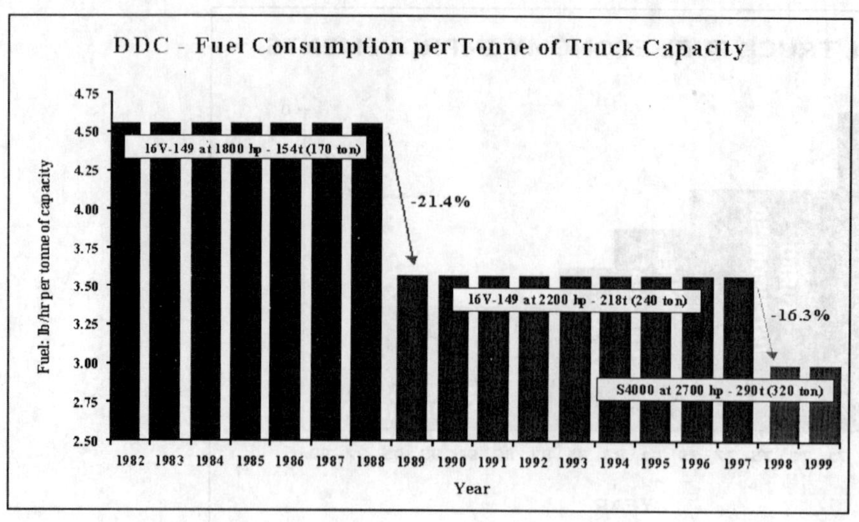

Figure 3: Fuel Consumption Improvements per tonne of Truck Capacity

Table 2: CAT Haul Truck Emission Data (CAT, 1998)

Caterpillar Data (98/10/30) ISO 8178; Type C1 8-Mode cycle data (weighted average)

CAT Machine: 777 (100 ton)
Engine: 3508 at a SCAC temp. of 56 degrees C.

Machine Model:	777C	777D	777D	777X
Engine Model:	EUI	B-Series	Tier I Reg.*	Tier II Reg.**
Rated Power (HP):	920	1000	1000	TBA
NOx (gram/HP-Hr):	10.06	7.48	6.90	4.80
CO (gram/HP-Hr):	1.65	0.53	8.50	2.60
HC (gram/HP-Hr):	0.28	0.24	1.00	N/A
PM (gram/HP-Hr):	0.13	0.14	0.40	0.15
***BSFC (gram/HP-Hr):	164.6	156.2	TBA	TBA
Maximum Payload (t):	86.2	91.0	91.0	91.0

CAT Machine: 785 (150 ton)
Engine: 3512 at a SCAC temp. of 56 degrees C.

Machine Model:	785B	785C	785C	785X
Engine Model:	EUI	B-Series	Tier I Reg.*	Tier II Reg.**
Rated Power (HP):	1380	1447	1447	TBA
NOx (gram/HP-Hr):	10.81	8.70	6.90	4.80
CO (gram/HP-Hr):	1.70	0.63	8.50	2.60
HC (gram/HP-Hr):	0.26	0.23	1.00	N/A
PM (gram/HP-Hr):	0.12	0.10	0.40	0.15
***BSFC (gram/HP-Hr):	155.9	151.8	TBA	TBA
Maximum Payload (t):	136.0	136.0	136.0	136.0

CAT Machine: 789 (195 ton)
Engine: 3516 at a SCAC temp. of 56 degrees C.

Machine Model:	789B	789C	789C	789X
Engine Model:	EUI	B-Series	Tier I Reg.*	Tier II Reg.**
Rated Power (HP):	1800	1904	1904	TBA
NOx (gram/HP-Hr):	10.85	9.61	6.90	4.80
CO (gram/HP-Hr):	1.04	0.55	8.50	2.60
HC (gram/HP-Hr):	0.28	0.25	1.00	N/A
PM (gram/HP-Hr):	0.14	0.08	0.40	0.15
***BSFC (gram/HP-Hr):	161.2	151.2	TBA	TBA
Maximum Payload (t):	177.0	177.0	177.0	177.0

CAT Machine: 793 (240 ton)
Engine: 3516 at a SCAC temp. of 65 degrees C.

Machine Model:	793B	793C	793C	793X
Engine Model:	EUI	B-Series	Tier I Reg.*	Tier II Reg.**
Rated Power (HP):	2160	2300	2315	TBA
NOx (gram/HP-Hr):	10.92	8.02	6.90	4.80
CO (gram/HP-Hr):	0.81	0.69	8.50	2.60
HC (gram/HP-Hr):	0.48	0.27	1.00	N/A
PM (gram/HP-Hr):	0.14	0.12	0.40	0.15
***BSFC (gram/HP-Hr):	160.6	160.9	TBA	TBA
Maximum Payload (t):	218.0	218.0	218.0	218.0

* Tier I and Tier II Regulation emission numbers are maximum permissible values. Actual CO, HC, PM values will be similar to B-series values.
** Tier II changes the convention by combining NOx and HC, therefore the Tier II value will represent NOx+ HC.
*** BSFC is associated with rated point.

Figure 3 shows the dramatic decrease in fuel consumption by moving to larger more productive equipment, and taking advantage of more efficient engines. Since the production of CO_2 is directly proportional to fuel consumption, the amount of CO_2 released into the atmosphere decreases by 21.4% when moving to a 218 tonne (240 ton) capacity haul truck, and another 16.3% to the 290 tonne (320 ton) truck. The total decrease per tonne of truck capacity could amount to over 34% with upgrades to equipment, in a totally unregulated environment. These decreases were driven by the economics of larger, more productive equipment coupled with more efficient and cost-effective engines. In addition, improved engine efficiency is responsible for reductions in HC, CO, and OOC emissions.

Data supplied from CAT was tailored to focus on haul trucks since it could show not only the engine evolution in terms of emissions and fuel efficiency, but also the equipment evolution in terms of size and productivity. Table 2 shows CAT data from four different sized haul trucks that all use 3500 series en-

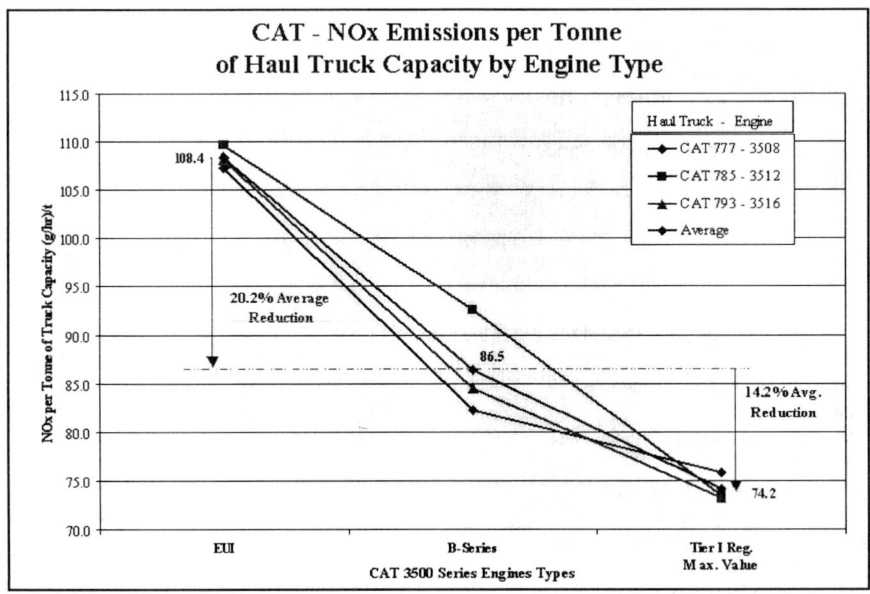

Figure 4: CAT NOx Emission Data per Tonne of Truck Capacity by Engine Type

gines in varying configurations. The technology covered is from the previous generation Electronic Unit Injection (EUI) engines to the current B-series engines that will, after final modifications, meet and exceed Tier 1 regulations in the year 2000. Tier 2 regulations are shown as the next goalpost for the year 2006. CAT believed it could meet Tier 1 regulations using the existing 3500 series engines with some modifications to electronics and injection timing. The EUI engines were then tested to establish base levels for comparing different emission reducing strategies. CAT will aim at being 15 to 20% lower than Tier 1 NOx emission levels in order to account for any variance from engine to engine. Emissions of HC, CO, and PM are already well below Tier 1 regulated levels, so the effort will focus on keeping the current emission levels from rising significantly.

Figure 4 shows the reduction in NOx emissions per tonne of truck capacity with the advances to engine design. The economics of acquiring new larger, more productive equipment, with improved engines that produce fewer emissions, is good for business as well as the environment. When the emissions data is tied to a truckload in the units of tonnes of material capacity, it is evident that emissions of NOx are reduced dramatically when equipment is upgraded.

4 CONCLUSIONS

The surface mining industry has done a good job over the years in using energy wisely. The operating cost savings when moving to larger, more productive equipment produce the added benefits of reduced fuel consumption and fewer diesel emissions on a unit moved basis. Fuel efficiency and CO_2 emissions have been reduced nearly 30% on a per horsepower basis over the last 30 years in a totally unregulated market. The marketplace drove the manufacturers to produce more fuel efficient and reliable diesel engines to power larger, more productive equipment. With efficiency comes reduced fuel consumption and reduced emissions of CO_2, CO, HC, and OOC.

Emission reductions of PM and NOx are also of concern, and engine manufacturers have met the challenge to reduce these components of diesel emissions while not allowing the other components to increase. Further reductions are regulated by the EPA with the high horsepower engines meeting Tier 1 targets in 2000. These non-road regulations have been phased in over several years starting with the higher volume engines having highway heavy-duty truck engine counterparts. Canada, through a memorandum of understanding, intends to follow the EPA regulations by requiring new engines for new equipment to be EPA certified before being imported. The concern is that the push to reduce NOx and PM will be at the expense of diesel engine efficiency, and that regulations will not allow time for technological advancements in engine design. Compliance may be accomplished through additional air cooling fans and other parasitics that will drive up fuel consumption and operating costs.

New capital is only employed when the economics are positive on a time-value basis. If the cost of new diesel engines is too high due to onerous emission standards, then upgrades to become more efficient will not occur. The economic incentive must remain for companies to strive for increased productivity, and fewer emissions per tonne of material moved in surface mining.

REFERENCES

CAT, 1998. Personal Communication and Statistical Data. Caterpillar Inc., Peoria, IL.

CUMMINS, 1998. Personal Communication. Cummins Engine Company Inc., Columbus, IN.

DDC, 1998. Personal Communication and Statistical Data. Detroit Diesel Corp., Detroit, MI.

EPA, 1998. Final Regulatory Impact Analysis: Control of Emissions from Non-road Diesel Engines. U.S. Environmental Protection Agency, Office of Mobile Sources. Report: EPA420-R-98-016, August. 127p.

JAQUES, A., 1992. Canada's Greenhouse Gas Emissions: Estimates for 1990. Environment Canada, Report EPS 5/AP/4

LEVELTON, 1999. Personal Communication: Review of Emission and other Characteristics for Heavy Duty Diesel Engines. Levelton Engineering Ltd., Calgary, AB

NAUSS, K., 1997. Diesel Exhaust: A Critical Analysis of Emissions, Exposure, and Health Effects. Summary of HEI Diesel Working Group Special Report, Health Effects Institute, Cambridge, MA.

NERA, 1997. Economic Evaluation of Regulations on Exhaust Emissions from Large Non-road, Compression Ignition Engines. National Economic Research Associates. Cambridge, MA. 38p.

Environmental Issues and Management of Waste in Energy and Mineral Production, Singhal & Mehrotra (eds)
 ISBN 90 5809 085 X

Disposal and utilization of mine waste rocks and flyash as a coal mine backfill

K. Matsui, T. Sasaoka, H. Shimada, M. Ichinose & S. Kubota
Kyushu University, Fukuoka, Japan

ABSTRACT: There is a big concern for the disposal and utilization of mine wastes and coal burning power plant waste materials. The dumping of mine waste and waste from power plants in the mined-out areas in underground mines has become increasingly important in the mining industry due to the lack of a dumping site.

This paper describes the trends of the Japanese coal mining industry and the importance of the backfilling system for underground coal mines in the world and discusses the potential for a range of coal mine waste rocks and flyash as a backfill for underground coal mines.

1 INTRODUCTION

In Japanese underground coal mines, waste rocks that are inevitably produced during mining activities are transported from underground to the surface and disposed of at a dumping site. Apart from the transportation problems, the disposal of waste rocks has a possibility of becoming a major environmental issue. Moreover, a great deal of refuse from coal burning power plants is produced every year. The amount of it is expected to be higher each year. The disposal and utilization of flyash is another issue.

A possible solution to these issues is to utilize them as backfill materials at underground coal mines.

In this study, the trends in Japanese coal industry and the importance of the study of the backfilling system in underground coal mines are described and the potential of a range of waste rocks and flyash as a backfill material for underground coal mines is discussed.

2 TRENDS IN THE JAPANESE COAL MINING INDUSTRY AND IMPORTANCE OF THE STUDY

Japan is dependent on overseas suppliers for 97% of the coal it uses. Japanese annual domestic production output stands at around 4 million tons (3%). The Japanese coal mining industry has already seen the closure of many of its collieries as part of the policy target envisaging a gradual contraction of coal production. At present, there are only two major underground coal mines (Ikeshima Colliery and Kushiro Colliery) in operation. They are now making extensive efforts to streamline in order to save costs. An appraisal can thus be made by saying that they play the role of serving as a nucleus for technical cooperation with overseas collieries and of balancing the economic burden placed on the public. As Japan continues to engage in cooperation with domestic advanced technology, such as safety technology, it plans to maintain and upgrade the coal supply potential of each country in the Asia region. This is a very important goal for Japan, in order to achieve greater stability in the supply of coal from overseas. Based on this philosophy, "New Coal Policy" principles have been announced. It is anticipated that the coal extraction conditions in the coal-producing countries will deteriorate in the future. In consequence, a "Five-Year Plan for Coal Technology Transfer" will be implemented from fiscal 2002 through 2006 in which the domestic collieries that are still in operation will play a key role. Under the five-year plan, coal technology will be transferred in an intensive and planned manner to coal-producing countries.

Joint research work at Ombilin Coal Mine,

Indonesia that has, since 1998, been conducted by Kyushu University, Japan, Bandung Institute of Technology, Indonesia and Institute of Sciences, Bandung, Indonesia is directed primarily towards the optimal support system and the development of optimal underground mining methods in Indonesia. Some results of the joint research have already been reported (Shimada et al., 1998, Anwar et al., 1998, Matsui et al., 1999, Anwar et al., 1999a and Anwar et al., 1999b).

In Indonesia, approximately 60 million tons of clean coal is produced annually. About 10 million tons is exported to Japan. Most of the produced coal is from surface mines. Only about 0.7 million tons is produced from underground mines at present. It is anticipated that most coal will have to be mined underground in the near future. For the period of transition from surface mining to underground mining, a highwall mining system should be considered.

The highwall mining system is to extract coal with an auger machine or a continuous miner. The continuous miner system includes the Addcar system and the Archveyor system. With the highwall mining system, there are many factors that have to be considered. Major factors are the coal recovery and the stability of the highwall.

A large amount of coal remains isolated and undeveloped as pillars due to previous indiscriminate mining operations performed by the use of the auger machine or the continuous miner. The former excavates holes up to 100 m long and 1.8 m diameter into the coal seams from highwalls. The latter excavates rectangular holes up to 300 m, and the size of the hole depends on the machines. Thus, the highwall mining causes highwall instability and also remains a precious coal reserve undeveloped as pillars.

The backfilling in the highwall mining would allow for a high coal extraction ratio while greatly reducing the threat of failure of the pillars and the highwall and damage caused by subsidence at the surface.

3 ENGINEERING ASPECTS OF BACKFILL

Waste disposal in underground mines is not a revolutionary concept. Waste rocks have been used in the world for many years as backfilling materials to provide additional support to underground excavations in mines. Recently, for the lack of a dumping site the use of backfill for regional and local support has been receiving increasing attention in the mining industry.

According to the previous research work (Afouz, 1994), backfilling presents the following merits to the underground coal mines:

1) Fills the excavated areas, promoting better support and ground control.
2) Provides a better environmental control on the waste rocks and coal preparation plant wastes.
3) Increases coal recovery, especially in room and pillar mining and highwall mining systems.
4) Reduces ventilation short-circuiting between adjacent mining sites.
5) Reduces cost of waste transportation to the mine surface dumping sites and tailing ponds, and associated up-keeping and monitoring of these facilities.

In the USA, increased opposition from environmental groups is severely restricting the operation and planning of large-scale surface mines. Some projects of mountaintop removal mining are having to be cancelled and downsized. In this case, as discussed in the previous section, the highwall mining system would be applicable and useful for protection of the environment and reclamation. The backfilling would increase the coal extraction ratio, keep the pillars and the highwall stable and control the subsidence at the surface.

4 BACKFILLING MATERIALS

Backfilling materials that are of concern to the Japanese coal mining industry can be broadly classified into the following three categories:

1) Waste originating from coal mines.
2) Waste originating from coal burning power plants.
3) Waste originating from other industries.

4.1 *Waste from coal mines*

The waste generating from coal mines is waste rocks from the underground sites and waste from preparation plants. The waste rocks that are inevitably produced during roadway drivage and repair work such as re-ripping and dinting are coal measures rocks, such as sandstones, shales, sandyshales, mudstones, conglomerates and so on. At Ikeshima Colliery, Japan, about 1.2 million tons of clean coal is produced and over 0.4 million tons of waste rocks is transported from the underground sites to the surface dumping sites.

Moreover, a large amount of wastes from the preparation plant is produced. The properties of the wastes vary depending on the mineralogical contents of the mother rock in which the coal is embedded. The waste quality depends on the method of mining and cleaning. The waste mainly consists of clays, quartz, carbonaceous materials, mica, pyrites and so on. At Ikeshima Colliery, about 0.5 million tons of wastes is produced from the preparation plant every year.

When using waste rocks, some problems for backfilling arise from the deterioration of the rocks due to weathering and slaking, loss of compacted strength and the eventual formation of acid water as a product of the leeching of the acid producing ingredients with in the waste, such as pyrite. There are other problems with using coal refuse, namely, the liberation of methane and the possibility of sudden ignition due to the amount of carbonaceous materials within the wastes.

4.2 *Waste from power plants*

The wastes generating from coal burning power plants are flyash, scrubber sludge, ash, spent bed, and fluidized bed wastes. In Japan, these types of wastes are generally called "coal-ash" and classified into two groups: flyash and clinker-ash (JFA, 1995). In Japan, over 130 million tons of coal is imported and about 40 million tons of coal is burned annually for generating electricity. Approximately 15 percent of all the coal burnt is transformed to coal-ash of which 85 to 95% is flyash and 5 to 15% is clinker-ash. The size of clinker-ash is larger than that of flyash. Therefore, it is crushed for utilization.

About 55% of coal-ash is now utilized and the remaining is disposed of at the disposal sites. However, the life of the disposal site is limited. It is difficult to find a new disposal site. It is requested that the percentage of the utilization of the coal-ash has to be increased in every field in Japan. However, the quality of flyash, depending on the coal type, restricts the extensive use of flyash.

The predominately spherical particle shape of the flyash provides a smooth flow of the mix by acting as a lubricant that increases the flowability of the mix and reduces the wear and tear of the equipment. The particle size range is from less than 0.1 up to 1 mm, the majority (90%) being less than 0.1 mm.

The addition of flyash also increases the specific surface area per unit weight and produces beneficial aspects on: segregation and rate of settlement, permeability, bleeding and pozzolanic activity.

Moreover, the low heat properties of flyash effectively reduce the temperature of hydration in mixes containing Portland cement by as much as 30%. In Japan, the mixes are called flyash cement and there are three types of Portland flyash cement: class A contains 5 to 10% flyash, class B; 10 to 20% and class C; 20 to 30%. The mix cement contributes to a cost saving of concrete mix without adversely affecting the long-term strength of the concrete.

The chemical composition of the flyash is shown in Table 1.

Table 1. Chemical composition of flyash.

Coal / Composition	Domestic coal	Oversea coal
SiO_2 (%)	50~55	40~75
Al_2O_3 (%)	25~30	15~35
Fe_2O_3 (%)	4~7	2~20
CaO (%)	4~7	1~10
MgO (%)	1~2	1~3
K_2O (%)	0~1	1~4
Na_2O (%)	1~2	1~2

4.3 *Waste from other industries*

The waste generating from other industries is industrial waste, such as soil from construction sites and waste concrete from destroyed old buildings and domestic garbage. A great deal of combustion residue is produced from domestic refuse incineration plants every year. The amount of this type of waste is expected to be higher each year. The public is less accepting of the storage of these materials in the surface dumping sites. In the past, the possibilities of depositing these materials in underground stable excavations where accessibility can be guaranteed have been investigated intensively. However, in Japan, the law strictly regulates the use of waste as backfilling material at underground mines.

5 BACKFILLING TECHNIQUES

5.1 *Backfilling systems*

The backfilling system depends on a variety of factors, the most important being the ground conditions, mining methods, constituents and the state of the mineral to be backfilled in the mined-out areas.

Backfilling operations can be categorized into either cyclic or delayed placement. In the cyclic placement, the backfilling materials are placed as part of the mine production cycle. On the other hand, in the delayed placement, the backfill is placed after the depletion of the mining.

The backfilling technique is also called stowing or packing. It is practiced mainly to backfill horizontal to vertical seam excavations such that exists in longwall, shortwall, room and pillar and highwall mining. There are four principal modes of placement of the backfilling materials to the underground sites: gravity, mechanical, hydraulic and pneumatic stowing.

5.1.1 *Gravity stowing*

The backfill placement is done by the force of gravity. This technique is economically feasible in workings steeper seams.

5.1.2 *Mechanical stowing*

This is also called conveyor stowing, in which the backfilling materials are transported by a conventional conveyor to the mined-out area. It is then transferred to a jet or high-speed conveyor and is thrown to the mined-out area.

5.1.3 *Pneumatic stowing*

The stowing material is transported in pipes by compressed air, and thrown to the backfilled area. The mine should have an ample supply of compressed air. The pipes are lined inside with anti-wear material to minimize the wear.

5.1.4 *Hydraulic stowing*

The waste rock is mixed with cement, flyash and water. It is the most advanced backfilling system in mining. The material is pumped into the mined-out area through pipeline. It requires a preparation plant at the mine surface or near the mining site to be stowed, a fill pump, pipelines, sump and ditches and a water pump to direct excess water to the preparation plant for recycling. The stowing material must be small enough to be pumpable and yet large enough not to stay suspended in water.

5.2 *Laboratory experiments*

In this experiment, different combinations of waste rocks, flyash and Portland cement were tested as a backfill mixture.

5.2.1 *Strength test*

As waste rocks, shale taken from Ikeshima Colliery was used. Mechanical properties of shale deteriorate significantly due to water and some types of shale exhibit a slaking phenomenon (Matsui et al., 1990). This shale is weak and shows the slaking effect in water. The mechanical properties (unconfined compressive strength; UCS, Brazilian tensile strength; BTS and tangent Young's modulus at 50% of UCS; E_{50}) decrease with increasing water content as shown in Figure 1.

Three types of shale aggregate size were considered in the experiment. The Ikeshima shale blocks were crushed with a hammer and classified into the following three sizes: they are the size of 1-5mm, 5-20mm, and 20-40mm. The shale

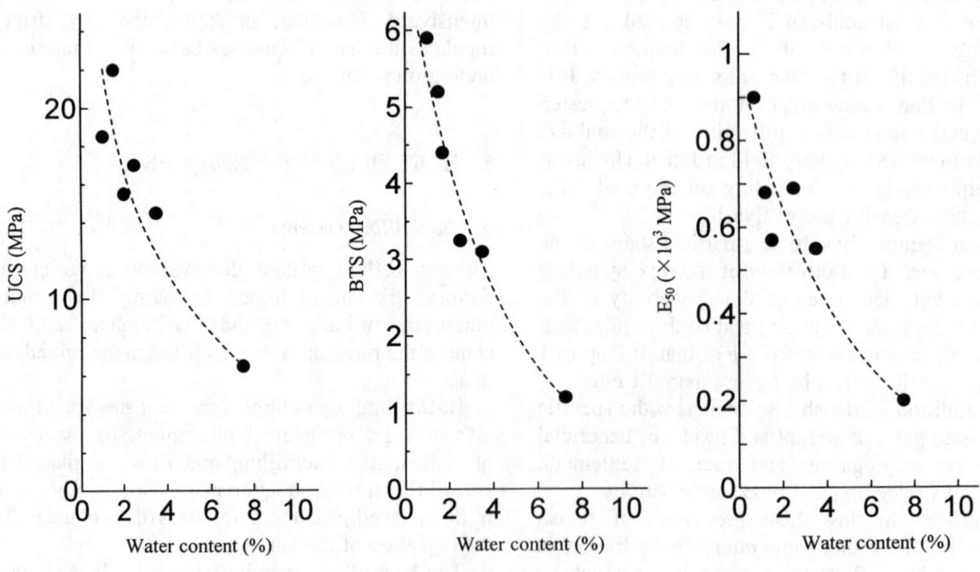

Figure 1. Relation between mechanical properties and water content for Ikeshima shale.

aggregates were cured at room temperature for four weeks.

The weight ratio of the flyash and cement slurry was as follows:

flyash : cement : water = 27:3:20.

A square column mold (100 mm ×100 mm× 200 mm) was used to produce the backfill mixture specimen.

Four types of curing conditions were considered, i.e. 3, 7, 14 and 28 days in the curing boxes.

The first stage of the experiment was to make a molded specimen. First the shale aggregates were filled in the mold, and then the flyash and cement slurry was pored using a small vibration. The porosity of the shale aggregate specimen or the volume percentage of the slurry is 32% for the 1-5 mm shale, 50% for the 5-20 mm shale and 55% for the 20-40 mm size shale. Finally these molds were placed under dry conditions for two days. After that, the specimens were removed from the molds and placed into the water curing boxes for the described period. For comparison, only flyash and cement slurry specimens were made.

Unconfined compressive strength test was performed for the specimens after curing. Each strength test was run for 3 specimens and the test results were averaged to obtain the mean value of their properties.

Figures 2 and 3 show the relationship between UCS/E_{50} and curing time for specimens. The increase in curing time affects the increase in the properties. The addition of weak shale reduces the mixture strength and modulus. There is a clear difference between shale aggregate sizes. The reduction rate for the large shale aggregate specimen is the highest among the three aggregates, because the shale is weak itself. The less the shale size, the larger the strength of the shale mixture.

5.2.2 *Discussion*

The required strength of the backfill depends on the strata and mining conditions: cover depth, rock types and properties, mining method, etc. At a shallow mine, the required strength is not critical compared to that at a deep mine. According to the research work in South Africa, backfilling shallow underground room and pillar mines with pulverized flyash slurry could significantly improve the mine operating conditions by providing improved roof support and increased extraction ratio (Wagner et al, 1979). The research also suggests that the complete filling of the rooms, up to the roof level, is not necessary for improved roof control. Filling up to 70% of the pillar height provided adequate confinement that constrained the lateral expansion of the pillar under concentrated compressive stresses. Moreover, by providing confinement to the pillars, thick coal seams have been successfully mined with a subsequent reduction of pillar height to width ratio and improved load bearing capacity of the pillars.

An increase in the amount of Portland cement causes an increase in the compressive strength of the backfill mixture. For example, in the case of water-cement ratio of 0.4, the UCS of the 20-40 mm size shale sample is 8.0 MPa (Matsui et al., 1999). It is possible to use the water-cement ratio of down to 0.4 in practical placement. Moreover, it is useful to use stronger rock as aggregates instead of weak rock

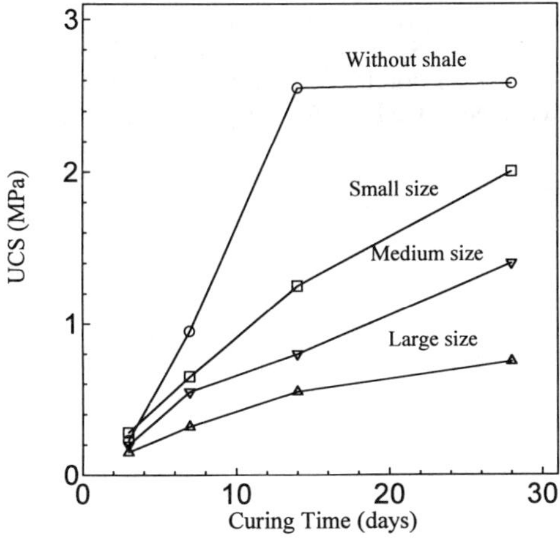

Figure 2. Relation between UCS and curing time.

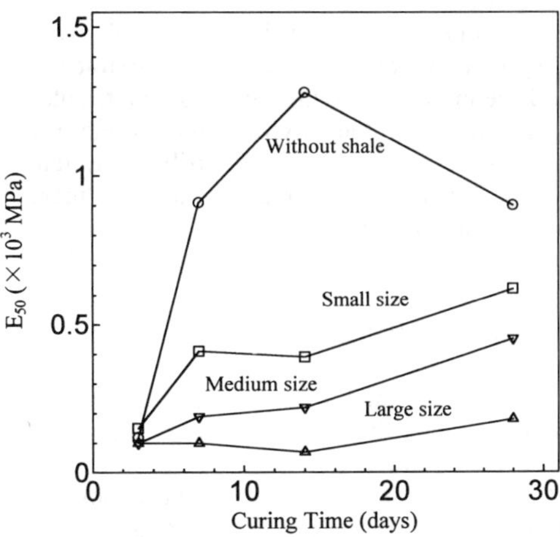

Figure 3. Relation between E_{50} and curing time.

in the cement slurry. The stronger the rocks, and the less the water-cement ratio, the larger the strength of the backfill mixture becomes. However, shrinkage up to 2% of the original volume occurs on drying. This means that there will be a gap between the backfill and the roof of the excavation. This is undesirable because it causes a reduction the supporting capability of the roof and the walls. Shrinkage can be reduced by the following measures:

1) Eliminating excess water through increasing the backfill slurry density.
2) Increasing the water drainage rate during the backfill placement. This will minimize the amount of the water left in the backfill after placement, which in turn minimizes the shrinkage.

Heat during the hydration reaction restricts the use in large volume in gassy mines.

Ideally, a backfill material, in addition to the strength requirements, should exhibit other properties, such as: quick setting and consolidation, non-toxic and properly sized for transportation to the underground sites. Another important property is the swelling characteristic of the backfill material. This would ensure the proper filling of voids up to the roof level and at the same time provide total confinement of the pillars.

6 CONCLUSIONS

An active use of mined-out areas of underground coal mines as a waste disposal site is very useful to solve the problem of the rapidly increasing production and stockpiling of wastes from power plants and mines. The backfilling in the highwall mining system not only increases the coal extraction ratio and the highwall stability, but also contributes to the protection of the environment around the mines. However, the optimal backfilling system needs more intensive investigation, because there are many factors to be considered.

ACKNOWLEDGMENTS

The authors are grateful to the managers and engineers of the Ikeshima Colliery for their assistance in this study.

All the opinions stated in this paper are those of the authors themselves and are not necessarily those of the colliery.

REFERENCES

Afrouz, A.A. (1994).Placement of backfill, *Mining Engineer*, February, pp. 205-211.

Anwar, H.Z., Shimada, H. Ichinose, M. and Matsui, K. (1998). Fundamental studies of the improvement of coal mine shales behavior, *Proc. of Regional Symposium on Sedimentary Rock Engineering*, Taipei, Taiwan, R.O.C., pp.45-50.

Anwar, H.Z., Shimada, H. Ichinose, M. and Matsui, K. (1999). Roof bolting application in longwall mining in Indonesia and Japan, *Proc. of 18th Int. Conference on Ground Control in Mining*, Morgantown, WV, USA, pp.256-262

Anwar, H.Z., Ichinose, M. Shimada, H. and Matsui K. (1999). Changes of mechanical properties of Indonesia coal measure rocks due to water, Hokkaido Geotechnics'99, No.10, pp.61-66, (in Japanese).

JFA (Japan Flyash Association). (1995). *Coal Ash Handbook*, (in Japanese).

Matsui, K., Shimada, H. and Ichinose, M. (1990). Characteristics of mechanical properties of coal measures rocks and coal and their application to mining, *Proc. 8th Symposium on Rock Mechanics*, Tokyo, Japan, pp. 375-380, (in Japanese).

Matsui, K., Shimada, H., Ichinose, M. and Anwar, H.Z. (1999). Reinforcement of slaking-prone mine tunnel floor by grouting, *Proc. Mine Planning and Equipment Selection 1999 & Mine Environmental and Economical Issues 1999*, Dnipropetrovsk, Ukraine, pp.253-260.

Simada, H., Matsui, K. and Anwar, A.Z. (1998). Control of hard-to-collapse massive roofs in longwall faces using a hydraulic fracturing technique, *Proc. 17th Int. Conference on Ground Control in Mining*, Morgantown, USA, pp.79-87.

Wagner, H. and Galvin, J.M. (1979). Use of hydraulically placed PFA to improve stability in bord and pillar workings in South African collieries, Symposium on the Utilization of Pulverized Fuel Ash, Pretoria, South Africa, Report No. CONF-7906215, pp.27.

Environmental Issues and Management of Waste in Energy and Mineral Production, Singhal & Mehrotra (eds)
© 2000 Balkema, Rotterdam, ISBN 90 5809 085 X

Subsidence prediction in Estonia's oil shale mines

J.R.Pastarus
Tallinn Technical University, Estonia

A.Toomik
Institute of Ecology, Estonia

ABSTRACT: This paper analysis the stability of the mining blocks in Estonian oil shale mines, where the room-and-pillar mining system is used. The pillars are arranged in a singular grid. The oil shale bed is embedded at the depth of 40-75 m. The processes in overburden rocks and pillars have caused the subsidence of the ground surface. The conditional thickness and sliding rectangle methods performed calculations. The results are presented by conditional thickness contours. Error does not exceed 4%. Model allows determining the parameters of spontaneous collapse of the pillars and surface subsidence. The surface subsidence parameters will be determined by conventional calculation scheme. Proposed method suits for stability analysis, failure prognosis and monitoring.

1 INTRODUCTION

The most important mineral resource in Estonia is a peculiar kind of oil shale. It is located in a densely populated and intensely farmed district. The structure of the productive oil-shale bed makes the rocks more difficult to break from the total massive. It is estimated that about 80-90% of the total underground oil-shale production is obtained by room-and-pillar method. The method is cheap, highly productive, easily mechanized, and relatively simple to apply. The area mined by this method reaches 100 km^2. It has become apparent that the processes in overburden rocks and pillars have caused unfavorable environmental side effects accompanied by significant subsidence of the ground surface. The horizontal bedding and small depth of oil shale seam enable the roof deformations to reach the land surface without essential reduction. The first spontaneous collapse of the pillars and surface subsidence took place in 1964. Up to the present, 39 failures in Estonian oil shale mines have been registered, which make up 9% from the total number of mining blocks and 2.5% from the mined out area.

Identification of the reasons of the surface subsidence, elucidation of the basic mechanism of this process and elaboration of the method of prognosis are the main aim of the present work.

The studies are based on the complex method, including:

- Investigations of in-situ conditions;
- Theoretical investigations;
- Modeling on PC.

For the feasibility study, stability problems were investigated. For the analysis the methods of conditional thickness and sliding rectangle have been used (Talve 1978). Visual Basic for application in Exel and MapInfo was used for numerical modeling. The results are presented by conditional thickness contours, which allows determining the failure and surface subsidence parameters. The method suits for stability analysis, prognosis and monitoring. It gives excellent results.

2 GEOLOGICAL CONDITIONS AND MINING METHOD

The commercially important part of oil-shale stretches from west to east for 200 km (from Rakvere to Luga), and from north to south for 30 km (from the coast of Gulf of Finland up to the shore of Lake Peipsi). The oil-shale bed lays in the form of a flat bed having a small inclination (2-3 m per km) in southern direction. The depth of the oil-shale bed varies: on the northern border of the field the bed lies under the Quaternary sediments, at the southern rim it lies even at the depth of 100-150 m. The thickness of the commercial oil-shale bed somewhat decreases west- and southward of the central part of the deposit. The reserves of oil shale in Estonia are estimated approximately at 4 thousand million tons (Estonian 1997).

The oil-shale seams occur among the limestone seams in Kukruse Regional Stage of the Middle Ordovician. It is a stratified sedimentary rock, rich in organic matter. The commercial oil-shale bed and immediate roof consists of oil shale and limestone seams. The main roof consists of carbonate rocks of various thickness.

The characteristics of the individual oil shale and limestone seams are quite different. The compressive strength of oil shale is 20-40 MPa and that of limestone is 40-80 MPa. The strength of the rocks increases in the southward direction. The volume density is 1.5-1.8 Mg/m^3, and 2.2-2.6 Mg/m^3 respectively. The calorific value of dry oil shale is about 7.5-18.8 MJ/kg depending on the seam and the area in the deposit.

In Estonian oil-shale mines the room-and-pillar mining system is used. The field of an oil shale mine is divided into panels, which are subdivided into mining blocks, approximately 300-350 m in width and from 600 to 800 m in length each. A mining block usually consists of two semiblocks. The oil-shale bed is embedded at the depth of 40-75 m. Its height corresponds to the thickness of the commercial oil-shale bed, approximately 2.8 m. The width of the room is determined by the stability of the immediate roof. The latter is very stable when it is 6-10 m wide. In this case, bolting must still support the immediate roof. The pillars are arranged in a singular grid. Actual mining practice has shown that pillars with a square cross-section suit best. The cross-sectional area of the pillars is 30-40 m^2, depending on the depth of the oil-shale bed.

The main operations carried out in rooms include bottom cutting, drilling of blastholes, blasting, loading of oil shale and supporting. A work cycle lasts for over a week.

3 PILLAR AND ROOF STABILITY PREDICTION METHOD

Pillar design in Estonian oil-shale mines is based on the tributary area theory using the Sheviakov's and Turners' calculation scheme (Room 1997). It is important to mark that in-situ condition pillar loads vary from place to place within a mining block. Consequently, the mining block stability depends on the real parameters of the pillar and roof. For the stability analysis the concept of critical width, methods of support coefficient, conditional thickness and sliding rectangle were used. They suit for modeling on PC.

The support coefficient and conditional thickness are presented by the following formulae (Pastarus 1982, Talve 1978):

$$K = \frac{S_p}{S_r}, C = \frac{H}{K} \qquad (1)$$

where K - support coefficient; C - conditional thickness, m; S_p - cross-sectional area of a pillar, m^2; S_r - roof area per pillar, m^2; H - thickness of the overburden rock, m.

Geometrical interpretation of support coefficient and conditional thickness is given in Figure 1. Conditional thickness represents the height of a prism whose cross-section equals the pillar cross-section area. Consequently, conditional thickness is related to the load on a pillar. If the load is too much for the pillars, a sudden failure is likely.

Pillar load depends on the width of the mining block, leading to the concept of the critical width.

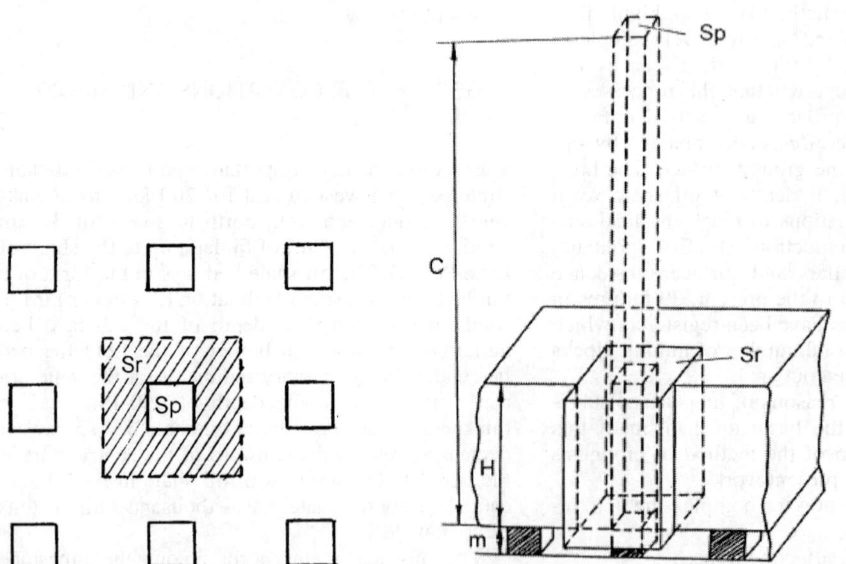

Figure.1. Geometrical interpretation of support coefficient (A) and conditional thickness (B) (Talve, 1978)

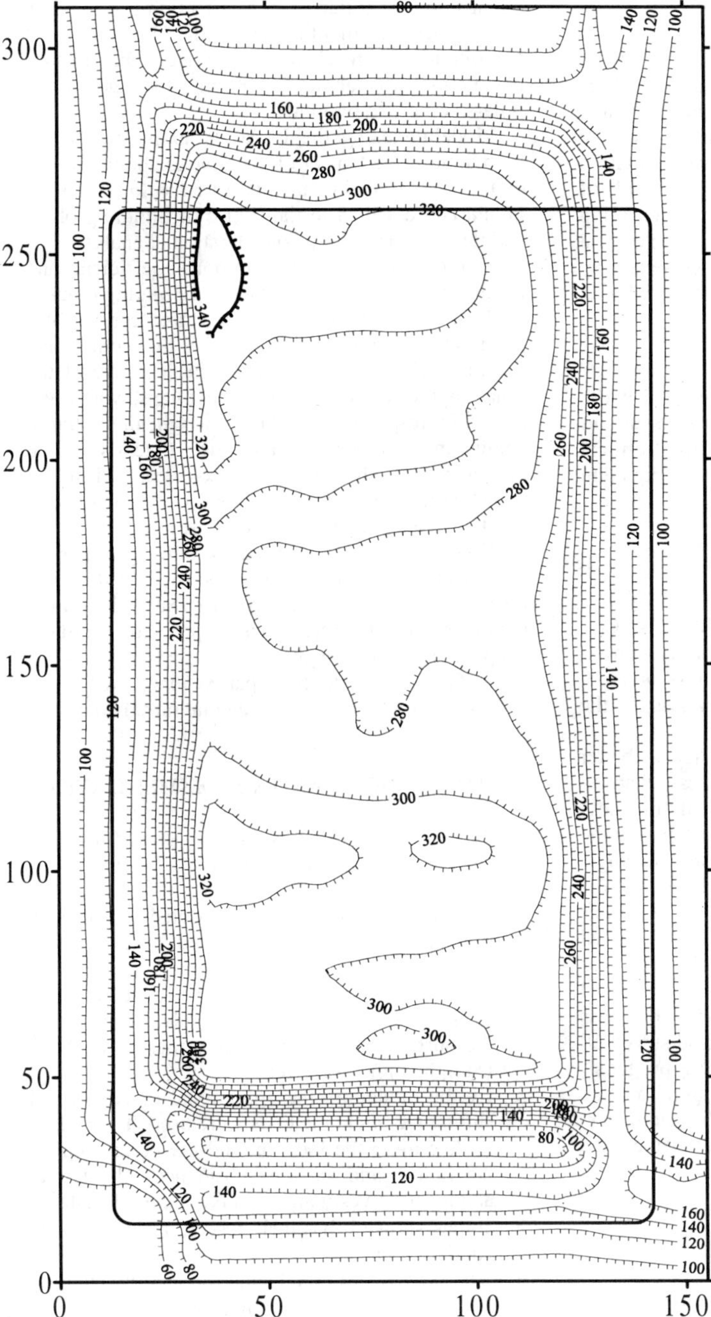

Figure.2. Support coefficient (A) and conditional thickness (B) contours of the left semiblock No.60 of the mine Ahtme

The critical width is the greatest width that the rock above the mine can span before its failure, or, if there are pillars, the width we must mine before the pillars accept the full weight of the overlying materials (Parker 1993). Typically, full load comes onto the pillars closest to the centre of a mining block. The load was determined using the two-dimensional Fast Lagrangian Analysis of Continua (FLAC) program. (FLAC 1993). FLAC has several built-in material behavior models and is particularly suitable for modeling non-linear, large-strain and physically unstable continuous systems. In fact, the best indicator of critical width in a given situation will be provided from old mine maps, by records of failures and surface subsidence, and from measuring roof-and-floor convergence in the mines. The critical width for Estonian oil shale mines is presented by the following formula (Room 1997, Stetsenko & Ivanov 1981):

$$L \geq 1.2H + 10 \tag{2}$$

In the three-dimensional case, the critical width transforms into the critical area. The average support coefficient and conditional thickness for a critical area can be expressed by the following equation (Pastarus 1982, Talve 1978):

$$K_C = \frac{\sum S_{pi}}{\sum S_{ri}}, C_C = \frac{H_a}{K_C} \tag{3}$$

where K_C - support coefficient for critical area; C_C - conditional thickness for critical area, m; S_{pi} - cross-section area of the i-th pillar, m^2; S_{ri} - roof area per the i-th pillar, m^2; H_a - average thickness of the overburden rocks for critical area, m.

Analysis showed that the critical area is better related to average conditional thickness than to average support coefficient. Conditional thickness contains more information and allows comparing the blocks in different depth. By the sliding rectangle method, the average conditional thickness of the critical area must be determined for all positions inside a mining block. For the modeling, the Visual Basic for application in Excel and MapInfo was used. The results are presented by conditional thickness contours. Error of conditional thickness contours does not exceed 4%. The method suits well for stability analysis, spontaneous failure prognosis and monitoring. The land deterioration on the mined-out territories is evaluated by the parameters of ground movement mechanism, using the conventional calculation methods.

4 RESULTS

Analysis by conditional thickness and sliding rectangle methods was made for 11 mining blocks (the mines Ahtme and Estonia). The investigation results of the mining block No.60 of the mine Ahtme are presented below.

The commercial oil shale bed of the thickness of 2.8 m is embedded at the depth of 50 m. Mining block is bordered by barrier pillars. A spontaneous collapse of the pillars in the left mining semiblock took place 8 months after the beginning of exploitation. It reached the surface. The area of destruction was about 26400 m^2, the age of the pillars being 1 to 6 months.

Modeling using the FLAC-program shows that the width of the barrier effect zone is about 1/4 -1/5 of the critical width. In this area the loads on the pillars are less than in the centre of a mining block and must be taken in to consideration at analysis.

Figure 2 presents the support coefficient and conditional thickness contours of the mining block.

Analysis shows that there are three centres of a potential collapse in the case of the conditional thickness C>340 m and C>320 m. The age of the pillars in the centre of a potential collapse was 1 and 2 months respectively. Likely enough, the collapse begins in these centres and then extends to the barrier pillars. The results of theoretical and in situ investigations in Estonian oil shale mines showed that they are close to the modeling results.

This analysis was based on the geometrical parameters of mining blocks. The method does not take into consideration the rheological and strength parameters of the rocks. These problems demand supplementary investigations.

Surface subsidence parameters will be determined by conventional calculation schemes.

5 CONCLUSIONS AND RECOMMENDATION

As a result of this study, the following conclusions and recommendations can be made.

1. Estonian oil shale mines are located in a densely populated and intensely farmed district where the room-and-pillar mining method is used. Pillar design is based on the tributary area method. The collapse of the pillars has caused the surface subsidence, which makes up 2.5% from the mined-out area.
2. By the modeling, the conditional thickness and sliding rectangle methods were used. They allow to determine the location and the parameters of the failure and surface subsidence. Theoretical and in situ investigations in Estonian oil shale
3. mines showed that their results are close to the modeling ones.
4. The proposed method suits for stability analysis, failure prognosis and monitoring.
5. Further investigations are aimed at the rheological and strength behavior of the rocks to determine the time and exact location of the collapse of pillars and ground surface.

The research was supported by Estonian Science Foundation, Grant No. 3651, 1999.

REFERENCES:

Borissov,A. 1980. Mechanics of Rock and Rock Masses., Moscow: Nedra (in Russian)

Estonian Statistics. 1997. 8(68), Tallinn (in Estonian)

FLAC,. Fast Lagrangian Analysis of Continua, 1993. Version 3.2, Vol.1. User's Manual. Minneapolis: Itasca Consulting Group, Inc.

Parker,I. 1993. Mine pillar design in 1993: Computers have become the opiate of the mining engineers Mining Engineering, July and August: 714-717 and 1047-1050. London

Pastarus, J. 1982. Study and analysis of the mining blocks situation in the mines of "Estonslanets" Transactions of Tallinn Polytechnic Institute 533: 3-14 (in Russian)

Room, pillars and safety zones calculation methods for underground oil-shale mining. 1997 Tallinn [in Estonian]

Stetsenko, V., Ivanov, G. 1981. Prognosis of the time-depending roof displacements for different span lengths. Oil-Shale11: 13-18 (EstNIINTI) (in Russian)

Talve, L. 1978. Check-up of real parameters of room-and-pillar mining in Estonia's oil shale mines. Ibid.451: 23-35 (in Russian)

REFERENCES

Brady, [illegible] 1985. Mechanics of Rock and Rock [illegible] (in Russian)

[illegible] 1997 [illegible] Institute [illegible] Estonia.

[illegible] AC [illegible] Analysis of Continua. 1991. Version [illegible] Manual. Minneapolis [illegible] Consulting Group, Inc.

[illegible] 1995. Mine pillar design in 1990. [illegible] the [illegible] mining [illegible] Mining Engineering. July and August [illegible] London.

[illegible] 1982. Stress analysis of the mining blocks [illegible] the mines of [illegible] Transactions of Tallinn Technical Institute [illegible] (in Russian).

[illegible] and safety [illegible] method [illegible] underground oil shale mining [illegible] Tallinn (in Estonian).

[illegible] G. 1981. [illegible] time [illegible] displacement [illegible] pillars [illegible] (in Russian).

[illegible] parameters [illegible] mining [illegible] shale [illegible] (in Russian).

Environmental Issues and Management of Waste in Energy and Mineral Production, Singhal & Mehrotra (eds)
© 2000 Balkema, Rotterdam, ISBN 90 5809 085 X

Environmental effects of increased blasting density in Appalachian limestone operations

Wm. R. Reed, E.C. Westman & C. Haycocks
Department of Mining and Minerals Engineering, Virginia Polytechnic Institute and State University, Blacksburg Va., USA

ABSTRACT: Studies have shown that an increased blasting density in Appalachian limestone quarries results in lower overall operational costs. Expanded blasting operations, however, exacerbate environmental side affects including ground vibrations, noise, flyrock, dust and increased nitrates in water discharged from mine sites. A study was made to determine if the environmental problems created were sufficient to negate the mining benefits of increased blasting.

1 INTRODUCTION

Blasting operations are performed to loosen and fragment the limestone in order to load, haul, and process the rock to its final product stage. Many studies have shown that as the blasting density is increased, there is a significant savings in both the primary and secondary crushing and grinding costs. One such study demonstrates that reducing the spacing and burden of a blast by 25% could result in a 10% savings for crushing and grinding costs (Fuerstenau, 1997). However, most studies that review the fragmentation of blasting operations generally do not evaluate the environmental effects of the increased blasting density. Therefore, a preliminary study to determine the environmental effects of blasting at limestone operations in the Appalachia area was performed. This study identified problem areas in blasting operations that may require further research.

2 METHODOLOGY OF SURVEY

A survey was conducted by interviewing quarry personnel who were knowledgeable about blasting-related environmental problems. All people contacted worked at operations located in Virginia. Initial contact was made by telephone, and later surveys concerning the blasting operations were completed by either personal interview or via telephone. Five persons, representing companies that owned or operated at many different quarry sites throughout the Appalachian region, agreed to participate with this survey.

The survey was developed to gather information concerning the blasting operations of the many different quarry sites, and to determine any negative environmental effects caused from blasting. Originally the survey was designed to gather specific information for each quarry site. However, preliminary interviews revealed that all individuals questioned worked at more than one quarry. The number of quarries worked at ranged from three to fifteen per individual. The time required to list site-specific information for this survey would have been unreasonable to expect from an interviewee. Thus, modifications were made to the survey in order to allow for the collection of multi-mine instead of site-specific data. Most questions were open-ended, requiring the interviewee to explain in greater detail their blasting operations.

3 SUMMARY OF BLAST DESIGN

After reviewing the results of the survey, a generalized blast design can be established for limestone quarrying operations. Table 1 shows the ranges for the different blasting design parameters. Many of the blast design parameters are generalized according to the individual discussions. It should be noted that these could vary widely, depending upon the different sites.

When the interviewee was asked whether the current design factors used by the individual operations were providing satisfactory fragmentation, the interviewee response was "yes" 80 percent of the time. Previous experiments at selected operations to improve fragmentation using improved blast design were carried out by making changes, such as reducing burden and spacing or increasing the powder factors. Qualitative analysis of the results showed an increase in blasting costs, with very little improvement in fragmentation.

Table 1. Typical design parameters for blasting operations.

Blast design category	Blast design
Blasthole diameter Millimeters (mm)	89 for steep topography 165 for flat benches
Bench height Meters (m)	12.2 – 18.3, some up to 27.4 Above 18.3 MSHA becomes concerned with bench height
Spacing & burden Meters (m)	Based on powder factor Typically 1.00 kg of explosive/metric ton of rock Typical of 4.3 x 4.9 to 4.9 x 5.8
Size of blast Kilograms (kg) of explosive	3175 – 6800 4535 average size
Type of explosive used	Ammonium Nitrate and Fuel Oil (AN/FO), dry conditions AN/FO Emulsion Blends, wet conditions
Frequency of blasting	Production rate: 1,800,000 metric tons of rock - 2 times/week 900,000 metric tons of rock - 1 time/week Smaller - 1 time/ 2-3 weeks

It is not known whether the fragmentation was evaluated from just a blasting perspective, or if the fragmentation was assessed by evaluating the entire operational costs, which would include crushing costs. In either case, the experimentation did not result in any permanent changes in the quarries blast designs. In all discussions, no formal documented experiments were completed, to show that increased blasting density results in savings of crushing and grinding costs.

4 BLASTING REGULATIONS

In reviewing the environmental regulations for blasting in the United States, the federal government does not regulate metal/nonmetal mining operations as stringently as they do coal mining operations. The Office of Surface Mining regulates blasting operations for environmental effects in coal mining operations, but these regulations do not apply to metal/nonmetal operations (National Archives and Records Administration, 1999). The regulation of effects of blasting is left to the individual states. These regulations can be different from state to state, but generally they follow the blasting regulations published by the Office of Surface Mining. The Virginia Department of Mining, Minerals, and Energy (DMME) regulates Virginia's quarrying operations.

The Mine Safety and Health Administration (MSHA) is responsible for enforcing safety regulations for all mining operations in the United States. MSHA has regulations for the transportation, storage, and safe use of explosives, but MSHA does not regulate that blasting operations control the use of explosives based on their environmental effects (National Archives and Records Administration, 1999).

In addition to the completed survey of the limestone operations, information concerning past blasting violations and blasting complaints from 1996 to 2000 was gathered for the state of Virginia from the DMME. The information is shown in Table 2 and Table 3. It should be noted that this only represents the number of complaints called to the DMME; it does not include any complaints called to the quarries.

Table 2. Type and number of violations against quarrying operations in Virginia.

Type of violations	Number of violations
Exceeded ground vibration limits	8
Exceeded airblast limits	13
Flyrock	1
Exceeded lbs of explosive/delay	9

Table 3. Type and number of complaints against quarrying operations in Virginia.

Type of complaints	Number of complaints
Ground vibration	139
Airblast	13
Flyrock	9
Dust	9
Subsidence	2
Loss of well-water	7

The number of complaints in the table is actually greater than the number of complaints logged by the DMME. Many of the categories listed above were mentioned in single complaints taken by the DMME. In those cases, the categories were separated out of the single complaint. For example, a complaint might state, "a blast shook my house, was very loud, and I saw a large dust cloud." In this case the DMME would log it in as one complaint. However, for the purposes of this study, this complaint would count as one ground vibration complaint, one airblast complaint, and one dust complaint.

It was interesting to note that eight blasts exceeded the ground vibration limit. While 139 complaints concerning ground vibrations from blasts at quarries were recorded. This shows that even though the quarries are generally within the legal limits there is still a noticeable effect to the public from blasting at the quarries and that the public perceives ground vibrations from blasting as the principle threat from limestone operations.

5 ENVIRONMENTAL EFFECTS OF BLASTING

5.1 *Ground vibrations*

Ground vibration from blasting operations was the most prevalent problem for blasting operations. The ground vibrations are measured by determining the peak particle velocity (PPV), using a seismograph. The PPV is used for establishing ground vibration limits. A PPV of 1 inch/second, [the ground vibration limit established by the DMME] is the limit used by most quarries. Table 4 shows the limits imposed by the state of Virginia (Virginia Department of Mines, Minerals, and Energy, 1998).

Table 4. Regulations for maximum allowable PPV for ground vibrations in Virginia.

Distance (D) from blasting site (feet)	Maximum allowable peak particle velocity for ground vibration* (inches/second)	Scaled distance (D_s) factor to be applied without seismic monitoring**
0 to 300	1.25	50
301 to 5,000	1.00	55
5,001 and beyond	0.75	65

* Ground vibration shall be monitored as the particle velocity. Particle velocity shall be recorded in three mutually perpendicular directions. The maximum allowable peak particle velocity shall apply to each of the three measurements.

** The following scaled distance equation shall be used whenever seismic monitoring is not used.

$$W = \left(\frac{D}{D_S}\right)^2 \tag{1}$$

Where W = maximum pounds of explosive per eight millisecond delay; D = distance to the nearest protected structure in feet; D_S = scaled distance factor from previous table.

The results of the survey indicate that the quarries generally did not have ground vibration problems with their blasting operations. Although they did acknowledge several complaints a year, they felt that ground vibrations from their operation were minimal.

Almost every blast was measured seismically. The measured PPV depends upon the geologic conditions of the site and the distance from the quarry to the nearest protected structure. Since these two factors are different for every site, it is impossible to get an average PPV for all operations. However, the range of PPV for blasting operations at the different quarries varied from 0.2 – 0.5 inch/second, and it was rare for a quarry to exceed the PPV limit of 1 inch/second when blasting.

The quarries have employed methods such as reducing the amount of explosive per delay or decreasing the duration of the blast to prevent ground vibrations. Decking the blastholes and having two initiations per blasthole at different delay intervals have reduced the amount of explosive used per delay. Reducing the number of blastholes per shot has decreased the duration of the blast, and has also resulted in lowering the amount of explosive used in the blast.

Complaints of ground vibrations from a blast were mitigated by meeting with the complainant to discuss the problem. Other measures to prevent complaints prior to blasting have been to notify the neighboring properties of the oncoming blast event prior to blasting and to offer tours of the operation to the neighbors. All of these measures have been helpful in reducing the number of ground vibration complaints.

5.2 *Flyrock*

DMME regulates the flyrock from a blast for coal mining by stating that flyrock cannot be thrown from the blast site, be thrown more than one half the distance from the quarry to the nearest dwelling or occupied structure, or be thrown beyond the permit boundary (Virginia Department of Mines, Minerals, and Energy, 1994). There is not a similar regulation for mineral mining, but this regulation is a good guide for the limestone operations to abide by.

Quarries rarely have a problem with flyrock, because they generally have an adequate distance between blasts and the nearest dwelling. Flyrock is also a visible hazard, and there is obvious damage whenever it occurs. This makes it harder for the quarries to deny such an incident.

Interviewees indicated that flyrock was relatively well controlled and not a problem at their quarries. Whenever flyrock does occur, it is generally an indication that the burden was not correct for the amount of explosive placed in the front row of blastholes. The incorrect burden causes flyrock to emanate from the blast.

Laser profiling the face of the blast can greatly reduce flyrock. Once completed, the profile can reveal areas where the burden is too excessive or too inadequate. In these areas, the amount of explosive placed in the blasthole can be adjusted accordingly to eliminate flyrock (Workman, 1999).

5.3 *Noise*

Noise from blasts is called air blast. The DMME regulates the amount of air blast allowable during a blast, as shown in Table 5 (Virginia Department of Mines, Minerals, and Energy, 1998). The amount of

Table 5. Regulations for maximum allowable decibel levels for airblast in Virginia.

Lower frequency limit of measuring system Hz (±3 dB)	Maximum Level Decibels (dB)
1 Hz or lower – flat response*	134 peak
2 Hz or lower – flat response	133 peak
6 Hz or lower – flat response	129 peak
C-weighted-slow response	105 peak dBC

* Only when approved by the DMME.

airblast allowable is measured in decibels (dBA) at the nearest dwelling or public building.

Airblast is measured simultaneously during the measurement of ground vibrations with the seismograph. Survey results indicated that the airblast was not a major problem. The amount of airblast for a blast ranged from 107 to 120 dBA.

Airblast has not been known to result in damage to any structure. It is more of a nuisance and often causes complaints. Generally when a complaint of airblast is received by the quarrying operation, the best method to mitigate it is to meet with the complainant and discuss the problem.

Air blast is known to be caused by the energy released from unconfined explosives, the energy release from improperly confined blastholes, and the movement of the burden and the ground surface (Virginia Department of Mines, Minerals, and Energy, 1994). Known measures used to prevent airblast include burying any unconfined explosives, such as detonating cord, to one foot or more, ensuring the amount of explosive in the front row of blastholes is correct for their corresponding burden, using stemming, such as crushed stone, which will not blowout during a blast, and avoiding blasting on low overcast days that indicate the presence of a temperature inversion (Virginia Department of Mines, Minerals, and Energy, 1994).

A temperature inversion causes the sound from the blast to be reflected back to the earth's surface instead of being dissipated in the atmosphere (Virginia Department of Mines, Minerals, and Energy, 1994). It is better to schedule blasts at noon or early afternoon when the chances of a temperature inversion existing are less. Blasting during these times is also better because most neighbors are not home or may be preoccupied with other activities, which may distract them from the blast.

5.4 *Dust*

After a blast occurs, generally a large cloud of dust forms. Dust is more pervasive when blasting during dry conditions. Currently, there are no known methods for reducing the amount of dust produced during a blast. Aside from any air quality regulations, there are also no known blasting regulations requiring the elimination of dust from a blasting operation.

A method to minimize complications from dust is to monitor the weather conditions when blasting. Windy days will cause the dust cloud to dissipate rapidly, but may move the dust cloud into unfavorable areas, depending upon the direction of the wind.

Some quarries go to great lengths to try to minimize dust from their entire operation, such as watering state highways at the entrances to their operations or purchasing a large tract of land so the dust dissipates before it leaves the property. While this does not eliminate any dust from blasting operations, it shows the quarry is proactive in addressing dust related problems, which may help public relations with their neighbors.

5.5 *Nitrates*

Nitrates in discharge water from mining operations have recently become a concern by the environmental regulatory agencies. Nitrates occurring in the discharge water are thought to come from the residue of incomplete detonation AN/FO in blasting operations or from spillage of explosive material, such as AN/FO, at the blast site (Revey, 1996). This problem is frequently present at mining operations that have significant groundwater flows. It is also possible to have nitrates in the stormwater effluent from mining operations, which would in turn be discharged from the mine site through the sites National Pollutant Discharge Elimination System (NPDES) permit.

In discussions with limestone operations in the Appalachia area, nitrates from blasting operations have not been a problem in the past. Many sites have to deal with the groundwater entering the quarry. Whenever samples of discharge water are taken, as required by NPDES permits, nitrates do not present themselves as a problem. In fact, one quarry that was required to test for nitrates in their discharge water found the amount of nitrates was in the lower range of parts per billion, which led to the conclusion that nitrates were not a problem at the site. Generally, the water discharged from limestone operations has no problems meeting the NPDES limits for discharge.

6 CONCLUSIONS

The results of this survey clearly demonstrate that, the limestone industry has few problems meeting the regulations for blasting as set by the government agencies. This may be due to the fact that most of the quarries have been in operation for many years and have accumulated vast knowledge and experience concerning the blasting operations at their site.

As far as mitigating any complaints or problems with the neighboring properties, the operations have been taking a proactive approach. They have been meeting with the public to inform them of the limestone operations and to educate them about the need for these quarries. They have also been working to improve their public image. Survey results indicate that these methods have been working well.

This study confirms that the most significant environmental effects of blasting in Appalachian limestone operations are ground vibrations. Vibrations generated from blasting are generally well below the legal limits set by DMME. However, the public still perceives the ground vibration from blasting as a threat. If blast vibrations are increased due to increased blast density, it is likely the public will react negatively, considering the current data on blasting complaints and violations.

Noise generated from blasting is generally below the legal limit, and therefore does not pose a problem except in a few cases. Flyrock is also generally not a problem and does not pose a threat, as most operations are located far enough away from neighboring properties. Dust can be a problem, but generally is not complained about. Nitrates in limestone operations do not seem to be a threat at all.

All of the environmental effects discussed could pose severe problems for the limestone operations if left unmanaged. But these operations have learned from past experience and are able to control their blasting operations to minimize any environmental effects. Overall, the limestone operations are managing the environmental effects from current levels of blasting. Therefore, the potential to reduce overall mining costs through increased blasting density becomes a viable proposition.

REFERENCES

Fuerstenau, M.C., Chi, G., Bradt, R.C., Ghosh, A., 1997. Increased Ore Grindability and Plant Throughput with Controlled Blasting. *Mining Engineering*, Vol. 49, No. 12: 70-75.

Revey, G.F., 1997. Practical Methods to Control Explosives Losses and Reduce Ammonia and Nitrate Levels in Mine Water. *Mining Engineering*, Vol. 48, No. 7: 61-64.

US National Archives and Records Administration, 1999. *Code of Federal Regulations, Title 30, Chapter I, Part 56, Subchapter N, Subpart E, Explosives*. US Government Printing Office.

US National Archives and Records Administration, 1999. *Code of Federal Regulations, Title 30, Chapter VII, Part 715.19, Use of Explosives.* US Government Printing Office.

Virginia Department of Mines, Minerals, and Energy, 1998. *Safety and Health Regulations for Mineral Mining 1998*, Commonwealth of Virginia.

Virginia Department of Mines, Minerals, and Energy, 1994. *Surface Blaster Certification Study Guide*. Commonwealth of Virginia.

Workman, L., 1999. Front Row Seat –Good Blast Performance, Safety, and Efficiency Begin in the Front Row. *Pit & Quarry*, Vol. 91, No. 11: 55-58.

Environmental Issues and Management of Waste in Energy and Mineral Production, Singhal & Mehrotra (eds)
© 2000 Balkema, Rotterdam, ISBN 90 5809 085 X

Analysis of Tunçbilek underground coal mine accidents based on risk analysis techniques

A.S. Selçuk
Department of Statistics, Middle East Technical University, Ankara, Turkey

C. Karpuz, H.Ş. B. Düzgün & M. Sari
Department of Mining Engineering, Middle East Technical University, Ankara, Turkey

ABSTRACT: Underground coal mine accidents result in many losses which directly reflects to the cost of mining. The randomness in the appearance of the accidents can be determined by frequency that considers how often a specific type of the accident may occur and the severity of the accident in which a financial loss arises resulted from the occurrence of the accident. Modelling these two basic variables statistically enables the researchers to estimate the probability of future losses and to determine the risk factor of the specific type of accidents. In this paper a risk assessment approach is considered to evaluate the risk profile in Tunçbilek underground coal mine in Turkey by using last four years of accident data.

1 INTRODUCTION

Underground coal mining has long been regarded as a hazardous activity everywhere in the world. The statistics presented by Turkish Social Security Institution (SSK) indicate how effective the coal mining losses are in Turkey on yearly base. Turkish coal mining industry ranks second following the construction industry in the occupational accidents, third in occupational disease, first in permanent injury and death (SSK 1991) and is the second after Korea in fatality rate according to International Labor Organisation statistics (ILO 1991). Underground coal mining is also considered as the most hazardous occupation in the United States (Coal Commission 1980).

Losses due to accidents have burden on the mining companies in terms of cost and administration. Production decreases as well as productivity as a result of accidents. Besides, the companies should struggle in a competitive market to survive. It has been shown by Bhattacherjee et al. (1992) that the total cost of accidents on average resulted in a loss of 1 million USD per year and an average loss of 5,000 USD per accident in a case mine. Therefore, main concern of mining companies should be to cut their expenses caused by accidents while increasing mine safety and their profit.

Risk assessment is one of the efficient methods which can be used in many fields for the identification, evaluation and control of risk. Since, it is an inter-disciplinary and practical method, this system is also to be applied in mining industry for the reasons stated above.

Implementation of risk assessment techniques in coal mining studies is limited in literature. However, many studies were done to diminish losses caused by accidents under the titles of job safety, worker health and job analysis (Golledge & Stokes 1985, Peay & Schaffer 1985, Butani 1988, Hayduk & Ritzel 1988, Leigh et al. 1991, Bozkurt 1993, Turin et al. 1995). These studies generally aimed to evaluate damages due to accidents/incidents and to take necessary precautions to prevent them before they occur. Methods for accident prevention analysis were developed by Gallagher et al. 1992, Grayson et al. 1992, Bhattacherjee et al. 1992, Staley & Foster 1996. Additionally, studies done on Practical Risk Assessment (Simpson & Moult 1995, Rowell 1996, Staley 1996, Van der Vyver 1997, Davies 1997) present the methods for identification of potential hazards, probability of occurrence and associated consequences of those hazards in the case of having a change in an existing operation or implementing a new system.

The objective of this study is to identify risk profiles of Tunçbilek Coal Mine based on several characteristics on past accident/injury data and to determine the statistical modelling of two basic variables, the frequency and severity of accidents, for the assessment of risk and evaluation of mine safety.

2 METHODOLOGY AND ANALYSIS

The coal mine considered here is one of the lignite reserves in Turkey. Tuncbilek mine is located in the south-west of Turkey. It is an underground mine and the mining method practiced is both classic and mechanized longwall mining.

From January 1996 to July 1999, a total of 599 incidents were reported. The accident data is categorised according to:

- name, birth date and job type of the injured
- the time, location and type of accident/incident
- body parts injured
- days off from work data (severity)

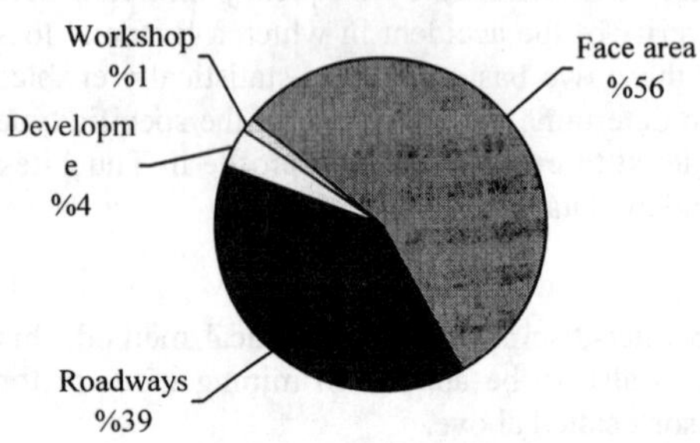

Figure 1. The distribution of accident locations

The accident/injury experience data were analysed initially by descriptive statistics techniques. For the location of accidents, the face area and roadways are the most hazardous places with 56% and 39% occurrences, respectively, as can be followed from Figure 1.

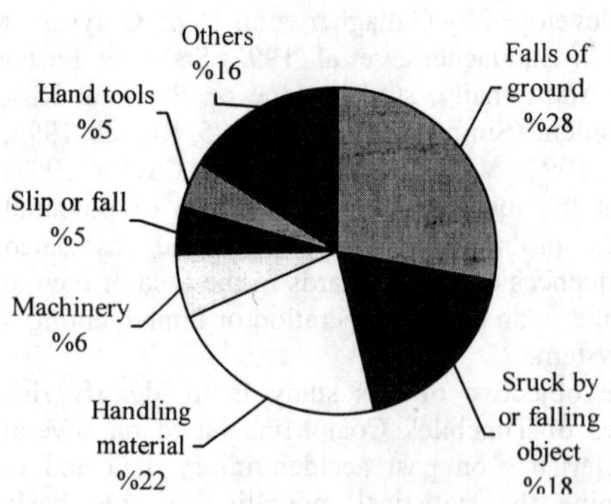

Figure 2. The distribution of accidents/injury types.

The analysis has also shown that the most common accident/injury types are falls of ground (roof, rock and coal), struck by or falling object and handling material, which altogether account for 70% of all injuries. Other remaining 11 accident/injury types constitute 30% of total injuries as presented in Figure 2.

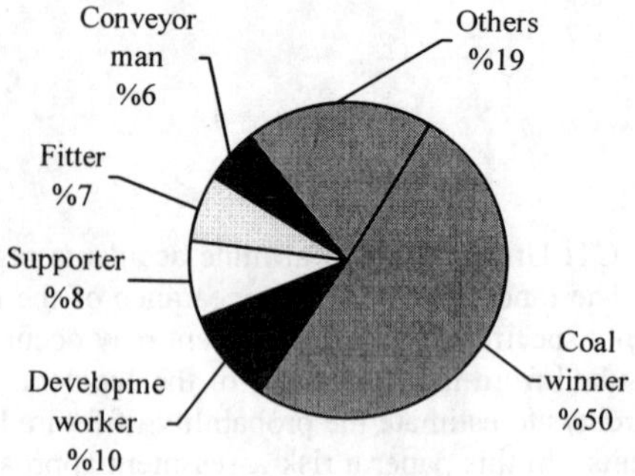

Figure 3. The distribution of job type in accidents

The classification by job type is presented in Figure 3. While coal winners experience almost half of the total accidents, the remaining incidents attributed to 16 other job types.

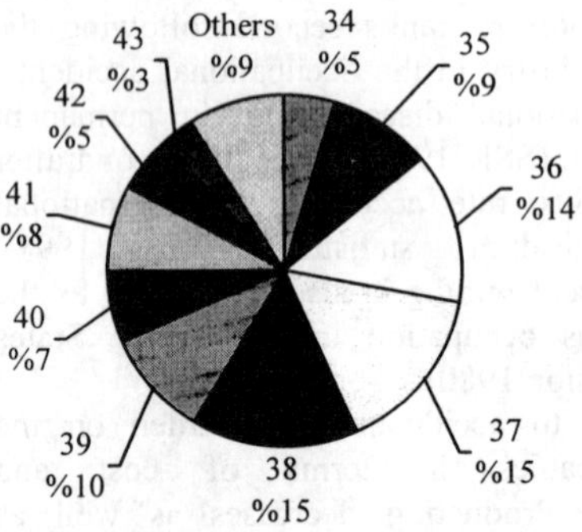

Figure 4. Distribution of age of the workers

The distribution of age of workers between 35-43 are presented in Figure 4. It can be concluded that experiencing an accident between ages 35-39 is more probable than the other age groups. In the analysis of mostly injured parts of body, hand, foot and main body take totally 77% of all injuries. This is shown in Figure 5.

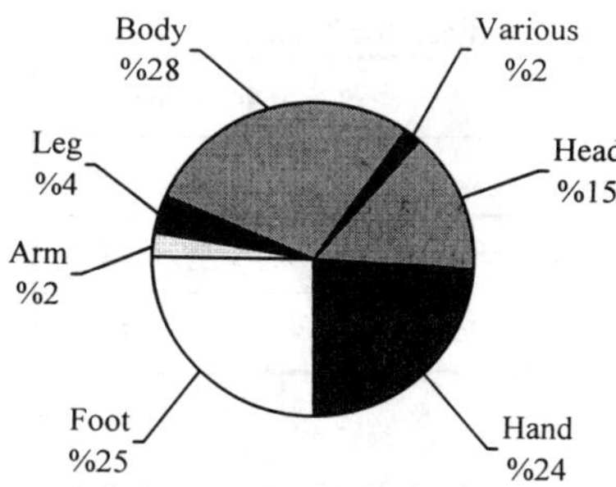

Figure 5. Distribution of injury based on body parts

At the second step of statistical analysis, distributional approach is considered. Determination of statistical distributions of frequency and severity for the basic variables leads the researches to evaluate the risk factor. This analysis is presented in the next section.

3 RISK ASSESSMENT

Risk is defined as uncertainty concerning occurrence of a loss (Rejda 1998). Mining activities require material, equipment, human resources and environment where the potential risk of catastrophic losses is very high. The uncertainty and variability in the occurrence of hazards necessitate the application of risk analysis techniques in mining practices.

Risk assessment consists of two distinct phases: a qualitative step of identification, characterizing, and ranking of hazards; and a quantitative step of risk evaluation which includes estimating the likelihood (e.g., frequencies) and consequences of hazard occurrence (Modarres 1993).

The quantitative analysis of accident and injury data for measuring the safety performance and identifying safety problems are usually done through Job Safety Analysis. Two basic safety indices, Accident Frequency Rate (AFR) and Accident Severity Rate (ASR), as follows (ILO 1962):

$$AFR = \frac{\text{number of injuries}}{\text{total man - hours of exposure}} * 10^6 \qquad (1)$$

$$ASR = \frac{\text{number of days - lost}}{\text{total man - hours of exposure}} * 10^3 \qquad (2)$$

These indices are calculated as the number of injuries for each million man-hours of exposure and total number of days-lost per thousand man-hours of exposure. Same approach has been adopted by the US Mine Safety and Health Administration (MSHA 1990) with changes in constants. In fact, these numbers can not exactly include uncertainty and variability inherent in the occurrence of accidents. On the other hand, in a study done by Kerkering & McWilliams (1987), appropriate statistical distributions were fitted by applying advanced statistical techniques on the mining accident data. Results showed that it was possible to measure mining safety and risk values in quantitative terms including the uncertainty and variability exposed by the data.

To quantify the risk, number of accidents (NOA) per month and days lost (DL) per month are considered. Probability distribution of these variables enriches the perspective of the researcher for future predictions. One method to determine the probability from past occurrences requires large number of data set to end up with reliable estimations (Vose 1996). There are many statistical distributions that can model the frequency histogram with varying degrees of success. The Poisson distribution is often the appropriate theoretical choices for modeling discrete data set in a given time interval.

Poisson distribution is expressed as

$$P(X = x) = \frac{e^{-\lambda}\lambda^x}{x!} \qquad x = 0,1,. \quad ;\lambda \geq 0. \qquad (3)$$

where x denotes the number of occurrences per given time and λ is the average rate of occurrence per given time (Benjamin & Cornell 1970).

Another method to detect Poisson distribution is to examine the statistical distribution of elapsed time between accidents. The elapsed time between two Poisson occurrences is assumed to have exponential distribution. Density function of Exponential (negative exponential) distribution with mean μ, is given in Equation 4.

$$f(x) = \frac{1}{\mu}\exp(-\frac{1}{\mu}) \qquad x \geq 0 \qquad \mu \geq 0 \qquad (4)$$

For implementation of the statistical analysis, the independency of Tunçbilek mine data is checked initially by performing a runs test (Daniel 1978). The result of this test shows that the number of accidents per week are independent of each other and random. Afterwards, to find the distribution type, a statistical software package, BestFit is executed. BestFit, evaluates all statistical distributions on the given data set and ranks the most appropriate distributions according to their statistical test criteria, Chi-square or Kolmogrov-Smirnov tests. The number of accidents per week is found to have a Poisson distribution with mean rate, $\lambda = 3.07$ per week. The Chi-square test statistics is significant enough to accept that the data set comes from proposed distribution. For the approval of this distribution, elapsed time between accidents are tested through BestFit and statistical distribution is searched. Exponential distribution is found to be the one of the proper distributions for the elapsed times supporting the assumption of Poisson distribution for the NOA. The frequency histogram of Poisson model fitted to the number of accidents per week in Tuncbilek mine is displayed in Figure 6. The plot of original data with the hypothetical distribution agrees each other.

Due to the flexibility of Poisson distribution, the analysis can be focused on daily basis. Table 1 summarises the safety indices for Tunçbilek coal mine data. It can be deduced from the table that hazard rate is the average number of accidents per day which is less than one accident per day. The probability of mine's accident-free operation for a day is 64% or equivalently, the probability of mine's accident-free operation for a week equals 5%, which is a quite small percentage. The risk of having an accident for any day and for one week equal to 0.36 and 0.95, respectively. That is, there is a 36% chance that there will be at least one accident on any given day and 95% probability that there will be at least one accident on a week. And finally, expected time between two consecutive accidents will be around 2.27 days.

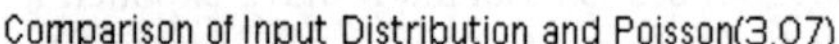

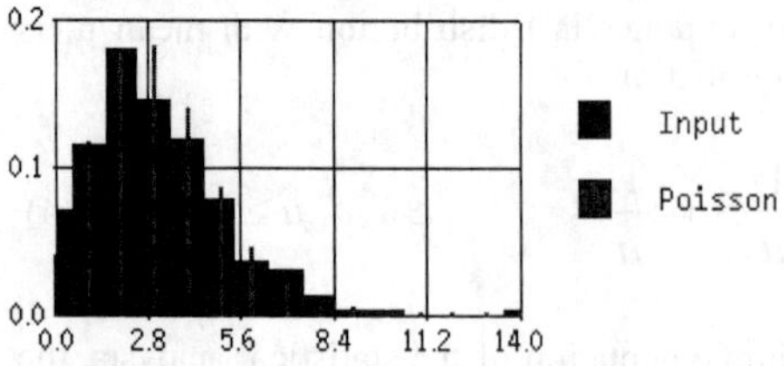

Figure 6. Distribution of NOA by BestFit.

Table 1. Mine safety indices for Tuncbilek mine

Hazard Rate	:	0.44 accident/day
Reliability or Safety	:	0.64
Risk or Accident Occurrence	:	0.36
Mean Time Between Accidents	:	2.27 days

The consequences that result in occurrence of accidents/incidents can be defined also in severity perspective. Since the number of days lost as a result of an injurious incident is the best available proxy measure of accident severity (Hull et al. 1996) and only days off from work data can be extracted from the database, days-lost due to accident are taken as the measure of injury seriousness. Similarly, using BestFit software package, the most appropriate distribution is checked for the data set and for Tuncbilek data, the days-lost data followed a one-parameter exponential distribution with mean rate of 10.8 days. This means that on average the days off from work because of an accident will be around 10.8 days. The typical frequency histogram of days-lost data is plotted in Figure 7.

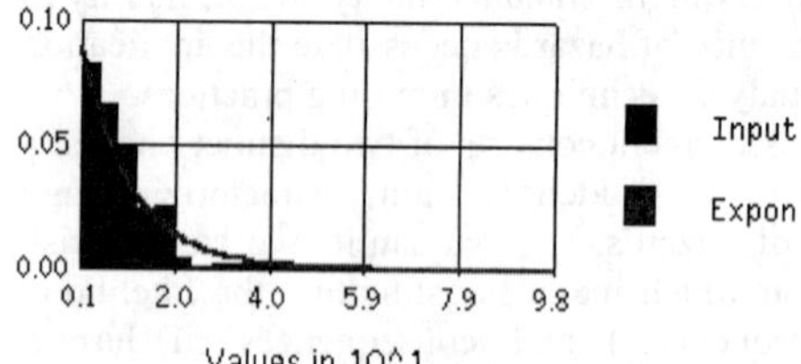

Figure 7. Distribution of days lost by BestFit.

4 RISK CLASSIFICATION SCHEMES

A simple, practical and schematic procedure to carry out risk assessment for the mine's accident/injury data will be presented in this section. Modification of the method proposed by Davies (1997), "Risk Classification Schemes" have been prepared for every considerable variable in the data set.

This technique presents the two way interaction between two variables, in which the one is replaced on x-axis and the other on y-axis. The farther get the coordinates from origin the higher

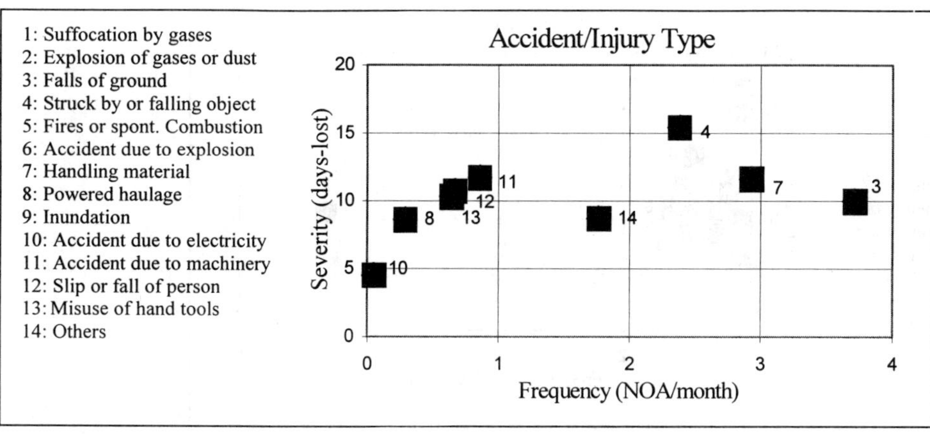

Figure 8. Risk classification scheme for Accident/injury type.

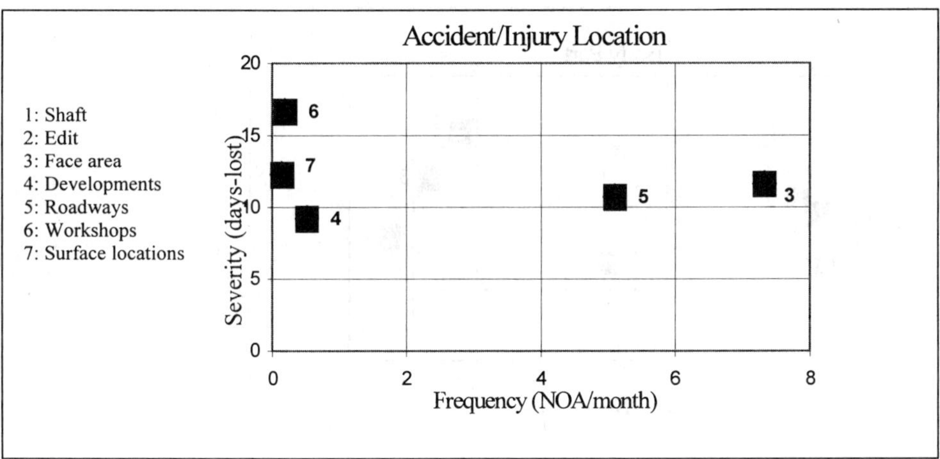

Figure 9. Risk classification scheme for accident/injury location.

gets the risk value. For evaluating the Tuncbilek mine data, risk classifications are performed for four variables. These are the accident/injury type, accident location, job types, and part of body injured on monthly base as illustrated in Figures 8-11. A careful examination of each scheme reveals much valuable information about level of risk in each component. High-risk groups can be easily distinguished from the lower ones. Therefore, these schemes are an excellent risk-screening tool. The procedure is quick, relatively repeatable and easily conveyed to others having little knowledge about risk evaluation.

From risk classification schemes, it can be concluded that, the most risky accident/injury types are falls of ground, struck by or falling object, handling material as shown in Figure 8. A similar situation is also seen in Figure 9. Face area and roadways are the risky places where most of accidents take place. In Figure 10, the coal winner as job type is easily distinguished from the others as having high level of risk. In the case of body part injured, hand, foot, and the main body are subjected to more injury risk than the other parts as can be followed from Figure 11. It can also be concluded

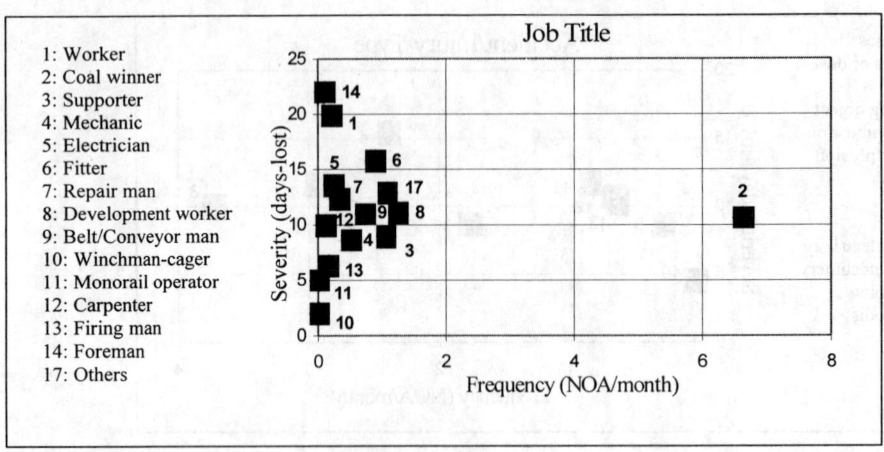

Figure 10. Risk classification scheme for job types.

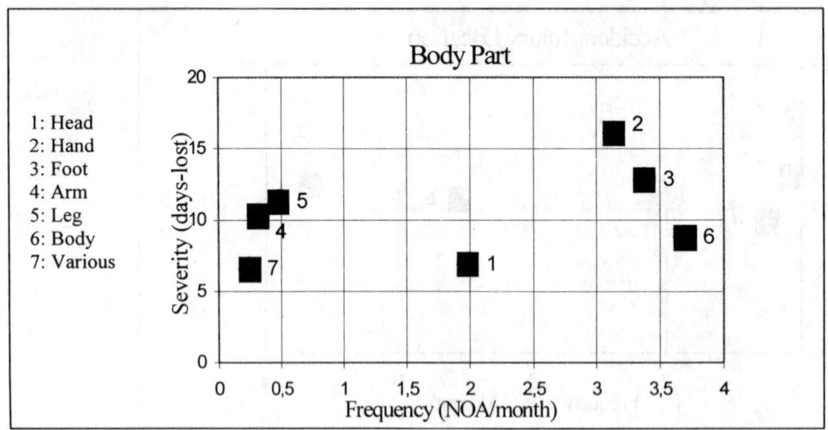

Figure 11. Risk classification scheme for the part of body injured.

from these charts that the frequency has more influence on the deviation of the risk than severity. Therefore, determination of risk levels as a function of severity and frequency can be assessed based on the behaviour of frequency distribution .

5 CONCLUSION

This study presents the evaluation of coal mine accident data by using risk assessment techniques. Randomness in the occurrence of accidents are incorporated into the model via statistical models. A real life accident data from Tunçbilek coal mine is used to show the implementation of the model. Two basic variables, frequency (NOA) and the severity (days lost) are expressed in terms of statistical distributions, Poisson and exponential distribution, respectively. Risk classification schemes are plotted to illustrate the effect of different characteristics on the variation of two variables. Risk assessment analysis enables researcher to predict the probability of an accident in any period for the purposes of being prepared for unexpected losses financially. Additionally, determination of high risk levels will improve the job safety and loss control programs.

As future study, risk retention levels for the mine management will be studied. The categorisation of risk levels through distributional approach for each variable will result in developing the most appropriate technique to eliminate the affect of hazards on coal mine enterprises.

REFERENCES

Benjamin, J.R. & Cornell, C.A. 1970. *Probability, Statistics and Decision for Civil Engineers*. New York: McGraw-Hill.

Bhattacherjee, A., Ramani, R.V. & Natarajan, R. 1992. Injury Experience Analysis for Risk Assessment and Safety Evaluation. *Mining Engineering*. 44 (12):1461-1466.

Bozkurt, R. 1993. *Application of the Modern Accident Prevention Techniques to OAL Mines (TKI)*. Master Thesis, METU, Ankara.

Butani, S.J. 1988. Relative Risk Analysis of Injuries in Coal Mining by Age and Experience at Present Company. *Journal of Occupational Accidents*. 10 (3): 209-216.

Coal Commission 1980. Coal Commision Finds Big Variations in Mine Safety. *The Energy Daily*, March 25: 2-3.

Daniel, W.W. 1978. *Applied Nonparametric Statistics*, Boston: Houghton Mifflin Company.

Davies, M.P. 1997. Potential Problem Analysis: A Practical Risk Assessment Technique for the Mining Industry. *CIM Bulletin*. 90 (1009): 49-52.

Gallagher, S., Bobick, T.G., & Unger, R.L. 1992. Reducing Back Injuries in Low-Seam Coal Mines Through Task Redesign. *New Technology in Mine Health and Safety*:310-323. SME, Littleton, CO.

Golledge, P. & Stokes, M.J. 1985. Development and Use of a Computer Based Lost-Time Injury Recording and Analysis System. *Proc. of 21st Int. Conf. on Safety in Mines Research Institutes*: 245-251.

Grayson, R.L., Layne, L.A., Althouse, R.C. & Klishis, M.J. 1992. Risk Indices for Roof Bolting Injuries Using Microanalysis. *Mining Engineering*. 44 (2):164-166.

Hayduk, J.L. & Ritzel, D.O. 1988. Diseases and Injuries among Coal Miners as Reported in a Rural Hospital. *Journal of Occupational Accidents*. 10 (3):77-93.

Hull, B.P., Leigh, J., Driscoll, T.R. & Mandryk, J. 1996. Factors Associated with Occupational Injury Severity in the New South Wales Underground Coal Mining Industry. *Safety Science*. 21 (3):191-204.

ILO 1962. Statistics of Industrial Injuries. *ILO Tenth International Conference on Labour Statisticians*, Geneva.

ILO 1991. Yearly Statistics. ILO, Geneva.

Kerkering, J.C. & McWilliams, P.C. 1987. Indices of Mine Safety Resulting from the Application of the Poisson Distribution to Mine Accident Data. USBM RI-9088, 31 pp.

Leigh, J., Mulder, H.B., Want, G.V., Farnsworth, N.P. & Morgan, G.G. 1991. Sprain/Strain Back Injuries in New South Wales Underground Coal Mining. *Safety Science*.14 (1): 35-42.

Modarres, M. 1993. *What Every Engineer Should Know About Reliabilty and Risk Analysis*.New York:Marcel Dekker Inc.

MSHA 1990. Injury Experience in Coal Mining. US Department of Labor, MSHA, Berkley, West Virginia.

Peay, J..M., & Schaffer, L. 1985. A Model Health and Safety Program for the Mining Industry. *Proc. of 21st Int. Conf. on Safety in Mines Research Institutes*: 259-262.

Rejda, G. E. 1998. *Principles of Risk Management and Insurance*, 6th Edition. Boston:Addison Wesley.

Rowell, W. 1996. Practical Risk Assessment. *Mining Technology*. 78 (899):184-188.

Simpson, G.C. & Moult, D.J. 1995. Risk Assessment as the Basis for the Introduction of New Systems. *Mining Technology*. 77 (891): 343-348.

SSK 1991. SSK Istatistik Yıllıgı (Statistics Bulletin of Social Security Institution). SSK Genel Müdürlügü, Ankara (in Turkish).

Staley, B.G. 1996. Risk Assessment for Busy Mine Managers. *Mining Technology*. 78 (899):201-204.

Staley, B.G. & Foster, P.J. 1996. Investigating Accidents and Incidents Effectively. *Mining Technology*. 78 (895): 67-70.

Turin, F.C. & Others 1995. Human Factors Analysis of Roof Bolting Hazards in Underground Coal Mines. USBM RI-9568, 25 pp.

Van Der Vyver, A.C. 1997. The Foundation to Practical Risk Assessment. *Journal of the Mine Ventilation Society of South Africa*. 50 (1): 29-31.

Vose, D. 1996. Quantitative Risk Analysis: A Guide to Monte Carlo Simulation Modelling. New York: John Wiley & Sons.

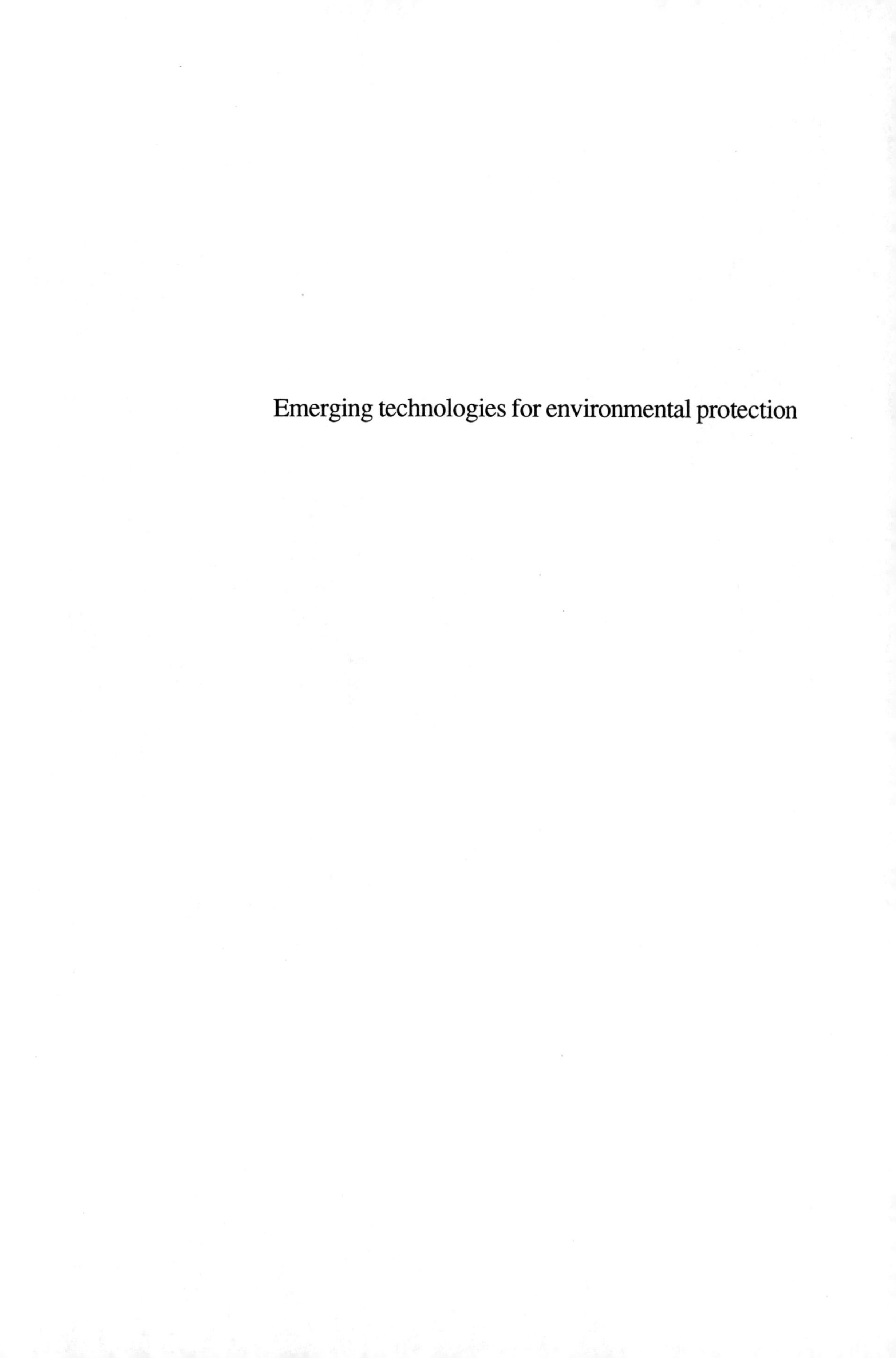

Emerging technologies for environmental protection

Environmental Issues and Management of Waste in Energy and Mineral Production, Singhal & Mehrotra (eds)
© 2000 Balkema, Rotterdam, ISBN 90 5809 085 X

Soil venting for site remediation after hydrocarbon pollution

G. Bonifazi, F. La Marca, P. Massacci & A. Matocci
Department of Chemical Materials, Raw Materials and Metallurgical Engineering, University of Rome 'La Sapienza', Italy

F. Luzzi
ERM Italia, Rome, Italy

ABSTRACT: Urban areas, utilized in past for stocking and distribution of oil products, inevitably presents phenomena of subsoil pollution. Such phenomena, also persisting after the end of industrial activities, involve conditions requiring interventions of environmental remediation. The present work analyzes the intervention procedures for the restoration of an area utilized as hydrocarbons deposit in an urbanized site of central Italy and of an area in the South of Italy, interested by a gasoline pollution due to underground storage tanks. Subsoil characteristics and its state of contamination have been preliminary investigated. Contamination has been located in two different layers, the first one extended in the vadose zone, the second one in the saturated zone. Specific interventions for the different delivery of pollution have thus been studied, that is:
- removal of the superficial contaminated soil;
- soil venting in the unsaturated zone of the subsoil;
- water pumping and treatment in the saturated zone.
The soil venting system resulted particularly important and complex, therefore the execution of pilot test has been made necessary. The results of the performed pilot test have been analyzed and the relevant parameters in the final plant design for the area restoration have been thus selected.

1 INTRODUCTION

Soil remediation procedures are particularly complex when the pollution originates from extended actions and when urban areas are interested.

The duration of the contamination effects involves, in fact, i) diffusion of pollutants far from the immission site and ii) distribution of pollutants by the different phase (adsorbed, residual, dissolved);

Urban areas involvement in pollution processes is increasing more and more, both because of the grow of cities toward areas originally designated to industrial activity, and to the risk linked to potentially pollutants storage and distribution inside areas characterized by a strong human presence (Visser, 1994; 1995).

In this work, the pollution related to hydrocarbon products storage and handling inside urban areas is examined, that is irreducible conditions of contamination due to spills and not detected or monitored losses, even if oil products distribution has been stopped from several years.

The knowledge of the polluted site condition represents the imperative starting point to design the remediation strategy (de Fraja Frangipane et al., 1994).

Suitable in situ sampling and appropriate analyses are essential to the evaluation of polluted site condition, which must be characterized by geological and morphological features and by past and present use description.

2 POLLUTED SITE CHARACTERIZATION

To completely describe a polluted site, it is necessary to define the following characteristics:

- geological characteristics, that is the identification of the geological structures typology, their lying referred to ground surface and eventually new civil buildings;
- petrographical characteristics, that is the identification of the mineralogical composition and the structure and texture of contaminated geological bodies, also enclosing or bordering;
- hydrogeological characteristics, that is the identification of the water table elevation, its feature and interaction with geological bodies, ground surface or discontinuity surfaces deriving from civil works. Groundwater flows and water table seasonal fluctuation have to be considered;
- morphological soil characteristics (topography) of geological and water bodies;
- networks structure (sewerage, water and energy distribution, telecommunications, road, etc.) eventually existing in the site;

- civil works and technological plants structure and settlement state on soil and subsoil;
- pollution characteristics concerning pollutants type, extension and concentration.

3 POLLUTED SITE INVESTIGATION METHODS

An investigation program addressed to characterize a specific site must be set up, following three different approaches (ASTM, 1994):
- exhaustive recognition of the characteristics, based on *in situ* surveying or on available documentation (e.g. civil works inventory, topography, etc.);
- indirect investigation by instrumentation able to provide spectral responses through application of specific stress (e.g.. seismic, radar, etc.) or surveying natural fields anomalies (e.g. electromagnetic and gravimetric fields, , etc.). Both the techniques can be applied utilizing instrumentation on the field (on the site to be investigated) or in remote positions (remote sensing);
- direct investigation by soil samples collection, monitoring wells installation and groundwater sampling.

Furthermore atmosphere transfer phenomena, releases and *ex situ* contaminants transport effects, through fluid medium (air and water) far from the polluted site, must not be neglected.

If the concentration of pollutants (dissolved hydrocarbons) in groundwater results higher than the limits of acceptance, a remediation process is necessary. In case of confined sites with possibility of easy aquifer interception, the "pump and treat" (P&T) method is revealed to be an effective method (EPA, 1996).

The P&T system is recommended for removal of the pollution in groundwater. The method is based on the set-up of a groundwater pumping system realized by means of couples of suitable wells (for injection and extraction). Such wells allow to maintain a constant drowdown of the water table, modifying the aquifer geometry and creating a capture zone so that fluids flow toward such wells, with consequent containment and recovery of dissolved phase.

Particular arrangements have to be considered for wells set-up. Wells must be complete with appropriate casing (high-density PVC pipe is suitable in case of low depths), micro-screened versus groundwater and unscreened through unsaturated horizons.

The cavity between casing pipe and borehole must be filled by naturally rounded fine siliceous gravel, 5 mm sized in correspondence of the screened section, to improve well hydraulic efficiency. In the superficial section, the well must be cemented by cement grout and closed upward by a walk-on manhole in cast iron.

The extracted water must be sent to a water treatment plant, by means of adsorption on activated carbon filters placed on a special metal container. After treatment, water must be used according to foreseen prescriptions by public control authorities.

4 INVESTIGATED AREAS

The former considerations have been applied with reference to two different polluted areas: the first one located in the Terni province, the second one in the South of Italy. In the following they will be referred as Test area 1 and Test area 2.

A general survey of the problems to face and the strategies to apply is reported for each of the two considered sites.

More detailed considerations have been developed for soil venting procedures, being this approach quite relevant in the scope of remediation of the studied contaminated areas.

5 SOIL VENTING PROCEDURE SET-UP

To define final and effective soil remediation strategies, a ventilation pilot test (Vent Test) can be performed.

This way, the pollutant removal velocity, the air flow and volatile organic compounds (VOC) concentration extracted from the polluted area and the radius of influence for each ventilation point have to be evaluated (EPA, 1996).

The mobile unity of the pilot system, utilized in this study, was composed by:
- a blower;
- an air flowmeter;
- a vacuum indicator;
- temperature indicators;
- an air/water separator;
- activated carbon filters;
- flow/vacuum control valves.

Furthermore, to perform the test it is necessary to install at least an extraction and/or injection well and at least two-three monitoring wells to measure the vacuum at different depths and distances from the extraction point.

Sampling devices should be installed to analyze the effluent air.

During each test, minimum 24 h, the extracted air flow (order of magnitude: hundred m^3/h) was realized adopting a depression of about 20÷100 mbar.

During the test, concentrations of several compounds have to be recorded continuously at blower exit, that is:
- VOC, in ppm;
- oxygen (O_2), in % volume;
- carbon dioxide (CO_2), in % volume.

At the same time, the induced vacuum in monitoring wells corresponding to three different flow step must be checked. The ventilation well radius of influence related to each flow value can then be valued.

It is important to perform, during the test, specific measurements, in order to gather data on hydrocarbon biodegradation in subsoil. Such a test consists in injecting in the ground, in the mostly VOC polluted area, air and a tracing inert gas. Through the injected air an oxygenation of the ground is produced, while the inert gas allows to estimate the system insulation. The injection is then interrupted and O_2 and CO_2 concentrations are monitored for about 24 h. Such concentrations are, therefore, compared with those measured before injection.

The biodegradation rates can be computed by the following relationship:

$$K_B = K_0 A D_0 C / 100 \quad (1)$$

where:

K_0 oxygen utilization rate;

K_B biodegradation rate (mg hexane for kg of soil per day)

A air volume (for soil mass);

D_0 oxygen density;

C stoichiometric hydrocarbon/oxygen mass ratio.

After the interruption of air injection, the evaluation of the carbon dioxide and oxygen concentration trend allows to recognize if an oxygen reduction occurs jointly to carbon dioxide production, according to the reaction, given for the toluene:

$$C_6H_5CH_3 + 9O_2 \rightarrow 7CO_2 + 4H_2O \quad (2)$$

6 TEST AREA 1

The area belongs to an ex petroleum bulk deposit, devoted to hydrocarbons storage and distribution, dismantled 15 years ago, and now interested by urban settlement, originally far away. It is a 5500-m^2 area, adjacent to the urban expansion area, Southward, and to an area predominantly for agricultural use, Northward.

From a geomorphological point of view, the studied area is located at 130 m above the sea level inside the Terni basin. The basin is characterized by two main stratigraphic groups, that is: i) formations of the triassic-miocenic sedimentary sea cycle, related to the Umbro–Marchigiano pelagic complex, and ii) formations belonging to pliocenic–quaternary sedimentary continental cycle.

From a lithological point of view, the area is extended on a layer of holocenic alluvial deposits composed by sand and sandy silt, with an over 10-m thickness. Below this formation, gravels are found, with a 20-30 m thickness, decreasing eastward. Finally, under the gravel layer, Villafranca's conglomerates are mainly found, with a 80-100 m thickness, alternating consolidated clayey silt.

The aquifer system of alluvial deposits, interacting with the investigated plant, is the gravel level, found below the sands, 10-15 m depth in Terni area and including the shallow aquifer of Terni basin. Under the gravel layer, the Villafranca's conglomerates, alternating consolidated clayey silt include relevant aquifers, due to the quite high permeability.

The superficial hydrography of the Terni Plain is dominated by a river (Nera) course, flowing in E-NE – W-SW direction.

The entire area was interested by hydrocarbons pollution of different entity interesting the superficial area (ground storage tanks for diesel fuel) and the underground (underground storage tank for gasoline) both as soil and water.

Furthermore, in the South-West area of the original oil storage, a water treatment plant is present, whose settling ponds still have superficial hydrocarbons evidence.

In order to survey the contamination distribution, the following data gathering has been carried on in the area, coherently with the general methodology above described:

- continuous and dry core drilling, up to 20 m depth;
- installation of monitoring wells;
- topographic survey;
- piezometric and fluids survey in all the installed wells;
- aquifer hydraulic test (Slug Test);
- collection of water, soil samples and related chemical analysis.

The informations obtained from such investigations have been:

- stratigraphy;
- depth to water table and product (hydrocarbon) and thickness of the "floating";
- pollutants characteristics.

Particularly the following data have been determined:

- VOC;
- total product hydrocarbons (TPH) (total hydrocarbons);
- aromatic organic solvents (BTEX);
- organic lead (Pb);
- MTBE.

The stratigraphy yield by the four surveys results as follows:

- from 0 to 2.0 m: landfill constituted by mixed-size and polygenic eterometric material;
- from 2.0 to 4.0 m: clayey silt, brownish/yellowish, with calcareous inclusions;
- from 4.0 to 12-13 m: coarse to fine size sand, yellowish , wet, slightly thickened;
- from 13 to 15 m: gravel in sandy matrix, saturated;

– from 15 to 17 m: marl clay, yellowish, very consistent, slightly wet.

An unconfined shallow aquifer has been detected corresponding to the gravel formation, about 1.5-2.0 m thick, with the bottom constituted by the malt clay level, slightly permeable (about 15 m depth from surface). The water table hydraulic gradient (i) has an average value equal to 0.0045 (0.45%), with a groundwater flow direction from North toward South.

Slug tests has been performed in the monitoring wells, developed according to Bower and Rice's method, measuring well recharge, following the extraction of a known quantity of water. The following equation was thus adopted:

$$Q = 2\pi Kd(ht/\ln(Re/rw)) \quad (3)$$

where:

K hydraulic conductivity;
Q lift water quantity;
Re radial distance to which drowdown is dissipated;
rw borehole radius;
d screened well length;
ht level at the time $t > t(0)$;

The effective groundwater flow velocity has been calculated applying the Darcy's formula:

$$v_{eff} = K * i / n_{eff} \quad (4)$$

where:

v_{eff} groundwater effective flow velocity;
K hydraulic conductivity;
i hydraulic gradient (dimensionless parameter);
n_{eff} effective porosity (dimensionless parameter).

Hydraulic conductivity values and time/drowdown curves have been determined for each well.

6.1 *Test area 1 pollution condition*

Sampling and analyses allowed to completely describe the subsoil quality, underlining the hydrocarbons presence, both as adsorbed product in unsaturated zone, both as dissolved in groundwater.

The adsorbed product presence in unsaturated zone has been extrapolated from the interstitial gas analyses, together to head space and gas-chromatography analyses performed on soil samples collected during the perforation phase.

The areas characterized by adsorbed products have been identified defining the lateral and vertical extension. The contamination level due to fluid losses from tanks and spill, in loading operations, has been evaluated.

The soil adsorbed products have been characterized on the base of hydrocarbon chains length and the presence of other compounds. It allowed to survey the light hydrocarbons presence in Test area 1, that, due to prolonged underground permanence (oil spill happened at least 10 years ago), produced a loss of volatile and biodegradable compounds mainly aromatic solvents, by natural attenuation.

The gas-chromatographic analyses, performed on water samples collected in the wells, revealed the presence of hydrocarbons dissolved in monitoring wells, the detected hydrocarbons being constituted by heavy aromatic solvents, presenting an hydrocarbon chain with more than 12 carbon atoms, and light solvents (BTEX). These results demonstrated as a chemical-physical transformation occurred, determining both the loss of short chain fractions and the enrichment in heavy compounds.

6.2 *Test area 1 remediation procedure design*

In order to design a safety system, two areas, characterized by structural difference in pollution, have been identified in the deposit:

– polluted soil in an unsaturated zone (South area);
– shallow polluted area (North area).

6.2.1 *South area*

This area was characterized by a level of contamination not reaching the "substitution level", as a consequence ventilation (soil ventilation) was assumed as correct remediation system.

Subsoil ventilation can be induced through suitable wells (vent wells), in depression by a Browler vacuum pomp. Inducing a depression in subsoil, the hydrocarbons equilibrium shifts toward the vapor phase, allowing a higher evaporation.

Extracted organic vapors in concentrations higher than the allowed limits, fixed by the legislation when released in the atmosphere, must be superficially treated by chemical-physical method, such activated carbon adsorption.

In order to optimize the process, it is important to perform a correct balancing of induced depressions of airflows and in respect of the radius of influence of each venting well, on the whole polluted area.

By forced ventilation, a subsoil oxygenation is realized, inducing aerobic hydrocarbons biodegradation processes. In Tables 1-3 the results of three pilot tests are reported. The VOC, O_2 and CO_2 concentration have been continuously recorded at bowler exit, related to different lift water quantity (Q).

The spatial position of venting wells must be established on the base of the estimated radius of influence obtained by comparison with similar subsoil and pilot tests results.

6.2.2 *North area*

Decontamination in shallow layer of soil, characterized by not elevated hydrocarbons concentrations, involves the mechanical ground removal up to the contaminated soil depth.

Table 1: VOC, O_2 and CO_2 concentration trends recorded at bowler exit, Q=128 m^3/h.

Time min	VOC ppm	O_2 %vol	CO_2 %vol
0	0	0.0	0.0
3	360	6.0	6.3
6	470	5.5	6.4
15	750	4.5	6.5
30	970	4.5	6.0
45	1530	4.4	5.7
60	1575	3.4	6.2
80	1390	2.7	7.2
100	1330	2.8	8.2
120	1300	2.8	7.9
140	1430	2.8	8.2
165	1575	2.9	7.3
185	1095	2.9	6.1
210	920	3.1	6.6
240	870	3.0	7.5

Table 2: VOC, O_2 and CO_2 concentration trends recorded at bowler exit, Q=160 m^3/h.

Time min	VOC ppm	O_2 %vol	CO_2 %vol
0	700	-	-
5	487	1.7	7.0
25	647	1.8	5.9
45	492	1.8	6.1
65	536	1.7	6.5
85	400	1.9	7.8
120	562	2.1	8.3
145	405	2.5	7.3
165	420	1.8	8.3
185	440	1.8	8.8
205	410	1.8	8.6
235	1030	1.9	8.3

Table 3: VOC, O_2 and CO_2 concentration trends recorded at bowler exit, Q=175 m^3/h.

Time min	VOC ppm	O_2 %vol	CO_2 %vol
0	0	-	-
5	735	1.3	9.6
25	505	1.8	8.5
45	862	2.3	7.9
65	1574	1.5	8.4
85	1136	1.5	8.1
105	1574	1.5	7.8
125	1426	1.8	7.8
155	1410	1.5	9.4
205	1135	1.7	8.1
225	1098	1.5	7.6
245	1180	2.0	6.9
265	1112	1.8	8.2
285	1210	2.0	7.8
305	1100	1.7	8.7
325	1075	1.8	8.6
345	985	2.0	8.0
365	1020	2.5	7.9
410	970	2.3	7.8

During ground removal operations, it is opportune to evaluate the quality of the removed soil by collecting representative samples and performing both quick field and laboratory analysis.

Comparing the concentrations resulting from the analyses and the maximum admissible values according to local legislation on soil quality standard for residential use, it is necessary to identify those portions of soil inside and outside the limits of acceptance. The first ones could be placed again on site, the second ones should be landfilled according to existing legislation on special waste transport and disposal.

7 TEST AREA 2

The area object of remediation, located in the South of Italy, was interested by a gasoline pollution (December 1995) due to the los of about 6000 kg of this fuel from two underground storage tanks.

In-situ Air Sparging (AS) and Soil Vapor Extraction (SVE) remediation technologies were used to clean up the soil and groundwater contaminated in major part by BTEX.

7.1 *Test area 2 pollution condition*

Site characterization was performed in two phases, included 40 point of Soil Gas Survey, installation of 6 monitoring wells with soil and groundwater sampling. The lithologic data indicate three different layers:

- the first layer, superficial, mainly constituted by silty clay ranging in thickness from 1 to 2.5 m;
- the second layer constituted by fine to medium sand ranging in thickness from 2.5 to 8 m;
- the last layer mainly composed by is sandy gravel ranging thickness from 8 to 15 m.

Groundwater table was located in sandy gravel unit at a depth of about 9 m below ground surface. Groundwater flow direction was from East to West, with hydraulic gradient of 0.002 (0.2%). Slug Test carried out in monitoring wells indicated an average hydraulic conductivity of $2\cdot10^{-3}$ m/sec.165

Concentration range of adsorbed BTEX detected in soil contaminated area was from 120 to 2.7 mg/kg and the concentration range of dissolved BTEX in groundwater was from 16.5 to 0.12 mg/l.

7.2 *Test area 2 remediation procedure design*

Remediation technology in accordance with the public local authority to clean up the contaminated area, was based on AS/SVE system.

All the process parameters are determined by AS/SVE pilot test.

Sparge and SVE wells were located to provide an area of pressure influence less than, and overlapped by the area of vacuum influence. Three Sparge/SVE wells were installed in the center of the BTEX plume, while five individual SVE wells were designed to cover the impacted area.

The SVE system is ran by two blower (4 kW) with a total flow rate of air extracted of 700 m^3/h.

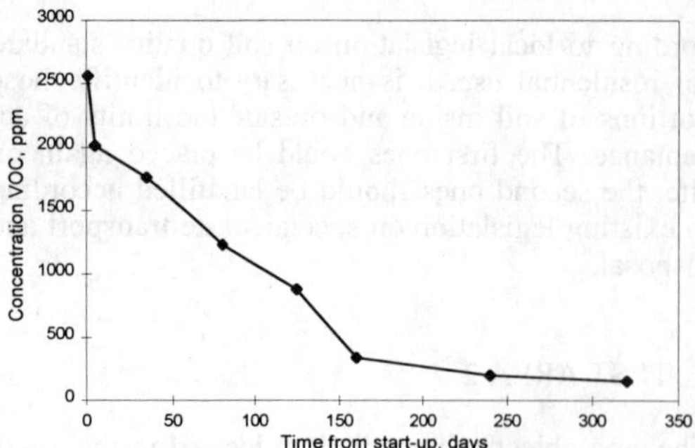

Figure 1: Off-gas (VOC) versus time.

while the AS system is ran by a compressor of 10 kW with a total flow rate of air injected of 95 m^3/h. A thermal/catalytic oxidizer has been used to treat the extracted vapor.

A monitoring program has been implemented including:

- process parameter (pressure, vacuum, temperature) monitoring;
- CO_2 off-gas and O_2 level detection;
- water table survey;
- air and water sampling;
- water chemical/physical parameter (redox potential, dissolved oxygen, temperature) control.

Prior to the system start-up, all monitoring wells were sampled and the water analyzed for TPH, BTEX, MTBE on March 1996, to establish the baseline conditions.

Remediation results were that, after 15 mounts of continuing application of AS/SVE system, the groundwater BTEX concentration, even in the source area (major contaminated zone) reduced from 16.5 mg/l to 0.025 mg/l. While the adsorbed BTEX concentration in major contaminated soil (near spilled tanks) were reduced from 120 to undetectable limit (<1.0 mg/kg). Out of the source zone, the concentration in groundwater and in vadose zone, were reduced lower to detectable limit.

Analysis of the parameter (VOC, O_2, CO_2, DO) detected during AS/SVE maintenance, demonstrated that for the first 6 months the most active process for remediation were the volatilization. Hydrocarbons recovered through volatilization were estimated from the TPH off-gas concentration and flow rate.

The volatilization recoveries decreased with time, increasing the biodegradation recoveries (monitored by CO_2 concentration) from 26% in the first 6 months, to 80% after 12 months (Figure 1).

8 CONCLUSIONS

The present work examined the pollution problems, and the related de-contamination strategies to adopt, related to hydrocarbon handling and storage. The information to gather and the methodologies to apply to reach this goal are outlined with the specific target to realize a full characterization of the site and to quantify the entity of the pollution. Starting from this set of information, suitable and reliable soil reclaiming procedures can be applied.

Different typologies of remediation have been investigated, according to different soil characteristics and different pollution grade. Considerations have been specifically developed to analyze the influence of contamination on the aquifer. Case studies concerning soil venting and hydrocarbon extraction monitoring, controlling and treatment procedures have been presented and discussed.

REFERENCES

ASTM, American Society for Testing and Materials, 1994. ASTM Standardization News, 22(5), May 1994.

de Fraja Frangipane, E., Andreottola, G. & Tatano, F., 1994. Terreni contaminati – Identificazione, normative, indagini, trattamento. C.I.P.A. Editore (in Italian).

EPA, 1996. EPA engineering sourcebook. Edited by J.R. Boulding, Ann Arbor Press, Inc.

Visser, W.J.F., 1994. Contaminated land policies in some industrialized country. 2nd Edition. The Hague.

Visser, W.J.F., 1995. Comparison of various international approaches to contaminated land. ISWA, Palermo 29 November – 1 December 1995.

Environmental Issues and Management of Waste in Energy and Mineral Production, Singhal & Mehrotra (eds)
 ISBN 90 5809 085 X

Control parameters in the sodium sulphate purification process

R.H.Cervera, L.V.Gutiérrez & N.J.Altamira
Mineral Processing Center, Servicio Geologico Minero Argentino, Argentina

E.J.Nolasco
Instituto de Technología Química, Universidad Nacional de San Luis, Argentina

ABSTRACT: In order to select the organic solvent and the operative variables that would allow the chloride ions to remain in a solution, the crystallization of sodium sulphate from sodium sulphate artificial brines was studied at laboratory scale, by adding methanol, ethanol and ethylene glycol, at different temperatures, evaluating its effect on the sodium sulphate recovery and the chloride content. The sodium sulphate recovery presents, in general, a tendency towards maximum values, with a temperature increase, independently of the solvent used, whereas the chloride ion content in the sodium sulphate crystals drops abruptly from 3000 ppm to less than 4ppm in the case of methanol, at a 30°C temperature. The sodium sulphate obtained may be anhydrous or hydrated, depending on the tested experimental conditions.

1 INTRODUCTION

This paper represents a step forward in the study being carried out on for the purification of sodium sulphate salts, contaminated by different salts, among them sodium chloride. Further data have been published recently (Cervera et al 1998).

There are numerous sodium sulphate depots in Argentina, (Angelelli et al 1983) as for example in the provinces of Salta, La Pampa and Santa Cruz, where one of the most frequent contaminant is represented by chloride ions, the contents of which vary between 5 and 20% of sodium chloride. During these brines' processing, there occurs a chloride content increase in the crystals obtained, forcing the production of drainings and the elimination of recirculating brine, with logical economic losses (Galli et al, 1978).

Sodium sulphate commercial concentrates containing 1-2% of sodium chloride are adequate for the detergent, glass and Kraft paper industries, but other markets, such as the chemical industry are more exacting, admitting maximum values of 10 ppm chloride. Figure 1 shows the solubility diagram for sodium chloride and for sodium sulphate in water, at a 1atm pressure and at a range of 0 to 100° C, diagramming the weight concentration of salts in percentage, as a function of the temperature.

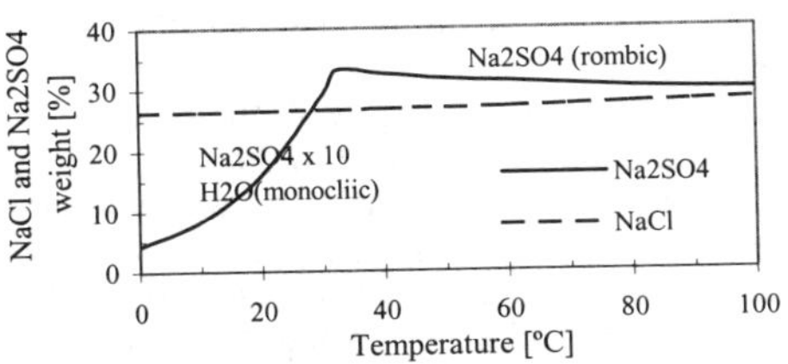

Figure 1. NaCl and Na2SO4 solubility in water.

This figure also shows that in an aqueous medium, the sodium chloride draws a solubility curve, with a slight positive slope, as a function of the temperature. The solid phase present is anhydrous sodium chloride, but the presence of a hydrated phase that disappears at 0.15° C is mentioned (Ullmann's 1993). In reference to sodium sulphate, the curves are clearly differentiated as from 32.4 °C. Below this temperature, it presents a direct solubility curve where the solid phase that coexists with this saturated solution is represented by decahydrous sodium sulphate and above this temperature, it decreases slightly and the solid phase is represented by anhydrous sodium sulphate.

The recovery of sodium sulphate from brines containing these salts in solution is quite a problem and it can be carried out taking advantage of the solubility difference between both. This can be done by water evaporation (with heat or vacuum), by diminishing the temperature, by adding an element that would compete with the water availability, such as a solvent miscible in water or another more soluble salt (Garside & Shah 1980, Kolthoff & Dendell 1956) or by using a combination of such methods (Gaska et al 1965).

Trials carried out in Canada (Mina-Mankarios &

Pinder 1991), describe an interesting process for crystallization of sodium sulphate, adding certain organic solvents to the acid brines.

With the objective of determining the most adequate parameters for the diminishing of the chloride ion content in the sodium sulphate crystals, same were tested in artificial sodium sulphate brines at 3000 ppm chloride, using the following solvents: methanol, ethanol and ethylene glycol, at 10, 20 and 30°C, at a unitary relation (solvent: artificial brine).

2 VALIDATION PROCEDURE

The equipment used for the purification of sodium sulphate brines can be divided into three systems:

The objective of the cooling and heating system is to supply the crystallizer with a flow of water at a constant temperature. It consists of a 20 l. thermic bath , a 700 cm3/min. submergible pump and a flowmeter connected to the crystallizer.

The crystallization system consists of a crystallizer, inside of which and held in place by a rubber ring there is a 70 ml flat-bottomed crystallizing cask which has a magnetic bar. Under the crystallizer there is a magnetic stirrer that regulates the stirring speed at 700 rpm. The temperature in the crystallizer is controlled by a 0.1°C resolution digital thermometer.

The filtering system is formed by a 0.5 HP vacuum pump with its accessories in the shape of a funnel and a container for collecting the filtered liquid.

The artificial brines are prepared by weighing the adequate sodium sulphate, at test temperature and at published values (Ullmann's 1993, Sehidell 1940), with less than 3% saturation values. The sodium chloride is added in solution form, to obtain a 3000 ppm chloride brine.

The crystals obtained by adding solvents are mineralogically characterized with a binocular magnifying glass in order to determine if same correspond to hydrated sodium sulphate (mirabilite) or anhydrous sodium sulphate (thenardite). They are then dried in a heater at 100-105°C for as long as it deems necessary, so as to reach a constant weight. The recovery is determined: as a percentage relation of the weight of the obtained sodium sulphate to the original one in the artificial brine. The chloride content is determined by colorimetry (ASTM D 512 1981), registering the absorption values in a Shimadzu UV-1601 Spectrophotometer.

3 RESULTS AND DISCUSSION

The influence of the type of organic solvent and the temperature will be analyzed for each solvent in particular.

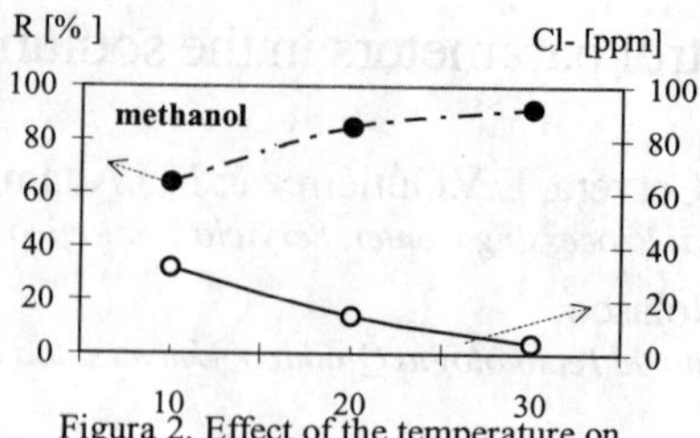

Figura 2. Effect of the temperature on the recovery and the choride content for methanol.

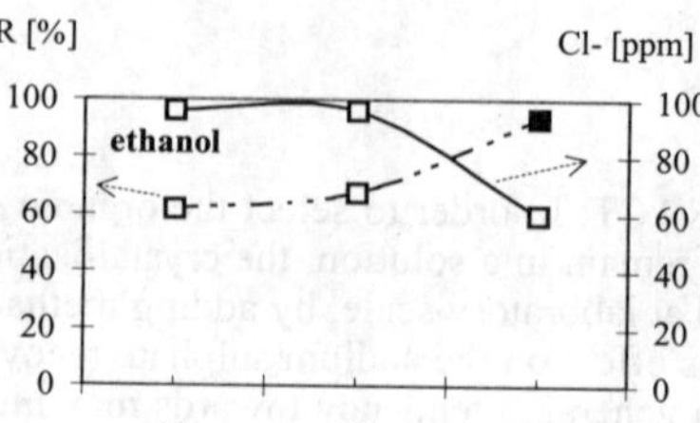

Figura 3. Effect of the temperature on the recovery and the choride content for ethanol.

methanol: In Figure 2, the recovery variation and the chloride content in the sodium sulphate crystals obtained as a function of temperature variation can be observed

Anhydrous sodium sulphate is crystallized between 10 and 30°C. The recovery curve increases rapidly, reaching maximum values of 91.5% at a 30 °C temperature.

The chloride ion content drops abruptly from 3000 ppm to values under 4 ppm, at a 30°C temperature.

ethanol: In Figure 3, the variation of the recovery and the chloride content in the sodium sulphate crystals obtained as a function of the temperature variation can be observed.

At a temperature ranging from 10 to 20°C the solid phase is represented by a decahydrous sodium sulphate, appearing anhydrous at 30°C The recovery curve is a little higher than that obtained with methanol, with a 93.6% maximum growth value at 30°C.

The chloride ion content is higher than that obtained with methanol, dropping to 60 ppm values, at 30°C.

ethylene glycol: In Figure 4, the recovery variation and the chloride content in the sulphate crystals obtained by temperature variation can be observed.

At 10°C no formation of crystals is observed, but at 20°C the solid phase is represented by decahydrous sodium sulphate, losing the crystallization water at 30°C. The sodium sulphate recovery with ethylene glycol is below that of ethanol and metha-

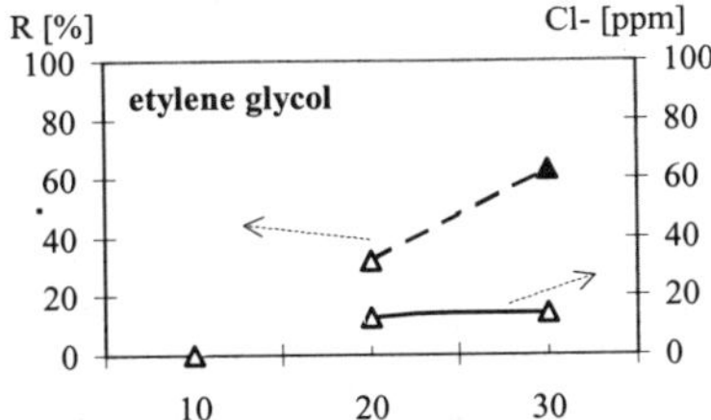

Figure 4. Effect of the temperature on the recovery and the choride content for ethilene glicol.

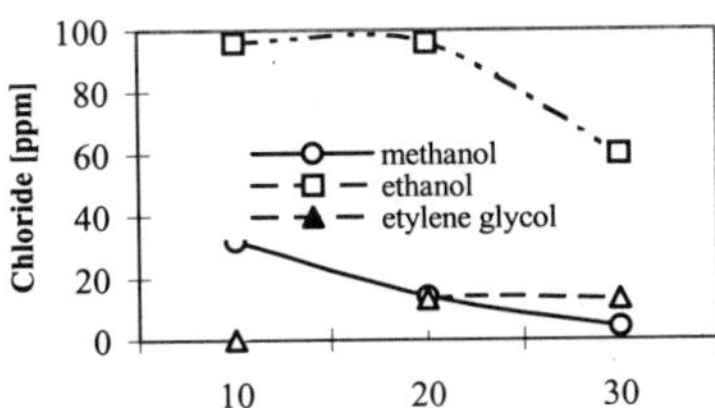

Figure 5. Effect of the temperature and of organic solvents on the chloride content for artificial brine.

nol, reaching maximum values of 62.8% at 30°C.

The chloride content at 20°C is below that of methanol and ethanol, remaining at 13 ppm until 30°C.

4 CONCLUSIONS

1. Due to the addition of organic solvents, the recovery of the sulphate recovery tends to increase its temperature, whereas the chloride content in the sodium sulphate crystals decreases abruptly from 3000 ppm to less than 4 ppm, in the case of methanol.
2. The sodium sulphate crystals obtained by using methanol as solvent, at a 30°C temperature, show a minimum chloride content of 4 ppm and a 91.5% sodium sulphate recovery.
3. The sodium sulphate crystals obtained with the use of ethanol as a solvent at a 30°C temperature show a chloride content of 60 ppm and a 93.6% sodium sulphate recovery.
4. The sodium sulphate crystals obtained using ethylene glycol as solvent at a 30°C temperature show a chloride content of 13 ppm and a 62,8% sodium sulphate recovery.
5. The results show that the greatest selectivity in the attenuation of the chloride ions in the sodium sulphate crystals is obtained with the use of methanol at a 30°C temperature, followed by ethylene glycol and ethanol, at the tested operating conditions (Figure 5).
6. The development of the present technology will allow for a simple, economical purification process, with a more positive environmental impact.

5 REFERENCES

Angelelli, V., Brodtkorb, M. K., Gordillo, C.. 1983.Las Especies Minerales de la República Argentina". Secretaría de Minería. Sta. Fé 1548 (1206) Buenos Aires

ASTM Standart test methods for choride ion in wa ter - Method C - Colorimetric method- 1981D 512 -81 pp. 396-398.

Cervera Y., R.H., Gutiérrez O., L.V., Crubellatti R., R.L. 1998. "Attenuation of contaminants in the process of obtaining a sodium sulfate at laboratory scale". Proceedings of the IV Intenational conference on clean technologies for the mining industry. Environment & innovation in mining and mineral technology. Universidad de Concepción. Chile. Tomo II.; pp. 901-909.

Galli, D. E., Viturro, C. I., Molina, A. C., Villa, W. C. y Zacur, A. 1978 "Análisis del contenido de cloruro de sodio en el proceso de purificación de sulfato de sodio". IV Jornadas de Ingeniería de Minas. Salta; pp. 229-247.

Garside, J. , Shah, M. B.. 1980 "Crystallization Kinetics from MSMPR crystallizers". Ind. Eng. Chem. Process Des. Dev. 19; pp. 509-514..

Gaska, R. A., Goodenough, R. D. and Stuart, G. A.. 1965. "Amonia as a solvent". Chemical Engineering Progress, 61 (1); pp. 139-144

Kolthoff, Y. M., Dendell, E. S. 1956. "Análisis químico cuantitativo. 6ta. Edición. Librería Editorial Niagar, S. R. L., Buenos Aires. pp. 132-162.

Mina-Mankarios, G., Pinder, K. L.. 1991. " Salting out of sodium sulphate"Cand. J. Chem. Eng., 69; pp. 308-324.

Sehidell, A.. 1940. "Solubilities of inorganics and metal-organic compound".3ra. Ed., Vol. 1, Van Nostrand. New York; pp. 1234-1265.

Ullmann's Encyclopedia of Industrial Chemistry. 1993.Vol. A24 . Sodium sulfates. Fifth Ed. USA.

Environmental Issues and Management of Waste in Energy and Mineral Production, Singhal & Mehrotra (eds)

Assessment of the reliability of tailings dam structures by centrifuge modeling

E. De Souza & A. P. Dirige
Department of Mining Engineering, Queen's University, Kingston, Ont., Canada

ABSTRACT: Tailings dam failures generally result in catastrophic ecological disasters. Failure of an embankment generally results in liquefaction of the unconsolidated tailings mass, which can flow many miles in the surroundings. Because of the potential hazard and severe environmental and safety impact created by mine tailings disposal, industry has taken leadership in developing extensive research towards improved tailings dam design and management techniques that address both the safety and the environmental issues of tailings disposal. New technology and risk-based design solutions have emerged, thus facilitating effective environmental impact assessments of planned, operating or after-use tailings impoundments. As a manifested contribution to this continuing effort, this paper introduces the centrifuge as an integral component of the design and verification process of mine tailings dams.

1 INTRODUCTION

The failure of a number of waste dumps and tailings dams over the last decade has directed public attention on the potential hazard and severe environmental and safety impact created by mine tailings disposal, and has renewed concerns on the need for more appropriate design, construction and management of tailings impoundments. Furthermore, the mining industry is more and more expected to achieve high environmental performance in their operations, and high design standards are required by government and regulators.

Although past engineering research efforts carried out within the mining industry have developed unique strategies that normally result in stable tailings dam conditions, the effective and economic design, construction and maintenance of safe mine tailings dams still constitute a major technical challenge to industry. A number of tools have, over the years, evolved to facilitate the design process and modern design methods and techniques continue to evolve.

This paper introduces the centrifuge as a modern and unique design and verification tool which may provide viable solutions to the continuing effort towards the construction of safe mine tailings dams. A number of case examples are provided to illustrate the effectiveness of the centrifuge as a modelling tool.

2 AN OVERVIEW OF MINE TAILINGS DAM ENGINEERING PRACTICES

2.1 *Dam Stability Issues*

As larger mining operations have become predominant in the last decade, larger tailings impoundments, containing billions of tons of tailings, and higher dam structures, reaching more than 200 meters, have evolved; the ecological impacts of such impoundments are being increasingly scrutinized by the general public. Recent massive impoundment failures have imposed new challenges to industry to not only guarantee the integrity of such structures over the life of a project but to minimize any ecological impact associated with tailings disposal.

Studies of incidents have indicated that most accidents and failures of tailings dams occur in active tailings ponds (USCOLD 1994). According to USCOLD the causes of failure of upstream tailings dams, the type of dam structure most susceptible to failure, are, in decreasing importance: slope instability, earthquake, overtopping, seepage, foundation and structural. Some of the most recent serious incidents include Aberfan in Wales, Buffalo Creek in USA, Bafokeng in South Africa, Stava in Italy, Merriespruit in South Africa (Brinsden & Lewis 1996), El Cobre in Chile (Zeng et al. 1998), Omai in Guyana (Vick 1996) and Los Frailes in Spain (Lindvall et al. 1999, Spooner 1999). It is generally agreed that most accidents are the result of a combination of faults, omissions, oversights and inappropriate management actions and that although such events are difficult to predict, appropriate management intervention could normally prevent most of the failures (Brinsden & Lewis 1996, Glos 1999).

2.2 *Environmental Impacts Associated With Tailings Dam Failures*

Other than the possible heavy loss of life and economic losses, environmental damaging impacts associated with structural failures in dams include (UNEP 1996):

- damage or destruction of valuable habitats and ecosystems;
- release of effluent from an impoundment may contaminate surface water;
- generation and release of acid mine drainage may occur;
- seepage of effluent throughout the base of the structure may contaminate groundwater;
- seepage may result in substantial mounding of the natural groundwater table and provoke soil salinity problems;
- dried tailings may be swept as dust by strong winds into neighbouring habitations or ecosystems;
- effluent in tailings impoundments may generate toxic gases that may poison birds attracted by water in dry regions;
- poorly designed and protected ancillary structures may trap or drown wildlife;
- landscape changes by major tailings impoundments may be unacceptable in areas of scientific or scenic significance.

Companies are now expected to have high quality standards to tailings dam construction and management. Any new mining project requires that extensive environmental impact studies, emergency preparedness studies, sustainable development studies, dam design risk assessment and reclamation programs be developed before site permitting and implementation.

2.3 *A Review of Tailings Dam Construction Methods*

A number of arrangements are required to be developed when constructing a tailings disposal system (UNEP 1996):

- a system for the delivery of the tailings to the disposal site;
- embankments to confine the tailings within the site;
- an arrangement for diverting natural runoff around or through the impoundment;
- a system for the deposition of the tailings within the impoundment;
- a facility for the evacuation of excess supernatant water from the impoundment;
- a system of measures to protect the environment from pollution.

Tailings dams can be classified based on the construction method either as water retention dams or raised embankments (ICOLD 1982). Water retention dams are constructed to their full height prior to impoundment while raised embankments are constructed in stages throughout the life of impoundment. Raised embankments are further classified according to the direction in which the embankment crest moves in relation to the starter dam: upstream, downstream and centreline dams.

The following more detailed classification of dams, based on the construction material utilized, has been suggested by UNEP (1996):

- conventional dam embankment;
- staged conventional embankment;
- staged embankment with upstream impermeable zone;
- embankment with tailings impermeable zone;
- embankment with tailings in structural zone (upstream, downstream, centreline construction);
- upstream construction using beach or paddock (spigot disposal method, sub-aerial disposal method, paddock disposal method).

Each method of construction has its inherited advantages and disadvantages. Water retention dams have a high initial cost while raised embankments have a distributed cost over the life of the project. The selection of the embankment type is a function of a number of factors including the characteristics and constraints of the site; earthquake considerations; the type of tailings; construction materials used; and cost considerations. The popular upstream raised embankment is the most susceptible to instability problems and appropriate design, construction and management practices must be carefully implemented.

A unique disposal system, referred to as the Thickened Tailings Disposal, has been developed to minimize the ecological problems associated with the operation of conventional tailings disposal systems (Robinsky 1999). It involves the process of thickening the tailings to a heavy slurry prior to disposal, thus creating a self-supporting deposit of tailings. Because the settling pond is eliminated, failure within the well consolidated tailings material would only result in local slump, without flow. It is also noted that the system can inhibit acid drainage. The fines in the slurry of the thickened tailings deposit promote high matric suction, thus maintaining the tailings containing sulphides in a saturated condition. Such saturated environment prevents the breakdown of sulphides into sulphuric acid which would otherwise dissolve other undesirable metallic elements that may contaminate the surrounding environment.

2.4 *A Review of Dam Design and Operation Practices*

The design process for a tailings disposal area involves the following steps (Taylor & D'Appolonia 1977):

- area selection;
- determination of environmental constraints;
- determination of subsurface conditions;
- determination of the layout limits of the disposal area;
- determination of embankment cross-sections;
- slurry retainage considerations;
- analytical considerations (stability, seepage);
- construction or operation considerations (equipment selection, haul road location, support facilities);
- cost considerations;
- environmental planning (reclamation, fugitive dust, abandonment).

The dam design process thus involve a complex optimization study involving geotechnical and engineering performance considerations and construction techniques integrated with site, environmental, operational and cost constraints.

The selection of the site for disposal is a crucial component of the dam design and planning process and involves consideration of a number of important factors (UNEP 1996):

- proximity to the plant: lower transportation costs;
- embankment recovery: select site with a large ratio of storage volume to embankment volume;
- topography: avoid site with very steep slopes;
- natural runoff: avoid site with a large catchment area;
- elevation: select site with elevation close to that of the plant;
- foundation conditions;
- potential mining sites: avoid sites likely to contain mineable resources;
- human presence: avoid populated centres and areas of human activity;
- visual intrusion: select sites hidden from public view;
- environmental sensitivity: avoid sites containing endangered species of flora or fauna;
- groundwater considerations: avoid sites which could result in excessive seepage from the impoundment into the groundwater;
- erosion: avoid sites prone to water or wind erosion.

The most commonly used methods for assessing dam structural stability include methods of limiting equilibrium analysis and stress-deformation analysis using numerical methods based on classical soil mechanics consolidation models. Such traditional design methods cannot properly describe the complex behaviour involving flow, settling and consolidation of tailings and may underestimate the structural performance of dams under different operational conditions. This could ultimately result in inappropriate designs where unstable structural conditions may arise during the operation of the tailings management project. Dynamic physical modelling is introduced in this paper as a modern design and verification tool which may provide viable solutions to the effective design of tailings dams. A combination of centrifuge and numerical modelling can form a powerful tool to enhance the design of safer and more economic dams.

The application of administrative controls and of risk-based assessments has been proposed for preventing dam failures during their operation. Brinsden & Lewis (1996) report a risk based approach in which the probabilities of all the faults that can progress to trigger failures are first estimated, and then are combined in a fault-tree analysis such that the probability of failure and associated losses can be determined and specific management actions to prevent faults developing can be established. Glos (1999) proposes the use of a four-part program of administrative controls for use in decreasing the risks and costs associated with tailings disposal facilities. The four proposed controls include the development of operating manuals to define the design criteria and construction procedures; the execution of quarterly operational inspections; the assessment of the seismic, hydrological and operational risk of the tailings facilities; and the provision of continuing technical advice to the line manager with respect to the technical aspects of the dam. This program is intended to provide the operational manager with detailed information to ensure that operation of the tailings facility follows the designer's intent, thus preventing facility misoperation and resulting failure.

3 THE CENTRIFUGE AS A TAILINGS DAM DESIGN TOOL

3.1 *Principles of Centrifuge Modelling*

Centrifuge modelling is a technique for reduced-scale physical modelling of gravity dependent phenomena, such as mine backfill and tailings dam behaviour. Centrifuge modelling is based on the fundamental concept that gravitational mass forces can be simulated by centripetal mass forces applied by the acceleration of a scaled model. The stress levels acting within an N-scaled model accelerated at N times the earth's gravity field are analogous to those acting on the prototype. For centrifuge modelling to be valid, similitude must exist between the accelerated scaled model and the prototype. Conditions of similitude are assessed using scale laws and dimensionless numbers. Scale factors and dimensionless numbers have been well established by many researchers for a multitude of engineering problems (Arulanandam et al. 1988, Schofield 1980).

Regardless of the type of application, the principles and design of scale model experiments are similar and have been well established. First, a geometrically-similar model, scaled down at 1/N, is formed with materials of identical properties as that of a prototype. The transformation of the prototype must ensure that the model will have the same relative position, proportions, values and structure. The model is, in essence, a small prototype. This transformation is followed by a dimensional analysis and the development of dimensionless numbers. Similitude between the prototype and the scale model is obtained when the same dimensionless numbers are obtained in the prototype and model. The scale model is then placed in the centrifuge strongbox and rotated at a desired angular velocity. A corresponding centripetal force acceleration is experienced by the model in flight, and this acceleration induces self-weight forces in the test material. Thus, if a 1/N scaled model is accelerated to N gravities, the inertial mass induced stresses at any point in the model will be identical to the gravitational mass induced stresses at any similar point in the prototype, providing that the boundary conditions are identical at all equivalent points. All stress-induced and stress-gradient induced events will, then, occur in the model as they would in the prototype, but model events depending on gradients will occur faster because the physical distances are reduced in the model. The model accelerated in the centrifuge thus experiences analogous behaviour (patterns of strain, deformation, distortion, arching, cracking, failure, etc.) as the prototype.

3.2 *Modelling Mine Tailings Dams*

Modelling tailings dams is based on the fundamental concept that gravitational mass forces can be simulated by centripetal mass forces applied by the acceleration of a scaled model. If the model dam is built from the same material as the prototype at a scale 1/N, and the centrifuge is run such that the centripetal acceleration acting in the model is N gravities, then the stress levels due to gravity at corresponding points in the prototype ($\sigma = \rho g z$) and in the model ($\sigma = \rho N g(z/N) = \rho g z$) will be identical. If model boundary conditions are made to correspond with the prototype, then the strain fields and hence deformations and behaviour will be similar. Modelling of dams using a centrifuge is particularly useful for cases involving three-dimensional problems and partly saturated conditions.

3.3 *The BMG Centrifuge Testing Facility*

A schematic of the Battle Mountain Gold Centrifuge Testing Facility is shown in Figure 1.

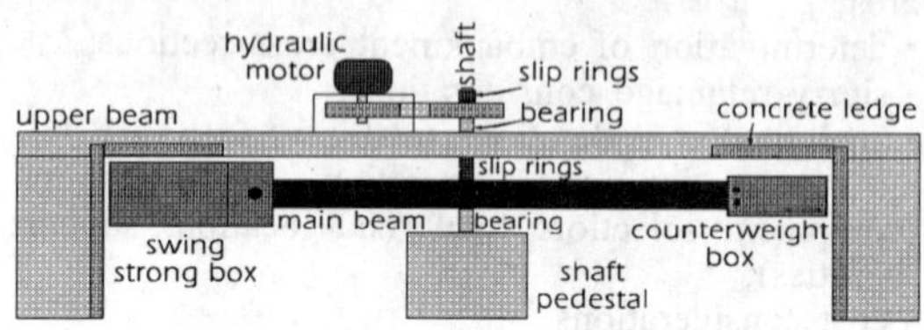

Figure 1. Schematic of the Battle Mountain Gold centrifuge testing facility.

The centrifuge is a 5.3 m diameter machine driven by a 50 kW hydraulic motor and speed control assembly. It is designed to rotate up to a maximum speed of 350 rpm and it is capable of exerting up to 330 gravities on typical models. It is rated at 30g-tonne, since it can accelerate a 0.1 tonne mass to 300 gravities. The main beam of the centrifuge is 5.19 m long, including strong boxes, and is fabricated from a 200 mm steel tube of 13 mm wall thickness. Swing strong boxes are available to house specific scaled models. These are attached to one end of the main beam by a 70 mm diameter steel pin. A counter-weight box is attached to the other end of the main beam by two 25 mm diameter steel pins. The beam rotates on a 76 mm diameter steel shaft which is powered by the 50 kW hydraulic motor. The speed is controlled by ramping mechanisms in the hydraulic power supply, and can be maintained at any level between 50 and 350 rpm. The centrifuge has a total of 12 slip rings and instrumented models are monitored via a high speed data acquisition system. An image processing system is also used for visualizing and digitizing the real time behaviour of models captured by a digital dynamic video monitoring system.

4 CASE APPLICATIONS

The progressive use of centrifuge dynamic testing

for modelling dam behaviour has been routinely presented in dedicated sessions of the centrifuge international conference series coordinated by the Technical Committee TC2 on Centrifuge Testing of the International Society of Soil Mechanics and Foundation Engineering and in a number of geotechnical conferences. Specific case applications are presented in this session.

4.1 *Earthquake Induced Failure in Tailings Dams*

Zeng et al. (1998) investigated the stability of embankments used for coal refuse impoundment employing the downstream, center and upstream methods of construction, typical of the mining areas of eastern Kentucky and West Virginia. Earthquake loading can generate excess pore pressures followed by liquefaction of the tailings mass, changing it into a heavy viscous fluid and producing a large pressure increase on the embankment which can induce failure.

The structure investigated was a tailings dam facility in eastern Kentucky constructed by the upstream method with a total height of 110 m. The stability analysis of the embankment under earthquake loading considered such factors as consolidation rate, drainage and pore pressures.

Three centrifuge tests were performed on a 1 m radius centrifuge, equipped with a servo-hydraulic shaker. Two models were tailings dams constructed by the upstream method and the third by the downstream method. The earthquake applied to the models was based on an earthquake recorded at the mine site, produced using the same frequency contents while scaling the peak amplitude. The models were fitted with accelerometers and pore pressure transducers. The downstream model was accelerated to 50g and then held at 50g for an equivalent of 30 prototype days before the earthquake was applied. The upstream models were built in two stages, the first stage was consolidated in the centrifuge for a prototype time of 150 days, and the second stage was allowed to consolidate for 30 prototype days before the earthquake was applied.

Figure 2 shows the profile of a downstream and of an upstream model before and after centrifuge earthquake simulation. Modelling has shown that a dam built by the downstream method has a much better seismic stability than a similar dam built by the upstream method. It was noted that the coal fine refuse can remain as a soft viscous material with low shear resistance even after many years of consolidation, and it has been recommended that, to improve the seismic stability of upstream dams, an improved drainage system for the coal fine refuse should be considered.

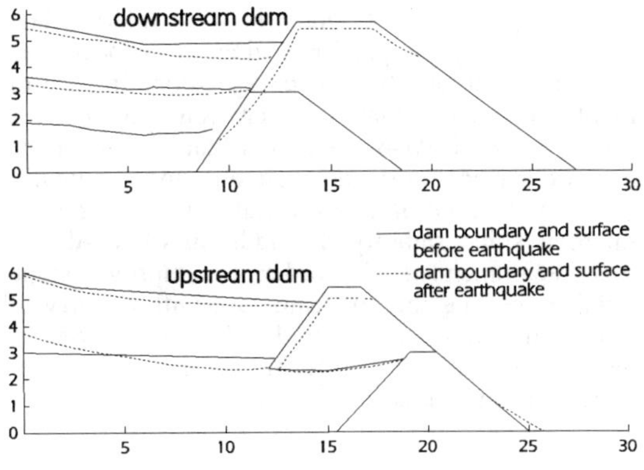

Figure 2. Profile of centrifuge models before and after earthquake test (after Zeng et al. 1998).

4.2 *Coal Waste Embankment Assessment Using Dynamic Centrifuge Modelling*

A series of centrifuge model tests were performed to investigate the stability of coal tailings embankments (Al-Hussaini et al. 1981). Eight model embankments, 254 mm high, 762 mm long and 152 mm deep, with various geometries, were instrumented and accelerated using an eight meter diameter centrifuge to 100-120 gravities. Corresponding prototype heights ranged from 18.6 to 29.6 m. The problems investigated included slope stability and seepage analysis. Coal waste from the western United States was used to construct the models. The materials had permeability values ranging from 10^{-4} to 10^{-8} cm/s, drained cohesive strengths ranging from 0.025 to 0.24 kN/m^2 and friction angles ranging from 26.0 to 34.7 degrees.

From the series of tests it was noted that slope stability calculations agreed well with observed model performance. Factors of safety ranged from 0.7 to 0.9 at the instant of failure for models with retrogressing slips and from 1.3 to 1.7 for stable embankments. Estimates of phreatic surface based upon Dupuit's parabolic assumption and finite element seepage analysis agreed well with phreatic surfaces observed in models with vertical upstream slopes. Models with more realistic sloping upstream surfaces demonstrated less favourable agreements. Figure 3 shows a schematic of the location of phreatic and observed and calculated failure surfaces for a tested model.

It was noted that, for embankments of cohesionless materials, permeability can govern the mode of failure. For materials of low permeability with high initial water content, as in hydraulically placed tailings, rate of construction may pose a stability problem. For materials of intermediate permeability, pore pressures due to seepage can cause mass instability and retrogression. In case of highly permeable materials, erosion may pose a stability hazard. Also noted was that exiting of the phreatic surface on the downstream slope generally leads to erosion or mass instability. In order to improve slope stability the addition of toe-drains was recommended. In order to improve slope stability, sealing the upstream face with a slurry of low permeability was found effective. However, inclusion of a highly permeable soil key was found to reduce embankment stability.

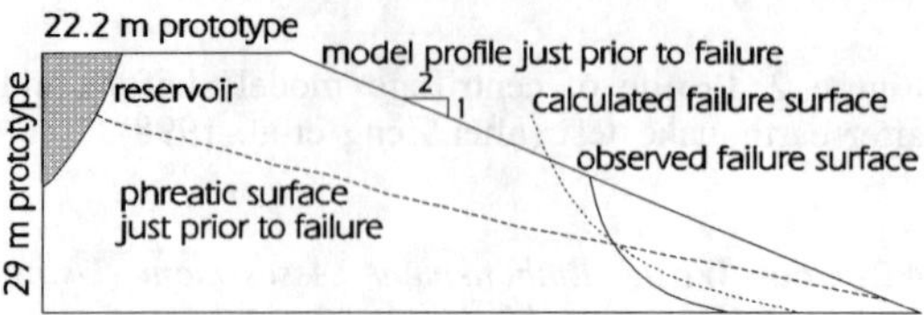

Figure 3. Location of phreatic and failure surfaces for a centrifuge model (after Al-Hussaini et al. 1981)

4.3 *Applications of centrifuge modelling to predict dam stability problems*

A series of centrifuge model tests, sponsored by the Bureau of Mines, were performed to investigate the safety of coal mine tailings embankments and to determine design criteria for tailings dams (McWilliams 1989). The problems investigated included slope stability, erosion and seepage, and the effects of compaction and weathering on tailings embankments. The centrifuge runs presented in this section were made at a 15 m diameter centrifuge installation capable of 240g.

A series of eight model tests were performed to determine the influence of packing density and material gradation on dam stability. Instrumented models were accelerated to a level of 100g and held for the duration of the test. The models which failed matched with predictions by a computer code used for slope stability analysis using the modified Bishop method of slices. However, the failure pattern does not necessarily follow the assumed classical slip circle shape. Two models in which the compaction state of the material was increased significantly, did not fail, thus demonstrating that, for the material tested, an increase in packing density can increase the stability of an embankment. Observations on the effect of gradation indicated that by increasing the amount of fines in the reservoir increased the stability of the structure. However, the increase in fines restrict the flow through the embankment, limiting the discharge from the reservoir and increasing the probability of overtopping the reservoir.

4.4 *Modelling Hydraulic Fracturing in Dam Cores*

A laboratory and centrifuge modelling study was carried out to examine hydraulic fracturing in earth dam cores due to differential settlement and stress arching (Mitchell & Abel 1995). The core of a composite earth dam, when designed with very high height to width ratios or where core trenches are relatively narrow with respect to trench depth, may be susceptible to hydraulic failures initiated by hydraulic fracturing. Although stress analysis techniques can adequately model the potential for hydraulic fracturing in earth dams, the centrifuge offers a simple and independent verification method of design.

In a simplified study, models of 200 mm wide and 120 mm wide cores were accelerated to 60 gravities to represent a 16.5 m high earth dam with a vertical compressible core and rigid shells. Models with core width to depth ratios of 1 and 0.6 were tested. A base total stress cell and a pore water pressure transducer were used to record pressure development. About one hour of centrifugation (150 prototype days) was allowed for settlements and arching to develop. At this stage the water level in the upstream was progressive raised to 250 mm (15 m prototype water head) in about 0.5 hours (2 months prototype) to achieve fill reservoir prototype pressures of 147 kPa. Rates of water flow through the core were monitored as the water level was raised. Increases in leakage rates were observed when the full reservoir condition was achieved, however such increases did not indicate a condition of hydraulic fracturing but one of reduced effective stress. The highest pressure monitored by the base transducer was just over 120 kPa in the 0.6 width to depth ratio models. The total stress cell indicated stress values consistently over 230 kPa in the 1.0 width to depth ratio models but noted values as low as 130 kPa in the 0.6 width to depth ratio models. Based on model results, the authors have suggested that the risk of hydraulic fracturing in dam cores under strongly arched stress conditions mat be independently examined by modelling in a centrifuge.

4.5 *The CANLEX Field Event*

The Canadian Liquefaction Experiment (CANLEX) involves the characterization of sand, based on laboratory and full scale field tests, in order to predict its static and dynamic liquefaction response

(List & Robertson 1995, Byrne et al. 1995a). In the field liquefaction event, at the Syncrude site at Fort McMurray, triggering of a liquefaction flow slide was based on the rapid construction of an embankment over a very loose saturated sand layer. The experimental study aimed at determining the level of disturbance required to trigger liquefaction, to determine if flow slide constitutes a prevailing response and to characterize the expected levels of deformations and associated geometry changes.

The initial phase of this experimental study, aimed at making predictions of response prior to the event and at facilitating the design and planning of the event, involved site characterization, laboratory characterization studies, centrifuge testing and numerical modelling. Details of the centrifuge modelling work (Byrne et al. 1995b, Phillips & Byrne 1994) and of the numerical analysis of the centrifuge tests (Cathro & Gu 1995, Chan et al. 1995) are presented to illustrate the use of centrifuge testing for assessing dam stability problems.

The centrifuge models were designed to simulate the planned field event which would comprise a 5-10 m thick saturated loose sand layer placed underwater and loaded by a stage-constructed embankment, 5-10 m in height. Liquefaction would be triggered by the rapid placement of the upper part of the embankment thus inducing high excess pore water pressure from an undrained loading condition.

The centrifuge models were prepared with Syncrude sand saturated with canola oil instead of water, to delay the rate of fluid dissipation, thus simulating near undrained conditions prior to load application. Because of the necessity of rapid load application, the upper part of the embankment was simulated by rapid placement of a weight rather than construction of a soil layer in flight.

The test work was carried in two stages. Stage 1 was successfully undertaken to verify if static liquefaction could be triggered under centrifuge loading. Testing comprised a scaled 13 m submerged slope loaded at the crest with a drop weight resulting in impact loading levels corresponding with applied pressures of 43 and 86 kPa. Stage 2 comprised four tests with a scaled 10 m target layer, a 5 m berm, and a controlled rapid loading assembly (Figure 4). Under an acceleration field of 50 g, the models were loaded in two stages of 60 kPa each, corresponding to approximately 3.5 m of soil placement per stage. Instrumentation included pressure and displacement transducers (PPPT and LDT, in Figure 4). The pattern of pore water response and the pattern of deformation after load application indicated that liquefaction failure could be successfully induced in the centrifuge. Test results have indicated that large deep-seated soil movements occur and most movements occur within seconds of load application. Measurements showed a sudden initial increase of excess pore pressure, followed by rapid dissipation. Centrifuge test results were significant for appropriate design verification of the planned field event.

A numerical simulation was also carried out using a finite difference solution (FLAC code) of the governing elastic-plastic stress-strain equations (modified Matsuoka Model). Numerically predicted and centrifuge measured pressure and displacement data were shown to be in reasonable agreement. Numerical simulations of the field event also indicated that placement of a 10 m embankment may trigger a liquefaction failure, under either undrained or transient conditions. A finite element analysis that combines liquefaction and consolidation analysis was also conducted to verify the centrifuge model results (Cathro & Gu 1995). Numerical model results confirmed that runaway liquefaction flow failures can occur under undrained conditions, but showed that the slope movements that occurred were stopped quickly by the effects of consolidation.

The field test was carried out during the summer of 1995 (Robertson et al. 1996). A flow slide was not initiated as the height of the embankment (8 m) and the direction of loading were not sufficient to trigger flow liquefaction within the loose sand foundation.

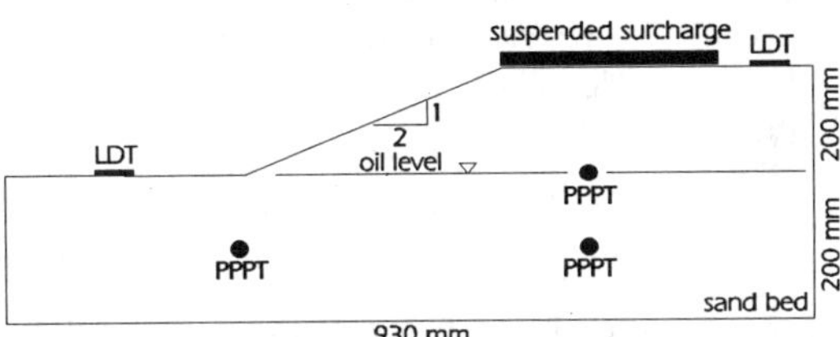

Figure 4. CANLEX centrifuge model test (after Byrne et al. 1995b)

5 CONCLUSIONS

Tailings dam failures generally result in devastating ecological disasters. This is because a conventional dam retains a massive amount of very loose unconsolidated tailings and large quantities of water and failure generally results in liquefaction of the unconsolidated tailings mass, which can flow many miles in the surroundings. Industry and regulators are aware of the need for improved tailings design and management techniques that address both the safety and the environmental issues of tailings disposal. This paper has demonstrated, through a series of case studies, the effectiveness of centrifuge modelling in assessing the stability of tailings dam structures and proposes that the centrifuge be adopted as an integral component of the design and verification process of mine tailings dams.

REFERENCES

Al-Hussaini, M.M., Goodings, D.J., Schofield, A.N. & Townsend, F.C. 1981. Centrifuge modeling of coal waste embankments. *Journal of the Geotechnical Engineering Division*. 107(GT4): 481-499. ASCE.

Arunalandam, K., Thompson, P.Y., Kutter, B.L., Meegoda, N.J., Muraleetharan, K.K. & Yogachandran, C. 1988. Centrifuge modeling of transport processes for pollutants in soil. *ASCE Journal of Geotechnical Engineering*. 114(2): 185-205.

Brinsden, W.K. & Lewis, P.A.P. 1996. Risk assessment in the design of tailings dams. *Journal of the South African Institute of Mining and Metallurgy*. 96(7):325.

Byrne, P.M., Robertson, P.K., Plewes, H.D., List, B. & Tan, S. 1995a. Liquefaction event planning. *48th Canadian Geotechnical Conference - Trends in Geotechnique*: 341-352. Canadian Geotechnical Society.

Byrne, P.M., Phillips, R. & Zeng, Y. 1995b. Centrifuge tests and analysis of CANLEX field event. *48th Canadian Geotechnical Conference - Trends in Geotechnique*: 353-365. Canadian Geotechnical Society.

Cathro, D.C. & Gu, W.H. 1995. Finite element analysis on CANLEX C-Core centrifuge test. *48th Canadian Geotechnical Conference - Trends in Geotechnique*: 367-374. Canadian Geotechnical Society.

Chan, D., Soroush, A. & Morgenstern, N.R. 1995. Numerical analysis of centrifuge modelling for the CANLEX experiment. *48th Canadian Geotechnical Conference - Trends in Geotechnique*: 375-382. Canadian Geotechnical Society.

Glos, G.H. 1999. Using administrative controls to reduce tailings-dam risk. *Mining Engineering*. 55(9): 31-33.

International Commission on Large Dams, ICOLD. 1982. *Manual on tailings dams and dumps*. Bulletin no. 45.

Lindvall, M., Oliva, A. & Erikson, N. 1999. The Aznalcollar tailings pond accident - environmental impact and recovery measures. *CIM Calgary 99. 101st Annual General Meeting*.

List, B.R. & Robertson, P.K. 1995. Canadian liquefaction experiment update - characterization of sand for static and dynamic liquefaction. *48th Canadian Geotechnical Conference - Trends in Geotechnique*: 49-58. Canadian Geotechnical Society.

McWilliams, P.C. 1989. Bureau of Mines geotechnical centrifuge research - a review. *Bureau of Mines Information Circular 9218*:19. United States Department of the Interior.

Mitchell, R.J. & Abel, A.C. 1995. Laboratory and centrifuge modelling on hydraulic fracturing in dam cores. *48th Canadian Geotechnical Conference - Trends in Geotechnique*: 839-846. Canadian Geotechnical Society.

Phillips, R. & Byrne, P.M. 1994. Modelling slope liquefaction due to static loading. *47th Canadian Geotechnical Conference*: 317-326. Canadian Geotechnical Society.

Robertson, P.K. et al 1996. CANLEX Phase III full scale flow liquefaction test: planning, objectives and conclusions. *49th Canadian Geotechnical Conference*: 567-578. Canadian Geotechnical Society.

Robinsky, E.I. 1999. Tailings dam failures need not be disasters - the thickened tailings disposal (TTD) system. *CIM Bulletin*. 92(1028): 140-142.

Schofield, A.N. 1980. Cambridge geotechnical centrifuge operations. 20th Rankine Lecture. *Geotechnique*. 30(3): 227-268.

Spooner, J. 1999. The tailings spill in southern Spain: sharing the experience. *CIM Calgary 99. 101st Annual General Meeting*.

Taylor, M.J. & D'Appolonia, E. 1977. Integrated solutions to tailings disposal. *Proceedings of the Conference on Geotechnical Practice for Disposal of Solid Waste Materials*: 301-326. ASCE.

United Nations Environment Programme, UNEP. 1996. *A guide to tailings dams and impoundments: design, construction and rehabilitation*. Bulletin 106.

USCOLD. 1994. Tailings dam incidents.

Vaid, Y.P., Sivathayalan, S., Uthayakumar, M., R. & Eliadorani, A. 1995. Liquefaction potential of reconstituted Syncrude sand. *48th Canadian Geotechnical Conference - Trends in Geotechnique*: 319-329. Canadian Geotechnical Society. Vancouver.

Vick, S.G. 1983. *Planning, design and analysis of tailings dams*. John Wiley & Sons.

Vick, S.G. 1996. Failure of the Omai tailings dam. *Geotechnical News*. 14(3).

Zeng, X., Wu, J. & Rohlf, R.A. 1998. Seismic stability of coal-waste tailings dams. *Geotechnical Earthquake Engineering and Soil Dynamics III. Proceedings of a Specialty Conference*: 950-961. American Society of Civil Engineers.

Environmental Issues and Management of Waste in Energy and Mineral Production, Singhal & Mehrotra (eds)

About methods of resource recovery from industrial waste

G.R.Gillich
'Eftimie Murgu' University, Reşiţa, Romania

ABSTRACT: Paper presents an important industrial activity to the mountain area of Banat Steel Factory Reşiţa - CSR , but also problems such as environment pollution, which is now given high concern. The current activity of the plant produces continuously waste as detailed above. This paper describes the way the waste is presently consumed, along with several proposals in order to intensify and improve the recycling activity.

1 INTRODUCTION

Steel Factory in Reşiţa-CSR, founded in 1771, is the oldest iron works in Balkans and one of the oldest in operation in the world. Building the first two furnaces has started in 1770, as ordered by Austrian Empress Maria Theresia. On July 3 1771, the furnaces „Franzikus" and „Josephus" have been fired up, running on iron ore from the mines nearby. The charcoal was produced in pits as the area was rich in wood. The whole project costs amounted 37,800 Guldens at that time.

Company's field has gradually become larger, as it started superior iron smelting and steel processing; new workshops have appeared, to produce chiefly railway equipment: tracks, engines, boilers, bridges, electrical motors, etc.

2 RECYCLING INDUSTRIAL WASTE IN CSR STEEL FACTORY

After the Second World War, the company has been split in several plants: mining companies, machine works, forest administration resulted thereof; the plant called CSR has taken over the iron and steel factories.

Presently, CSR operates an ore processing plant, furnaces, Siemens-Martin plants, electric furnaces, rolling mils, maintenance shops as well as power production and distribution facilities. Two projects started after 1989 are about to be commissioned: the electric steel plant and continuous casting plant.

The 229 years of industrial activity have brought fame and prosperity to the mountain area of Banat, but also problems such as environment pollution, which is now given high concern. Main issues on the topic are: solid waste disposed in dumps and air pollution due to fumes and dust emission. The waste disposed for years in dumps are:

- non-metallic waste (mainly slag and bricks) and
- metallic waste (both ferrous and non-ferrous).

The current activity of the plant produces continuously waste as detailed above. The use of these, both resulted from daily operation as well as dumped for years after years is a matter of prime importance, both economical and ecological. This paper describes the way the waste is presently consumed, along with several proposals in order to intensify and improve the recycling activity.

The processing of dump materials consist of the following operations:

- selection of ferrous, non-ferrous and non-metallic components;
- crushing;
- grinding;
- metallic particle selection from the slag mass.

The figure depicts the existing waste processing traces in full line and the proposed operations in dotted line.

3 CONCLUSIONS

1 The waste resulted from daily operation as well as dumped for years after years is a matter of prime importance, both economical and ecological.
2 The analysis in this field should be made also about every plant with technical details.
3 These recycling solution are under certain condi-

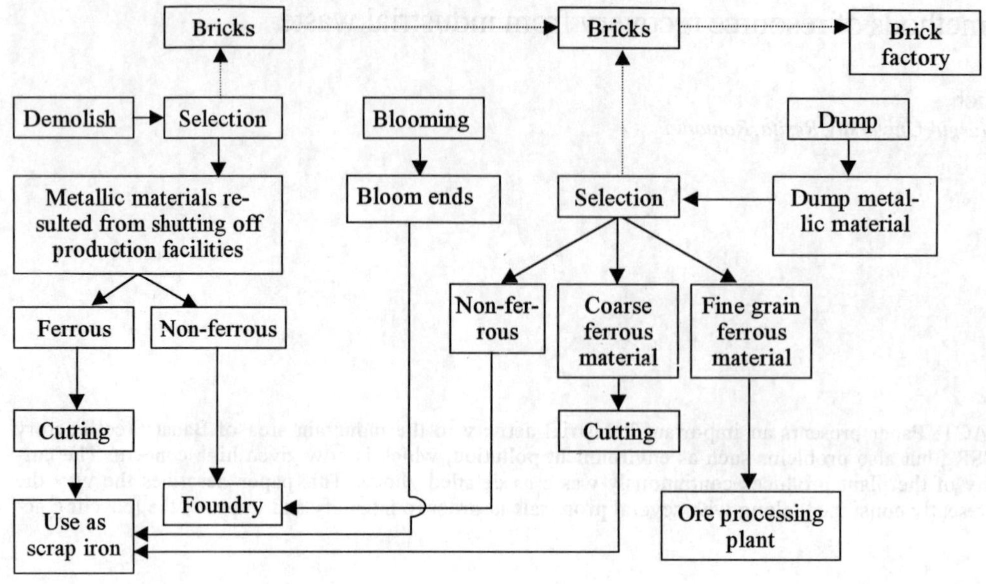

tion more economic that the stocking one.
4 Interest should be briefly drawn to the exploitation of secondary recycling energy resources.

4 REFERENCES

Belu, S., Ieremia, A. 1971.*Două secole de siderurgie la Reşiţa*, Cluj: Întreprinderea Poligrafică.
Oprescu, I., 1974. *Utilaj metalurgic-îndrumar de calcul şi proiectare,* Bucharest, Institutul Politehnic.
Pugna, I, Muţiu, C. *Relaţia om-maşină-mediu,* Timişoara, Editura Facla.
Nanu A., 1972. *Tehnologia materialelor,* Bucharest, Editura Didactică şi pedagogică.
Gillich, G.R. Reciklaza industrijskog otpada – izvor zastite okoline, *Eco-Conference 99, Novi Sad sept.1999, vol.1* : 259-260.

Environmental Issues and Management of Waste in Energy and Mineral Production, Singhal & Mehrotra (eds)
© 2000 Balkema, Rotterdam, ISBN 90 5809 085 X

The role of LCA in performance assessment in minerals processing – A copper case study

D.P.Giurco, M.Stewart & J.G.Petrie
CRESTA (Centre for Risk, Environment, Systems Technology and Analysis), Department of Chemical Engineering, University of Sydney, N.S.W., Australia

ABSTRACT: This paper demonstrates the use of Life Cycle Assessment (LCA) as a basis for assessing potential environmental performance in the minerals industry through a copper case study. The case study is sourced from a demonstration plant of a hydrometallurgical process that converts copper sulfides and oxides to LME Grade A copper using a leach / electrowinning technology. The life cycle approach is supported by a sequential leach extraction protocol to better quantify the potential mobility of metals and complementary anions from solid waste residues. Sequential leach tests showed that the majority of iron and sulfur, the main components of these residues, are unlikely to mobilize under waste deposit conditions. This information was used to modify the potential burdens predicted using conventional LCA practice, giving a more realistic view of the overall potential impact associated with this process. Certain environmental effect scores were sensitive to changing process conditions; the eco-toxicity indicator was affected by changing process feed composition, whilst the global climate change indicator (greenhouse effect) was affected by changes to the current density for electrowinning.

1 INTRODUCTION

A significant challenge facing the minerals industry is to improve its environmental reporting for new, as well as established processes. Whilst there is increasing awareness within the industry of the need to adapt to the call for 'triple bottom line' accountability, within which environmental and social impacts are identified explicitly, the availability of tools to assist in such an assessment is often thought to be lacking.

This paper brings together Life Cycle Assessment (LCA) and established solid waste characterization tests to provide a new combination of assessment tools which reduces the uncertainty of LCA assumptions with respect to potential mobility of heavy metals from solid waste. This adds value to LCA in the context of minerals processing by addressing the gross assumption made in LCA that all the metals in waste deposits will mobilize, which is often an unreasonable one.

1.1 *Environmental assessment*

Numerous approaches exist for assessing the environmental performance of a process. These include LCA, Environmental Impact Assessment (EIA), Materials Intensity per Unit of Service (MIPS), Cost Benefit Analysis (CBA) and Materials Flux Analysis (MFA).

Each technique has particular strengths. However, their application is dependent on the stage of the project's development (Stewart et al. 2000). In this work, we are interested particularly in looking at scale-up issues around new technologies, and how one might extrapolate findings on environmental performance gleaned from pilot and demonstration scale equipment to full scale plant, with no *apriori* knowledge of a particular site. Unlike EIA, LCA has the flexibility to be used without site-specific information. LCA provides information on the performance of a process relative to its potential to contribute to recognized environmental problems: e.g. Greenhouse Effect, Ozone Depletion, Acidification, Eutrophication, Human Toxicity, Eco-Toxicity and Smog. Definitions of these terms are given in Table 1.

Traditionally, LCA has been used in the comparison of the potential environmental impacts involved with manufacturing consumer products. The extension of LCA as a tool for analysing mineral processes highlights significant deficiencies which must be addressed (Stewart & Petrie 1996). In particular, the LCA methodology presupposes that all metals present in solid waste residues will leach (mobilize) from these residues and enter the environment. This

Table 1. LCA impact category descriptions

Impact Category	Description
Greenhouse Effect	Contribution to global warming due to the Greenhouse Effect.
Acidification	Increasing the pH of natural systems through mechanisms such as acid rain.
Eco-toxicity	In this study, Eco-toxicity means toxicity to aquatic ecosystems. It is reasoned that any terrestrial toxicity will eventually appear in groundwater and is covered under aquatic toxicity.
Human Toxicity	Based on established Human-Toxicological Classification Values. Not an indication of the potential of the process to kill people, rather how close the process approaches exposure limits set by, for example, the World Health Organisation. This impact category is not related to occupational health and safety.
Ozone Depletion	Potential to deplete the ozone layer.
Eutrophication	Excessive algal production in rivers and shallow water courses (algal bloom) caused by an environment rich in NO_3 and PO_4.
Smog	Atmospheric pollution in the form of smog.

assumption takes no account of the thermodynamic stability of the metallic compounds in solid wastes. There is a need for waste characterization tools to augment environmental assessment tools such as LCA in order to give a more realistic description of impact potential associated with metals' release from solid waste deposits.

1.2 *Aims of the paper*

The aims of this paper are to:

1. Demonstrate the effectiveness of LCA for environmental assessment in minerals processing and the circumstances under which it is most applicable.
2. Show how Sequential Leach tests can be used in conjunction with LCA to provide a better informed environmental assessment.
3. Demonstrate that this approach can highlight the sensitivity of the environmental assessment to changing process conditions. This is demonstrated for the particular case of a novel copper electrowinning process.

2 COPPER PROCESS DESCRIPTION

The copper process in the case study uses an innovative leach solution to extract copper from a copper sulfide and copper oxide concentrates. The solution is purified without using solvent extraction, as impurities are precipitated into a solid waste residue. Electrowinning at high current densities produces a LME Grade A copper product.

A demonstration plant for the process has been successfully operated and plans for a full-scale plant are being developed.

3 LIFE CYCLE ASSESSMENT

3.1 *Introduction*

LCA is an environmental analysis tool which can be used to determine the potential environmental performance of a service, product or process in a transparent and objective manner (SETAC 1992). It is based on a rigorous flowsheeting approach to process modeling whereby the environmental impacts associated with resource consumption and waste generation are identified explicitly (Clift & Longley 1995). This explicit identification of the drivers behind environmental impacts enables LCA to provide a link between economics and the environment by quantifying the manner in which waste leads to pollution (Guinee et al. 1993a ,b).

The significance of LCA in environmental management has been highlighted by its inclusion in the ISO 14000 series of standards for environmental management. ISO 14041 through to ISO 14043 focus specifically on LCA (ISO 1997a, b, c). In the case study, LCA is applied at demonstration plant scale. This provides an excellent opportunity to use information gained from the LCA to guide strategic design and management decisions to ensure optimal performance at the full-scale plant. In addition, it provides a sound basis for the process to achieve compliance with ISO 14000, the specific standard for environmental management systems.

3.2 *Generic LCA methodology*

LCA involves four stages. The first deals with the definition of the system and its boundaries. A key feature of LCA is that it expands the boundary of the investigation beyond the manufacturing facility itself to include consideration of the environmental impacts associated with resource extraction, the process itself, use and re-use of the product through to final disposal (Clift & Longley 1995). The impacts of the product are tracked from 'cradle-to-grave'.

In this way, a system under investigation could include an analysis of copper ore extraction and concentration in, say, Europe, coupled with production of copper product in Australia and the use and disposal of the copper product in North America. Consequently, LCA can begin to apportion the burden of environmental degradation to both the users of the product as well as to the manufacturing process and its sub-sets (Ayres & Simonis 1994). This is not seen in any way to be a dilution of environmental responsibility. Rather, by making explicit the total impacts associated with the product, LCA provides a platform from which social intervention can be used to effect improvement in environmental performance (Petrie 1998, Petrie & Clift 1994).

Establishing Life Cycle Inventories (LCIs) is the second stage of LCA. A Life Cycle Inventory is the quantification of inputs to the system and wastes generated by the system under consideration. These can be based on mass and energy balances for the process.

The third stage of LCA, impact assessment, links the process inputs and wastes to a recognized set of environmental impacts already described. Furthermore, the relative importance of the different environmental impacts can be ranked in a specific context to allow social values to be taken into consideration (de Oude 1993). This relative valuation of impacts is not trivial. It is necessary to present the information from the analysis in a way that can be interpreted by stakeholders in the decision making process. This makes it possible to elicit information on the preferences of the stakeholders which can be incorporated into the LCA (Cowell 1998, Meittinen & Hamalainen 1997).

The most notable feature of LCA is the way in which it generates an environmental perspective based on rigorous process analysis. Critical impacts are identified, as are the waste streams that give rise to them. Likewise, the steps within the process which generate these wastes can be highlighted. This leads to the fourth stage of LCA which is improvement analysis. Improvement analysis targets specific parts of the process where changes will have the greatest effect on reducing adverse environmental impacts.

In this case study, the internationally acceptable LCA software tool and database Simparo® was used (Product Ecology Consultants 1999).

4 COPPER CASE STUDY

4.1 *LCA in minerals processing*

LCA as an environmental assessment tool applied in the minerals' context has certain deficiencies. In general, it does not take explicit account of the temporal and spatial effects of the environmental impact. These are important elements for any environmental assessment of minerals technologies, where effects are often site-specific, and may present over considerable periods of time, even beyond plant closure. In effect, LCA deals with 'impact potentials' in a worst-case analysis, within which it is assumed that all materials contribute to environmental impacts in direct proportion to their mass in any product or waste stream, irrespective of their true mobility, or bio-availability.

The mobility of metallic species will depend on the mineralogy of the residue, and the type of waste management practices employed. To address the first issue, the stability of the solid waste residue from the copper case study was studied using sequential leach tests adapted from Tessier et al. (1979). A link is made between sequential leach tests and waste management practices in the industry. This choice of sequential leach and the methodology itself is discussed in section 5.

4.2 *LCA system boundary*

In the absence of site-specific information, as in the copper case study involving a new copper refining process, the system boundary was reduced from the 'cradle-to-grave' analysis conventionally used in LCA, to a 'cradle-to-gate' analysis covering the initial mining of copper ore up until it leaves the factory as copper ingot. Conventional impact categories were employed in this assessment. The use and disposal of copper product was excluded because it is very difficult to assess, is common to all copper producing technologies and not within the control of the operating company.

4.3 *Analysis of strategic scenarios*

In the present study, four different feed concentrates and one concentrate blend were investigated to see how they changed the environmental profile of the copper process. The concentrates were from ore bodies in Australia and Portugal.

The sensitivity of the environmental impact to increasing the nominal current density used for electrowinning copper was also examined.

4.4 *LCA results*

This section discusses the results of the LCA study.

4.4.1 *Direct vs. indirect effects*

The results of the copper case study showed that contributions to the impact categories of Greenhouse Effect, Ozone Depletion, Acidification, Eutrophication, Smog, Eco-toxicity and Human Toxicity can be attributed to so-called 'direct' and 'indirect' effects of the copper process. 'Direct' effects arise from the emissions of the copper process itself, predominantly the solid waste residue. 'Indirect' effects are those arising from the manufacture of inputs to the process such as the preparation of concentrate and manufacture of electricity to be used by the process. Downstream impacts of the process associated with the use, re-use and disposal of the product would also be regarded as 'indirect'. However, the 'cradle-to-gate' boundary definition used in this case study excludes consideration of these effects.

Contributions to Acidification, Greenhouse Effect, Ozone Depletion, Eutrophication and Smog are 'indirect' effects, mainly resulting from the burning of coal to generate electricity used it the process. It must be noted that the electricity supply for this case study was modeled as coming from the burning of coal which is accompanied by a significant contribution to the Greenhouse Effect.

As stated, Acidification is an 'indirect' effect. Unlike pyrometallurgical processes which can emit significant quantities of sulfur dioxide gas, this hydrometallurgical process has sulfur leaving as a stable compound in its solid waste residue and hence does not contribute a 'direct' effect to Acidification. This is an advantage for the hydrometallurgical process which provides no on-site contribution to Acidification.

The impact categories of Eco-toxicity and Human Toxicity are predominantly 'direct' effects of the process. An operating company would have most control over these effects, and can manage its process to reduce these potential impacts - for example, by running the process in a manner which creates a more stable waste. The adoption of best practice in waste management should ensure that the release of mobile species into the environment is limited.

4.4.2 *Sensitivity to nominal current density for electrowinning*

Increasing the nominal current density in the electrowinning cells would reduce the capital cost of the plant by reducing the number of cells required, but increase the operating costs by increasing the electrical requirement.

In this case study, the result of increasing electricity consumption is a proportional increase in contribution to the Greenhouse Effect impact category.

This finding can be used to guide company policy around optimizing energy management, and the promotion of 'greener' energy sources as opposed to coal.

4.4.3 *Sensitivity to feed concentrate*

The effect of changing the feed concentrate to the process on each impact category in the case study is outlined in Table 2.

Changing the feed to the process had most effect on potential Eco-toxicity, which is a 'direct' effect. The choice of feed concentrate changed the potential

Table 2. Coefficient of variation for each impact category, showing the variability of that category with change in feed.

Impact category	Coefficient of variation* (%)
Eco-toxicity	33
Eutrophication	26
Greenhouse Effect	22
Human Toxicity	22
Acidification	15
Ozone Depletion Potential	12
Smog	12

* The coefficient of variation is equal to the standard deviation divided by the mean expressed as a percentage.

Eco-toxicity for the process by up to a factor to three. The main elements contributing to this potential toxicity were iron and sulfur. This is illustrated in Figure 1. The 'indirect' effects associated with inputs to the process were less affected by changing the feed.

4.4.4 *Focus on Eco-toxicity*

The previous section described how potential Eco-toxicity was sensitive to the process feed which influences the composition of the solid waste residue. Being a 'direct' effect and hence one which the company can most control, it is informative to study Eco-toxicity in more depth and place in context the usefulness of the sequential leach tests. The same argument holds for Human Toxicity which is mostly a 'direct' effect too.

The potential risk for Eco-toxicity is a function of a metal's 'Exposure' to the natural environment and the 'Ecological Effect' caused in the natural environment by the metal (US EPA 1996). The Simapro® database was used to quantify the 'Ecological Effect' of metals. No further research in this area was undertaken. This paper focuses on providing a better indication of 'Exposure', rather than the default approach in LCA which assumes that all metals within the solid waste will mobilize and impact on the natural environment. Sequential leach tests were used to characterize the mobility of the solid waste. Such tests provided an indication of intrinsic 'Exposure' potential even in the absence of specific knowledge of waste management practice.

5 SEQUENTIAL LEACH TESTS

5.1 *Choice of sequential leach over TCLP*

Sequential leach tests were chosen as a quick, relatively simple method for better understanding the mobility of the solid waste residue from the copper process. Employing these tests seeks to address the first part of the gross assumption made in LCA that:

1. all metals in the solid residue mobilize and
2. all mobile metals enter the natural environment.

Historically, the Sequential Leach protocol was developed to assess the age of strata in a soil profile. The protocol exposes the sample to increasingly more aggressive leaching solutions in a step-wise sequence (Tessier et al. 1979). In so doing, it leaches the least stable materials first and the more stable materials in later extraction steps.

Analysis was performed on selected metal species in order to ascertain stability trends for each metal. The ultimate aim of the sequential leach tests was to provide an association between the aggressiveness of sequential leach steps and solid waste management scenarios. This is discussed later in this paper.

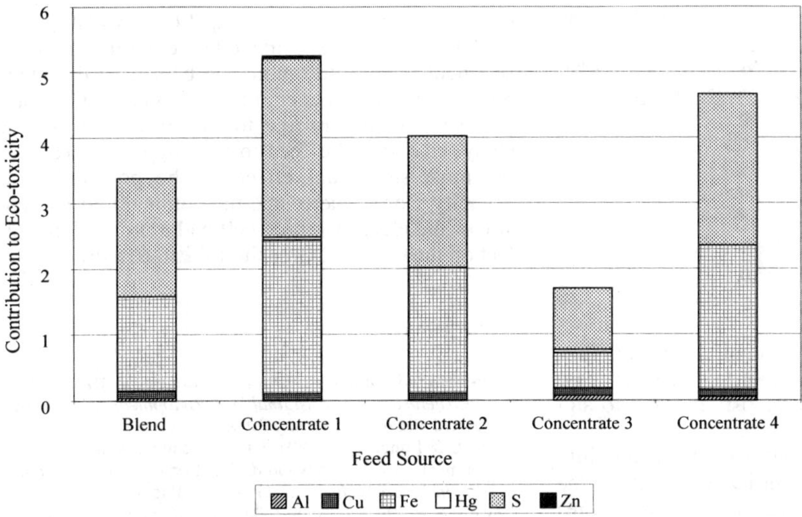

Figure 1. Comparison of Eco-toxicity (in relative units) with variation in feed concentrate

Other methods which can be performed to characterize the nature of solid wastes include the standard TCLP and amended TCLP (Peterson 1998). Sequential leach tests are preferred to these other protocols as discussed below.

The TCLP is a batch test developed by the US EPA to determine whether the co-disposal of different domestic wastes represented a significant environmental risk. Its use has been extended to quantifying risks associated with solid waste management in general. TCLP tests are often regulatory requirements.

The TCLP test involves the aggressive mixing of the sample in acetic acid (standard test) or artificial rain (amended test). After a fixed time, analysis is performed to determine which metals have leached from the sample.

The disadvantage of the TCLP test is that the mobility of metals cannot be correlated with likely mobility under waste management conditions. Mixing conditions and leachants are not similar to those likely under waste management conditions and the test only provides one data point making it impossible to establish trends for metal behaviour.

The sequential leach tests use a series of leachants that span a range above and below that of likely conditions arising in a managed solid waste facility and are hence more informative than a TCLP tests.

5.2 *Sequential leach methodology for copper case study*

The sequential leach protocol outlined by Tessier et al. (1979) includes five stages. For the case study at hand, a modified sequential leach protocol using three stages was employed. The authors developed this protocol in conjunction with the Australian Nuclear & Technology Organisation (Lowson 1999). The first stage uses NH_4Cl (0.1 M) to remove soluble salts and exchangeable metals. The second stage uses Tamm's Acid Oxalate to remove amorphous oxides and the third stage uses Shuman's Reagent to remove crystalline oxides.

Metals mobilizing at the first sequential leach stage will also mobilize if water were allowed to infiltrate the deposit.

The second stage of sequential leach represents relatively harsh environmental conditions. Metals mobilizing at this stage would only do so in an uncapped deposit that was exposed to significant acid rain.

Stage three of the sequential leach requires the mobilization of metals bound up in a crystalline matrix e.g. goethite and hematite. This is highly unlikely to occur in a waste management facility.

For these reasons, the second stage of sequential leach was chosen to represent a conservative practical limit for the potential mobilization of metals from the solid waste deposit.

5.3 *Sequential leach results*

Results showed that a significant amount of material did not mobilize from the solid waste residue.

The results in Table 3 show that while iron and sulfur make up over half of the solid waste residue, only 10% of these metals mobilize after stage two of the sequential leach. In practical terms, it is unlikely

Table 3. Mass percent of selected metals mobilized after the second stage of sequential leach.

Metal	Metal fraction in solid residue	Mass percent of metal fraction mobilized
Sulfur	26.5	10
Iron	25.0	10
Copper	6.6	99
Magnesium	2.6	69
Zinc	0.2	57
Lead	0.015	2
Arsenic	0.005	10

that more than 10% of iron and sulfur will mobilize from the solid waste residue in the copper case study, even in an uncapped deposit exposed to significant rainfall.

Feeding the revised estimates of metal mobility back into the LCA has a significant impact on the potential Eco-toxicity and Human Toxicity scores which are influenced by the metals in the solid waste residue.

Potential Eco-toxicity decreases by a factor of 5.5 and potential Human Toxicity decreases by a factor of 3.

The amount of metals leached at the second stage of sequential leach was generally higher that that leached in the TCLP test. This allows the second stage of sequential leach concordance with legislative requirements. In addition it provides a conservative but realistic upper threshold for potential metal mobility from a waste deposit independent of waste management practices.

6 CONCLUSIONS

1. Through a case study, this paper has shown that LCA can be used to support the assessment of environmental performance of a minerals process. LCA is an especially useful tool when site specific information is not available.

2. Combining sequential leach tests with LCA gives a better understanding of the potential mobility of solid wastes, an issue LCA does not currently address. In the copper case study, the sequential leach tests showed that 90% of the iron and sulfur in the sold waste residue was in a stable form. This is an especially significant finding, as these two elements and their compounds are major contributors to both Eco-toxicity and Human Toxicity impact categories.

3. The methodology described here permits an assessment of the effect of changing process parameters. Two examples were explored here – the effects of changing feed quality and current density in the electrowinning cells. Interestingly, such changes suggest that improvements to process performance are generally associated with a shift in impacts between 'direct' and 'indirect' impact categories. This shift is identified clearly using LCA, whilst it may well be overlooked using other environmental assessment tools. Little more can be said at this stage other than that designers of new minerals technologies should be sensitive to this fact. In this case study, changing the feed to the copper process had the most significant effect on the potential Eco-Toxicity scores. Increasing the nominal current density of the electrowinning cell had a proportional effect on the potential Greenhouse Effect score.

REFERENCES

Ayers, R.U. & Simonis, U.E. 1994. *Industrial Metabolism - Restructuring for Sustainable Development.* New York: United Nations University Press.

Clift, R. & Longley, A. 1995. Clean Technology and the Environment. In R.I. Kirkwood & A.J. Longley (eds), *Introduction to Clean Technology.* Glasgow: Blackie.

Cowell, S.J. 1998. *LCA and Decision-Making: Revised Summary of Meeting 2 of SETAC working group on LCA and Decision Making.* Guildford: SETAC.

de Oude, N. 1993. Clean Production Strategies. In T. Jackson (ed.), *Product Life-Cycle Assessment.* London: Lewis.

Guinee, J.B., Udo de Haes, H.A. & Huppes, G. 1993a. Quantitative Life Cycle Assessment of Products. 1: Goal definition and inventory. *J Cleaner Production* 1(1): 3-13.

Guinee, J.B., Udo de Haes, H.A. & Huppes, G. 1993b. Quantitative Life Cycle Assessment of Products. 2: Assessment. *J Cleaner Production* 1(2).

International Standards Organisation (ISO). 1997a. *Environmental Management - Life Cycle Assessment - Principles and Framework*; ISO code 14 040.

International Standards Organisation (ISO). 1997b.; *Environmental Management - Life Cycle Assessment - Goal and Scope Definition*; ISO code 14 041.

International Standards Organisation (ISO). 1997c. *Environmental Management - Life Cycle Assessment - Life Cycle Impact Assessment*; ISO code 14 043.

Lowson, R.T. & Rajaratnam, G. 1999. *TCLP and Sequential Leach Tests*, Australian Nuclear Science and Technology Organisation Internal Report, Lucas Heights, NSW, Australia.

Lowson, R.T. 1999. Australian Nuclear Science & Technology Organisation. Private Communication.

Meittinen, P. & Hamalainen, R.P. 1997. How to benefit from decision analysis in environmental Life Cycle Assessment (LCA). *European Journal of Operational Research*, 102: 279-294.

Petersen, J. 1998. *Assessment and Modelling of Chromium Release in Minerals Processing Waste Deposits*; PhD Thesis, Department of Chemical Engineering, University of Cape Town, South Africa.

Petrie, J.G. & Clift, R. 1994. Life Cycle Assessment. *MERN Research Bulletin*, No 7, December.

Petrie, J.G. 1998. The use of LCA in effective environmental decision making, its interface to the Impact Assessment process and its role in Environmental Management Systems. *Paper presented at Life Cycle Assessment for Asia Pacific regions - UNEP/APEC/NEDO/AIST Symposium; Tsukuba, Japan.*

Product Ecology Consultants 1999. *SimaPro 4.0 The Life Cycle Assessment tool.* Retrieved January 20, 1999 from the World Wide Web <http://www.pre.nl/simapro.html>.

Society for Environmental Toxicology and Chemistry (SETAC) 1992. *Guidelines for Conduct of LCA Peer*

Review Draft Paper. Pensacola: SETAC.

Stewart, M. & Petrie, J.G. 1996. Life cycle assessment for process design- the case of minerals processing. in M.A. Sanchez, F. Vergara & S.H. Castro (eds), *Clean Technology for the Mining Industry.* Concepcion.

Stewart, M., Basson, L., Alexander, B. & Petrie J.G. 2000. Allied tools to LCA in the process industries. *Paper presented at 2nd National Conference on Life Cycle Assessment, Melbourne, 23-24 February 2000.*

Tessier, A.., Cambell, P.G.C. & Bisson M. 1979. Selective extraction procedure for the speciation of particular trace metals. *Anal Chem.* 51(7): 844-851.

US Environmental Protection Agency 1996. *Proposed Guidelines for Ecological Risk Assessment*, Risk Assessment Forum. 1:129.

Environmental Issues and Management of Waste in Energy and Mineral Production, Singhal & Mehrotra (eds)
© 2000 Balkema, Rotterdam, ISBN 90 5809 085 X

Design and reliability engineering of aquifer CO_2 disposal in Alberta

Akihiro Hachiya & Samuel Frimpong
University of Alberta, Edmonton, Alb., Canada

ABSTRACT: Disposal of carbon dioxide (CO_2) in land aquifers is a potentially viable option for Alberta in dealing with long-term CO_2 emissions. Studies have confirmed the existence of appropriate aquifers in the Alberta Basin for CO_2 storage over a long geological period. In this study, the authors have carried out a detailed review of the geology of the land aquifers in the Alberta Basin and the stability of CO2 in these aquifers. In particular, the authors have reviewed the phase dynamics and the trapping mechanisms of CO_2 in aquifers, the porosity, permeability and fracture pressure properties of the host aquifers. A rigorous design of an aquifer disposal of CO_2 in Alberta from coal-fired power plants is carried out based on review results. Analytical designs are carried out for the liquefaction-transportation-injection networks based on the reaction kinetics and phase dynamics of CO_2 under different temperature and pressure conditions. These models are validated using flue gas data from the 546 MW Wabamum Power Plant in Alberta. The design results show that the respective energy requirements for capturing, liquefying, transporting and injecting CO_2 in the Glauconitic aquifer are 68.5MW kW, 48.77 MW, 251 kW and 1500 kW. Reliability engineering analysis also shows that CO_2 will be trapped safely in the aquifers by chemical and hydrodynamic trapping.

1. INTRODUCTION

At the Third Conference of the Parties to the United Nations Framework Convention on Climate Change (COP3), the global community made a commitment to reduce GHG emissions to a selected target by the year 2012. Canada made a commitment to reduce greenhouse gas emissions by 6 percent below 1990 levels by the year 2012 at the Kyoto conference [Gunter, et al., 1998]. However Canada is expected to emit about 560 million tonnes of CO_2 in the year 2012 which is 22% above the 1990 level [Gunter et al., 1998]. It is generally considered that Canada find it difficult to reduce the future greenhouse gas emissions to that level in spite of new energy technologies and new energy sources.

The growth of industry and population will continue to, therefore energy consumption, especially fossil fuels, will grow and this will cause CO_2 emission growth more than reduction. In order to achieve these targets, CO_2 storage or disposal options have been proposed. The option includes CO_2 storage or disposal into geological sinks such as an aquifer, depleted oil reservoir, coal bed or the deep ocean. Bachu, et al. (1996) states that the ocean disposal option still has numerous hazards associate with its implementation by recent studies [Bachu, et al, 1996] Disposal into depleted oil, gas reservoirs and salt beds have limited capacity [Bachu, et al., 1996]. Coal beds disposal has not been tested, therefore, CO_2 aquifer disposal option is the most feasible option for the immediate solution for this problem [Bachu, et al., 1996]

2. DESIGN OF AQIFER DISPOSAL SYSTEM

2.1 *Aquifer CO_2 Disposal*

Naturally occurring aquifer are geological formations of traps bordered by layers of sandstones and limestones that can contain water. Studies in the geological formation of Alberta have confirmed the existence of appropriate aquifers with potential for CO_2 disposal and storage over a considerable long geological period [Bachu, et al., 1996]. These aquifers indicate the Glauconitic and Nisku aquifers within the Alberta Basin. CO_2 is an ideal candidate for aquifer disposal because of its high density and solubility in water at relatively high pressures. Discussions and analysis in this chapter will focus on CO_2 disposal in aquifer, phase diagram and behavior of CO_2, and trapping mechanisms for CO_2 disposal. The geology and stability of the aquifer structures will also be discussed and analyzed technical and safety basis for the disposal system.

2.2 *The Required Conditions of CO_2 Aquifer Disposal System*

Aquifer CO_2 disposal and storage require the following conditions [Bachu, et al., 1996]:

- The top of the aquifer must be below 800m from surface to keep CO_2 in the super critical state.
- The aquifer should be capped by impermeable (sealing) layers, regional aquitard.
- The aquifer should have enough porosity and adequate permeability. The near well permeability should be high (above 100md) to allow good injectivity, but the regional permeability should be low (under 100md) to yield long CO_2 residence time.
- The injection site should be close to CO_2 emission site.

CO_2 is in a super-critical state at over 87.98° F (31.1 °C) and 1069.4 psia (7.38MPa). At this state, CO_2 behaves like gas with liquid density. Alberta's aquifers below 1000m have higher temperatures and pressures than the critical point so that CO_2 must be sent to the aquifer in the super-critical state [Bachu, 1995, Bachu, et al.,1996].

In a super-critical state, the solubility of CO_2 in water is much higher than in the gaseous state [Gerrard, 1980]. The reactivity with minerals contained in the formation water is also higher because some of the CO_2 dissolved in the water. This CO_3^- reacts with minerals and become trapped, for example as $CaCO_3$. Also in the super-critical state, CO_2 has a higher density so that the volume rate is more efficient. The target aquifer has a top pressure of 12.5MPa and temperature of 50°C, which show that, under these conditions, CO_2 is in the super-critical phase.

2.3 *CO_2 Trapping Mechanism in Aquifers*

The injected CO_2 travels in the aquifer in both dissolved and immiscible phases depending on the aquifer permeability. The dissolved CO_2 will travel in an aquifer with extremely low velocity, and residence time in the order of 1 million years. The immiscible CO_2 will also travel with residence time in the order of millions of years [Bachu, Gunter and Perkins, 1994]. Therefore, CO_2 can be trapped in an aquifer for long geological period in the liquid phase. This phenomenon is referred to as hydrodynamic trapping. Formation waters range in composition from pure water to brine. Thus when liquid CO_2 is injected into an aquifer, there is the possibility that it will react with the elements in the formation water or the minerals comprising aquifer rocks [Bachu, Gunter and Perkins, 1994]. By reacting, CO_2 can be trapped as different substances in the aquifer. This phenomenon is referred to as chemical or mineral trapping.

2.4 *Hydrodynamic Trapping*

When CO_2 is injected into an aquifer at an appropriate pressure, it moves away from the injection well and flows within the natural flow regime. Once outside the injection-well radius of influence, the flow of immiscible CO_2 will travel at the same speed as the formation water in the regional flow system. In Alberta, there are many suitable aquifers in the Alberta Basin for hydrodynamic traps of CO_2.

2.5 *Chemical (Mineral) Trapping*

The chemistry of the formation water and rock mineralogy also increases the potential for CO_2 disposal through chemical reactions. Chemical reactions in a carbonate aquifer immobilise CO_2 as another carbonated substance [Perkins and Gunter, 1995]. Preliminary study shows that aluminosilicate minerals could sequester injected CO_2 in siliciclastic aquifers in two forms depending on the dominant cations. When the dominant cation is Na^+ or K^+, the concentration of bicarbonate is built up in the aqueous phase and forms bicarbonate brine. When the dominant cations are Ca^{++}, Mg^{++} and Fe^{++}, the concentration of bicarbonate is built up because of high solubility of sodium and potassium and the low solubility of calcite, dolomite and siderite which form precipitates [Perkins and Gunter, 1995]. Fracture pressure is the maximum injection pressure allowable below this pressure. Injection causes no rock fracture or disruption in the aquifer. The fracture pressure in the Glauconitic aquifer is about 33.5MPa. Therefore, injection pressures lower than 33.5MPa are theoretically safe for CO_2 disposal into this aquifer. Injectivity simulation studies, conducted with an assumed maximum injection pressure of 90% of fracture pressure, have concluded that injection of CO_2 is safe under this condition [Bachu, et al.,1996]. CO_2 is a very good solvent in water, and thus, it is possible to exist in the liquid phase with water and be stored in an aquifer under proper conditions. CO_2 can thus be trapped in an aquifer for a long geological time scale by this hydrodynamic trapping when the outer permeability of the injection zone is low [Bachu, et al.,1996]. When CO_2 is injected into an aquifer at an appropriate pressure, it moves away from the injection well and flows with the natural flow regime. Beyond the radius of influence of the injection well, the flow of immiscible

CO_2 will travel at the same speed as the formation water in the regional flow system. The Glauconitic aquifer is suitable for CO_2 disposal because the for mation water flows downward and are caught in the hydodynamic regime of the formation water [Bachu and Undershultz, 1995].

The Glauconitic aquifer has suitable conditions for CO_2 chemical trapping [Perkins and Gunter, 1995]. In the Glauconitic aquifer, the dominant cations are Ca^{++}, Mg^{++} and Fe^{++}, resulting in equations (1.1) and (1.2).

$$H_2O = H^+ + OH^- \qquad (1.1)$$

$$H_2O + CO_2 = HCO_3^- + H \qquad (1.2)$$

Some of the CO_2 exists as bicarbonates or as bicarbonate ion with the proton ion at any pressures. The proton results in acidic condition in the aquifer which can affect the silicate minerals in the aquifer. This results in free Ca^{++} ions as in equation (1.3). The fastest chemical reaction is the precipitation of calcium carbonate in equation (1.4).

$$H_2O + CaAl2SiO8 + 2H^+ = Ca^{2+} + Al2Si2O5 \bullet (OH)_4 \qquad (1.3)$$

$$Ca^{2+} + HCO_3^- = CaCO_3 + H^+ \qquad (1.4)$$

CO_2 can eventually be stored permanently as a solid $CaCO_3$. Gunter and Perkins (1995) estimated that the capacity of CO_2 by this chemical trapping is about 0.5 Mt per 1 square kilometer in the Glauconitic aquifer. By the chemical trapping simulation, there will be complete equilibrium in 820 years and that 6.2 moles of carbon dioxide will be trapped as calcite per kg of formation water in the Glauconitic aquifer [Perkins and Gunter, 1996]. As a result of the above reasons, siliciclastic aquifers are prime targets for mineral trapping of CO_2. In Alberta, the Glauconitic Aquifer has suitable minerals for CO_2 chemical trapping [Perkins and Gunter, 1995].

2.6 *Geology of the Alberta Basin*

The Western Canadian Sedimentary Basin consists of two basins called the Alberta Basin and Williston Basin with rich coal, oil and gas as shown in Figure 1. The Alberta basin is an accumulation of marine and near-shore sedimentary rocks. The basin geometry and structure are the result of two major phases of basin development. The first passive margin phase was dominated by carbonate deposition on the continental margin. The second phase began in the Middle Jurassic with the onset of convergent tectonism and the formation of a foreland basin dominated by clastic sedimentation of sandstone and shales [Bachu, Gunter, and Perkins, 1994]. The Glauconitic and Nisku Aquifers in the Alberta Basin qualify as suitable aquifers for CO_2 disposal and storage. The Glauconitic has been selected as the target aquifer for this study. The Glauconitic aquifer is capped at the top by the Grand Rapid Formation Aquitard and at the bottom by the Ostracod Beds aquitard.

Grand Rapids Formation capping the Glauconitic Aquifer is divided into two layers. The lower layer on the top of the Glauconitic Aquifer has a continuous basal shale zone about 10m thick. The upper layer contains thin interbedded siltstones, shales and limestone [Bachu, et al.,1996]. Shale is any mudrock that exhibits lamination or fissility or both. Siltstone is also mudrock with 50% or more silt-sized material [Prothero and Schwab, 1996]. The Ostracod Beds aquitard is relatively uniform in its lithology and thickness with an average thickness of 18m. This contains black mudstones with abundant shale beds. There are occurrence of quartz sandstone and siltstones [Bachu, et al.,1996]. Sandstones are major reservoirs of groundwater and petroleum [Prothero and Schwab, 1996]. The Glauconitic Aquifer has an average thickness of 14m. From bottom to top, the Glauconitic Sandstone consists of argillaceous sandstone grading upward into thin, stacked cycles of fine- to medium-grained, porous, salt-and-pepper sandstone. Detailed layers are identified by a cross-bedded to massive sandstone base grading upward into bioturbated sandstone at the top. This aquifer is capped by a medium-grained sandstone that grades upward into a white siltstone [Bachu, et al.,1996].

2.7 *Rock Properties*

The relevant rock properties for CO_2 aquifer disposal are porosity, permeability and mineralogy. The Glauconitic Aquifer is classified as a mature to submature litharenite. The sandstone is fine-midium grained, sub-angular to sub-rounded, moderately well sorted, and has good porosity for injection. The ratio of quartz, feldspar and rock fragments ranges from 55:4:41 to 40:3:57. Monocrystalline and polycrystalline quartz grains are very clean. A few samples contain kaolinite coatings, with dolomite and calcite crystal growth along grain contacts. Rock fragments include chert, glauconite, mudstone, and dolomite. The high proportion of clay in the Glauconitic Aquifer is due to the presence of glauconite that ranges in size from sand-size grains to clay-size grains [Bachu, et al.,1996]. Porosity data for this study area were measured by Bachu, et al., (1996) using core analysis, and the results are provided in Table 1.1. From the analysis, generally there is no vertical trend in porosity values in any unit.

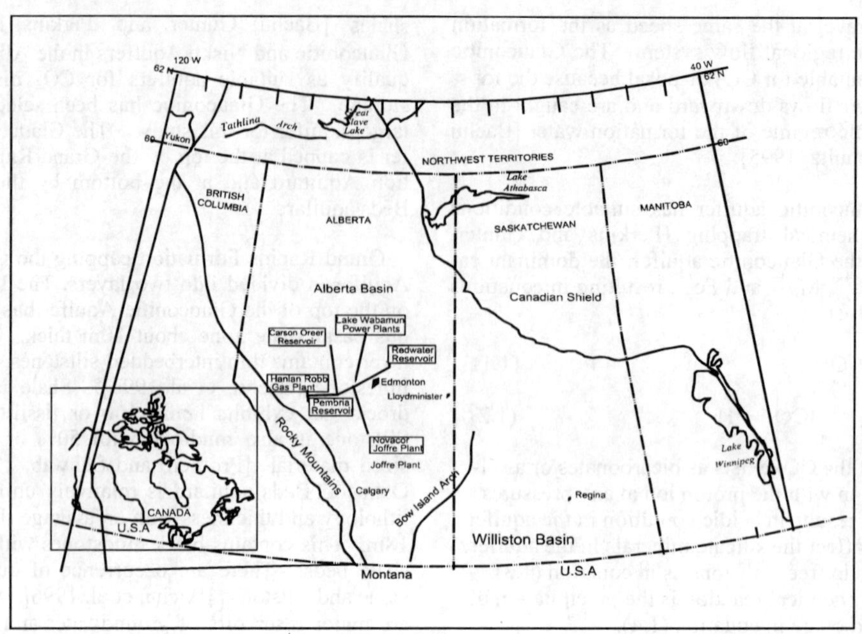

Figure 1 Location of Alberta Basin

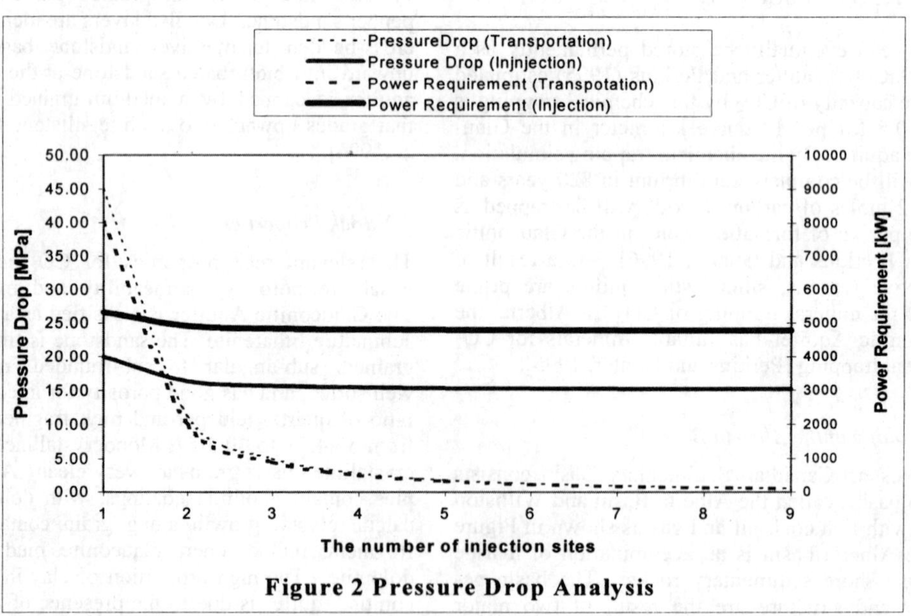

Figure 2 Pressure Drop Analysis

As shown, the porosity of the Mannville strata group of the Grand Rapids Formation, Glauconitic and Ostracod Beds is quite variable, but the average is higher for the Glauconitic Aquifer. The Grand Rapids Formation has only half of the porosity of the Glauconitic Aquifer. The high porosity in the Glauconitic aquifer increases its capacity for CO_2 disposal and storage.

Table 1.1 Porosities of the Mannville Strata Group [Bachu, et al.,1996]

Stratigraphic Unit	Min [%]	Average [%]	Max [%]
Grand Rapids Formation	5.60	6.10	6.60
Glauconitic	11.80	11.90	12.00
Ostracod Beds	1.30	7.80	17.10

Rock permeability analysis provided by Bachu, et al (1996) also shows that there is no vertical trend of permeability values as shown in Table 1.2. The data shows that the Glauconitic Aquifer has significantly higher permeability values compared to that of the capping aquitards.

Table 1.2 Permeability of the Mannville Strata Group [Bachu, et al.,1996]

Stratigraphic Unit	Min [md]	Average [md]	Max [md]
Grand Rapids Formation	0.01	0.10	1.00
Glauconitic	13.40	14.15	14.95
Ostracod Beds	0.01	1.87	212.73
Ellerslie Member	0.03	4.06	201.93

As a result of the porosities and permeabilities, the aquitards confine the flow to the Glauconitic aquifer. Generally, water flows in the horizontal direction because of the higher permeabilities. No vertical water flow occurs due to the lower permeabilities. In the Grand Rapid Formation, there exist frequent coal layers. This layer acts as the cap for the aquifer. As a result of the hydrodynamic trapping mechanism and permeability distribution, the injected CO_2 can have a long residence time in the aquifer. Erosional rebound leading to reverse flow from aquifers into shaley aquitard with lower permeability was observed by Neuzil (1993) for Williston Basin. A similar result was observed in a sub-Andean foreland basin in Colombia [Villegas 1994, Bachu, et al 1996]. From these results, it can be concluded that other aquifers in the world have similar characteristics and mechanisms of formation water flow system, which enhance the advantage of CO_2 aquifer disposal [Bachu, et al., 1996]. The carbon dioxide is confined to the aquifer even if there are no chemical reactions to form minerals in the aquifer [Bachu, et al.,1996].

2.8 *Fracture Pressure*

The stability and residence of injected CO_2 in the target aquifer also depend on the fracture pressures in this aquifer. Fracture pressure is the maximum pressure for injection, under which the rock is theoretically stable. The fracture pressure in the Glauconitic aquifer is about 33.5MPa. Therefore it can be concluded that the injection pressure lower than 33.5MPa is theoretically safe for CO_2 disposal into Glauconitic Aquifer. Alberta Research Council has done injectivity simulations, to study the stability of CO_2 in the Glauconitic aquifer. They assumed maximum injection pressure to be 90% of estimated fracture pressure for CO_2 stability [Bachu, et al., 1996]. Bachu, et al., (1996) carried out this injectivity study under the following conditions.

- The aquifers are homogeneous
- The thickness of the aquifer is constant
- The small dip of the aquifer is ignored
- CO_2 is in the super critical state in the aquifer and is treated as single phase fluid.
- Capillary pressure effects are negligible
- The relative permeability curves for the carbon dioxide-water were not measured

By injecting CO_2 into high permeability zone, fracture will be avoided and the efficiency of injection will be higher [Bachu, et al.,1996]. In addition, if maximum pressure is set as 90% of fracture pressure that would be 30.12MPa. They concluded that a high injection pressure results in high injection rate. About 50% more carbon oxide can be injected when injection pressure increases from 25.15MPa to 30.12MPa [Bachu, et al, 1996]. Changes in porosity had minimized effect on injection rate. The carbon dioxide might propagate farther in the case of the lower porosity. The permeability has a very significant effect on the capacity of CO_2 disposal. The total amount of carbon dioxide is more than 15 times greater when permeability changes from 6.2md to 100md [Bachu, et al.,1996]. Here, injection pressure is the pressure at the bottom of well.

3 DESIGN OF AQUIFER DISPOSAL SYSTEM

3.1 *System overview*

The aquifer CO_2 disposal system comprises technologies for capturing, liquefaction, transportation and injection. Figure 3 indicates the system scheme. After flue gases come out from greenhouse gas emission site, CO_2 is captured to make disposal effi-

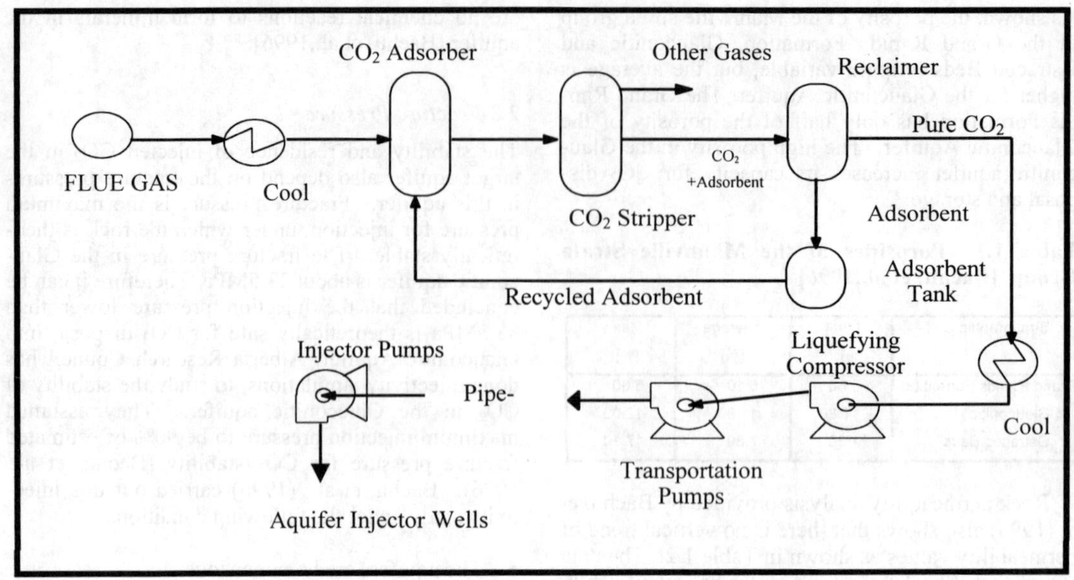

Figure 3 Aquifer CO_2 Disposal System (ACDS)

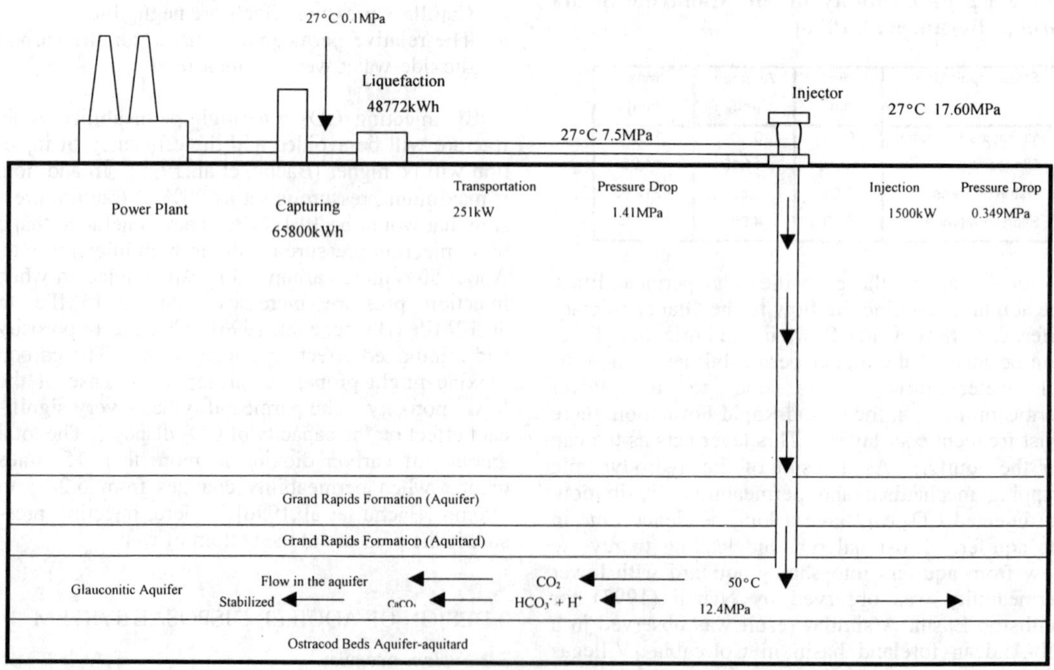

Figure 4 Energy Requirements of System

ciency high. After capturing, CO_2 is liquefied and sent to injection site. Before injection, CO_2 is more pressurized to meet the injection pressure requirement of the aquifer. In this section, the design of each procedure is discussed and case study is carried out for the Wabamun Thermal Power Plant of TransAlta Corporation.

Flue gases are cooled down before capturing because the adsorbent, K-S, captures CO_2 at about 60 degree Celsius. After adsorption, the gases are sent to a stripping tower. The gases are removed subsequently from the top to the bottom of the tower depending on the difference in densities from lighter substances to heavier. The CO_2 has the greatest density among gases

by absorbent. The CO_2 remain at the bottom of the tower and is recovered at reclamer. The CO_2 with K-S is heated up to be decomposed with K-S at a temperature of 120°C. The absorbent is sent back to its storage tank to be recycled and CO_2 is sent to the liquefaction plant after cooling down to room temperature 27°C. At this point, 90% of CO_2 is recovered with a purity of 99%. The pure CO_2 is liquefied by pressurizing with 7.5 MPa of pressure in the liquefaction compressor at 300K. The liquefied CO_2 is sent to injection sites through the pipelines by pumping with the pressure equivalent to the pressure loss between pumps and injection sites to keep the liquefied pressure. Figure 4.4 indicates procedure of injection. The liquid CO_2 is injected in aquifer by pumping through the injector well. CO_2 is pressurized with pressure, which is greater than the aquifer pressure and pressure drop occurring in the injector well. The pressure of the Glauconitic aquifer is 12.4 MPa. Pressure drop must be estimated depending on the system design and environment of the injection sites.

3.2 *Mathematical Models of the Disposal System*

After flue gases released from greenhouse gas emission source, CO_2 is separated from other flue gases before injection. To make disposal efficiency high, CO_2 must be separated from flue gas as much as possible because if there exist another substances, system requires more extra energy while liquefaction. Kansai Electric Power Company of Japan and MHI, Ltd., of Japan have developed high efficiency CO_2 capturing system. This technology is applied as capturing system for this study. In the system, energy requirement is given by following equation.

$$PR_{cap} = UP_{cap} * T_{CO_2} \qquad (1.5)$$

Liquefaction is required to change CO_2 from a gas phase to liquid phase because liquid CO_2 has high density so that it can be sent with high efficient flow rate to the injection sites. The CO_2 is compressed by a compressor with pressure of over 7.4MPa. Extremely high power compressor is required for this step. The required energy model for liquefaction is:

$$PR_{liq} = UPl_{iq} * T_{CO_2} * R_{CO2} \qquad (1.6)$$

PR_{liq} indicates the total power requirement for liquefaction system. UP_{liq} indicates the unit power requirement kW per a tonne of CO_2. T_{CO2} is the amount of CO_2 emission per hour. R_{CO2} is recovery ratio of CO_2 from the capturing system. Liquefied CO_2 is transported to injection site through pipelines by pumps. Pumps must push CO_2 with pressures greater than pressure losses given by the equation (1.3) between liquefaction station and injection site. Otherwise, CO_2 changes its phase to gas again. The required power is given by following equation.

$$\Delta P = 4\lambda \left(\frac{L}{D} \right) \left(\frac{\rho < u >^2}{2} \right) \qquad (1.7)$$

λ is the friction loss coefficient [-] given by Reynolds Number which is given by following formulas:

$$Re = \frac{D < u > \rho}{\mu} \qquad (1.8)$$

is viscosity of CO_2 [Pa s] ε is average roughness of the pipe is the roughness of surface of inside of a pipe. L is the length of the pipeline [m]. D is the diameter of the pipeline [m] ρ is the density of CO_2 [kg/m^3]. <u> is average velocity of CO_2 [m/s]. This average velocity rate <u> is shown as following equation:

$$<u> = V / \pi * n * R^2 \qquad (1.9)$$

The variables are indicated as V, the volumetric flow rate {m3 / s], n is the number of injector wells.

$$V = \frac{CO_2 * 1000}{\rho} \qquad (1.10)$$

The volumetric flow rate V can be shown as following;

$$\Delta P = 4\lambda \left(\frac{L}{D} \right) \left(\frac{CO_2^{\,2} * 10^6}{2\rho (n\pi R^2)^2} \right) \; [Pa / site] \qquad (1.11)$$

CO_2 is CO_2 flow rate after purification [t/sec]. The power of injection can be estimated by;

$$p\,[Pa] * V\,[m^3/s] = p * V\,[W] \tag{1.12}$$

Therefore, the power requirement for the pipeline is:

$$PR_{trns} = \Delta P * V = \lambda\left(\frac{L}{D}\right)\left(\frac{\rho * CO_2^2 * 10^6}{2(np\pi R^2)^2}\right)\left(\frac{CO_2 * 1000}{\rho}\right) \tag{1.13}$$

The liquefied CO_2 sent to injection site is injected into underground thorough the injector well that the same type of the petroleum production well. The well is consist of tubing in which liquid goes and casing which covers and protects tubing and injection pump/compressor. The required pressures and energy for injection can be calculated by following formulas.

CO_2 must pressurized to recover pressure drop between starting point and injection site. The pressure drop is discussed below.

The injection pressure at the surface must satisfy following equations (1.14) and (1.15)

$$P_{total} \geq P_{aqu} + \Delta P_{total} \tag{1.14}$$

$$\Delta P_{total} = \Delta P - \Delta Pg \tag{1.15}$$

Where P_{aqu} is aquifer pressure [Pa], ΔP is pressure drop between surface [Pa] and well head and ΔP_g [Pa] is given by $\Delta P_g = \rho gL$; and this pressure is occurred by gravity. ρ is density of CO_2 [kg/m^3], g is gravity constant [m/s^2] and h is the length of the well [m]. This value can be estimated by the equations used in the pipeline design part below.

$$P_{total} \geq Paqu + 4\lambda\left(\frac{h}{D}\right)\left(\frac{\rho * CO_2^2 * 10^6}{2(np\pi R^2)^2}\right) - \rho gh \tag{1.16}$$

$$P_{total} = P_{bot} + 4\lambda\left(\frac{h}{D}\right)\left(\frac{\rho * CO_2^2 * 10^6}{2(np\pi R^2)^2}\right) - \rho gh \tag{1.17}$$

Let the pressure at the bottom of well P_{bot} . This is set greater than aquifer pressure. For injection, power in equation (1.18) is required. Total power requirement for the system can be estimated by adding PRcap, PRliq, PRtrans, and PRinj.

$$PR_{inj} = \Delta P_{total} * V = \left(P_{bot} + 4\lambda\left(\frac{L}{D}\right)\left[\frac{P * (CO_2)^2 * 10^6}{2[np\pi R^2]^2}\right] + pgl\right) * \left[\frac{CO_2 * 10^3}{p}\right] \tag{1.18}$$

3.3 *Case Study for Wabamun Thermal Power Plant*

In this study, Wabamun Thermal Power Plant (Trans Alta Corporation) is selected as the CO_2 emission site for the case study of Aquifer CO_2 Disposal System (ACDS). The Wabamun Thermal Power Plant is located at Wabamun, 65km west of Edmonton, Alberta. In this plant, 2.8 million tonnes of coal are burned annually to generate 548 MW of electricity. The flue gases from this power plant are 4,584 kt of CO_2, 15,400 kt of N_2, 1,357,918 H_2O, and 992,451 of O_2 annually [TransAlta Co., 1998]. According to the data from Kansai Electric Power Company and MHI, Ltd., the capturing process requires total electricity 35.5 MW to separate 253t of CO_2 [Iijima, et. al, 1998]. This indicates that 140kW of electricity is required to capture 1 tonne of CO_2. This conversion factor is applied to the Wabamun case. At the Wabumn plant, 470.77 t of CO_2 is recovered per hour. Equation (1.2) is used to estimate the power requirement for liquefaction. The liquefaction system changes CO_2 from the gases phase to the liquid phase. The CO_2 is compressed in a compressor. The required power for liquefaction is estimated by using equation (1.2) as 0.1036 kwh/kg or 103.6 kwh/t at 300 K (27° C) and 0.1Mpa (atmosphere) [Pak, Nakamura and Suzuki, 1997] The above data are used for all cases because the capturing and liquefaction systems are the same for all cases.

The length of well, the radius of well are fixed with 1490m, 3 inches,respectively. In this study, the length of the well depends on the depth of the Grauconitic aquifer, which length is 1480m. The radius of the pipelines, density of CO_2 and the CO_2 flow rates are fixed at 3 inches [Bachu, et al.,1996]. Flow Rate of Wabamun Power Plant after capturing is estimated as 130.77 kg/s or 470.77 t/h. These data are used as input data for calculation. The required pressures and power requirement for pipeline for transporting liquid CO_2 can be calculated from equations (1.7) and (1.8). The number of service injection site in shown as Table 4.2. The length of pipeline is 5000 m for first six sites because CO_2 moves away from injector well up to 5km [Bachu, et al.,1996]. The permeability within 5km from the injection sites is assumed as 100md. The outer zone's permeability is assumed much lower. This means the CO_2 forms the circle, which has maximum radius of 5km and thickness of 13m. Average roughness of the pipe is assumed as 0.06 inch. The liquefied CO_2 is transported at a pressure of 7.5MPa and a temperature of 300 K. Top injection pressure and power requirement of the injector well can be estimated using equations (1.11) and (1.12).

3.4 *Result*

Figure 2 shows the pressure drop for pipeline and injector well and the power requirement for both pipeline and injector well in the case the injection pressure at the aquifer is 30.12MPa. The results show that as the number of injection sites increases, the pressure drop decreases. Especially, the pressure drop and power requirement for transportation shows very steep decreasing as the number of sites increases.

The maximum pressure drop for transportation is 45.02MPa and power requirement is 8020kW for only one injection site. This is sixteen times as big as that of 4 wells. On the other hand, the pressure drop analysis does not show the steep decrease like pipeline. This is because the injector well involve the pressure drop caused by gravity and this is much bigger than the pressure drop occurred by friction in the pipe. About 15.28 MPa of pressure drop is occurred by gravity for all case. The pressure drop occurred by the friction in the pipe varies from 4.71 MPa to 0.059MPa. This is much smaller than pressure drop occurred by the friction in the pipe so that the results do not show big difference like pipeline.

The energy requirement is different from pressure drop. For various pipelines, the energy requirement shows almost the same tendency but for injectors the results are different. This is because injection pressure is required to be at least over 12MPa at the aquifer. As long as results show, more than four sites are the optimum number for injection sites to save the energy requirement. The power requirements are estimated for four injection sites. The power requirements are 68500 kW for CO_2 capturing plant and 48772 kW for liquefaction. The transportation and injection system require 251kW and 1500kW, respectively. Total energy requirement for the system is 116 MWh. This is about 22% of the total generated energy at the Wabamun plant. Because the capturing system re quires about 68.5 MW of power and liquefaction system requires extremely high power, which is 48.7 MW.

Figure 4 shows the design of aquifer CO_2 disposal system for the Wabamun plant and the physical state of the CO_2 and the energy requirements for each system.

CONCLUSION

A detailed study is carried out on the geology of the Alberta Basin, the safeability of the host aquifer, and the physico-chemical behavior of CO_2. Technical feasible disposal systems are designed based on injection pressures and local aquifer permeabilities at injection site. These models are validated using the Wabamun coal-fired power plant. The Glauconitic aquifer in the Alberta Basin is suitable for CO_2 disposal and storage. The average depth and thickness of the aquifer are 1480m and 13m respectively. The permeability is between 6.2 and 100md. Poros
ity is between 6 and 12%. The top pressure is 12.4 MPa. Capacity of CO_2 in the Glauconitic aquifer are between $2.8*10_6$ and $2.2*10^7$ tonnes over a period of 30 years. The energy requirement for the disposal system for Wabamun plant is 116 MWh. This is about 22 % of the total generated energy at Wabamun plant.

REFERENCES

1. Bachu, S., Gunter, W.D., Law, D., and Perkins, E.H., 1996, *Aquifer Disposal of Carbon Dioxide*; © Hitchon Geochemical Service Ltd., Edmonton, Canada:
2. Bachu, S., Gunter, W.D. and Perkins, E.H., 1994, *Aquifer Disposal of CO_2:Hydrodynamic and Mineral Trapping*; Energy Conversion Management, Vol.35 No.4; © Elsevier Science Ltd; 269-279
3. Bachu, S., 1995,"Synthesis and Model of Formation-Water Flow, Alberta Basin, Canada"; *AAPG Bulletin, Vol.79 No.8*; © AAPG:1159-1178.
4. Bachu, S. and Gunter, W.D., 1998, "*Storage Capacity of CO2 in Geological Media in Sedimen tary Basins with Application to The Alberta Basin*"; Energy Conversion Management © Elsevier Science Ltd
5. Prothero, D.R.and Schwab, F., 1996 *Sedimentary geology : an introduction to sedimentary rocks and stratigraphy Prothero,* ©W.H. Freeman, New York,
6. Gunter, W.D., Wong, S., Cheel, D.B., and Sjostrom, G., 1998, Large CO_2 Sinks: Their Role in the Mitigation of Greenhouse Gases from an

International, Canadian and Alberta Perspective; *Unpublished Report*, Alberta Research Council, Edmonton

7. Neuzil, C. E., 1993,*"Low Fluid Pressure Within the Pierre Shale: A Transient Response to Erosion"*; Water Resources Research,Vol.29, No.7; ©The American Geophysical Union:
8. TransAlta, 1999, "*Heat Efficiency Progress via e-mail*"; 4 Feb 1999; TransAlta Ltd., Calgary, Alberta.
9. Iijima, M., 1998, "A Feasible New Flue Gas CO_2 Recovery Technology for Enhanced Oil Recovery"; © SPE (Paper 396

Environmental Issues and Management of Waste in Energy and Mineral Production, Singhal & Mehrotra (eds)
© 2000 Balkema, Rotterdam, ISBN 90 5809 085 X

Fine coal recovery utilizing landfill-derived liquids

S.T. Hall
School of Chemical, Environmental and Mining Engineering, University of Nottingham and LLAFF Technical Services, Nottingham, UK

ABSTRACT: Liquid effluents that arise in the landfilling of municipal solid wastes, both leachate and gas condensates, can replace some or all of the conventional coal flotation reagents. The waste management industry currently must treat these effluents, to destroy contained organics, prior to their discharge to the environment and hence they are available at no cost, or even with "dollars attached". These landfill-derived liquids contain valuable short-chain fatty acids (e.g. valeric and caproic acids) that act as both frother and collector in flotation processes. They can be further concentrated by membrane filtration to reduce the transport costs of such liquids and the ease of their use in coal recovery systems. This paper will discuss how these liquids are formed in landfills, their typical compositions and how they can be employed to recover fine coal. Data from both conventional flotation feeds (-0.5 mm) and column flotation feeds (-0.1 mm) is presented.

1 INTRODUCTION

Landfill leachates are highly polluted wastewaters resulting from the ingress of water to landfill wastes and the degradation and leaching of materials from the wastes. They are highly variable in composition between sites and with time. They contain biodegradable compounds, inorganic salts and trace recalcitrant pollutants. As a minimum they require treatment for the reduction of they biochemical oxygen demand (BOD), chemical oxygen demand (COD) and ammoniacal nitrogen contents prior to release to the environment.

Currently most landfills in the UK discharge leachates to foul sewer. The landfill operator paying the local water company significant amounts of money calculated on a modified trade effluent formula. This formula accounts for the quantity of leachate (plus gas condensate where condensate is collected and returned to the waste or leachate lagoon) and the toxicity when compared with domestic sewage (chemical oxygen demand and suspended solids). Other components of concern to the water companies are ammonia concentration and dissolved methane. The high chemical oxygen demand (COD) values of acetogenic leachate – mostly due to the volatile fatty acid components (predominant in the early years of landfill operation) – means that the disposal costs can easily be in excess of £10/m^3 with a large landfill discharging several hundred cubic metres per day.

Several companies have installed some form of pre-treatment, with the most common method being the aerated lagoon (Ashbee and Fletcher, 1993). Here bacterial action promoted by aeration reduces the COD/BOD levels over a period of time (typically 10 – 20 days); however the biomass generated (0.5 g per g of COD removed) must routinely be removed for disposal. This is currently deposited into the landfill, not only occupying valuable void space, but also contributes to the biodegradable input to the landfill which the recent EU landfill directive seeks to reduce over future years. Capital costs for such lagoons are estimated at up to £10,000/m^3 of installed daily capacity with operating costs of £1 – 5/m^3 (power for aeration accounting for a significant proportion of operating cost).

It has recently been demonstrated in laboratory experimentation that landfill leachate can act as the source of effective froth flotation reagents (Hall, 1999). It is believed that the volatile fatty acids (VFAs) and alcohol content of the leachate accounts for the frothing and collecting action in the flotation process. The adsorption of the VFAs onto the coal particle surfaces reduces the COD and BOD in the filtrate from the flotation process. Ammonia is believed to adsorb onto the surfaces of the fine clay minerals, comprising the high-ash shale residue, and is removed from the process filtrate.

However, it is unlikely to be economic to transport the large volumes of relatively dilute leachates to employ them in existing (and numerically decreasing) coal and mineral processing facilities in the UK. However, landfill gas condensates, possible in combination with membrane filtration (reverse osmosis), might offer an economic source of organic chemicals to utilise in such processes.

Landfill gas condensates arise whenever landfill gas is collected for flaring or power generation. This is becoming the normal requirement given the increased concern over methane emissions and their contribution to global warming. Many gas extraction systems are now being installed with in-built facilities to fit leachate-pumping equipment. LFG is usually saturated with water when it enters the collection pipework. The temperature at which it is formed is 30 – 40 °C or higher. The gas emerges from the body of the fill, is drawn along the collection pipelines, where it cools and condenses. The typical method of controlling condensate in pipelines is to install a drum (a water "knockout pot") at low points in individual lines. These will drain through a liquid seal ("syphon"), which allows continuous discharge of condensate back to the landfill (estimated to be around 20 tonnes/hour at a large landfill site). Alternately, condensates are discharged to sewer after pH adjustment.

Gas condensates offer a possibly concentrated and purified source of volatile fatty acids (VFAs) transferred from the body of the waste (the VFA content of two UK gas condensate samples is given in Table 1). It has been generally concluded that landfill gas condensate from sites in the UK that have received primarily household wastes is not necessarily a hazardous material. Landfill leachates are an additional potential source of VFAs, but are

Table 1. Volatile fatty acid content of two UK landfill gas condensate samples.

Volatile Fatty Acid (ppm)	Gas Condensate A	Gas Condensate B
Acetic acid	32	6.8
Propionic acid	146	43
Iso-butyric acid	86	37
n-butyric acid	470	170
Iso-valeric acid	60	27
n-valeric acid	180	76
Iso-caproic acid	9.3	3.6
n-caproic acid	410	196

lower in concentration and contain many other substances of environmental concern.

The innovation is in the use of such liquids, possibly concentrated by membrane filtration, as flotation reagents. Figure 1 indicates the froth generation properties of solutions of UK gas condensate sample B, solutions of reagent-grade caproic acid and 100 and 1000 ppm solutions of MIBC.

2 EXPERIMENTATION AND RESULTS WITH COAL FLOTATION FEEDS

A coal flotation feed sample was collected prior to the addition of flotation chemicals from an operating UK coal preparation plant. A landfill leachate sample was collected from a landfill site receiving municipal solid wastes in the UK (Table 2). Allowing the solids to settle and decanting off the supernatant water thickened the coal flotation feed sample. Laboratory flotation tests were conducted by intro-

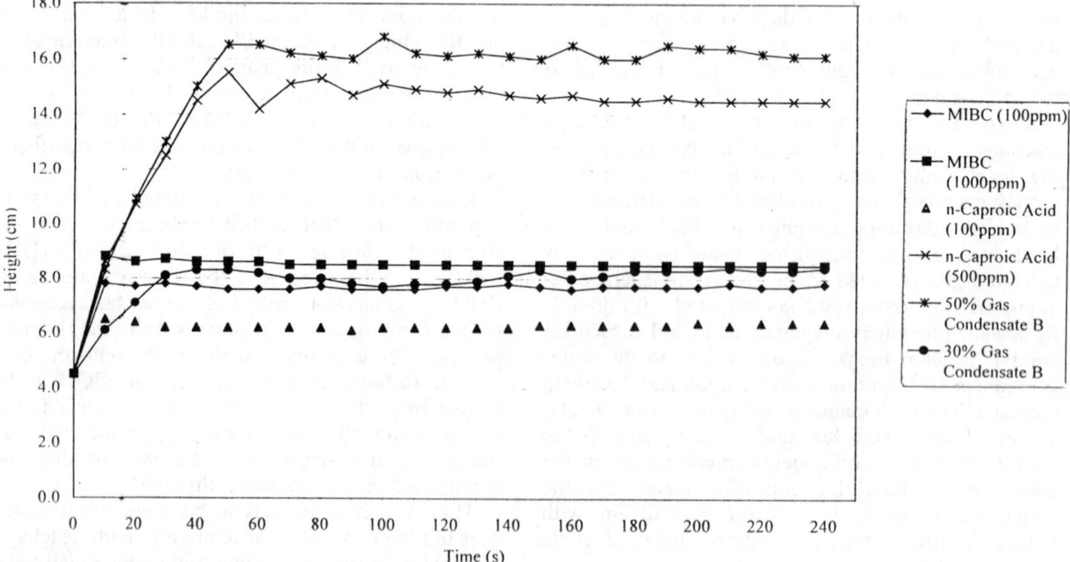

Figure 1. Froth height vs. time

ducing 40% by volume of either tap water (control experiment) and/or landfill leachate to make up the necessary volume (4.5 litres) in a Denver laboratory batch test cell.

A 200 litre sample of the leachate was concentrated by reverse osmosis (RO) to generate approximately 10 litres of concentrate. A similar flotation test was conducted employing only 2% by volume of RO concentrate.

Conventional flotation reagents were added to the control experiments (MIBC frother and kerosene collector) in similar dosages to those employed commercially (up to 1 kg/tonne). Concentrates were collected incrementally and all samples, including final tailings were filtered and dried. All dry samples were weighed and sub-sampled for ash analysis. Tailings samples were retained in a moist form for vegetation growth experiments. Table 3 presents the results of these flotation tests.

Similar flotation tests were conducted employing conventional reagents (MIBC and n-dodecane), reagent-grade caproic acid and quantities of the two gas condensate samples (with and without addition of MIBC and n-dodecane), (Table 4).

In addition, two tests were performed with samples of column flotation feed (-150 microns) with gas condensate only (Table 5).

3 RECOVERY OF COAL LAGOON MATERIALS – LEACHATE PRETREATMENT

The removal of organics (expressed as BOD and COD) and ammonia in the flotation process enables it to act as a leachate pretreatment operation. The volume of leachate to be treated limits the required process capacity, however filtrate recycle (prior to eventual discharge) will have possible benefits of further organics and ammonia removal, plus the generation of additional clean coal byproduct. The operation is now conducted at the landfill site with coal transported to the process.

One potential material for separation employing leachate at landfill sites is lagoon coal. Coal lagoons are a source of valuable coal from historical or current coal mining activities. Coal lagoons contain fine particles of coal and shale, generally untreated prior to surface disposal. Such lagoons have considerable negative environmental impact, including CO_2 emissions from coal oxidation and possible spontaneous combustion. The recovery of the contained coal by froth flotation alone is often uneconomic, however it can become economically attractive by incorporating the material in a leachate pretreatment scenario, especially as the former lagoon site might subsequently become available as a landfill void.

Table 2. Indicative analysis of landfill leachate sample used in flotation testing

Approximate Age (Years)	2
pH	7.4
BOD**	64
COD**	1200
TOC*	84
Chloride*	1100
Ammoniacal nitrogen*	370

Units; *mg/l, ** mg 0_2/l.

Table 3. The results of coal flotation with MIBC and kerosene, landfill leachate; and RO concentrate of landfill leachate

Chemical Addition	Yield (wt%)	Concentrate Ash (%)	Tailings Ash (%)
Conventional Reagents (MIBC + kerosene)	58.3	9.0	91.4
Leachate (40% by volume)	54.8	6.6	88.0
Leachate RO Conc. (2% by volume)	55.7	12.2	77.4

Table 4. Response of UK conventional coal flotation feed sample using landfill gas condensate.

Test No.	Chemical Addition	Yield (Wt%)	Concentrate Ash (%)	Tailings Ash (%)
1	440 g/t n-dodecane + 200 ppm MIBC	48.5	7.2	64.9
2	440 g/t caproic acid	55.5	5.9	90.9
3	50 ml gas condensate A	21.2	5.8	59.7
4	50 ml gas condensate A + 220 g/t n-dodecane + 100 ppm MIBC	46.4	6.5	84.4
5	100 ml gas condensate A	58.6	7.8	85.1
6	100 ml gas condensate A + 220 g/t n-dodecane + 100 ppm MIBC	61.9	8.6	89.9
7	200 ml gas condensate A	63.5	10.1	90.7
8	200 ml gas condensate A + 220 g/t n-dodecane + 100 ppm MIBC	64.4	10.7	91.5
9	50 ml gas condensate B	21.2	7.1	54.8
10	50 ml gas condensate B + 220 g/t n-dodecane + 100 ppm MIBC	46.4	7.1	77.2

Table 5. Response of UK column coal flotation feed sample using landfill gas condensate samples in conventional flotation laboratory testing.

Test	Chemical addition	Yield (Wt%)	Concentrate Ash	Tailings Ash
A	25 ml gas condensate A	46.1	12.0	88.3
B	25 ml gas condensate B	56.0	11.3	85.5

Table 6. Results of flotation testing employing landfill leachate and conventional process flotation reagents where necessary.

	Landfill Leachate	Effluent from two stage 20% solids flotation
Chemical Oxygen Demand (COD) (mg/l)	2800	110
Ammoniacal Nitrogen as N (mg/l)	1190	71
Suspended Solids SS (mg/l)	159	5
Indicative Trade Effluent Disposal cost	2.55	0.42*

* @ 200m^3/day a saving in typical trade effluent charges of £440/day (£132,000/year)

A coal lagoon sample (80% - 100 microns; 37.5%) was slurried in a landfill leachate at 20% solids by weight and flotation conducted to completion. Small additions of conventional reagents were added in the later stages of flotation. The products were filtered and the filtrate recycled to a second-stage of flotation, conducted under similar conditions. The leachate and second-stage filtrate analysis is reported in Table 6, together with an indication of the savings in trade effluent disposal costs.

4 VEGETATION GROWTH TRIALS

The potential for the contained nutrients in the coal flotation tailings to promote the growth of grass was studied by seeding 1 litre pots with 50 perennial rye grass seeds. Four duplicate tests were conducted on conventional coal flotation tailings, leachate flotation tailings, leachate flotation tailings plus 5% topsoil and leachate flotation tailings plus 15% topsoil. After nine weeks of growth, the biomass produced was harvested and dried. The ammonia concentration in the tailings and the conversion to nitrate was monitored.

In the growth trials, the tailings from leachate-containing flotation produced significantly more biomass over the 9-week period than the conventional tailings (Table 7).

In addition, the ammonia concentration within the tailings decreased rapidly during the first few days of experimentation to less than 1 ppm at the end of the incubation period.

Table 7. The mount of biomass produced (dry weight) after nine week growth trials in coal flotation

Growth Substrate	Biomass (g)
Conventional flotation tailings	0.033
Leachate flotation tailings	0.566
Leachate tailings + 5% topsoil	0.833
Leachate tailings + 15% topsoil	0.833

5 DISCUSSION

It has been shown that volatile fatty acids (VFAs) can act as effective coal flotation reagents, providing both frothing and collecting actions. The presence of volatile fatty acids, plus short-chain alcohols, in landfill-derived liquids makes both leachates and gas condensates potential sources of flotation chemicals. For efficient flotation performance, especially in the recovery of coarser coal particles, their action might require the addition of supplementary conventional reagents.

Landfill gas condensates, offering a purified and concentrated source of VFAs, might provide a source of flotation reagent for use in conventional coal preparation. The VFAs are biodegradable and their source offers considerable environmental merit in promoting their utilisation.

The use of froth flotation at landfill sites offers an innovative leachate pre-treatment process. The removal of organics, ammonia, suspended solids, colour and dissolved methane in flotation and product dewatering reduces the subsequent disposal cost of the process filtrate. The separation of fine coals from lagoon materials offers a valuable by-product for combustion, or possibly co-combustion in waste-to-energy incineration of wastes. The ammonia content of the high-ash shale residue is appropriate to promote rapid vegetation of an otherwise sterile substrate making it suitable for site rehabilitation. In addition, this concept can potentially remove coal lagoons, a significant pollution source in areas of current or historical coal mining activity, creating possible sites for future landfilling (Honaker, Mohanty and Patwardhan, 1998).

Other materials, such as industrial minerals, contaminated soils and recycled materials (deinking of paper, plastics segregation), can potentially be treated using landfill-derived liquids (leachates, gas condensates or products generated from these sources).

6 CONCLUSIONS

The volatile fatty acids and short-chain alcohols in landfill-derived liquids can act as useful collectors and frothers in froth flotation.

Fine coal can be recovered from conventional and column flotation feeds using landfill-derived liquids.

Fine coal feed materials, including coal lagoon materials, can act as the "reagents" in leachate pre-treatment at landfill sites, providing a clean coal concentrate by-product.

The ammonia in leachates is adsorbed onto the high-ash shale tailings in concentrations appropriate to stimulate rapid vegetative growth.

The innovative utilisation of landfill-derived liquids in froth flotation has identified a number of significant commercial opportunities in materials separation.

7 REFERENCES

Ashbee & Fletcher, 1993. Reviewing the options for leachate treatment. *Wastes Management, August:* 32–33.

Hall, S.T. 1999. The Use of Landfill Leachate for the Recovery of Fine Coal by Froth Flotation. *Coal International, January:* 22–23.

Honaker, R.Q. Mohanty, M.K. & Patwardhan, A. 1998. High quality coal extraction and environmental remediation of fine coal refuse ponds using advanced cleaning technologies. Pasamehmetoglu & Ozenoglu (eds): 573-579. Balkema.

Environmental Issues and Management of Waste in Energy and Mineral Production, Singhal & Mehrotra (eds)
 ISBN 90 5809 085 X

Biofiltration: A cost-effective technique for controlling methane emissions from sub-surface sources

J.P.A. Hettiaratchi, V.B. Stein & G. Achari
Engineering for the Environment, Department of Civil Engineering, Faculty of Engineering, University of Calgary, Alb., Canada

ABSTRACT: Focus of this paper is a new technique to control the escape of methane from sub-surface sources into the atmosphere. This innovative, but simple, technique involves manipulation of surface soil layers to oxidize methane to carbon dioxide. It is based on optimizing the growth of naturally occurring methanotrophic bacteria in soils. Conditions required for optimum growth of methanotrophic bacteria include soil moisture content, soil type, temperature and source strength of methane. The application of microbial oxidation for controlling CH_4 emissions from soil migration at well sites, as well as landfills is possible, and is suited for sites with CH_4 flux rates as high as 166 $g*m^{-2}day^{-1}$.

1 INTRODUCTION

The major environmental issues identified with the emission of methane (CH_4) and non-methane volatile organic compounds (NMVOCs) into the atmosphere are global warming, formation of lower atmospheric or tropospheric ozone, and human exposure to toxic compounds. Methane has 21 times the global warming potential of carbon dioxide (CO_2) over a 100-year time horizon (IPCC, 1996). Therefore, the reduction of CH_4 emissions, in addition to CO_2 emissions, could produce significant greenhouse gas control benefits. Since CH_4 has relatively shorter residence time in the atmosphere as compared to other greenhouse gases, such as CO_2 or nitrous oxide (N_2O), any reduction in CH_4 emissions may quickly produce benefits.

Industry is seeking cost-effective methods to minimize greenhouse gas emissions. In this paper, a simple but innovative technique that involves biological conversion of methane to carbon dioxide is evaluated for potential application in upstream oil and gas industry operations and in sanitary landfills. Initially, background information on methane emissions from these sources is provided. Next, the process of biological methane oxidation is described, along with its primary controlling factors. Following that, a brief description of the research being undertaken at the University of Calgary in this field is provided, along with some recent results. The paper is concluded with brief descriptions of potential methods for applying methane biofiltration in the field.

2 BACKGROUND

Two of the most important sub-surface sources of methane emissions in Alberta, namely upstream oil and gas operations and sanitary landfills, are described below.

2.1 *Methane Emissions from Upstream Oil and Gas Industry Operations*

The upstream oil and gas industry has been identified as one of the major contributors of CH_4 to the atmosphere. Carbon isotope measurements indicate that about 20% of the total annual global methane emissions are related to the production and use of fossil fuel (IPCC, 1996).

Within the oil and gas industry sector, conventional oil production is the topmost contributor of total hydrocarbons (32.4%), closely followed by gas production (29.6%) and heavy oil production (17.7%). In upstream oil and gas operations, the practice of process venting is the primary source of CH_4 emissions. An estimated 1230 kilotonnes of hydrocarbons was emitted to atmosphere via process venting in 1989 (Holford, 1998). This accounted for almost 0.8% of the total hydrocarbons produced that year, and comprised of 785 kilotonnes of CH_4 and 445 kilotonnes of NMVOCs. The practice of process venting contributed 27% of the total hydrocarbons emitted by oil and gas industry, and of which 57% came from venting of casing gas at heavy oil well sites.

The upstream oil and gas operators of Alberta have joined the Voluntary Climate Change Challenge Programme (VCCCP) to support Canada's commitment to the international community for stabilizing greenhouse gases as agreed upon within the framework of the Kyoto Protocol. The oil and gas industry has targeted the reduction of CH_4 emissions as the key activity towards achieving greenhouse gas stabilization goals. Furthermore, in Alberta, the environmental liabilities of the oil and gas industry have increased. According to the "zero tolerance" standards, any well site can not be reclaimed, or abandoned, until all indications of CH_4 gas leakage through the surface casing and/or in the soil are completely eliminated.

One of the emission sources of oil and gas sector that has received recent attention is from the production of heavy oil. This source can be sub-divided into; 1. Surface casing vent flow and soil migration gas, 2. Production casing vent flow / solution gas.

2.1.1 Surface casing and soil migration gas

Gas leakage from petroleum or gas wells can manifest itself outside the outermost casing (either production or surface casing). In these cases, the escaping gas is known to migrate in soils and escape into the atmosphere. These CH_4 emissions have not been accurately quantified, due to the lack of reliable well site emission monitoring data and effective reporting mechanisms. However, studies have been carried out to quantify the CH_4 gas migration from heavy oil production sites near Lloydminster, Alberta (Erno and Schmitz, 1996, Schmitz et al., 1994). Erno and Schmitz (1996) investigated in situ flow rates of surface casing vent flows and soil gas migration in the Lloydminster area. They reported that 23% of the wells had surface casing vent flows and 45% of the wells had detectable soil gas migration in the immediate vicinity of the well. For the majority of wells, surface casing vent flow rates ranged from $0.01 m^3$/day to 100 m^3/day, and less than 5% of wells had leakages greater than 100 m^3/day. Soil gas migration rates were mostly less than 0.01 m^3/day, and no well exceeded migration rates of 60 m^3/day.

During migration of CH_4 in soils adjacent to these wells, some quantities of methane could potentially be oxidized by indigenous microbial populations. Although this phenomenon is known to oil and gas operators, soil methanotrophy has not been used as a technique to control methane gas emissions at well sites.

2.1.2 Production casing vent flow / solution gas

Solution gas is the hydrocarbon gas associated with oil that is produced from reservoirs. Because it is under the increased pressure found in oil reservoirs, it is kept "in solution" with the liquid hydrocarbon phase. During the production of oil, the pressure is reduced to atmospheric pressure and these gases are liberated from the liquid phase to form a separate vapour phase. In Alberta, most of this gas (about 92%) is collected and processed at gas processing plants and sold. The remainder is flared or vented (Holford, 1998).

2.2 Methane Emissions from Sanitary Landfills

In sanitary landfills, organic matter decays anaerobically to produce large quantities of CH_4 and CO_2. Emissions from landfills account for almost 20% of the worldwide anthropogenic CH_4 emissions into the atmosphere (Nozhevnikova et. al. 1993). Extraction, with or without energy recovery, is the most common method currently employed to control such emissions. However, this method is not adaptable in most landfills, especially in small landfills. Even in landfills with gas extraction, available techniques are capable of capturing only 40 to 90% of the produced gas (Augenstein and Pacey, 1996). The rest escapes into the atmosphere.

Traditionally, landfill cover systems are designed to minimize water infiltration with the specific aim of minimizing leachate generation. All layers, except the gas collection layer, are selected to minimize water infiltration into waste, by maximizing surface run-off (suitable slope and vegetation), evapotranspiration (vegetative layer), storage (root zone) and sub-surface run off (drainage layer and low-permeable layer). Although, not recognized by most designers, natural bacteria ("methanotrophic bacteria or methanotrophs") in top-most layer of a cap are known to oxidize methane to carbon dioxide.

Methane oxidation by methanotrophs has been observed at oil well sites and sanitary landfills. The challenge is to provide suitable conditions for these organisms to proliferate and optimally oxidize methane. It is necessary to determine the extent to which biological oxidation in soils can serve as a technique for reducing methane emissions, and to determine how to select or manipulate soils in a manner which maximises their CH_4 oxidation potential.

3 SOIL METHANOTROPHY: A LITERATURE REVIEW

3.1 Methanotrophic Bacteria

Methyltrophs are microorganisms that can use one-carbon compounds that are more reduced than carbon dioxide (e.g. CCl_4, CH_4). Methanotrophic bacteria are the subset of methyltrophs that possess the specific enzyme methane mono-oxygenase which enables them to utilise methane as a sole source of energy and as a major source of carbon (Haber, et al., 1983). They oxidise methane through methanol to formaldehyde, which they then either assimilate for the synthesis of cell material or further oxidise to carbon dioxide (Topp and Hanson, 1991).

Methanotrophic bacteria are present in small numbers in most soils. Long term exposure of soils to high levels of methane encourages the growth populations of methanotrophs with a high capacity for methane oxidation (i.e. a high V_{max}, where V_{max} is defined as the rate of CH_4 oxidation when CH_4 is not limiting). These populations are also characterised by a low affinity for CH_4 (i.e. a low K_m, where K_m is defined as the concentration of CH_4 which results in a CH_4 oxidation rate equal to one half of V_{max}) (Bender and Conrad, 1994).

3.2 Factors Affecting Soil Methanotrophy

Soil methanotrophy has been reported most commonly in agricultural soils, forest soils, tundra, bogs and landfill cover soils. Once established, the methanotrophs survive under intermittent CH_4 flow conditions (Kightley et al., 1995). Intermittent flow conditions are common in landfills in northern climates. The soil moisture content (MC) affects both physics and biology of soil methanotrophy. Parameters such as soil type and structure, hydraulic conductivity and MC are inter-related, and could be influenced by weather conditions. According to field and laboratory observations, optimum MC for sandy-loam and sandy-clay varies between 10% and 20% (Whalen et al., 1990). The greatest methanotrophic activity has been observed when particle diameters range between 0.5 and 2 mm.

Among the conditions required for optimum growth of methanotrophic bacteria, oxygen concentration in soil and nutrients are two of the parameters that can be manipulated by selecting the proper type of soil and long-term operational practices. Temperature is a critical factor in northern climates. These three factors are discussed below.

3.2.1 Oxygen Concentration

Methane oxidation by methanotrophs occurs only in environments where CH_4 and O_2 occur simultaneously. Methane oxidation has been observed to decrease at O_2 mixing ratios of less than 3%, but is only slightly sensitive to changes in O_2 concentration at mixing ratios above 3% (Czepiel et al., 1996; Stein and Hettiaratchi, 2000).

De Visscher et al. (1999) also observed the depth of the highest V_{max} in their soil columns to occur where the O_2 concentration was close to K_{O2}, which in their case was 1.23%. For some reason the presence of O_2 in concentrations exceeding K_{O2}, limits the maximum size of a methanotrophic microbial community.

3.2.2 Nutrients

Amending soil with ammonium (NH_4) has been found to substantially reduce oxidation of CH_4 at atmospheric concentrations (Steudler et al., 1989; Schnell and King, 1994). Schnell and King (1994) found that NH_4^+ also inhibits CH_4 oxidation at higher CH_4 concentrations. However, de Visscher et al. (1999) recently contradicts this finding, suggesting that it applies only to soils that have not been exposed to historically high CH_4 concentrations. Kightley et al. (1995) conducted experiments to determine the effects of NH_4NO_3 (2 g N * Kg soil^{-1}), K_2HPO_4 (100 mg P * Kg soil^{-1}) , and anaerobically digested sewage sludge (100 mg of N and P * Kg soil^{-1}) on CH_4 oxidation. The sewage sludge increased CH_4 oxidation by 26%, whereas NH_4NO_3 inhibited CH_4 oxidation and K_2HPO_4 had no effect. However, de Visscher et al. (1999) found that an addition of either NH_4Cl or $NaNO_3$ (25 mg of N * Kg soil^{-1}) resulted in a doubling of the CH_4 oxidation rate. The seeming contradiction in these researchers' results may be due to a difference in the resulting ionic concentrations of their soils. In the experiments of Kightley et al. (1996), the CH_4 oxidation may not have been inhibited specifically by NH_4^+, but instead by an increased ionic concentration. This might also explain why they observed sewage sludge to increase the CH_4 oxidation rate, as the ionic concentrations resulting from its addition may have been below amounts that cause an ionic inhibitory effect.

Clearly, nitrogen amendment experiments in which the effects of ionic inhibition are controlled, should be performed on a variety of soils and at a variety of nitrogen concentrations. However, based on the work of de Visscher et al. (1999), the possibility of significantly increasing CH_4 uptake by a controlled release of mineral N into the soil seems very promising.

3.2.3 Temperature

The response of microbial CH_4 oxidation to varying temperature has been quantified by manipulating temperature during soil sample incubations (Czepiel et al., 1996; Stein, 2000). This temperature response can be described by the Arrhenius relationship, in which oxidation increases exponentially to a distinct maximum, and then decreases with continued temperature increase.

The exponential constants seem to depend on the temperature at which the population was grown. For example, in a collaborative study undertaken with the Asian Institute of Technology (Visvanathan et al., 1999), we have observed that methanotrophs grown at an average temperature of 32°C exhibited a CH_4 oxidation rate that is six times higher at 22°C than at 10°C , whereas methanotrophs grown at a temperature of 22°C exhibit only a 2.4 fold increase in CH_4 oxidation when the temperature is increased from 10 to 20°C (Stein and Hettiaratchi, 2000). Nozhevnikova et al. (1992) observed an even smaller response to temperature increase in a soil taken from a Moscow landfill. For their soil, increasing the temperature from 6°C to 25°C resulted in the CH_4 consumption rate increasing only by a factor of 2.5.

4 METHANE BIO-OXIDATION RESEARCH AT THE UNIVERSITY OF CALGARY

Almost all of the work performed to date investigating the factors which influence microbial methane oxidation in soils have relied on batch experiments in which soil is placed in jars which are then injected with methane. This approach does not simulate the reduction in the areal extent of oxygen penetration caused by its advective displacement by methane and by its consumption due to microbial methane oxidation. Also, the batch experiments performed to date haven't allowed the microbial populations to reach their potential and so do not indicate the maximum amount of methane that can be oxidized in soil systems. To adequately simulate these mass transfer limitations soil column experiments were used in our investigations.

The use of soil columns provides a realistic simulation of soil methanotrophy as encountered in field situations.

Batch experiments do not account for the reduction in the areal extent of O_2 penetration that might be caused by its advective displacement by CH_4. Simultaneous column experiments were undertaken at the University of Calgary and Asian Institute of Technology (Bangkok, Thailand). Details of our laboratory set-up for column studies are described elsewhere (Stein and Hettiaratchi, 1997).

The columns are constructed from Plexiglas tubes with an inner diameter of 5.5 inches. Methane is fed through the bottom of the columns, simulating the range of fluxes encountered at heavy oil well sites. Air is passed across the top of each column through ports in the head caps. This permits measurement of the methane flux from the soil surface, and also maintains natural oxygen concentrations within the soil.

The concentration of methane and carbon dioxide exiting the column is measured using infrared gas analysers. Gas sample ports are placed at 10-cm intervals down the column, and are plugged with silicone septa. These permit analysis of gas concentration depth profiles for determining the depth of oxygen penetration and regions of maximum CH_4 concentration. Methane nitrogen, oxygen and carbon dioxide are measured using gas chromatography. A few different soil types were tested, including topsoils, a sandy loam, and sedge peat.

5 RESULTS AND DISCUSSION

The following is a brief summary of the key findings of our research and those of other researchers regarding biofiltration as a technique for controlling CH_4 emissions.

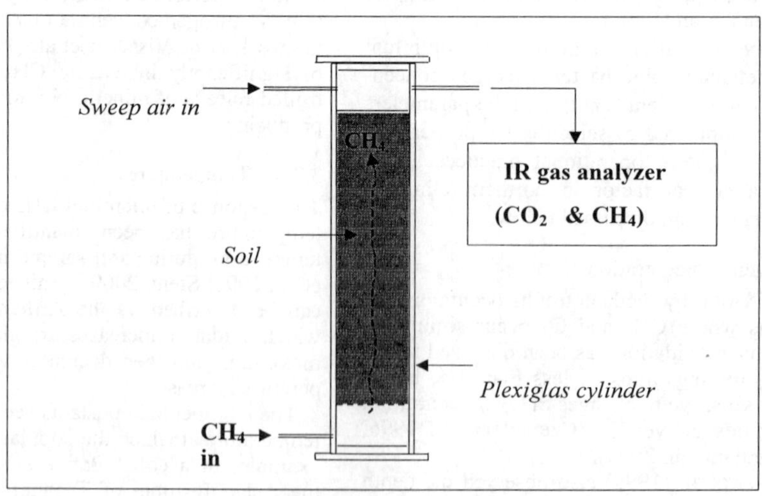

Figure 1. Soil Microcosm Schematic

5.1 Efficiency of Methane Oxidation

All of the soils tested developed significant capacities to oxidise CH_4, provided that their moisture content was not too far from the optimum (which was typically 10-15% on a dry weight basis). Some of the soils that were obtained from the field had moisture contents that were sub-optimal for biological CH_4 oxidation. The oxidation activity of these soils would increase by as much as an order of magnitude when their moisture content was increased to an optimal level.

The highest rate of methane oxidation observed in laboratory-scale soil microcosms was in a sedge peat, for which CH_4 consumption rate was 186 $g*m^{-2}*day^{-1}$. A comparably high oxidation activity was also observed in a sandy loam. The oxidation efficiencies observed for these two soils were similar when the columns were performing optimally (Figure 2).

After operating for 74 days, the oxidation efficiency of the sandy loam decreased significantly (Figure 3). The top half of the soil columns were observed to have dried somewhat, and were operating at sub-optimal moisture contents. However, based on experiments that were performed to investigate the effect of moisture content on CH_4 oxidation rate, the decrease in oxidation activity after 74 days can not be entirely explained in terms of a decrease in moisture content.

A possible explanation for the observed decrease in oxidation efficiency is that the biofilters' kinetics shifted from growth kinetics to nutrient limited kinetics. The cessation of growth in a biofilter can be caused by nutrient depletion in the bed, which forces the bacteria into maintenance metabolism, which is of a relatively lower rate compared to substrate consumption during the active growth of the acclimation phase. This hypothesis is under further investigation.

Similar results were obtained from soil column experiments in Thailand. When simulated landfill gas, containing 60% CH_4 and 40% CO_2, at a volumetric rate of 5 mL/min was passed through a Type I soil (consisting of 70% sand, 15% clay and 15% silt), about 65% of methane was oxidized at steady state. When the flow rate was increased to 9 mL/min, the immediate result was a decrease in percentage oxidation (to about 35%). However, methane oxidation rate increased from 61.74 g/m^2.day to 120.64 g/m^2.day (a 95% increase). In Type II soil (consisting of 70% sand, 25% clay and 5% silt), the change in flow rate decreased the percentage oxidation from a high of about 51% to about 20%, whereas the oxidation rate increased from 48.88 to 87.29 g/m^2.day (a 44% increase). These results show the pronounced impact of the composition of soil, and flow rate on oxidation. An ideal soil layer should contain mostly sand and silt and small amounts of clay.

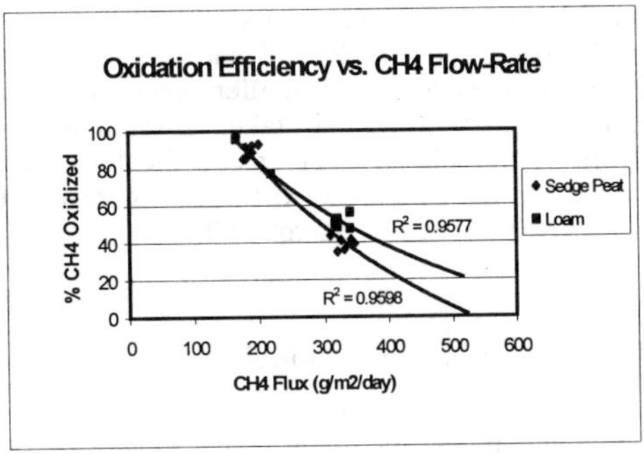

Figure 2: CH_4 biofilter efficiency (optimal)

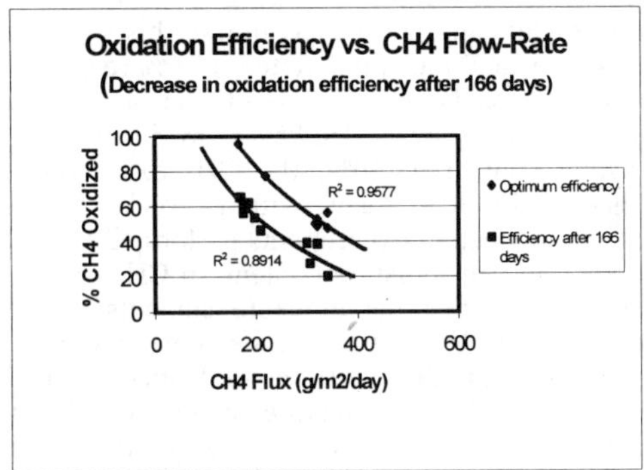

Figure 3: Decline in biofilter efficiency

5.2 Effect of Oxygen Concentration and Temperature

We found the O_2 half-saturation constant (K_{O2}) to be 1.14%. Interestingly, we also observed that methanotrophs seem to thrive in micro-aerobic environments. The highest V_{max} within our soil columns occurred at the depth at which the O_2 concentration was between 0.75% and 1.5%, i.e. close to K_{O2}. At this depth, we observed the V_{max} value to be approximately four times higher than at the depths where the O_2 concentration was higher. But because the effects of O_2 on the CH_4 oxidation rate also exhibit Michaelis-Menten characteristics, the actual oxidation rate at these depths was typically only two to three times the rate at depths with higher O_2 concentrations, because O_2 was at a sub-saturating concentration.

In the collaborative study undertaken with the Asian Institute of Technology (Visvanathan et al., 1999), we have observed that methanotrophs grown at an average temperature of 32°C exhibited a CH_4 oxidation rate that is six times higher at 22°C than at 10°C , whereas methanotrophs grown at a temperature of 22°C exhibit only a 2.4 fold increase in

CH_4oxidation when the temperature is increased from 10 to 20°C (Stein, 2000). Nozhevnikova et al. (1992) observed an even smaller response to temperature increase in a soil taken from a Moscow landfill. For their soil, increasing the temperature from 6°C to 25°C resulted in the CH_4 consumption rate increasing only by a factor of 2.5.

5.3 *Maximum CH_4 Oxidation Rates*

Soil column experiments were performed with a variety of soil types, including a landfill cover loam, two top-soils, and an organic soil (sedge peat) at CH_4 flux-rates ranging from 160 to 320 $g*m^{-2}*day^{-1}$. In all of the soils tested, the steady-state oxidation rates were between 92 and 120 $g*m^{-2}*day^{-1}$ at field moisture contents. One exception to this was a top-soil with a field moisture content of 6% (dry weight basis). Its steady-state CH_4 oxidation rate was less than 10 $g*m^{-2}*day^{-1}$. However, by increasing its moisture content to 10%, the CH_4 oxidation rate climbed to 124 $g*m^{-2}*day^{-1}$ within a week.

Batch incubation experiments performed to evaluate the effect of moisture content on CH_4 oxidation rate indicated that the moisture contents of these soils were sub-optimal. For example, the rate of CH_4 oxidation in the landfill cover loam could be increased by 60% if its moisture content of 9.4% was increased to 16%. This would result in an oxidation rate between 160 and 192 $g*m^{-2}*day^{-1}$ (depending on the CH_4 flux-rate). Such an oxidation rate is comparable to the maximum rate observed by Kightley et al. (1996) in a coarse sand (166 $g*m^{-2}*day^{-1}$) and to the maximum rate observed by deVisscher et al. (1999), which was 172 $g*m^{-2}*day^{-1}$ in an agricultural soil with a 16.5% moisture content.

Moisture content is seen to be a critical variable in determining a soil's CH_4 oxidation potential. The type of soil selected for an oxidative cover will influence the moisture content, with the optimal choice depending on climatic variables such precipitation, temperature, solar flux, average wind speed and the type of vegetative cover. Hydrologic models such as BROOK90 can be used to predict a soil's moisture content as a function of these climatic variables and the soil's physical properties.

However, one cannot use moisture content as the sole criterion in selecting the optimal soil for CH_4 oxidation. Its oxidation potential also depends upon the depth of O_2 penetration, which is influenced by at least three factors: the soil's air-filled porosity, the rate of displacement of the normal soil atmosphere by the upward movement of CH_4, and the rate of microbial CH_4 oxidation on a volumetric basis. All of these parameters can be easily estimated, with the exception of the volumetric CH_4 oxidation rate. This depends on the microbial population's kinetic parameters, which are discussed in the next section.

5.4 *Reaction Kinetics*

The CH_4 oxidation rate is a function of the CH_4 concentration, and exhibits typical Michaelis-Menten characteristics (Bender and Conrad, 1992; Czepiel et al., 1996; Stein, 2000) and is described by the following equation:

$$\nu = \frac{V_{max} \times [S]}{K_{app} + [S]}$$

where:

ν = CH_4 consumption rate ($g*day^{-1}*g$ dry weight^{-1})
[S] = CH_4 mixing ratio [ppmv] in air
V_{max} = maximum CH_4 consumption rate ($g*day^{-1}*g$ dry weight^{-1})
K_{app} = half saturation constant [ppmv]

King (1992) reported that the V_{max} parameter correlates well with methane flux rates. This suggests that the supply of methane to the zone of oxidation may determine V_{max}. However, it has yet to be determined whether there exists a predictable relationship between V_{max} and the rate of CH_4 flux among diverse sites. Czepiel et al. (1996) observed the following linear relationship between V_{max} at the depth of maximum oxidation and the in situ CH_4 concentration:

$$V_{max} = 50 * C_{CH4}$$

where:

V_{max} = maximum CH_4 oxidation rate (nmol * $hour^{-1}$ * g dry soil weight^{-1}) at the depth of maximum oxidation
C_{CH4} = in situ soil gas CH_4 mixing ratio at a depth of 7.5 cm

In the column experiments we performed using a soil similar in texture to that investigated by Czepiel et al. (1996), we found that this equation accurately predicted the V_{max} values at the zone of maximum oxidation. Kinetic parameter values from our laboratory experiments, together with values generated by others, are presented in Table 1.

However, examination of some of the V_{max} values for CH_4 oxidation reported by others reveals that this relationship is applicable only to certain soil types. For example, Nozhevnikova et al. (1992) report a V_{max} of 25000 nmol * $hour^{-1}$ * g $d.w.^{-1}$ in a sandy clay landfill cover. The above equation would imply that their soil was exposed to a CH_4 concentration of 500%, which is impossible. The highest CH_4 oxidation V_{max} value observed in the U of C study was 1944 $g*m^{-2}*day^{-1}$.

Once a relationship between a soil's microbial kinetic parameters and its physical properties and gas concentrations can be established, it will be possible to quantify the amount of CH_4 a given soil cover can

Table 1: Kinetic parameters for methane oxidation

Soil origin and type	V_{max} (nmol*h^{-1}*g dry soil^{-1})	K_{app} (ppm)	Reference
Forest soil above natural gas source in Switzerland	44500 ± 12	100 000 ± 21	Bender and Conrad, 1992.
New Hampshire landfill Sandy clay loam	40 – 2594	195-5847	Czepiel et al., 1996.
Essex, UK landfill Coarse sand	998 ± 58	3793	Kightley et al., 1995.
Moscow landfill Sandy clay mixture	5000 – 25000	n/a	Nozhevnikova et al., 1993.
Calgary Springbank landfill Sandy loam	1802	7759	U of C Study

oxidize for a given methane flux-rate and set of climatic variables (by using a reactive-transport model in conjunction with the BROOK90 hydrologic model). One can then select the optimal soil type and thickness for an oxidative cover or biofilter.

6 APPLICATIONS OF SOIL METHANOTROPHY

6.1 *Biofilter Medium*

Our column experiments, and research by others, indicate that significant quantities of CH_4 can be oxidized in soil, and therefore methanotrophy provides a means of treating casing gas at many oil well and landfill sites, instead of venting the gas directly to the atmosphere. While all of our research has been on soils as a biofilter media, three groups of European researchers have been conducting experiments to assess the efficacy of compost as methane biofiltration media.

Straka et al. (1999) recently reported CH_4 oxidation rates in a passively aerated compost biofilter that are two orders of magnitude higher than those observed in the soils tested in the U of C study. Based on O_2 mass transfer limitations and the maximum V_{max} values for CH_4 oxidation reported in literature, their reported oxidation rate of 23,760 g*m^{-2}*day^{-1} seems physically impossible.

Humer and Lechner (1999) have also reported higher CH_4 oxidation rates in a compost biofilter than in natural topsoil or in a humic garden soil biofilters. They observed methane oxidation rates that were between 272 and 362 g*m^{-2}*day^{-1} in a well decomposed waste compost. These rates are three times higher than the rates observed in the U of C study.

Finally, Dammann et al. (1999) also demonstrated that a compost biofilter media results in a high CH_4 oxidation rate, having observed molar oxidation rates that were an order of magnitude higher than those observed in the U of C soil or peat biofilters (Stein and Hettiaratchi, 1999). However, their compost biofilters were treating gas streams with inlet CH_4 concentrations that were less than 1/20 of the inlet concentrations in the U of C biofilters.

6.2 *Soil Migration and Surface Casing Gas*

Based on the results of research reported to date, including our soil column experiments, it appears that control of CH_4 emissions arising from soil and surface casing migration may be limited to sites where the CH_4 flux rates are lower than 120 g*m^{-2}day^{-1}. However, steady state CH_4 oxidation values as high 166 g*m^{-2}day^{-1} in a coarse sand have been reported by others (Kightley et al., 1996), demonstrating that a significant improvement in CH_4 oxidation efficiency can be attained through the selection of an optimal soil texture.

While biofiltration of CH_4 is highly efficient only under low flow rates, the obvious advantages to such systems would be their relatively low capital cost and low operating expense.

6.3 *Solution Gas*

Conceivably, solution gas could be treated by diverting it through a soil biofilter. Based on the above results, a passively aerated 10m x 10m biofilter might be capable of oxidising as much as 25 m^3 of CH_4 per day for 8 months of the year.

However, some researchers who have studied the kinetic parameters of soil methanotrophs have observed V_{max} values as high as 44,500 nmol*h^{-1}*g dry soil^{-1} (Bender and Conrad, 1992), which is significantly greater than the highest V_{max} observed in this study (i.e. 1800 nmol*h^{-1}*g dry $soil^{-1}$). This suggests that much higher CH_4 removal rates may be attainable in a bio-filter. However, at such a high CH_4 consumption rate the depth of O_2 penetration would be extremely shallow, reducing the areal extent of methane oxidation, as was observed in computer simulations of the soil biofilters. But if such a biofilter were somehow aerated, then the overall oxidation efficiency might be increased by an order of magnitude.

6.4 *Sanitary Landfills*

The landfill final cover at sanitary landfills play an important role in minimizing infiltration into waste layers, thereby is crucial for controlling leachate production. A modified cover incorporating biofiltration should ensure leachate control function is not jeapordized. A cap consisting of a vegetative top soil layer and a sub-surface support layer could ensure both optimum biofiltration and minimum rainwater infiltration. A thick top soil layer containing a good growth of vegetation will oxidize methane as well as promote evapotranspiration. The primary differences between this final cover and the typical cover design used by landfill engineers are; elimination of the low permeable barrier layer and increased thickness of the top soil layer. The elimination of the barrier layer ensures even flow of gas across the landfill cover.

7 CONCLUSIONS

The application of microbial oxidation for controlling CH_4 emissions from soil migration at well sites and landfills is possible, and is suited for sites where the CH_4 flux rates are as high as 166 g*$m^{-2}$$day^{-1}$. For optimum results, the top soil layer should be selected and maintained to provide ideal moisture content, oxygen penetration etc. However, such conditions could be site specific. For example in a climate that is quite droughty, a soil that is more capable of retaining water is required.

Conceivably, solution gas and casing vent flow gas could be treated by diverting it through a soil biofilter, provided that its flow rate is low enough. For example, a 10m x 10m x 1m soil biofilter could treat a CH_4 flow of up to 25m^3/day (for 8 out of 12 months of the year). However, if such a biofilter could somehow be aerated, it might be possible to increase this amount by an order of magnitude.

A landfill cap should be modified to ensure a thick layer of top soil and good growth of grass are maintained. For even flow of gas through the cover, the barrier layer should be eliminated.

8 ACKNOWLEDGEMENTS

Funding for this research is being provided by the National Science and Engineering Research Council of Canada, Husky Oil Operations Limited, and the Canadian Association of Petroleum Producers. The authors would also like to thank Ron Schmitz of Husky Oil Operations Limited for his contribution to this research.

9 REFERENCES

Augenstein, D. And J. Pacey. 1996. Landfill Gas Energy Utilisation: Technology Options and Case Studies. EPA-600/R-92-116. US Environmental Protection Agency. Office of Air and Radiation.

Bender, M. and R. Conrad. 1994. "Methane oxidation activity in various soils and freshwater sediments: Occurrence, characteristics, vertical profiles and distribution on grain size fractions." J. Geophys. Res. 99(D8):16531-16540.

Czepiel, P.M., B. Mosher, P.M. Crill, and R.C. Harriss. 1996. "Quantifying the effect of oxidation on landfill methane emissions." J. Geophys. Res. 101:16721-16729.

Erno, B. and R. Schmitz. 1996. "Measurements of soil gas migration around oil and gas wells in Lloydminster area." Journal of Canadian Petroleum and Technology. 35(7):37-45.

Haber, C.L., L.N. Allem, S. Zhao and R.S. Hanson. 1983. "Methyltrophic bacteria: biochemical diversity and genetics." Science. 221:1147-1153.

Holford, M. 1998. "An Assessment of Alternatives to Flaring", M. Eng. Thesis, University of Calgary, Calgary, Alberta.

IPCC (Intergovernmental Panel on Climate Change). 1996a. "Summary for Policy Makers." In Climate Change 1995: The Science of Climate Change. Eds. J.T. Houghton, L.G. Meira Filho, B.A. Callander, N. Harris, A. Katternberg and K. Maskell. Cambridge University Press, UK.

Kightley, D., D. Nedwell, and M. Cooper. 1995. "Capacity for methane oxidation in landfill cover soils measured in laboratory-scale microcosms." Appl. Environ. Microbiol. 61(2):592-601.

Nozhevnikova, Alla N., A.B. Lifshitz, V.S. Lebedev, and G.A. Zavarzin. 1993. "Emission of methane into the atmosphere from landfills in the former USSR." Chemosphere. 26:401-417.

Schmitz, R., D.V. Stempvoort, and B. Erno. 1994. "Gas migration research—working towards risk-based management." Paper presented at the 4th Annual Symposium on Groundwater and Soil Remediation. Sept. 21-23, 1994. Calgary, AB.

Stein, V.B. and J.P.A. Hettiaratchi. 1997. Methane Oxidation in Soils as a Tool for Reducing Greenhouse Gas Emissions. In proceedings Emerging Air Issues for the 21st Century; an International Speciality Conf., ASWMA/ASPB/APEGGA, Calgary, Alta, Sept. 1997

To, N.M., J.P.A. Hettiaratchi, J.W. Shaw, C. Chomsurin, and C. Visvanathan. 1997. A Methodology for Determining Gas Source Strength. In proceedings of Emerging Air Issues for the 21st Century; an International Speciality Conference, ASWMA/ASPB/APEGGA, Calgary, Alta, Sept. 1997

Topp, E. and R.S. Hanson. 1991. "Methane Oxidising Bacteria" In Microbial Production and Consumption of Greenhouse Gases: Methane, Nitrogen Oxides, and Halomethanes. Eds. John E. Rogers and William B. Whitman. American Society for Microbiology, Washington D.C.

US EPA 1993. Solid Waste Disposal Facility Criteria: Technical Manual. US Environmental Protection Agency. EPA530-R-93-017

Whalen, S.C., W.S. Reeburgh, and K.A. Sandbeck. 1990. Rapid Methane Oxidation in a Landfill Cover Soil. Applied Environmental Microbiology. 56(11): 3405-3411.

Environmental Issues and Management of Waste in Energy and Mineral Production, Singhal & Mehrotra (eds)
© 2000 Balkema, Rotterdam, ISBN 90 5809 085 X

Kinetics and spectroscopy of Pb immobilization on rock phosphate at constant pH

R. Melamed
CETEM / CNPq, Center for Mineral Technology, Rio de Janeiro, Brazil

P. Self & R. Smart
IWRI-Ian Wark Research Institute, University of South Australia, Adelaide, S.A., Australia

ABSTRACT: We investigated the mechanisms of Pb immobilization on rock phosphate with the aim to utilize this material as a coating at disposal sites or as a filter for effluent treatment. The research involved kinetic studies of Pb reaction with the rock phosphate, at constant pH, in nitrogen atmosphere. Spectroscopy of the products of the reactions was carried out by SEM and XPS. The results indicate that at the natural rock phosphate pH (8.7), when no dissolved phosphate ions (P) are present, the immobilization of Pb is relatively slow, but efficient, and is related to surface complexation and nucleation mechanisms. At a pH value of 3.0, rock phosphate dissolves and approximately 3 mmol/L of P are present in solution. At this pH, the immobilization is instantaneous owing to the chemical precipitation of $PbHPO_4$, as indicated by the speciation program *MINTEQA2.* At a pH value of 5.7, the efficiency is relatively lower. The mechanism of immobilization at this pH involve a transition between chemical precipitation and surface complexation. When sulfate ions are present in solution, at low pH (pH 3.0), as in the case of acid mine drainage, the immobilization of Pb occurs via preferential chemical precipitation of anglesite, while P does not take part in the reaction.

INTRODUCTION

More attention has been given to the hazard that toxic elements can impose to the environment and to man. Thus, it became imperative that anthropogenic activities comply with environmental regulations. For instance, a big challenge for the mining-metallurgical industry has been the transport of toxic chemicals through soils, at disposal sites, resulting in the contamination of surface and ground waters. In this framework, "*in situ*" immobilization techniques are of particular interest because they are more cost-effective. One of these techniques consists of the application of phosphate rocks, at waste disposal sites, or in effluent treatment, for the immobilization of heavy metals.

The mechanisms of immobilization that control the transport and fate of any heavy metal consist basically of chemical precipitation and sorption at mineral surfaces, decreasing its bio-availability. While chemical precipitation depends on the complexation in solution and on the solubility of the compound formed, the sorption of heavy metals is a complex phenomena, in which mechanisms such as adsorption (which includes outer- or inner- sphere surface complexation), surface precipitation or co-precipitation, and intra-particle diffusion occur concurrently or simultaneously. The distinction between these mechanisms is of critical importance because they dictate the reactivity and mobility of any metal species.

An analyses of the industrial materials for use as a chemical and geochemical barrier for Uranium (Morrison & Spangler, 1992) indicated that hidroxyapatite was very efficient in Uranium extraction from the liquid effluent, while the rock phosphate tested was not. Ma et al. (1995) showed that the efficiency of phosphate rocks in the immobilization of Pb in soils depends on its solubility. Other studies suggested that an important retention mechanism for Pb (Suzuki et al., 1981; 1982; 1984), Zn (Ingram et al., 1992), and Cd (Christoffersen et al., 1988; Dalas & Koutsoukos, 1989) by hidroxyapatite occurs via ionic exchange with Ca. However, the results of Ma et al. (1995), above mentioned, implies a mechanism in which the phosphate ion (P) precipitates with Pb, forming a pyromophite type mineral.

Phosphate rocks are also efficient in the mitigation of acid mine/rock drainage (Kalin et al., 1995). In this respect, the addition of phosphate rocks to pyritic tailings results on a coating of $FePO_4$ which hinders pyrite oxidation (Evangelou, 1994; 1996; Georgopoulou et al., 1995).

In the case that the immobilization mechanism involves a sorption phenomenon, the retention of metals increases as the system pH is increased, producing a sigmoidal function, referred as "adsorption edge", which reflects the affinity of the metal specie for the mineral surface (Hingston et al., 1981; Sposito, 1984). It is expected that under acid mine drainage conditions, when metals are prompt to move due to leaching-induced low system pH, the phosphate anion will decrease metal mobility by increasing the pH of the system via ligand exchange with OH functional groups of mineral surfaces (Melamed et al., 1995), and by increasing the net negative surface charge of oxide surfaces (Rajan, 1976, 1979; Wann & Uehara, 1978; Bolan & Barrow, 1984; Sposito, 1984; Barrow, 1987; Melamed et al., 1994). In fact, it has been shown that the phosphate anion enhances the mobility of anions which form outer-sphere surface complexes such as Br^- e o NO_3^-, via electrostatic repulsion (Melamed et al., 1994), and increases drastically the mobility of other oxy-anions which form inner-sphere surface complexes such as arsenate, via competitive sorption (Peryea, 1991; Melamed et al., 1995). On the other hand, the phosphate anion decreases the mobility of K (Wann & Uehara, 1978), of Cd (Nicholson et al., 1994) and of Hg (Melamed & Villas Bôas, 1998).

This paper elucidates the immobilization mechanisms of Pb by a natural phosphate rock, at constant pH, emphasizing the kinetics of the reactions and the spectroscopy of the products.

MATERIALS AND METHODS

The goethite used in the batch sorption experiment was prepared following the methodology of Schwertmann & Cornell (1991).

The major components of the rock phosphate utilized in this study are given in Table 1. Its particle size distribution shows 95.5% of the material passing through the 325 mesh sieve (35 μm).

Batch sorption experiments of Pb sorption on the pure iron oxide system (goethite) were designed to show the effect of pH on the sorption of Pb in the absence of P. The reaction was carried out in centrifuge tubes, containing 0.2 g of goethite. KNO_3 was used as background electrolyte at a concentration of 0.1 M, and 0.1 M HNO_3 was used to adjust the system to the desired pH. To each tube it was added 0.1 mL of 4.83 mmol Pb/L, resulting on an initial Pb concentration of 86 μmol/L. The final solid to solution ratio was 1:28. Tubes were agitated for 48 hours. After this reaction period, the suspensions were filtered through a 0.22 μm filter paper and the supernatants were analyzed by ICP-AES. The amount sorbed was calculated from the difference between the amount added and the amount remaining after the reaction.

The kinetic studies of Pb immobilization were conducted in a batch reactor cell, constantly stirred for 24 hours, in nitrogen atmosphere, at constant pH, by using an automatic titrator. For each experiment, 2 g of rock phosphate were equilibrated with 0.1 *M* KNO_3. The pH values studied were 3.0, 3.7, 5.7 and 8.7, using 0.1 *M* HNO_3 and 0.1 *M* KOH for adjustments. At the lowest pH studied, Na_2SO_4 at a concentration of 200 mg S/L was also used as the background electrolyte to simulate acid mine drainage conditions. In this case, 0.1 *M* H_2SO_4 and 0.1 *M* NaOH were utilized for pH adjustments. When equilibrium of the system pH was established, a 1 mL solution of 1.6×10^{-4} M or 1.1×10^{-3} M Pb was added, at which point, the experiment was initiated. The volume of solution, at the beginning of the experiment was 300 mL, resulting in an initial 1:150 solid to solution ratio. However, the solid to solution ratio was recalculated for each run according to the volume of acid added for pH adjustments. During the entire reaction period, Eh values were monitored by a combination electrode. Samples of the suspension, with 15 mL each, were withdrawn by a syringe, at 0, 5 minutes, 15 minutes, 30 minutes, 1 hour, 2 hours, 6 hours and 24 hours after the reaction was initiated, and immediately filtered through 0.22 μm porous filter paper. After separation, the supernatant was analyzed for Pb, P, Ca, Fe and Al by ICP-AES and the solid material was analyzed by SEM and XPS.

Table 1: Chemical analyses of the Rock Phosphate

	%
P_2O_5	34.91
CaO	46.53
Fe_2O_3	4.63
SiO2	0.66
MgO	0.13
$BaSO_4$	4.27

RESULTS AND DISCUSSION

Figure 1 shows the effect of pH on Pb sorption at the goethite surface. These results indicate that the ability of cationic species to form inner-sphere surface complexes is related to the ability of the specie in solution to form hydroxides. The analogy between surface complexation and solution complexation was demonstrated by Hayes & Katz (1996) with the following reactions:

$$Me^{2+} + H_2O = Me(OH)^+ + H^+ \quad (1)$$

$$Me^{2+} + \equiv OH = \equiv OMe^+ + H^+ \quad (2)$$

where Me represents the cationic metal and ≡ represents the surface site.

In fact, it has been pointed out (Kinniburgh et al., 1976; McKenzie, 1980) that the affinity of the α-FeOOH surface for cations agrees with the first hydrolysis constant (pK_{Cu}=7.7, pK_{Pb}=7.7, pK_{Zn}=9.0, pK_{Co}=9.7, pK_{Ni}=9.9, pK_{Cd}=10.1), following the order: Cu = Pb > Zn > Ni > Co > Cd

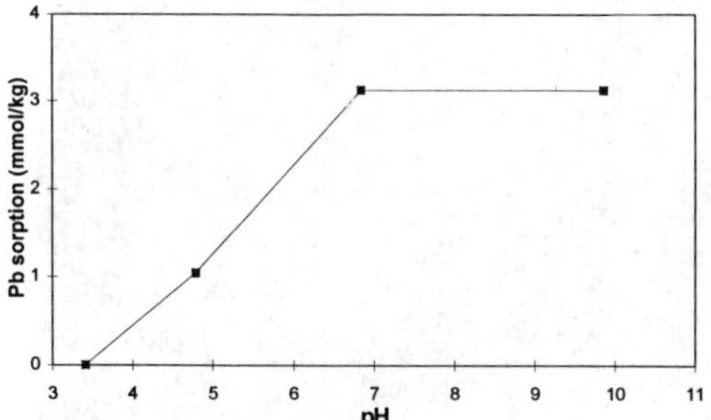

Figure 1. The sorption of Pb on the goethite surface at different pH values

Table 2. Ca and P dissolution from rock phosphate at different pH values

pH	Ca	P
	mg/L	
3.0	200	90
3.7	155	60
5.7	20	6
8.7	3	0.5

Table 2 shows the solubility of the phosphate rock at different pH values. The results indicate that the dissolution of P and Ca is non-stoiquiometric with a Ca:P ratio which decreases with pH.

The kinetics of the Pb immobilization on the rock phosphate, at pH 3.0, utilizing 0.1 M KNO_3 as background electrolyte and at a Pb initial concentration of 1,6x10^{-4} M is shown in Figure 2 and at a Pb initial concentration of 1,1x10^{-3} M in Figure 3. No matter the amount of Pb added, all Pb was immobilized by the rock phosphate at pH 3.0, reflecting the efficiency of this material at low pH. The results indicate also a very fast reaction rate.

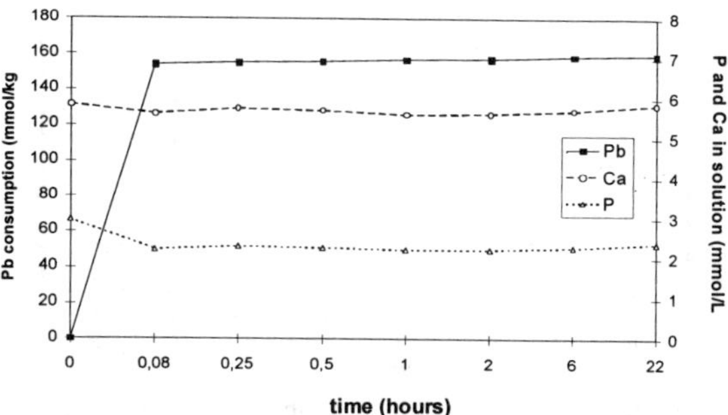

Figure 3. Kinetics of Pb immobilization on the rock phosphate at pH 3. [Pb]=1,1x$10^{-3}$$M$ and 0.1 M KNO_3

Figure 3, for instance, shows that after 5 minutes of reaction, approximately 95% of the total Pb added were consumed. Figures 4, 5 and 6, show the kinetics of Pb immobilization at pH values 3.7; 5.7 and 8.7, utilizing 0.1 M KNO_3 as background electrolyte and Pb initial concentration of 1.1x10^{-3} M. In general, no matter the equilibrium pH, the concentration of P in solution decreases and stabilizes concurrently with the consumption of Pb. However the concentration of Ca varies with the consumption of Pb. The concentration of Ca may decrease, remain constant or even increase. The results in Figures 3, 4, 5 and 6 indicate differences in the mechanisms of Pb immobilization due to the dissolution of P associated to the equilibrium pH of the system. At pH values of 3.0 and 3.7, chemical precipitation of Pb in solution occurs in an instantaneous reaction. However, at pH values of 5.7 and 8.7 the reaction is relatively slower with large differences in reaction rate at 6 hours and at 24 hours, and little P in solution, indicating Pb sorption at the surface of the rock phosphate.

Figure 7 summarizes the consumption of Pb by the rock phosphate, at all values of pH studied, indicating the high efficiency of the rock phosphate in Pb immobilization at low pH values (3.0 and 3.7)

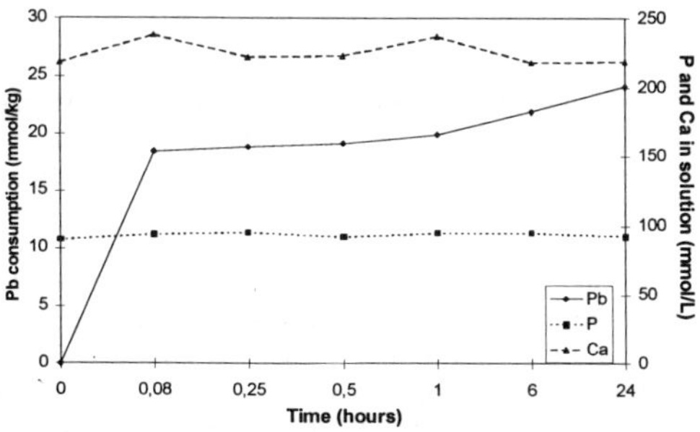

Figure 2. Kinetics of Pb immobilization on the rock phosphate at pH 3. [Pb]=1,6x$10^{-4}$$M$ and 0.1 M KNO_3

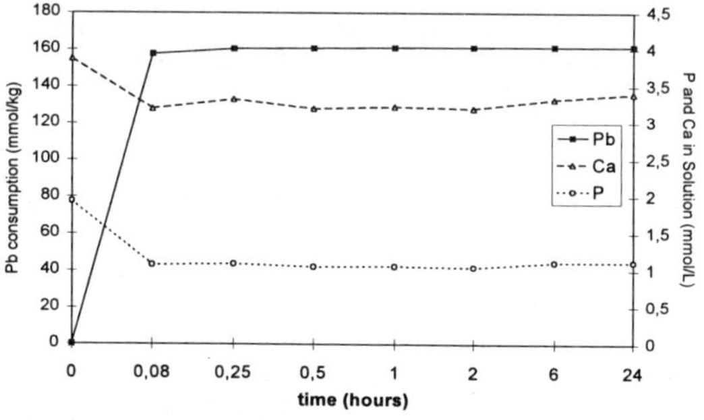

Figure 4. Kinetics of Pb immobilization on the rock phosphate at pH 3.7. [Pb]=1,1x$10^{-3}$$M$ and 0.1 M KNO_3

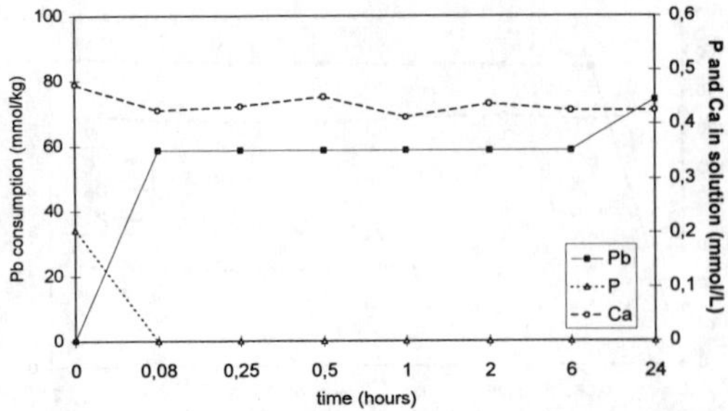

Figure 5. Kinetics of Pb immobilization on the rock phosphate at pH 5.7. [Pb]=$1{,}1\text{x}10^{-3}$ *M*, 0.1 *M* KNO_3

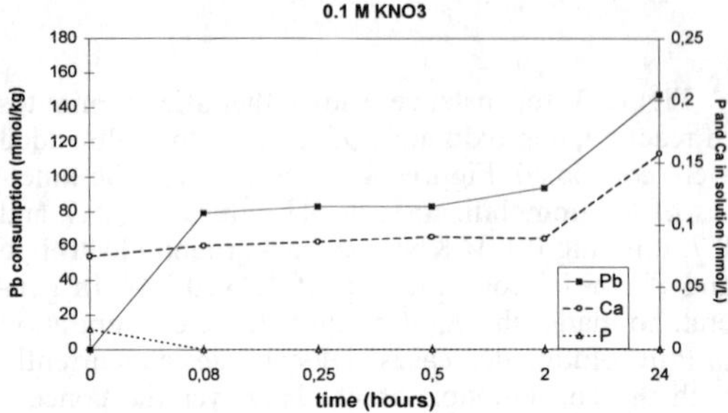

Figure 6. Kinetics of Pb immobilization on the rock phosphate at pH 8.7. [Pb]= $1{,}1\text{x}10^{-3}$ *M*, 0.1 *M* KNO_3

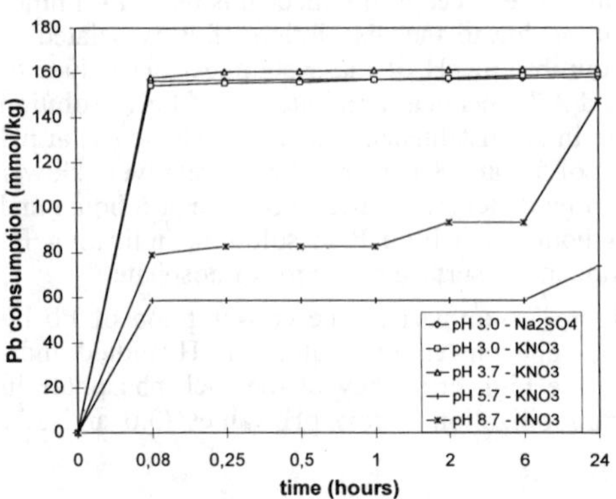

Figure 7. Kinetics of Pb immobilization on the rock phosphate at pH values of 3, 3.7, 5.7 and 8.7, with 0.1 *M* KNO_3.[Pb]= $1{,}1\text{x}10^{-3}$ *M.*

and the differences in mechanisms. At pH 5.7, the rock phosphate has a low efficiency reflected by low chemical precipitation and low sorption, as expected, based on the results of Figure 1.

The SEM images of the solid samples (Figure 8 - pH 3.0; Figure 9 - pH 3.7; Figure 10 - pH 5.7; Figure 11 - pH 8.7) indicate very different morphologies, with the formation of crystals with balloon shape at pH values 3.0 and 3.7, and the formation of crystals

Figure 8. SEM of Pb on Rock Phosphate, pH 3

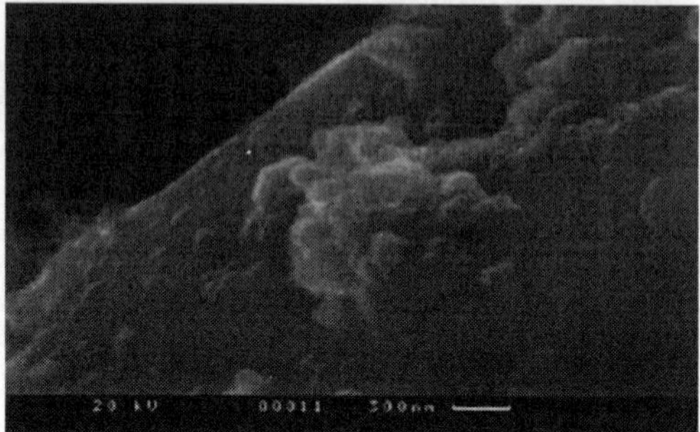

Figure 9. SEM of Pb on Rock Phosphate, pH 3.7

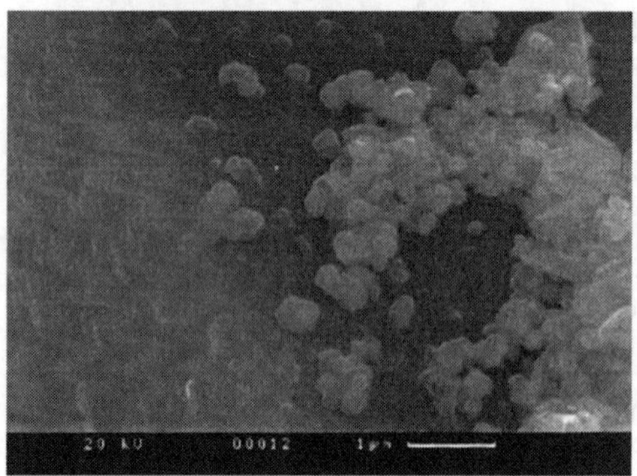

Figure 10. SEM of Pb on Rock Phosphate, pH 5.7

with hexagonal orientation at a pH value 8.7. At pH 5.7, the crystals have a less defined morphology which is attributed to a transition between chemical precipitation and sorption.

The SEM images and the respective EDX resulting from the addition of Pb to samples of solutions equilibrated at different pH values, without the solid material indicate that at pH values of 3.0 and 3.7 the crystals formed are composed exclusively by Pb and P. Ca did not co-precipitate in the reaction. The results of the speciation program *MINTEQA2* indicate the formation of $PbHPO_4$ at pH values of 3.0 and

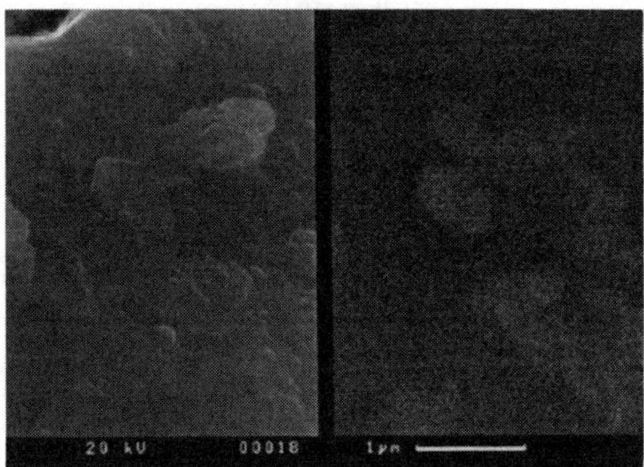

Figure 11. SEM of Pb on Rock Phosphate, pH 8.7

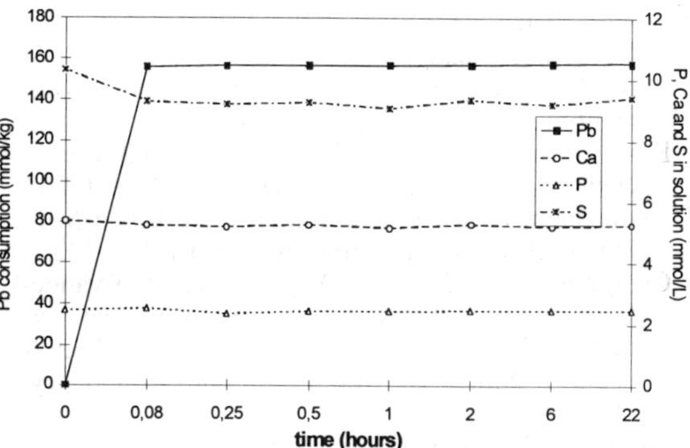

Figure 12. Kinetics of Pb reaction with rock phosphate. PH 3. Na_2SO_4 at 200 mg S/L [Pb]= $1,1x10^{-3}$ *M*

3.7; formation of $Pb_3(PO4)_2$ at pH 5.7 and $Pb(OH)_2$ and hydrapatite at pH 8.7. ICP-AES analysis of the liquid samples indicate the absence of chemical precipitation at pH 8.7, suggesting surface precipitation of Pb as proposed in Equation 2. These results are in agreement with the XPS analysis which indicate a high surface concentration of Pb, at pH 8.7, and a decrease in the surface concentration of Pb as the system pH decreases.

The quantity of acid or base added during the pH stabilization and adjustment after the addition of Pb at different pH values reveals that at all pH values studied, at the moment of Pb addition to the system, the pH decreases rapidly and KOH is delivered to maintain the system pH at its equilibrium value. At pH values of 3.0 and 3.7, a dissolution-precipitation of the rock phosphate occurs with the precipitation of Pb and P in solution in the form of a pyromophite like mineral (Ma et al., 1995), in accordance to the following reactions:

$$Ca_{10}(PO_4)_6X©+12H^+ \xrightarrow{\text{dissolution}} 10Ca^{2+}+6H_2PO_4^-+2X^- \quad (3)$$

$$10Pb^{2+}+6H_2PO_4^-+2X^- \xrightarrow{\text{precipitation}} Pb_{10}(PO_4)_6X_2(C)+12H^+ \quad (4)$$

or

$$Ca_{10}(PO_4)_3(CO_3)_3F(OH)(C)+6H^+ \xrightarrow{\text{dissolution}} 10Ca^{2+}+3H_2PO_4^-+3(CO_3)^{2-}+F^-+OH^- \quad (5)$$

$$10Pb^{2+}+3H_2PO_4^-+3(CO_3)^{2-}F^-+OH^- \xrightarrow{\text{precipitation}} Pb_{10}(PO_4)_3(CO_3)_3F(OH)(C) + 6H^+ \quad (6)$$

When the kinetic studies involved the stabilization of Pb in the presence of sulfate, at pH 3.0, for the simulation of acid mine drainage, there was no P consumption, but the consumption of S from solution (Figure 12). The crystals formed have different morphology and composition as compared to the crystals formed in the absence of sulfate. The analogous results of SEM and respective EDX indicate the formation of crystals that do not contain P. These results are confirmed by the speciation program *MINTEQA2* which shows the formation of anglesite, preferentially. It is interesting to point out that a very low decrease in system pH occurred concurrently to the formation of anglesite.

CONCLUSIONS

This study shows the efficiency of rock phosphate on the immobilization of Pb at low pH values where phosphate ions are present in solution and the differences in reaction mechanism according to the solubility of the phosphate rock or to the presence of S in the system.

ACKNOWLEDGEMENTS

The main author is thankful to CNPq for the support given and to IWRI for the support, time spent and availability of equipment utilized in this study.

REFERENCES

Bolan, N.S. & Barrow, N.J. 1984. Modelling the the effect of adsorption of phosphate and other anions on the surface charge of variable charge oxides. J. Soil Sci. 35: 273-281.

Christoffersen, J., Christoffersen, M.R., Larsen, R., Rostrup, E., Tingsgaard, P., Andersen, O. & Grandjean, P. 1988. Calcif.Tissue Int. 42:331-339.

Dalas, E. & Koutsoukos, P.G. 1989. J. Chem. Soc., Faraday Trans. 1. 85:3159-3164.

Evangelou, V.P. 1994. Potential microencapsulation of pyrite by artificial inducement of $FePO_4$ coatings. Paper presented at the International Land Reclamation and Mine Drainage Conference and the Third International Conference on the Abatement of Acidic Drainage, Pittsburg, PA.

Evangelou, V.P. 1996. Pyrite oxidation inhibition in coal waste by PO_4 and H_2O_2 pH-buffered pretreatment. International Journal of Surface Mining, Reclamation and Environment 10: 135-142.

Georgopoulou, Z.J., Fytas, K., Soto, H., & Evangelou, V.P. 1995. Pyrrhotite coating to prevent oxidation. Paper presented at Sudbury'95, Conference on Minig and the Environment, Sudbury, Ontario.

Ingram, G.S., Horay, C.P.& Stead, W.J. 1992. Caries Res. 26: 248-253.

Kalin, M., Fyson, A., & Smith, M.P. 1995. Reduction of acidity in effluent from pyritic waste rock using natural phosphate rock. Paper presented in 27th CMP.

Ma, Q.Y., Logan, T.J. & Traina, S.J. 1995. Lead immobilization from aqueous solutions and contaminated soils using phosphate rocks. Environ. Sci. Technol. 29: 1118-1126.

Melamed, R., Jurinak, J.J & Dudley, L.M. 1994. Anion exclusion-pore water velocity interaction affecting transport of bromine through an Oxisol. Soil Sci. Soc. Am. J. 58: 1405-1410.

Melamed, R., Jurinak, J.J., & Dudley, L.M. 1995. Effect of adsorbed phosphate on transport of arenate through an Oxisol. Soil Sci. Soc. Am. J. 59: 1289-1294.

Morrison, S.J.& Spangler, R.R. 1992. Extraction of uranium and molybdenum from aqueous solutions: a survey of industrial materials for use in chemical barriers for uranium mill tailings remediation. Environ. Sci. Technol. 26: 1922-1931.

Nicholson, F.A., Jones, K.C., & Johnston, A.E. 1994. Effect of phosphate fertilizers and atmospheric deposition on long-term changes in the cadmium content of soils and crops. Environ. Sci. Technol. 28: 2170-2175.

Rajan, S.S.S. 1976. Changes in net surface charge of hydrous alumina with phosphate adsorption. Nature. 262: 45-46.

Rajan, S.S.S. 1979. Adsorption of selenite, phosphate and sulphate on hydrous alumina. J. Soil Sci. 30: 709-718.

Schwertmann, U. & Cornell, R.M. 1991. Iron oxides in the laboratory. Preparation and Characterization. WCH-PO box 101161, D-6940 Weinheim. Weinheim. New York. Basel. Cambridge.

Sposito, G. 1984. The surface chemistry of soils. Oxford Univ. Press, New York.

Suzuki, T., Hatsushika, T., Hayakawa, Y. 1981. J. Chem. Soc., Faraday Trans. 1. 77:1059-1062.

Suzuki, T., Hatsushika, T., Miyake, M. 1982. J. Chem. Soc., Faraday Trans. 1. 78:3605-3611.

Suzuki, T., Ishigaki, K., Miyake, M. 1984. J. Chem. Soc., Faraday Trans. 1. 80:3157-3165.

Xu, Y., Schwartz, F.W., & Traina, S.J. 1994. Sorption of Zn^{2+} and Cd^{2+} on hydroxyapatite surfaces. Environ. Sci. Technol. 28: 1472-1480.

Environmental Issues and Management of Waste in Energy and Mineral Production, Singhal & Mehrotra (eds)
© 2000 Balkema, Rotterdam, ISBN 90 5809 085 X

Silica micro encapsulation: An innovative commercial technology for the treatment of metal and radionuclide contamination in water and soil

P. Mitchell
Klean Earth Environmental Company, Penryn, UK

J.T. Rybock & A.L. Anderson
Klean Earth Environmental Company, Lynnwood, Wash., USA

ABSTRACT: Klean Earth Environmental Company (KEECO) has developed the Silica Micro Encapsulation (SME) technology to treat heavy metals and radionuclides in water and soil. Unlike conventional neutralization/precipitation methods, SME encapsulates the contaminants in a permanent silica matrix resistant to degradation under even extreme environmental conditions. Encapsulated metals and radionuclides are effectively immobilized, minimising the potential for environmental contamination and impacts on human or ecosystem health. The effectiveness of SME has been proven through independent reviews, laboratory and field trials and commercial contracts, and the technology can be used to control and prevent acid drainage and the transport of soluble metals from mine sites, tailings areas, landfills and industrial sites. Successful demonstrations in the treatment of sediments and in brownfield redevelopment, treatment of metal-finishing wastewaters, and control of hazardous, low-level, and mixed waste at DOE/DOD sites and commercial nuclear power plants have also been undertaken. This paper describes the reactions involved in the SME process, the methods by which SME chemicals are introduced to various media, and recent project applications relevant to the cost effective remediation and prevention of environmental problems arising from energy and mineral production.

1 INTRODUCTION

Metals contamination and pollution is a widespread and growing problem in the natural and built environment. Industrial activities of various kinds have issues relating to both the chronic (long-term) and acute (short-term) release of metals. In addition to the ongoing consequences of operational-related releases, metals have also accumulated from past industrial activity at concentrations that may be harmful to humans and natural resources. In the absence of fission or fusion reactions, metals cannot be destroyed or transmuted. Therefore, where feasible, industrial practices should be modified to prevent or minimize the quantity or toxicity of metal releases (e.g. through the implementation of improved "housekeeping" measures, better environmental management systems, or clean and cleaner technologies). In many cases, however, damage to the natural and built environment has already occurred, while in others the implementation of better management or cleaner technologies still results in the generation of significant residual wastes that are contaminated by toxic metals.

Whereas metals cannot be destroyed, the potential to control their toxicity, bioavailability and mobility does exist, and a number of technologies exist which offer varying degrees of success in this regard. Unfortunately, conventional technologies have not performed well in either prevention or control of metals contamination. For example, approximately 75% of USA Superfund sites contain metals as a form of contamination with lead the most common, followed by chromium, arsenic, zinc, cadmium, copper and mercury (Evanko & Dzombak 1997). The most common remedial approaches to metals contamination at both Superfund and brownfield sites are simple containment (capping) or excavation and disposal at an offsite landfill (EPA 1999). In some cases, soils are stabilized with cement, fly ash, blast furnace slag or an organic binder prior to disposal. The result is that most of the metals are retained in their original state of contamination, although a portion may be temporarily immobilized by the transient precipitation of chemically unstable hydroxides within the solid matrix (Shively et al. 1986). Over time, as the cap weathers and degrades, and the liners fail through physically-, chemically- or biologically-mediated processes (as eventually will any human construction that is not subject to regular maintenance), the pH and redox characteristics of the soil will change, leading to a real risk of metal remobilization from the unstable precipitates, with all the toxicological and liability issues that such an event brings.

Acid rock drainage (ARD) is another metals-related problem – in fact, one of the largest and most intractable problems facing the non-ferrous metal and coal mining sectors. Once initiated, the cycle of chemically and biologically mediated and catalyzed reactions that lead to ARD generation is often difficult, if not impossible, to stop. Experience in historic mining regions of the world indicates that the problem may persist for centuries at certain sites. Continuing global depletion of non-acid generating oxide ore deposits and the increasing dominance of sulfide ores can only lead to an increased potential for ARD incidents unless improved preventative and treatment options are developed (Mitchell & Potter 1999).

Yet ARD treatment has been dominated for many years by the application of lime in a number of basic and more sophisticated guises. The drawbacks and deficiencies of lime-based treatment systems vary according to site-specific factors but generally include high capital, operating and maintenance costs, chemical instability and poor handling characteristics of treatment sludge, limited potential for resource recovery, and long-term liability associated with disposal of the sludge (Mitchell et al. 1999). Other conventional approaches to ARD treatment - including additions of caustic soda or magnesium hydroxide - generally fare no better than lime in terms of environmental performance and total project cost.

To an ever increasing extent, the public, politicians and regulatory agencies are expressing their dissatisfaction with conventional means of addressing society's ongoing problems with metal contamination and pollution in the built environment. Whether the situation involves lime-based treatment of ARD, cement-based treatment of metals-contaminated sludges, or a simple "dig and haul" procedure, the "solution" is often little more than a temporary fix to a large and growing problem. If standard conventional technologies continue to dominate, stakeholders in developed and developing countries face the real prospect the risks and liabilities associated with metal contaminants returning in the future, at an environmental and financial cost that few will wish to bear. Furthermore, few if any of the conventional standards meet the tenets of sustainable development and the concept of not passing environmental problems to future generations.

2 SILICA MICRO ENCAPSULATION: A BETTER ALTERNATIVE

In response to the known drawbacks of conventional metals treatment systems, Klean Earth Environmental Company (KEECO) of Lynnwood, Washington (USA) has developed a high-performance, low-cost technology for the prevention and treatment of metals-contaminated waters and soils. Silica Micro Encapsulation (SME) does precisely what its name implies – it encapsulates metals in an impervious microscopic silica matrix (essentially locking the metals in very small sand-like particles) which prevents the metals from migrating or otherwise adversely affecting human health or the environment. The solution to metals contamination in the environment may reside in one of the most common, durable and inert substances on earth – silica.

SME normally achieves control of contaminants in a single step, without the need for pre-treatment with chemicals or post-treatment flocculation or filtration. The reaction begins with a pH adjustment that initiates the precipitation of heavy metals from water (including pore water in solid media). An electrokinetic reaction promotes metal hydroxyl formation and condensation polymerization, resulting in the chemisorption of the metals into a three-dimensional structure, or matrix, composed primarily of silica. These microscopic matrices contain no fissures or fractures, completely surround the metal precipitates, and continue to strengthen with time. The bound metal precipitates are in the form of a fast-settling, sand-like product, which is environmentally benign and resistant to degradation under even extreme environmental conditions. Due to the coarse size and high density of the particles (relative to precipitates resulting from liming, for example), no flocculent is required to induce settling, which occurs naturally at a high rate. Contrary to conventional treatment sludges that can seriously degrade over time, the SME silica matrix continues to strengthen, further isolating contaminants from the environment.

SME is a very robust technology that has been demonstrated to work effectively on heavy metals (such as chromium, copper, lead, mercury and zinc), metalloids (such as arsenic), and radionuclides (such as uranium). It can be applied to surface and groundwaters, wastewaters, sediments, sludges, soils, mine tailings, and other complex media. In addition to the control of metals, SME chemicals have been shown to reduce dissolved solids (such as sulfates) and to degrade hydrocarbons (such as gasoline and fuel oil) and other organic chemicals through a high-energy oxidation process.

3 COMPONENTS OF THE SME TECHNOLOGY

3.1 *Chemical products and their applications*

KEECO has produced three SME reagents, each a unique proprietary calcium/silica-based formulation, manufactured as a powder and applied in dry or slurried form. The names and applications of each of these chemical reagents are as follows:

KB-1™ for aqueous media:
- Neutralization of ARD and mine process water
- Treatment of metal plating and other industrial wastewaters
- Treatment of stormwater
- Polishing of municipal and industrial water supplies
- Treatment of landfill leachate
- Recovery of metal resources

KB-SEA™ for solid media:
- Onsite remediation of contaminated soil
- Stabilization of sediments
- Stabilization of industrial and municipal sludges
- Silica coating to prevent or control future acid generation

META-LOCK™ for radionuclides and difficult-to-treat wastes:
- Conversion of mixed waste to rad waste
- Volume reduction of rad waste
- Reuse/recycling of treated wastewaters
- Combined metals encapsulation and hydrocarbon oxidation

This paper emphasizes the application of KB-1™ and KB-SEA™ to waste management in the energy and mineral production industries, although META-LOCK™ has considerable potential in mitigating problems associated with the generation and management of wastes from uranium mining operations.

3.2 *Delivery and mixing systems*

For delivery of SME chemicals to aqueous waste streams in slurry form, KEECO developed the K-250 shear-mixing unit. This machine blends chemical powder with a bleed stream from the main wastewater flow. Efficient mixing is accomplished through the blending action of a high shear mixer to maximize chemical reactivity of the slurry and the use of compressed air to inject the slurry into the main waste stream. The sheared chemical slurry is dispersed through a chemically resistant, ventilated plastic manifold to facilitate rapid dispersion of the reagent throughout the stream causing an immediate reaction between dissolved metals and the SME reagent. A single K-250 has been demonstrated in field trials to treat a flow of 2,000 gallons per minute (gpm), and a parallel configuration of units could treat flows many times higher. This capability significantly reduces the capital outlay that is often associated with systems that must respond to seasonal or production-related changes in flow.

Another advantage of the K-250 unit is that it can be mobilized to a site and made operational in a matter of a few days. For many situations involving long-term, continuous flow of wastewater, however, a slurry system may not be the best method for delivery of SME chemicals. Not only do the silica-based chemicals cause wear on certain moving parts and require regular maintenance, but the chemical feed rate using a slurry system is usually much higher than stoichiometric calculations would indicate. To overcome these two issues, KEECO developed and patented the K-series mixers for high-efficiency, dry chemical injection.

The K-10™ mixing system is a portable device designed to treat wastewater flows of 15-20 gpm, while the K-500 ™ system is much larger and can handle flows of approximately 500 to 800 gpm. Both the K-10™ and the K-500 ™ use little power (<0.4 and <2 horsepower, respectively) and, with minimal moving parts, require very little maintenance. In both systems, untreated water enters the top of an inverted hemisphere mixing chamber through a spring-loaded valve and forms a thin film along the inner surface of the chamber. Chemical is moved from the reagent hopper by a pH-controlled variable speed auger onto a rotating powder beater in the center of the mixing chamber, projecting the chemical as a fine smoke-like dust into the thin film of water. With the high surface area and the fine pulverization of the chemical, contact between the wastewater and SME chemical occurs immediately and the silica encapsulation process begins in the chamber. Upon leaving the unit, a retention time of only a few minutes is usually sufficient for the encapsulated metals to settle out before the treated water can be safely discharged.

For applications of SME chemicals to soils, sediments and sludges, delivery and mixing methods will depend on the particular situation. Where relatively dry materials have been excavated, mixing is easily accomplished in a standard pug mill or ribbon blender – after a reaction period of one to two days, the treated material can spread back onto the site. Where the contamination is relatively shallow, SME chemicals can be applied and mixed in-situ using tillers and other conventional land farming equipment. Where the contamination is deeper and excavation is not feasible, chemicals could be introduced in-situ using auger mixers or pressure injectors. If the sediment or sludge has a high water content, it may be advantageous to treat the waste using a K-series mixer (described above for aqueous wastes).

4 PROJECT EXPERIENCE

KEECO believes that SME represents the first serious alternative to conventional technologies for preventing and treating metals contamination. The interest in this technology and its robust nature to address a wide range of media, site and contaminant conditions is confirmed by an upsurge in the last few years of successful field trials and commercial projects, as summarized below.

4.1 *Bunker Hill Mine*

The Bunker Hill Mine in Idaho was one of the largest lead and zinc mines in the United States and has been in virtually continuous operation for over 100 years. With 31 levels, over 150 miles of drifts, 6 miles of major inclined shafts and a total volume of disturbed ground reaching approximately 5 cubic miles (Trexler, 1975; Eckwright, 1982), the mine is an archetypal example of the long-term consequences of failing to manage and plan for the prevention and control of ARD from the outset and also of the difficulties of retrospectively implementing preventative measures in underground workings once the process of acid generation is under way. Preventative measures have also been hindered by the infiltration and percolation of incipient rainfall through several thousand feet of sulphide-bearing rock overlying the mine workings. The mineralogy of the mine is complex and gives rise to a broad range of metals and metalloids in the acid discharge, including lead, zinc, iron, copper, manganese, cadmium, arsenic and antimony.

In 1991 the New Bunker Hill Mining Company purchased the mine and shortly thereafter the US Environmental Protection Agency (EPA) began to remediate the site under the Superfund program. An acid stream of approximately 500 gpm at a pH of 1.96 was emanating from a large underground working.. The new owner came under heavy pressure from EPA to improve the quality of water being discharged from the mine and thereby reduce the load on the central treatment facility. Standard treatment processes had proved unsuccessful or uneconomic (including the use of lime-based systems), and subsequently KEECO was invited to undertake trials at the mine.

Results of the trials were exceptional. A K-250 shear-mixing plant was set up and running within hours of its arrival at the mine site. At a dose rate of 2.0 grams per liter (g L^{-1}), the SME-treated waters achieved discharge criteria and were sufficiently clean to satisfy the drinking water criteria for arsenic, cadmium, chromium, lead and zinc (Table 1). The silica-encapsulated sediments also passed EPA's Toxicity Characteristic Leaching Procedure (TCLP), confirming the non-hazardous nature of the waste and eliminating the expense of hazardous waste disposal. It was considered feasible to return the encapsulated sediments to worked-out areas of the mine, removing the need for a surface disposal site.

Table 1. Bunker Hill Mine Water Quality (pH in standard units, metals in mg L^{-1})

Parameter	*Untreated Water*	*Treated Water*	*DW Standard*
pH	2.0	9.0†	6.5 – 8.5
Aluminum	0.585	0.027	0.05
Cadmium	0.401	<0.001	0.005
Copper	0.555	0.002	0.15
Iron	146.029	<0.001	
Lead	1.291	0.004	0.015
Zinc	199.645	<0.001	5.0

† Following liquid-solid separation, the supernatant pH returns to approx. 7.

KEECO and the New Bunker Hill Mining Company evaluated the possibility of recovering a zinc product from the contaminated water as an integral part of the treatment process. Using a primary treatment step with the addition of 1.5 g L^{-1} of KB-1™ to raise the pH to 5.5, and a secondary step where a further 0.5 g L^{-1} was added to bring the pH to 8.5, zinc-depleted and zinc-rich sediments were separated. Typical analysis of the primary and secondary stage sediments is shown in Table 2. The secondary stage sediment was composed of approximately one-third zinc, with the other major components being calcium, iron and manganese. Based on the chemical stability of the sediment, it is likely that a pyrometallurgical process would be the most appropriate route for metal recovery from SME-treated waters. While this approach would not be suitable for every operation, given the fact that no additional KB-1™ is required, it has tremendous potential at certain sites to generate additional revenue to offset the cost of treatment.

Table 2. Sediment Analysis by SEM-EDS (Wt%)

Element	*Primary Stage*	*Secondary Stage*
Aluminum	2.48	2.09
Cadmium	9.10	6.16
Iron	31.07	3.60
Magnesium	1.40	3.68
Manganese	1.36	7.72
Oxygen	37.98	32.47
Sulfur	4.80	6.58
Silicon	3.94	4.38
Zinc	7.44	33.05
Total	99.56	99.72

4.2 *Leviathan Mine*

Utilizing the same method of slurry injection described above for the Bunker Hill Mine application, the company executed an emergency response water treatment project at the Leviathan Mine, an abandoned mine site located in the Sierra Nevada Mountains of California. The project was completed for ARCO, which was obligated under a Consent Order by the EPA to treat several million gallons of ARD. The waters were collected throughout the winter and spring in lined containment ponds that, upon overflow, would discharge into an adjacent freshwater stream.

The acid waters collected within the lined ponds were generated from a number of sources including underground workings, seeps and surface runoff. Although they contained a host of metals and metalloids, in addition to a significant concentration of sulfate, the metals of most concern were aluminum, arsenic, copper, iron and nickel. Furthermore, the stability of the sludge generated from the treatment process was important as the expense of transporting and disposing the sludge off-site was prohibitive.

The site posed significant logistical challenges that made the possibility of mobilization and execution of traditional active treatment approaches unfeasible. The site lacked infrastructure and power supply, offered only limited access roads and staging areas for chemical storage and system layout, and presented extremely harsh weather conditions. Due to the compact size of the KEECO K-250 slurry injection unit, the small footprint of the water treatment and sludge handling systems, and the low power requirements, the SME technology offered a viable alternative to active chemical treatment at a site that previously had few options to economically accomplish such treatment objectives.

After treatment with KB-1™, the treated water was discharged into a sequence of three 20,000-gallon tanks. The tanks served as retention zones for sludge settling and were connected to allow the treated decant water to weir over from tank to tank, providing a total of approximately 45 minutes retention time measured from the inflow point on tank 1 to the exit portal on tank 3 (most of the sludge settled out in tank 1). Chemical analysis was performed on water from tank 3.

As shown in Table 3, SME treatment successfully reduced the concentrations of the key metals by over 99.5%, most below detectable limits. Furthermore, the sulfate concentration was reduced by as much as 89%.

Table 3. Leviathan Mine Water Quality (pH in standard units, dissolved metals in mg L^{-1})

Parameter	*Raw Water*	*Sample #1*	*Sample #2*	*Sample #3*
pH	2.5	9.5	8.8	7.5
Aluminum	1000	1.1	4.6	4.2
Arsenic	38	0.019	0.017	0.018
Copper	8.4	<0.02	<0.02	<0.02
Iron	2100	<0.1	<0.1	0.27
Nickel	22	<0.1	<0.1	0.1
Sulfate	14000	1500	NA	1800

Several sludge samples were collected from the three tanks. Each sample passed the TCLP action limits for all regulated metals, thereby offering the client a number of disposal options that are fiscally reasonable, eliminating the requirement to transport the sludge out of state for disposal at a hazardous waste repository.

4.3 *Wheal Jane Mine*

The sudden release of 10 million gallons of highly acidic, metal contaminated ARD from Wheal Jane Mine in Cornwall, UK was a significant event for the mining industry, regulators and other stakeholder groups, both in the UK and in a wider context. Following the release, and after two years of evaluating over 400 UK and global companies, the National Rivers Authority (since 1996, the Environment Agency) allowed KEECO to demonstrate the effectiveness of KB-1™ at Wheal Jane Mine. A K-500 dry feed injection plant applied the KB-1™ to the waters at a flow rate of 500 gpm.

Results from the project were exceptional (Table 4). All evaluated metals were removed to levels well below project criteria. The SME treatment process reduced levels of cadmium -- the metal of greatest concern -- to below the detection limit of 1.0 µg L^{-1}. Additionally, the sediments produced from the treatment process settled out within a four-minute residence time without the addition of flocculents or polymers and successfully passed TCLP for all evaluated metals, demonstrating that the sediment would require no further treatment and would not be considered a hazardous waste.

Table 4. Wheal Jane Mine Water Quality (mg L^{-1})

Parameter	*Untreated Water*	*Treated Water*	*Proposed Criteria*
Arsenic	0.279	0.018	0.05
Cadmium	0.066	<0.001	0.001
Copper	0.948	0.006	0.028
Iron	260	0.283	1.0
Zinc	104.2	0.123	0.5

4.4 *Meldon Quarry*

Located in southwest England, the Meldon Quarry is one of a number of quarries generating ARD with high concentrations of dissolved metals. The existing water treatment system relies on caustic soda that not only consistently fails to achieve the discharge criteria, but is potentially hazardous to workers and produces an ultrafine sludge with extremely poor settling characteristics. Using KEECO's K-10 dry-feed system in a small-scale field demonstration, the drainage was successfully treated with KB-1™ as shown in Table 5.

Table 5. Meldon Quarry Water Quality (pH in standard units, metals in mg L^{-1})

Parameter	*Untreated Water*	*Treated Water*	*DW Standard*
pH	3.51	6.2	6 - 9
Aluminum	37.12	0.11	0.5
Cadmium	0.01	<0.01	0.01
Copper	0.35	0.1	0.2
Iron	0.67	<0.01	1.1
Nickel	2.81	1.65	2.3
Zinc	3.13	0.56	2.5

The SME water treatment was followed by sludge separation. KEECO's process produced a dense, rapidly settling precipitate that, like the Wheal Jane sludge, settled in about four minutes without recourse to further chemical additives. As shown in Table 6, leach tests agreed with the Environment Agency demonstrated that the sludge would not require the Agency's consent for disposal and could be safely deposited on-site.

Table 6. Meldon Quarry Sludge Leach Testing (pH in standard units, metals in mg L^{-1})

Parameter	*Deionized Water Leach (pH 7)*	*Deionized Water and Sulfuric Acid Leach (pH 3.7)*
pH	9.74	9.80
Aluminum	14.24	14.19
Cadmium	<0.01	<0.01
Copper	<0.01	<0.01
Iron	<0.01	<0.01
Nickel	<0.01	<0.01
Zinc	<0.01	0.02

4.5 *Metal Finishing Wastes*

A metal finishing company in Everett, Washington had been treating its wastes using sodium hydroxide for pH adjustment to 9.0 and alum as a precipitant. However, treatment consistency was variable and permit violations occurred. KEECO developed a treatment process that included a sodium hydroxide pretreatment step to raise the pH from 1.7 to 7.0. KB-1™ was then added at 1.3 g $L^{-1}$to raise the pH to 7.25, and the supernatant was analyzed. As shown in Table 7, this treatment program successfully met the discharge criteria.

This project demonstrated that KB-1™ can replace alum in the metal finishing treatment process with a KB-1™ utilization rate at one-sixth of the alum dosage, with an additional reduction in sodium hydroxide consumption. Because the unit costs of alum and KB-1™ were similar and KB-1™ provided better treatment, the facility subsequently switched to the SME treatment process.

Table 7. Metal plating solution treatment with KB-1™ (pH in standard units, metals in mg L^{-1})

Parameter	*Raw Conc.*	*Treated Conc.*	*Criteria*
pH	1.7	7.25	5 – 10
Chromium	377	0.48	1.71
Copper	115	0.09	2.07
Nickel	31	0.50	2.38
Zinc	23	0.07	1.48

4.6 *Industrial Stormwater Runoff*

KEECO was asked to demonstrate its SME technology on stormwater runoff at a former metal smelting site in western Washington. Contaminants of concern were arsenic, copper, lead, and zinc. The pilot-scale demonstration treated 15 gpm using META-LOCK™ and a K-10 mixer.

The treatment system consisted of ferric sulfate pretreatment (co-precipitant), META-LOCK™ treatment, sludge separation, filtration, and acid neutralization. A pH of approximately 9.5 was required to achieve zinc precipitation. Acid neutralization was required to meet the effluent pH requirements for discharge. The metal contaminants were reduced to levels well below those set by the discharge criteria (Table 8). Also, the treatment sludge passed TCLP tests, with arsenic 0.9 mg L^{-1} and lead <0.02 mg L^{-1} (criteria for both are 5.0 mg L^{-1}).

Table 8. Water Quality Test Results (mg L^{-1})

Metal	*Raw Water Conc.*	*Average Conc.*	*Discharge Criteria*
Arsenic	9.2	0.03	0.1
Copper	0.71	<0.02	0.1
Lead	0.070	<0.05	0.1
Zinc	3.0	<0.02	0.1

4.7 *Sutter Gold Mine Tailings*

Sutter Gold Mining Company in California requested KEECO to evaluate the feasibility of amending mine tailings with KB-SEA™ prior to their disposal to reduce or prevent future acid generation. The contaminant of concern was arsenic.

In a bench-scale study, samples of whole tails and the ⁻325 mesh fraction were analyzed under an Aqua Regia Digestion and EPA's Synthetic Precipitation Leaching Procedure (SPLP) for total and leachable arsenic concentrations. Following raw sample analyses, the samples were treated with varying amounts of SME product KB-SEA™ to reduce the leachable arsenic concentration to below 50 micrograms per liter ($\mu g\ L^{-1}$). As shown in Table 9, KB-SEA™ successfully achieved a reduction in the leachable arsenic concentration to below the stated goal in both the whole tails (using ≥2% KB-SEA™ by dry weight of the tails) and in the ⁻325 mesh fraction samples (using ≥3% KB-SEA™ by dry weight of tails).

Table 9. KB-SEA™ Amended Sample Analyses

KB-SEA™ Added (% dry weight)	*Whole Tails*		*⁻325 Mesh Tails*	
	Arsenic SPLP ($\mu g\ L^{-1}$)	% Reduction	Arsenic SPLP ($\mu g\ L^{-1}$)	% Reduction
0% - Untreated	96.8	N/A	95.5	N/A
1%	79.1	-19%	130	+27%
2%	31.6	-68%	73	-24%
3%	12.8	-87%	40	-42%
4%	4.6	-95%	NP	N/A

N/A = Not applicable. NP = Not performed due to insufficient sample quantity.

4.8 *Coeur d'Alene River Sediments*

Prior to the 1920s, mine and ore processing wastes from the Silver Valley – one of the largest producers of silver, lead and zinc in the world – were discharged directly into the South Fork of the Coeur d'Alene River in Idaho. Consequently, the Lower Coeur d'Alene River received tens of millions of metric tons of mine wastes that are now intermixed with sediments and deposited in the riverbed, riverbank and floodplain. Due partly to seepage of groundwater through the riverbanks, high loadings of lead, zinc and cadmium into the river present a threat to agricultural, recreational and ecological uses of the system.

With funding from EPA, the Idaho Water Resources Research Institute (IWRRI) at the University of Idaho conducted a demonstration program to compare remediation technologies for controlling seepage from these fluvially-deposited mine wastes. KEECO's SME technology was one of only four technologies that completed the program; the others were phosphate, apatite mineralization, and Apatite II technologies.

IWRRI sampled and analyzed test plot soils for metals leachability initially in April 1998 and then one month and eight months following the August 1998 field treatment program. The results are summarized in IWRRI's October 1999 report titled "Demonstration of *in-situ* metals-fixation technologies on the Coeur d'Alene River, Idaho."

The four technologies were compared according to six measures of their ability to control metals migration through the riverbanks. The SME technology was ranked as the sole "best performer" in two of the six categories and "among the best performers" in three others, resulting in a "best performer" ranking in 83% of the categories. The other three technologies were ranked among the "best performers" in 17%, 33% and 50% of the categories. The study also found that SME had no adverse effect on soil fertility parameters (N, P, K and organic matter).

4.9 *Tributyltin Study*

Cornwall College (UK) conducted a bench-scale laboratory treatability test in May 1999 evaluating the ability of various reagents (two unidentified organic substrates, vermiculite, KB-1™ and KB-SEA™) to treat tributyltin oxide (TBTO). Ten grams of each reagent were placed in 250 ml conical flasks with 100 ml of 10 mg L^{-1} TBTO solution for 24 hours under three different conditions - aerated, shaken, and undisturbed. The solutions were then filtered and analyzed for total tin by inductively coupled plasma-mass spectrometry.

As shown in Table 10, KB-1™ outperformed the other reagents under all three treatment protocols and reduced the tin in the aerated sample to less than the detection limit (0.1 mg L^{-1}). This test demonstrated the effectiveness of KB-1™ in treating dissolved TBT in aqueous media. The next phase of the study will involve direct application of KB-1™ or KB-SEA™ to TBT–contaminated sediments. The long-term effectiveness of the silica encapsulation in preventing future remobilization of the tin will also be evaluated.

Table 10. Treatment of a Tributyltin Oxide Solution

	Total Tin (mg L^{-1})		
Reagent	Aerated	Shaken	Undisturbed
Blank	13.4	14.6	15.5
KB-1™	<0.1	0.6	1.3
KB-SEA™	1.1	0.8	2.2
Organic # 1	2.8	2.9	5.9
Organic # 2	1.0	1.8	1.6
Vermiculite	8.2	9.6	8.9

4.10 *Delaware Sediments*

Historical deposits of lead contaminated sediments were encountered at an upland riverfront in a highly industrialized area of Delaware in November 1999. TCLP lead concentrations averaged 21 mg L^{-1} with a high value of 83 mg L^{-1}. EA Engineering, Science and Technology evaluated a range of treatment technologies and compared costs of three viable alternatives.

1 Off-Site disposal in a RCRA Subtitle C (hazardous waste) landfill ($750,000)
2 Off-site treatment and disposal in a RCRA Subtitle D (non-hazardous waste) landfill ($300,000)
3 On-site reuse of sediments treated with KEECO's SME technology ($200,000)

The Delaware state regulators reviewed the EA/KEECO treatability study and work plan and agreed that, if post-treatment sampling proved that the lead had been effectively fixed in silica and prevented from exceeding the 1.0 mg L^{-1}leach criterion, treated sediments could be returned to the site as "clean fill".

The selected SME chemical product, 300 ton-per-hour capacity pug mill, front-end loaders and ancillary equipment were mobilized to the site after the sediments were pre-screened to three inch minus. The loaders were used to both apply the chemicals directly into the piles of excavated sediment and then feed this material into the pug mill for more thorough mixing. The entire 2,000 cubic yards of sediments was treated in two-and-a-half days. Of the 40 post-treatment samples collected, 38 yielded undetectable TCLP lead values (at a detection limit of 0.4 mg L^{-1}) and 2 measured 0.7 mg L^{-1} or less. In addition to immobilizing the lead, SME treatment reduced moisture levels within the sediments from near 20 percent to less than 1 percent, which made the material much more suitable as fill. Based on the successful sample results, the treated sediment was approved to remain on-site.

5 CONCLUSIONS

The SME technology is unique in its permanent encapsulation of metal contaminants, which greatly reduces or eliminates the need for costly hazardous waste disposal and the environmental liabilities associated with future remobilization of metals. Under most conditions, the capital, operating and life cycle costs of SME treatment are lower – and sometimes by an order of magnitude or more – than conventional methods for addressing metals contamination (Anderson & Rybock 2000).

The energy and mining industries are relatively conservative when investing their funds in new technologies, which reflects in part the low profit margin of many companies in these sectors. However, as more stringent regulatory criteria are imposed and public awareness of environmental impacts of energy and mining operations increases, new technology must be incorporated into project design and closure plans. Evidenced by the successful results of several large-scale field applications, the SME technology represents an opportunity to meet the ever-increasing demand for better treatment of metal-contaminated wastes. It represents a substantive stepping-stone for industry to use in the course of strengthening its commitment to more environmentally responsible energy production, mining and waste management practices.

REFERENCES

Anderson, A. & J. Rybock 2000. Life cycle cost considerations in ARD control: SME versus lime and other conventional approaches. SME 2000, Salt Lake City, Utah.

Eckwright, T. S. 1982. Patterns of water movement in the Bunker Hill Mine. Unpublished M.Sc. thesis, University of Idaho.

EPA 1999. Treatment technologies for site cleanup: annual status report (ninth edition). Environmental Protection Agency, Office of Solid Waste and Emergency Response, EPA-542-R99-001.

Evanko, C. R. & D. A. Dzombak 1977. Remediation of metals-contaminated soils and groundwater. Ground-Water Remediation Technologies Analysis Center, Technology Evaluation Report TE-97-01, 53 pp.

Mitchell, P. & C. Potter 1999. The future of waste treatment in the mining industry. Proceedings of the Global Symposium on Recycling, Waste Treatment and Clean Technology (REWAS '99), San Sebastian, Spain.

Mitchell, P., J. Rybock & A. Wheaton 1999. Treatment and prevention of ARD using silica micro encapsulation. 1999 Proceedings, Mining and Reclamation for the Next Millennium, American Society for Surface Mining and Reclamation.

Mitchell, P., C. Potter & M. Watkins 2000. Quarrying and ARD: developing cost-effective solutions in a low profit margin arena. International Conference on Acid Rock Drainage (ICARD 2000), Denver, USA.

Shively, W., P. Bishop, D. Gress & T. Brown 1986. Leaching tests of heavy metals stabilized with portland cement. J. Water Pollution Control Federation, 38:234-241.

Trexler, B. D. 1975. The hydrogeology of acid production in a lead-zinc mine. Unpublished Ph.D. thesis, University of Idaho.

Environmental Issues and Management of Waste in Energy and Mineral Production, Singhal & Mehrotra (eds)
© 2000 Balkema, Rotterdam, ISBN 90 5809 085 X

Effects of recycling on sustainable supply for base metals

G. Mogi, T. Adachi & J. Yamatomi
The University of Tokyo, Japan

S. Mismi
Ube-Mitsubishi Cement Research Laboratory, Japan

ABSTRACT: Amount of potential resource and the rate of recycling as well as limitation of use by substitution are the most important keys for sustainable development and utilization of metals. Several factors are competing each other to sustain the demand, which basically depend on population and economical environment. The relative advantage of recycling compared to virgin metal depends on the technical development, which usually is hard to predict, to increase the efficiency of separation while reducing the cost.

In this system dynamics model, the local recycling system is embedded in the global supply-demand structure of base metals (copper, lead and zinc) and the effect of recycling to the supply of each base metal is analyzed for over the next fifty to hundred years. The effect of substitution for sustainable supply of copper is also evaluated.

1 INTRODUCTION

Since the report for the Club of Rome's project on the predicament of mankind "the limits to growth" (Meadows et. al. 1972) was published, a lot of efforts have been invested to point out the condition for sustainable development of the human society. As designated in "the global 2000 report to the president" (US Government 1980) or later also in "beyond the limits" (Meadows et. al. 1992), the keys for sustainability in raw materials are the rate of recycling, improvement in efficiency, extension in life of products and limitation of use. In this study, we concentrated to evaluate the effects of recycling and partly also the limitation of use due to backstop and substitution on sustainable supply for base metals as copper, lead and zinc.

2 OUTLINE OF THE MODEL

2.1 *Demand*

Presumed demand or consumption for each metal is exogenously given by applying the historical trend of the consumption against the world population to the predicted one by the United Nations. Figure 1 shows the historical trend of demand against world population for lead, zinc and copper. For lead, a case, in which the growth of its demand deviates negatively from the growth of population, is also considered.

We also made up a model, in which the metal will be substituted depending on its extraction cost. Even though metals in general are substituted with other metals or several different materials, we only considered a single type of material for substitution and set only one backstop price at which the substitution takes place. The purpose of this model is mainly to compare the difference in various demand structures.

2.2 *Supply*

Two sources of supply, namely primary production, which means extraction from the run-of-mine ore, and secondary production, which means extraction mainly from the scrap by recycling, are considered in the model. Figure 2 shows major factors influencing the competence of primary and secondary productions.

In the case of copper, twelve groups of primary producer countries, classified according to regional difference of the deposits such as physical potential

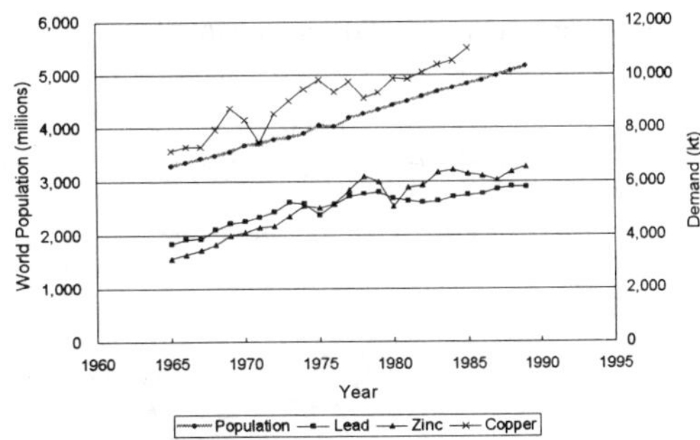

Figure 1. Historical trend of demand against world population for lead and zinc.

or economical environment, are competing each other based on extraction cost, which is the sum of the operating costs for mining and refinery. In case of lead and zinc, those are subdivided only into three groups due to their relatively small areal diversity.

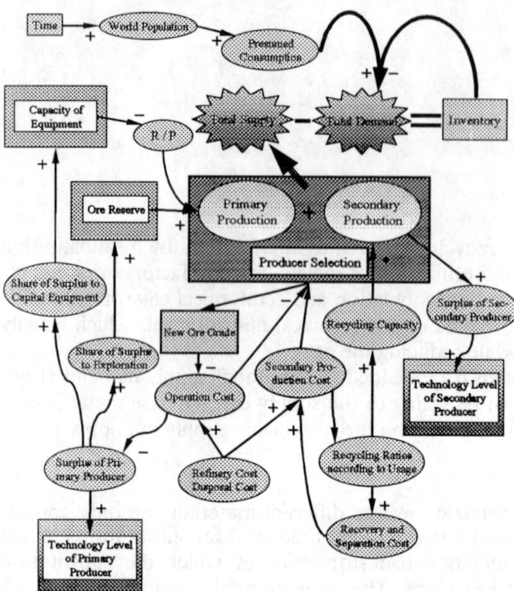

Figure 2. Major factors influencing the competence of primary and secondary productions.

Twelve groups for copper are followings:

1) Papua New Guinea and Asia except Japan
2) Australia
3) Canada
4) Chile and other American countries
5) Europe and Japan
6) Mexico
7) Peru
8) South Africa
9) US
10) Former USSR
11) Congo
12) Zambia and other African countries

Three groups for lead and zinc are followings:

1) North and South America
2) Asia and Oceania
3) Europe, Africa and ex-communist countries

Once the demand is presumed at a certain period, each primary producer group and the secondary producer are competing with each other and fulfill the demand, unless it exceeds the total production capacity. In Base Model 2, however, secondary production is intentionally promoted in consideration of present situation in industrialized countries and priority is given to secondary producer to fulfill the demand up to its relatively limited production capacity.

Figure 3 shows the principal flows and stocks of the model. Metal in the undiscovered reserves stock flows to identified reserves stock according to the discovery rate and flows further to the market according to the production or supply rate. The consumed metal flows to the disposal plant stock after certain delay time in the market. Then a part of it will be recycled and flows again to the market but this time as a secondary product. The rest will be disposed as waste and flows out as final disposal.

Primary production cost tends to rise according to the decrease in average grade of ore. The historical relationship between the weighted average ore grade and the cumulative production is used to estimate the average grade at certain point of time. Figure 4 shows the average ore grade-cumulative production relationship of copper in Canada. Lower limit of cut off grade is set as 0.1%, 1.0% and 1.0% for copper, lead and zinc, respectively.

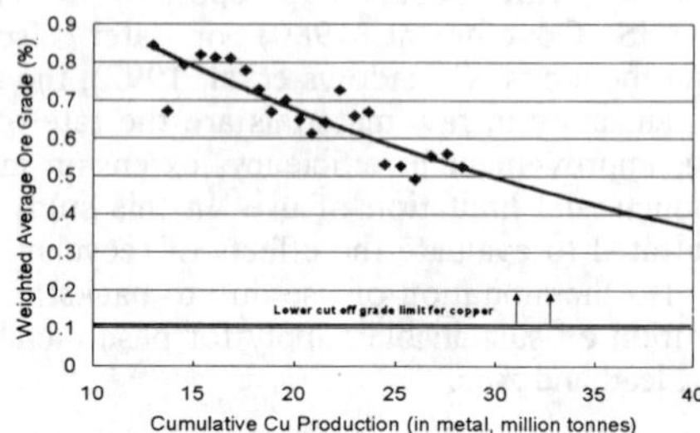

Figure 4. Average ore grade-Cumulative production relationship of copper in Canada.

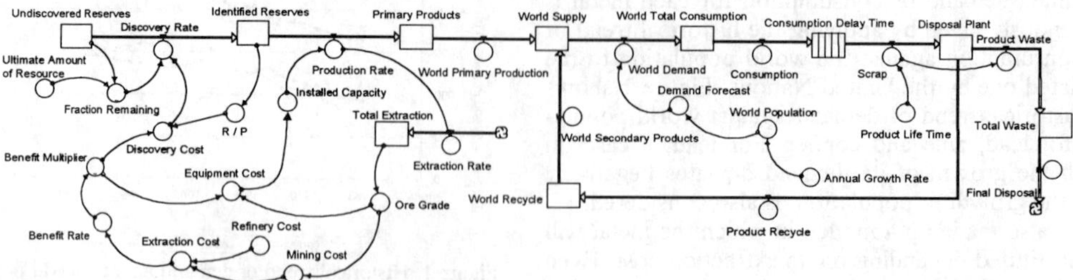

Figure 3. Principal flows and stocks of the system dynamics model.

On the other hand, technological development in mining and smelting, which depends on the cumulative investment in research, tends to curtail production cost (Behrens III 1973). The relationship between cumulative investment in R&D and technological level is expressed as a logistic curve in the model. Outcome of the research is assumed to spread over the world instantaneously and will be shared. The amount of investment in R&D depends on the amount of profit and capital as well as exploration investments of primary producers. Some of the R&D investment for smelting is assumed to come also from the secondary producers, who also intend to increase their competence.

2.3 *Recycling*

A part of the used metal will be salvaged from the market and recycled as secondary product. Secondary production cost consists mainly of salvage-separation cost and smelting cost. Smelting cost for secondary production is assumed to be equal to that for primary production. Technological development or investment in R&D either in salvage-separation sector and smelting sector also causes reduction in cost and increase the competitiveness of the secondary producers. The amount of investment in R&D mainly depends on the amount of their profit.

On the other hand, it is obvious that free competition is not the only motivation for recycling. Spirit of environmental responsibility, which is usually hard to evaluate economically, is definitely another driving force to promote recycling. On this account, reclamation cost, which will be added to the LME price of the metal to promote recycling, is introduced in the model.

2.4 *Substitution*

Another key factor for sustainability in raw material supply, in addition to promoting recycling, is limitation of use due to substitution. We made up a model in which a backstop for producing cost is taken into consideration. In this model, primary and secondary producers also compete with each other and the presumed demand will be fulfilled by those most cost efficient producers. But if once the marginal cost of the supplier exceeds the set backstop cost, substitution occurs and the rest of the demand will be supplied by the third source.

3 INDICATED SCENARIOS

3.1 *Lead*

Figures 5a, 5b show the results of the executions for lead, of which 5a is for the Base Model 1, where free competition takes place between primary and secondary producers and 5b for Base Model 2, where primary production is promoted under limited production capacity considering the present situation.

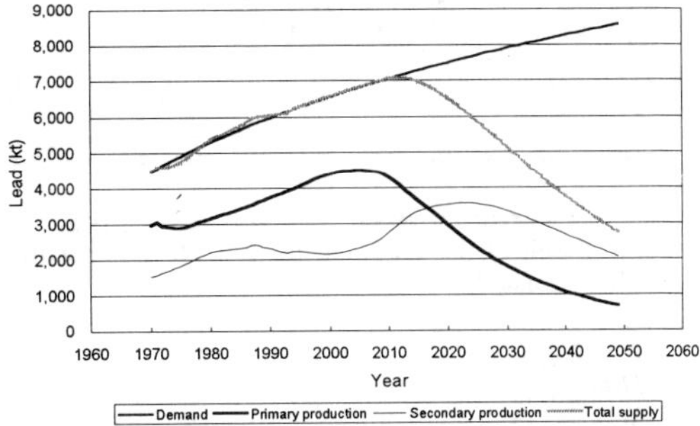

Figure 5a. Result of the execution for lead (free competition).

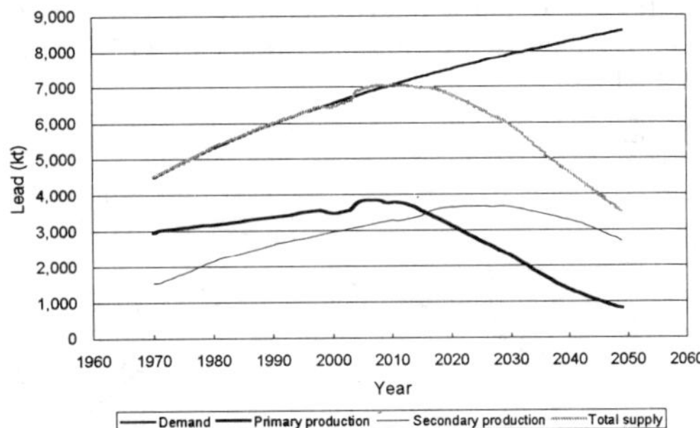

Figure 5b. Result of the execution for lead (recycling promoted).

Lead is starting to deplete at around year 2010 in either case due to the relatively scarce amount of its ultimate resources. Even though depletion is assumed to occur more gradually if recycling is intentionally promoted, the effect is quite limited as the present recycling rate of lead is already relatively high compared to other two metals and its remarkable growth cannot be expected even under promoting environment. Practically, at the time when depletion becomes obvious and the commodity get scarce in the market, the price will rise and the demand diminish to the level that even with declined production and any kind of probable substitution the market can reach the equilibrium.

The recent trend of growth in lead consumption within developed countries partly shows a negative deviation from the trend of population growth. Such trend is reflected in the model, whose result of execution is shown in Figure 6. In this case, the timing of the onset of scarcity is apparently postponed for more than twenty years, compared to the case in Figure 5a. But this scenario could be impractical as lead is usually accompanied by zinc, so that the level of its primary production also depends on the process of depletion of zinc.

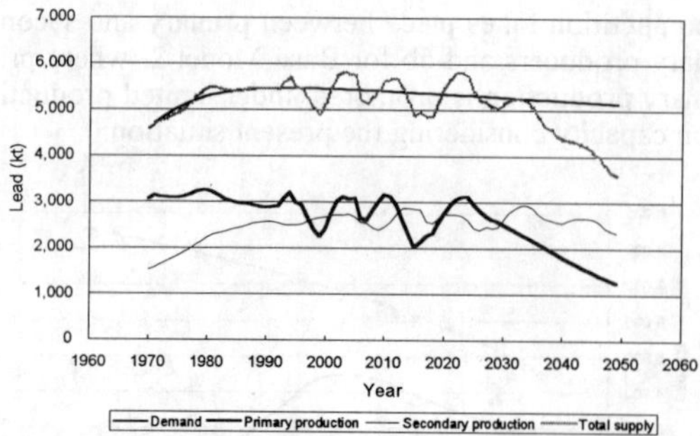

Figure 6. Result of the execution for lead (free competition, diminishing demand).

3.2 *Zinc*

The distinctive feature of zinc is the technological difficulty of its recycling as its main usage is plating. Figures 7a, 7b show the results of the executions for zinc, of which 7a for the Base Model 1 and 7b for the Base Model 2.

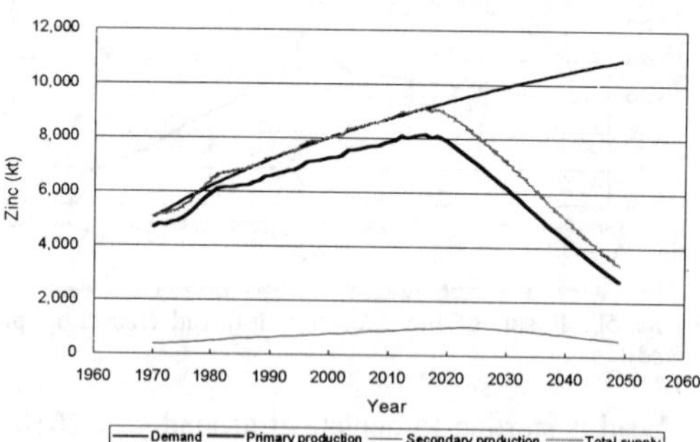

Figure 7a. Result of the execution for zinc (free competition).

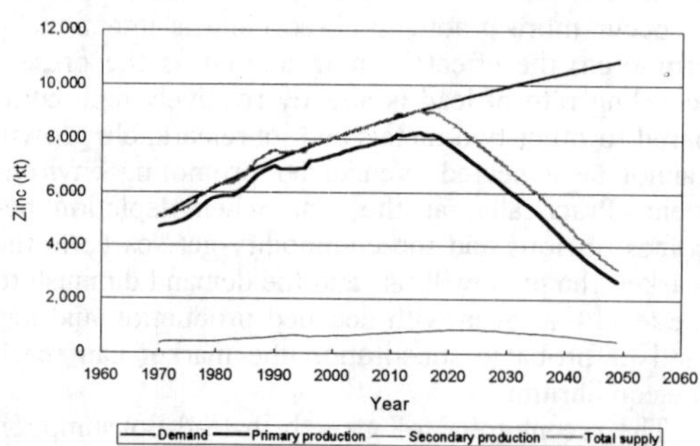

Figure 7b. Result of the execution for lead (recycling promoted).

Zinc also is starting to deplete shortly after year 2010 in either case. There is practically no difference in results between the two models, as recycling only plays a very limited role in case of zinc and the risk of depletion is most urgent among the three metals. However, a risk to fall into scarcity crisis may be quite low as the reserves of its substitutes, for example aluminum for galvanized steel plate for car manufacturing etc., are considered being quite abundant.

3.3 *Copper*

Potential resources of copper is relatively abundant compared to other two base metals. Also a systematical recycling already takes place in the society, even though the present recycling rate is not as high as lead. But unlike other two metals, the resources is unevenly distributed and nearly 40% of the recent potential reserves is assumed to exist in Chile alone.

Figures 8a and 8b show the results of the executions for copper, of which 8a for the Base Model 1 and 8b for the Base Model 2. Recycling is developing fairly smooth, even under free competition environment, compared to other two metals and helps to increase the sustainability of the metal supply. It is noteworthy to say that the scarcity only sets in at last around year 2060. But the effect of promoting secondary production is also quite limited at least over the long-term point of view.

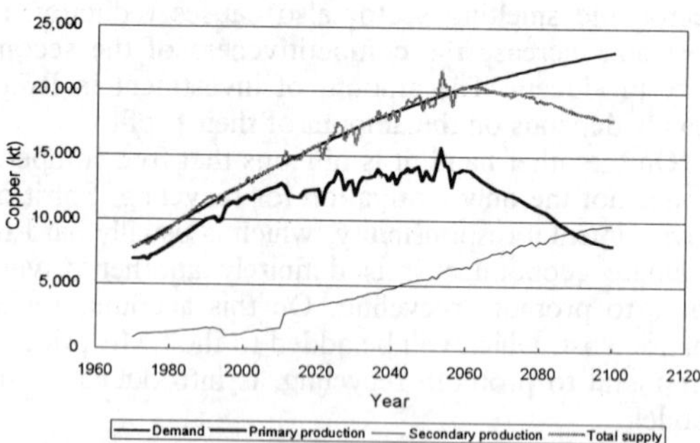

Figure 8a. Result of the execution for copper (free competition).

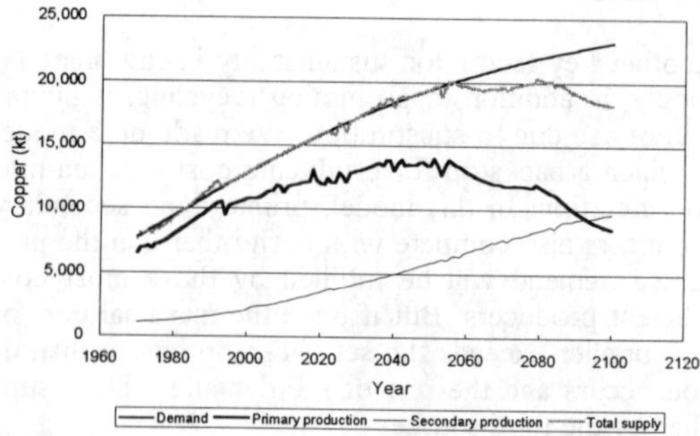

Figure 8b. Result of the execution for copper (recycling promoted).

A third source of supply with constant production cost and unlimited capacity is introduced to the model to evaluate the effect of backstop and substitution. Figures 9a and 9b are results of executions of

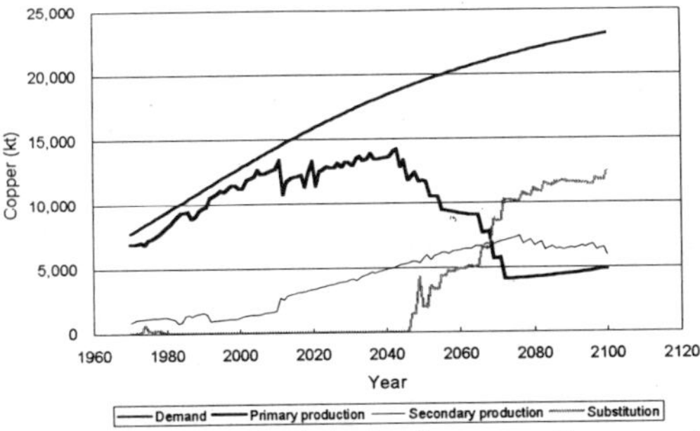

Figure 9a. Influence of substitution for copper (free competition, low backstop cost).

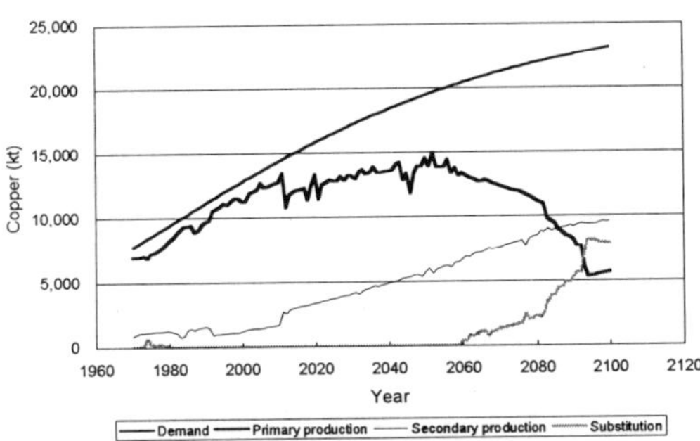

Figure 9b. Influence of substitution for copper (free competition, high backstop cost).

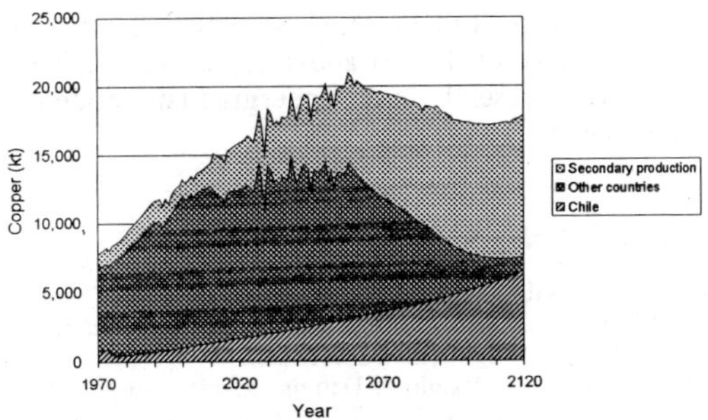

Figure 10a. Changes in shares of copper supply (free competition, without substitution).

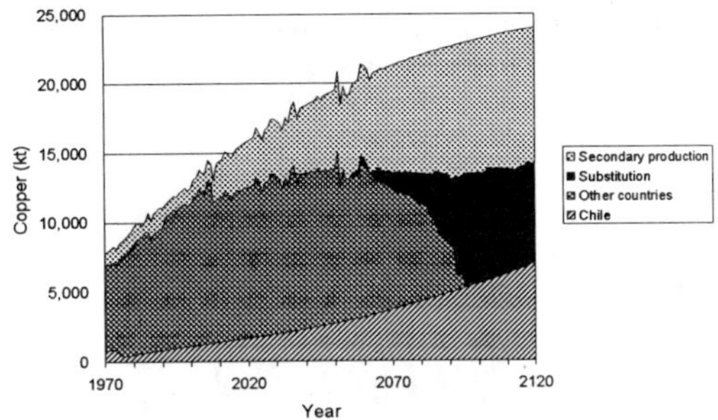

Figure 10b. Changes in shares of copper supply (free competition, substitution at high backstop cost).

such cases with different backstop costs. The production cost of the third source for 9b is set 35% higher than for 9a. Onset of substitution occurs in between year 2040 and 2060. Primary production from countries other than Chile and later on also a part of secondary production will be substituted in both cases. Consequently, considering a time span of over hundred years, substitution is definitely necessary to keep the sustainability of copper supply.

Figures 10a and 10b show the changes of shares in copper supply. Figure 10a is without substitution, where primary production of countries other than Chile will be gradually taken over by Chile and secondary production. Practically of course, as we also mentioned before, at the time when the commodity get scarce in the market, the price will rise and the demand diminish to the level that even with declined production the market can reach the equilibrium. Especially under the assumption that substitution can take place, primary production from countries other than Chile will rapidly taken over by substitutes just after the onset of scarcity in the market. Consequently the share of Chile in copper production including secondary production will rise from ca.15% just before the onset of substitution to ca. 40% at the beginning of the twenty-second century.

4 CONCLUSIONS

For lead and zinc, a sustainable supply for the next hundred years only by help of recycling is uncalculated. Onset of substitution, for example aluminum for galvanized steel plate or several kind of unleaded batteries, at early stage is definitely necessary. But in spite of their scarcity, the risk of future crisis through lack of those resources is supposed to be quite low, mainly due to various options and possibilities for substitution and relatively homogeneous distribution of the resources.

In case of copper, by help of steady development of secondary production, sustainable supply for the next fifty years will be easily attained compared to other two metals. But due to the recent upward revision of growing demand, scarcity could occur in the last half of the next century. Introduction of substitution is inevitable to assure the sustainability throughout the next century. Unlike the other two metals, copper is geographically unevenly distributed. Consequently, the importance of Chile in the copper production is supposed to further increase with time.

5 ACKNOWLEDGEMENT

The authors acknowledge the contribution of Mr. N. Hashimoto, Dr. S. Murakami and Mr. T. Kurita, who are all graduates of laboratory of mining engineering, the University of Tokyo and also of Mr. T. Nakayama, who presently is an undergraduate student of the above lab.

6 REFERENCES

Behrens III William 1973. The dynamics of natural resource utilization. *Toward global equilibrium: Collected papers.* Cambridge: Wright-Allen Press, Inc.

Meadows Donella, Meadows Dennis, Randers Jørgen & Behrens III William 1972. *The limits to growth.* New York: Universe Books.

Meadows Donella, Meadows Dennis & Randers Jørgen 1992. Beyond the limits. Vermont: Chelsea Green Publishing Company.

US Government 1980. *The global 2000 report to the president – Entering the twenty-first century-.*

Environmental Issues and Management of Waste in Energy and Mineral Production, Singhal & Mehrotra (eds)
© 2000 Balkema, Rotterdam, ISBN 90 5809 085 X

A design of cylindrical holes penetrating through the electrode used to prevent diffusion of hazardous heavy metal ions from waste landfills

N. Nakayama, T. Komai, Y. Ogata & M. Utagawa
National Institute for Resources and Environment, Tsukuba, Japan

ABSTRACT: Among the subjects to be considered for improving the safety of waste landfills, inhibition of the diffusion of heavy metal ions to the soil or groundwater is one of the most important. For this purpose, the authors propose an electrode through which a large number of cylindrical holes penetrate. This electrode, the electrochemical potential of which is kept constant, is planned to be installed around the landfills or across the water channels in soil. Heavy metal ions diffusing into the electrode holes were reduced to metal and trapped on the inner surface of the hole. An equation with which to design the optimum shape of the holes (radius and length) so as to inhibit the diffusion effectively is proposed and the validity of it is examined through experiments using electrodes with thus-designed holes. Diffusion was inhibited by the electrodes and experimental results agreed well with those predicted by the calculation using the equation.

1 INTRODUCTION

Recently, the possible risks of soil and groundwater contamination in the environment, and above all, around waste landfills have been pointed out, and the development of advanced techniques are required to more safely manage hazardous chemicals such as heavy metals in waste landfills. Among the subjects to be considered for improving the safety of waste landfills, inhibition of the diffusion of heavy metal ions to the environment (soil and groundwater) is one of the most important. In Japan, the diffusion of such ions is presently inhibited by covering the soils with thin resin sheets on which the waste is deposited, or by containing the waste in concrete cells (Honma 1995). In these systems, the ions may possibly leak and diffuse into the environment, for example, when the sheets or concrete cells are partially damaged due to collision with sharp waste materials, during an earthquake, or due to the deterioration of the resin or concrete. Therefore, it is desirable to implement another means of inhibiting the diffusion of the ions around the resin sheet or concrete barrier to avoid environmental pollution in the emergency mentioned above.

One strategy is to install electrodes around the resin sheets or concrete walls of the cells, with which to remove the chemicals. For this purpose, we propose an electrode with a large number of cylindrical holes that penetrate through the electrode material. Such electrodes, the electrochemical potential of which is kept constant, are planned to be installed around landfills, close to the sheet or the concrete wall, or across water channels in soil. Heavy metal ions diffusing into the electrode holes are reduced to metal and trapped on the inner surface of the hole. Although electrochemical reactors with outer and inner cylindrical electrodes have been reported (Bisang 1990), electrodes with cylindrical holes used to inhibit the diffusion of hazardous chemicals in the environment seem not to have been studied. In this paper, an equation with which to design the optimum shape of holes (radius and length) to inhibit the diffusion effectively is proposed and the validity of the equation is examined through experiments using a solution containing heavy metal ions and electrodes made of gold, according to the results of calculation using the equation.

2 DESIGN OF THE CYLINDRICAL ELECTRODE HOLE

The optimum radius and length of the cylindrical hole were designed using a diffusion equation. The concentration displacement of the heavy metal ions in the electrode hole was calculated using a diffusion equation. A schematic diagram of the cylindrical hole electrode is shown in Figure 1. The ions diffuse from the left side of the hole into the electrode hole. Circular cylindrical coordinates (r,z) were used in this calculation, where the origin was the center of the inlet (left side) of the cylindrical hole.

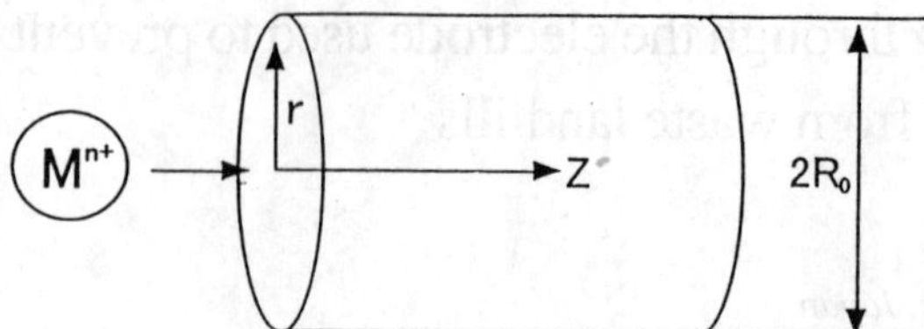

Figure 1. Schematic diagram of the cylindrical hole electrode.

The concentration C(r,z) is governed by the following Laplace equation in cylindrical coordinates in dimensionless form.

$$\frac{\partial^2 C}{\partial r^2} + \frac{1}{r}\frac{\partial C}{\partial r} + \frac{\partial^2 C}{\partial z^2} = 0 \qquad (1)$$

This equation was solved under the following assumptions.

(1) The concentration was equal to 0 on the surface of the electrode.

(2) The concentration was equal to 0 at the infinite hole length.

(3) The concentration was equal to that of bulk solution in the electrolyte contained in the inner vessel at the center of the hole inlet .

The following boundary conditions reflect the model assumptions.

$$C(R_0,z)=0 \qquad (2)$$
$$C(r,\infty)=0 \qquad (3)$$
$$C(0.0)=C_0 \qquad (4)$$

where R_0 = hole radius; C_0 = concentration of ions in the electrolyte contained in the inner vessel. By solving equation (1) for C(r,z) yields

$$C(r,z)=C_0\ J_0\left(\frac{-2.4048r}{R_0}\right)\exp\left(\frac{-2.4048z}{R_0}\right) \qquad (5)$$

where J_0 = the Bessel function of the first kind, with zero order. Equation (5) is applicable to all heavy metal ions since it is independent of the parameters characteristics of the ions such as diffusion constants. This equation shows that the concentration displacement along z is determined by the ratio of z to R_0 and the concentration decreases exponentially as z increases. The calculated ratios of C(0,z) to C_0 or C(0,0) are shown in Figure 2 and Table 1 as a function of z/R_0. The radius and the length of the hole could be determined using the equation, such that the concentration at the outlet could be decreased to values less than those regulated in corresponding standards or laws. The validity of the equation was examined through the experiments described in the next section.

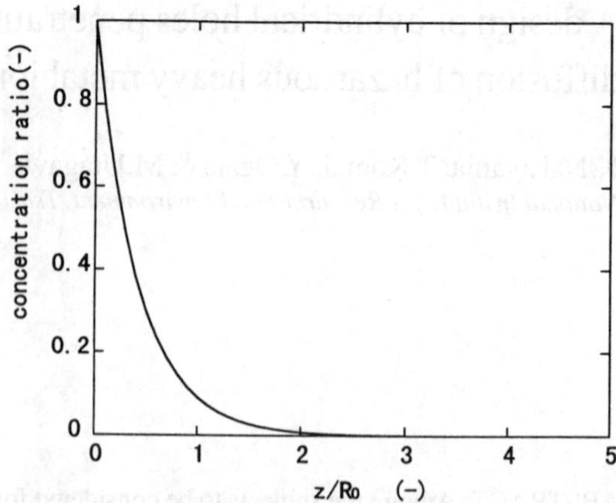

Figure 2. Calculated ratios of C(0,z) to C(0,0).

Table1. Calculated ratios of C(0,z) to C(0,0).

Z/R_0	C(0,z)/C(0,0)
0	1.0
1.0	9.0×10^{-2}
2.0	8.2×10^{-3}
2.9	9.4×10^{-4}
3.0	7.4×10^{-4}
3.9	8.5×10^{-5}

3 EXPERIMENTAL

Figure 3 shows a schematic diagram of the electrodes used in this study. Two types of electrodes were used; a column-type electrode (electrode A) and a disk-type electrode (electrode B). The column type electrode was made of two semicylindrical gold electrodes. These two electrodes were put together and a column-shaped electrode (4.5 mm of outer radius, 0.5 mm of inner radius, 16 mm of length) was assembled. The disk-type electrode was made of a gold disk (10.5 mm of radius, 5mm of thickness) with 101 cylindrical holes (0.5 mm of radius) penetrating through it within the radius of 15 mm. The values of z/R_0 for the column- and the disk-type electrode are 32 and 10, respectively, and the calculated concentrations of the ions at the outlet of the electrodes using equation (5) are 1.4×10^{-67} and 1.3×10^{-21} mol/dm^3, respectively. Based on these values, the concentrations at the electrode outlet were expected to be negligible. The former and latter electrodes were installed in electrode holders A and B, respectively, as shown in Figure 4, and were attached to the electrolyte reservoir shown in Figure 5. The reservoir was composed of two components, or an inner and an outer vessel. The electrode holders and the vessels were made of PTFE (polytetrafluoroethylene). The assembly of the electrode and the attachment of the electrode holders were

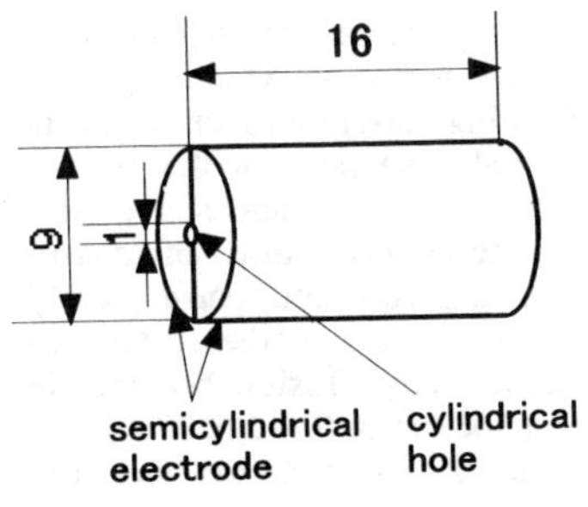

(A) column-type electrode

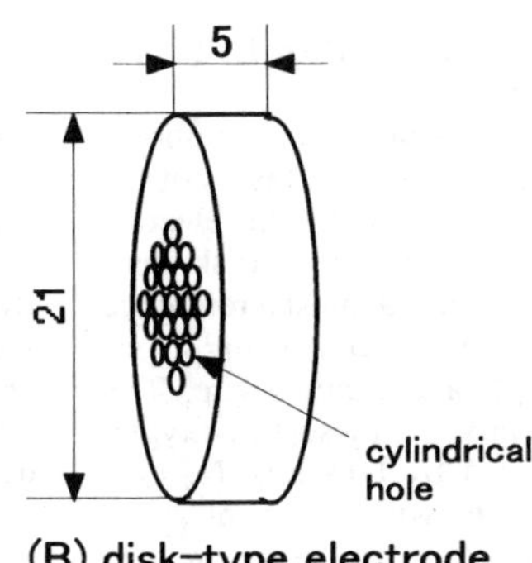

(B) disk-type electrode

Figure 3. Schematic diagram of the electrodes.

55
Pt wire
41
glass filter
Au electrode
glass filter

(A) column-type electrode holder

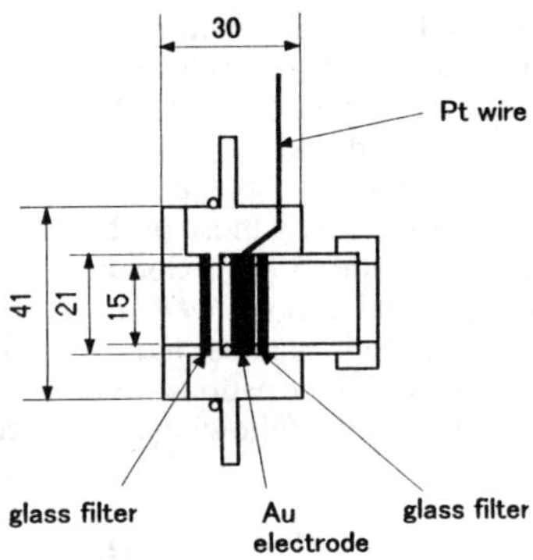

(B) disk-type electrode holder

Figure 4. Schematic diagram of the electrode holders.

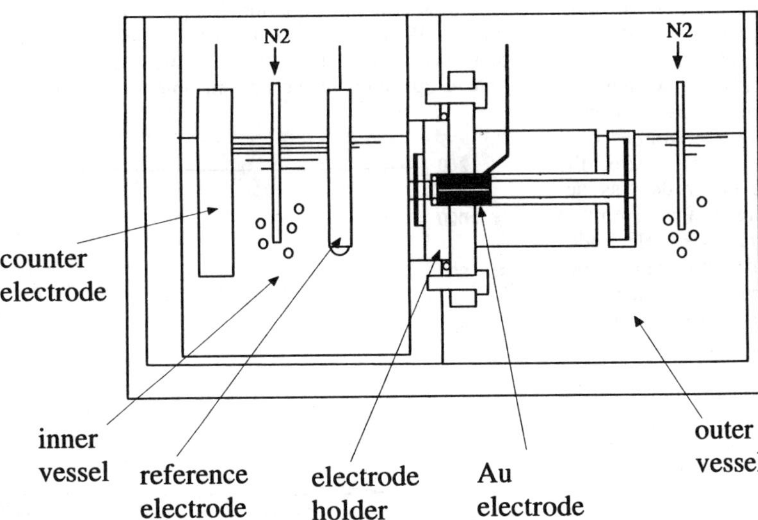

Figure 5. Schematic diagram of the electrolyte reservoir.

carried out in the electrolyte (0.2mol/l of KCl solution) beforehand filled in the reservoir (or inner and outer vessel) to avoid the accumulation of air bubbles in the electrode holes and to equalize the levels of the electrolyte surface in the two vessels. The electrochemical potentials of the column and the disk electrodes were kept at constant values against a reference electrode (saturated calomel electrode), using a potentiostat connected through a platinum wire. The electrolytes were stirred mildly and dissolved oxygen in the electrolyte was purged by bubbling N_2 gas. Plugging of the electrode holes with N_2 bubbles and the advection of ions contained in the electrolyte into those were prevented by attaching glass filters (thickness: 0.26mm, pore size: 1.6 μ m) to both sides of the holders, consequently, the ions passed through the electrode hole only by diffusion.

Zinc cations were selected for examination because they are classified as a harmful element (the concentration is regulated as 1 ppm or 1.5×10^{-5} mol/dm^3 in Japan for drinking water) (Nihonkagakukai 1986) and the reduction potential of zinc cations is comparatively low among those of the harmful heavy metal cations (Vetter 1967). Consequently, most of the heavy metal cations were reduced simultaneously with zinc cations under the reduction potential of the zinc cation. Zinc cations were reduced to zinc metal acoording to the following reaction.

$$Zn^{2+} + 2e^- \rightarrow Zn \qquad (6)$$

Zinc cation was supplied as $ZnCl_2$. After the potentials of the column- or the disk-type electrodes were kept at constant values, $ZnCl_2$ powder was added into the inner vessel until the concentration of zinc cations became 2.1×10^{-3} mol/dm^3. The cations diffused through the cylindrical holes and were reduced on the hole inner surfaces. The inhibitory effects of the holes of the electrodes on zinc cation diffusion were evaluated by the measurement and comparison of the concentrations of zinc cations in the electrolytes contained in the inner and outer vessels. After the measurements, the column-type electrode was detached from the holder and separated again into two semicylindrical electrodes. After these electrodes were washed with pure water and dried, the distributions of zinc on the inner surfaces of the electrodes were measured by EPMA (electron probe microanalyzer) along the grooves corresponding to the electrode hole.

The potentials of the column- and the disk-type electrode to be maintained were determined using the current-potential curve measured in the same $ZnCl_2$-KCl solution as mentioned above, with a column-shaped gold electrode (9.0mm of diameter, 5mm of length). In this measurement, the electrode was polarized cathodically from the rest potential with N_2 gas bubbled into the electrolyte, and the current-potential curve was obtained. The potentials of the electrodes were kept at those in the region in which a plateau corresponding to a limiting current was shown in the curve. At these potentials, the rates of the electrochemical reactions on the electrode increased to those sufficient to decrease the concentrations of ions on the electrode surface to almost 0 and, consequently, the diffusion rates of the ions onto the surface increased to a maximum; the diffusion became the rate-determining step (Bockris & Reddy 1970). Under this condition, boundary condition (2) was satisfied.

Electrodes were polarized and kept at the constant potential using the HOKUTO DENKO HZ3000 system. The concentrations of zinc cations were determined using a SEIKO DENSHI SPS4000 inductively coupled plasma emission spectrometer. A NIHON DENSHI JSM5300LV (with a Philips EDAX 9800 system) electron probe microanalyzer system was used for the measurements of the zinc distribution on the electrode surfaces.

4 RESULTS AND DISCUSSION

Figure 6 shows the current-potential curve obtained for the 0.2 mol/dm^3 of KCl solution containing 2.1×10^{-3} mol/dm^3 of $ZnCl_2$. Cathodic current increased rapidly at potentials higher than about -1000 mV. This increase in the current was probably related to the reduction of hydrogen cations (electrolysis of water). A plateau appeared around -1300 mV. This plateau was attributable to the limiting current for the reduction of the Zn cation according to equation (6), since the standard electrode potential in this solution was calculated to be about -1080 mV vs SCE using the Nernst equation and the reported standard electrode potential for equation (6) (-763 mV vs SHE) (Vetter 1967). The potential of the electrodes to be kept was selected from those in the region that the plateau appeared in the current-potential curve. Ac-

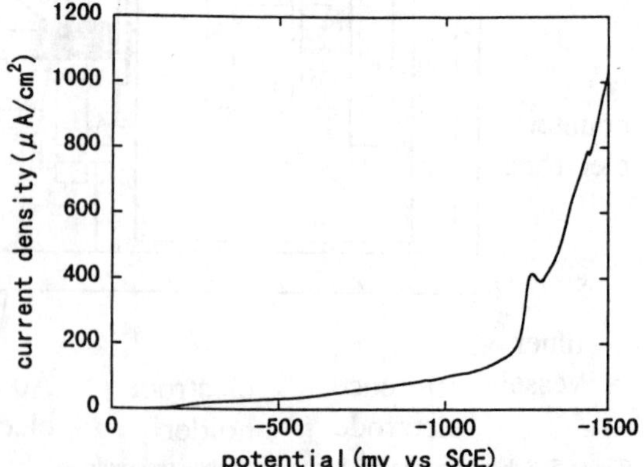

Figure 6. Current-potential curve obtained for the 0.2 mol/dm^3 of KCl solution containing 2.1×10^{-3} mol/dm^3 of $ZnCl_2$.

cording to the Nernst equation, the electrode potential shifted to a lower potential by 70mV when the concentration decreased to 1.5×10^{-5} mol/dm^3 which corresponds to the value regulated for drinking water as is mentioned above. The electrode potential to be maintained was determined to be -1400 mV, considering this shift of the potential.

Figures 7 and 8 show the changes in the concentrations of zinc cations in the electrolytes contained in the inner and outer vessels, for the column- and the disk-type electrode, respectively. The changes in the concentrations in the electrolytes contained in the inner vessel were small and almost the same as the initial values through the measurements, while the concentrations for the outer vessel were lower than the detection limit (about 7.7×10^{-8} mol/dm^3) as long as the electrodes were kept at -1400 mV, and increased after the circuit was opened. This showed that the diffusion of zinc cations was effectively inhibited by these electrodes as was predicted in the calculation.

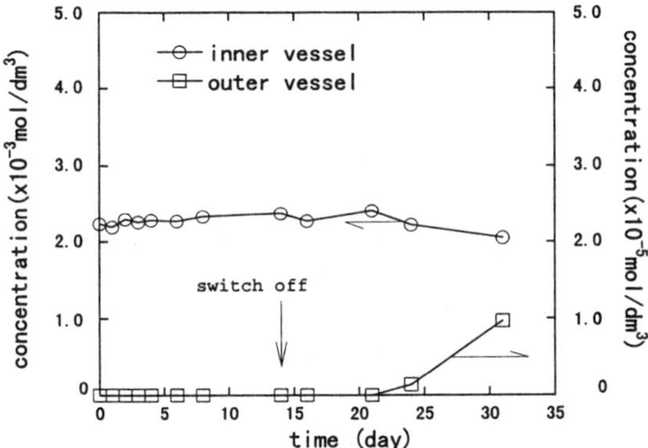

Figure 7. Changes in the concentrations of zinc cations in the electrolytes contained in the inner and outer vessels.(column-type electrode; 0.2 mol/dm^3 of KCl solution containing 2.1×10^{-3} mol/dm^3 of $ZnCl_2$.)

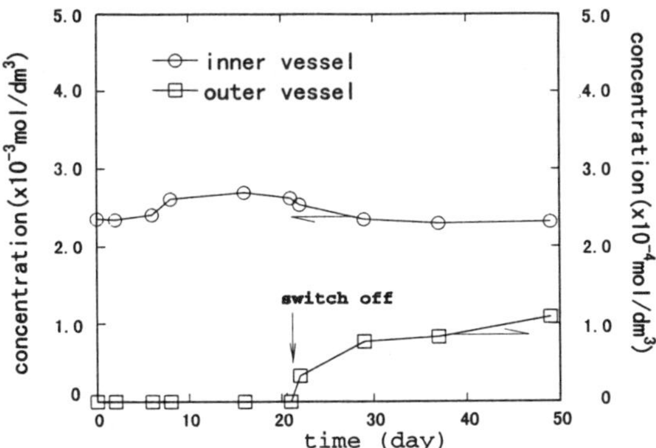

Figure 8. Changes in the concentrations of zinc cations in the electrolytes contained in the inner and outer vessels.(disk-type electrode; 0.2 mol/dm^3 of KCl solution containing 2.1×10^{-3} mol/dm^3 of $ZnCl_2$.)

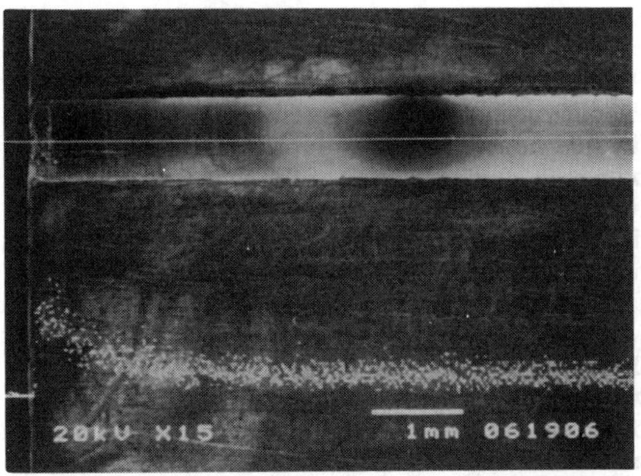

Figure 9. Typical SEM micrograph and a result of EPMA line analysis of the inner surface of the column-type electrode.

Figure 9 shows a typical SEM micrograph and a result of EPMA line analysis of the inner surface of the column-type electrode obtained after the measurement of the change in the concentration mentioned above. A groove corresponding to the cylindrical hole in the electrode is shown in the upper part of this micrograph and the line analysis was performed along the white line shown in the groove. The left side of the micrograph corresponds to the side of the inner vessel and was an inlet of the cylindrical electrode hole for the diffusion of zinc cations. Zinc cations diffused from the left side to the right side. A trace of metal deposition was found on the left side around the inlet of the hole. The white dotted line on the lower side of the micrograph shows the result of the line analysis of elemental zinc. Zinc concentration on the surface was maximum around the edge of the inlet and decreased abruptly as the length from the edge increased. The zinc concentration became almost negligible at length more than about 1.3mm. This result indicated that most of the zinc cations were reduced around the inlet of the electrode. The calculated ratios of $C(0,z)/C(0,0)$ shown in Figure 2 and Table 1 were about 0.01 and 0.001 at the z/R_0 ratios of 2 and 2.9 which corresponded to the lengths of 1.0 and 1.45mm, respectively. This agreed closely with the result observed by EPMA, indicating that equation (5) is applicable to the prediction of the concentration displacement in the cylindrical holes. This equation is thought to be useful for the design of electrodes with holes to inhibit the diffusion of heavy metal ions if the diameters of the holes are comparatively small and the reduction of the ions is performed at potentials at which the limiting currents are obtained.

5 CONCLUSIONS

An equation used to design an electrode that inhibits the diffusion of the heavy metal ions by means of cylindrical holes penetrating through the electrode, was proposed, and the validity of the equation was examined through experiments using a column- and a disk-type gold electrode which were designed based on the equation (hole radius: 0.5mm). The results obtained are summarized as follows.

(1) The concentration displacement of the ions along the hole length was calculated using the equation that was represented by the product of the Bessel function of the first kind with zero order as a function of the hole radius and the exponential function as the function of the length from the inlet of the hole. The concentrations of the ions at the outlet of the hole were determined from the ratio of the hole length (the thickness of the electrode) to the hole radius.

(2) The diffusion of zinc cations in the $2.1x10^{-3}$ mol/dm^3 of $ZnCl_2$ solutions was effectively inhibited by the electrodes. EPMA analyses of zinc displacement on the inner surface of the electrode hole showed good agreement between the observed results and those predicted by the calculation using the equation. The proposed equation is considered to be useful in designing the electrode.

REFERRENCES

Bisang, J. M. 1990. Simultaneous solution of the potential equations for the metal and solution phases in cylindrical electrochemical reactors. *J. Appl. Electrochem.* 20: 723-727.

Bockris, J. O'M. & Reddy, A. K. N. 1970. *Modern Electrochemistry.* New York: Plenum Press.

Honma, S. 1995. *Data Guide Chikyu Kankyo.* Tokyo: Aoki Shoten.

Nihonkagakukai 1986. *Kagakubinran Oyoukagaku hen I*. Tokyo: MARUZEN.

Vetter K. J. 1967. *Electrochemical Kinetics.* New York San Francisco London:. ACADEMIC PRESS.

Environmental Issues and Management of Waste in Energy and Mineral Production, Singhal & Mehrotra (eds)
© 2000 Balkema, Rotterdam, ISBN 90 5809 085 X

Evaluation of gas emission from closed mines surface to atmosphere

Z. Pokryszka & C. Tauziède
Institut National de l'Environnement Industriel et des Risques (INERIS), Verneuil en Halatte, France

ABSTRACT: The closed mines are likely to release into atmosphere polluting or/and dangerous gases. The detection and hazard evaluation of those emissions are a complex problem. In order to quantify and qualify these gas emissions and check gas migrations special measuring methods are required. The research done at INERIS resulted in a reliable and appropriate methodology using a flux chamber. This methodology is intended for detecting, quantifying and qualifying gas discharges, but can also be applied to verify the effectiveness of the preventive means used for reducing emissions (sealing operations, gas drainage, etc.).The method has already been applied in an operational manner for diagnosis and appraisal on a number of sites (old mines, landfills, etc.). Although originally designed for measuring methane flows, it is fully useable in or transposable to any similar situation of gas flows emitted by the ground, subject possibly to some specific modifications..

1 SURFACE EMISSION OF FIREDAMP

Experience in different coalfields has shown that gas that is released and accumulates in old underground workings tends to migrate towards the surface. This gas can represent a significant hazard because it may be both flammable and suffocating.

A number of incidents and even accidents resulting from gas rising to the surface have already been noted in different countries (Burrel et al., 1996; Kral et al., 1998).

These events were caused by the following risk situations usually encountered:

- the isolated and concentrated release of firedamp through mine shafts, faults, cracks, etc. providing a direct link between old workings and the surface;
- the build-up of gas in confined or semi-confined spaces (cellars or basements, even dwellings or buildings themselves, but also underground networks, and so on);
- the venting of contained firedamp to the atmosphere (for example where drilling or civil engineering works are carried out near old workings);
- enclosure of a gas-emitting surface (for example, by constructing a building).

2 AIMS AND PRINCIPLES OF THE METHOD

For concentrated emissions or when gas builds up in a confined space, simple measurements of gas concentrations in the atmosphere can be sufficient to detect and evaluate the hazards.

The approach is much more complicated where there is diffuse emission from the ground, because the mere fact that measurable concentrations of gas are present does not always mean that a truly significant flow exists. Moreover certain gases, such as carbon dioxide and methane, may be present in the soil naturally.

Accordingly, in order to be able to detect, quantify and qualify gas discharges of this kind, as well as to evaluate the effectiveness of the preventive means used, it is indispensable to have reliable and proven methods for measuring the flow of gas emanating from the surface of the ground.

With this in view, INERIS has been developing and validating such methods since 1992 (Pokryszka et al., 1995). These methods, which were originally intended for landfills and contaminated ground, quickly found an application in mines.

The latest measurement technique developed is based upon a principle of the accumulation chamber. This is a local and direct type of measurement, that does not depend on assumptions about how gases are emitted into the atmosphere and dispersed. The method involves using a chamber to cover a certain area to avoid perturbing the environment too much (see figure 1). The gases given off from the covered area then build up in the chamber. In this way it is possible to monitor how the atmosphere becomes enriched in methane. A sample of the mixture is fed to an analyser and then returned to the chamber, so that the gases are recirculated. By monitoring the rate at which the recirculated mixture is enriched in methane, it is possible to deduce the local methane flow at the point in question.

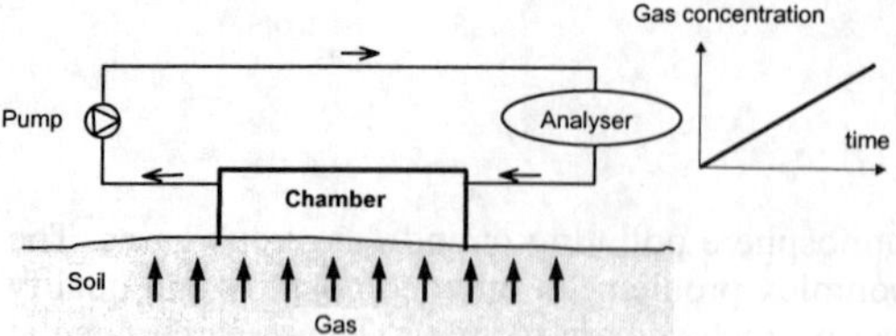

Figure 1. INERIS flux chamber diagram.

The dimensions of the chamber and the operating parameters of the method were optimised at the design stage on a test rig (figure 2) using known gas flows. At this stage it was also possible to determine the influence on the results of the measuring system itself, the nature of the soil, the wind and many other factors, so that due allowance could be made in the calculations.

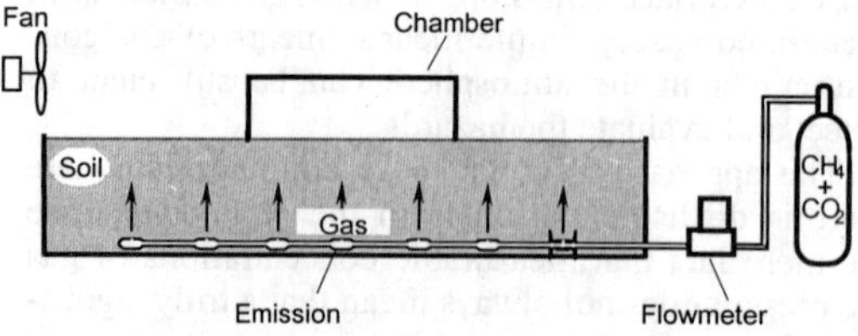

Figure 2. Diagram of test rig for chamber validation.

Following the laboratory validation phase, the method was fine tuned under real conditions. This stage confirmed the relevance of the operating principles of the method defined in the laboratory.

The measurement system used is relatively simple to operate. The total time necessary for an individual measurement is of the order of 5 to 10 minutes, so that a large number can be made in a day (from 40 to over 60 points according to the difficulties of the site).

The flow densities that can be determined depend on the capabilities of the analyser used for measuring the concentration of a given gas in the chamber. For methane, measured with a resolution of 1 ppm, it is possible to measure flows ranging from under 0.1 up to 4000 Ncm³/minute/m².

If it is necessary to determine the total flow from a given surface area, the point measurements are made using a spatial sample approach adapted, on a case-by-case basis, to the local emission conditions and to the desired precision of estimation.

The local values measured are then interpolated and extrapolated using specific methods (for example, geostatistics), in order to estimate the total flow and to supply other necessary data, such as surface mapping, the spatial variability of the flow, and so on.

The exact procedures involved in this method are protected by a European patent (No. 96-05996, filed on May 14th 1996 and entitled "Measurement of gas flows through surfaces").

3 TYPICAL APPLICATIONS

The method has been applied in a operational manner for diagnostics and appraisals at a number of sites (old mines, waste dumps, landfills, and so on). Two typical cases are presented here.

3.1 *Emission of firedamp from an old coal mine*

Measurements made for preventive purposes on land located above an old coal mine revealed methane concentrations in the soil exceeding 40% in volume. Since the site was accessible to the public, INERIS was called in to evaluate the risk.

Using the chamber method, significant gas flows were measured, that were capable of creating dangerous accumulations on the surface (figure 3).

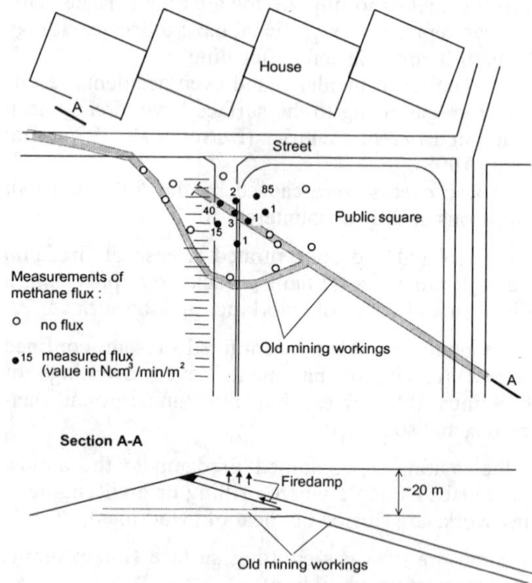

Figure 3. Measuring of firedamp emission on the surface of a closed coal mine.

3.2 *Emission of gas from a MSW landfill*

Municipal Solid Waste (MSW) landfills are substantial source of biogas emissions, with methane as a major component. In certain cases these emissions can represent a fairly considerable risk, notably as regards explosion.

As part of an investigation for the French Ministry for the Environment, INERIS conducted a measurement serie to quantify the flow from a dump covering 8 hectares Emission mapping was also done, involving 400 measurements over the area studied (figure 4).

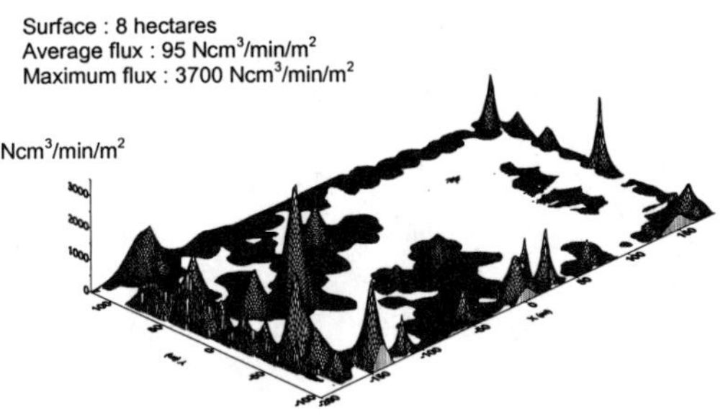

Figure 4. Mapping of methane emission from a landfill.

4 CONCLUSIONS

The research done at INERIS resulted in a reliable and appropriate methodology for quantifying the complex phenomenon of the emission of gas flows from a land area.

The methodology is intended for detecting, quantifying and qualifying gas discharges, but can also be applied to verify the effectiveness of the preventive means used for reducing emissions (sealing operations, drawing off gas, etc.).

The method has already been applied in an operational manner for diagnosis and appraisal on a number of sites (landfills, old mines, etc.). Although originally designed for measuring methane flows, it is fully useable in or transposable to any similar situation of gas flows emitted by the ground, subject possibly to a few modifications of detail.

The method is protected by European patent No. 96-05996.

5 REFERENCES

Burrel R., Friel S. 1996. The effect of mine closure on surface gas emission. *Conference on the Environmental Management of Mine Operations*, Proceedings, IBC eds, London.

Kral V., Paletnik M., Novotny R. 1998. Methane from closed-down mines in the soil. *International Conference on Coal-Bed Methane Technologies of Recovery and Utilisation,* Proceedings, GIG eds, Katowice.

Pokryszka Z., Tauziède C., Cassini Ph. 1995. Development and validation of a method for measuring biogas emissions using a dynamic chamber. *5th International Landfills Conference - Sardinia'95*, Proceedings vol III, CISA eds, Cagliari.

Environmental Issues and Management of Waste in Energy and Mineral Production, Singhal & Mehrotra (eds)
© 2000 Balkema, Rotterdam, ISBN 90 5809 085 X

Thermal water distribution monitoring system for Baile Herculane resort

I. Ruja, N. Potoceanu & S. Szabo
'Eftimie Murgu' University, Reşiţa, Romania

ABSTRACT: The paper presents the main features of Baile Herculane Spa constituting an important touristic sight and bathing resort. We present several solutions for a system of monitoring the temperatures of the thermal springs and the possibility of a real time acquisition of the ions in the air and the centralization of results in the dispatcher's room.

1 FOREWORD

Baile Herculane is a resort renowned for its sulfurous thermal and radioactive springs, known for over 2000 years.

The radioactive minerals have healing properties together with the hypo-thermal springs are recommended in the treatment of rheumatism, peripheral nervous system diseases, skin diseases, chronic professional intoxication, digestive diseases, chronic hepatitis etc.

1.1 *Features of the biological active aero-ions*

In Baile Herculane, at an altitude of 160 meters one breathes the same air as at 3000 meters due to the natural negative ionization, witch constitutes a source of bio-energy for the organism. This phenomenon acts upon the central nervous system, simulates the growth and acts against stress. Table 1 presents the measurements of the concentrations of positive and negative ions taken in the representative points on a surface of 2 square kilometers.

1.2 *Characteristics of the thermo-mineral springs*

Characteristic of the thermo-mineral springs
Thermo-mineral waters are found in the chalk and granite rocks over 1000 meters deep. The mineralization of the thermal waters is caused by the dissolution of soluble compounds of rocks and their radioactivity is determined by the radioactive disintegration occurring in the depths of the Earth.

The chemical composition of the springs is very stable in time. It is presented in table 2.

Table 1. Concentration / Measurement points

Concentration	n+ ion/cm^3	n- ion/cm^3	n+/n-
Point 1	574	648	0.9
Point 2	428	418	1.03
Point 3	811	980	0.87
Point 4	897	1006	0.9
Point 5	696	806	0.95
Point 6	602	728	0.89
Point 7	657	860	0.82
Point 8	733	875	0.86

Table 2. Total mineralization

Parameter		Spring				
		7 izvoare	Hercules	Diana	Neptun	Traian
Q	[l/s]	7	1.67	1.67	5	5.6
T	[degC]	53.5	56	62	55.9	60.5
Mineralization	[mg/l]	1149	3460	5853	7006	7540
Ca^{++}	[mg/l]	80.6	526	740	934	1020
Mg^{++}	[mg/l]	70.5	48.6	92.4	49.1	67.3
$Na^{+}K^{-}$	[mg/l]	487.5	732	10.4	1492	1270
Cl^{-}	[mg/l]	5349	2516	4029	4982	4767
$HCO3^{-}$	[mg/l]	390.4	183	209.4	143.4	158
Br^{-}	[mg/l]	0	0	8.0	8.0	12.0
H_2S	[mg/l]	14.05	12	66.56	66.25	72.0

The figures in the table represents maximum values.

2 THE MONITORING SYSTEM

In order to perform a real time measurement of the spring's main parameters: temperature, debit (flow), pressure, we implement a date acquisition system with teletransmission which allows the registration of values in the dispatcher's room.

Figure 1 shows the block diagram of the acquisition system.

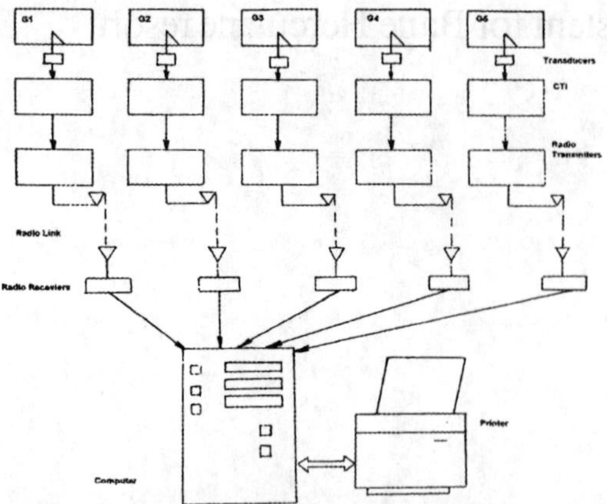

Figure 1. Block diagram of the acquisition system

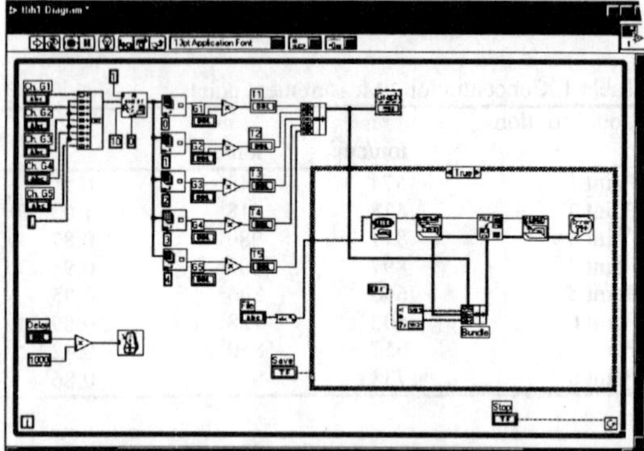

Figure 2. Block diagram for the virtual instrument

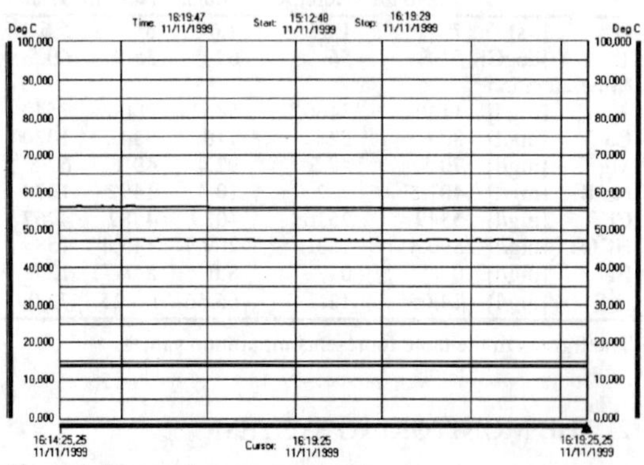

Figure 3. Temperature measurements

The main elements are : G1-G5 springs, C computer, CTI voltage-current converters, RE radio emitter, RR radio receiver.

The data acquisition system uses the acquisition card AT-MIO-16E1 ant the LabVIEW software. The transmission frequency of the radio link is in the 938-958 MHz band.

3 EXPERIMENTAL RESULTS

The virtual instrument used for monitoring the springs is presented in figure 2.

Figure 3 presents the real time acquisition of temperatures from 2 sources.

4 CONCLUSIONS

- The partial results obtained show that the pro posed solution is correct.
- In order to acquire the flow rate and pressure, it is necessary to prepare and build the proper transducers and adapters specific to the measurement of the resort thermal waters.
- The mineralization of thermal waters require special measures for the precise working of the transducers.
- The real time acquisition of the concentration of aeroions constitutes a future goal in completing the resorts' monitoring.

REFERENCE:

National Instruments Corp. 1998. *LabVIEW User Manual*. Austin, TX, USA

National Instruments Corp. 1998. *Virtual Bench. User Manual*, Austin, TX, USA

Cristescu, I. 1996. *The Cerna Miracles*, Herculane, In Hercules Friends, Romania

Potoceanu, N. & Chincea, I. & Cornean, A.M. 1999. Elements of Medium Conservation in Caras-Severin County. *Analele U.E.M., F.S.A*, ISSN 1453-7394, pages 304-311, Romania

Ruja, I. & Szabo, S. & Liuba G. 1998, Applications of virtual instrumentation in technical measurements, *Virtual Instrumentation News,* in romanian ISSN 1453-8059, volume 4: pages 169-173, Romania

Ruja, I. & Szabo, S. 1997, Virtual Temperatue Regulation, *Instrumentation Horizons*, volume 5, pages 39-42

Tolle, H. 1968. *Messverfahren fur warme- und maschinentechnische untersuchungen*, Leipzig : VEB Deutscher Verlag fur Grundstoffindustrie, Germany

Environmental Issues and Management of Waste in Energy and Mineral Production, Singhal & Mehrotra (eds)
© 2000 Balkema, Rotterdam, ISBN 90 5809 085 X

Mining waste: Transforming mining systems for waste management

Malcolm Scoble, Bern Klein & W. Scott Dunbar
Department of Mining and Mineral Process Engineering, University of British Columbia, Vancouver, B.C., Canada

ABSTRACT: Mining environmental management tends to focus on concerns over the impact of waste disposal on surface, primarily in the form of tailings and waste rock structures. This relates to the fact that traditionally waste products have only been returned to the mining void in limited quantities and surface disposal technologies have been paramount. Mining systems need to be reengineered, based on a new paradigm that mining is a business whose success is fundamentally dependent upon waste management. Strategies are available to shift this paradigm and to minimize the surface impact of waste disposal. Firstly, mining methods need to be developed that are more selective, i.e. leave waste in place. The eventual, more futuristic outcome of this strategy may well be the implementation of in situ or solution mining. The adoption of pre-concentration and even the full integration of mineral processing within the mine workings could be important in this strategy in the interim. Secondly, technologies should be further pursued that enable the return of waste securely to the void. This paper considers the design of new mining methods that minimize waste output. It then reviews technologies for in situ pre-concentration and mineral processing. Suitable processing technologies with a small footprint, that require limited infrastructure and can be used for coarse particle processing are considered. Finally it addresses the issues associated with the return of waste to mined voids.

1 INTRODUCTION

Awareness of the potential environmental impact of mine waste disposal has intensified recently, fostered by the increasing scale of mining, and the evolution of environmental technologies, legislation and ENGO's. The power of the media and the sequence of tailings dam failures in recent years have heightened this awareness.

The paradigm has been proposed that "mining companies are waste management companies" (Moss 1999). The growth in larger, surface-mined deposits at very low grade has prompted this viewpoint. It is possible that many new surface mines will find it difficult in the future to gain permitting to operate without agreeing to backfill their mined voids.

In this paper, waste is considered in a mineralogical context. It is the geological material encountered in a mining operation that is below the cut-off grade. The implication therefore is that waste may become ore if commodity prices rise, or vice versa. It is possible to seek new marketable uses for what exists currently as waste, for example, witness the efforts expended in Europe to find new uses for waste materials from coal mining. The disposal of marginal waste, that is encountered as mining proceeds, would therefore best be undertaken so that the material is still accessible in the future in the event of a rise in commodity prices. This is an example of the need to build in operational flexibility within waste disposal. The technology and markets for waste utilization can also change with time. Deposits of past waste, whether from mining, other industrial activity or domestic refuse will likely be commonplace orebodies of the future. Mining will be an integral part of the process of recycling in sustainable development. Responsibility for more efficient and total recovery of ore deposits will increase as global populations and expectations expand in the future. Sustainable development principles are being applied increasingly by mining companies.

The disposal of waste will also become an increasingly challenging issue for underground mines. Several of Canada's underground mines are mature and face the need to mine at much greater depths, (Scoble, 1994). In such circumstances, there is a major incentive to leave waste in the mine rather than suffer the costs of transporting it to surface. In underground mines there is at least the opportunity to dispose of the waste back in the mined voids during the life cycle of the mine. This may also be the case in open-caste or strip mining on surface, but in open pit mines this is only the case at the end of the mining operation. Many of the older open pits are now looking to continue as underground operations. This is possible only where low-cost, bulk mining is feasible, generally by using a caving method of working. Unfortunately such methods do not facilitate the underground disposal of waste.

An obvious solution to many of these issues is not to mine so much waste in the first place, i.e. the practice of selective mining. This means the minimization of waste production but may not necessarily correspond to maximizing the recovery of the mineral resource.

A further step to minimize waste generation is to employ mineral pre-concentration or concentration close to the mining face, so that significant waste is removed from the ore and disposed into the void rather than transported out from the mine, processed and then disposed on surface. Comminution of development and stoping waste *at the face* presents the opportunity for pre-concentration as well as for slurrification and hydraulic transportation in pipelines either to back-filling sites or the mill. The pre-concentration requirements as well as the hydraulic transport constraints would determine the degree of comminution.

Unfortunately, fine processed waste is often reactive and contains water. The evolution of paste back-fill and co-disposal technologies may well overcome some of these issues.

Heap leaching has been more successful in surface mines than underground. But on surface this still entails large movements of waste and liquors. There is much to be gained by the integration of the mine and mill through a new paradigm "processing at the face". The ultimate extension in a more futuristic form of mining will be in situ mining, where only the metal/mineral of value is extracted from the deposit.

Parsons and Hume (1997) place the following waste management practices in order of decreasing preference:

- waste avoidance, i.e. practices that prevent the production of waste;
- waste reduction, i.e. practices that reduce waste;
- waste re-use, i.e. direct re-use of waste materials;
- waste recycling or reclamation, i.e. using components of waste in other processes;
- waste treatment, i.e. to reduce hazard, usually at the site of generation;
- waste disposal.

Mining companies use the last three options extensively. Other industries, notably the chemical industry, have made use of the first three options by modifying or changing processes. The problem with application of the first three options to mining wastes is that in the current mining paradigm, as ore grades become lower, then more waste is produced. Major rethinking of mining, mineral processing and waste management technology and its use is necessary for the mining industry to take advantage of these more preferable waste management options.

2 SELECTIVE MINING

The ability to mine selectively is strongly dependent upon the knowledge of the variability of the mineralization and its host rock-mass. The orebody complexity is dependent upon the variability in grade distribution, morphology, and geotechnical characteristics. The quality of the initial exploration program and the subsequent delineation phase governs the degree of selectivity attainable. A recognition of this has prompted significant current research into geophysics: more traditional gravity, magnetic, reflection seismic, electromagnetic and radiometric methods, plus sonar, crosshole seismics, seismic tomography, resistivity tomography, borehole radar, mine seismic profiling, radio tomography and ground penetrating radar. This has been accompanied by the realization of the need to integrate data into multidimensional Spatial Information Systems, capable of integrating disparate data sources, database operations, visualization and spatial analysis (Francis et al., 1998).

Open pit mining does not facilitate the same degree of selectivity as in underground mining. There is reduced flexibility in terms of the ability to change sequencing and mine in areas of choice within the orebody. Selective mining in low grade, disseminated mineralization brings much less benefit than in situations where ore is high grade with clear boundaries. The degree of selectivity is dependent upon the morphology of the orebody: irregular outline orebodies require more precise mining methods, based upon more detailed delineation. The beneficial impact on reducing dilution by employing selective mining systems will be proportionately greater in narrower orebodies. Bulk, high productivity mining tends to be applied to large ore-bodies where selectivity has less impact. Selective methods generally entail smaller volumes of ore excavation in order to follow complex ore outlines more closely, tending to reduce productivity.

Selective mining methods can also employ segregation of ore and waste in the excavation process, itself. This is the principle of the resueing method of stoping, where waste and ore are broken in separate cycles to segregate ore and waste right from the outset within the stope. The waste can then be left within the stope. This enables waste minerals or those deleterious to downstream processing in zones within orebodies e.g. arsenic-rich zones in base metal deposits, to be broken and dealt with separately. Selective mining can also be important for segregating and dealing with potential contaminants, such as reactive sulfides, that propagate acid rock drainage. This entails selective waste disposal,

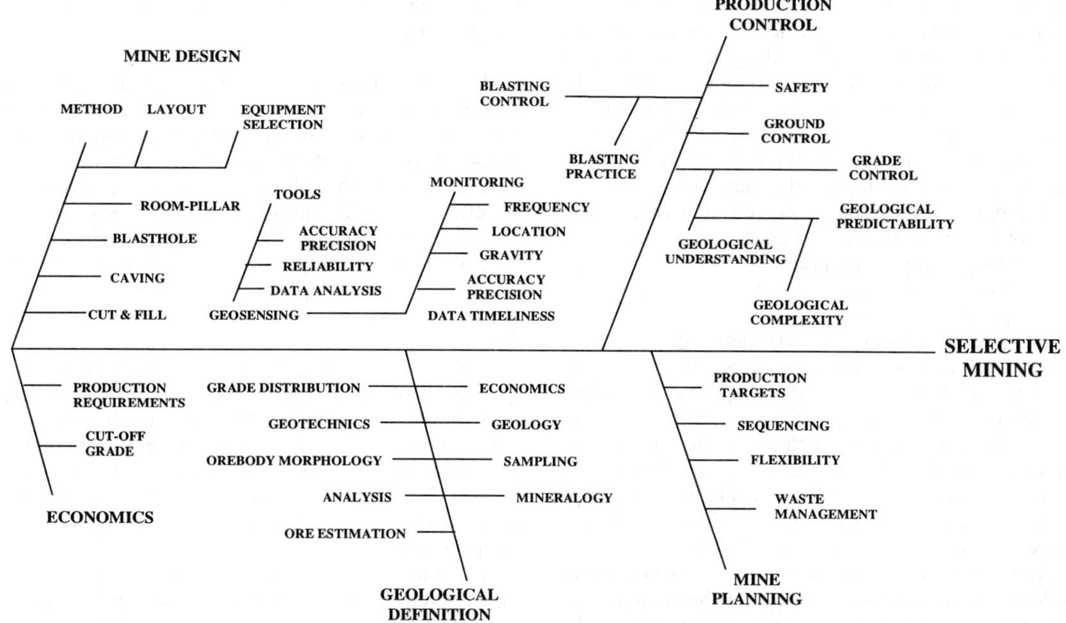

Figure 1. An overview of the factors controlling the implementation of selective mining.

which is a sensible practice when ranges of reactive waste are to be handled. Selective mining then presents the opportunity to be proactive in anticipating waste management issues.

Selectivity governs mining losses and dilution. Knissel et al, (1995), defined selectivity as the measure of the recovery (relative losses) and cleanliness achieved (relative dilution) within the mining process.

The preceding paragraphs describe only some of the factors that control the implementation of selective mining. A more detailed overview of the factors that need to be considered (Francis, 2000) is shown in Figure 1.

3 MINERAL PROCESSING

3.1 *Processing underground*

Ore is transported from the mine to the mineral processing plant where valuable minerals are separated from tailing waste. The valuable minerals usually comprise only a small portion of the mined ore, leaving the bulk to be transported to waste dumps, tailing ponds or returned to underground workings. From a mineral processing perspective, waste management objectives are to:

- Minimize the amount of reactive ARD generating wastes,
- Minimize the production of fine waste, and
- Minimize handling and transport of waste.

The extent to which these objectives can be achieved is related to the geographical setting, the geology and mineralogy of the ore deposit and several other logistical factors that are specific to each ore deposit.

Integrated mining and recovery systems, such as underground mineral processing/ pre-concentration, can help to achieve the waste management objectives. Processing the ore underground would reduce the costs of bringing ore to the plant and of returning backfill waste to the mine workings. As mines become deeper, cost savings from underground processing become more significant. While the reduced transport costs are clearly an incentive, there are constraints upon the construction and operation of an underground processing plant (Lloyd, 1978):

- The physical volume of the processing facility should be as small as possible,
- The waste should be suitable for backfill with regards to its physical properties and volume,
- The process should be robust with respect to being capable of treating ore at a range of feed rates and grades while maintaining high metal recoveries.

Underground crushing facilities are common, however there are few examples of other underground processing facilities. The most noteworthy is Codelco Chile's Andina Mine, where the mineral processing plant was built underground because of the extreme climatic conditions (Brewis, 1995). The processing plant was constructed in two large underground chambers, one housing the grinding section and the other the copper flotation plant. Ore is transported from open pit and underground workings to the plant where it is processed at a rate of approximately 33,000 tpd. While the Andina Mine is an extreme example, it represents a precedent for the design and construction of other underground processing plants.

A more conservative underground mineral processing facility would pre-concentrate the ore to reduce the amount of material transported to surface and leave a significant portion of the waste behind for backfill. The packing volume of the broken rock limits the amount of waste rock that can be returned to the excavation. Based on the packing density, approximately 60% of the tonnage mined can be replaced in the excavations (Lloyd, 1978).

The main technology used for pre-concentration is dense media separation, which has been applied to the separation of metal bearing sulfides from siliceous gangue for particles ranging in size from 0.25 to 500 mm. An underground processing facility would include crushing, screening, dense media separation and a dense media recovery circuit. The products are essentially dry (with the exception of surface moisture) for transport.

For ores that require further particle size reduction to achieve liberation, underground grinding and coarse particle flotation circuits have been proposed (Lloyd, 1978). In this case, pumping would be required to transport a bulk concentrate to the surface and to return the tailings to the underground excavations.

3.2 *Pre-concentration*

The objective of pre-concentration is to reject barren waste at as coarse a particle size as possible. The ability to achieve this objective is determined by the liberation characteristics of the ore. The potential benefits of pre-concentration include:

- Reduced environmental concerns, since the coarse barren waste product can be disposed of separately and less fine reactive (sulfide) waste is produced for disposal in tailings ponds or as backfill;
- Smaller footprint for processing facility at the surface;
- Reduced capital and operating costs resulting from reduced material handling, and smaller grinding and downstream processing facilities;
- Reduced operating cost per ton of ore processed, by lowering grinding work index and abrasion as a result of rejecting siliceous material prior to grinding.

For pre-concentration, dense media separation (DMS) is the most suitable process due to its high processing capacity and ability to make sharp separations at coarse particle sizes. Schena et. al. (1990) developed a methodology to assess the economic benefits of pre-concentration. Other coarse particle processing technologies that are being developed and may be applied to pre-concentration include dry high intensity magnetic separation using rare earth magnets and optical sorting (Arvidson and Reynolds, 1995).

The dense medium consists of an aqueous suspension of finely ground particles (typically magnetite, ferro-silicon or galena) that has an effective density between the densities of gangue (siliceous) and valuable mineral (sulfides) components. The medium imparts a buoyant force on particles causing low-density particles to float while allowing high-density particles to sink.

There are two basic classes of dense media separators referred to as static separators and dynamic separators. Static separators, such as drums, vessels and baths, are suited for particle sizes ranging from approximately 2mm to 500 mm and have a unit capacity of up to 300 t/h. Dynamic separators, such as the DSM Cyclone, Dyna-Whirlpool and Tri-Flo utilize centrifugal acceleration to achieve separation. Large diameter separators (up to 1 m) can process particles up to 100 mm at a rate of up to 400 t/h. Through optimization of dense media rheology, the size limit that can be effectively separated has been lowered to 0.25 mm. Dynamic separators are more compact than static separators and the centrifugal flow produces a sharper separation.

Despite the potential benefits of pre-concentration, it is often overlooked during flowsheet design. An example is the McKinnon Creek lead, zinc, gold, silver prospect in British Columbia, where metallurgical studies conducted over a twenty-year period had neglected to evaluate pre-concentration. In 1997, simple heavy liquid float sink testing revealed that crushing to minus 2 inches followed by heavy media separation would reject 40% of the plant feed while recovering greater than 97% of the metals. The coarse reject contained low sulfide grades, which allowed for easy disposal or the possibility to be sold as aggregate. Pre-concentration significantly improved the economic viability of the project, partially due to the reduced waste disposal and transport costs. An operating scenario that involved pre-concentration indicated that the project was economic with an internal rate of return of 18% based on gold $350/oz, silver $6.00/oz, zinc $0.55/lb and lead $0.30/lb (Trafford, 1998a,b).

Pre-concentration has been used at Cominco's Sullivan lead zinc mine in British Columbia since 1949 (Winckers, 1981). Crushed +3/16 inch ore is processed at a rate of 500 tph with two static dense media baths to separate siliceous minerals from sulfide minerals. The dense media (S.G. 2.95) is prepared from fine galena, which is produced in downstream flotation. DMS rejects approximately 30% to 35% of mill feed weight as a low sulfide waste product.

Dense media cyclones have been used in a similar application at the Mount Isa lead zinc silver mine in Australia since 1982. The plant rejects 30-35% of the run-of-mine ore while maintaining 96-97% metal recovery (Munro et al, 1982). More recently, AMT have elected to install dense media separation for pre-concentration at its Copper Creek project near San Manuel, Arizona. Crushing to -13 mm followed by DMS will reject 75-80% of the feed while maintaining copper recoveries of 93-95% (McCulloch et al, 1999). Although the primary incentive for these installations is increased plant capacity and reduced capital and operating costs, benefits are also realized for waste management.

For most ores, liberation of metal bearing sulfides requires further particle size reduction that can be achieved only from grinding. One innovative process was designed by the Chamber of Mines of South Africa Research Organization for underground processing of gold ores from the Witwatersrand (Lloyd, 1978). The process design was based on consideration of underground space limitations for a processing facility, mineralogical/liberation properties of the ore and the capacity to return waste backfill to excavations. The raw ore is ground in a centrifugal mill and screened at 3-mm. The minus 3-mm fraction is floated in specially designed flotation cells for use with very coarse particles to recover gold and gold bearing sulfide minerals. The flotation tailings are classified using hydro-cyclones, which return coarse gangue, moderately coarse valuable minerals and locked valuable minerals to the mill. The cyclone over flow is a waste product, which is thickened for backfill. The estimated gold recovery is 98% while rejecting 60% of the raw feed.

Peters et al (1999) addressed factors governing the feasibility of underground pre-concentration:

- The orebody characteristics are key parameters for the selection of the most suitable mining method. While bulk mining methods have the lowest operating costs, their application typically increases ore dilution. Selective mining methods, on the other hand, limit ore dilution, but do necessarily result in considerably higher costs (typically increased by a factor of 1.5 to 2.0). The ability of a pre-concentration/concentration plant to effectively eliminate most of the gangue minerals prior to hoisting of the ore could therefore have an impact on the selection of the most suitable mining method.
- Geology. The characteristics of the rock-mass hosting the mineral processing plant play an important part in controlling the stability of the excavation and need for long-term support. Faults in particular have to be identified and stability analysis has to consider the ground support required maintaining the excavation security. It is advisable to install the processing plant at some distance from the mining activities to limit the risk of damage caused by seismic events.
- Distance of mine from concentrator. A concentrator located in close proximity to the mine permits the utilization of both conventional handling systems as well as pumping systems. As the distance of the concentrator increases, then a pipeline application becomes less feasible due to the associated high capital requirements for its construction.
- Backfill requirements. The backfill requirements are primarily a function of the selected mining method. Apart from caving methods, every mining method can facilitate the use of backfill material. Backfill strength requirements, however, vary with the applied mining sequence and the deposit geology. In order to ensure the backfill strength, geotechnical strength tests have to be conducted on samples of the processing plant rejects. Rejects from gravity separation circuits are the preferred backfill materials, since long-term chemical reactions are not to be expected due to the absence of reagents. The potential for oxidation of sulfide minerals in the backfill and its impact should also be assessed.

3.3 *ARD Prevention: Segregation of waste products*

The legacy of past mining practices is large quantities of acid generating waste rock and tailings. In 1995, estimated acid producing and potentially acid producing mine wastes in Canada were 1,877.7 million tonnes of tailings and 738.9 million tonnes of waste rock. The corresponding estimates of acid producing liability were $1.52 billion and $0.40 billion, respectively (Feasby and Tremblay, 1995). Although grossly over simplified, these translate into costs of $0.81/tonne acid generating tailings and $0.54/tonne acid generating waste rock. The cost/tonne values are significantly under-estimated, as the estimated liability does not include waste stored sub-aqueously.

A specific example is the former Britannia Mine in British Columbia, where oxidation of sulfide minerals in underground mine excavations produces acid rock drainage that flows into Howe Sound. A comparison of alternative treatment scenarios indicated that the most cost-effective option would involve high-density sludge treatment and off-site storage of the sludge.

For a facility treating 500 m^3/h, the capital cost was estimated to be $4.66 million and the annual operating cost was $0.96 million (O'Hearn and Klein, 2000). It is believed that the effluent would need to be treated in perpetuity.

A pro-active approach to waste management can reduce the ARD liability. Feasby and Tremblay (1995) describe the role of mineral process engineers in the prevention of acid generation from sulfide-containing wastes.

For waste rock, acid formation can be prevented for the short term by adding alkaline materials such as lime or limestone. The acid generating was rock can be either placed in a dump with bedded layers of alkaline material or blended with the alkaline material.

For selective flotation of base metal sulfides, lime is added to depress pyrite and pyrrhotite. The addition of lime contributes to the alkalinity of tailings. Based on sulfide mineral contents of the tailings, more lime may be added to reduce the acid generating potential.

Sulfide removal from tailings allows for several disposal options. The isolated sulfide material can be stored below the surface of the tailings to limit the potential for oxidation. Alternatively, separate underwater disposal in a lined containment has been proposed.

As discussed above, pre-concentration of metal bearing sulfides using dense media separation isolates the sulfide minerals from non-reactive gangue prior to grinding.

At several existing plants, high sulfide cleaner flotation tailings, are often combined with low sulfide scavenger tailings. In this case, the high sulfide cleaner tailings can be isolated for disposal by installing a separate pumping system. Such steps can substantially reduce the acid generating potential of the bulk of the tailing material.

Sulfide removal can be achieved by froth flotation however; flotation of gangue sulfide minerals in base metal sulfide processes may not be trivial. In base metal plants, selective flotation against pyrite is commonly achieved by adding lime to increase the pH. Subsequent pyrite flotation from the tailings may require the addition of acid to lower the pH.

An alternative approach would involve bulk sulfide flotation to recover all sulfide minerals followed by depression of the gangue sulfides during selective flotation. This approach would involve redesign of flotation circuits. For some ores, depressing gangue sulfide minerals that had been activated for flotation may be problematic.

At the University of British Columbia, research is underway to evaluate a continuous centrifugal gravity concentrator for the separation of sulfides from tailings. These types of concentrators are a relatively new technology that may have application in this area. Most noteworthy are the Falcon C, Knelson CVD, Kelsey Jig and Mozley Multi-gravity Concentrator. They represent a clean non-chemical technology that could separate sulfides from the tailing stream and would not interfere with upstream metal recovery processes. The potential for these technologies may be limited, since sulfide particles in tailings are often finer than the siliceous gangue particles.

Oxidation of sulfide minerals in a controlled environment using autoclaves can eliminate the acid generating potential. Autoclaves are used at some gold mines to oxidize gold bearing sulfide minerals (refractory gold) prior to cyanidation. Treatment of waste is in this manner is not economic, however, controlled oxidation may be justified in some situations. At the Miramar Con Mine, arsenic trioxide waste from roaster operations that accumulated over decades from mining in the area is presently being treated in this manner (Young and Allan, 1994).

"Tailings rheology engineering" involves thickening of the tailings prior to placing in a pond, which helps to maintain a high water table in tailing ponds by capillary action. If the water table can be kept above sulfide layers, the rate of oxygen diffusion and sulfide oxidation will be limited. The addition of lime to the tailings can increase the apparent viscosity, which reduces particle size segregation, improves the moisture retaining capacity and decreases oxygen diffusion. Similarly, non-reactive tailings with low permeability can be used for layered covers.

4 PASTE AND CO-DISPOSAL OF WASTES

Paste backfill is a reconstituted form of tailings formed by separating the coarse and fine-grained fractions of the tailings in a hydrocyclone. The coarse-grained fraction (the cyclone underflow) is allowed to drain naturally and the fine-grained fraction (the cyclone overflow) is de-watered in thickeners or by filtration. The two size fractions are then re-combined together with cement and other coarse-grained materials may be added to enhance strength and decrease water retention. The result is a viscous material that can be molded relatively easily into a desired form.

The fundamental problem in the design of a paste production system is to reduce the water content of the tailings sufficiently so that the size fractions do not segregate and so that the paste can be transported in a pipeline. These are conflicting requirements but an optimum amount of de-watering can be found for a broad range of tailings materials by simply varying the particle size distribution. It is desirable to reduce the content of particles of size 20 microns or less since electrostatic charges surrounding these particles attract water molecules thus increasing moisture retention. The usual design goal is to reduce the <20-micron fraction to a minimum of 15% to 20% by weight. (Brackebusch, 1994; Cincilla et al, 1998).

Paste is commonly considered for use as backfill support in underground voids and has been considered for use to reshape surface landforms near mining sites. If paste backfill is used as support to facilitate further mining, the paste production operation requires very close integration with the mine plan. The paste must be available when and as needed; there is no opportunity to "stockpile" paste. As a paste production operation involves several types of equipment including hydrocyclones, filters, mixing tanks, and storage silos, the issue of system reliability arises. To some extent this can be dealt with via redundancy or by designing equipment so that it is easily repaired, if necessary.

An equally important issue is the design of a paste production system that can adapt to changes in particle size distribution, mineralogy, or moisture content. Relatively small variations in these properties can result in large changes in paste viscosity and the ability to transport the paste and in the long-term stability of the paste backfill. Sensor and control systems are currently available to handle changes in moisture. Further development of such systems, improvements in pumping or transporting technology and in additives to improve flow are all necessary so that paste production plants can be designed to deal with a wide variation in tailings properties.

Paste technology potentially allows for a considerable amount of innovation in mine waste management. Co-disposal of tailings and waste rock, where waste rock would be used as part of the coarse fraction of the paste, is one such possibility. Alternatively, a waste rock and tailings paste could be deposited on land. Combinations of these alternatives could be considered to provide more flexibility. Currently these possibilities suffer from the uncertainty associated with the long-term chemical and physical stability of tailings/waste rock combinations.

Paste can be considered as a form of waste reduction since it reduces the amount of water to be managed in a tailing pond. Although, the water extracted from the tailings may be used for processing, it must be treated and disposed of after the mine is closed. This is likely a better alternative than the long-term liability associated with having water remain in a tailings pond after mine closure where it can generate acid or high metal concentrations in surface or groundwater for some years after the mine is closed.

Capital investment for a paste production plant is high compared to other backfill systems. In most cases, this if offset by lower operating costs and increased productivity. Even more important is the avoidance of long-term liabilities associated with tailings ponds and waste rock dumps. Although difficult, it is important to take these factors into account when evaluating waste management systems.

CONCLUSION

A more holistic approach to the design of mining and milling systems in mines is needed which accounts for the different aspects of environmental and social costs and benefits. This should optimize the engineering design of excavation, transport and processing in terms of more global economic factors. Waste management costs would then be transparent and taken into account in the mining and milling design. Technology development should be focussed on integrated selective and flexible mining systems.

REFERENCES

Arvidson, B., Reynolds, M.S., New Photometric Ore Sorter for Conventional and Difficult Applications, *Processing for Profit Conf., Industrial Minerals,* Amsterdam, 1995.

Brackebusch, F. W., 1994. Basics of paste backfill systems. *Mining Engineering*, October, pp 1175-1178.

Brewis, T., 1995. Andina Develops for the Future. *Mining Magazine*, London, February 1995, pp.78-87

Cincilla, W., Landriault, D., Newman, P., and Verburg, R., 1998. Paste Disposal. *Mining Environmental Management*, May, pp.11-15.

Feasby, D.G., G.A. Tremblay, Role of Mineral Processing in Reducing Environmental Liability of Mine Wastes, *27th Annual CMP Meeting,* Ottawa, 1995.

Francis, H., Internal Ph.D. Report, Department of Mining and Mineral Process Engineering, University of British Columbia, 2000.

Francis, H., Salter, D. and M. Scoble, 1998. Intelligent Ore Delineation: Integrating Geosensing and GIS. *100th Ann. General Meeting, Can. Inst. Min. Metall.*, Montreal, CD ROM.

Knissel, W., Jurgen, H. and M. Fahlbusch, 1995. Significance of Selective Mining Exploitation for Economical and Environmentally Beneficial Underground Ore Mining. *Int. Jnl. Min. Resources Engineering, Imp. Coll. Press*, 5, 2, pp. 165-174.

Lloyd, P.J.D., The potential of Integrated Mining and Extraction Systems on the Witwatersrand, *Proc. 11th Int Commonwealth Mining and Metallurgy Congr. IMM,* Hongkong, 1978.

McCulloch, W.E., R.B.Bhappu, J.D. Hightower, Copper Ore Pre-concentration by Heavy Media Separation for Reduced Capital and Operating Costs, *Copper 99*, Phoenix Arizona,1999.

Moss. A., Looking into the crystal ball: the industry in the 21st Century. *Global Mining Conference, Pricewaterhouse-Coopers*, 1999.

Munro, P.D., I.S. Schache, W.G. Park, R.M.S. Watsford, The Design, Construction and Commissioning of a Heavy Media Plant for Silver-Lead-Zinc Ore Treatment - Mount Isa Limited, *XIV Int. Min. Proc. Congr.*, VI-6.1-20, Toronto, 1982.

O'Hearn, T., Klein, B. A Comparison of Acid Rock Drainage Treatment Scenarios at the Former Britannia Mine, *Sixth Int. Symposium on Environmental Issues and Waste Management in Energy and Mineral Production*, Calgary, 2000.

Parsons, A. S. and Hume, H. R., 1997. The contribution of new technology to improved environmental performance in the mining industry. *UNEP Industry and Environment*, October-December, pp 38-43.

Peters, O., Scoble, M. and T. Schumacher, 1999. The technical and economic potential of mineral processing underground. *Annual General Meeting, Can. Inst. Min. Metall.*, Calgary, CD ROM.

Schena, G.D., R.J. Gochin, G. Ferrara, "Preconcentration by Dense-Medium Separation - An Economic Evaluation", *Trans. IMM, Section C*, 99, 1990, pp. C21 - C31.

Scoble, M., 1994. Competitive at Depth: Re-Engineering the Hard Rock Mining Process. Keynote Lecture, *1ST North American Rock Mechanics Symposium*, Austin, Balkema, Rotterdam.

Trafford, G.T., Company Receives Scoping Study Weymin Mining Corporation, *News Release* October 26, 1998.

Trafford, G.T., Results Nearly Double Grade to 13 Grams Gold Per Tonne, *News Release* Weymin Mining Corporation, January 20, 1998.

Winckers, A.H., The Use of Rubber Mill Liners at Cominco's Sullivan Concentrator" *13th CMP Annual Meeting,* Ottawa, 1981, pp 300-310.

Young, P.D., M. Allan, Commissioning and Early Operation of the Con Mine Pressure Oxidation Plant, *26th CMP Annual Meeting*, Ottawa, 1994.

Environmental Issues and Management of Waste in Energy and Mineral Production, Singhal & Mehrotra (eds)
© 2000 Balkema, Rotterdam, ISBN 90 5809 085 X

Application of thermal drilling machine to waste processing

S.Shimada
Institute for Environmental Studies, Graduate School of Frontier Science, The University of Tokyo, Japan

ABSTRACT: Jet piercing machine was originally developed for the purpose of thermal drilling and cutting of hard rocks. This machine can be used for waste processing. A flame jet expelled from the thermal drilling machine (jet burner) can crush, grind, disintegrate and dry materials. A flame jet expelled from a combustion chamber through the nozzle of a jet burner was used to process sludge, which had high water content and flocculated easily in a wet condition, and oily sludge, the mixture of oil, water, sand and others. A system for processing flocculated wastes and oily wastes by using a thermal drilling machine has been developed. Flocculated wastes with a high water content were dried and at the same time crushed (or disintegrated) in the processing vessel of the system. Oily wastes were cracked into lighter oils and solids were separated. The drying- disintegration and /or recycling of aluminum hydroxide sludge and crude oil sludge were processed for the recycling use and the disposal. Firstly, the component of the processing system and the mechanism for generating the flame jet, are reported. Secondly, the experimental results of the processing of aluminum hydroxide sludge, oil sand and crude oil sludge are reported. Finally the processing and recycling of many kinds of wastes by this system are proposed.

1 INTRODUCTION

As a flame jet generated by a thermal drilling machine (or, jet burner) has high temperature and pressure, it has been used for many processes such as thermal drilling, cutting, crushing, disintegration, drying, evaporation and thermal cracking. Originally, the flame jet was developed for the purpose of drilling crystalline hard rocks in the 1940's. However, it is now used for heat generator in chemical plants. The flame jet is generated in a combustion chamber by burning fuel and oxidizing agent. The flame jet can be used to dry and disintegrate flocculated materials. These wet flocculated materials become lumpy by conventional drying methods. Therefore, they need to be disintegrated after drying, to form a powder. The drying - disintegration system using a flame jet combines these two processes.

This paper reports the outline of the "Flame Jet Processing System", which is using the principle of thermal drilling machine as a heat generator, and some processing results on materials with a high water content and oil.

2 JET BURNER OPERATING SYSTEM

2.1 *Jet burner*

A jet burner with a water-cooled combustion chamber (hereafter, denoted C.C.) was used (Fig. 1). The water used for cooling the C.C. was circulated between the jet burner and a cooling tower.

The fuel was kerosene. The kerosene was supplied to the burner by a kerosene pump and atomized at an injector (8). The compressed air supplied by a compressor flowed through a pipe into the C.C. (2) through a swirler (3). The swirler promoted the mixing of the kerosene and the air and at the same time protected the inner surface of the C.C. by cooling. The air and the atomized kerosene composed a combustible mixture. The combustible mixture was ignited by the ignition plug (7). The burned gas was expelled through a nozzle throat (5) in a supersonic condition. A flame detector (4) installed in the burner detected extinguishment of the flame in the C.C.. The cooling water was circulated through the annulus pipe (6) by a water pump in order to protect the C.C. from damage.

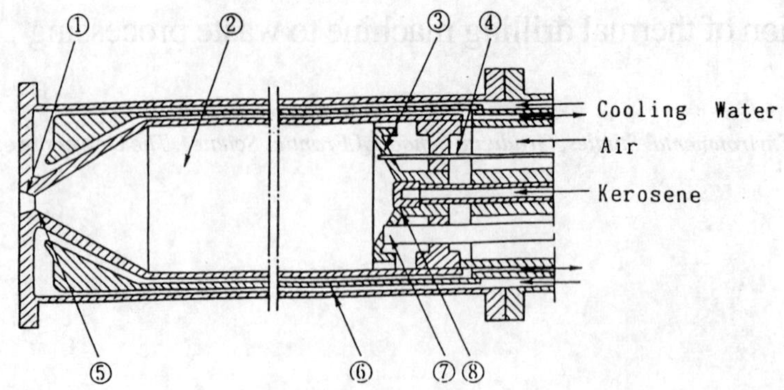

Fig. 1 Jet burner of water cooling type

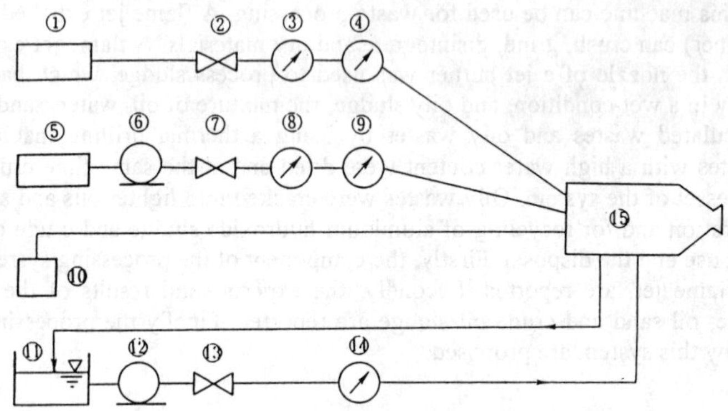

Fig. 2 Operating system of jet burner

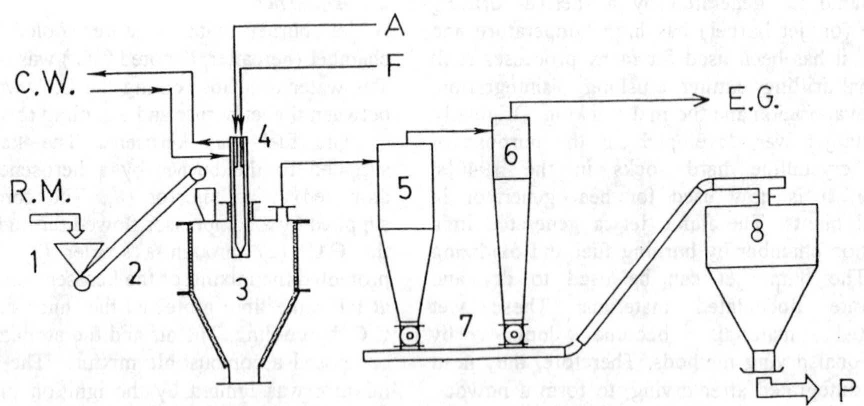

Fig. 3 Flow sheet of flame jet processing system
A: compressed air F: fuel C.W.: cooling water
R.M.:raw material E.G.:exhaust gas P:product

2.2 *Operating system*

Fig. 2 shows the operating system. The compressed air was supplied by a compressor (1) to a burner (15) through a control valve (2), a pressure gauge (3) and a flow meter (4). Kerosene was supplied by a kerosene pump (6) from a kerosene tank (5) to the burner through a control valve (7), a pressure gauge (8) and a flow meter (9). The mixture of fuel and air was ignited by an ignition plug, to which the power is supplied by an electric power source (10). The cooling water was supplied by a water pump (12) from a cooling tower (11) to the burner through a control valve (13) and a flow meter (14).

3 CHARACTERISTICS OF A FLAME JET

The thermo-dynamic properties of the flame jet influence its processing efficiency. Measurement of the flame characteristics is essential for applications of a flame jet. Temperature, pressure and velocity were measured by Pitot tubes and thermocouples. The principle and the details of the measurement are described in the references. The results are summarized below. The details of the results are published in the references

1) Temperature: The maximum temperature of the flame jet was about 1,500K. The calculated temperature in the C.C. was 2,000K.

2) Pressure: The maximum pressure of the flame jet was about 80% of that in the C.C.. The pressure in the C.C. generally ranges 5.0 - 6.0 atms in gauge pressure.

3) Velocity: The maximum velocity was 1,300 m/sec at the nozzle exit.

4) Distribution: Pressure, temperature and velocity fluctuated in the range of a supersonic region due to the generation of oblique shock waves. The length of a supersonic region was about 10 times nozzle throat diameter from the nozzle exit.

5) Heat transfer coefficient: The maximum heat transfer coefficient of a flame jet, when it impinged onto a flat surface was 6,000 $W/m^2/K$ at a stagnation point.

4. PROCESSING SYSTEM

Fig. 3 shows the basic flow sheet of a processing system. The pressure and flow rates of the compressed air, kerosene and cooling water were controlled by a control panel.

The raw material (sludge etc.) in a hopper (1) was fed into a processing chamber (3) by a conveying mechanism (2) and processed (dried and disintegrated) by the flame jet expelled from the jet burner (4). The temperature in the processing chamber was controlled by adjustment of the feed rate of the material. The dried and disintegrated material (product) was collected in a gravitational separator (5) and cyclone (6). Other particle collection equipment, such as bag-filter can be used additionally depending on the particle diameter of the product. The product recovered from the bottom of the collecting equipment was transported to a storage hopper (8) by a conveyor (7). Other letters in the Fig.3 means that, C.W.: cooling water, A: compressed air, F: fuel (kerosene), R.M.: raw material, E.G.: exhaust gas, P: product.

The weight and water content of the product were measured. The temperature in the processing chamber was measured by a pair of thermocouples installed at the exit of gas flow in the drying chamber. When the resistance of the separators was high, suction was provided by installing a blower. Several types of processing chamber were used. The maximum inner diameter and the height of the processing chamber of a standard type were 800 mm and 1,500 mm, respectively. The distance between the nozzle exit and the bottom of the chamber was about 600 mm. The inner volume of the processing chamber was $0.3m^3$. The raw material was fed into the chamber through the feed inlet near the jet flame. The raw material collides against the bottom of the chamber due to the action of the jet. The raw material was crushed and disintegrated in this collision. The disintegrated particles flowed up to the duct with the exhaust gases, meanwhile they were ground by collision between particles, and were fed into the separators.

Fig. 4 shows the flow sheet for processing the oily materials. Additionally to Fig. 3, an oil condenser, a recovery oil tank and a deodorizer were installed to recover the oil and to minimize the odor emitted by the reaction of sulfur contained in the oil.

5. CHARACTERISTICS OF THE SYSTEM

The processing system using a flame jet is very useful for drying- disintegration of materials, such as sludge wastes. The processed wastes can be recovered as powder and recycled or used for other purposes. The followings are the expected advantages using jet burner for processing sludge and oily materials.

1) High heat transfer coefficient: As mentioned before, the heat transfer coefficient is high. This

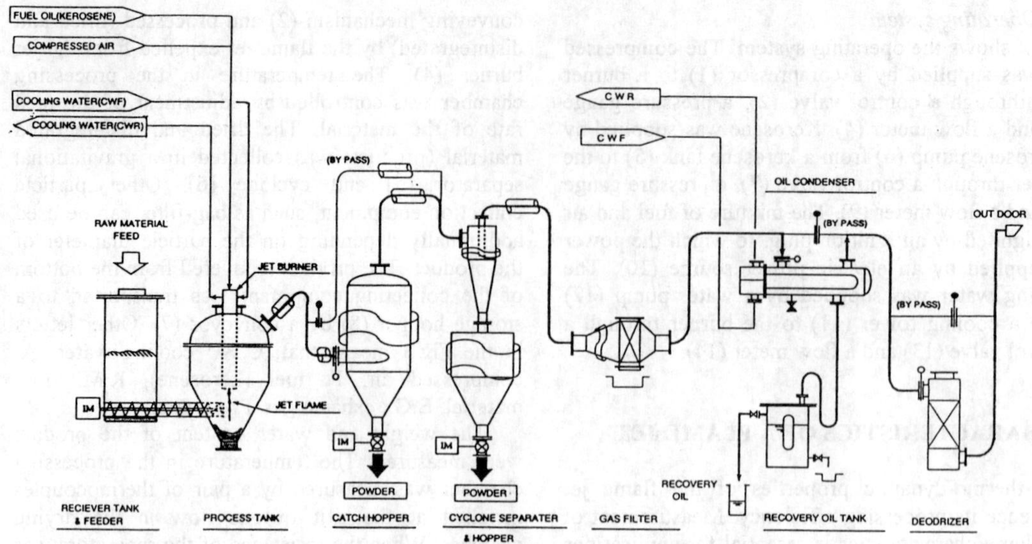

Fig. 4 Flow sheet of crude oil sludge processing system

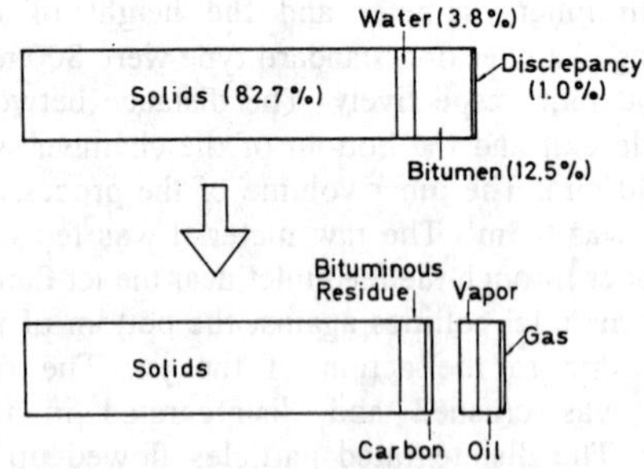

Fig. 5 Weight change of components of oil sand (pre- and post-processing)

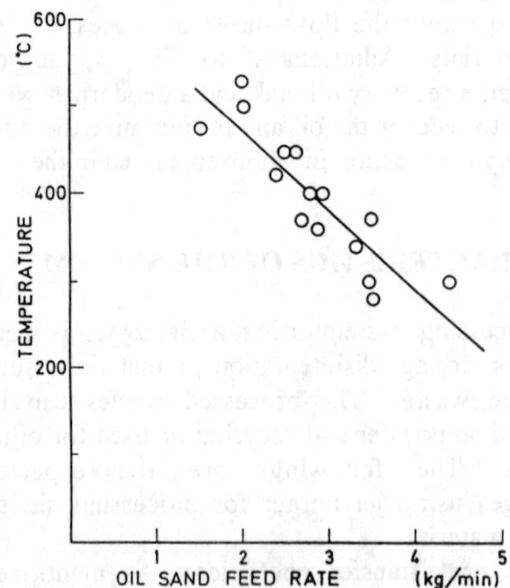

Fig. 6 Relationship between processing rate and processing temperature

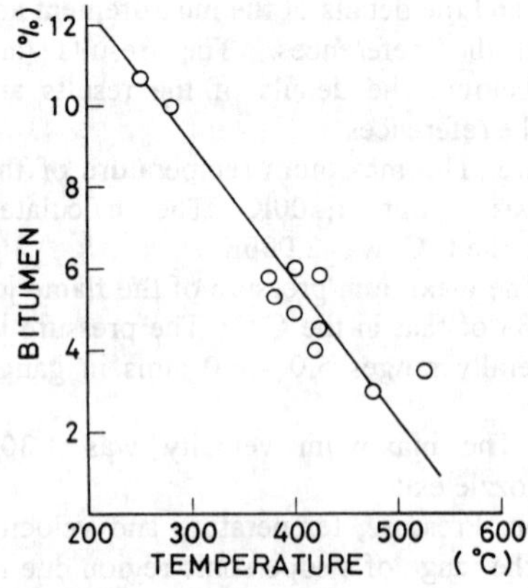

Fig. 7 Relationship between processing temperature and weight of residual bitumen

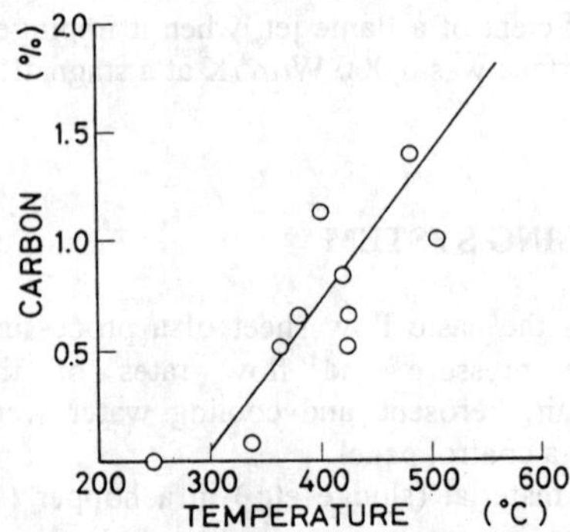

Fig. 8 Relationship between processing temperature and carbon weight

allows the drying chamber to be small.

2) Inert gas: The burned gas has lean oxygen in it. Accordingly, though the gas has a high temperature, the wastes being processed do not burn. The processed wastes are recovered and re-utilized. This implies that the oily materials are processed without combustion in this system.

3) Direct heating: As the heating method of this system is a direct method, the thermal efficiency is expected to be high.

4) Jet with a pressure: Owing to the high pressure of the flame jet, it can be operated in liquid materials. The heat is exchanged directly between the flame and the liquids.

5) Combined process with crushing and/or disintegration: The crushing (grinding) and/or disintegration of the materials ensures that the total surface area of the resultant particles is large. Therefore, the drying process is completed in a short time. Moreover, because of the disintegration of the surface of the materials, the new surface with high water content is exposed, and wastes such as sludges are dried in a short time. Many kinds of sludges with water contents of more than 80% are dried to water contents of from 5% to 40% in one drying process.

6) Prevention of particle adhesion on chamber wall: Owing to the high velocity of jet flowing on the inner wall surface of the drying chamber, the particles, though they are sticky, does not adhere on the wall, which adhere on it by a conventional drying method,

7) Pneumatic transportation: As the materials are transported by the stream of burned gas, the system needs no special transportation and discharge mechanism.

6 RESULTS

6.1 *Aluminum hydroxide sludge*

The water content of the aluminum hydroxide sludge was about 80%. The water content of dried one was between 16 - 22 %. The thermal efficiency was less than 50%. Thermal efficiency here was defined as (heat used for making steam) / (heat of combustion of kerosene). This reason of the low efficiency was that the heat taken by the exhaust gas was not recovered. The recovery of the heat of exhaust gas is necessary in practical plants. The maximum volumetric heat transfer coefficient (VHTC) was 600 W/m^3/K.

The standard operating condition of the jet burner and the processing rate are shown in Table 1.

Table 1 : Operating condition of jet burner for processing aluminum hydroxide sludge

Pressure in C.C. :	0.6 MPa
Flow rate of air :	2.55 m^3/min
Flow rate of kerosene:	0.17 l/min
Processing temperature:	300-375 °C
Water content (wet base)	
Pre- processing :	77 %
Post- processing :	17 %
Processing rate :	1.4 kg/min

6.2 *Oil Sand*

The oil sand is like a sludge containing oil. Athabasca oil sand of Alberta, Canada was used to process and recover oil by the Flame Jet Processing System.

Fig. 5 shows the pre- and post-processing weight change of components of oil sand. The operating condition of the jet burner was shown in Table 2.

Other results are shown in Fig. 6 – Fig. 9. Fig. 6 shows the relationship between processing rate and processing temperature. Fig. 7 shows the relationship between processing temperature and weight of residual bitumen, which was not processed and remained in the sand. Fig. 8 shows the relationship between processing temperature and the carbon weight. With the increase of processing temperature, the weight of residual bitumen decreases and the carbon weight increases. Due to the some errors on the operation of condenser, all components of oil could not recovered at the condenser. Therefore, the expected weight of recovered oil was calculated from the following equation.

$$12.5 - (\text{Residual bitumen} + \text{Carbon}) \quad (\%)$$

The processing cost of oil sand by Flame Jet Processing System calculated by using the above equation is about US$40/bbl. Fig. 9 shows the property difference of oils recovered from oil sand by Flame Jet Processing and solvent extraction. This figure shows that the lighter oil is recovered by Flame Jet Processing System.

Table 2: Operating condition of jet burner for processing oil sand

Pressure in C.C. :	0.6 MPa
Flow rate of air :	1.31 m^3/min
Flow rate of kerosene:	0.1 l/min
Excess air ratio :	1.22

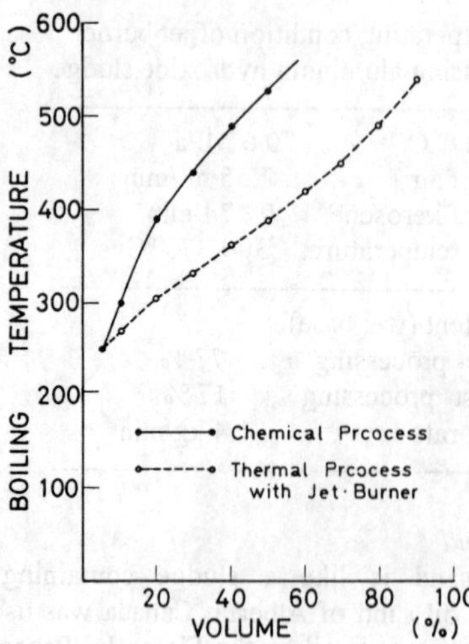

Fig. 9 Property difference of oils recovered from oil sand by Flame Jet Processing and solvent extraction

Table 3 : Percentage of residual oil in solids

Processing temp. (°C)	Percentage (%)
200	6.08
250	5.56
275	0.32
300	0.44
350	0.85

6.3 *Crude Oil Sludge*

Crude oil sludge exhausted from crude oil tankers was processed by this processing system. Crude oil sludge is a mixture of solids (steel and sand), water and oil. The percentage of these components varies widely in the range of 20-60% (in weight) for oil, 15-40% for water and 15-60% for solids. The crude oil sludge is hard to handle, for it forms slurry, mud and cake, depending on the percentage of the components.

Table 3 shows the percentage of oil residue remained in the recovered solids after processing. If the disposal standard of the solids is less than 1% for remained oil, the crude oil sludge must be processed at the temperature higher than 275 °C.

7. CONCLUSION

The outline of the Flame Jet Processing System and some processing results were reported. Flame Jet Processing System is a very useful processing method to dry sticky flocculated materials, which is rather difficult to dry by a conventional drying method, and to process oily waste.

Acknowledgment

The author expresses his hearty gratitude to Hitachi Techno Engineering Co., Ltd. for the cooperation in promoting this study.

Environmental Issues and Management of Waste in Energy and Mineral Production, Singhal & Mehrotra (eds)
 ISBN 90 5809 085 X

An integrated process for utilization of gypsum and pyrite wastes

D.Tao
Department of Mining Engineering, University of Kentucky, MMRB, Lexington, Ky., USA

M.Abdelkhalek, S.Chen & B.K.Parekh
Center for Applied Energy Research, University of Kentucky, Lexington, Ky., USA

M.T.Hepworth
Department of Civil Engineering, University of Minnesota, Minneapolis, Minn., USA

ABSTRACT: Pre-combustion coal cleaning and post-combustion flue gas desulfurization (FGD) using lime or limestone are the primary methods employed to reduce sulfur dioxide (SO_2) emission from coal combustion. These processes generate voluminous pyrite and gypsum solid wastes that are usually landfilled, occupying thousands of acres of land and creating serious land and water pollution problems due to the release of acids and toxic substances. An integrated process has been developed for the combined utilization of gypsum and pyrite wastes by converting them into useful products including lime, sulfur, and direct reduced iron (DRI). The process includes four major steps. The first step is to concentrate pyrite from coal tailings using flotation or gravity separation. The second is the thermal decomposition of pyrite to pyrrhotite and sulfur. This is followed by the reduction of pyrrhotite with carbon in the presence of lime to produce iron and calcium sulfide. The fourth step involves the reaction of calcium sulfide with gypsum to regenerate lime. The chemical reactions involved in the process were studied using thermogravimetric analysis (TGA) and X-ray diffraction (XRD) techniques. Process variables studied included reaction temperature, time, reactant composition, and purity of reactants.

1 INTRODUCTION

The 1990 Clean Air Act Amendments (CAAA) require coal-burning utilities reduce sulfur dioxide (SO_2) emissions to 1.2 lb per million Btu by the year 2000. Pre-combustion coal cleaning and post-combustion flue gas desulfurization (FGD) processes using lime or limestone are the primary means used to achieve the compliance. Coal cleaning produces cleaner and higher-calorie coal products for utilities by reducing pyritic sulfur and other ash forming minerals. FGD processes prevent sulfur dioxide formed during coal combustion from releasing to air and forming acid rain.

Coal cleaning and FGD processes produce millions of tons of gypsum and pyrite wastes annually in the coal-related energy industry. The American Coal Ash Association (1992) reported that about 20 million tons of FGD by-product gypsum ($CaSO_4$) was produced in 1991. U.S. Environmental Protection Agency (1988) predicts that about 50 million tons of FGD by-products will be produced by the year 2000 from burning high sulfur coal for generation of electricity using Clean Coal Technology (CCT). FGD processes also consume millions of tons of limestone and lime. General Accounting Office (1977) estimated that 40% lime or 70% limestone produced in the U.S. is used for FGD processes. Calcination of limestone releases tremendous amounts of carbon dioxide into atmosphere as a greenhouse gas, which is believed to cause global warming. The binding treaty signed by the U.S. government on the 1997 International Conference on Global Climate Change in Kyoto, Japan requires a drastic reduction in carbon dioxide emissions in the U.S.

Pyrite (FeS_2) is ubiquitous in the earth's crust and exists in significant quantities in many coal and mineral (zinc, lead, copper, uranium, gold, silver, etc.) deposits. The Commonwealth of Kentucky produces about 40 million tons of high sulfur coal that contains about 7% pyrite. About half of pyrite in the coal or 2.8 million tons is discarded into tailing ponds. Other five major high sulfur coal producing states, i.e., Illinois, Indiana, Ohio, Pennsylvania, and Virginia, generate similar amounts of pyrite wastes (Feeley et al., 1990). It is estimated that about 10 million tons of pyrite is discarded every year from high sulfur coal production in these six states. Mining and processing of other minerals also produce huge amounts of pyrite wastes.

Gypsum and pyrite wastes are currently landfilled although efforts have been made to use FGD by-products in road construction, acid soil neutralization, material production, etc. (Shainberg et al., 1989). Pyrite contains heavy metals such as arsenic, cobalt, copper, lead, nickel, and zinc (Wewerka, et al., 1982). When pyrite waste is exposed to air and ground water or rain in impoundments, it readily oxidizes, forming large amounts of acid and releasing soluble toxic substances that pollute land and water sources. The coal mining industry alone spends about half billion dollars a year to neutralize acids produced from pyrite oxidation (Wildeman, 1991). It is estimated that pyrite-containing wastes pollute over 12,000 miles of river and streams and 180,000 acres of lakes and reservoirs in the U.S. (Kleinmann, 1989). Similarly, impoundment of gypsum wastes imposes the risk of continuous leaching of environmentally undesirable constituents into the groundwater and streams. Landfilling of pyrite and gypsum wastes also occupies thousands of acres of land. The average size of a U.S. landfill encompasses 173 acres and has a potential capacity of six to seven million tons (Hite, et al., 1994). This suggests that landfilling of 50 million tons of FGD by-products alone will consume more than 1236 acres of land per year. An environmentally sustainable approach to disposing of pyrite and gypsum wastes is critical for continued existence and growth of coal-based energy industries.

The present investigation was intended to develop a novel integrated thermochemical process to utilize millions of tons of gypsum and pyrite wastes generated annually by the U.S. energy industries and reduce the emission of millions of tons of greenhouse gas carbon dioxide. This was accomplished by converting gypsum and pyrite wastes to marketable products such as lime, direct reduced iron (DRI), and sulfur products and obviating the need to calcine millions of tons of limestone for use in utility scrubbers. Figure 1 shows the developed process which primarily consists of four steps:

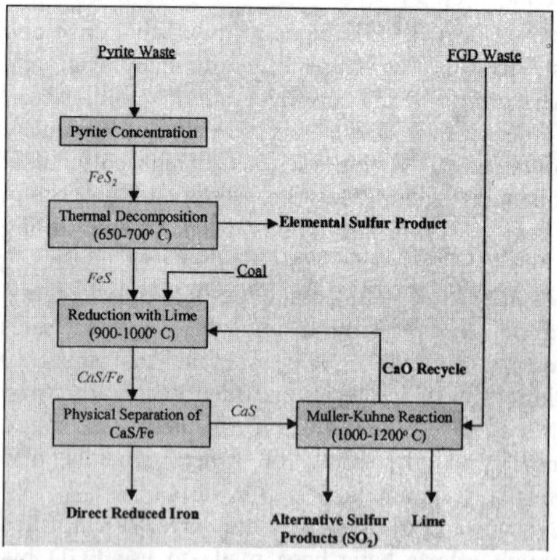

Figure 1. Thermochemical process for utilization of pyrite and gypsum wastes.

1. Pyrite (FeS_2) is concentrated from tailings discharged from coal-cleaning operations. This is achieved by physical separation processes such as gravity separation and flotation.

2. Pyrite is thermally decomposed to pyrrhotite (FeS) and elemental sulfur:

$$FeS_2 = FeS + S. \tag{1}$$

This reaction requires a non-oxidizing atmosphere and a temperature of 650-700° C for favorable kinetics. The labile sulfur vapor from the decomposition of pyrite is removed by condensation. The composition of pyrrhotite produced depends on the reaction temperature, retention time, and particle size of pyrite. Impurities such as arsenic, lead, and selenium in pyrite evaporate and report with elemental sulfur during this process.

3. Pyrrhotite is reacted with lime in a reducing environment to form calcium sulfide and direct reduced iron (DRI):

$$FeS + C + CaO = Fe^o + CaS + CO. \tag{2}$$

Steps 2 and 3 can be accomplished as a two-stage fluidized-bed process with reaction (1) as the first stage and reaction (2) as the second stage. The hot off-gas, rich in CO from the second stage would provide the enthalpy for the first stage of reactions.

4. Calcium sulfide is employed as a reductant to react with gypsum wastes to produce lime and sulfur dioxide:

$$CaS + 3CaSO_4 = 4CaO + 4SO_2 \quad (Muller-Kuhne\ reaction). \tag{3}$$

The lime can be recycled to the FGD process that produces gypsum waste while concentrated sulfur dioxide can be sold directly or used to produce sulfuric acid. The overall reaction is:

$$FeS_2 + 3CaSO_4 + C + 1/2O_2 = Fe^o + 3CaO + 4SO_2 + S + CO_2 \tag{4}$$

Reaction (4) indicates that one mole of FeS_2 (120 g) will react with three moles of $CaSO_4$ (408 g) to produce one mole of Fe (56 g), three moles of CaO (168 g), four moles of SO_2 (256 g), and one mole of elemental sulfur (32 g). CaO will be fed back to the

FGD process while Fe, SO_2, and elemental sulfur are all salable products.

2 EXPERIMENTAL

The thermogravimetric analysis (TGA) of chemical reactions involved in the process was investigated using a thermal gravimetric analyzer Model TGA 7 from Perkin Elmer. A sample of approximately 1000 mg reactant(s) was used in each TGA test. Nitrogen gas of dry grade was passed through the reaction tube at a flow rate of 200 cm^3/min to create an inert atmosphere. During the reaction, the weight of the sample was recorded at 30 second intervals into a data file which was processed later to generate the "percent completion vs. time" curve. A tube furnace (Type 59544) from General Signal was used to study the thermal reactions on relatively larger scales under the optimum conditions determined from the TGA characterization.

A Denver D-12 flotation cell and a Falcon Concentrator SB4 were employed for pyrite concentration from tailings based on the difference in surface hydrophobicity and specific gravity, respectively. A Davis tube magnetic separator was used for the separation of directly reduced iron from calcium sulfide.

Samples of pyrite waste and FGD gypsum waste were obtained from the pyrite rejection machine and flue gas scrubber, respectively, of an LG&E midwestern power plant. The pyrite sample was crushed and screened to different size fractions prior to its use. Pure pyrite and pyrrhotite samples were acquired from Ward's Geology, NY. Lime of analytical grade (98% purity) was dehydrated and decarbonated at 1000°C. Calcium sulfate of analytical grade (-45 μm) was dehydrated at 650°C. Pure pyrite, pyrrhotite, lime, and activated carbon in the size fraction of +45-74 μm were used in experiments, unless specified otherwise.

3 RESULTS AND DISCUSSION

Figure 2 shows the –65+325 mesh (-0.210+0.044 mm) pyrite concentration results obtained using the Falcon Concentrator. Pyrite can be upgraded from about 54% to 69% at about 70% recovery. Direct froth flotation using Aero 633 as coal depressant and ethyl xanthate as pyrite collector produced more than 90% grade pyrite at reduced recovery.

Figure 3 shows the TGA data on decomposition of pure pyrite at reaction temperatures of 600, 650, and 700°C, respectively. The reaction completion rate was strongly dependent on the reaction temperature. For example, only 12% completion was achieved at 600°C after 2 minutes. The completion increased to 35% at 650°C and more than 90% at 700°C within the same reaction time. Obviously, higher reaction temperature was favored for increased reaction rate.

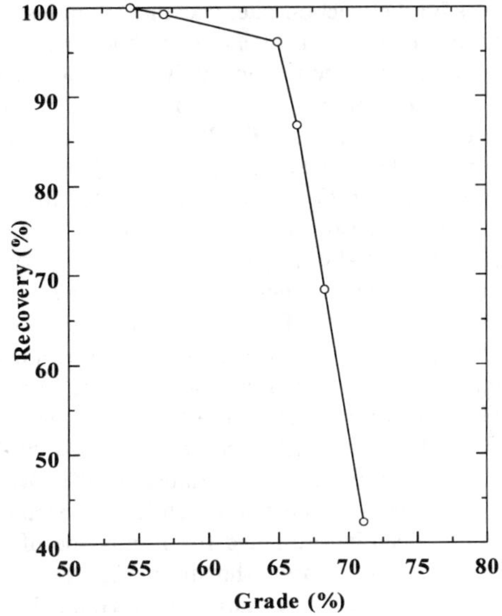

Figure 2. Pyrite concentration using Falcon Concentrator

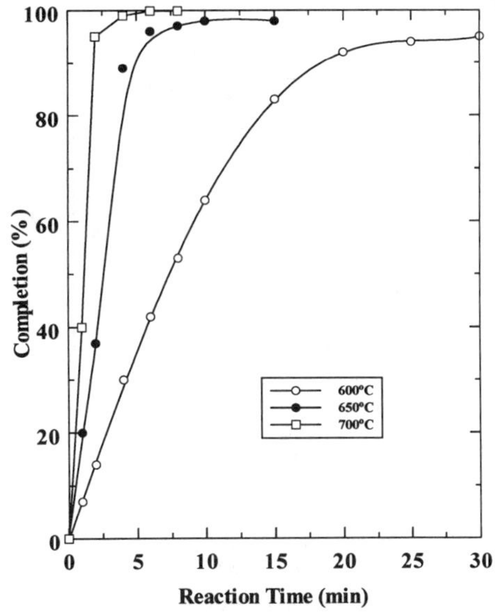

Figure 3. TGA results of pure pyrite decomposition at different temperature

This is in agreement with the fact that Reaction (1) is an endothermic process with a free energy of 43500-44.85T cal (Turkdogan, 1980). The same behavior was observed with 90%, 70%, and 55% grade pyrite samples although less pure pyrite showed slightly lower reaction rate.

To investigate effects of surface area of reactant particles on the reaction rate, decomposition tests of pyrite powders in different size fractions were performed in the tube furnace at 800°C. The results are shown in Figure 4 for –50 mesh (-0.30 mm), -100 mesh (-0.15 mm), and –200 (-0.074 mm) size fractions. The 50% reaction completion was achieved at about 23, 7.5, and 4.5 minutes for –50 mesh, -100 mesh, and –200 mesh fraction, respectively. Smaller particle size or larger surface area increased the decomposition reaction rate.

Figure 5 shows the TGA data on the kinetics of reaction (2) using pure FeS at varying temperatures. As temperature (T) increased from 950°C to 1100°C, reaction rate (k) increased more than eight times. A 50% completion was achieved at approximately 77, 26, 13, and 9 minutes at temperatures of 950, 1000, 1050, and 1100°C, respectively. The rate constant of the reaction can be estimated from the slope of the linear part of the curve shown in Figure 5 and correlated with the temperature T in Equation (5):

$$\ln k = -2.752\times10^4 / T + 22.14. \qquad (5)$$

The activation energy was determined to be approximately 229kJ/mol, which is in reasonabble aggrement with the result reported by Jha and Grieveson (1992). Figure 6 shows the X-ray diffraction (XRD) pattern of the reaction product obtained after reaction at 1000°C for 65 minutes. The pattern consists of strong intensity peaks corresponding to calcium sulfide (C) and α-iron (D) and weak peaks of unreacted pyrrhotite (E).

Figure 7 shows the effect of relative surface area abundance on the rate of reaction of mixtures with different FeS:CaO:C mole ratios at 1000°C. A stoichiometrical excess of CaO and/or C increased reaction rate. Comparison of curves 3 and 5 in the figure indicates that the lime content in the initial mixture has more significant effects on the rate of the reaction than carbon content. The 50% completion time was decreased from 40 minutes to 27 and 22 minutes by doubling the initial carbon content and lime content, respectively. Similar

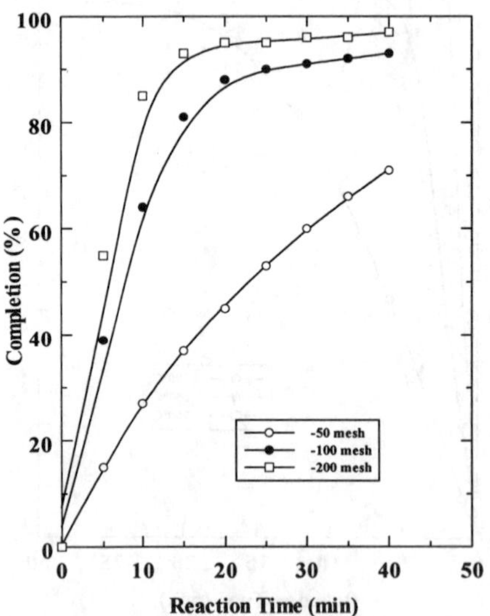

Figure 4. Effects of particle size on pyrite decomposition in tube furnace.

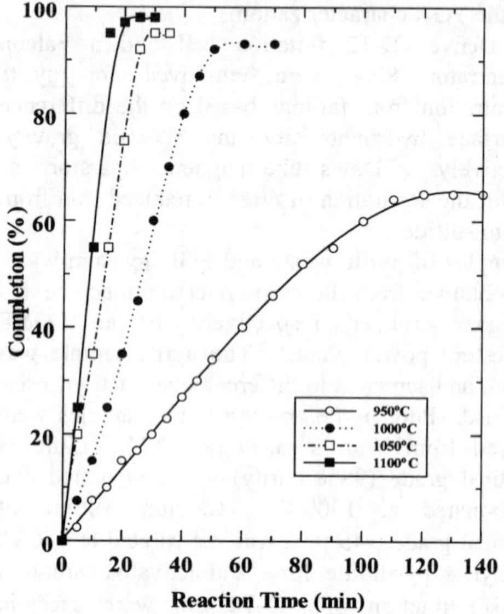

Figure 5. TGA results of reaction of pyrrhotite and lime at different temperature

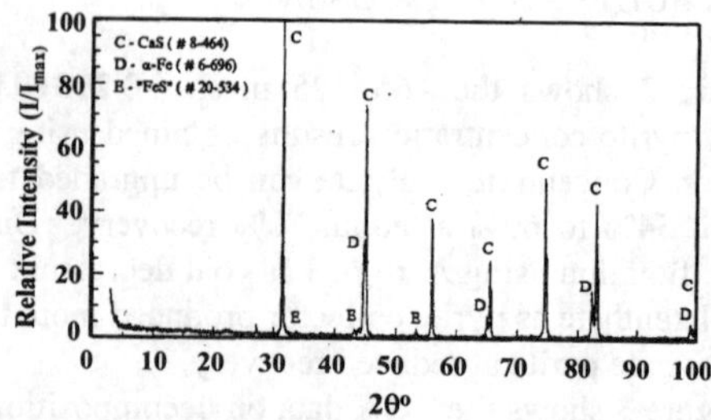

Figure 6. XRD (Cu Kα1 radiation, λ=0.15406 nm) pattern of reaction products.

results were generated with 88% and 70% grade FeS although less pure pyrrhotite showed somewhat slower reaction kinetics. Increase in lime content also helps to reduce sulfur content in iron, as demonstrated in Figure 8 where sulfur content in direct reduced iron (DRI) separated magnetically from other species is plotted as a function of current intensity. This was because the desulfurization of iron with lime was controlled by the solid state counterdiffusion of oxygen and sulfur through calcium sulfide, which built up and surrounded lime (Boyd et al., 1975). By increasing the lime content in the initial mixture, the lime surface area was increased. Therefore, thinner calcium sulfide layers were produced during the reaction and less time was necessary for oxygen and sulfur to diffuse through these layers. The XRD analysis shows that the excess lime did not react with iron and remains in the mixture as a pure phase.

The effective separation of DRI produced in reaction (2) is necessary for the process to proceed. Magnetic separation is one of the techniques that can be used to separate iron from calcium sulfide due to their difference in magnetic susceptibility. The composition of the separation product depends on the reaction temperature. With the reaction product obtained at 1000°C, a single stage of magnetic separation under the optimum operating conditions produced the magnetic product that recovered 97.5% iron and rejected 86.7% calcium and 88.7% sulfur.

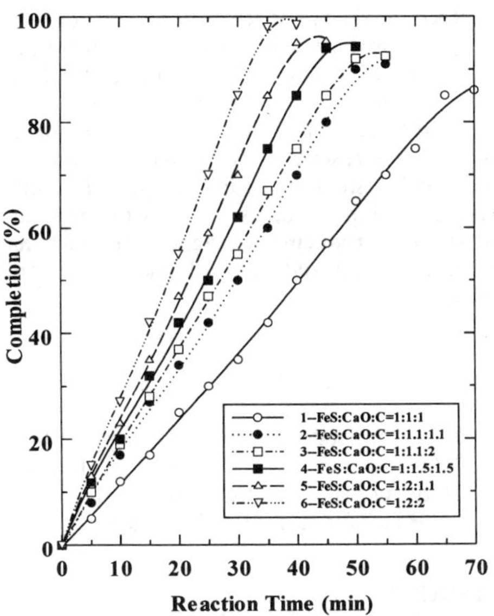

Figure 7. Effects on pyrrhotite reduction rate of relative surface area abundance of reactants.

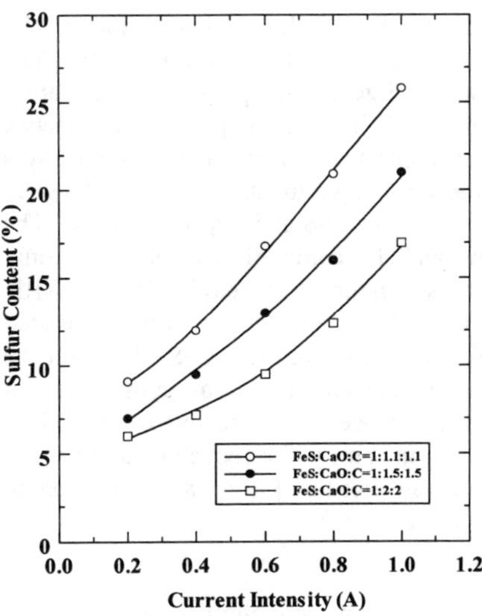

Figure 8. Effect of current density of magnetic separator on DRI sulfur content produced under different conditions.

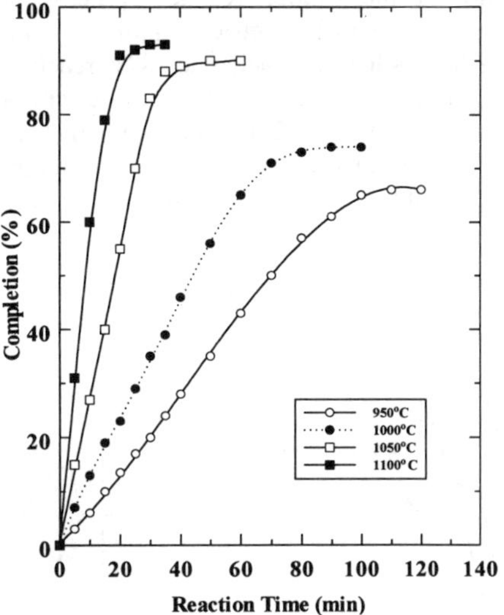

Figure 9. TGA results of Muller Kuhne reaction at different temperature.

The high sulfur content of DRI shown in Figure 8 was caused, in large part, by unreacted pyrrhotite. Since the magnetic susceptibility of pyrrhotite is about 20times greater than that of calcium sulfide, essentially all unreacted pyrrhotite reported together with DRI to the magnetic fraction. More work is needed to reduce sulfur in DRI to an acceptable level.

The kinetics of the Muller Kuhne reaction was studied as a function of temperature. The TGA results shown in Figure 9 for pure CaS and $CaSO_4$ indicate that at 1150°C 90% completion was reached in about 20 minutes and slower reaction kinetics was observed at lower temperatures. Similar results were obtained with impure CaS and gypsum wastes. The reaction rate and the completion extent were only slightly decreased in the presence of CaO. The XRD analysis showed that the product mixture consisted of CaO and traces of $CaSO_4$ and CaS. It should be noted that complete conversion of gypsum is not necessary in this process since recovered lime does not require a high purity for recycling to flue gas scrubber. The rate constant k for reaction (3) can be described in Equation (6):

$$\ln k = -2.660\times10^4 / T + 20.417, \qquad (6)$$

which resulted in an activation energy of 221 kJ/mol, which is in excellent agreement with other investigators' data (Turkdogan and Vinters, 1976).

Figure 10 shows the effect of surface area or relative abundance of CaS and $CaSO_4$ on the reaction rate of the Muller Kuhne process at 1100°C. The stoichiometric mixture with $CaS:CaSO_4=1:3$ mole ratio reached the 87% reaction completion within 34 minutes. The mixture with a stoichiometric excess of $CaSO_4$ (mole ratio of CaS and $CaSO_4$ smaller than 1:3) showed faster reaction rate. For example, the mixture with a 2 time stoichiometric excess of $CaSO_4$ (with $CaS:CaSO_4=1:6$ mole ratio) reached the same reaction completion within about 15 minutes. The mixture with a stoichiometric excess of CaS showed similar but less significant effects. The results indicate that the excess of the reactants, particularly $CaSO_4$, enhances the degree of reaction completion, which represents a favorable economic situation for the recovery of lime through the Muller-Kuhne reaction.

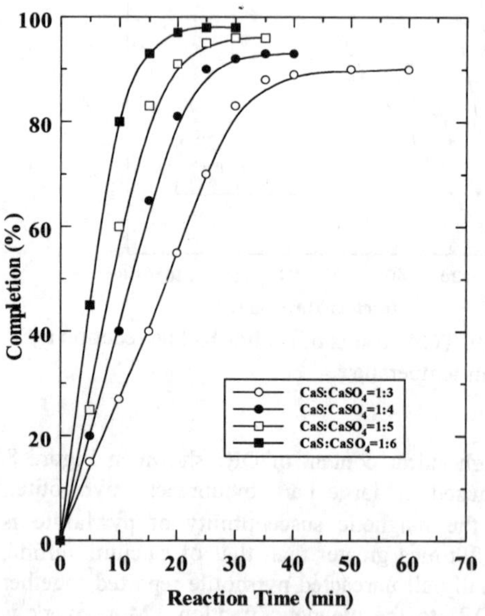

Figure 10. Effects on Muller Kuhne reaction rate of relative surface area abundance of reactants.

4 CONCLUSIONS

The present TGA and tube furnace study has demonstrated that pyrite and gypsum wastes can be converted into iron, lime, and sulfur products using a thermochemical process. The overall reaction can be represented by:

$$FeS_2 + 3CaSO_4 + C + \tfrac{1}{2}O_2 = \\ Fe^o + 3CaO + 4SO_2 + S + CO_2. \qquad (7)$$

This reaction can be accomplished in three consecutive steps that included pyrite decomposition to pyrrhotite, reduction of pyrrhotite to iron, and the Muller Kuhne reaction. Higher temperature increased the rate of all thermochemical reactions involved. A temperature of 650°C seemed to be appropriate for pyrite decomposition. The reduction of pyrrhotite with carbon in the presence of lime had a favorable kinetics above 1050°C. The Muller Kuhne reaction required a temperature above 1100°C for sufficient reaction rate. A stoichiometric excess of lime and carbon in initial mixtures enhanced the rate of the iron production and increased the degree of reaction completion. The Muller-Kuhne process was effective in converting gypsum to lime that can be recycled to FGD processes. The excess of $CaSO_4$ significantly enhanced the kinetics and degree of completion of the Muller Kuhne reaction. Higher reaction rate was achieved with smaller reactant particle size. Magnetic separation can be used to separate iron and calcium sulfide, producing a magnetic product that recovers 97.5% iron and only 13.3% calcium and 11.3% sulfur.

5 ACKNOWLEDGEMENT

The authors acknowledge the financial support from the U.S. DOE through Federal Energy Technology Center (Pittsburgh) Contract No. DE-FG26-98FT40114, Bob Pattern, Project Manager.

REFERENCES

American Coal Ash Association (ACAA), 1992. "*1991 Coal Combustion By-Product—*

Production and Consumption." ACAA, Washington, DC, 1992.

Boyd, D.C., Phelps, W.C., Jr., and hepworth, M.T., 1975. Met. Trans. B, 6B, pp. 87-93.

Feeley, T.J., III, Walsh, J.D., Gala, H.B., and Sniegicki, J.L., 1990. *Proc. Of Processing and Utilization of High-Sulfur Coals III*, edited by R. Markuszewski and T.D. Wheelock, Elsevier Science Publishers B.V., Amsterdam, pp. 99-107.

General Accounting Office, 1977. "*U.S. Coal Development – Promises, Uncertainties*," Report to Congress, EMD-77-43, pp. 6.20,21 and 6.50,51.

Hite, D., Chern, W.S., and Fredrick, J.H., 1994. *Proc. Of Eleventh Annual International Pittsburgh Coal Conference*, Pittsburgh, PA, Sep. 12-16, pp.431-436.

Jha, A. and Grieveson, P., 1992. *Scan. J. Metall.*, 21, pp. 50-62.

Kleinmann, R.L.P., 1989. *E&MJ*, July, p. 161.

Shainberg, I., Sumner, M.E., Miller, W.P., Farina, M.P.W., Pavan, M.A., and Fey, M.V., 1989. *Advances in Soil Science*, edited By B.A. Stewart, Vol. 9, Springer-Verlag, New York, pp. 1-111.

Turkdogan, E.T., 1980. *Physical Chemistry of High Temperature Technology*, Academic Press, New York.

Turkdogan, E.T. and Vinters, J.V., 1976. Trans. Inst. Min. Metall., Sect. C, 85, pp. C117-C123.

U.S. Environmental Protection Agency (USEPA). "*Wastes from the Combustion of Coal by Electric Utility Power Plants.*" EPA/530-SW-88-002, USEPA, Washington, DC, 1988.

Wewerka, E.M., Williams, J.M., Wagner, P., 1982. *The Use Of Multimedia Environmental Goals To Evaluate Potentially Hazardous Trace Elements From High Sulfur Coal Preparation Wastse,* Los Alamos National Laboratory Report LA-9189 MS UC-901.

Wildeman, T.R., 1991. *Gelogy in Coal Resource Utilization*, edited by D.C. Peters, Techbooks, Fairfax, VA, p. 499.

Environmental Issues and Management of Waste in Energy and Mineral Production, Singhal & Mehrotra (eds)
© 2000 Balkema, Rotterdam, ISBN 90 5809 085 X

Fluff separation by image analysis

C.Vergati, G.Bonifazi, F.La Marca & P.Massacci
Department of Chemical, Materials, Raw Materials and Metallurgical Engineering, University of Rome 'La Sapienza', Italy

ABSTRACT: In demolition processes of the motorcar after dismounting, shredding, metallic materials sorting and after an appropriate venting and air classification, a special fraction is obtained, conventionally named fluff. Fluff is a light fraction composed by organic material, rubbers, tissues, normally used for the stuffing, the insulation, the gaskets and the tapestry of a car. Although such a fraction is not able to be discriminated by specific weight based methods, the separation in its main constituents is required to get recycled products of high commercial value. Particularly it is necessary to get an organic fraction, composed exclusively by plastics and rubber and deprived as far as possible of contamination due to other materials in the fluff mixture. Therefore an optical selection technique based on image processing, aided by proper software of recognition, has been adopted. The system of image survey is based on digital black and white (B&W) image acquisition of the objects that flow and are properly lighted. A suitable number of parameters are obtained for each object.
From the gained data, a feature vector of representative objects has been extracted in order to achieve objects selection in 2 classes according to the market specification for recycled materials. The classification of the objects in different classes of destination has been carried out using a classification algorithm adopting various experimental modalities. Satisfactory results in terms of correct recognition (>80%) have been achieved.

1 INTRODUCTION

The aim of the paper is to point out suitable procedures able to characterize the fluff components and separate the organic fraction (plastics and rubber) from the other (foam rubber, textiles, electric wires and metals), both resulting from car dismounting and shredding. Products obtained after separation have to be useful for material and/or energy recovery. In fact, the organic fraction has a high-embodied energy and a high calorific value. On the contrary, the other fraction has to be dumped because to day it cannot be economically recovered. Image analysis was chosen as a proper technique to investigate fluff characteristics. It's rather a new technique, quite cheap and fast, comparing to traditional separation methods (Bonifazi & Massacci, 1996).

2 DATA SET STRUCTURE

In the case of fluff either shape- and texture-features have been considered: 17 shape parameters and 14 texture parameters, related to each object identified in each image, have been computed. These 31 parameters were stored originating a matrix characterized by a number of rows equal to the number of the analyzed objects and a number of columns equal to the number of extracted numerical parameters. Another column was finally added, containing the class target of each object of the fluff sample considered to a previous established criterion of material assignment.

A feature vector of 32 elements was thus associated to each object.

3 FEATURE EXTRACTION BY IMAGE PROCESSING

3.1 *Image acquisition: lighting conditions*

The first step in the analysis was the image acquisition. It was realized at the Delft University of Technology (The Netherlands).

B&W images of fluff were acquired in static conditions. A white background was used because plastics and rubber were characterized by very low average gray level values. The main problem was the choice of a right lighting system to enhance texture and avoid formation of shadows. An indirect lighting was adopted.

The acquisition set up was realized embedding the system constituted by the camera, the lights and the samples in a box, internally covered with shuffled aluminum sheets to obtain a suitable light diffusion. As a consequence, optimal acquisition conditions were realized for fluff, but unfortunately not for the metals that were confused by the image recognition procedures, during the following processing stages. Metals, in fact, according to their high degree of reflectance, were badly recognized, being "confused" with the background. Such a procedure, if not correct in terms of imaging procedures, has practically not influenced in this study, being the target the recognition and the classification only of the light fractions (fluff).

3.2 *Segmentation and labeling*

B&W images (Figure 1) were transformed into binary images by segmentation adopting a threshold method. A program was implemented to choose the threshold value considering both the average gray level of the whole image and the total area of each object. In this way, it was possible to recognize the nature of the different objects being the particles of metals and wires very clear (high average gray level value) and covering very small areas in comparison with those of fluff (Figure 2).

Labeling was carried out to recognize each object in the binary image: 8-connected neighborhoods were adopted. A noise reduction procedure was also applied to eliminate objects smaller then 100 pixels, representative of noise-fragments (metal or background surface anomalies) generated by segmentation.

3.3 *Morphological parameters*

The following morphological parameters were computed for each segmented particle on the image:
- perimeter (p): sum of pixels belonging to the boundary of a particle (a pixel has been considered a perimeter pixel if it was an on pixel and if at least one of its neighborhood pixels was off);
- area (A): sum of pixels having the same label value;
- circularity (p^2/A): indicating an elongated shape;
- Feret diameter (d_F): the longest projection of the particle's profile in any direction (it was approximated to the diagonal of the bounding box, that is the smallest rectangle that could contain the object dominion);
- eccentricity (e): the eccentricity of the ellipse that had the same second moments as the object dominion;

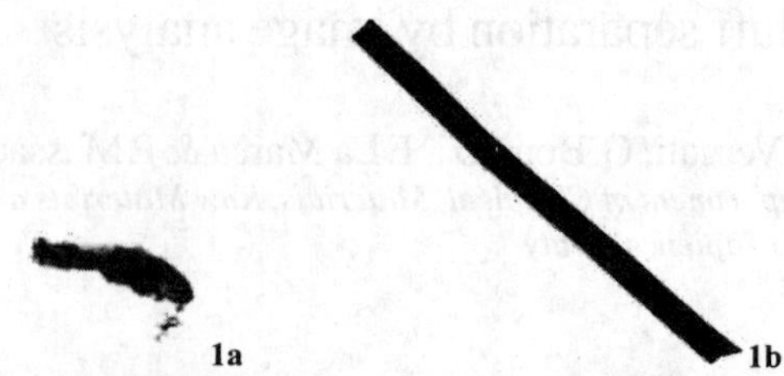

Figure 1: B&W images of fluff (1a) and wire fragment (1b).

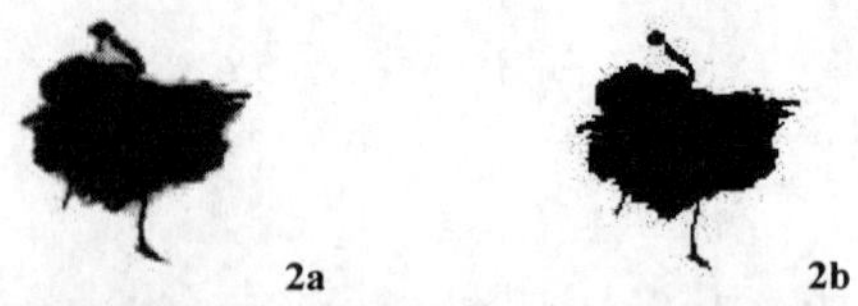

Figure 2. Fluff object: as acquired (2a) and as resulting from segmentation (2b).

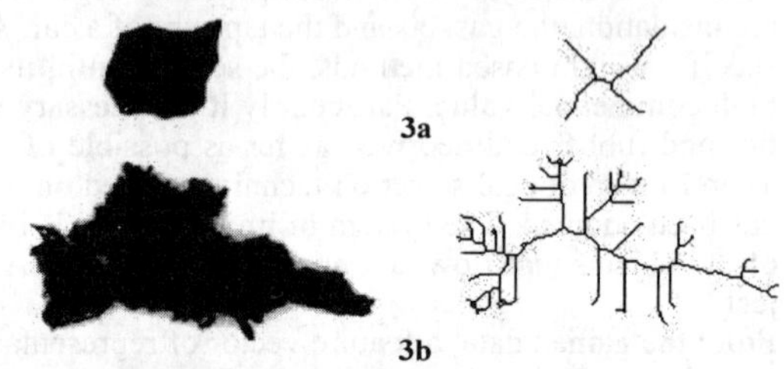

Figure 3. Skeletons (on the right) of two different fluff particles (3a: plastics, 3b: textiles) forming complex "dendritic" structures.

- orientation (o): the angle (in degree) between the x-axis and the major axis of the ellipse that had the same second moments as the object dominion;
- Euler number (E): the number of the objects in the image field minus the number of holes in the objects in the same field (E = 1 if there were no holes);
- holes area (A_H): difference between the area of each object dominion with all holes filled in and the net area of the same dominion;
- equivalent diameter (D_{eq}): diameter of a circle with the same area as the object dominion;
- solidity (s): ratio between the area of an object dominion and the area of the smallest convex polygon that could contain the same dominion;
- extent (ext): ratio between the area of an object dominion and the area of the bounding box, that is the smallest rectangle that could contain the same dominion.

3.4 *Skeleton analysis*

The skeletonization is a morphological operation that reduces each object in an image to lines, without changing the essential structure of the object (see Figure 3).

After having observed the differences between skeletons related to objects of different nature, a program to compute other two morphological parameters, i) skeleton length and ii) number of branches in the skeleton, was implemented.

3.5 *Fractal analysis*

The irregularity of the boundary of each object was studied by fractal analysis (Mandelbrot, 1983). The method of perimeter equispaced exploration was chosen. It is based on the approximation of the object boundary length to the perimeter of a polygon whose sides are the chords of different lengths that join boundary pixels far a fixed number of steps each other (Kaye et al., 1986).

Changing the number of steps between boundary pixels, a polygon family was built. The smallest adopted step was equal to 1 pixel and the largest one was equal to 1/3 of the Feret-diameter always expressed in pixels. As the perimeter evaluation was influenced by the exploration starting point, the boundary exploration was repeated ten times so that the starting point changed at every turn. After ten turns, the segment closing the boundary was measured and the boundary length was divided to ten.

Different boundary lengths related to different steps, both normalized in respect to Feret diameter, were plotted on a Richardson plot to evaluate the fractal dimension (Kaye, 1993).

3.6 *Texture analysis*

First of all, the mean gray level, the variance and the standard deviation of the gray level of each object in the image were computed.

Texture analysis was performed adopting the co-occurrence matrix method (or spatial-tone dependence matrix method). According to this method, a relationship may be defined between the gray level value of a pixel and that of pixels at known distance (*d* pixels) and direction (*k*) (Haralick et al., 1973).

Four main directions were chosen (0°, 45°, 90° and 135°) and the distance *d* was fixed equal to 1 so that it was possible to recognize every changing in textural structure.

Four matrices were built corresponding to each direction: the general element (*l*, *m*) represented the number of pixel pairs whose distance along the fixed direction was *d*, and whose gray level values were *l* and *m*, respectively:

$$M_{0°}(l, m, d) = \#\{[(p, q), (r, s) \in (L_y L_x)\ (L_y L_x)]\ p-r = 0, |q-s| = d, I(p, q) = l, I(r, s) = m\} \tag{1}$$

$$M_{45°}(l, m, d) = \#\{[(p, q), (r, s) \in (L_y L_x)\ (L_y L_x)]\ (p-r = d, q-s = -d) \text{ or } (p-r = -d, q-s = d)\ I(p, q) = l, I(r, s) = m\} \tag{2}$$

$$M_{90°}(l, m, d) = \#\{[(p, q), (r, s) \in (L_y L_x)\ (L_y L_x)]\ |p-r| = d, q-s = 0, I(p, q) = l, I(r, s) = m\} \tag{3}$$

$$M_{135°}(l, m, d) = \#\{[(p, q), (r, s) \in (L_y L_x)\ (L_y L_x)]\ (p-r = d, q-s = d) \text{ or } (p-r = -d, q-s = -d)\ I(p, q) = l, I(r, s) = m\} \tag{4}$$

The four matrices were normalized by dividing to the number of considered pixel pairs. Using the normalized matrices, a series of eleven texture parameters were derived:

Angular Second Moment (ASM)

$$f_1 = \sum_l \sum_m \{M_k^*(l,m,d)\}^2 \tag{5}$$

Contrast

$$f_2 = \sum_{n=0}^{N_g-1} n^2 M_k^*(l,m,d) \quad |l-m|=n \tag{6}$$

Correlation (COR)

$$f_3 = \frac{\sum_l \sum_m [(lm) M_k^*(l,m,d)] - \mu_x \mu_y}{\sigma_x \sigma_y} \tag{7}$$

Sum of Squares (SSQ)

$$f_4 = \sum_l \sum_m (l-\mu)^2 M_k^*(l,m,d) \tag{8}$$

Inverse Difference Moment (IDM)

$$f_5 = \sum_l \sum_m \frac{1}{1+(l-m)^2} M_k^*(l,m,d) \tag{9}$$

Sum Average (SAV)

$$f_6 = \sum_{l=2}^{2N_g} l p_{x+y}(l) \tag{10}$$

Sum Entropy (SUE)

$$f_7 = -\sum_{l=2}^{2N_g} p_{x+y}(l) \log[p_{x+y}(l)] \tag{11}$$

Sum Variance (SUV)

$$f_8 = \sum_{l=2}^{2N_g} (l-f_7)^2 p_{x+y}(l) \tag{12}$$

Entropy (ENT)

$$f_9 = -\sum_l \sum_m M_k^*(l,m,d) \log[M_k^*(l,m,d)] \tag{13}$$

Difference Variance (DVA)

$$f_{10} = \text{variance of } p_{x-y} \tag{14}$$

Difference Entropy (DEN)

$$f_{11} = -\sum_{l=0}^{N_g-1} p_{x-y}(l)\log[p_{x-y}(l)] \qquad (15)$$

The final value of each parameter was computed as a mean of the four values related to each matrix. In this way, the co-occurrence matrix method could be considered a rotation independent method.

4 CLASSIFICATION

Classification of the objects in the reference classes was carried out using the software T.R.A.C.E. (Total Recognition by Adaptive Classification Experiments) pointed out by Patrizi (1979).
The algorithm, starting from the data-set built after image analysis, worked in two phases (Massacci et al., 1993; Nieddu & Patrizi, 1998):

- a training phase: to identify the set of reference vectors characteristic of each target class;
- a classification phase: to assign each objects to one of the target classes on the basis of the minimum distances between the feature vector of each object and the reference vectors.

A set of experiments was carried out in order to evaluate:

- the contamination error for the class targeted as plastic and rubber: expressed as the fraction of objects incorrectly classified in this class, belonging to the other one targeted as foam rubber, textiles, electric wires and metals;
- the assignment error: expressed as the fraction of plastics and rubber incorrectly classified in the other one, that is the loss fraction.

The aim of the experiment was to recognize the significance of the features adopted for object characterization after image analysis.

4.1 *Classification of fluff in two classes*

Taking into consideration all the extracted features, the results of classification in two classes (class A = textiles, foam rubber, electric wires and metals and class B = plastics & rubber) are given in Table 1.

4.2 *Image acquisition sensitivity*

Some experiments were conducted on a series of images characterized by different contrast between the white background of a conveyor belt and the dark objects constituting the fluff lying on the same conveyor.

These tests have been carried out made in order to recognize the sensitivity of the image acquisition procedure in front of variation of regulating parameters during the sorting process.

The results related to the different image acquisition set-up are shown in Table 2.

After comparison of assignment and contamination errors related to the classification results obtained for plastics and rubber recognition it appeared the better regulation of the sorting process competes to the adoption of less contrast in image acquisition. In fact, adopting lighting conditions, realizing a strong contrast between the white background and dark objects in the images, shadows appeared around the objects with the consequence to introduce some noise in segmentation; as a consequence differences between shapes and background were not well enhanced.

Table 1. Fluff classification in two classes: plastics and rubber (53.65% wt) and textiles, foam rubber, electric wires and metals (46.35% wt). All features adopted for object characterization. Low contrast was selected for image acquisition.

	Target class		Error	
Assignment class	textiles, etc.	plastics & rubber	assignment	contamination
textiles, etc.	98.64%	5.23%	1.36%	5.78%
plastics & rubber	1.36%	94.77%	5.23%	1.22%
Total	100.00%	100.00%		

Table 2. Fluff classification in two classes: plastics and rubber (53.65% wt) and textiles, foam rubber, electric wires and metals (46.35% wt). All features are adopted for object characterization. Low or high contrast was alternatively selected for image acquisition.

	Low contrast	High contrast
Assignment error in the two classes	5.23%	5.48%
Contamination error for plastic & rubber	1.22%	5.48%

4.3 *Classification in three classes*

The contribution of shape and texture parameters on correctness of classification was studied comparing the results obtained by the algorithm T.R.A.C.E. alternatively using the shape parameters and the texture parameters. The results are shown in Table 3. The best results have been obtained considering both shape and texture features.

Table 3. Fluff classification in two classes: plastics and rubber (53.65% wt) and textiles, foam rubber, electric wires (46.35% wt). Shape and texture features are alternatively adopted for object characterization.

	Shape features	Texture features	Shape and texture features
Global correctness of classification	96.05%	93.48%	97.84%
Contamination error for plastic & rubber	1.44%	6.87%	1.43%

As the tailing class appeared to be very heterogeneous, three classes were considered:

- textiles and foam rubber (tail);
- plastics and rubber (product);
- electric wires and metals (tail).

Table 4. Fluff classification in three classes: plastics and rubber (52.33% wt), textiles and foam rubber (32.28% wt) and electric wires and metals (15.39% wt). Shape and texture features are together adopted for object characterization.

	Target class			Error	
Assignment class	textiles & foam rubber	plastics & rubber	electric wires & metals	assignment	contamination
textiles & foam rubber	98.60%	5.99%	0.00%	1.40%	8.96%
plastics & rubber	1.40%	94.01%	0.23%	5.99%	0.98%
electric wires & metals	0.00%	0.00%	99.77%	0.23%	0.00%
Total	100.00%	100.00%	100.00%		

The results are shown in Table 4. They are very close to those obtained for the classification of the materials in two classes (Table 1).

5 NUMBER OF FEATURES

A set of experiments was carried out adopting a procedure for increasing the number of features. Every column of the original data set was thus multiplied for all the other columns. A new feature and a new expanded data set was thus obtained. It could happen an easier classification in front of more numerous parameters available to compare the objects.

This procedure was applied for fluff classification into two classes (plastic and rubber) and (rubber, textiles, electric wires and metals). The results are shown in Table 5.

While the assignment error decreased in comparison with the previous experiments, the contamination error was higher. The lower correctness (%) in classification was due to the introduction of redundant parameters.

6 CONCLUSIONS

From the results obtained by the applied imaging procedures it is evident as classification is strongly influenced by experimental and data processing adopted strategies, such as sampling procedure, acquisition set-up and feature extraction. More specifically the classification algorithm has to be suitably trained through the selection of a set of features that is really able to describe the morphological, morphometrical and textural attributes of the particles (fluff) object of investigation. Furthermore, in order to reduce the level of contaminants (misclassified particles) in the final product, a consistent training sample set has to be selected.

Particularly, suitable training phase of the classification algorithm has to be adopted taking into consideration convenient features related to well defined classes. In order to obtain low levels in contamination in the marketable class of objects it needs to adopt a consistent training sample.

Table 5. Fluff classification in two classes: plastics and rubber (41.93% wt) and textile, foam rubber, electric wires and metals (58.07% wt). Shape and texture features are together adopted for object characterization. Multiplication procedure was adopted.

	Target class		Error	
Assignment class	textiles, etc.	plastics & rubber	assignment	contamination
textiles, etc.	95.12%	4.05%	4.88%	2.98%
plastics & rubber	4.88%	95.95%	4.05%	6.58%
Total	100.00%	100.00%		

ACKNOWLEDGMENTS

The authors would thank Prof. G. Patrizi and Dr. L. Nieddu for the precious help and suggestions in T.R.A.C.E algorithm utilization. The authors wish to dedicate an imminent paper to point out its applications and versatility.

REFERENCES

Bonifazi, G. & Massacci P., 1996. Particle identification by image processing. *KONA Powder and Technology*. No. 14.

Haralick, R.M., Shanmugan, K. & Dinstein, I., 1973. Textural Features for Image Classification. *IEEE Transaction on Systems, Man, and Cybernetics*. 3: 6.

Kaye, B.H., 1993. *A Random Walk through Fractal Dimension*. VCH - Verlagsgesellschaft, Weinheim.

Kaye, B.H., Clark, G.G., Leblanc, J.E. & Trottier, R.A., 1986. Image Analysis Procedures for Characterising the Fractal Dimension of Fine Particles. *1st World Congress on Particle Technology* - Nuremberg.

Mandelbrot, B.B., 1983. *The fractal geometry of nature*. New York : W.H. Freeman and Co.

Massacci, P., Bonifazi, G. & Patrizi, G., 1993. Alternative feature selection procedures for particle classification by pattern recognition techniques. *XVIII Mineral Processing Congress*, Sidney.

Nieddu, L. & Patrizi G., 1998. Formal methods in pattern recognition: a review. European Journal of Operational Research.

Patrizi, G., 1979. Optimal Clustering Properties. *Ricerca Operativa*: 10: 37-60.

Environmental Issues and Management of Waste in Energy and Mineral Production, Singhal & Mehrotra (eds)
© 2000 Balkema, Rotterdam, ISBN 90 5809 085 X

Self-cleaning acoustic screen filter system for mine waste water

J.R.Woolsey
The University of Mississippi, University, Miss., USA

S.Z.Shkundin
Moscow State Mining University, Russia

ABSTRACT: The polystaged process of mine waste water purification includes the stage of fine solid particle separation from waste water. Known screen filters have a common disadvantage: since the media being filtered typically includes fine grain and adhesive substances, the screen very quickly becomes clogged and must be removed or cleaned. The technology implemented in the acoustic screen filter system (ASF) concept uses acoustic energy, coupled hydraulically through the subject liquid media and fine mesh screen to perform: 1) a separation of solid suspended particles from a host liquid. 2) enhanced filtration of the liquid fraction from the solids fraction to specified limits. and 3) continuous cleaning of the screen, preventing screen blinding. The basic filter system consists of a tank body equipped to introduce the subject liquid with contained suspended solids, house a fine micron mesh screen, support a suitable energy source, permit removal of filtered liquid, and promote removal of suspended solids concentrate.

1. BACKGROUND

Introduction: The acoustic screen filter (ASF) system concept, which formulates the basic technology of this study, is the product of a cooperative project between the Mississippi Mineral Resources Institute (MMRI) and the Moscow State Mining University, Moscow, Russia. It involves the use of low frequency acoustic energy to effect continuous separation and filtration of fine suspended particles from a host liquid with continuous flow of filtered liquid and continuous removal of solids concentrate. The primary advantages over existing fine micron filter technology are the elimination of the need for replaceable elements and requirements for periodic disruption of filtering during filter backwash cycles, all of which add to overall cost.

Concept: The concept of the subject study involves the application of acoustic energy, coupled hydraulically through liquid media, and a fine mesh screen to perform the following:

1 separation of solid suspended particles from a host liquid through the breakdown of surface tension;
2 enhanced filtration of the liquid fraction from the solids fraction to specified limits, retaining a specified size range of solids; and
3 continuous cleaning of the screen, preventing screen blinding and promotion of a continuous filtration process.

Description: The basic system consists of a tank body with a means of: 1) introducing the subject liquid with contained suspended solids; 2) housing a fine micron mesh screen; 3) mounting of a suitable energy source; 4) removal of filtered liquid; and 5) removal of suspended solids concentrate (Figures 1 and 2).The process by which the system works involves the supply of energy to a low frequency pulsing device which creates hydraulic pressure waves throughout the liquid media, acting on and through the micron mesh screen. The action of the hydraulic pressure waves is best described as positive and negative semiperiods of the basic wave form.

The actual physical processes involved may be considered in two phases. The first relates to the process of separation where hydraulic pressure waves induce vibro-agitation (threshold of cavitation) which serves to breakdown surface tension between the solid particles and suspending liquid (i.e. water) and the screen mesh and passing liquid. The second phase relates to a process of filtration whereby the separated solid particles are filtered out from the liquid fraction which then passes through the pore spaces of the screen. The hydraulic pressure waves serve to drive the separation/filtration process

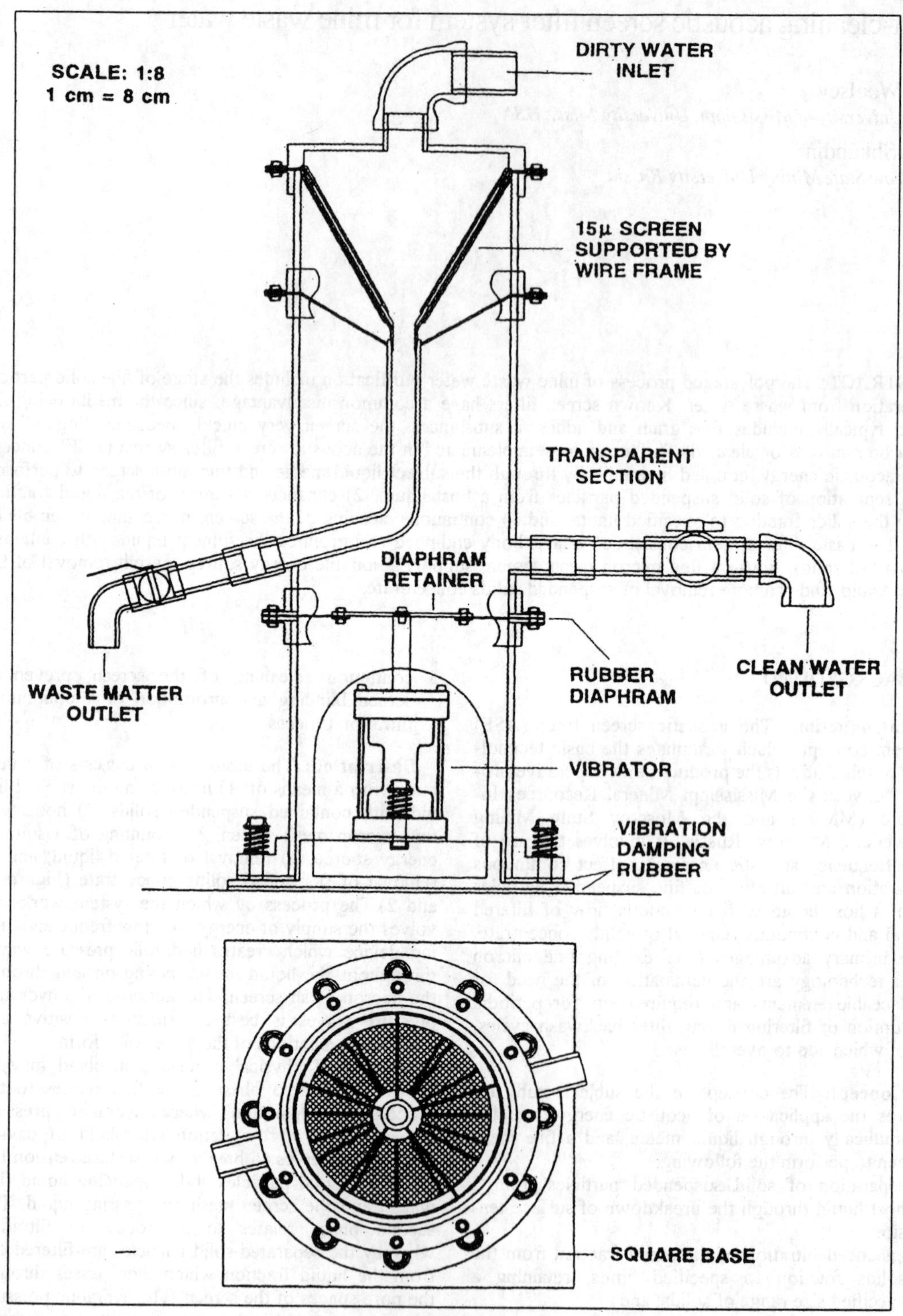

Figure 1. Acoustic Screen Filter - Prototype Model.

and to continuously clean the screen. This may be most simply described as a process by which the pressure waves act on a given section of screen surface at a given instant of time and progressively and cyclically affect the entire screen in unit time. In examination of the wave action, the negative pressure semiperiod assists a static pressure head in enhancing liquid flow through the pore spaces of the screen, followed in turn by the positive (back) pressure semiperiod which assists in the dislodging of particles that tend to become jammed in the screen pore spaces. Further, this positive pressure semiperiod serves to force the increasing concentration of solids back from the screen surface further dewatering the solids concentrate by hydraulic compression. The filtered material then progressively and continuously gravitates down the screen surface to an exit port with the aid of redundant vibration. In unit time, the cyclic positive and negative semiperiod pressure waves tend to form a boundary layer Tipple effect progressing cyclically over the entire surface of the screen. This action further serves to provide a continuous cleaning effect assisting in the continuous flow of liquid filtrate through the screen and continuous removal of concentrated, dewatered solids.

A secondary unique beneficial effect noted in early ASF tests is the observed killing and/or stressing of harmful bacteria which may pass through the screen with the filtered liquid.

Potential Application: ASF technology has a wide span of potential applications ranging from mine waste water cleanup to pre- and post-cleaning of industrial water with ultimate benefits to the environment and savings in energy costs. Also of interest is, in a sense, the reverse of the process, by which a desired fine solid particulate is dewatered and recovered. Early interest in the ASF concept stemmed from a need to recover very fine grained (70-80 micron) titanium oxide sand particles (limonite), which were being lost from a hydrocyclone in the overflow liquid. Also of interest was the increasing environmental concern to reduce/remove suspended solids (silt) from sand and gravel process water. A variety of similar environmental cleanup and mineral processing applications can be anticipated with the increasing need to conserve mineral resources, including water, and the utilization of otherwise high grade mineral fines formerly discharged to tailings dumps. In certain cases the reworking of tailings dumps in concert with environmental cleanup, utilizing ASF technology may have economic advantages.

2.SYSTEM TESTS

ASF Prototype Model Tests: A series of tests were established to evaluate the ASF concept and related components within the scope of the proposed objectives. The first test involved the ASF prototype model with the cone screen configuration (Figure 1), driven first by an external pneumatic vibrator, and second by an electrically powered transducer. Using an internally mounted directional accelerometer, frequency and amplitude were monitored for both devices. The purpose of the test was twofold: first to see if previous preliminary results could be duplicated within the specified low frequency/high amplitude range, i.e., 10 to 20 Hz and 0.2 to 0.8mm respectively; and second, to compare these results with a system driven by a somewhat higher frequency/lower amplitude of 3000 Hz and 0.01 to 0.03mm. Both systems were rated at similar levels of effective energy consumption: approximately .20 KW, and processing a throughput of 150 liters per minute. All other variable factors were held constant for the test; i.e. screen mesh, 20 microns, and volume of test water, 600 liters, with approximately 50 parts per thousand suspended solids.

Results from this test demonstrated the effectiveness of the low frequency/high amplitude system over the alternative high frequency (3000 HZ) device. On breakdown and examination of the screen of the former system, it was found to be clean; whereas, in the latter case, approximately 10% of the screen surface was found to be blinded by fine particulate solids.

The ASF prototype model test further demonstrated the concept by which relatively high amplitude/low frequency (and tow energy consumption) was capable of establishing a cyclic field of flux acting on and through the screen to dislodge fine particles that would otherwise blind the screen, as was indeed the case when amplitude was diminished with the higher frequency device.

A second beneficial effect was also noted. It was feared that suspended fine diameter organic filaments (less than 15 microns) would become aligned with the short axes perpendicular to the screen and thus result in their passing or becoming threaded in the screen mesh. On close examination, it was discovered that the elongated filaments and fibers tended to align themselves with their long axes perpendicular to the direction of the secondary force field, which is, in effect, perpendicular to the surface of the screen at any point on the screen. This orientation then facilitates filtration. As noted earlier, a boundary layer is created, possibly by this same cyclic field of flux, acting in a direction normal to the screen, thereby preventing full and continuous contact of the cloud of particulate solids with the screen

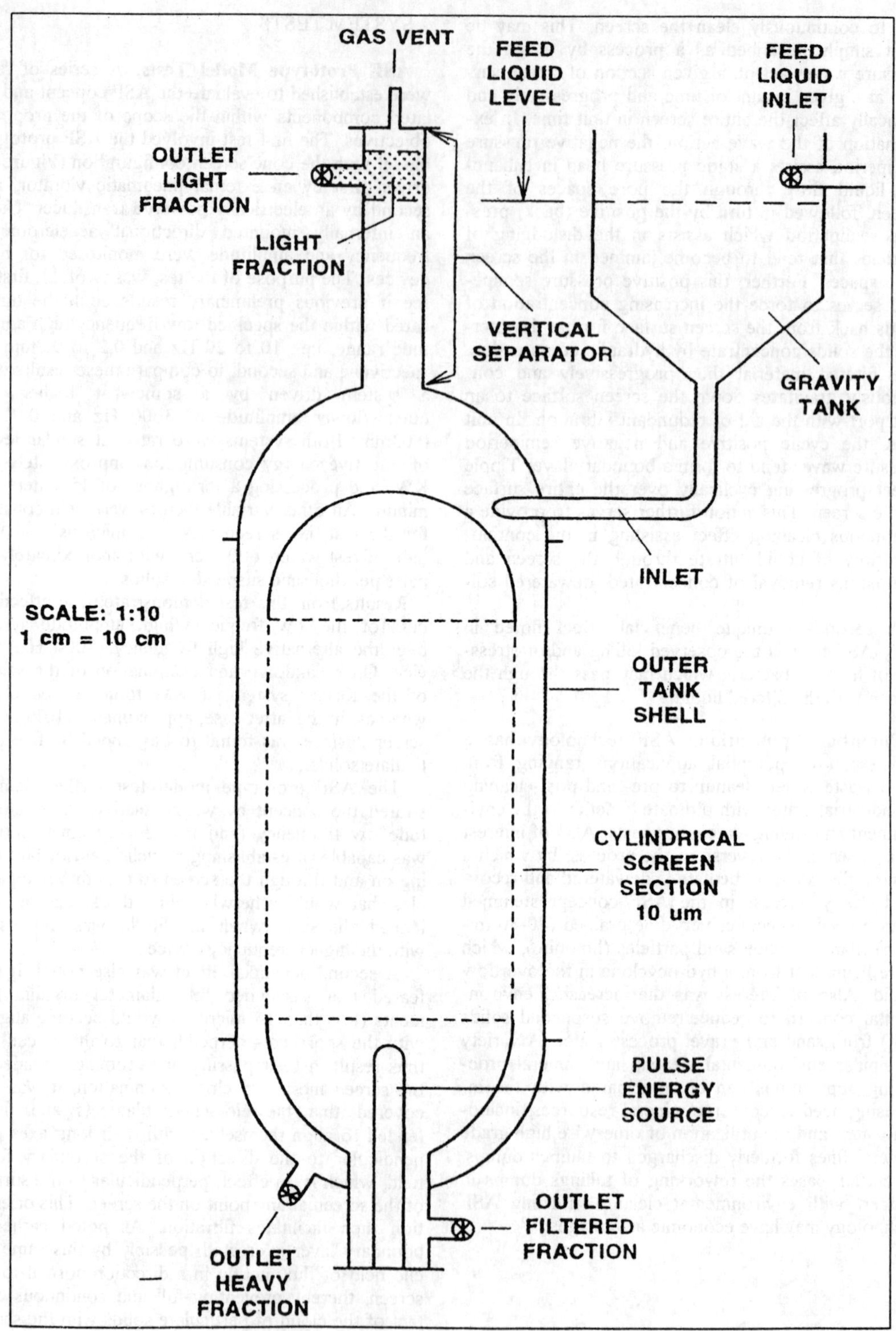

Figure 2. Acoustic Screen Filter - Production Test Model.

surface, which in turn enhances the continuous self cleaning effect of this unique ASF process.

3.CONTINUED DEVELOPMENT

Objectives: Based on the success of previous tests, it was decided that a more comprehensive investigation would be made of the ASF concept. A production test unit was built. Primary objectives were:

1 modification of the basic ASF prototype design to more efficiently handle a variety of solid particles ranging in density from less than to greater than that of water;
2 investigation/evaluation of several acoustic generators, utilizing various drives and several forms of energy (hydraulic, pneumatic, and electric)
3 with the purpose of evaluating efficiency (energy consumption per unit volume of the liquid/solid mixture processed); and
4 evaluation of fine mesh screen size of 10 microns.

Equipment and Test Apparatus: The ASF production test model (Figure 2) is a modified version of the ASF prototype model (Figure 1) designed to handle organic and inorganic solids of variable density and concentrations. The desired throughput flow rate is approximately 1000 liters per minute which is somewhat of a standard unit cell volume capacity for many mining and industrial applications.

The ASF production test model (Figure 2) was designed to accommodate various acoustic generators and driving mechanisms as part of the test. The original prototype used a diaphragm mechanism driven by a pneumatic vibrator. Testing of the production test model will compare the efficiency of the diaphragm unit (driven by various external drives) with two different internal energy sources (relative to the water chamber). One internal source is a hydraulic (water) gun, which works by driving a piston shuttle at high velocity to displace water at selected volumes and rates. The water gun to be used in the test is an S-15 supplied by Seismic Systems, Inc. (SSI) of Houston, Texas. The other source is a more or less conventional sparker system (also internally mounted), which provides pulse energy by high voltage capacitor discharge. The energy of the sparker can likewise be selected to a desired level of intensity and firing rate. Through the respective adjustments, both systems can deliver a desired acoustic pulse spectrum producing hydraulic waves within a selected range of frequency and amplitude for the optimization study.

Test Protocol: The first phase of the test (simulated mixture) is designed to model the various energy type applications for the purpose of focusing on energy efficiency relative to the basic operational effectiveness. The second phase will be an applied test for the purpose of evaluating the viability of ASF technology to an actual mining/industrial waste water problem.

Design Considerations and Modifications: Various design modifications to the ASF prototype model were considered to deal with the manifold problems posed by waste water applications. The approach selected for the ASF production test model (Figure 2) was designed for removal of the light fraction (lighter than water) via a vertical separator fitted to the top of the outer filter body tank. The separator plumbing, centered in the upper portion of the hemispherical head should receive and pass "lights" concentrated in the cyclonic vortex of the upper, liquid entry section of the outer chamber. Lofting of the "lights" into and through the vertical separator is assisted by the system's hydraulic head, with exit controlled by valving

Solid particles that are neutral or heavier than water will be separated and filtered as described by the ASF process. Considering the sparker pulse generator configuration, compressed air may be introduced to the near vicinity of the arc, thereby enhancing both the energy of the pulse as well as the generation of ozone. An additional beneficial effect of this method would be the flotation of particulate organic solids by the disseminated gas bubbles, further enhancing the lofting of "lights" to the separator.

Key components incorporated in the ASF production model are: 1) a 15 micron stainless steel wire mesh screen with approximately 2.8 square meters of working surface area; and 2) a tungsten electrode sparker, firing in air/oxygen saturated water at two times per second. The pretested pulse energy level is approximately 1000 joules per second. The anticipated throughput, as previously mentioned, is 1000 liters per minute. Power consumption is expected to approximate one KW/hr.

Accessory components will include such items as the necessary inlet/outlet valves for control of fluid levels. Initially they will be manually operated (with the exception of a float valve on the inlet feed for the gravity tank). Depending on the continuity of the suspended solids ratio (or lack thereof), the final valve arrangement will be controlled either by timing or by acoustic monitoring.

4. THE THEORETICAL APPROACH TO THE DESCRIPTION OF THE FILTERING PROCESS.

Depending on the concentration of solid in liquid we have to consider two cases:

1 rarefied monodispersive mixture motion and
2 close packed dispersive mixture motion.

Let's imagine the internal part of the filter like a cylinder and introduce the following designation.

r - distance from the axis of the filter chamber; radius of the tube .
v - the velocity of the water .
u - the velocity of the dispersion phase;
c - concentration of the particles;
S - tube -section area;
b - radius of the pores in filter;
ε - porosity of the filter.

Equations for rarefied monodispersive mixture motion can be obtained in the following way.

First, it's necessary to determine all forces, acting upon the particle.

Force $\vec{f}$, acting from carrier phase (water) side to the spherical particle with radius R, combines from four components [1,2,3]:

$$\bar{f} = \bar{f}_\mu + \bar{f}_A + \bar{f}_m + \bar{f}_B \quad \text{, where:}$$

force of viscous friction (Stocks):

$$f_\mu = -6\pi\mu R(U - V) \tag{1}$$

μ - viscous of carrier phase (water)
V - the velocity of the water.
U - the velocity of the dispersion phase

Buoyancy force:

$$\bar{f}_A = \frac{4}{3}\cdot\pi\cdot R^3\cdot\rho\cdot\left(\frac{d}{dt}\bar{V} - \bar{g}\right) \tag{2}$$

where
g - free fall acceleration ,
ρ - the density of the carrier wave;

Associated masses force:

$$\vec{f}_m = \frac{2}{3}\cdot\pi\cdot R^3\cdot\rho\cdot\left(\frac{d}{dt}\vec{V} - \frac{d}{dt}\vec{U}\right) \tag{3}$$

Basse force:

$$\vec{f}_B = 6\cdot R^2\cdot\sqrt{\pi\cdot\rho\cdot\mu}\cdot\int_{-\infty}^{\infty}\frac{d}{d\tau}\left(\vec{V}-\vec{U}\right)d\frac{\tau}{\sqrt{t-\tau}} \tag{4}$$

Moreover, the gravity force acts on this particle. The interacting forces applied to the particles in rarefied system are being neglected.

Then the particle motion law will be:

$$\frac{4}{3}\cdot\pi\cdot R^3\cdot\rho_o\frac{d}{dt}\vec{U} = \frac{4}{3}\cdot\pi\cdot a^3\cdot\rho_o\cdot\vec{g} + \vec{f} \tag{5}$$

where:
ρ_0- the density of the particles;
The equation of the carrier phase motion:

$$(1-c)\cdot\rho\cdot\frac{d}{dt}\vec{V} = \tag{6}$$

$$= -\frac{d}{dx}\vec{P} + \frac{d}{dr}F + \frac{F}{r} + (1-c)\cdot\rho\cdot\vec{g} - n\cdot\vec{f}$$

Where:

$F = \mu\cdot\frac{d}{dr}V$ - tangential stress in the viscous carrying liquid.

$n = \frac{c}{\left(\frac{4}{3}\cdot\pi\cdot R^3\right)}$ - number of particles in the unit volume.

$n\cdot f$ - сила, действующая со стороны частиц на единицу объема несущей жидкости.

The velocity should fulfill the condition of adhesion on the boundary $r=a$:
r - distance from the axis of the filter chamber;
a - radius of the tube.

$$V(a) = 0 \tag{7}$$

Equations for close packed depressive mixture motion can be obtained in the following way. The equation of dispersive phase differs from (6) by presence of pressure σ, which is transferring from one particle to another. The friction between particles can be taken into account trough tangential stress.

Therefore we have:

$$\rho_o\cdot c\cdot\frac{d}{dt}\vec{U} = -\frac{d}{dx}\vec{\sigma} + \frac{d}{dr}\tau_0 + \frac{\tau_0}{r} + \tag{8}$$

$$+c\cdot\left(-\frac{d}{dx}\vec{P} + \frac{d}{dr}\vec{\tau} + \frac{\vec{\tau}}{r}\right) + \rho_o\cdot c\cdot\vec{g} + n\cdot(1-c)\cdot\vec{f}_\theta$$

For the granular medium the tangential stress τ_o will be determine from the Coulomb`s law:

$\tau_0 = k\cdot\sigma$ - on the sliding surface, and when

$\tau_0 \le k\cdot\sigma$ - deformation in a medium is absent.

where : k - friction coefficient,

σ - the internal pressure in a particles because of contact interaction .

For the carrier phase we can use (6), where the interaction with particles is being determined by the force f_θ, and in tangential stress τ only the friction with the tube wall (face) is taken into account.

$$\rho\cdot(1-c)\cdot\frac{d}{dt}\vec{V}=(1-c)\cdot\left(-\frac{d}{dx}\vec{P}+\frac{d}{dr}\vec{\tau}+\frac{\vec{\tau}}{r}+\rho\cdot\vec{g}-n\cdot\vec{f}_\theta\right)$$

$$\vec{\tau}=\mu\cdot\frac{d}{dr}\vec{V} \qquad (9)$$

The force f_θ may be found from the experiment of flowing by the dense grain layer.

For the theoretical definition of force f_θ we have to solve the equation of liquid carrier phase in a complicated realm with conditions of adhesion on the tube body, and on a particles either.

$$\rho\cdot\frac{d}{dt}\vec{V}=-\frac{d}{dx}\vec{P}+\frac{d}{dr}\vec{\tau}+\frac{\vec{\tau}}{r}+\rho\cdot\vec{g} \qquad (10)$$

Then the force f_θ will be determined as a total stress of liquid on the particles surface.

For estimation of force f_θ, by the order of magnitude one can replace occupied by the liquid complicated region by the pores structure with pores (canals in a form of cylinders), which radius is comparable with radius of particles R.

Then the task will be transformed to the calculation of the hydraulic resistance in a cylinder channel.

REFERENCES

1. Landau L. , Lifshits E. Hydrodynamics. M.: Science, 1986, p. 130-132
2. Nigmatoulin R. Dynamics of multiphase mediums. part. I. – M.: Science, Head red. phis. math. liter., 1987, -464 p.
3. Petrov A. Variational methods in dynamics of incompressible liquid. Moscow State University, 1985, p. 72-77.

Environmental Issues and Management of Waste in Energy and Mineral Production, Singhal & Mehrotra (eds)
 ISBN 90 5809 085 X

Modeling of methane emission control from heavy oil wells

M.Yang & A.K.Mehrotra
Department of Chemical and Petroleum Engineering, University of Calgary, Alb., Canada

J.P.A.Hettiaratchi
Engineering for the Environment, Faculty of Engineering, University of Calgary, Alb., Canada

ABSTRACT: This paper presents the results of a study of the means of controlling casing gas (methane) emissions from heavy oil wells in the Lloydminster area in Canada. The study was aimed at providing a better understanding of the current casing gas methane emissions in the area and evaluating the potential for reducing these emissions by alternative technologies. Current casing gas emissions in the area were evaluated based on the best available information, and a database was established. Short-term trends of casing gas availability were forecasted. A Geographic Information System (GIS) was employed to simulate a multi-level gas clustering process. An optimization model was developed to evaluate potential technology applications and to find the least-cost solutions for reducing methane emissions in the study area.

1 INTRODUCTION

Methane emission into the atmosphere has become one of the major environmental concerns because of several reasons:

1) Methane, molecule for molecule, is a significant greenhouse gas with a global warming potential (GWP) of 21 on a 100-year time scale. This high GWP value makes methane the second largest greenhouse gas in the atmosphere just after carbon dioxide.

2) The concentration of methane in the atmosphere has increased by 145% over the last 250 years, while the concentration of carbon dioxide has increased by only 27% (IPCC, 1996).

3) A significant portion of methane emissions could be avoided by converting methane into carbon dioxide. Conversion of one unit mass of methane is equivalent to the reduction of 18 unit mass of carbon dioxide.

Based on these considerations, reducing methane emissions has become an intensively discussed subject since the Canadian Association of Petroleum Producers (CAPP) launched the Voluntary Climate Challenge Program.

Heavy oil production is one of the major sources of methane emissions from the up-stream oil and gas industry in the province of Alberta (CAPP, 1992). Identified sources of methane emissions in heavy oil fields are production casing venting, surface casing venting, storage tank leaks, soil gas migration and transportation facility leaks (Hettiaratchi & Kuang, 1997). The largest single emission source is the gas vented from production casings, which is called the casing gas. In general, the casing gas from heavy oil wells is sweet and contains mainly methane (CAPP, 1993). At present, it is used as fuel for engines and storage tank heaters at some well sites, but most of the casing gas is vented into the atmosphere. Traditionally, it has been believed that, since the volume of casing gas venting is very small, it is difficult to use or treat the gas by current technologies.

In recent years, investigations have been initiated to identify potential technologies to reduce methane emissions and to develop new technologies that can operate with small gas volumes. The technologies potentially suitable for dealing with the casing gas from heavy oil wells include the use of casing gas for power generation and the conversion of methane into carbon dioxide and water by either catalytic oxidation or bio-oxidation (CAPP, 1993). In regard to the small and unstable gas flow rate, a proposed option of combining gas flows from adjacent sources, or "clustering", to obtain larger gas volumes has attracted attention (Holford & Hettiaratchi, 1998). However, no study has been conducted on the feasibility of applying gas clustering and gas utilization technologies in heavy oil fields.

In order to study the potential and options to reduce methane emissions from heavy oil wells, 1351 active heavy oil wells in the Lloydminster area were chosen to investigate the situation of casing gas emissions and the possibility of gas clustering and utilization. The Lloydminster area is one of the major heavy oil producing areas in Western Canada. It covers the area from Township 42 Range 17 Meridian West 3 in Saskatchewan to Township 58 Range 10 Meridian West 4 in Alberta.

The principal objective of the study was to provide a better understanding of current methane emissions from heavy oil wells and to identify feasible solutions to reduce the emissions. This included: 1) evaluating current casing gas emissions from heavy oil wells and analyzing the trends of future casing gas availability, 2) establishing a casing gas database, 3) exploring the potential and method of gas source clustering, and 4) developing an optimization model and recommending least-cost solutions for reducing casing gas emissions from heavy oil wells.

2 METHODOLOGY

2.1 *Data acquisition and casing gas evaluation*

The first step of the study was data acquisition and verification. Data used in this study were obtained from three sources: existing databases, previous research and industry sources. Based on the data obtained from these sources, current casing gas venting from wells in the study area was evaluated. The casing gas data were then sent back to the operations personnel working in the area for verification. Operations personnel in the oil fields verified the casing gas data and re-tested gas venting for some high emission wells. This process ensured that the best available information was adopted in the research. A database including casing gas volumes and the geographic locations of these emission sources was then established.

Future casing gas availability was forecasted by applying production decline analysis combined with the observation of production gas oil ratios (GOR). Exponential declining was assumed in the production decline analysis for 12 selected pools (Lyons, 1996). Average yearly declining rates of oil production for wells in the 12 pools were evaluated and applied to predict future oil production. An overall average decline rate was applied to predict oil production for wells in other pools (Yang, 1999). Other factors, such as oil demand and oil prices in the international oil market, may also affect oil production rates, but these factors were not considered in detail in this study.

2.2 *GIS modeling*

A Geographic Information System (GIS)□IDRISI □was employed to analyze the distribution of gas sources and to explore the possibility and the potential of gas clustering in the area. The advantages of the GIS are its abilities of input, storage, manipulating, integrating, analysis and visualization of spatially related information for features on the earth surface and link the spatial information with attribute data of these features. These advantages allow us to simulate the real situation of the well location, the volume of casing gas associated with these wells, and the geographic conditions surrounding these wells, all of which make the gas clustering modeling more realistic.

To cope with the high deviation of gas flow rates from the studied wells, a multi-level clustering model was designed. The rationale used was: To transmit small gas flows over a short distance and large gas flows over a longer distance. The distance of gas transmission is a function of gas flow rate, transmission cost, and gas prices. In this study, gas clustering radii were determined according to the relationship of gas flow rate and economic transmission distance derived by Holford and Hettiaratchi (1998). The derivation of the relationship was based on low-pressure gas gathering technology, and considered both pipeline and compressor costs.

2.3 *Optimization modeling*

In order to evaluate potential technologies and to recommend feasible solutions for reducing methane emissions from the study area, an optimization model was developed. Linear programming method was applied in the optimization modeling. The objective function in the model was to minimize the total system cost, subjected to the planned target for greenhouse gas reduction. Other constraints included gas availability at each site and minimum gas requirements for each technology applications. A mathematical programming software LINGO5.0 was applied to solve the linear programming problem.

The optimization model is a multi-period linear programming model (Yang, 1999). A 4-year period was considered. The input of the gas availability at 191 gas sites with flow rates greater than 200 m^3/day was provided by the GIS simulation model. Three technologies were evaluated in the model, which are catalytic oxidation, power generation using mini-turbines and power generation using gas turbines. Six scenarios were designed for the reduction of greenhouse gas (GHG) emissions, varying from 60% reduction to 85% reduction. The economic performance of the discussed technologies was evaluated at different power prices in the range from 2 ¢/kWh to 4 ¢/kWh.

Gas royalty is also a factor affecting the economic performance of power generation technologies. It is not certain whether royalties would be collected on casing gas used for power generation. In this study, the situations when gas royalty is not applied and when gas royalty is at \$0.20/GJ and \$0.40/GJ, respectively, were discussed.

3 RESULTS AND DISCUSSIONS

3.1 *Status and forecasting of casing gas availability*

An investigation of the current casing gas venting was carried out for the 1351 active wells in the study area. Of these wells, 516 wells are located in the province of Alberta and 835 wells are located in the province of Saskatchewan. Regulations for oil industry operations are slightly different in these two provinces. Volumes of casing gas production from oil wells are required to be reported in the province of Alberta, but not in the province of Saskatchewan. Therefore, casing gas data for wells in Alberta are fairly up-to-date, while gas data for wells in Saskatchewan are not always up-dated and sometimes not available. Of the 1351 wells in our study, casing gas data are not available for 398 wells, with most of these wells located in Saskatchewan.

At present, casing gas is used as fuel for engines or storage tank heaters at 356 well sites. Gas from 79 wells was completely consumed by engines or heaters. Excluding these 79 wells and the 398 wells without casing gas data, the remaining 854 wells vented 393.6×10^3 m^3 of gas daily in 1999, with an average gas venting of 461 m^3/(well.day). On well-by-well basis, the volume of gas venting is highly deviated, varying from 1 m^3/day to 25.6×10^3 m^3/day. Although two-thirds of the 854 wells vented casing gas at less than 300 m^3/day, the other one-third of wells with gas venting greater than 300 m^3/day contributed 89% of the total gas emissions from the 854 wells. This implies a great potential for gas utilization in terms of the amount of methane emission reduction.

Forecasting of gas availability for the next five years was based on the forecasting of future oil production rate and an assessment of GOR trends. Considering the uncertainty associated with oil production, a five-year period forecasting was attempted. Production decline analysis was applied on pool bases to predict oil production rates. A conservative forecast was considered in this study. The analysis predicted production decline rates of between 1% to about 5% a year for studied pools (Yang, 1999).

There are other factors that could also affect gas availability in the future. For instance, future oil production will also depend on the situation of the international oil market. If the demand for oil and the price of oil decrease, oil production would decline, which may lead to less casing gas production. On the other hand, at higher oil demand and higher oil prices, the production of oil would increase, which may bring more casing gas production.

3.2 *Gas source clustering simulation*

Gas clustering process was simulated in a GIS based modeling system. For very small gas streams, gas transmission should not be considered as an option due to economic reasons. In this study, wells with gas flow rate less than 200 m^3/day were not included in gas source clustering. The remaining 345 wells were put into the GIS for a two-step clustering simulation.

The first level clustering dealt with wells with gas rates between 200-300 m^3/day. Clustering radii were set at 500 m, and the wells falling in this category were examined individually. When there were other wells located with a distance less than 500 m to the examined well, the examined well was clustered to the closest well. When there was no well with a distance less than 500 m to the examined well, the examined well was not clustered. After the first level clustering, some wells originally with gas rate between 200-300 m^3/day were clustered to form larger gas sites. These sites as well as the wells originally with gas rates greater than 300 m^3/day were considered for second level clustering. The wells with gas rates less than 300 m^3/day after the first level clustering were ignored in the second clustering.

For the second level clustering, a number of wells were chosen as gas gathering sites due to their favorable locations (generally, these wells are located in the center of a cluster of wells). The clustering radii were determined at 1100 m. The simulation process was designed to ensure the satisfaction of two conditions: (1) all clustered wells were connected to gas gathering sites at a distance less than 1100 m, and (2) when there were more than one gas gathering sites close to the examined well, the well was clustered to the nearest gathering site.

Comparisons of gas sites before and after clustering are presented in Table 1 and Table 2. After the two-level clustering, the 345 wells originally with gas rates greater than 200 m^3/day were clustered into 151 gas sites with gas rates greater than 300 m^3/day and 40 not clustered wells with gas rates between 200–300 m^3/day. The most significant impact of the gas clustering is the increase in percentage of gas sites with larger gas flows. Before clustering, the number of gas sites with gas flows greater than 1000 m^3/day accounted for 31% of the total number of sites. After clustering, this group of sites accounted for 48% of the total number of sites. The share of combined gas of this group of sites increased even more, from 70% to 89% of the total gas volumes from the 345 wells considered in the clustering simulation. The overall average gas rate for the 345 wells was 1064 m^3/(well.day) before clustering, which increased to 1921 m^3/(site.day) for the 191 gas sites after clustering.

The impact of gas clustering on potential technology applications is demonstrated in Figure 1 and Figure 2. The percentage of gas sites with potential for gas turbine applications increased from 3% before clustering to 11% after clustering, while the combined gas volume from this group of sites increased from 25% to 46% of the total gas volume from the studied wells. At the same time, the share of gas sites with potential for mini-turbine applications decreased from 76% before clustering to 68% after clustering. The combined gas volume of this group of gas sites decreased from 71% to 51% of the total gas volume. These observations indicated that part of gas sites with potential for applying mini-turbines were upgraded to gas sites with potential for gas turbine applications. Since gas turbines are more energy efficient in power generation than mini-turbines, this improvement would increase the overall energy efficiency of the gas utilization system.

The overall average gas volume of clustered gas sites was significantly higher than that before clustering. There were 263 wells with gas rates between 300-4000 m^3/day before clustering, which are potential sites for mini turbine applications. Average gas rate of the 263 wells was 984 m^3/(well.day). After clustering, 130 gas sites fell in the group of potential sites for mini-turbine applications. The average gas rate of the 130 gas sites reached 1451 m^3/(site.day), which is about 50% higher than that before clustering. Although mini-turbines can be operated at gas rate as low as 300 m^3/day, they would achieve higher efficiency at larger gas flows. It is obvious that gas clustering not only could improve the potential for using larger power generators, but could also improve the efficiency of smaller power generating units.

Table 1: Distribution of gas sites before and after clustering

Group by gas rate (m^3/d)	Before clustering		After clustering	
	Number of wells	Share (%)	Number of Sites	Share (%)
>10000	2	0.6	5	2.6
4000-10000	9	2.6	16	8.4
1000-4000	95	27.5	71	37.2
300-1000	168	48.7	59	30.9
200-300	71	20.6	40	20.9
Total	345	100.0	191	100.0

Table 2: Combined gas volume before and after clustering

Group by gas rate (m^3/d)	Before clustering		After clustering	
	Combined gas(m^3/d)	Share (%)	Combined gas(m^3/d)	Share (%)
>10000	2	0.6	5	2.6
4000-10000	9	2.6	16	8.4
1000-4000	95	27.5	71	37.2
300-1000	168	48.7	59	30.9
200-300	71	20.6	40	20.9
Total	345	100.0	191	100.0

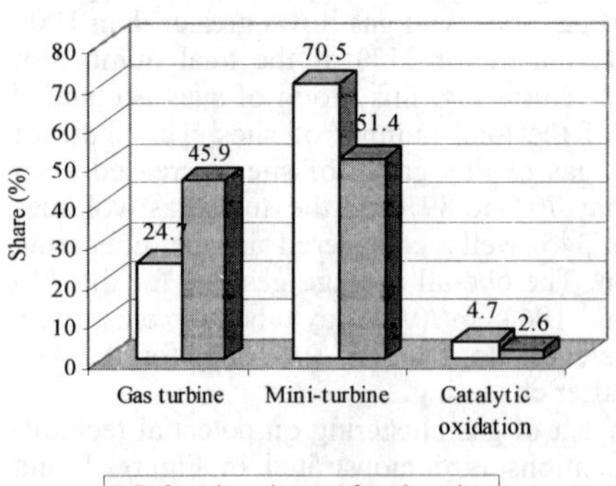

Figure 1. Comparison of number of gas sites for potential technology applications before and after clustering.

3.3 *Optimization results*

As described in Section 2.3, the optimization model considered 191 gas sites, a four-year period, three technologies, six scenarios for greenhouse gas (GHG) emission reduction and different power prices. Although the model was used for evaluating the situations when the royalty is not applied on gas used for power generation and when the royalty is at \$0.20/GJ and \$0.40/GJ, respectively, only the results for no gas royalty applied and royalty at \$0.40/GJ are presented in this paper.

Figure 3 demonstrates the value of total system cost under various conditions. If the gas royalty is applied, at power prices of above 3 ¢/kWh, the total system cost was negative for all GHG reduction scenarios, which means that the system could actually make profit from power generation. The total system cost increased to a certain level when the gas royalty is collected on the gas used for power generation. For the gas royalty of \$0.40/GJ, the system cost became positive at a power price of 3 ¢/kWh. However, at a power price of 4 ¢/kWh, the system would be profitable even at gas royalty of \$0.40/GJ. It is also observed that the system would not make profit

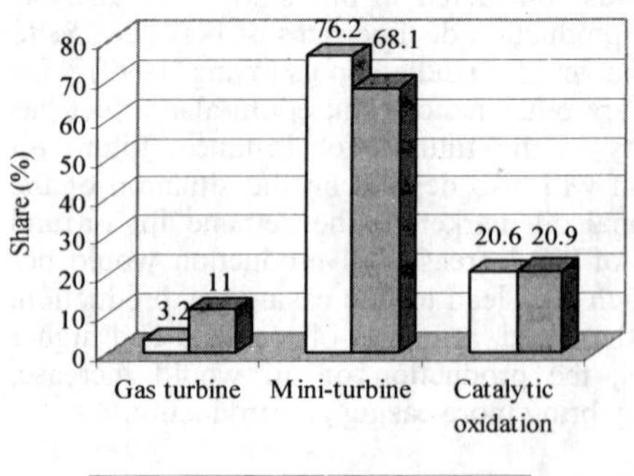

Figure 2. Comparison of combined gas for potential technology applications before and after clustering.

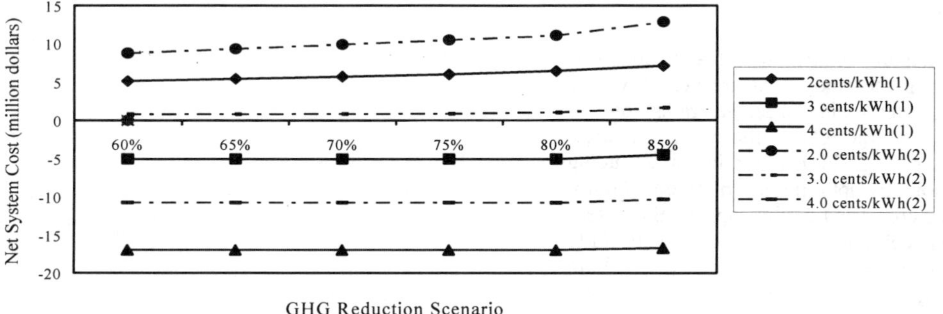

Figure 3. Net system cost versus greenhouse gas reduction scenarios.
(1): No royalty is applied. (2): Gas royalty is at $0.40/GJ.

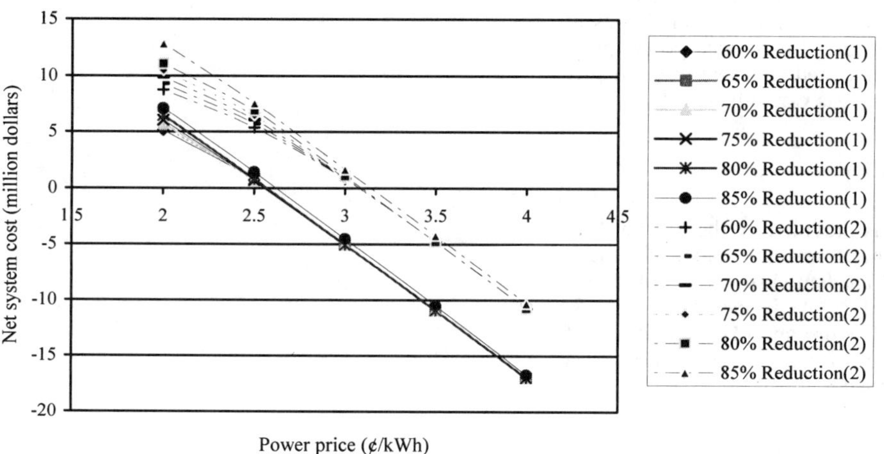

Figure 4. Net system cost versus power price.
* (1): No gas royalty is applied. (2): Gas royalty is at $0.40/GJ.

at a power price of 2 ¢/kWh, even if the gas royalty were not applied.

Figure 4 is another presentation of the optimization results, in which the minimum power price required for the system to be profitable can be observed. When no gas royalty was collected on the gas used for power generation, the system became profitable at power price of 2.6 ¢/kWh. The curves of system cost would shift upward under all GHG reduction scenarios when the gas royalty was at $0.40/GJ. The minimum power price required for the system to be profitable increased to about 3.1 ¢/kWh.

The results from optimization modeling suggested that two power generation options were more competitive. Catalytic oxidation was considered by the model only when very strict GHG emission constraint was imposed on the model. For GHG reduction scenarios below 80% level, catalytic oxidation was not considered by the model. An obvious consequence associated with power generation options is the impact of power price and gas royalty. As power price decreases, the minimum gas rates required for applying the two power generation technologies increases. At a power price of 4 ¢/kWh, the minimum gas requirement was 5900 m^3/day for gas turbine applications and 370 m^3/day for mini-turbine applications (gas royalty is not applied). However, when power price decreased to 2 ¢/kWh, the minimum gas required for gas turbine applications increased to 9400 m^3/day and for mini turbine applications increased to 590 m^3/day. It is observed that at very low power prices, mini-turbine would be the most attractive option for reducing GHG emissions in the study area.

4 CONCLUSIONS

Although methane emission rates from most heavy oil wells in the study area are very low, the small number of wells with larger gas rates contributed the majority of the total emissions in the area. These gas sites possess high potential for reducing methane emissions by applying alternative technologies. Gas source clustering could improve the potential for various technology applications, especially the potential for power generation in terms of both the amounts of gas to be utilized and the overall energy efficiency of the gas utilization system. The two power generation technologies, namely gas turbines and mini-turbines, are more competitive than others in reducing methane emissions in the study area. Power price and gas royalty policy could affect the application of power generation technologies in the area, however, at fairly reasonable power price or if part of power generated could be used on site, the utilization of casing gas would be feasible.

ACKNOWLEDGMENT

Financial support was provided by the Natural Sciences and Engineering Research Council of Canada (NSERC) and the Canadian Association of Petroleum Producers (CAPP). Assistance from Mr. R. Schmitz, Mr. M. R. Holford and Ms. A. Mah, all of Husky Oil Operations Ltd., Calgary, Canada, is gratefully acknowledged.

REFERENCES

Canadian Association of Petroleum Producers (CAPP), 1992. A Detailed Inventory of CH_4 and VOC Emissions from Upstream Oil and Gas Operations in Alberta (Report), Calgary, Canada.

Canadian Association of Petroleum Producers (CAPP), 1993. Options for Reducing Methane and VOC Emissions from Upstream Oil and Gas Operations—Technical and Cost Evaluation (Report), Calgary, Canada.

Hettiaratchi, P. & Kuang, E. C., 1997. Control of Methane Emissions from Heavy Oil Production (Report), University of Calgary, Calgary, Canada.

Holford, M. R. & Hettiaratchi, P., 1998. *An Evaluation of Potential Technologies for Reducing Solution Gas Flaring in Alberta*. MEng Thesis, University of Calgary, Calgary, Canada.

Intergovernmental Panel on Climate Change (IPCC), 1996. *Climate Change 1995: The Science of Climate Change*. Cambridge Press, UK.

Lyons, W. C., 1996. *Standard Handbook of Petroleum & Natural Gas Engineering*. Gulf Publishing, Houston, TX, USA.

Yang, M., 1999. *Controlling Methane Emission from Heavy Oil Wells: Gas Clustering Simulation and Optimization Modeling*. MSc Thesis, University of Calgary, Calgary, Canada.

Environmental Issues and Management of Waste in Energy and Mineral Production, Singhal & Mehrotra (eds)
© 2000 Balkema, Rotterdam, ISBN 90 5809 085 X

A review of emerging technologies for sustainable use of coal for power generation

T.M.Yegulalp
Henry Krumb School of Mines, Columbia University, New York, N.Y., USA

K.S.Lackner & H.J.Ziock
Los Alamos National Laboratory, N.Mex., USA

ABSTRACT: Concerns about climate change and environmental consequences of increased levels of atmospheric CO_2 will require the power generation industry to reduce CO_2 emissions from current levels. Unfortunately, for reductions to have the desired effect they will have to be large. While the schedule of the Kyoto Protocol may appear daunting, in the long term mere compliance with the protocol will hardly change the rate of increase of atmospheric CO_2. There are, however, technical, economical and practical choices that will allow the industry to meet the requirements of the protocol and ultimately stop the increase in atmospheric CO_2 without eliminating coal from the fuel mix. In this paper, we present a review of current and emerging technologies for CO_2 sequestration. We provide a summary of the underlying scientific principles and discuss the practical and economic aspects of sequestration technologies, which will allow continuing use of global coal resources with minimum or no impact on CO_2 levels in the atmosphere.

1 INTRODUCTION

In this paper, we focus on coal as a major source of energy and emerging technologies that will allow coal to be used in power generation with minimal or no accumulation of CO_2 in the atmosphere. The problem of continuously increasing the CO_2 content of the Earth's atmosphere requires solutions, which will allow sustainable power generation at increased levels. Within the last few years the consensus has been steadily growing, that excess carbon dioxide will cause a significant change in climate that will have repercussions on a wide variety of human activities. Most predictions are, however, based on complex climate models that still cannot fully capture all the physical effects that interact in leading to a change of climate. Consequently, there remains a considerable amount of uncertainty.

The carbon dioxide content of the atmosphere has increased dramatically since the beginning of the 19th century. It has risen by about 30%, from 280 ppm to 360 ppm. This change is well documented (Siegenthaler & Oescher, 1987, Keeling et al., 1995). The increase recorded during the last 40 years (from 315 ppm to 360 ppm) accounts for more than 50% of the total increase during the last two centuries. The general consensus is such that the increase of CO_2 in the atmosphere is due to human activities, primarily the combustion of fossil fuels. It has been pointed out (Keeling *et al.*, 1995) that the rate of increase closely tracks the growth in the generation of CO_2 from fossil fuels and cement production. At present, about 60% of the CO_2 thus generated remains in the atmosphere. The 6 Gt of carbon that are emitted annually (EIA, 1998) amount to 1% of the 570 Gt of carbon that naturally would reside in the atmosphere as CO_2.

It has also been shown that the change of atmospheric carbon dioxide is accompanied by a corresponding drop in the atmosphere's oxygen level, which implicates combustion processes, rather than, for example, a decrease in the rate of photosynthesis, which would also increase CO_2 levels. Furthermore, precision measurements of concentration gradients in CO_2 and O_2 point to the Northern Hemisphere as the source of the excess CO_2 (Keeling et al., 1996). Considering the current and projected future fossil carbon consumption and the available fossil carbon resources, it is conceivable that in the distant future the rate limiting resource is oxygen from the air rather than fossil carbon from the ground. Long before this point is reached, CO_2 levels would have reached intolerable levels. Thus, CO_2 sequestration is an important step to maintain access to fossil fuels for centuries to come.

Here, we are not concerned with the straightforward approach of avoiding CO_2 production by either foregoing energy production or by using other forms of energy. These methods can and will contribute to the reduction of CO_2 emissions. Nevertheless, energy conservation and energy efficiency will fall far short from what would be required. Alternative forms of energy that don't produce CO_2 are currently far too expensive to compete. Unless there is a major and unexpected technological breakthrough for a carbon free energy resource, one cannot expect to remove the largest contributor to today's energy mix from a growing world energy market.

2 COAL'S STRENGTH

Among raw energy resources, coal is a strong competitor. Coal is available on every continent and in virtually every country. Mining technologies are well advanced and coal has by far the lowest cost of all fossil fuels. However, energy extraction from coal requires processing which adds to the cost. Energy needs to be put in the form of electricity or in the form of chemical energy carriers that could be readily distributed and consumed without undue environmental impact. The low price of coal allows for substantially more processing than would be acceptable for crude oil or raw natural gas.

As chemical energy carriers become more distinct from the raw resources, raw resources become interchangeable. For electricity the transition to interchangeable energy sources has already occurred. Electricity is generated from oil, gas and coal as well as from non-fossil sources. Local conditions prevailing at a particular time determine which resource is cheapest and will be used. Coal is not a likely source of gasoline or natural gas, as they are too close to crude oil and raw natural gas, respectively. Among the energy carriers contemplated for the future, the most prominent is hydrogen. Methanol also may find use in the transportation sector. Both are sufficiently different from all raw energy resources that their production from coal could be considered. The higher processing cost must be offset by the much lower cost of coal.

Managing the carbon cycle is made easier if the energy is distributed in a carbon free format, for example as hydrogen or as electricity. For carbon based energy carriers a solution needs to be found that can recover carbon at the end of the distribution chain, or one needs to look at other sources of carbon that are sequestered instead.

Hydrogen production from coal would transfer the energy that is stored in coal into a clean hydrogen gas that is most likely derived from reducing water. In transferring the oxygen that is attached to hydrogen to carbon, one transfers the energy from a carbon-based carrier to hydrogen.

3 SUSTAINABLE COAL

Sustainability is usually claimed by technologies relying on renewable resources. However, sustainability or a state near sustainability is achievable for other technologies as well. A technology should be considered "sustainable" if the intended or unintended consequences of its use do not force an abandonment of this technology.

Resource size limits any technology's sustainability. Even renewable resources are not infinite. For practical purposes we consider a technology as "sustainable" if resource depletion is not an issue on a time scale for which humans can reasonably plan.

Coal consumption for the foreseeable future is not limited by resource availability. Worldwide consumption (including lignite) is on the order of 4 to 5 Gigatons (Gt) per year (United Nations, 1997). Accessible coal is estimated to exceed 10,000 Gt and currently recoverable proven reserves of coal are in the order of 1000 Gt (United Nations, 1997). With the current consumption rate, the proven coal reserves would last at least two centuries, and the entire coal resources would last 20 centuries. Coal's availability virtually guarantees that energy will not run out for many generations. The challenge to coal's sustainability lies in the environmental impact of consuming such a staggering volume of material. Small concentrations of impurities in the coal, multiplied by a large rate of consumption, amount to large emissions of problematic materials. Many of these environmental impacts have been addressed for some time. With regard to impurities and undesirable combustion byproducts, technology is already moving towards zero emission coal. The one serious challenge to sustainability of coal that so far has not received adequate attention is the emission of CO_2.

4 THE CARBON DIOXIDE PROBLEM

The safe and permanent disposal of carbon dioxide is the most difficult challenge to the sustainability of coal., The industry shares this problem with all other fossil energy sources. Coal is carbon intensive, it generates more CO_2 per unit of energy than most other energy resources. There are however, natural gas wells with such high CO_2 content that the raw gas is even more carbon intensive than coal. For example, the Natuna field, which is the largest natural gas field in the world, contains 71% CO_2. The largest carbon disposal operation, at Sleipner field off the shore of Norway, sequesters CO_2 from a natural gas well (Herzog et al., 2000).

In order for carbon not to accumulate in the atmosphere, or for that matter elsewhere in the carbon surface pool, future energy technologies need to provide the means of capturing the oxidized carbon and dispose of it in a safe and permanent manner. We emphasize the need for disposing of the oxidized carbon, i.e. CO_2, or carbonates, since all reduced form of carbon carry an amount of energy that is comparable to the energy carried per unit of carbon by fossil fuels.

To set the scale for the problem, consider the fact that a century of CO_2 emissions at the current rate amounts to 600 Gt. This is more than the natural carbon content of the atmosphere. It matches the entire living biomass on the surface of the earth and it is 40% of the carbon found in soil and detritus. At 39,000 Gt of carbon, the ocean is the biggest reservoir of mobile carbon, mainly in the form of bicarbonate ions. However, using mobile carbon in the ocean as a yardstick is misleading because a change of about 1,000 Gt is sufficient to decrease the ocean pH everywhere by 0.3. Such a change in acidity would have substantial environmental effects. If we follow the trend set during the last century, the potential output during the 21st century could be 4 to 6 times larger than the initial rate of output at the beginning of the century suggests. If energy production stays on its current path, likely problems in the future are not limited to climate. Excess CO_2 in the atmosphere results in physiological changes in plants and animals and more subtle changes in ecological systems. The concomitant acidification of the ocean will pose its own problems to an ecological system, which seems to be quite sensitive to the pH of the water. Corals in particular are sensitive to such changes (Kleypas, 1999). A continuous rise in atmospheric CO_2 levels is not sustainable and steps must be taken to stop a further rise. While the debate of what level of CO_2 is tolerable is still ongoing, hardly anyone would advocate levels that exceed twice the preindustrial value. At this point the surface ocean pH would have changed by 0.3. Current worldwide increases in emissions suggest that this point is not far away and is likely to be reached by the middle of this century.

In order to stop the increase of atmospheric CO_2 one has to drastically reduce CO_2 emissions. At present, the oceans absorb CO_2 because the continuous rise in atmospheric levels maintains a gradient. As the increase is arrested, the gradient is no longer maintained and fluxes out of the atmospheric reservoir are greatly reduced. Model calculations suggest that even after doubling the CO_2 from preindustrial levels the steady state emissions must be exceedingly small. To maintain a constant CO_2 level, emissions must be reduced. Even at twice the preindustrial level of CO_2, sustainable emissions could rapidly fall to 30% of 1990 emission rates.

The consequence of these physical constraints is that as atmospheric CO_2 approaches the maximum tolerable level, the annual worldwide emission allowance will rapidly shrink. Dividing the world emission allowance evenly among 10 billion people would lead to a per capita allowance which is only about 3% of the current US per capita emission rate.

5 TECHNOLOGY FOR SUSTAINABLE COAL

To avoid the built-up of carbon in the surface pool, carbon needs to be shepherded through the entire energy cycle. In the case of coal, the carbon comes out of the ground, essentially as $CH_{0.8}$. The mixture also contains ash and impurities like sulfur, nitrogen, and heavy metals. A well-designed plant will combine the collection of CO_2 with the cleanup of all other pollutants. CO_2 generation is unavoidable; in order to extract the energy, either as electricity or as a chemical energy carrier free of carbon, the carbon needs to be oxidized. Carbon is either combusted with the oxygen from air or it is used to reduce the oxide of some other element, which then carries the energy in a carbon free form. Combustion with air complicates the CO_2 recovery because it carries a large amount of nitrogen through the combustion process. Not only does this greatly increase the potential for undesirable nitrogen compounds, it also dilutes the CO_2 by roughly a factor of four. To complicate matters further, CO_2 and N_2 are both gases suggesting a difficult separation problem. One approach to the problem is to separate the oxygen from the air and perform the combustion in either pure oxygen or a mixture of flue gas CO_2 and O_2. In either case the resulting combustion products will be dominated by CO_2 and water, which are easily separated.

The other option is not to bring in air at all, but shifting the oxygen from another compound to the carbon. A potential candidate here is water, which is reduced to hydrogen. One problem with this approach is that the stoichiometry and energy balance typically do not match and additional oxygen is needed to complete the transition while balancing the energy of the reaction. For example, in transferring the oxygen in water to carbon and forming CO_2 an additional amount of energy, about 40% of the heat of combustion of carbon, is needed to complete the reaction.

After CO_2 has formed, it needs to be separated out in a concentrated stream that can be readily disposed of. At issue is the nature of the gas formed. The admixture of particulates, sulfur, nitrogen etc. can make for a corrosive mixture that rules out some of the more advanced ways of separation. For example, such an environment will challenge membrane technologies. Absorption of CO_2 during the reaction provides a most convenient approach. Carbon dioxide acceptor processes fall into this category.

6 ANAEROBIC HYDROGEN PRODUCTION FOR ZERO EMISSION COAL

With the exception of biomass sequestration, which takes CO_2 directly out of the air, sequestration will require a concentrated input stream of CO_2. Thus, sustainable coal technology not only needs CO_2 disposal options but also CO_2 collection options. Retrofitting existing power plants for collection of CO_2 is possible but it is a costly proposition. A better approach for the long term is to reoptimize a power plant with the new design constraints in mind. New designs will look quite different, since the new optimum is not likely to near current designs.

One approach to zero emission coal is the CO_2 acceptor process, which has been pioneered by Consolidated Coal in the sixties and seventies. The basic idea is to assist the reforming shift reactions, which make hydrogen from water by turning carbon into CO_2 and which use CaO to remove the CO_2 from the reaction products. Not only does the reaction with lime remove one of the reactants and thus let the reaction proceed further, but it also provides in the exothermic carbonation reaction the energy that is required to complete the shift reaction. It is an accident of nature that the reaction

$$CaO + C + 2H_2O \rightarrow CaCO_3 + 2H_2 \quad (1)$$

is balanced in enthalpy. Starting with liquid water the reaction is exothermic by a mere 0.6 kJ/mole. The heat of combustion of carbon which sets the scale is 393kJ/mole (Figure 1.).

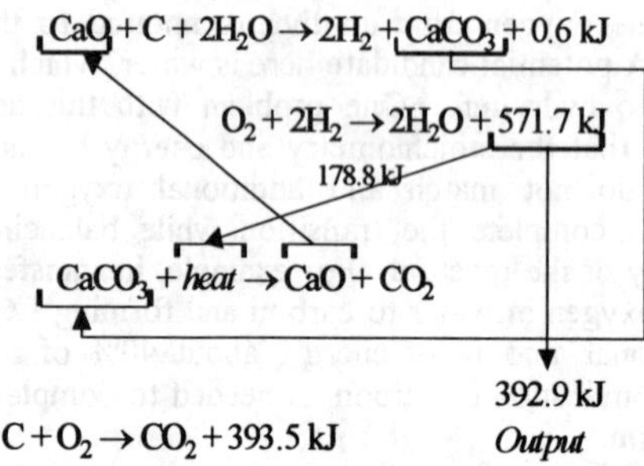

Figure1. Energy balance in hydrogen production from coal

Ziock and collaborators are developing an anaerobic process for hydrogen production from coal (Lackner et al., 1999b) in which coal, water and lime are used to form hydrogen and limestone as an intermediary. The hydrogen when it is combusted releases an amount of energy, which combines the heat of combustion of coal with the heat of carbonation of CaO. A fraction of this combustion heat can be used to calcine the CaO and thus extract the CO_2 from the process. The remaining energy in the hydrogen than matches the heat of combustion of the coal. If all the hydrogen is run through a high temperature solid oxide fuel cell it generates an amount of electricity that nearly matches the heat of combustion of coal. The thermodynamically unavoidable waste heat is of a sufficient quality that it can be used to calcine the calcium carbonate. Thus the effective efficiency of the fuel cell is boosted by the ratio of the heat of combustion of hydrogen to the heat of combustion of carbon. In an indirect manner, this is a carbon-based fuel cell The theoretical efficiency of a carbon burning fuel cell would be 102% based on the ratio of Gibbs free energy to enthalpy. The process of making intermediate hydrogen is, however, not reversible and consequently the theoretical limit of this specific approach at 93% is somewhat lower. Practical implementations will fall short of this ambitious goal. Nevertheless this approach has the potential for extremely high efficiencies (Figure 2.).

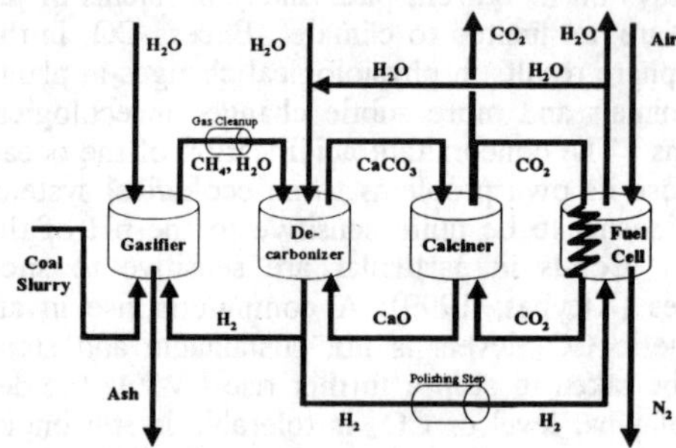

Figure 2. A block diagram of CaO driven hydrogen production system

The economics of this process look very promising. The process steps are not that different from pressurized fluidized bed technologies which are used in upgrading conventional coal fired plants for merchant plant installations. As a means of producing hydrogen this process may prove very cost efficient. Ultimately, of course, it relies on the price difference between coal and natural gas. It is worthwhile pointing out that nearly all attempts of utilizing coal have simpler counterparts for natural gas. Thus the future of coal will depend on the availability and price of natural gas. The advantage of coal is in its lower starting cost.

7 SEQUESTRATION TECHNOLOGIES

Sequestration in its broadest sense is any technology that keeps CO_2 out of the atmosphere. Thus, sequestration technologies can prevent the accumulation of CO_2 in the air without curtailing the use of fossil fuels. Sequestration can be accomplished in a variety of ways. CO_2 could be collected at the point of combustion or later taken from the air. CO_2 could be stored in gaseous form or could be chemically transformed before it is disposed as waste. It has been suggested that some of it is recycled back into the economy.

Currently, only biomass generation extracts CO_2 from air. The 1:3000 dilution has made this approach look too difficult. On the other hand the CO_2 in a volume of air represents an amount of heat of combustion that exceeds the kinetic or wind energy of the same amount of air by two orders of magnitude (Lackner et al. 1999a). Since wind energy appears close to economically viable, CO_2 extraction from air may prove to be attractive. Extraction from air would eliminate the need for a dedicated infrastructure for the transport of CO_2 and no change would be needed in present combustion technology. Extraction from air would be particularly useful in counteracting the emissions of small, distributed and often mobile sources of CO_2, which together amount to about half of all CO_2 emissions.

The other alternative would be to collect CO_2 at the site of combustion. While difficult for mobile sources, it is quite easy for large stationary sources. It is the obvious method of choice if the flue gas is sufficiently rich in CO_2. In this case, chemical scrubbing or membrane separation techniques can be used (Meisen & Shuai, 1997). Retrofitting existing power plants with low concentration of CO_2 around 10% to 15% in the flue gas make this a costly option raising the cost of electricity by 30% to 40%. Integrating CO_2 collection into an integrated gasifier combined cycle plant is much easier because pressures are higher and the exhaust is much richer in CO_2. In the above mentioned plant designed to produce hydrogen, collection of CO_2 comes naturally and is achieved with hardly any incremental cost. Cost estimates vary from an increase by 30% to 40% of electricity costs in retrofits, to very small cost increments for hydrogen producing power plants (Lackner et al., 1998). Transporting and shipping CO_2 will not stop sequestration. Pipeline technology for transporting carbon dioxide is already in place and costs have been estimated at \$0.01/ton km (Audus et al., 1995). Recent studies for very large pipelines have arrived at even lower costs. However, the options for turning CO_2 into valuable products are extremely limited considering the fact that the volume of CO_2 is very large. CO_2 emissions in the US alone amount to about 20 t/year per person (EIA, 1998). This is more than three times per capita crushed stone consumption in the US. Considering the value of crushed stone today, transportation costs alone would make this option very unattractive.

We can easily rule out carbon recycling in the form of plastics (Halman, 1993, Arresta, & Tommasi, 1997) or other organic compounds. The synthesis of nearly all carbon rich products which require substantial amounts of energy are usually far in excess of what has been extracted at the power plant.

7.1 *Biomass Sequestration*

Biomass generation has been considered as a method of sequestration. Photosynthesis is a natural process and extracts CO_2 from air to form starch or similar organic materials by adding H_2O and sunlight. Biomass can accumulate in standing forests or other green areas. This however is a means of collecting energy, which ultimately will be wasted. It is difficult to store the perpetual accumulation of carbon as biomass. A mature forest will lose about as much biomass as it generates. Furthermore, because collection rates are very small (Ranney & Cushman, 1992), one needs to dedicate unrealistic amounts of land or ocean to use this option as the sole means of sequestration of CO_2 (Lackner et al. 1998, Sedjo & Solomon, 1989). The annual collection of carbon on an acre of land at best compensates for a couple of minutes worth of CO_2 from a one GW power plant (Ranney & Cushman 1992). Natural forests would fall far short of this number.

7.2 *Underground Injection*

Instead of releasing it to the air, CO_2 can be injected into suitable geological formations for permanent storage. This idea is already practiced at a limited extent for various purposes. For example, because of high carbon tax in Norway (\$55/t of CO_2) CO_2 stripped from natural gas is injected into an aquifer 1000 m under the sea floor in the North Sea (Kaarstad & Audus, 1997). It is also a common practice in crude oil and natural gas production to inject CO_2 to increase production rates. In another case, CO_2 is injected to recover methane from deep coal seams (Gunter et al., 1997). Since CO_2 is adsorbed much more strongly than methane, CO_2 exchanges places with the adsorbed methane, which can then be produced and CO_2 remains behind fixed in place.

7.3 *Ocean disposal*

Ocean disposal has been extensively studied (Herzog et al., 1997). There are various forms, which differ in how and where CO_2 is introduced into the ocean. CO_2 is transported in an undersea pipeline from the shore, or it is introduced from a ship that carries it to a deep part of the ocean. CO_2 is introduced as a compressed gas at great depth, or injected as a water clathrate. It can be introduced as dry ice or bubbled into intermediate depth where it dissolves in the water. While still many questions remain, one could consider this option as available on a small scale. Very deep storage has the advantage that the CO_2 becomes denser than water and forms a lake on the bottom of the ocean, which only gradually dissolves into the surrounding water (Herzog et al., 2000).

Ocean circulation guarantees that over time the highly soluble CO_2 is mixed into the ocean as a whole. The allowable change in pH sets an upper limit on how much can be stored in the ocean. Approximately 1000 Gt of carbon added as bicarbonate ions to the ocean would change the overall pH by 0.3. Time constants for exchange with the air are estimated between 500 years to a few thousand years. Oceans are a natural sink that is much larger than the atmosphere. In that sense, this is an accelerated natural process.

7.4 *Carbonate Disposal*

Except biomass generation, all sequestration methods consider disposing CO_2 in gas form. A newly emerging technology suggests that it is not only feasible but also quite advantageous to dispose of CO_2 in the form of carbonates (Lackner, et al. 1998). This idea was first suggested by Sefritz (Seifritz, 1990). The reaction of CO_2 with common mineral oxides to form carbonates like magnesite or calcite is exothermic and thermodynamically favored under ambient conditions. Consequently it is possible to dispose of CO_2 as a solid mineral carbonate (Lackner et al., 1995, Lackner et al., 1997a). The resulting waste product is environmentally safe and thermodynamically stable. There is no shortage of raw materials to bind with CO_2. In fact, the known resources for such raw materials are more than what is needed for even the most optimistic estimates of fossil energy reserves (Lackner et al., 1997a). The reaction is well known to geologists because it occurs spontaneously on geological time scales.

The carbonation reaction can be shown by the simple reaction of binary oxides, MgO and CaO. These reactions are exothermic.

$$CaO + CO_2 \rightarrow CaCO_3 + 179 \text{ kJ/mole} \qquad (2)$$

$$MgO + CO_2 \rightarrow MgCO_3 + 118 \text{ kJ/mole} \qquad (3)$$

Even compared to the heat released in the combustion of carbon (394 kJ/mole), these reactions release substantial heat. In nature, however, calcium and magnesium are rarely available as binary oxides. They are found typically as calcium and magnesium silicates. The carbonation reaction is still exothermic for common calcium and magnesium bearing minerals. In such cases however, the heat release is reduced. As an example consider the carbonation reactions of forsterite and serpentine. For forsterite and serpentine respectively:

$$1/2Mg_2SiO_4+CO_2 \rightarrow MgCO_3+1/2SiO_2 +95\text{kJ/mole} \qquad (4)$$

$$1/3Mg_3Si_2O_5(OH)_4+CO_2 \rightarrow MgCO_3+2/3SiO_2 +2/3H_2O+64\text{kJ/mole} \qquad (5)$$

Both of these reactions are favored at low temperatures. In nature magnesite and silica are common in serpentinized ultramafic rocks. Their formation is due to natural CO_2 –rich fluids percolating through mineral deposits. Magnesite is stable and not likely to release the bound CO_2 again.

One can accelerate the process by injecting concentrated CO_2 into underground formations which are likely to react with it. These include limestone reservoirs that would form bicarbonates, or silicate rocks that can form carbonates. The advantage of forming carbonates in underground injection is that it solves the problem of long-term stability. Once carbonates have been formed the CO_2 can no longer escape to the surface. The concern about seepage is heightened by the large volumes of CO_2 that need to be injected underground. At a nominal density of water a one Gigawatt coal fired power plant of 33% conversion efficiency would generate a volume of CO_2 that would have to raise the ground over a 10km by 10km area by 7 cm.

An alternative route to forming stable mineral carbonates is an above ground industrial process. In this case the rock is mined, ground up and reacted with CO_2 to form solid mineral carbonates. The advantage of the above ground process is that the resulting material is much better controlled. Rather than relying on circumstantial evidence that the carbonate indeed has been formed or will form in the foreseeable future one has a product in hand and can determine its quality.

Lackner et al. (Lackner et al., 1997b) have been studying this reaction over the last few years with the purpose of developing an economically viable above ground process that accelerates the natural reaction rate so that it can be performed cost-effectively. Since the reaction generates heat, the aim is for an implementation without an external supply of energy. At present, this process is still in an early research phase. The goal is to achieve a cost of about \$20/t of CO_2 but this still needs to be demonstrated. The long-term stability of the waste product makes the process appealing.

Figure 3. shows a 2 GW power plant fed from a 9 kt/day coal mine by pipeline at a serpentine mine by pipeline at a rate of rate of 18 - 20 kt/day coal slurry. The plant produces 2.6 kt/day of hydrogen, which is then combusted in a fuel cell. CO_2 emissions are captured and compressed and sent to a carbonation plant where the carbonation process takes place. Calcium and magnesium carbonates are solid which is desirable in above ground disposal. The materials formed can be stored at the serpentine mine as landfill and will not leave the disposal site. Magnesium proved more attractive since there are large deposits of magnesium rich minerals. Peridotite, and serpentinized peridotite rocks can have an MgO content between 35% and 40% by weight, whereas abundant calcium silicates rarely have more than 12 % to 15%

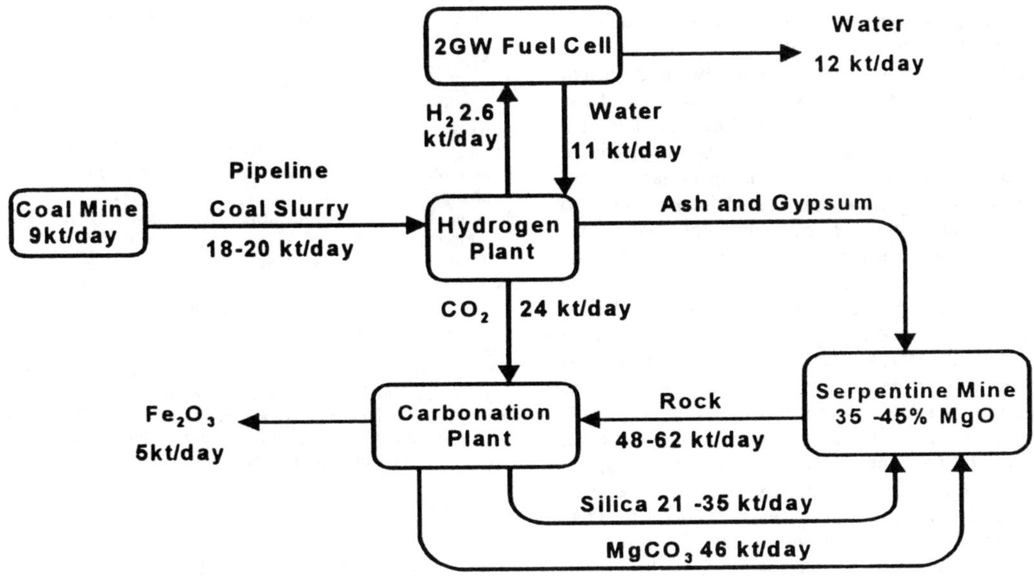

Figure 3. A block diagram of zero emission system

of CaO by weight. In addition, the magnesium silicates are more reactive and are therefore more suitable for above ground carbonation.

The process implies a large mining effort, but the areal extent of the mine is small compared to the coal mine that produces an equivalent amount of coal. Overburden on serpentinite rock is generally insignificantly small and the minerals occur in thick layers rather than thin seams. Nevertheless, the mass of material required is larger by a factor of six than the mass of coal that is used as fuel As a result, the formation of carbonate will have to be performed at the mine site, and the resulting silica and carbonates will be stored in the mine. Since volumes increase in the process, some modification of the local terrain's profile is unavoidable.

Mining costs appear to be quite low. The mining is similar to copper mining and the amount of peridotite required for a GW power plant is small compared to the amount of ore mined in a large copper mine. Cost estimates, based on other mining operations suggest a cost of about \$8 per ton of CO_2.

Future research efforts will have to focus on the details of the chemical processing. Direct carbonation of the mineral rock is feasible but at this point still too slow to be economic. An aqueous process using hydrochloric acid that is completely recovered within the process has been demonstrated to be fast. However, the number of steps is too large and the large amount of steam generation makes the process energy inefficient even though the overall reaction is exothermic (Lackner et al. 1997a, Goff & Lackner 1998). A very similar process was suggested in the past (Houston, 1945, Barnes et al., 1950) to overcome Mg shortages. A variation of this process which uses molten $MgCl_2$ salts to dissolve the rock is similar to a process Noranda will use in a commercial plant that utilizes serpentine mine tailings to produce metallic magnesium. The molten salt process for forming carbonates is currently under investigation. While the thermodynamics has been demonstrated, experiments verifying the kinetics are still underway. In summary the feasibility of carbonate formation has been demonstrated, but a more streamlined implementation will be required to be cost effective. However, reasonable assumptions about thermodynamically efficient processes suggest costs for the entire disposal process, including rock mining, crushing and milling, of about \$20/t of CO_2. For a 66% efficient power plant this would add less that 1 cent to the cost of a kilowatt hour.

8 CONCLUSIONS

It is clear that the CO_2 problem needs to be addressed in a rational and systematic way. It is also clear that energy saving strategies, forestation and alternative fuels alone will not solve the problem simply because of huge quantities of CO_2 production. In the foreseeable future, however, technologies are expected to be on line for zero emission processes which will make coal, the most abundant fuel, become a sustainable resource for at least 2 centuries. These technologies will also help solve the

CO_2 problem associated with other fossil fuels by capturing CO_2 at the source of its generation.

Economically, coal has a strong advantage in that the raw resource is much cheaper. Coal cost at the minemouth is about \$0.50 to \$1.00 per GJ. The cost for natural gas is \$2 to \$3 per GJ by comparison. The low cost of coal is in part offset by larger handling cost and larger cleanup cost. When it comes to carbon sequestration, the lower cost of coal allows one to budget a certain amount for CO_2 disposal without losing the competitive edge. For example the price difference between nuclear and coal suggest a buffer of about \$40 to \$60 per ton of CO_2 (Herzog et al., 1997). Relative to renewable options like solar energy the margin is even higher. Another comparison is relative to natural gas. For every dollar difference in price between natural gas and coal, the cost of CO_2 capture and disposal can go up by \$22/ton of CO_2. This scenario assumes that the price difference between coal and gas is reduced because coal needs to pay an additional cost for removing that amount of CO_2 by which it exceeds the output from natural gas power plants. For the sake of this discussion we assume comparable efficiencies and no additional CO_2 in the natural gas.

Processing of coal to produce hydrogen even without sequestering of CO_2 will pay in the short run to eliminate the problems associated with sulfur and nitrogen oxides emissions. It is conceivable that while for efficient and economic technologies are being developed for CO_2 sequestration a transition to hydrogen production can be implemented as a first step to zero emissions energy production systems.

REFERENCES

Arresta, M. & Tommasi, I. 1997. Carbon dioxide utilization in the chemical industry. *Energy and Conservation Management*, 38 Suppl, S373–S379.

Audus, H., Riemer, P. W. F. & Ormerod, W. G. 1995. Greenhouse gas mitigation technology results of CO_2 capture & disposal studies. In B. Sakkestad(ed), *Proceedings of the 20th International Technical Conference on Coal Utilization & Fuel Systems.* Clearwater Florida, March 20–23, 1995. 349–358.

Barnes, V. E., Shock, D. A., & Cunningham, W. A. 1950. In *Utilization of Texas serpentine*. 5-52. Austin TX: University of Texas Publication No. 5020.

Energy Information Administration (EIA) 1998. *International Energy Outlook 1999* . DOE/EIA, Washington DC.

Goff, F. & Lackner, K.S. 1998. Carbon dioxide sequestering using ultramafic rocks. *Environmental Geosciences*, 5 (3) 89-101.

Gunter, W. D., Gentzis, T., Rottenfusser, B. A. & Richardson, R. J. H. 1997. Deep coalbed methane in Alberta, Canada: a fuel resource with the potential of zero greenhouse gas emissions. *Energy and Conservation Management,* 38 Suppl, S217–S222.

Halman, M. M. 1993. *Chemical Fixation of Carbon Dioxide.* CRC Press, Boca Raton.

Herzog, H., Drake, E. & Tester, J. 1993. A research needs assessment for the capture, utilization and disposal of carbon dioxide from fossil fuel-fired power plants. *Energy Laboratory, Massachusetts Institute of Technology*, Cambridge, MA 02139-4307.

Herzog, H., Eliasson, B. & Kaarstad, O. 2000. Capturing greenhouse gases. *Scientific American*. Feb. 2000. New York.

Herzog, H., Drake, E., & Adams,E. 1997. CO_2 Capture, reuse, and storage technologies for mitigating global climate change." *DOE Order No. DE-AF22-96PC01257.*

Houston, E. C. 1945. *Magnesium from olivine*. New York: American Institute of Mines and Materials Engineering, Tech. Pub. 1828 Class D No. 85.

Kaarstad, O. & Audus, H. 1997. Hydrogen and electricity from decarbonised fossil fuels. *Energy and Conservation Management*. 38 Suppl, S431–S436.

Keeling, C.D., Whorf, T. P. Wahlen, M. &. van der Plicht, J. 1995. Interannual extremes in the rate of rise of atmospheric carbon dioxide since 1980. *Nature,* 375, 666–670.

Keeling, R. F, Piper, S. C. & Heimann, M. 1996. Global and hemispheric CO_2 sinks deduced from changes in atmospheric O_2 concentration. *Nature*, 381, 218–221.

Kleypas, J.A., Buddemeier, R.W., Archer, D. Gattuso, J.P., Langdon, C., Opdyke, B.N. 1999. Geochemical consequences of increased atmospheric carbon dioxide on coral reefs, *Science,* 284**,** 118–120.

Lackner, K.S., Wendt, C.H., Butt, D.P., Joyce, E.L. & Sharp, D.H. 1995. Carbon dioxide disposal in carbonate minerals. *Energy 20*, 1153–1170.

Lackner, K.S., Butt, D. P., Wendt, C. H., Goff, F. & Guthrie, G. 1997a. Carbon dioxide disposal in mineral form: Keeping coal competitive. *Los Alamos Report LAU-UR* -2094.

Lackner, K.S., Butt, D.P., & Wendt, C.H. 1997b. Magnesite disposal of carbon dioxide. *The Proceedings of the 22 nd International Technical Conference on Coal Utilization & Fuel Systems,* March 16–19, 1997, Clearwater, Florida, U.S.A., 419–430.

Lackner, K. S., Butt, D. P. & Wendt, C. H. 1998. The need for carbon dioxide disposal: A threat and an opportunity. *Proceedings of the 23 rd International Conference on Coal Utilization & Fuel Systems*, Clearwater Florida, March 1998, 569–582

Lackner, K.S., Ziock, H.J., Grimes, P. 1999a. Carbon dioxide extraction from air: Is it an option? *Proceedings of the 24th International Technical Conference on Coal Utilization & Fuel Systems,* Clearwater Florida, March 1999, 885–896.

Lackner, K.S. Ziock, H.J., Harrison, D. 1999b. Calcium oxide assisted hydrogen production from fossil fuels & water. Los Alamos Report, LA-UR-994291, (1999).

Meisen, A & Shuai, X. 1997. Research and development issues in CO_2 capture. *Energy and Conservation Management*. 38 Suppl, S37–S42.

Ranney, J. W. & Cushman, J. H. 1992. Energy from biomass. In R. Howes & A. Feinberg (eds.), *The Energy Sourcebook*: 299-311. American Institute of Physics, New York.

Sedjo, R. A., & Solomon, A. M. 1989. Climate and forests. In N. Rosenberg, W. E. Easterling III, P. R. Crosson & J. Darmstater (eds.), *Greenhouse Warming: Abatement and Adaptation.* : 135-129. Proceedings of a workshop held in Washington DC, June 14-15, 1988 Resources for the Future, Washington DC.

Seifritz, W. 1990. CO_2 disposal by means of silicates. *Nature*, 345, 486.

Siegenthaler, U. & Oeschger, H. 1987. Biospheric CO_2 emissions during the past 200 years reconstructed by deconvolution of ice core data. *Tellus,* 398**,** 140–154.

United Nations 1997. 1995 *Energy Statistics Yearbook*. New York.

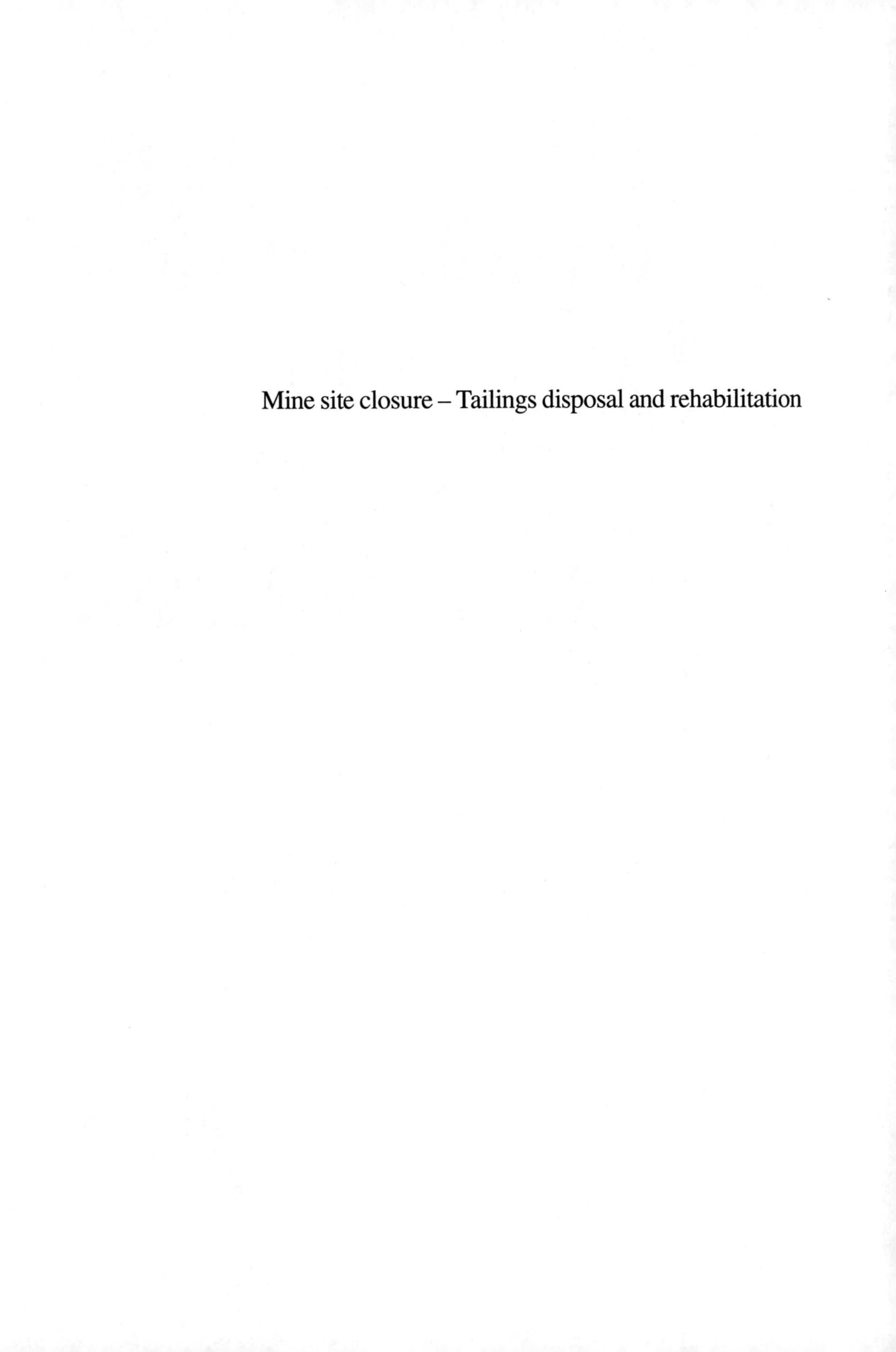

Mine site closure – Tailings disposal and rehabilitation

Environmental Issues and Management of Waste in Energy and Mineral Production, Singhal & Mehrotra (eds)
© 2000 Balkema, Rotterdam, ISBN 90 5809 085 X

Natural regeneration of vegetation on the mining waste sites

C. Biro & M. Georgescu
Department of Mining, University of Petrosani, Romania

ABSTRACT : Natural regeneration of vegetation is the way by which nature heals the wounds caused by human activity to the environment. Our paper intends to present the lows and dynamics of this phenomenon on the waste sites resulted from the underground and surface mining activity. Once we know the laws and their dynamism we can decode nature's secrets and apply them in an efficient way to solve some ecological, artificial reconstruction issues. The paper also states that, using some minor artificial intervention, this phenomenon can be oriented to bring as major results in vegetal regeneration at mining wastes sites. Our paper presents the observations results made during 10 years. We have discovered that there are 25 species at grass and 3 species at trees which can grow on the wastes. We tried to make some link between the demands at the studied species and the potential of the habitat.

1. INTRODUCTION

The degradation of the natural environment due to various human activities can be considered a major issue of the third millenium. Among the many existing degradation phenomenon, the present paper is framed in the problems which result from the enclosed Figure 1.

Anthropogenic degradation of the environment → Many
↓
Our goal
↓
Degradation caused by the mining industry → Many
↓
Our goal
↓
Degradation caused by the waste deposition
→ Social
→ Of the soil
→ Of the landscape
→ Hydrologic
→ Climatic
→ Geomorfological
→ Of the air

Figure 1. Target our goal

The solution of this problem is difficult in Romania at present. Although there is the necessary legislation, materialized by Law No.137/1995, Law of the Environment Protection, its carrying out is difficult due to the acute lack of funds for these activities, as well as owning to some obscure economical interests. The studied phenomenon affects about 4000 ha in Romania, from which 270 ha in the Jiu Valley coal basin, region which constitutes the target of our research.

Because the obscure economic interests concerning the protection of the environment are difficult to be controlled in this period, through this paper we look for less expensive solution in order to solve the above mentioned issue; solution which aim, first of all, the decoding of some natural regularities which conducted in a proper way, with minimum investments should offer maximum results. The solutions are foreseen to be difficult because the vegetal soil was not retrieved and the stability projects were not observed at the construction of the waste dumps from Jiu Valley.

The challenge is major because it will be necessary to work on the waste sites without soil with nude rock above ground and on sites with advanced instability.

The proposed goal of the paper is to bring back in the economical circuit these waste dumps and to solve concomitantly the aspects related to the protection of the environment. The general accepted and applied methods have been limited so far to the artificial regeneration of these either agricultural or forest surfaces - investments funds consumer works. But from the practical experience and from the undertaken studies results that these surfaces can be naturally regenerated only by applying some organizational measures, by adopting some management rules of these surfaces.

From the point of view of the protection of the environment it has to be mentioned the fact that the coal mining activities in the Jiu Valley have been restricted and they will dramatically limited due to the unprofitableness of the coal worked off in such conditions, so the affected surfaces will be more restricted in the future.

The investigations were made on three waste dumps, namely: one from the West side of the Jiu Valley, having a surface of 9.6 ha ; one in the central zone with a surface of 10,0 ha; one from the East side with a surface of 7,8 ha. We mention that the waste dumps are old and inactive for at least 10 years.
Throughout the paper we shall try to decode the regularities which rule this natural regeneration phenomenon, its dynamics and the offered solutions.

2. WORKING PRINCIPLES

2.1 *Expeditionary research*
2.2 *Stationary research*
2.3 *Laboratory analysis*
2.4 *Analysis and correlation*

3. UNDERTAKEN INVESTIGATION

3.1 *The study of the climatic conditions*

The ecological factors.	The values of ecological factors.
Annual average temperature (° C)	5.1
Annual average precipitation	822 mm

3.2 *Geological investigation*

It has been made an attentive investigation at the rock deposits from the waste dumps which has led to the following conclusions:
- the waste dumps consist at the following rock : clays, marls, gristone, microconglomerates, coal.
- from the mineralogical point of view in the composition of the above mentioned rocks of there are various minerals, as:

The clay consists in clayey minerals, respectively in order of preponderence kaolinite, illite, montmorrilonite. Beside of these clayey minerals can be mentioned fine nice powders quarz, feldsporssa as well as animal and vegetal organic substance resulted from the decomposition of some plants pemainders and lamelliobranchiate and gestoropoda shells. There are (absorbant iron , hydrolisis (limonite)), too.

The marl has the same minerological composition as the clay, at which a 40% CaCO content are added and bituminous substances.

The gristones consists of less than 2 mm quarz granules, mica (muscovite, biotite) feldspar, amphiobite united by cement which has clayey, marl, limstonecharacter and more rarely, siliceous; there is also animal and vegetal organic substance.

Microconglomeration consisted at meta morphic rocks, especially at quartzites, phyllites, slates and amphobolites united by a bounding cement having clayey and marl character.

Due to the structure and texture coming in contact with the atmopheric agents, there are desintegrated very soon in milimeter and ever submilimeter fragments. The proportion of the rough skeleton due to this fact is of the 40-50%, encouraging the development of a considerable useful phisiological volume.

3.3 *Pedological investigation*

Because the waste dumps are old, due to the physomechanical desintegration and the chemical alteration, respectively to the herbaceous and arborescent vegetation which have been installed, it has been found the existence of a soil formation process which generated a little advanced soil with an A-C profile; A-represents the very thin humic horizont, with a humus content below 1,0%; C-represents the diverse desintegrated and alterated waste material.The humus is framed in the moder oligotrophic type. The acidity varies between
ph = 6-8 and the content of nitrous substances is much inferior to the zonal type of soil. The identified texture is sandy - clayey. The soil is compact moderated, specific feature given mostly by the waste underlayer. The edafic volume is medium owing to the diverse desintegrated and alterated underlayer. The degree of saturation in the alkalies is about 60%. The total potential trophicity frames the type of the soil at the oligotrophicity degree. The pedohydrological regime is that of precipitations

without phreatic contribution and it does not affect the trophicity of the soil. Under the influence of the general and zonal climate the soil formation processes have the tendency to develop in the direction of the formation of a zonal soil typically. (Tîrziu 1997)

3.4 *Geomorphological investigation*

Regarding the altitude, the waste dumps are situated between 500-800 m and they were built due to the limited surfaces on the general directions of the valley, namely East-West. On that account many slopes were generated with North and South orientation, with obvious impacts on the thermal and hydral regime from the created slopes. The steps of the wastes and the slopes have, in general an inclination of 45% and a height between 10-25 m, ending in a levelled plateau, having also a different thermal and hydral regime.

The manner of the construction of the wastes generates many vegetation situations and conditions which are constituted in microstations and they are analized as such. In general the relief of the wastes is not divided into fragments, the slopes and the plateaux being continue.

3.5 *The study of the natural degradation*

As the studied wastes have been built in a long period of time having a considerable age - because they were not the subject of some rehabilitation actions - they undergone to degradation as a result of climatic factors and they were affected especially by the pluvial erosion, above ground and in depth. The surfaces which did not regenerated naturally are the most exposed. In the case of the surfaces affected by the erosion in depth the only remedy remain the adaptation to the rehabilitation solution, some interventions with special vegetal supporting works which will be introduced artificially.

3.6 *The investigation of the zonal vegetation*

The studies surfaces are situated, from the point of view of the bioclimatic zone, in the forest zone underzone FD4 +FM1, mountain and premountain storey of beech forests, storey characterized by a temperate continental climate. The main species of the storey is the *Fagus sylvatica* which forms vaste wooded massives in zone. But our interest is aimed toward the pioneer species specific to this storey, namely *Betula verucosa, Almus incana and Populus tremula*, arborescent species with high dissemination capacity and with reduced exigency given the environmental conditions.The mentioned species have succeeded to install themselves in a natural way on the studied waste dumps. In order to understand this mechanism it necessary to be made a correlation between the previously described stational potential and the exigencies of the existing which are reproduced in the ecological record cards shown further on (Chiriţă 1977).

Ecological record card (Stănescu 1979)
Species: *Betula verrucosa*

No. crt.	Ecological factors	Ideal states of the ecological factors
1.	Annual average temperature (C)	5 - 8
2.	Annual avarge precipitations (mm)	700 - 1100
3.	Altitude (m)	600 - 1200
4.	Exposition	sunny semisunny
5.	Degree of saturation in alkalies (V%)	25 - 65
6.	Soill acidity (pH in the water)	6 - 4.4
7.	Humus type	Moder oligotrophic
8.	Depth (cm)	60 - 140
9.	Edafic volume (m/m)	0.45 - 0.90
10.	Global trophicity potential	50 - 140
11.	Pedohydrological regime	precipitation
12.	Soil compactness	loose
13.	Soil texture	sandy - claye

It can be noticed the fact that there is almost a perfect correlation between the potential of the station and the exigencies of the presented species.

Ecological record card (Stănescu 1979)
Species: *Alnus incana*

No. crt.	Ecological factors	Ideal states of the ecological factors
1.	Annual average temperature (^{0}C)	6 - 11
2.	Annual avarge precipitations (mm)	400 - 800

3.	Altitude (m)	300 -1000
4.	Exposition	semishady semisunny
5.	Saturation degree in alkalies (V%)	60 - 90
6.	Soil acidity (pH in water)	6 - 6.8
7.	Humus type	moder oligotrophic
8.	Depth (cm)	50 - 140
9.	Edafic volume (m^3/m^2)	0.45 - 0.90
10.	Global potential trophicity	50 -140
11.	Pedohydrological regime	precipitations
12.	Soil compactness	compact
13.	Soil textures	luteos - clayey

The previous conclusions can be repeated in the case of this species, too.

The third mentioned species *Populus tremula* will not be analysed because it appears only occasionally, in the studied surface and we consider that it has no distinct practical interest.

Besides the studied arborescent species, 25 species of grass were identified which are, actually, the first that tookhold of the dump sites, constituting a premise for the release of the soil formation processes and the installation of the arborescent species. The identified species of grass are : *Centaurea rhenana, Horcus lanatus, Gentiana ciliata, Calamagrostis epigeios, Plantago media, Scaleiosa ochroleuca, Lontodom autumnalis, Euphrasia stricta, Trifolium pratense, Chrysantemum leucanthenum, Tussilago farfara, Cirisum arvense, Lotus corniculatus, Tripleuspermum inodorum, Eqvisetum arvense, Polygonum minus, Clematis vitalba, Parnassia palustris, Melilotus albus, Daucus karota, Trighochin palustris, Poa triviotis, Poa pratensis, Agrostis stolonifera, Centaurea austriaca* (Simionescu 1973). The multitude of the species of grass is amazing.

It has to be mentioned that the regenerated species are found again in the immediate vicinity of the studied surfaces as natural autochtonous species.

It has to be highlighted that from among the species of grass *Tussilago farfara* develops a very advanced radicular system, reaching a depth of about 50 cm in the disintegrated and alterated waste, exploring an impressive mineral support volume.

3.7 *The administrative conditions of the studied sites*

These sites belong to the mining unites which created them and owing to the restriction of the activity, the unites left them in fallow. They have not been prepared in any way for reintegration in the economic circuit. The existing vegetation has been installed naturally but it is much disturbed by the abusive grazing practiced in the respective perimeter.

We consider that by a simple circumstance which could ensure the peace in those perimeters, the integral natural regeneration of the studies surfaces could succeed.

year: 1989

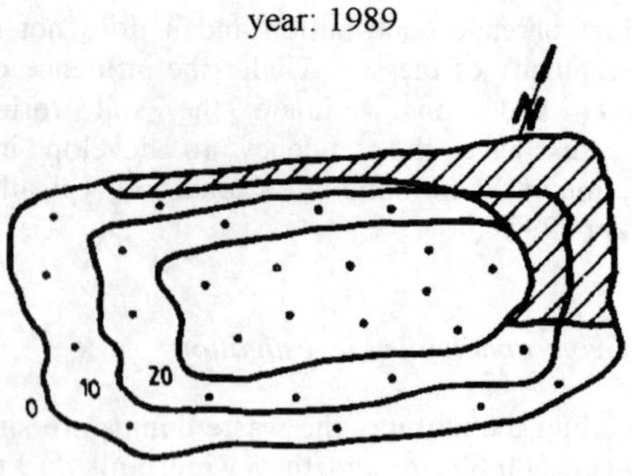

0.1S //// Regeneration *Alnus incana* (Brush regeneration; diameter =1 cm; height = 0.8m)

0.9S ░ Regeneration *Tussilago farfara* (average density 3 samples/m^2)

year: 1999

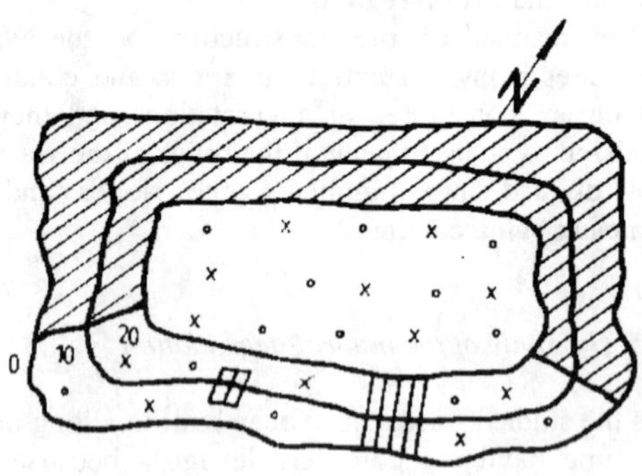

0,6S //// Regeneration *Alnus incana* (Brush regeneration ; diameter = 5 cm, height = 4,0 m)

0,4S ░ Regeneration *Tussilago farfara* and other species enumerated in text (average density 68 samples/m^2)

Figure 2 The dynamics of the regeneration on the waste dump number 1

year: 1989

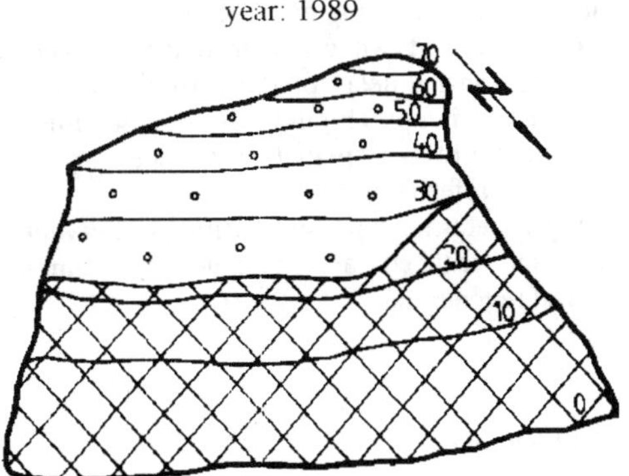

0,6S ////Artificial regeneration with *Hopophae rhamnioides*
0,4S ░ Regeneration *Tussilaga farfara* (average density 1 samples/ m^2)

year : 1999

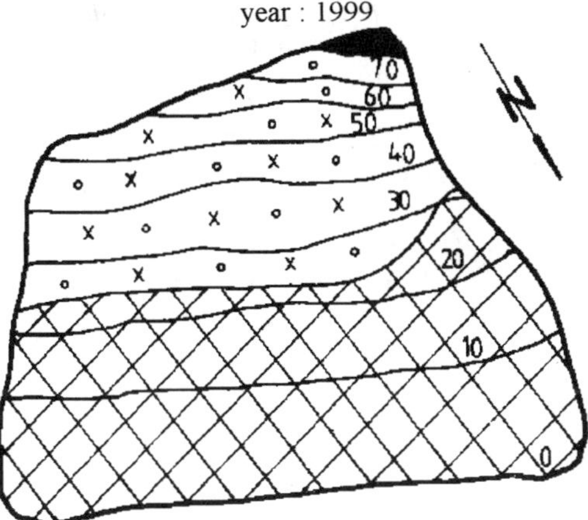

0,4S ░ Regeneration of all species of grass enumerated in text
(average density 57samples/m^2)

Figure 3 The dynamics of the regeneration on the waste dump number 2

year: 1989

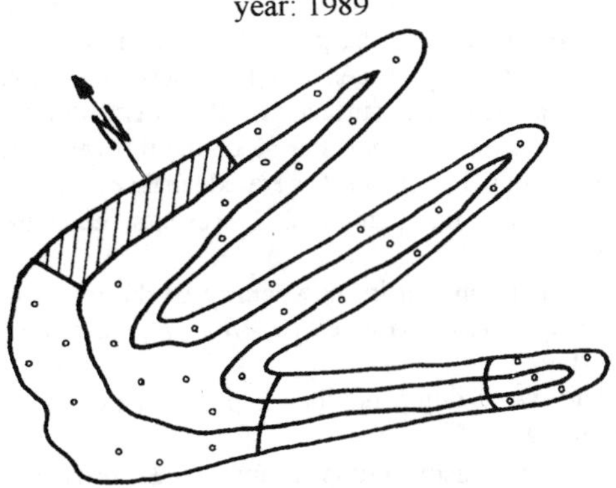

0,1S ////Regeneration of the *Betula verucosa* (brush regeneration; diameter = 1 cm, height =0.6 m)
0,7S ░ Regeneration of the *Tussilago farfara* and of other species, in a little extent (average density 3 samples/m^2)
0,2 □ Active surface of the waste.

Year: 1999

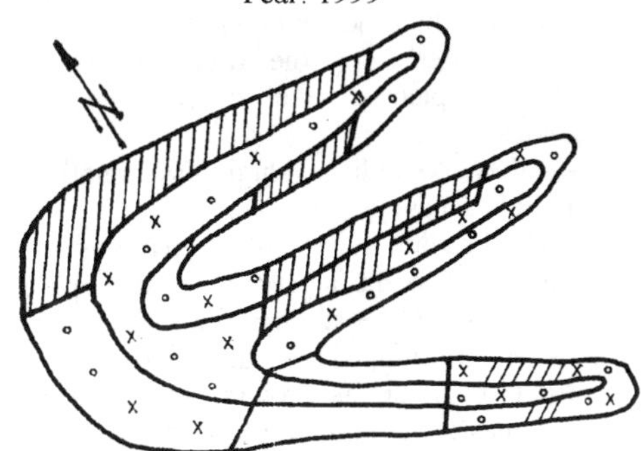

0,4S ////Regeneration of the *Betula verucosa* (brush diameter=3 cm; height=2-3 m)
0,5S ░ Regeneration of the *Tussilaga farfara* and other species in a little extent (average density 50 samples/m^2)
0,1S □Active surface of the waste.

Figure 4 The dynamics of the regeneration on the waste dump number 3.

4.THE MONITORING AND THE DYNAMICS OF THE NATURAL REGENERATION

The monitoring of the regeneration has been made by successive raising in plane of the regenerated surfaces from 2 to 2 years, concomitantlly picking up dentrometrical data (diameter on the level of ground, height) in the case of the arborescent vegetation and for the herbaceous vegetation it has been established the density of the samples on 1 m^2 in the most characteristic points (representative, averages)

The evolution of the natural regeneration went on in different ways on those three studies waste dumps, both from the point of view of the species which regenerated themselves and from the point of view of the realized degree of covering.

In the next figures there are shown the situations of the surfaces, at the beginning and at the end of the studies period.

The draft solutions (with level curves) have the mission to suggest the percentage of the regenerated surfaces and the regenerated species. The notation "S" represents the surfaces of the waste dumps and the index on the front, the % of natural regeneration (example: 0.1S=10% regenerated surface).

The waste dump number 1 presents the dynamics from Figure 2.

We mention that in the vecinity of the waste dump there are nature samples of *Alnus incana* along the Jiu river.

The waste dump number 2 presents the dynamics from Figure 3.

The waste dump number 3 presents the dynamics from Figure 4 .

We mention that in the vicinity of the waste dump there are meadows with brushes of *Betula verucosa* at the fructification age.

5. CONCLUSION

The regeneration of the waste dumps following the coal mining is possible naturally. The possibility can be realized in the following conditions :

a. The existence on the waste dumps of some species of plants which meet the following conditions:

- they should have a high dissemination power (many light seeds) ;
- they should have low pretentious given the environmental conditions (rustic species - pioneer)
- they should be autochtonous, from the vegetal sites where are situated the studied sites;
- they should be fast growing in order to finish the state of massive as soon as possible.

b. The mine waste do not require a preliminary preparation, excepting those with unstable balance.

c. In general it has been noticed the installation of the vegetation on the slopes with Northern, North-Eastern and North-Western exposition with a more favorable temperature and humidity regime.

d. The integral natural regeneration of the wastes with arborescent species have not succeeded during 10 years, but the mixed regenerations has been fulfilled plenarily.

e. As concerns the dynamics of the regeneration, it has been found a very tight correlation with the peace from the studied perimeters.

As the waste dumps are situated in general in zones inhabited by rural people whose main preocupation is the animal breeding, the grazing in the main obstacle for the natural regeneration of the waste dumps.

Through the erodation of the sprouts and the consumption of the herbaceous layers, respectively through the ramming and the disintegration of the mineral support we consider that this issue is the main reason for delay of the natural regeneration of the waste dumps.Through the surrounding of the surface could be created that necessary peace at which we referred to previously.

In prospect the expensive artificial intervention can be much reduced, and used just to complete the natural regeneration.

REFERENCES

Chiriţă, C. s.a. 1977. *Staţiuni Forestiere*. Editura Academiei R.S.R. Bucureşti

Simionescu, I. 1973. *Flora României*.Editura Albatros. Bucureşti.

Stănescu, V. 1979. *Dendrologie*. Editura Didactică şi Pedagogică. Bucuresti.

Târziu, D. 1997. *Pedologie şi staţiuni forestiere*. Editura Ceres. Bucureşti.

Environmental Issues and Management of Waste in Energy and Mineral Production, Singhal & Mehrotra (eds)
© 2000 Balkema, Rotterdam, ISBN 90 5809 085 X

Greenhouse growth trials of perennial ryegrass in coalmine spoils

B.K.C.Chan & A.W.L.Dudeney
T.H.Huxley School, Royal School of Mines, Imperial College of Science, Technology and Medicine, London, UK

ABSTRACT: As part of a European Union research programme on the rehabilitation of coal tips, a greenhouse growth trial was carried out with samples of spoil from two tips in Germany. Perennial ryegrass *Lolium perenne cv Prohpet* was sown in 1 kg spoil samples of particle size – 6.7 + 2.3 mm, with and without addition of ammonium nitrate and calcium phosphate in a greenhouse at 23°C, with 19 hours light per day for 39 days. Potash was expected to be available in the spoil. Nutrient uptake, growth height and root penetration were examined. From the results, all the plants grew and took up N, P and K although differently according to the fertilizer present and the particular spoil characteristics. It was found that the roots adhered to the surface of the spoil and penetrated into cracks. From height measurements, correlation was observed between potassium uptake from the spoil and plant growth.

1 INTRODUCTION

Coal was the major source of energy for industry and domestic use in Europe for much of the 20th century. Due to the lower demand for coal on the global market, many coal mines have been closed in recent years. This led to increasing problems of costly rehabilitation of waste tips and derelict land. Many research programmes have been carried out, especially in agriculture and forestry since they are the most common land uses for reclaimed colliery spoil (Hester & Harrison 1997). As sufficient soil was not normally available, different materials were used on top of the spoil as growing media and substitutes for topsoil, e.g., bottom ash and lime (Tedesco et al. 1999); arbuscular-mycorrhizal fungi (AMF) and composited sewage sludge (Thorne et al. 1998); sewage sludge and limestone (Joost et al. 1987) and pulverized refuse fines (PRF) (Chu & Bradshaw 1996).

Industrial wastes, such as sludge from water and waste water purification plants, fly ash from power stations, red mud from aluminium production and flotation tailings from coal preparation plants, can cause major environmental problems and need to be safely disposed of. In this project, which is centred in Germany and has contributors in the Ukraine and UK, the overall objective was to use designed mixtures of sewage sludge and other industrial wastes to form an artificial top soil (ATS) to cap the waste-dumps and at the same time create a fertile substratum for revegetation. Such ATS addition can contribute to the solution of the environmental problems: safe disposal/utilization of industrial residues and rehabilitation of waste dumps resulting from the mining industry. As a preliminary study prior to field application of ATS, the coal spoil was to be tested as a growing medium. The specific objectives were to evaluate the plant growth, nutrient uptake and root penetration in the mine spoil through greenhouse experimentation. An understanding of the above would assist in the management of successful reclamation efforts, in particular promoting root growth beneath the ATS layer.

2 SOURCE OF MATERIAL

Samples of coal spoil were taken from two tips located near Saarbrücken in Germany.

Reden (Grube Reden): the tip is approximately 0.8 x 0.9 km with elevation about 60 - 80 m above the surrounding area. It is situated on an impermeable (carboniferous) base. Mining operations ceased in this area in 1996, and ATS (80% spoil, 10% sewage sludge and 10% wood residues) was emplaced on most of the surface by SaarMontan to a depth of 2 m (Neu 1997).

Duhamel (Bergwerk Ensdorf mine): the tip is approximately 0.72 km in diameter at the base with elevations about 130 - 140 m above the surrounding area and resting on a sandstone base. It is an active site and large areas have yet to be completed and rehabilitated.

A 35 – 40 kg sample of dump material was taken from two locations on each tip for laboratory testing.

Reden (1): nearly flat area near the centre of the top of the dump.

Reden (2): steeply sloping area on the north side of the tip, which had not been previously rehabilitated.

Duhamel (1): steeply sloping area on the north side of the top of the tip.

Duhamel (2): steeply sloping area on the south side about half way up the tip.

The locations were representative of sites selected for future larger scale rehabilitation testwork.

3 METHODS

3.1 *General procedures*

The samples were analysed for moisture content, visual characteristics and elemental and qualitative mineralogical composition prior to growth trials.

Figure 1 shows the flowchart of experimental procedures. The 40kg samples were crushed to -6.7 mm and representative samples were taken, after coning and quartering, for the analyses indicated. Elemental analysis, by Inductively Coupled Plasma spectrophotometry (ICP), was carried out after grinding to –106 μm. The sample was digested in nitric, perchloric and hydrofluoric acids. Sink-float tests were carried out with 500 g samples previously crushed to - 3.35 + 1.18 mm to determine the content of coal and shales using a zinc chloride solution (s.g. 1.57).

3.2 *Laboratory growth trials*

Representative samples (- 6.7 mm) from the four sites were sieved to - 2.36 mm. The – 6.7 + 2.36 mm fractions were used for pot experiments. The -2.36mm fraction was not used due to problems of aeration and water logging. 5-inch pots were used with 1 kg of spoil in each pot. Perennial ryegrass *Lolium perenne cv Prohpet* was chosen due to its exceptionally persistency and high yield character under wide ranging conditions, winter hardiness and good crown rust disease resistance.

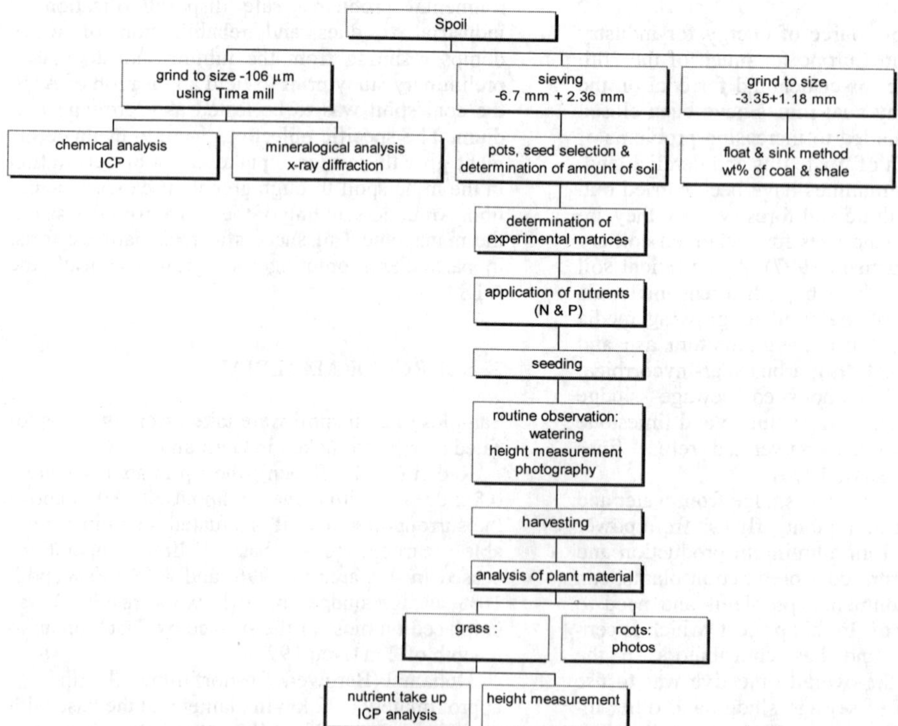

Figure 1 Schematic diagram of experimental procedures

Three pots were used for each site, one as a control without any nutrient, the second with nitrogen and the third with nitrogen and phosphorous. Two pots of John Innes No. 1 compost were also used as controls. In order to observe if the roots were able to penetrate through the spoil, four transparent plastic pots were used with addition of nitrogen.

The application rates of nutrients were based on field recommendations (Rowell 1994). In the experiments, 0.34 g NH_4NO_3 for nitrogen and 0.12 g $Ca_3(PO_4)_2$ for phosphorus were applied per pot.

The agricultural rate of application of ryegrass is 32 kg/ha, i.e. 3.2 g/m^2 (Rowell, 1994). For 5 inch pots, the seed requirement was 0.2 g per pot. For the square 9 x 9 cm clear plastic pots, the requirement was 0.12 g. Seeds were sown 10 mm below surface. 60 ml of deionised water was added from a syringe each day. The greenhouse was at 23°C with 19 hours light per day. The beginning of seeding was 2nd February 1999; the last day of cutting was on 12th March 1999. The total growth period was 39 days.

4 RESULTS AND DISCUSSION

4.1 *Spoil characteristics*

Table 1 shows a summary of the properties of the four waste samples. From initial observations, the majority was shale. After crushing, the three aged samples gave a greyer appearance (as unoxidised surfaces were exposed) while for the new spoil sample (Duhamel (2), south side,) the sample was grey before and after crushing. The Reden samples were much more friable than those from Duhamel, thus reflecting the longer period over which weathering processes had been in progress.

Table 1. Summary of the properties of the spoils

	Reden (1)	Reden (2)	Duhamel (1)	Duhamel (2)
Colour	Brown	Brownish grey	Grey	Grey
Observation	easily broken to pieces	easily broken to pieces	hard	hard
Moisture content (wt%)	3.45	0.39	0.00	0.62
Float (wt%)	4.21	1.52	1.58	1.08
Woodchips (wt%)	0.99	0.00	0.02	0.00
Minerals (qualitative)	Quartz Clinochlore Illite	Quartz Clinochlore Illite	Quartz Clinochlore Illite-2 Muscovite	Quartz Clinochlore Illite-2 Muscovite Kaolinite

The table illustrates that the samples contained just a few percent coal and, in the case of the rehabilitated Reden sample, about 1% woodchips originating from the ATS.

From mineralogical analysis by x-ray diffraction, the main constituents were seen to be quartz and clay minerals. No pyrite was detected by XRD or light microscopy. This result was consistent with the field observation of a non-ochreous leachate and confirmed that acid run-off was not to be expected from these tips.

The results of ICP analysis are shown in Table 2. The main constituents determined were aluminium (110-114 mg/g), iron (38-48 mg/g) and potassium (22-27 mg/g). Silicon was also abundant but not measured by this procedure. Sulphur (separately determined) was low at 2-4 mg/g, indicating the lack of pyrite. There were no significant concentrations of toxic metals.

Table 2. ICP analysis of elements in the spoils

Elements	Reden (1) mg/g	Reden (2) mg/g	Duhamel (1) mg/g	Duhamel (2) mg/g
Li	0.09	0.09	0.13	0.10
Na	2.22	2.39	1.46	1.80
K	23.60	26.90	22.75	22.20
Be	0.003	0.003	0.003	0.003
Mg	8.7	9.89	8.36	8.44
Ca	5.98	1.67	3.16	4.00
Sr	0.14	0.10	0.10	0.11
Ba	0.59	0.48	0.53	0.49
Al	113.0	114.0	113.0	110.0
La	0.05	0.04	0.04	0.04
Ti	4.53	5.03	4.60	4.50
V	0.12	0.13	0.11	0.11
Cr	0.11	0.11	0.12	0.11
Mo	0.009	0.007	0.009	0.01
Mn	0.65	0.48	0.77	0.98
Fe	38.30	39.50	42.05	47.60
Co	0.02	0.01	0.12	0.02
Ni	0.06	0.06	0.07	0.06
Cu	0.07	0.05	0.04	0.04
Ag	0.001	0.0007	0.0005	0.0009
Zn	0.18	0.10	0.24	0.30
Cd	0.002	0.001	0.001	0.002
Pb	0.07	0.06	0.10	0.13
P	0.79	0.26	0.35	0.33
S	4.28	2.12	2.44	1.84

The waste materials from the two tips were mutually similar from the point of view of elemental analysis and original mineralogy. In principle, there was sufficient potash, phosphate and trace nutrients in the spoils to encourage plant growth.

4.2 *Growth trials*

Many methods for evaluating soil fertility are based on observation or measurements on growing plants. There are many environmental factors known to influence plant growth. The following are probably the most important (Tisdale and Nelson 1975).

Temperature	Soil reaction
Moisture supply	Oxygen in the atmosphere
Radiation energy	
Composition of the atmosphere	Compactness of the soil
Gas content of the soil	Toxic components
Supply of mineral nutrient elements	

In this work, only the effect of nutrient addition to the spoils was studied assuming that plants could take up potash from the spoil.

From the greenhouse experiment, the seeds started to germinate after five days in all 18 pots. Figure 2 shows the different heights of the grass up to 37 days. No clear correlation was observed between plant height and time. However, there was generally rapid growth in the first ten days in all cases due to the readily available nutrient in the seeds. Thereafter, rapid growth continued up to 20 days in compost (the grass reaching 20 cm in height) with much slower growth in the other cases. In Reden (1), the growth was intermediate in rate, and there was no significant difference in grass height whether N or N and P were added. Enough nutrients were present but clearly the rooting conditions were less favorable than in compost. The final height was 13 - 16 cm. In the other cases, the growth was slow after the first ten days and the grass height only reached 10 - 12 cm. In Figure 3(a), without nutrient addition and Figure 3(b) with addition of nitrogen, ryegrass in Reden (1) was stronger compared with the other 3 sites because of the presence of old ATS. However, growth was healthy in all cases. After the addition of N and P (Figure 3(c)), there was no difference in height. From this, it was deduced that potash was readily available in the spoil.

Table 3. Potassium uptake by plants

Spoils		K in spoils (mg/g)	K uptake by plants (mg/g)
Reden (1)	No nutrient	23.60	24.90
	+ N		22.70
	+ N&P		18.50
Reden (2)	No nutrient	26.90	26.20
	+ N		24.70
	+ N&P		21.00
Duhamel (1)	No nutrient	22.75	19.10
	+ N		15.70
	+ N&P		19.10
Duhamel (2)	No nutrient	22.20	13.10
	+ N		10.10
	+ N&P		14.40
Compost	Pot 13		28.60
Compost	Pot 14		25.60

In order to examine the quantity of nutrient the grass had taken up, ICP analyses of ryegrass samples were carried out. The samples were analysed by using the standard methods of nitric acid and perchloric acid attack. As expected (Table 3), all the plants took up potassium from the spoil. Only 4.11 mg/g of potassium (20 % of the total) was found in the ryegrass seeds themselves. In Reden(1) and Reden (2), potassium uptake was higher than those in Duhamel (1) and Duhamel (2), since potash is more readily available in aged spoil. There was a correlation of potassium uptake with the height of the plants, but this was not unequivocal (Figure 4). Figure 5 shows the height of grass on day 37 against the uptake of total nutrient. In this case, there was also a correlation between nutrient uptake and the grass height.

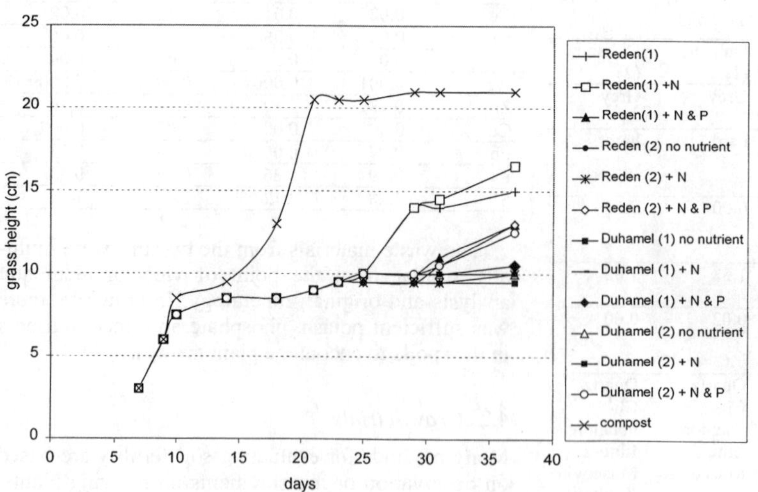

Figure 2. Recorded grass growth

(a) (no nutrient) 1 2 3 4 5

(b) with N addition 1 2 3 4 5

(c) with N & P addition 1 2 3 4 5

Figure 3. Comparison of the height of plants in 4 different locations and compost after 37 days
pot 1 - Reden (1), pot 2 - Reden (2), pot 3 - Duhamel (1), pot 4 - Duhamel (2) and pot 5 - compost

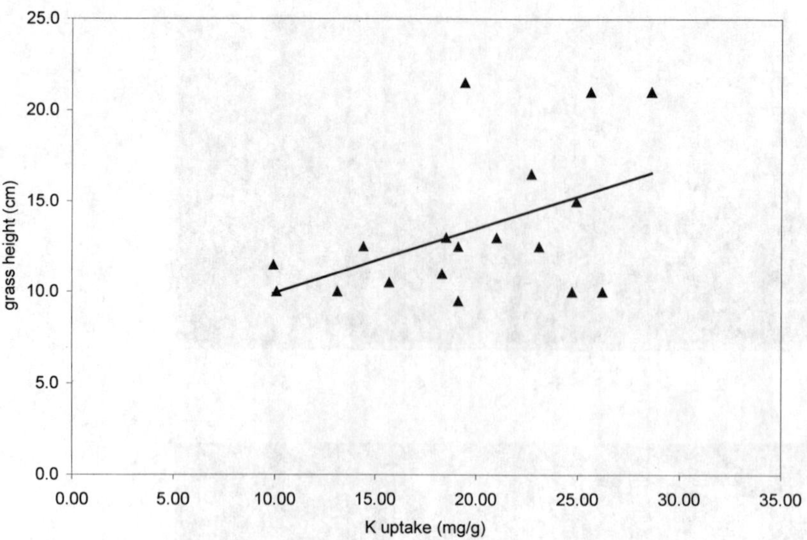

Figure 4. Recorded height of grass as a function of potassium uptake on day 37.

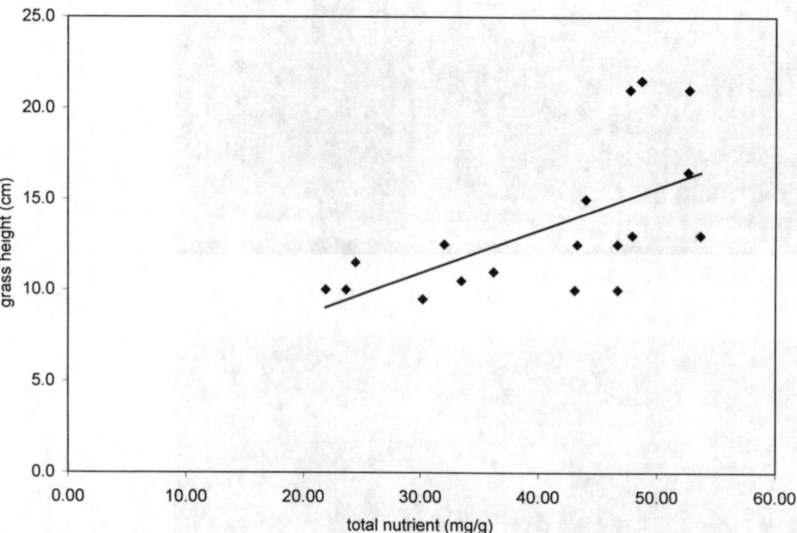

Figure 5. Recorded height of grass as a function of total nutrient uptake on day 37.

The phosphorus uptake (Figure 6) in each pot correlated strongly with the appearance of the grass which was healthier and more luxuriant when the nutrient was added. Thus the phosphorus in the spoil was not readily available.

In order for the plants to grow, the roots had to reach out and penetrate the spoil particles to assimilate potassium. In this study, it was important to observe if this process could sustain plant growth. It was found that the roots did in fact tend to adhere to surfaces and cracks of the shales (Figure 7). They also adhered to the woodchips (Reden 1)

5 CONCLUSION

- **The mineralogy is fairly constant in all the sites but Reden (1) had residues of previous rehabilitation which, even after four years, still promoted growth.**

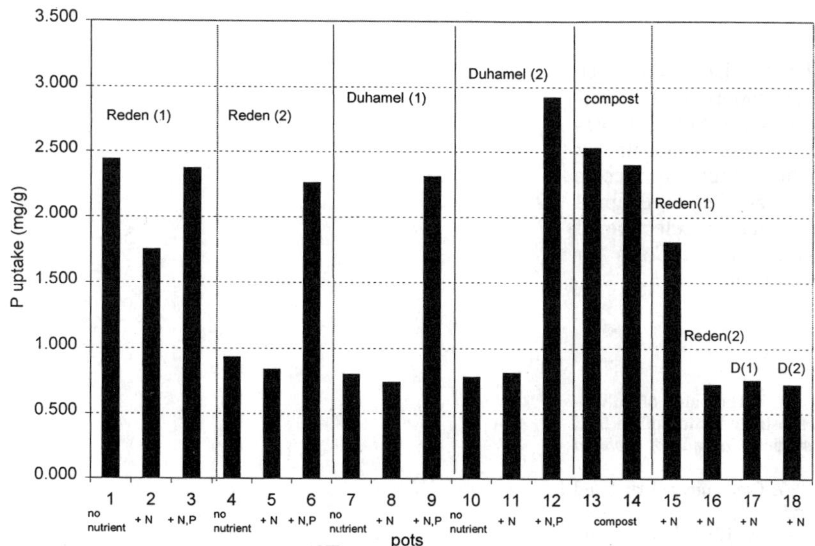

D(1): Duhamel (1), D(2): Duhamel (2)
Figure 6. Phosphorus uptake

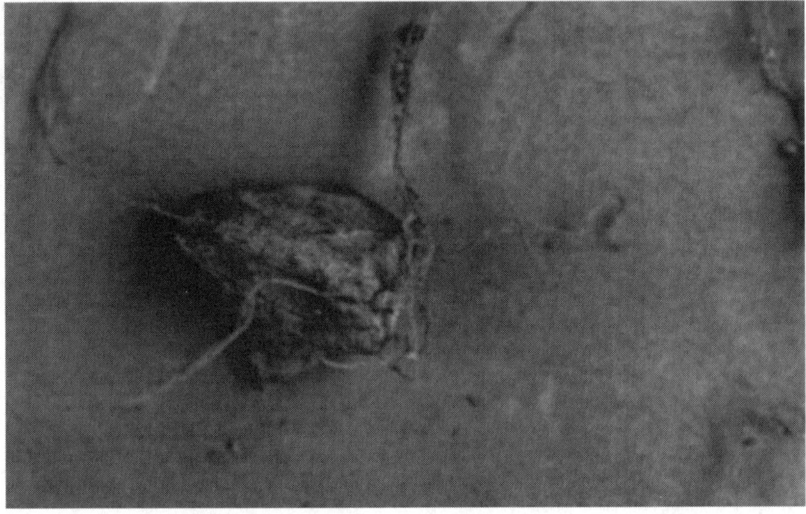

Figure 7. Photograph showing root penetration into the spoil

- Under the specified experimental conditions, ryegrass could grow in the spoil samples taken from the four sites even without nutrient addition. However, growth was strongest where ATS had previously been applied.
- All the plants took up K from the spoil although differently according to the fertilizer present and the particular spoil characteristics. A positive correlation existed between potassium uptake and the height of the grass; and total nutrient with the height of the grass.
- Plants with nitrogen and phosphorus fertilizers added grew more strongly than those with nitrogen fertilizer only. The higher the phosphorus uptake, the stronger the plants.
- Root hairs adhered to the surface of the spoil and penetrated into the cracks to assimilate potassium.

ACKNOWLEDGEMENT

The authors would like to thank European Commission for the financial support and the project partners of their advice and assistance, in particular T. Neu of SaarMontan Gesellschaft für bergbaubezogene Dienstleistungen mbH for access to sites and provision of spoil samples, D. Brignall of Wardell Armstrong for advice on selection of seeds, I. Tarasova, O. Demin and B. Coles for laboratory work at Imperial College.

REFERENCE

Chu, L.M. and Bradshaw, A.D. 1996. The value of pulverized refuse fines (PRF) as a substitute for topsoil in land reclamation. II Lysimeter studies. *Journal of Applied Ecology*. 33:858-865.

Hester R.E. & Harrison, R.M. 1997. Contaminated land and its reclamation. Thomas Telford.

Joost, R.E., Olsen, F.J. and Jones, J.H. 1987. Revegetation and minesoil development of coal refuse amended with sewage sludge and limestone. *J. Environ. Qual.* 19(1):65-68.

Rowell, D.L. 1994. Soil Science – methods and applications, Longman Scientific & Technical.

Tedesco, M.J., Teixeira, E.C., Medina, C. and Bugin, A. 1999. Reclamation of spoil and refuse material produced by coal mining using bottom ash and lime. *Environmental Technology*. 20:523-539.

Thorne, M.E., Zamora, B.A. & Kennedy, A.C. 1998. Sewage sludge and Mycorrhizal effects on Secar Bluebunch Wheatgrass in mine spoil. *J. Environ. Qual.* 27:1228-1233.

Neu, T. 1997. Die Rekultivierung der Bergehalde Reden als Beispiel für die erfolgreiche Gestaltung einer Bergbaufolgelandschaft. *Erzmetall.* 50: Nr.6, 390-395.

Tisdale, S.L. & Nelson, W.L. 1975. Soil fertility and fertilizers, 3rd edition, Collier Macmillan Publishers, London.

Environmental Issues and Management of Waste in Energy and Mineral Production, Singhal & Mehrotra (eds)
 ISBN 90 5809 085 X

The environmental effects of rock drains

Murray Fitch & Mike Thompson
Piteau Engineering Associates Limited, Calgary, Alb., Canada

Bill Kovach
Luscar Limited, Line Creek Mine, Sparwood, B.C., Canada

ABSTRACT: The efficient disposal of waste rock at coal mines is of significant economic importance and may mean the difference between a viable and uneconomic mine. When blast rock is disposed of in a valley bottom through which a watercourse passes, the base of the dump is referred to as a "rock drain". The use of formal rock drains to convey significant streamflows at mines in the mountainous regions of western Canada dates from 1980. Since then, rock drains have come into use at the majority of Rocky Mountain coal mines in Alberta and British Columbia.

The Rock Drain Research Program was the first comprehensive study of the physical and flow-through characteristics of rock drains and their environmental effects. The program involved analysis and intensive field data collection (over a four-year period) at Manalta Coal Ltd.'s Line Creek Mine in southeastern British Columbia.

The project comprised investigations into the physical, flow-through, and environmental characteristics of rock drains. Potential environmental issues included rock drain effects on suspended solids and bedload, water temperature, water chemistry, aquatic invertebrates, and fish.

This paper will present an overview of the "environmental effects" investigations of the program. These effects, though measurable in most cases, were found to be much less significant than initially anticipated. Apart from the physical reduction of aquatic habitat due to the presence of the drain, the impacts of the drains were found to be quite localized.

1 INTRODUCTION

The Rock Drain Research Program was the first comprehensive study of the physical and flow-through characteristics of rock drains and their environmental effects. Field data were collected at Manalta Coal Ltd's Line Creek Mine in southeastern British Columbia (Figure 1).

The program was formally initiated in 1992 with final reporting in 1997. The program was jointly funded by Manalta Coal Ltd., the governments of Canada and British Columbia, and Komex International Ltd.

The Line Creek mine is located in a subalpine valley near the town of Sparwood, in southeastern British Columbia, Canada. The drainage basin of Line Creek ranges from approximately 1300 to 2300 m, and lies just to the east of the continental divide. Average annual precipitation at the site was 533 mm (1983-1997), and mean monthly air temperatures range from a low of -11°C in January to a high of 13.7°C in August.

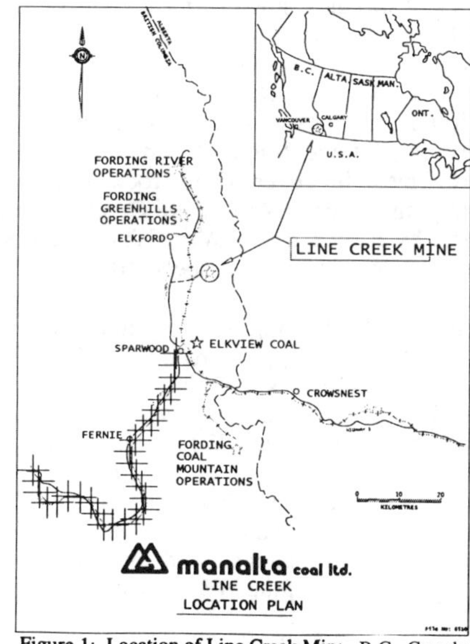

Figure 1: Location of Line Creek Mine B.C., Canada

The primary focus of the study were the two drains on Line Creek, called the "Main Rock Drain" (MRD) and the "Rock Drain Extension" (RDE) (Figure 2). The MRD was the original drain on Line Creek, dumping having commenced in 1990. Construction of the RDE, lying upstream of the MRD, was constructed in 1994. The changing drain configurations presented significant problems for data analysis, but also the opportunity to examine the effects of drain length on a variety of environmental variables.

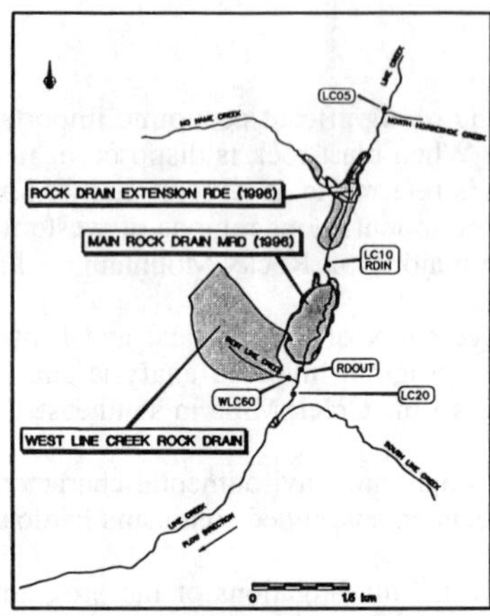

Figure 2: Line Creek Site Plan

The Line Creek drainage is comprised of a number of tributary streams, the most significant of these being Line Creek and South Line Creek, which join downstream of the rock drain. Other tributaries include No Name Creek, which now discharges into the RDE, and West Line Creek, which enters downstream of the MRD. A small portion of No Name Creek is covered by rock drains, while virtually the entire West Line Creek drainage is covered by a rock drain. West Line Creek was of particular interest with respect to water quality concerns.

Line Creek is a high-energy mountain stream, with a bed slope of approximately 2-3% through the areas occupied by rock drains. The streambed is relatively coarse (typically gravel and larger). The mean annual flow at the Water Survey of Canada (WSC) gauging station (drainage area =138 km^2) is 2.1 m^3/s, with mean monthly flows varying between 0.5 m^3/s in February and 7.9 m^3/s in June. Average flows at the downstream end of the MRD are approximately half of the WSC discharges.

2 WATER TEMPERATURE

The thermal regime of a stream has significant ecological effects on aquatic biota. Cold-blooded organisms such as aquatic invertebrates and fish are particularly sensitive to temperature change. Depending on the length and severity of the temperature change, an organism's metabolism, growth, reproduction, and distribution may be affected. Temperature changes upstream can alter the existing aquatic community downstream. If the resulting temperature changes are beyond the existing organisms' range of tolerance, their numbers will decrease and more-tolerant organisms may colonise the area.

It is reasonable to assume that temperatures within the rock drain are less variable than ambient temperatures. While the drain is highly porous to air, the huge mass of rock within the drain probably reacts slowly to changes in outside temperature. Consequently, one would expect a reduction in short-term variations in water flowing through the drain, but an increase in longer-term thermal effects. Lastly, when water flows through the drain during the day, it does not receive heat inputs from the sun, resulting in a relative reduction in downstream water temperatures.

2.1 *Methods*

From 1989 through 1994, stream temperatures were monitored at 15 minute intervals upstream (LC10) and downstream (LC20) of the MRD. In 1994, a monitoring station was added upstream of the RDE (LC05), thereby permitting an examination of temperature effects of both drains.

2.2 *Results*

Mean Daily Water Temperatures: The historical water temperature data indicate that both mean daily water temperature and the magnitude of diurnal variations in water temperature have decreased at the downstream site since the construction of the MRD.

Prior to the construction of the MRD (1989-1990), the average temperature increase between LC10 and LC20 was between 0.80 and 0.96°C. This rise in temperature was likely due to heating directly from the sun or from residual heat in the streambed and banks at night. By 1992, the temperature increase had been reduced to 0.18°C and by 1993, there was actually a very slight decrease in mean daily temperature between LC10 and LC20. The previous heating, which had occurred in the stream prior to rock drain construction, was effectively eliminated and by 1993, mean daily temperatures at the two sites were comparable.

Diurnal Variations in Water Temperature: The 1989 data, prior to rock drain construction, indicates that water temperature at both sites followed a cyclic pattern with the lowest temperature early in the morning and a peak temperature around mid-day. Upstream water temperatures were consistently lower than downstream temperatures and both showed a maximum average diurnal variability of 3 to 4°C.

In 1992, after two years of rock drain operation, upstream temperatures followed the same cyclic pattern as seen prior to rock drain construction, while downstream temperatures fluctuated less. For example, the maximum variation in water temperature in June was approximately 2.5°C at LC20.

The 1993 and 1994 average hourly water temperatures show a trend similar to the 1992 data. Downstream variation in water temperature was less than variation upstream. The temperature variation in LC05, upstream of the RDE, exhibited a slightly greater diurnal variation than LC10 and LC20.

2.3 *Discussion*

The thermal regime of a stream is actually a composite of absolute temperature, amplitude of temperature changes, and rates of change that vary over time and with position in the water course. A variety of hydrological, topographical, and meteorological factors affect the thermal pattern seen in undisturbed stream systems. The aquatic invertebrate and fish community in Line Creek has evolved life histories consistent with the local thermal regime. An altered thermal regime has been shown to adversely affect aquatic communities.

Temperature changes of the magnitude found in the temperature study are not likely to result in measurable impacts on the aquatic community for a number of reasons. The magnitude of the changes in water temperature is small. The average temperature reduction of approximately 1°C across the MRD is simply not significant enough to cause substantial ecological effects. As well, prior to construction of the drain, there was a natural increase in temperature between LC10 and LC20 of approximately 1°C. Although the drain has eliminated this natural temperature increase, it has not shifted the mean temperature outside the range that was previously found. Diurnal variability has been reduced but, once again, the temperatures are within the ranges previously found within this section of the creek. Lastly, temperatures in the creek have been reduced, rather than increased. Studies have found that salmonid species are more likely to be adversely affected by significant temperature increases, rather than decreases.

3 WATER CHEMISTRY

Concerns regarding water quality effects of rock drains are focused primarily on (i) acid-rock drainage, and (ii) nitrogen additions. A potential for acid-rock drainage occurs when sulphide-rich rock is mined. Under aerobic conditions, sulphur can be converted by bacteria to sulphuric acid. Acidic water promotes leaching of metals from rocks. A second water quality concern is the contamination of water by nitrogen residues from ammonia nitrate based explosives used in blasting. The waste rock, which forms the rock drain, has surficial blasting residues, and concern has been expressed that these residues may leach nitrogen compounds into surface water. Nitrogen is expected to be primarily in the form of nitrate, although both ammonia and nitrite forms are also possible. Nitrite and ammonia are more toxic to aquatic life than is nitrate, but react quickly with oxygen to form nitrate. It is unlikely that streams flowing through a rock drain will pick up enough nitrogen residue to cause an impact under normal flow conditions for two reasons.

1. The large rocks that form the base of the drain, have a low surface-area-to-volume ratio. Therefore, they are covered with a relatively small proportion of the total-nitrogen residue load.
2. Only a very small proportion of the total dump is contacted by the stream.

Nitrogen loading of the streams may be expected to increase under flood conditions when higher water levels in the drain result in water contacting "fresh" rock within the dump, which may have nitrogen residues. Water seeping through the dump surface (where the fine particles are concentrated) may also be elevated in nitrate. These effects are expected to be short-lived and will continue only until the blasting residue is leached from the rock surface.

3.1 *Methods*

Water samples were taken regularly at a number of locations on the mine site, as a part of license requirements, and analyzed for several parameters. Data from the mine from 1980 through 1995 were incorporated into the water chemistry database. Most of the RDRP-specific water quality data were collected in 1993, with ten samples collected at the upstream and downstream ends of the MRD, and a variety of samples collected at other locations (such as the No Name and West Line Creek drains). The range of chemical analyses included major ions, metals, and some organics (e.g. phenols and DOC). Results are presented only for the 1993 data set.

3.2 *Results*

Major Ions and Indicator Parameters: The major ions (calcium, magnesium, sodium, potassium, bicarbonate, chloride, and sulphate) typically constitute the majority of dissolved ions in both surface and groundwaters. All sites but WLC60 (the outlet from West Line Creek) showed similar major ion chemistry. The two WLC60 samples were significantly higher in sulphate and more highly mineralized than the samples from the remaining sites, perhaps reflecting the long residence time within the drain.

Nitrate-N: WLC60 has much higher concentrations of nitrate-N than the other sites, though the concentration is below the Canadian Drinking Water Guideline of 10 mg/L. A small increase in nitrate occurred as water passed through the MRD - the median increase was approximately 0.15 mg/L, with an average outlet concentration of 0.3 mg/L.

Total Dissolved Solids (TDS): As was the case for nitrate and sulphate, the TDS concentrations at WLC60 were substantially higher than at the other sites. The median TDS concentration at WLC60 is near the drinking water aesthetic objective of 500 mg/L, while the other sites are well below this level. The data shows quite a small, but relatively consistent, increase in TDS across the drain, which is of little concern.

Trace Parameters: Typically, metals and other trace parameters are introduced into water by leaching under acidic conditions. Line Creek, No Name Creek, and West Line Creek are all somewhat alkaline. Slight increases in concentrations across both the No Name and Line Creek Rock Drains were noted for most parameters, but the levels were typically well below the BC water quality criteria for aquatic life.

3.3 *Discussion*

The Line Creek drains have a small but measurable effect on the water chemistry of Line Creek. Nitrate levels were found to increase a small amount across the drain; dissolved solids concentrations differed little across the drain; and the pH is alkaline, suggesting good buffering capacity and few concerns about acid rock drainage. There may be some flushing of nitrogen compounds during high flow events, but it would appear that if this were to occur, the effects would be very short-lived.

4 AQUATIC INVERTEBRATES

This program investigated the effects of the Line Creek rock drains on the invertebrate drift, benthic invertebrates, transport of organic carbon in the form of coarse particulate organic matter (CPOM), and the potential effects on the invertebrate prey of resident fish.

4.1 *Methods*

In 1993, two invertebrate drift sites (upstream and downstream of the MRD) were assessed for the variability in basic ecological parameters. In 1994, a third sampling site was added 70 m downstream from the toe of the MRD. In 1995, benthic invertebrate samples were also collected. Invertebrate drift was not assessed, although drift samplers were used to collect coarse particulate organic matter (CPOM). Stomachs of cutthroat trout were also collected in 1995 to determine the preferred invertebrate prey taxa of these fish. A comparison of upstream and downstream abundance of these taxa was conducted.

For each invertebrate sample, the following basic ecological parameters were assessed: total invertebrate abundance, number of invertebrate taxa, total abundance and number of taxa of the particularly sensitive orders (Ephemeroptera, Plecoptera, and Trichoptera (EPT index)), and the Shannon Diversity Index.

4.2 *Results*

Invertebrate Drift Abundance, Number of Taxa, and Diversity: In 1993, drift samples were taken from upstream and downstream of the MRD. Statistically, there was no significant difference in abundance between the two sites. The average number of taxa and EPT counts were similar to that of total counts showing larger numbers downstream, but no significant difference between the two sites. The calculated diversity values showed a more even distribution of individuals in the taxa at the upstream site, but the difference was not statistically significant.

In 1994, drift samples were taken from upstream, downstream, and at the toe of the MRD. The downstream samples had the lowest total invertebrate count while the toe site had the largest. A statistically significant difference occurred between the toe and downstream sites. The highest number of invertebrate taxa was found in the upstream drift samples and the fewest in the downstream samples. A significant difference occurred between the upstream and downstream sites. The Shannon Diversity Index values showed a more even distribution of individuals in the taxa at the downstream site than at the toe site. A significant difference occurred between the upstream and toe sites. EPT counts were lowest for the downstream site and highest at the toe. The upstream site had the largest number of EPT taxa while the toe and downstream sites had fewer. Statistical analysis showed no significant difference between any sites for the 1994 total EPT counts, and significant differences between upstream and toe sites and upstream and downstream sites for the number of EPT taxa.

Benthic Invertebrate Abundance, Number of Taxa, and Diversity: Total benthic invertebrate abundance was lowest at the downstream site and highest at the toe of the MRD. A significant difference occurred between the toe and downstream sites. The abundance of EPT in the benthic samples was lowest upstream and highest at the toe of the MRD. Significant differences occurred between the upstream and toe sites, as well as upstream and downstream sites. The largest number of invertebrate taxa in the benthic samples was found at the toe of the MRD while the least was found at the downstream site. Statistical analysis indicated no significant difference between sites. The Shannon Diversity Index values calculated for the downstream and toe sites showed a more even distribution than the upstream site. A significant difference occurred between the upstream and downstream sites, as well as the upstream and toe sites.

Coarse Particulate Organic Matter Analysis: The largest amount of organic matter collected in drift samples came from the downstream site and produced a significant difference between it and the other two sites. There was no significant difference between the upstream and toe sites.

Cutthroat Trout Gut Content Analysis: There were significant differences in the benthic counts of several invertebrate prey taxa between the upstream and downstream sites.

4.3 *Discussion*

Drift and Benthic Invertebrates: If the rock drains had an adverse effect on invertebrate drift or the benthic invertebrate community, it would be expected that the site immediately downstream of the MRD would have a lower abundance of invertebrates, fewer number of taxa, and lower diversity than the upstream site. These trends were not seen in the invertebrate drift data collected for this program and a statistically significant difference in the total number of invertebrates in the benthic community was found only between the toe and downstream sites. This suggests that any influence of the MRD on these basic ecological parameters is within the range of that found in the upstream reach.

The number of individuals in the EPT taxa was smaller upstream than downstream of the MRD. If the rock drain was the cause of an adverse effect, the numbers at the toe should be less than the other two sites. This suggests that differences between the benthic invertebrate communities may be related to factors other than the rock drain.

Coarse Particulate Organic Matter: If the rock drain had an effect on the transport of CPOM, it would be expected that the toe site would have the least CPOM due to a filtering effect. There was significantly *less* CPOM in the drift upstream of the MRD than at the toe and downstream sites. The results suggest that any effect of this rock drain on the mass of the transported CPOM is negated by the input from riparian vegetation immediately downstream of the rock drain.

Cutthroat Trout Stomach Contents: If the MRD influenced the availability of invertebrate prey, it was expected that the total abundance and number of these taxa would be less at the toe and increase further downstream. It was found, however, that the preferred prey species were significantly more abundant in benthic samples at the toe site than further downstream. Furthermore, several of the species found downstream did not exist in the upstream samples. In 1993 and 1994, there was no significant difference in the abundance of these preferred prey species in the drift samples collected. It is important to note, however, that the volume of food in the cutthroat trout stomachs was very small and was collected only once. The data must therefore be viewed with caution.

General Findings: In this program, the MRD did not appear to significantly affect invertebrate drift and the benthic community in the manner expected. There were no significant differences in any of the ecological parameters assessed between invertebrate drift upstream and downstream of the rock drain. While significant differences between benthic invertebrate populations were found upstream and downstream of the MRD, the rock drain itself did not seem to be the cause of these differences. These results suggest that any influence of the rock drain on the invertebrate drift and benthic invertebrates is small and does not exceed the natural variation of populations within the stream. Any effect of the rock drain on the transport of CPOM was rapidly ameliorated downstream by the input of additional leaf litter.

5 FISHERIES

Line Creek supports indigenous populations of bull trout (*Salvelinus confluentus*), westslope cutthroat trout (*Oncorhynchus clarki lewisi*), and mountain whitefish (*Prosopium williamsoni*) (B.C. Research 1977, and Allan 1987). The cutthroat trout population is largely resident (Allan 1995), whereas the whitefish population is present in the stream only during spring/summer (Allan 1987 and 1995). The bull trout population consists of resident rearing juveniles and an adult spawning cohort that inhabits

Line Creek from mid summer to September (Allan 1987, 1991a, b, 1993).

Minor changes in density and distribution of cutthroat and juvenile bull trout have been attributed to natural variation, with the exception of those losses incurred by burial of habitat beneath the Line Creek rock drains. In fact, perhaps as a result of fishing restrictions on the creek, fish populations doubled in 1991 and again in 1995 (Allan 1991, 1995), and continue to remain healthy and viable.

6 CONCLUSIONS

Despite their huge size and appearance, the environmental impacts of the two Line Creek rock drains are not as substantial as had been initially anticipated. Small changes in temperature and water chemistry were observed across the drains, but contrary to expectations, there were minor (if any) effects on the benthic invertebrate community and the downstream fishery.

REFERENCES

Allan, J.H. 1987. *Fisheries Investigations in Line Creek - 1987*. Report of Pisces Environmental Consulting Services Ltd. to Crows Nest Resources Limited, Sparwood, BC.

Allan, J.H. 1991a. *Fisheries Investigations in Line Creek - 1991*. Report of Pisces Environmental Consulting Services Ltd. to Crows Nest Resources Limited, Sparwood, BC.

Allan, J.H. 1991b. *Spawning and Emergence of Bull Trout and Cutthroat Trout in Line Creek*. Report of Pisces Environmental Consulting Services Ltd. to Line Creek Resources Ltd., Sparwood, BC.

Allan, J.H. 1993. *Fisheries Investigations in Line Creek and South Line Creek - 1994*. Report of Pisces Environmental Consulting Services Ltd. to Line Creek Resources Ltd., Sparwood, BC.

Allan, J.H. 1995. *Fisheries investigations in Line Creek - 1995*. Report of Pisces Environmental Consulting Services Ltd. to Line Creek Resources Ltd., Sparwood, BC.

BC Research 1977. *Stage 2 Environmental Study of the Line Creek Project*. Volume 1. Report of BC Research to Crows Nest Industries, Fernie, BC.

Environmental Issues and Management of Waste in Energy and Mineral Production, Singhal & Mehrotra (eds)
© 2000 Balkema, Rotterdam, ISBN 90 5809 085 X

Revegetation: Invasion and succession above the 45th parallel

H.G.Gerbrandt
Department of Environmental Engineering, Montana Tech of the University of Montana, Butte, Mont., USA

P.T.Sawyer
Department of Biology, Montana Tech of the University of Montana, Butte, Mont., USA

ABSTRACT: Montana Resources is currently reclaiming old and new waste dumps at several sites in the vicinity of its Butte copper and molybdenum mine. In order to improve seed mixes and plantings on future dump reclamation, Montana Tech's Environmental Engineering and Biology Departments have just completed five years of monitoring at two sites seeded in the springs of 1992 and 1994 respectively. Tech's analysis determines which seeded and planted species survive and flourish, and which species, both desired and unwanted, move on to the site naturally. Montana Resources is using this information to modify future seed mixes and plantings.

1 INTRODUCTION

Revegetation of disturbed soils above the 45th parallel must take into account not only soils, slopes, elevation, and aspect; but also climate, short growing seasons, and scarcity of frost-free days. In the western United States and Canada, semi-arid conditions further limit the plant choices. Therefore, use of native species and tested cultivars becomes mandatory, and use of species with a successful track record imperative.

In order to minimize reclamation maintenance (reseeding and re-fertilizing) and maximize the "bang for the (seeding) bucks", Montana Resources (MR) along with Montana Tech of The University of Montana have been monitoring revegetation at several reclaimed waste rock dumps at MR's open pit copper and molybdenum mine in Butte, Montana, USA. This paper describes the monitoring program and analyzes five consecutive years of data at the Woodville waste dump.

2 SITE DESCRIPTION AND HISTORY

This site is 25 acres in area, at an elevation varying from 6000 to 6200 feet above sea level. The site has a mildly sloping area at the top of a hill, with semi-spherical slopes below the top, running from west around to northeast. Slopes range from flat on top to 20% on the sides.

This area is a waste rock dump site. After grading to current slopes, the site was covered with approximately 4 feet of subsurface material. MR personnel describe this material as "non-acid generating, oxidized leached capping". On the top of the hill, 24 inches of topsoil was placed on top of 24 inches of the capping material described above. This represents the best soils for revegetation. The north-facing slope material is characterized as the poorest.

After placing and grading the cover and topsoil, 200 lbs. per acre of commercial fertilizer was applied and chiseled in. Mulch was applied and crimped, and the site was seeded in April and May of 1994. The site is not fenced, and vehicular access is available, however, the remoteness of the site minimizes site degradation from recreational vehicles. The site is accessible to herbivorous wildlife, but cattle are absent.

The seed mix was prepared by Grassland West of Clarkston, Washington. The mix is presented in part in the first two columns of Table 1, and contained eight grasses, four forbs, and three shrubs. The application rate was 29.5 lbs. per acre (Mustoe, 1994). Soil tilling, mulching, crimping, fertilizing, and seeding were performed by Bernie Jennson, a local contractor. Additionally, 4000 Lodgepole pines and 150 Douglas fir were planted in May of 1994 by Bitteroot Native Growers and 2500 Aspen added in May of 1995 by MR personnel. These trees were not spaced uniformly across the site, but concentrated in specific areas.

Butte's climate can be severe, with occasional winter temperatures of -40° F and cool short summers. Precipitation is approximately 13 inches per year. This climate will support cool season grasses, forbs, and shrubs on dryer slopes (south-facing and west-facing). Wetter slopes, higher elevations, and wetlands will support trees as well as wetland varieties.

3 MONITORING

The Woodville site measures approximately 25 acres. It would be nearly impossible to scrutinize every square foot of surface at the site during the monitoring. The approach taken was to monitor a small but statistically significant portion of the site and extrapolate the data collected to the site as a whole. Current practice is to monitor a transect (designated monitoring area) approximately 100 square meters (1000 square feet) for each ten acres of distinct plant community (plants that share nearly identical climate, soil, elevation, solar aspect, wildlife, etc.). Using this criteria, three transects would be required at the Woodville site. At MR's request, the number of transects was reduced to two. One transect was located on the wetter, north-facing slope and one on the drier west-facing slope. Monitoring methods are described below.

3.1 Monitoring Methods.

Three methods were used to monitor plant species:

(1) Belt Transects were used for counting and inventorying trees. A belt transect is a rectangular area six feet wide by two hundred feet long, which is marked by driving steel posts (called monuments) into the ground at the center of the short sides (see Figure 1a). Each tree falling within the 6' by 200' rectangle was identified and counted during the yearly monitoring.

(2) Daubenmire quadrats (Daubenmire, 1959) were used for inventory of grasses. Daubenmire quadrats are a series of 40 frames, each 0.1 square meters (20 cm x 50 cm), laid out on the long sides of the belt transect, as shown in Figure 1b. Species frequency, density (number of plants per area), percent canopy cover (shadow of the plant projected onto the ground surface) and importance value (Daubenmire, 1968) were determined and reported. In order to clarify the terms used above, definitions are provided:

Frequency. The percentage of Daubenmire plots containing at least one plant of a given species.

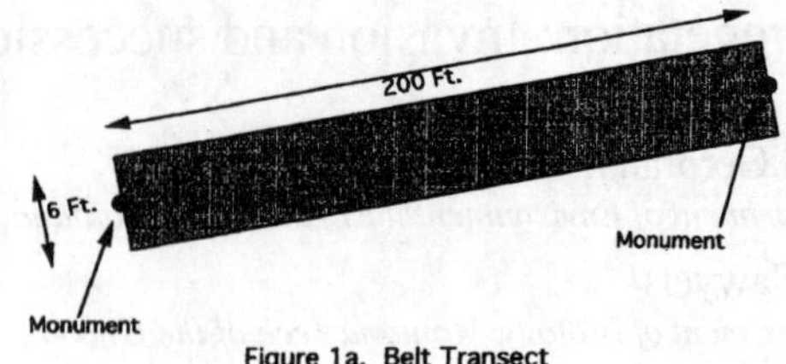

Figure 1a. Belt Transect

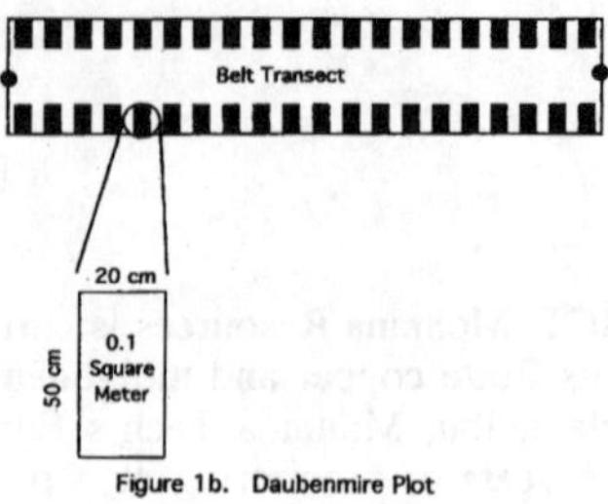

Figure 1b. Daubenmire Plot

Density. Average number of plants per area. In this report, density is reported at the average number of plants per square meter as identified in the transects (trees and shrubs) or plots (grasses and forbs).

Percent Cover. Percent of the plot covered by the shadow of the plant projected on to the ground surface at high noon.

Importance Value. Indicates the relative importance of each species compared to other species at the site. The Importance Value can mathematically be written as:

$$IV = \frac{SpeciesFrequency}{TotalSpec.Frequency} * 100 + \frac{Species\%Cover}{TotalSpec.\%Cover} * 100 + \frac{SpeciesDensity}{TotalSpec.Density} * 100$$

The actual Importance Value has little meaning in itself, only in relation to Importance Values of other species. Importance values are often used to determine the dominance of species.

(3) In addition to the transect and Daubenmire plot monitoring, a species list was prepared, showing all species of trees, shrubs, forbs, and grasses noted at the site.

3.2 Frequency of Monitoring.

Monitoring was performed yearly for a period of five years, with an optional follow-up investigation ten years after seeding. Monitoring was performed in June or July after seed heads appeared, assuring accurate identification.

TABLE 1. West-facing transect importance values.

SPECIES	Pounds Per acre	1995	1996	1997	1998	1999
Alfalfa	**1**	55.0	21.8	81.5	63.3	81.9
Alsike Clover		0.0	3.3	1.8	0.0	0.0
Arrowleaf Balsam-root	**2**	0.0	0.0	0.0	0.0	2.7
Black Medic		0.0	0.0	10.4	0.0	0.0
Bluebunch Wheatgrass	**2**	8.0	3.5	0.0	1.4	2.4
Canadian Bluegrass	**1**	1.2	3.3	7.9	2.8	7.6
Collomia		0.0	0.0	0.0	0.0	9.0
Covar Sheep Fescue	**2**	62.0	39.5	88.9	144	98.8
Crested Wheatgrass		0.0	0.0	0.0	0.0	1.7
Dandelion		0.0	0.0	0.0	0.0	1.3
Fanweed		23.0	45.1	13.7	0.0	18.8
Common Salsify		0.0	0.0	0.0	0.0	1.3
Hoary Aster		0.0	0.0	0.0	1.7	0.0
Knapweed		4.0	4.7	3.5	3.1	10.8
Lambs Quarter		0.0	0.0	0.0	0.0	1.7
Lewis Flax	**2**	14.0	8.8	12.4	22.6	7.0
Loesel Tumble-weed		36.0	18.7	4.9	0.0	19.8
Michaux Safebrush		3.0	2.6	4.2	5.7	5.8
Pompelly Brome		1.2	3.8	2.2	0.1	1.4
Sanfoin	**0.5**	6.0	3.7	2.9	0.0	0.0
Silverleaf Phacelia		0.0	3.4	1.7	0.0	2.8
Streambank Wheatgrass		0.0	3.8	0.1	0.0	0.0
Thickspike Wheatgrass	**4**	5.0	10.1	0.1	0.0	1.7
Tumble Mustard		6.0	17.8	0.7	0.0	2.8
Wheat		7.0	14.2	0.0	0.0	0.0
Yarrow		38.0	62.8	63.2	55.7	23.6
Yellow Sweet Clover		18.0	11.0	0.0	0.0	0.0

TABLE 2. Northeast-facing transect importance values.

SPECIES	Pounds Per acre	1995	1996	1997	1998	1999
Alfalfa	**1**	79.2	75.4	111	124	124
Alsike Clover		1.8	4.0	4.0	0.0	0.0
Autumn Willow-Herb		0.0	0.0	0.8	0.0	0.0
Bluebunch Wheat-grass	**2**	0.0	5.3	0.0	0.0	0.0
Canadian Bluegrass	**1**	0.0	1.6	0.0	0.0	7.1
Michaux Sagebrush		1.8	2.1	5.2	3.7	3.1
Covar Sheep Fescue	**2**	55.5	63.0	99.2	154	107
Fanweed		18.5	51.0	0.0	0.0	10.3
Kentucky Bluegrass		0.0	0.0	0.0	0.0	1.8
Knapweed		7.6	0.0	4.6	2.3	0.2
Lewis Flax	**2**	10.1	16.3	9.8	1.5	0.0
Loesel Tumble-weed		40.8	0.2	1.3	0.0	11.4
Penstemon		0.0	0.0	1.0	0.0	0.0
Pompelly Brome		0.0	0.9	1.0	0.0	0.5
Sowthistle		0.0	3.6	0.0	0.0	0.0
Stream-bank Wheatgrass		2.2	2.4	0.8	0.0	0.0
Thickspike Wheat	**4**	0.0	0.0	0.0	0.0	2.2
Tumble-Mustard		0.0	6.7	0.0	0.0	0.0
Western Wheat-grass	**4**	0.0	0.8	0.0	0.0	0.0
Wheat		2.0	0.8	0.0	0.0	0.0
White Camphia		0.0	0.0	1.3	0.0	0.0
Yarrow		43.4	51.0	58.3	14.2	32.0
Yellow Sweet Clover		33.0	0.0	0.0	0.0	0.0

4 MONITORING RESULTS

Individual trees growing within transects were counted, and results are presented in Section 4.2. Frequency, density, percent cover, and importance value (IV) of each grass or forb species were determined. While frequency, density, and percent cover provide valuable information, comparative values between species may often be deceiving. For instance, a frequency of one hundred Fanweed plants would provide very little cover, while a single mature Alfalfa plant might provide cover of 100% within a Daubenmire frame. Therefore, vegetation data in this report is presented in the form of importance values. Importance values, in themselves, have no actual meaning, but do reflect frequency, density, and percent cover; and are useful in comparisons with other species. Actual frequencies, densities, and cover values are available from the authors.

4.1 *Grass and Forb Importance Values*

From five years of density, frequency, and cover data, IV's for all species found within the Daubenmire plots were calculated. IV's from the West-facing transect are presented in Table 1, with values for the North-facing transect appearing in Table 2. Seeded species are bolded, while invaders (non-seeded species) are in regular font style. The seeding rate is listed in the second column, followed by the IV's.

4.2 *Tree Counts*

No trees were detected within the transects during the five year period. Believing that the transects were not reflective of actual tree densities at the site, a wider monitoring effort was performed by the 1996 Ecology class at Montana Tech. In October of that year, 32 circular plots, each encompassing 375 square meters (70 feet in diameter), were surveyed for Aspen, Lodgepole pine, and Douglas fir (see Figure 3. Additionally, vigor classes were identified for each tree:

Vigor Class	**Condition**
1	Dead
2	Not Healthy
3	Healthy
4	Very Healthy

The 1996 data summary is presented in Table 3. The data shows 117 identified Lodgepole pines within the study areas, at a density of 39 trees per acre in the surveyed areas. 22 of these were considered very healthy, 60 were found to be healthy, 33 appeared "not healthy" and 2 dead trees were found. 10 Douglas firs were found, ranging in vigor from not healthy to very healthy. Only one unhealthy Aspen was identified.

5 DISCUSSION OF RESULTS

In order to discuss the "success" of the revegetation effort, an understanding of the reclamation objec-

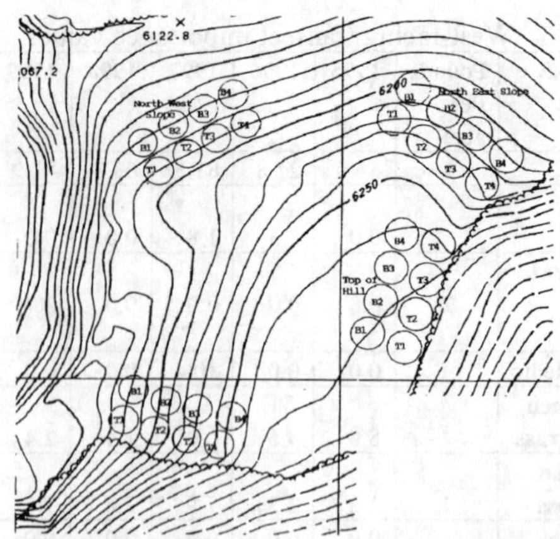

Figure 3. Circular Plots on Woodville Hill.

tives is essential. For instance, if food production for livestock is the primary objective, then success would be based on the dominance and yield of high food value grasses. For this site, the following objectives are assumed:

- Erosion Control. Reduce erosion and sediment leaving the site.
- Site Stability. Achieve long term structural sta bility of the slopes.
- Reduction of Infiltration into Waste Rock. Uptake by the vegetation should prevent substantial amounts of precipitation from reaching the underlying waste material, some of which may be acid-producing.
- Asthetics. The site should be aesthetically pleasing to the general public.

An additional Woodville site objective is the creation of wildlife habitat. However, neither livestock nor wildlife food production is an important objective of the reclamation at the site. With these objectives in mind, a data analysis is presented below.

5.1 *Grasses and Forbs*

Both transects at the Woodville site sustain healthy vegetation. Growth is so dense and lush that it is difficult to conduct the survey. The seeded species Covar sheep fescue and Alfalfa dominate the site (see IV's on Tables 1 and 2). Seeded Lewis flax, Canadian bluegrass, and Bluebunch wheatgrass are also growing in both transects. Of the non-seeded species, Yarrow is the most important, with IV's of 9 and 8 at the North- and West-facing transects in 1999. Of the non-seeded species growing at the site, Michaux sagebrush is perhaps the most significant,

Table 3. Tree Monitoring, 1996, Woodville Site, Montana Tech Ecology Class. Number of seedlings found in each vigor class.

	Lodgepole Pine				**Aspen**				**Douglas Fir**			
Vigor Class	1	2	3	4	1	2	3	4	1	2	3	4
North East Slope												
1 Top		2	4									
1 Bottom			3									
2 Top			4	1								
2 Bottom			3	1								
3 Top	1	2	3	1						1	1	
3 Bottom			1	2								
4 Top		1	3	2								
4 Bottom			1									
Top of Hill												
1 Top												
1 Bottom			3									
2 Top												
2 Bottom												
3 Top		1										
3 Bottom		1								1		
4 Top		3	1	1								
4 Bottom		1	1	2								
West Slope												
1 Top				3								
1 Bottom		1				1						
2 Top		10	10									
2 Bottom		10	9	2								
3 Top			3	1								
3 Bottom												
4 Top											1	
4 Bottom												
NorthWest Slope												
1 Top		1	1	1							3	
1 Bottom												
2 Top			4	3							2	1
2 Bottom												
3 Top	1		3	2								
3 Bottom												
4 Top			3									
4 Bottom												
Total	**2**	**33**	**60**	**22**	**0**	**1**	**0**	**0**	**0**	**2**	**7**	**1**

as it serves as forage for deer and elk. Of the seeded species, three (Indian ricegrass, Rubber rabbitbrush, and Wild rose) have not been identified on the site. It is believed that seeds from these species did not produce viable plants.

1999 saw a marked increase in diversity at the Woodville west-facing transect. Slenderleaf collomia, Silverleaf phacelia, Lambs quarter, Arrowleaf balsam-root, Common salsify, Dandelion, and Hoary aster were identified within the transect for the first time. While these species are present with small IV's, diversity is considered beneficial to the revegetation objectives.

Spotted knapweed, a State of Montana Category 1 noxious weed (most serious of three categories), is present in the north and west transects at importance values of 0.1 and 6 in 1999. 1999 showed an increase of IV for this weed in the west-facing transect and a decrease in the north-facing, as shown in Tables 1 and 2. Other weeds found on these two sites include Curly dock, Common mullein, Butter and Eggs, and Dalmation toadflax.

Bare ground cover values at the Woodville transects (not presented in the Tables) are 4% and 4%, Woodville in 1995.

Thus, the vegetation cover on the site has increased every year since seeding.

Weedy species are diminishing in importance as grasses and forbs take hold. This is evidenced by a decrease in IV's of Fanweed, Loesel tumbleweed, Knapweed, and Tumble mustard.

Importance values of most species have declined in the last two years. This may be an indication that initial and subsequently applied fertilizer is being used up by the plants, and periodic application may be necessary to maintain 1997 vegetation levels. However, the reclamation objectives are still being met under the current densities. There is no evidence of erosion or slope instability, and the site is certainly aesthetically pleasing. No acid seeps have been identified at the toes of the dump slopes.

5.2 Tree Plantings

In 1996, a survey of 2.8 acres of the original 25 identified 126 live trees out of the roughly 6500 planted trees. It is obvious the the grasses and forbs, especially the Alfalfa, have out-competed the trees. The faster growing grasses and forbs shaded out the trees, killing most of them. Only after five years of growth are the remaining trees starting to appear above the canopy of the Alfalfa.

While no trees were found in either Woodville transect, Lodgepole pines are growing on the site, as evidenced by the 1996 survey. Visual inspection of the forest surrounding the dump shows conifers invading and taking over sites previously dominated by Aspen. It is the opinion of the authors that in spite of the severe competition of the Alfalfa for light, enough conifers will survive on the dump site to serve as seed sources at a later date, when the vigor and vitality of the Alfalfa has declined. Furthermore, because of the trend of conifer takeover of Aspen habitat in the site vicinity, it is doubtful that planting of Aspen would result in a long-term establishment of this species.

6 CONCLUSIONS

Two transects were established at the Woodville site and inventoried as part of a five year monitoring effort to determine successful and unsuccessful revegetation seed mix designs. The following findings summarize the most salient points of interest:

- Alfalfa and Covar sheep fescue dominate the site.
- Weedy species are diminishing in importance as
- Diversity is increasing, a trend which helps to meet the reclamation objectives.
- Of the fifteen seeded grass and forb species, only four are established on the site.
- Grasses and forbs out-competed the planted tree seedlings during the initial five years, killing most of them. Enough trees will survive to serve as seed stock for the future.
- Noxious weed density, cover and frequency are low, but should be monitored.
- Planting of Aspen is considered a waste of resources for the Woodville site. Conifers are more likely to achieve success, and will dominate the Aspen given enough time.
- Desired reclamation species such as Yarrow, Alsike clover, and Common sagewort moved into the Woodville site without seeding. Their successes point to possible inclusion in future seed mixes and plantings.
- Use of 24" of topsoil at the Woodville site may be an important factor in the success of vegetation at the site.

At the discretion of either MR or Montana Tech, a follow-up inventory may be performed in 2004 to gain a ten-year perspective. A tenth year inventory is strongly suggested.

7 REFERENCES

Daubenmire, R. 1959. A canopy-coverage method of vegetational analysis. Northw. Sci. 33:43-64

Daubenmire, R. 1968. Plant communities. Harper & Row, Publishers, New York.

Dorn, Robert D., 1984. Vascular plants of Montana. Mountain West Publishing, Cheyenne, Wyoming

Mustoe, M. 1994. Personal correspondence to John Burke, Manager of Engineering, Montana Resources.

Soil Conservation Service, US Department of Agriculture, 1980. Standardized plant names for Montana. Bozeman, Montana.

8 APPENDIX

SCIENTIFIC NAME

Trees

Pinus contorta Dougl. ex Loudon
Populus tremuloides Michx.
Pseudotsuga menziesii (Mirb.) Franco

Shrubs

Chrysothamnus nauseosus (Pall.) Britt.

Grasses

Agropyron cristatum (L.) Gaerth
Agropyron dasystachyum (Hook) Scribn.
Agropyron riparium Scribn. & Smith
Agropyron smithii Rydb.
Agropyron spicatum (Pursh) Scribn. & Smith
Bromus inermis Leyss.
Bromus pumpellianus Scribn.
Festuca ovina L.
Festuca ovina var duriuscula (L.) Koch
Oryzopsis hymenoides (Roem & Schult.) Rycker exAper
Poa compressa L.
Poa prenensis L.
Triticum aestivum L.

Forbs

Achillea millefolium L.
Artemisia michauxiana Bess.
Balsamorhiza sagittata (Pursh) Nutt.
Centaurea maculosa Lam.
Chenopdium album L.
Collomia linearis Nutt.
Epilobium paniculatum Nutt.
Linaria dalmatica (L.) Miller
Linaria vulgaris Miller
Linum lewisii Pursh
Lychnis alba Mill.
Machaeranthera canescens (Pursh.) Gray
Medicago lupulina L.
Medicago sativa L.
Melilotus alba Desr.
Melilotus officinalis (L.) Lam.
Onobrychis viciaefolia Scop.
Penstemon sp.
Phacelia hastata Dougl.

SCIENTIFIC NAME

Rhumex crispus L.
Sisymbrium altissimum L.
Sisymbrium loeselii L.
Sonchus sp.
Taraxacum officinale Weber
Thlaspi arvense L.
Tragopogon dubius Scop.
Trifolium hybridum L.
Verbascum thapsus L.

Scientific names from Dorn, 1984

COMMON NAME

Lodgepole pine
Quaking aspen
Douglas fir

Rubber rabbitbrush

Crested wheatgrass
Thickspike wheatgrass
Streambank wheatgrass
Western wheatgrass
Blue bunch wheatgrass
Smooth brome
Pumpelly brome
Covar sheep fescue
Hard sheep fescue (Durar Hard Fescue)

Indian ricegrass
Canada bluegrass
Kentucky bluegrass
Wheat (Pregreen)

Western yarrow
Michaux sagebrush
Arrowleaf balsamroot
Spotted knapweed
Lambsquarter
Sleanderleaf collomia
Autumn willowherb
Dalmation toadflax
Butter-and-Eggs
Lewis' flax
White cockle
Hoary aster
Black medic
Alfalfa
White sweet-clover
Yellow sweet-clover
Sainfoin
Penstemon
Silverleaf phacelia
COMMON NAME
Curly dock
Tumblemustard
Loesel tumblemustard
Sowthistle
Dandelion
Fanweed
Common salsify
Alsike clover
Common mullein

Common names from "Standardized Plant
Names for Montana, 1980

Environmental Issues and Management of Waste in Energy and Mineral Production, Singhal & Mehrotra (eds)
© 2000 Balkema, Rotterdam, ISBN 90 5809 085 X

High value tree reclamation research

Donald H.Graves, James M.Ringe & Matthew H.Pelkki
Department of Forestry, University of Kentucky, Lexington, Ky., USA

Richard J.Sweigard
Mining Engineering, University of Kentucky, Lexington, Ky., USA

Richard Warner
Biosystems and Agricultural Engineering, University of Kentucky, Lexington, Ky., USA

ABSTRACT: A multidisciplinary group of researchers at the University of Kentucky initiated a diverse study to evaluate the effects of soil compaction on the survivability and growth of high value tree species. This study was established on the Starfire Mine owned by Cyprus-Amax in Perry County, Kentucky. The team of researchers encompassed the various expertise areas in the Department of Biosystems and Agricultural Engineering, Forestry, Mining Engineering, and Geology. Since its initiation, a weather station has been established, over 57,000 trees have been planted, a passive dewatering system initiated, and a fertigation study constructed. The trees, consisting of seven species, have been planted on areas that were compacted, lightly compacted, and uncompacted. Bulk density and penetrometry data have been gathered on all the planted sites to determine the relative compaction of each planting area. Compaction is already a contributing factor to either or both tree survival and growth. Additional studies being conducted include the effects of alternative ripping methods on reclaimed areas that were compacted in the normal manner prior to the concept of a new grading standard. These plots were mulched and planted in the spring of 1999. Additional studies will be established depending upon the availability of funds.

1 INTRODUCTION

Mine soil grading standards adapted with the enactment of the federal surface mining law (PL 95-87) resulted in compaction of the soil material to the point that tree roots were unable to penetrate for proper tree growth and development. This has resulted in high mortality rates for seedlings which precluded the development of forests with high value species.

Research in Illinois has proven that uncompacted pre-Law mining sites have resulted in some of the most productive areas in the state for the growth of such species as yellow-poplar, white oak, and black walnut (Ashby et al, 1978). Recent studies at Virginia Polytechnic Institute and State University have verified such results when mine soil is loose dumped without grading and planted to various tree species (Burger and Torbert 1992). Research at the University of Kentucky and elsewhere has confirmed that surface mines reclaimed since the enactment of PL 95-87 have proven to be hostile to the re-establishement of trees to mined areas. Usually, only pioneer (early invaders), intolerant species such as cottonwood, black locust, sycamore, Virginia pine and other such species can be established with any degree of expected success.

During the reclamation process, soil is removed ahead of the pit and is either stockpiled to wait for replacement, or is hauled immediately to graded spoil areas behind the pit. One consequence of soil transportation includes compaction. Generally, compaction produces an undesirable change in the physical properties of a soil. It is a common and potentially serious problem on reclaimed mine lands. A highly compacted soil reduces root growth, which in turn reduces vegetation success. The degree of soil compaction is a function of applied pressure, soil structure, and soil characteristics. The equipment used to transport the soil applies pressure to the surface, which results in an increase in soil density (Dollhopf and Postle 1988).

In addition, several soil properties change as a result of soil compaction. Soil density increases as the soil particles are more closely together. The mechanical resistance to penetration increases, depending on moisture content and size of soil particles. The hydraulic conductivity is reduced due to an increase in compaction (Barnhisel 1988). These changes occur as a consequence of soil movement and restoration. Mine soils general show little or no structure throughout the profile. Structure may begin to appear in the mine spoil within 50 years (Dollhopf and Postle 1950).

Several options exist for the mine operator when treating compacted soils. Deep tillage has been used frequently to correct physical problems in dense, massive soil (Barnhisel 1988).

The overall success of trees grown on relocated compacted soil depends primarily upon its physical condition (Barnhisel, Powell, and Hines 1996). This is determined by the physical properties of soil such as structure, bulk density, and resistance to mechanical penetration. In an unstructured soil, aggregates are separated by planes of weakness that are represented by open spaces. These macro pores allow for air and water movement and root exploration. The removal and stockpiling of soil, necessary for reclamation, breaks up these aggregates so that the original structure of the soil is destroyed.

Upon removal and replacement, the soil is compacted with earth moving equipment to the extent that the original structure is completely modified. A crust is typically found on the surface layers due to a final grading and leveling of this soil. There are no apparent lines of weakness with the exception of dessication cracks near the surface due to wetting and drying. This reduces water infiltration and transmission through the soil. Layers of extremely compacted soil develop at various levels in the profile due the intensity of traffic. The pore space is greatly reduced by the loss of macro pores. These factors combine to produce a soil with poor physical properties that inhibit root growth (Dollhopf and Poste 1988). Soil compaction is the unsuitable result of soil transportation due to the break-up of its structure.

Various methods have been proposed to circumvent the problem of excessive compaction. These methods have been developed primarily for lands being reclaimed for row crops in the Midwest. One method is to use dump trucks to haul and dump subsoil material with truck wheel traffic being limited to graded spoil beneath the subsoil. This material is then graded by "low ground pressure" dozers and followed by placement of topsoil, either directly by scrapers or in "windrows" of confined scraper traffic and then spread onto graded subsoil via light dozers.

A more effective, yet expensive method is the placement of subsoil by a large belt conveyor equipped with a spreader boom. This method requires virtually no traffic during spoil placement. Dunker et al (1992) showed significantly greater corn and soybean yield for this method versus scraper-placed material for each of seven consecutive years. Hooks et al (1992) also reported that truck placement resulted in higher yields than with scraper placement.

There remains, however, significant material handling difficulties associated with alternatives to removal and replacement of soil with scrapers. Scraper hauling is more time efficient and devising practical means of segregating, placement, and grading of topsoil (A horizon) material without significant compaction of the subsoil material is very difficult.

This research project was needed to quantify the environmental impact associated with reforestation with a low / no compaction mine soil placement. It is anticipated that the use of selective reforestation methods on low / no compaction spoil material combined with organic soil modifiers will not only provide a better micro climate or environment, but will also achieve reduced storm runoff and lower sediment loading of streams. The instrumentation and data acquisition system is documenting the reduced impact from this multi-disciplinary research project.

2 METHODOLOGY

Nine (9) berm enclosed cells were constructed. Each comprised 2.5 acres inside the berm. Each cell contains 21 one-tenth acre plots with 10 foot alleyways. Three of the bermed cells were compacted using the normally accepted spoil grading systems. Three were "back-dumped" using Euclids and lightly graded using small dozers. The final three were back dumped with Euclids and not further disturbed except to grade alley ways for later access.

One cell in each spoil placement method was mulched with 40 tons per acre of processed hardwood bark. One was mulched with 40 tons per acre of ground straw and manure. One cell in each set was left unmulched to determine the effects of organic material on spoil modification.

Three areas of 2.5 acres (without berms) were deep cross ripped with a D-9 bulldozer to a depth of at least 3 feet. Another three areas of 2.5 acres without berms were cross ripped with a tractor and a conventional farm subsoiler. Very little penetration was achieved with the second subsoiling method. The six areas were mulched the same as on the nine bermed cells.

Each 2.5 acre research area contains 21 one-tenth acre plots (seven tree species replicated three times). Each plot was initially planted with 121 tree seedlings on a 6 foot spacing. All plots were seeded with a slow establishing, non-aggressive mixture of grasses and legumes.

Loose fill in the bermed areas have produced a wide range of mico-sites. Instrumentation has been placed in bermed and non-bermed areas to characterize micro-climate variables related to tree growth and development. Soil moisture, soil temperature, bulk density, air temperature and relative humidity have been recorded. A fully automated weather station monitors precipitation inputs and other basic climate variables in the study area.

To enhance data gathering in the cells, a 16 acre light to ungraded area or drag line dumped material was selected for planting. This area was hydro-seeded with a native non-invasive mixture of grass and legume seed like the rest of the research area and planted with the same seven species of tree seedlings. The tree species include yellow-poplar, black walnut, white pine, white ash, red oak, white oak, and paulownia.

A deep profile, constant-rate recording penetrometer was built by Southern Illinois University and University of Illinois Cooperative Reclamation Research Station for this project. The device measures soil strength over a depth of approximately four feet. The penetrometer has been mounted on a tractor where a hydraulic cylinder is used to force a steel rod, with a metal cone attached, into the top of the spoil. The machine records the resistance to penetration of the entire range of the hydraulic ram.

A set of background data is being obtained for the constructed cells. Additional background data is being obtained for typical reclaimed land that has not received any special treatment. This permits a comparision between the initial soil strength of the constructed cells and the surrounding reclaimed land. It is anticipated that the initial measurements of penetration resistance for both types of areas would be quite high and would consist primarily of refusal at each location. It was also anticipated that, over time, a more normal soil strength and profile would develop.

An instrument network was designed to monitor soil water and temperature conditions in the experimental cells. One hundred soil water tension access tubes were constructed and installed. An electronically data-logged time domain reflectometry system continuously monitors soil moisture. Twelve stainless steel pan lysimeters were fabricated and installed during spoil placement.

An passive dewatering sediment control system was constructed in connection with the three compacted cells. Flumes were designed and placed to direct the water runoff from the compacted cells into a sediment pond. A specially constructed dam with a slow sand filter and storm data acquisition devices was constructed.

Cells were constructed to study high-value tree species survival and growth under conditions of trickle irrigation in mine spoil. The area consists of

Table 1. Average survival, height, and growth on drag line spoil, 1996-1998.

Drag line spoil	Species	1996 average survival %	1997 average survival %	1998 average survival %	1996 average heigh (cm)	1997 average height (cm)	1998 average height (cm)
uncompacted	white pine	51	48	50	35	53	18
	white ash	92	89	91	67	92	24
	black walnut	92	77	77	54	55	1
	yellow-poplar	57	47	44	43	51	7
	white oak	81	62	64	33	36	3
	northern red oak	88	83	82	45	64	19
	Total	**77**	**68**	**68**	**46**	**58**	**12**
compacted	white pine	67	27	27	27	30	3
	white ash	93	100	87	42	49	7
	black walnut	93	53	60	50	43	-7
	yellow-poplar	67	13	13	32	30	-2
	white oak	47	33	33	27	32	5
	northern red oak	67	53	33	24	31	7
	Total	**72**	**47**	**42**	**34**	**36**	**2**

38 20 ft. by 20 ft. plots. Tipping buckets constantly monitor the flow of infiltrated water into all 38 plots.

3 RESULTS

This project is expected to continue for 20 years or longer. It is now in its infancy, but beginning to yield some interesting results. In the uncompacted drag line spoil (Table 1), after three years, white ash had an average survivability of 91 percent and averaged 24 cm of height growth. Yellow poplar was the least successful survivor with an average of 44 percent and walnut has averaged only 1 cm of height growth.

In the compacted dragline spoil, white ash still had a survivability rate of 87 percent. White ash and northern red oak have averaged 7 cm of height growth. Yellow-poplar averaged only 13 percent survivability and walnut averaged a negative 7 cm of height growth during the three years. Negative height growth is a result of dieback, and in some cases, browse damage.

Survival in the loose dumped cells after three years is very good (Table 2). White oak averaged 97 percent survival, but is closely followed by white ash at 86 percent and northern red oak at 85 percent. Paulownia averaged only 40 percent survival which may be attributed to the very small planting stock. The paulownia that survived averaged 49 cm of height growth. White pine averaged 25 cm of height growth and white ash 19 cm. Black walnut averaged a negative 13 cm of growth after 3 years. The overall survivability of all tree seedlings averaged 76 percent with 13 cm of height growth.

White ash was the leading survivor in the rough graded cells at an average of 85 percent, followed by northern red oak at 70 percent. The lowest survivability was again paulownia with an average of 44 percent. Paulownia height growth averaged 29 cm for the three years and white ash averaged 17 cm. Black walnut averaged a negative 9 cm of height growth and white pine averaged 11 cm and yellow-poplar 10 cm. The overall survival averaged 64 percent and overall height growth averaged 7 cm.

White ash was the leading survivor in the compacted cells averaging 58 percent. The lowest survival percentage was for white pine and paulownia at 11 percent each. The surviving paulownia height growth was 26 cm and for white pine height growth averaged -1 cm for the three years. Black walnut averaged a negative 20 cm in height increment over the same period. The average survival of these compacted cells was 39 percent and average height growth was - 5 cm. Only white ash and paulownia exhibited positive height growth.

When we look at the average survivor from 1996 to 1998 in the loose drag line spoil, we see that northern red oak, yellow-poplar, and black walnut are slowly declining. White oak, white ash, and white pine are showing increases from resprouting after dieback or browse. On the compacted drag line spoil, only black walnut is increasing after the initial dieback or browse damage. White ash resprouted and then declined in the following year. Northern red oak is decreasing each year, while white oak, yellow-poplar, and white pine are remaining constant after initial mortalities.

Survival is decreasing in the loose dumped cells for all species except paulownia and black walnut. In the rough-graded cells, all species are declining except for white oak. Survival of all species is declining in the compacted cells except white oak and paulownia. White ash still has a survival rate of 85 percent in these cells.

The average height growth was greater for all species in the uncompacted dragline spoil with the exception of white oak. The compacted cells resulted in a decreased height growth for all species. White pine, yellow-poplar, northern red oak, and white oak appear to perform better in the loose dumped cells.

Table 3 presents the 1999 summary of soil properties for each cell in the Starfire Mine High-Value Tree Reclamation Project. It identifies the average depth to refusal, average penetration resistance and average dry bulk density as related to surface conditions in the cell. Table 4 presents the tree survival rate associated with each cell's surface conditions.

Figure 1 appears to initially verify that tree survival rate is closely associated with depth to refusal. Table 5 presents the average survival rate as a function of depth to refusal. There appears to be some overlap in survival rates between loose dump cells and those that were loosely compacted (strike-off cells). Compacted cells yield results markedly lower than loosely compacted cells. Many factors affect the characteristics of compacted cells, but primary among them are operator experience and size and weight of the equipment (dozer) used to grade the spoil.

It is hoped that one or a combination of the above measurements will eventually allow the determination of productivity of a site when it is prepared for tree planting. This could be determined on a species specific basis to assist in selecting trees which would maximize economic returns on a reclaimed site with certain soil conditions.

The three compacted 2.5 acre plots (7, 8, and 9) have been generating storm event data since the spring of 1998 for the surface runoff and erosion portions of this study. Initially, this was a primary focus of the research effort. Coupled with this is the routine maintenance and inspection of the instrumentation with each site visit. Site visits are conducted after a rainfall event greater than 0.3 inches in the nearby towns of Jackson and

Quicksand, Kentucky, as recorded on the University of Kentucky Agricultural Weather Site on the Internet. It has been necessary to replace two data loggers due to moisture problems, and three liquid level actuators have been replaced due to cable damaged caused by animals or human vandalism. In addition, vegetation growth has been periodically removed from the inlet and outlet of the flumes so that flow stages through the flumes would not be affected by backwater.

The sediment control enhancement by passive dewatering systems is linked to the surface runoff project in that the runoff generated from the three compacted spoil sites is directed to a pond with influent and effluent samplers and data loggers. The focus has been to collect storm samples and flow data and monitor the sealing effort on the pond floor. The performance of the sand filter system in the pond has been the focus of data acquisition to date, and the bulk of the effluent data has been generated for the passive dewatering control system. Other outflow data has been obtained from the perforated

Table 2. Average survival, height, and growth in constructed, bermed cells (1997-1999).

	Species	1997 average survival %	1998 average survival %	1999 average survival %	1997 average height (cm)	1998 average height (cm)	1999 average height (cm)	1998 average growth (cm)	1999 average growth (cm)
loose dumped spoil	white pine	92	85	82	36	40	66	4	25
	white ash	95	88	86	40	51	71	11	19
	black walnut	64	74	76	80	79	66	-1	-13
	yellow-poplar	93	86	77	24	32	48	8	16
	royal paulownia	35	37	40	10	56	105	47	49
	white oak	88	76	87	28	30	39	1	10
	northern red oak	99	86	85	28	31	41	4	10
		81	**76**	**76**	**35**	**46**	**59**	**11**	**13**
rough graded spoil	white pine	87	79	51	34	36	46	2	11
	white ash	98	86	85	41	54	71	13	17
	black walnut	86	92	61	82	72	63	-10	-9
	yellow-poplar	94	63	61	24	29	40	5	10
	royal paulownia	52	48	44	11	63	92	52	29
	white oak	73	76	78	28	31	33	3	2
	northern red oak	96	71	70	29	28	31	-1	3
		84	**74**	**64**	**35**	**45**	**52**	**9**	**7**
compacted spoil	white pine	37	27	12	34	33	32	-1	-1
	white ash	91	87	85	40	43	45	3	2
	black walnut	41	39	34	79	51	31	-28	-20
	yellow-poplar	59	50	30	27	25	21	-3	-4
	royal paulownia	21	11	12	23	56	82	33	26
	white oak	49	38	49	27	24	16	-3	-7
	northern red oak	82	66	51	27	20	17	-7	-3
		54	**45**	**39**	**37**	**36**	**31**	**-1**	**-5**

riser when a storm event was large enough to raise the pond stage up to the elevation of the perforations (3 ft./min.) while still dewatering through the sand filter. These storm events facilitate documenting the combined performance of this control system. A pressure transducer stage recorder has been installed in the pond so a stage-discharge relationship can be ascertained for the spillway configuration.

Tipping bucket data acquisition in the field determination of infiltration has been continued since the Spring of 1998 and will continue for the duration of the project. The tipping buckets are read after storm events and/or upon each site visit, generally every two or three weeks. All data are tabulated on a spreadsheet showing incremental and cumulative infiltration volumes and rates. There has been considerable variation in infiltration volumes. The high variability is likely due to the presence of large rocks creating preferential flowpaths and the location of depressions created during spoil dumping and grading operations.

Twenty of the infiltration plots are exhibiting excellent infiltration data. The remaining eighteen plots have infiltration rates that are suspect. An intense field investigation will be conducted to determine why flows are so low on the 18 plots. The construction of 38 plots was based on the expected loss of 14 plots. The remaining plots will provide an excellent experimental design. Loss of plots was expected since spoil was dumped from 85-ton trucks with some rocks being larger than automobiles.

There has been noticeable weathering of rocks over time on these sites. Weathering and exposure has broken down a significant amount of shale and sandstone, making it easier for trees to increase and spread their root mass. The trees have shown promising growth patterns.

Equipment maintenance has consisted of tipping bucket inspections and repair. Tipping bucket counter screws have become loose or fallen out on some buckets, rocks occasionally fall and block the tipping area, and other counters have become inactive due to broken springs or weathering or other malfunctions. All repairs are made and documented at the time of the site visit to minimize the interruption of data collection.

Trees have been planted throughout the area of the 38 infiltration cells. The trickle irrigation system have been designed. All areas have been mulched. Intallation of the irrigation system awaits the

Table 3. Summary of soil properties for each cell in the Starfire Mine High-Value Tree Reclamation Project.

Cell	Average depth to refusal (m)	Avg. Penetration Resistance (psi)	Avg. dry bulk density (pcf)	Surface condition
1	5.8	624	101.7	Srike-off
2	8.7	662	93.4	Loose dumped
3	11.0	549	102.5	Loose dumped
4	11.3	731	97.5	Loose dumped
5	6.2	769	104.5	Strike-off
6	12.8	699	109.2	Strike-off
7	3.8	657	110.9	Compacted
8	5.4	861	113.4	Compacted
9	4.6	811	114.4	Compacted
10	8.2	491	75.8	Dozer ripped
11	13.3	569	96.4	Dozer ripped
12	7.9	590	86.6	Tractor ripped
13	8.8	625	83.6	Tractor ripped
14	10.7	504	98.3	Dozer ripped
15	8.7	575	103.2	Tractor ripped
Spoil Pile	6.3	579	103.2	Drag line spoil

Table 4. 1999 tree species survival rate for the Starfire Mine High-Value Tree Reclamation Project.

Cell	Tree Species Survival Rate (%)							Average Survival Rate (%)	Surface Condition
	Black Walnut	Royal Paulownia	Northern Red Oak	White Ash	White Oak	White Pine	Yellow-poplar		
1	57	37	66	78	61	47	47	56	Strike off
2	76	44	82	89	85	87	80	78	Loose dumped
3	80	58	87	77	89	86	80	80	Loose dumped
4	73	20	88	91	86	72	71	72	Loose dumped
5	53	34	61	85	87	43	63	61	Strike off
6	74	63	871	93	85	62	73	76	Strike off
7	1	9	31	71	8	1	4	18	Compacted
8	31	19	58	96	57	7	48	45	Compacted
9	71	7	64	88	83	30	37	54	Compacted
10(1)	---	---	---	---	---	---	---	---	Dozer ripped
11(1)	---	---	---	---	---	---	---	---	Dozer ripped
12(1)	---	---	---	---	---	---	---	---	Tractor ripped
13(1)	---	---	---	---	---	---	---	---	Tractor ripped
14(1)	---	---	---	---	---	---	---	---	Dozer ripped
15(1)	---	---	---	---	---	---	---	---	Tractor ripped
Spoil Pile(1)	---	---	---	---	---	---	---	---	Drag line spoil
Survival Rate (%)	57	32	69	85	71	48	56	60	

Notes: (1) Cells planted in 1998. First year growth data not collected.

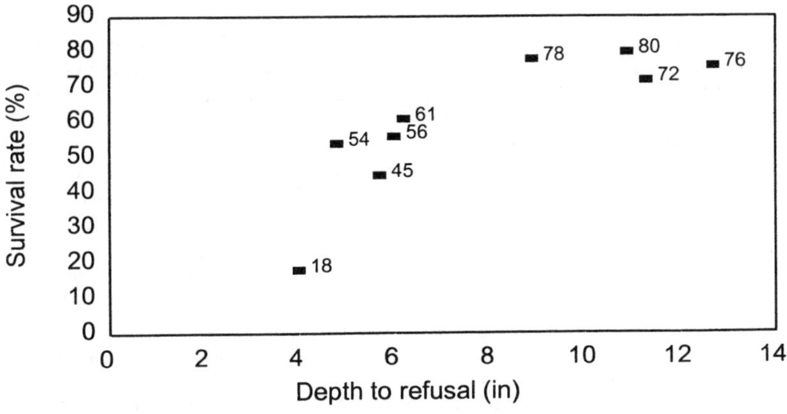

Figure 1. 1999 Tree Survival rate as a function of depth to refusal.

Table 5. 1999 tree survival rate as a function of depth to refusal in the Starfire High-Value Tree Reclamation Project.

Cell	Average Depth to Refusal	Average Survival Rate (%)	Surface Condition
1	5.8	56	Strike off
2	8.7	78	Loose dumped
3	11.0	80	Loose dumped
4	11.3	72	Loose dumped
5	6.2	61	Strike off
6	12.8	76	Strike off
7	3.8	18	Compacted
8	5.4	45	Compacted
9	4.6	54	Compacted
10	8.2	---	Dozer ripped
11	13.3	---	Dozer ripped
12	7.9	---	Tractor ripped
13	8.8	---	Tractor ripped
14	10.7	---	Dozer ripped
15	8.7	---	Tractor Ripped
Spoil Pile	6.3	---	Drag line spoil

construction of the water supply pond. Installation of the irrigation system will be accomplished withing three days of the pond completion. Data acquisition will commence upon completion of these two tasks.

4 CONCLUSION

It is important to understand that this is a work in progress and final conclusions cannot be drawn at this point. There are some very positive trends being observed in the data being collected from this study. There is little doubt that compaction has a very negative effect on both survival and growth for the species selected for this study. We are seeing at this time that light compaction is not detrimental to some species at this stage of their development. We also see that no compaction is beneficial to those species not affected by light grading. Time will be the determinant of which system works best, but evidence from past research indicates that growth, yield, and soil formation are increased by having little or no compaction from heavy equipment, and in areas of no compaction, growth and yield of trees exceeds that of undisturbed (background) forests since uncompacted spoil allows greater root development than is found in natural forest stands.

REFERENCES

Ashby, W. C., C. A. Kolar, M. L. Guerke, C. F. Persell and J. Ashby. 1978. Our Reclamation Future With Trees, Coal Extraction, and Utilization. Research Center, Southern Illinois University at Carbondale. 101 pp.

Barnhisel, R. I., J. L. Powell, and D. H. Hines. 1986. Changes in Chemical and Physical Properties of Two Soils in the Process of Surface Mining. Transactions, American Society for Surface Mining and Reclamation.

Barnhisel, R. E. 1988. Correction of Physical Limitations to Reclamation. Reclamation of Surface-Mined Lands. L. R. Hossner, editor. Vol. 1. CRC Press, Inc. Boca Raton, Florida, pp. 192-211.

Burger, J. A. and J. L. Torbert. 1992. Restoring Forests on Surface-Mined Lands. Virginia Polytechnic and State University Pub. # 460-123. 16 pp.

Dollhopf, D. J. and R.C. Postle. 1988. Physical

Parameters that Influence Successful Minesoil Reclamation. Reclamation of Surface-Mined Lands. L. R. Hossner, editor. CRC Press, Inc. Boca Raton, Florida, pp. 93-99.

Dunker, R. E., C. L. Hooks, S. L. Vance, and R. G. Darmody. 1992. Rowcrop response to high traffic vs. low traffic soil reconstruction systems. Proceed. of 1992 Natl. Symp. on Prime Farmland Reclamation, St. Louis, MO.

Hooks, C. L., R. E. Dunker, S. L. Vance, and R. G. Darmody. 1992. Rowcrop response to truck and scraper hauled root media systems in soil reconstruction. Proceed. of 1992 Natl. Symp. on Prime Farmland Reclamation, St. Louis, MO.

Environmental Issues and Management of Waste in Energy and Mineral Production, Singhal & Mehrotra (eds)
 ISBN 90 5809 085 X

Stabilisation of erodible soil by fly ash and blast furnace slag

B. Indraratna & W. Salim
Civil Engineering Division, University of Wollongong, N.S.W., Australia

ABSTRACT: Residual soils are found on large terrains which are often subjected to considerable erosion. This paper presents a detailed laboratory evaluation of the effect of fly ash and blast furnace slag on two different soils: (a) colluvium (sandy loam) from New South Wales, Australia and (b) erodible dispersive clay from northeast Thailand. Geotechnical tests were conducted to determine the compaction characteristics and the compressive and shear strength properties of the blended and natural soil specimens. The effect of the above mentioned industrial wastes on the rate of erosion and on the associated pH levels is also investigated. Fine grained fly ash is found to be useful as a void filler if used in substantial quantities, whereas self-hardening milled slag is more effective in terms of improving the internal friction angle of the treated soil.

1. INTRODUCTION

Erodible soils have sometimes caused massive soil loss from unprotected land areas and have contributed to the collapse of foundations and embankments in certain parts of the world. In New South Wales (NSW), Australia, fine-grained colluvium of low shear strength (often called sandy loam) occupies large terrains. In the absence of effective ground protection, these soils are subjected to considerable erosion and mass movement. Graham (1989) estimated that about 12% of available land is potentially vulnerable to moderate gully erosion, while a further 3% has already been affected by moderate to severe rill and sheet erosion. Landslides involving such soils and the transport of eroded sediments in selected waterways have also been documented (Chowdhury and Young, 1987). Although hydrated lime is conventionaly used in stabilising these deposits (Fordham, 1993), the effectiveness of fly ash and milled slag (blast furnace) in soil stabilisation has not been properly investigated in Australia.

Highly dispersive soil is found in the Khorat plateau of Thailand, and the effectiveness of using fly ash in stabilising this silty clay was studied. Fly ash is produced as a by-product of coal (lignite) combustion, and was obtained from Mae Moh power plant. Fine-grained silty soils can be rapidly eroded by the segregation of individual particles (dispersed in suspension) even under mild hydraulic gradients. In the northeast part of Thailand, several irrigation dams have been affected by surface and internal erosion (Cole et al., 1977). These structures often show pronounced rill and gully marks. Highly erodible and often dispersive silty clays are also found as flood plain and terrace deposits (Fernando, 1979), and their total clay fraction is generally less than 30%.

The aim of this paper is to investigate the effect of additives on the stabilisation of the two above mentioned soils of highly erodible nature. The emphasis was given on the use of industrial wastes such as fly ash and blast furnace slag as additives. The use of waste materials in soil stabilisation is economically and environmentally attractive, because of the direct benefits that can be derived in waste management schemes through reducing the volume of spoil tips (eg. slag embankments) and the area of disposal sites. Currently, the use of these by-products is limited. For example, coarse slag is used as road base and as lightweight aggregates in concrete, whereas milled slag (pulverized) is used as a partial replacement for Portland cement, and for brick making. In the steel manufacturing industry of NSW, the output of fly ash and blast furnace slag is overweighs than its consumption in industry re-use.

Finely pulverized fly ash is the major by-product of coal combustion plants, whereas slag is a by-product of the steel industry, where it is separated from molten iron in the blast furnace. The environmental implications of industrial by-products should not be overlooked. If fly ash is disposed as raised embankments, potential air pollution due to strong winds and the risk of embankment failures under heavy rainfall cannot be ignored. More than 9 million tonnes of fly ash and 3.5 million tonnes of slag are produced annually in the state of New South Wales alone, of which less than 10% is used in the construction industry. It should be noted here that in Australia, fly ash and blast furnace slag do not contain significant levels of toxic materials or acid forming compounds such as pyrites and sulphur. Therefore, the utilisation of these materials for soil improvement will not pose a risk of groundwater contamination, or increasing soil acidity.

1.1 Properties of erodible soils

1.1.1 Colluvium

This soil was collected near the the town of Picton, and is termed as the 'Picton group' of high erosion potential. The dominant landform consists of steep concave upper slopes and irregular lower slopes with colluvial benches. The soil is a finely textured, red or brown sandy loam of porous structure with a very small component of angular gravels (2 - 5mm). The crests and upper slopes are generally covered with up to 80 cm of this sandy loam, except where severe erosion had reduced its depth to 30 - 40 cm. Hazelton and Tille (1990) have estimated the annual soil loss to be around 300 tonnes/ha on the steep slopes, and up to 170 tonnes/ha for the exposed sub-soils. Throughout the terrain, sheet, rill, gully and tunnel erosion marks are clearly observed.

Fine-grained soil samples were obtained at 0.5m depth, where significant gully erosion had occurred. The liquid and plastic limits of the soil were determined to be 37% (Casagrande apparatus) and 20%, respectively, giving a plasticity index of 17%. Linear shrinkage was not more than 12.5% of the specimen length, although conventional laboratory testing indicated a shrinkage limit of 46%. Shrinkage in the field is clearly evident where cracking patterns and marked depressions are noticed on dried out surfaces. This high shrinkage level (with openings up to 20 mm) in itself contributes to the erosion problem where the 'cracked' soil is more susceptible to disintegration.

The specific gravity of the soil was estimated to be 2.68, which is higher than that of fly ash (2.15-2.45) and milled slag (2.3-2.4). As a result, the blended soil is expected to have a specific gravity smaller than that of the natural soil. As the natural soil was also found to be acidic (pH of 4.9-5.1), alkaline additives such as milled slag would assist in raising the pH levels of the blended matrix. This is beneficial because a neutral or slightly alkaline soil does not make corrosion of buried metal elements and concrete members in the ground.

The particle size distribution (PSD) of a typical specimen is illustrated in Fig. 1, where the fines below 75 microns have been obtained by the particle size analyser and hydrometer tests (Indraratna, 1996). The PSD curves of milled slag, fly ash and conventional lime (hydrated) are also plotted for comparison, which indicate that the dispersed soil particles are clearly smaller than the grain sizes of additives. This erodible loam consists predominantly of silt (49%) and clay (39%) sized particles with 12% fine to medium sand. According to the Unified Soil Classification System, the soil falls into the category of an inorganic clay of low plasticity (CL), and it is indeed such soils that are prone to signifcant erosion (Goldman et al., 1986).

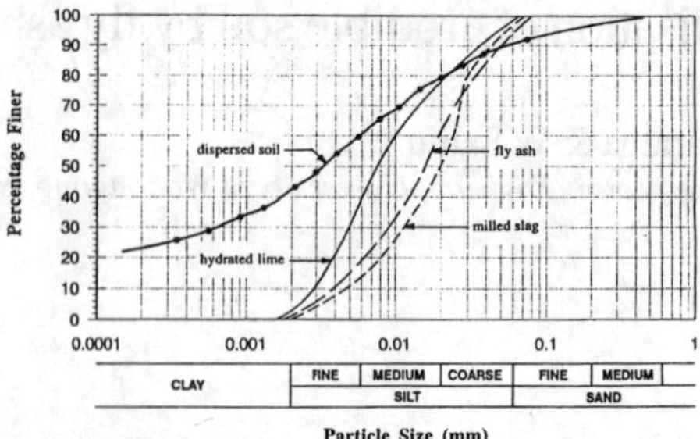

Figure 1. Particle size analysis of colluvial loam in comparison with the grain size distribution of additives (Fordham, 1993).

1.1.2 Properties of dispersive silty clay

The Khorat area in northern Thailand is an undulating plateau interspersed with isolated hills and wide sloping valleys. The extremely high rate of erosion of the clayey formations are associated with their natural dispersive chracteristics, attributed to the high percentage of exchangeable sodium ions. The dispersiveness of these soils is a result of sodium adhering to montmorillonite than the presence of Na+ in the pore water. The soil obtained from the Khorat plateau has an average exchangeable sodium percentage (ESP) greater than 60%, and the cloudiness associated with stagnant water pools is definitely indicative of the dispersive nature of these soils (Indraratna et al., 1991a). A large number of small irrigation dams and road embankments in the area have been affected with significant erosion, and the gully and rill marks are often pronounced on these earth structures. Embankments and foundations composed of dispersive soils are particularly vulnerable to failure initiated by piping.

The high erosion rate of these dispersive soils is also associated with their low plasticity, having a typical liquid limit of 38%, a plastic limit of 16% and a shrinkage limit of 43%. The considerably cracked, dry parts of the terrain emphasise the high shrinkage level of the soil. The specific gravity of the soil was found to vary between 2.65 and 2.68, and a near-neutral pH level of 6.7 was observed. The particle size distribution of this soil is similar to the Australian sandy loam described earlier, characterised by a silt content of 47% and a clay content of 43%, the remainder (10%) being fine to medium sand. This soil is classified as 'CL' in the Unified Soil Classification System as a silty clay of low plasticity.

2. WASTE MATERIALS AS ADMIXTURES

Use of lime and cement as soil admixtures is popular in surface treatment and deep mixing. The lime column technique employed in foundation engineering has been effective in enhancing bearing capacity (Broms, 1982; Honjo et al., 1991; Yamanouchi et al., 1982). The use of hydrated lime in deep mixing has been successful in Japan (Saitoh et al., 1985). Soft clay studies in Southeast Asia have shown that consolidation settlements can be reduced by lime treatment (Balasubramaniam et al., 1989). In contrast to lime treatment, limited studies have been conducted on the use of industrial waste materials such as fly ash, milled slag and mine tailings in ground improvement (Indraratna, 1996). For instance, the properties of fly ash with regard to strength and compaction characteristics has been discussed by Natt and Joshi (1984) and Indraratna et al. (1991b). For a uniform sand found in New South Wales, Hausmann (1990) has shown that the porosity and the unconfined compressive strength can be optimized by using a high content of fly ash approaching 20%. In Australia, the utilization of blast furnace slag (from steel industry) in soil improvement is hardly documented.

Mae Moh fly ash shows slightly pozzolanic and self-hardening characteristics (ASTM Class C) because of its higher free lime content (1.5-2.0%). In comparison, NSW fly ash is non-pozzolanic (ASTM Class F: free lime content < 0.2%). While the latter was characterised by a low specific gravity of 2.15, the former showed a significantly higher specific gravity of 2.45. This was due to the presence of a larger quartz content and a lesser carbon content. Milled (pulverised) slag supplied by the Australian Steel Mill Services, NSW indicated significant self-hardening with time. The unconfined compressive strength of milled slag increased with time at a decreasing rate (Fig. 2) for specimens compacted at their natural moisture content of 10%. This self-hardening property is due to the presence of free lime and gypsum in milled slag, which undergo hydration in the presence of moisture.

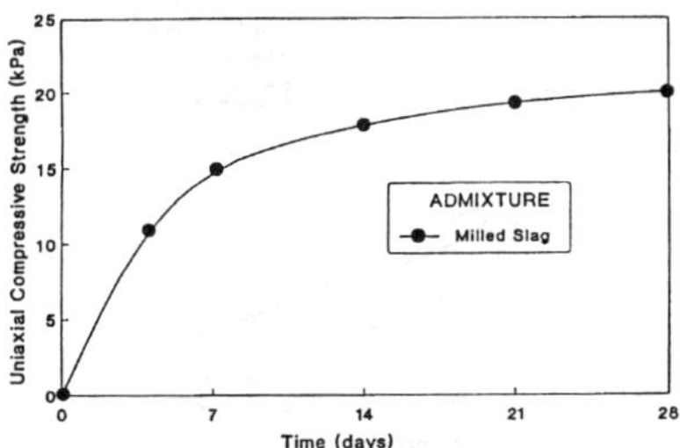

Figure 2. Time dependent increase in uniaxial compressive strength of milled slag due to self-hardening (Indraratna, 1996).

3. BEHAVIOUR OF BLENDED SOIL

The geotechnical behaviour of treated soil with fly ash and milled slag is discussed in this section. In particular, reference is made to compaction, unconfined compression strength and triaxial shear strength characteristics.

3.1 Alkalinity

Colluvium (sandy loam) from NSW, Australia is naturally acidic (pH = 4.9). Therefore, it is relevant to investigate the effect of additives on the pH level of the treated soil. The addition of fly ash and milled slag increases the pH of blended soil (Fig. 3), although the effect is not as marked as conventional lime treatment. It may be noted that the use of hydrated lime exceeding 3% (pH > 10) can be detrimental to some vegetation. At pH levels exceeding 12, certain geotextiles and reinforcement grids are prone to alkaline attack. Fly ash shows the least effect on pH levels, whereas 2-5% of milled slag makes the soil slightly alkaline.

3.2 Compacted Density

Compaction tests were carried out on blended specimens to study the effect of fly ash and milled slag on the dry density, at varying moisture content. Figure 4 illustrates the role of milled slag and fly ash on the dry density of sandy loam. For all treated specimens, the dry density decreased with an increase in additives. This is partly because of the lower specific gravity of the additives in comparison

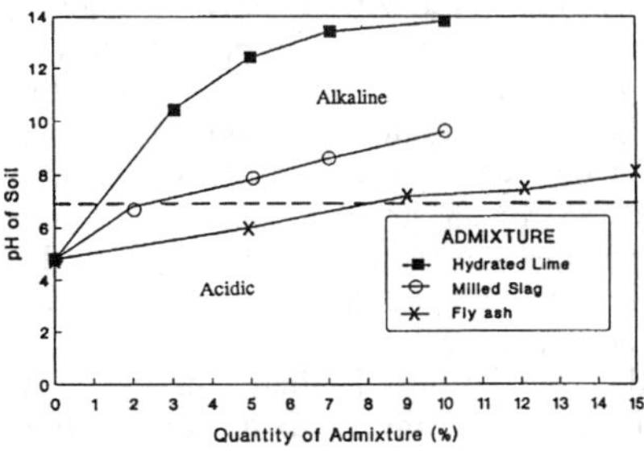

Figure 3. Effect of additives on pH levels of colluvial loam (Indraratna, 1996).

with that of the soil. Milled slag consumes water during the hydration process, thereby increasing the optimum moisture content. This can be considered as an advantage when working with wet fills. The specimens blended with milled slag behave differently from the fly ash treated specimens. The compaction curves move closer towards the zero air void line (wet side of the optimum) as the milled slag content is increased. The decreased dry density of the blended soil can be considered as beneficial in reducing mass movements, because at limit equilibrium, the factor of safety is inversely related to the bulk density. The NSW fly ash is not very effective in improving the compactability of the colluvium. This is because the maximum dry density continues to decrease and deviate further from the zero air void line as the fly ash content is increased.

The effect of lignite fly ash on the dry density of the dispersive soil from Thailand is shown in Fig. 5. These curves indicate the opposite effect, where the increase in fly ash content generally increases the dry density of the treated soil, while reducing the optimum moisture content. Increasing dry density can be obviously attributed to the higher specific gravity of the Mae Moh fly ash (2.45) in contrast to the NSW fly ash (2.15). It is also clear from Fig. 5 that increasing the fly ash content above 10% has insignificant effect on the dry density.

3.3 Compressive strength

The maximum unconfined compressive strength (UCS) of treated sandy loam is related to the percentage of fly ash and slag (Fig. 6). The effect of conventional lime addition is also shown for comparison in Fig. 6. Increasing the slag content beyond 2% is associated with a gradual drop in UCS. Although, the effect of slag is not as dramatic as lime, milled slag shows a 20-25% increase in strength at 2% slag content. This is economically impressive, considering the relatively low cost of milled slag compared to commercial hydrated lime. The effect of fly ash on increasing the uniaxial compressive strength is not very significant. On the contrary, as the fly ash content exceeds 6%, the strength begins to decrease substantially. It may be noted that the NSW fly ash is non-pozzolanic (no self-hardening). In contrast to milled slag, the addition of about 5% hydrated lime almost doubles the apparent UCS, while further increase in lime content causes a drop in strength. Excessive lime can be detrimental in soil stabilisation if the available silica in the soil is consumed during hydration (Van Impe, 1989).

Lignite fly ash from Thailand is slightly pozzolanic. Its effect on the uniaxial compressive strength of the Khorat dispersive clay is considerable (Fig. 7). Even though a substantial gain in

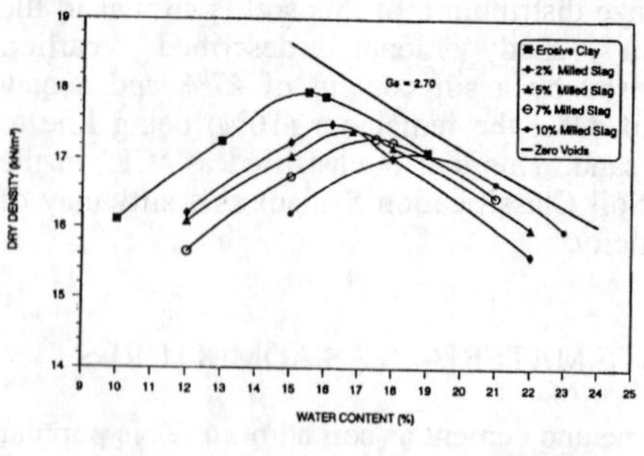

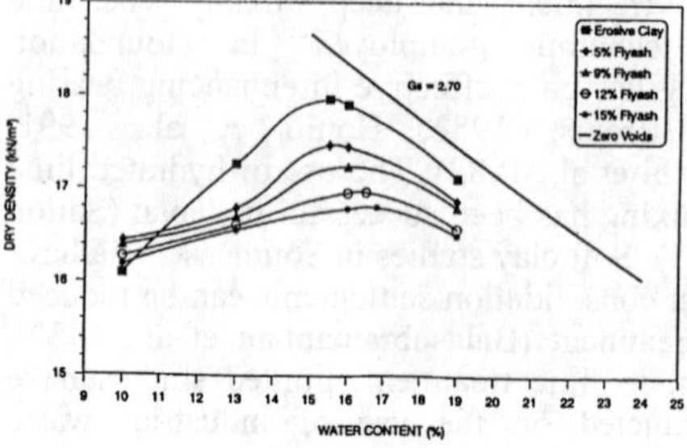

Figure 4. Variation of dry density with moisture content of colluvial loam blended with additives (Fordham, 1993).

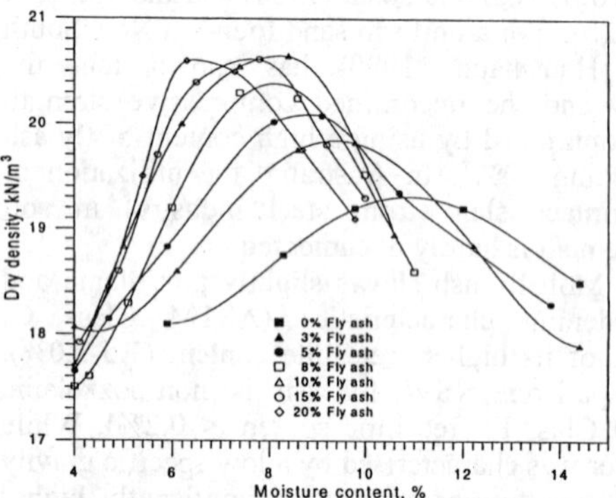

Figure 5. Effect of lignite fly ash on the dry density of dispersive soil.

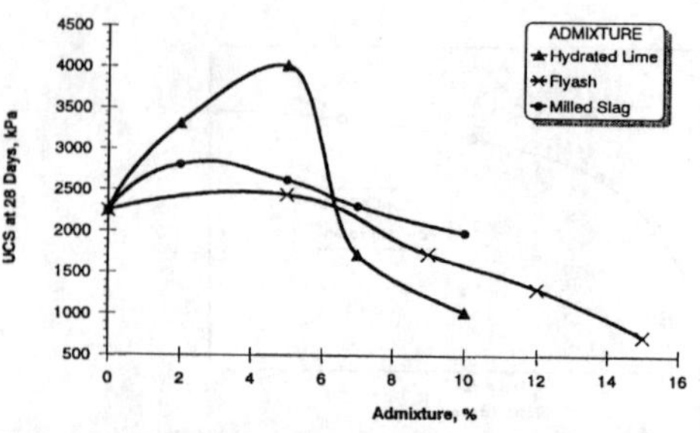

Figure 6. Effect of additives on compressive strength of colluvial loam (Indraratna, 1996).

compressive strength and initial stiffness is associated with the increase in fly ash content, the increase in fly ash content increases the brittleness of the treated silty clay (Fig. 7). Increased brittleness is not desirable if the stabilised soil is to be utilised as a landfill or as a structural fill. For instance, in the case of a dam core, enhanced brittleness causes tensile cracking, and increases the internal permeability. As Mae Moh fly ash is self-hardening, the effect of curing time on the compressive strength of the treated soil was also studied (Fig. 8). The gain in strength after three weeks of curing is not significant.

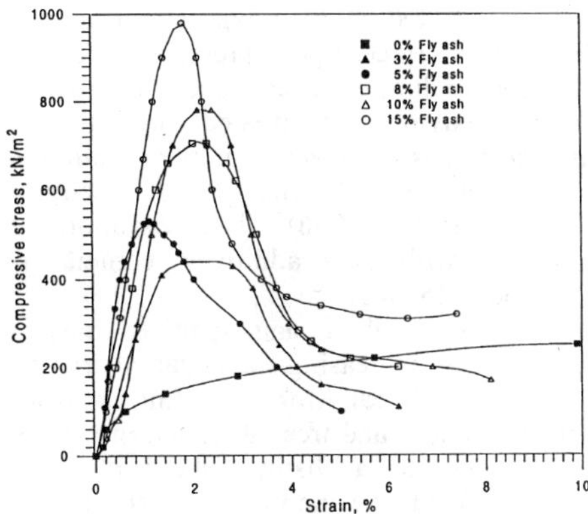

Figure 7. Effect of Mae Moh fly ash on compressive strength of Khorat dispersive clay.

3.4 *Erosion of fly ash blended soil*

In the laboratory, pinhole tests were conducted on Khorat dispersive soil samples (natural and treated) to assess how fly ash treatment affects the rate of erosion. As Mae Moh fly ash is pozzolanic, it was anticipated that fly ash treatment would increase the overall cohesion, thereby enhancing the resistance to erosion. Standard cylindrical test specimens were prepared (38mm in diameter with a 1 mm co-axial hole along the length), which were subjected to hydraulic heads of 50, 180 and 380 mm. The time-dependent flow quantities through the specimens were monitored, and the effluent turbidity was recorded. The enlargement of the pinhole diameter corresponding to the applied hydraulic heads, and the resulting seepage rate through the pinhole were considered to be indicative of the degree of erosion. In order to quantify the effect of fly ash on the erosion rate, the dispersive silty clay was mixed with various proportions of fly ash, ranging from 3% to 20% by weight.

The untreated soil is highly erodible due to its natural dispersive behaviour, and falls in the ASTM category of D1 or D2. Preliminary tests revealed that the addition of a small amount of fly ash (3-4%) is enough to categorise the treated soil to the less erodible classes. With fly ash contents exceeding 5%, the treated soil was characterised with significantly reduced flow rates through the pinhole. The stabilised soil could be classified as non-dispersive and non-erodible (ND1). However, at large fly ash contents (> 10%), the flow rates increased again, which can be attributed to the decrease in overall cohesion due to excessive fly ash in the blended matrix. A fly ash content of about 8% gave the optimum resistance to erosion, as verified by the measured turbidity of the pinhole effluent. The variation of flow rate with the fly ash content under three different hydraulic heads was measured (Fig. 9). Based on this data, the optimum fly ash content is around 8%, which is in agreement with the turbidity measurements.

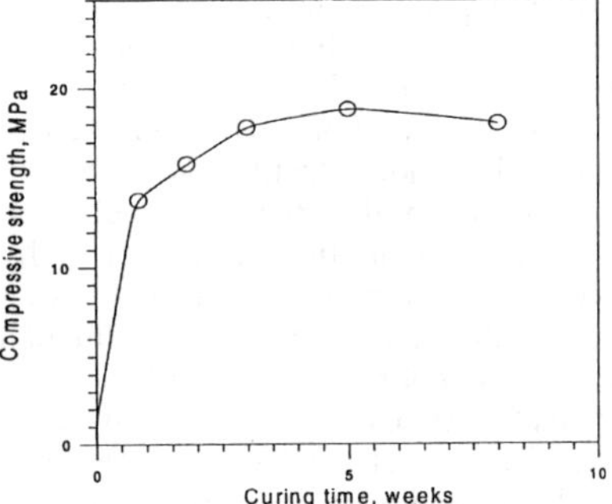

Figure 8. Effect of curing time on the compressive strength of treated dispersive clay (Indraratna et al., 1991a).

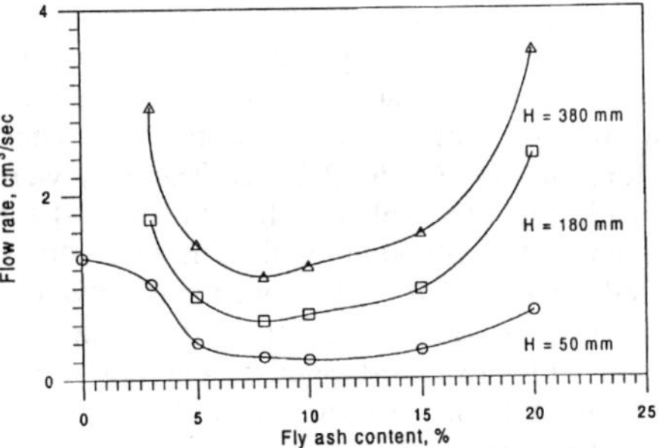

Figure 9. Variation of measured flow rate with fly ash content in the pihole apparatus (Indraratna et al., 1991a).

3.5 *Shear strength characteristics*

A large number of consolidated-drained triaxial tests (38mm x 76mm specimens) were carried out on compacted specimens to investigate the effect of fly ash and milled slag on the drained shear strength of the treated colluvium. Specimens were subjected to confining pressures of 100, 200 and 400 kPa, respectively, and tested at small axial strain rates of 0.01-0.03% / min. Excess pore pressures were not allowed to develop during the slow shearing process. Stress-strain behaviour of treated and natural colluvium specimens compacted at their optimum moisture content of 15% has been discussed elsewhere (Indraratna, 1996). The variation of friction angles with the additives content is discussed in the following.

The fly ash and milled slag additives have a significant effect on increasing the apparent friction angle, based on the linear Mohr-Coulomb analysis. All specimens (natural and treated) tested in triaxial compression indicated a distinct shear plane at failure. As expected, the inclination of the failure plane was influenced by the confining pressure and the quantity of additives. Unlike the slag-treated specimens, a more brittle behaviour was observed for fly ash-treated specimens, where a post-peak sudden drop of strength was evident at confining pressures exceeding 200 kPa. Irrespective of the fly ash content, these specimens failed within a narrow range of axial strains (3-5%). The slag-treated specimens indicated a more ductile nature during shearing. More significantly, they showed well developed shear planes at failure. A number of fly ash-treated specimens also indicated tensile cracking in addition to the ultimate shear failure. Although tensile cracking is often attributed to end friction, it is anticipated that pozzolanic materials such as hydrated lime and milled slag which undergo self-hardening would provide greater resistance to tensile splitting.

The behaviour of conventional lime-treated specimens was also studied in triaxial compression for comparison with the fly ash and milled slag treated samples (Indraratna, 1996). In comparison with fly ash-treated soil, specimens blended with lime and milled slag showed a ductile, strain-softening response in traixial compression. Their axial failure strains (3% to 9%, see Table 1) were more sensitive to the applied cell pressure and the additives content. Although smaller in peak stress, the milled slag-treated soil specimens revealed slightly larger plastic strains prior to failure, in comparison with the lime-treated counterparts. The linear Mohr-Coulomb criterion was employed to represent the failure envelopes of all specimens, both natural and treated. As shown in Fig. 10, all stabilised samples were characterised by enhanced friction angles (ϕ) in comparison with the natural soil. A maximum friction angle (47°) was obtained for the 5% slag-treated specimens, while the corresponding friction angle for the fly ash treated specimen was 42°. With regard to the internal friction angle, the optimum milled slag content is in the vicinity of 5% (similar to hydrated lime). The optimum fly ash content, however, is considerably higher at about 12%.

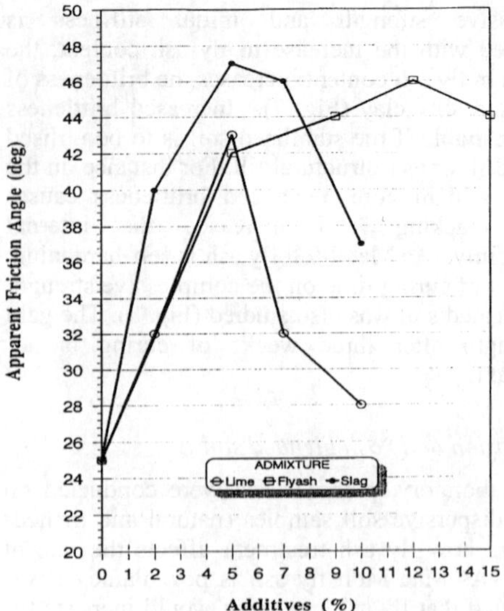

Figure 10. Effect of additives on apparent friction angle of treated soil (Indraratna, 1996).

Table 1. Effect of lime and slag on failure strains of blended specimens.

Type of Specimen	Failure Strain, %		
	$\sigma_3 = 100$ (kPa)	$\sigma_3 = 200$ (kPa)	$\sigma_3 = 400$ (kPa)
Natural Soil	3.4	3.9	5.2
2% lime	3.6	4.2	5.8
slag	3.8	4.8	6.2
5% lime	3.3	3.9	4.2
slag	2.7	4.3	5.9
7% lime	3.5	4.4	5.4
slag	4.4	5.7	9.0
10% lime	3.8	5.2	6.4
slag	4.4	4.6	6.5

Table 2. Effect of Additives on Cohesion Intercept.

	Apparent Cohesion Intercept, kPa							
	Content of additives, %							
Additive	0	2	5	7	9	10	12	15
Hydrated lime	150*	400	360	420	-	460	-	-
Flyash	150*	-	320	-	250	-	250	200
Milled Slag	150*	340	140	150	-	250	-	-

*Natural soil compacted around 15% moisture content.

Considering the effect of these additives on the shear strength of the stabilised soil, it may be concluded that milled slag is the most effective at low content of additives. At relatively high proportions of additives exceeding 12%, fly ash is effective partly due to the silica content that enhances the internal shearing resistance (mechanical interlocking of particles) by reducing the porosity of the blended mix. The improvement of internal shear strength by milled slag is, however, attributed to the combined effect of cementation bonding and void ratio reduction (Indraratna, 1996). As shown in Table 2, the apparent cohesion (c) of the treated specimens does not reveal a specific relationship to the type or content of the additives. The untreated (natural) soil has a cohesion intercept of 150 kPa, and has the lowest friction angle of 25°. According to the data, none of the industrial wastes seem to increase the apparent cohesion. This suggests that the role of the additives discussed in this paper is mainly a function of increasing the shear strength through the internal friction angle. Further studies are necessary to quantify the effect of these additives on the cohesion intercept.

4. ECONOMIES OF SOIL TREATMENT

The cost of soil treatment by waste materials can be quantified on the basis of the (a) uniaxial compressive strength and (b) shear strength of the blended matrix. On one hand, based on the uniaxial compressive strength (Fig. 6), the optimum contents of milled slag and fly ash are close to 2% and 5%, respectively. With regard to the internal friction angles (Fig. 10), the optimum additive contents for slag and fly ash are close to 5% and 12%, respectively. Without doubt, these results imply that a substantially larger quantity of fly ash is required to obtain similar benefits associated with milled slag. Even though milled slag (A$ 100-110/tonne) is slightly more expensive than fly ash (A$ 65-75/tonne), it is clear that milled slag is still more effective and economical than fly ash for treating colluvium. Nevertheless, both milled slag and fly ash are considerably cheaper than commercial lime treatment, where the cost of hydrated lime is greater than A$145 per tonne.

As shown in Fig. 10, it was milled slag that provided the greatest friction angle (47°) for treated specimens. The choice of 5% milled slag treatment will provide the maximum shearing resistance as well as the lowest cost. In the case of slope stabilization and improvement of the bearing capacity of foundations, the friction angle is more important than the uniaxial compressive strength. In terms of improving the uniaxial compressive strength, neither milled slag nor fly ash is comparable to conventional lime treatment (Fig. 6).

Considering the high cost of hydrated lime, it may be concluded that conventional lime treatment is the most suitable for highway embankment construction, where both the compressive strength and the shear strength are equally important in preventing failure under heavy traffic loads.

5. CONCLUSIONS

In this paper, the effects of milled slag and non-pozzolanic fly ash on the behaviour of two different soils were studied, namely: (a) colluvial soil from NSW, Australia, and (b) erodible dispersive soil from Thailand. Milled slag, like hydrated lime, improves the strength and resistance to erosion of soil as a result of hydration induced cementation. In contrast, non-pozzolanic fly ash which has little or no free lime content (ASTM Class F) shows limited effectiveness in improving the properties of the colluvial loam. However, Mae Moh fly ash (ASTM Class C) used to stabilise the dispersive silty clay of Thailand was effective in enhancing its compressive strength and shear strength, apart from its resistance to erosion.

Milled slag increases the shear strength and uniaxial compressive strength substantially at the optimum percentage of say 5%. Non-pozzolanic fly ash does not significantly increase the time-dependent compressive strength, although it decreases the porosity of the blended soil. Therefore, it can still be regarded as a useful void filler which improves internal friction. A small percentage of milled slag (about 5%) is sufficient in achieving the optimum friction angle of the blended specimens (47°). In comparison, a 5% fly ash (Class F) gives an apparent friction angle of 42°, which is still a marked improvement of the natural loam which would otherwise have a friction angle of 25°.

In view of relative costs, in the stabilisation of the colluvial loam from NSW, milled slag offers the most attractive option, where the improvement of shear strength is imperative. Conventional lime treatment is most appropriate where marked improvement of both compressive and shear strengths is a criterion (eg. road embankment construction).

Naturally, the results of this study are directly influenced by the properties of the two soils tested, namely, colluvium and the dispersive silty clay. Other types of soils are expected to behave differently, depending on their plasticity (clay content), initial degree of compaction, and the grain size distribution among, other factors. Moreover, properties of both milled slag and fly ash can vary significantly from place to place. As a result, direct extrapolation of these data to other soil types should not be exercised without caution.

ACKNOWLEDGMENT

The laboratory assistance given by a number of technical staff at University of Wollongong (Alan Grant, Ken Cunningham and Ian Laird) is gratefully appreciated. The support provided by former students, Glen Fordham and N. Kuganenthira is much appreciated.

REFERENCES

Balasubramaniam, A.S, Bergado, D.T, Buensuesco, B.R., and Wang, W.C., 1989. Strength and deformation characteristics of lime-treated soft clays, *Geotechnical Engineering* (SEAGS), Vol. 20, No. 1, pp.49-65.

Broms, B., 1982. Lime columns in theory and practice, *Proc. Int. Conf. Soil Mechanics*, Mexico, pp 149-166.

Chowdhury, R.N. and Young, A.R.M., 1987. Field guide to slope instability in the Wollongong-Picton-Nattai region of New South Wales, *Report prepared for the 5th Int Conf and Field Workshop on Landslides*, Wollongong, 62p.

Cole, B.A., Ratanasen, C., Maiklad, P., Liggins, T.B., and Chirapuntu, S., 1977. Dispersive clay in irrigation dams in Thailand. *ASTM Special Technical Publication No. 623* (Sherard, J.L. and Decker, R.S., editors), pp. 25-41.

Fernando, M.J., 1979. Distribution and properties of dispersive soils in the vicinity of Lam Sam Lai dam site and their effects on embankment dam construction. *M.Eng thesis, Asian Institute of Technology*, Bangkok, Thailand.

Fordham, G., 1993. Utilization of waste products in geotechnical projects. *B.E. thesis, University of Wollongong*, NSW, Australia.

Goldman, S.J., Jackson, K. and Bursztynsky, T.A., 1986. Erosion and sediment control handbook, International Edition, McGraw-Hill.

Graham, O.P., 1989. Land degradation survey of NSW: 1987-1988, *Technical Report No 7, Soil Conservation Service of NSW*, pp. 1-30.

Hazelton, P.A. and Tille, P.J., 1990. Soil landscapes of the Wollongong - Port Hacking, *Soil Conservation Service of NSW*, Sydney, pp. 50-53.

Hausmann, M.R., 1990. Engineering principles of ground modification, McGraw-Hill, 632p.

Honjo, Y., Chen, C.H., Lin, D.G., Bergado, D.T. and Okamura, R., 1991. Behaviour ofimproved ground by the.deep mixing method, *Proc Kozai Club Seminar*, Bangkok, Thailand.

Indraratna, B., Nutalaya, P., Kuganenthira, N., 1991a. Stabilisation of a dispersive soil by blending with flyash, *Quarterly Journal of Engineering Geology*, No. 24, pp. 275-290.

Indraratna, B., Nutalaya, P., Koo, K.S. and Kuganenthira, N., 1991b. Engineering behaviour of a low carbon, pozzolanic fly ash and its potential as a construction fill, *Canadian Geotechnical Journal*, Vol. 28, pp. 542-555.

Indraratna, B., 1996. Utilization of lime, slag and fly ash for improvement of a colluvial soil in New South Wales, Australia. *Geotechnical and Geological Engineering Journal*, Chapman & Hall, Vol. 14, pp. 169-191.

Natt, G.S. and Joshi, R.C., 1984. Properties of cement and lime-fly ash stabilized aggregate, Transport Research Record 998, *Transportation Research Board*, National Research Council, pp. 32-40.

Saitoh, S., Suzuki, Y. and Shirai, K., 1985, Hardening of soil improved by deep mixing method, *Proc. 11th Int. Conf. Soil Mech. & Found. Eng.*, San Francisco, Vol. 3, pp. 1985-1988.

Van Impe, W.F., 1989. Soil Improvement Techniques and their evolution, Balkema Publishers, Rotterdam.

Yamanouchi, T., Miura, N., Matsubayashi, N. and Fukuda, N., 1982. Soil improvement with quicklime and filter fabric, *J. Geotechnical Engineering*, ASCE, Vol. 108, pp. 935-965.

Environmental Issues and Management of Waste in Energy and Mineral Production, Singhal & Mehrotra (eds)
© 2000 Balkema, Rotterdam, ISBN 90 5809 085 X

Benefication of Emet colemanite tailings by ultrasonic sound waves

H. Ipek & V. Bozkurt
Osmangazi University, Eskisehir, Turkey

ABSTRACT: Boron is widely used in many areas. More than sixty percent of the world's boron deposits is in Turkey. In Emet colemanite plant, it is estimated that approximately 1.5 million tons of tailings are stored at the tailing ponds. B_2O_3 grade of the tail is 18-24 % and its particle size is -0.2 mm.

In this study, the concentration of Emet colemanite tails is investigated by using ultrasonic sound waves. Ultrasonic sound waves are between the 13 - 100 kHz and these waves cause a vibration in liquid. This effect is called cavitation.

A pre-concentrate with 34.49 % B_2O_3 and 92.62 % recovery was obtained. This pre-concentrate was further concentrated using magnetic separator and multi gravity separator (MGS). A final concentrate with 37.15 % B_2O_3 and 88.53 % recovery was obtained. Accordingly, tailing ponds expenses will have reduced and the environment will have been protected.

1 INTRODUCTION

Turkey has a considerable potential of boron minerals which are widely used in many fields of industry such as glass, ceramic, agriculture, chemistry, and detergent. With the development of technology, the areas where boron is used are increasing. In previous studies, it was determined that more than 60 % of the world's boron deposits are Turkey and Turkey is the second boron producer of the world preceded by the US. Boron is produced in four different areas in Turkey. One of them is Emet colemanite plant.

At all plants in Turkey, the concentration of boron minerals is carried out by removing clays with water and classifying these after washing or dispersing and washing (Aytekin & Sonmez 1991). However, with these methods clays that are located between the crystals and at the surfaces of boron minerals can not be removed completely. Consequently, better quality concentrates can not be obtained. In addition, these methods produce much more tailings which cause storage and environmental problems.

Sounds over the audio-frequency are called ultrasound (13-100kHz). Ultrasound waves have been used in industry for cleaning metal surfaces. Fine clay particles on ore grains were removed by the pull a part effect of ultrasonic sound waves. It was investigated whether or not boron minerals can be enriched with ultrasound waves. Therefore, the purpose of this work is to investigate the possibility of concentrating boron using ultrasonic sound waves.

2 MATERIAL AND METHODS

2.1 *Material*

The sample used in this experimental study was taken from an old tailing pond of Etibank Emet colemanite mine. In this plant, colemanite concentrates in three different size fractions are produced size fraction; (-100+25 mm, -25+3 mm, and –3+0.2 mm). –0.2 mm is dumped to the tailing pond. Tailing grades vary between 20 – 24 % B_2O_3. Up to now approximately 1.5 million tons tailings have been accumulated.

The sample was reduced to 50 kg by coning and quartering. The complete chemical and screen (wet) analysis of the sample are given in Table 1 and Table 2, respectively. According to XRF analysis, the sample consists of colemanite, dolomite, montmorillonite, calcit and quartz.

Table 1. Complete chemical analysis of the sample.

Substance	Contents, %
B_2O_3	22.39
SiO_2	20.10
CaO	19.75
MgO	10.32
R_2O_3	5.40
K_2O	1.50
Na_2O	0.37
As	0.20
Loss of ignition	20.83

Table 2. Screen analysis and B_2O_3 distributions of the sample.

Particle Size (mm)	Weight %	B_2O_3 %	Distribution B_2O_3 %
+ 0.300	18.02	34.93	28.21
- 0.300 + 0.212	16.56	28.62	21.24
- 0.212 + 0.150	15.46	24.72	17.13
- 0.150 + 0.106	17.80	23.05	18.40
- 0.106 + 0.053	9.45	19.75	8.37
- 0.053 + 0.038	6.52	16.13	4.71
- 0 038	16.19	2.68	1.94
Total	100.00	22.31	100.00

2.2 *Methods*

In the first group of experiments was used a stainless steal tank with 12 transducers in 400 x 300 x 450 mm dimensions and power supply was 500 W. Basically in the ultrasonic bath an electrical current produced with high frequency an voltage by applying to crystals with piezoelectric properties, it is obtained that high frequency vibrations. In the experiments the tank was filled with water up tol the determined level and it was run for 10 minutes to reduce the amount of solved air and then the sample was added.

In the second group of experiments, the sample which had treated by ultrasonic waves was passed through 0.038 mm screen. Deslimed and dried sample was through a high field intensity, permanent magnetic separator and Multi-Gravity-Separator.

3 EXPERIMENTAL STUDIES

3.1 *Ultrasonic treatment experiments*

The following experiments were carried out in order to determine the effect of ultrasonic treatment periods and pulp densities.

3.1.1 *Effect of ultrasonic treatment periods*

The following experiments were carried out at different time intervals in order to determine the effect of ultrasonic treatment on the removal distribution of clays. Experimental conditions and results were listed below (results of 15, 30, 60 and 120 minutes are given in tables 3, 4, 5 and 6, respectivly).

Experimental Conditions

Particle size	: - 0.500 mm
Material weight	: 250 g
Pulp density	: % 20 w
Heat of water	: 20 °C
Times of ultrasonic baths	: 30, 60, 90, 120 min

Table 3. Effect of ultrasonic bath period ; 15 minutes

Particle Size (mm)	Weight %	B_2O_3 %	Distribution B_2O_3 %
+ 0.300	17.10	35.64	27.39
- 0.300 + 0.212	14.58	31.47	20.62
- 0.212 + 0.150	14.31	30.43	19.57
- 0.150 + 0.106	13.93	25.29	15.84
- 0.106 + 0.053	9.40	21.01	8.89
- 0.053 + 0.038	5.72	17.27	4.44
- 0 038	24.96	2.90	3.25
Total	100.00	22.25	100.00

Table 4. Effect of ultrasonic bath period ; 30 minutes

Particle Size (mm)	Weight %	B_2O_3 %	Distribution B_2O_3 %
+ 0.300	14.40	37.36	25.23
- 0.300 + 0.212	11.72	36.12	19.85
- 0.212 + 0.150	10.56	33.52	16.60
- 0.150 + 0.106	11.48	29.33	15.80
- 0.106 + 0.053	9.24	27.72	12.01
- 0.053 + 0.038	4.12	24.22	4.68
- 0 038	38.48	3.23	5.83
Total	100.00	21.32	100.00

Table 5. Effect of ultrasonic bath period ; 60 minutes

Particle Size (mm)	Weight %	B_2O_3 %	Distribution B_2O_3 %
+ 0.300	13.20	37.64	23.56
- 0.300 + 0.212	10.72	36.79	18.70
- 0.212 + 0.150	9.92	34.24	16.10
- 0.150 + 0.106	10.84	33.03	16.98
- 0.106 + 0.053	7.96	31.88	12.03
- 0.053 + 0.038	4.00	37.70	5.25
- 0 038	43.36	3.59	7.38
Total	100.00	21.09	100.00

Table 6. Effect of ultrasonic bath period ; 120 minutes

Particle Size (mm)	Weight %	B_2O_3 %	Distribution B_2O_3 %
+ 0.300	13.00	37.98	23.41
- 0.300 + 0.212	10.23	36.99	17.94
- 0.212 + 0.150	9.42	34.11	15.23
- 0.150 + 0.106	10.96	34.26	17.81
- 0.106 + 0.053	7.51	33.03	11.76
- 0.053 + 0.038	4.31	28.16	5.75
- 0 038	44.57	3.83	8.10
Total	100.00	21.09	100.00

Overall results of + 0,038 mm size fractions versus time is given in Table 7.

Table 7. + 0.038 mm overall results versus time

Time Minutes	Weight %	B_2O_3 %	Distribution B_2O_3 %
0	83.81	26.10	98.06
15	75.04	28.68	96.75
30	61.52	31.29	90.61
60	56.64	34.49	92.62
120	55.43	34.97	91.91

When the experiment results were examined, it was found that there was no change in % B_2O_3 grades after a 60 minutes long ultrasonic bath period.

3.1.2 *Effect of pulp density*

The following experiments were performed at different pulp densities in order to determine the effect of pulp densities on the distribution of clays. Results of 40 % pulp density is given in Table 8.

Experimental Conditions

Particle size	: - 0.500 mm
Material weight	: 250 g
Pulp density	: 20 % w, 40 % w
Heat of water	: 20 °C
Times of ultrasonic baths	: 60 min

Table 8. Effect of pulp density ; 40 % w

Particle Size (mm)	Weight %	B_2O_3 %	Distribution B_2O_3 %
+ 0.300	15.42	35.43	25.48
- 0.300 + 0.212	12.46	32.16	18.69
- 0.212 + 0.150	11.64	29.57	16.06
- 0.150 + 0.106	14.32	27.62	8.45
- 0.106 + 0.053	8.98	23.45	9.82
- 0.053 + 0.038	5.73	25.78	6.89
- 0 038	31.45	3.14	4.61
Total	100.00	21.44	100.00

Table 9. + 38 mm overall results versus pulp density

Pulp Density weight %	Weight %	B_2O_3 %	Distribution B_2O_3 %
20	56.64	34.49	92.62
40	68,55	29,83	95,39

When worked with 20 % pulp density, higher grade was obtained than that of 40 % pulp density.

From the above results, it is seen that a pre-concentrate which is 34.49 % B_2O_3 and 92.62 % recovery can be obtained.

3.2 *Magnetic separation experiment*

The mineralogical and petrografic studies indicated that clay minerals from Emet formation contained considerable amount of iron and traces of minerals having magnetic properties, such as biotite, epidot, muscovite and turmaline. Also it was determined that susceptibility of magnetic product, which was produced by high field intensity wet magnetic separation applied to –1 mm material, increased to 1.32 x 10^{-5} cm^3/g (Sonmez, Ozdag & Savas 1996). Magnetic beneficiation studies of deslimed tailings were in the light of the data obtained.

Deslimed and dried pre-concentrate was fed to a high field intensity, permanent magnetic separator (PERMOLL®).

The best grade of concentrate was obtained with experimental conditions listed below.

Experimental Conditions

Particle size	: - 0.500 + 0.038 mm
Material weight	: 140.5 g
Content	: 34,41 % B_2O_3
Belt speed	
1. stage	: 100 rpm
2. stage	: 100 rpm
Splitter position	
1. stage	: 30°
2. stage	: 30°

Table 10. Magnetic separation experiment

Products	Weight %	B_2O_3 %	Distribution B_2O_3 %
Concentrate	89.14	37,15	96.24
Midd. + Tailing	10.86	11,91	3,76
Pre-Concentrate	100,00	34,41	100,00

When the concentrate was related to the original feed

Table 11. Magnetic separation experiment

Products	Weight %	B_2O_3 %	Distribution B_2O_3 %
Concentrate	50.52	37.15	88.53
Tailing 2	5.68	11.91	3.19
Tailing 1	43.80	4.01	8.28
Feed	100.00	21.20	100.00

These results reveals that a salable product with 37.15 % B_2O_2 grade and 88.53 % recovery could be obtained.

3.3 *Multi gravity separation experiment*

It is known that, Multi-Gravity-Separator (M.G.S.) is used at the minus 500 microns particle sizes ores successfully.

In this study M.G.S. was tried alternatively to magnetic Separator because of cheaper and lower operating cost.

Experimental Conditions

Particle size	: - 0.500 + 0.038 mm
Material weight	: 141.3 g
Content	: 34,13 % B_2O_3
Wash water flow	: 6 lt/min.
Pulp density	: 33 % w
Drum inclination	: 4°
Drum rotation speed	: 210 rpm
Amplitude	: 10.5 mm
Frequency	: 5.7 cps

Table 12. MGS experiment

Products	Weight %	B_2O_3 %	Distribution B_2O_3 %
Concentrate	67.37	36,18	71,42
Tailing	32.63	29,90	28,58
Pre-Concentrate	100,00	34,13	100,00

When the concentrate was related to the original feed

Table 13. MGS experiment

Products	Weight %	B_2O_3 %	Distribution B_2O_3 %
Concentrate	38.08	36.18	65.50
Tailing 2	18.44	29.90	26.21
Tailing 1	43.48	4.01	8.29
Feed	100.00	21.03	100.00

As can be seen from the results, a salable product with 36.18 % B_2O_3 grade and 65.50 % recovery could be obtained. However, 65.50 % recovery is not good enough despite 36.18 % B_2O_3 is saleable grade because 36 % B_2O_3 is acceptable grate to sale.

4 CONCLUSIONS

At the light of above results, the following results can be produced from the study

1. When ultrasonic time period increases , it was obtained cleaner concentrate. However, the increase the time period, the increase dissolution of boron minerals. After the 60 minutes ultrasound bath periods this dissolution increasing stayed.

2. The pre-concentrate which was 34.49 % B_2O_3 and 92.62 % efficiency was obtained by the ultrasonic bath. But this pre-concentrate can not sale directly or it has to be mixed with the 3 – 25 concentrates.

3. It was tried that obtained pre-concentrate was enrichment by magnetic separator. After magnetic separation, a concentrate which was 37.15 % B_2O_3 and 88.53 % efficiency was obtained. This is a concentrate which is salable easily.

4. Alternatively, it was tried that this pre-concentrate was enrichment by multi gravity separator. However, it could not obtain a acceptable concentrate because of gravity nearness between the boron and gangue minerals.

REFERENCES

Aytekin, Y. & Sonmez, E. 1991. The investigation of concentration possibilites of Kırka Tincal ore and concentrate. PhD. Thesis. Dokuz Eylül University. Izmir, Turkey.

Ozdag, H., Bozkurt, R. & Ucar, A. 1988. The recovery of boron from Kestelek boron washing tailings. Proceeding of 2th International Mineral Processing Symposium, 238-249. Izmir, Turkey.

Sonmez, E., Ozdag, H. & Savas, M. 1996 Benefication of Emet tailings by water absorption + mechanical attrition + magnetic separation. Proceedings of 6th International Mineral Processing Symposium,143-148. Izmir, Turkey.

Sonmez, E., Ozdag, H. & Savas, M. 1997 An investigation on the use of ultrasonic wasves for the enrichment of colemanite tailings. Proceedings of 15th mining congress of Turkey, 319-323. Ankara, Turkey.

Environmental Issues and Management of Waste in Energy and Mineral Production, Singhal & Mehrotra (eds)
© 2000 Balkema, Rotterdam, ISBN 90 5809 085 X

Identification of long-term environmental monitoring needs at a former uranium mine

R.C.Lee, R.Robinson, M.Kennedy & S.Swanson
Golder Associates Limited, Calgary, Alb., Canada

M.Nahir
Public Works and Government Services Canada, Edmonton, Alb., Canada

ABSTRACT: The former Rayrock uranium mine in the Canadian Northwest Territories (NWT) was abandoned in 1959 and remediated in 1996. Canadian government agencies were interested in developing short- and long-term monitoring plans for the site to satisfy Atomic Energy Control Board (AECB) requirements and to ensure that potential human health risks from ionizing radiation remain minimal. This study determined whether potential risks are likely to remain minimal, identified data gaps, and recommended a long-term plan. A high-level analysis was conducted in conjunction with risk estimations to determine sensitivity of risks to uncertainties associated with radionuclide concentrations or radiation levels in exposure media including gamma radiation, radon, soil, water, and consumption of wild game/fish. Potential human health risks under current conditions are minimal under any conceivable exposure scenario. Recommendations for long-term monitoring focus on the exposure pathways of greatest concern, combined with the feasibility of further data collection.

1 INTRODUCTION

The former Rayrock Mine is located approximately 145 km northwest of Yellowknife, NWT. Underground mine exploration and development activity began in 1955. The mine was abandoned in operable condition in 1959. During the operation of the mine and associated mill, 70,903 tonnes of unneutralized tailings were discharged. The tailings piles are weakly acid generating, and are potential sources of radionuclides of the uranium-thorium decay series and heavy metals to the surrounding area.

The Rayrock site underwent a remediation program in 1996. The goals of the remediation program were to stop further mobilization of contaminants from the site into the surrounding environment, and to reduce hazards to humans potentially visiting the area. The remediation was not intended to remove contaminants and restore pristine conditions in the immediate area of the mine. Remediation included sealing the mine adit and ventilation shafts, removal of radioactive material from the dump and disposal of the dump material across the tailings ponds, and thick-capping of tailings ponds with silty-clay material borrowed from an old airstrip near the mine site.

A short-term environmental monitoring program was designed (Golder Associates Ltd. 1996) and implemented between 1996 and 1998 by the Department of Indian Affairs and Northern Development (DIAND), Public Works and Government Services Canada (PWGSC) and the Low Level Radioactive Waste Management Office (LLWRMO) of the Atomic Energy Control Board (AECB).

The results of the short-term monitoring program (LLWRMO 1999) indicated that the remedial activities were successful in reducing gamma radiation and radon emissions (the main radiation sources) to acceptable regulatory limits. The results of multiple-pathway radiation dose estimations in several different exposure scenarios indicated that even in extreme exposure scenarios doses were minimal and well below regulatory dose limits.

The present analysis includes an evaluation of the relative impact of exposure routes on total doses, identification of areas of uncertainty that are likely to be important to the proposed long-range monitoring plan at the site, recommendations regarding long-range monitoring, and identification of possible risk management strategies at the site.

2 EXPOSURE AND UNCERTAINTY ANALYSIS

The LLWRMO (1999) found that external gamma radiation is the most potentially important source of ionizing radiation at the Rayrock site, although the total dose is minimal. However, uncertainty associated with the doses was not assessed. In the following analysis, the dependency of dose on time is examined, as well as the uncertainty associated with contaminant concentrations/activity. Doses resulting from multiple simultaneous exposure routes (e.g. gamma plus water plus food) are not assessed here because the LLWRMO determined that there is no unacceptable residual risk remaining at the site even in an extreme worst-case scenario. The purpose of the present analysis is to guide further data-gathering efforts and possible further mitigation strategies.

2.1 *Exposure Models*

The general form for the exposure model is:

$$D_i = (C_{i:s} - C_{i:b}) \times IR_i \times ED \times CF \quad (1)$$

where D_i = site-related incremental radiation dose from exposure route i; $C_{i:s}$ = site media concentration/activity from exposure route i; $C_{i:b}$ = background or control concentration/activity from exposure route i; IR_i = intake or exposure rate associated with exposure route i; ED = exposure duration; and CF = unit and/or dose conversion factor(s).

Specific models that represent the exposure routes of interest are listed in Table 1.

Table 1. Radionuclide dose equations. All equations are based on LLWRMO (1999). Variable definitions are listed in Table 2.

Exposure Pathway	Dose Equation (Incremental)
Ingestion of meat or fish	$D_{m/f} = (C_{m/f:s} - C_{m/f:b}) \times IR_{m/f} \times DCF_1 \times ED$
Ingestion of water	$D_w = (C_{w:s} - C_{w:b}) \times IR_w \times DCF_2 \times ED$
Inhalation of dust	$D_d = (C_{d:s} - C_{d:b}) \times InhR \times TSP \times DCF_3 \times UCF_1 \times ED$
Inhalation of radon	$D_r = (C_{r:s} - C_{r:b}) \times DCF_4 \times ER \times UCF_2 \times ED/WM$
External exposure to gamma radiation	$D_g = (DR_{g:s} - DR_{g:b}) \times UCF_2 \times UCF_3 \times ED$

Mitigation decisions are normally based on the incremental dose because it is not possible to mitigate background ionizing radiation exposures. AECB is in the process of adopting the International Commission on Radiation Protection (ICRP) annual dose limit for the public of 1 mSv, which is incremental to background (AECB 1998). This dose limit is used here for comparative purposes.

The variables in the exposure models are heterogeneous across space or a population ("variable"), and the parameters (e.g. mean, variance) of these distributions are uncertain (inaccurate and/or imprecise). Variability is an inherent feature of space or populations and is irreducible; uncertainty stems from inadequate knowledge of a variable and is reducible given more data and/or more representative information. It is not possible, given the current data, to adequately characterize the true variability associated with potential doses at the Rayrock site. The LLWRMO analysis covered a wide range of potential exposure scenarios, and thus addressed at least some of the variability in dose that is likely to exist. The current assessment examines uncertainty because reduction of uncertainty is one goal of long-term monitoring.

Each of the variables in the models above can be represented as a distribution of values (other than the unit conversion factors, which are deterministic by definition). These variables are expected to be independent, and thus there is no need for a defined correlation structure in exposure modeling. Uncertainty associated with physiological variables is minimal, and is not quantitatively evaluated here (mean point estimates are used instead). Estimates of these variables are listed in Table 2. Development of contaminant concentration/activity uncertainty distributions is described below.

2.2 *Contaminant Concentration/Activity Uncertainty Distributions*

The uncertain variables that are of interest in terms of long-term monitoring are exposure media concentrations/activity of radionuclides. Media concentration/activity distributions used here represent the uncertainty associated with doses from defined exposure media to an individual who wanders at random through the site.

Table 2 contains the distributions used for media concentration/activity variables along with the sources of data. In all cases, uniform distributions are defined with a lower bound of zero (a physical limitation) and an upper bound of the maximum measured concentration/activity as determined in the LLWRMO (1999) short-term monitoring study. Uniform distributions are used, in accordance with the principle of maximum entropy, when knowledge is limited to the bounds of a distribution (Lee & Wright 1994). Use of uniform distributions implies that a true average and/or statistical spread is not known for site contamination, and this distribution maximizes uncertainty within reasonable bounds.

Direct measures relating to potential human exposure such as consumption of wild meat and/or fish and surface water are used in this assessment rather than attempting to model food chain or indirect uptake pathways. Groundwater is not a viable direct exposure pathway at this site. It is possible that direct exposures to sediments could occur, perhaps in fishing activities, but it is unlikely that receptors would receive appreciable doses from this route in terms of dermal absorption or gamma radiation exposure.

Table 2. Radionuclide dose variables and factors. Distributions are represented as U(x,y), where U = uniform distribution, x = lower bound, and y = upper bound.

Variable or Factor	Definition	Value	Units	Reference	Note
$C_{m/f:s}$	Current site-related activity of uranium in meat and/or fish	U(0,0.0007)	Bq/g	LLWRMO 1999, Table 5.2	1
$C_{m/f:b}$	Current background activity of uranium in meat and/or fish	U(0,0.058)	Bq/g	Swanson 1996, in Golder 1996, Table 6	2
$C_{w:s}$	Current site-related concentration of uranium in surface water	U(0,263)	μg/L	LLWRMO 1999, Table 4.3	3
$C_{w:b}$	Current background concentration of uranium in surface water	U(0,0.6)	μg/L	LLWRMO 1999, Table 4.3	4
$C_{d:s}$	Current site-related concentration of uranium in dust	U(0,65)	μg/g	LLWRMO 1999, Table 4.12	5
$C_{d:b}$	Current background concentration of uranium in dust	U(0,16)	μg/g	LLWRMO 1999, Table 4.12	6
$C_{r:s}$	Current site-related activity of radon	U(0,794)	Bq/m³	LLWRMO 1999, Table 4.13	7
$C_{r:b}$	Current background activity of radon	U(0,135)	Bq/m³	LLWRMO 1999, Table 4.13	8
$DR_{g:s}$	Current site-related dose rate of gamma radiation	U(0,1.6)	μSv/hr	LLWRMO 1999, Table 4.14	9
$DR_{g:b}$	Current background dose rate of gamma radiation	U(0, 0.1)	μSv/hr	LLWRMO 1999, Table 4.14	10
$IR_{m/f}$	Ingestion rate of meat and/or fish	Mean = 270, variable 0 to 1500	g/d	O'Connor 1997	11
IR_w	Ingestion rate of water	Mean = 1.5, variable 0 to 4	L/d	O'Connor 1997	12
InhR	Inhalation rate	Mean = 16	m³/d	O'Connor 1997	13
TSP	Concentration of total suspended particulates	50	μg/m³	LLRWMO 1999	
ED	Exposure duration	Variable (0 to 20 days)	d	Professional judgement	14
DCF_1	Dose conversion factor 1 (ingestion of meat/fish)	0.0025	mSv/Bq	LLRWMO 1999	15
DCF_2	Dose conversion factor 2 (ingestion of water)	3.2×10^{-5}	mSv/μg	LLRWMO 1999	15
DCF_3	Dose conversion factor 3 (inhalation)	0.0017	mSv/μg	LLRWMO 1999	15
DCF_4	Dose conversion factor 4 (radon)	0.0011	m³×mSv/Bq	LLRWMO 1999	15
ER	Equilibrium ratio (radon daughters)	0.1	--	LLWRMO 1999	16
WM	Working month	170	hr	LLRWMO 1999	17
UCF_1	Unit conversion factor	0.000001	g/μg	LLWRMO 1999	--
UCF_2	Unit conversion factor	24	hr/d	LLWRMO 1999	--
UCF_3	Unit conversion factor	0.001	mSv/μSv	LLRWMO 1999	--

Table 2 General Note: Maximum value represents maximum at any location on the site or background locality according to referenced data unless otherwise noted.

Table 2 Specific Notes:

1 Maximum value represents organ levels in a pike caught from Sherman Lake.

2 Background concentrations not available from LLRWMO (1999). Historical maximum value represents average tissue activity in white suckers taken from lakes near Uranium City (^{210}Pb activity).

3 Sample taken from Lake Alpha. A conversion was applied to measured uranium levels by the LLWRMO (1999) to account for greater activity of some of the uranium series radionuclides.

4 Sample taken from Maryleer Lake.

5 See text regarding "dust" assumptions. Surface sample taken from downwind of north tailings pile. A conversion was applied to measured uranium levels by the LLWRMO (1999) to account for greater activity of some of the uranium series radionuclides.

6 See text regarding "dust" assumptions. Surface sample taken from near Maryleer Lake.

7 See text regarding breathing zone assumptions. Average concentration west of the south tailings area.

8 See text regarding breathing zone assumptions. Average concentration at Maryleer Lake. Emil River samples not assessed by LLWRMO due to uncertainties in data (see LLWRMO 1999).
9 Measurement at north tailings area.
10 Measurement at Maryleer Lake.
11 Mean represents average wild game ingestion rate for Native Canadian adults of both sexes (includes Inuit populations). Maximum value represents the 99th percentile of a lognormal distribution of wild game ingestion rates for Native Canadian adults of both sexes (includes Inuit populations). Note that approximately 70% of respondents in the original survey stated that they ate no wild game. Ingestion rates are lower for fish consumption.
12 Mean represents average water ingestion rate for Canadian adults of both sexes. Maximum value represents the 99th percentile of a lognormal distribution of water consumption rates for Canadian adults of both sexes.
13 Represents the mean of a lognormal distribution of 24-hour inhalation rates for adults of both sexes.
14 Maximum represents the longest reasonable period that a person would spend continuously at the site, based on professional judgement.
15 See LLWRMO 1999 for detailed explanation of dose conversion factor.
16 ER represents the ratio of radon daughter products to radon gas.
17 Standard radiological protection assumption.

Use of the distributions defined here may introduce *a priori* conservative biases (toward safety) in the following ways:

- Because statistically designed sampling has not been performed at this site, these distributions do not represent spatial variability in contaminant concentrations/activity. Rather, they represent uncertainty associated with exposure media concentrations/activity to which a random individual may be exposed. Due to site characteristics and topography, it is less likely that the individual will be exposed to relatively high contamination areas than to lower contamination areas. The assumption of uniform probability over the range of values specified is a conservative assumption.
- Exposure concentrations/activity of dust and ambient radon are highly dependent on season and weather conditions. A simple soil-to-dust conversion is used here (the same as used in LLRWMO 1999), and measured concentrations of radon are assumed to be the same as in the breathing zone of the receptor (the same assumption as used in LLWMO 1999). Both of these assumptions are highly conservative, particularly in the winter months when snow covers the ground.
- The maximum surface water values used reflect sampling results from Lake Alpha, a shallow lake near the mine. This lake is not considered to be a desirable potable water source in comparison to the much larger, more distant, and less contaminated Lake Sherman, which is connected to Lake Alpha.

2.3 *Methodology*

The point estimates and distributions defined in Table 2 are combined in the appropriate equations listed in Table 1. The dose conversion factors used here were calculated by the LLWRMO by application of standard health physics assumptions (LLWRMO 1999). Monte Carlo simulation (using the Excel® add-in Crystal Ball®) is used to subtract background media concentration/ activity distributions from site-related media concentration/activity distributions. Ten-thousand random draws are used for each simulation; this number of iterations allows a reasonable degree of confidence to be placed in upper percentiles of the output distributions (a greater than 99% confidence level in the 99th percentiles) without requiring an excessive amount of computer time (IAEA 1989). A Boolean statement is used to convert negative values resulting from the subtraction operation to zero. Thus, if a large number of subtractions result in negative values, the distribution will incorporate a "spike" of zero values. Percentiles of the "net" site-related concentration/activity distributions are incorporated into exposure algorithms; the results are distributions of possible incremental doses that can result from particular exposure routes.

Depending on the exposure route, different variables can be used as independent variables to provide an indication of the point when the AECB/ICRP annual dose limit of 1 mSv is attained. Exposure duration (ED) is used as an independent variable here. Calculations are performed for a series of time steps up to 20 days (a reasonable maximum ED; see Table 2). The median, 5th percentile, and 95th percentile of the uncertain cumulative dose for each exposure route are then calculated. Intake rate (IR) is also used as an independent variable in the cases of water and meat and/or fish ingestion. Dose calculations are performed for daily intake rates up to a reasonable maximum for each route (4 L per day for water, and 1500 g per day for wild meat and/or fish, see Table 2).

2.4 *Simulation Results*

Table 3 presents the results of the simulated site-related concentration/ activity distributions from exposure media, and Table 4 and 5 present the results of the simulated doses associated with those media.

There are differences between the doses as calculated by the LLWRMO (1999) and the doses calculated here due to subtraction of background concentrations/activity in this analysis, differences in intake values, and differences in exposure durations. However, the general conclusions are approximately the same as those made by the LLWRMO. The following ranking of pathways applies to daily doses across percentiles and exposure durations:

1 External exposure to gamma radiation (highest);
2 Ingestion of surface water from water bodies near the mine site;
3 Inhalation of radon;
4 Inhalation of dust; and
5 Ingestion of wild meat or fish (lowest).

It was not possible to perform traditional statistical analyses (due to the lack of data from a statistically-designed sampling plan) to address the question of whether site contamination in particular media exceeded that of background concentrations/activity to be expected in a mineral-rich area. However, it is informative to examine the percentiles of the simulated site-related distributions. For example, approximately 10% of the dust concentration distribution is below zero, and 95% of the wild meat/fish distribution is below zero. This provides an indication that the site-contamination-related wild meat/fish distribution is no different than background. This conclusion is uncertain due to the limited number of samples, but nonetheless may indicate that it is likely that little if any appreciable off-site contamination is presently occurring that is reflected in native foods.

The coefficients of variation (which provide a means of comparing relative uncertainty) of site-related radionuclide concentration/activity distributions are similar (Table 3). The coefficient of variation for the meat/fish concentration distribution is noninformative due to the large number of zero values.

2.5 *Discussion*

Gamma radiation accounts for approximately 3 times the dose from the next most important pathway (water ingestion). It would take almost a month of continuous exposure to reach the proposed incremental annual dose limit of 1 mSv for a 95th percentile exposure to gamma radiation (Table 4). Using average exposure assumptions, it would take 17 days of continuous simultaneous exposure to the 95th percentile of all exposure routes evaluated to reach the dose limit.

As an extremely conservative worst-case scenario, total doses (as opposed to incremental doses) were calculated using maximum values for all exposure media (without subtracting background levels), as well as maximum values for water (4 L/d) and wild meat/fish ingestion (1500 g/d). It would take 11 days to reach the exposure limit under this scenario (Table 5), which is virtually impossible. These results indicate that there is minimal remaining risk to humans from radionuclide exposure at the Rayrock site under any conceivable circumstance under present conditions.

The majority of the minimal potential risk from radionuclide exposure at the Rayrock site is associated with pathways that are directly associated with on-site contamination (e.g., gamma radiation from capped tailings). Potential off-site exposure pathways such as consumption of migratory wild game or consumption of water from off-site sources will result in much lower doses than on-site pathways. Therefore, there is a high degree of certainty that exposure from off-site sources would pose even lower potential risk than the minimal risk from on-site exposure.

There are a number of considerations in interpreting these results:

- The results are indicative of current conditions only;
- The calculations of site-related doses are dependent on limited data, especially for indicators of potential off-site exposure such as fish;
- Maximum values used here may not be representative of true maximums at the site;
- Appropriate background values are uncertain due to the natural heterogeneity of radionuclides in a mineral rich area;
- Realistic exposure patterns and behaviour of potential receptors are uncertain; and
- Uncertainty is associated with a number of standard health physics assumptions in dose conversions.

As indicated above, not all sources of uncertainty were quantitatively addressed in this analysis. However, quantitative characterization of these sources of uncertainty would not be expected to change the conclusions of this report appreciably. Further characterization of background concentrations is not informative under current conditions because both incremental and total potential doses at the site are minimal. The long-term monitoring recommendations are targeted toward assuring that mitigation efforts performed to date continue to keep the potential for human radionuclide exposure at a minimal level.

3 LONG-TERM MONITORING RECOMMENDATIONS

A long-term monitoring plan is designed to focus on the media that offer the most appreciable potential for unacceptable exposures and that have appreciable associated uncertainty, balanced with the feasibility of sample collection based on the experience gained in the short-term sampling.

Table 3. Results of contaminant concentration/activity modelling.

	Gamma (μSv/hr)	Radon (Bq/m³)	Dust (μg uranium/g)	Water (μg uranium/L)	Meat/Fish (Bq/g)
Statistic					
Mean	0.72	382	25	132	0
Standard Deviation	0.44	230	18	76	N/A
Coefficient of Variation	0.61	0.60	0.72	0.58	N/A
Percentile					
5%	0.03	24	0	14	0
25%	0.33	182	8.2	66	0
50%	0.72	385	24	131	0
75%	1.10	580	40	197	0
95%	1.40	740	54	250	0

Table 4. Estimated exposure-route-specific incremental doses using average values for exposure assumptions (see Table 2). All doses given in mSv.

Exposure Route	Percentile	Exposure Duration (days) *1*	*5*	*10*	*15*	*20*
External exposure to gamma radiation	50th	1.7×10^{-2}	8.6×10^{-2}	1.7×10^{-1}	2.6×10^{-1}	3.5×10^{-1}
	95th	3.4×10^{-2}	1.7×10^{-1}	3.4×10^{-1}	5.1×10^{-1}	6.7×10^{-1}
Inhalation of radon	50th	6.0×10^{-3}	3.0×10^{-2}	6.0×10^{-2}	9.0×10^{-2}	1.2×10^{-1}
	95th	1.2×10^{-2}	5.7×10^{-2}	1.2×10^{-1}	1.7×10^{-1}	2.3×10^{-1}
Inhalation of dust	50th	3.3×10^{-5}	1.7×10^{-4}	3.3×10^{-4}	5.0×10^{-4}	6.6×10^{-4}
	95th	7.4×10^{-5}	3.7×10^{-4}	7.4×10^{-4}	1.1×10^{-3}	1.5×10^{-3}
Ingestion of surface water	50th	6.3×10^{-3}	3.1×10^{-2}	6.3×10^{-2}	9.4×10^{-2}	1.3×10^{-1}
	95th	1.2×10^{-2}	6.0×10^{-2}	1.2×10^{-1}	1.8×10^{-1}	2.4×10^{-1}
Ingestion of meat and/or fish	50th	0	0	0	0	0
	95th	0	0	0	0	0

Table 5. Estimated exposure-route-specific incremental doses using extreme values for ingestion exposure assumptions (4 L/d for water and 1500 g/d for meat/fish), and highest measured values for exposure media concentrations/activity (without subtracting background concentrations) (see Table 3). Note that this is a *highly unrealistic* scenario. All doses given in mSv.

Exposure Route	Exposure Concentration	Daily Dose	Days to Reach 1 mSv Annual Incremental Dose Limit
External exposure to gamma radiation	1.6 μSv/hr	3.8×10^{-2}	26
Inhalation of radon	794 Bq/m³	1.2×10^{-2}	83
Inhalation of dust	65 μg uranium/g	8.8×10^{-5}	11,400
Ingestion of surface water	263 μg uranium/L	3.4×10^{-2}	29
Ingestion of meat and/or fish	0.0007 Bq/g	2.6×10^{-3}	384
Simultaneous exposure to all routes and associated concentrations/activity as defined above	N/A	8.7×10^{-2}	11

As indicated earlier, the ranking of media according to potential for radiation exposures is:

1 External exposure to gamma radiation;
2 Ingestion of surface water from water bodies near the mine site (e.g. Lake Alpha);
3 Inhalation of radon;
4 Inhalation of dust; and
5 Ingestion of wild meat or fish.

The uncertainty associated with contaminant concentrations/activity is similar across the exposure media as indicated by the respective coefficients of variability (Table 3). The exceptions are levels in meat and fish, which are highly uncertain due to lack of data. It is not possible to estimate a coefficient of variability for this pathway due to the apparent lack of difference between site-related concentrations and background. Due to the nature of caribou movements, it is unlikely that these animals would accumulate sufficient incremental doses from environmental media at the Rayrock site so that the meat would pose a potential hazard. However, fish have more direct and food-chain contact with potentially contaminated media, and it is possible that the true variability in radionuclide levels was not captured in the short-term program.

The feasibility and cost of collecting additional information vary dramatically across the exposure media. The following criteria can be used to score the different media as to feasibility of additional data collection in a remote Northern area:

- Necessity for use of helicopters as opposed to float planes for access;
- Necessity for use of other transport equipment such as all-terrain vehicles or snowmobiles as opposed to access on foot;
- Necessity for use of large sampling equipment such as drilling rigs;
- Time of year that samples should be collected;
- Objective hazards associated with sampling (e.g., necessity of traversing hazardous terrain, potential for adverse weather, etc.); and
- Number of person/hours necessary for sample collection.

Based on discussions with the site manager and with environmental professionals who have experience in different types of sample collection schemes, qualitative "feasibility scores" were assigned to the different exposure media according to the criteria above.

"Dose scores" that reflect the 50th percentile of daily risks (Table 4) were assigned to appropriate media. Table 6 is a matrix of estimated dose vs. feasibility of sampling scores.

The long-term sampling plan should optimally focus on those media that pose the greatest potential for doses along with ease/low cost of sample collection. The media that satisfy these criteria include gamma radiation, surface water, radon and fish.

Gamma radiation is associated with the highest potential doses, and it is relatively straightforward to collect these data. Also, gamma radiation measurements provide a measure of the integrity of the existing tailings caps. Therefore, collection of gamma measurements is important to the long-range sampling plan. Personal dosimetry of sampling personnel is an important occupational safety measure.

Surface water ranks relatively high in terms of potential doses. The hydrology of the site is currently uncertain due to limited information regarding transport pathways (i.e., surface flow vs. groundwater input). Therefore, it is useful to perform surface water sampling, in both summer and winter, as well as sampling of groundwater from an important groundwater well downgradient from the mine to determine time trends and relative contributions from transport media to surface water bodies. Surface water sampling in winter will eliminate the contribution of surface water flow to total contaminant load, and will allow the possible contribution of groundwater flow to surface water bodies to be assessed.

Radon ranks high in terms of potential doses in the current assessment. However, the highly conservative assumption was made that radon measurements reflect breathing zone concentrations. Although radon is unlikely to be associated with appreciable exposures under current conditions, radon measurements provide a means of monitoring integrity of caps on shafts and adits.

Table 6. Matrix of estimated doses and feasibility of sample collection.

	Feasibility		
Dose Exponent	*Low*	*Medium*	*High*
$>10^{-2}$			Gamma
$10^{-2} - 10^{-3}$		Radon	Surface water
$10^{-3} - 10^{-4}$			
$10^{-4} - 10^{-5}$	Dust, meat	Fish	

Although fish appear to pose minimal risks, the large uncertainty associated with true site-related concentrations and the relative feasibility of collecting additional data contribute to this media's inclusion in the list. The fish sampling program outlined here includes provision for sampling fish in two lakes: a nearby reference lake that is unaffected by potential contamination from the Rayrock mine site; and Sherman Lake, which is a likely source of fish used for consumption.

Additional sampling of the remaining media will not be informative nor feasible. Further characterization of background concentrations will not be informative under current conditions because both incremental and total potential doses at the site are minimal, and extensive characterization of background heterogeneity is not feasible. Because wild meat appears to be associated with minimal exposures, further sampling of intermediate media such as lichens and macrophytes will also not be informative. Surface water and fish (direct exposure pathways) sampling should be sufficient to characterize the water environment. Sediment sampling will not be required unless significant increases in radionuclide levels in surface water and/or fish over time are evident.

The recommended plan includes periodic monitoring of the following components for a time scale of years:

1 Visual inspection of tailings and adit/shaft cap integrity;
2 Gamma radiation surveys and dosimetry in tailing areas;
3 Radon measurements in the vicinity of adits/shafts;
4 Surface water sampling in Lake Alpha and in the southwest arm of Sherman Lake; and
5 Groundwater sampling in a monitoring well downgradient from the mine.

Additionally, the following one-time data collection and interpretation exercise is recommended to improve the data upon which further recommendations may be based:

6 Fish collection in a reference lake and Sherman Lake.

There are several approaches for assessment of whether contaminant levels are increasing appreciably or significantly in a statistical sense over time.

A dramatic increase (e.g., 100%) in a measured value in a particular medium over the previous measured value would be an obvious trigger for further investigation. Assuming that trends in contaminant levels are likely to be more gradual over a long period of time, a nonparametric statistical trend test such as the Mann-Kendall test (Gilbert 1987) can be employed. Use of such a test is recommended after the first 5 and 10 years, and every 10 years thereafter, to determine whether significant trends in contaminant concentrations are occurring. Statistically significant upward trends in contaminant levels over time will trigger re-evaluation of the monitoring and risk management plan.

Potential exposure media that are not listed above do not appear to pose appreciable risks, and reduction of uncertainty associated with exposures to these media will not be informative. However, further sampling of these media may be necessary if the recommended inspections and sampling indicate a change in current physical conditions. Changes in conditions associated with the monitored media, such as significant upward trends in surface water radionuclide concentrations, may be difficult to assess without an extensive statistically designed monitoring plan. However, a consistent increasing trend in surface water concentrations may indicate that a change in the physical condition of the site is occurring, and that the monitoring plan should be revised. The monitoring plan should be revised in any event in 100 years due to long-term uncertainties and possible changes in land use.

4 RISK MANAGEMENT

Both the LLWRMO (1999) and current assessments indicate that the potential for unacceptable radiation exposures at the Rayrock site is very low as long as the existing physical barriers remain intact. The long-range monitoring plan will provide information regarding the continued integrity of these low risks, and risk management will consist of ensuring that existing remedial approaches stay intact. Any compromise of existing physical barriers should be assessed and mitigated in an expedient fashion.

Additional recommendations can be made for management of non-radiation-related risks that may exist at the site. Long-term liability management is important at a site such as Rayrock where wastes have been left *in situ*. Implementation of legal land use restrictions, after consultation with any affected stakeholders, is recommended. It appears unlikely that the site can be released for unlimited public use in the near future, due to the potential for disturbance of remedial measures and for exposure to non-radiation physical hazards. Erection of "No Trespassing" signs and maintenance of caps on adits and shafts is important to avoid the potential for accidents.

5 REFERENCES

AECB (Atomic Energy Control Board), 1998, *Assessment and Management of Cancer Risks from Radiological and Chemical Hazards*, Minister of Public Works and Government Services Canada, Ottawa.

Gilbert, R.O., 1987, *Statistical Methods for Environmental Pollution Monitoring*, Van Nostrand Reinhold, NY.

Golder (Golder Associates Ltd.), 1996, *Environmental Monitoring Program for Assessing Remediation Efforts at the Rayrock Uranium Mine, Northwest Territories*, prepared for Public Works and Government Services Canada, October.

IAEA (International Atomic Energy Agency), 1989, *Evaluating the Reliability of Predictions Made Using Environmental Transfer Models*, Safety Series No. 100, Vienna.

Lee, R.C. & Wright, W.E., 1994, Development of Human Exposure-Factor Distributions Using Maximum-Entropy Inference, *J. Exp. Anal. & Env. Epid.* 4:329-341.

LLWRMO (Low-Level Radioactive Waste Management Office), 1999, *Short-Term Environmental Monitoring Program, Rayrock Uranium Mine: Revision 1*, prepared for Indian and Northern Affairs Canada, March.

O'Connor (O'Connor Associates Environmental Inc.), 1997, *Compendium of Canadian Human Exposure Factors for Risk Assessment*, Ottawa.

Environmental Issues and Management of Waste in Energy and Mineral Production, Singhal & Mehrotra (eds)
© 2000 Balkema, Rotterdam, ISBN 90 5809 085 X

Water quality issues associated with implementation of composite tailings (CT) technology for managing oil sands tailings

M.D.MacKinnon, J.G.Matthews, W.H.Shaw & R.G.Cuddy
Syncrude Canada Limited, Edmonton, Alb., Canada

ABSTRACT: In the oil sands industry, Composite Tails (CT) is an attractive tailings management approach. In the CT process, the fines or clay fraction of extraction tailings are mixed with the coarse sand fractions, such that, with the addition of a coagulant aid, a non-segregating deposit is produced. Various coagulant aids have been tested and have been shown to destabilize suspensions of fines and clays. After treatment, the fines aggregate and with the stress or load of the coarse sand fraction will dewater, with the fines remaining trapped with the coarse fraction, and particle-free water being released. The rate and degree of water release determine when the resulting deposit can be classed as a solid or dry landscape for reclamation. The coagulation of the clays can be accomplished with a variety of chemicals, both organic and inorganic, that initiate clay destabilization and affect water quality. Depending on the coagulant aid, changes in pH, salinity, cation and anion levels, buffering capacity and toxicity have been observed. The impacts on water quality with the use of several chemical treatments are described and potential influence of the CT release waters on operational aspects and reclamation issues are discussed.

1 INTRODUCTION

Large quantities of tailings are produced during separation of bitumen from oil sands at the two currently operating commercial operations, Syncrude Canada Ltd. and Suncor Energy Inc., in northeastern Alberta. An aqueous digestion method is used in extraction of the bitumen (FTFC 1995). This results in large volumes of tailings (sand, fines, unrecovered hydrocarbons, and water with dissolved components). The tailings are transported hydraulically, and when deposited, they segregate (MacKinnon 1989). While the coarse sand fraction settles quickly to form beaches, most of the fines (silts and clays) enter settling basins to form a stable suspension that requires a long time to fully consolidate (Mikula et al 1996). As this suspension densifies, it is referred to as mature fine tails (MFT). Reclamation of the MFT has been examined using both "wet" (retention of fluid character in aquatic ecosystems) and "dry" (deposits suitable for terrestrial vegetation) approaches (Gulley & MacKinnon 1993). The retention of this material as a fluid will require stable secure holding areas, while terrestrial landscapes may allow for easier reclamation. To this end, a CT process, known as composite tailings at Syncrude and consolidated tailings at Suncor, has been developed to answer this challenge to the oil sands industry. Under present plans, over 50% of the extraction tailings will be handled in the CT process.

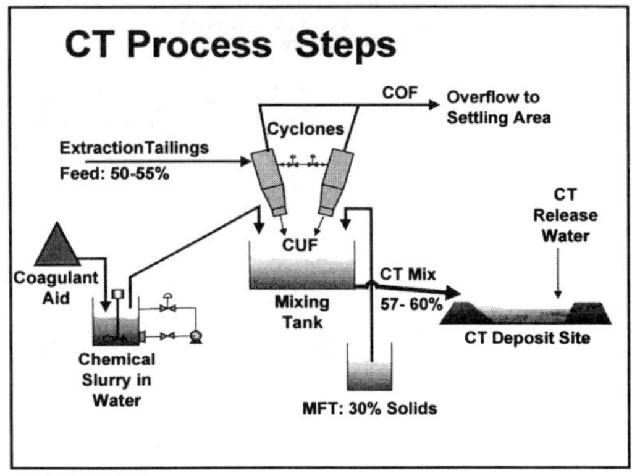

Figure 1: Schematic of the CT Process. Solids contents expressed as wt%.

2 CT OR NON-SEGREGATING TAILINGS

In the CT process, the extraction tailings (ET) stream is densified using a hydrocyclone (Fig. 1). The densified cyclone underflow (CUF) is combined with mature fine tails (MFT) and a coagulant aid to form a non-segregating slurry (FTFC 1995). Unlike the conventional extraction tailings, when the CT slurry is deposited, the fines remain with the coarse sand fraction, and particle-free water is released. The CT process enhances the permeability of the mix and the sand acts as a stress or load, which enhances the dewatering of the deposit (Matthews et al. 2000).

The goal of the CT process is to retain and consume MFT with the coarse solids so that it will:

- allow for the development of a sustainable and acceptable landscape, and
- produce release water with no detrimental effects on the ongoing operations, or the performance of end-of-lease landscape components.

The fines content in the CT mix is expressed as the SFR, the ratio of sand (>44*u*m) to fines (<44*u*m) fractions. The SFR is controlled by the MFT addition to the CUF during the mixing stage (Figure 1). The segregation boundary of CT depends on:

- density (solids content) of mix (wt%),
- fines content (SFR, >44*u*m/<44*u*m), and
- type and dosage of coagulant aid.

In the present paper, the role of the coagulant aid on CT-release water chemistry will be discussed.

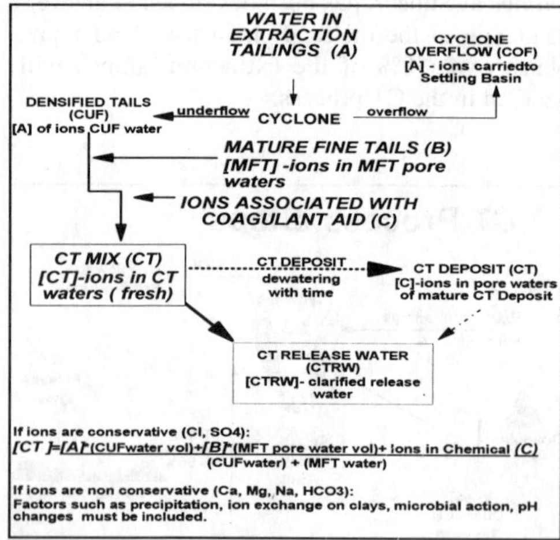

Figure 2: Ionic content of CT-release water (CT) as a function of extraction tails (A), MFT (B) and a coagulant aid (C). The nature of ionic species will affect the predicted levels.

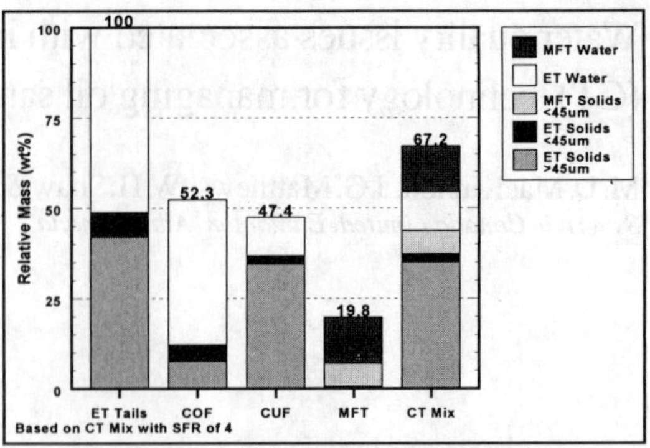

Figure 3a: Relative mass of components in CT process and the CT product. Normalized to 100t of extraction tails (ET) delivered to hydrocyclone.

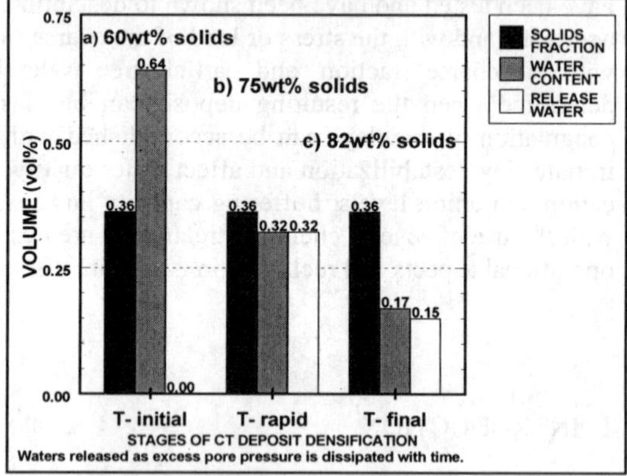

Figure 3b: Predicted fate of water (vol%) in CT Mix (initial) after deposition, in the short term (rapid) and long term (final).

The composition of the CT-release water depends on the feed streams and the coagulants used (Fig. 2):

- extraction process water: composition reflects water management, process chemicals, and leachable components from the oil sands ore,
- MFT pore water, and
- coagulant dissolution, ion exchanges on clays, and reactions.

2.1 *Distribution of Solids and Water in CT Process*

In producing CT, various tailings streams are combined (Fig. 1 & 2). In Figures 3a and 3b, a distribution of the water and solids components from these streams in the resulting CT deposit is shown. About half the water in the CT mix is derived from the MFT that is added to the densified extraction tails (CUF) to produce the desired sand to fines ratio (SFR) (Fig. 3a). During hydrocycloning, most of the extraction water and fines (solids less than 44*u*m) are carried over in the overflow stream (COF) to a conventional settling basin. As a result, most of the fines and water in the CT come from the MFT. This is an important factor when predicting CT release water quality (Fig. 2).

During CT production, the recipe followed will determine the source of the waters and fines in the deposited material (Fig. 3b). Upon deposition, water begins to be released relatively rapidly from the deposit (Pollock et al. 2000). This enhanced permeability and dewatering makes the CT process an attractive tailings disposal option.

As the CT deposit densifies (from 60% to 75% solids) over this initial period (days to weeks), about 50% of initial water can be expected to be released. This particle-free release water is available for recycle and use in the extraction process (Fig. 4). After deposition, further dewatering proceeds as the excess pore pressure of the CT deposit is dissipated. The rate and extent of this dewatering depends on deposit composition, density, rate and thickness of deposition, treatment efficiency and permeability (Pollock et al. 2000). In a scenario where the deposit is contained, another 20-30% of the CT water will be released. As a result, the final contained CT deposit volume would be reduced by over 40% of the initial CT mix.

The integration of CT into tailings management systems (Fig. 4) will affect water management in an oil sands operation. The CT release water must be suitable for recycle and not limit reclamation options. The impact of the chemical used must be considered. In addition, changes in water quality will occur after deposition through exchanges with the atmosphere, interaction with clays, dilution with other waters, and *in situ* changes within the deposit (Fig. 4).

In the initial deposition area, the CT release water will be held briefly, prior to transfer to a larger settling basin from which the recycle water needs of the plant are drawn. Within the settling pond, a combination of longer retention times and commingling with other non-CT water streams will further amend water properties (Mikula et al 1996).

2.2 *Water Quality Changes in CT Process*

Chemical amendments in CT production can lead to shifts in pH, alkalinity, and salinity, as well as reactions and exchanges with the clay components, and the addition of conservative ions associated with

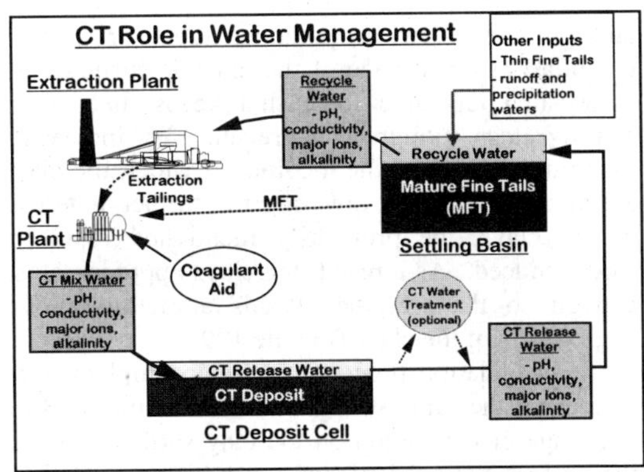

Figure 4: Schematic of pathway of CT release water and impact on recycle water quality.

the coagulant aid itself. After deposition, further changes from pH buffering with atmosphere, equilibria between the water and solid phases, and microbiological activity as the CT deposit goes from aerobic to anaerobic conditions will affect water properties. The overall effects of these combined changes are not well understood.

During the development of CT, various coagulant aids have been evaluated. Many chemicals have been shown to produce non-segregating mixtures with desired geotechnical properties over the operating ranges expected (Matthews et al. 2000). Some of these include acid, lime, acid/lime combinations, gypsum, alum, and polymers. Ongoing work is proceeding at modifying treatments through various combinations (i.e. lime/CO_2, gypsum/lime, CO_2). Results from these treatments will not be discussed in the present paper.

While the various chemicals will produce CT at suitable coagulant dosages, there are significant differences in the composition of the resulting CT water. Some of the issues raised about this CT release water (CTRW) include:

- increasing salinity,
- scaling potential from increased Ca^{+2} content,
- changes in pH,
- corrosion potential ,
- potential effects on oil sand processibility,
- impact on recycle and reclamation.

2.3 *Coagulation Step in CT Process*

The chemicals identified as effective coagulant aids initiate the coagulation process through changes in the ionic strength or character of the medium in which the clays are found or through the exchange of cations on the surface of the clays. In non-treated tailings, the electrical potential (zeta or surface potential) at the surface of the clays is sufficient to establish a repulsive force that keeps them from coming close enough to aggregate. By increasing the ionic strength of the medium in which the clays are dispersed, and by lowering the surface potential of the clays, the previously mentioned repulsive force reduced. As a result, the clays approach close enough so that van der Waals attraction allows aggregation of the clays (Levine 1993).

The coagulation process can be accomplished by increasing the ionic strength of the medium and by exchange or adsorption on the clay surfaces so that the surface potential of the clays (degree of effect depends on the cation) is reduced. As the repulsion associated with the double layer of the clay is reduced, the attractive forces of the clays are able to overcome the repulsive forces (Hiemenz 1977, Everett 1988,). With aggregation, the effective size of the clays will be increased sufficiently so they can be entrained within the sand fraction in the CT mixes.

The ionic strength required for this to occur is defined as the critical coagulation concentration (CCC). The actual concentration needed to reach the CCC is dependent on the charge or valence (z) of the counter ions. In the Schultze-Hardy rule, the effectiveness of the counter ion, in destabilizing clays, is proportional to z^6 (Na^+ <Ca^{+2}<<Al^{+3}, relative effective coagulation effect, Everett 1988).

In successful CT treatments, the dosage of an inorganic coagulant aid must be sufficient to increase the ionic strength of the electrolyte above CCC levels for tailings clays (mix of kaolinite and illite) (Mikula et al 1996). By adding coagulants, the concentrations of polyvalent cations are increased in the medium, and will be available for exchange on the clay surfaces. If cation exchange occurs, the displaced ion from the clays will be seen in the CT water. At the critical concentration for coagulation, the overall salinity of the CT water, both from the ions associated with the added chemical and those exchanged from the clays, will influence the CT release waters.

3 CT RELEASE WATER QUALITY

The impacts of various chemical treatments in the CT process on the properties, salinity, and levels of specific ions have been determined. Through both small-scale laboratory experiments and large-scale field tests of the viability of the CT process, the properties of the waters have been examined. In the following sections, the release water properties for some of the successful CT coagulant aids are presented. For the purposes of this, the composition of the waters in the CT components (CUF and MFT waters, Fig. 2) are considered the same for all treatments. In this comparison of water qualities, it is assumed that we are treating a CT mix with a solids content of about 60% (wt%) and a SFR of 4. In each treatment case, a dose response of the changes in major properties in the CT mix is given.

3.1 *Treatment Options for CT Production*

For coagulation to be initiated in the CT process, both salinity and availability of polyvalent cations must be increased. At Syncrude, the waters in the mixes are already relatively saline (conductivity in the 3- 4mS/cm range), so the addition of coagulant aid will increase the overall salinity, but only by 10-50%. However, in many of the cases, shifts in pH, increases in polyvalent cations or the release of available cations for exchange on the clays (Ca^{+2}, Al^{+3}) will be sufficient to allow coagulation. Some of these changes for several coagulant aids in the Syncrude context will be described.

3.1.1 ***Acid (H_2SO_4)***

Sulphuric acid is a effective coagulant aid that is readily available chemical. It has been shown to produce non-segregating deposits over a wide operating range (FTFC 1995). However, the resulting changes in pH and ionic content of the CT release waters have made this an unattractive option.

As acid is added, the pH is buffered by the high alkalinity of the tailings water (Fig. 5). Beyond dosages of about 20meq/L (650g of concentrated H_2SO_4, density of 1.84g/ml, per m^3 of CT mix), the rate of change in pH and of the other variables becomes more pronounced. The initial alkalinity (expressed as dissolved inorganic carbon, DIC) of the water in the CT mixes prior to treatment is about 20meq/L. The initial acid addition is buffered by this alkalinity. Once the alkalinity is consumed, a rapid decrease in pH was observed. Salinity increases were evident, with ions such as Na^+ and Ca^{+2} being leached from the clays and SO_4^{-2} being added into the water from the acid (Fig. 5). In addition other ions such as Mg, Al and Fe were elevated in the CT waters.

With acid, non-segregating character was observed at dosages in the 1000-1200mg/L (FTFC 1995). At this dosage, there is an excess of calcium ions present in the mix (Fig. 5). Coagulation of the clays could be initiated by this and other polyvalent cations in the aqueous medium or through its exchange on the clays for Na^+. On a positive note, during the acidification step, there was a removal of dissolved organics (naphthenic acids) and when pH was returned to the neutral range, the acute toxicity (based on Microtox bacterial bioassay) was removed.

3.1.2 *Lime (CaO or $Ca(OH)_2$)*

During the development of CT as a tailings option, lime (quicklime-CaO or hydrated lime-$Ca(OH)_2$) was shown to be a robust and effective chemical for the production of non-segregating deposits over a wide range of operating conditions (FTFC 1995, Caughill et al. 1993). In lime treatment, the pH is elevated as lime dissociates and releases OH^- ions. As the pH in the CT mix increases, the bicarbonate-carbonate (HCO_3^--$CO_3^{=}$) equilibrium shifts towards $CO_3^{=}$.

The carbonate reacts with the Ca^{+2} ion added with the lime to produce a calcite ($CaCO_3$) precipitate. This loss of inorganic carbon species (DIC) lead to a reduction of the conductivity in the CT water (Fig. 6). As a result, even though large quantities of Ca^{+2} are added to the CT mix, no free Ca^{+2} is observed until $Ca(OH)_2$ dosages exceeded 1250g/m^3 (about 35meq/L). Prior to treatment, the CT mix had about 20meq/L of inorganic carbon (or about 40meq/L of the CO_3^{-2} species) to react with the Ca^{+2} ion. After removal of the CO_3^{-2} as calcite (>1250mg/L of lime), a more rapid increase in pH is seen with increasing dosages (Fig. 6). However, not as much free Ca^{+2} ion was evident in the produced water as expected. Rapid increases in conductivity and Na^+ are observed. This suggests that divalent Ca^{+2} is exchanging with Na^+ on the clays in the mix (Fig. 6).

For similar mixes, it was determined that the optimum dosages necessary for making a good performing non-segregating mix are in the 1500g/m^3 range (Matthews et al., 2000). This means that the resulting produced waters would have high pH and salinity, with low alkalinity. The change in ionic content, the increase in Ca^{+2} in the water and its exchange on the clay surfaces, appear to push the mix beyond the CCC for the clays. In lime treatment, the resulting high pH of the deposit will lead to the dissolution of the organic acids (naphthenic) into the water phase. Some of these will be lost as the Ca+2 salts but an increase in the CT water will occur. As a result, the lime-produced CT waters displayed high levels of acute toxicity (Microtox bioassay). Risks associated with recycle of this water back into the process, as well as potential terrestrial and aquatic reclamation impacts must be addressed.

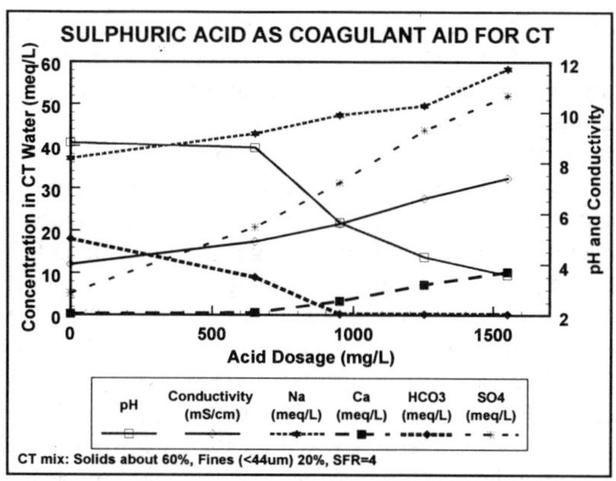

Figure 5: Change in water properties as a function of sulphuric acid dosage in a CT mix of about 60% and a SFR of 4.

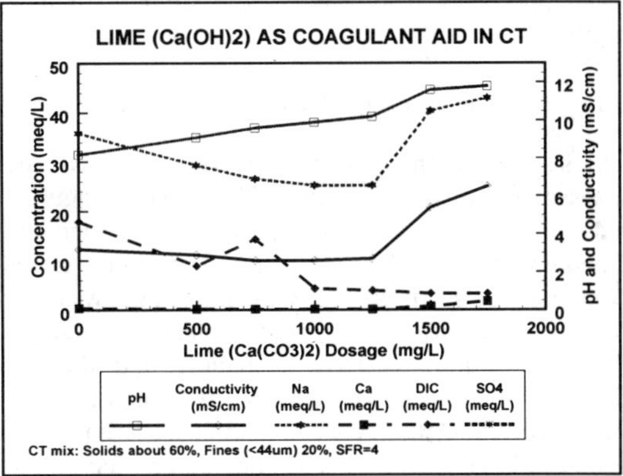

Figure 6: Change in water properties as a function of lime ($Ca(OH)_2$) dosage in a CT mix of about 60% solids and a SFR of 4.

3.1.3 *Acid/Lime (H_2SO_4/CaO)*

As was seen for the acid and lime options, some issues surrounding water quality are evident (Fig. 5 & 6). In lime treatment, most of the chemical aid is actually wasted in the formation of calcite. An option to minimize this inefficiency involved adding acid (H_2SO_4) initially to reduce the pH (6-6.5) and the alkalinity (300-500mg/L), followed by the addition of lime (CaO) (FTFC 1995, Liu et al. 1996). This approach results in more effective release of Ca^{+2} ion from the lime into the electrolyte, as well as allowing exchange on the clays. In the first stage, the pH drops to about 6, while in the lime step, the pH returned to the 8.5-9.0 range (Fig. 7). At this pH, the loss of Ca^{+2} as calcite is not favoured. As a result, free Ca^{+2} was seen (40-60 mg/L) in the produced waters. In Figure 7, changes in water properties for a range of acid/lime mixes is shown. Competing reactions between the acid and lime stages are occurring. Based on the development of non-segregating deposit characteristics, dosages of about 700/500g/m^3 (acid/lime) were found effective over an expected operating range. The produced CT displayed enhanced dewatering rates and good depositional characteristics. The use of the acid resulted in an overall increase in the conductivity of the CT release waters, primarily through the addition of SO_4^{-2} from the acid. The loss of alkalinity was significant (>50%) but was less than with either of the chemicals individually. A decrease in acute toxicity (Microtox bioassay) was seen in this treatment. With the drop in pH during the acid step, there was a removal through adsorption of some of the naphthenic acids, that remained tied up even when pHs were readjusted to the slightly alkaline region.

3.1.4 *Gypsum ($CaSO_4.2H_2O$)*

The success of the acid/lime treatment lead to the assessment of gypsum ($CaSO_4.2H_2O$) as a coagulant aid. Gypsum proved to be a robust chemical amendment over wide operating conditions (solids contents, SFRs). It is readily available as a waste of other industries (fertilizer plants, flue-gas desulphurization), is easy to handle and the CT deposit performed comparably to that of lime-produced CT (Matthews et al. 2000). It has been tested in large field tests and is being applied to full scale operations (Shaw et al., 1996, Pollock et al

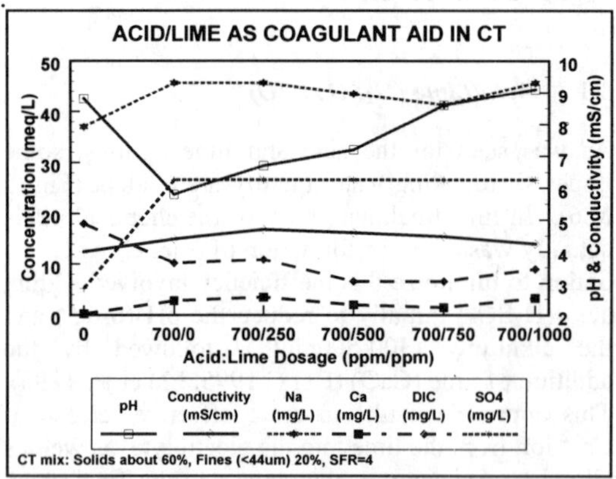

Figure 7: Change in water properties as a function of acid/lime (H_2SO_4/CaO) dosage (g/m^3) in a CT mix of about 60% and a SFR of 4.

2000). At Syncrude, gypsum has been chosen as the base case for planning and implementation of composite tailings.

Gypsum is an acid salt, which will slightly reduce the pH of the CT mix during treatment from >8.5 to about 8.0 (Fig. 8). This pushes the carbonate cycle towards bicarbonate, so precipitation of Ca^{+2} as calcite is not favoured. This allows for greater levels of "free" Ca^{+2} in solution (Mikula, et al., 1996, Shaw et al., 1996). The possible competing reactions of the calcium have been shown to be a combination of cation exchange on the clays present (subsequent release of Na^{+}), precipitation of carbonates (results in reduction in alkalinity), and dissociation (increase in Ca^{+2} in the electrolyte) (Fig. 8). With gypsum treatment, the mechanism for the coagulation step will be a combination of increasing total electrolytic salinity and the divalent cation levels, and shifting clay surface character through the exchange of sodium with calcium. The resulting aggregation is quite evident as the viscosity of the CT mix increases dramatically when the optimum CCC is met. Gypsum proved over a number of laboratory and field tests to be easy to use and handle and robust over a wide range of operating conditions (Shaw et al., 1996).

For gypsum, optimum dosages for reliable CT production were in the 1000-1200g/m^3 range (Matthews et al. 2000). With the aid of mass balances, the fate of the added Ca^{+2} was assessed. There appear to be three main sinks for the added calcium from the gypsum:

- about one third exchanges on the clays (release of Na^{+}),
- one third is lost as the calcite precipitate (reduction in alkalinity), and
- one third is present as the Ca^{+2} in solution.

This means that produced CT waters will have Ca^{+2} contents in the 70-120mg/L level. This depends primarily on pH and clay content of the mixtures.

The other significant change in the gypsum treatment is the addition of the counter ion, SO_4^{-2}, into the CT release waters. At optimum dosages, this will result in an ionic loading to the CT release waters of about 1000mg/L per treatment cycle. Most of the increase in salinity associated with the gypsum-treatment approach results from this addition of the relatively conservative sulphate ion.

The particle-free waters released after gypsum treatment affect both aquatic and terrestrial toxicity. With the slight decrease in pH in gypsum treatment and the initiation of coagulation of the clays, a loss of 20-30% of the naphthenic acids levels from the pre-treatment levels was observed (Fig. 8a). This removal results in the reduction of the acute toxicity

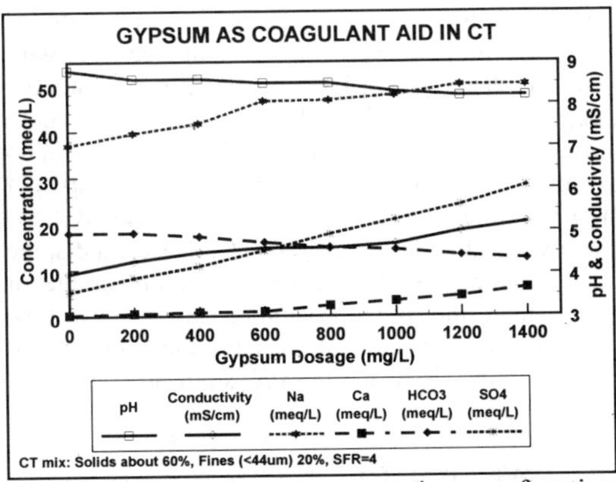

Figure 8a: Change in water properties as a function of gypsum ($CaSO_4.2H_2O$) dosage (g/m^3) in a CT mix of about 60% and a SFR of 4.

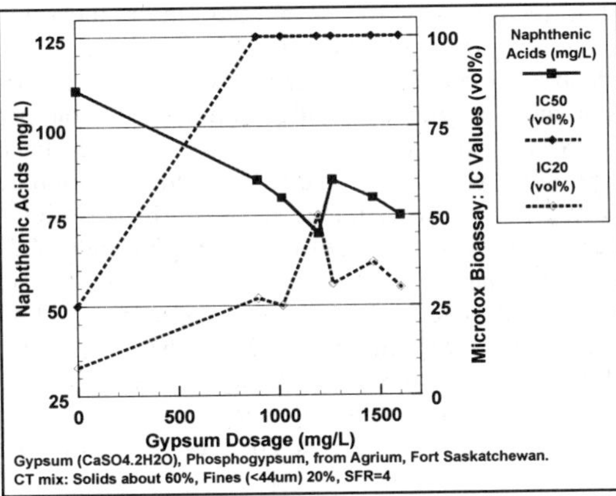

Figure 8b: Change in naphthenic acid content and acute toxicity in gypsum-CT release water from Syncrude's 1997 prototype field test. Toxicity is reported as IC_{50} and IC_{20} (vol%).(Microtox bioassay).

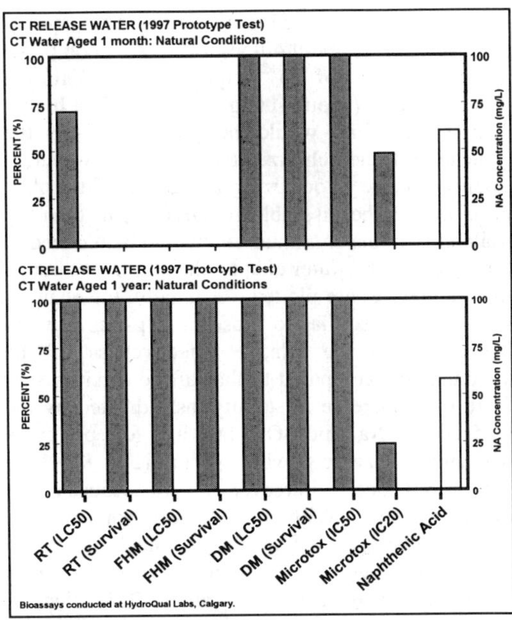

Figure 8c: Change in bioassay results for gypsum-CT release waters from Syncrude field test after 1 month and 1 year of ageing under natural conditions. RT- rainbow trout, FHM- fathead minnow, DM- *Daphnia magna*, Microtox- bacterial bioassay.

(Microtox bioassay) of the release waters (Fig. 8b). Chronic or sub-lethal toxicity of the gypsum-produced CT waters improves dramatically with time. After only one year of ageing under natural conditions, the CT release water from gypsum treatment passed most of the available acute and chronic bioassays (Fig. 8c). In studies of potential terrestrial impacts of gypsum CT waters on vegetation, it has been shown the major factor to be addressed with respect to gypsum CT water will be salinity (Renault et al. 1998). Little evidence of trace metal or organic impacts has been indicated. Ongoing studies are being undertaken in this area.

3.1.5 *Alum ($Al_2(SO_4)_3.14.3H_2O$)*

The large increase in salinity, particularly the free-Ca^{+2} levels, in gypsum-CT waters prompted an investigation of potential alternative coagulant aids to gypsum. The intent was to find a chemical that would be as effective as gypsum, but have less impact on water quality. One chemical that seems to meet this is alum (papermakers alum, a 48% solution of $Al_2(SO_4)_3.14.3H_2O$). In alum treatment, there is a complex series of reactions involving the dissociation of the aluminum salt to release free-Al^{+3}, which then undergoes a rapid hydrolysis to various charged hydroxide intermediates (Amirtharajah & Mills, 1982). The effect of the Al^{+3} species in the electrolyte on the clay interactions should follow the Schultze-Hardy z^6-rule. The required concentration of the Al^{+3} ion should be less than 10% of that of a Ca^{+2} ion. However, at the pH of the CT mixes (8-9), alum effectiveness is not optimized. In addition, cation exchange on clays in the resulting high-density slurries is possible.

From laboratory and flume testing, the effective concentration of alum (48% solution) for production of CT over expected operating ranges (solids and

SFRs) is in the 1000-1250g/m^3 range. This equates to the addition of about 40-55g/m^3 of Al^{+3} compared to about 250g/m^3 of Ca^{+2} required at the optimum gypsum dosage (about 1000g/m^3). This is a higher alum dosage than would be predicted from the application of the Schultze-Hardy rule. However, if cation exchange is occurring and some of the Al^{+3} ion is lost as the insoluble $Al(OH)_3$ form, then the availability of the aluminum ion would be reduced.

In Figure 9, the water quality changes over a range of alum dosages are plotted. Generally, the changes with alum are similar to those of gypsum, but the degree reflects the relative effectiveness of the aluminum salt compared to that of the calcium salt. There is an increase in salinity, as indicated by the conductivity, Na^+ and SO_4^{-2} but, it is less per cycle of treatment than seen with gypsum (Fig. 8). Free calcium is a concern in recycle waters because of the scaling potential. With alum, there is some release of Ca^{+2}, although not from the chemical added, but through cation exchange. The increase in Ca^{+2} from the alum treatment will be about 25% that of gypsum. Alum is acidic, so the drop in alkalinity (expressed as DIC) seen results from the buffering reaction of the bicarbonates in the mixes (Fig. 9). Since alum dosages are lower, the resulting drop is less than with gypsum. The composition of the alum-produced CT water will show an increased salinity, but the overall changes will be less and operational risks would be lower than with gypsum. However, until the chemical has been fully tested under operational field conditions to assess its robustness and the depositional properties of the CT it is premature to recommend this as an alternative treatment approach.

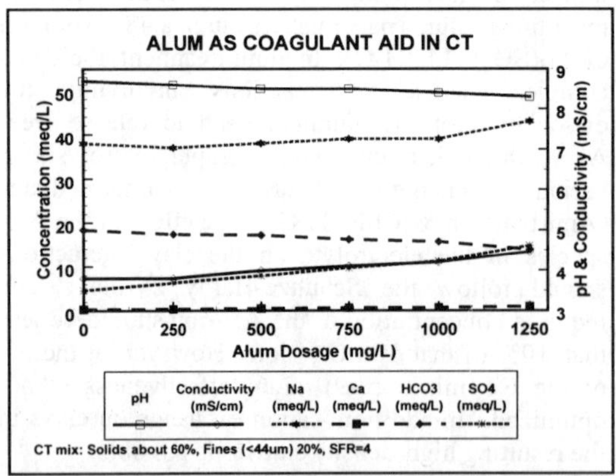

Figure 9: Change in water properties as a function of alum ($Al_2(SO_4)_3$. $14.3H_2O$) dosage (g/m^3) in a CT mix of about 60% solids and a SFR of 4.

The toxicity of the alum release waters is reduced in the CT process, but to a lesser degree than with gypsum. There have been concerns over potential bioavailability of Al from the alum. Studies on alum usage in water treatment have shown that there is no evidence of pathways for delivery of biologically significant levels of aluminum in natural systems (Stauber et al. 1999). In alum-CT release waters, the Al^{+3} content was less than 1mg/L (comparable or lower than the other coagulant aids). To help confirm this, ongoing studies on the aquatic and phytotoxicity associated with alum-produced CT waters are underway.

3.1.6 *Polyelectrolytes*

The success of the CT process requires the coagulation of the clays. A number of inorganic chemical will achieve this clay destabilization. Unfortunately, as we have seen, these inorganic coagulants affect water quality. Ideally, the use of an organic chemical, that performed as well as the inorganic coagulants, could eliminate the salinity concerns. Flocculation of fine tails has been reported with high molecular weight anionic polymers (polyacrylamides such as Allied Colloids Percol LT27A) at dosages in a 10-30g/m^3 range (Xu & Cymerman 1999). When used as a coagulant aid for CT, dosages of 50-100g/m^3 were required to make non-segregating mixes (Matthews et al 2000). While the impact on water quality was negligible, its deposition and strength performance for the produced CT were not as good as seen with the inorganic coagulants.

Currently, no organic coagulants have been identified as acceptable replacements for the inorganic chemicals discussed earlier. Polymers will have negligible impact on water quality, but unless they can produce deposits with good geotechnical properties and with acceptable dewatering and stability characteristics, they cannot be considered as viable alternatives to inorganic coagulants. While the concept of organic coagulants is still a goal, the present screening work has not indicated that they are robust enough for consideration in commercial application.

4 SUMMARY OF WATER QUALITY EFFECTS

The production of a non-segregating mixture with acceptable handling, transport, depositional and post-depositional properties over an operational range of CT mix solids contents (45-65wt%) and fines content (SFR of 3-5) has been demonstrated with several coagulant aids (FTFC 1995, Matthews et al 2000). As can be seen in Figure 10a, coagulant aids will impact the composition of the water released from the resulting deposits. They can directly add ions through dissolution, release ions into solution through leaching or cation exchange reactions, or lead to precipitation or shifts in equilibria established in the CT mixes.

In Figure 10b, the relative changes in water quality for each coagulant aid, at what is considered an effective dosage (acid- 1100g/m^3, lime 1500g/m^3, acid/lime- 700/500g/m^3, gypsum- 1000g/m^3, alum- 1250g/m^3, polymer- 75g/m^3) for an operational mixture (solids- 60%, SFR- 4) are shown. In this plot, the impacts of treatments are expressed as expected changes per treatment cycle. For each variable, the changes are reported relative to non-treated release water. The general trends in water quality for each of the CT chemicals are summarized in Table 1. In most cases, an increase in salinity will occur, with special attention needed regarding the increase in divalent cations in the release waters. The addition of CT waters into the recycle water system (Figure 4) means that impacts on plant operations must be considered. The increasing salt loading and the presence of higher levels of ions such as Ca^{+2} raise issues regarding:

- the potential for scaling in process units,
- the potential increase in corrosion of susceptible metallurgy, and
- the interference with efficiency of the bitumen recovery or froth treatment in extraction.

These are sensitivities that will need to be monitored closely during the implementation of the CT process at full-scale rates starting in 2000. Over the longer term, aspects pertaining to the eventual reclamation of the CT influenced areas (both aquatic and terrestrial) must be considered. The general increase

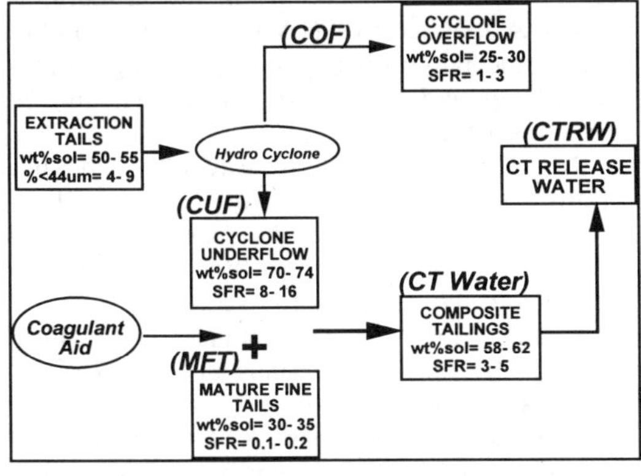

Figure 10a: Schematic of composite tailings (CT) production and resulting release water.

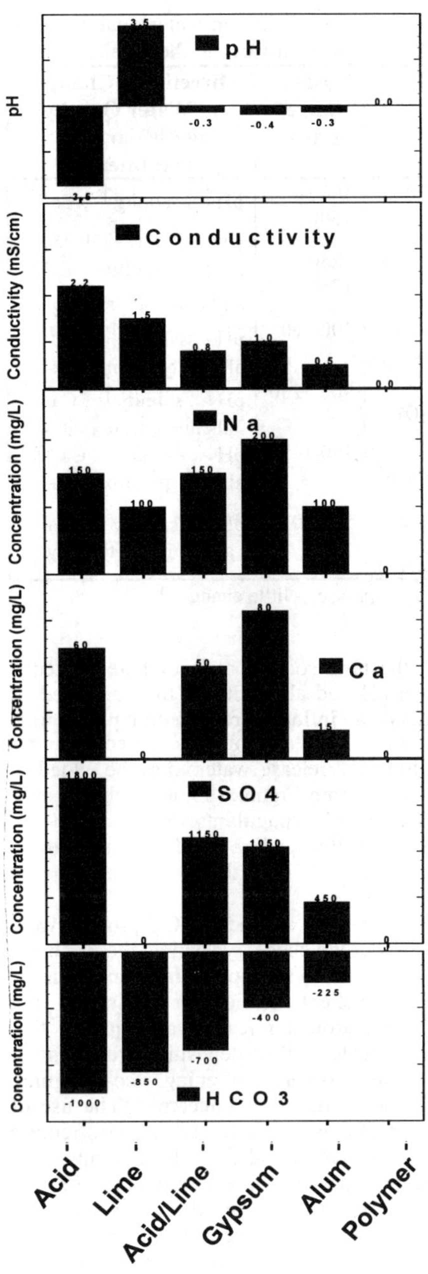

Figure 10b: Changes in CT release water properties per treatment cycle under similar operating conditions (solids- 60%, SFR- 4) using acid (H_2SO_4- 1100g/m^3), lime($Ca(OH)_2$-1500g/m^3),acid/lime (H_2SO_4/CaO- 700/500g/m^3),gypsum ($CaSO_4.2H_2O$- 1000g/m^3), alum ($Al_2(SO_4)_3.14.3H_2O$- 1250g/m^3), and polyelectrolyte (LT27A- 75g/m^3).

Table 1: Summary of effects on water quality in the CT process expected with various chemicals.

Chemical	Dosage Range (g/m^3)	Directional Changes in Water Quality with Various Treatments*
Acid (H_2SO_4)	1000-1500	pH↓ , salinity↑ , Ca^{+2}↑ , alkalinity↓, toxicity↓
Lime (CaO or $Ca(OH)_2$)	1250-1750	pH↑ , salinity – , Ca^{+2}– , alkalinity↓ , toxicity↑
Acid/Lime (CaO)	700/500	pH– , salinity↑ , Ca^{+2}↑ , alkalinity↓ , toxicity↓
Gypsum ($CaSO_4.2H_2O$)	900-1250	pH↓ , salinity↑ , Ca^{+2}↑ , alkalinity↓ , toxicity↓
Alum ($Al_2(SO_4)_3.14.3H_2O$)	750-1250	pH↓ , salinity↑ , Ca^{+2}↑ , alkalinity↓ , toxicity –
Polyelectrolytes (anionic polyacrylamides)	50-100	pH – , salinity – , Ca^{+2}– , alkalinity – , toxicity –

* ↓ decrease,↑ increase, – little change

in salinity, the high sodium and sulphate contents, and the final pH and alkalinity of the contained or release waters may influence reclamation planning.

Depending on the treatment, specific components in the resulting CT release waters may be added or lost (Fig. 10b). From Figure 10b and Table 1, It is obvious that organic coagulants demonstrated the least impact on water quality. Unfortunately, these coagulants produced CT that appeared to be undesirable.

When all the factors around the CT process were considered, Syncrude chose gypsum as the preferred coagulant for the initial startup and implementation of CT. It is a excellent chemical for making CT, but there are issues around release water quality that need to be evaluated. Post-deposition treatment of the CT release water (softening, co-mingling, mixing) may alleviate water concerns. The use of alternative chemicals, such as alum, will require field-testing to ensure that they perform comparably to gypsum and produce a CT product with the desired properties. In addition, the water produced from these alternatives must be shown to be superior for both recycle and reclamation to that of gypsum if they are to be considered a viable replacement.

5 ACKNOWLEDGEMENTS

This project required the support of many colleagues who have helped in the lab and field components of this study. Project Development and Operations groups were associated with many aspects of this study. Excellent support for the analytical aspects of this study was provided by Syncrude's Operations Lab at site and the Research Lab in Edmonton Research facility. In particular, the authors thank Betty Fung for her support and analytical expertise.

6 REFERENCES

Amirtharajah, A. & K.M. Mills 1982. Rapid-mix design for mechanisms of alum coagulation. J. AWWA. 74(4): 210-216.

Caughill, D.L., N.R. Morganstern, & J.D. Scott 1993. Geotechnics of nonsegregating oil sand tailings. Can. Geotech. J. 30: 801-811.

Caughill, D.L., J.D. Scott, Y. Liu, R. Burns, & W.H. Shaw 1994. 1993 Field Program on Non segregating Tailings at Suncor Inc., *Proceedings 47th Canadian Geotechnical Conference,* Halifax, September, 1994.

Everett, D.H. 1988. *Basic Principles of Colloid Science*. London: Royal Society of Chemistry.

FTFC (Fine Tailings Fundamentals Consortium) 1995. In: *Advances in Oil Sands Tailings Research*, Alberta Department of Energy, Oil Sands and Research Division. 1995.

Gulley, J., M.D. MacKinnon 1993. Fine Tailings Reclamation Utilizing a Wet Landscape Approach, Proceedings Fine Tailings Symposium. *Proceedings Fine Tailings Symposium. Oil Sands - Our Petroleum Future Conference, Edmonton, April, 1993*. Paper F23 pp24.

Hiemenz, P. C. 1977. *Principles of Colloid and Surface Chemistry*. New York: Marcel Dekker.

Levine, S. 1993. Mathematical modelling of the microstructure present in oil sands fine tailings. *Proceedings Fine Tailings Symposium. Oil Sands - Our Petroleum Future Conference, Edmonton, April, 1993*. Paper F8 pp.43.

Liu, Y.B., D.L. Caughill, W. Shaw, E. Lord, R. Burns, & J.D. Scott 1996. Volume reduction of oil sands fine tails utilizing nonsegregating tailings. In *Tailings and Mine Waste '96.* Rotterdam:1995 Balkema. pp73-81

Lo, R.C., E.R.F. Lord, J.M. Coward 1994. Oil Sands Tailings Reclamation Involving Sand-Fines Mixtures. In: *Proceedings First International Congress on Environmental Geotechnics, Edmonton, July 1994*. pp. 519 - 524.

MacKinnon, M.D. 1989. Development of the tailings pond at Syncrude's oil sands plant: 1978-1987. AOSTRA J. Res. 5: 109-133.

Matthews, J.G., W. H. Shaw, R.G. Cuddy & M.D. MacKinnon 2000. Development of composite tailings technology at Syncrude Canada Ltd. In:

Proceedings of SWEMP 2000, Calgary Alberta, May30-June 2, 2000.

Mikula, R.J., K.L. Kasperski, & R.D. Burns 1996. Consolidated tailings release water chemistry. In: *Tailings and Mine Waste '96.* Rotterdam:1995 Balkema. pp459-468.

Mikula, R.J., K.L. Kasperski, R.D. Burns, & M.D. MacKinnon 1996. Nature and fate of oil sands fine tailings. In: *Suspensions: Fundamentals and Applications in the Petroleum Industry.* Washington, DC: American Chemical Society. 677-723.

Levine, S. 1993. Mathematical modelling of the microstructure present in oil sands fine tailings. In: *Proceedings Fine Tailings Symposium. Oil Sands - Our Petroleum Future Conference,* Edmonton, April, 1993. Paper F8. pp1-43.

Pollock, G.W., E.C. McRoberts, G. Livingstone, G.T. McKenna, & J.G. Matthews 2000. Consolidation behaviour and modelling of oil sands composite tailings in the Syncrude CT prototype. In: *Tailings and Mine Waste '00.* Rotterdam: 2000 Balkema. *Fort Collins, January 2000.* pp121-130.

Renault, S. , C. Lait, J.J. Zwiazek, & M. MacKinnon 1998. Effect of high salinity tailings produced from gypsum treatment of oil sands tailings on plants of the boreal forest. Environmental Pollution. 102: 177-184.

Scott, D., Y. Liu, & D.L. Caughill 1993. Fine Tailings Disposal Utilizing Non segregating Mixes, *Proceedings Fine Tailings Symposium. Oil Sands - Our Petroleum Future Conference, Edmonton, April, 1993.* Paper F18 pp. 1 - 19.

Scott, J. D., Y. Liu, & D.L. Caughill 1993. Fine Tails Disposal Utilizing Nonsegregating mixes. In: *Proceedings of Oil Sands — Our Petroleum Future Conference*, Edmonton, Alberta, Canada. Apr. 4 – 7, 1993. F18.

Shaw, W., G. Cuddy, G. McKenna, G., & M. MacKinnon 1996. *Non-segregating Tailings: 1995 NST Field Demonstration.* Edmonton, 96-2: Syncrude Canada Ltd. 4 vols.

Shaw, W., B. Livingstone, J. Beck 1993.. Disposal of Fine Tailings Utilizing Hydrocyclone Techniques. In: *Proceedings Fine Tailings Symposium. Oil Sands - Our Petroleum Future Conference, Edmonton, April, 1993.* Paper F19 pp. 1 - 20.

Stauber, J.L., T.M. Florence, C.M. Davies, M.S. Adams, & S.J. Buchanan 1999. Bioavailability of Al in alum-treated drinking water. J. AWWA. 91(11): 84-93..

Xu, Y. & G. Cymerman 1999. Flocculation of fine oil sand tails. In: *Proceedings of 3rd UBC-McGill Bi-Annual Symposium on Fundamentals of Mineral Processing, Quebec City, August, 1999.* 15pp.

Environmental Issues and Management of Waste in Energy and Mineral Production, Singhal & Mehrotra (eds)

Development of composite tailings technology at Syncrude Canada

J.G. Matthews, W.H. Shaw, M.D. MacKinnon & R.G. Cuddy
Syncrude Canada Limited, Edmonton, Alb., Canada

ABSTRACT: During extraction of bitumen from the Athabasca Oil Sands, an aqueous fines suspension, called mature fine tails (MFT), is produced. The geotechnical characteristics of MFT demand long term storage in geotechnically secure containment areas. The composite tailings (CT) process involves mixing a coarse tailings stream with a MFT stream and adding a coagulant to form slurry that rapidly releases water when deposited and binds the MFT in a coarse tailings/MFT deposit. Thus, more of the fines can be stored in a geotechnical soil matrix, which reduces the inventory of fluid-fine tails and enables a wider range of reclamation alternatives. CT process optimization, coupled with public and regulatory consultation at key milestones, has led to a corporate commitment to implement this technology. This paper reviews key aspects of the evolution of the CT process at Syncrude, including segregation, depositional and geotechnical characteristics of CT mixes that are formed with various chemical aids. Some of the treatments discussed include those based on the use of acid, lime, gypsum, alum, and organic polymers as the coagulant aids.

1 INTRODUCTION

Processing of Athabasca Oil Sands using a water-based extraction process at the current commercial oil sands operations, Syncrude Canada Ltd. and Suncor Energy Inc., generates an increasing volume of clay-rich fine tailings (Fig. 1). In response to the challenge of fine tailings management, Syncrude plans to begin making composite tailings (CT) on a commercial scale in 2000. Initial placement of CT will be in the mined-out area of the present base mine east of Hwy 63 on Lease 17.

This paper will describe the composite tailings process, the benefits of CT, and outline some of the research and development work carried out at Syncrude Canada Ltd. to develop this tailings management technique.

1.1 The Composite Tailings (CT) Process

The CT process involves addition of a chemical coagulant to a slurry, comprising of sufficient solids content and clay minerals fraction, that there is a shift in the coarse-fines segregation boundary. As a result of the treatment, the slurry mix becomes non-segregating during transport, discharge, and deposition (Fig. 2). The product is a material containing the coarse- and fine-grained tailings solids from which a particle-free water is rapidly released upon deposition. The solids are retained within a homogenous and uniform deposit. While these deposits are initially soft and require containment, an increase in strength to allow a relatively rapid reclamation is anticipated.

The water released from the CT deposit is essentially suspended solids-free. It will comprise a significant portion of the Plant recycle water used in Syncrude's bitumen extraction process and other plant operations (MacKinnon et al 2000).

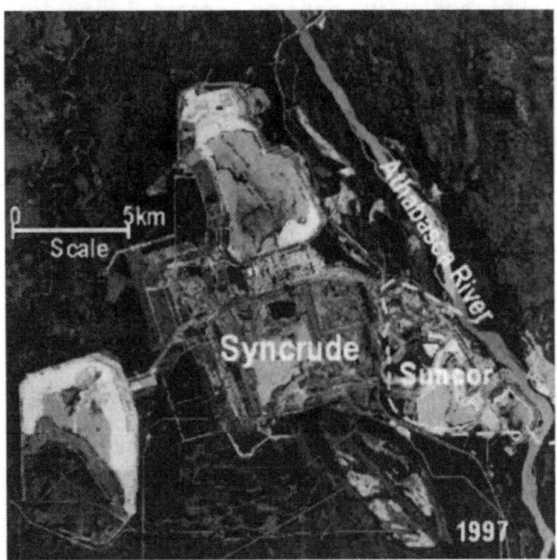

Figure 1: Aerial photo of Syncrude and Suncor oil sands plants.

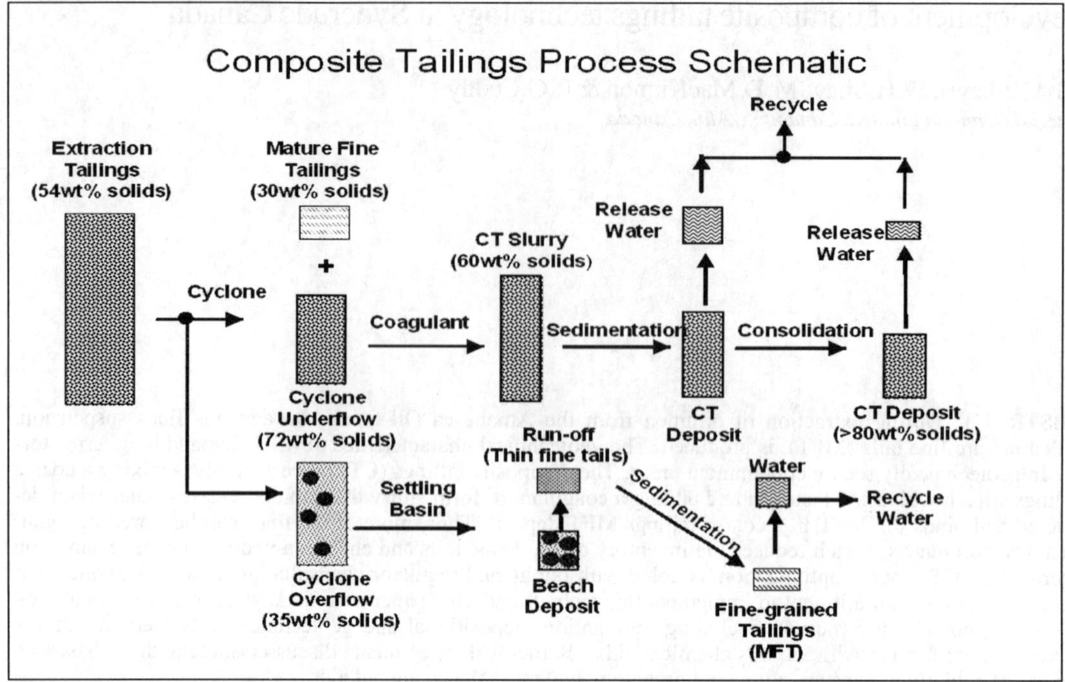

Figure 2: Outline of the CT Process.

1.2 *Benefits of Composite Tailings*

Syncrude, through the implementation of CT, will accrue a number of benefits:

- Existing volumes of fine tailings (MFT) on Leases 17/22 will be maintained or reduced,
- Disturbed areas will be reclaimed faster,
- The percentage of dry landscape at lease closure will be increased — a positive response to public and regulatory concerns expressed regarding the long term management of fluid fine tailings, and
- Tailings management and storage costs will be reduced. Increased reclamation and mine planning flexibility may also provide opportunities for further production cost savings.

2 BACKGROUND

Current tailings management practices at Syncrude involve discharging the tailings slurry onto beaches or into cells. The coarse fraction (>44μm) of the mixture settles rapidly to form a loose water-saturated sand deposit (beaches, dykes). Approximately 50 to 70 percent of the fines (< 44μm) in the extraction tailings slurry segregate from this deposit and wash into a settling basin along with the process-affected waters released during discharge and deposition. This thin slurry of fines (6–10wt% solids) eventually develops into MFT (>30wt% solids).

Without changes to the current tailings management strategy, it is estimated that Syncrude's Leases 17/22 MFT inventory could increase from the current 360 million cubic metres to one billion cubic metres by 2030. This provides incentive to create tailings that do not require fluid containment.

2.1 *Previous Research Efforts in Tailings*

Over the past three decades, the chemical amendment of whole tailings to initiate coagulation of the fines has been assessed. This research has included dry tailings filtration (Liu et al 1977; Coward 1996), centrifugation (Cymerman 1991), and other mechanical enhancements. In the 1980s, work focused on applying the benefits of coagulation to tailings deposition (FTFC 1995; Shaw et al 1996).

Research in the early 1980s indicated that Syncrude coarse extraction tailings could be made non-segregating with the addition of lime (FTFC 1995). Though not developed to commercial implementation, this work set the stage for further development of chemical amendment processes in support of creating non-segregating tailings (CT).

In 1990, lime was added to a Syncrude extraction tailings stream to create a non-segregating beach deposit, but it was very soft and had low beach angles (Cuddy et al 1991). Subsequent laboratory work at the University of Alberta confirmed that SCL extraction tailings could be made non-segregating through the addition of lime (Caughill 1992). Laboratory testing of the deposits created in this work indicated acceptable long-term geotechnical characteristics that would allow a "dry" landscape for reclamation, but they would require containment during placement and dewatering.

These findings coincided with an in-house mine plan review in 1993 and an opportunity was identified to place non-segregating tailings into below-grade mined-out pits. The decision to pursue this opportunity prompted a more comprehensive laboratory program. Combined with these positive technical results and identified opportunities, a number of significant achievements since 1993 supported continued evaluation and development of the CT process including:

- Successful field pilot trial of CT at SCL in 1995 (Shaw et al 1996),
- Suncor initiated commercial operation of CT process in 1996, and,
- Successful full-scale prototype demonstration of CT at Syncrude in 1997/98.

3 CT TECHNICAL ASPECTS

Through extensive research and development initiatives CT has evolved through a number of process changes (FTFC 1995). Factors identified as influencing segregation of CT mixes include:

- fines content (fraction of total solids),
- total solids content (density),
- particle size gradation,
- mineralogy of fines fraction, and
- water chemistry).

These factors can be manipulated individually or in combination to shift the segregation boundaries. The segregation character can be further manipulated with chemical amendments.

3.1 *Segregation Mechanics*

Two segregation boundary curves are given in Figure 3. One curve is for Syncrude tailings without chemical amendment and the other is for gypsum treated tailings (1000g $CaSO_4 \cdot 2H_2O$ per m^3 of CT slurry). The curves represent segregation curves under laboratory testing conditions and a shift in these boundaries would be expected under full-scale discharge and deposition conditions. All %fines - %solids combinations plotting above the segregation line (Fig.3) are non-segregating, those falling below the line will be segregating.

Three basic means of slurry manipulation have been evaluated in the course of the CT development:

1. Increased solids content, primarily by hydrocyclone densification,
2. Increased fines content, primarily through enrichment with settled fine tailings, and,
3. Chemical adjustment by addition of coagulants.

The effect of each of these factors is depicted in Figure 3. Initial slurry with 23% fines and 63% solids (*point A*) plots below the segregation curve for untreated Syncrude tailings and is a segregating mix. This slurry is made non-segregating using three basic approaches above:

In the first case, the slurry is densified (as with hydrocyclones or a thickener) to 68% solids without a change in the %fines in the tailings slurry (*point B*). This shifts the slurry composition above the segregation curve for untreated Syncrude tailings and the slurry becomes non-segregating.

Alternately, the mix can be made non-segregating by enriching the slurry with fines (relative to coarse solids and without a change in slurry solids content). In this case the fines content is increased to 28% (*point C*) and again the new slurry "state" is on the non-segregating side of segregation boundary curve.

Finally, the mix can be made non-segregating by addition of a coagulant (*point A*). In this case, the slurry composition does not shift in %solids-%fines space but instead the appropriate segregation curve for reference dramatically shifted. This example depicts a gypsum-treatment curve. With gypsum, point A is well above the segregation boundary curve and the slurry is non-segregating.

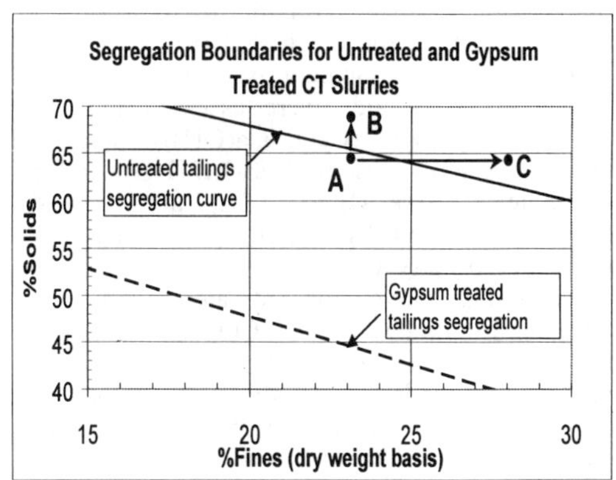

Figure 3: Segregation curves for "raw" and gypsum treated CT slurry.

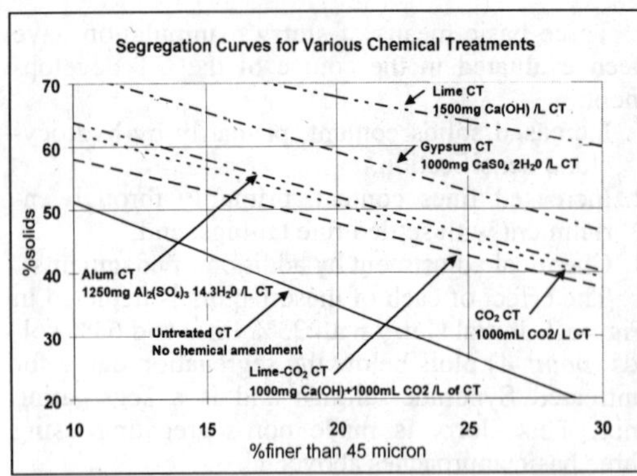

Figure 4: Segregation curves for various coagulant treatments

3.2 *Role of coagulants*

Addition of a coagulant to the CT mixture (coarse tailings-MFT mixture) is an essential component of the CT process. The coagulant produces changes in the properties of the clays in the CT mixture. Coagulation of clays can be initiated by pH adjustment, salinity or specific cation adjustment, or through cation exchange (MacKinnon et al 2000).

Without addition of a coagulant, coarse tailings slurry exhibits a gap-graded particle size distribution; it is the gap-graded nature of the PSD that leads to segregation of the fines from the slurry during discharge and deposition (FTFC 1995). Addition of a coagulant causes the colloidal particles to aggregate. This has the effect of changing the particle size distribution of the solids in the slurry to provide a more uniformly graded mixture that can suppress or eliminate segregation. With the addition of a coagulant, the coarse tailings slurry can be made non-segregating during transport, discharge, and deposition (Scott et al 1993).

During the development of the CT process, various coagulant aids were assessed including acid (H_2SO_4), lime (CaO, $Ca(OH)_2$), acid-lime (H_2SO_4 - CaO), gypsum ($CaSO_4 \cdot 2H_2O$), sodium aluminate ($Na_2Al_2O_3$), alum (48% $Al_2(SO_4)_3 \cdot 14.3H_2O$), lime-$CO_2$, CO_2, and organic polymers (polyacrylamides).

Segregation curves for untreated, lime-, gypsum-, alum-, lime-CO_2, and CO_2-treated tailings slurry are given (Fig. 4). Each curve represents the segregation boundary for a given coagulant treatment and coagulant dosage in %fines-%solids space.

3.2.1 *Coagulant Impact on Water Chemistry*

In MacKinnon et al (2000), the impact of coagulant on the quality of CT waters is discussed. Laboratory and field studies were undertaken to evaluate these impacts.

CT made using inorganic coagulants exhibited the greatest effect on water quality, generally with a resulting increase in overall or specific ionic loading and a shift in the relative ionic composition of the CT release waters. While CT produced using organic flocculants showed no net increase in overall ionic loading of produced waters, the performance of the resulting CT was not deemed acceptable based on its geotechnical properties.

3.2.2 *Coagulant Screening Process*

Following Syncrude's 1990 Field Trial using lime treatment of extraction tailings streams (Cuddy et al 1991), acid and lime were considered as candidate coagulants. As part of this comparison, the release water quality from lime, acid, and acid-lime and gypsum produced CT mixes was assessed (MacKinnon et al 2000). Though all coagulant treatments were found to produce acceptable non-segregating slurries at appropriate dosages (Fig. 4), significant differences were noted in the CT pore- and release-waters quality. As well, the segregation characteristics, deposit dewatering and consolidation rates, and economics were factors in coagulant selection.

Based on on-going research efforts (FTFC 1995), gypsum was proven to be a robust, effective, easy to handle, and readily available alternative to lime and acid-lime treatment. As a result, Syncrude decided to use gypsum as the coagulant for a 1995 field demonstration and the 1997/98 prototype programs. In these programs, slurries that were non-segregating during transport, discharge and deposition and which displayed suitable geotechnical characteristics were produced under operational conditions. Through this evaluation process, gypsum was selected as the coagulant for the commercial application of CT at Syncrude.

Though gypsum has been demonstrated as an affordable and effective coagulant for the CT process, work continues in screening alternative coagulants. Key factors assessed in screening

Figure 5: Syncrude's 1997/98 CT Prototype trial.

alternatives include supply cost, dosage, effects on pore and release water, segregation susceptibility, depositional performance and geotechnical performance.

3.2.3 Gypsum Dosages at Field Scale

In the development of the gypsum-based CT process, differences were noted in dosages required to prevent segregation of the CT solids when lab-scale performance was compared with field-scale performance. Gypsum dosages required for full-scale operation have been found to be as much as twice the dosage recommended based on laboratory findings. Laboratory testing on gypsum treated CT was carried out at the University of Alberta to identify minimum and recommended gypsum dosages for use in the 1995 Field Demonstration program. Through this study, a dosage of 900 grams gypsum per cubic metre of CT was selected for the program. When this dosage was tried during the initial stages of the 1995 Field Demonstration, it was found to be inadequate to prevent segregation of most of the CT slurry produced. In response to this, gypsum dosages were incrementally increased and a minimum field dosage of 1200 grams gypsum per cubic metre of CT slurry was identified. To ensure effective production of non-segregating slurry, the field dosage was boosted further to 1400 grams gypsum per cubic metre of CT slurry. Some of this difference reflects the moisture content and debris in the available (agricultural grade from Domtar).

3.3 Densification of the Coarse Tailings Stream

In Scott et al (1993), it was shown that Syncrude coarse tailings streams can be made non-segregating without a densification step. These predictions are made based on results from laboratory standpipe testing programs. In practice, most Syncrude coarse tailings slurry cannot be made non-segregating without some degree of densification as part of the CT process (Shaw et al 1996).

The process step of coarse tailings densification provides a number of process benefits including reduced treatment costs and an opportunity to consume MFT inventory.

3.3.1 Reduced treatment costs.

Densification increases the solids content of slurry. For a fixed fines content, increasing the solids content reduces the likelihood for the slurry to segregate. Therefore, if the densification process is implemented and dosage rates are left unchanged, the slurry is less susceptible to segregation for a given fines content.

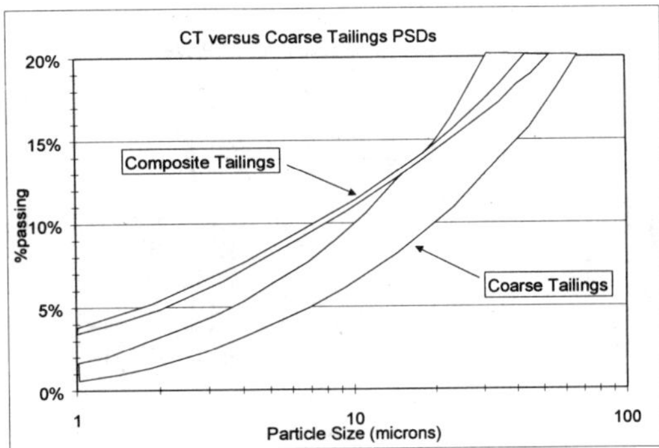

Figure 6: Fines consumption with CT

3.3.2 Opportunity to consume MFT inventory.

Densification of the coarse tailings stream, followed by the addition of MFT, creates an opportunity to consume existing MFT inventory. When CT, with a sand-to-fines ratio (SFR) of 4:1 (< 44μm basis) is produced it has almost twice as much of the <5.5μm fines fraction as an "un-enriched" extraction coarse tailings stream with a SFR of 5:1 (Fig. 6). Since the finest mineral particles are the primary MFT-formers (FTFC 1995), this difference has a significant impact of MFT production. More MFT can be consumed through the densification of the coarse tailings stream, followed by addition of MFT, than can be achieved by capturing all available fines in the extraction coarse tailings slurry alone.

3.4 Addition of Mature Fine Tailings

One of Syncrude's biggest challenges is management of fluid fine tailings inventory. In 1997, Syncrude carried a base mine MFT inventory of approximately 360 million cubic metres. PSDs for typical Syncrude tailings streams are given in Figure 7. CT provides an opportunity to manage the fines content of the CT slurry produced, providing a means of engineering the solids storage efficiency and geotechnical performance of the deposit(s) created. A primary incentive of CT is to create non-segregating slurries with higher fines contents and a resulting consumption of MFT-forming fines. In contrast, controlling the fines-enrichment factor must be considered to optimize tailings storage. The aim of designing the CT slurry is to ensure that these interests are balanced to provide optimum tailings storage efficiency and acceptable geotechnical performance, while maintaining mine planning flexibility.

Factors which were considered in support of fines enrichment included dosage efficiency, MFT content and long-term and depth dependence of the geotechnical integrity of the deposit. Factors that

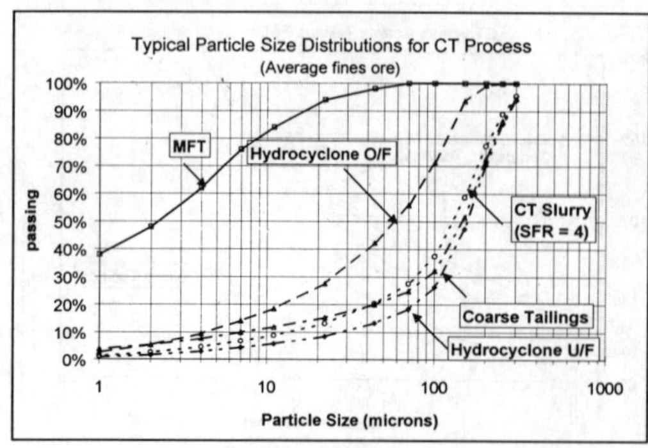

Figure 7: Typical PSDs for CT Process

may limit the fine content of the CT product included slower dewatering and consolidation rates, reduced hydraulic conductivity and reduction in short term/shallow deposit integrity.

The field scale demonstration and prototype programs were important in increasing understanding of the CT geotechnical performance.

Laboratory work done in advance of the field programs was extensive, but it was unable to account for all of the processes at play at full-scale implementation. Almost all of the geotechnical models required re-calibration after the field trials to account for significant differences in process rates and effects between the laboratory and field scale. The current consolidation and permeability models for Syncrude CT are discussed below. These models were initially developed from carefully controlled laboratory testing programs and then re-calibrated based on observed field behaviour from the 1995 and 1997/98 field programs (Pollock et al 2000).

3.4.1 Fines Content and CT Compressibility

Compressibility curves for various sand-to-fines ratios (SFRs, >44μm/<44μm) are given (Fig. 8). These curves represent the state of knowledge regarding anticipated stress-strain performance of the CT deposits based on all work done to date (including laboratory and field data, with precedence given to performance from the 1995 and 1997 field trial programs). Although these curves depict the current state of knowledge, careful monitoring will be necessary with commercial start-up to verify the model parameters and account for a number of significant influences including; larger deposition areas, winter CT deposition, extraction tailings stream variability, and deeper deposit thickness. It may be seen that current models predict increased CT deposit compressibility with increased fines contents (Fig. 8).

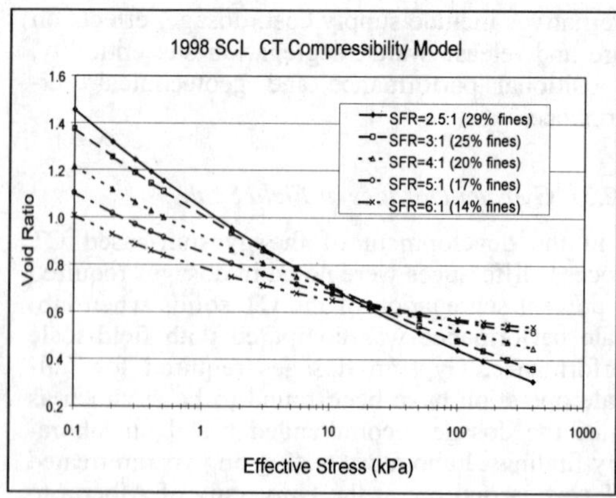

Figure 8: CT Compressibility

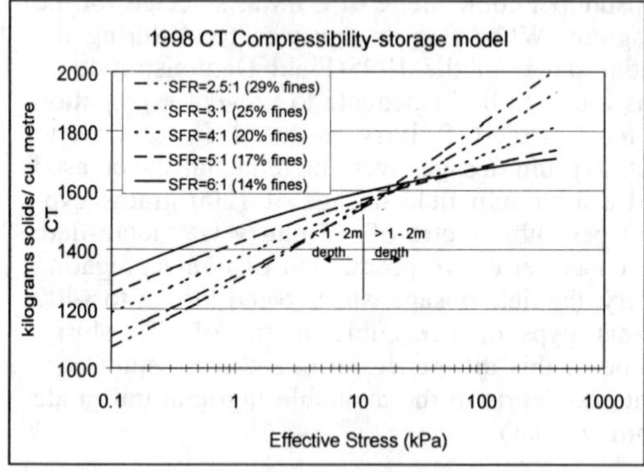

Figure 9: CT Solids Storage Efficiency

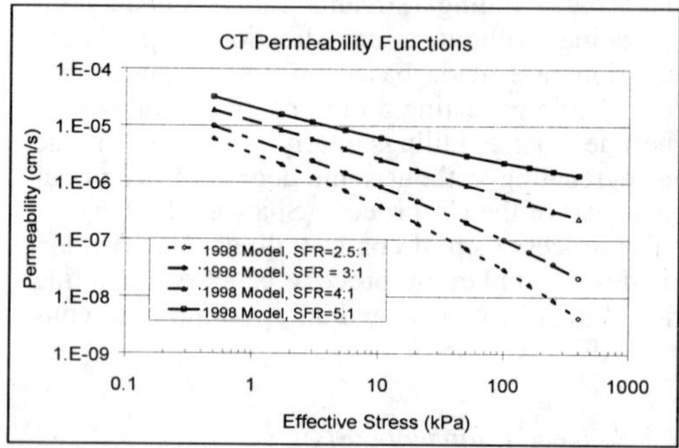

Figure 10: CT Permeability

Storage efficiency of CT can be represented in terms of tonnes of dry solids stored per cubic metre of CT deposit across a range of effective stress conditions (Fig. 9). Model analyses indicate that greater storage efficiency is associated with higher fines contents at depths in excess of approximately two metres (i.e. effective stresses in excess of approximately 20 kPa).

3.4.2 *SFR Influence on CT Permeability*

The permeability of CT at a given void ratio decreases with increased fines contents (Fig. 10). In the plot of the permeability-fines void ratio, there is a single curve describing the permeability of CT across the range of SFRs evaluated (<44μm basis). After deposition and completion of initial sedimentation and consolidation, the permeability of the CT appears to be almost wholly influenced by the fines (<44μm) content of the deposit (Pollock 2000).

3.4.3 *SFR for Commercial Implementation*

A target SFR of 4:1 (<44μm basis) has been chosen for commercial implementation in 2000. This choice was influenced by precedent established through research and development efforts and by a desire to optimize the rate of consolidation. Most Syncrude field experience included a ratio of 4:1, providing confidence that this SFR will provide adequate geotechnical performance and satisfy planning needs.

3.5 *CT Implementation and MFT Inventory*

While the CT process consumes existing MFT, thin fine tails (TFT) continue to be generated from the hydrocyclone overflow stream and from other tailings streams (Fig. 2). The new TFT will densify and continue to produce MFT in the active settling basins. With the implementation of CT in 2000, it is anticipated that net MFT production will be eliminated at Syncrude's base mine operation. The immediate benefit to Syncrude will be reduced fluid containment requirements. This plan assumes production of CT slurry at 58wt% solids with a SFR of 4:1 produced with 95 percent efficiency in creating non-segregating slurries. Sensitivities factored into this prediction include the assumed hydrocyclone solids and fines capture efficiency and CT plant operating efficiency. Should deposit performance warrant it, there is potential to increase MFT consumption through increased fine solids enrichment.

4 FUTURE CHALLENGES FOR R & D

Certain technical issues require continued evaluation and development, including:

- Assessment of water quality impacts on extraction processibility, plant operations, and reclamation,
- Continued assessment of alternative coagulants,
- Monitoring and evaluation of environmental and reclamation aspects, including substrate design, vegetation selection, and landscape alternatives, and
- Monitoring and prediction of depositional and geotechnical performance including ongoing optimization of sand-to-fines ratio (i.e. fines consumption versus geotechnical integrity and performance).

4.1 *Water Impacts*

Water chemistry changes will be associated with implementation of the CT process. Issues to continue monitoring and addressing include effects of changes on ore processibility and plant operations and reclamation design. Due regard must be given to these issues to ensure that potential detrimental effects of water chemistry changes are mitigated (MacKinnon et al 2000).

4.2 *Reclamation Design*

Reclaimed composite tailings deposits will comprise a significant part of Syncrude's final landscape. It is anticipated that approximately two thirds of the CT deposits will be capped with tailings sand and will be reclaimed using established reclamation practices. The remaining one third of the deposits will not be capped with sand but will be amended with peat or capped with reclamation materials and vegetated. Most of this uncapped landscape will be found in lowlands areas and will be developed to support wetland communities. CT reclamation research is currently focussed on:

- Evaluating the effects of CT water on plant growth, survival, and propagation,
- Selection of salt tolerant species for vegetation,
- Determination of most effective capping strategies (including materials, depth, timing) to mitigate effects of CT release water on root zones, and,
- Evaluation of methods to develop and improve soils on CT surfaces.

4.3 *Geotechnical Performance*

Development of effective consolidation models for a range of CT mixtures allows assessment of various deposition and volume-balancing scenarios in support of optimized mining and extraction tailings objectives. With a focus on maximizing the depletion of mature fine tailings (MFT) reserves, there is a potential to create non-segregating slurries with lower sand-to-fines ratios. Balancing this potential is the recognition that as the fines content goes up, the rate of water release and consolidation goes down, potentially adding more pressure on site tailings and water storage capacity needs. Continued refinement of consolidation models for CT, based on field performance at increasing scales, will offer increased confidence regarding the prediction of future performance and consequential identification of planning sensitivities.

5 SUMMARY

Syncrude Canada Ltd. is committed to continuous improvements in the areas of productivity and environmental performance. The company is also committed to restoring the land to an acceptable and sustainable use.

One challenge facing Syncrude is fine tailings management. To meet the challenges associated with fine tailings management and to support its commitment for continuous improvements, Syncrude maintains a strong research and development focus on tailings management alternatives.

Over recent years, composite tailings (CT) technology has been developed to become an important part of the tailings management solution. Much of the CT development process has taken place through the concerted and collaborative efforts of the oil sands industry, consultants, academics and regulatory agencies. The technology began initially as a simple process of adding lime to a coarse tailings stream and has developed through a number of key modifications to become a robust and commercially viable tailings management technology.

Commercial implementation of the CT process provides Syncrude an opportunity to improve tailings management efficiency and reduce the volumes of fluid fines to be managed in the long term.

More research and development work is planned to continue improving the CT process. In particular, challenges remain in assessing reclamation alternatives, assessing potential negative effects of CT release water on bitumen recovery and plant operations efficiency, seeking continuous process improvements (including alternative coagulants), and optimization of storage plans through manipulation of the CT mixtures.

6 CONCLUSIONS & RECOMENDATIONS

A process has been developed which will enable Syncrude to achieve improved fine tailings management performance. The process was developed through programs of progressive and increasing scale and scope, from bench scale testing to the creation of a prototype scale deposit. This effort has been an example of successful partnering of university and industry research initiatives and close cooperation and support from development and operational groups. The composite tailings (CT) process, which emerged from these ongoing research and development improvements, has been tested and found to be robust and effective in producing a non-segregating tailings slurry through a wide range of operating conditions with favourable economics.

Though much has been learned about the CT process, there are a number of areas requiring further investigation. Specific aspects to monitor with full-scale production will include field deposit performance (consolidation and water release rates) and release water chemistry changes. As well, consideration will be given to potential differences between CT produced from Clark Hot Water Extraction (CHWE) tailings and CT made at Aurora using Low Energy Extraction (LEE) tailings.

AKNOWLEDGEMENTS

The authors wish to graciously acknowledge the efforts of everyone that has contributed to the successful development of the composite tailings process. Many dedicated people worked to make this project a success including: the geotechnical engineering group at the University of Alberta, research partners at CANMET (Devon), industry partners at Suncor Energy Inc., consultant expertise from Golder Associates and AGRA Earth and Environmental, and the many Syncrude employees and contract support staff. Collectively, many hours of effort have gone into the development of the CT process. Congratulations to all involved for a job well done.

REFERENCES

Caughill, D.L. 1992. *Geotechnics of Non-segregating Oil Sand Tailings*. University of Alberta, M.Sc. Thesis. June 1992.

Coward, J. 1996. *Feasibility and Economics of Whole Tailings Filtration*. Edmonton, Syncrude Research Department, Progress Report, 25 (1) pp. 80 – 124, 1996.

Cuddy, G., R. Lahaie, & W. Mimura 1991. *1990 Sand-fine tails field program*. Edmonton, Syncrude Research Dept. Research Report, 91-5, 3v., 1991.

Cymerman, G. 1991. *Fine Tails Compaction Tests at Bird Laboratory*. Edmonton, Syncrude Research Department, Progress Report, 20 (1) pp. 83 – 113, 1991.

FTFC (Fine Tailings Fundamentals Consortium) 1995. In: *Advances in Oil Sands Tailings Research*, Alberta Department of Energy, Oil Sands and Research Division. 1995.

Liu, J. K. 1977. *Dry Tailings Disposal — Filtration Demonstration Run*. Edmonton, Syncrude Canada Ltd. Research Department, Progress Report, 6 (7) pp. 84-91, 1977.

MacKinnon, M.D., J.G. Matthews, W.H. Shaw, &

R.G. Cuddy 2000. Water quality issues associated with implementation of composite tailings (CT) technology for managing oil sands tailings. In: *Proceedings of SWEMP 2000*, Calgary, Alberta, May 30 – June 2, 2000.

Pollock, G.W., E.C. McRoberts, G. Livingstone, G.T. McKenna, & J.G Matthews 2000. Consolidation behaviour and modeling of oil sands composite tailings in the Syncrude CT prototype. In: *Tailings and Mine Waste '00*. Rotterdam: 2000 Balkema. *Fort Collins, January 2000*. pp121-130..

Scott, J. D., Y. (Bill) Liu, & D.L. Caughill 1993. Fine Tails Disposal Utilizing Nonsegregating mixes. In: *Proceedings of Oil Sands — Our Petroleum Future Conference*, Edmonton, Alberta, Canada. Apr. 4 – 7, F18, 1993.

Shaw, W., G. Cuddy, G. McKenna, & M. MacKinnon 1996. *Non-segregating Tailings: 1995 NST Field Demonstration*. Edmonton, 96-2: Syncrude Canada Ltd. 4 vols.

Environmental Issues and Management of Waste in Energy and Mineral Production, Singhal & Mehrotra (eds)
© 2000 Balkema, Rotterdam, ISBN 90 5809 085 X

Environmental planning considerations for the decommissioning, closure and reclamation of a mine site

D.M. Mchaina
Boliden Limited, Etobicoke, Ont., Canada

ABSTRACT: This paper discusses environmental management and technical considerations for the decommissioning, closure and reclamation of a mine site. The management and technical aspects discussed include planning; decommissioning, closure and reclamation design basis and performance measures; implementation; monitoring; and financial assurance. Other aspects such as stakeholder consultation and basis for cost estimates for financial assurance are beyond the scope of this paper.

1 INTRODUCTION

The mining industry is one resource-based sector to know in advance that its operations have a finite duration. Historically, many mine sites were abandoned and not decommissioned and reclaimed. Environmental awareness on the part of the mining sector, governments, regulatory authorities, and financial institutions has prompted decommissioning, closure and reclamation aspects to be considered as integral parts of the mining cycle. As such, environmental planning for mine closure is now a worldwide occurrence, even where there is no jurisdiction requiring companies to submit closure plans prior to the granting of a license to develop or operate a new mine. Most of the mining companies now design their works in readiness for closure. This planning approach is known as “designing for closure”. Designing for closure requires the mining, milling and environmental teams to work together so that reclamation, as part of mine closure planning, forms an integral part of the planning process which includes feasibility studies and environmental impact assessments for new mines. Closure is an increasingly important aspect in the management of existing operations. The primary aim of mine closure planning is to ensure that the decommissioning and reclamation of the site can be successfully achieved, whilst satisfying the following objectives:

- To allow a productive and sustainable post-mining use of the site which is acceptable to all stakeholders.
- To protect public health and safety
- To alleviate or eliminate environmental damage and as a result encourage environmental sustainability
- To conserve valuable attributes; and
- To minimize adverse socio-economic impacts.

This paper discusses environmental management and technical considerations for the decommissioning and closure of mine workings.

2 CLOSURE PLANNING CONSIDERATIONS

Closure planning requires an understanding of statutory requirements, standards to be achieved, application and appropriateness of performance criteria; financial assurance and bonding requirements; consultation (local community and regulators) and communication and reporting requirements. Historically, final closure planning has been done during the final years of the mine's operational life after a strategic decision had been made concerning the planned closure date for the mine. In recent years, mine owners have been developing initial closure plans at a mine feasibility stage.

Closure planning should include scenarios that take into account eventualities which may include as temporary suspension and states of inactivity and long-term care and maintenance that may be brought about by factors such as pre-mature closures. Premature closures are usually due to aspects such as low commodity prices, reduced resource grade, smaller than anticipated reserves, increased production costs, technical and political constraints.

The site management should also continuously evaluate the operations impact on the receiving environment with the objective of minimizing future environmental impacts and rehabilitation costs. Overall, the closure and decommissioning plan should demonstrate how the site can be closed in a controlled manner.

2.1 *Closure Design Basis*

In order to close a mine in a proper and acceptable manner, it is imperative to develop a number of design criteria. These are used to define the specific performance requirements to be met during the implementation of the closure plan. The actual performance criteria tend to vary from one jurisdiction to another and are site specific and will also depend upon the overall closure objectives and prevailing regulation. The design considerations should focus on issues such as physical stability, chemical stability, biological stability, hydrology and hydrogeology, geographical and climatic influences, local sensitivities and opportunities, land use, financial assurance, bonding and socio-economic considerations. Some of these considerations are discussed below.

The selection of the appropriate design criteria requires an understanding of the failure modes of structures, their construction materials, extreme events such as high precipitation and floods, landslides and avalanches, earthquakes, drought and fires. Perpetual disruptive forces such as wind erosion, sheet erosion, physical weathering, freezing and thawing, chemical weathering, seasonal frost penetration and microbial action may induce other failure modes. Full account must be given to these aspects in the design periods and factors of safety proposed. Specific design criteria have been developed by several agencies outlined in the list below:

- International Committee on Large Dams (ICOLD) for those sites with tailings and/or water retention dams.
- US Committee on Large Dams (USCOLD) for those sites with tailings and/or water retention dams.
- Canadian Dam Association (CDA) Safety Guidelines
- Canadian Mining Association (MAC) Guidelines to Management of Tailings Facilities.
- Acid Rock Potential – MEND and B.C. Acid Rock Drainage Guidelines.
- Federal, state, provincial and local regulatory requirements – health, safety and reclamation.

Design criteria ensure that the closure meets overall objectives of mine site reclamation are met. The objectives can be considered under the following site conditions:

2.1.1 *Physical Stability*

Physical structures such as crown pillars, pit slopes, underground openings, tailings storage facilities (TSF), spillways, surface openings, waste rock dumps, leach dumps and water retaining dams must meet the following:

- Be physically stable and designed in accordance with acceptable design criteria
- Pose minimal hazard to the public health and safety as a result of failure or physical deterioration
- Continue to perform the function for which they were designed

2.1.2 *Chemical Stability*

The reclaimed mine site and associated structures must be chemically stable. This means surface waters and groundwater must be protected against adverse environmental effects resulting from discharges. In addition, these discharges should not endanger public health and safety, nor result in unacceptable deterioration in environmental resources. Aspects to be monitored closely in closed mines include short and long term changes in tailings geochemistry, seepage from tailings storage facilities (TSF), mine workings, waste rock dumps, and chemistry of surface water draining from the site. Effects due to acid rock drainage (ARD), leaching of metals, and flushing of process reagents or other chemicals must be mitigated. Control and mitigation measures must be specific to the source and contaminant type.

2.1.3 *Biological Stability*

The biological stability of a closed site and potential effects on the surrounding environment are closely related to methods of reclamation, the end land use and the physical and chemical characteristics of the site. The success of reclamation and re-vegetation of the site can influence its chemical and physical stability. It may be required to monitor and manage reclaimed areas until the vegetation is self-sustaining and meets the requirements of the landowner, or until their management can be integrated into the management of the surrounding area. Biological stability also applies to other environments such as an aquatic habitat.

2.1.4 *End Land Use*

The end land use of a closed mine site is generally determined by a number of factors including current land use surrounding the site, viability of re-using site infrastructure and facilities, extent of any environmental impacts and extent of soil contamination. Other considerations may include the need to safeguard against physical, chemical and biological hazards, the location, zoning conditions, availability and real estate value, regulatory standards and other requirements. In general, the mine site should be reclaimed so that the ultimate land use and morphology of the site are compatible with either the current land use in the surrounding area, or with the premining environment. The area could be maintained as an industrial or commercial site if it is appropriate.

2.1.5 *Climatic and Geographical Conditions*

Regional and local climatic information is normally used to resolve questions concerning aspects such as hydrology and hydro - geology. The effect of climate on mine closure measures include: precipitation and extreme events such as floods, landslides and high precipitation. Precipitation affects the overall water balance of the site and the ingress of water to tailings or waste rock dumps, and hence influences the chemical and physical stability of the site together with its contaminant transport parameters. Extreme events influence erosion and weathering forces and subsequently the physical and chemical stability of the site.

The effects of geography on mine closure measures include proximity of local population and resource users downstream of the mine; and proximity of surface and groundwater, which will influence their susceptibility to contaminants of concern released from the closed mine.

2.1.6 *Reclamation Planning*

Reclamation planning is normally dictated by legal requirements hence varies from jurisdiction to jurisdiction. Before the objectives of a decommissioning and reclamation plan are stated, it is important to define the three commonly used terms to describe Restoration, Reclamation and Rehabilitation.

The term "Restoration", the repair of land disturbed by mining, is normally used for situations where disturbed land is left in exactly the same condition as it was prior to disturbance. This requires returning the land to the same pre-mining contours, surface and groundwater patterns and plant and animal ecosystems. This implies that there is no difference between pre-mining and post-mining land conditions. In practice there have been few instances where 100% restoration would be possible. Generally, there will always be differences between pre-mining and post closure land configuration.

"Reclamation" is a term commonly used in the Americas and Europe. The term implies the return of a site that approximates a pre-mining condition and that is inhabited by the ecosystem organisms nearly depicting the same composition and density as those prior to mining.

In general, Reclamation is interpreted as including every process which enhances soil conservation and productive land use to transform land which has been impacted by industrial use back into the flow of normal changes of the landscape (Goodman, 1972).

The term "Rehabilitation" is used to describe the return of a disturbed site to a form and productivity level that conforms to a defined land use. This term suggests that alternative end land uses have been examined, and their potential for reaching each of the land uses has been assessed. This term is the most commonly used in Australia.

Many jurisdictions in the Americas issue reclamation permits that stipulate conditions that meet the reclamation and rehabilitation definitions.

It is important to note that irrespective of which terminology is used, they are all part of the mining closure planning process. In this paper, the term reclamation includes rehabilitation.

The concept of "reclamation" combines all measures needed to make surface mined landscapes productive and visually attractive again.

3 CLOSURE IMPLEMENTATION FOR MAJOR SITE COMPONENTS

Closure plan should include a sequence of events starting from cessation of mining and processing and milling operations through decommissioning, demolition, reclamation and long term care and maintenance if required. Figure 1 outlines a typical sequence of events for mine closure.

The mine components that are important to closure may be divided into the following functional components:

- Underground mine facilities

Main sub components may include surface openings such as adits, boreholes, shafts, declines and portals. The main concerns with these components are safety and physical, chemical and biological stability issues.

- Open pit

Typical primary problems and hazards associated with abandoned open pits include unstable pit walls and slopes; illegal access, and illegal dumping; pollution of surface water and/or groundwater and visual impacts. To mitigate these problems, closure plan for open pits should take into consideration aspects such as pit integrity, water balance, quality and quantity of the surface water and groundwater, geotechnical stability and potential for flooding of the pit. The main concerns associated with pit closure include safety, physical stability, chemical stability and biological stability.

- Waste rock dumps and overburden piles

Potential problems and hazards associated with rock and overburden piles may include unstable slopes, illegal access, ARD, pollution of water regimes, illegal dumping and dust. Again, just like the above components, the issues associated with waste rock dumps include safety, physical stability, chemical stability and biological stability.

- Tailings Storage Facilities (TSF)

The potential problems and hazards associated with closed TSF include unstable slopes, foundation failure, overtopping resulting in failure, ARD, seepage or leakage of tailings water leading to down-

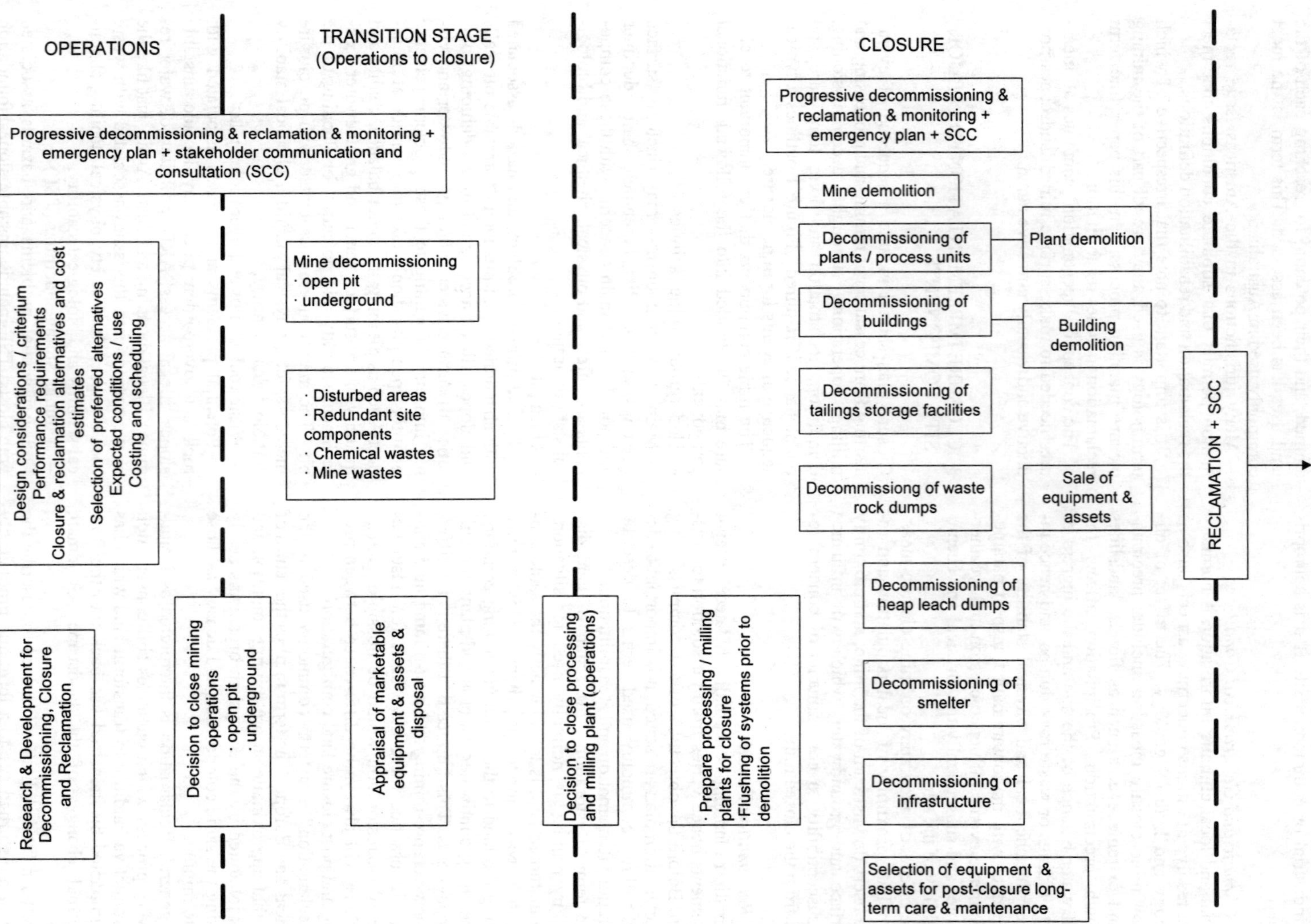

Figure 1: Decommissioning, Closure & Reclamation Implementation Sequence

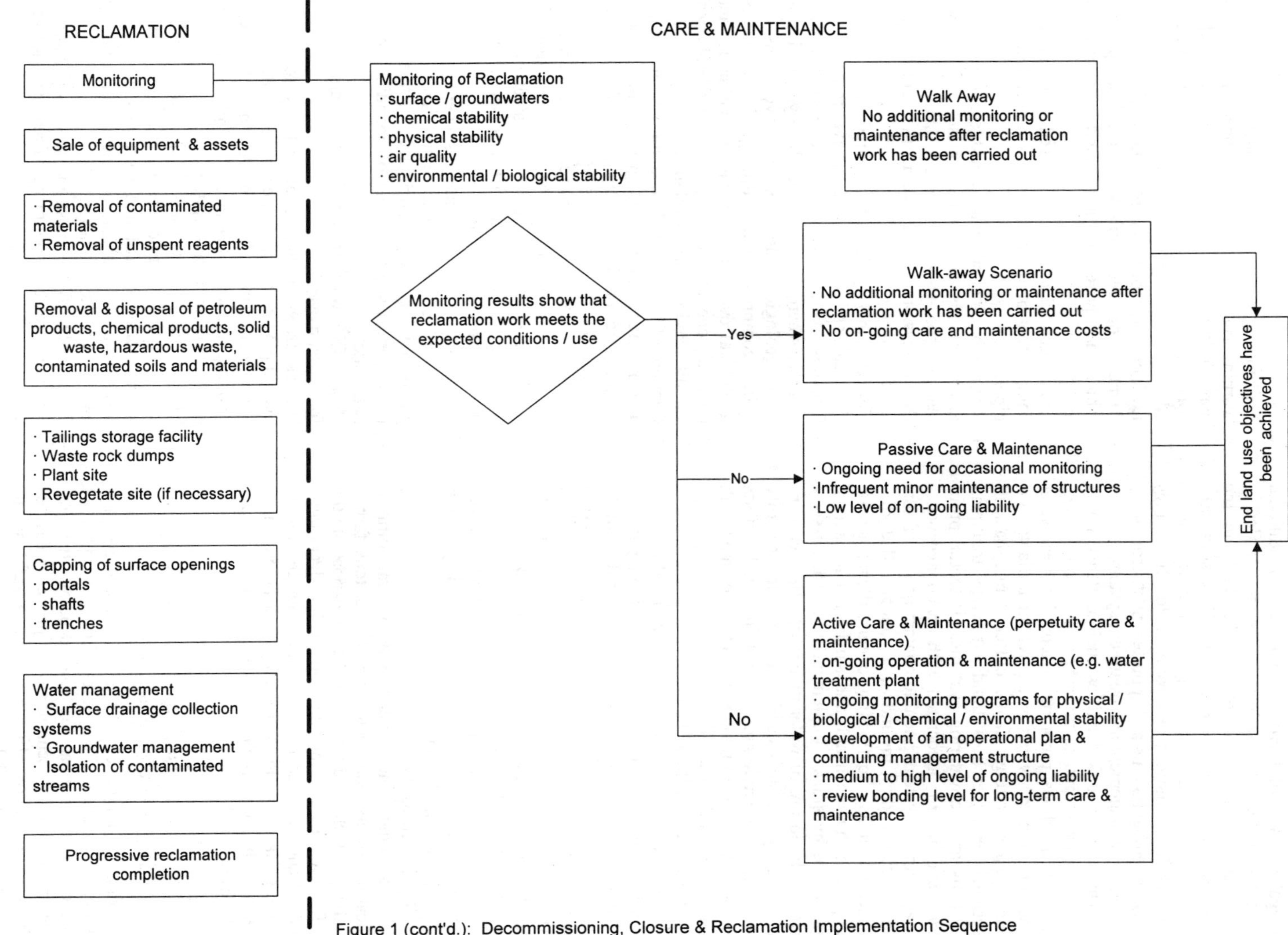

Figure 1 (cont'd.): Decommissioning, Closure & Reclamation Implementation Sequence stream contamination, pollution of water regimes, illegal access, wind erosion and dust. Additional potential triggers and failure modes are detailed in the Mining Association of Canada Guide to the Management of Tailings Facilities. The issues associated with TSF are safety, physical stability, chemical stability, biological environment for the tailings surface, embankments and appurtenant structures.

– Water management and treatment systems

This may include facilities such as dams, reservoirs, spillways, intake structures, diversion ditches, culverts, pipelines, pumping systems for fresh water and contaminated water, water treatment plants, settling ponds, tailings dams, coffer dams, backfill systems, engineered wetlands, sewage treatment systems and de-watering systems. Potential problems and hazards associated with abandoned water management systems include contamination of surface water and/or groundwater, uncontrolled discharges that lead to flooding events, illegal access, impact on human health and safety and impact on fauna and livestock. Implementation issues associated with closure of water management systems include safety, physical stability and chemical stability and biological/environmental stability. All efforts must be made to minimize impacts associated with the closure of these systems.

– Process and milling plants

This includes the entire process and milling flow sheet. Potential concerns include safety, physical stability of remaining structures, chemical stability associated with residual chemicals spilled during processing and biological/environmental stability. The closure plan should ensure that these concerns are dealt with and their impacts are minimized.

– Buildings and equipment

All buildings and equipment should be removed unless a specific function is necessary during or after closure implementation.

– Service Infrastructure

This includes the power supply system, communication towers, water supply pipelines, tank farms, railways, tracks and roads. As much as possible the infrastructure components should be decommissioned and removed from the site. The decommissioning plans should integrate any reusable components of the infrastructure into post mining land use wherever possible.

3.1 *List of Selected Potential Decommissioning, Closure and Reclamation Measures*

Decommissioning, closure and reclamation programs are entirely site specific and may consist of any or all of the measures detailed below. It is important to employ technologically proven measures that are simple, cost effective, practical, safe and environmentally sound. Detailed below are some of the general measures that can be employed depending on prevailing circumstances and conditions at the site.

- Fence using wire mesh or boulders, berming and warning signs to restrict unauthorized access to components such as open pits, open holes, TSF, waste rock dumps, leach dump and associated facilities.
- Lock out and isolate potential dangers.
- Remove or decommission those components which may pose danger to and/or environment and public.
- Discharge and remove all unused and redundant components of the site such as power lines and pipelines.
- Remove or block all unused bridges, access roads to components such as TSF, waste rock dumps, leach dumps and water retaining dams.
- Decommission and remove all fuel and oil storage facilities.
- Collect, label as legally required and pack all chemicals and reagents and remove from the site
- Flooding of open pits is a suitable rehabilitation measure where there are favorable hydrologic and hydro-geologic conditions. However, this requires comprehensive hydrological and hydrogeological evaluations to ensure that safety and physical, chemical and environmental/biological stability environs are maintained.
- Capping of waste rock dumps and TSF to control and minimize surface and groundwater impacts resulting from metal leaching and ARD.
- Decontaminate all reagent systems.
- Identify contamination and remediate, or remove contaminated material to an approved disposal facility or remediate in situ to a satisfactory level.
- Maintain necessary infrastructure as in operational phase of the mine.
- Restore natural drainage patterns where possible.
- Decommission treatment plant and process plants; remove all equipment and structures.
- Display/erect signs warning of potential dangers at the site.
- Design TSF, waste rock and water retaining dams for extreme events.
- Increase freeboard and install spillway, to prevent overtopping erosion by an extreme event.
- Supplement an ARD control measure with flooding to control reactions, or collect and treat using active or passive systems.
- Design tailings and waste rock dumps surface and/or diversion structures to minimize surface water entering the TSF and waste rock dumps.
- Establish a self – sustaining vegetation cover for long-term site stabilization.
- Evaluate the effectiveness and success of the closure measures by monitoring of performance criteria.

4 MONITORING

The monitoring program should address the management, operation, maintenance, surveillance measures, impact mitigation and method of record keeping for all structures and facilities that will remain on the site after closure and reclamation. Figure 1 shows the monitoring program timeline for all phases of closure, reclamation and long-term care and maintenance. The monitoring program should be designed to address safety, physical (e.g. static and dynamic scenarios), and chemical and environmental/biological stability to verify whether expected conditions and uses have been achieved. The closure plan should include the purpose and objective of the monitoring program; type and methods of monitoring to be conducted; monitoring locations and equipment or instrumentation; and frequency of monitoring and expected duration of the program. Additional requirements may include Quality Assurance and Quality Control procedures; methods to be used to inspect, record and evaluate data; parameters to be measured; analytical detection limits; data storage and security; reporting; and procedures for verifying achievement of the expected conditions described within the closure plan.

Site-specific strategies for monitoring physical stability and chemical stability issues should be developed for all mine components.

Monitoring programs are entirely site specific and may consist of any or all of the following methodologies to provide qualitative and quantitative data.

- Inspections – qualitative and quantitative
- Surveying
- Sampling

Monitoring programs for mine components such as tailing dams or covers over acid generating wastes can be expected to continue for the long term to demonstrate that adverse conditions have not developed.

4.1 *Biological Monitoring*

The purpose of the biological systems monitoring program is to provide an additional check on the effectiveness of the closure measures in terms of overall environmental effects and to verify the sustainability of the reclaimed area with particular attention to vegetation. The biological monitoring program should be designed to determine aspects such as the effects of mine effluents and others discharges on fish and other aquatic organisms and aquatic habitat; effectiveness of remedial action taken; and the adequacy of existing control measures and the need for follow-up measures. It is recommended to plan a bio-monitoring program to depict changes during and following each stage of closure to assist in the evaluation of the effectiveness of the reclamation measures.

5 FINANCIAL CONTROL ASPECTS OF CLOSURE

It is now a common practice for mining companies to set aside adequate financial resources for progressive reclamation, closure, decommissioning and reclamation programs. In most cases these are estimated costs for each phase of the closure plan together with costs associated with monitoring and a long-term site management program, if applicable. The acceptability of a financial assurance is a determination that rests with the regulatory authorities. The requirements for financial assurance vary from one jurisdiction to another. Financial assurance can be in various forms such as "Trust Fund", "Performance Bond", Irrevocable Letter of Credit", "Insurance Policy", "Parent Company Guarantees", "Title Deeds – Pledging of Assets", "Mutual Fund" and "Company Stocks". However, the most commonly used forms of financial assurance are bonds and letters of credit. The details pertaining to the development of accurate estimation of reclamation costs is beyond the scope of this paper.

6 SUMMARY

Decommissioning, closure and reclamation planning has in recent years become a legal necessity, a sound business approach, and an environmental responsibility. Most mining companies view reclamation planning as an integral part of the operating plan.

A decommissioning, closure and reclamation plan must consider the long-term physical and chemical effects on both aquatic and land elements, including surface and groundwater, due to components remaining at the site after mining operations have ceased. The key to any plan is to achieve the expected end use conditions through establishing performance criteria. Reclamation planning is a dynamic process and must be flexible with the objective to adapt to site-specific situations.

ACKNOWLEDGEMENTS

Thanks to Ms. B. Parker for the help rendered to organize the paper.

REFERENCES

Box, T.W., 1979. "Surface Mine Rehabilitation – Some Experiences in the United States". *In Proceedings of a Workshop on Management of Lands Affected by Mining,* (Editors,

R.A. Rummery and K.M.W. Howes), Kalgoorlie, May 28 – June 1, 1979. pp. 55-67, CISRO Division of Land Resources Management, Perth.

Goodman, G.T., and Bray, S., 1972. *Ecological Aspects of the Reclamation Derelict and Disturbed Land.* University College of Swansea, Department of Botony. Manuscript, 1972.

Mining Association of Canada (MAC), 1998. 'A Guide to the Management of Tailings Facilities". The Mining Association of Canada, Ottawa, Canada.

Ontario Ministry of Northern Development and Mines, 1994. "Rehabilitation of Mines Guidelines for Proponents". Mine Site Reclamation Section, Mining and Land Management Branch, Mines and Minerals Division. Ontario, Canada.

Environmental Protection Agency, Australian Federal Environmental Department, 1995. "Rehabilitation and Revegetation". One Module in a Series on Best Practice Environmental Management in Mining. June 1995.

Marcus, J.J. (ed.), 1997. "Mining Environmental Handbook: Effects of Mining on the Environment and American Environmental Controls on Mining". Imperial College Press, Imperial College, London, U.K.

Ripley, E.A., Redmann, R.E., and Crowder, A.A., 1996. "Environmental Effects of Mining". St. Lucie Press, Delray Beach, Florida, USA

Ibbotson, B., and Phyper, J.D., 1996. "Environmental Management in Canada". Mcgraw-Hill Ryerson Limited, Whitby, Ontario, Canada.

Gunn, J.M. (ed.), 1996. "Restoration and Recovery of an Industrial Region: Progress in Restoring the Smelter-Damaged Landscape near Sudbury, Canada. Springer-Verlag, New York Inc.

British Columbia Technical and Research Committee on Reclamation, 1999. Proceedings of the Twentieth Annual British Columbia Mine Reclamation Symposium. Kamloops, British Columbia, June 17 to 20, 1996.

Environmental Issues and Management of Waste in Energy and Mineral Production, Singhal & Mehrotra (eds)
© 2000 Balkema, Rotterdam, ISBN 90 5809 085 X

Development of an environmental impact and mitigation assessment program for a tailings storage facility stability upgrade

D.M. Mchaina
Boliden Limited, Etobicoke, Ont., Canada

S. Januszewski
Boliden Westmin (Canada) Limited, Myra Falls Operations, Canada

R.L. Hallam
Hallam Knight Piésold Limited, Canada

ABSTRACT: This paper summarizes an environmental impact and mitigation assessment plan developed for a tailings storage facility stability upgrade at a Canadian base-metal mine. The plan details environmental aspects relevant for channel modification involving the narrowing of a creek channel width. Planning for a tailings storage facility stability upgrade is site-specific dependant; however, this paper details a number of aspects that may be taken into consideration when planning for a project of a similar nature. The aspects include: evaluation of environmental effects and their significance, analyses of cumulative effects, in-situ fish habitat characterization and enhancement program, regulatory requirements, provisions for accidents and malfunctions and fisheries compensation requirements. These considerations together with the proposed monitoring activities for the on-going construction activities are reviewed and discussed in this paper. These aspects can be applied in similar situations after considering factors such as topography, geology, climate, biophysical setting, availability of construction materials, mode of construction, reclamation options and applicable laws.

1 INTRODUCTION

A tailings storage facility is an essential part of the entire mining process and requires to be managed in a safe and environmentally responsible manner through the full life-cycle approach. Tailings storage facilities are site-specific and complex, involving unique biochemical settings and physical –characteristics. Their effective management depends on applying both managerial and technical expertise. As such, a comprehensive tailings storage facility upgrade program requires evaluation of potential impacts to ensure that the proposed project is environmentally sound. This implies that the effects of the project over its projected life do not unacceptably degrade the environment.

This paper discusses the environmental impact, cumulative effect analysis and mitigation assessment for a proposed tailings storage facility stability upgrade.

The proposed tailings storage facility seismic upgrade design will incorporate a number of key components:

- narrowing of the adjacent creek by foundation widening
- foundation compaction
- earth-fill buttress
- improved outer drain groundwater collection system which collects seepage from the facility
- continued development of one or more clean fill borrow areas to provide construction material
- expansion of quarry operations located east of the tailings storage facility to provide construction material
- relocation of a pumphouse return pipeline and backfill sand pipeline
- additional monitoring instrumentation
- closure planning
- contingency planning including an earthquake preparedness plan.

Based on the above key components, a systematic evaluation of potential effects and their significance; cumulative effects, mitigation measures was undertaken.

2 ENVIRONMENTAL EFFECTS OF THE PROJECT AND THEIR SIGNIFICANCE

2.1 *Criteria for Evaluating Adverse Environmental Effects*

In order to determine whether or not the adverse environmental effects were considered significant or not, five criteria were taken into consideration:

Table 1: Environmental Impact Review. Significance Evaluation and Ranking

Activity	Projected Impact	Significance Evaluation: Magnitude Severity	Geographic Extent	Duration / Frequency	Reversible / Irreversible	Ecological Context	Overall Significance Ranking	
Construction Phase I								
Earth and Rock Quarry Screening Plant	Surface Disturbance, Soil Erosion, Stream Sedimentation, Noise Dust	Low	Minor/Several	Medium Term	Reversible	Mine Site/ Immediate Area (IA)	Terrestrial Aquatic	Not Significant
Deployment of Aquadam	Instream Works, Temporary Hydraulic Constriction of Stream	Low	Limited	Short Term	Reversible	Fisheries In-Stream	Aquatic	Not Significant
Removal and Stockpiling Rip-Rap	Surface Disturbance, Soil Erosion, Stream Sedimentation	Moderate	Limited	Short Term	Reversible	Fisheries In-Stream	Aquatic	Significant
Construction of Rip Rap Apron	Instream Works, Stream Sedimentation	High	Limited	Short Term	Reversible	Fisheries Habitat	Aquatic	Significant
Construction of Rip Rap Apron	Permanant Hydraulic Constriction of Stream, Loss of Fish Habitat	High	Moderate	Permanent	Irreversible	Fisheries In-Stream	Aquatic	Significant
Raise Berm Fill 2 m	Soil Erosion, Stream Sedimentation	Low	Limited	Short Term	Reversible	Mine Site/IA	Terrestrial	Not Significant
Upstream Raise of Tailings Embankment	Topographic Modifications	Low	Limited	Short Term	Irreversible	Mine Site/IA	Terrestrial	Not Significant
Construction / Operation Phase II								
Continued Earth and Rock Quarry Screening	Surface Disturbance, Soil Erosion, Stream Sedimentation	Low	Minor/Several	Medium Term	Reversible	Mine Site/IA	Terrestrial Aquatic	Not Significant
Decommissioning Exisiting Drainge System	Loss of Contamination to Groundwater into adjacent creek	Moderate	Moderate	Short Term	Reversible	Fisheries Habitat	Aquatic	Significant
Dynamic Compaction	Loss of Contaminated Pore Water to adjacent creek	Low	Limited	Short Term	Reversible	Fisheries Habitat	Aquatic	Significant
Dynamic Compaction	Noise, Vibration	Low	Limited	Short Term	Reversible	Mine Site/IA	Terrestrial	Not Significant
Placement of Compacted Fill	Soil Erosion, Stream Sedimentation	Moderate	Moderate	Medium Term	Reversible	Fisheries Habitat	Aquatic	Significant
Installation of Cement-Bentonite Cutoff Wall	Loss of Fines to adjacent creek, Stream Sedimentation	Moderate	Moderate	Medium Term	Reversible	Fisheries Habitat	Aquatic	Significant
Installation of Seepage (Bio-Polymer) Collection Pipe	Loss of Fines to adjacent creek, Stream Sedimentation	Moderate	Moderate	Medium Term	Reversible	Fisheries Habitat	Aquatic	Significant
Installation of Embankment Fill	Soil Erosion, Stream Sedimentation	Low	Limited	Short Term	Reversible	Mine Site/IA	Terrestrial	Not Significant
Installation of Rip-Rap Reventment	Soil Erosion, Stream Sedimentation	Low	Limited	Short Term	Reversible	Mine Site/IA	Aquatic	Significant
Pipe Bridge and Support Modifications	Instream Works, Soil Erosion, Stream Sedimentation	Low	Limited	Short Term	Reversible	Fisheries Habitat	Aquatic	Significant
Installation of Armoured Spillways	Topographic Modifications, Visual	Low	Limited	Short Term	Reversible	Mine Site/IA	Terrestrial	Not Significant
Installation of Right Bank Rip-Rap Reventment	Surface Disturbance, Soil Erosion, Stream Sedimentation	High	Moderate	Permanent	Irreversible	Fisheries In-Stream	Aquatic	Significant
Installation of Fisheries Habitat Enhancement	Instream Works, Stream Sedimentation	Low	Limited	Short Term	Reversible	Fisheries In-Stream	Aquatic	Not Significant
Complete Upstream Tailings Embankment	Topographic Modifications, Visual	Low	Limited	Short Term	Reversible	Mine Site/IA	Terrestrial	Not Significant
Installation of Instrumentation	No Projected Impacts	Low	Limited	Short Term	Reversible	Mine Site/IA	Terrestrial	Not Significant
Construction / Operation Phase III								
Raise Berms to El 3372 m	Topographic Modifications, Visual	Low	Limited	Short Term	Irreversible	Mine Site/IA	Terrestrial	Not Significant
Extention of Armoured Spillways	Topographic Modifications, Visual	Low	Limited	Short Term	Irreversible	Mine Site/IA	Terrestrial	Not Significant
Installation of Instrumentation	No Projected Impacts	Low	Limited	Short Term	Reversible	Mine Site/IA	Terrestrial	Not Significant
Install New Pump Station	No Projected Impacts	Low	Limited	Short Term	Reversible	Mine Site/IA	Terrestrial	Not Significant
Upstream Raise of Tailings Embankment	Topographic Modifications, Visual	Low	Limited	Short Term	Reversible	Mine Site/IA	Terrestrial	Not Significant
Construction / Closure Phase IV								
Raise Berm to Final Crest Elevation	Topographic Modifications, Visual	Low	Limited	Short Term	Irreversible	Mine Site/IA	Terrestrial	Not Significant
Complete Emergency Spill Way	Topographic Modifications, Visual	Low	Limited	Short Term	Irreversible	Mine Site/IA	Terrestrial	Not Significant
Capping Tailings Surface/Swales, Wetland Habitat.	Component of Closure Plan	-	-	-	-	-	-	-
Installing New Lynx Diversion Channel	Component of Closure Plan	-	-	-	-	-	-	-

Legend:
TSF = Tailings Storage Facility
AI = Immediate Area
AC = Adjacent Creek

- magnitude (severity) of the adverse environmental effect,
- geographical extent of the adverse environmental effect,
- duration of the adverse environmental effect
- reversibility nature of adverse environmental effect, and
- ecological context.

It is clearly acknowledged that this form of evaluation is somewhat subjective and includes evaluator biases. However, the inherent subjectivity in this evaluation can be partially eliminated by setting out definitions under each criteria as to degree.

The definitions and ranking used in this evaluation of significance is set out in the following:

- Magnitude of the adverse environmental effect, where magnitude refers to severity. Minor or inconsequential effects may not be significant, but effects that are major or catastrophic will be significant. The ranking of magnitude in this evaluation is given as *Low, Moderate, or High* depending on the perceived severity of the impacts and is a subjective ranking.
- Geographic extent of the adverse environmental effect. Localized adverse effects may not be significant, but widespread effects may be significant. The ranking of geographic extent in this evaluation is ranked as follows:
  - *Minor* if the works involve only 100 to 200 m of embankment at a time,
  - *Moderate* if the works involve the entire length of the tailings embankment at one time, or
  - *Extensive* if the impacts of the works extend beyond the battery limits of the project;
- Duration and frequency of the adverse environmental effect are ranked as follows:
  - long term and/or frequent adverse effects may be significant,
  - short term and/or temporary nature may not be significant.
- A ranking of short term is given if the duration of the work is only for a few weeks, medium term if the work is projected to last for extended periods of time, and permanent if the works are fixtures as in the case of a constructed embankment or a restriction of the creek.
- Degree to which the adverse effect is reversible or irreversible. Reversible adverse environmental effects may be less significant than effects that are irreversible. A ranking of reversible is given if the immediate effects cease when the work ceases, and irreversible if the effects are long lasting. For example, the effect of works in and about a stream which is likely to cause sedimentation but would cease as soon as the work ceases and would be ranked reversible. The effects on sedimentation such as impacts on fish redds or attached algae are not ranked even though they may continue after the work had ceased. The ranking of irreversible is given when the works are intended to be permanent such as constriction of the stream.
- Ecological context of the adverse environmental effect. The adverse effects of projects may be significant if they occur in areas or regions that have not already been adversely affected by human activities and/or are ecologically fragile and have little resilience to imposed stresses. All of the proposed work will be carried out at the Project Site, thus all of the impacts occur within an area of important ecological context. However, they also all occur within an area that is already subjected to mine development. Consequently, the effects are divided into two categories:
  - those that may have an effect on the aquatic environment or fisheries habitat and
  - those that may have an effect on the terrestrial environment or park topography (i.e. public visibility and wilderness experience).

The potential impacts on the aquatic environment and fisheries habitat of the adjacent creek are judged to have a more important ecological context than those on the terrestrial environment only because, the proposed works are modifications to existing disturbances and as such will be incorporated into the final reclamation and closure plans.

2.2 *Significance Evaluation*

The results of the significance evaluation are presented in Table 1. In overall terms, those activities or works that have an aquatic aspect or a potential to effect the water quality in, or the fish habitat capability of, adjacent creek were judged as being Significant, even though they may be minor in magnitude, of very limited geographical extent, of short duration, and completely reversible.

Those activities or works that are essentially terrestrial or topographical in nature, such as berm raises that have a potential effect on the project area environment (visual and wilderness experiences) were judged as being Not Significant, since these activities and works are modifications to, extensions of, or improvements to existing disturbances and are to be incorporated into the overall project reclamation and closure plans. Again it is stressed that this analysis was not intended to examine a wider range of possible minor indirect and secondary effects of the works or activities, such as wildlife avoidance.

Positive impacts resulting from the implementation of this project include an improved outer drain groundwater collection system, improved aquatic habitat, installation of additional monitoring instrumentation, improved tailings storage facility seismic stability and design of tailings storage facility for closure.

Table 2: Potential Mitigation Review and Cumulative Effects Analysis

Activity	Proposed Mitigation and Control Measures and Analysis of Cumulative Effects		Cumulative Impacts
Construction Phase I			
Earth and Rock Quarry Screening Plant		Settling ponds and total recycle of wash water. Surveilance /	Not Applicable
Deployment of Aquadam	Limited length (150 m), limited use during low flows, restricted to fisheries window.	System itself is mitigation. Installation effects minimal.Remove aquadam under high flows.	Not Applicable
Removal and Stockpiling Rip-Rap	Work done behind aquadam under low flows. Apply sediment control as required.	Cease work under high flows.	Not Applicable
Construction of Rip Rap Apron	Work done behind aquadam under low flows. Apply sediment control as required.	Mitigation appears adequate. Cease work under high flows.	Not Applicable
Construction of Rip Rap Apron	Fisheries habitat enhancement program proposed.		Not Applicable
Raise Berm Fill 2 m	Work done behind aquadam under low flows. Apply sediment control as required.	Mitigation appears adequate. Cease work under high flows.	Not Applicable
Upstream Raise of Tailings Embankment	Work done behind aquadam under low flows. Apply sediment control as required.	Surveillance / Construction Monitor.	Not Applicable
Construction / Operation Phase II			
Continued Earth and Rock Quarry Screening		Settling ponds and total recycle of wash water. Surveilance /	Not Applicable
Decommissioning Exisiting Drainge System	Groundwater from existing outer drain will be pumped to Pump Station	Surveillance / Construction Monitor / Water Quality Monitoring.	Not Applicable
Dynamic Compaction	Pumping groundwater to tailings storage facility.	Surveillance / Construction Monitor / Water Quality Monitoring.	Not Applicable
Dynamic Compaction	Vibration limited to 20 m, no mitigation proposed.	Surveillance / Construction Monitor.	Not Applicable
Placement of Compacted Fill	Work under low flow, behind aquadam. Dirty water to TSF. Sediment control.	Cease work under flood flows. Surveillance / Construction Monitor.	Not Applicable
Installation of Cement-Bentonite Cutoff Wall	Work under low flow, behind aquadam. Dirty water to TSF. Sediment control.	Cease work under flood flows. Surveillance / Construction Monitor.	Not Applicable
Installation of Seepage (Bio-Polymer) Collection Pipe	Work under low flow, behind aquadam. Dirty water to TSF. Sediment control.	Cease work under flood flows. Surveillance / Construction Monitor.	Not Applicable
Installation of Embankment Fill	Work done behind aquadam under low flows. Apply sediment control as required.	Cease work under high flows.	Not Applicable
Installation of Rip-Rap Reventment	Work done behind aquadam under low flows. Apply sediment control as required.	Cease work under high flows.	Not Applicable
Pipe Bridge and Support Modifications	Sediment control as required.	Sediment control works may be required.	Not Applicable
Installation of Armoured Spillways	Apply sediment control as required.	Surveillance / Construction Monitor.	Not Applicable
Installation of Right Bank Rip-Rap Reventment	Work done behind aquadam under low flows. Apply sediment control as required.	Cease work under high flows.	Not Applicable
Installation of Fisheries Habitat Enhancement	Sediment control as required.	Sediment control works may be required.	Not Applicable
Complete Upstream Tailings Embankment	Work done behind aquadam under low flows. Apply sediment control as required.	Surveillance / Construction Monitor.	Not Applicable
Installation of Instrumentation - Not in Creek	Apply sediment control as required.	Surveillance / Construction Monitor.	Not Applicable
Construction / Operation Phase III			
Raise Berms to El 3372 m	Works above adjacent creek. Apply sediment control as required.	Surveillance / Construction Monitor.	Not Applicable
Extention of Armoured Spillways	Works above adjacent creek. Apply sediment control as required.	Surveillance / Construction Monitor.	Not Applicable
Installation of Instrumentation	Works above adjacent creek. Apply sediment control as required.	Surveillance / Construction Monitor.	Not Applicable
Install New Pump Station	Works above adjacent creek. Apply sediment control as required.	Surveillance / Construction Monitor.	Not Applicable
Upstream Raise of Tailings Embankment	Works above adjacent creek. Apply sediment control as required.	Surveillance / Construction Monitor.	Not Applicable
Construction / Closure Phase IV			
Raise Berm to Final Crest Elevation	Works above adjacent creek. Apply sediment control as required.	Surveillance / Construction Monitor.	Not Applicable
Complete Emergency Spill Way	Works above adjacent creek. Apply sediment control as required.	Surveillance / Construction Monitor.	Not Applicable
Capping Tailings Surface/Swales, Wetland Habitat.			
Installing New Diversion Channel	Apply sediment control as required.		

2.3 *Cumulative Effects*

This section evaluates cumulative environmental effects that are likely to result from a project as a consequence of its combination with other projects or activities that have been or will be carried out within the project area. The cumulative impacts from future projects that are likely to be carried out are the focus of this analysis.

The temporal boundaries for this assessment were assumed to encompass any likely projects or activities that would occur during the Project Site mine life. On the basis of current ore reserves the project is expected to continue for between 10 and 15 years. The object of the reclamation and closure program is to restore the original vegetation of the valley, once all mining activities have ceased. The area would then be returned to park use.

The geographic boundary of the project for this cumulative effects review is assumed to include the entire watersheds of two creeks and the southern part of a lake that lies entirely within the park.

There are currently no other projects permitted within the Project Site and it is extremely unlikely that the governments will authorize any additional projects, other than those projects. To the knowledge of the Company, there are no other proposed or likely projects within the defined geographic boundaries. As such, there are no projected cumulative effects. Cumulative effects analysis is given in Table 2.

3 PROPOSED MITIGATIVE MEASURES

The following physical design principles will be reasonably used to avoid adverse impacts of channel modifications.

- where possible, channel straightening and steepening will be avoided
- bank stability will be promoted by judiciously placing riprap. Riprap material will be large, dense, and angular rock
- emulating nature in design channel form by creating appropriately spaced pools and riffles
- accumulation of woody materials along the shoreline will be encouraged to create pools, increase structural complexity, provide fish cover, form substrate for invertebrates and trap gravel for spawning and invertebrate production.

Activities or works that have an aquatic aspect or a potential to effect the water quality in, or the fish habitat capability of the adjacent creek were judged as being Significant, regardless of their magnitude, geographic extent, duration, or frequency.

Potential effects on the aquatic environment can be grouped into three principle categories:

- release of contaminated groundwater to adjacent creek during construction and operation
- loss of sediment to the adjacent creek during construction and operation
- physical disruption or loss of aquatic habitat due to construction and operation.

3.1 *Groundwater Control*

Environmental protection considerations have been included in the construction plans to intercept contaminated ground water during the construction process. Groundwater from the outer drain and the pore water from the dynamic compaction process will be picked up at each of the outer drain access chambers and pumped to a pumphouse, while the new cement-bentonite cutoff wall and replacement bio-polymer drain are being constructed. Contaminated surface and groundwater from the construction of the new cement-bentonite cutoff wall and replacement bio-polymer drain will be pumped to the tailings storage facility. Dynamic compaction will progress from upstream to downstream and gradually closer to the outer drain and access chambers to minimize the effects of compaction on the outer drains functioning time. Piezometers are to be installed in advance to monitor pore pressures. In addition to the proposed mitigative measures, the following will be implemented:

- construction of the new cement-bentonite slurry cutoff wall and replacement bio-polymer drain will cease during flood flows down the creek,
- the environmental monitor be responsible for monitoring groundwater migration to the adjacent creek, via the proposed piezometer installations, and
- the environmental monitor will be responsible for monitoring water quality in adjacent creek for evidence of groundwater seepage.

3.2 *Sediment Control*

Provisions have been included in the construction plans for a sediment control program for both in-stream and above-stream activities. Sediment control for in-stream and near-stream works such as removal of the rip-rap, placement of the rip-rap toe, placement of the 2-m lift, and installation of the new cement-bentonite cutoff wall and replacement bio-polymer drain will be carried out, in limited distances, by the use of an Aquadam. Sediment control for above-stream works, such as berm construction and raises, will consist of ditches, sumps, sump pumps, silt fences, and other works as required. In addition to the above mitigative measures, the following will be incorporated into the sediment-control program.

- a sediment control program will be incorporated into the soil/gravel screening and rock quarrying processes, and include containment and total recycle of wash water using settling ponds.

- the Environmental Monitor will be responsible for monitoring soil/gravel screening and rock quarrying processes.
- the works be suspended during sudden periods of high or flood flows when the effectiveness of the aquadam can not be assured.

3.3 *In-situ Aquatic Habitat Enhancement*

Aquatic habitat enhancement considerations have been included in the future operation of the proposed seismic upgrade to provide locations for fish refuge during both high and low flows by including a low flow channel, boulder groupings, and riffle heads. Although the current reach in the adjacent creek that lies adjacent to the tailings storage facility does not contain fish, reasonable efforts will be made for it to provide a conduit for any fish migrating to the lower portions of the creek or downstream environment and may contribute to fish productivity.

To enhance stream alteration the design considers the interplay among aquatic habitat factors and physical factors of stream velocity, water quality, depth and substrate. Proposed structures to enhance stream habitat take those factors into account. The proposed structures to enhance stream habitat include a low-flow thalweg (notched channel along entire length), boulder clusters for water velocity easing and aeration, larger rip-rap and filter material to minimize risk of erosion, sediment generation and also to provide additional crevice space in the wetted channel for refuge and cover for fish. The proposed enhancement program will produce a pool/riffle environment required for aquatic habitat. For example, in a pool-riffle environment, riffles function as food production and spawning areas.

Riffles exhibit relatively shallow depths, higher than average velocity, and coarser substrate than pools. Velocity is the primary parameter describing the distribution of aquatic invertebrates. Velocity in riffles governs the rate of oxygen transfer to properly sized substrate, i.e., rubble, boulders, cobbles; thereby supplying oxygen and removing metabolic waste products from intergravel areas. Water velocity increases the exchange rate thereby enhancing respiration and food acquisition. Optimal velocity is subject to debate, but the range for riffle segments for good stream productivity is between 0.5 and 3 fps (0.15 and 0.9 mps). The design targets such velocity ranges to minimize impacts.

Benthic invertebrates decrease in number and diversity as substrate is changed from rubble to coarse gravel to fine gravel and sand. Rubbles appear to play a key role in the riffle environment. It provides a broad surface for invertebrates to cling to and functions to protect insects from high velocities. Velocity also functions as the vehicle for drift, which is the movement of organisms downstream by current. Drift supplies the mechanisms to acquire food, which advances increased population densities and diversity. Efforts will be made to minimize impact on substrate.

The parameters of velocity, depth and substrate combine in the riffle environment to provide an optimal habitat for aquatic invertebrates. The repetitive pool-riffle succession creates an excellent habitat for food production, spawning, cover, and resting.

Other structures for stream habitat enhancement that may be considered include current deflectors and low profile dams, and selected placement of logs. Depending on how they are placed, these devices can locally increase velocity, create pools, provide gravel-trapping areas, remove silt from spawning areas, protect stream banks, enhance pool-riffle sequencing, aerate the water, reduce or increase water temperature, and generally create enhanced habitat and simultaneously provide bank stabilization.

Detailed design for construction and placement of the structures in and near the creek will be done in consultation with the respective authorities before project implementation. Follow-up studies to assess the effectiveness of structures installed in the creek will be conducted in consultation with the responsible authorities. Detailed mitigative measures are given in Table 2.

3.4 *Surveillance Programs*

During the construction phase a number of other incidents and malfunctions could require the deployment of mitigative measures and potentially emergency response teams. The Project Site's Surface Emergency Response Program (ERP) will be used to respond to accidents and malfunctions associated with the system when required. The ERP is an established program with a mandate, coordinator, trained teams with monthly training sessions, and a regularly updated response manual. Specific details of environmental aspects that will need to be addressed during construction, together with their respective response plans, will be prepared for regulatory review prior to implementation of construction work.

4 PROPOSED MONITORING ACTIVITIES

The purpose of monitoring is to collect data of known quality to verify projected impacts, evaluate the effectiveness of mitigation measure, and determine the effectiveness of habitat enhancement efforts. Regulatory agencies and the Company will review the results of all monitoring activities. If environmental changes varied substantially from those predicted, the regulatory agencies, and the Company, will determine what actions, if any, the Company will need to implement to reduce or eliminate project-related effects. Table 1 lists project related environmental effects.

A detailed environmental monitoring plan will be developed by the Company in cooperation with the government agencies, specifically for surface water, groundwater and fisheries. This monitoring plan will combine the following elements:

- purpose, goals, monitoring objectives, and contingency plans
- environmental monitoring network design
  - station location
  - constituent selection
  - sampling frequency
  - sample collection, handling, and shipping
- Field Sampling
  - protocols
  - field quality control
  - constituents
- Laboratory Procedures
  - analytical techniques
  - quality control and assurance procedures
  - data recording standards
  - required lab quality control sample and frequency
- Chain of Custody
- Data Handling
  - data screening and verification
  - database maintenance
  - data reporting and distribution
- Data Analysis Needs
  - trend analysis
  - compliance with water quality criteria
  - best management practice effectiveness
- Monitoring Evaluation
  - establishment of evaluation criteria that would invoke implementation of additional mitigation measures
  - quarterly and annual reviews of program for update and/or modification of program and plan
- Follow-up

5 CONCLUSIONS

Managing tailings storage facility upgrade works adjacent to a water course can be highly complex involving the input of many agencies and groups, as well as biological and technical concerns. To address the concerns, a systematic approach was developed to determine the environmental effects and their significance together with cumulative effects and potential mitigation measures.

Upgrade work is proceeding as planned. The design aspects of the project are satisfactory and constructible from a practical standpoint. The company continues to optimize the construction process and refine the mitigation measures, both to reduce environmental impacts and streamline the process.

Proper project planning that encompasses environmental aspects assures that project implementation is done correctly, and in an efficient and environmentally sound manner. The type of management philosophy and commitment to the environment will meet the long-term vision of creating a stable, self-sustaining environment through an improved outer drain groundwater collection system, aquatic habitat, tailings storage facility seismic stability and design of the tailings storage facility for closure.

REFERENCES

Klohn, E.J. 1979. Seepage Control for Tailings Dams. *Proceedings of 1st International Mine Drainage Symposium*, Denver, Colorado.

Kohler, C.C., and Hubert, W.A., 1993. Inland Fisheries Management in North America. *American Fisheries Society*, Bethesda, Maryland, USA 1993.

Lo, R.C., and Klohn, E.J., 1996. Design Considerations for Tailings Dams. *Proceedings of the Third International Conference on Tailings and Mine Waste 1996.* [1]A.A. Balkema / Rotterdam / Brookfield / 1996.

Marcus, J.J. (Editor), 1996. *Mining Environmental Handbook: Effects of Mining on the Environment and American Environmental Controls on Mining.* Published by Imperial College Press, 516 Sherfield Building, Imperial College, London.

Mining Association of Canada, 1998. *A Guide to the Management of Tailings Facilities.* 1105-350 Sparks Street, Ottawa, Ontario, Canada, September 1998.

Portfors, E.A. 1981. Environmental Aspects and Surface Water Control. *Proceedings of Seminar on Design and Construction of Tailings Dams.* Colorado School of Mines. July 1981, pp. 99-117.

Environmental Issues and Management of Waste in Energy and Mineral Production, Singhal & Mehrotra (eds)
© 2000 Balkema, Rotterdam, ISBN 90 5809 085 X

Effect of a one-time application of municipal biosolids on plant growth at a metal mine tailings impoundment, five years of results

R.L. McNearny
Montana Tech of the University of Montana, Butte, Mont., USA

ABSTRACT: Vegetation growth is often limited in metal mine spoils under semi-arid climatic conditions. This study was conducted to determine whether a one-time biosolids application, limestone and wood residue addition could enhance the establishment of vegetation on metal mine spoil under these conditions. Municipal sewage sludge (as biosolids) were applied at a rate of 0 Mg/ha (control), 22.4 Mg/ha, 44.8 Mg/ha, and 67.2 Mg/ha, to the Kennecott Utah Copper Corporation tailings impoundment located near Magna, Utah. Five test sites on the impoundment were monitored for five years, beginning in 1994. Above ground biomass production, percent cover of individual plant species, and the concentration of heavy metals in plant tissue were evaluated. A conclusion of the study is that biosolids application enhanced the plant growth capability of the tailings significantly. The increase in biomass and percent cover was proportional from the control rate to the 67.2 Mg/ha application rate.

1 INTRODUCTION

A large number of disturbed lands and unproductive areas exist across the United States that needs to be reclaimed and revegetated (Webber, 1992). At the same time, the volume of municipal sewage sludge, containing the necessary important soil nutrients for land reclamation, is being produced in large quantities throughout the country. This sludge, also termed "biosolids," could be used, and is being used, to reclaim these areas.

Surface mining has already disturbed over 1.6 million ha in the United States. In Utah, U.S.A., the surface area that is disturbed and should be reclaimed is about 6,073 ha (Sopper, 1993). The types of disturbed land where sewage sludge might be used beneficially are copper mines, phosphate mines, zinc and lead smelters, and other metal and coal waste contaminated areas (Schroeder, 1995). However, there is one disadvantage of biosolids application: the content of elements found in wastes from industrial sources, particularly heavy metals, may be introduced into the environment and consequently the human food chain. The U.S. Environmental Protection Agency (EPA) has thus developed guidelines and regulations regarding this issue (Norton-Arnold, et al., 1995).

The potential use of biosolids in the reclamation of disturbed lands is tremendous. Biosolids add not only organic matter, which is one of the most important components in the improvement of soil physical properties, but also biosolids are considered to be very helpful in the process of soil pH stabilization and fertility (Sopper, 1993). Land application of biosolids has been shown to promote biodiversity (Pichtel, et al., 1994). Biosolids for land application must be aerobically or anaerobically processed, composted, or lime treated to reduce pathogens and must contain low levels of potentially toxic heavy metals, as required by federal regulations (EPA, 1994).

In addition to potentially phytotoxic chemical properties, metal mine tailings often have physical characteristics that may make them inhibitory to plant life in other ways. Tailings often lack organic matter and microorganisms, since ores are in shortage of organic materials in their original state, and little is added in the extraction process. Nutrient cycles cannot be established until organic material has accumulated and microorganisms are present and active. This deficiency causes a lack of revegetation success. But biosolids are a valuable addition to this type of spoil. It often adds the organic matter necessary to improve root zone properties and may also add the microorganisms needed to establish healthy nutrient cycles (Munshower, 1994).

In Utah, very few studies utilizing biosolids as an effective soil conditioner and fertilizer had been conducted (Nielson and Peterson, 1972; Sabey, et al., 1990). The project discussed in this paper is one of the first studies on the land application of biosolids implemented under the semi-arid field condi-

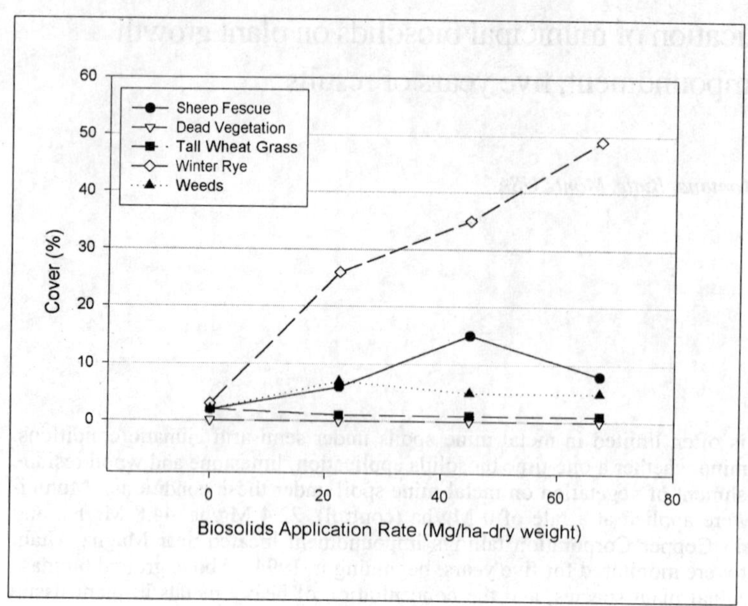

Figure 1-1995 Cover and Biodiversity

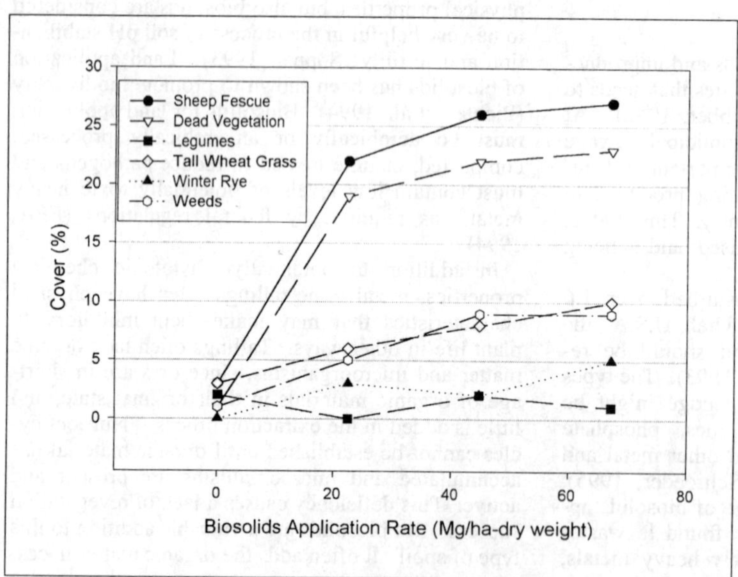

Figure 2-1996 Cover and Biodiversity

tions typical of the state of Utah. This project utilizes almost 33 ha of land as test sites. This paper reports improvement in plant growth after five growing seasons, as a result of the addition of biosolids, limestone, and wood residue.

The possible benefit of biosolids amendment in mitigating problems of vegetation establishment under semi-arid climatic conditions has not been fully examined, therefore the objectives of the study reported here were to determine: (1) the effect of the different treatments on the above-ground biomass production and the percent cover of individual species, and (2) the extent of the concentration of heavy metals from the tailings-biosolids mix in the plant tissue of test species as a result of the different treatments.

2 EXPERIMENT AND MATERIALS

The project was begun in July 1994. Five test sites were established along the north to northwest side of the KUCC tailings impoundment. The tailings impoundment is approximately 2300 ha, and is scheduled to be expanded considerably over the next few decades. The experimental design consisted of a single-factor linear design that utilized randomized complete plots for all the test sites. Each site was divided into 16 0.4-ha test plots. The test plots were approximately 30 m wide to facilitate spreading of the biosolids. Biosolids were added to the test sites at the four application rates of 0 Mg/ha (as a control), 22.4 Mg/ha, 44.8 Mg/ha and 67.2 Mg/ha. Each application rate was repeated four times. Coarse sand-sized limestone particles were added at the rate of 13.4-Mg/ha to test site No. 1 to evaluate the potential for raising the pH of the tailings. Wood residue was added at the rate of 67.2-Mg/ha to test site No. 3 to evaluate the potential for improving the texture of the tailings as well as supply an additional source of organic material. Both amendments were added prior to the addition and incorporation of the biosolids. Test site No. 3B received aerobically treated biosolids from the Magna, Utah, city wastewater district. Test sites No. 1, 2, 3, and 4 received anaerobically treated biosolids from the Central Valley Wastewater Reclamation Facility (CVWRF) located in the Salt Lake metropolitan area. Thus, the treatments consisted of (1) the addition of anaerobically digested biosolids plus limestone to test site No. 1, (2) the addition of anaerobically digested biosolids only to test sites No. 2 and 4, (3) the addition of anaerobically digested biosolids plus wood residue to site No. 3, and (4) the addition of aerobically treated biosolids only to test site No. 3B. Test sites No. 1 through 3B are northwest facing, while test site No. 4 is north facing. The test sites were arranged in this manner to evaluate differences in slope aspect. On initial analysis of the test results, no difference could be found between test sites No. 2 and 4, thus the results from these test sites were combined for this paper.

3 METHODS

Baseline data collection prior to application of the biosolids was conducted July through August 1994. Tailings samples were taken from each plot at a depth interval of 0-15 cm. The samples were collected by driving a clean-wall tube sampler into the soil and putting the resulting samples into lidded plastic sample buckets. Composite samples were made by mixing equal parts of the four samples collected according to application rate. The Utah State University Soil Laboratory performed testing for agronomic properties of the samples.

Application of the biosolids, followed by disking to a depth of 30 cm, was completed in the fall of 1994. Each test plot received one of the four different treatments of biosolids. Analysis for the regulated metals, as required by 40 CFR 503 (U.S. Code of Federal Regulations), was performed on a continuing basis by CVWRF during the time of application of biosolids to the impoundment. At no time were regulatory limits exceeded. All sites were resampled for agronomic properties in the December of 1994 and again in June of 1995 through June 1999.

In the spring of 1995, Kennecott personnel planted all test sites with a rangeland seed drill. The seed mix of winter rye (*Secale cereale*) was plant as a nurse crop at a rate of 45 kg/ha. A perennial mix of yellow sweetclover (*Melilotus officinalis*), tall wheatgrass (*Agropyron elongatum*), and sheep fescue (*Festuca ovina*) was planted at a rate of 3.4 kg/ha, and alfalfa (*Medicago sativa*) was planted at a rate of 5.6 kg/ha.

Vegetation enhancement success was measured based on comparison with the control plots of each test site as a function of measured above-ground biomass production, species diversity and percent cover as measured along one transect across the plot by the line intercept method as described by Chambers and Brown (1983). For biomass production, plant species were harvested from the center of the test plot, dried and weighed to obtain total above-ground dry biomass results. A total of 80 plant tissue samples were collected for metal analysis. The plant tissue samples were represented by three different species; namely winter rye, sheep fescue, and tall wheatgrass.

Biomass and cover data were analyzed statistically using the STATGRAPHICS (Manugistics, 1994) and SIGMASTAT (Jandel Scientific, 1994) statistical packages. Biomass data were evaluated according to the following: analyses of plant growth within each test site according to the four different

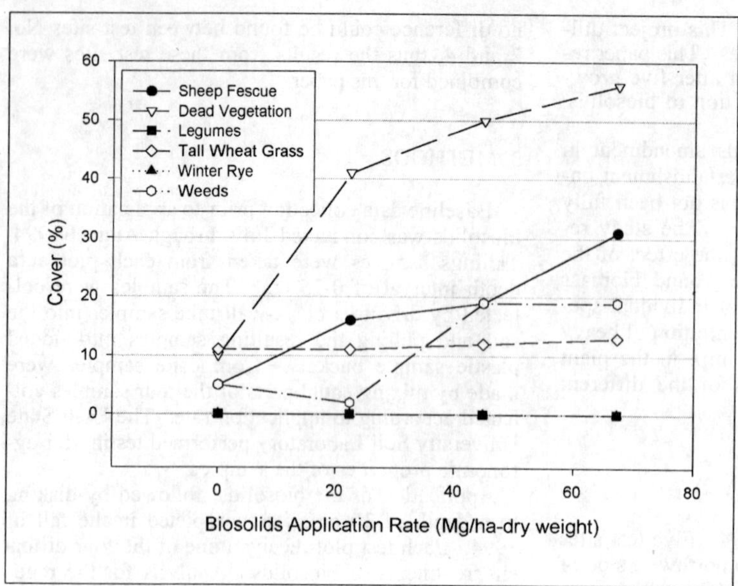

Figure 3-1997 Cover and Biodiversity

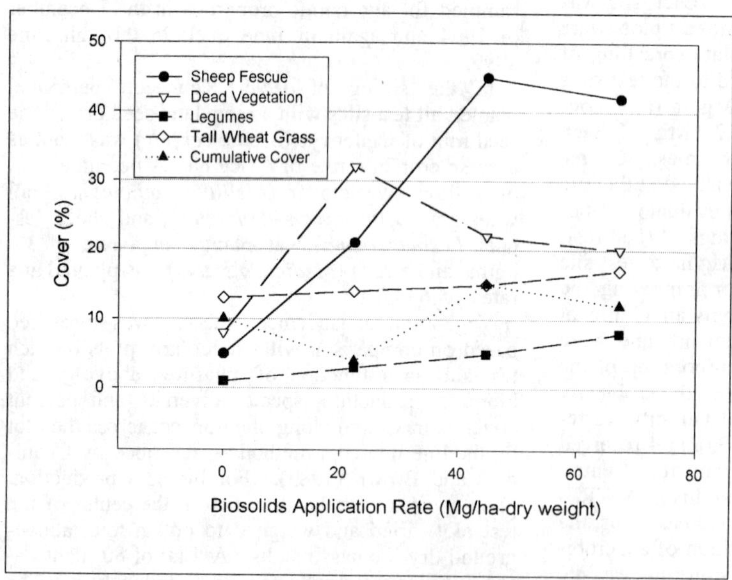

Figure 4-1997 Cover and Biodiversity

Table 1. Plant tissue copper levels (mg/kg).

Plant Species	Biosolids Application Rates			
	0 Mg/ha	10 Mg/ha	20 Mg/ha	30 Mg/ha
Winter Rye	28.8	21.8	18.9	20.7
Sheep Fescue	54.9	45.1	46.1	39.2
Tall Wheat Grass	101.7	135.4	53.4	35.3

treatments were performed using analysis of variance (ANOVA) or one-way analysis on ranks, depending on the type of data distribution. Each site was analyzed and evaluated separately. Mean separation was conducted using the Student-Newman-Keuls multiple range test. The Kruskal-Wallis test for non-parametric distribution was also applied when the data failed the test for normality. Correlations between the biosolids application rate and biomass results were performed using either the Pearson or Spearman correlation procedures for identifying trends. Percent cover for each species found on the test plots were analyzed using one and two-factor Kruskal-Wallis tests and the Student-Newman-Keuls test for mean separation.

4 DISCUSSION OF VEGETATION RESPONSE

A positive growth response of planted test species, namely winter rye, sheep fescue, tall wheatgrass, yellow sweet clover, and alfalfa to biosolids addition was apparent within the first two months after planting and throughout the life of the study. Statistically significant differential biomass responses were observed on all test plots where biosolids were added when compared to the control plots. Biomass production increased significantly with each increase in the biosolids application rate.

Biosolids amendment increased dramatically the growth of winter rye in the first year (Figure 1). It exhibited the greatest overall mean percent cover (29 %), followed by sheep fescue (7 %), and kochia (*Kochia* sp.) (5 % as an invasive weed). There was a statistically significant increase in biomass production with increasing biosolids application rate for winter rye, sheep fescue, and kochia. In contrast, other test species did not display any statistically significant difference in growth, including halogeton (*Halogeton glomeratus)*, also an invasive weed. These results may be explained in part by the plant growth dominance in the first two years of the study by winter rye, which was planted at a rate that exceeded the planting rates of the other test species by a factor of eight.

All test species were examined for any kind of correlation between biosolids application rate and percent cover using either the Pearson product moment correlation or the Spearman rank order correlation test. For example, a positive trend was established for winter rye, which initially produced the greatest biomass and percent cover of all the species planted. Winter rye production increased significantly with increasing biosolids application rate, but dropped off significantly in the next three years (Figures 2 through 4) and had essentially disappeared from the test plots by 1998 (Figure 5).

The plant tissue concentrations of metals, except for Cu, in the three species tested (winter rye, sheep fescue, and tall wheatgrass) were within normal limits for plants grown on western soils (Logan and Chaney, 1983, and Mendel and Kirkby, 1987). The plant tissue Cu concentrations can be attributed to the level of Cu in the tailings rather than to the addition of biosolids. The highest levels of Cu were seen in the tall wheat grass, followed by sheep fescue, then by winter rye. Table 1 lists the mean levels of Cu for the three test species. Testing for plant tissue metal concentrations were dropped after the first year since the metal levels were considered normal.

Percent cover increased rapidly for the 44.8 and 67.2 Mg/ha biosolids application rates, attaining 100 % cover by 1997 (Figure 6). The 22.4 Mg/ha biosolids application rate caused a linear increase in cover from the beginning of the study, attaining a 97% cover in 1999. The percent cover for the control plots never increased above 38%. Most of this increase was due to growth of weeds.

Biomass production increased dramatically in the first year of the study. This increase was again due to the rapid growth of winter rye, which was planted as a nurse crop to protect the sites from erosion (Figure 7). The rapid decrease in biomass was also

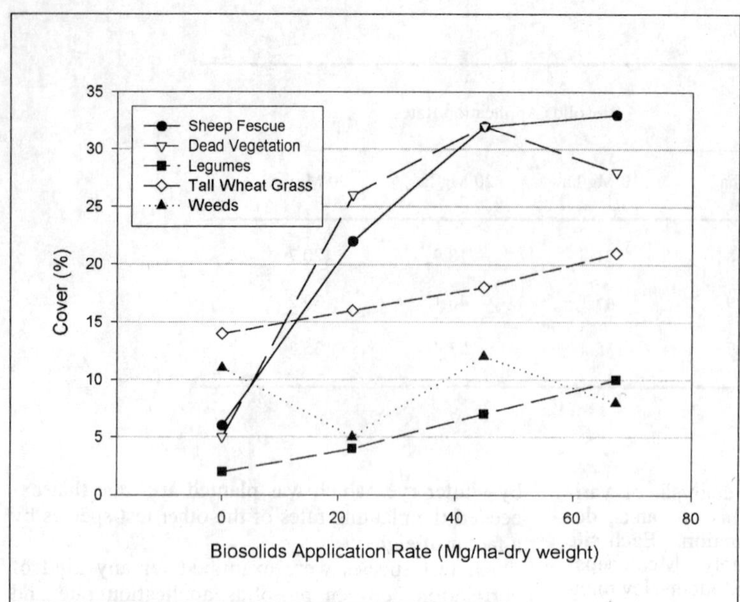

Figure 5-1999 Cover and Biodiversity

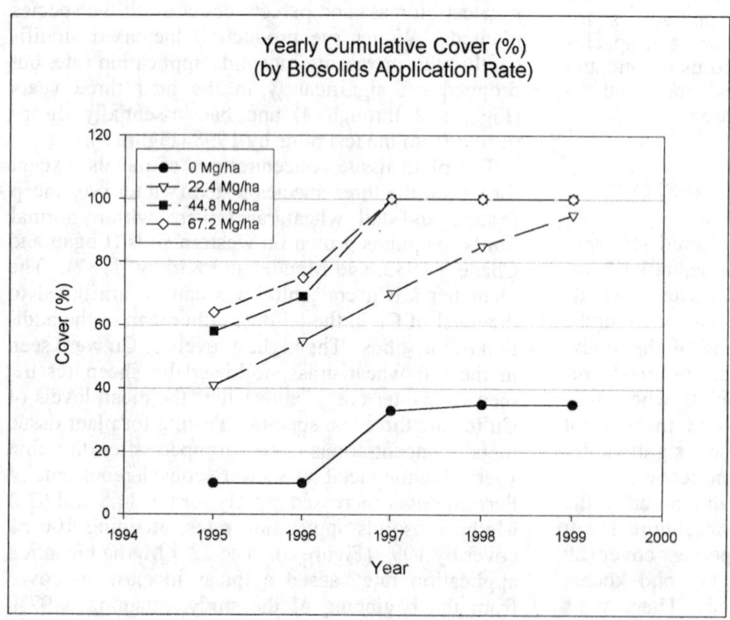

Figure 6-Yearly Cumulative Cover

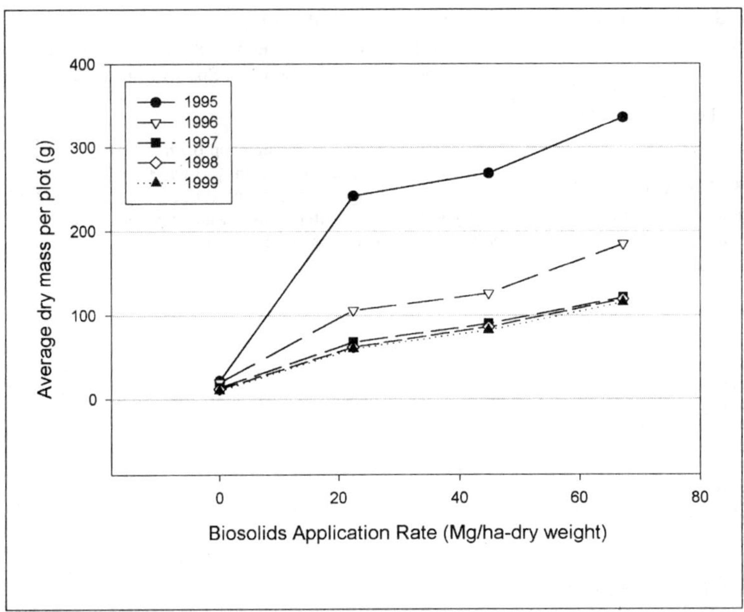

Figure 7-Yearly Biomass Production

due to winter rye, which essentially did not reseed itself for the following growing seasons, as would be expected for a nurse crop. However, the dead vegetation left by the winter rye did provide cover for additional erosion protection. This dead plant matter was subsequently mineralized and recycled over the remaining three growing seasons. It now appears that the amount of annual biomass production for all the test plots, and biosolids application rates, has now since stabilized. For example, for the 67.2 Mg/ha application rate, it appears that the annual average biomass production rate has stabilized at 110 g/plot. This is significant as no additional amendments or irrigation of these plots were used.

5 CONCLUSIONS

It was evident that the municipal sewage sludge used in this study as biosolids generally enhanced the plant growth capability of the tailings throughout the life of the experiment. It caused a considerable initial response in growth of winter rye, followed by a more gradual increase of more desirable reclamation species, such as the legumes. The highest biosolids application rate of 67.2 Mg/ha caused the greatest growth both in terms of aboveground biomass production and an increase in percent cover. One hundred percent cover had been achieved by the third year of the study. And, very importantly, it was shown by this study that biosolids addition did not influence the uptake of plant-available heavy metals.

6 ACKNOWLEDGEMENTS

Kennecott Utah Copper Corporation and the State of Utah, through the Department of Environmental Quality, funded this research.

7 REFERENCES

Chambers, Jeanne C. and Ray W. Brown 1983. Methods for Vegetation Sampling and Analysis on Revegetated Mined Lands. General Technical Report INT-151, Intermountain Forest and Range Experiment Station, Ogden, Utah.

EPA (U.S. Environmental Protection Agency), 1994. Land Application of Sewage Sludge: A Guide for Land Appliers on the Record Keeping and Reporting Requirements of the Federal Standards for the Use of Disposal of Sewage Sludge, 40 CFR 503. Office of Wastewater Enforcement and Compliance. EPA/831-B-002b. June 1994.

Jandel Scientific, 1994. Sigmastat, Statistical Software for Windows, Jandel Scientific, San Rafael, CA.

Lindsay, W. L., and W. A. Norvell, 1978. Development of a DTPA soil test for zinc, iron, manganese and copper. Soil Sci. Soc. Am. Proc. 42: 421-428.

Logan T. J. and R. L. Chaney, 1983. Utilization of municipal wastewater and sludge on land-metals. In A. L. Page (Ed.), Utilization of municipal wastewater and sludge on land, University of California, Riverside, pp. 235-326.

Manugistics, Inc., 1994. Statgraphics Plus for Windows, Rockville, MD

Mendel K. and E. A. Kirkby, 1987. Principles of plant nutrition, 4th Ed. International Potash Institute, Switzerland, pp.123-125.

Munshower F.F., 1994. Site preparation. In: Practical Handbook of Disturbed Land Revegetation. Lewis Publishers, Boca Raton, Florida, pp.57-72.

Nielson, Rex F. and H. B. Peterson, 1972. Treatment of Mine Tailings to Promote Vegetative Stabilization. Bulletin 485, Agricultural Experiment Station, Utah State University, Logan, Utah. June 1972.

Norton-Arnold, M., L., Fitzhugh, and V. Fischer, 1995. Long Range Planning for Biosolids Management. Water/Engineering and Management 2: 142.

Pichtel, J. R., W. A. Dick, and P. Sutton, 1994. Comparison of Amendments and Management Practices for Long-Term Reclamation of Abandoned Mine Lands. J. Environ. Qual. 23: 766-772.

Sabey, B. R., R. L. Pendeleton, and B. L. Webb, 1990. Effect of Municipal Sewage Sludge Application on Growth of Two Reclamation Spoils. J. Environ. Qual. 19: 580-586.

Schroeder, S. A., 1995. Topographic Influences on Soil Water and Spring Wheat Yields on Reclaimed Mineland. J. Environ. Qual. 24: 467-471.

Segal, W. and R. L. Mancinelli, 1987. Extent of regeneration of the microbial community in reclaimed spent oil shale land. J. Environ. Qual. 16: 44-48.

Sims, J. R., and Jackson G. D., 1971. Rapid analysis of soil nitrate with chromotropic acid. Soil Sci. Soc. Am. Proc. 35:603-606.

Sopper, W.E., 1993. Municipal Sludge Use in Land Reclamation. The Pennsylvania State University Press, University Park, PA, pp. 45-56.

Watanabe, F. S., and Olsen S. R., 1965. Test of an ascorbic acid method for determining phosphorus in water and $NaHCO_3$ extracts from soil. Soil Sci. Soc. Am. Proc. 28: 677-678.

Webber, M. D., 1992. Canadian approach for limiting metals on land from municipal sludges. In: C. Lue-Hing (Ed.), Municipal Sewage Sludge Management: Processing, Utilization and Disposal, Vol. 4, Water Quality Management Library, Technomic Publ. Co., Lancaster, PA, pp. 643-653.

Environmental Issues and Management of Waste in Energy and Mineral Production, Singhal & Mehrotra (eds)
© 2000 Balkema, Rotterdam, ISBN 90 5809 085 X

Tailings revegetation – On non-ameliorated substrate

J. M. Osborne
School of Environmental Biology, Curtin University of Technology, Perth, W.A., Australia

The Westonia open pit gold mine, 320 km east of Perth, Western Australia (31°18'S, 118°42'E) was decommissioned in July 1991. Normandy Mining Pty Ltd seeded in May-June 1993, the entire surface of the tailings storage facility. The rehabilitation objective was to stabilise the tailings structure. Tailings slurry had been pumped into three Cells at different periods throughout mining operations. The near surface layer of tailings is most significant from a rehabilitation perspective, as it is where plants are growing.

Vegetated areas have lower salinity than areas lacking vegetation (8.23 compared with 47.39 dS m^{-1} ECe). Revegetation density averages 1.734 stems m^{-2} in October 1999 and foliage cover is 30 percent. Nine chenopods were recorded in October 1999 with *Maireana brevifolia, Atriplex undulata* and *A. lentiformis* more prevalent. Grasses continue to show accelerated establishment. A cryptogamic cover prevalent over the upper surfaces of "raw" tailings appears to be enhanced under established vegetation where moisture levels and the soil nutrient status are inherently higher.

The tailings profile in the centre of the third cell (vegetation absent) showed similar changes in salinity to previous samplings. The surface sample in April 1999 (102 dS m^{-1}) was less saline than the April 1998 sampling (225 dS m^{-1}). The layer between 0.1 m and 0.4 m remains extremely saline, *i.e.* range 83-150 dS m^{-1} ECe. Around the perimeter of his cell the tailings is coarser, and sampling of the profile to 0.3 m depth gave salinities from 5 to 20 dS m^{-1} ECe.

1 BACKGROUND

The Westonia open pit gold mine, 320 km east of Perth, Western Australia (31°18'S, 118°42'E) was decommissioned in July 1991. The initial success of revegetation trials established on non-ameliorated tailings over the storage facility by Curtin University, prompted Normandy Mining Pty Ltd to seed the entire surface (40 ha) of the tailings storage facility using a suite of chenopod species (saltbushes and bluebushes). Seeding was at 2 kg ha^{-1} or 4 kg ha^{-1}. The rehabilitation objective was to stabilise the tailings structure. Tailings slurry had been pumped into three Cells (A, B, and C) at different periods throughout mining operations. Cells A and B were used for initial deposition beginning in 1986, and Cell C was used in the last twelve months of operations. The process water used at Westonia was saline (TDS 35,000 ppm, approximate ECe 56 dS m^{-1}). A layer of precipitated salt mined a few metres above the water table further increased water salinity inside the processing plant. Most of this tailings had been pumped into the upper layer of Cell A and Cell B. Stockpiled low grade ore, stripped from the surface at the commencement of operations, was processed late in the mine life. This material was ground coarser than previous materials, and was deposited into the upper layer of Cell C. This near surface layer of tailings is most significant from a rehabilitation perspective, as it is where plants are growing.

To date a total of 12 assessments have been completed over the Westonia tailings storage facility, commencing in April 1994 (Table 1). The most recent assessment was in October 1999, 75 months after the large scale seeding in May – June 1993.

Table 1: Timing of revegetation assessments over the Westonia tailings storage facility, and revegetation age.

Assessment Date		Revegetation Age
April	1994	10 months
April	1995	22 months
September	1995	27 months
February	1996	32 months
April	1996	34 months
September	1996	38 months
April	1997	45 months
September	1997	50 months
April	1998	57 months
September	1998	62 months
April	1999	69 months
October	1999	75 months

2 DATA AND COMMENTS

2.1 *Rainfall*

The Westonia Mine tailings storage facility lies within the eastern wheatbelt region of Western Australia. Annual rainfall is relatively low (320 mm average per annum) but sufficiently reliable to allow successful cropping of wheat, oats and barley (Fig. 1). At conclusion of mining and processing operations tailings was extremely saline. Consistent rainfall during the years since decommissioning has facilitated movement of salts down the profile, and enhanced plant establishment. A vegetation cover has lowered surface temperatures over the tailings storage facility, and further assisted salinity reduction by interrupting the capillary rise of salts (Fig. 2).

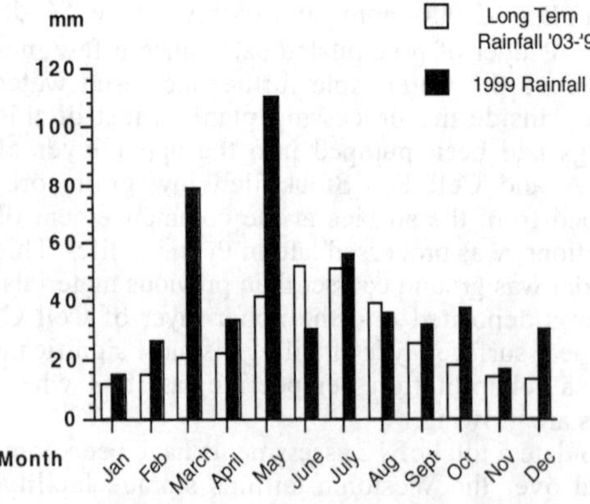

Figure 1: January - December 1999 rainfall (mm) and long term averages, for Merredin (40 km south-west of Westonia).

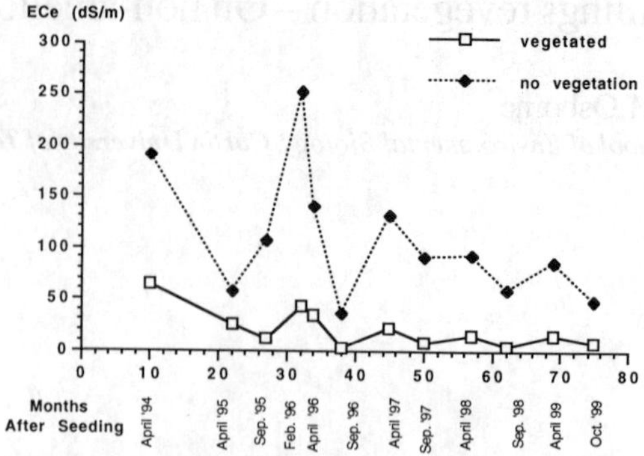

Figure 2: Mean salinity values from vegetated (n = 78) assessment transects, and areas lacking vegetation (n = 52): A comparison over time.

2.2 *Analysis of Consolidated Tailings*

Mean surface salinity over vegetated areas of the tailings storage facility is lower in October 1999 than was sampled in April 1999 (8.2 and 15.0 dS m^{-1} ECe respectively). After winter rains, salinity decreases with an increase during hotter drier months (see Fig. 2). Over a five year period September samplings have given a salinity range of 2.8 - 10.6 dS m^{-1} ECe.

Salinity over non-vegetated areas of the tailings decreased between September 1998 (58.7 dS m^{-1} ECe) and October 1999, now averaging 47.4 dS m^{-1} (Table 2). Following summer surface salinities are higher, *e.g.* 85.5 dS m^{-1} in April compared with 47.5 in October 1999.

Table 2: Non-vegetated areas of the tailings storage facility, salinity (dS m^{-1} ECe) from April 1998 to April 1999 samplings.

CELL	Sampling Period (units dS m^{-1} ECe) April '98	Sep '98	April '99
A n= 14	144.7	86.4	131.4
B n =11	159.4	102.4	154.8
C n = 27	37.1	26.5	33.4

Sampling points located in the lower lying, central area of the cells continue to show elevated salinity, *e.g.* 202.4 dS m^{-1} from one point on Cell C. However, there has been a general marginal decrease on lower lying areas over the twelve months. The 24 upper surface non-vegetated samples from Cells A and B gave 17 samples greater than 100 dS

m^{-1} ECe in April 1999 compared with 21 samples in April 1998.

On vegetated areas, salinities in October 1999 are similar to those in September 1998, significant decreases are not apparent (see Fig. 3). It is noted values on the vegetated areas are low, reflecting salts have not risen to the surface; on Cell C salinity was 2.6 dS m^{-1} ECe in September 1998 and 1.8 dS m^{-1} ECe in October 1999. There is extensive uptake of water by the plant roots that have now extended into the tailings.*

*Further field and laboratory studies continue – refer Prof J.M. Osborne supervisor, with Dr A. Schatral, of PhD study (1998-2000) on chenopod growth on tailings; and *Atriplex nummularia* and *Maireana brevifolia* Honours investigation (1999 – Westonia study site).

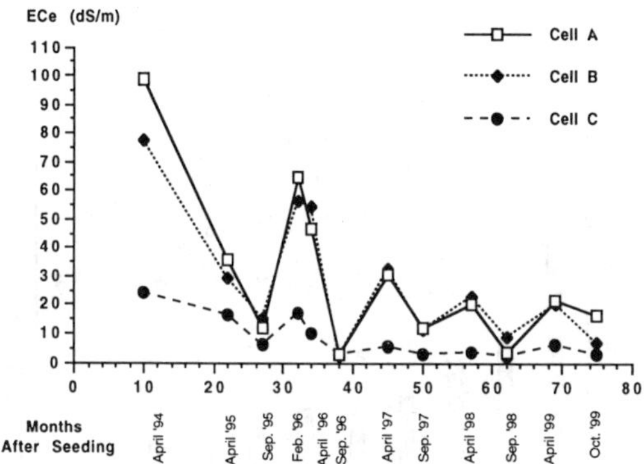

Figure 3: Mean salinity values from vegetated assessment transects, comparing each Cell, over time.

2.3 *Revegetation Cover and Density*

Revegetation density (number of stems per m^2) was greater in April 1999 than September 1998, encouraging after a summer period and likely facilitated by good rains in January to April (see Fig. 1); 2.44 stems m^{-2} in April 1999 compared with 2.35 m^{-2} in September 1998. Sampling in October 1999 gave a decrease in plant numbers (now 1.74 stems m^{-2}) (Fig. 4). However, plant cover over the tailing surface has remained at a highest level since seeding. On average, 30 percent of the vegetated area is covered by revegetation, and October 1999 gives the highest of the five post-winter readings (Fig. 4).

2.4 *Revegetation Composition*

Nine (9) chenopod species were recorded over the tailings storage facility in October 1999, along with

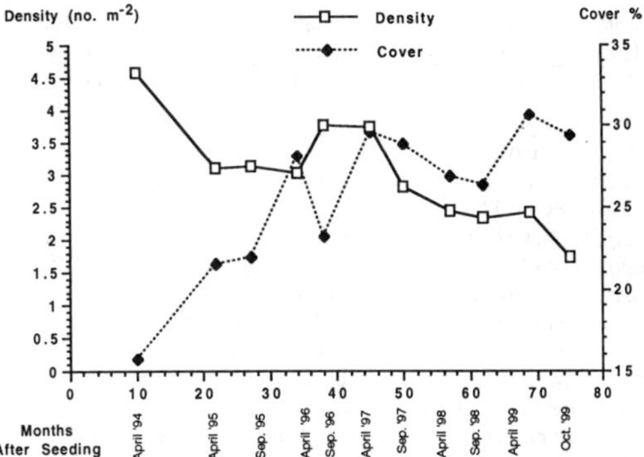

Figure 4: Change in average revegetation density (per m^2) and cover (%) values (Cells A, B, and C combined) over time.

two introduced grasses, and members of the families Asteraceae and Zygophyllaceae (Table 3). Considering all 26 assessment transects density averaged 1.74 plants m^{-2} and there was foliage over 29 percent of the areas sampled (noting random sampling with stratification of vegetated and non-vegetated areas).

Maireana brevifolia (mean IVI of 101), *Atriplex undulata* (mean IVI = 53) and *Atriplex lentiformis* (mean IVI = 40) are prevalent on the tailings impoundment after the summer period and there dominance is reflected in numbers present and percent of the foliage cover.

The prevalence of the introduced grasses (*Avena fatua* {wild oats} and *Bromus* sp. {brome}) has continued to increase (IVI = 47 - considering cover only). Both were well established at the base of dead *Atriplex lentiformis* stems, or in areas of apparent higher soil moisture. At the base of plant stems surface temperatures are lower, there is moisture accumulation, and nutrients from decaying dead foliage provides a more amenable media for growth.

2.5 *Cryptogamic Cover*

A change noted over the upper surface of "raw" tailings in September 1998 was the prolific and recent appearance of a cryptogamic cover under the revegetation. This was first recorded in September 1998 when along each of the twenty-six permanent assessment transects cryptogamic cover was scored between 0 (not present) and 5 (100 percent cover) (Table 4). Cryptogams are more common around the base of plants. Similar values were shown when recordings scored after winter were compared with those taken post summer (April 1999) (see Table 4).

Table 3: Westonia tailings revegetation species recorded October 1999 over all three cells. Mean density, cover, heights and IVI values from twenty-six 20 m by 1 m transects

FAMILY Species	Common Name	Density (m^2)	Cover (%)	Height (cm)	IVI
ASTERACEAE					
Sohchus oleraceus	common sowthistle		0.398		1.17
CHENOPODIACEAE					
Atriplex amnicola [a]	river saltbush	0.052	0.594	69	8.82
Atriplex bunburyana	silver saltbush	0.123	3.292	65	28.37
Atriplex lentiformis [a]	quail bush	0.267	1.824	56	39.74
Atriplex nummularia	old man saltbush	0.050	0.610	94	7.34
Atriplex undulata [a]	wavy leaf saltbush	0.387	2.232	38	53.06
Atriplex vesicaria	bladder saltbush	0.042	0.417	34	7.48
Enchylaena tomentosa	ruby saltbush	0.010	0.096	15	1.43
Maireana brevifolia [a]	small leaf bluebush	0.794	3.948	14	100.89
Sclerolaena diacantha	copper burr	0.010	0.025	11.5	2.43
POACEAE					
grasses [b]			15.867		47.34
ZYGOPHYLLUM					
Zygophyllum aurantiacum	shrubby twinleaf	0.002	0.075	9.75	1.48
	TOTAL	1.737	29.378		300.00

[a] direct seeded species
[b] includes *Avena fatua* (wild oats) and *Bromus* sp. (brome)

Table 4: Cryptogamic cover (scored between 0-5) along each of the 26 permanent assessment transects - September 1998 and April 1999.

Cell A		Cell B		Cell C	
Sept. '98	April '99	Sept. '98	April '99	Sept. '98	April '99
0.0	1.0	3.0	3.0	3.0	3.5
4.0	3.5	0.0	0.0	3.0	1.5
1.0	2.5	0.5	0.0	0.5	0.5
2.0	1.0	0.5	1.0	1.5	0.5
3.0	3.0	2.5	3.0	1.5	2.5
0.2	0.5	0.5	0.5	1.0	2.5
0.3	0.0	3.0	0.0	0.0	0.0
0.5	0.5	0.0	0.0	0.0	0.0
				3.0	1.5
				1.0	1.0
1.38	1.50	1.25	0.94	1.45	1.35

2.6 *Batter Revegetation - Northern and Southern Aspect*

There was foliage cover over 40 percent of the southern batter surfaces of the tailings facility in September 1998 and October 1999 (Fig. 5). Vegetation continues to establish slowly over the northern batters of the tailings storage facility. In September 1998 plant density averaged 0.98 plants m^{-2} with a ground cover of 2 percent, and 12 months later comparable values are 0.75 plants m^{-2} and 7 percent.

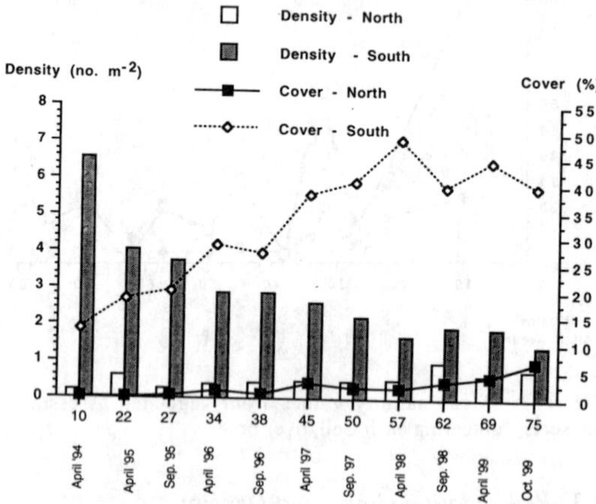

Figure 5. Average revegetation density (per m^2) and cover (%) recorded on the northern and southern batters of the tailings impoundment between April '94 - October '99. Months after seeding given.

Soil salinity along the northern and southern batters increases over the summer period. April 1998 and 1999 values however are comparable, *i.e.* 14.63 and 15.26 dS m^{-1} ECe respectively for the northern batter and 4.43 and 5.70 dS m^{-1} ECe

2.7 *Tailings Profile*

The salinity profile in the non-vegetated centre of Cell C follows a different pattern to that encountered around the vegetated perimeter of the cell (compare Figure 6 showing non-vegetated profile with Figure 7 profile from vegetated perimeter). In the central region, the tailings consist of a high proportion of fine material, which when moist becomes "Plasticine-like".

The April 1999 profile taken from the centre of Cell C (vegetation absent) showed similar changes in salinity to the previous samplings (see Fig. 6). The surface sample in 1999 was less saline than the April 1998 sampling (102 compared with 225 dS m^{-1}). Between 0.1 - 0.4 m depth salinity ranged from 83 - 150 dS m^{-1}, but it is noted samples below this depth (down to 1 m) are considerably less saline, *e.g.* 25 dS m^{-1} at 0.5 m depth and 37 at 1.0 m (Fig. 6). Salinity decreased lower in the profile.

Around the perimeter of Cell C the raw tailings is much coarser, and hence leaching potential greater. Vegetation has successfully established. Previous profile samplings have indicated a peak in salinity at 0.5 m depth. In September '98 this peak was not observed (10 dS m^{-1}). However in April 1999, at 0.4 m to 0.6 m depth salinity had increased, 37 and 36 dS m^{-1} respectively and showing similar levels to those of April 1997 and 1998. Salinity then decreased with depth, from 17 dS m^{-1} (0.7 m) to 15 dS m^{-1} (1.0 m) at the April 1999 sampling (see Fig. 7).

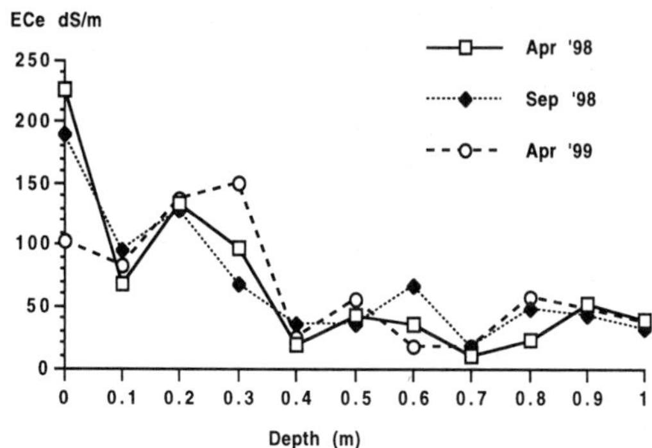

Figure 6. ECe (dS m^{-1}) of tailings samples collected to 1 m at 0.1 m intervals at a non-vegetated site in Cell C; April '98 to April '99.

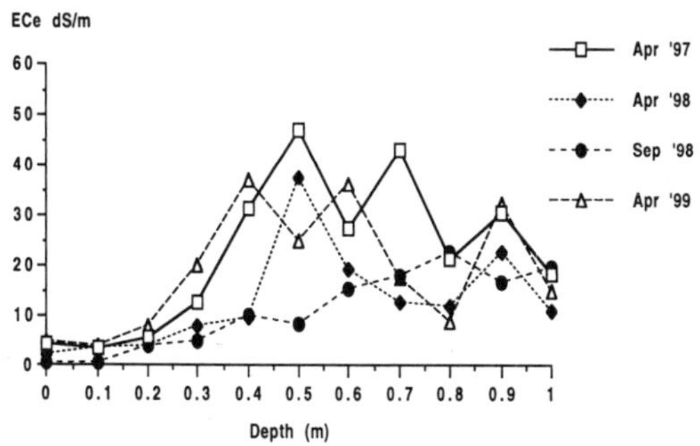

Figure 7. ECe (dS m^{-1}) of tailings samples collected to 1.0 m at 0.1 m intervals at a vegetated site in Cell C; April '97 to April '99.

3 BIBLIOGRAPHY

Brearley, D.R. & Osborne, J.M. 1999. Ecosystem Development of Revegetation on tailings at Westonia - September 1998. Prepared for the Normandy Mining Ltd pp 24.

George, P.R. & Wren, B.A 1985. Crop tolerance to soil salinity. Western Australian Department of Agriculture Technote 6/85.

Green, J.W. 1985. (2nd edition) *Census of the Vascular Plants of Western Australia*. WA Herbarium, Department of Agriculture, Western Australia.

Green, J.W. 1987. *Census of the Vascular Plants of Western Australia Supplement No. 7*. WA Herbarium, Department of Agriculture, Western Australia.

Magurran, A.E. 1988. Ecological Diversity and Its Measurement. University Press, Cambridge, Great Britain.

Mueller-Dombois, D. & Ellenberg, H. 1974. Aims and Methods of Vegetation Ecology. Wiley and Sons, New York.

Osborne, J.M. & Brearley, D.R., 1994. Revegetation on tailings at Westonia. Prepared for the Normandy Poseidon Group pp 58.

Osborne, J.M. & Brearley, D.R. 1995. A revegetation option for saline tailings in Western Australia. In: R.K. Singhal, Mehrotra, A.K., Hadjigeorgiou, J. & Poulin, R.. (eds). Proceedings: Fourth International Symposium on Mine Planning & Equipment Selection. A.A. Balkema, Rotterdam, Netherlands. pp 1011-1016. ISBN 90 5410 5590

Osborne, J.M. & Brearley, D.R. 1999. Ecosystem Development of Revegetation on tailings at Westonia - April 1999 Summary. Prepared for the Normandy Mining Ltd pp 20.

Osborne, J.M., Brearley, D.R. Fox, J.E.D. 1993. Assessment procedures for revegetation on waste rock dumps in semi-arid Western Australia. In: Proceedings: Gold fields International Conference on Arid Landcare. pp 129–150.

Osmond, C.B., Bjorkman, O., & Anderson, D.J. 1980. *Physiological Processes in Plant Ecology. Towards a Synthesis with* Atriplex. Springer Verlag, Berlin, Heidelberg, New York.

Rayment G.E. & Higginson, F.R. 1992. *Australian Laboratory Handbook of Soil and Water Chemical Methods.* Inkata Press, Australia.

Rogers, N. 1997. Salinity tolerance of two selected chenopods from a rehabilitated minesite and natural habitat in the Goldfields of Western Australia. Honours Dissertation. Curtin University of Technology. pp 61.

Zar, J.H. 1996. *Biostatistical Analysis*. (3rd edition) Prentice-Hall, Inc. New Jersey.

Environmental Issues and Management of Waste in Energy and Mineral Production, Singhal & Mehrotra (eds)
© 2000 Balkema, Rotterdam, ISBN 90 5809 085 X

Completion criteria – Case studies considering bond relinquishment and mine decommissioning: Western Australia

J.M.Osborne & D.R.Brearley
School of Environmental Biology, Curtin University of Technology, Perth, W.A., Australia

ABSTRACT: The Western Australia Government Agencies controlling environmental bond assessment and land relinquishment after mine decommissioning advocate proactive environmental management. Meeting environmental objectives ensures return of bond monies and facilitates release of the Company from environmental obligations. Completion criteria that confirm establishment of a self-perpetuating and resilient vegetation cover over mine waste are being developed. Post-mining land use generally reverts to maintaining the natural ecosystem or rangeland. Within this climatic zone very distinct vegetation communities are closely linked to soil substrate classification. Matching surface waste characteristics from a mine waste dump with those of surrounding analogue communities is the first step in determining an achievable vegetation cover. Examples of the methodology as applied to one Western Australian mine site, a semi-arid location (annual rainfall 320 mm) (Westonia - 31°18'S, 118°42'E) is presented. There has, at Westonia, been establishment of Eucalypt Woodland on mildly saline waste surfaces. The combination of direct seeding at the appropriate sowing time and good topsoil management, has led to successful revegetation of a suite of species with all life forms represented.

1 BACKGROUND

1.1 *Completion Criteria*

Operators of open cut gold and nickel mines in arid/semi-arid Western Australia are putting an increasing effort into rehabilitation of their waste rock dumps. Following the recent good wet seasons in the interior of Western Australia there are many examples of extremely successful vegetation establishment on dumps. This has highlighted the situation that neither the operators nor the regulators know what to expect in the way of sustainability of the resultant ecosystems. No one knows what ecosystem objectives should be set for waste dump rehabilitation. Indeed there have been instances where the long term rehabilitation objectives appear to have been generated by the personal opinions of individuals. This is not surprising, as currently we are unable to answer basic questions such as:

- what constitutes success in waste dump rehabilitation
- is the vegetation likely to live to maturity
- will the ecosystem regenerate and be sustainable
- what is an achievable objective for dumps in this region
- is this dump heading towards success or towards failure
- how do we know whether there will be success or failure

A Minerals and Energy Research Institute of Western Australia project was designed to provide these answers. Companies need to be able to answer these questions to plan their rehabilitation programs, and companies and regulators need to be able to set criteria for lease completion conditions. Data and methodology exists in some large-scale Australian rehabilitation operations, *e.g.* Western Australia northern sand plain mineral sands, Western Australia and northern Queensland bauxite; but not for arid zone waste rock dumps (after all, waste rock dump is very different to input back filling).

The senior author had accumulated sufficient data for waste rock dumps to give preliminary indications of revegetation development. The current project is consolidating a pre-existing knowledge base, and using it to develop rehabilitation protocols and assessment procedures. Attainable criteria for progressive revegetation, and lease completion conditions (completion criteria) are being generated.

The Western Australia Government Agencies controlling initial environmental bond assessment and land relinquishment after mine decommissioning advocate proactive environmental management approaches by the mining operator. Meeting environmental objectives ensures return of bond monies and facilitates the release of the Company from environmental obligations. The reported paper is part of a long term study supported by Government and seven gold and (or) nickel mining companies (*via* the Minerals and Energy Research Institute of Western Australia).

The project aims to develop completion criteria which confirm the establishment of a self-perpetuating and resilient vegetation cover, over mine waste, at the earliest point in the development of the rehabilitated ecosystem. If a vegetation cover is both self-perpetuating and resilient a number of ecological processes will be occurring. The quantitative assessment of outcomes is the basis for the development of the completion criteria, and include plant cover, revegetation density, species richness, and reproductive potential.

Before developing completion criteria the end land use must first be defined. All the sponsoring Company field sites are located in arid (desert: summer but more effective winter rainfall) and semi-arid (semi-desert: Mediterranean) regions of Western Australia on Crown Land, with all but one being on pastoral (rangeland) leases. Post-mining land use in these areas generally reverts to maintaining the natural ecosystem or rangeland. Rangeland areas within this climatic zone comprise very distinct vegetation communities that are closely linked to the substrate type. Matching surface waste characteristics from a mine waste dump with those of surrounding analogue communities is the first step in determining what vegetation cover is achievable. "It is not possible to rehabilitate to self-perpetuating vegetation without reference to the surrounding ecosystem on which the constructed one will ultimately depend for stability" (Biggs 1996).

Completion criteria are being developed at each mine site using quantitative data gathered from:

i) baseline studies of analogue communities - distinct vegetation units not impacted on by mining activities; and

ii) long term ecosystem studies of rehabilitation research trials (research trials have been regularly monitored {annual or biannual}, currently up to nine years following revegetation by direct hand sowing of seed).

This project recognises the importance of developing completion criteria prior to, or during the early phases of the mining operation. The completion criteria developed are compatible with the end land use, the substrate being dealt with, and the limitations of the arid/semi-arid climatic regime. The methodology for developing completion criteria is easily applied and economical.

Presented below is the methodology as applied to one Western Australian mine site, a semi-arid location (annual rainfall 320 mm) (Westonia - 31°18'S, 118°42'E). There has, at Westonia, been establishment of Eucalypt Woodland on mildly saline waste surfaces. Two other analogues sites are illustrated. The combination of direct seeding at the appropriate sowing time and good topsoil management, has led to successful revegetation of a suite of species with all life forms represented.

1.2 *The Site*

The Westonia open-cut goldmine is located 320 km east of Perth, in the eastern wheatbelt of Western Australia. Eucalyptus Woodland is the dominant native vegetation cover in the region, with *Eucalyptus salubris* (gimlet), *E. salmonophila* (salmon gum) and *E. longicornis* (morrell) the common mallee and tree species. The understorey composition and structure is variable in response to changing soil conditions, however typical associations are low chenopod shrubs or mid-tall acacia/melaleuca shrubs.

The conclusion of mining operations at Westonia in 1988 marked the commencement of rehabilitation activities the following year. Our assessments focussed on the four waste dumps (A-D) which comprise saline weathered waste rock from the open-cut.

Salinities ranged from 4.2 dS m^{-1} ECe (mildly saline) to 37.9 dS m^{-1} ECe (extremely saline), equivalent to 2,714–24,256 ppm salts in solution. Surface soil over the waste dumps was typically of basic reaction, however localised areas were initially acidic (pH range 4.9-9.2). The topsoil cover spread over the waste rock materials was non-saline, ranging from 1.20 to 2.56 dS m^{-1} ECe (768 - 1,639 ppm salts in solution equivalent).

A mean annual rainfall in the order of 320 mm falls largely during winter months. Extremes of temperature are typical. Species selected for revegetation must tolerate a low and sometimes unreliable rainfall regime and extremes of temperature. In early May 1990 the surfaces and slopes of four waste dumps at the Westonia operation were direct sown with a suite of local species local. Seed mixtures included salt tolerant varieties specifically adapted to the saline medium initially present, and a variety of species found in the adjacent woodland (Table 1).

Rehabilitation strategies likely to enhance field establishment on saline rock materials were introduced prior to seeding. These included the addition of a topsoil layer, deep ripping, ploughing or harrowing of the topsoil and underlying rock materials, and application of fertiliser. Fletcher (1993) found deeper ripping (to 1 m) was beneficial to the establishment of acacias. Eucalypts were assisted by the presence of ponding zones. Applying fertiliser at an appropriate regime to waste rock materials lacking topsoil enhances chenopod growth (Oliver *et al.* 1992, Stewart & Osborne 1993, Osborne & Sneesby in prep.). The response of eucalypts and woody perennials such as acacias to different fertiliser application rates is less well studied.

Revegetation has now been assessed on eight occasions between September '90 (5 months after seeding) and September '98 (101 months - 8.5 years after seeding). For Waste Dumps 1A, 1B and 1C data from permanent assessment transects on the upper flat surfaces and battered slopes were considered separately. For Waste Dump 1D only the upper surfaces have been monitored, as slopes were left at angle of repose to reduce the clearing of woodland along the adjacent road side.

2 RESULTS AND COMMENTS

2.1 *Waste Dumps*

The upper surfaces and battered (to ≤ 200) slopes of the four waste dumps have shown similar change with respect to salinity and pH over the nine year assessment period. Initially soils were saline (ECe 25 dS m^{-1}) and alkaline (pH 8.2). Salinity has decreased over time averaging less than 1.4 dS m^{-1} in September '98 for all surfaces, with the exception of Waste Dump 1C - upper (6 dS m^{-1}). Soil pH remains alkaline in September '98, averaging between 8.3 and 8.6 (September '98). A general decrease in pH was recorded for most surfaces by April '96, seventy-two months after seeding. During the same year foliage cover over the same surfaces decreased. Biological processes involved with litter decomposition were likely responsible for the increased soil acidity.

Relatively high seeding rates (Table 1) and good quality seed resulted in abundant seed germination, with subsequent high seedling densities recorded five months after seeding; for both the upper and batter surfaces. Increased competition between establishing plants contributed to progressive decreases in average density over the course of the initial four assessments (Fig. 1).

For Waste Dump 1A a greater response in plant density and revegetation cover was recorded over upper surfaces, in comparison to the batters. For the upper surfaces plant density has ranged between 0.27 and 0.65 stems m^{-2} since April '96. In comparison plant density over the battered surfaces has ranged between 1.23 and 1.64 stems m^{-2} for the same period. Revegetation cover was typically 15 percent higher over the upper surfaces, in comparison to the batters. Ground cover has shown an increasing trend, averaging 44 percent and 26 percent over upper and batter surfaces respectively in September '99. Average ground cover decreased for both areas in April '95 and April '96, in response to lower rainfall experienced during 1994.

Both vegetation density and cover on Waste Dumps 1B and 1C (Fig. 1) showed similar trends to Waste Dump 1A. In September '98 revegetation cover over the upper and batter surfaces of Waste Dump 1B averages 48 and 43 percent respectively; Waste Dump 1C - 40 and 30 percent respectively.

Following high initial plant density recorded during the first growing season on upper surfaces of Dumps 1B and 1C (5.61 stems m^{-2} and 4.78 stems m^{-2} in September '90), density has decreased over time to average 0.25 stems m^{-2} and 0.47 stems m^{-2} in September '98 (Fig. 1).

The structure and composition of revegetation over Waste Dump 1D, chenopod shrubland dominated by oldman saltbush (*Atriplex nummularia*), differs from that establishing over the other three waste dumps (*Eucalyptus, Acacia* dominated).

2.2 *Species Composition*

Atriplex semibaccata (creeping saltbush) was an important coloniser species with high numbers present during the early stages of ecosystem development (Fig. 1). The prostrate spreading habit of this species is of considerable benefit, facilitating surface stability initially, but was then out-competed by other desirable longer lived species. Sixty months after seeding *Atriplex semibaccata* was only recorded in very low numbers, and by September '98 (101 months after seeding) was present on only one Waste Dump (1C) (Fig. 2).

Table 1. Species sown over the upper surfaces of Waste Dump 1C (4 ha)- Westonia waste dump; seeded 5 May 1990.

FAMILY Species	Common Name	Seed Rate (kg ha^{-1})
CAESALPINIACEAE		
Senna filifolia	desert cassia	0.100
CHENOPODIACEAE		
Atriplex nummularia	oldman saltbush	0.800
Atriplex semibaccata	creeping saltbush	0.800
Enchylaena tomentosa	ruby saltbush	0.100
Maireana brevifolia	5-winged bluebush	0.500
MIMOSACEAE		
Acacia acuminata	jam wattle	0.200
Acacia hemiteles	tan wattle	0.200
Acacia nyssophylla		0.180
MYRTACEAE		
Eucalyptus loxophleba	York gum	0.100
Eucalyptus salubris	gimlet	0.300
Eucalyptus sargentii	salt river gum	0.200
Melaleuca pauperiflora	boree	0.200
PITTOSPORACEAE		
Pittosporum phylliraeoides	native willow	0.500
PROTEACEAE		
Hakea multilineata		0.062
SAPINDACEAE		
Dodonaea viscosa	hop bush	0.300
TOTAL		4.542

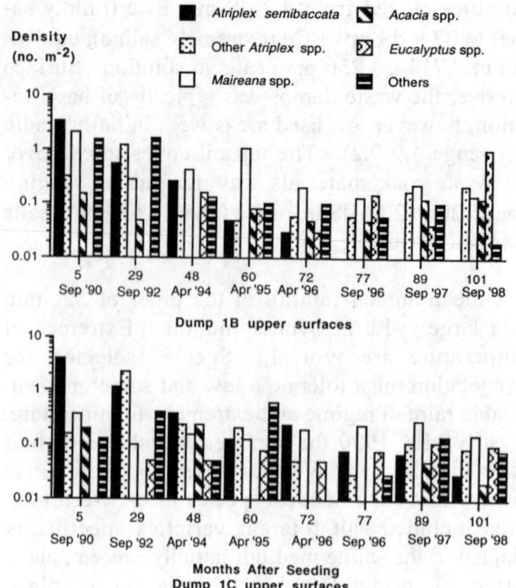

Figure 1. Plant density (log_{10} scale) for dominant taxa on Dumps 1B and 1C upper surfaces, showing changes over time - September '90 to September '98.

Other *Atriplex* spp. established gradually, replacing *A. semibaccata* in the revegetation by April '95 (60 months after seeding). Saltbushes, especially *Atriplex nummularia*, are most prevalent over Waste Dump 1D, providing 25 percent ground cover in September '98. The taxa has also established well over Waste Dump 1A, with 9 percent ground cover over upper and batter surfaces (September '98)For Waste Dumps 1B and 1C where taller eucalypts and acacias have successfully established (see Fig. 2), *Atriplex* spp. are less dominant, providing approximately 6 percent and 2 percent ground cover respectively.

Maireana spp. (bluebushes) are a minor component of the revegetation on all four waste dumps. In September '98 the taxa approximates 2 percent ground cover for the upper and batter surfaces of Waste Dump 1C. For the other surfaces assessed, it provides less than 0.6 percent cover.

Grasses, were volunteer colonisers 29 months after seeding, when surface salinity had decreased to

Table 2. Trends in revegetation development from quantitative observations (over time) on 35 rehabilitation areas located between Westonia (North Eastern Wheatbelt), Laverton (North Eastern Goldfields) and Kalgoorlie (Eastern Goldfields) - Western Australia.

Ecosystem Development
1-2 Years Growing Season: Prolific seed germination and subsequent establishment of high seedling densities. Low foliage cover during the initial growing season, increasing during the second year. Soils are typically of highest salinity during the early stages of revegetation. Annual species (especially *Atriplex* spp.) are more prominent than perennial life forms. This is more evident with increasing salinity.
3 years and beyond growing seasons: Competition between establishing plants leads to predictable changes in density and ground cover. Plant density decreases over time before stabilising; revegetation cover increases before also stabilising. The point at which both density and cover stabilise is determined by the soil resource supply. Variability in environmental conditions affects the stability of both parameters. Revegetation parameters are often synonymous with analogue communities of similar vegetation structure and composition.
For Low Chenopod Shrublands a small number of taxa establish themselves as dominants within the revegetation (keystone species). Common examples include *Atriplex vesicaria, A. bunburyana, A. stipitata* and *Maireana pyramidata.* For revegetation where a number of distinct strata are establishing (*Eucalyptus* above *Acacia* above *Atriplex*), the major taxa are usually more evenly represented.
Significant decreases in density, cover and richness occur in response to disturbance *e.g.* drought, overgrazing. If the revegetation is resilient, parameters will return to pre-disturbance levels within 1 - 2 growing seasons, dependent on rainfall. A response in revegetation cover often lags one season behind the plant density response.
Significant decreases in vegetation parameters recorded on rehabilitated sites may be in response to the senescence and death of a dominant species. Maternal plants are all seeded during the same year and subsequently influenced by similar environmental factors, providing like life spans, resulting in high numbers of the same species dying during the same year. This can trigger a change in species dominance within the rehabilitation.
Maireana spp. seed often requires an after-ripening period, and can take up to three years before germinating and establishing in the revegetation. Abundant seed germination for this taxa is observed following heavy summer rainfalls.
Exotic weed species are more likely to influence rehabilitation performance when occurring in the soil seed bank prior to the seeding of desirable rehabilitation species, or when an established vegetation cover is disturbed.
Low Shrublands are a particularly successful vegetation cover over waste dumps throughout the arid zone, due to the halophytic and xerophytic nature of the dominant chenopod component.
Heavy summer rainfall from tropical cyclone events is very important to overall revegetation development, and may be of particular importance for the establishment of *Acacia* and *Senna* spp. as a pre-cursor for breaking seed dormancy. Summer rainfall also extends the length of the growing season.
A vegetation cover decreases surface temperature over the dry summer months, reducing the capillary rise of salts. Salts are progressively leached from the upper profile. There is now scope to trial the remedial seeding of woody perennial species (acacias, sennas, eucalypts) when surface salinities have declined to tolerable levels.

tolerable levels. They have persisted at ground level providing average ground cover up to 56 percent. In September '98 average ground coverage provided by grasses ranged between 10 and 34 percent on the dumps.

Mid-stratum and upper storey species including *Acacia* and *Eucalyptus* were first recorded in the revegetation 29 months after seeding (September '92). Both taxa displayed initial slow development, up to September '97, when significant increases in ground cover were evident. The sudden increase in productivity coincided with the first flowering by many species and subsequent production of seed. *Acacia* spp. and *Eucalyptus* spp. have established on the batter surfaces of Waste Dump 1A (2 and 6 percent ground cover respectively), but are absent from the upper surfaces of the dump. On Waste Dumps 1B and 1C *Eucalyptus* is better established over the upper flat surfaces (14 and 20 percent cover respectively - see Fig. 2), in comparison to batters (6 and 3 percent cover respectively). *Acacia* spp. are successfully established over upper and batter surfaces of Waste Dump 1B (7 and 10 percent cover), but have been less successful on Waste Dump 1C (1 percent for both surfaces). Differences in topsoil quality (storage and handling practices) during waste dump construction has likely affected the subsequent establishment of hard seeded species by impacting the soil seed store.

2.2.1 <u>Eucalyptus</u> *Woodland over Tall Scrub*

The *Eucalyptus* Woodland over Tall Scrub analogue community occurred north east of the waste dumps. The ground coverage of this community averaged 44 percent in September '97 and 52 percent in April '99 (Fig. 3). The dominant upperstorey eucalypts (20 percent) were the greatest contributors to overall foliage cover, along with grasses (15 percent) and acacias (5 percent; see Fig. 3). The openness of the understorey was reflected by low plant density, between 1.9 stems m^{-2} (September '97) and 0.9 stems m^{-2} (April '99). Higher density in September 1997 was an abundance of the short lived annual

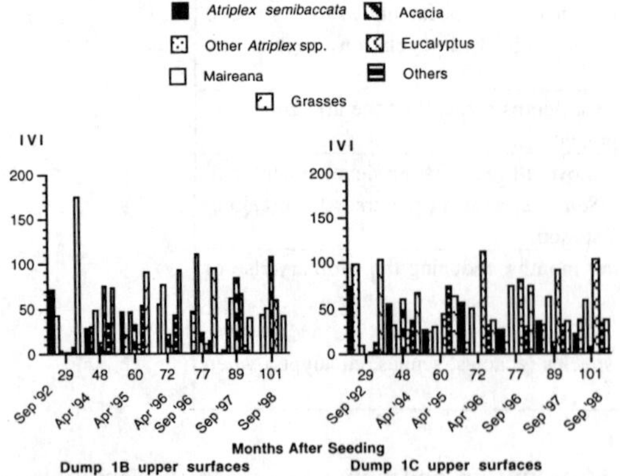

Figure 2. Importance Value Index (IVI), which considers frequency, density and cover, for dominant taxa on Dumps 1B and 1C upper surfaces showing changes over time - September '92 to September '98.

Ptilotus exaltatus (grouped into the "other species" - Fig. 3) which proliferated following late unseasonal rainfalls. Soils are non-saline (1.0 dS m^{-1}) and alkaline (pH range 7.3 - 7.8).

2.3 *Analogue Sites - Westonia*

Three analogue communities were sampled in the Westonia region, each differing in overstorey and understorey structure and composition. Dumps 1B and 1C are compared with the *Eucalyptus* woodland over Tall Shrub (Fig. 3). Also sampled were *Eucalyptus* Woodland over *Atriplex stipitata* Low Shrubland, and *Atriplex vesicaria* Low Shrubland.

3 DISCUSSION

Eight and a half (8.5) years following seeding, the rehabilitated site was comparable to the analogue site with respect to a number of vegetation parameters, including plant density, vegetation cover (Fig. 3) and species richness (see also Fig. 4). Established plants are setting viable seed that is recruiting, and vegetation has survived through different, and very variable, rainfall regimes. There is now modelling to obtain a generalisable measure, that considers broad plant groupings.

Rehabilitation is often looked at as being a technical procedure, with the "crude" objective of stabilising the surface by obtaining any "sort of" permanent vegetation cover as quickly and cheaply as possible. In reality, the rehabilitation of a functional ecosystem requires a thorough understanding about how the stable, mature ecosystem functions and maintains itself. Currently this knowledge base is incomplete. Before disturbance occurs there should be an understanding of the ecological relationships within the functioning ecosystem, and the potential impacts the mining operation will have on the components of that ecosystem. Rehabilitation is effectively an attempt to artificially overcome the factors considered to restrict ecosystem development. The actual rehabilitation operations will often be dictated by engineering or financial considerations, but their underlying logic should be ecological (Bradshaw 1994).

Surveys undertaken in the arid interior of Western Australia indicate the recurrence of distinct vegetation communities, which have evolved in association with the underlying geology and soils. Although plant species richness within each community is

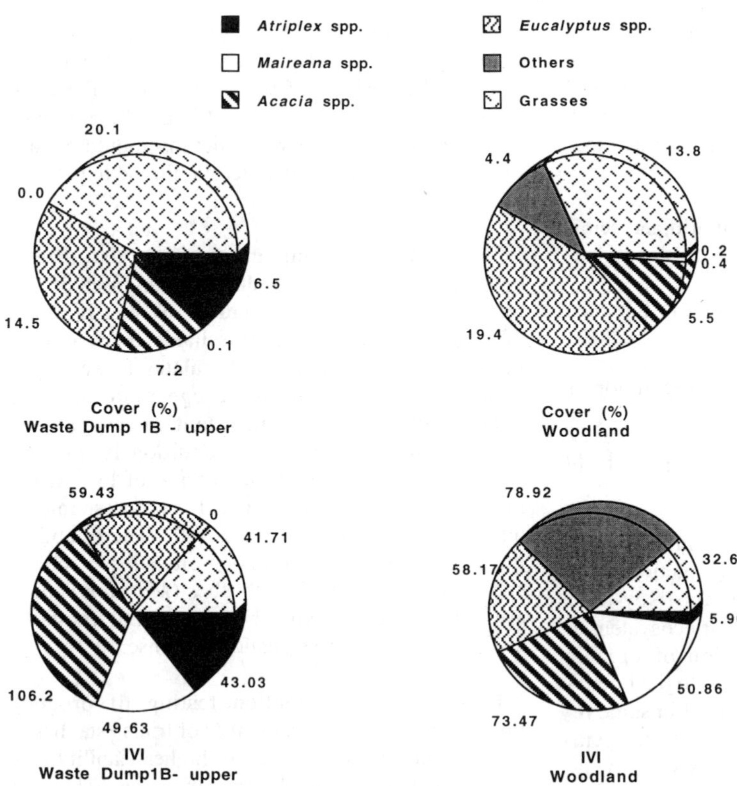

Figure 3. Waste Dump 1B and Eucalypt Woodland, April 1999 data giving percentage foliage cover and Importance Value Index (IVI - maximum 300, considering frequency, density and cover).

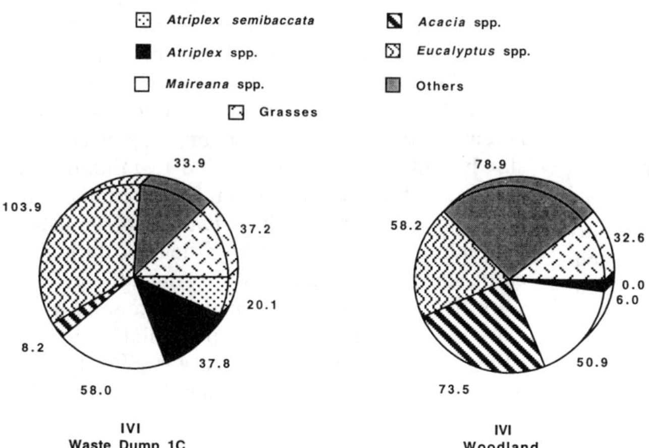

Figure 4. Waste Dump 1C and Eucalypt Woodland, April 1999 data providing Importance Value Index (IVI - maximum 300, considering frequency, density and cover).

typically low, overall diversity is increased due to the large spatial heterogeneity throughout the rangeland (Noy-Meir 1981). It is the undisturbed vegetation communities (analogue sites) on which site specific rehabilitation scenarios should be modelled.

Three characteristics are typical of arid and semi-arid ecosystems:

- a single factor, water, is limiting for most biological processes;
- vertical stratification of the vegetation is poorly developed, and
- species diversity at the community level tends to be low.

To understand an ecosystem one must first observe the dynamic behaviour and interaction of its components over at least a few years (Noy-Meir 1981). Processes within a functional ecosystem at any one time include decomposition of organic matter, nutrient assimilation, microbiological activity, and plant production, all determined in some respect by environmental variables (rainfall, temperature) which cannot be controlled. Alternatively, variables associated with the rehabilitation plan (waste characterisation, topsoil cover, batter angles, deep ripping), which also effect ecosystem development, are under our jurisdiction. Likewise, seed mixture composition and individual species sowing rates have a long term impact on ecosystem development. Seed mixtures can be tailored to suit different environmental constraints and are easily manipulated in response to required ecosystem outcomes.

Processes within an ecosystem are difficult to measure and quantify. Outcomes from these processes are easier to measure, for example commonly represented by a number, weight, cover, or concentration (see summary Table 2). For minesite rehabilitation, revegetation parameters are logical and valid assessment criteria. Data are easily and efficiently recorded, and information can be accurately related back to the rehabilitation plan for immediate implementation of any appropriate modifications.

Moisture is the primary limiting resource in arid and semi-arid Western Australia, with most biological processes occurring in "pulses" induced by rainfall events (Bridges *et al.* 1972, Westoby 1972). The quantity of available water is not only affected by precipitation inputs, but also by soil characteristics and the position of the ecosystem in the landscape (Ludwig and Tongway 1996). The construction and contouring of waste dumps should always focus on the retention of water within the upper profile.

The season of precipitation inputs also affects both the growth and reproductive responses of different species in the revegetation. Winter rains are of lower intensity but more reliable than summer rainfall. Summer rainfall stimulates a more pronounced response from the revegetation, however winter rainfall remains important to revegetation maintenance. Revegetation condition is immediately affected by an extended period of low winter rainfall, however recovery occurs almost spontaneously as tropical summer rains fall. The establishment of hard seeded species (*e.g. Acacia* spp., *Senna* spp.) and a number of bluebushes (*Maireana* spp.), are stimulated following heavy summer rainfalls, evoking an overall revegetation response.

Quality and germination testing of progeny revegetation seed from a number of trials established seed was generally of similar or higher viability than that of the maternal seed originally sown. This was confirmed by field responses one year after a drought, when very high plant densities were recorded following rainfall.

Rangeland assessments (Ludwig and Tongway 1996) have found annual life forms to be fast growing, reaching maximum production within two months following significant rainfall. Annuals also established at higher density in more open areas. Perennial species, in comparison, responded more slowly and over a longer period to rainfall, with maximum production within four months. We found similar trends, with *Atriplex* spp. annuals in particular, establishing at high density during the early stages of rehabilitation. Annual life forms were also more prominent when surface salinity was extreme. However, once a perennial vegetation cover is established, the annual component within the rehabilitation becomes less significant.

4 BIBLIOGRAPHY

Biggs, B. 1996. Guidelines for Mining in Arid Environments. Mining Operations Division, Department of Minerals and Energy Western Australia. ISDN 0 7309 7802 8.

Bradshaw, A.D. 1994. Alternative end points for reclamation. In: J. Cairns (ed.) *Rehabilitating damaged ecosystems.* (2nd ed.). Lewis Publications, Boca Raton, p. 425.

Bridges, K.W., Wilcott, C., Westoby, M., Kickert, R. & Wilken, D. 1972. *Nature: a guide to ecosystem modelling.* Presented at the Ecosystem Modelling Symposium, American Institute of Biological Sciences Meeting, Minneapolis.

Fletcher, D.L. 1993. Factors affecting woody perennial establishment on a saline mullock dump in semi-arid Western Australia. *MSc Thesis*, School of Environmental Biology, Curtin University of Technology.

Ludwig, J.A. & Tongway, D.J. 1996. A landscape approach to rangeland ecology. In: J.A. Ludwig, D.J. Tongway, D.O. Freudenberger, J.C. Noble, & K.C. Hodgkinson (eds.) *Landscape Ecology, Function and Management: Principles from Australia's Rangelands.* CSIRO Publishing, Melbourne.

Ludwig, J.A. & Whitford, W.G. 1981. Short-term water and energy flow in arid ecosystems. In: D.W. Goodall & R.A. Perry (eds.) *Arid Land Ecosystems.* 271-299. Cambridge University Press, Cambridge, United Kingdom.

Noy-Meir, I. 1981. Understanding arid ecosystems: The challenge. In: D.W. Goodall & R.A. Perry (eds.) *Arid Land Ecosystems.* 447-449. Cambridge University Press, Cambridge, United Kingdom.

Oliver, K.E., Fox, J.E.D. & Osborne, J.M. 1992. Growth of chenopod and woody perennial species on fertilised waste rock material from Mt. Keith. *Report* to Western Mining Corporation. October 31st, 1992. pp 44

Osborne J.M. & Sneesby, J.E. (in preparation) Revegetation studies at the tailings dam site at Westonia Gold Mine.

Osborne, J.M., Brooks, D.R. & Carey, L.J. 1985. Sampling strategies: A quantitative study on rehabilitated minesites, Eneabba, WA. *Proceedings AMIC Environmental Workshop*, pp. 293-315. AMIC, Townsville.

Osborne, J.M., Fox, J.E.D. & Mercer, S. 1993. Germination response under elevated salinities of six semi-arid bluebush species. In: H. Lieth and A. Al Massoum (eds.) *Towards Rational Use of High Salinity.* Kluwer Academic Publishers. Netherlands. Vol. 1, pp. 323-338.

Osborne, J.M., Stewart, L.S. & Bouwhuis, E. 1992. Experimental design: research approaches to semi-arid problems. *Proceedings of Workshop on Rehabilitation of Arid and Semi-arid Areas.* May 14 - 15th 1992, Kalgoorlie.

Schatral, A., Osborne, J.M. & Brearley, D.R. 1999. Weeds threatening rehabilitation efforts on mine sites in the goldfields of Western Australia: Preliminary studies on the germination ecology of *Rumex vesicarius* (Family Polygonaceae). In: C.J. Asher and L.C. Bell (eds.) *National Workshop on Native Seed Biology for Revegetation.* Perth, Australian Centre for Minesite Environmental Research, Queensland. ISBN 09585366 43

Stewart, L.S. & Osborne, J.M. 1993. Germination characteristics and growth of chenopod and woody perennial species on saline oxidised mine waste materials from Kalgoorlie. A *Report* to Kalgoorlie Consolidated Gold Mines and Kaltails Mining Services. 21.7.93. pp 157.

Westoby, M. 1972. *Problem orientated modelling: a conceptual framework.* Presented at the Desert Biome Information Meeting, Tempe.

Environmental Issues and Management of Waste in Energy and Mineral Production, Singhal & Mehrotra (eds)
© 2000 Balkema, Rotterdam, ISBN 90 5809 085 X

From abandoned pit to trailhead: One community's success story

Susan B. Patton
Montana Tech of the University of Montana, Butte, Mont., USA

Bruce Hall
Bonner Development Group, Milltown, Mont., USA

ABSTRACT: In the spirit of community, the Bonner Development Group took upon the task of reclaiming an abandoned gravel pit with very limited resources. The Bonner Development Group of Milltown, Montana is a proactive organization of community representatives who work cooperatively to promote growth that balances the environ-ment with industrial development. The group was the enthusiastic recipient of a dona-tion of an abandoned gravel pit on the banks of the Milltown reservoir near the conflu-ence of two rivers, the Blackfoot and the Clark Fork. Because the mine was developed in the mid-1960's the reclamation bond was not sufficient to cover event the cost of seeding. Through the efforts of over 40 separate entities, public, and private, the site has successfully undergone closure well under the budget of $250,000. This paper out-lines the engineering design and creative financing required to reclaim the site and cre-ate a recreation and wildlife viewing area to serve as the trailhead for the Kim Wil-liam's Trail

KEYWORDS: Reclamation, closure

1 BACKGROUND OF THE SITE

The abandoned Champion International Sand and Gravel Pit is located along the northeastern border of the Milltown Res-ervoir in Milltown/Bonner Montana. The community lies 11 km east of Missoula, MT along Interstate Highway 90. Refer to Figure1 for a location map. The confluence of the two rivers, the Blackfoot and Clark Fork with their distinctive character and environment is the area's most dominant natural feature. Highlights are the wetlands and the run-ning waters downstream of the Milltown Dam and up the Blackfoot. The steep valleys and hillsides accent the rivers and riparian open spaces. Preserving these resources and associated wildlife and utilizing them to enhance the area's economy and quality of life is essential. (Klapwyk and Patterson, 1997).

The gravel pit site is 6 hectares in size. The acreage are known as uplands with the sand and gravel pit occupying 3 of the hectares; the remaining 3 hectares of the uplands are comprised of well vegetated benches with thick stands of Douglas Fir and Ponderosa Pine.

The gravel pit was occupied for a few years in the mid-1960's. The sand and gravel from the pit was used in the con-struction of the new Interstate Highway 90. Approximately 155,000 m3 of material was removed from the site. Prior to the pit opening, the land was used for grazing; from the existence of the rela-tively young stand of trees, it seems that the land may have been cultivated for grazing and since the land has been al-lowed to rest the trees have started to grow.

2 HISTORY OF BONNER DEVELOPMENT GROUP

In January of 1995 the formation of a local group in opposition of the placement of a pre-release facility in the community was the seed to the incorporation of the Bonner Development Group (BDG) as a not-for-profit corporation in May of 1995. The mission of the BDG is to work cooperatively to promote the kinds of growth that will achieve a balance between the native beauty of their community environment and the commercial, residential, and industrial development that brings employment, prosperity, and infrastructure support. In December of 1995 BDG received a land donation of 16 hectares from the Champion International Corporation.

The property is situated at the eastern end of the Milltown reservoir having acreage on both the north and south shores of the Clark Fork River. The property covenants stated that it may only be used as "open space" and according to the Environmental

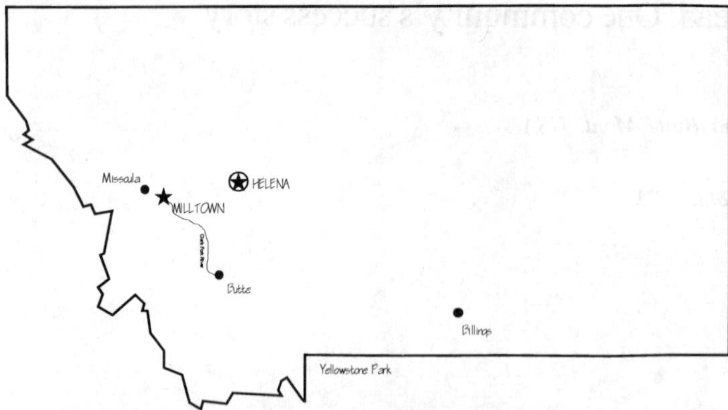

Figure 1. Location map of Milltown/Bonner Montana.

Protection Agency (EPA) is expected to be a high quality recreation area.

The EPA is involved with the site because of the superfund classification of the reservoir. The reservoir was formed through the construction of a dam in 1907. This dam is located at the junction of the Clark Fork and Blackfoot Rivers. Since its creation, the reservoir has served as a recreational outlet for both the Bonner and Missoula communities. The north fork of the reservoir, which constitutes the end of the Blackfoot River, has been kept open to both motorized and non-motorized recreational activities; the South Fork of the reservoir-Clark Fork River-has been restricted to non-motorized recreation only. The construction of the dam for power generation not only provided for an increase in recreational opportunities but also provided for an increase in ecological diversification. Eighty five different species of waterfowl, songbirds, and raptors including the Bald Eagle have been observed near the reservoir (BDG, 1999). Over the years since the construction of the dam, sediment layers have been deposited. In addition to creation of unnatural wetlands, the dam has also created a sort of tailing's repository. Mine tailings, which originated in Butte and Anaconda, have been deposited in the reservoir. The reservoir currently contains an estimated 4.6 million cubic meters of sediment carried from upstream activities including mining, logging, agriculture and highway construction. As a result of this the Milltown Reservoir has been included as part of the Clark Fork Corridor and subject to reclamation. Of great concern to the residents of the area is maintaining their relationship with the reservoir.

Community forums followed the formation of BDG with the outcome a published community action plan. In August of 1996, the BDG received the first of a series of Rural Community Assistance (RCA) grants for a feasibility study of re-spanning the reservoir and thereby connecting the north and south shore of the reservoir. This connection would allow access to the Kim William's Trail system throughout Missoula County. Figure 2 presents a visual overview of the community vision. In July of 1997 BDG was awarded a second RCA grant for an engineered design of the north shore trailhead.

3 DESIGN OF TRAILHEAD

Students of Mining Engineering at Montana Tech of the University of Montana completed the reclamation plans. The reclamation plan tasks included the identification of site features, definition of design criteria, and the design itself. The first task, identification of site features, included the inventory of the site to determine all features that required reclamation efforts. This task also included the research and identification of recreation opportunities. During this phase the site was documented using photography.

Community forums were held to help identify recreation opportunities. The local community, including the school children, was heavily involved at this stage. Representatives of the local Audubon Society and Montana Fish, Wildlife & Parks were also consulted as to appropriate recreational use of the site.

Task two incorporated the site topographic survey and mapping. A records search was conducted to develop a site history. The third and final task was that of design. The planning and design stage was performed to identify and optimize viable options on the site. The design provides for an aesthetically pleasing landscape design with emphasis on low maintenance. In addition to landscape design, the following items were incorporated into the reclamation design:

Hiking and cross country skiing trails
Wildlife viewing access
Interpretive signage
Picnic shelters

Access to the site is by non-motorized only; more specifically, the design provides for pedestrian,

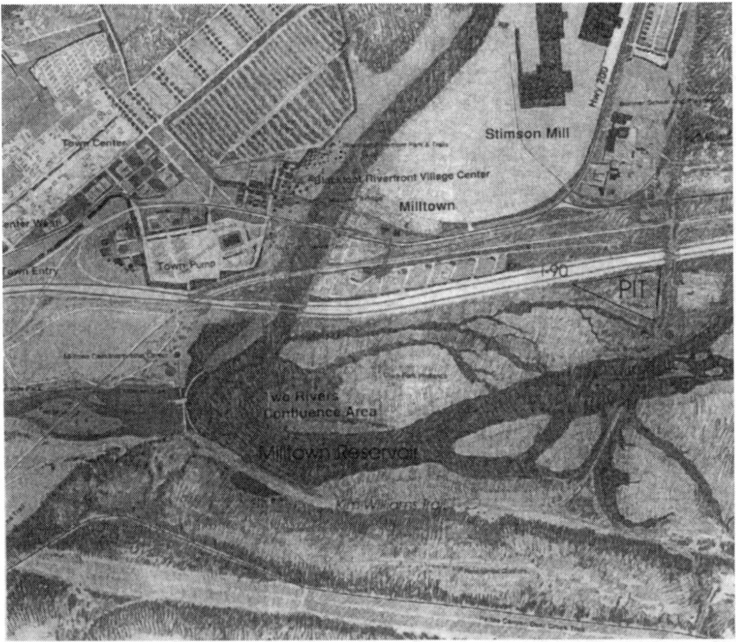

Figure 2. Overview of community action plan showing link to Kim William's Trail.

equestrian and bicyclist only.

The final design for the trailhead had four components, three re-grading with topsoiling phases and one trail construction. Phase 1 includes a considerable amount of fill material to be brought in to bring a steep portion of the pit to a 3:1 slope. The fill would then be covered with topsoil and prepared for seeding. Fill is required rather re-sloping to save existing trees. The site lies near the interstate and trees help filter the road noise. Funding permitting, further fill could be brought to the site in phases 2 and 3 to enhance the aesthetics. This essentially provided for creating pond like depressions and small mounds to break up the linearity of the pit floor. The final phase included the construction of a trail system with shelters and viewing stands. The estimated cost of this project was $250,000.

4 FINANCING THE RECLAMATION

The next problem facing the BDG was how to fund the reclamation phase. The trail construction was funded by a third RCA grant. Breaking the project into smallest components possible made the project more manageable and more realistic for in-kind and volunteer donations. The largest cost was that of earthwork and topsoil at an estimated $100,000. The Montana Natural Resource and Reclamation Program provides $2 million each biennium for rehabilitation of lands damaged by natural resource development. Montana Tech in conjunction with BDG was successful in obtaining $57,000 toward the purchase and placement of topsoil.

Once the site was topsoiled, the trails could be constructed and seeding take place. The majority of the maintenance at the site will be control of noxious weeds. Under the RDGP grant, any land under bond (regardless of the size of the bond) could not be reclaimed. There is an active bond on two acres of the property. The side slopes were left without resloping and topsoil to satisfy a portion of this requirement. Additional funding and services was realized from over 40 other entities. The story of the entities is the success of how businesses, corporations, governmental agencies, and individuals caught the dream of this grassroots community effort.

The 40 entities recognized at the dedication ceremony are too numerous to list here but among them were large companies, individuals, state, federal and local agencies. A large construction firm working on the nearby interstate highway. They provided fill material from demolition of interstate spans in the area. The re-bar was removed, compacted in place, covered with a gravel layer, regraded and topsoiled. In this way a portion of the gravel had come back "home". This work had a value over $40,000. Others recognized included local residents and businesses that donated countless hours of their time, materials, equipment and labor. A local construction firm prepared the parking lot for the county to pave. A local church group assisted with bench placement. The Montana Conservation Crew worked in the rain on trail construction using manual labor only. The Montana Conservation Crew is a service learning based agency in which students earn college credit while working on community projects. The county correctional facility provided labor to thin trees to promote healthy tree growth. The end result is a community trail at a 25% lower cost than the estimate.

Even without the access to the south shore available yet, the site experiences a number of users. The caboose that graces the entrance to the park has been used to exhibit local artists work, a canoe workshop was held on site to over 40 participants, numerous local residents walk the trail for exercise.

5 RECENT DEVELOPMENTS

BDG recently received word that they were successful candidates to receive a bridge from Silver Star, MT. The bridge is planned for demolition and came up for possible adoption. BDG was successful in obtaining the bridge from the Montana Department of Transportation. It is a 60 m span that will provide the approach for the bridge spanning the reservoir. Fundraising underway for the proposed rustic pole and beam construction picnic pavilion. The estimated cost is $90,000 with over half that raised at this time.

6 REFERENCES

BDG, 1999, *Milltown Wildlife Inventory*, Montana Fish, Wildlife & Parks and Five Valleys Audubon Society, BDG, Milltown, MT.

Klapwyk, G. and J. Patterson, 1997, *Engineering Design of the Reclamation Plans for the Champion Internation Gravel Quarry to a North Shore Trail Head for the Kim Williams Trail*, Montana Tech of the University of Montana, Butte, MT, October.

Environmental Issues and Management of Waste in Energy and Mineral Production, Singhal & Mehrotra (eds)
© 2000 Balkema, Rotterdam, ISBN 90 5809 085 X

Modeling of sub-aerial tailings deposition and optimum deposition design in arid regions

Yunxin Qiu & Dave C.Sego
Department of Civil and Environmental Engineering, University of Alberta, Edmonton, Alb., Canada

ABSTRACT: Mineral and energy production produces large volumes of tailings. Most tailings are disposed of on the earth surface. How to minimize the environmental impact and reduce the storage volume of the tailings has become a major concern of mining operations. In addition, the pressure on the optimal use of water resources especially in arid regions is a critical issue. Therefore, how to maximize recycling of the processing water is an important concern in these regions. The sub-aerial deposition technique is a promising method for dealing with these problems. In this paper, the criterion for the optimum deposition is presented. A one-dimensional model of the sedimentation, consolidation and desiccation of subaerially deposited tailings is described. Desiccation cracks are incorporated in the design model which is applied to simulate the sub-aerial deposition of two typical tailings. The modeling results are presented and discussed.

1 INTRODUCTION

Tailings are commonly deposited as a slurry into an impoundment formed by embankments or dykes. The sub-aerial disposal method is a promising tailings disposal method. It involves systematic discharging and deposition of tailings in a thin layer from one or more points along the perimeter of the impoundment area. As a slurry flows, the tailings settle, drain, partially air dry and increases in density. The released water drains to a low point where it is recycled into the processing system, decanted to a separate containment facility for reuse or evaporation, or allowed to evaporate in place. Subsequent layers are then deposited, and the disposal cycle is repeated.

In general, mines produce and discharge tailings continuously. To maintain continuous deposition in the sub-aerial method, the tailings empoundment is divided into several deposition cells to allow for tailings to be continually deposited into different cells with time.

Increasing stringent environmental regulation requires that the environmental issues be considered for new facilities during the operation, as well as for closure of the facility. Many factors affect the behavior of the sub-aerial deposited tailings during the operation of the facility, such as deposition frequency, sub-layer thickness, climate conditions, desiccation cracks, etc. (Qiu and Sego, 1998a). After closure, the minimum volume of the tailings and higher tailings placement density are expected to contribute to its long-term safety and assist with reclamation. Therefore, a rational method to predict the response and behavior of the sub-aerial deposited tailings is needed to aid in developing design criteria for the long-term performance of these facilities.

After a brief review of existing predictive models, this paper presents the theory used to model sedimentation, consolidation and desiccation cracking. Then, numerical solution of the model is described. Furthermore, the model was applied to simulate the overall sedimentation, consolidation and desiccation process of the sub-aerial deposited tailings. The modeling results are described and conclusions are presented.

2 BRIEF LITERATURE REVIEW

Lachenbruch (1961,1962) studied the development of shrinkage cracks in permafrost soils and presented a solution for crack depth and crack spacing. To estimate the depth and spacing of tension cracks one must evaluate the problems of initiation, propagation, and arresting of a tension crack in an extended medium.

Bronswijk (1988, 1989, 1990, 1991) studied shrinkage, displacement and cracking of a clay under desiccation, and developed theoretical relationships between vertical soil movements and water-content changes in clays undergoing desiccation. Bronswijk (1990) concluded that in-situ shrinkage was isotropic. Therefore, the measured one-dimensional surface subsidence of the soil was converted into a three-dimensional volume change and to crack volume. However, the model can not be used to predict the spacing of the cracks.

Morris et al. (1992) reviewed the occurrence and morphology of cracks in dry-climate regions of Australia and Canada, and developed theoretical relationships between the depth of cracks, the properties of the soil and a given suction profile.

Abu-Hejleh and Znidarcic (1995) studied desiccation theory for soft cohesive soils. A key element in their model is to consider the principle of effective stress in the slurry to relate total stress induced by a condition of zero lateral strain during shrinkage to the suction and the effective stress path followed by any soil element during consolidation. The soil starts cracking during one-dimensional shrinkage when the total lateral tensile stress at any point reaches the tensile strength of the soil.

Recently, Konrad and Ayad (1997) have presented an idealized framework for the prediction of the spacing between primary shrinkage cracks in cohesive soils undergoing desiccation. In their model, the linear elastic fracture mechanics (LEFM) theory was used for determination of the ultimate depth of a crack under a given lateral stress field. The theory of linear elasticity was used to compute the stress redistribution around the crack in order to determine the extent of the stress relief zone, which, in turn, relates directly to the spacing of these primary cracks. This model can not predict the width of the tension cracks.

3 THEORY

3.1 *Governing equations*

The governing equations were derived based on the equilibrium and continuity of the mass of the tailings and Darcy's law. The governing equation for sedimentation and consolidation are:

$$\frac{\partial \sigma}{\partial z} + \gamma_w [\theta(1+e) + G_s] = 0 \tag{1}$$

and

$$\frac{\partial}{\partial z}\left\{\frac{k}{(1+e)}\left[\frac{\partial h}{\partial z} + (1+e)\right]\right\} = \frac{\partial[(1+e)\theta]}{\partial t} \tag{2}$$

where σ is total stress in the soil, z is solids coordinate which is defined as the volume of tailings particles per unit area lying between the datum plane and the point being analyzed, γ_w is the unit weight of water, θ is the volumetric water content of the soil, e is the void ratio of the soil, G_s is the specific gravity of the tailings particles, k is the hydraulic conductivity of the soil, h is the pore pressure head and t is time.

The governing equation was changed into an unique one-variable-dependent form

$$f\frac{\partial h}{\partial t} = \frac{\partial g}{\partial z}\left\{\frac{\partial h}{\partial z} + (1+e)\right\} + g\left(\frac{\partial^2 h}{\partial z^2} + \frac{\partial e}{\partial z}\right) \tag{3}$$

with $\quad f = f(h,e) = e\dfrac{dS}{dh} + S\cdot f_e \qquad$ (3a)

and

$$f_e = f_e(h,e) = -\gamma_w \frac{de}{d\sigma'}\left\{h\frac{d\chi}{dS}\frac{dS}{dh} + \chi\right\} \approx \frac{de}{dh} \tag{3b}$$

and $\quad g = g(h,e) = \dfrac{k(S(h),e)}{1+e} \qquad$ (3c)

where, S is the degree of saturation and χ is the Bishop parameter which equals unity for saturated soils and decreases with the degree of saturation of the soils.

3.2 *Boundary conditions*

3.2.1 *Upper boundary conditions*

During sedimentation, the effective stress is assumed to be zero. Hence, a prescribed surface flux,q_z=H_z, condition is applied at the upper surface as follows:

$$q_{z=H_z} = -\frac{k(1-G_s)}{1+e} \tag{4}$$

During saturated consolidation, if surface settling dominates the surface flux, once the surface void ratio reaches the settled e_{set} it is held at this value until evaporation becomes the dominate boundary condition, i.e. the surface flux due to self-weight consolidation is less than the potential evaporation rate.

When evaporation dominates the surface flux, i.e., the surface flux due to self-weight consolidation is less than the potential evaporation rate, the upper boundary flux is equal to the actual evaporation rate,

$$-\frac{k}{1+e}\left\{\frac{\partial h}{\partial z} + (1+e)\right\} = E_a \tag{5}$$

3.2.2 *Lower boundary conditions*

Impermeable boundary condition requires that

$$\frac{\partial h}{\partial z} = -(1+e) \quad (6)$$

When there is a bottom drainage system in the sub-aerial tailings disposal facility. The boundary condition for the permeable bottom can be considered as "free drainage". The lower boundary condition in this situation can be expressed according to Swarbrick (1992) as:

$$\left.\frac{\partial h}{\partial z}\right|_{z=0} = 0 \quad (7)$$

3.3 *Design criterion and schemes*

In the sub-aerial tailings deposition, since bottom drainage exists, the best way to collect recyclable water is through downward drainage to the collection system. The design criterion for sub-aerial tailings deposition in the arid regions is to maximize the amount of water available for recycle combined with minimizing the tailings volume in the tailings facility, which in turn, requires that a high hydraulic conductivity in each tailings layer be maintained.

There are two choices to meet the design criterion. The first scheme is to keep the tailings fully saturated at all time, i.e., all deposited layers should be placed to maintain full saturation, thus the hydraulic conductivity of the tailings is not affected by desaturation. The other scheme is to allow sufficient drying to crack each layer. The equivalent hydraulic conductivity of the cracks is very high.

3.4 *Crack initiation, propagation and crack width*

The condition for crack initiation at the tailings surface is when the total lateral stress reaches the tensile strength of the tailings, i.e.

$$\sigma_h = -\sigma_t \quad (8)$$

where σ_h is the lateral stress and σ_t is the tension strength of the tailings.

After its initiation, the crack usually propagate downward. Assume that the theory of LEFM applies, then the ultimate depth of the cracks, b, should be calculated based upon the tensile stress distribution applied on the sides of the crack and the value of the stress intensity factor, κ, which changes with crack length.

Assume that the actual tensile stress distribution applied on the side of the crack can be approximated by a trapezoidal distribution, and can be considered as the superimposition of a uniform distribution of σ_t and a triangular distribution with a maximum value of (σ_t - σ_3(b)), thus, the stress intensity factor corresponding to the trapezoidal tensile stress distribution, κ_{trap}, can be obtained

$$\kappa_{trap} = \kappa_{unif} - \kappa_{tria} = \kappa_c \quad (9)$$

where κ_c is the tailings' fracture toughness.

Substituting Lachenbruch's (1961) solution to equation (9), then rearranging yields(when b = a):

$$b = \left[\frac{\kappa_c}{0.44\sigma_t + 0.68\sigma_3(b)}\right]^2 \quad (10)$$

where b is the depth of the crack, σ_t is the tensile strength of the tailings, σ_3(b) is the horizontal stress at the depth of b and κ_c is the tailings fracture toughness.

After each crack has reached its ultimate depth b, the surface tensile stress in its vicinity will be reduced and the horizontal stress on the crack walls must vanish. Based on the principle of superimposition, when a perturbation stress with a value of $-\sigma_3$ is added to the existing stress field, the stress condition on the crack surfaces is satisfied (Lachenbruch 1961). The stress relief caused by the crack can be calculated by numerical methods. Assume that the second crack initiates at the point where the tensile stress equals 95% of the tensile strength. According to this assumption and the surface lateral stress relief, the crack spacing, B, can be determined. Based on the assumption for the tensile stress distribution, an alternative method can be used to calculate the crack spacing, i.e., directly use the calculated stress relief for a semi-infinite medium (Lachenbruch 1961).

The crack width must be evaluated to account for the effect of the cracks on the macro hydraulic conductivity. Consider an element with a width and length the same as the spacing of the crack. Assume that shrinkage of the tailings is three dimensional and isotropic. Total volume of the element at full saturation (V_0) is

$$V_0 = B^2 \cdot D_0 \quad (11)$$

where B is the spacing of cracks and D_0 is the layer thickness at full saturation.

Then the change in volume of this element is (Bronswijk 1989 and 1991)

$$\Delta V = \left[1 - \left(1 - \frac{\Delta D}{D_0}\right)^3\right] \cdot V_0 \quad (12)$$

where ΔV is the decrease in volume of the tailings matrix due to shrinkage, ΔD is the decrease in layer thickness due to shrinkage from fully saturation state to when the crack propagates, D_0 is the layer thickness at full saturation, V_0 is the volume of the tailings matrix at saturation.

The total crack volume is (Bronswijk 1989)

$$V_{cr} = \Delta V - B^2 \cdot \Delta D \quad (13)$$

where V_{cr} is the total crack volume.

Based on geometry,

$$V_{cr} = 2 \cdot B \cdot b \cdot \upsilon \quad (14)$$

where υ is the width of the crack.

Therefore the width of the crack can be obtained

$$\upsilon = \frac{\Delta V - B^2 \cdot \Delta D}{2B \cdot b} \quad (15)$$

3.5 *Recycle water amount*

Water balance components during the operation of the sub-aerial tailings disposal can be summarized as follows (Figure 1):

- Initial input water during discharge (Q_1)
- Evaporation loss of water (Q_2)
- Drainage through the free drainage zone (Q_3)
- Drainage through the bottom sand filter blanket (Q_4)
- Discharge due to consolidation of tailings (Q_5)
- Recycle water from drainage collection (Q_6)
- Precipitation (rain or snow) (Q_7)
- Storage water in the tailings (Q_8)

The amount of recyclable water per unit area can be determined using the following equation.

$$Q_6 = Q_1 + Q_7 - Q_2 - Q_8 \quad (16)$$

The evaporation and the storage can be calculated using:

$$Q_2 = \int_0^{t_d} E_A(t)dt \quad (17)$$

where $E_A(t)$ is the actual evaporation rate, it is a function of the time when the potential evaporation is constant, and t_d is the exposure time at the surface of tailings facility.

$$Q_8 = \int_0^{H(t)} \theta(z,t)dz \quad (18)$$

where H(t) is the total thickness of the tailings layer at time t, $\theta(z, t)$ is the volumetric water content, and a function of both the depth and time.

4 NUMERICAL SOLUTION

Because it is difficult to find the analytical solutions to the governing equations, a numerical method is used to find the solutions to the governing equations.

One of the best methods for solving one-dimensional problems is to use the Douglas-Jones predictor-corrector method (Gilding 1983). Assume the total solids height is H_z and total time is T. To apply the Douglas-Jones method (Douglas et al. 1963) to equation (3), the solids height H_z is discretized into R portions of equal length, $\Delta z = H_z/R$. The point $z_i = i\Delta z$ denotes computational points for i =1, 2, 3, ..., R. The time t is discretized into N portions of equal length, $\Delta t = T/N$, $t_n = n\Delta t$. Let the numerical solution for h at point z_i and time t_n be denoted by h_i^n. Based on the Douglas-Jones method, the finite difference equation for the predictor step takes the form:

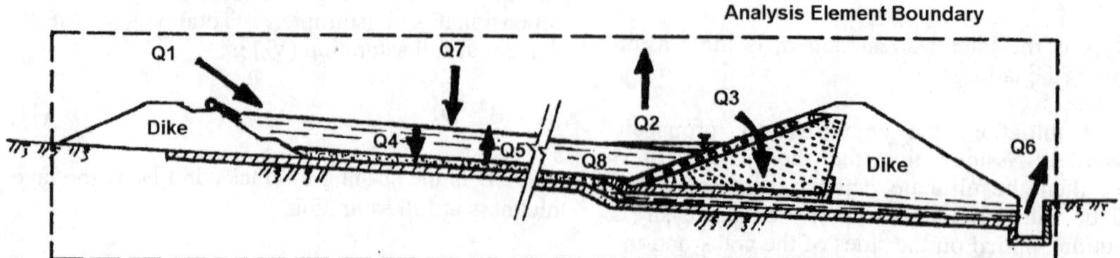

Figure 1. Water balance components during the tailings pond operation

$$f_i^n \frac{h_i^{n+\frac{1}{2}} - h_i^n}{\frac{1}{2}\Delta t} = \left(\frac{g_{i+1}^n - g_{i-1}^n}{2\Delta z}\right)\left\{\frac{h_{i+1}^n - h_{i-1}^n}{2\Delta z} + \left(1 + e_i^n\right)\right\} + g_i^n\left[\frac{h_{i+1}^{n+\frac{1}{2}} - 2h_i^{n+\frac{1}{2}} + h_{i-1}^{n+\frac{1}{2}}}{(\Delta z)^2} + \frac{e_{i+1}^n - e_{i-1}^n}{2\Delta z}\right] \tag{19}$$

with

$$f_i^n = f\left(h_i^n, e_i^n\right) \tag{19a}$$

and

$$g_{i+1}^n = g(h_{i+1}^n, e_{i+1}^n) \tag{19b}$$

and

$$g_{i-1}^n = g\left(h_{i-1}^n, e_{i-1}^n\right) \tag{19c}$$

The corrector equation follows the predictor with the form:

$$f_i^{n+\frac{1}{2}} \frac{h_i^{n+1} - h_i^n}{\Delta t} = \frac{g_{i+1}^{n+\frac{1}{2}} - g_{i-1}^{n+\frac{1}{2}}}{2\Delta z}\left\{\frac{h_{i+1}^{n+\frac{1}{2}} - h_{i-1}^{n+\frac{1}{2}}}{2\Delta z} + 1 + e_i^{n+\frac{1}{2}}\right\} +$$
$$+ g_i^{n+\frac{1}{2}}\left[\left(\frac{h_{i+1}^{n+1} - 2h_i^{n+1} + h_{i-1}^{n+1}}{2(\Delta z)^2} + \frac{h_{i+1}^n - 2h_i^n + h_{i-1}^n}{2(\Delta z)^2}\right) + \frac{e_{i+1}^{n+\frac{1}{2}} - e_{i-1}^{n+\frac{1}{2}}}{2\Delta z}\right] \tag{20}$$

Table 1. The basic physical parameters of mine tailings (modified from Qiu and Sego, 1998b)

Tailings	Copper	Coal
Specific Gravity, G_s	2.75	1.94
Compression parameters ($e=A(\sigma')^B$), A	0.9233	1.4444
Compression parameters ($e=A(\sigma')^B$), B	-0.0491	-0.1536
Hydraulic conductivity parameters ($k=Ce^D$), C (m/s)	0.0001	0.000002
Hydraulic conductivity parameters ($k=Ce^D$), D	3.0892	4.0686
Tensile strength, σ_t (kPa)	20	10
Fracture toughness, κ_c(kN/m$^{1.5}$)	3.5	3.75
Original solids contents, S_0	35%	41.5%
Potential evaporation rate, E_p x 10^{-8}(m/s)	3.0	3.0
Deposited thickness (m)	0.1	0.1

Equation (19) and (20) are general equations for internal nodal points, and can be reduced to the general form:

$$a_i h_{i-1} + b_i h_i + c_i h_{i+1} = d_i \tag{21}$$

Applying equation (21) to each node in both the predictor and corrector results in a system of equations which can be solved efficiently by the Thomas algorithm (Gilding 1983).

5 DISCUSSION OF MODELING RESULTS

Two types of tailings, i.e. copper and coal tailings were selected. The model was applied to the tailings. The basic properties of the tailings and parameters used in the model are summarized in Table 1. The modeling results are presented and discussed in the following sections.

5.1 *Void ratio change*

The void ratio profile changes in the copper tailings from sedimentation to desiccation are shown in Figure 2. It is shown that during the early stage of sedimentation the void ratio decreases rapidly. From consolidation to desiccation cracking, the void ratio changes much slower. It is shown that the void ratio at the bottom decreases more quickly than at the surface, this may be due to the drainage system existing beneath this layer.

The void ratio profile changes with time in coal tailings is illustrated in Figure 3. It is shown that the lower part of the profile changes more quickly than the upper part. This indicates that the sedimentation and consolidation occurs more quickly in the lower part than the upper part. This difference in the coal tailings is more significant than was observed in the copper tailings.

5.2 *Settlement*

The predicted settlement of the copper tailings is shown in Figure 4. During sedimentation, the copper tailings settles quickly, by the end of sedimentation, the accumulated settlement reaches about 6.6 cm. During the early stage of consolidation, the copper tailings still settles with an observable rate. 6 hours after deposition, the copper tailings continue to settle but slowly. The figure indicates that 6 hours after deposition is sufficient for the layer, i.e. the next layer of copper tailings can be deposited since the majority of the settlement is complete. Figure 5 presents the settlement of the coal tailings. Compared to the copper tailings, the settlement of

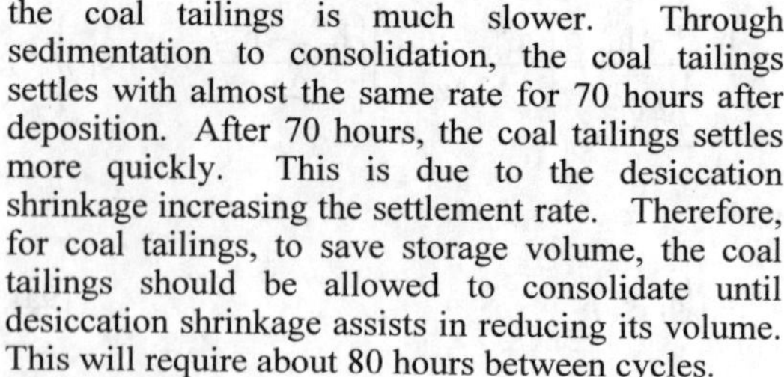

the coal tailings is much slower. Through sedimentation to consolidation, the coal tailings settles with almost the same rate for 70 hours after deposition. After 70 hours, the coal tailings settles more quickly. This is due to the desiccation shrinkage increasing the settlement rate. Therefore, for coal tailings, to save storage volume, the coal tailings should be allowed to consolidate until desiccation shrinkage assists in reducing its volume. This will require about 80 hours between cycles.

5.3 *Recyclablé water*

Generally, mineral processing requires large amounts of water. Hence, the demand on water resources is becoming a critical issue, especially in arid regions. The process water needs to be recycled as much as possible. Normalized recyclable water amount changes versus time in the copper tailings is shown in Figure 6. The figure shows that after the first 5 hours, the recyclable water increases quickly to 90% of the initial input water. After that, it increases very slowly. Therefore, from the point view of maximizing the recyclable water amount,

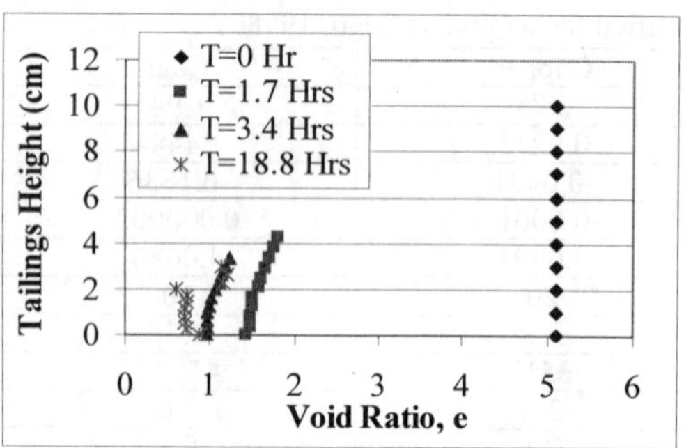

Figure 2. Copper tailings void ratio profile changes

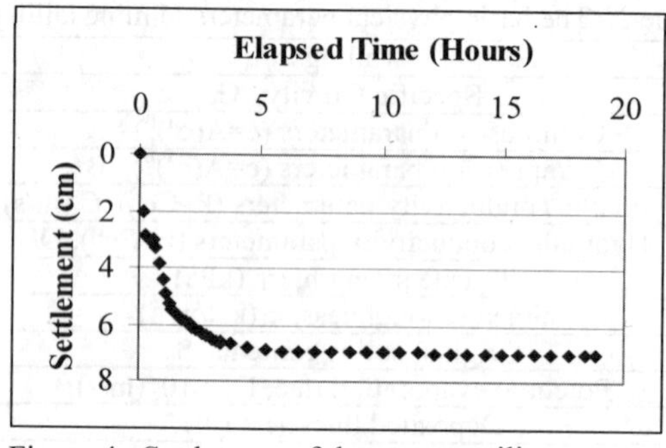

Figure 4. Settlement of the copper tailings

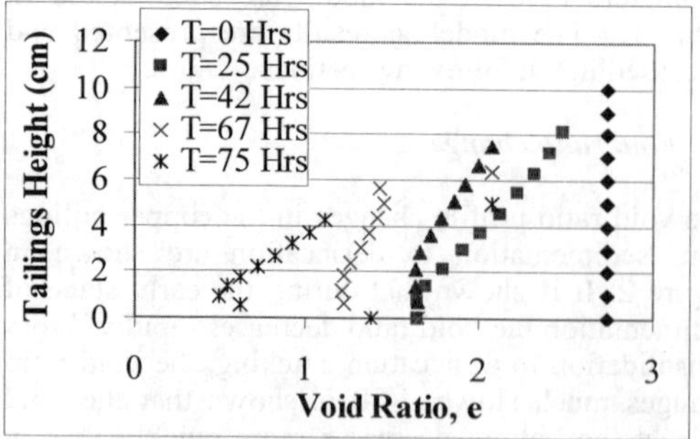

Figure 3. Coal tailings void ratio profile changes

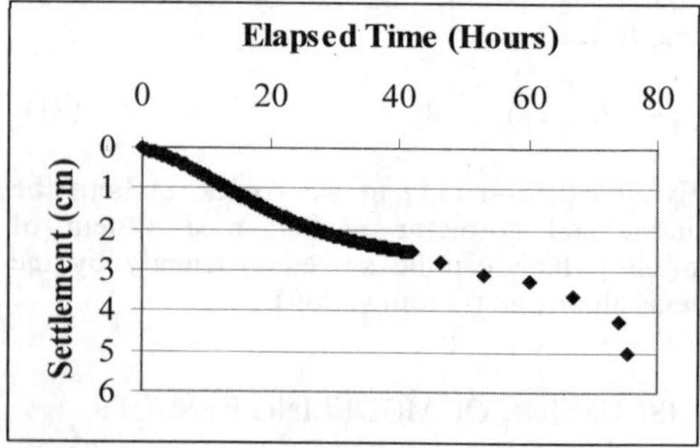

Figure 5. Settlement of the coal tailings

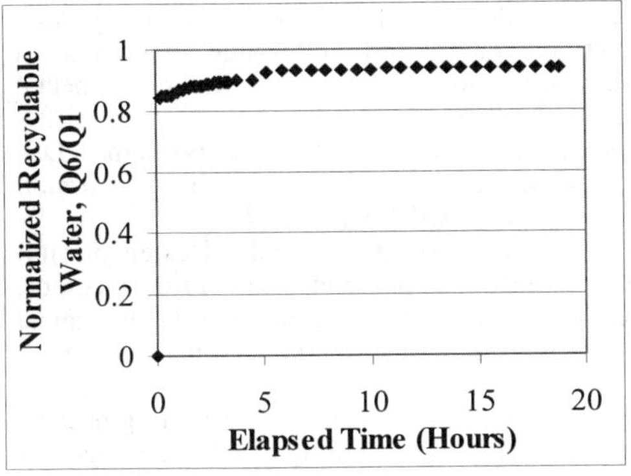

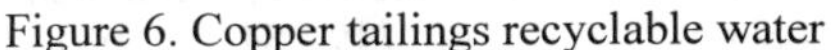

Figure 6. Copper tailings recyclable water

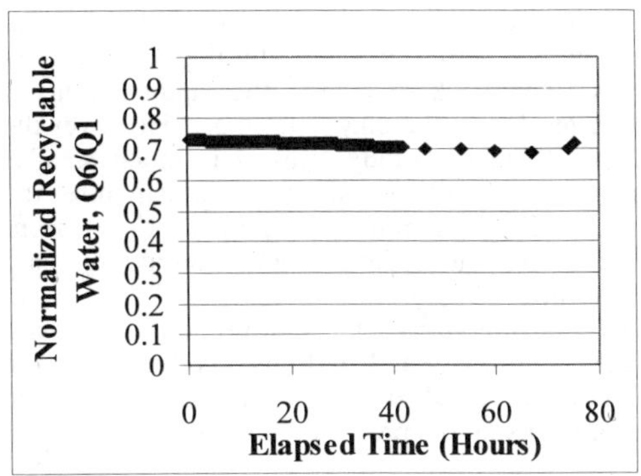

Figure 7. Coal tailings recyclable water

Table 2. Predicted tensile crack dimensions

Deposition Thickness (cm)	Copper Tailings			Coal Tailings		
	Crack Spacing (cm)	Crack Width (cm)	Total Time (Hours)	Crack Spacing (cm)	Crack Width (cm)	Total Time (Hours)
10	149	0.07	19	125	0.26	75
20	165	0.02	29	134	0.18	149

Note: the crack depth is equal to the layer thickness at cracking.

allowing the copper tailings to settle about 5 hours after deposition should be sufficient to produce substantial recycle water from the deposited layer.

Figure 7 presents the normalized recyclable water in the coal tailings. The predicted data shows that the recyclable water decreases slightly from 73% to 69% during the sedimentation and saturated consolidation, and then rebound to 72%. Therefore, from sedimentation to desiccation to cracking, the recyclable water from the coal tailings undergoes little change.

5.4 *Tension cracks*

With the progress of desiccation, internal horizontal tensile stresses build when tailings are exposed to the air. When the horizontal tensile stress exceeds the local tensile strength, a tensile crack occurs locally and may propagate through the layer. In this model, the whole process from deposition to desiccation is simulated until the tensile crack extends through the layer thickness. The predicted tensile crack dimensions and the total time required from deposition to cracking through the whole profile for 0.1m and 0.2m deposition thickness are listed in Table 2.

The predicted data indicate that the crack spacing of the copper tailings is larger than that of the coal tailings, while the crack width of the copper tailings is smaller than that of the coal tailings. In addition the total time required to form a crack in copper tailings is less than one day, while for the coal tailings it is more than 3 days for a thickness of only 10cm. The data also shows that the thicker the deposited layer the larger the crack spacing and the longer total time for it to develop. However, the thicker the deposited layer the narrower the crack width. Therefore, in the practice, based on the available deposition segments and the area of each segment, the optimum deposition thickness can be determined to desiccate and crack the deposited layer just before a new layer is deposited.

6. CONCLUSIONS

From this numerical study, the following conclusions can be drawn:

The physical processes of the sub-aerial tailings disposal method are sedimentation, consolidation and desiccation. These processes can be modeled using a unified theory which can be solved using numerical techniques.

The sandy tailings settles more quickly than the clayey tailings and most settlement of the sandy tailings occurs during sedimentation. However, clayey tailings settle slowly with a consistent rate until desaturation and volume shrinkage begins.

Desiccation causes tension cracks to initiate and propagate in both the sandy tailings and clayey tailings. The crack spacing in the sandy tailings is larger than that in the clayey tailings. However the crack width in the sandy tailings is much smaller than one in the clayey tailings. Meanwhile, the clayey tailings require a longer time from sedimentation to the crack initiation and propagation.

Based on the size of the tailings facility, in situ climate conditions and tailings productivity, the optimum sub-aerial deposition parameter can be determined using the presented model.

ACKNOWLEDGEMENT

The authors would like to thank the Natural Sciences and Engineering Research Council of Canada (NSERC) for providing the financial support for this research. The support of Luscar Sterco (1977) Ltd (Coal Valley Mine) and Kennecott Mining (Salt Lake) by supplying tailings used in this research is appreciated.

REFERENCES

Abu-Hejleh, A.N., and Znidarcic, D. 1995. Desiccation theory for soft cohesive soils. *Journal of Geotechnical Engineering*. ASCE, 121(6): 493-502.

Bronswijk, J.J.B. 1988. Modelling of water balance, cracking and subsidence of clay soils. *Journal of Hydrology*. 97: 199-212.

Bronswijk, J.J.B. 1989. Prediction of actual cracking and subsidence in clay soils. *Soil Science*. 148(2): 87-93.

Bronswijk, J.J.B. 1990. Shrinkage geometry of a heavy clay soil at various stresses. *Soil Science Society of America Journal*. 54: 1600-1602.

Bronswijk, J.J.B. 1991. Drying, Cracking, and Subsidence of A Clay Soil in A Lysimeter. *Soil Science*. 152(2): 92-99.

Douglas, J. And Jones, B.F. 1963. On predictor-corrector methods for nonlinear parabolic differential equations. *Journal of Society, Industry and Applied Mathematics*. 11(1): 195-204.

Gilding, B.H. 1983. The soil-moisture zone in a physically-based hydrologic model. *Advances in Water Resources*. 6:36-43.

Konrad, J. -M. and Ayad, R. 1997. An idealized framework for the analysis of cohesive soils undergoing desiccation. *Canadian Geotechnical Journal*. 34: 477- 488.

Lachenbruch, A.H. 1961. Depth and Spacing of Tension Cracks. *Journal of Geophysical Research*. 66(12): 4273-4292.

Lachenbruch, A.H. 1962. Mechanics of thermal contraction cracks and ice-wedge polygons in permafrost. Geological Survey of America, Special Paper No. 70. 69p.

Morris, P.H., Graham, J., and Williams, D.J. 1992. Cracking in drying soils. *Canadian Geotechnical Journal*. 29: 263-277.

Qiu, Y. and Sego. D. C. 1998a. Design of sub-aerial tailings deposition in arid regions. Proceedings of 51st Canadian Geotechnical Conference. Edmonton, Alberta, October 4-7.Vol. 2. pp. 615-622.

Qiu, Y. & Sego, D.C. 1998b. Engineering properties of mine tailings. Proceedings of 51stCanadian Geotechnical Conference. Edmonton, Canada, October 4-7. Vol. 1. pp. 149-154.

Swarbrick, G.E. 1992. Transient unsaturated consolidation in desiccating mine tailings, PhD thesis presented to the School of Civil Engineering, University of New Southern Wales.

Environmental Issues and Management of Waste in Energy and Mineral Production, Singhal & Mehrotra (eds)
© 2000 Balkema, Rotterdam, ISBN 90 5809 085 X

Revegetation of alpine mines and quarries – An interdisciplinary research project

A. Sanak-Oberndorfer & T. Oberndorfer
Department of Mining Engineering, University Leoben, Austria

R. Schaffer
Agrar-Consulting und Handelsunternehmen, Austria

ABSTRACT: This paper presents a 3-years research project on reclamation in alpine conditions. After the description of the general situation and the goals of the project in the first part the experimental tests conducted and the results gain are described. The second parts deals with observations of reclamation results from a series of mines visited. Finally in the last part the importance of an integral assessment of mining and reclamation, in particular in respect of safety, is highlighted.

1 INTRODUCTION

Due to the high population density and the high portion of mountainous landscape mining of Austria and population have to coexist close together. By number and material produced quarrying of building material is the dominant sector of mining in Austria, complemented by mining of industrial minerals and very few metal mines. Typical quarry operations might vary between 100.000 to 300.000 t/a.

Because of the generally noted increasing sensibility of the public against visual impacts and emissions caused by mining, in particular open-pit mining and quarrying operations face increasing problems gaining mining permissions. This tendency is also reflected by increasingly restrictive laws. It has become standard acknowledgment that mining has to eliminate the damage done as soon and thoroughly as possible, most preferable already during mining activity. However, in the current situation it is not sufficient to state the intention of doing so, but it is necessary to prove already in the state of planning and approval that the methods proposed for reclamation will show the expected effect.

Reclamation is not a new field. There exists a lot of experience in making mining areas available for agriculture, forestry or habitation. However, in alpine areas some special circumstances must be considered, above all the high ranking the aspect "nature" is likely to have (see Table 1). This means, that in alpine areas the potential conflict between mining and habitants in the vicinity is extended by a potential conflict with nature, or - to be more precise - what human mankind thinks nature is. Unfortunately these opinions on "how nature has to be" differ from time to time and as well as between peoples involved.

This means that the final goal of reclamation of mining in alpine areas can be very different from standard cases, e.g. in terms of diversity of flora and fauna or creating living space for threatened specimen of wildlife. Accordingly standard techniques of reclamation are not appropriate, both for ecological and topographical reasons.

Table 1. Aspects for reclamation of mining in alpine areas

Mining
- Economy
- Safety during and after operation
Habitants
- minimizing impact on the living space of humans. Objective and subjective aspects.
- appearance of landscape (during/after mining), design of landscape
- immediate and immediately visible reclamation
- minimized visibility of mine or quarry
- minimization of emissions (noise, dust, water, waste)
- (economy of mining, working places, etc.)
Nature
- adequate appearance, ranging from bare and steep rock walls to forest and alpine meadows
- diversity of flora (no single-specimen vegetation)
- living space for wildlife, threatened or even "exotic" specimen
- high quality standards on water resources
- sustainability

The aspects exemplified in Tab. 1 are characterized by very interlaced relationships.

Usually, the costs of reclamation are not an economical threat for mining (which dose not mean that it should not been tried to make reclamation as cheap as possible). More serious is that certain requests and requisitions from ecological representatives have an effect on mining methods and mining design, and in this way also an effect on economy and above all safety, which is the more frequent source of conflict. For alpine mining operation in addition it is likely that contradictions between the demands of local habitants and "nature" (representatives) arise. A very simple example for this is the opposition between rapid and sustainable reclamation.

2 PROJECT

The project "Rekult" presented in this paper was launched 1997 and ran 3 years. It was funded by the Austrian government with a total of 18 Mio ATS (2 Mio Can $).

The goal of the project was to develop and define up-to-date and adequate reclamation measures for mines in alpine areas in order to comply with ecological and economical requirements. According to the interlaced relationships between aspects involved the project was not restricted to the reclamation process itself, but was considered as an integral part of mining operation.

In the first place the project is intended as an assistance for mining companies in terms of finding the most appropriate and cost effective method, and – even more – helping to avoid failures. But the concept aimed also towards an assistance for authorities involved, in terms of making the interrelations between mining and reclamation evident. This seems necessary because due to the current legislative situation some authorities involved are not familiar with specific problems of mining.

Roughly speaking the project is divided into two parts. One part deals with experimental investigations, which were conducted on a huge testing area of an open pit mine in Austria. This part aimed towards the development of reclamation techniques and finding limits of application. The second part comprises theoretical considerations accomplished by numerous mine visits and discussions with authorities involved. The intricate and interdisciplinary character of the problem was well respected in the project team, which consists of experts in the fields of biology, forestry, geology and mining.

Summarizing the following tasks were investigated:

- which parameters are essential for the development of new vegetation (success / failure)
- what methods are applicable and promising
- can any change in mine design assist the success of reclamation (decrease reclamation costs by in-time preparation of necessary steps)
- does a proposed reclamation method adversely effect mining operation in terms of economics and safety

Figure 1. View of mine site VA-Erzberg

3 EXPERIMENTAL TESTS AT VA-ERZBERG.

VA-Erzberg is a open-pit mining on siderite and with 4 Mio t/a the biggest mining operation in Austria. It is located in Styria within mountainous area (Figure1). The altitude ranges between 700 – 1500 m (2000 – 4500 ft) and the mine site has an overall extension of approximately 600 ha, with 140 ha being waste dumps.

Several aspects make this mine site extremely problematic from the point of view of reclamation and hence preferable for the tests, the most important being:

- High, steep and lengthy waste dumps (impossible to cover with soil, problem of erosion)
- Carbonatic iron ore with high diversity of mineral components (schist, limestone, siderite, hydroxide)
- Alpine climate, with a precipitation averaging 1200mm per year, high temperature differences during one day, etc.
- Very huge area – making a micro-climate of its own

3.1 *Methods applied*

The methods tested were designed according to the general goal, which can be characterized by: assisting nature to build its own vegetation. This method is meanwhile widely accepted as the favorable method for abandoned mining areas in alpine areas. The advantages of this approach are sustainability, applicability under difficult conditions and economy. The main disadvantage is that such type of reclamation takes some time.

The tests were executed using hydro-seed. This well-established method has some particular advantages at alpine conditions, notably: no coverage with soil necessary or even counterproductive, less risk of erosion, flexible in respect of seed regime and economic application even on steep slopes. The limited working distance and the necessity of preferable weather during working are the main disadvantages.

Generally speaking two groups of tests can be distinguished:

- Initiation of vegetation: this is applied at completely bare areas. The goal was to develop in-situ a A_i layer (soil) by weathering of rock (source of mineral-nutrient) and by the biomass of the plants. This was tried to achieve by seeding (grass, trees), adequate preparation of ground, and additional nutrient supply for preferable conditions for growth of seeds. Much care was taken to select plants in harmony to the local vegetation.
- Acceleration of growth by supporting genuine processes: this is applied at areas where already isolated vegetation or pioneer vegetation exists, but very slow growth can be observed due to extreme shortage of nutrient. In these cases the focus was on supplying additional and specifically adjusted nutrient in order to improvement vitality, while no or only scarcely additional seed is used. Special care was taken that no existing species are negatively effected by the growth of on specific plant.

About 200 tests totaling more than 50 ha were conducted under different conditions and using different methods, which were all accompanied by an extensive monitoring program, covering:

- Documentation of the initial geological, geometrical and biological status of the area.
- Lysimeter measurements of water quality and geo-chemistry (in particular after significant events such as heavy rain fall.
- Weather station (influence of weather events on parameters investigated)
- Biological monitoring in year or half-year intervals, covering: number of germs, portion of species, diversity of species, coverage by seeded and autochthonous plants, soil analysis (incl. micro-biological parameters). These parameters are compared with the typical local vegetation and define the success of the reclamation.

The analysis focused on the investigation of the dependence of reclamation-success on factors such as:

- exposition, steepness, fragmentation (fine or blocky ground)
- geological and lithological conditions
- regime of seeds
- fertilization and refertilization variants
- soil glueing and strengthening
- integration of wood-sticks (combined methods)

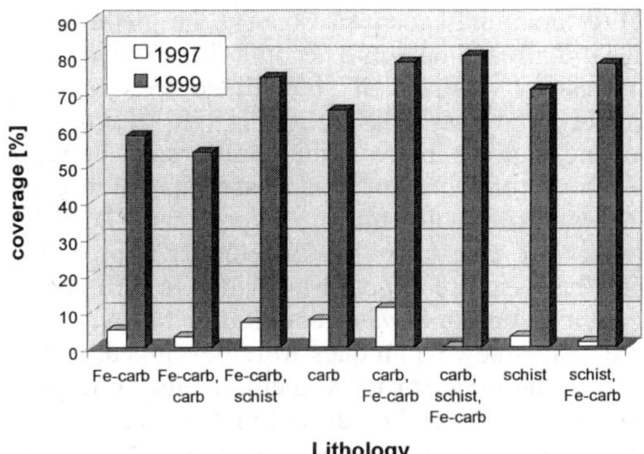

Figure 2. Comparison of coverage of initial state and after 2 years - by distinct lithological conditions

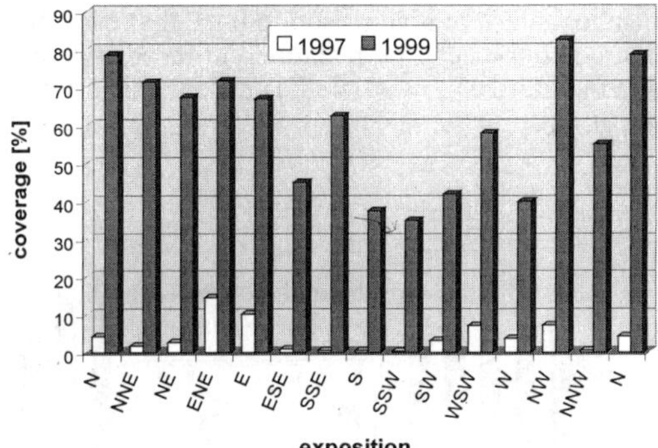

Figure 3. Comparison of coverage of initial state and after 2 years – by exposition

3.2 *Results*

A prime aspect was the interdisciplinary investigation of the data monitored and compiled independently by distinct experts. From the very beginning the data were entered into a database, giving full access to the information gained to all partners involved. The database allows analysis and comparison of data and their interdependencies, even for none-computer experts. Failures could be quickly detected, and necessary additional measures could be initiated. This lead to a series of improvement of the methods applied.

Summarizing it can be stated that the reclamation test were quite successful. Figure 2 and 3 show a comparison of the initial state at the beginning and after the first two-year period in terms of coverage by vegetation. There is only a slight variation of the reclamation-success for distinct lithologies (Figure 2), while there are more obvious differences for different expositions (Figure 3). It is important to properly interpret these figures. It means that it succeeded to find adequate methods for all situations, and not that e.g. lithology does not significantly effect reclamation, because of course the methods applied did already take into account these specific circumstances. This can be illustrated by the monitoring results after the first year (Figure 4), showing a much more evident difference between the more preferable north and east exposition compared with the south and west exposition. These differences are caused by the different length of sunshine the areas are exposed to, which in turn effects the temperature and water conditions. Motivated by these results new techniques were developed. Figure 5 shows the effect of these methods after one year, in this case expressed by the amount of biomass. It is plain to see that the positive effect is only relevant under less preferable conditions.

This example indicates also that there is a notable potential of cost savings, if the appropriate method is applied. The promising results of the new technique mentioned above give reason to hope that even difficult conditions can be reclaimed at reasonable cost, and therefore is now offered as a standard service of the group.

According to the general goals of reclamation there are many more success parameters. Figure 6 shows the number of species, which is regarded as an important key-figure in particular in respect of sustainability of the reclamation. The increase of species shown confirms the proper selection of the seed and fertilizer regime. The proof of such a development is a perfect argument for ecological reclamation.

Very interesting results were gained also in respect of the influence of the fragmentation of the material. It turned out that very coarse and also fine mono-sized materials are very problematic for reclamation. The interesting point is that the investigations have shown, that in many cases it is more cost-effective to adjust the way a waste dump is built rather than applying cost-intensive reclamation methods afterwards – another example that reclama tion and mining operation must not be separated for proper evaluation.

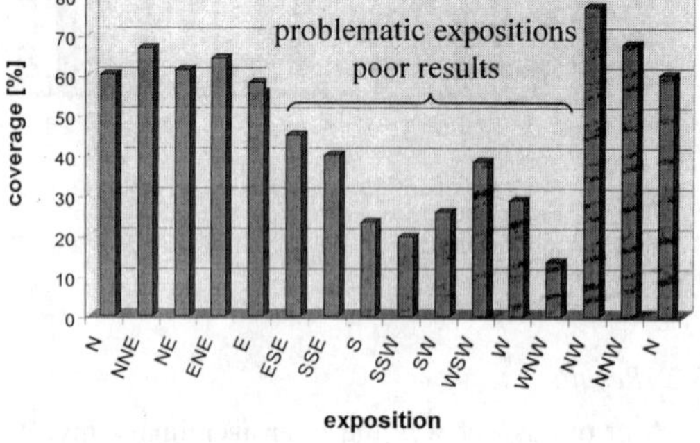

Figure 4: Coverage after one year – showing expositions with poor reclamation results

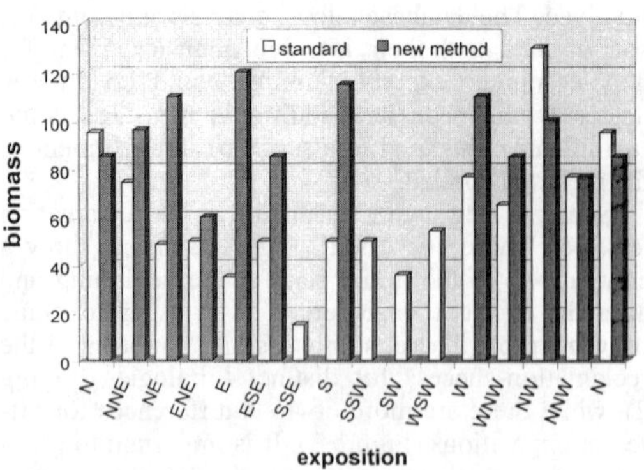

Figure 5. Comparison of the standard and new method after one year.

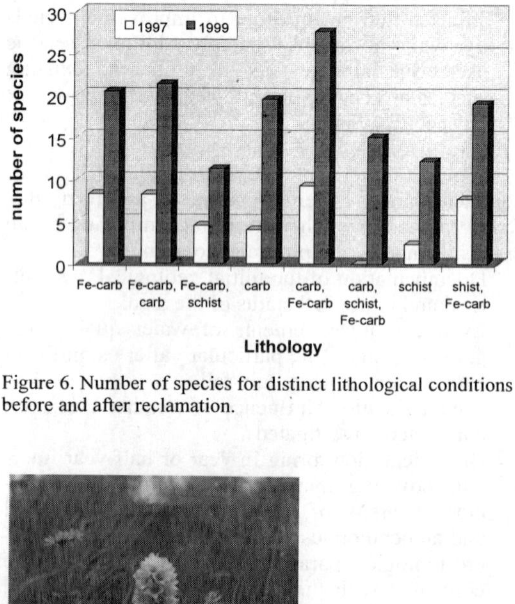

Figure 6. Number of species for distinct lithological conditions before and after reclamation.

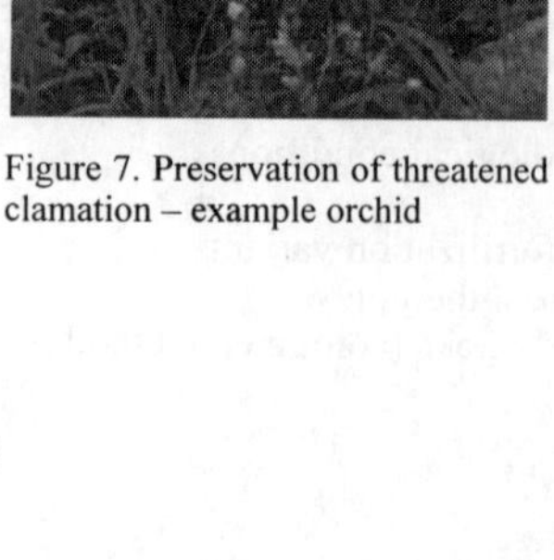

Figure 7. Preservation of threatened species by considerate reclamation – example orchid

4 EXTERNAL AREAS

Despite the variety of conditions available at the test site, as a matter of fact not all conditions could be covered. For this reason numerous mine sites (in operation and abandoned) were visited and documented. For financial reasons no monitoring was possible. The advantage of this additional approach is that the result of reclamation (induced or autochthonous) could be observed after much longer time periods than possible within the project, giving important indications in terms of sustainability.

Minerals of prime interest for Austria are limestone, quarzite and diabas. The focus of the documentation was the differences of short-term and mid-term development under theses conditions. In addition many other criteria expected to be of relevance are included.

Again all data are compiled in a database. In this case the data focus on qualitative and descriptive data, which are less likely to be analyzed statistically but serve as a source of information.

These systematical observations and investigations prove the complexity of the problem. For example the abandoned mine-site "Radmer" on siderite with completely identical geological and geographical conditions as the test site at VA-Erzberg shows very successful autochthonous vegetation, in contrast to what can be observed at VA-Erzberg. This can be explained by the rather small mining area of "Radmer", with completely different mechanisms in revegetation (Figure 8). This exemplifies that we must proceed with caution if distinct sites are compared and conclusions are drawn.

It is obvious that the real value of this database will turn up with increasing number of mines covered. For this reason the database is supposed to be continued by the department of mining engineering. It is expected that in future the database will be of great help for mining companies in order to verify proposed methods or falsify allegations from opponents.

5 INTEGRAL CONCEPT

According to the goal of the project, reclamation is always regarded as an integral part of mining operation with mutual dependencies.

Two distinct mining methods are applied in quarrying in Austria. The distinction is mainly by the direction of advance, and in consequence by the type of haulage used (Figure 9). "Wall-mining" proceeds in horizontal direction with rather small berms. The blasted material is usually pushed down to the deepest level and mucked and hauled there. Contrary bench mining proceeds form top downwards, having only one or two benches in operation and requiring haulage to the lowest level, usually by truck or glory hole.

Figure 8. Autochthonous vegetation of bench face after 12 years (Radmer)

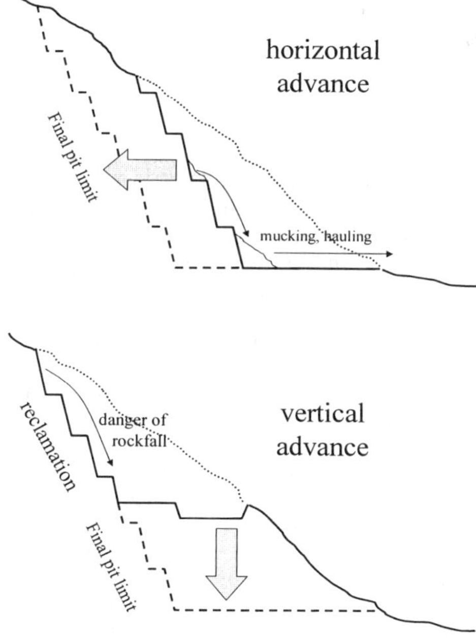

Figure 9. Sketch of two principle mining methods applied in Austrian quarrying

By historical reasons wall-mining is dominant for smaller quarries. However, there is a obvious preference of ecological representatives towards bench mining. This is mainly due to the limited visual impact and the option for simultaneous reclamation during mining.

A main aspect of the investigations was on the effect of this preference on mining activity. Although the stated advantages are out of question, there are also some disadvantages, in first place concerning safety. A consequence of simultaneous reclamation is that the wall has to be brought into the final shape already during mining. Ecological representatives prefer the elimination of the berms and benches to provide a "natural look" of the remaining wall. This approach has some serious consequences on safety, because the berms have obviously a protection function against rock fall, in this case both during and after mining.

By means of analytical simulation techniques it was tried to find the most critical parameters of the shape of the wall on rock fall. The findings are:

- berm width is an essential parameter
- Irregularity plays a very important role, because of the increased chance of very disadvantageously rebouncing angles
- Vegetation has – at least in the beginning – no relevant protection function
- Height of benches and total wall and also the total slope angle (in typical ranges) have less significant effect
- Loose material and well developed vegetation can help to stop rocks at the berms

The conflict between visual appearance and safety is unsolvable, and form the safety aspect the complete abundance of berms is not acceptable. However, there are some measures to decrease this discrepancy. One is that adequate preparation of the wall can reduce the risk that rocks start falling down the wall. Blasting with pre-splitting is one way to achieve this, but unfortunately completely opposed to a "natural" appearance. More adequate is the working of the wall using hydraulic hammers. It is important to note that experiences with pre-splitting show that the problem can be moderated but not solved. This means number of berms can only be decreased. Damping material on the remaining safety berms and protection damps against working area are helpful. In terms of mine design reduction of overall slope angel is a promising measure, giving sufficient space for effective berm width. Unfortunately this leads to an increase of the extension of the mine, which is – falsely – regarded as unecological. Quite contrary, larger mine extension and the resulting slow down of advance downwards has also some positive ecological effects. Reclamation has more time to develop, hence sustainable methods can be applied, and the emerging vegetation can develop protection function. In addition mining technology can be applied which is more beneficial form ecological point of view, e.g. belt conveying in combination with in-pit crushing.

6 OUTLOOK

The experimental test have detected and confirmed the interrelation between reclamation and geometrical and geological conditions. Improvements of the methods applied have increased the potential of success at acceptable costs. It is planned to continue the monitoring program in order to gain more information in respect of the sustainability of the methods.

In addition, the thorough concept of the project proved that the reclamation success depends on many more parameters, e.g. geographical and topological situation. It will be tried to enlarge the cases of observations in order to provide even more reliable conclusions.

In particular the mutual influence of mine design and reclamation is highlighted. Mine safety must have highest priority. The clear description of these interactions is necessary to bring the complexity of the problem into the consciousness of all groups involved, although this is expected to be a lengthy process.

Environmental Issues and Management of Waste in Energy and Mineral Production, Singhal & Mehrotra (eds)
© 2000 Balkema, Rotterdam, ISBN 90 5809 085 X

In-pit tailings disposal management facility, Fosterville Gold Project, Australia: A case study

A.J.Trevenen & K.Biswas
School of Engineering, University of Ballarat, Vic., Australia

V.Langdale
Perseverance Exploration Pty Limited, Australia

ABSTRACT: The issue of permanent disposal of mine tailings is a contentious one, raising concerns from both mining companies and their environmental counterparts. The Fosterville Operation, owned and operated by Perseverance Exploration Australia Pty Ltd., is currently tackling the problem of tailings disposal in relation to its proposed expansion of the Fosterville Sulphide Project. Following extensive research and consultation, it was decided that the most suitable method of disposal in this case, was in-pit disposal. Subsequently, the Fosterville Pit, which was successfully mined out in 1993, has become the focus of a substantial backfilling and reclamation process. This paper intends to provide a case study, detailing the chosen methods of deposition and rehabilitation, suitable for the application of the Fosterville Operation.

1 INTRODUCTION

1.1 Background

Located in the Central Victorian Goldfields, is the Fosterville Gold Project, owned and operated by Perseverance Exploration Australia, Pty Ltd. The current surface operations produce gold from weathered oxide ore, which is typically processed using cyanide leaching techniques. Deeper, unweathered sulphide deposits underlying the current operations have been explored and confirmed as a suitable resource for further mining. With the discovery of a new ore type, comes the inevitable addition of mining methods, ore treatment, and water and waste management (Perseverance Exploration Pty Ltd, 1996).

1.2 Ore Processing Techniques

The proposed processing technique utilises traditional flotation and leaching methods, with the addition of bacterial oxidation techniques, as developed by Minsaco Pty Ltd. of South Africa. This BIOX® process utilises a combination of bacteria to liberate the trapped gold, in preparation for subsequent cyanidation and processing. The flotation tailings and the counter-current decantation (CCD) neutralisation residue are to be used in the fill medium for the

Fosterville Open Pit. The tailings from the carbon-in-pulp/carbon-in-leach circuit contain cyanide levels that render them unsuitable for effluent recirculation through the processing plant. These tailings are to be disposed of in a separate tailings compound.

2 WASTE CHARACTERISATION

The Fosterville Sulphide Project is currently undergoing financial evaluation and is not yet operational. Therefore, tailings samples were sent to various metallurgical and geochemical laboratories for processing and property analysis.

2.1 Physical Properties

Limited information was available on the physical properties of both the flotation tailings and the CCD neutralisation residue. However, a screen analysis revealed that approximately 80% of the material sampled passed a 150μm screen aperture and the elements found in the samples provide some indication of the possible physical properties of the material.

2.2 Chemical Properties

A thorough geochemical analysis was undertaken by Environmental Geochemistry International (EGI) Pty Ltd., to determine the multi-elemental composition of both the liquid and solid fractions of the tailings samples, and to identify potential toxic elements of environmental concern (EGI, 1996).

The flotation tailings were found to have a low sulphur content, moderate acid neutralisation capacity (ANC)[*], and were classified as non-acid forming (EGI, 1996). The solids contained within the flotation tailings were significantly enriched with arsenic and antimony, while the liquor was found to contain relatively high levels of soluble salts, whilst maintaining a relatively neutral pH of 6.5 (EGI, 1996). Also detected within the liquor was relatively high soluble arsenic levels (EGI, 1996).

The CCD neutralisation residue was essentially found to be precipitated calcium sulphate or gypsum, with a high arsenic content of 3.9% (EGI, 1996). Although the soluble arsenic concentration was lower than that of the flotation and CIP/CIL tailings, it was still high enough to cause environmental

[*] The Acid Neutralising Capacity (ANC) is measured by the amount of acid consumed by a sample, during a simple chemical reaction.

concern (EGI, 1996). Concentrations of other environmentally significant heavy metals were low and generally less than detection (EGI, 1996).

Following the mixing of the flotation tailing liquor and the CCD neutralisation residue, chemical changes resulted in a reduction of arsenic concentration and increases in concentrations of calcium, magnesium and sulphate (EGI, 1996).

3 METHOD OF DEPOSITION

The method of disposal was pre-determined by the availability of the mined-out Fosterville open pit. The actual method of deposition, however, required greater research. Following an approximated economic comparison between several proposed methods, a conventional wet deposition method with a slimes liner was decided upon.

3.1 Wet Deposition Method

The wet deposition method is a relatively conventional method of tailings deposition, which utilises the properties of classified hydraulic fill. Classified hydraulic fill refers to the predominantly sand fraction of the tailings material, suspended within a hydraulic slurry.

The finer fractions of tailings are removed using a thickener/hydrocyclone combination, leaving the more coarse fractions better suited to tailings disposal. The finer fractions are then stored for use in the slimes liner (refer section 3.3) or deposited in the purpose designed tailings dam. With the removal of the slimes fractions the permeability is increased, thus allowing the tailings to settle faster and the liquid to be decanted from the surface. Generally, the tailings are thickened to between 40 - 65% solids depending on the type and method of thickener used.

The cyclone underflow would then be pumped, via the tailings pipeline, to the Fosterville open pit. The peripheral discharge method allows the tailings to beach from the periphery inward (Ritcey, 1989). Subsequently, a pond forms in the centre of the pit, which allows for relatively simple decantation using floating pumps. After a completed layer of tailings is achieved and sufficient time is allowed for the tailings to consolidate, the discharge points are rotated ready for the placement of the next layer.

As limited information was available on the physical characteristics of the tailings material, it is difficult to assess the behaviour of the tailings over time. The tailings height would be raised by a series of lifts, to a height of no more than five metres, utilising the coarse fill material. Sufficient time would be allowed in between the placement of each lift. This allows for maximised dewatering and consolidation of the tailings material.

Careful consideration needs to be given to the time allowed between the placement of each layer of tailings. Sufficient time is needed to remove the decant water from the pit. However, the longer the tailings surface is exposed to free oxygen, the greater the risk is of the contained pyrites oxidising. This creates a separate issue of dealing with the acidic conditions created in that environment.

3.2 Dewatering

In this section, the term water control not only takes into account the surface and groundwater requirements, but also the control of effluent removed from the tailings material.

In an in-pit tailings disposal facility, water control consists of two primary areas; groundwater management and tailings dewatering. These two areas combined, are critical in ensuring the tailings consolidate sufficiently and remain a stable body.

When hydraulic fill is discharged into a tailings impoundment, the liquor gradually separates from the solids. The particle size distribution of the material determines the rate at which the two mediums separate. Following separation, the liquid can be collected and recirculated through the process plant. The remaining solids are then left to consolidate and stabilise.

The most suitable method of dewatering is dependent on the characteristics of the tailings and the pit. In this case, a combination of dewatering methods is proposed to obtain the best results.

The use of cyclones to separate the finer tailings fractions from the remaining coarser fractions (refer section 3.3), aids in the dewatering process. Combined with a floating pump system and the dewatering occurring due to self-weight consolidation, the tailings are given sufficient chance to dewater before the next layer is placed on top. The decant water is pumped from the top of the settled tailings and recirculated through the processing plant.

Regional groundwater flow in the region is controlled via a system of dewatering wells located around the perimeter of the Fosterville pit and adjacent areas. These wells currently minimise the inflow of groundwater in the pit area.

3.3 Slimes Liner

Basically, there were two main options available for the control of contaminated water in tailings impoundments; preventing it from leaving the impoundment or capturing after it exits the impoundment (USEPA, 1994). Aside from wanting to minimise the amount of contaminants leaving the tailings impoundment, controls were also required to minimise the amount of groundwater influx into the pit area. As the groundwater in the Fosterville region is saline, a maximum groundwater concentration of 7% is enforced within the water recirculation system. This is to ensure the bacterial oxidation process runs effectively. For this reason, a liner/buffer was deemed the most suitable method of ensuring the tailings and tailings effluent remains fully contained within the Fosterville pit.

Unlike synthetic liners, tailings slimes liners are relatively low cost, and have the added advantage of disposing of the finer tailings fractions in the process. Seepage reduction is achieved by virtue of the low permeability of the finer fractions of tailings which slows down, and depending on the thickness of the layer, prevents any leachate from entering or exiting the tailings impoundment (Vick, 1983).

For a slimes liner to be constructed, slimes must constitute a considerable fraction of the total mill tailings (Vick, 1983). Due to the nature of the proposed processing method of the Fosterville Sulphide Project, it is anticipated this requirement will be met. The finer particles are to be separated through the use of cyclones, which rely on strong centrifugal forces for separation. These finer fractions are then stored for use in the liner.

It is proposed that the slimes be mixed with a binder, namely Portland cement, prior to disposal around the walls of the pit. As the liner is required to prevent groundwater flows into the pit, a height of at least RL 140 is to be reached along the walls, with the liner. This is due to the location of the groundwater table at RL 135. It is also recommended that in order to further prevent the mixing of saline and decant waters, the Fosterville pit be dewatered for its initial life. Once the tailings become established and exert their own hydrostatic pressure, further dewatering may not be required.

It is anticipated that the most appropriate method of distribution would be through the use of shotcrete equipment, in which the slimes/cement binder mixture is sprayed on the inside walls of the Fosterville pit. The material would then be left to cure for an appropriate length of time, before the remaining fractions of tailings were deposited on the surface. Careful consideration must be given to the length of time required, as too long a time will lead to total drying, causing localised cracking and reduced effectiveness of the slimes liner.

Once the slimes liner is established, the remaining finer tailings fraction are to be deposited into a nearby tailings dam. This conventional tailings dam is to be established to contain the continuous flow of finer particles resulting from the treatment process.

4 REHABILITATION

'The ever-increasing need for metals, and the ability of modern mining and processing methods to develop low grade ore-bodies economically, has placed increased strain on the environment at a time when the demands for high environmental standards are also increasing' (United Nations, 1992).

The primary aim of rehabilitation is to restore land to its original form. For once forested areas, gradual re-introduction of indigenous flora and fauna species allows the native ecosystem to be re-established in the mining affected area.

The area incorporating the Fosterville Pit was originally an established Box-Ironbark forest. The dominant flora species in this forest were eucalypts, including Grey Box, Red Box and White Box (Perseverance Exploration, 1996). Extensive surveys carried out in the area displayed considerable biodiversity of both flora and fauna.

4.1 Revegetation Constraints

The physical, chemical and biological properties of the tailings material render it unsuitable for plant growth, when used alone. The small and relatively uniform particle size distribution, unfavourable porosity, aeration, water infiltration and percolation properties all combine to make the tailings a harsh material for plant growth (Johnson, Cooke & Stevenson, 1994). Added to this is the concentration of essential trace elements, which even at low concentrations can be toxic to plants (eg. copper & zinc) (Johnson, Cooke & Stevenson, 1994). Other non-essential elements, which may be less toxic to vegetation, may be harmful to animals that feed on the plants in the area (Johnson, Cooke & Stevenson, 1994). Therefore, the need for capping is integral in ensuring adequate plant growth on the Fosterville pit site.

Another primary constraint relevant to the Fosterville Operation is the presence of significant quantities of iron pyrites (FeS_2). Pyrite-bearing wastes that are disposed of at neutral or slightly alkaline pH can degrade within months or years to produce extreme acidity (Johnson, Cooke & Stevenson, 1994). The rate at which the pyrites oxidise is greatly influenced by the surface area available for weathering, the native carbonate content of the material, and the size, morphology and type of pyrite present (Johnson, Cooke & Stevenson, 1994). The presence of carbonates in the Fosterville Sulphide Project tailings indicates that the pH of the material may vary over time, the effect of which is yet to be determined.

The quantity of essential plant nutrients contained within the tailings material needs also to be considered. Commonly, nitrogen and phosphorus levels are relatively low, and deficiencies in potassium, calcium and magnesium may also occur (Bradshaw & Johnson, 1992). However, in this case the tailings from the flotation process are enriched in potassium, calcium and magnesium. This will provide some benefits when re-establishing a native cover.

4.2 Revegetation Objectives

Essentially, the primary rehabilitation objective is to re-establish the original ecosystem to the mining affected land. The more site specific objectives include long-term stability of the land surface, development of sufficient drainage system, reduction of leaching throughputs, and minimisation of potentially toxic elements released into the surrounding environment.

4.3 Revegetation Technique

Revegetation of the Fosterville Pit site can be divided into three separate areas; stabilisation of the exposed surface, direct seeding of vegetation, and planting of indigenous tube stock. Following the completion of these stages, continuous monitoring is required to ensure the vegetation is given the best possible chance of success.

4.3.1 Stabilisation of Exposed Surfaces

To prevent, or at the least reduce, erosion, the exposed tailings surface requires covering. By providing a suitable cover, the growing conditions for the vegetation are also greatly improved.

Capping requires various layers of material to ensure an adequate cover is established (*refer figure 1*).
The first layer is constructed using compacted clay, which provides an impermeable layer between the tailings and the vegetation. This layer is aimed at reducing the upward movement of metals and inhibiting infiltration, reducing possible leachate generation (Minerals Council of Australia, 1999). In most cases, an impermeable layer of at least 300mm depth is required (Johnson, Cooke & Stevenson, 1994).

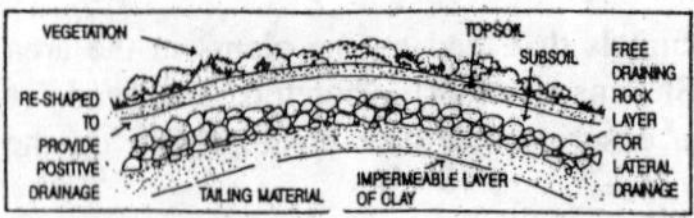

Figure 1. Various Layers Required for Capping (Minerals Council of Australia, 1998)

The second layer is a free draining rock layer for lateral drainage (Minerals Council of Australia, 1998). Landforming and re-shaping is required to aid the productivity of this layer, ensuring surface runoff is controlled. Positive drainage allows the runoff to be collected in drains and directed to storage areas for further use. Again, a 300mm layer of porous material is recommended (Johnson, Cooke & Stevenson, 1994).

The top two layers are constructed using subsoil and topsoil respectively. The primary aim of these layers is to provide a rooting medium, which allows establishment and growth of stabilising vegetation and storage of soil moisture (Johnson, Cooke & Stevenson, 1994). A combined depth of 300 - 500mm is regarded as suitable for root growth (Johnson, Cooke & Stevenson, 1994).

4.3.2 Direct Seeding

Direct seeding is the most straightforward and common method of revegetation. It involves the identification and collection of native seed from nearby indigenous vegetation. The seed is collected in the summer months then distributed over the required areas. To alleviate compaction, the land should be contour ripped to a depth of 300mm, as loose soil tends to hold seed in soil cavities and provide protection as the seeds germinate and establish themselves (Perseverance Exploration, 1996). As the long-term growth of vegetation requires an adequate supply of nitrogen, the use of fertilizers may be required depending on the condition of the topsoil material (Johnson, Cooke & Stevenson, 1994).

4.3.3 Indigenous Tube Stock

To establish an effective ground cover, it is often necessary to use a combination of revegetation methods. By planting tube stock, of both indigenous grasses and trees/shrubs, some form of ground cover can establish prior to the commencement of direct seeding. At the Fosterville Pit site, the tube stock should be planted in autumn for best results (Perseverance Exploration, 1996).

5 CONCLUSION

The environmental effects of a mining operation can never be erased. However thorough research into suitable reclamation techniques, following de-commissioning, can ensure the original ecosystem is restored and the land is returned to nature.

REFERENCES

Bradshaw AD & Johnson MS, *Minerals, Metals and the Environment.*, Institute of Mining & Metallurgy, London, 1992

Environmental Geochemistry International Pty Ltd, *Interim Report: Geochemical Characterisation of CIP/CIL Tailings, Flotation Tailings & Neutralised CCD Thickener Residue, Fosterville Gold Project.*, EGI Pty Ltd, Balmain, 1996

Johnson MS, Cooke JA & Stevenson JKW, *Mining and its Environmental Impact: Revegetation of Metalliferous Wastes and Land After Metal Mining.*, Royal Society of Chemistry, Cambridge, 1994

Minerals Council of Australia, *Mineral Rehabilitation Handbook 2nd Edition.*, Minerals Council of Australia, ACT, 1998

Perseverance Exploration Pty Ltd, *Fosterville Gold Project: Project Upgrade – Environmental Effects Statement.*, Perseverance Exploration Pty Ltd, Fosterville, 1996

Ritcey GM, *Process Metallurgy 6, Tailings Management: Problems and Solutions in the Mining Industry.*, Elsevier Science Publishers BV, Amsterdam, 1989

United Nations, *Mining and the Environment: The Berlin Guidelines.*, United Nations Department for Technical Co-operation and German Foundation for International Development, Mining Journal Books, London, 1992

United States Environment Protection Authority – Office of Solid Waste, *Special Waste Branch Technical Report: Design & Evaluation of Tailings Dams.*, USEPA, Washington DC, 1994

Vick SG, *Planning, Design & Analysis of Tailings Dams.*, John Wiley & Sons, Canada, 1983

Environmental Issues and Management of Waste in Energy and Mineral Production, Singhal & Mehrotra (eds)
© 2000 Balkema, Rotterdam, ISBN 90 5809 085 X

Sediment re-suspension in shallow water under wind-induced waves and current

Y.Yang, A.G.Straatman & E.K.Yanful
The University of Western Ontario, London, Ont., Canada

ABSTRACT: Sediment re-suspension in shallow water covers is an important problem in mine waste management, where shallow ponds are used to store tailings. In the present study, a wind-wave tank is used to investigate the resuspension process using both inert materials (glass beads) and actual mine tailings. The wind speed above the tank is monitored using a hot wire and the wave height using a wave meter. The wind induced motion and sediment concentrations in the water are measured using an Acoustic Doppler Velocimeter (ADV). Experiments have been conducted to determine sediment resuspension for the following conditions: water depth ranging from 6.35 cm to 11.43 cm; wind speed ranging from 5 m/s up to 15 m/s. Attempts are made to correlate the sediment concentration with wind speed and water depth.

1 INTRODUCTION

The use of shallow water covers to prevent mine tailings from oxidation is very common in mine waste management. Large quantities of mine waste are deposited into tailings ponds every year. If wind generated motions in the pond are strong enough, the sediment may resuspend into the pond water and react with dissolved oxygen and possibly generate acid, thereby polluting the surrounding environment. The concentration level of the resuspended sediment in the pond water determines the satisfactory performance of these ponds.

The resuspension and entrainment of sediment particles in a shallow tailing pond is initiated by bed shear stress induced by wind generated motions in the pond. On investigation of sediment resuspension in a shallow pond, the following factors should be considered: (a) wind conditions; (b) wind generated wave; (c) wind generated current; (d) wave and current interactions; (e) sediment properties. In the literature, most studies on suspended sediment transport have considered open channel flow (Rouse 1938, 1939, Einstein & Chien 1952, Yalin 1977, Van Rijn 1984). For these cases, the bottom shear stresses are generated mainly by current. Some other studies considered the effects of both wave and current motion on the initiation of sediment suspension (Van Rijn et al 1993, 1995, Williams et al 1999). In a shallow tailing pond, wind generated flow is countercurrent flow, so the bottom shear stress is actually generated by wind wave as well as countercurrent flow.

Based on the empirical relationship between wind speed and shear stress, Rodney & Stefan (1987) proposed a conceptual model for wind-generated sediment resuspension for fine particles in a shallow pond. Lawrence et al. (1991) developed a model to express the depth of water needed to prevent the wind-wave-induced suspension of sediments in mine tailings ponds for deep-water conditions. The depth is expressed as a function of four factors: the threshold velocity, the wind velocity, the fetch over which the wind blows, and the wavelength. Mohamed et al (1996) presented a minimum depth of water cover needed to prevent tailings resuspension based on a set of wave tank experiments.

The most effective method for studying resuspension of particles and wind-induced motions is to construct a well-controlled laboratory experiment. While it is extremely difficult, if not impossible, to develop a properly scaled model, the main motions and the substantial characteristics of the flow field can be studied using a small-scale wind-water tunnel. The purpose of the present study is to experimentally investigate the sediment resuspension under wave and current conditions in shallow water using both inert material and real tailings. The experiments were carried out for a variety of wind speeds, water depth and sediment sizes and densities. The preliminary experimental results are presented in this paper. The objective is to relate the wind speed and water depth to the local sediment resuspension induced by wind-generated waves and countercurrent flow. The results from this experiment can be used to validate numerical modeling.

The long-term goal is to extend a numerical model to the full-scale prediction of sediment transport in real tailing ponds.

2 EXPERIMENT SETUP

Experiments are carried out in a closed-circuit wind tunnel with a wave tank, which is located in the Boundary Layer Wind Tunnel Laboratory (BLWTL), the University of Western Ontario. The water tank is 5.22 m long, 0.4 m wide and 0.2 m deep, and the wind tunnel is the same width and 24 cm high, as shown in Figure1.

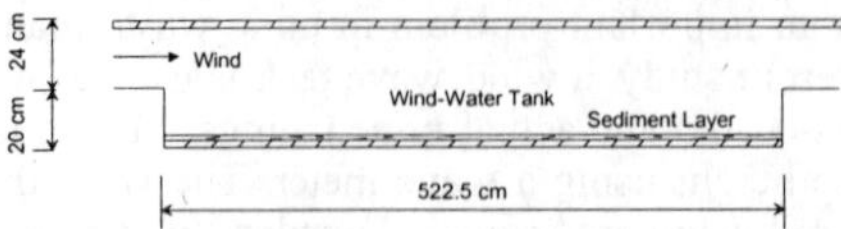

Figure 1. Wave and countercurrent flow is generated in a closed-circuit wind tunnel with a wave tank

The test section is located from 2.0 m to 2.5 m from the inlet of the wind tunnel, where there is a glass window. The flow pattern and sediment resuspension can be observed through the window.

Two sediment materials were used in this study. The first sediment material is soda-lime; a kind of solid glass sphere. The mean particle size is 20 μm in diameter. The specific gravity is 2.5. The second sediment material is fresh mine tailings from Heath Steele Mine Ltd., New Brunswick, which has a specific gravity 4.0. The particle size distribution is shown in Table 1. The finer (%) given in the table refers to the percentage of particles that are finer than the diameter, also given in the table.

Table 1

Finer (%)	80	50	20
Diameter (mm)	0.045	0.021	0.006

A 2-mm thick layer of sediment materials was carefully placed at the bottom of the water tank. To reduce the end effects of the wave tank, the sediment covered the entire bed except for the end portions of the tank. Tap water was then used to fill the tank. Once the water tank was filled, the wind was blown for a short time to get the air bubbles out of the sediment. The sediment was then allowed to settle before an experiment was started.

When an experiment was started, the wind speed accelerated very quickly to reach its desired value. The wind speeds ranged from 5 m/s to 15 m/s and water depth was varied from 6.35 cm to 11.43 cm.

A Pitot tube was used to monitor the wind speed through the test. The wind velocity profile was measured using a hot wire. A wave meter was used to measure the wave height.

Instantaneous velocity and suspended sediment concentration values were measured above the bed using a Sontek three-dimensional Acoustic Doppler Velocimeter (ADV). The sampling rate for all measurements was 25 Hz.

The experimental conditions are shown in Table 2

Table 2

Wind	Water depth (cm)				
(m/s)	11.43	10.16	8.89	7.62	6.35
15.0	-	-	-	+	+
11.07	-	-	-	*	*
10.11	-	-	+	+	+
9.04	-	+	+	+	*
7.82	-	+	+	+	-
6.39	-	+	+	+	*
4.95	+	+	+	+	-

+ Performed. * Using both materials. - Not performed.

3 RESULTS AND DISCUSSION

The *u, v* and *w* velocities and concentration (c) of suspended sediment were measured at two cross-sections, 2.0 m and 2.5 m from the upstream edge of the water tank, respectively. The recording of instantaneous velocity and concentration signals was started before the wind was turned on and completed when a constant concentration of suspended sediments was reached.

Sediment resuspension occurred under all experimental conditions when the glass beads were used. However, in only two cases, wind speeds of 11.07 m/s for water depth of 6.35 cm and 7.62 cm, did mine tailings go into resuspension. This is due to the high specific gravity of mine tailings.

Figure 2 shows a typical time series of velocity in the stream-direction x, lateral direction y and vertical direction z at a distance z=1.0 mm above the bed and at x = 2.0m. Figure 3 shows the time series of the concentration of suspended sediment at the same point and over the same time period as Figure 2. It is clear that the resuspension occurred after the wind was turned on. When the maximum concentration level was reached, the averaged concentration remained virtually the same. That indicates that the suspended sediment reached the equilibrium state at that point in the water. During the period that the concentration value was increasing, a slope can be found that shows the incremental rate of increase of the suspended sediment.

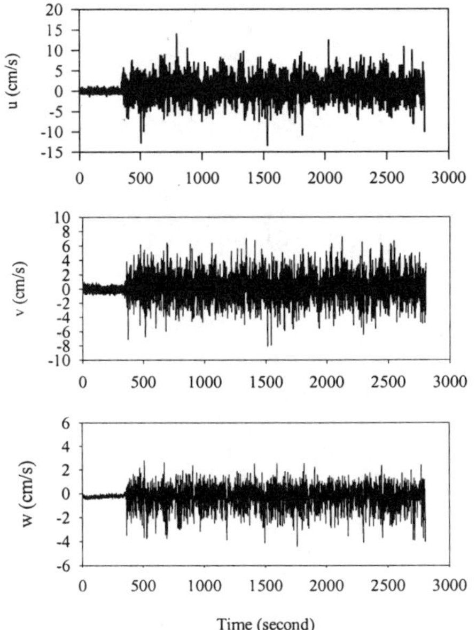

Figure 2.Instant velocity u, v and w at cross-section x=2.5 m for wind speed = 11.069 m/s.

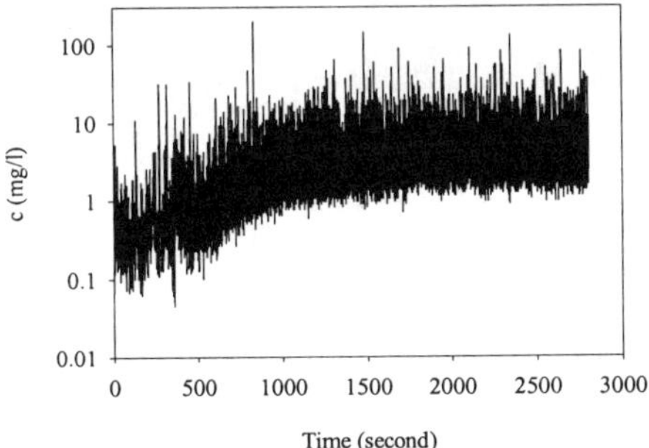

Figure 3.Instant concentration at cross-section x=2.5 m for wind speed = 11.069 m/s.

The vertical distributions of the suspended sediment were measured after the maximum concentration was reached. From the experiments, it was found that the higher a wind speed, the higher the concentration of the suspended sediment obtained for a constant water depth, as expected. Figure 4 shows several profiles obtained at x=2.5 m, for a water depth h=7.62 cm under different wind speeds. The concentration is much higher near the bed and decreases when the distance from bed increases. Apart from the near bottom range, the suspended sediment concentration is nearly uniform with depth. These observations are similar to those made in open channel flow for fine sediments (Rouse, 1938).

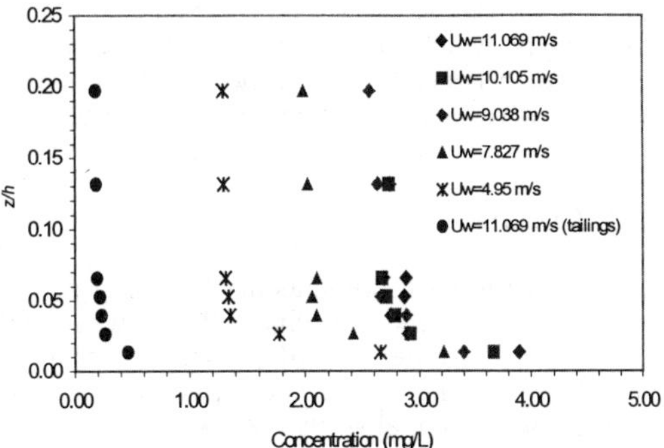

Figure 4. Concentration profile X=2.5 m, water depth h=7.62 cm under different sind speeds

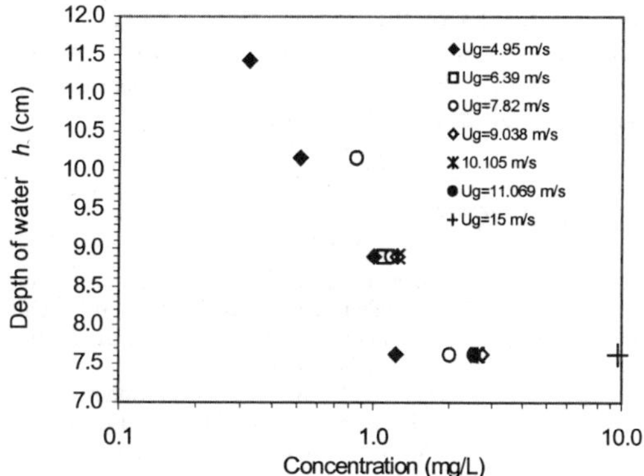

Figure 5. Depth averaged concentration vs. water depth at x=2.0 m under different wind speed

Depth averaged concentrations, excluding the first measurement point above the bed, are shown in Figure 5. The depth-averaged concentration indicates the suspension capacity of sediment for the shallow water, which is important for estimating the total suspended sediments. It is clear that the concentration of suspended sediment increases with decreasing water depth and increasing wind speed, again, as might be expected. Using Figure 5, if a particular concentration level was desired and the wind speed was known, the required water depth could be obtained. Thus, Figure 5 characterizes our wind-water tunnel facility for the experiments performed to date. A plot of this type would be of enormous benefit in industry, since it could be used to characterize tailings ponds in terms of wind, water depth and particle concentration. Unfortunately, due to scaling issues, the data cannot be transferred directly from the scaled model to the field. However, information from Figure 5 can be used to tune validate computational models, which can then be used to predict the relationship between concentration, particle size, wind speed and water depth in real tailings pond.

4 CONCLUSION

A set of laboratory experiments has been carried out to explore the influence of wind-induced wave and current on the resuspension of particles in shallow water. The experiments were performed in a small-scale wind-water tunnel so that the wind speed, water depth and particle size could be carefully controlled. Experiments were carried for several variations of wind speed, water depth and particle size. The main results of this study are the characterization of the facility in terms the mentioned parameters. The results of this study are to be used in the development and validation of a numerical model, which can then be used to explore the same phenomena in full-scale tailings pond.

5 REFERENCES

Einstein, H.A. & N. Chien 1952. Second approximation to the solution of the suspended load theory. Series 47, Issue No.2, *Institute of Engineering Research*, Univ. of California, Berkeley, Calif., Jan. 31, 1952.

Lawrence, G.A., Ward, P.R.B. & M.D. MacKinnon 1991. Wind-Wave-Induced Suspension of Mine Tailings in Disposal Ponds – a Case Study. Can. *Journal of Civil Engineering*. 18. 1047-1053.

Mohamed, A.M.O., Yong, R.N., Caporuscio, F. & R. Li 1996. Flooding of a mine tailing site: Suspension of solids-Impact and prevention. *Int. J. of Surface Mining, Reclamation and Environment*.10, 117-126.

Rodney, M.W. & H.G. Stefan 1987. Conceptual model for wind-generated sediment resuspension in shallow ponds. *Proc. 1987 National Symposium on Mining, Hydrology, Sedimentology, and Reclamation*, Univ. Of Kentucky, Lexington, Dec. 7-11,

Rouse, H. 1938. Experiments on the mechanics of sediment suspension, *Proc. 5th Inter. Congress of Applied Mechanics*, Vol. 55, John Wiley and Sons, Inc., New York, N.Y.

Rouse, H. 1939. Analysis of sediment transportation in light of fluid turbulence. *Soil Conservation Service Report No.* SCS-TP-25, United States Department of Agriculture, Washington, D.C.

Van Rijn, L.C. Martin W. C. Nieuwiaar, Van der Kaar, T, Eelco Nap & A. Van Kampen 1993. Transport of fine sand by currents and waves, *J. of Waterway, Port, Coastal, and Ocean Engineering*, 119 (2), 123-143.

Van Rijn, L.C. & F.J. Havinga 1995. Transport of fine sand by currents and waves. II, *J. of Waterway, Port, Coastal, and Ocean Engineering*, 121 (2), 123-133.

Yalin, M.S. 1977. *Mechanics of sediment transport*. 2nd Ed., Pergamon Press, Oxford, England.

Mine site closure – Acid rock drainage

Environmental Issues and Management of Waste in Energy and Mineral Production, Singhal & Mehrotra (eds)
© 2000 Balkema, Rotterdam, ISBN 90 5809 085 X

Bioactivation and bioaugmentation of a passive reactor for acid mine drainage treatment

S. Beaulieu, G.J. Zagury, L. Deschênes & R. Samson
NSERC Industrial Chair in Site Remediation and Management, École Polytechnique de Montréal, Qué., Canada

ABSTRACT: Interest in passive remediation technologies used for acid mine drainage (AMD) treatment has been increasing considerably in the last decade. Although passive systems involving sulfate reducing activity offer many advantages, treatment performance still needs to be improved. Therefore, the application of bioactivation and bioaugmentation techniques on these systems was investigated. Bioactivation of a granular peat moss (GPM) was performed in 0.75 l laboratory batch reactors with a sulfate-reducing bacteria (SRB) consortium. Bioactivation increased the average sulfate reduction rate from 69 to 167 mg/l.d. Three 9 l columns were filled with a mixture of organic substrates, inorganic materials and a culture medium. The first column was inoculated with the bioactivated consortium, the second with a non-bioactivated sediment and the third was a control. Synthetic AMD was fed in a continuous mode. Results obtained after 60 days show similar sulfate and metal removal in all reactors with slightly higher performance in the inoculated and bioactivated columns.

1 INTRODUCTION

Acid mine drainage (AMD) is a persistent environmental problem at many active and abandoned mining sites. Sulfide minerals in contact with oxygen and water, combined with the presence of oxidizing bacteria, will be oxidized to produce AMD. AMD is characterized by high acidity and high concentration of metals and sulfate. Because of the negative effects of AMD on streams and waterways, it usually requires treatment before being released in the environment.

Considering their minimal operation and maintenance requirements, passive system technologies involving sulfate-reducing activity are very promising remediation techniques (Wildeman & Updegraff 1997). The main mechanism taking place is performed by SRB, that can be found in natural environments where anoxic conditions prevail. SRB consume organic matter, produce HCO_3^- that raises the pH and reduces sulfate present in AMD to sulfide under anaerobic conditions. Sulfide then combines with metal cations to form insoluble metal sulfides. However, a certain amount of time (in most cases 3 to 6 weeks) is needed for the establishment of bacterial sulfate reduction, and subsequent drop in dissolved metal concentrations.

The use of bioactivation and bioaugmentation techniques should minimize the lag phase and improve the performance of passive systems involving SRB activity. Bioactivation consists in stimulating the growth of an SRB consortium on an appropriate support and increasing its density in a short period of time. Once the microbial consortium is stimulated, bioaugmentation of the reactor is carried out by inoculating the media with the activated consortium. Batch experiments performed previously with two different SRB consortia demonstrated that the bioactivation of granular peat moss could considerably increase sulfate reduction activity (Beaulieu et al. 1999).

The objective of this study is to minimize the lag phase and to improve the performance of passive systems involving SRB activity using bioactivation and bioaugmentation techniques.

Results of bioactivation experiments and preliminary results of bioaugmentation experiments (2 months) in laboratory pilot scale reactor columns are presented.

2 MATERIALS AND METHODS

2.1 *Bioactivation set-up*

Bioactivation was performed in 750 ml glass reactors sealed with teflon lined screw caps. Cubes of granular peat moss (1-2 cm^3) were saturated with Postgate medium B (Postgate 1984) for 24 h. NaOH 5N was added to adjust the pH to 5.5. The SRB source was a sediment collected in an inactive mine area in Eastern Townships (Quebec). The solid portion of a 250 ml centrifuged suspension (11 000 g) of this sediment was then added to the reactor as the SRB source. Nitrogen was sparged (1 l/min) into the reactor until the oxido-reduction potential (ORP) reached –100 mV. Mininert® valves were used for sample collection. The reactors were incubated at room temperature (22 ± 1°C). All batch experiments were conducted using 4 replicates with an abiotic control. $HgCl_2$ (0.5 g/l) and NaN_3 (0.5 g/l) were added to the abiotic control and an additional amount of $HgCl_2$ was added as and when necessary.

During bioactivation, the reactors were spiked periodically with fresh lactate, sulfate and nutrients. When sulfate concentration was lower than 500 mg/l, a sulfate spike was added. Sulfate salts ($CaSO_4$, $FeSO_4$ and $MgSO_4$) were added with lactate when necessary. Essential nutrients, ammonium chloride and potassium phosphate, were added with every spike. Reactor preparation and sample collection were performed in an anaerobic glove box filled with nitrogen.

2.2 *Bioaugmentation set-up*

Bioaugmentation experiments are currently being performed in three laboratory pilot-scale (9 l) continuous column reactors filled with a reactive mixture containing various organic (wood chips, leaf compost, poultry manure) and inorganic materials (sand, gravel, limestone). The composition of the reactive mixtures, which is based on Cocos et al. (2000), is shown in Table 1.

Table 1. Composition (wt %) of column reactive mixtures.

	Bioactivated reactor	Inoculated reactor	Control reactor
Wood chips	18	18	24
Leaf compost	18	18	24
Poultry manure	16	16	22
Urea	3	3	4
Sand	6	6	8
Limestone	3	3	4
Gravel	10	10	14
Bioactivated GPM	26	-	-
Anoxic sediment	-	26	-

The first column was inoculated with the bioactivated GPM, the second with an anoxic sediment collected at the mine site described previously while the third consisted of a control reactor (reactive mixture without inoculum). A synthetic acid mine drainage was used as the influent. AMD composition was based on the chemical characterization of an effluent leaching from the mine site where the SRB consortium was collected. A fresh solution of synthetic AMD was prepared with distilled water and sulfate salts twice a week. Average metal and sulfate concentrations were 1 mg/l copper (Cu), 15 mg/l zinc (Zn), 300 mg/l iron (Fe) and 1 500 mg/l sulfate (SO_4^{2-}). The solution had a pH of 3.8-4.0. To ensure reducing conditions in the reactive mixture, the columns were saturated with Postgate B culture medium and incubated at room temperature for 1 week. AMD was then fed to the columns at a constant flowrate with a peristaltic pump. To exclude air, the columns were completely saturated with liquid. The flowrate was adjusted to ensure a minimum residence time of 12 hours in the columns. Results obtained after 60 days of operation are presented.

2.3 *Sampling and analysis*

During bioactivation and bioaugmentation experiments, pH (Orion ROSS 8103 BN electrode), redox potential (Accumet electrode), dissolved sulfide (methylene blue method), and ferrous iron (phenanthroline method) were analyzed immediately after sample collection. Filtered samples (0.45 µm cellulose acetate filter paper) for lactate and sulfate determinations were stored at 4 °C before analysis. A spectrophotometer (HACH DR/2010) was used for ferrous iron, lactate, sulfate and sulfide analysis (HACH Procedure Manual 1998). Calcium, copper, total iron and zinc concentrations were determined by atomic absorption spectrophotometry (AAS) on filtered (0.45 µm) and acidified (HCl) samples.

In the first week of bioactivation, samples were collected every 2 to 4 days. Then, as the sulfate reduction rate increased, samples were collected every day for pH, lactate and sulfate analysis. Iron concentration was analyzed before and after each spike of sulfate salts.

During the bioaugmentation experiment, feed and effluent samples were collected every week. Ferrous and total iron, sulfate, calcium, zinc, copper, pH and ORP analysis were performed on every sample. Sulfide concentration was analyzed in effluent samples only.

3 RESULTS

3.1 *Bioactivation*

Figure 1a shows a typical sulfate reduction pattern observed in one of the four batch reactors (reactor 2). In this reactor, the initial sulfate reduction rate was 67 mg/l·d. After each spike of sulfate, it increased and reached a maximum of 233 mg/l·d. Figure 1b presents the evolution of sulfate concentration in the abiotic control (reactor 5). Unlike other reactors, sulfate was not depleted and its concentration increased after each spike. $HgCl_2$ had to be added periodically in order to maintain abiotic conditions in the control reactor.

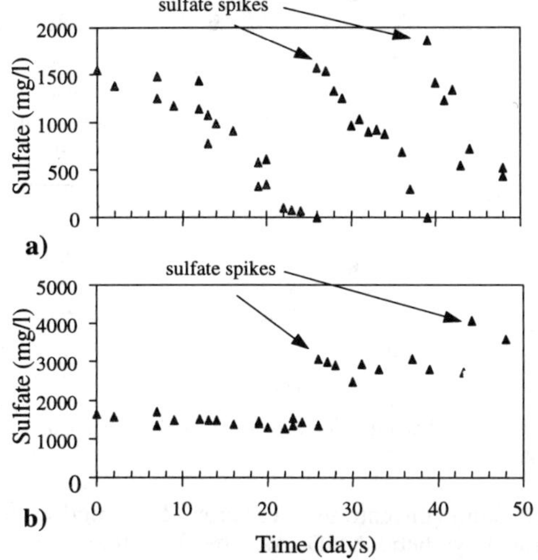

Figure 1. Sulfate concentration versus time in batch bioactivation reactor : a) reactor 2; b) reactor 5 (abiotic control).

Table 2 shows initial and maximum rates of sulfate reduction, number of sulfate spikes, and time needed to obtain the maximum rate for each reactor. Sulfate reduction rates were estimated with linear least square regression performed on the experimental curves. Average coefficient of determination (r^2) was 0.80. Higher sulfate reduction rates were obtained in reactors 2 and 4, although all reactors contained the same inoculum and media. The inoculum used for bioactivation experiments was a creek sediment. Its heterogeneous characteristics could explain the differences observed. The inoculum used in reactor 1 and 3 could have had a lower sulfate reducing bacteria density, resulting in a lower sulfate reduction activity.

Table 2. Sulfate reduction rates reached in the reactors after bioactivation.

Reactor	Sulfate reduction rate *		Time	Sulfate
number	initial	maximum	(days)	spikes
1	55	120	35	1
2	67	233	41	2
3	71	131	51	2
4	82	185	25	1
5**	0	0	-	2

* sulfate reduction rates in mg/l·d.
** abiotic control

After 26 days, pH in the reactors (1, 2, 3 and 4) was stable at 6.9. This typical increase in pH due to HCO_3^- formation was expected (Postgate 1984; Gibson 1990). Quite the reverse was observed in the abiotic control (reactor 5), where an average pH of 4.8 was measured. This acidic pH in addition with sulfate accumulation in the abiotic reactor strongly suggest that sulfate reduction and pH increase in other reactors (1, 2, 3 and 4) is the result of bacterial activity.

3.2 *Bioaugmentation*

After two months of continuous operation, results show a similar behavior for the three column reactors. The pH of synthetic AMD increased from 3.9 to more than 7.5. Negative ORP values in the effluent show that reducing conditions were maintained in the three column reactors (Table 3). Average pH and ORP values for the control reactor (mixture without inoculum) show that the reactive mixture provides a high reducing and alcaline environment favorable for the growth of an SRB population.

Table 3. Average pH and ORP of feed and effluent of the three columns after 56 days of operation.

	pH		ORP (mV)	
Column	in	out	in	out
Bioactivated	3.9	7.6	275	- 97
Inoculated	3.9	7.9	275	- 118
Control	3.9	7.8	276	- 150

Figure 2 shows sulfate concentration in the three columns, influent and effluent. A significant sulfate drop was observed after 30 days of operation in the three columns. Sulfate reduction was observed in the control column even though it was not inoculated. The sulfate reduction in addition to metal removal and ORP values at which the SRB activity is optimum (Prasad et al. 1999) suggest that the sulfate was utilized following bacterial activity. An SRB consortium could have been provided by the manure or the compost of the reactive mixture.

On day 56, sulfate reduction increased in the inoculated column where more than 280 mg/l was removed, compared to 205 mg/l and 193 mg/l in the bioaugmented and control reactor column respectively. A lag phase associated with bacterial sulfate reducing activity was observed in the three reactors. This adaptation period could therefore not be avoided in the bioaugmented column.

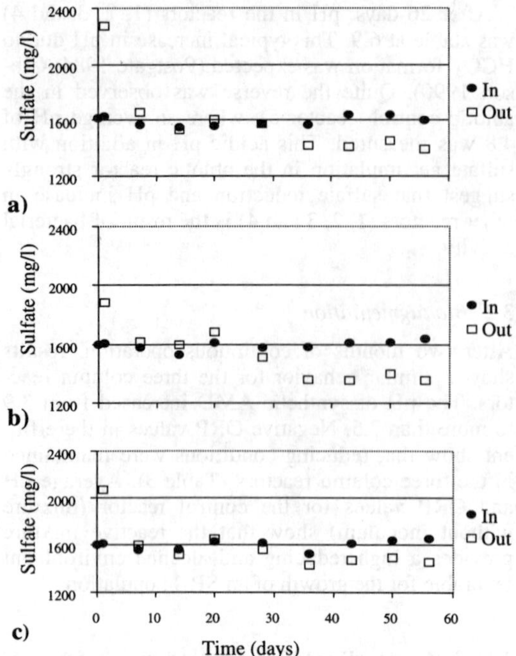

Figure 2. Sulfate concentration in reactor columns influent and effluent: a) bioaugmented column; b) inoculated column; c) control column.

Bioactivation was performed with lactate as a carbon source. An adaptation period could have been necessary for the SRB consortium to be able to use the new carbon sources of the reactive mixture. A direction for future research could be to assess bioactivation with complex carbon sources, similar to those contained in the reactive mixture.

Figures 3a and 3b present copper and zinc concentrations in column effluents. Average concentrations of copper and zinc measured in the influent are shown on the figures.

The dotted lines in Figures 3a and 3b represent respectively 97 % and 95 % copper and zinc removal efficiencies. After more than 40 days of operation, Cu removal reached 97 %. This treatment efficiency was preceded by a stabilization period where Cu removal was erratic. This phenomenon could be attributed to Cu specific affinity with soluble organic matter and colloids. Copper complexation with soluble organic matter and its strong adsorption with humic acids has been reported before (Zagury et al. 1999; Jurinak & Tanji 1993).

Figure 3b shows that zinc removal started rapidly in the three reactors. After 7 days, Zn removal reached 95 % for both inoculated and bioactivated column reactors while removal was lower in the control column.

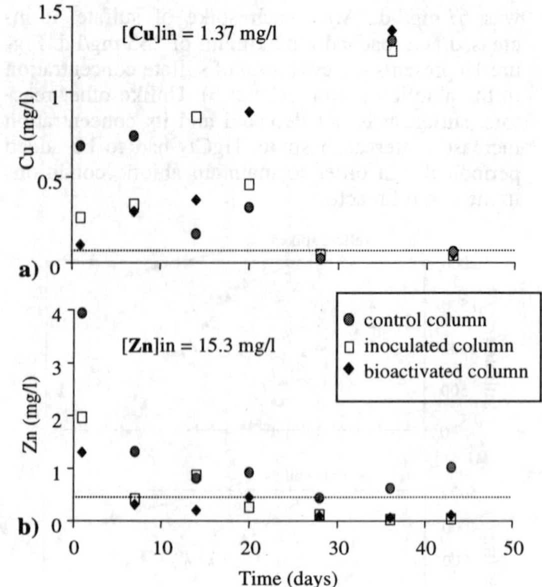

Figure 3. Metal concentrations in column effluents: a) Cu; b) Zn.

Calcium concentration was around 0.3 mg/l in the influent synthetic AMD. As a result of the presence of limestone in the reactive mixture, average calcium concentration in the effluent increased to 54, 37 and 41 mg/l for the bioaugmented, inoculated and control column respectively. During the course of the experiment, hydrogen sulfide levels in all column effluents remained under the detection limit. Black deposits in the effluent tubes and low levels of hydrogen sulfide indicate that formation of metal sulfides was occurring in the three columns.

Figure 4 shows effluent ferrous iron concentration and pH of the inoculated column. Initial pH of the effluent was 9 and it decreased to reach a stable value just under 7 after one month of operation. A similar effluent pH and Fe^{2+} concentration behavior was observed in the bioactivated column and in the control column.

Ferrous iron concentration behavior can be divided in two phases. Fe^{2+} concentration first started to decrease as pH decreased. This phenomenon can be attributed to ferrous hydroxide/Fe^{2+} equilibrium which is strongly pH dependant. It can also be attributed to Fe^{2+} adsorption on organic matter. The following phase was a Fe^{2+} concentration increase in the column effluent. This increase was associated with a pH stabilization around 7. Organic matter cations adsorption capacity is lower at low pH values (McLean & Bledsoe 1992).

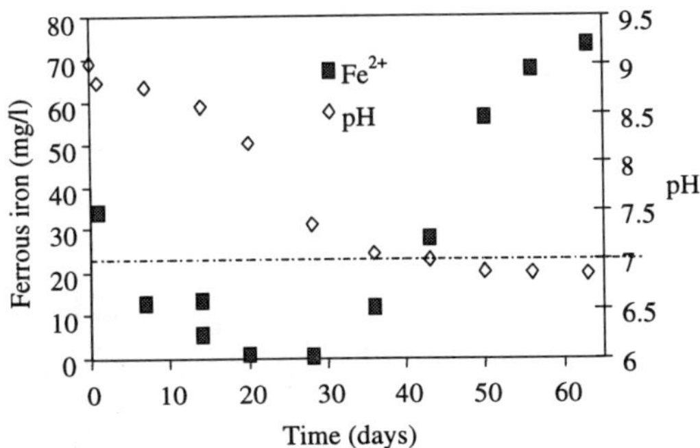

Figure 4. Effluent Fe^{2+} concentration and pH (inoculated column).

However, as observed for Cu, the system is probably not fully stabilized and the expected third phase should be ferrous sulfide formation.

4 CONCLUSIONS

Up to this point, the following conclusions can be drawn from the bioactivation experiments and the bioaugmentation preliminary results.

- Bioactivation increased the average bacterial sulfate reduction rate from 69 to 167 mg/l·d. The maximum sulfate reduction rate was 233 mg/l.d.
- Similar sulfate removal was observed in the three reactor columns with slightly higher performances in the inoculated and bioactivated reactor columns.
- Inoculation of the bioreactor column with a bioactivated GPM did not help in avoiding or minimizing the lag phase.
- The reactive mixture itself provided a high reducing and alcaline environment favorable for the establishment of an SRB activity without any inoculation.
- Zinc removal was higher than 95 % for both inoculated and bioactivated column reactors while removal was lower in the control column.
- After a stabilization period, Cu removal reached 97 % in all column reactors.

REFERENCES

Beaulieu, S., G.J. Zagury, L. Deschênes & R. Samson 1999. Use of bioactivation and bioaugmentation techniques for treating acidic metal-rich drainage. *The Fifth International In Situ and On-Site Bioremediation Symposium* 5 : 211-216, San Diego, California

Cocos, I.A., G.J. Zagury, & R. Samson 2000. Statistical design for reactive mixture selection in mine drainage treatment. In *SWEMP 2000*. Rotterdam:Balkema.(in press).

Gibson, G.R. 1990. A review : Physiology and ecology of the sulphate-reducing bacteria. *Journal of Applied Bacteriology* 69 : 769-797.

HACH Procedure Manual 1998. DR/2010 Spectrophotometer Handbook, 4ed, HACH Company, USA.

Jurinak, J.J. & K.K. Tanji 1993. Geochemical factors affecting trace element mobility. *Journal of irrigation and drainage engineering*. 119(5): 848-867.

McLean, J.E. & B.E. Bledsoe 1992. Behavior of metals in soils. *Ground Water Issue* United States Environmental Protection Agency. EPA/540/S-92/018.

Postgate, J.R. 1984. *The sulphate reducing bacteria* 2nd ed. Cambridge University Press, 208 p.

Prasad, D., M. Wai, P. Bérubé & J.G. Henry 1999. Evaluating substrates in the biological treatment of acid mine drainage. *Environmental technology* 20: 449-458.

Wildeman, T., & D. Updegraff, 1997. Passive bioremediation of metals and inorganic contaminants. *Perspectives in environmental chemistry*, Oxford University Press, pp.473-495.

Zagury, G.J., Y. Dartiguenave & J.C. Sétier 1999. Electroreclamation of heavy metals contaminated sludge: Pilot scale study. *Journal of environmental engineering*. 125: 972-978.

Environmental Issues and Management of Waste in Energy and Mineral Production, Singhal & Mehrotra (eds)
© 2000 Balkema, Rotterdam, ISBN 90 5809 085 X

Electrochemical prevention of acid mine drainage

J.H.R. Brousseau – *ENPAR Technologies Inc., Orleans, Ont., Canada*
L.P. Seed & G.S. Shelp – *ENPAR Technologies Inc., Guelph, Ont., Canada*
M.Y. Lin – *Geotechnical Research Centre, University of Western Ontario, London, Ont., Canada*
J.D. Fyfe – *Falconbridge Limited, Sudbury Operations, Onaping, Ont., Canada*

ABSTRACT: One of the most serious problem facing the mining industry is the oxidation of sulphide rich mine tailings and sulphide rich orebodies that ultimately results in the release of metal-rich acidic waters. Enpar Technologies patented electrochemical protection provides a new cost effective engineering solution to the acid drainage problem. This new approach can complement conventional strategies or it can be used on its own to mitigate AMD/ARD. By using electrochemistry, the thermodynamic stability of sulphide rich minerals is significantly improved. This is possible because sulphide rich minerals frequently exhibit semiconductor properties. Electronegative polarization is used to prevent the oxidation of sulfide rich mineral deposits in a manner similar to that used in cathodic protection of steel structures. The resulting cathodic polarization leads to *in-situ* electrochemical reduction of oxygen at the tailings/soil interface and a shift of the redox potential of the sulphide minerals to a stability field where oxidation is not thermodynamically favoured. Results from a field installation at a mine tailings site near Sudbury is presented in this article along with a discussion on future field application systems.

1 INTRODUCTION

Acid mine drainage (AMD) resulting from the disposal of mining waste containing sulphide minerals is considered one of the largest environmental problems facing the mining industry today. AMD is formed as a result of the oxidation of sulphide minerals that ultimately results in the release of metal-rich acidic waters. The oxidation of iron sulphides is catalyzed by bacteria and proceeds according to the following reaction:

$$2FeS_2 + 7.5O_2 + 7H_2O \rightarrow 2Fe(OH)_3 + 4H_2SO_4 \quad (1)$$

where FeS_2 represents pyrite.

The conventional method of dealing with the AMD problem involves treatment by liming. More recently geotextile liners and engineered soil covers have been developed to prevent the infiltration of oxygen to the tailings to reduce the oxidation rate of the tailings.

ENPAR Technologies Inc., based in Guelph, Ontario, Canada, has developed and patented a novel electrochemical technology for the prevention of AMD from tailings deposits. This new approach involves negatively polarizing the tailings/electrolyte interface, thus, reducing dissolved O_2, *in situ*, at the surface of the tailings. The end result is to stop the oxidation of the tailings. A modified design (patent pending) has been developed for low level sulphide wastes including waste rock.

2 SCIENTIFIC BASIS OF THE TECHNOLOGY

The scientific principles behind implementation of the technology were establish through a series of laboratory experiments. Sulphide minerals are known to behave as semi-conductors; high-sulphide bedrock has been shown to act as an electrode in an electrochemical cell. Laboratory studies and field tests conducted by ENPAR demonstrated that high sulphide tailings can act as an electronic conductor. Therefore, large bodies of sulphide rich tailings are amenable to electrochemical manipulation. Basically, the tailings deposit can be converted into the cathode of an electrochemical cell.

Figure 1 illustrates a schematic of an electrochemical system utilizing tailings as the cathode. Electrons are transferred to the tailings through the electrical circuit, negatively polarizing the tailings/electrolyte interface, reducing the redox potential of the tailings.

Oxygen is reduced, *in situ*, at the tailings/electrolyte interface through the electrochemical reduction of O_2 as described by the following reaction:

$$2H_2O + 4e^- + O_2 \rightarrow 4OH^- \quad (2)$$

The rate at which the electrochemical reduction reaction proceeds is controlled predominantly by the applied current density which is related to the rate at which oxygen diffuses through the soil cover or overburden to the tailings. As well as removing dissolved oxygen, hydroxyl ions will be generated at the tailings, therefore inhibiting the oxidation of sulphide minerals and associated release of metal ions. The negative polarization of tailings also changes the thermodynamic properties of the iron sulfide minerals in the tailings to more stable conditions.

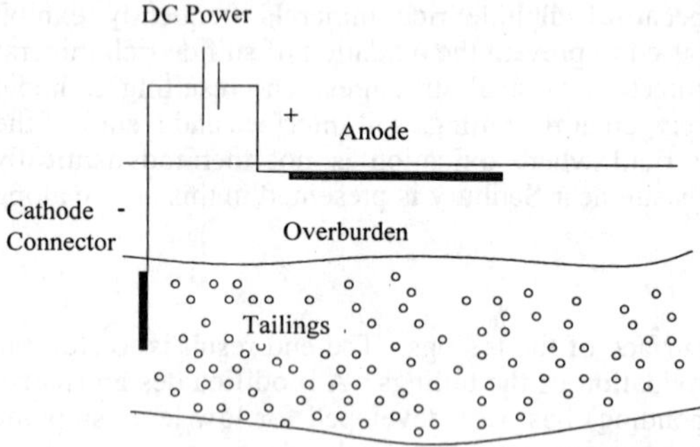

Figure 1: Schematic of Electrochemical Protection System

2.1 *Monitoring Oxygen, In Situ*

The aforementioned electrochemical principles can be exploited as a method of determining O_2 concentration at the tailings/overburden interface. Since the tailings is acting as a single large conductor, it is possible to survey the impact of the electrochemical treatment on the tailings by conducting half-cell potential surveys.

A half-cell potential survey consists of a series of field potential measurements between the conductive tailings body (which behaves as a metallic structure) and a reference electrode placed in contact with the earth directly above the tailings under investigation. The reference electrode commonly used for such surveys is a copper-copper sulphate reference electrode. It consists of a copper rod in a saturated solution of copper sulphate crystals. This reference cell has a stable, constant potential against which the potential of the tailings is measured. A schematic of how the measurement is recorded is presented in Figure 2. The potential that is recorded is a measure of the redox potential of the tailings which is an indirect measure of O_2 concentration.

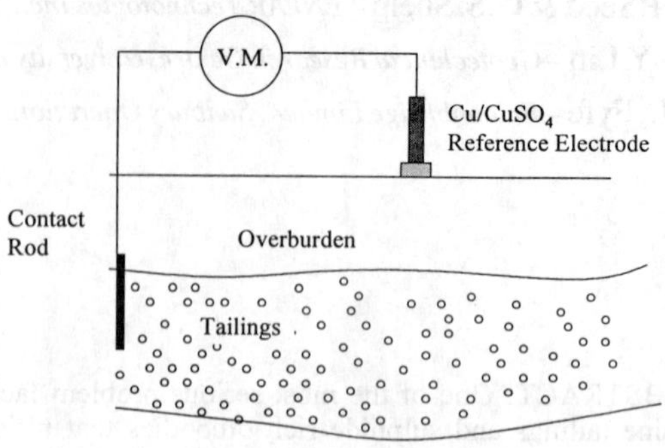

Figure 2: Half-Cell Survey of Tailings

3 FIELD TESTING OF ELECTROCHEMICAL SYSTEM

3.1 *Installation*

In the spring of 1998 ENPAR installed a pilot-scale electrochemical treatment system at a tailings deposit in the Sudbury basin (Ontario, Canada). The tailings deposit encompasses a 5 ha area and consists primarily of pyrrhotite at a concentration of approximately 60-70%. The tailings is overlain with a 2-2.5 m deep cover comprised of glacial fluvial sands and gravel. Oxidation of the tailings and subsequent contamination of the groundwater with acidic leachate were encountered prior to installation of the treatment system.

After initial testing and characterization of the tailings pond, a 1.2 ha plot was selected for use as a test area for the system. The remaining area of the tailings deposit acts as the control and was also monitored during the pilot test. Anodes were installed in the overburden, contact was made to the tailings body, and then DC power was supplied to the system using a rectifier.

3.2 *Results*

A first static half-cell potential survey of the tailings site was performed prior to installation of the electrochemical treatment system in the spring of 1999. The results presented in Figure 3 show that the redox potential (unpolarized potential) of the tailings was in a positive potential range for most of the site.

After the electrochemical treatment system was energised for approximately one month, half-cell potential measurements were recorded immediately after current interruption "instant off" (Figure 4).

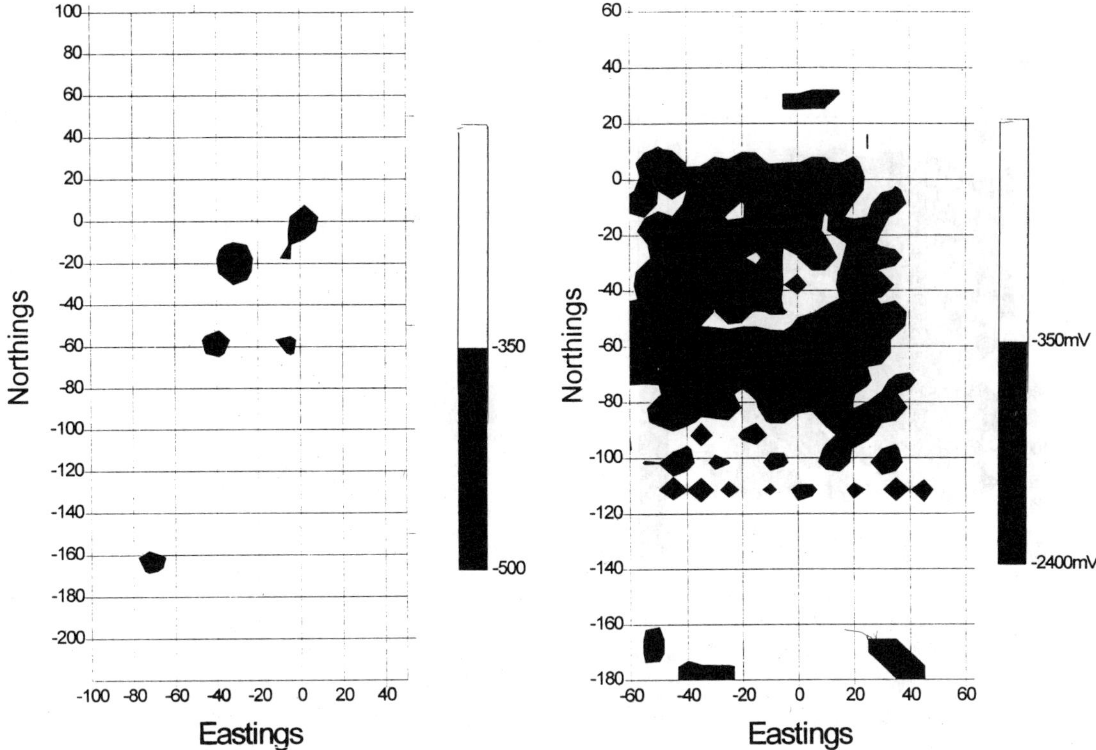

Figure 3: Static Measurement - May 22, 1999

Figure 4: Instant Off Measurement - Aug. 4, 1999

The potentials in the treatment area (bounded by 0 to100 south and 60 east to 40 west) are much more electronegative than the original static potentials and the untreated tailings area. The half-cell potential measurements of the tailings indicated that oxygen has been depleted at the tailings (by greater than 99.9%) and that the tailings are thermodynamically more stable (i.e. less likely to oxidize).

Another electrochemical indication that we have successfully induced cathodic polarization on the tailings is presented in Figure 5 which shows the difference in potential between the instant off potential and the fully depolarization survey which was recorded 18 hours after current interruption. In the treated area most of the potential has depolarized by more than 50 mV which shows that we have successfully induced cathodic polarization to the tailings and promoted the beneficial reactions discussed in the introduction section.

3.3 *Leachate Monitoring*

Lysimeters were installed in the unsaturated zone of the tailings throughout the control and treatment area. Changes in pore water chemistry are expected to be detected gradually once the contaminated pore water is displaced within the treatment zone. Preliminary leachate results sampled 3 months after start-up indicated an increase in porewater pH in the treatment zone of 0.7 units to a pH of 4.14. Leachate at the test site will be sampled during the following year.

3.4 *Power Requirements*

The test site is currently operating at a power consumption of approximately 9.6 kW. However, the system is installed at a site with an overburden characterized by high electrical resistivity (gravel and sand) of greater than 500,000 ohm-m. Future installations at sites undergoing the development of decommissioning plans will include a low resistivity cover (less than 10,000 ohm-m), such as silt, clay or silt loam. The lower resistivity overburden (i.e. soil) would decrease power costs significantly. An electrochemical system can be designed to utilize solar panels to provide DC power. As an alternative, under certain conditions, a galvanic system requiring no external power source can be designed and implemented. The design of an electrochemical system is site specific and depends primarily on site characteristics.

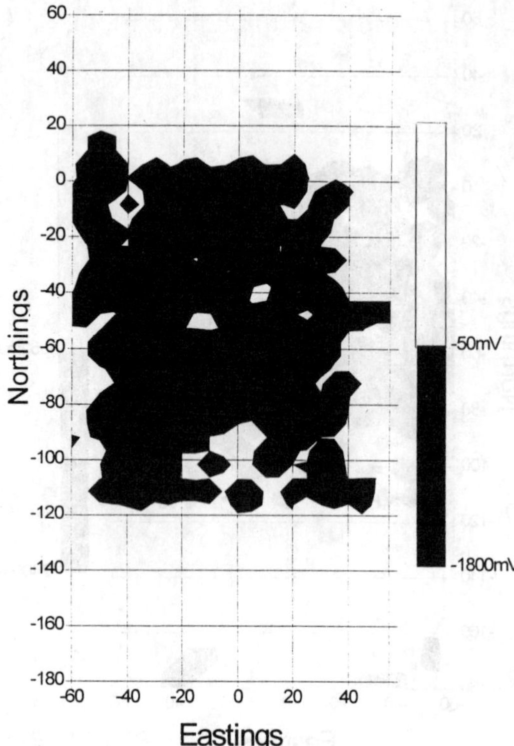

Figure 5: Polarization Based on 18 h Off

4 FUTURE DIRECTIONS

The field site will be continued to be monitored during the following year. An electrochemical treatment system has been developed for the remediation of low sulphide tailings and waste rock and is progressing to pilot-scale testing.

5 SUMMARY

A novel electrochemical treatment system to prevent the formation of acid mine drainage is currently undergoing pilot-scale testing at a tailings deposit. The electrochemical technology has been successful in removing dissolved oxygen and lowering the redox potential at the tailings/electrolyte interface. Also, by negatively polarizing the tailings, a thermodynamically stable environment is created within the tailings. By removing oxygen *in situ*, the technology tackles the problem at the source, thereby, shutting down the formation of AMD. The technology is robust and modular and can be used in conjunction with a low cost soil cover to remediate and revegetate a tailings deposit.

Environmental Issues and Management of Waste in Energy and Mineral Production, Singhal & Mehrotra (eds)
© 2000 Balkema, Rotterdam, ISBN 90 5809 085 X

Statistical design for reactive mixture selection in mine drainage treatment

I.A.Cocos, G.J.Zagury & R.Samson
NSERC Industrial Chair in Site Remediation and Management, École Polytechnique de Montréal, Qué., Canada

ABSTRACT: Generation of acid mine drainage (AMD) results from the biological and physico-chemical oxidation of sulfide minerals. The use of a sulfate-reducing reactive wall in the path of AMD contaminated groundwater is recognized as a cost-effective alternative method. However, a rigorous method for reactive mixture selection still needs to be investigated. The aim of this study was to select the most appropriate reactive mixture using a multiple factor design based on sulfate reduction rate. Mixture proportions (wt %) were settled for wood chips (0-30) poultry manure (10-20) leaf compost (0-30) and for two types of porous support: oxidized tailings (0-8) and sand (0-8). The reactivity of 17 mixtures was assessed using a synthetic AMD. Results indicate that within 41 days, sulfate concentrations decreased from initial concentrations of 2000-3200 mg/l to < 90 mg/l. Sulfate reduction rates ranged from 35.6 mg/l·d to 156.3 mg/l·d. The most efficient reactive mixture in terms of sulfate reduction rate was selected.

1 INTRODUCTION

Acid mine drainage (AMD) is a major pollution problem associated with the mining industry. Atmospheric oxidation of sulfide ores in tailings impoundment involves biological and physico-chemical oxidation processes which lead to the generation of an acidic leachate plume adversely affecting both surface and ground waters (Gray 1998). Moreover, AMD can cause widespread and often intermittent pollution (Kelly 1988).

Current techniques for AMD treatment include minimizing sulfide oxidation, collection and treatment of the acidic leachate, and preventing the migration of tailings pore water. Biological approach to AMD treatment, which is a possible alternative to chemical treatment, intend to induce sulfate-reducing conditions particularly by adding an organic waste product to stimulate the activity of sulfate reducing bacteria (SRB). Under anaerobic conditions, SRB oxidize simple organic compounds (CH_2O) with sulfate and thereby generate hydrogen sulfide and bicarbonate ions (Dvorak et al. 1992, Blowes et al. 1995):

$$SO_4^{2-} + 2CH_2O \xrightarrow{SRB} H_2S + 2HCO_3^- \quad (1)$$

Furthermore, the dissolved sulfide binds with most metals released from AMD to form insoluble metal sulfide according to the reaction:

$$H_2S + Me^{2+} \longrightarrow MeS + 2H^+ \quad (2)$$

The heavy metal confinement technique based on sulfide precipitation is more commonly used for AMD treatment over the conventional hydroxide precipitation because of the highest degree of metal removal at low pH (pH 3 to 6). Meanwhile, the sludge characteristics are improved i.e. chemically more stable, denser and less voluminous sludge (Peters & Ferg 1987).

The use of a sulfate-reducing reactive wall, which is installed below the ground surface in the path of contaminated groundwater, is being more and more recognized as an alternative and cost effective method (Benner et al. 1997, 1999). Waybrant et al. (1998) assessed the reactivity of 8 organic-carbon reactive mixtures containing four components: an organic source, a bacterial source, a neutralizing agent, and a porous medium. The reactivity of the mixtures varied with those containing several organic sources being most reactive. However, a rigorous and methodical selection of the most reactive mixture, and the influence of each component on the reactivity still needs to be investigated.

The objective of this study was to select the most appropriate reactive mixture using a multiple factor design based on sulfate reduction rates. For this purpose, mixture proportions were settled for three different organic substrates and for two porous media, while other component proportions (bacterial source, neutralizing agent, and urea) remained constant.

2 MATERIALS AND METHODS

2.1 *Experimental set-up*

A synthetic acid mine drainage was used to ensure constant quality of water throughout the study. Simulated AMD composition (Table 1) was based on the chemical characterization of a groundwater plume discharging from the study mine site (Abitibi, Quebec). Required quantities of sulfate salts of Fe, Cu, Ni, Zn, Co, Mn, and Na were added in a 1 litre volumetric flask and made up to volume with deionized water. 6N NaOH was slowly added to buffer the solution until the final pH was around 5.5-6.

Table 1. Composition of synthetic acid mine drainage.

Element	Desired concentration (mg/l)	Chemical used	Amount of chemical added per litre (mg)
Fe	800	$FeSO_4 \cdot 7H_2O$	3971
Cu	0.08	$CuSO_4$	0.2
Zn	7.3	$ZnSO_4 \cdot 7H_2O$	32
Ni	0.8	$NiSO_4 \cdot 6H_2O$	3.5
Co	0.89	$CoSO_4 \cdot 7H_2O$	4.2
Mn	47	$MnSO_4 \cdot H_2O$	144.5
SO_4	2940	Na_2SO_4	1093
pH	5.5-6.0	-	-

Among the carbon sources already studied for bacterial activity proliferation, two main groups were retained: cellulosic wastes (wood chips, sawdust, bark, maple leaves), and organic wastes (manure, dehydrated whey, starch). Results of previous research (Kuyucak & St-Germain 1994) indicated that cellulosic waste alone would not sustain SRB growth. Consequently, a mixture of cellulosic and organic wastes was considered. Wood chips, leaf compost, and poultry manure were selected for their ability to support sulfate reduction. Moreover, their different composition in terms of complex and simple organic compounds should ensure shorter and longer periods of time for this degradation. Wood chips and leaf compost were obtained from a local provider, while a local poultry farm delivered the manure. The upper and lower percentages (wt %) for these components in the mixtures were:

$$0 < \text{wood chips } (X_1) < 30 \tag{3}$$

$$0 < \text{leaf compost } (X_2) < 30 \tag{4}$$

$$30 < X_1 + X_2 < 40 \tag{5}$$

$$10 < \text{poultry manure } (X_3) < 20 \tag{6}$$

Two types of porous support (reactive or inert) were used in order to assess their different influence on bacterial activity. The reactive porous support (oxidized tailings), was obtained from an inactive mine, and clean silica sand was obtained from a local stone quarry. The constraints (wt %) imposed for oxidized tailings and silica sand were:

$$0 < \text{oxidized tailing } (X_4) < 8 \tag{7}$$

$$0 < \text{silica sand } (X_5) < 8 \tag{8}$$

$$5 < X_4 + X_5 < 8 \tag{9}$$

SRB source used was a creek sediment (the presence of SRB was suspected by the black color and the H_2S odor of the sediment) sampled from the anaerobic zone of an AMD-affected stream in an inactive mine area. Enumeration of SRB in the sediment using Most Probable Number technique (ASTM 1990) showed a SRB concentration of 30×10^3 cell./100 ml. The added percentage of bacterial source was 37 wt %, and was constant for all mixtures.

A neutralizing agent (2 wt % limestone) was added to each mixture in order to maintain near neutral conditions (Blowes et al 1995).

Urea, a chemical which has a high percentage of nitrogen (65 %), was added to each mixture in order to reach an appropriate C: N: P ratio of 110:7:1 to support high SRB activity (Gerhardt 1981). The percentage of urea was constant (3 wt %) for all mixtures.

After the mixtures were added, all the flasks were filled with 300 ml of deoxygenated synthetic mine drainage.

The 5 factors mixture design yielded 17 mixtures to be tested (Table 2), after all constraints on mixture proportions were considered. Consequently, 17 batch reactors were run in duplicate in order to assess the efficiency of each of the reactive mixtures in terms of sulfate reduction and heavy metal precipitation.

2.2 *Sampling and analysis*

Laboratory batch experiments using 500 ml glass reaction flasks were conducted at room temperature ($22 \pm 1^{0}C$). Reaction flasks sampling ports were fitted with Teflon-lined septa so that substances could be removed by piercing with a syringe. Moreover, sampling sessions were conducted in an anaerobic glovebox.

Mine drainage samples were collected on days 0, 5, 9, 13, 17, 21, 25, 33, and 41. To obtain representative samples from water phases on day 0, reactive mixtures were allowed to settle for approximately 12 h before sampling. Measurements of pH (ORION ROSS 8175 BN electrode), redox potential (ACCUMET electrode), alkalinity (2N H_2SO_4 titration), dissolved sulfide (Methylene Blue Method) and Fe^{2+} (1.10 Phenanthroline Method) were determined immediately after sample collection. The pH electrode was calibrated using pH 4.0, 7.0 and 10.0 buffer solutions, and the ORP electrode performance was confirmed using pH 4.0 and pH 7.0 quinhydrone standard solutions.

pH and ORP were analyzed directly in the sampling solution according to Standard Methods (APHA 1985a) while alkalinity, dissolved sulfides and Fe^{2+} concentrations were determined on a filtered (0.45µm cellulose acetate filter paper) sample.

Filtered samples were used for Ni, Cu, Zn, and SO_4^{2-} determinations. TOC concentration was determined using a Total Carbon Analyzer by the Persulfate-Ultraviolet Oxidation Method (APHA 1998b) and nitrogen was determined by the Total Kjeldahl Method (APHA 1998b) on a H_2SO_4 preserved sample.

Cu, Ni, and Zn concentrations were determined by atomic absorption spectrometry (AAS) and SO_4^{2-}, iron, and sulfide concentrations were analyzed by spectrophotometry using standardized substances and an Hach digital spectrophotometer DR/2010 (HACH Procedure Manual 1998).

3 RESULTS

3.1 *Sulfate-reduction rates*

Batch experiments were run for 41 days. During the course of the experiment, sulfate concentrations decreased from 2020 mg/l - 3254 mg/l to < 90 mg/l depending on mixture. Sulfate-reducing conditions developed rapidly in all mixtures after an initial acclimation period, which ranged from 0 to 21 days (Fig. 1).

Sulfate-reduction rates were calculated using linear least-square regression analysis for the entire period of study, $T_{0\text{-}41}$ (Table 3). The average sulfate removal rate for the 17 reactors was 70.7 mg/l·d and the coefficients of determination (r^2) ranged from 0.82 to 0.98.

A second sulfate-reduction rate, $T_{modified}$, was calculated using linear least-square regression dis-

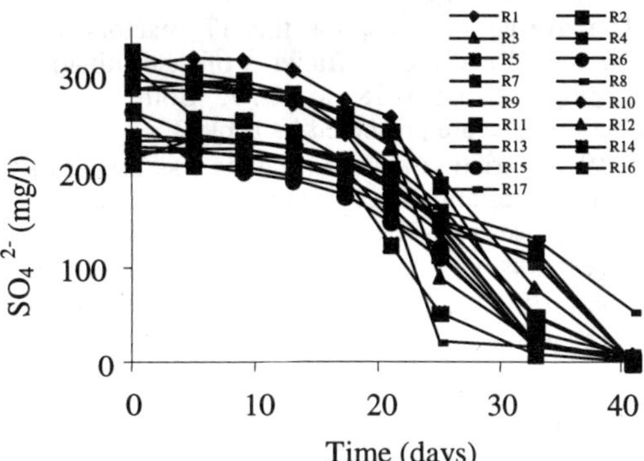

Figure 1. Sulfate concentration versus time for batch mixtures.

Table 2. Composition of reactive mixtures (R_1 – R_{17}) used in batch experiments (wt %).

Variable components*	R_1	R_2	R_3	R_4	R_5	R_6	R_7	R_8	R_9	R_{10}	R_{11}	R_{12}	R_{13}	R_{14}	R_{15}	R_{16}	R_{17}
Cellulosic wastes																	
Wood chips	30	30		-	30	10	30	10	30	30	3	3	30	30	10	10	17.87
Leaf compost	-	-	30	30	10	30	10	30	3	3	30	30	10	10	30	30	17.87
Organic wastes																	
Poultry manure	20	20	20	20	10	10	10	10	20	20	20	20	13	13	13	13	15.75
Oxidized mine tailings	8	-	8	-	8	8	-	-	5	-	5	-	5	-	5	-	3.25
Silica sand	-	8	-	8	-	-	8	8	-	5	-	5	-	5	-	5	3.25

*Constant proportion components: bacterial source 37 wt %, neutralizing agent 2 wt % and urea 3 wt %.

Table 3. Calculated sulfate reduction rates (mg/l·d) for the 17 batch reactors and their duplicates.

Reactor	T(0-41)	Tmodified	Reactor	T(0-41)	Tmodified
1.1	85.8	115.1	1.2	77.9	104.2
2.1	86.5	100.3	2.2	87.6	104.9
3.1	85.4	128.2	3.2	70.8	97.2
4.1	78.0	74.6	4.2	79.8	83.0
5.1	53.2	82.3	5.2	57.8	78.9
6.1	66.5	68.2	6.2	64.6	64.3
7.1	63.1	63.1	7.2	58.4	79.7
8.1	35.6	45.9	8.2	52.0	82.6
9.1	77.7	100.7	9.2	79.1	118.1
10.1	84.8	132.2	10.2	84.0	125.5
11.1	81.6	129.4	11.2	77.0	119.9
12.1	83.6	153.5	12.2	84.3	156.3
13.1	61.6	95.8	13.2	64.0	97.2
14.1	71.7	131.8	14.2	67.8	102.7
15.1	63.4	93.1	15.2	63.0	96.1
16.1	58.8	92.6	16.2	58.8	107.0
17.1	70.7	150.2	17.2	67.6	150.3

regarding early-time data, which may have been affected by acclimation periods and adsorption, and late-time data which may have been sulfate limited (Waybrant et al. 1998).

The average $T_{modified}$ for the 17 reactors was 103.7 mg/l·d and the coefficients of determination (r^2) ranged from 0.78 to 0.99. Both sulfate-reduction rates are presented in Table 3.

A 10 % variation between values was used as a criterion to establish lag-phase and final sulfate-limited phase, considering that spectrophotometer precision was 9.8% for sulfate analysis. Both rates will be further used in the statistical interpretation of the results.

3.2 *Heavy metal removal and change in pH*

Dramatic decreases were observed for the assessed metal (Cu, Ni, and Zn) concentrations. Results showed that around 90 % of total metal removal was effected within 17 days. Average heavy metal removal percentages calculated for all mixtures were 96 % for Cu, 69 % for Ni and 88 % for Zn.

After 17 days, pH was stable near 7.9 for all mixtures while final average ORP was -353 mV suggesting strong sulfate-reducing conditions.

Reduction of the initial concentration of sulfate (≈ 2020 mg/l) should generate, according to equation 1, a theoretical amount of alkalinity of 2567 mg/l as HCO_3^-, and a theoretical amount of H_2S of 715 mg/l. The final measured average alkalinity for the reactors was 187.2 mg/l $CaCO_3$, and the final average dissolved sulfide concentration was 4.5 mg/l. However, according to equation 2, dissolved sulfides bind with most metals released from AMD to form insoluble metal sulfides. This could explain the low dissolved sulfide concentration measured. Sulfate-reduction rates in addition with metal removal percentages and ORP values, at which the SRB activity is optimum (Prasad et al. 1999) strongly suggest that the substrate was utilized following bacterial activity.

3.3 *Preliminary statistical analysis*

A preliminary statistical analysis of the results was carried out using Quality Control and Experimental Design operating module of STATISTICA software (StatSoft, Inc. 1995). Imposed variables to produce the X-bar charts (Fig. 2) were sulfate reduction rates, either $T_{0\text{-}41}$ or $T_{modified}$, while variables with sample identifiers were the 17 mixtures.

Sulfate-reduction rates varied between 35.6 mg/l·d and 87.6 mg/l·d for $T_{0\text{-}41}$, and between 45.9 mg/l·d and 156.3 mg/l·d for $T_{modified}$ depending on mixture composition (Table 3). Repeatability between duplicates was good with an average range of 4.1 mg/l·d for $T_{0\text{-}41}$ and an average range of 11.7 mg/l·d for $T_{modified}$. Reactors 3 and 8 had a lower repeatability with ranges between duplicates slightly exceeding upper control limit (13.3 mg/l·d) for $T_{0\text{-}41}$ sulfate-reduction rate.

As shown on Figure 2, the three most reactive

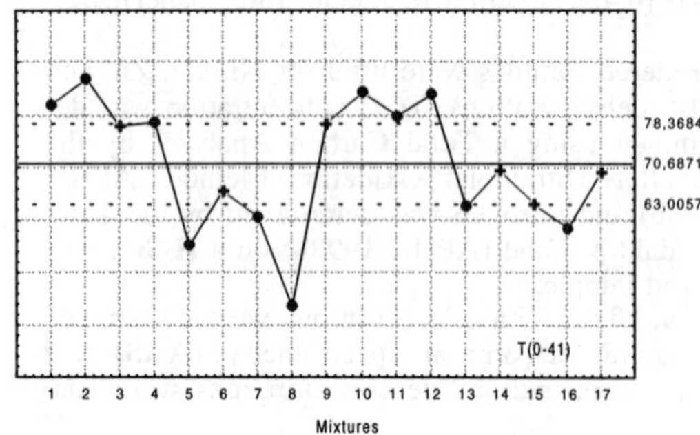

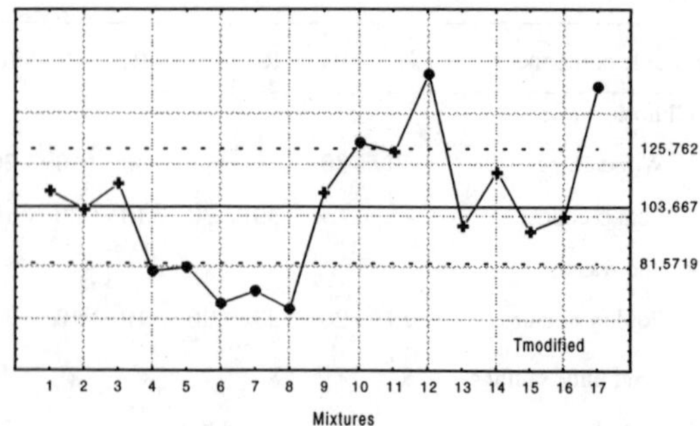

Figure 2. Mean values of sulfate reduction rates (mg/l·d).

mixtures in terms of sulfate reduction for T_{0-41} were R_2 (87.0 mg/l·d), R_{10} (84.4 mg/l·d) and R_{12} (83.9 mg/l·d). For $T_{modified}$, the most reactive mixture was R_{12} (154.9 mg/l·d). Since the second calculated sulfate-reduction rate better reflects the sulfate reduction process by excluding initial and final phases, and since R_{12} reactor had a greater reactivity for both calculated rates, it was considered as the most reactive mixture. Mixture proportions of R_{12} are 3 % wood chips, 30 % leaf compost, 20 % poultry manure and 5 % silica sand. These component percentages are in accordance with preliminary results of statistical analysis, which reveals that the most important component for sulfate reduction is poultry manure. R_{12} mixture has the maximum percentage (20 wt %) of poultry manure and the lowest percentage of porous support tested in this study (5 wt %).

Statistical analysis is currently under process, and extensive results will be further published.

4 CONCLUSIONS

These results support the basic assumption of the study that mixture composition (especially component proportions) has an important impact on its reactivity.

All mixtures tested were efficient in terms of sulfate reduction and metal precipitation. Final concentrations for sulfates were lower than 90 mg/l and the average rate of sulfate-reduction was either 70.7 mg/l·d for T_{0-41}, or 103.7 mg/l·d for $T_{modified}$. However, sulfate–reduction rates varied between 35.6 mg/l·d and 87.6 mg/l·d for T_{0-41}, and between 45.9 mg/l·d and 156.3 mg/l·d for $T_{modified}$ depending on mixture composition.

The most reactive mixture in terms of sulfate reduction was R_{12}, which contains 3% wood chips, 30% leaf compost, 20% poultry manure (maximum proportion tested in the study) and 5% silica sand as a porous support.

The influence of each component on sulfate-reduction rate is currently under study. Modelling of sulfate reduction-rate is also under process and results will be published in the near future.

ACKNOWLEDGEMENTS

The authors acknowledge the financial support of the Chair partners: Alcan, Bell Canada, Canadian Pacific Railway, Cambior, Centre d'expertise en analyse environnementale du Québec, Elf-Aquitaine, Hydro-Québec, Petro-Canada, Natural Science and Engineering Research Council (NSERC), Solvay, and Ville de Montréal.

The authors also wish to thank Pr. Bernard Clément, Department of Mathematics and Industrial Engineering, Ecole Polytechnique de Montréal, for his help and useful advice during statistical design and interpretation of results.

REFERENCES

APHA 1985a. *Standard methods for the examination of water and wastewater.* 16th edition. American Public Health Association, Washington D.C.

APHA 1998b. *Standard methods for the examination of water and wastewater,* 20th edition, American Public Health Association, Washington D. C.

ASTM 1990. *Standard test method for sulfate reducing bacteria in water and water-formed deposit.* D4412-84, pp.533-535.

Benner, S.G., D.W. Blowes & C.J. Ptacek 1997. A full-scale porous reactive wall for prevention of acid mine drainage. *Ground Water Monitoring Remediation* Fall 1997: 99-107.

Benner, S.G., D.W. Blowes, W.D. Gould, R.B. Herbert & C.J. Ptacek 1999. Geochemistry of a permeable reactive barrier for metals and acid mine drainage. Environmental Science and Technology 33: 2793-2799.

Blowes, D.W., C.J. Ptacek & J.G. Bain 1995. Treatment of mine drainage water using *in situ* permeable reactive walls. *Mining and the environment*, Sudbury.

Christensen, B., M. Laake & T. Lien. 1996. Treatment of acid mine water by sulfate-reducing bacteria; results from a bench scale experiment. *Water research* 30: 1617-1624.

Dvorak, D.H., R.S. Hedin, H.M. Edenborn & P McIntire 1992. Treatment of metal contaminated water using bacterial sulfate reduction: results from pilot scale reactors. *Biotechnology and bioengineering* 40: 609-616.

Gerhardt, P. 1981. *Manual of methods for general microbiology.* Washington D.C.ASM Publication.

Gray, N.F. 1998. Acid mine drainage composition and the implication for its impact on lotic systems. *Water research* 32: 2122-2134.

HACH Procedure Manual 1998. DR/2010 Spectrophotometer Handbook, 4ed, HACH Company, USA.

Kuyucak, N. & P. St-Germain 1994. In situ treatment of acid mine drainage by sulfate reducing bacteria in open pits: scale-up experiences. *International Land Reclamation and Mine Drainage Conference*, Pittsburgh.

Kelly, M.D. 1988. Mining and the freshwater environment. London: Elsevier Applied Science.

Peters, R.W. & J. Ferg 1987. Dissolution/ leaching behavior of metal hydroxide/metal sulphide sludges from wastewaters. *Hazardous waste and hazardous materials.* 4:325-331.

Prasad, D., M. Wai, P. Bérubé & J.G. Henry 1999. Evaluating substrates in the biological treatment of acid mine drainage. *Environmental technology* 20: 449-458.

StatSoft, Inc. 1995. STATISTICA for Windows [Computer program manual]. Tulsa, OK: StatSoft, Inc., 2300 East 14th Street, Tulsa, OK, 74104-4442.

Waybrant, K.R., D.W. Blowes & C.J. Ptacek 1998. Selection of reactive mixtures for use in permeable reactive walls for treatment of mine drainage. *Environmental science and technology* 32: 1972-1979.

Widdel F. 1988. Microbiology and ecology of sulfate- and sulfur-reducing bacteria. pp. 469-586. In: A.J.B. Zehnder (ed.), *Biology of anaerobic microorganisms.* Wiley, New York.

The authors also wish to thank Dr. Bernard Clément, Department of Mathematics and Industrial Engineering, École Polytechnique de Montréal, for his help and useful advice during statistical design and interpretation of results.

REFERENCES

APHA 1985. *Standard methods for the examination of water and wastewater*. 16th edition. American Public Health Association, Washington DC.

APHA 1998. *Standard methods for the examination of water and wastewater*. 20th edition. American Public Health Association, Washington DC.

ASTM 1990. *Standard test method for [illegible] ... Annual Book of ASTM Standards*, [illegible].

Benner, S.G., Blowes, D.W. & Ptacek, C.J. 1997. A full-scale porous reactive wall for prevention of acid mine drainage. *Ground Water Monitoring & Remediation* [illegible]: 99–107.

[illegible], C., [illegible], W.D., [illegible], R.R., [illegible] & [illegible] 2000. [illegible] of a permeable reactive barrier for metals and acid mine drainage. *Environmental Science and Technology* 35: 2470–2476.

Blowes, D.W., Ptacek, C.J. & [illegible] 1995. Treatment of mine drainage water using in situ permeable reactive walls. *[illegible]*.

[illegible], [illegible], M., [illegible] & [illegible] 199[illegible]. Treatment of acid mine water by sulfate reducing bacteria: results from a bench scale experiment. *Water Research* [illegible].

Dvorak, D.H., Hedin, R.S., Edenborn, H.M. & McIntire, P.E. 1992. Treatment of metal contaminated water using bacterial sulfate reduction: results from pilot scale reactors. *Biotechnology and Bioengineering* 40: 609–616.

[illegible], R. 1991. *[illegible]*. [illegible]: CANMET publications.

[illegible], K.P. 199[illegible]. [illegible] drainage composition and the [illegible] for its impact on lake systems. *[illegible]*: [illegible].

[illegible] 1997. *[illegible]* [illegible] Company, USA.

[illegible], [illegible] & [illegible] 199[illegible]. [illegible] of acid mine drainage by sulfate reducing bacteria. [illegible]. [illegible] and [illegible]. [illegible].

[illegible], M.T. [illegible]. *[illegible] and [illegible]*. London: Elsevier Applied Science.

[illegible], R. & [illegible] 199[illegible]. [illegible] of metals and sulphate by sulphate-reducing bacteria [illegible]. *[illegible]*: [illegible].

[illegible], M. [illegible] 199[illegible]. [illegible] *[illegible] and [illegible]*. [illegible].

[illegible] 199[illegible]. [illegible] ... [illegible] ... Tokyo [illegible].

[illegible], K.R. [illegible] 199[illegible]. [illegible] of sulfate reducing bacteria [illegible]. *[illegible] and [illegible]*.

[illegible] 1988. [illegible] for sulfate reduction [illegible]. [illegible].

mixtures, in terms of sulfate reduction. For T_{10} it was R_3 ([illegible] mg/l/d), R_2 ([illegible] mg/l/d) and R_5 ([illegible] mg/l/d). For T_{20}, the most effective mixture was R_3 ([illegible] mg/l/d). Since the sulfate-reducing bacteria populate better in the sulfate reduction process by exceeding initial and final phases, this mixture has a greater reactivity for both calculated rates, it was considered as the most reactive mixture. Mixture proportions of R_3 are: 30% wood chips, 30% leaf compost, 20% poultry manure, and 20% silica sand. These component proportions are in accordance with preliminary results of statistical analyses, which reveal that the most important component for sulfate reduction is poultry manure. R_3 has the highest percentage (20 wt %) of poultry manure and the lowest percentage of porous support used in this study (20 wt %).

Statistical analysis is currently under process and extensive results will be further published.

CONCLUSIONS

The results support the basic assumption of the study that mixture composition (especially component proportions) has an important influence on its reactivity.

All mixtures used were efficient in terms of sulfate reduction and metal precipitation. Final concentrations for all metals were lower than [illegible] mg/l, and the average rate of sulfate reduction was either [illegible] mg/l/d for T_{10} or [illegible] mg/l/d for T_{20}. However, the rates of sulfate reduction varied between [illegible] mg/l/d and [illegible] mg/l/d for T_{10}, and between [illegible] and [illegible] mg/l/d for T_{20}, depending on mixture composition.

The most reactive mixture in terms of sulfate reduction was R_3, which contains 25% wood chips, [illegible] leaf compost, 20% poultry manure, the highest proportion used in the study, and 20% silica sand as porous support.

The influence of each component in sulfate reduction is currently under study. Modeling of sulfate reduction rate is also being processed and results will be published in the future.

ACKNOWLEDGEMENTS

The authors acknowledge the financial support of the [illegible] partners: Alcan, Bell Canada, Canadian Pacific Railway, Cambior, [illegible], [illegible] environmental, [illegible] Québec, [illegible], [illegible] Québec, [illegible], and Natural Sciences and Engineering Research Council (NSERC). [illegible] and [illegible].

Environmental Issues and Management of Waste in Energy and Mineral Production, Singhal & Mehrotra (eds)
© 2000 Balkema, Rotterdam, ISBN 90 5809 085 X

The use of basic additives to tailings in layered co-mingling to improve AMD control

S. Fortin, A. Lamontagne & R. Poulin
Department of Mining and Metallurgical Engineering, Laval University, Sainte-Foy, Que., Canada

N. Tassé
INRS-GeoResources, Sainte-Foy, Que., Canada

ABSTRACT: This paper presents the results of a laboratory study done to verify the concept of co-mingling of waste rock with layers of compacted tailings including a 10% mass fraction of alkaline additives. The use of basic material should help to control the generation of acidity. Co-mingling is expected to retard the onset and reduce the magnitude of acid mine drainage (AMD) production in the waste rock. The results obtained show that the co-mingling method would be efficient to mitigate AMD by maintaining nearly water-saturated conditions in compacted tailings layers. Moreover, the use of alkaline residues would improve the efficiency by providing a larger reservoir of alkalinity for neutralisation, thus ensuring prevention of AMD on a longer term.

Keywords: co-mingling, acid mine drainage, prevention/control methods, cement kiln dust, red muds

1 INTRODUCTION

Acid mine drainage is perhaps the most serious environmental problem facing the mining industry. At open pit mines, tons of waste rock have to be disposed of on mine sites. The waste rock is coarse and, unless submerged, are stored in large, partially water-saturated, porous heaps or piles. When sulphides are present, acid mine drainage (AMD) can be produced shortly after placement. Sulphide oxidation is catalyzed by bacteria and produces acidic effluent with heavy metals in solution. The oxidation of sulphides (generally pyrite) is exothermic, leading to high temperatures within the piles. Transfer of oxygen is enhanced by thermal convection, in addition to simple diffusion. When the process of AMD has begun, it is very difficult, even impossible to stop it. It is thus of great interest to prevent the onset of AMD.

This paper proposes a method aiming to prevent the apparition of AMD in a waste rock pile. The final goal is to facilitate the closure of the mine in a safe and economical way. This method proposes to limit the transfer of oxygen and water by interlayering some compacted fine material layers in a waste rock pile during its construction. The fine material considered is mine tailings which also need to be disposed of at mine sites.

In this study, the tailings used are net acid generators. Highly alkaline material was blended with the tailings in order to form a layer able to neutralize oncoming acidity and to stabilize metals through precipitation/adsorption mechanisms. The basic materials used were cement kiln dust (CKD) produced by the Portland cement industry and red muds issued from the Bayer process used in the aluminum industry. The recycling of these industrial residues in a useful manner such as the layered co-mingling promotes their valorization in the context of sustainable development.

Test columns have been carried out and this paper presents the results issued from the characterization of each material used in the columns as well as the geochemical trends identified after 9 months of monitoring.

2 ACID MINE DRAINAGE PROCESSES

The most important process within waste rock dumps is the oxidation of the pyrite present in the rocks. Oxidation involves oxygen consumption, even though the direct oxidant is often ferric iron (Fe^{3+}). A supply of oxygen is thus required to sustain pyrite oxidation. The reaction is catalyzed by bacteria such as *T. ferrooxidans* that grow upon iron and sulphur consumption at low pH values. It follows that keeping near neutral pH values and low oxygen concentrations would keep bacterial activity at a minimum, thus lowering AMD production.

Heat production also occurs since pyrite oxidation is strongly exothermic (11.7 MJ/kg of pyrite oxidized). The release of heat can drive temperature up, as high as 70°C in the heart of a pile (Gélinas et al. 1992). This increase in temperature is important since it totally modifies the mechanisms responsible for oxygen transport to oxidation sites. Initially, diffusion is the main process providing oxygen within waste rock accumulations. Diffusion is driven by gradients in oxygen concentration between the atmosphere and the partially gas filled pores. In materials of low permeability, such as in mine tailings, diffusion is believed to remain the only means of oxygen transport. However, in higher permeability materials, following an initial increase in temperature, temperature-driven convection currents are generated (Lefebvre 1994, Kuo & Ritchie 1999). The resulting advection is a much more efficient oxygen transport process than diffusion and sustains higher oxidation rates.

Finally, water infiltration occurs through the unsaturated porous material and picks up oxidation products to form an acidic leachate containing high concentrations of sulphates and metals that can contaminate both surface and groundwater.

3 DESCRIPTION OF THE LAYERED CO-MINGLING METHOD

Co-mingling is a method whereby waste rock is co-disposed of with tailings, the basic premise being that the modified waste material has a lower bulk permeability. Layered co-mingling refers to the disposal of waste rock with the addition, at predefined construction stages, of compacted layers of mine tailings. This results in an interlayered material whose bulk properties are different from those of each layers. The layered co-mingling process requires dewatering of the tailings and some means of transporting them to the disposal area. Dewatering is a prerequisite for compacting the tailings layers, which is necessary to obtain the capillary barrier effect.

Laboratory investigations have shown the potential benefits of the layered co-mingling (Poulin et al. 1996). Potential advantages over other disposal methods are discussed by Lamontagne et al. (1999). It is of particular interest to enhance the neutralization potential of the fine layers for stopping the migration of acid away from its production site. The use of other industrial wastes showing a strong alkaline character seemed well suited to attain this goal and was to be tested in the laboratory.

4 DESCRIPTION OF MATERIALS

The tailings used in the study are those from Les Mines Selbaie, Abitibi, Québec. Selbaie mine is a copper and zinc mine but silver and gold are also exploited. These tailings were chosen for the analysis because of their typical overall characteristics. The physical characteristics have been broadly described elsewhere (Poulin et al. 1998) and are no longer discussed in this paper. Mine tailings were fresh and non oxidized when placed in the experimental columns.

One of the basic material used was cement kiln dust (CKD). This is a waste residue primarily composed of oxidized, anhydrous, micron-sized particles generated as a by-product of Portland cement fabrication. CKD are collected from electrostatic precipitators during the production of cement clinker at high temperatures. Upon contact with water, high concentrations of potassium, sulphates and caustic alkalinity are leached out, with other constituents to lesser levels (Duchesne & Reardon 1998). CKD is composed of a melange of oxides, aluminosilicates, sulphates and chlorides. Some of these phases such as lime (CaO), arcanite (K_2SO_4) and sylvite (KCl) are unstable or highly soluble under atmospheric conditions. When put into contact with water, these phases will either dissolve completely or precipitate under more stable secondary phases. The CKD used for this study was provided by Ciment St-Laurent, Joliette, Canada.

The other basic material used were red muds, which are a clay-like chemical waste issued from the alkaline extraction of the alumine contained in bauxite through the Bayer process. They are so called "red" because of their high content in iron oxides, primarily as hematite. During the Bayer process, a certain fraction of a caustic liquor (NaOH) is added to the bauxite and some of it remains even after washing, along with a liquor composed of sodium carbonate and sodium aluminate (Fortin 1991).

The red muds used for this study are those from the ALCAN Vaudreuil plant, Jonquière, Canada.

The waste rock used come from the South Dump of La Mine Doyon, Abitibi, Québec. This waste rock has a proven high reactivity (Gélinas et al. 1992), that is an advantage for laboratory tests where time is an important factor. The waste rock had been excavated for an unknown period of time and was partially altered when placed into the columns.

Table 1 summarizes the main chemical, mineralogical and physical parameters relevant to the present study, while Figure 1 presents the grain size distribution for the fine materials. Standard acid base accounting (ABA) test was conducted on materials and results clearly show that mine tailings are acid generating (Table 1). However, they are planned to be used as water saturated layers, thanks to their capillary properties. In these conditions, oxidation should not proceed at a significant rate and no acidic drainage should be produced. Other interesting observations from Table 1 are the high natural pH of both CKD and red muds, and the considerable sulphur content of the CKD, probably mostly as sulphate within arcanite. Also, the red muds possess a great specific surface area, which quality favors a high reactivity from this material. The primary mineralogical phases observed are in agreement with those described in the literature.

Table 1: Main characteristics of the materials used in the experimental columns.

Material	pH	S_T*	NNP*	S.S.A.*	1st phase under XRD
Waste rock	4.2	4.12	- 49.1	---	Quartz, clinochlore, illite, pyrite.
Tailings	5.6	4.43	- 132.1	0.98	Quartz, illite, clinochlore, muscovite.
CKD	12.6	5.10	721.0	2.25	Calcite, lime
Red muds	11.4	0.36	260.0	15.47	Hematite, sodalite

*S_T: Total Sulfurs (%); S.S.A.: Specific area (m^2/g); NNP: Net Neutralization Potential (kgH_2SO_4/t).

5 EXPERIMENTAL COLUMNS

Five columns were prepared for the leaching tests. The first (reference) contains only waste rocks from La Mine Doyon. Column 2 has been constructed with tailings free of basic additives, following the scheme of layered co-mingling of waste rock (Poulin et al. 1996). Columns 3 and 5 tested the layered co-mingling technique with the addition of a 10 % mass

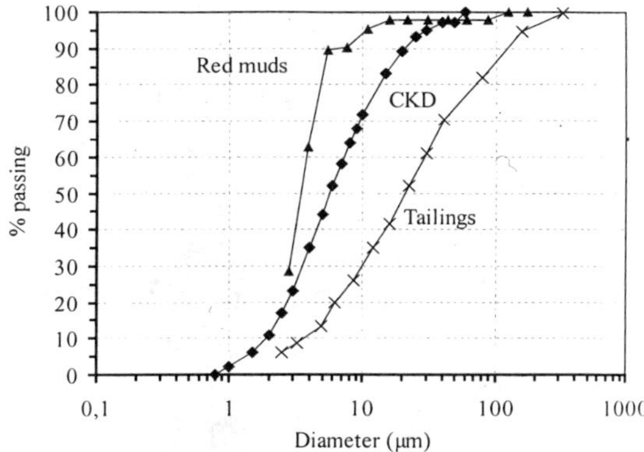

Figure 1: Grain size distribution of the fine materials

fraction of alkaline additive mixed to the tailings layers.

The clear Plexiglas columns were opened at the top and fed with deionized water at a controlled rate of 115 mL/d by a peristaltic pump.

Reference column contains only waste rock. The two other columns involved two 0.5m layers of waste rocks alternating with three 0.1 m layers of compacted tailings (Fig. 2). The thickness ratio of waste rock to tailings and additives was 5:1. Proctor curves have been done for both tailings-additives mixes. The fine materials were compacted on the wet side of the Proctor curve (Lapierre et al. 1990). They were introduced in three sub-layers in order to reach a thickness of 0.1m/layer.

6 INSTRUMENTATION, SAMPLING AND ANALYSIS

6.1 *Monitoring water content*

Time-domain reflectometry (TDR) probes were installed in each compacted layers for water content monitoring. TDR is based upon the dielectric properties of a medium (Topp et al. 1980). Readings were taken bi-weekly and weekly for the first three months of the experiments and monthly for the remainder of the test. Water contents did not change significantly in the layers during the experiment. The layers were almost fully water saturated in the beginning of the experiments (about 90 %) and they kept the same degree of saturation. These values were confirmed at the end of experiment by measuring the water content using the oven method.

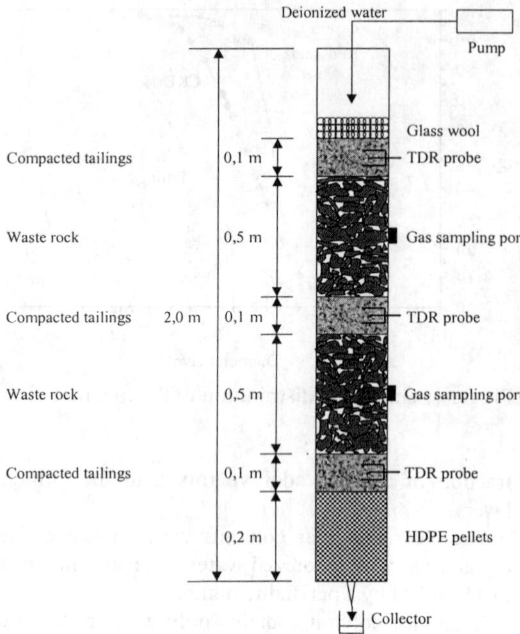

Figure 2: Experimental columns.

6.2 *Monitoring gas composition*

Rubber septums were installed at each levels of the waste rock layers. Gases were sampled twice a week for the first few weeks and once a week afterward. The absence of O_2 demonstrates that the capillary barrier effectively prevents O_2 replenishment. The evolution of gas through time shows that for both columns, oxygen is completely consumed within 60 days in the experiment, thus suggesting that the diffusion rate of oxygen through the tailings was much lower than the rate of oxygen consumption within the columns.

6.3 *Monitoring chemical composition*

Leachates were collected weekly and analyzed for the following parameters: pH, alkalinity and metal concentrations. In order to compare the pH of the solution at the exit of the circuit with that of the water *in situ*, interstitial water was sampled by squeezing the layer material with a hydraulic press after dismantling the columns (Tassé & Germain 1994). All samples were filtered on a 0.45 μm aceta-cellulose membrane, acidified with 1M nitric acid and kept at 4°C until analysis by ICP-AES for the following elements : Al, As, Ba, Ca, Cd, Co, Cr, Cu, Fe, K, Mg, Mn, Na, Ni, P, Pb, S, Si, Sr, Zn. Sulphur was considered as sulphates, and iron as ferrous iron, as demonstrated by comparative determinations in similar acid drainage (Tassé & Germain 1994). Acidity was assumed to be controlled mainly by ferrous iron concentration. X-ray diffraction (XRD) and scanning electron microscopy (SEM) were used to help identify both the primary and secondary minerals present. Finally, in instances where the small amounts of the phases present prevented detection by XRD and/or SEM techniques, mineral saturation index calculations were used to help identify probable mineralogical controls. Saturation indexes were calculated with U.S.G.S. PHREEQC code (Parkhurst 1995).

7 GEOCHEMICAL RESULTS

7.1 *pH, iron and sulphates variations*

Figure 3 (a, b, c) shows evolution of pH and of iron and sulphate concentrations through time in the three columns of interest, *i.e.* reference, co-mingling with CKD and co-mingling with red muds. Results from column 2 (co-mingling with tailings only) are detailed elsewhere (Lamontagne et al. 1999). For the reference column, initial pH is acidic at about 2.5 and remains strongly acid near 2.0 over the entire course of the experiment. In column including 10 % CKD, pH is maintained at strong alkaline levels for the whole experiment, despite a drop around day 60. Duchesne & Reardon (1998) also noted rapid stabilization of pH and alkalinity when CKD are put in contact with water. The pH of column with 10 % red muds is somewhat less regular than the above. After some 10 days of leaching, the pH rises quickly from its initial value at 6.0 toward a peak near 8.0. It remains stable for 25 days before slowly decreasing to a bottom value around 4.8 after 80 days, before steadily increasing for 50 days to reach a stable plateau at neutral pH for the remain of the test. Comparison of pH profiles to outflow rates do not show any relationship. Leachate alkalinity was relatively low for both columns with CKD and red muds, with values averaging 60 and 25 ppm $CaCO_3$ respectively. These results suggest that the high pH values do not involve automatically high alkaline capacities.

The concentrations of the main oxidation products, iron and sulphates, vary with ups and downs and plateaus. In the reference column, concentrations drop for both ions the first 25 days, then increase to reach plateaus at 3200 and 11,400ppm, respectively. In the co-mingled columns, both leachates show grossly analogous profiles. Regarding sulphates, the general trend is a step-wise decrease with concentrations plateaus. For the CKD column, concentrations decreased from a 22,500ppm plateau between 0 and 80 days to 17,400ppm between 115-165 days. In the red muds column, the plateaus occurred in the periods between 30-60 and 75-125 days at 17,700 and 16,500ppm sulphates. Both columns seemed to stabilise at 3000 and 4500ppm sulphates from 200 and 150 days until the end.

Acid water readily recovered from the reference column at the very beginning of the experience shows that the altered waste rock used to fill the columns were acid-producing when they were sampled. Oxidation continued in the lab without needs to overcome eventual neutralisation from other host rock minerals, with sulphide minerals easily available for further alteration, and little needs for incubation of oxidising bacteria, contrary to fresh material. Air was free to diffuse from the top, so O_2 did not go below 18 % in the pore spaces. A sharp drop of iron and sulphates concentrations in the first three weeks suggests that labile sulphides were first oxidised. Leaching of soluble iron minerals such as melanterite and/or other hydrated iron-sulphate minerals is also possible. Concentrations then rose until full stabilization of the oxidizing system. The concentration plateaus that prevailed for more than 200 days until the end of the experience testify that the concentrations of dissolved iron and sulphates were controlled by geochemical equilibrium. Iron hydroxides and gypsum are potential solubility controls, and geochemical calculations are in agreement with their precipitation. At 2.0, pH is somewhat lower than what would be expected in an iron hydroxide buffered system (Blowes & Ptacek 1994). However, the low pH in the leachate do not totally reflect the geochemical conditions prevailing in the column, since it is not in equilibrium with the solid phases. Jarosite could control the solubility of both iron and sulphate, but calculations do not support jarosite precipitation.

For the CKD column, the generally high pH reflects the lime content of the CKD. The punctual drop in pH observed after 65 days corresponds precisely to the period of complete O_2 consumption, *i.e.* to the maximum oxidation rate. Co-mingled columns both became depleted in oxygen within 40-65 days, meaning that oxygen consumption by oxidation was faster than oxygen replenishment under dissolved or gaseous form, and that availability of oxygen became the rate limiting factor for sulphide oxidation. Despite a much lower oxidation rate and lower O_2 in pore spaces, sulphate concentrations are significantly higher than for the reference column, but constantly decreasing. This can best be explained by the initial composition of CKD with 5.1 % sulphur, mostly as arcanite which is readily leached out. Also, it is thought that an important fraction of the sulphates produced came from the leaching of secondary phases linked to the previously altered waste rocks. However, the sulphates produced were rapidly stabilised through precipitation mostly as gypsum, as proposed by mineral saturation index calculations and SEM observations.

Besides a peak of iron of 60 ppm at 60 days, Fe concentrations have remained very low at all time in the CKD column. It is likely that the high pH values meant a high availability in –OH groups so that iron ions could easily form stable minerals such as goethite (FeOOH) and iron hydroxydes ($Fe(OH)_3$). Visual observations agree with PHREEQC calculations, which predict supersaturation. Unlike the reference column, leachate from the CKD column seemed to be in chemical equilibrium with the solid phase when compared to *in situ* concentration and pH values (Table 2).

Table 2: Comparison of pH and alkalinity values in leachate and in tailings pore water at the end of test.

Column	Layer	Layer pH	Layer ALK*	pH (end)	ALK* (end)
1	---	---	---	1.85	---
3	1	8.75	43		
(CKD)	2	8.57	40	9.41	0
	3	9.27	53		(nil)
5	1	8.01	53		
(red	2	7.73	75	6.94	9
muds)	3	n.a.	n.a.		

*ALK: Alkalinity (ppm $CaCO_3$)

Later decrease in sulphate concentrations from the CKD column is likely related to lower oxidation rates and to arcanite depletion, since the variation curve for K deviates from that of sulphates after a certain time. About 1.2 % CO_2 were observed in the pore space of waste rock at the top, testifying for carbonate neutralisation of acid, which is the source of Ca^{2+} for gypsum precipitation. The lower CO_2 concentrations measured at the end of test confirm the lower oxidation rates.

In the red mud column, the sharp initial rise in pH corresponds to the leaching of the caustic solution (NaOH) trapped in the red muds that increase pH to basic values near 8.0. The pH fall that follows is likely attributed to the complete consumption of O_2 (between 37-45 days), *i.e.* the period of highest oxidation rate observed in the column as proposed by the sulphates peak. The system reacts quickly by neutralising acidity with carbonates dissolution so that pH rise and remain at neutral levels. Since initial red muds contained no sulphate-bearing minerals, the rather high sulphate concentrations observed must be produced through the oxidation of waste rock and/or tailings and also through the leaching of sulphate-bearing secondary phases attached to the waste rock. Resulting sulphates are even higher at some point than reference column, even though O_2 values in the pore spaces are much lower. However, sulphate levels rapidly decrease to low values. Neither PHREEQC calculations nor microscopic analysis support the formation of a secondary phase such as gypsum that could stabilise sulphates, so that the drop in sulphates varies accordingly to a strong decrease in oxidation rates. Moreover, the calcium variation curve do not support gypsum precipitation.

The iron concentrations in the red muds column remain very low (maximum observed at 12ppm) during the entire experiment, thus suggesting a strong solubility control by the solid phase. Mineral index calculations favour the formation of many iron compounds such as iron hydroxides, geothite and hematite. The latter two have also been observed under SEM. Iron sulphates are another possible phase stable under the conditions observed. Sampling of interstitial water seem to reveal that the leachate attained equilibrium conditions with the solid phase (Table 2).

Between 1.8-3.1 % CO_2 have been measured in the pore spaces at the middle of experiment, attesting for acid neutralisation by carbonate dissolution. The lower CO_2 values measured at the end of testing (between 1.0-1.9 %) suggests lower oxidation rates. This, again, was proposed by the decrease in sulphate concentrations that was observed (Fig. 3c).

7.2 *Other elements*

Variations of Zn, Cu and Al are not illustrated here. Zinc is observed in slightly higher concentrations in the reference column than in the two others but concentrations are too low to express any trend. On the other hand, copper concentrations are much higher in the reference column than in the other columns. Cu variations are grossly similar to those of iron and sulphates in the reference column, with, however, a slow decreasing trend rather than a plateau. However, Cu concentrations are too low to show any pattern in co-mingled columns. In addition, aluminium concentrations are higher by two to three orders of magnitude in the drainage water from the reference column. Again, aluminium concentrations are too low to show any trend in both columns 3 and 5.

Initial tailings and waste rock composition partly explain those behaviours. Zinc concentrations are almost the same in both tailings and waste rock. However, the tailings are the end product of the extraction processes of a base metal ore, whereas the waste rocks are the unprocessed sterile rejects of rocks surrounding gold ore. Mineralogical and textural properties of zinc minerals are doubtless quite different, with tailings showing a much higher specific surface area. While Zn should react less in the reference column, the Zn that is leached out from co-mingled columns is stabilised through co-precipitation and/or adsorption mechanisms with the solid phase.

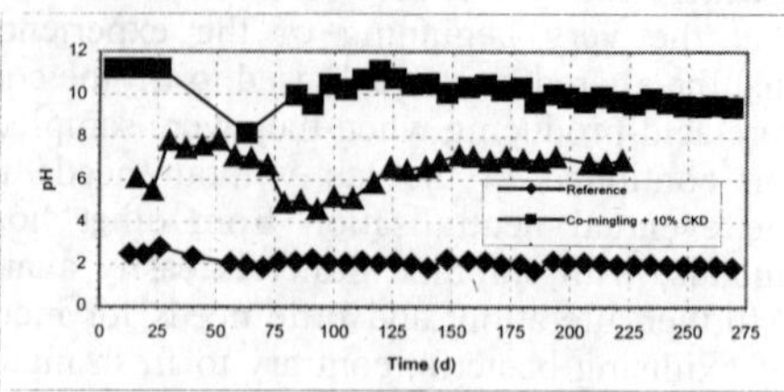

a) pH Vs. Time

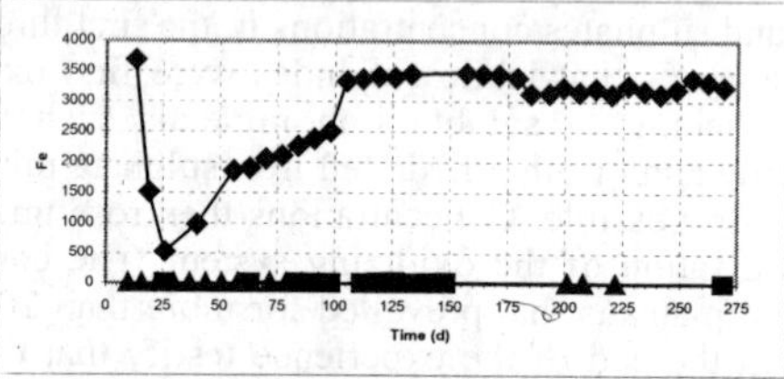

b) Iron concentration (ppm) Vs. Time

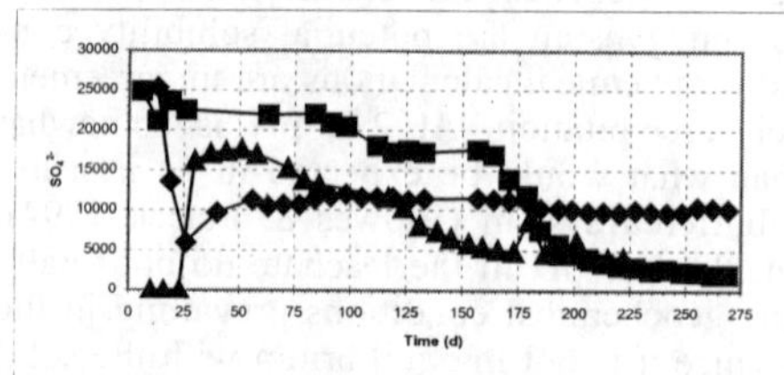

c) Sulphates concentration (ppm) Vs. Time

Figure 3: Evolution curves of pH and elements.

Initial copper concentration is higher in the tailings than in the waste rock. Differences in mineralogical and textural properties between tailings and waste rock may be more important for copper than it was for zinc. However, valence of copper varies, as iron, with likely different behaviour in reference-column compared to co-mingled columns. *In situ* geochemical conditions in tailings and basic additives are not known well enough to digress on the mineral form under which copper might be trapped within the layers, but it is. In the reference column, copper is leached as iron is, but absence of plateau suggests that it was not subject to a solubility control.

Initial Al concentrations are quite similar in tailings and waste rock. Here, the very large difference between the reference column and the co-mingled columns are entirely related to solubility controls by pH in the reference column and tailings-additives layers in co-mingled columns. Aluminium is leached from the dissolution of aluminosilicates and kept dissolved in the very low pH waters of the reference column. In the co-mingled columns, the higher pH values of the layers (Table 2) promote the precipitation of aluminium as boehmite (AlOOH) and gibbsite ($Al(OH)_3$), according to PHREEQC calculations. The fine materials layers thus act as a sink for metal ions, as indicated by the strong decrease in concentrations that can be up to one thousand-fold in the case of aluminium concentrations.

8 BACTERIAL ACTIVITY

Biological oxidation of metal sulphides is a highly significant process in AMD production, since pyrite oxidation by atmospheric oxygen alone is a process too slow to be of any significance. Thirty years ago, Singer & Stumm (1968) pointed out the strong catalytic effect of bacteria to oxidize Fe^{2+}. Among the numerous micro-organisms accompanying and paving the way to massive acid production, *Thiobacillus ferrooxidans* is probably the most efficient species for oxidizing iron sulphides such as pyrite. A bacterial count for *T. ferrooxidans* was thus performed in order to compare the controlling effects of co-mingling and basic additives not only on the pH, but also on the bacterial activity which indirectly controls oxidation rates.

8.1 *Sampling and analytical method*

The top and base of the reference column and each five layers of co-mingled columns were sampled at the end of the experiments. Micro-organisms were isolated using standard micro-biological procedure.

Bacterial enumeration was done following the Most Probable Number method (MPN; Koch 1981) using a procedure adapted by Lafleur et al. (1993). This method is of particular interest for this experiment, because it allows counting in cases where kinetics of growth are highly variable, because it is a selective method permitting to isolate the species of interest and because it is not altered by the presence of solidifying material. However, the method is not statistically robust and provides only indicative answers.

8.2 *Results*

MPN was determined from published tables and were compared to a reference sample fed with pure ferrous sulphate. Results are summarised in Table 3.

These numbers clearly show that the compacted layers (tailings + basic additives) control efficiently bacterial activity when compared to the reference column. All compacted layers including red muds did not record any bacteria, while layers from the CKD column are lower by two to three orders of magnitude relative to the reference column. It is concluded that bacteria had difficulty to adapt to the media as pH of the CKD columns (around 9.0) and red muds column (around 7.8) are much higher than the optimum range for *T. ferrooxidans* growth.

After 40-60 days of testing, the only O_2 available was that dissolved in water that was close to saturated with respect to O_2 with an average concentration of 8 ppm (less than 0.05 %). However, as O_2 was consumed and pH decreased, anoxic conditions were developed where partial pressure of O_2 was low and temperature were probably quite stable. At this stage, microbial activity became more important due to oxidation of elemental sulphur combined with ferric iron reduction in acid conditions. Although oxygen availability is a limiting factor to bacterial observation, it was observed by Guay (1994) that activity could continue in total absence of oxygen as ferric iron (Fe^{3+}) is becoming the primary electron acceptor (oxidant). From these considerations, even a cover preventing oxygen diffusion could not totally stop microbial activity as long as ferric ion is present in the pore water. A cover could thus only reduce or control the oxidation process.

Table 3: Bacterial counting results from the modified MPN method.

Source	Layer	MPN (bacteria/mL)
Reference	Column 1-A	13,000
	Column 1-B	27,500
Co-mingled column with CKD	Tailings-top	792
	Waste rock-top	74
	Tailings-middle	231
	Waste rock-base	no sample available
	Tailings-base	275
Co-mingled column with red muds	Tailings-top	0
	Waste rock-top	no sample available
	Tailings-middle	0
	Waste rock-base	no sample available
	Tailings-base	0

9 CONCLUSION

Acid mine drainage (AMD) is the largest environmental problem facing the mining industry today. Open pit mining generates large volumes of waste rock that are problematic when sulphides are present. Results from this paper have shown that the addition of basic industrial residues to layered co-mingling can mitigate AMD. The presence of fine material limits the mobility of water and air in waste piles. The presence of basic residues increase the alkaline reservoir required to neutralise acidity at the source and keep the pH at near-neutral values. This study, using relatively reactive waste rock, demonstrates the benefits of co-mingling involving mixing of tailings with basic materials.

The fine material layers act as a trap for different metals favouring precipitation. This sink of metals operates because of the relatively high pH of tailings-additives pore water. Sulphate and iron leachate concentrations against time are consistent with this explanation. Bacterial activity is greatly reduced in the co-mingled set up including the addition of a 10% mass fraction of alkaline residues because of lack of O_2 and higher pH, both conditions unfavourable to strong levels of bacterial activity.

Previous modelling of waste piles (Lamontagne et al. 1999) based on these results shows that optimisation of design is possible with a tailings to waste rock ration different than 1:5, with the actual materials of this experiment a ratio around 1:10 could be sufficient. Finally, the use of alkaline industrial residues offer a promising option to improve the efficiency of tailings layers in the layered co-mingling method in cases where the tailings would be acid-producing.

REFERENCES

Blowes, D.W. & C.J. Ptacek 1994. Acid-neutralization mechanisms in inactive mine tailings. In : Short course handbook on environmental geochemistry of sulfide mine-wastes : 271-292, Mineralogical Association of Canada.

Duchesne, J. & E.R. Reardon 1998. Determining controls on element concentrations in cement kiln dust leachate. *Mine Waste Management*, in press, June 1998.

Fortin, J. 1991. Caractérisation minéralogique et chimique de boues rouges et leur mise en végétation. Mémoire de maîtrise, Faculté des sciences de l'agriculture et de l'alimentation. 98 p.

Gélinas, P., Lefebvre, R. & M. Choquette 1992. Characterization of acid mine drainage production from waste rock dumps at La Mine Doyon, Quebec. Second Int. Conf. on Environm. Issues and Manag. of Waste in Energy and Mineral Prod., Calgary, Sept. 92.

Guay, R. 1994. Diversité microbiologique dans la production de drainage minier acide à la halde sud de la Mine Doyon. MEND Report 1.14.2. March 1994, 64 p. and app.

Koch, A.L. 1981. Growth measurement , Manual of methods for general bacteriology: 179-207. Editors P. Gerhardt and R.N. Costilow (Washington D.C., American Society for microbiology ,1981).

Kuo, E.Y. & A.I.M. Ritchie 1999. The impact of convection on the overall oxidation rate in sulfidic waste rock dumps, Mining and the Environment II, Sudbury'99, pp.211-220.

Lafleur, R., Roy, E.D., Couillard, D. & R. Guay 1993. Determination of iron oxidising bacteria numbers. Biohydrometallurgical technologies: 433-441. Vol.2, *Fossil energy materials, bioremediation, microbial physiology*. Edited by A.E. Torma, M.L. Appel and C.L. Brierley.

Lamontagne, A., Lefebvre, R., Poulin, R. & S. Leroueil 1999. Modelling of acid mine drainage physical processes in a waste rock pile with layered co-mingling, 52nd Canadian Geotechnical Conference, oct. 25 to 27, Saskatchewan.

Lamontagne, A., Fortin, S., Poulin, R., Tassé, N. & R. Lefebvre 2000. Layered co-mingling for the construction of waste rock piles as a method to mitigate acid mine drainage – laboratory invectigations. 5th Int. Conf. On Acid Rock Drainage. May 21-24, 2000, Denver, Colorado.

Lapierre, C., Leroueil, S. & J. Locat 1990. Mercury intrusion and permeability of Louisville clay, *Can. Geotech. J.* 27: 761-773.

Lefebvre, R. 1994. Caractérisation et modélisation numérique du drainage minier acide dans les haldes de stériles. Ph.D. thesis, Department of geology and geological engineering, Université Laval, 375 p.

Parkhurst, D.L. 1995. User's guide to PHREEQC – A computer program for speciation, reaction-path, advective transport, and inverse geochemical calculations. U.S. Geological Survey. *Water-Resources Investigations* Report 95-5227.

Poulin, R., Lamontagne, A. & J. Hadjigeorgiou 1998. Layered co-mingling, multiple-barrier ARD control – Laboratory trials. Proceedings SWEMP'98, Turkey 18-20 May : 451-456.

Poulin, R., Hadjigeorgiou, J. & R.W. Lawrence 1996. Layered mine waste co-mingling for mitigation of acid rock drainage, *Trans. Inst. Min. Metal.* (105): A55-A62.

Singer, P.C. & W. Stumm 1968. Acid Mine Drainage : The Rate-Determining Step. *Science* 167: 1121-1123.

Tassé, N. & D. Germain 1994. Caractérisation géochimique du parc à résidus miniers Canadian Malarctic, Rapport final. Service du développement minier, Natural Res., 204 p.

Topp, G.C., Davis, J.L. & A.P. Annan 1980. Electromagnetic determination of soil water content: Measurements in coaxial transmission lines. *Water Reso. Rese.* 16 (3): 574-582.

Environmental Issues and Management of Waste in Energy and Mineral Production, Singhal & Mehrotra (eds)
 ISBN 90 5809 085 X

AMD generation at sub-zero temperatures

R.C.Godwaldt, K.W.Biggar, D.C.Sego & J.Foght
University of Alberta, Edmonton, Alb., Canada

ABSTRACT: This paper discusses several factors which may affect the long-term integrity of structures constructed from frozen sulfidic material given the exothermic nature of the sulfide oxidation and the predicted ongoing oxidation of sulfides in sub-zero temperatures. The exothermic nature of the sulfide oxidation reactions may lead to the development of microclimates within the structure, leading to the creation of unfrozen water zones. Although the chemistry and kinetics of AMD generation are well understood at moderate temperatures, few studies have explored the process at sub-zero temperatures.

1 BACKGROUND

With exploration and mining activity recovering in the arctic, and due to intense environmental scrutiny, attention is again being focussed on waste treatment issues in the north. Decades of careless resource management have left a legacy of abandoned mine sites, waste dumps, and ongoing cleanup requirements. Over 75 sites in the Canadian territories have been identified by Indian and Northern Affairs Canada (1994a, b) as requiring investigation into their potential for acid mine drainage, a condition in which sulfide material reacts with oxygen and water to generate acid through a combination of chemical and biological activities.

The resulting acidic conditions are responsible for increased heavy metal concentrations in the water, which are toxic to plant and animal life. These sulfides may be present in a mineral or material of economic interest, waste generated from mining operations, or from disturbed material on a construction site. In any case, once the natural sulfides have been disturbed, it is the responsibility of the operator to manage the resulting contamination.

One of the recommended strategies for control of acid mine drainage in northern regions is the use of permafrost to permanently encapsulate the reactive material (GEOCON 1993; Davé et al. 1996; Dawson & Morin 1996). Preliminary results at Nanasivik Mine, NWT, indicate that this method has been successful in continuous permafrost regions, although no long-term data or performance monitoring have been published (LeDrew, pers. comm.) While permanent storage may be achievable in extreme cold regions for some material, it is necessary to examine the factors that determine whether a specific material remains inert *in a given environment*.

The questions that are considered in this paper include: 1) at what temperature does the biological component of AMD becomes negligible, and 2) what are the factors influencing unfrozen water content within the sulfide rich tailings? As mentioned previously, encapsulation of potentially acid forming material in a permanently frozen state has shown promising results under certain conditions; however, this paper will demonstrate that despite experiencing success for some conditions, the presence of sub-zero temperatures does not necessarily preclude the formation of AMD. This study examines the potential employing highly sulfidic coarse grained material in an unsaturated zone.

2 REVIEW OF ACID MINE GENERATION

2.1 *Role of bacteria in acid generation*

Even at moderate temperatures, the biological sulfide oxidation processes are relatively poorly understood, but there are several well-documented summaries available describing the catalyzing effects of aerobic sulfide oxidizing bacteria (Ingledew 1990; Morin et al. 1991; Otwinowski 1994). ‘Acidophiles’ are organisms that have the ability to grow and reproduce in an acid environment (Ingledew 1990), some at pH 1.0 and lower. In these low pH environments, life is based on sulfate as the predominant anion, as most other acids cannot support acidophilic habitats. Reports indicate that the oxidation catalysis by bacteria has a significant effect on the reaction rates (Davé et al. 1996), though recent publications challenge this claim (Morin 1999). That the bacteria are able to survive in these extreme environments is attributed to the presence of a cytoplasmic membrane (Ingledew 1990). This membrane acts as an osmotic barrier, and is capable of maintaining a neutral pH inside, protecting the cell. It is this ability of bacteria to modify their environment that is of further importance to this study.

A key assumption in reviewing the effects of cold temperature on microbial sulfide oxidation catalysis is that microbial growth depends on the presence of liquid water. It is likely incorrect, however, to assume that all microbial activity ceases at 0°C, although two models described in the literature make this assumption (GEOCON 1993; Davé et al. 1996). This is reflected by the sudden change in the reaction rate gradient as shown in the figures below. In Figure 1, the data do not actually extend to 0°C, yet the assumption is made that the curve passes through the origin. Figure 2, however, shows data for a biological oxygen demand (BOD) tests, demonstrating biological activity, in which, at the 0°C intercept, the growth rate is greater than zero. If this curve is valid for the oxidation of sulfides, the previous models must be questioned.

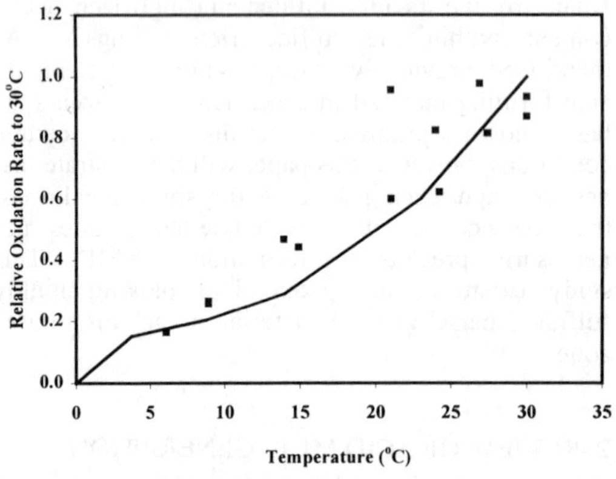

Figure 1. Computer model prediction (Modified from GEOCON 1993)

The presence of psychrotrophs (cold tolerant bacteria) is well documented (Vincent 1988; Berthelot et al. 1994; Russell & Hamamoto 1998) and demonstrates that bacteria are capable of existing in extreme cold environments. It is thus important to determine to what degree the sulfide oxidation reaction is catalyzed at sub-zero temperatures. It is possible that at lower temperatures, the catalyzing effect is still present, though reduced. Given that early studies in food science have demonstrated bacteria are capable of germinating at -6°C (Halvorson et al. 1961), an examination into microbial activity at sub-zero temperatures within sulfide rich materials is in order.

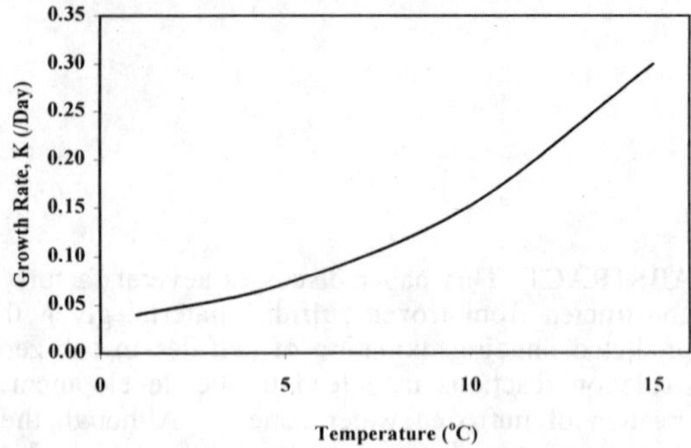

Figure 2. Biological oxygen demand in fresh water (Modified from Heroux & Rowney 1987)

2.2 *Effects of Cold Temperatures on Microorganisms*

The slowing of the physiological processes, and therefore the catalyzing effects, in cold temperatures is due to 'gelling' of a fluid layer of lipids within the cell membrane (Vincent 1988). As the temperature decreases, the membrane loses its fluidity and impairs the activity, or activation energy, of certain proteins. The rate at which the bacterium slows its physiological processes due to the cold has traditionally been described by the Arrhenius equation (Equation 1), where k is the rate constant, E_a is the activation energy and T is the temperature (in degrees Kelvin). A is a constant and R is the universal gas constant.

$$k = A \cdot \exp\left(\frac{-E_a}{RT}\right) \quad (1)$$

This equation, however, does contain certain inherent errors, as it does not consider systems containing more than one enzyme, nor complex systems in which a single enzyme may have several stages (Vincent 1988).

A second model that has been used to predict microbial activity is the square root model (Equation 2), proposed by Ratkowski et al. (1982), which utilizes the temperature difference between the optimum temperature for growth and the actual temperature.

$$\sqrt{k} = b \cdot (T - T_{min}) \qquad (2)$$

This relationship applies for the temperature range between the minimum and optimum temperatures for growth, where k is the specific growth rate, b is the slope of the regression line, T is temperature of concern, and T_{min} is the temperature, in Kelvin, where the regression line cuts the temperature axis. Figure 4 shows the effect of plotting the Arrhenius data with a square root model. The data shown are for mesophilic bacteria (those that predfer moderate temperatures). Psychrophilic (cold-loving) bacteria demonstrate the same curve, though shifted to colder temperatures (Russell & Hamamoto 1998).

The square root model is an improvement over the Arrhenius model as it incorporates a changing activation energy, E_a, and has met with success in the food industry at temperatures from -2°C to 15°C.

Although no reference has been found listing specific sulfide oxidizing bacteria that grow in the sub-zero temperature ranges, it is recognized that many species of bacteria remain to be identified. One recent study of note (Meldrum 1998) has shown

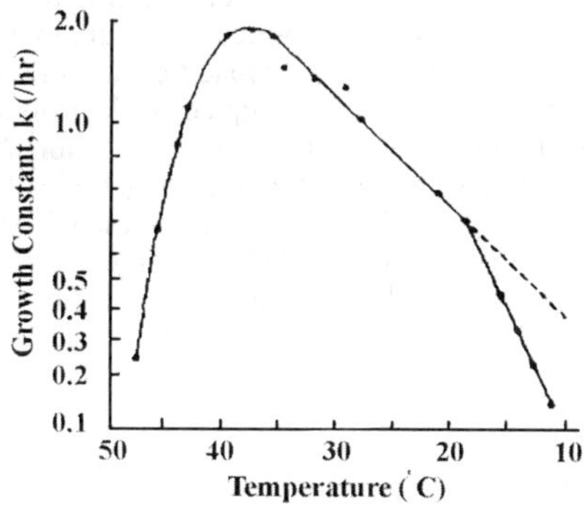

Figure 3. Arrhenius plot of growth rate constant (Modified from McMeekin et al. 1988)

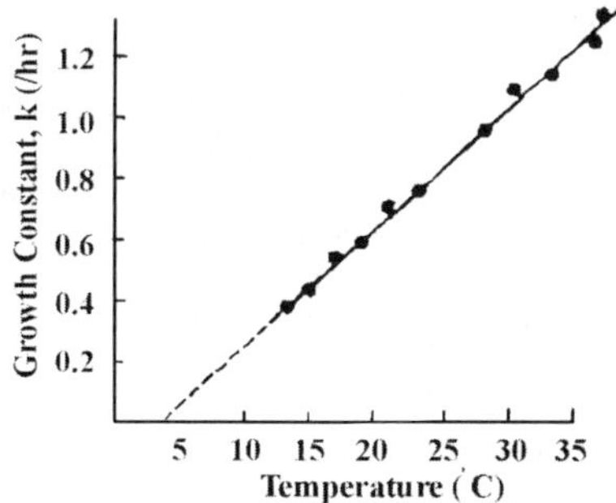

Figure 4. Plot of data from Figure 3 according to Equation 2. (Modified from McMeekin et al. 1988)

oxygen being consumed in a sulfidic environment at -2°C.

That the author attributes this oxygen consumption to bacterial activity indicates the potential for the existence of psychrotrophic acidophiles. Unfortunately, each species of bacteria has different optimum, preferred and lethal temperature ranges, and the well-known sulfide oxidizing species may not be able to survive at sub-zero temperatures. As such, research into isolating these bacteria and characterizing their growth temperatures represents an important step in predicting acid generation at sub-zero temperatures.

2.3 *Microbial Adaptation to Cold Temperatures*

Although it is assumed that complete freezing does preclude biological activity and acid generation, it must be proven that bacteria are capable of resisting the extreme temperatures and can survive in super-cooled environments (Vincent 1988). Bacteria are surrounded by cell membranes which contain very small aqueous channels. As the external environment cools, the external water begins to crystallize; however, the ice growth is restricted as it reaches the channels. According to surface tension / capillary pressure theory, the size of the ice crystal is proportional to its freezing point. Vincent (1988) reports super-cooling in bacteria to at least -10°C, which corresponds to an ice crystal radius, and therefore membrane channel, of 30nm.

As the temperature continues to decrease, ice crystals form in the water and the remaining solute concentration increases, resulting in an osmotic pressure gradient across the plasma membrane. To maintain equilibrium, the bacterium rejects water, increasing its internal solute concentration, thereby lowering the freezing point of the water within the bacterium further. As the concentration of solutes reach the solubility level, certain compounds will precipitate, and the eutectic temperature will be reached. For further detail, the reader is directed to Hayashi (1991). At this point, complete freezing is assumed to preclude further biological activity.

3 UNFROZEN WATER CONTENT

Another issue associated with the determination of the minimum temperature of acid generation relates to the presence of unfrozen water. This section does not propose new models for predicting unfrozen water content with temperature or freezing point depressions, but discusses the factors that could enhance the acid generating environment. For the purposes of this paper, the hysteretic effects of temperature on unfrozen water content are ignored and the temperatures are assumed to be reached by cooling. This section will review the significant factors including temperature, soil suction and the presence of solutes and charged surfaces. Research into predicting unfrozen water content is extensive, with many models employed for the various factors involved; however, a review of the literature did not reveal any all-encompassing relationships.

3.1 *Matrix Suction*

As mentioned previously, matrix suction, or surface tension, is thought to have a significant influence when considering the bacterial component. It has been presented by Williams (1991) as a function of the Clausius-Clapeyron equation (Equation 3), where T is the temperature in Kelvin, P_w is water pressure, P_i is ice pressure, V_w and V_i are specific volumes of water and ice, and L_f is the latent heat of fusion.

$$T - T_0 = \frac{(P_W \cdot V_W - P_i \cdot V_i) \cdot T}{L_f} \qquad (3)$$

It is necessary here, to clarify the difference between capillary space and adsorption space. Capillary space, as used in this paper, is the zone in which soil water is only affected by the laws of surface tension, as in the case of sands. Adsorption space is the zone in which soil water is strongly affected by the charged surfaces of the colloidal material (Black 1990), whether mineral or bacterial.

3.2 *Effect of Solutes*

The factor that has received the most attention is the effect of solutes on the unfrozen water content. The freezing point depression, or increases in the solubility of ice, is related to water activity. By definition, water activity is the availability of water in a solution. According to solvate theory, as described by Chen & Nagy (1986), 'part of the water can combine with the dissolved substance (non-electrolytes) or ions (electrolytes) to form hydrates'. This water no longer acts as a solvent. Calculations of freezing point depression due to solutes are still very rough as little is known about the thermodynamic properties of electrolytes in solution below 0°C, with the exception of sodium chloride. It is known though, that at high concentrations of calcium chloride, brines can remain entirely unfrozen at -50°C (Thurmond & Brass 1987).

The equations for predicting the freezing point depression vary significantly in complexity, the simplest considering only the effects of dilute solutes. In dilute solutions, the freezing point depression can be calculated using the molarity of the solutes, derived from the Van't Hoff equation (Equation 4) which can be simplified in an ideal setting to:

$$\Delta T \approx -1.860 \cdot (v \cdot m_B) \qquad (4)$$

where m_B is the molarity of the solute, T is temperature (Kelvin), and v is the number of aqueous species resulting from the dissolution of the solute (Marion 1995). This model is insufficient for the purposes of this discussion due to the effects of solute redistribution, or solute exclusion, which can concentrate the solution at the freezing front to 80 times the original concentration, provided that there is no solute inclusion or salt precipitation (Marion 1995). As the ice crystals reject the solute, the concentration in the remaining unfrozen water increases, further decreasing the freezing point. Given the high concentrations of metals in solution associated with acid mine drainage, one can anticipate the potential for significant freezing point depression. Freezing point depression in these environments has not been well addressed.

At higher solute concentrations, additional parameters are necessary (except for simple solutions, ie. NaCl). The complex systems, as modeled by the Pitzer equations, relate ion activity coefficients and osmotic coefficients to enthalpies, entropies, Gibb's energies, heat capacities and molal volumes of highly concentrated aqueous electrolyte solutions, to temperatures below -50°C (Marion & Grant 1997). The Pitzer equations are algebraically extremely complex and have been simulated by computer (Mironenko et al. 1997), but are limited with respect to chemical interactions. According to Marion & Grant, this theory cannot be applied to frozen ground until:

1. Sufficient heat capacity and density data are collected for supercooled electrolyte solutions at sub-zero temperature, and
2. The interaction of ionic solutes with layered silicates can be incorporated.

The relationship between unfrozen water content and solute content is further complicated due to electrolyte effects on capillary pressures of frozen soils. Additional capillary pressure is generated as the solute distorts the ice-liquid interface (Marion & Grant 1997). The consequence of this is that the ability to predict the temperature at which sulfide-oxidizing solutions will completely freeze is indeterminate.

3.3 *Charged Surfaces*

The final factor affecting the presence of unfrozen water is the presence of charged surfaces in the form of clay minerals or bacteria (Cullimore 1998). The charged surfaces contain a region known as the diffuse double-layer, in which water, a dipole molecule, is sorbed to the surface in proportion to the strength of the surface charge (Marion 1995). Empirical results, though now dated, have shown that the relationship can be represented by a simple power curve (Equation 5),

$$w_u = \alpha \cdot \theta^\beta \tag{5}$$

where α and β are characteristic soil parameters and θ is the temperature (C degrees below zero). Over 30 sets of values have been compiled by Andersland & Ladanyi (1994); for the soils not listed, the liquid-limits (Equation 6) have been used as an indirect method for determining unfrozen water content (Tice et al. 1976).

$$w_{u,\theta=1} = 0.346 \cdot w_{N=25} - 3.01 \tag{6}$$

where N is the number of blows required to close the standard groove in the liquid limit test (ASTM Standard D 4318-84); however, these equations are based on tests only to -2°C. Figure 5 shows the power law curve for a series of soils; however, while these curves are effective for material that has been previously tested, no direct relationship has been established between the strength of the charged surface and the unfrozen water content with temperature. This relationship would be very helpful when dealing with complex/composite materials.

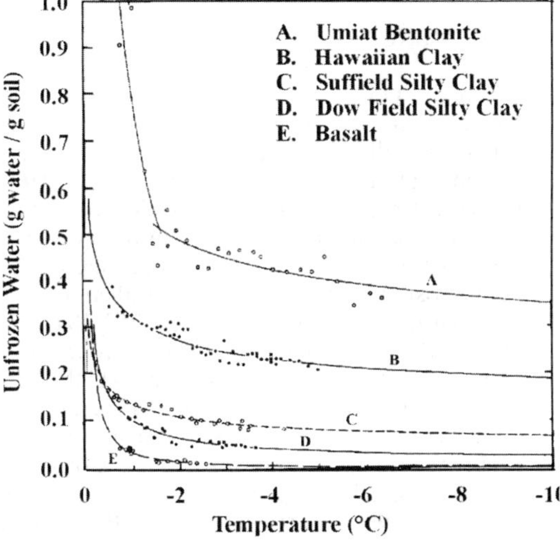

Figure 5. Unfrozen Water Content with Temperature for Five Soils (Modified from Anderson & Morgenstern 1973)

3.4 *Oxygen Availability*

The final consideration for acid generation is the availability of oxygen. In addition to the oxygen transport by advection through the unsaturated zones, the dissolved oxygen in the pore water must also be considered. A review of the literature did not locate any data for oxygen levels in sub-zero water, but it is known that the levels of dissolved oxygen increase as the temperature decreases, up to 14.5 mg/L at 0°C (Davé et al. 1996). Studies would have to be performed to determine if at these concentrations, sufficient oxygen is present to prevent the environment from becoming anaerobic.

4 CONCLUSIONS

The natural progression of the argument presented above is to question the long-term behavior of sites that rely solely upon freezing temperatures for the disposal of acid generating mine tailings; however, if the question of acid generation was strictly limited to mass balances, there would be little effort required. The final consideration is that the reaction is exothermic, giving off energy as the pyrite is oxidized. The following iterative steps describe a self-propagating oxidation environment (modified from GEOCON 1993):

Step 1. A grain of sulfide mineral is oxidized by elemental oxygen or another oxidizing agent, such as Fe^{3+}, releasing energy. 22500 kJ/kg of pyrite is released at a rate of $1.2x10^{-2}$ J/s at 30°C, and $5.5x10^{-4}$ J/s at -2°C (Meldrum 1998). These values vary with sulfide type, mineral grain properties and other mineral conditions. [Note: An earlier study (Dawson & Morin 1996) listed 0.012 kJ released per kg of pyrite oxidized. This is a calculation error, and as this was one of the first reports examining acid generation at sub-zero temperatures, it may be the reason that subsequent research has been scarce.]

Step 2. The heat is either absorbed by the proximate pyrite grains or is conducted away by non-reactive grains and/or pore water. The amount of heat that is conducted away from the grain is dependent on the thermal conductivity of the surrounding mass. This conductivity varies with moisture content, material type, and temperature.

Step 3. The heat that is not conducted away from the area either increases the temperature locally or is used in other energy consuming processes, such as melting ice. 334 kJ are required to melt 1 kg of ice. The melting of ice will result in the release of available dissolved oxygen and exposure of additional sulfide grains.

Step 4. If the temperature of the local area increases, the rate of further oxidation is increased (as the rate of oxidation is a function of the temperature). If additional oxidizing agents and sulfide minerals remain, the process repeats Step 1 at a higher rate of oxidation.

Step 5. If the thermal gradient surrounding the oxidizing zone is sufficiently great, the heat will dissipate into the surrounding mass and no change in the oxidation reactions will occur. If the heat lost to the mass is greater than the heat generated in the zone at any time, the local temperature will decrease, thereby decreasing the oxidation rate.

Hypothetically, if high pyrite concentrations are present and the super-cooled solution has elevated electrolyte concentrations in addition to oxygen availability, a zone may develop in which the rate of heat loss to the surroundings is low enough to maintain elevated local temperatures and promotes further melting and oxidation. Once these warmer microclimates are established, the integrity of the structure may be at risk due to free water lenses that decrease the strength of the material (Anderson & Morgenstern 1973).

Facilities constructed for permanent disposal should consider ongoing observations of the site to ensure that the bacterial and chemical activities remain very low. It is conceivable that over long periods, if the unfrozen water develops in a low pH, well aerated environment, the increase in bacterial activity due to the preferred acidic environment could offset the decrease in activity due to cold temperatures. This could lead to an eventual risk of failure of the containment system.

5 FURTHER RESEARCH

Several factors need to be better understood for accurate prediction of acid generation from mine tailings at sub-zero temperatures:

1. The behavior of supercooled electrolyte solutions
2. The relationship between charges surfaces and unfrozen water content
3. The existence and behavior of psychrophilic acidophiles
4. The solubility of oxygen at sub-zero temperatures

Analysis of cold weather sulfate oxidation is currently being conducted at the University of Alberta on samples of unweathered tailings that were retrieved from a mine in the Yukon, containing approximately 70% pyrite. The laboratory component of the testing program involves using humidity cells (Modified ASTM Standard D5744-88) at temperatures of 20°, 3° and 1°, with minor modifications for temperature and particle size effects, and static batch reaction tests at 20°, 10°, 3°, 1.5°, 0°, -1.5°, -3°, and -20°C. The intent is to develop a correlation between the humidity tests (accepted standard testing) and the batch tests at 20°, 3° and 1°, then extrapolate these results to the batch tests being conducted at colder temperatures.

The humidity cell tests will run for 20 weeks, analyzing the leachate for metals, sulfate, acidity, conductivity and redox potential at regular intervals. The batch test program is comprised of eight sets of cells with approximately 500g of unsaturated, unweathered tailings. Each cell is maintained at a constant temperature, with continuous air feed. At regular intervals, the frozen cells are thawed and the water is collected. These solutions will then be tested for the same parameters as the humidity cells. These tests will run for at least 1 year.

6 ACKNOWLEDGEMENTS

The authors wish to thank the Environmental Services, Public Works and Government Services Canada, the Department of Indian and Northern Affairs for their financial support, and to Michael Nahir, who brought this project to our attention. Thanks also to Gerry Cyre, Christine Hereygers, and Steve Gamble for their work in the manufacturing and operation of the humidity and batch cells.

7 REFERENCES

Andersland, O.B. & B Ladanyi 1994. An Introduction to Frozen Ground Engineering. Chapman Hall, New York, 352p.

Anderson, D.M. & N.R. Morgenstern 1973. Physics, Chemistry, and Mechanics of Frozen Ground: A Review. *In* North American Contribution 2nd International Conference on Permafrost, Yakutsk, U.S.S.R. Washington D.C.: National Academy of Sciences, pp. 257-288

Barthelot, D., L.G. Leduc & G.D. Ferrone 1994. The Absence of Psychrophilic Thiobacillus Ferrooxidans and Acidophilic Heterotrophic Bacteria in Cold, Tailings Effluents from a Uranium Mine. Canadian Journal of Microbiology, 40:60-63.

Black, P.B. 1990. Three Functions that Model Empirically Measured Unfrozen Water Content Data and Predict Relative Hydraulic Conductivity. CRREL Report 90-5. 7p.

Chen, C.S. & S. Nagy 1986. Prediction and

Correlation of Freezing Point Depression of Aqueous Solutions. Presentation at 1986 Winter Meeting – American Society of Agricultural Engineers, Chicago, Illinois, December 16-19. 17p.

Cullimore, R. 1998. Personal communication.

Davé, N., E. Blanchette & E. Giziewicz 1996. Roles of Ice, in the Water Cover Option, and Permafrost in Controlling Acid Generation from Sulphide Tailings. Report for Department of Indian and Northern Affairs Canada. MEND Report 1.61.1. 61p.

Dawson, R.F. & K.F. Morin 1996. Acid Mine Drainage in Permafrost Regions: Issues, Control Strategies and Research Requirements. Report for Department of Indian and Northern Affairs Canada. 59p.

GEOCON 1993. Preventing AMD by Disposal of Reactive Tailings in Permafrost. Report for Department of Indian and Northern Affairs Canada. MEND Report 6.1. 98p.

Halvorson, H.O, J. Wolf & V.R. Srinivasan 1961. Initiation of Growth at Low Temperatures *In* Proceedings - Low Temperature Microbiology Symposium, Campbell Soup Company, Camden. pp. 27-40.

Hayashi, Y. 1991. Micro-macro Freezing of Biological Substances. Third International Symposium on Cold Regions Heat Transfer, June 11-14, University of Alaska Fairbanks. p. 47-56.

Heroux, J. & A.C. Rowney 1987. Biodegredation Kinetics in Arctic Waters *In* Cold Regions Environmental Engineering – Proceedings of the Second International Confernce, Edmonton, Mar 23-24. pp. 198-209.

Indian and Northern Affairs Canada 1994. Acid Rock Drainage Potential in the Northwest Territories, NMEND #2. 47p.

Indian and Northern Affairs Canada 1994. Acid Rock Drainage Potential in the Yukon Territory, NMEND #3. 39p.

Ingledew, W.J. 1990. Acidophiles *In* Microbiology of Extreme Environments. Edwards, C.A. (ed), Open University Press. pp. 33-54.

Konrad, J. –M. 1989. Unfrozen Water as a Function of Void Ratio in a Clayey Silt. Cold Regions Science and Technology, 18:49-55.

Marion, G.M. 1995. Freeze-Thaw Processes and Soil Chemistry. CRREL Special Report 95-12. 23p.

Marion, G.M. & S.A. Grant 1997. Physical Chemistry of Geochemical Solutions at Subzero Temperatures. CRREL Special Report 97-10. pp. 349-356.

McMeekin, T.A., J. Olley & D.A. Ratkowsky 1988. Temperature Effects on Bacterial Growth Rates *In* Physiological Models in Microbiology, M.J. Bazin & J.I. Prosser (eds.), CRC Press, Boca Raton, Vol. 1, pp. 76-89.

Meldrum, J.L. 1998. Determination of the Sulphide Oxidation Potential of Mine Tailings from Rankin Inlet, Nunavut, at Sub-zero Temperatures. MSc. Thesis, Queen's University, Kingston, Ontario. 115p.

Mironenko, M.V., S.A. Grant & G.M Marion 1997. FREZCHEM2 – A Chemical Thermodynaic Model for Electrolyte Solutions at Subzero Temperatures. CRREL Report 97-5. 9p.

Morin, K.A. & N.M Hutt 1998. Contribution of Bacteria to Sulfide-Mineral Reaction Rates in Natural Environments. Internet Case Study for November 1998. http://www.mdag.com/cs11-98.htm.

Morin, K.A., Gerencher, E., Jones, C.E. & D.E. Konasewich 1991. Critical Literature Review of Acid Drainage from Waste Rock. Report for Department of Indian and Northern Affairs Canada. MEND Report 1.11.1. 175p.

Otwinowski, M. 1994. Quantitative Analysis of Chemical and Biological Kinetics for the Acid Mine Drainage Problem. Report for Department of Indian and Northern Affairs Canada. MEND Report 1.51.1. 76p.

Otwinowski, M. 1995. Scaling Analysis of Acid Rock Drainage. MEND Report 1.19.2. 67p.

Ratkowski, D.A., R.K. Olley, T.A. McMeekin & A. Ball 1982. Relationship Between Temperature and Growth Rate of Bacterial Cultures, J. Bacteriol, 149,1.

Russell, N.J. & T. Hamamoto 1998. Psychrophiles *In* Extremophiles: Microbial Life in Extreme Environments. K. Horikoshi & W.D. Grant (eds.), Wiley-Liss, New York, pp. 25-45.

Thurmond V.L. & G.W. Brass 1987. Geochemistry of Freezing Brines – Low Temperature Properties of Sodium Chloride. CRREL Report 87-13. 11p.

Tice, A.R., D.M. Anderson & A. Banin 1976. The Prediction of Unfrozen Water Contents in Frozen Soils from Liquid Limit Determinations. U.S. Army Cold Regions Research Laboratory. CRREL Report 76-8. 9p.

Vincent, W. 1988. Microbial Ecosystems of Antarctica. Cambridge University Press, Cambridge. 291p.

Williams, P.J. 1964. Unfrozen Water Content of Frozen Soils and Soil Moisture Suction. Geotechnique 14(3): 213-246.

Williams, P.J., 1991. Thermal Properties and the Nature of Freezing Soils. Third International Symposium on Cold Regions Heat Transfer, June 11-14, University of Alaska Fairbanks. p. 57-67.

Environmental Issues and Management of Waste in Energy and Mineral Production, Singhal & Mehrotra (eds)

Treatment of acid drainage from a uranium mine by means of a passive system

S.N.Groudev, P.S.Georgiev, I.I.Spasova & A.T.Angelov
University of Mining and Geology, Sofia, Bulgaria

K.Komnitsas
National Technical University of Athens, Zografos, Greece

ABSTRACT: Acid drainage waters from an uranium mine were treated by means of a pilot-scale passive system consisting of an anoxic cell and a constructed wetland arranged in a series. The anoxic cell was filled with a mixture of biodegradable solid organic substrates (spent mushroom compost, sawdust and cow manure) and was inhabited by a mixed microbial community consisting of sulphate-reducing bacteria and other metabolically interdependent microorganisms. An efficient removal of the pollutants was achieved in the anoxic cell. The microbial dissimilatory sulphate reduction and the biosorption were the main processes involved in this removal. The effluents from the anoxic cell were enriched in dissolved organic compounds. These compounds were efficiently degraded in the constructed wetland, which was characterized by an abundant water and emergent vegetation and a diverse microflora.

1 INTRODUCTION

The uranium ore deposit Curilo located in Western Bulgaria, about 35 km north from Sofia, was a site of intensive mining activities for a long period of time. Both open-pit and underground mining techniques were used to recover the uranium ore. Apart from uranium, the ore contained some copper minerals such as chalcopyrite, covellite and chalcocite and was rich in pyrite. After the exhaustion of the high-grade ores the "classical mining" was stopped but later an operation for in situ leaching of some low-grade ores was started. Leach solutions consisting of diluted sulphuric acid were injected through numerous boreholes to irrigate the ore mass and dissolved uranium. The pregnant solutions were collected by a system of drainage boreholes and were treated by ion exchange for recovering uranium. In 1990 this leaching operation was stopped due to a complex of economical, political and environmental reasons.

Since 1990 the abandoned mine sites are sources of acid drainage waters. These waters are generated as a result of the bacterial oxidation of the residual amounts of pyrite and other sulphide and uranium minerals. The waters have a low pH (usually in the range of 2.5 – 3.5) and contain radioactive elements (uranium, radium, thorium), toxic heavy metals (copper, zinc, cadmium, manganese, iron), arsenic and sulphates in concentrations much higher than the relevant permissible levels for waters intended for use in the agriculture and/or the industry.

Currently, the polluted waters are collected together and treated by ion exchange and partial chemical neutralization. However, these "active techniques" are very expensive. The most attractive from both technological and economical points of view is the possibility to treat the polluted waters by means of a passive system (Gazea et al. 1995, Cambridge 1995, Gusek 1995, Groudev et al. 1999). This assessment is based on the composition of these waters, the character of the geological, hydrogeological and climatic conditions in the area the availability of sufficient ground surface and a suitable landscape and on the type of the indigenous vegetation. In 1998 a pilot-scale passive system consisting of an anoxic cell and a constructed wetland was put into operation to check the above-mentioned possibility. Data about this operation are presented in this paper.

2 EXPERIMENTAL

The waters being treated in this study had a pH in the range of 2.8-3.5 and the concentration of the above–mentioned pollutants were 2-5 times higher than the relevant levels for waters intended for use in the agriculture and/or the industry (Table 1).

Table 1. Data about the drainage waters before and after their treatment by the natural wetland

Parameters	Before treatment	After treatment	Permissible levels for waters used in agriculture and industry
pH	2.8 – 3.5	7.1 – 7.3	6 - 9
Eh, mV	530 - 680	230 – 310	-
Dissolved O_2, mg/l	4.1 – 6.4	4.2 – 6.2	2
Total dissolved solids (TDS), mg/l	914 - 1450	370 - 745	1500
Solids, mg/l	32 - 91	23 - 82	100
Oxidativity (by $KMnO_4$), mg/l	3.7 - 14	8.0 - 32	40
Chemical oxygen demand (COD), mg/l	6.4 - 35	17 - 77	100
Biological oxygen demand (BOD_5), mg/l	1.7 – 8.2	4.1 - 18	25
Sulphates, mg/l	515 - 974	204 - 410	400
Uranium, mg/l	0.91 – 2.84	<0.10	0.6
Radium, Bq/l	0.23 – 0.60	<0.05	0.15
Total β-activity, Bq/l	0.99 – 2.35	<0.15	0.75
Copper, mg/l	0.68 – 3.52	<0.05	0.5
Zinc, mg/l	0.44 – 4.15	<0.05	10
Cadmium, mg/l	0.03 – 0.10	<0.004	0.02
Manganese, mg/l	0.71 – 5.10	<0.10	0.8
Iron, mg/l	180 - 347	<1.0	5
Arsenic, mg/l	0.18 – 1.07	<0.02	0.2

The quality of the passive system influents and effluents was monitored at least once per week in the period June 1998 – July 1999. The parameters measured in situ included: pH, Eh, dissolved oxygen, total dissolved solids and temperature. Elemental analysis was done by atomic absorption spectrophotometry and induced coupled plasma spectrophotometry in the laboratory.

The isolation identification and enumeration of microorganisms were carried out by described elsewhere (Karavaiko et al. 1988, Groudeva et. al. 1993)

The anoxic cell represented a pond with a volume of 9.5 m^3 (3 m x 2 m x 1.5 m) constructed in the ground. The cell was filled with a mixture of biodegradable organic substrates (spent mushroom compost, sawdust and caw manure) and was inhabited by a mixed microbial community consisting of sulphate-reducing bacteria and other metabolically interdependent microorganisms (Table 2). Sulphates and nutrients needed for the microbial growth were partially provided by the spent mushroom compost, which contained large amounts of sulphates, phosphates and various microelements.

The constructed wetland was a pond covering an area of 140 m^2. The impermeable rock bottom of the wetland was covered by a 50 cm layer consisting of a mixture of soil with a high content of organic matter, spent mushroom compost, silt and sand. The watercourse in the wetland proceeded in several shallow channel formed within the above-mentioned layers. The length of these channels was about 20 m each. The wetland was characterized by an abundant water and emergent vegetation and a diverse microflora. Typha latifolia and Typha angustifolia as well as different algae were the main plant species in the wetland. Different aerobic heterotrophic bacteria were the prevalent microorganisms in the microbial community. However, various anaerobic microorganisms were well present in the bottom zone (especially in the bottom sludge) of the wetland. Various protozoa, insects and other invertebrate organisms were also permanent inhabitants of the wetland.

The water treatment was carried out during the different climatic seasons and at different temperatures. The flow rate varied in the range of 2 to 6 m^3/24 h.

It was found that an efficient removal of pollutants from the waters being treated was achieved (Table 1), particularly during the warmer months of the year (from April to October). Under such conditions, even at dilution rate as high as 0.5 h^{-1}, the concentrations of pollutants were

Table 2. Microflora composition of the anoxic cell

Microorganisms	Cells/ml
Sulphate-reducing bacteria (related to the genera Desulfovibrio, Desulfobulbus, Desufococcus, Desulfobacterium, Desulfotomaculum, Desulfosarcina, Desulfobacter)	10^6 - 10^8
Cellulose-degrading microorganisms	$10^4 – 10^6$
Denitrifying bacteria	$10^2 – 10^4$
Methane-producing bacteria	$10^2 – 10^3$
Anaerobic heterotrophic bacteria related to other physiological groups	$10^1 – 10^3$

decreased below the relevant acceptable levels. The removal of pollutants was efficient even during the cold winter months when the temperatures were close to 0°C. However, the residence times were much longer.

The microbial dissimilatory sulphate reduction and biosorption were the main processes involved in this removal. The uranium was precipitated mainly as the mineral uraninite (UO_2). The toxic heavy metals and arsenic were precipitated mainly as the relevant insoluble sulphides. Most of the radium as well as portions of the uranium, heavy metals and arsenic were removed by adsorption on the organic matter in the anoxic cell. The pH of the waters being treated was increased to levels about the neutral point. This increase of the pH was connected not only with the alkalinity produced during the microbial dissimilatory sulphate reduction but also with the chemical neutralisation caused by alkaline components present in the spent mushroom compost. The efficiency of the microbial dissimilatory sulphate reduction depended to a great extent on the digestibility of the solid organic substrates in the anaerobic cell. The concentration of dissolved organic compounds, which were products from this digestion and were used by the sulphate-reducing bacteria as sources of carbon and energy, decreased in the course of time. However, it was possible to use efficiently the anoxic cell under continuous – flow conditions for a period of about 6 months without any change of the solid substrates.

The effluents from the anoxic cell were enriched in dissolved organic compounds. The concentration of ammonium ions in the effluents was also increased considerably due to the ammonification of the organic matter. The concentration of phosphate ions was also increased and this was connected with the solubilization of some phosphate present in the spent mushroom compost. The effluents still contained high concentrations of hydrogen sulphide.

In the constructed wetland the intensive growth and activity of different heterotrophic microorganisms resulted in a significant decrease in the concentration of dissolved organic compounds in the waters. The ammonium ions were removed by their oxidation to nitrates by the chemolithotrophic nitrifying bacteria as well as a result of their assimilations by the plant and microbial cenoses in the wetland. The phosphates were removed as a result of both precipitation and assimilation processes. No pathogenic microorganisms were detected in the product water.

3. CONCLUSIONS

1. The anoxic sulphate-reducing and alkalinity producing cell achieved an efficient removal of radioactive and toxic heavy metals and arsenic from waste waters, decreasing the concentrations of these pollutants below the relevant permissible levels for waters intended for use in the agriculture and/or the industry.
2. The microbial dissimilatory sulphate reduction was the main process involved in the removal of pollutants. The efficiency of these process depended to a great extent on the digestibility of the solid organic substrates present in the anoxic cell.
3. Uranium was precipitated in the anoxic cell mainly as the mineral uraninite (UO_2), while the toxic heavy metals and arsenic were precipitated mainly as the relevant sulphides. Most of the radium as well as portions of the above-mentioned pollutants were removed by adsorption on the organic matter in the anoxic cell.
4. The effluents from the anoxic cell were enriched in dissolved organic compounds. These compounds were efficiently removed by degradation caused by heterotrophic microorganisms inhabiting the constructed wetland.
5. The data from this study revealed that passive system consisting of anoxic cells and constructed wetlands arranged in series can be efficiently applied in commercial scale to treat acid drainage waters.

ACNOWLEDGEMENT

A part of this study was funded by the National Science Fund (Research Contract № TH-803/98).

REFERENCES

Cambridge, M. 1995. Use of passive system for the treatment and remediation of mine outflows and seepages. *Miner. Industry Int.* May 1995 : 35-42.

Gazea, B., K. Adam & A. Kontopoulos 1995. A review of passive systems for the treatment of acid mine drainage. Paper presented at the *Minerals Engineering'95 Conference, St. Ives, England, 14-16 June 1995.*

Groudev, S.N., P. S. Georgiev, K. Komnitsas, I. I. Spasova & A. T. Angelov 1999. Treatment of drainage waters from a flotation tailings pond by a natural wetland. In I. Gaballach, J. Hager & R. Solozabal (eds), *Global Symposium on Recycling, Waste Treatment and Clean Technology, San Sebastian, 5-9 September 1999*: vol. III, 2133 --2140. Warrendale, PA: TMS, The Minerals, Metals & Materials Society.

Groudeva, V.I., I.A. Ivanova, S.N. Groudev & G.C. Uzunov 1993. Enhanced oil recovery by stimulating the activity of the indigenous microflora of oil reservoirs. In A.E. Torma, M.L. Appel & C.L. Brierley (eds), *Biohydrometallurgical Technologies*, vol. II, 349-356. Warrendale, PA: TMS Minerals, Metals & Materials Society.

Guzek, J.J. 1995. Passive-treatment of acid rock drainage: What is the potential bottom line? *Mining Engineering*: *March 1995*: 250-253.

Karavaiko, G.I., G. Rossi, A.D. Agate, S.N. Groudev & Z.A. Avakyan 1988. *Biogeotechnology of Metals. Manual.* Moskow: GKNT International Projects.

Environmental Issues and Management of Waste in Energy and Mineral Production, Singhal & Mehrotra (eds)
© 2000 Balkema, Rotterdam, ISBN 90 5809 085 X

Use of electric utility wastes for control of acid mine drainage

A.G.Kim
US Department of Energy, Pittsburgh, Pa., USA

ABSTRACT: Placement of fly ash in abandoned, reclaimed or active surface coal mines is intended to reduce the amount of acid mine drainage (AMD) produced. Water quality changes have been monitored at three surface mines where fly ash grout was injected after reclamation. Also, a laboratory column leaching study exposed samples of fly ash to AMD surrogates for 30 to 180 days. Changes in acidity and potential release of heavy metals were primary areas of interest. Both field and laboratory studies indicate that fly ash may be an economical reagent for ameliorating acid mine drainage without adverse environmental effects.

INTRODUCTION

Over 50 % of electricity produced in the U.S. is generated by coal-fired power plants, most of which burn pulverized coal (PC). Fluidized bed combustion (FBC) boilers which burn a mixture of coal or coal waste and limestone, constitute a small proportion of the power generation facilities.

Of the 726 million mt of coal used annually to generate electric power, approximately 10 to 15 % is recovered as coal combustion by-products (CCB) (Tyson 1992). According to the American Coal Ash Association, 92 million mt of CCB were generated in 1996, including fly ash (60 %), bottom ash (15 %), flue gas desulfurization sludge (23 %) and boiler slag (2 %) (ACAA 1996). FBC units generated 8.5 million mt of products (FBCP) in 1995 (CIBO 1997). Because of the added sorbent, an FBC boiler burning coal generates approximately 110 kg of by-product per metric ton of coal, approximately 40 % more than a conventional PC boiler (79 kg/mt).

To reduce costs and to limit the need for off-site disposal, many power producers are actively marketing CCB for use in a variety of applications, including cement/concrete, fill, road base, waste stabilization and agriculture. On the average, less than 25 % of all CCB is utilized. Less than 1 % of all CCB produced is used in coal mine remediation. Approximately 75 % of FBCP is beneficially used, and in 1995, 3.3 million mt was utilized in coal mine remediation (CIBO 1997).

Acid mine drainage (AMD), a legacy of coal mining in the U.S., has had an impact on surface streams since the eighteenth century. A 1967 survey estimated that over 16,000 km of streams in Appalachia were degraded by drainage from mines and that over 8,000 km were acidic as a result (FWPCA 1969). The Federal Water Pollution Control Act of 1972, as amended by the Clean Water Act of 1977, and the Surface Mining Control and Reclamation Act, require that water from active mines be treated, in perpetuity if necessary, to meet National Pollution Discharge Elimination System (NPDES) standards for discharge to surface streams. The compliance with these discharge standards is largely responsible for a significant improvement in stream water quality over the past 25 years. Nationwide, the cost of treating AMD from coal mining operations is estimated at over $1 million/day (Kleinmann 1989). Preventive measures during reclamation or closure that inhibit the formation of AMD can significantly reduce this economic burden.

The formation of AMD requires pyrite, air and water to produce ferrous ions and sulfuric acid:

$$2FeS_2 + 2H_2O + 7O_2 \rightarrow 4H^+ + 2Fe^{2+} + 4SO_4^{2-} \qquad \textbf{(1)}$$

The oxidation and hydrolysis of the ferrous iron produces hydrated iron oxide and additional acidity.

$$2Fe^{2+}+\frac{1}{2}O_2+2H^{+}\rightarrow 2Fe^{3+}+H_2O \quad (2)$$

$$2Fe^{3+}+4H_2O\rightarrow Fe_2O_3\cdot H_2O+6H^{+} \quad \mathbf{(3)}$$

The acid lowers the pH of water, and the iron compounds form an unsightly precipitate that coats stream beds, destroying habitat for aquatic life. As the pH of the system decreases (<3) ferrous iron is oxidized by *Thiobacillus ferrooxidans,* and the ferric ions oxidize pyrite to produce additional acidity:

$$FeS_2+14Fe^{3+}+8H_2O\rightarrow 15Fe^{2+}+2SO_4^{2-}+16H^{+} \quad \mathbf{(4)}$$

To limit the formation of AMD, pyrite must be isolated from air or water, which may not be feasible in a large natural system. However, if acid formed can be neutralized and the pH maintained above 3, the effect of ferric ion oxidation and the activity of the iron-oxidizing bacteria can be minimized, significantly reducing the amount of AMD formed.

In mine reclamation, CCB may impact the discharge of AMD through four possible mechanisms: neutralization, bacterial inhibition, pyrite encapsulation and water diversion (Kim and Cardone, 1997). If the CCB is alkaline, it can neutralize acidic groundwater; and, at a higher pH, bacterial activity is inhibited. A pozzolanic CCB can encapsulate pyrite, isolating it from air and water and preventing the formation of AMD. The deposition of CCB or a CCB grout can also reduce the permeability of mine strata, diverting water away from acid-forming materials. Because of the high concentration of calcium and the formation of cementitious compounds, FBCP may be a superior reagent for AMD control in inactive or abandoned mines.

In evaluating the potential beneficial use of combustion products at mine sites, two fundamental points should be considered: the effect of the alkaline material on the net acid production and the potential release of trace metals from the combustion by-products to surface streams or groundwater.

FIELD STUDIES

To evaluate the effect of fly ash injection on AMD, water quality was monitored at three reclaimed surface mines (Bognanni, Fran and Pierce) where a fly ash grout was injected into subsurface areas believed to be zones of acid production (Ackman et al. 1996). At these three sites, when mining was completed, overburden spoil was covered with a variable thickness of soil, then planted with grass. However, water, either from precipitation or infiltrating groundwater, apparently reacted with buried pyrite to produce AMD. The injection of the alkaline fly ash grout was tried as a relatively simple method to neutralize existing acid and to reduce the rate of acid formation. The objective at all three sites was to determine if this method was applicable as a single permanent treatment for AMD control.

Water samples were collected at outflows, from boreholes within injection areas, and from boreholes in ungrouted control areas for a year or more before the injection of the grout, and for a year or more after grout injection. In addition to the before and after comparison, inflow and discharge samples were also compared, as were samples from the grout injection areas and from ungrouted control areas.

Site Descriptions

The three sites described below were all surface coal mines that had been reclaimed with conventional methods; all had developed acid seeps after reclamation.

The Bognanni site is a 14.6 ha reclaimed strip mine in Greene County, PA (Kim & Ackman 1994). The depth of buried spoil ranges between 5 and 15 m, averaging about 10 m, under 2 m or less of vegetated cover. If it is assumed that the porosity of the material is 20 %, the pore volume in the 10 m of spoil above the pit floor is estimated as 7,500 m^3. Terrain conductivity (Ackman & Cohen 1994) indicted that groundwater flow at the site was from east (the area of a buried highwall) to west through the reclaimed area to a seep approximately 70 m south of the site boundary. Thirty-four wells were used for injection, with the majority of these located in a 1.2-ha central section of the reclaimed area. Nine monitoring wells are located in the grouted area and four wells are in spoil areas that were unaffected by grout injection. Additional monitoring points are the Inflow and the discharge point, labeled the Seep. The seep is outside the target area and drains portions of the reclaimed mine that were unaffected by grout injection.

The Fran site is a 15 ha reclaimed strip mine in Clinton County, northern Pennsylvania. Discrete piles of tipple refuse or pit cleanings are believed to be buried beneath spoil of pyritic shale and sandstone (Schueck et al. 1996). Geophysical techniques were used to estimate the location of the buried refuse and the direction of water flow. Infiltrating precipitation is believed to flow through the area and then leave the site either through a seep or through fractures in the pit floor.

Forty-two monitoring wells were drilled on and adjacent to the site. Water samples were also collected at surface discharges 66 m south of the site. For this site, values are averages of all samples for the ungrouted area (Spoil), for the area where grout was placed (Injection), for the area that would have been affected by water from the injection area (Downdip), and for the seep (Discharge 1) and fractured area (Discharge 2).

The Pierce site is a 6 ha section of a 32 ha reclaimed surface mine in Upshur County, West Virginia (Hawkins et al. 1991). The spoil contained part of the Lower Kittanning coal seam, gray carbonaceous shale and a gray sandstone. Within the injection area, 15 monitoring wells were drilled and cased with PVC pipe. Water samples were also obtained at a discharge point at the edge of the reclaimed area. At this site, monitoring wells are grouped as Inflow, Injection for the grouting area, Downdip for the area receiving water from the grouted area, Spoil for the ungrouted area, and Seep for the discharge. The values for each set are averaged for: 1989, pre-grouting; 1990, immediate post-grouting; and 1995, later post grouting.

Materials

At the Bognanni site, the grout was prepared with water from the AMD treatment system and AMD treatment sludge or lime waste (which consists of about 30 pct unused lime). Three fly ashes were used, two from conventional power plants and one FBC ash. The amount of fly ash injected into a single hole varied between 0.4 and 41 m^3 with an average of 5.5 m^3. On a volume basis, fly ash constituted less than 6 pct of the injected grout, and the 192 m^3 of fly ash is equivalent to less than 5 pct of the estimated pore volume in the 1.2 ha section. The fly ash grout was pumped to refusal at a maximum pressure of 3.5 kg/cm^2.

At the Fran site, a grout of fluidized bed combustion (FBC) ash and water (1 m^3/800 L) was injected into pods of refuse in order to divert water from the acid-producing materials. The same grout mixture, which formed a low-strength cement, was used to cover the piles of buried coal refuse and to seal fractures in the pit floor. Grout was pumped to refusal in 650 holes, averaging 3 to 9 m in depth. The total of 3440 m^3 of injected grout is the equivalent to 4 % of the total volume of voids.

The grout used at the Pierce site was a mixture of 375 g of cement and 750 g of fly ash per L of water. A total of 380 m^3 of grout was injected through 62 cased wells, primarily in two areas near the buried highwall. The volume of injected grout is comparable to less than 1% of the estimated void volume at the site.

Water Quality

Although water samples were collected at all three sites, before and after grouting, the most comprehensive data set is available for the Bognanni site. The water entering the Bognanni site had an average pH of 7 and contained 2 ppm total iron, 1 ppm aluminum, 35 ppm sulfate, less than 1 ppm barium, and approximately 0.1 ppm each cobalt, chromium, nickel, antimony and zinc (Kim & Ackman, 1995). The pH of the water decreased to 5 where it entered the reclaimed area (Inflow), then decreased to almost 3 within the spoil. Prior to grouting, the water at the seep had a pH of 3.2. In 3 years after grouting, the pH of water at the seep increased to 3.3, slightly higher than the pH in the ungrouted area (Fig. 1).

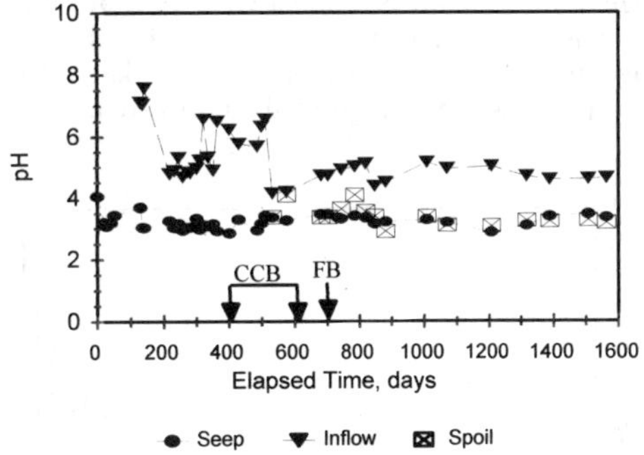

Figure 1. Change in pH with time at inflow, outflow and control area, Bognanni site.

Immediately after grout injection, the pH in the area of injection increased. Over the three year post-injection period, the average pH at wells in the grouted area increased 0.5 pH units (Fig. 2).

The concentrations of ferrous and total iron, calcium, magnesium, aluminum, sodium and manganese tended to be higher in the grouted area than in the control area. The increase in the concentrations of calcium, magnesium and aluminum in the spoil area after grouting may be due to the presence of lime and AMD sludge in the grout. In the grouted area, the average total iron concentration remained almost constant before and after grouting, but the proportion of ferrous iron decreased. At the seep, the concentrations of ferrous and total iron decreased after grouting, but the proportion of ferrous iron remained constant. The concentrations of ferrous and total iron, magnesium, aluminum, and manganese at the seep all decreased by approximately 35% after grouting.

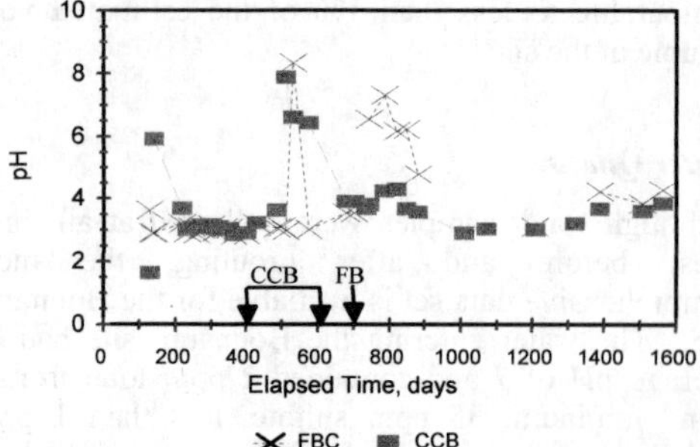

Figure 2. Change in pH at monitoring wells in areas grouted with fly ash from conventional (CCB) and fluidized bed (FBC) power plants, Bognanni site.

Trace element analysis indicated that barium (Ba), present in the incoming water, was not detected at the seep (Fig. 3). Only the concentrations of cobalt (Co), nickel (Ni), selemium (Se)and zinc (Zn) were greater than 1 mg/L. The difference in the concentrations in areas grouted with CCB or FBC fly ash was less than 0.5 mg/L. At the seep, the concentration of these metals were less than average concentrations in the ungrouted control area.

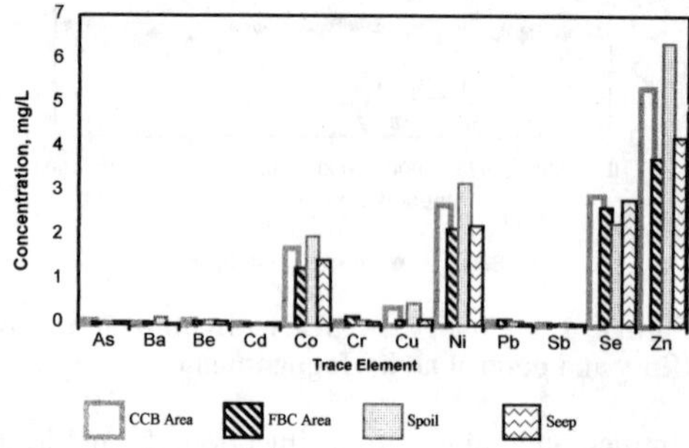

Figure 3. Average concentrations of trace metals in grouted area, control area and outflow, Bognanni site.

At the Pierce site, water samples were obtained monthly prior to grouting (1989) and immediately after grouting (1990). During 1995, water samples were collected at 3 month intervals. The average pH of the water in the injection area decreased in the year after injection, but had increased when the water was sampled five years later (Fig. 4). Similar changes in average pH were observed in the samples from areas downdip of the injection area and at the discharge point. However, the water entering the site (inflow) and in the ungrouted spoil area also had a higher average pH in 1995.

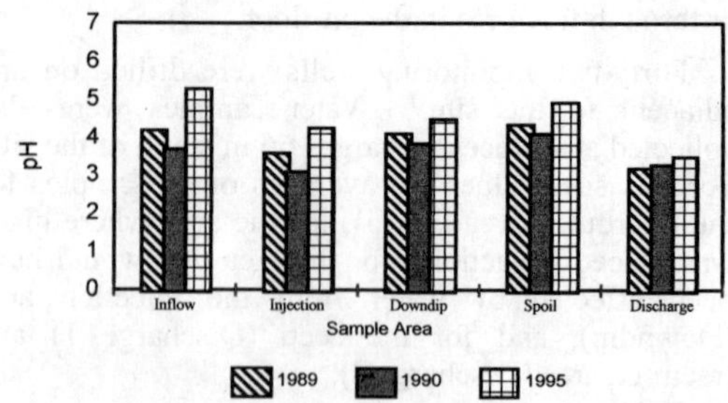

Figure 4. Average pH before(1989) and after injection of fly ash grout, Pierce site.

Acidity, measured as ppm of $CaCO_3$, decreased in the injection area, as well as in the inflow and discharge samples. In the untreated spoil area, acidity increased. Although water quality improved at the discharge point, a pH less than 4 and an acidity of approximately 100 ppm indicate continued release of AMD.

Trace element concentrations, determined in 1995, were higher in the injection, downdip and discharge samples than in inflow or ungrouted spoil samples (Fig. 5). Only the concentrations of cobalt, nickel and zinc exceeded 0.2 mg/L; all values were less than freshwater aquatic life criteria.

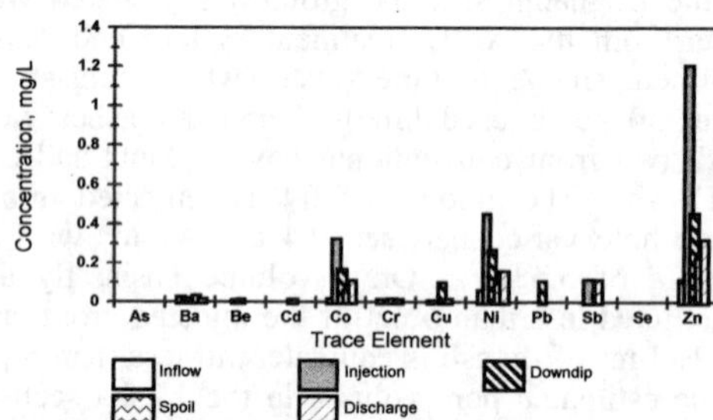

Figure 5. Comparison of post-injection trace element concentrations at Pierce site.

At the Fran site, water samples were collected on a fairly regular basis before and after grouting. These are grouped as before (1990), immediately after (1992) and more recent (1994). The pH increased in samples from the injection area, the downdip areas, and from two discharge areas (Fig. 6). The pH of water in the untreated spoil also increased immediately after injection, but then decreased. The acidity decreased in all areas, except in the untreated spoil. The pH of the discharges at less than 3 and acidity exceeding 2000 ppm indicate that the site continues to produce a significant amount of AMD.

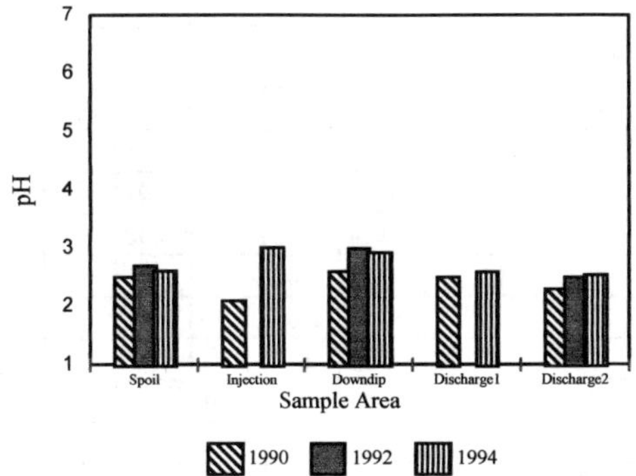

Figure 6. Average pH in untreated and treated areas before (1990) and after grout injection, Fran site.

In 1994, average concentrations of arsenic, cobalt, copper, nickel and zinc were higher in the injection area than in background areas. (Fig. 7). However, the concentrations in downdip and discharge samples were closer to those in the ungrouted spoil.

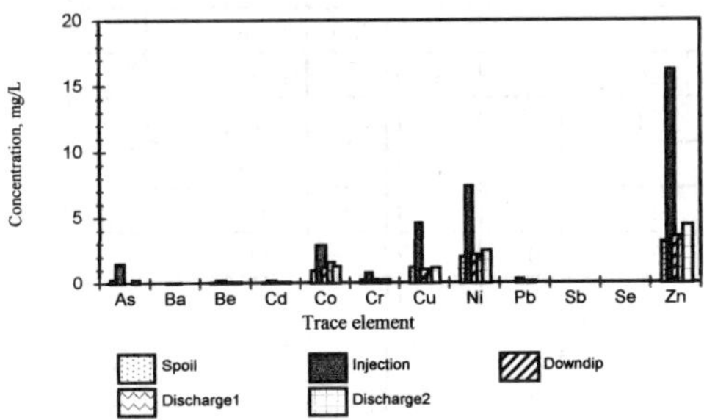

Figure 7. Concentration of trace metals in treated and untreated areas 2 years after grout injection, Fran site.

At all three mine sites, the injection of fly ash grout produced minimal improvement in water quality at the discharge point. The magnitude of the change may be related to the relatively small volume of material introduced into the area of acid generation and to the small proportion of the site treated.

LEACHING STUDY

When deposited in a mine environment, the fly ash could be exposed to a variety of fluids, acidic, neutral and basic, possibly containing high levels of ferric iron. The potential release of trace elements is one of the primary concerns in placement of fly ash underground. In order to characterize the potential effects, the US Department of Energy is conducting a long term column leaching study of the effects of various leachants on fly ashes.

Laboratory Methods

One kg samples of fly ash are placed in 5 cm by 1 m acrylic columns. The leachant solution flows through the column at a rate between 130 and 275 mL/d. Leachate samples, collected at 2 to 3 day intervals, are analyzed for pH, acidity and/or alkalinity, ferrous iron, total iron, aluminum, manganese, magnesium, calcium, sodium, potassium, sulfate and the heavy metals arsenic, barium, beryllium, cadmium, cobalt, chromium, copper, nickel, lead, antimony and zinc. The entire system was designed for the simultaneous leaching of four different fly ashes by seven leachants using a total of seven reservoirs and 28 columns (Kim & Sharp 1995, Kazonich & Kim 1997).

Although seven leachants are used in the study, this discussion focuses on the AMD surrogates sulfuric acid (H_2SO_4), pH = 1.2; and ferric chloride ($FeCl_3$ with HCl), pH = 1.95. The results with deionized water (H_2O), pH = 7 are included as the baseline standard. The length of the leaching test varies between 30 and 180 days. Of environmental interest is the potential release of RCRA heavy metals: arsenic (As), barium (Ba), cadmium (Cd), chromium (Cr), lead (Pb), and selenium (Se).

Results

The leaching study indicates that fly ashes generate alkalinity when exposed to an acid leachant. Although the initial effect is most obvious, the pH of the outflow sample remains significantly higher than that of the leachant solution (Fig.8).

The average release of arsenic, beryllium, cadmium, cobalt, lead, antimony and selenium is less than 5 mg per kg of fly ash in water and the AMD surrogates (Fig. 9). The average cumulative amount of chromium, copper and nickel leached is less than 10 mg/kg. The total amount of barium and zinc leached exceeds 10 mg/kg.

Most elements tend to be more soluble in sulfuric acid and ferric chloride than in water. The average release of heavy metals from combustion by-products is less than 10 % for all trace metals except beryllium and copper in sulfuric acid. Only the solubility of zinc exceeds 20 % of the amount in the sample (Fig. 10).

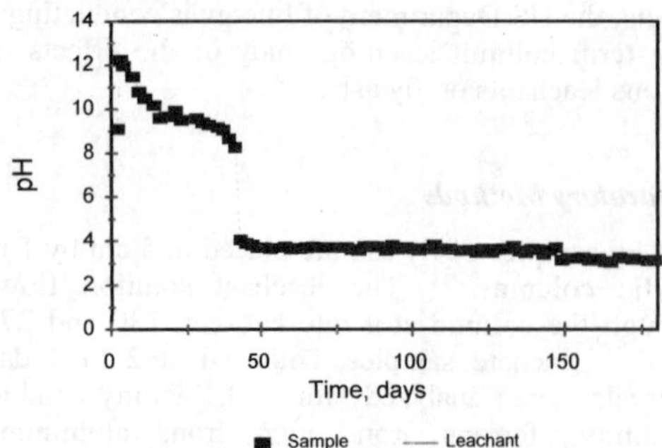

Figure 8. Comparison of pH of leachate samples to pH of leachant, Fly Ash # 25 in .1N H_2SO_4.

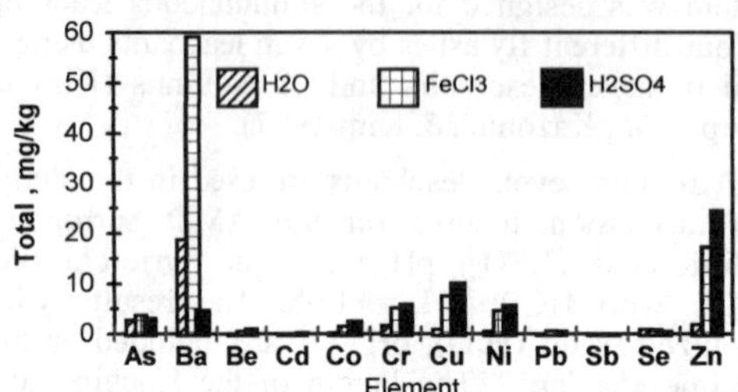

Figure 9. Average cumulative amount of trace metals leached in water, ferric chloride, and sulfuric acid.

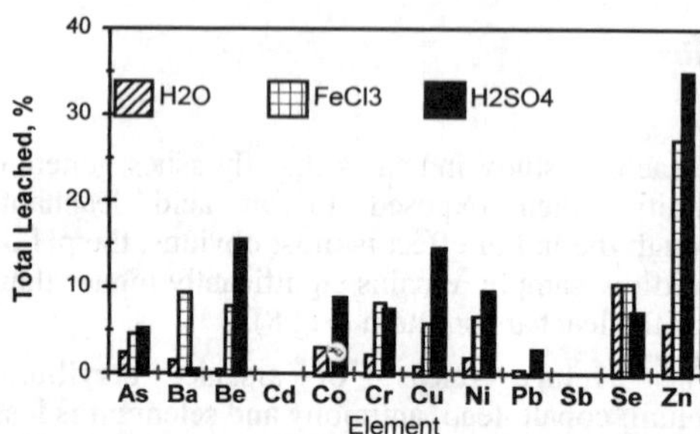

Figure 10. Relative amount of trace metals leached from fly ash in water, ferric chloride and sulfuric acid.

When individual samples are considered, the release of RCRA elements in water is generally much less than the RCRA standard for toxicity (Table 1). Of 192 potential values, almost 50 % are less than 0.01 mg/L, and only 2 values exceed the RCRA standard. When leached with ferric chloride, average concentrations tend to be higher for some samples, but almost 50 % are less than 0.01 mg/L. Only 3 average values exceed the RCRA standard (Table 2). In 0.1 N sulfuric acid, 2/3 of the average values are greater than 0.01 mg/L, but only one value exceeds the RCRA standard (Table 3).

Table 1. Average concentration of RCRA elements in water, mg/L.

Sample #	As	Ba	Cd	Cr	Pb	Se
RCRA STD	5	100	1	5	5	1
2		0.01	0.00	0.01	0.01	
3	0.06	0.14	0.01	0.05	0.08	
4		0.13	0.01	0.42	0.02	
5		0.36	0.01	0.16	0.02	
6	1.28	0.01				0.02
7	1.60	0.01		0.01		0.07
8		0.26		0.14		
9		0.78	0.01	0.02		
10		38.12		0.02		
11	0.67	0.03		0.02		
12	1.18	0.02		0.01		0.12
13	0.97	0.03	0.01	0.01		
14		0.14		0.14		
15	0.11	0.36	0.04	0.63		0.69
16	1.86	0.18		0.01		
17		0.13	0.01	0.03		
18	0.02	0.24		0.06		
19		0.13	0.01	0.06		
20		18.99		0.01		
21		0.14		0.14		
22		5.24		0.28	0.00	
23	0.01	2.89		0.06		0.03
24	0.23	0.02		0.25		
25		0.07		0.02		0.01
26		0.24		0.01		0.09
27	0.25	0.03		0.31		1.52
28	0.33	0.09	0.03	0.66	0.01	0.06
29	0.81	0.14	0.04	0.05		2.06
30	0.61	0.02		0.06		0.22
31	0.28	0.08	0.01			0.09
32		0.11		0.52		
33		0.59		0.38		

SUMMARY

At three field sites, differences in water quality between areas injected with fly ash and areas of untreated spoil indicated localized effects, particularly a decrease in acidity and an increase in pH. The field studies indicate that injection of CCB in areas of AMD produces no significant release of trace metals.

Comparison of the water quality between the spoil and the discharge indicated that the fly ash grout produced a slight but measurable change at the discharge point. At all three sites, the amount of fly

Table 2. Average concentration of RCRA metals in $FeCl_3$ leachate, mg/L.

Sample #	As	Ba	Cd	Cr	Pb	Se
RCRA STD	5	100	1	5	5	1
2	0.21	2.13		1.62		
3	0.42	1.80		1.14	0.31	
4		0.29				
5		0.38				
6	0.61	12.86		1.49	0.40	0.12
7	0.60	15.69		1.40	0.74	0.02
8		0.43				
9		8.12				
10		18.33				
11	0.82	14.31	0.08	1.01	0.14	1.31
12	0.75	15.12	0.05	0.36	0.09	
13	0.91	20.63	0.01	1.26	0.46	
14	1.05	1.47	0.02			
15	2.02	16.60	0.04		0.41	1.85
16	2.23	15.23	0.03			
17	1.12	1.76	0.02			
18	0.47	3.71		0.55		
19		0.29	0.04			
20		15.02				
21		0.27				
22	0.01	24.09		0.95		
23	0.01	7.34				
24	0.24	5.41	0.02	3.91	0.93	
25		0.36	0.00			
26		0.60	1.05			0.10
27	0.74	6.81	0.02	3.74	0.73	0.53
28	1.31	9.32	0.05	4.89	0.67	0.28
29	0.53	1.37	0.03			0.29
30	0.76	20.84	0.02	2.15		0.11
31	1.89	11.65		2.51	0.58	0.03
32		0.50	0.00	0.86	0.06	
33	0.03	2.34		1.14	0.11	

Table 3. Average Concentration of RCRA metals in H_2SO_4, mg/L.

Sample #	As	Ba	Cd	Cr	Pb	Se
R C R A STD	5	100	1	5	5	1
2	0.42	0.04	0.01	0.60	0.05	
3	1.46	0.03	0.00	0.36	0.57	
4		0.17	0.01	0.48	0.02	
5		0.10	0.01	0.45	0.07	
6	0.51	0.12	0.02	0.52	0.08	
7	1.53	0.15	0.01	0.51	0.07	
8	0.09	0.12	0.01	0.59	0.08	
9		0.22	0.01	0.01		
10		0.20	0.01	0.03		
11	0.61	0.12	0.04	0.53	0.09	
12	0.43	0.09	0.04	0.43	0.06	
13	0.52	0.12	0.05	0.46	0.01	
14	0.20	0.16	0.02	0.49		
15	0.36	0.09	0.02	0.47	0.02	2.10
16	0.95	0.09	0.05	0.79		
17	0.14	0.13	0.04	0.39		
18		0.19	0.01	0.39		
19		0.15	0.04	0.41		
20		11.35		0.00		0.06
21		0.15	0.03	0.28		
22		5.62		1.08		
23	0.00	0.87		0.11	0.00	0.04
24	0.19	0.07	0.01	1.14	0.01	
25	0.14	0.10	0.02	0.51	0.01	
26		0.14		0.01		
27	0.02	0.11	0.04	1.79		0.12
28	0.73	0.10	0.07	3.13		0.10
29	0.03	0.11	0.04	0.41		0.69
30	0.05	0.12	0.03	0.51		
31	0.02	0.09	0.08	0.73	0.04	
32		0.12	0.24	0.89		
33		0.14	0.01	0.58		

ash injected and the volume of subsurface area it could have affected are relatively small. . Engineering design, the distribution of material and applied volume have a significant effect on results, particularly at abandoned or inactive mines. These studies demonstrate that fly ash can be used in passive AMD treatment, but that long term control of AMD will require larger volumes of CCB and a more efficient method of application

The laboratory leaching studies indicate that trace metals, particularly barium and zinc, may be released from CCB upon exposure to AMD. However, the amount is relatively small. For the 32 samples studied, only 1% of average leachate concentrations of the RCRA metals exceed the standard.

The data from the leaching tests also indicate that the mobilization of trace elements from CCB is not a simple function of the concentration in the solid. For all of the trace elements, except zinc, less than 20 % of the amount present in the ash is extracted by long-term leaching.

The results of both laboratory and field studies support the use of CCB as an inexpensive reagent to control the formation of AMD.

REFERENCES

ACAA. 1996. Coal Combustion By-Product - Production and Use. American Coal Ash Association, Alexandria, VA, 1p.

Ackman, T.E. and K.K. Cohen. 1994. Geophysical Methods Remote Sensing Applied to Mining Related Environmental Problems. Proc. International Land Reclamation and Mine Drainage Conference, 1994, BuMines Special Pub. SP06D-94, v. 4, pp. 208-217.

Ackman, T.E., J.R. Jones and A.G.Kim. 1996. Water Quality Changes at Three Reclaimed Mine Sites Related to The Injection of Coal Combustion Residues. 13th Annual Pittsburgh Coal Conf University of Pittsburgh, Pittsburgh, PA, USA, pp. 1055-1060.

Council of Industrial Boiler Owners (CIBO). 1997. Report to the U.S. Environmental Protection Agency on Fossil Fuel Combustion Products from Fluidized Bed Boilers. CIBO Special Project on Non-Utility Fossil Fuel Ash Classification.

Hawkins. J.W., T.E. Ackman and W.W. Aljoe. 1991. The Effect of Grout Injection on Mine Spoil Groundwater Hydrology. Presented at the National Meeting of the American Society for Surface Mining and Reclamation, Durango CO., May 14-17, 199 1.

FWPCA. 1969. Stream Pollution by Coal Mine Drainage in Appalachia. Federal Water Pollution Control Administration, U.S. Dept. of Interior, 1967, revised 1969, 271 pp.

Kazonich, G. and A.G. Kim. 1997 Leaching Fly Ash with Environmental and Extractive Lixiviants, 12th International Symposium on Coal Combustion By-Product Management and Use, American Coal Ash Association, Orlando, FL,, p 11-18.

Kim, A.G. and T.E. Ackman. 1994. Disposing of Coal Combustion Residues in Inactive Surface Mines: Effects on Water Quality, Proc. International Land Reclamation and Mine Drainage Conference, 1994, BuMines Special Pub. SP06D-94, v. 4, pp, 228-236.

Kim, Ann G. and Carol Cardone. 1997. Preliminary Statistical Analysis of Fly Ash Disposal in Mined Areas. Proc: 12th International Symposium on Coal Combustion By-Product Management and Use. American Coal Ash Association, v. 1, pp. 11-1 - 11-13.

Kim, A.G. and F.A. Sharp. 1997, Leaching Coal Combustion By-Products with Acidic, Basic, and Neutral Liquids, *1995 International Ash Utilization Symposium,* CAER, Lexington, KY, 6 pp.

Kleinmann, R.L.P. 1989. Acid Mine Drainage. Engineering and Mining Journal. July 1989, Chicago, IL Vol. 190, No. 7, pp. 16I-16N.

Schueck, J., M. DiMatteo, B. Scheetz and M. Sillsbee. 1996. Water Quality Improvements from FBC Ash Grouting of Buried Piles of Pyritic Materials on a Surface Coal Mine. Proc. 13th Annual Mtg. American Society for Surface Mining and Reclamation, May 18-23, 1996, Knoxville, TN, pp.308 - 320.

Tyson, S. S. 1992. Guidance for Coal Ash Use Codes and Standards Activities for High-Volume Applications, Ninth Annual Pittsburgh Coal Conf, University of Pittsburgh, Pittsburgh, PA, pp.815-817.

Environmental Issues and Management of Waste in Energy and Mineral Production, Singhal & Mehrotra (eds)
 ISBN 90 5809 085 X

Development of a process to prevent acid generation from waste rock and mine tailings

R. Mehta
University Center for Environmental Sciences and Engineering, University of Nevada, Reno, Nev., USA

S. Chen & M. Misra
Department of Metallurgical and Materials Engineering, University of Nevada, Reno, Nev., USA

ABSTRACT: A process to prevent acid generation from waste rock and mine tailings is currently under development at the University of Nevada, Reno (UNR). The UNR process is modification of the existing DuPont Process where an effort is underway to replace the use of potassium permanganate by other inexpensive coating agent such as magnesium oxide. In addition, other advantages of the UNR process are that it is applicable in a relatively lower pH range (9-10) as opposed to pH (12) and is easy to deploy in the field. In this paper, we will present the passivation results of pure pyrite, waste rock and mine tailings.

INTRODUCTION

Acid mine drainage (AMD) poses a serious threat to the environment due to its strong acidity (pH <2). This condition occurs as a result of the oxidation of pyrite such as, iron and other sulfides present in waste rock and tailings exposed in the environment. The oxidation of the pyrite results in the production of sulfuric acid. According to some estimates, the oxidation of one mole of pyrite by oxygen can produce two moles of sulfuric acid (Nordstrom, 1982).

$$FeS_2 + H_2O + 2O_2 = Fe^{2+} + 2SO_4^{2-} + 2H^+$$

Government Regulations prohibit the discharge of AMD into environment as it can change the ecological system balance. Federal water quality standards in the U.S. require that the pH of water discharge be between 6.0 and 9.0, and total iron and manganese concentration should be less than 3.2 mg/l and 2.0 mg/l respectively.

According to the US Bureau of Mines estimates, the abandoned coal and metal mines, and the associated piles of mine waste adversely affect over 12,000 miles of rivers and streams, and over 180,000 acres of lakes and reservoirs in the US (Kleinmann, 1989). In the U. S., the mining industry spends over $1 million per day to treat AMD (Evangelou, 1995). Therefore, finding an economic and effective way to prevent AMD is essential for the mining industry.

The various prevention methods currently in use include neutralization, inundation, and encapsulation. However, these methods have some drawbacks. For example, neutralization method only neutralizes the surface acidity. Because of the partial oxidation of pyrite, it is difficult to achieve total neutralization in a specified time scope. Inundation method though limits the access of oxygen by maintaining the waste in a saturated or submerged condition, the passivation lacks due to the fluctuations of water levels. The effectiveness of this method is further limited by the existence of ferric ion, which is an alternate oxidizer. The encapsulation method, on the other hand works by depositing a stable coating on the surface of pyrite, thus preventing the exposure of pyrite surfaces to outside oxidizers. If successfully employed, this technology can provide an effective and economic way to prevent AMD.

DuPont has invented a passivation process to prevent AMD (De Vries, 1996). The process works by contacting the potassium permanganate to the sulfide containing minerals at pH 12. This forms a stable coating on the sulfides. The process has produced beneficial results, but has some problems. First, the required pH value (12.0) needed for this process is prohibitively high and secondly, the dosage level of potassium permanganate required to form the stable coating makes it expensive for the treatment of wastes rocks and tailings.

The purpose of this research is to develop a passivation process, which would utilize a lower pH value, and which would use other inexpensive coating agents, that are more economically viable.

In this research, pure pyrite samples were passivated and the hydrogen peroxide oxidation method as suggested by the DuPont was used to determine the acid generation potential of the passivated samples. The method is inexpensive and a quick way to screen the desired operating conditions.

EXPERIMENTAL

Samples and Reagents:

The four samples were used in the passivation work: pure –325 mesh pyrite, the Hecla sample (waste rock), the Newmont sulfide tailings sample, and the Brohm's Ruby Gulch sulfide tailings sample. As received waste rock and tailings samples were crushed and -10 mesh samples were used in the tests. Other chemicals used were: magnesium oxide (94.1%) from Backer Chemical and calcium oxide (99.2%) from MC/B.

Passivation Tests Procedure:

Three types of tests were conducted.

Blank: 5 gms of pyrite/waste rock/tailings was placed in a 250 ml beaker and 20 ml of deionized water was added. No other chemical reagent was added. pH was monitored for 3 hours and the sample was filtered and washed with 200 ml deionized water.

Control: 5 gms of pyrite/waste rock/tailings was placed in a 250 ml beaker and 20 ml of deionized water was added. Then 20 mg of calcium oxide was added. pH was monitored for 3 hours and the sample was filtered and washed with 200 ml deionized water.

UNR Process: 5 gms of pyrite/waste rock /tailings was placed in a 250 ml beaker and 20 ml of deionized water was added. Then 20 mg of magnesium oxide and 20 mg of calcium oxide was added. pH was monitored for 3 hours and the sample was filtered and washed with 200 ml deionized water.

Hydrogen Peroxide Test Procedure:

The washed samples from the above tests were subjected to 85 ml deionized water and 15 ml of 30% hydrogen peroxide. The slurry was stirred gently and the solution pH was monitored for 24 hours at specific time internals.

RESULTS AND DISCUSSION

The solution pH changes during the hydrogen peroxide test are shown in Figure 1. The blank sample was quickly oxidized, giving a final pH of 2.5 after 1500 minutes. The control sample also showed no passivation, because the pH dropped to below 3.0. The UNR Process sample, however, exhibited passivation since the final pH of 7.46 was observed.

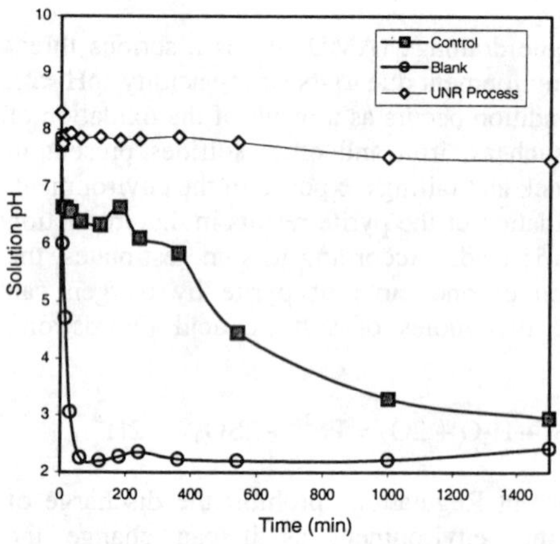

Figure 1. Acid Generation Potential of Passivated Pyrite Sample

The results show that by the addition of specified amount of magnesium oxide, passivation can be achieved at a pH around 10. It is postulated that a the reaction between sulfide minerals and the excess magnesium oxide forms stable cementing phases such as magnesium oxy-sulfates which are precipitated on the pyrite surfaces. In order to examine the formation of coatings on the surface of passivated pyrite surfaces, Scanning Electron Microscopy (SEM) was used. Figure 2 and Figure

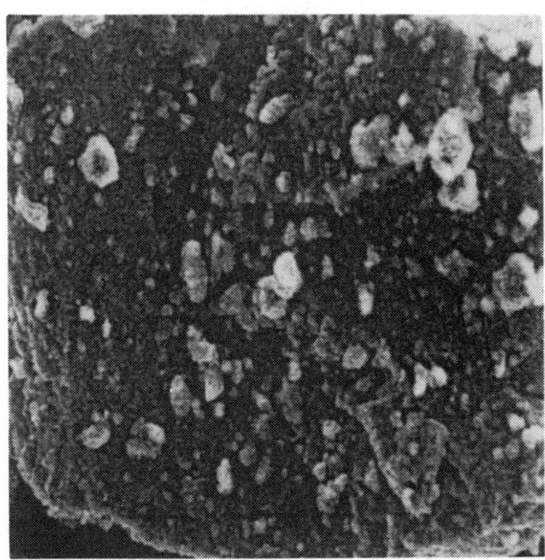

Figure 2. Pyrite Before Passivation

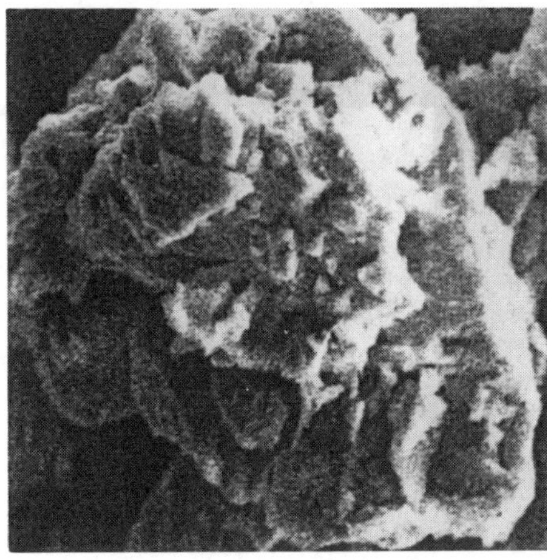

Figure 3. Pyrite after H_2O_2 Test

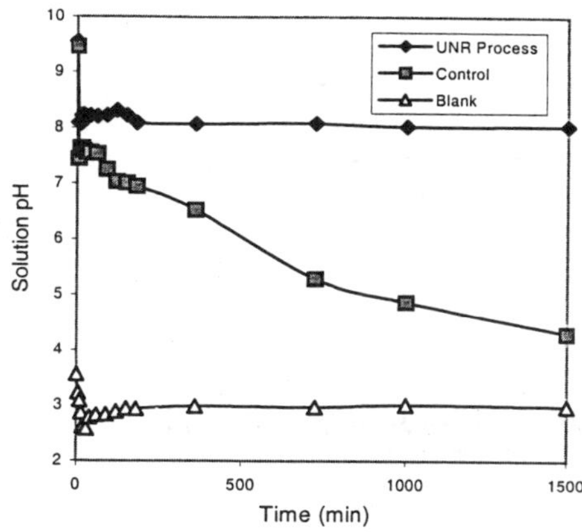

Figure 4. Acid Generation Potential of the Passivated Hecla Rock Sample

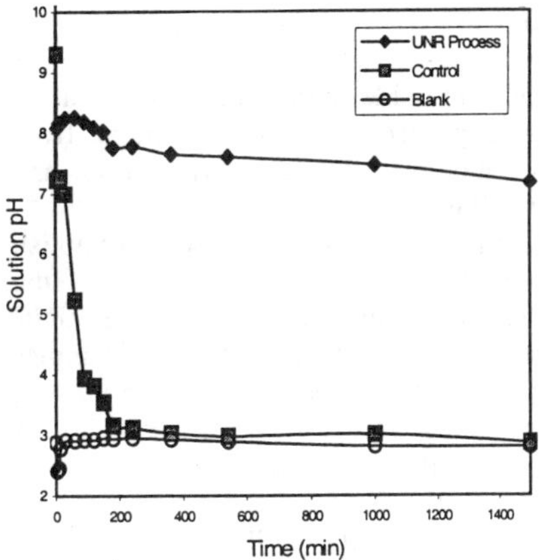

Figure 5. Acid Generation Potential of the Passivated Brohm's Ruby Gulch Sample

3 · are respectively the pyrite surface before passivation and after hydrogen peroxide oxidation.

The initial pH of the Hecla sample was around 3.5. The surface acidity of Hecla sample was not very high because when 20 mg of calcium oxide was added the solution pH increased to 12. The acid generation test results are presented in Figure 4 which shows that the blank dropped to pH below 3.0, the control dropped to pH below 4.3, however, the UNR Process sample kept the pH above 8.0.

The pyrite content of the Brohm's Ruby sample was high. As shown in Figure 5, the solution pH of the blank sample was as low as 2.8. The control sample again did not increase the pH, which remained at 2.87. However, the UNR Process sample passivated the sample, which was evident, as the pH of the solution did no fall below 7.0.

The Newmont sample showed the highest acidity. The results are plotted in Figure 6. The pH of the blank sample was as low as 2.25. The pH of the control sample again dropped below 3.0. However, the pH of the UNR Process sample maintained itself above 7.22. It should be noted that due to the high acidity of the tailing sample, an extra 10 mg of MgO was added.

The results show that the use of magnesium oxide resulted in passivating all of the sulfides samples. The coating formed remained stable under strong oxidizing conditions when subjected to hydrogen peroxide treatment.

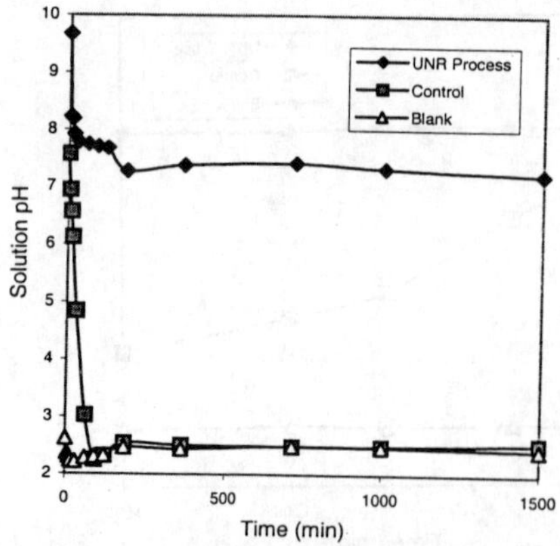

Figure 6. Acid Generation Potential of the Passivated Newmont Sample

CONCLUSIONS

1. Magnesium oxide is an effective passivating agent to prevent acid mine drainage from iron-containing sulfides in waste rock and mine tailings.
2. Further test work is underway to establish the long-term weathering behavior of the passivated samples treated by the UNR Process. Initial tests show that the samples passivated by the UNR Process are passing all standards except that relatively higher levels of sulfates are observed even at low acidity levels (pH >6).
3. The future efforts will be placed on developing methods and reagents to maintain the stability of oxy-sulfates.

REFERENCES

1. De Vries, N.H.C., Process for Treating Iron-containing Sulfidic Rocks and Ores, US patent Number 5,587,001, Dec. 1996.
2. Evangelou, V. P., Pyrite Oxidation and its Control. CRC Press, 1995
3. Kleinmann, R. L. P., Acid Mine Drainage: U.S. Bureau of Mines Researches and Develops Control Methods for Both Coal and metal Mines, E&MJ. July 1989
4. Lawrence, R.W., Laboratory Procedure for the Predication of Long Term Weathering Characteristics of Mining Wastes, *Acid mine Drainage: Design for Closure,* BiTech Publishers Ltd., 1990
5. Nordstrom, D. K., Aqueous Pyrite Oxidation and the Consequent Formation of Secondary Iron Minerals, in Acid Sulfide Weathering, Geochemistry and Relationship to Manipulation of Soil Minerals, Hossner, L. R., et al, (Eds.), Soil Science Society of American Press, Madison, WI, 1982

Environmental Issues and Management of Waste in Energy and Mineral Production, Singhal & Mehrotra (eds)
© 2000 Balkema, Rotterdam, ISBN 90 5809 085 X

A comparison of acid rock drainage treatment scenarios at the former Britannia Mine

T.O'Hearn
British Columbia Research Incorporated, Vancouver, B.C., Canada

B.Klein
University of British Columbia, Vancouver, B.C., Canada

ABSTRACT: In the absence of passive acid rock drainage treatment processes, effluents must be treated to remove metals and neutralize acid. Presently, the High Density Sludge (HDS) lime treatment process is the main technology that is used. This paper compares alternatives to the HDS process: a modified HDS process that utilizes pulp mill ash wastes rather than lime; and, the novel Bio-sulfide process. Results from pilot scale studies on the Britannia Mine effluent are used as the basis for comparison of technical aspects relating to effluent quality and sludge disposal as well as for comparison of capital and operating costs. The results indicate that, despite the high risk associated with new technologies, alternative ARD treatment processes deserve consideration.

1 BRITANNIA MINESITE: BACKGROUND

The Anaconda Britannia Mine was operated from 1905 until final closure in 1974 during which time approximately 48 million tonnes of ore and waste rock were mined for the recovery of silver, gold, cadmium, copper, lead, and zinc. Infiltration of precipitation through the mine workings, which intersect the surface, react with the sulfide mineralization to generate acid rock drainage (ARD). This ARD is discharged from two locations: the 2200 portal, in the upper workings, which discharges intermittently to Britannia Creek; and the 4100 portal bulkhead located near Britannia Beach, which is subsequently piped to discharge at depth in Howe Sound.

Anaconda sold the property to Copper Beach Estates (CBE) in 1979 and currently CBE is held in receivership. The BC government is attempting to sell off the site, with the condition that moneys must be posted to cover the capital expenditure of a permanent water treatment facility. The quality and quantity of ARD at Britannia have been measured sporadically since 1974 with more frequent monitoring occurring recently (Table 1). Cu, Zn and acidity (pH) are the main contributors of toxicity at the site. The 2200 and 4100 discharges would need to be combined and treated, yielding an average flow rate of approximately 500 m^3/h. It is expected that treatment of Britannia ARD will be required over a very long time period. For design, operation, and cost purposes; this study considers that treatment will be required in perpetuity.

Three types of ARD treatment processes have been pilot tested at Britannia: high density sludge (HDS), a modified version of HDS and the Bio-sulfide processes. This paper compares these processes, firstly, on a technical basis and then on a cost basis, to allow recommendation of the most suitable ARD treatment process.

Table1: Britannia ARD quality and flow

Parameter (ppm)	4100 Portal Low	4100 Portal High	2200 Portal Low	2200 Portal High	Ave.	Regs*
Copper	12	28	0.3	115	23	0.3
Iron	2	34	0.40	55	9	
Lead	0.07	0.25	0.04	0.28	0.13	0.2
Zinc	12	28	3	48	26	0.5
pH	3.10	4.50	2.40	4.60	3.4	6.0
Sulfate	1140	1900	200	1950	1530	
Flow (m^3/h)	170	1700	0	420	500	

*Mean Metal Mining Effluent Regulations, Fisheries Act, Canada Gazette Part II, Vol. III, No 5, monthly mean values.

2 TECHNICAL EVALUATION

2.1 *HDS and Modified HDS Processes*

Environment Canada and Cominco Engineering Services Ltd. (CESL) conducted pilot scale testing of the HDS process at the Britannia site. The purpose of this testing was twofold: to demonstrate the conventional HDS process for treatment of Britannia ARD, and to research the potential of using alternative reagents, consisting of pulp mill ash wastes (modified HDS).

Preliminary bench scale testing with conventional hydrated lime indicated that an effluent with low levels of metals could be produced. Pilot testing was conducted at site during April-June 1997. Most testing was performed on the 4100 discharge with the exception of a two-day trial conducted on the 2200 discharge.

In the HDS process, the co-precipitation of iron and other base metal hydroxides upon the surfaces of recycled sludge particles allows for effective removal of base metals from the ARD (Kuit, 1980). The HDS process is normally operated at pH 9.0 to 9.5 as most metals precipitate quickly and thoroughly below this range. The by-products of the neutralization stage are metal hydroxide and calcium sulfate precipitates.

For the modified HDS process precipitator catch and top ash, which are waste products from the nearby Howe Sound Pulp and Paper pulp mill (Port Mellon, B.C.) were used in place of lime. The precipitator catch and top ash were prepared to 25 - 75% -200 mesh (75 μm) and -20 mesh (750 μm), respectively.

Pilot testing consisted of a series of test runs using the three alkali reactants. Conditions were varied according to the degree of underflow recycle rate, retention time and pH. The test conditions for the best HDS treatment results are listed in Table 2.

Table 2: HDS test conditions for best treatment

Parameter	Level	
Feed rate	1198	ml/min
U/F recycle rate	87	ml/min
Lime addition rate	0.44	kg/m^3
Flocculant addition rate	2.17	kg/m^3
Reactor 1 pH	9.5	
Reactor 2 pH	9.5	
Clarifier O/F pH	9.5	
Retention time	41	Min
Recycle ratio	23:1	
Solid generation rate	0.50	g/L
Clarifier U/F S.G.	1.11	
Clarifier U/F % Solids	15.6	

The maximum attained density of the clarifier underflow was 15.6% solids by wt, which is lower than the typical range of 25 - 35% solids. The low solid content is mainly due to the relatively low iron concentration (1-10 ppm), but also due to the presence of zinc and aluminum, which act to retain water within the sludge. For ARD with low metals concentrations, several weeks of operation may be required to attain the maximum sludge density. For this reason, it is likely that longer operation of the process would produce a denser sludge.

The sludge density for the precipitator ash and top ash tests peaked at approximately 37% and 35%, respectively. CESL considered these sludge densities to be near maximum and that increasing the amount of underflow recycled would not increase the sludge density significantly.

At pH 9.0, the consumption of precipitator ash and top ash were 22 and 16 times greater than the lime consumption, respectively. However, the rates of wet sludge generation for precipitator catch and top ash were only 3 times greater than that for lime.

Vacuum filtration tests conducted on the HDS product increased the solids content to 31%. Precipitator catch and top ash sludges were de-watered to 53% and 65% solids, respectively.

Toxicity testing on the effluents generated from the lime and top ash neutralization showed these effluents to be non-lethal at 100% concentration. However, precipitator catch effluent caused 50% rainbow trout mortality after 96 hours at 100% concentration. The toxicity was attributed to dioxin and furan concentrations in the effluent, which ranged from 10-11 ρg/L and 100-130 ρg/L, respectively. Current federal regulatory limits for discharge of dioxins and furans are 15 ρg/L and 50 ρg/L.

2.2 *Bio-sulfide Process*

The Bio-sulfide process was developed to adapt the sulfide reduction process to the treatment of acidic mine drainage. Batch and continuous laboratory tests conducted using Britannia ARD samples indicated that both copper and zinc could be quickly removed to below target levels. In May, 1995 NTBC Research constructed a pilot scale Bio-sulfide process plant at the Britannia mine-site, which they operated through to October, 1996 (Rowley et al, 1997).

With the Bio-sulfide process, raw ARD enters the chemical circuit and reacts with the hydrogen sulfide produced in the biological circuit. Only a fraction of the treated ARD is directed through the biological circuit to be used as a sulfate source. Metal sulfides are precipitated separately by controlling pH and sulfide concentration. Alkalinity from the biological circuit or alkali reagents can be used to provide the stepwise pH adjustments. The biological circuit produces alkalinity through the biological reduction of sulfate to sulfide, which produces HCO^{-3}.

The Britannia Bio-sulfide pilot plant consisted of two interconnected yet independent circuits, the biological circuit and the chemical circuit (Rowley et al, 1997). Pilot scale testing demonstrated that each circuit could be operated independently (i.e. while the other circuit was shut down).

The Bio-sulfide pilot plant at Britannia was operated as a series of specific plant runs intended to test particular aspects of the plants operation such as metal removal, product precipitate quality, sustainable treatment, effects of seasonal flow changes, and effects of operation upsets.

Due to the extremely low solubility of metal sulfides, it was expected that the Bio-sulfide sulfide precipitation process would achieve very low levels of metals in solution. During continuous operation, metal concentrations in the process discharge of <0.01 mg/L Cu, <0.05 mg/L Zn and <0.001 mg/L Cd were routinely achieved. For some runs, levels as low as <0.001 mg/L Cu and <0.001 mg/L Zn were obtained.

The production of saleable Cu and Zn sulfide products is considered essential to achieve a cost advantage over other ARD treatment processes. Table 3 shows the concentration of selected elements in the Cu and Zn products.

As the pH of the ARD was below 3.0, it contained very low levels of dissolved ferric iron, however there was a small amount of ferric hydroxide, which settled out with the Cu sulfide. The CuS concentrate was considered to have an acceptable grade for sale to a smelter.

The ZnS concentrate contained a considerable quantity of Cu. However, it is expected that full scale operation under optimized conditions would yield a product with a higher Zn grade and a lower Cu grade. Based on laboratory testing by NTBC Research, Zn grades approaching 60% Zn should be realized.

Five days of continuous pilot plant operation demonstrated that the levels of alkalinity and sulfide supplied by the biological circuit were sufficient for continuous and efficient operation of the chemical circuit. Over this period, the plant operated continuously and displayed no identifiable trends or upsets.

Table 3: Bio-sulfide process product assays

Product	Cu (%)	Zn (%)	Fe (%)	As (ppm)	Ag (ppm)	Tot S (%)
CuS	41.2	0.36	7.71	2800	21	26.2
ZnS	22.6	26.1	3.11	6	2.1	27.9

2.3 *Discussion of pilot scale results*

The best overall effluent quality associated with each of the treatment processes, as well as the federal discharge criteria, are shown in Table 4.

These results shows that the Bio-sulfide process is the only treatment which does not comply with the federal criteria due to its low pH effluent and thus, is considered insufficient for treatment. All other treatment processes comply fully with the federal regulations. However, due to the organic toxicity associated with the effluent from the modified HDS process using precipitator ash, this option was considered unsuitable.

As there are two ARD treatment processes, which were demonstrated to be sufficient, it is reasonable to proceed to an evaluation of cost effectiveness of these methods to allow for selection of a single process or treatment scenario. Although the Bio-sulfide process was inadequate, consideration was given to a treatment scenario consisting of the Bio-sulfide process to remove metals followed HDS treatment to neutralize the effluent.

Table 4: Best effluent produced from treatment

Process	PH (ppm)	Cu (ppm)	Zn (ppm)
Bio-sulfide	4.0	<0.01	0.10
HDS	9.5	<0.05	0.05
Mod. HDS – PC ash	9.5	0.10	0.08
Mod. HDS – Top ash	9.0	<0.05	0.09
Federal Regs.	6.0	0.3	0.5

*Average effluent quality for 5-day run conducted July, 1996.

3 COST ANALYSIS

3.1 *Basis for cost estimate*

The following treatment scenarios were chosen for evaluation of cost effectiveness.

A. HDS – stand alone
B. Modified HDS – utilizing top ash
C. Bio-sulfide coupled with HDS

Wherever possible, data from the pilot scale tests were used for estimation of costs. Measured reagent consumption rates were used for determining reagent costs. Measured sludge generation rates and densities were used to determine sludge volumes for de-watering and disposal cost estimation.

Some capital and operating costs from a report by H.A. Simons (1998) were used as a basis for estimation. MEND-derived cost estimation models (1994) were used for estimation of HDS capital and operating costs. The MEND models are based on realized costs from seven HDS operations in Canada. Capital cost estimation is made by extracting cost information from a graphical model using only the known ARD in-flow rate and net acidity level. Operating cost estimation is made by choosing the closest model in terms of in-flow rate and acidity loading and prorating the costs where applicable. The capital costs include indirect costs and a contingency of 25%. Operating costs include a contingency of 10%.

A cost estimate, containing a 15% contingency factor, prepared by NTBC Research in 1997 for a full-scale plant at Britannia was used for Bio-sulfide costing. This estimate was checked by

H.A. Simons and was found to be reasonable, though somewhat conservative (Rowley, 1998).

Quotations for specific equipment, such as pressure filters, and sludge haulage were obtained for verification purposes or to cost certain scenarios which had little or no existing cost information available. The most recent unit costs were selected from the information sources noted above.

3.2 *Reagent and sludge cost estimation*

For the HDS pilot testing, a sludge density of 15.6% solids was measured, yielding a sludge generation rate of 14,100 wet t/yr based on flow rate of 500 m^3/h. The modified HDS process using top ash gave a denser sludge at 35.0% solids yet a significantly higher rate of sludge generation at 46,400 wet t/yr. The HDS and modified HDS pilot tests indicated flocculant consumption rates of 2.2 and 0.7 kg/m^3, respectively. These findings were for test runs that provided the best quality effluents.

3.3 *Estimated capital costs*

Estimated capital costs for each treatment scenario are presented in Table 5, which provides estimates for treatment and de-watering facilities for each treatment scenario.

Table 5: Estimated capital costs of treatment

Scenario	Treatment Plant	De-watering Facility	Total Cost
HDS	$3.4	$1.26	$4.66
Mod. HDS	$4.1	$3.60	$7.70
Bio-sulfide & HDS	Bios.: $2.8 HDS: $3.4	$1.08	$7.28

*Note: All costs in millions of dollars.

The capital cost for the HDS (stand-alone) treatment plant is based on that of the MEND estimation models. Costs for the modified HDS was based on an expectation that a 20% increase in cost, over that of HDS, would be realized due to required scale up of certain equipment. NTBC Research provided the cost for the full scale Bio-sulfide plant. An HDS plant used to treat Bio-sulfide effluent would have a required capacity equal to that of the stand alone HDS plant and thus, equal cost.

De-watering facility capital costs were based on quotations received from Larox Corp. for pressure filtration equipment. The size and type of equipment required were based on the quantities and densities of sludge to be filtered, the expected final sludge densities and the expected composition of the sludge. A 50% contingency was added for costs of peripheral equipment such as pumps, piping, electrical, etc. as suggested by Larox Corp. The de-watering costs also include a 100% markup for associated infrastructure and a 20% contingency factor.

3.4 *Estimated operating costs*

The items considered for estimation of treatment costs are process reagents, operating and maintenance labor, utilities, operating capital and maintenance capital. Sludge costs include costs for de-watering and disposal.

Total labor consists of operating and maintenance labor. Operating labor is estimated to be 67.7% of the total labor as per MEND. Operating capital is the capital required for replacement of significant equipment and includes monitoring and analytical costs. Maintenance capital covers the replacement of parts and consumables. De-watering operating costs include labor, utilities, and operating/maintenance capital. Disposal costs includes sludge handling, construction of a landfill, and trucking to and from the landfill site.

Estimated operating costs for each treatment scenario are presented in Table 6. Lime consumption rates include a 20% contingency. Total labor and power consumption rates were based on the MEND estimation models. For the modified HDS treatment scenario, it is expected that ash would be provided at no cost. Labor and utility components for the HDS and modified HDS scenarios are assumed to be equal. Power consumption rates include a 15% contingency.

Sludge production rates and thus, reagent consumption rates were estimated to be 66.7% of the stand-alone HDS treatment scenario. These rates are based on the relative quantity of sludge associated with the Cu and Zn removed through the recovery of these metals by the Bio-sulfide process.

3.5 *Sludge Disposal Scenario*

In order to estimate the overall sludge operating costs, a disposal scenario was assumed. Consultation with Environment Canada regarding the possibility of disposal of sludge to Howe Sound indicated this scenario would not be permitted. Disposal of the sludge underground was considered risky due to the potential for re-solubilization of sludge metals by the acidic conditions within the workings. On-site disposal to a secure landfill was viewed as a temporary option due to the limited amount of suitable terrain.

Table 6: Summary of operating costs for ARD treatment options

Item	Unit Cost	Estimated Consumption Rate	Estimated Cost per Annum*
		HDS Treatment	
Lime	$110/t	0.53 kg/m^3 or 2,320 tpy	$256
Flocculant	$3,500/t	11.5 tpy	$41
Oper. & Main. Labor	$25/h	8,700 h/yr	$218
Power	$0.06/kW-h	817,000 kWh/yr	$49
Operating Capital	35% of operating labor		$51
Maintenance Capital	2% of capital cost		$68
			Total = $683
		Modified HDS Treatment	
Flocculant	$3,500/t	3.6 tpy	$13
Oper. & Main. Labor	$25/h	8,700 h/yr	$218
Power	$0.06/kW-h	817,000 kWh/yr	$49
Operating Capital	35% of operating labor		$51
Maintenance Capital	2% of capital cost		$82
			Total = $413
		Bio-sulfide/HDS Treatment	
Bio-sulfide Plant			
Nutrients	$0.48/kg	70,800 kg/yr	$34
Propane	$0.19/L	289,000 L/yr	$55
Oper. & Main. Labor	$25/h	5,460 h/yr	$137
Power	$0.06/kW-h	333,000 kWh/yr	$20
Operating Capital	35% of operating labor		$32
Maintenance Capital	2% of capital cost		$56
			Sub-total = $334
HDS Plant			
Lime	$110/t	1,550 tpy	$171
Flocculant	$3,500/t	7.7 tpy	$27
Oper. & Main. Labor	$25/h	8,700 h/yr	$218
Power	$0.06/kW-h	817,000 kWh/yr	$49
Operating Capital	35% of operating labor		$51
Maintenance Capital	2% of capital cost		$68
			Sub-total = $584
			Total = $918

*Estimated Cost Per Annum in Cdn$1,000

Table 7: Estimated annual sludge de-watering/disposal costs

Process	Est. Volume of Sludge (wet t/yr)	Est. Volume of Dewatered Sludge (wet t/yr)	Estimated Dewatering Cost	Estimated Disposal Costs	Haulage Costs	Est. Total Sludge Costs
HDS	17,000 t	8,600 t	$146,000	$35,000	$97,000	$278,000
Mod. HDS	55,700 t	30,000 t	$317,000	$120,000	$338,000	$775,000
Bio-sulfide & HDS	11,400 t	5,800 t	$123,000	$24,000	$66,000	$213,000

H.A. Simons' considered 15 years worth of sludge could be stored on site, however this report is more concerned with the long-term sludge disposal. Rail haulage from the site to a smelter is possible with a subsidy but would require negotiations and thus, is difficult to cost. Off-site disposal of sludge is another possible option, which is currently being utilized by some mining operations in Canada. This option is viable and acceptable to regulating authorities and was therefore chosen as the disposal option for this report.

The nearest area with suitable terrain for such sludge disposal is the Squamish town-site situated approximately 12 km to the north on Highway 99.

It is estimated, that trucks would make a round trip of 25 km in order to haul the mechanically de-watered sludge to a secure landfill. The estimated costs associated with this sludge disposal scenario are shown in Table 7.

The de-watered sludge from the Bio-sulfide/ HDS treatment scenario is assumed to have the same density as from the stand alone HDS process (i.e. 31% solids). Therefore the de-watering costs were assumed to be the same with the exception of the maintenance labor, which was prorated according to the quantity of sludge to be de-watered. Disposal costs were based on MEND information at $4/t, which includes on-site haulage and handling costs. Haulage costs were based on a quotation from Trimac Transportation Services Inc. of $0.35/tonne/km. Table 8 summarizes the operating costs of each treatment scenario.

3.6 *Summary of cost analysis*

Table 9 summarizes the estimated annual capital and operating costs for each treatment scenario. For the Bio-sulfide scenario, operating revenue from the expected sale of the concentrates was included. Net present value (NPV) was calculated for each scenario over a 100-year period of treatment based on an annual discount rate of 3%.

Revenue for the Bio-sulfide concentrates was based on the value of the Cu and Zn contained within the Britannia ARD as estimated by H.A. Simons according to 1998 metal prices.

3.7 *Discussion*

The estimated capital and operating costs for the stand alone HDS process compared well to other estimates (H.A. Simons, 1998, SRK, 1991). The estimated Bio-sulfide capital and operating costs are considered to be as reliable and somewhat conservative. As these costs are difficult to estimate, due to a lack of precedence, there is considerably less confidence placed upon them in comparison to cost estimates for industrially proven processes.

Capital costs for Modified HDS treatment are estimated to be much higher than those for the stand alone HDS treatment, partially due to the significantly higher volumes of neutralizer required but mainly due to the relatively high quantities of sludge that is produced. As a result, associated de-watering costs are 290-330% higher than the costs for other options. Capital costs for the Bio-sulfide/ HDS scenario, are relatively high as would be expected.

The treatment costs for the Modified HDS option, were the lowest, mainly due to significantly lower reagent costs. Bio-sulfide/ HDS treatment costs were the highest, as expected, at nearly 35% more than that of the stand alone HDS.

The net operating costs for the Bio-sulfide with HDS treatment scenario is slightly lower than the HDS stand alone scenario. Modified HDS treatment has the highest estimated total operating costs, approximately 17% higher than the Bio-sulfide/ HDS scenario. A much different situation arises with respect to estimated sludge costs: Bio-sulfide/ HDS treatment costs are nearly 24% lower than for the stand alone HDS process; and the Modified HDS sludge costs are nearly 280% higher than for the stand alone HDS process.

Combining the offsetting treatment and sludge costs tends to decrease the difference between the Modified HDS and Bio-sulfide/ HDS options. Total operating costs are within 5%, with the Modified HDS being more costly. Stand alone HDS, with intermediate treatment and sludge costs, provides the lowest total operating cost. However, once the revenue for the Bio-sulfide Cu and Zn concentrates are taken into consideration, this scenario yields the lowest net operating cost; approximately 5% lower than the stand alone HDS option.

NPV for the stand-alone HDS and the Bio-sulfide/ HDS scenarios are within 3% over a 100 year period. The NPV for the Modified HDS process is highest, at 29% greater than that of stand alone HDS.

For mine sites with less acidic ARD, (higher pH in the range of 5 to 6), stand alone Bio-sulfide treatment may be more cost effective than stand alone HDS. The cost advantage would be due to significantly lower treatment costs and no associated sludge costs.

Table 8: Summary of operating costs

Scenario	Treatment Plant Costs	Sludge Costs	Total Cost
HDS	$683,000	$278,000	$961,000
Mod. HDS	$413,000	$775,000	$1,188,000
Bio-sulfide & HDS	$918,000	$213,000	$1,131,000

Table 9: Summary of cost analysis

Process	Capital Cost	Operating Cost Per yr.	Operating Revenue	Net Operating Cost Per yr.	NPV (100 yr.)	Relative Cost Comparison
HDS	$4.66	$0.961	—	$0.961	$35.0	1.00
Mod. HDS	$7.70	$1.188	—	$1.188	$45.2	1.29
Bio-sulfide & HDS	$7.28	$1.131	$0.220	$0.911	$36.1	1.03

*Note: All costs are millions of Canadian dollars.

4 CONCLUSIONS

On a technical basis, excluding effluent quality, all three processes have characteristics, which indicate that they are suitable to treat Britannia ARD. From a treatment perspective, the HDS and modified HDS (using top ash) processes have demonstrated successful treatment of Britannia ARD yielding effluent specifications that are within the federal discharge limits. At Britannia, the Bio-sulfide process yields an effluent, which easily meets the federal criteria with respect to metals concentrations however, it fails to raise the pH sufficiently.

HDS and Bio-sulfide/ HDS treatment scenarios are the most cost effective of the three alternatives examined at Britannia. HDS and Bio-sulfide/ HDS treatment scenarios have a near equal NPV over a 100-year period of treatment. The Modified HDS scenario has significant lower cost effectiveness.

To compare the merits of the stand alone HDS and the Bio-sulfide/ HDS scenarios, the underlying risks associated with each option must be considered. As the Bio-sulfide process has not been proven industrially, more risk accompanies this option. Therefore, HDS is considered to be the preferred process option of the three and is recommended for use at the Britannia site.

5 REFERENCES

Cominco Engineering Services Ltd. Pilot Scale Testing of the High Density Sludge Process, Britannia Mine, AMD Treatment, Britannia Beach, BC. August, 1997.

H.A. Simon Ltd. Treatment of Acid Drainage at the Anaconda - Britannia Mine, Britannia Beach, B.C. , March, 1998.

Kuit, W.J., Mine and Tailings Effluent Treatment at the Kimberley, B.C. Operations of Cominco Ltd, CIM Bulletin, December 1980, pp.148-176.

MEND Program. Acid Mine Drainage - Status of Chemical Treatment and Sludge Management Practices. MEND Report 3.32.1, June 1994.

Rowley, M.V., D.D. Warkentin and V. Sicotte. Site Demonstration of the Bio-sulfide Process at the Former Britannia Mine. Fourth International Conference on ARD Proceedings, Volume 3, pg. 1533-1547. 1997.

SRK (B.C.) Inc. and Gormely Process Engineering. Evaluation of ARD From Britannia Mine and the Options For Long Term Remediation of the Impact on Howe Sound. November, 1991.

Environmental Issues and Management of Waste in Energy and Mineral Production, Singhal & Mehrotra (eds)
 ISBN 90 5809 085 X

Alternatives for mitigation of acid mine drainage in a coal mine

P.S.Soares, L.S.Borma & V.P.Souza
CETEM, Centre for Mineral Technology, Rio de Janeiro, Brazil

E.Van Huyssteen
CANMET, Canada Centre for Mineral and Energy Resources, Ottawa, Ont., Canada

J.P.S.Schultze
CRM, Companhia Riograndense de Mineração, Porto Alegre, Brazil

ABSTRACT: The impact of mining on the quality of surface and ground water is a hotly debated issue of great interest worldwide. The most significant among the several problems that can affect the quality of these waters, is the generation of acid mine drainage (AMD). Unmanaged it can render the use of impacted waters totally unsuitable for recreational, agricultural or drinking purposes, and also seriously damage the eco-system in the receiving environment. The Brazilian mining industry is striving to adopt sustainable environmental practices. As part of a contribution to these efforts CETEM and CANMET have put forward a joint project on mine rehabilitation. CRM, a coal mining company, is one of the partners actively involved in the Project through its Candiota coal mine located in southern Brazil. The initiative includes a systematic survey of existing local environmental data, supported by laboratory and fieldwork to identify and evaluate different suitable rehabilitation technologies. The project began in January 1999 and will extend over a period of three years. This paper reports the Project's main achievements during its first year and describes the approach adopted by the partners - CETEM/CANMET/CRM – in dealing with acid mine drainage taken as one of the issues within an integrated mine rehabilitation concept.

1 INTRODUCTION

The growing importance of environmental issues has been a major source of environmental pressure on the mining industry worldwide. One can say that particularly in countries where this industry plays an important economic role, managers and researchers will certainly face increasing challenges in this new millennium to produce metals and mineral products according to the principles of sustainable development.

This is the case of Brazil where mineral production generates about US$ 7.5 billion. Taking into account the metallurgical and mineral transformation industry these figures reach US$ 53 billion which is equivalent to 7.0% of the Country's domestic gross product. These values can also be added on to others, related to a number of services and small businesses levered by mining and metallurgical industries such as maintenance and equipment, spare parts, servicing, commerce and so on.

The great international significance of the Brazilian mining industry can also be measured by its capacity to attract foreign investments. Presently Brazil is ranked third in the world among the most attractive for investments in mining.

In Brazil, as in other countries the use of environmentally sustainable technologies has unfortunately not always been the practice of the whole mineral industry. On the contrary the public image of this industry has been associated with pollution of rivers, air and land as well as the depletion of non-renewable natural resources. As a consequence, it has been continuously subjected to environmental pressures.

The main sources of these pressures have been the improvement of the national regulation and enforcement system, the stakeholders' interests and the increasing awareness of the society of the importance of environmental quality.

Within this complex scenario the Brazilian mineral industry has also to overcome the challenges of a global market where commodity prices are depressed and international non-tariff barriers based on environmental issues can jeopardise its leading commercial position.

A possible response to these challenges is the adoption of sustainable mining practices in Brazil which encompass, among other aspects, a comprehensive mine rehabilitation plan throughout the mining cycle and a technically sound decommissioning stage at the end of the mine life.

As part of their contribution to the efforts of the Brazilian mining industry as it strives to adopt and develop environmentally sustainable practices, CETEM and CANMET have put forward a joint project on mine rehabilitation.

The technology transfer elements of this initiative are partially sponsored by CIDA - Canadian International Development Agency. The overall initiative includes a systematic survey of local environmental data as well as laboratory and fieldwork to evaluate different rehabilitation technologies over a period of three years, beginning in January 1999.

The project has been carried out by building partnerships with various Brazilian mine operators who are willing to make their sites available for demonstration purposes. CRM, a coal mining company was one of the first partners actively involved in the Project through its Candiota mine located in southern Brazil. The choice of Candiota as a partner was based on the fact that it is a major coal producer in Brazil and its management is strongly committed to the improvement of the company's environmental performance.

This paper reports the Project's main achievements during its first year and describes how the partners - CETEM/CANMET/CRM - have been dealing with acid mine drainage taken as one of the issues within an integrated mine rehabilitation concept.

2. THE MINE SITE

2.1 *Location*

The Candiota Mine is operated by CRM, a state owned company and it is located in the state of Rio Grande do Sul, the most southern Brazilian state in a subtropical region close to the border with the Republic of Uruguay as shown in Figure 1.

The operation is headquartered in the village of Candiota which is about 2000 kilometres away from Brasilia and 410 kilometres from Porto Alegre, Rio Grande do Sul's capital city. The local population is 12,000 people and the main economic activities in the region are coal mining, electricity generation due to the 400MW thermal-power station that is fed by the mine production, cattle breeding and rice cropping.

The local landscape is mainly dominated by small hills locally called *coxilhas* which are usually not higher than 100 m and covered by pastures.

2.2 *General climate data*

Temperatures vary over a wide range in Candiota during the year. From a mild and rainy winter (– 4^0C in July) to a hot summer (+40^0C in January). The average temperature is about 17^0C though. Even in severe winters, snow is very rare and when temperatures drop below zero degrees Celsius, heavy frosts occur.

The cumulative annual rainfall is 1400 mm. The rains are more intense from July to October. The relative air humidity is slightly higher in winter (83%) than in summer (73%) and the region has about 2.440 hours of sun/year.

Figure 1. Location of Candiota Mine (Mina Candiota)

2.3 The *ore*

Coal reserves represent a significant fraction of the non-renewable energy resources in Brazil. The largest deposits are located in the southern state of

Rio Grande do Sul. They contain 28 billion tons or 89,3% of Country's total reserves. Candiota is the state's larger deposit, containing 44 % of its reserves or 38% of the Country's total reserve (DNPM, 1995).

Candiota Mine produces about 100,000t coal per month and the company has 15 claims in the region, which are called *malhas* numbered 1 to 15. The whole area is approximately 65km by 25km encompassing a coal reserve of almost 12 billion tons (MME 1985). The operation started back in 1961 and since then different areas have been mined. The present target is a 1.9 million ha area called Malha 4.

The Candiota coal is part of the Rio Bonito geological formation which is composed by sediments of variable grain sizes laid in seams. It occurs in two seams known as upper and lower seam which are 5.0 m thick in total and separated by a sandstone layer and a non economical coal seam 0.8 to 1.5m thick. The lower seam is about 12 to16 m deep (Fiedler & Solari, 1988; Schultze, 1998).

The ash content of the Candiota coal is particularly high averaging 50.3% (dry basis) in the run of mine (ROM) and its sulphur content is 2.4%. Moreover, heavy metals are also detected in the ore body.

3. THE PROJECT APPROACH

3.1 *General Considerations*

A balanced site reclamation plan requires an appreciation of the global aspects of a site. It starts with creating an inventory of the various site components, ranking the areas of concern, and only then focusing on the more troublesome issues. This approach is cost effective because it avoids one falling into the trap of collecting more data than is appropriate for remediating a site (vanHuyssteen, 1997).

One can say that mine rehabilitation operations on strip mined areas generally involves re-contouring, landscaping, and emplacing soil on the mined area and revegetating it. A site rehabilitation which focuses exclusively on these topics though, will almost certainly fail, as a multidisciplinary approach is definitely necessary.

All potential impacts of a mine operation on the various environmental components must be taken carefully into consideration and the most cost-effective options to mitigate these impacts should be implemented for a successful rehabilitation project.

Some of these impacts are summarised as in Table 1.

Table 1 – The impact of Mining & Processing on the Various Environmental Components (van Huyssteen, 1998)

Environmental Components	Impact	Manifestation
AIR	Gas and particulate transfer	• Particulate dispersion • Change in air composition • Noise
TERRESTRIAL ENVIRONMENT	Solids transformation and moving	• Point sources of contamination • Waste piles • Change in vegetation cover • Ground instability • Change in surface morphology • Underground cavities
SURFACE WATER ENVIRONMENT	Soluble element transfer and suspended solid load change	• Chemical contamination • Change in suspended load • Destabilisation of banks • Diversion of water courses • Creation of new water bodies
GROUND WATER ENVIRONMENT	Soluble element transfer	• Chemical contamination • Change in water table level • Change in aquifer properties

The work started by implementing a systematic stepwise approach, according to the following topics:

- Conducting a preliminary survey on the mining and waste management practices on site.
- Assessing the existing data and reports on the environmental impacts of the operation.
- Conducting on - site training of the mine's technical staff.
- Recommending and executing a complimentary diagnostic laboratory test work program.
- Recommending further investigative fieldwork.

These topics were broken down into activities and relevant information was systematically collected. The progress of the work is summarised as follows.

3.2 *Mining and waste management practices on site*

The ore in Candiota is extracted by strip mining using a 38 cubic yard walking dragline. The waste to ore ratio is 1.6 m^3/ton and the "run of mine" extracted coal seam is jaw crushed and conveyed to the thermal - power station four kilometres away from the mine site. No further beneficiation facilities are available on site.

A large amount of wastes (overburden, sandstone and a non-economical intermediate coal seam) is produced in the process.

Mined areas are usually re-contoured to a shape as close as possible to the local *coxilhas* and subsequently covered with top soil about 0.20m thick. This soil comes from adjacent more recently mined areas. Various types of grasses and in some cases trees, such as *acacias* or *eucalyptus* are presently being used for re-vegetation.

In the older mined areas no topsoil was kept for rehabilitating the mined areas. Re-vegetation was then usually done just after re-contouring the wastes, straight onto mine wastes without a soil cover. Consequently, different degrees of success were achieved.

A small portion of the ash produced in the thermal-power station is sold to local cement factories. Most of it goes back to the mine and fills some of the empty spaces created in mined areas. These ash deposits correspond to about 1% of the mine site area A schematic view of the process is shown in Figure 2.

Acid generation from the coal wastes is significant despite the fact that the sulphur content of the waste material can be as low as 1%. This is because the volume of the wastes is significant and the waste is widely spread all over the site in a manner which promotes weathering and AMD generation.

3.3 *Previous data about the operation*

Many Brazilian researchers have studied the environmental impacts of the Candiota Mine and the adjacent thermal-power station.

Andrade (1985), Fiedler (1987) and Zanella (1988) characterized the environmental impacts associated with the coal combustion and the ashes generated in the thermal-power station. The authors mention problems associated with heavy metals in gaseous emissions and discuss the practice of disposing ashes in the mine and dust generation.

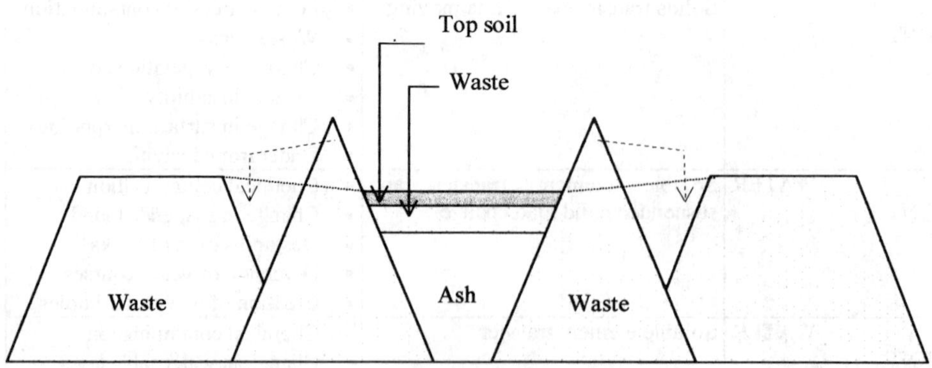

Figure 2: Back-filling procedure with ashes as practised at Candiota

Soares (1995, 1998) studied the metal mobility and kinetics of pyrite oxidation in Candiota. Schultze, (1998) assessed some alternatives for mine rehabilitation focusing on slope stability and re-vegetation issues. In addition to these reports, the official Environmental Impact Assessment (EIA, 1998) study commissioned by the company also presents a comprehensive collection of information about the operation and its impact on the surrounding environment.

References about the impact of the mining on the quality of surface waters of the region and AMD generation date back to 1988 (Fiedler & Solari, 1988). These authors analysed samples taken from Arroio Poaca and its tributary Arroio Carvoeira. These rivers collect mine site effluents although the spring of Arroio Poaca can be considered as a reference point (or baseline) as mining does not influence it.

Results as shown in Table 2 revealed AMD generation. Recent data collected at the same points confirm previous evidence as also shown in Table 2. As observed, results do not do not meet CONAMA (Brazilian Environmental National Council) standards.

3.4 *Technical training and recommended complimentary work on AMD.*

To address the generation of AMD, the Project team began firstly by organizing training sessions at the mine site aimed at giving Candiota's technical staff a comprehensive overview of all elements involved in developing an integrated rehabilitation plan, focussing particularly on AMD prediction, prevention and mitigation.

Secondly, since AMD generation and water availability on site are closely related, climate and topographical data have been systematically compiled and analysed to enable a site water balance to be calculated and examined in conjunction with the water drainage pattern of the mine property.

The use of dry covers to mitigate AMD has been considered, as water resources in the region are apparently relatively scarce. (evaporation > precipitation for many months).

Simultaneously, a series of supporting laboratory tests have been recommended and are being performed at CETEM. Generated data will be used to support the choice of the best rehabilitation plan and particularly the most adequate AMD management option for the mine property.

Tests on materials available at the mine site include grain size analysis, moisture retention curves, chemical and mineralogical characterization.

Laboratory scale AMD simulation tests are being performed to give the Project team an insight on various chemical processes involved during the weathering of the coal wastes and the rate of oxidation rate of sulphide minerals present in Candiota wastes.

Table 2 – Water samples from Candiota (apud Fiedler & Solari, 1988)

Parameter ($mg\ L^{-1}$)	Mine drainage	Carvoeira Creek		Poacá Creek			Standard Values CONAMA
				Spring (baseline)	After mine		
pH	2.3 - 3.0	3.1 - 3.6	1.88*	6.0 - 7.7	3.2 - 4.6	2.9*	6.0 - 9.0
Fe	10 – 237	3.9 - 6.3	4.8*	0.3	0.9	1.5*	0.3
Zn	0.35 - 1.72	0.14 - 0.24	0.18*	0.02	0.04	0.04*	0.18
Co	0.28 - 0.63	0.07 - 0.09	-	0.009	0.016	-	0.2
Ni	0.26 - 1.17	0.10 - 0.15	-	0.011	0.028	-	0.025
Mn	2.80 - 12.3	1.15 - 1.33	4.0*	0.15	0.59	2.3*	0.10
Cd	0.004 - 0.01	0.002 - 0.004	-	0.002	<0.001	-	0.001
Cr	0.01 - 0.03	<0.01	-	<0.01	<0.01	-	0.5
Cu	0.01 - 0.10	0.009 - 0.016	-	0.004	<0.002	-	0.02
Pb	0.036 - 0.082	0.024 - 0.030	-	0.026	0.02	-	0.03

* samples taken in 1999.

Complimentary laboratory tests have also been performed to evaluate the possibility of using power station ash as neutralising agent for AMD.

Lysimeters have been assembled in the laboratory with wastes from Candiota, to simulate and study the possible use of geochemical barriers *in situ* as a means to stop acid generation.

On site fieldwork has been also recommended to assess and position the local water table. A preliminary drilling campaign involving solids sampling and piezometers installation has been indicated to examine how the local water table depth varies according to rainy or dry seasons as this can be crucial to understand how AMD generation varies in the property throughout the year

During this campaign water and solid samples will also be collected which will contribute to evaluate the stage of the AMD process in the field. Ground water quality, and the direction and rate of its movement, in the waste will be determined.

4 CONCLUSIONS

Generally speaking, any mine rehabilitation plan is heavily site specific. In other words it is not possible to design a rehabilitation plan which would universally fit all mine sites regardless of their characteristics. Each property has different location, geology and surrounding area. This determines different geographical (weather, topography, rivers, vegetation, etc), socio-political (local policy, neighbour communities, borders, etc), geological and economic constraints which will demand different rehab solutions.

On the other hand, an AMD management and control plan can only be successful if it is design as one of the elements of a more comprehensive mine site rehabilitation Project.

This Project should be a multidisciplinary exercise encompassing various different issues as location of the mine property, local climate and hydrology, engineering materials available on site, local vegetation, slope stability, erosion, and so on.

The joint Project carried on by CETEM, CANMET and CRM has been addressing AMD in Candiota within this perspective.

ACKNOWLEDGEMENTS

The authors wish to thank the Canadian International Development Agency (CIDA) & Agência Brasileira de Cooperação (ABC) for sponsoring and participating in this program.

REFERENCES

Andrade, A 1985. *Characterisation of Flying Ashes from Candiota Coal. Porto Alegre, RS.* M.Sc. Thesis. Rio Grande do Sul Federal University (*in Portuguese)*.

National Department for Mineral Production (DNPM) 1995. *Coal Industry Annual Report.* DNPM, Brasília (*in Portuguese)*.

EIA 1998. *Environmental Impact Assessment/ Candiota Mine – Malha VII. Companhia Riograndense de Mineração/CRM.* Vol. I. CIENTEC (*in Portuguese)*.

Fiedler, H.D. 1987. *Characterisation of Candiota Coal and Environmental Consequences of its reprocessing.* M.Sc. Thesis. Rio Grande do Sul Federal University (*in Portuguese)*.

Fiedler H.D. & A.J. Solari 1988. Environmental Impact of Candiota Mine on Regional Surface Waters. *In: III National Meeting on Mineral Treatment and Hydrometalurgy*. pp. 483-498 (*in Portuguese)*.

Ministry of Mines and Energy (MME) 1985. *Coal Industry Annual Report*, Brasília (*in Portuguese)*.

Schultze, J.P.S. 1998. *Candiota Mine: Rehabilitation Studies.* Internal Report UNISINOS University (*in Portuguese)*.

Soares E.R. 1995. *Heavy Metal Mobility in Materials from Candiota.* M.Sc. Thesis 58p. Viçosa Federal University (*in Portuguese)*.

Soares E.R. 1998. *Kinetics of Pyrite Oxidation and Heavy Metals Release in Sediments of Candiota - RS.* DSc. Thesis, Viçosa Federal University (*in Portuguese)*.

van Huyssteen, E. (1998). CANMET INTEMIN Overview of Environmental Baselining.

Published by CANMET/MMSL ISBN 0-660-17565-7. Issued on CD-Rom entitled Baselining – A reference Manual obtainable from CANMET/MMSL or INTEMIN

van Huyssteen,E.(1997):*The Multi -disciplinary Aspects of Mine Site Rehabilitation*, published in Mining and Environment, edited by Roberto C. Villas Boas, Peres Barbosa, Robert Hargreaves – Rio de Janeiro: CANMET/CETEM, ISBN 85-7227-104-X published by CETEM.

Zanella R. 1988. Investigation on Environmental Issues Related to Coal Burn up and Exploration in Candiota. M.Sc. Thesis. Santa Maria University (*in Portuguese*).

Environmental Issues and Management of Waste in Energy and Mineral Production, Singhal & Mehrotra (eds)
© 2000 Balkema, Rotterdam, ISBN 90 5809 085 X

Arsenic problems in mine tailings

C.A.Taschereau
Omai Gold Mines Limited, Guyana

K.Fytas
Department of Mining and Metallurgy, Laval University, Québec City, Qué., Canada

ABSTRACT : Arsenic is an element frequently associated to mining operations around the world. Its long term stability in mine tailings may be a problem due to its particular chemical properties. It has been frequently observed that arsenic goes into dissolution in conditions where other metals are stable. This paper presents a case study of a tailing site where arsenic has been periodically released. Data collected over a nine-year period are presented and analysed. The paper identifies the causes underlying the periodic arsenic release and defines the treatment methods in order to correct the situation.

1 INTRODUCTION

The association of arsenic with gold or gold-copper orebodies is well known and documented (Boyle et Jonnasson, 1973). This is explained by a similar solubility of gold and arsenic in the hydro-thermal solutions.

Arsenic ore is generally considered refractory as it interacts with gold in most extraction processes and causes lower recoveries (Vaughan and al., 1992). However, it is more because of its complicated chemical stability that arsenic constitutes a serious environmental problem for the mineral industry around the world.

The history of arsenic's high toxicity has led to a very strict control of it in the mine effluents. Moreover, the arsenic has a complex chemical behaviour, it can bond with many different elements and reacts differently depending on the physico-chemical properties of the water, making the prediction of its stability rather difficult (Frost, 1967).

2 THE CHIMO CASE

2.1 *Mine overview*

The Chimo mine is located 50 km East of Val d'Or in Northern Quebec. The mine has been in operation in different time periods from 1936 until its final closure in 1997. Between 1989 and 1997 the mine was owned and operated by Cambior Inc.

A cyclic increase of arsenic concentrations in the effluents during the summer months was observed at Chimo Mine. In order to explain and control this phenomenon that forced the mill to temporarily interrupt its effluent discharge in the summer in order to comply with the regulated maximum arsenic limits, a thorough investigation was initiated. The findings brought a new explanation to the arsenic dissolution occurring in the tailings pond.

During the studied period, from 1994 to 1997, the underground mine was mining ore composed of quartz lenses associated with pyrite and arsenopyrite mineralisations with an average grade of 3-4 g Au/tonne. The mining method used was the long-hole cut-and-fill. Mining was carried out on 19 levels down to a depth of 865m. The production rate was of about 1500 tonnes per day.

The Chimo mill started its operation in 1994. Previously, the ore was being hauled to Béliveau mine for processing. At the closure of that mine in 1993, the mill was moved to the Chimo mine site.

2.2 *Process description*

The grinding process used a jaw crusher and a 20 foot SAG mill. The mill was producing a gravity concentrate which accounted for about 30% of the gold content and a flotation concentrate of pyrite and arsenopyrite containing the balance of the gold. The concentrate is was then shipped to Yvan Vézina mill for CIP recovery. The gold overall recovery is about 92%.

The flotation was carried out at natural pH without lime or acid addition. The reagents used were a Xanthate as collector, a froth promoter in flotation and a flocculant in the thickener.

Table 1. Composition of Streams

	Ore	Concentrate	Tails
Ag	<1 ppm	3 ppm	<1 ppm
As	8800 ppm	>9999 ppm	530 ppm
Au	3 ppm	100 ppm	<1 ppm
Cu	190 ppm	310 ppm	31 ppm
Fe	4.30%	12%	2.50%
Pb	<1 ppm	22 ppm	<1 ppm
Zn	210 ppm	950 ppm	80 ppm

2.3 *Tailings*

About half of the tailings were returned underground as paste backfill. The rest was pumped to a 170,000 m^2 tailings pond with a capacity of 300,000 m^3. The tailings pond was constructed on the slope of an esker on one side and composed of three dams of an average height of 2m (maximum 3.5m) on the other sides.

As the sulphides were associated with gold, the tailings were mainly composed of feldspaths, quartzite, calcite and hornblende. They contained more than 500 ppm of arsenic and 2.5% of iron. We should note that since no roasting or hydrometallurgical process had been used, the arsenic was still present as arsenopyrite (FeAsS).

The tailings disposal was partly sub-aquatic, about 50% of the tailings being immersed underwater. Because of their high carbonate and low sulphur content the tailings were not acid generating.

The average size of the particles in the tailings is 80%<75μ (200 mesh). The density of the dry tailings is about 1.5 tonnes/m^3.

The mill was using the water pumped from the mine as process water. Consequently, there was no water recirculation from the tailings to the mill.

3 OBSERVATIONS

The tailings pond was put in operation in May 1994 at the start-up of the mill. During the first month of operation, lime was required at the discharge of the tailings pond to increase the pH above the regulated maximum limit of 6.5. This was due to a very low natural pH of the groundwater in the tailings area. Samples indicated a natural ground pH below 3. This low-pH period provoked a limited increase of iron, copper and zinc in the effluents but not of arsenic.

During June and July of 1994 the pond slowly filled up and at the end of July, the pH stabilised around 7.5 (Figure 3). In August, it was noticed that algae and larvae began to multiply in the tailings. The algae density went high enough to increase the turbidity in the water and make it almost opaque. Simultaneously, the pH rose up to 9.5. Finally, in late September, the phenomenon reduced by itself and the algae disappeared, followed by a return to a normal pH.

During the winter months of 1995, everything remained normal in the tailings pond with clear and neutral pH effluents.

In April a light increase of the aquatic life was noticed, corresponding to the spring increase in temperature. This increase was followed by an increase of the arsenic concentrations. Starting from a range of 0.1 ppm to 0.2 ppm observed previously, it became an average of 0.27 ppm by the end of the month.

In the beginning of May very hot weather with many sunny days were experienced. This time quickly followed by a new algae bloom. This time the pH rose to 9.6 and the arsenic rose to 0.53 ppm.

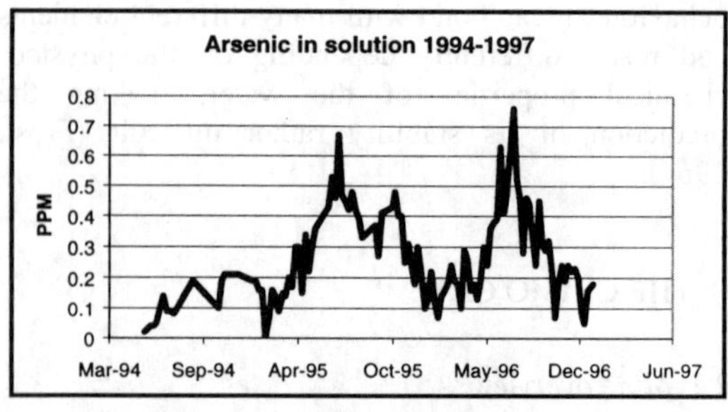

Figure 1. Dissolved arsenic variation through time

Finally, in late June a reduction of temperature caused a simultaneous reduction of the algae and the arsenic to lower levels. However, during the rest of the summer (until October) the algae remained but at lower levels compared to those of May-June.

From July to October, the level of arsenic stayed at an average of 0.4 ppm and it was only in November that the average went back under 0.3 ppm.

From December 1995 to April 1996, the concentration of arsenic in the effluent was 0.2 ppm.

However, in May 1996, for a third time, the now expected algae bloom reappeared, again accompanied by an increase of pH and arsenic. The latter even reached 0.63 ppm in June.

We should note that only the arsenic showed an important increase during the whole life of the tailings pond. Figure 2 shows the dissolved iron variation between 1994 and 1997.

The main peaks of iron were observed at start-up of the mill in May 1994 when the acidity of the water probably dissolved part of the metals. The iron also peaked in May 1996 when the addition of ferric sulphate was tested, without positive results, at the discharge of the tailings.

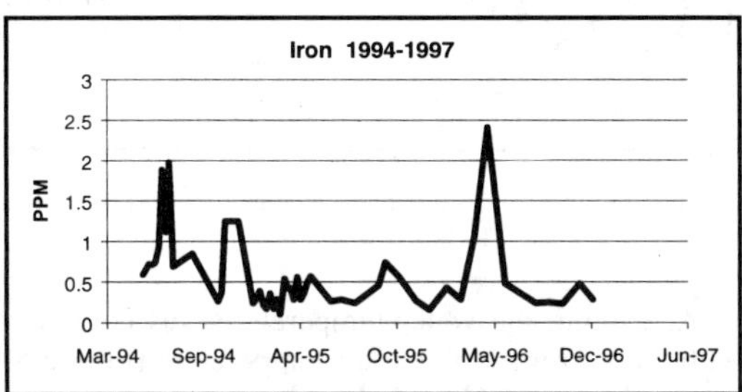

Figure 2. Dissolved iron variation through time

The other metals present like copper, zinc and lead increased only during the low pH period to subsequently fall below the detection level.

If we sum up, the characteristics of the situation were:

- Water temperature increase;
- pH increase;
- Increase of biota (aquatic life);
- Increase of arsenic in solution;
- No increase of other elements (Fe, Cu, Zn, Pb, etc);
- Return to normal when temperature decreased.

4 HYPOTHESES

To explain the phenomenon of periodic arsenic release, many hypotheses were examined, but globally we can regroup them in 2 categories:

1. The arsenic was usually precipitating in the tailings pond, but due to unknown reasons (ex. short residence time or short circuits in its path) it was not precipitating anymore during certain periods.
2. There is some dissolution of the arsenic already present in the tailings due to a variation of the physico-chemical parameters in the pond water or pore-water.

The first hypothesis was ruled out by sampling the feed and the discharge of the tailings pond and assaying for arsenic in July 1996. It was shown that the arsenic concentration in the tailings discharge was more than twice the arsenic concentration in the feed (a mean value of 0.25 ppm of arsenic in the feed vs. 0.6 ppm in the discharge.

A study of the arsenic charge of the effluent demonstrated that it did increase by many times during the summer months. However, we were able to prove that the observed increase in arsenic level was not only related to a reduction of the flow rate due to reduced precipitations during the summer.

In order to find the origin of the arsenic we tried to determine the form of the arsenic measured in the effluent. Arsenic could have been in one or many of the following forms (Moore, 1990):

- Solid in suspension;
- Organic in micro-organisms;
- In solution as Arsenite (As^{3+});
- In solution as Arsenate (As^{5+}).

Our chemical analyses indicated that all of the arsenic was dissolved in the solution as arsenite, which is known to be much easier to dissolve than the arsenate but it is only formed in anaerobic-reducing conditions (Moore, 1990).

Additional sampling showed that most of the arsenic feeding the tailings was in solid form being removed by a 0.45 μm filtration prior to analysis.

It became therefore clear that the problem was a dissolution of the arsenic already contained in the sediments of the tailings.

Starting from this information we examined the following hypotheses as possible causes of the arsenic dissolution:

- Variation of the ore composition
- Variation of the tailings composition
- Effect of pH
- Effect of Temperature

We supposed that there might have been some variations of the chemical composition or of the nature of the arsenic in the mined ore. Even if the seasonal nature of the phenomenon was an argument against this explanation, it could be alleged that some mining areas contained a different ore liberating more arsenic in solution and causing an increase of the pH of the water.

However, a number of tests, assays and geologic studies contradicted this hypothesis.

The second hypothesis examined was a possible change of the tailings composition. As an example the exposure to air of a wider zone of the tailings during the summer, might have caused oxidation and liberation of some arsenic.

However, once again observations of the tailings, samples taken from different zones and assaying allowed rejection of this hypothesis.

At first glance the most important parameter suspected to be associated with the arsenic increases was the pH that increased with the cyanide during the algae blooms. However, the graph of Figure 3 indicates that the arsenic is not closely correlated with the pH.

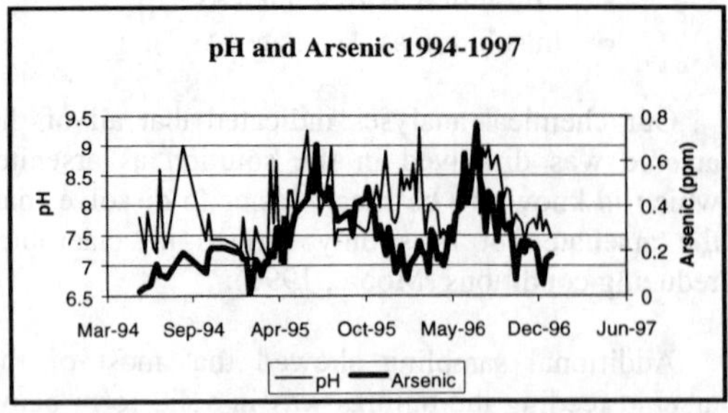

Figure 3. Variation of pH and dissolved arsenic

As we can see the peaks are not corresponding perfectly and most important the pH increased during the winter 1996 without affecting the arsenic concentration at all.

The only parameter that strongly correlates with the As is the temperature of the water which follows almost perfectly the arsenic dissolutions (Figure 4).

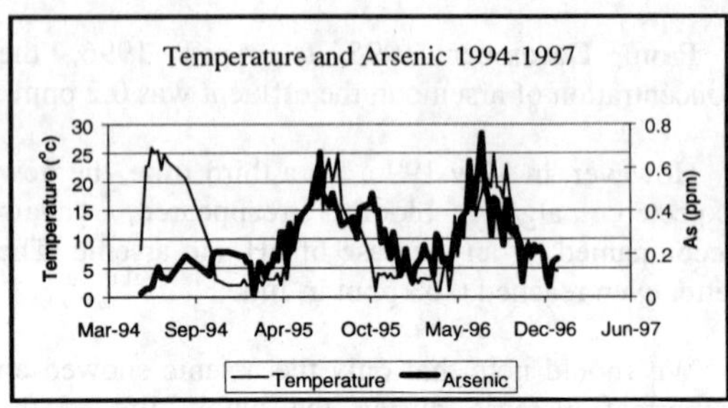

Figure 4. Variation of temperature and dissolved arsenic

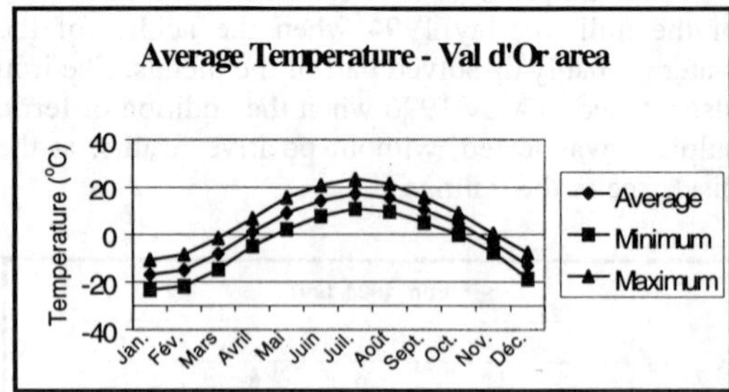

Figure 5. Variation of average temperature over the year

At Chimo, the water temperature was consistent with the average ambient air temperature (Figure 5), and was also affected by the duration of sun exposure.

The effect of the temperature on the sorption and desorption of trace elements is not well known.

The relationship seems complex and is generally not considered in the range of temperature generally observed (Laxen, 1983).

However, many chemical reactions are highly affected by the temperature, as an example, the O_2 solubility is reduced by half when the temperature raises from 0^oC (14.6 ppm) to 30^oC (7.6 ppm).

Therefore, the water temperature was closely associated to the arsenic dissolutions but the mechanisms were not clear yet.

Some literature examples indicate an increase of the arsenic in solution during the summer. Aggett in 1988 showed that the ratio of arsenite/ arsenate was passing from 1:20 in the winter to 1:1 in the summer. He attributed this observation to the bacterial activity but without proving it. McLaren in 1994 had observed a seasonal cycle of the arsenic concentration in a river in New-Zealand. Once again, the temperature was suspected without however proving it.

The other interesting factor in our case is the algae bloom, which also correlated with the arsenic increase.

The possibility of the arsenic existing in an organic form was also considered since it had been observed in similar cases (Bowell and al., 1994). However, this hypothesis had already been rejected by the testing of the arsenic form.

5 EXPLANATION

As far as it can be determined, no single factor explains the phenomenon. But by considering the tailings pond as a dynamic system we can explain the periodic arsenic dissolution as it follows.

The tailings pond could be considered as a shallow lake with a low agitation. In these conditions, the effect of the sun creates distinct layers in the pond.

During the summer, the surface layer (epilimnion) is heated by the sun and is floating over the lower layer (hypolimnion) because of its lower density. An intermediary layer (thermocline) separates both. This phenomenon is common for this type of stretch of water and is called thermal stratification.

At Chimo, the tailings pond water proved to have the ideal conditions for algae and plants growth: neutral pH, warm temperature, absence of toxin and a high quantity of nitrates (coming from partially detonated ANFO explosives) acting as a fertiliser.

This created an ideal environment for the quick multiplication of algae and the different forms of aquatic life observed, up to a very high density. The nutriments which usually limit the biota multiplication are here in high concentration as nitrates.

Consequently we obtain a high density of algae in the sun-exposed epilimnion. The photosynthesis of the algae creates a high level of oxygen in the top layer and aerobic conditions. The photosynthesis of algae can be simplified as (1):

$$6CO_2 + 6H_2O + h\nu \rightarrow C_6H_{12}O_6 + 6O_2 \qquad (1)$$

The photosynthesis also provokes some chemical reactions in the water, including:

$$CO_3^{2-} + H_2O + h\nu \rightarrow CH_2O + OH^- + O_2 \qquad (2)$$

where CH_2O = biomass

This reaction and the liberation of the ion OH^- is responsible for the increase of the pH of the water.

But in the lower layer the dead algae are accumulating and the bacterial activity caused by their organic degradation leads to anaerobic and reducing conditions.

These reducing conditions and the high pH of the water led to the formation of highly mobile arsenite.

Consequently the reaction is as follows:

Nitrate (from explosives) in the water
+
High Temperature and a lot of sun exposure
↓
Biota Multiplication (algae)
↓
Increase of the pH
+
Formation of a Reducing zone at the bottom of the tailings
↓
Remobilization of Arsenic

6 SOLUTIONS

The solution adopted at Chimo in order to comply to the regulated maximum arsenic limits in the effluents was the use of ferric sulphate, causing a precipitation of the Arsenic as ferric arsenate. This chemical form is considered relatively stable in neutral pH conditions (Krause & Ettel, 1988).

The ferric sulphate was added directly in the second half of the tailings pond, after the sedimentation of most solids. Trying to add it directly in the mill or at the discharge of the tailings proved to be inefficient.

This method allowed the mine to comply to the maximum arsenic limits in the effluents defined by the environmental legislation during the period where the arsenic dissolution was observed.

Other solutions considered were the reduction of the nitrates at the source or the use of inverse osmosis. But considering the impending closure of the mine at that time, no long term plans were developed.

7 CONCLUSIONS

The behaviour of arsenic in a tailings pond is more difficult to anticipate than for most of the other contaminants due to its complicated chemical stability.

As we have seen, a seemingly trivial event like an algae bloom may initiate an arsenic dissolution by raising the pH and creating a zone of reducing conditions.

This interaction between the nitrates, algae, pH increase and reducing zone could help to explain other cases observed in tailings pond or mine water ponds.

In the Chimo case, the true origin of the whole situation was the presence of nitrates in the water. To avoid a similar situation, nitrates need to be controlled directly in the mine and the presence in the ore of any non-detonated explosives should be avoided. Nitrates could have more complex effects than it is usually assumed and their effects on water ecosystems downstream of the mines need to be studied further.

8 ACKNOWLEDGEMENT

The authors gratefully acknowledge the mining company Cambior Inc. and especially Serge Vézina, director on research and environment for his kind permission to publish the data from the Chimo Mine.

REFERENCES

Aggett, J., 1988, Current understanding of the arsenic cycle in Waikato hydro-lakes. Trace elements in New Zealand : Environmental, human and animal. *Proc. NZ trace elements group conf.*, 30 Nov-2 Dec, Lincoln college, Canterbury, N-Z

Boyle, R.W. and I.R. Jonasson, 1973, The geochemistry of arsenic and its use as an indicator element in geochemical prospecting, *J. Geochem. Explor.*, 2 : 251-296

Bowell, R.J., N.H. Morley et V.K. Din, 1994, Arsenic Speciation in soil porewaters from the Ashanti Mine, Ghana, *Applied Geochemistry*, Vol. 9, pp. 15-22

Brannon, J.M., and W.H. Patrick, 1987, Fixation, transformation and mobilization of arsenic in sediments, *Environmental Science and Technology*, 21 : 450-459

Frost, D.V., 1967, Arsenic in biology – retrospect and prospect, *Fed. Proc. Fed. Am. Soc. Exp. Biol.*, 26 : 194-208

Krause, E., and V.A. Ettel, 1988, Solubilities and stabilities of ferric arsenate compounds, *Hydrometallurgy*, 22 : 311-337

Laxen, D.P.H., 1983, Cadmium adsorption in freshwater – a quantitative appraisal of literature, *Sci. Total Environ.*, 30:129-146

Manahan, Stanley E., 1993, *Fundamentals of Environmental Chemistry*, Lewis Publishers, 844p.

McLaren, S.J. and N. D. Kim, 1994, Evidence for a Seasonal Fluctuation of Arsenic in New-Zealand Longest River and the Effect of Treatment on Concentration in Drinking Water, *Environmental Pollution*, Vol. 90, No 1, pp. 67-73

Mok, W.M. et C.M. Wai, 1990, Distribution and mobilization of arsenic and antimony species in the Coeur D'Alène River System, Idaho, *Environ. Sci. Technol., 24: 102-108*

Moore, James W., 1990, *Inorganic Contaminants of Surface Water*, Springler-Verlag, New-York, 334p.

Taschereau, C.A., 1998, *Le problème de l'arsenic dans les résidus miniers – Le cas de la mine Chimo*, M.Sc. Thesis, Université Laval, 225p.

Vaughan, J.P., R.C. Dunne and S. Bacigalupo-Rose, 1992, Mineralogical Aspects of the Treatment of Arsenical Gold Ores, *in Innovations in Gold and Silver Recovery*, Randol, p.3011-3012

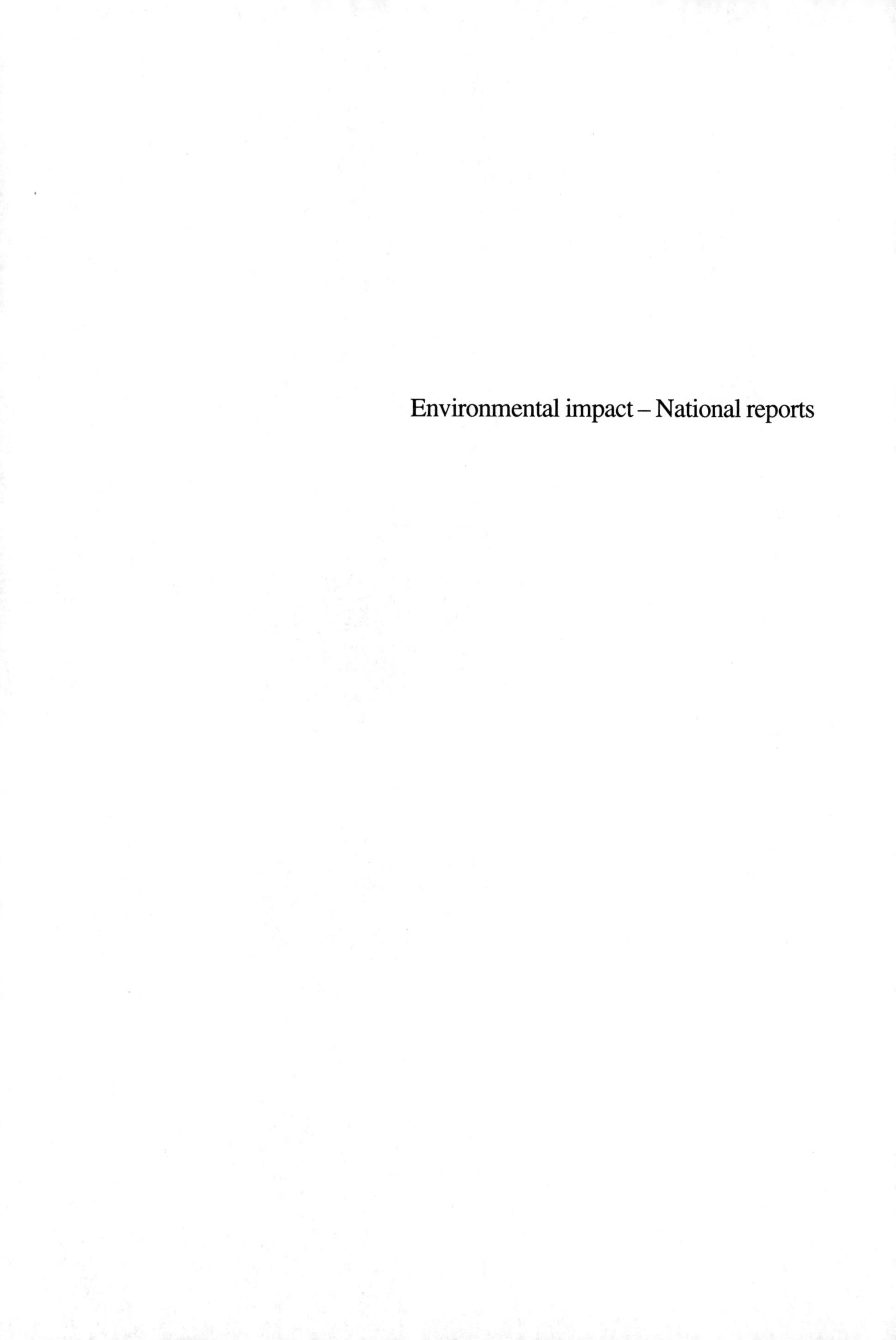

Environmental impact – National reports

Environmental Issues and Management of Waste in Energy and Mineral Production, Singhal & Mehrotra (eds)
© 2000 Balkema, Rotterdam, ISBN 90 5809 085 X

An approach to the environmental baseline chart in Argentina

J.C.Avila
Tucumán University, San Miguel de Tucumán, Argentina

ABSTRACT: This paper deals about the environmental baseline status in Argentina at the present times. The global characterization of the environment is a first step in quantifying impacts on nature that should be produced by a mining project. The quantitative characterization of environmental resources and targets which may be influenced by the focused activity is an important stage that perhaps takes a long period of time according to the knowledge of environmental database. Although the nation has an important tradition in the field of natural sciences, the data are not organized regarding environmental aspects. The environmental baseline have been performed only in few mining projects. In order to mitigate this deep lack of knowledge the National Geological Service is developing since 1997 the program named Environmental Baseline Chart of Argentine Republic (scale 1:250.000). The program addresses the different thematic items that must be taken in account. Mains are Lithology, Geomorphology, Soils, Climate, Superficial hydrology, Vegetation and use of the land and Natural and cultural patrimony.

1 INTRODUCTION

The knowledge about the environmental baseline of a region is the main start point in order to develop any kind of project that could impact in the environment.

Argentina has an important background in the field of natural sciences but, in general, data are not organized regarding environmental aspects. Only in few mining projects this type of studies have been performed (Lavilla & González, 1999)

In order to mitigate this deep lack of knowledge the National Geological Service is developing since 1997 the program named Environmental Baseline Chart (EBC) of Argentine Republic scale 1:250,000 (SEGEMAR, 1997).

The main objective of EBC is to obtain a complete and homogeneous information about the environment of argentinian territory in order to be used as an inventory of the natural resources.

Other targets of EBC are to optimize the use and protection of natural resources, to establish uniform symbols for the different environmental elements, and to define sensitive zones that must be carefully studied and protected.

2 CONTENT OF EBC

In order to satisfy main objectives, EBC must contains the following thematic items: Lithology, Geomorphology, Soils, Climate, Superficial hydrology, Vegetation and use of the land and Natural and cultural patrimony.

2.1 *Lithology*

This is the basic element in an environmental cartography. It should be developed from geological maps previously done in various scales.

The geological map must be properly simplified grouping formations that have similar physical and chemical characteristics that define permeability and geotechnical parameters of the rocks.

The map should contains main structural features as faults, folds, *etc.*

Main units to be represented are as follows:

- plutonic rocks (acid, intermediate, basic and ultrabasic);
- volcanic rocks (acid, intermediate, basic and ultrabasic);
- metamorphic rocks (slates, phyllites, schists, cornubianites, migmatites, milonites) and

- sedimentary rocks (clastic, organic and chemical origin).

Three ranges of permeability should be defined: high (> 10^{-4} m/s); media (between 10^{-4} and 10^{-6} m/s) and low (< 10^{-6} m/s). The permeability is related to porosity, intergranular, fissure and karst features.

Consistency, with reference to the lithification grade of the rock, should be defined in three categories: high, media and low.

Alteration in this case means the weathering grade of rocks. It is also defined in three levels (not or low altered, media and high altered).

The map should contains also detailed mining data about active and inactive mines (ore, gangue, reserves, mining and treatment methods, production rates, tails disposal, *etc.*).

2.2 *Geomorphology*

The geomorphological units should be grouped in relation to morphogenetic systems.

Main morphogenetic systems are:

- structural modelling;
- volcanic modelling;
- fluvial modelling;
- poligenic modelling;
- glacial and periglacial modelling;
- karstic modelling;
- eolic modelling;
- lacustral modelling;
- slope modelling;
- shore modelling and
- anthropogenic modelling.

Physiographic position and chronology of the units must also be indicated.

Permeability should be defined as in item 2.1 and the materials should be described using the "Earth Manual", Bureau of Reclamation, U.S. Department of the Interior.

In this item also a semiquantitative relief character (relief amplitude) must be defined. The relief amplitude is related to the difference of level between highest and lowest point in the terrain.

The relief amplitude is classified as:

- plain (gap less than 50m);
- low ridge (gap between 50 and 200m);
- hill (gap between 200 and 1000m) and
- mountain (gap more than 1000m).

2.3 *Soils*

The soil map from an pedologic point of view regarding also productive potential must be done. Soil types must be defined using "Soil Taxonomy" from Soil Survey Staff, U. S. Department of Agriculture (U.S.D.A.). Taxonomic categories should be: Order; Suborder; Great group and Subgroup.

Each soil type description should be complemented with the following characteristics:

- slope (flat to rugged);
- rock outcrops (absents to abundants);
- stony (absent to dominant);
- texture (clayey to sandy);
- depth (slight deep to deep);
- structure (lamellar to granular);
- hydromorphy (well drained to bad drained);
- flooding risk (low to high);
- water availability for plants (low to high);
- hydric erosion (low to high);
- wind erosion (low to high);
- vulnerability related to pollution (low to high);
- productivity uses and index (urban, agricultural, industrial, mining, *etc*);
- chemical properties (organic content, pH, soluble salts, *etc.*)

2.4. *Climate*

In an environmental research, studies about the climate are absolutely necessaries. Specially those related with precipitation, temperature and winds.

The climate map should have information about meteorological station regarding available equipment.

Annual isotherms (interval: 1°C) annual rainfall curve (interval: 100 mm) and wind rose should be detailed.

Climate types should be defined using Köppen (1948) classification.

The climate map must have also the following information:

- maxim rainfall in 24 hours;
- maxim instant rainfall;
- monthly media rainfall;
- annual media rainfall;
- monthly media temperatures;
- annual media temperatures;
- annual maxim temperature;
- annual minimum temperature;
- number of freezing days
- annual potential evapotranspiration
- climatic index of aridity (Thornthwaite & Mather, 1957)
- frequency, velocity and intensity of winds.

2.5 *Superficial hydrology*

Regarding the extreme importance that water means for biota and human beings, EBC must contain a hydrological superficial map.

This item should address the follows features:

- complete drainage including temporary streams (flows);
- hydraulic infrastructure (channels, damps, *etc.*, volume and surface)
- lakes, ponds, glaciers, *etc.* (volume and surface);
- main divortium acuaria (basins);
- gauging station (physical and chemical features of water)
- water pollutant points (physical and chemical features of water)

2.6 *Vegetation and use of the land*

Vegetation is one of the most important element related with environment. In this item the use of the land is also included because in a great part of Argentina natural vegetation has been eliminated or modified due to anthropic activity.

With regard to natural vegetation main phytogeographic units are:

- High grassland
- Dense shrubsland
- Medium grassland with scattered shrubs
- High andean grassland
- Bogs
- Gallery forest
- Low shrubsland with scattered columnar cacti.

In each unit followings features should be expressed:

- Covering grade. From grade 1 (5%) to grade 5 (100%).
- Slope grade. Flat (<25%) to rugged (>45%)
- Alteration grade (low to high)
- Abundant species
- Botanical endemia
- Faunal contained

With respect to the use of the land the follows parameters should be described:

- Urban use (population, social structure, health, economic activity)
- Industrial use (type and characterization of industry)
- Mining use (type and characterization of mine and plant)
- Forestal use (type of wood, reforestation plains)
- Agricultural use (type of cultivation, techniques, fertilizers)
- Cattle use (type, quantity, quality)
- Land used by aboriginal communities (ethnic, population, economic activities).

2.7 *Natural and cultural patrimony*

A well developed EBC have to include an inventory of natural and cultural inheritance that exist in the territory.

There are several categories of protected areas. Main are:

- I Scientific reserve
- II National park
- III Natural monument
- IV Natural reserve
- V Protected landscapes
- VI Resources reserve
- VII Natural – cultural reserve
- VIII Multiple use reserve
- IX Biosphere reserve
- X World patrimony sites

There are also sites of special scientific interest (geological, biological, cultural, *etc.*)

2.8 *Explicative text*

The explicative text should contain the following chapters:

- Abstract
- Introduction
- Thematic chart explanation
- Environmental explanation
- Conclusions
- References

3 DISCUSSION

The EBC program in Argentina has been briefly explained.

From my point of view some thematic items should be modified. There are:

- *Hydrology*: This chapter must include not only superficial water but also underground aquifers that have an important role in the environment specially in desert regions.
- *Vegetation*: This item should be recalled *Biodiversity* regarding that Fauna y Flora have identical weight in the study of the biota.
- *Use of the land*: This item has sufficient importance that must be separately discussed.

4 CONCLUSION

Nevertheless different points of view about some items, the really important notice is that Argentina is now in the correct way in order to know about the environmental baseline of the territory, specially regarding that mining industries are going to develop very fast in the next years.

REFERENCES

Köppen, W. 1948. Climatología. Fondo de Cultura Económica. México, 478 pp.

Lavilla, E. O. & González, N. 1999. Biodiversity of Agua Rica, Catamarca, Argentina. BHP, Fundación Miguel Lillo. Tucumán, 619 pp.

SEGEMAR, 1997. Carta Línea base Ambiental de la República Argentina. Normativa de realización.(unpublished).

Thornthwaite, C. W. & Mather, J. R. 1957. Intructions and tables for computing potential evapotranspiration and the water balance. Centerton, New Jersey, USA.10(3): 181-312p.

Environmental Issues and Management of Waste in Energy and Mineral Production, Singhal & Mehrotra (eds)
© 2000 Balkema, Rotterdam, ISBN 90 5809 085 X

Impact of small-scale gold mining on water quality

Rosa Cidu
Dipartimento Scienze della Terra, Università di Cagliari, Italy

ABSTRACT: The impact of gold exploitation on the aquatic environment has been evaluated by monitoring the waters in the area downstream from the tailings basin. Mining at Furtei (Sardinia) started in 1997 involving open cut over 2 km^2. Though the ore processing via cyanidation includes solution recycling and cyanide recovery, extreme CN, Hg and Cu dissolved contents occur at the tailings impoundment, but they do not seem to have affected the chemical composition of the water over the past three years, compared to the period before mine operations started. A leachate rich in CN, Hg and Ag was observed downstream from the impoundment after the solution level in it increased significantly. Afterwards, the contaminated waters were pumped back to the tailings basin in order to contain the dispersion of toxic components. With periodic monitoring it was possible to pinpoint the contaminant plume and prevent the dispersion of contaminants.

1 INTRODUCTION

Mining is a particular industry since large amounts of rock need to be processed to produce very small amounts of usable material, often generating considerable disruption in the physical environment. Mining itself may only occur for a few years, but its legacy in wastes is likely to persist much longer. Plans for modern mining exploitation, especially in developed countries, involve the assessment of the environmental impact during the exploitation, reclamation works, and all the requirements associated with the cessation of mining operations (Marcus 1997). Within this context, the potential impact of mining works on the quality of water resources is a primary concern in the local communities. A balance between mining interests and public awareness must therefore be achieved.

Mining operations may intercept groundwater, and the water quality may deteriorate due to the significant increase in specific surface of material in contact with the water. In mining areas, the wastes from exploitation represent the most important, long-term hazard for water quality. Mine waters usually show low pH, carry high amounts of metals, and sometime cause impressive visual impacts (Bowen et al., 1994). However, methods need to be implemented to assess how the mining will affect the surrounding groundwater, and how long its adverse effects will persist. This study was undertaken to monitor the water quality during small-scale gold exploitation in a populated area that hosts important water resources.

2 THE STUDY AREA

The study area covers about 30 km^2 near the village of Furtei (population: ~ 2,000) in southern Sardinia. It is a hilly area with a maximum elevation of 374 m a.s.l. and the vegetation is scanty with cultivated valleys at an altitude of ~100 m. The morphology was altered by decades of kaolin exploitation which left many abandoned quarries.

Oligocene calc-alkaline volcanic rocks, mainly andesites including explosive products (Assorgia et al. 1994), with marked NW-SE trending lineaments, are dominant in the area (Fig. 1). The volcano-sedimentary sequence underwent important alteration processes related to epithermal systems. Concentric patterns with an alteration intensity that increases towards the eruptive centers are observed. Marine, carbonate sediments of Miocene age (Pecorini 1966), with a maximum thickness of 30 m, outcrop at the borders of the area (Fig. 1), overlying the volcanic rocks.

A high-sulfidation, epithermal gold deposit was found related to advanced argillic alteration in the volcanic rocks. The pyroclastic and breccia units were more intensely affected by hydrothermal

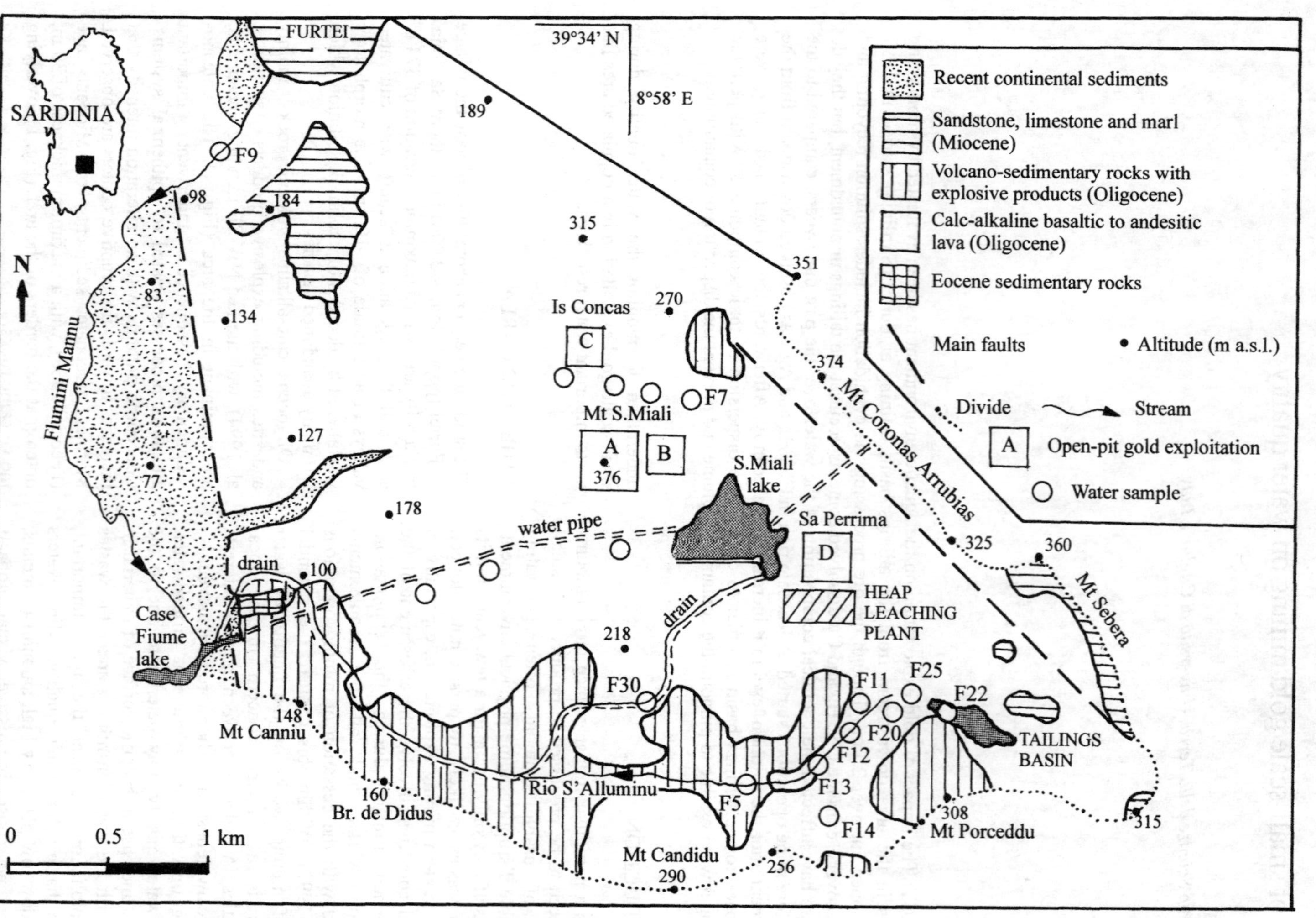

Figure 1. Schematic geological map of the Furtei area showing the location of water samples.

alteration than massive andesites. Four alteration types were recognized: silicification with massive and residual silica, and chalcedony bodies; advanced argillic alteration consisting of kaolinite, dickite, and residual silica; argillic alteration distinguished by montmorillonite; and propylitization with quartz, calcite, chlorite, and pyrite (Ruggieri et al. 1997, and references therein). The ore stage postdates hydrothermal alteration. Total ore reserves are 2,150,000 tons at a grade of 2.82 Au g/t (Madau et al. 1996). The ore deposit comprises an oxidized zone, and a deeper primary sulfide zone characterized by a vertical zoning in mineral assemblages: pyrite-enargite-luzonite-gold at higher levels, accompanied by tennantite and tellurides at depth; native gold, containing 0.7-7.8 wt % silver, occurs as blebs in enargite and luzonite (Ruggieri et al., 1997). The supergene alteration assemblage includes jarosite, gypsum and other sulfates, iron-hydroxides and covellite (Garbarino et al. 1990).

The Furtei Project (Sardinia Gold Mining S.p.A., SGM) involves open cut mining of four ore bodies (A, B, C, D in Fig. 1) located over an area extending 2 km^2. The project, which started in 1997, processes the oxide and transition ore at a rate of 270,000 t/y (tons per year) via a conventional cyanidation and CIP (carbon-in-pulp: adsorption on carbon from the leach pulp) circuit. Gold and silver are eluted using the AARL (Anglo American Research Laboratory) system. When the sulfur content of the ore increases, additions will be undertaken to incorporate a sulfide flotation circuit. The process plant comprises recycling of the solutions discharged into the tailings sedimentation basin, and recovery of cyanide.

In the Furtei area, the long-term mean rainfall is 540 mm/y distributed over 55 rainy days, with almost dry summers. Rainfall was higher in 1996 (730 mm), lower in 1998 (350 mm), and close to it in 1997 and 1999. The mean annual temperature is 17°C, and evapotranspiration is high. In the area around the mine two dams play an important role in the regional water supply. The S. Miali lake (Fig. 1) is filled by a water-pipe from an upper lake of the Flumendosa hydrologic system, far outside the study area. The lake water is renewed weekly and a drain allows water-transport to the lower Case Fiume lake, from where water is distributed over a large area in southern Sardinia and used in agriculture and for drinking.

The hydrogeology is mainly controlled by the altered volcanic rocks: fracturing and vuggy silica zones allow the groundwater to flow, and promote mineral oxidation. In contrast, clay zones inhibit the groundwater flow and allow sulfide minerals to persist near the surface. The water circulation is relatively close to the surface, usually less than 20 m below ground level, and becomes deep only rarely and close to regional structures. The water table is very near the surface along the Rio S'Alluminu valley. The water table is drained by intermittent streamlets and springs, with flow rates that strongly decrease over the dry periods. The water circulation in the carbonate rocks is limited by their reduced thickness in the area.

3 SAMPLING & ANALYTICAL METHODS

Since the area downstream from the tailings basin and the heap leaching plant is potentially subjected to a higher water contamination risk, sampling was carried out mainly at the Rio S'Alluminu valley. Based on previous records (Cidu et al. 1997), additional samples were considered as the exploitation progressed. The considered samples include waters from streams, springs, ponds, lakes and drillholes. The location of the water samples is shown in Figure 1. The S. Miali Lake (F30) is fed with water, mostly drained from Paleozoic metamorphic rocks located far outside the study area. The Flumini Mannu river (F9) drains a large hydrologic basin mostly made up of Tertiary marine sediments. These samples represent waters not interacting with the ore deposit.

Periodic monitoring was chosen according to the plan of the SGM Company (Madau et al. 1996), but it was flexible as more information became available. Some factors of a potentially higher risk, such as storm events and the increase in the hydrostatic pressure at the tailings basin, were taken into account. In January 1997 sampling was carried out immediately before exploitation, following an extended rainy season, while in November 1997 it was immediately after a three-day heavy rain event, and in November 1999 after a storm event that produced 92 mm of rainfall in 20 hours. Other sampling periods represent conditions from low rain to dryness.

The waters were filtered in situ through a 0.4 μm pore-size membrane filter, and collected into acid-cleaned polyethylene bottles. Sample stabilization using suprapure acid was carried out in situ upon filtration: 1% HNO_3 for most metals, 0.2% HCl for As and Sb, and 0.2% H_2SO_4 plus $KMnO_4$ for Hg.

Temperature, pH, redox potential (Eh), conductivity, and alkalinity were measured at the sampling site. Free cyanide by spectrophotometry was determined on non-filtered samples within three hours after sampling, with a 10 μg/l detection limit. Turbid waters were filtered before cyanide analysis.

Table 1. Ranges of chemical components in the Furtei waters over three years exploitation of gold (1997-1999).

No.		pH	Eh	TDS	Ca	Mg	Na	K	HCO_3	Cl	SO_4	SiO_2	Al	B	Li	Ba	Sr	Fe	Mn	Zn	Cd	Pb	Cu	Ni	Co	As	Hg	Ag	CN
			V	g/l	mg/l	mg/l	mg/l	mg/l	mg/l	mg/l	mg/l	mg/l	mg/l	µg/l	µg/l	µg/l	mg/l	mg/l	mg/l	mg/l	µg/l	µg/l	µg/l	µg/l	µg/l	µg/l	µg/l	µg/l	µg/l
F30 a	min.	8.0	0.44	0.3	41	14	25	2.0	116	29	57	2	0.1	25	1	28	0.1	0.1	0.03	0	<0.1	0.2	1	1	0.1	1	0.3	<0.1	<10
	max.	8.4	0.46	0.5	64	22	41	2.1	138	52	152	4	1	32	4	39	0.3	0.5	0.5	0.01	0.1	2	50	2	0.7	4	0.7	<0.1	<10
F9 b	min.	7.5	0.38	0.7	71	28	83	7	240	83	100	5	0.02	140	9	40	0.6	0	0.01	0.01	<0.1	1	2	1	0.1	1	0.3	<0.1	<10
	max.	8.2	0.45	1.1	124	45	136	9	370	170	190	9	0.3	280	30	70	1.3	0.3	0.1	0.02	0.1	2	10	2	4	8	2.6	<0.1	13
F14 c	min.	7.2	0.35	0.7	36	34	160	5	234	143	173	13	0.01	115	1	10	0.3	0.1	0.05	0.01	0.1	0.2	1	1	0.1	0.3	0.2	<0.1	<10
	max.	7.6	0.48	1.6	108	71	280	16	328	264	510	23	0.1	150	4	38	0.7	0.9	0.8	0.04	0.2	3	10	2	3	7	1.9	<0.1	14
F7 d	min.	5.5	0.26	4.7	440	120	890	12	70	1060	2000	9	0.1	530	2000	8	3.8	13	2.6	0.2	0.1	1	1	1	0.4	200	0.1	<0.1	<10
	max.	5.8	0.35	4.9	470	150	1000	17	160	1170	2160	14	4	740	3000	15	4.3	50	3.2	0.6	0.7	10	45	10	20	380	1.1	<0.1	<10
F20 d	min.	6.4	0.15	3.6	230	43	880	12	460	770	1400	10	0.01	450	70	8	5.0	1.7	0.1	0.01	0.1	1	3	2	0.2	11	0.2	<0.1	<10
	max.	7.1	0.22	4.9	320	63	1180	14	500	1080	1800	16	0.3	530	90	10	6.6	3.1	0.2	0.5	2	11	8	7	3	25	1.3	0.4	<10
F25 f	min.	6.4	0.16	5.6	450	240	750	8	210	1500	2400	19	0.1	330	20	17	2.3	0.5	4	0.03	0.4	1	5	30	340	0.2	0.5	<0.1	<10
	max.	7.7	0.61	7.0	770	320	1120	14	610	1900	2700	30	12	460	120	27	3.7	27	28	2.1	9	7	90	60	540	48	1.5	13	12000
F11 e	min.	4.3	0.37	2.6	300	120	400	8	<10	660	1080	6	0.2	230	45	21	1.4	0.3	4	0.5	4	1	8	16	40	0.2	0.2	<0.1	<10
	max.	6.4	0.59	6.1	700	310	990	13	120	1820	2740	29	48	500	150	46	3.4	1.6	20	4.5	39	35	220	90	370	0.8	0.7	6	19
F12 f	min.	4.3	0.25	3.6	370	150	580	11	<10	970	1400	6	0.6	340	30	16	1.9	0.5	9	0.5	2	1	3	18	50	0.1	0.3	<0.1	<10
	max.	7.3	0.59	6.9	690	360	1040	14	240	2070	2900	14	27	710	90	31	4.2	6.5	36	4.7	16	8	84	90	350	0.4	1.6	<0.1	23
F13 f	min.	4.7	0.29	5.5	520	270	870	9	20	1490	2190	6	0.02	450	15	15	3.6	0.1	0.04	0.03	1	1	5	5	2	0.3	0.1	<0.1	<10
	max.	7.1	0.42	7.8	720	470	1280	16	430	2260	2870	16	16	700	50	43	6.0	0.4	26	3.7	18	4	50	80	200	2	0.6	<0.1	<10
F5 f	min.	7.6	0.34	5.9	610	320	930	11	400	1500	2300	12	0.02	380	22	33	4.4	0	0.1	0.01	0.3	1	1	2	1	0.7	0.3	<0.1	<10
	max.	8.0	0.41	7.9	740	480	1370	14	500	2290	2800	16	0.1	600	55	50	6.3	0.7	4	0.04	2.5	3	13	19	19	1.8	0.5	<0.1	<10
F22 g	min.	9.4	0.34	2.5	200	3	680	10		360	1100	21	0.1	70	32	12	0.5	0.1	0.01	0.05	0.2	1	15000	180	60	170	300	7	70000
	max.	10.4	0.39	3.9	450	10	1540	32		540	1500	48	0.2	350	64	36	0.9	3	0.3	0.5	5	2	140000	340	400	4800	3000	600	270000

a: S.Miali lake; b: Flumini Mannu river; c: Mt Porceddu spring; d: drillhole; e: S'Alluminu pond; f: S'Alluminu streamlet; g: tailings impoundment.

Anion species were determined by ionic chromatography on filtered, non-acidified samples. Analysis of metals was by GFAAS, ICP-OES and ICP-MS. Mercury, antimony and arsenic were determined by ICP-MS after flow-injection Hg-vapor or hydride generation.

Precision, usually 6% and 11% at the mg/l and μg/l level, respectively, was evaluated by random replicate analysis and duplicate samples. Accuracy, estimated by analyzing certified standard material, and using different techniques, was similar to precision. The potential contamination of the samples was monitored by field blank solutions treated and analyzed following the same procedure as the samples. Concentrations of elements in field blanks were always below detection limits, calculated on the basis of 5 times the standard deviation of the blank mean value for each analytical sequence.

4 RESULTS AND DISCUSSION

The chemical characteristics of the waters in the Furtei area prior to the beginning of gold mining are reported in Cidu et al. (1997). The ranges of chemical components of the most representative waters at Furtei over three years exploitation are reported in Table 1. These waters show pH values from acidic to alkaline, total dissolved solids (TDS) in the 0.3-7.8 g/l range, different chemical composition, and a wide range of metal concentration. The different chemical characteristics strongly depend on the interaction of the water with the altered volcanic rocks and ore minerals from the Furtei area.

The concentrations of the chemical components in some representative waters immediately before the gold exploitation are reported in Figure 2. Bicarbonate prevails on sulfate and chloride in the low salinity waters, while sulfate and chloride dominate at high salinity. Chloride, sodium and magnesium are closely correlated with TDS. In the high-salinity waters, sulfate is controlled by the equilibrium with respect to gypsum, while calcite equilibrium is attained at intermediate TDS. High concentrations of Al, Fe, Mn, and Zn occur at acidic condition in waters interacting with altered volcanic rocks.

Samples from the Santu Miali lake (F30) and the Flumini Mannu (F9) are waters not interacting with the rocks in the mining area. The lake water at F30 shows very low salinity (≤0.5 g/l), and sulfate up to 150 mg/l. The river water F9 shows salinity up to 1.1 g/l, and sulfate 190 mg/l. In both waters, low concentrations of trace metals occur.

Spring F14 receives water interacting with the kaolinized volcanic rocks of Monte Porceddu. It shows a balanced chemical composition, TDS up to 1.6 g/l and low concentrations of metals related to the ore deposit. At this site, the water level decreased in the summer season, and the spring went dry from July to October 1999.

Samples F11, F12, F13, and F5 represent the emergence of the water table at decreasing height along the Rio S'Alluminu valley. These waters are characterized by high salinity, a dominant sulfate composition, and high concentration of Sr, B, Li, and Mn, while metals related to the gold ore (such as Cu, As, Hg) are usually low or below detection limit. In these waters, which are strongly influenced by rainfall, high concentrations of Al and Fe, and increased concentrations of Zn, Cu, Co, and Ni, are observed at acid pH. Whitish deposits, mostly made up of sulfate minerals, are often observed at the F11, F12, and F13 sampling sites at pH's greater than 5.5.

The drillhole water F20 (and the drillhole F7) shows high concentrations of sulfate, Fe, Mn, and As. Reddish deposits, mostly made up of iron oxy-hydroxides, were always observed at emergence. Samples F20 and F7 did not seem directly affected by rainfall, and represent the groundwater interacting with disseminated minerals related to the gold deposit.

The analytical results during the monitoring period show that the above characteristics are very similar to those observed in 1994, long-before mining (Cidu et al., 1997), and that they are roughly maintained at all water sites over three-years exploitation. A higher pH, and a marked decrease in metal concentrations due to a dilution process, was observed in surface waters and in the shallow groundwater after rainy periods.

Though in contact with the atmosphere, the water at pond F11 may be considered representative of the shallow groundwater in the Rio S'Alluminu valley. The behavior of some metals at F11 during monitoring is shown in Figure 3. It can be observed that the highest concentrations of metals occur at an acid pH (4.1-4.6), while a dilution effect is clearly observed after heavy rain. Metal contents after three years' exploitation are close to those measured prior to the beginning of mining, with the exception of Mn that shows a marked increase (from 5 to 20 mg/l).

The solution at the tailings impoundment (F22 in Table 1) shows extremely high concentrations of Hg, As, Cu, and about 200 mg/l of free cyanide. It therefore poses a potential hazard to the aquatic

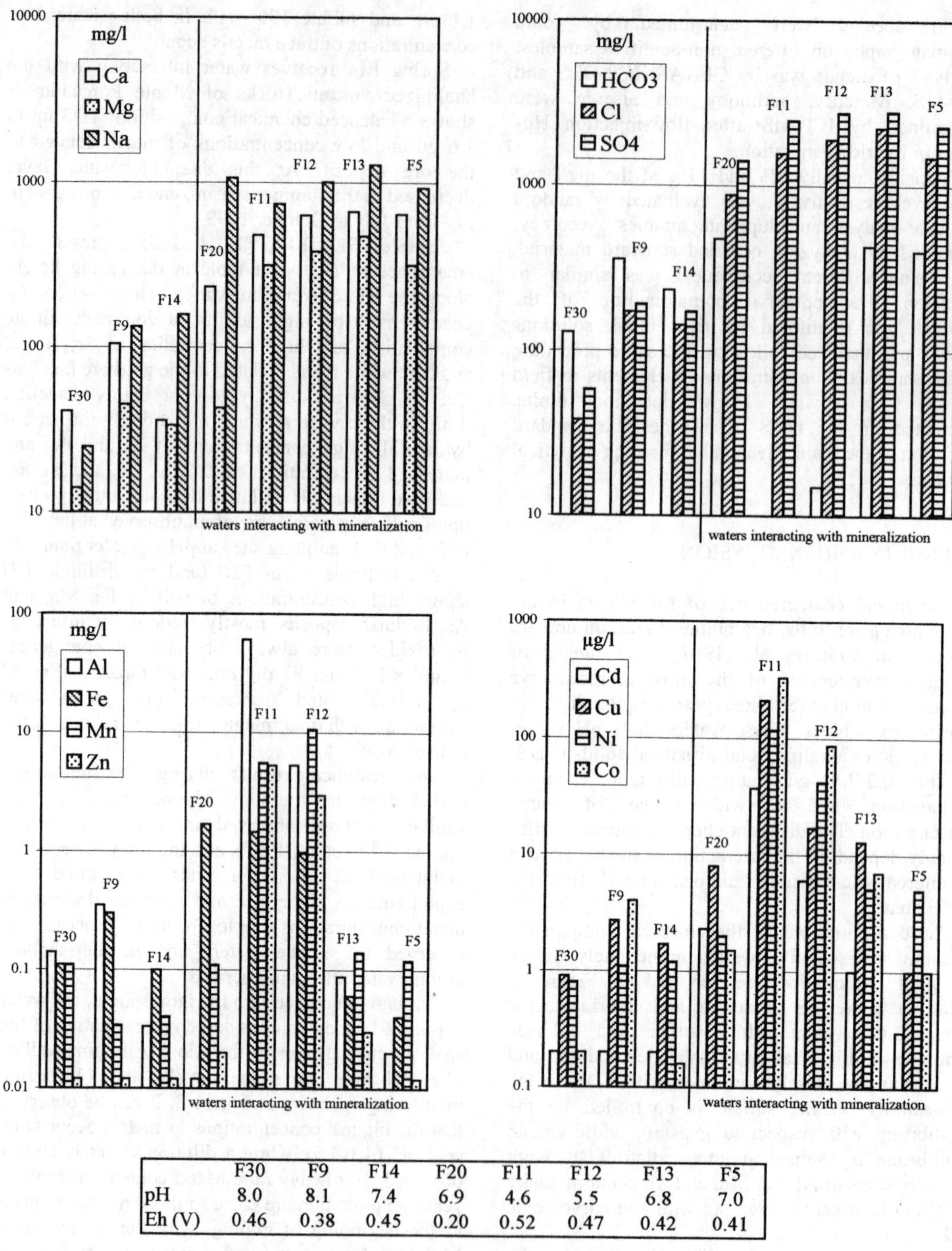

	F30	F9	F14	F20	F11	F12	F13	F5
pH	8.0	8.1	7.4	6.9	4.6	5.5	6.8	7.0
Eh (V)	0.46	0.38	0.45	0.20	0.52	0.47	0.42	0.41

Figure 2. Concentrations of dissolved chemical components in some waters from the Furtei area (water sampling on January 1997, immediately before gold exploitation).

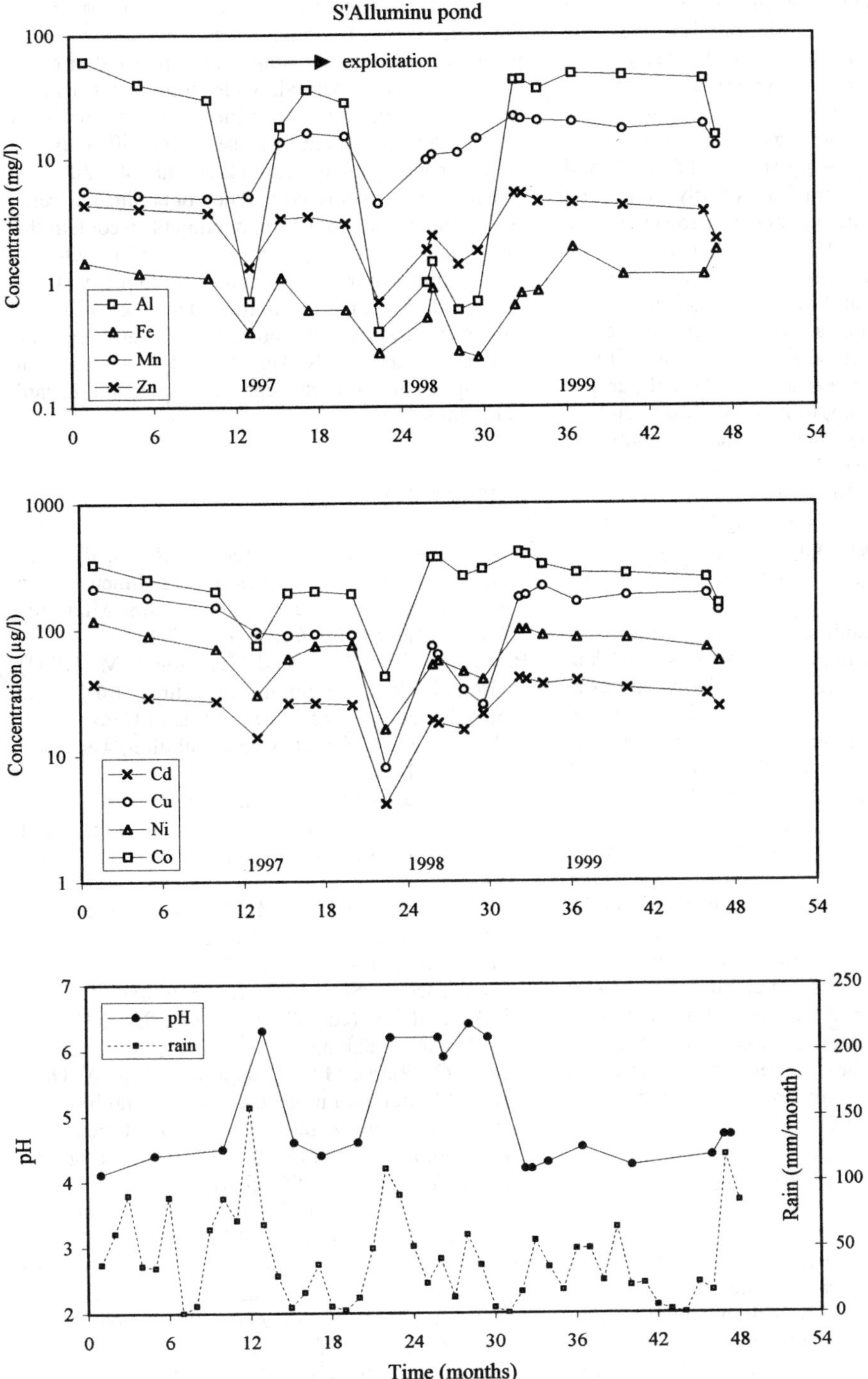

Figure 3. Concentrations of some metals, pH, and monthly rain at F11 over the 1996-1999 period.

environment. No leaching from the tailings basin was observed until the solution level at the impoundment has been kept low. In February 1998, when the hydrostatic pressure increased significantly, water flowed out immediately downstream from the tailings basin. This water (F25) contained CN (120 μg/l), Hg (2.5 μg/l), and Ag (2 μg/l) at concentrations clearly indicating seepage from the impoundment. However, the concentrations of CN, Hg and Ag decreased by about 50 m downstream, and were found very close to background level at F11, about 250 m further downstream. Sampling a week later showed a similar situation. On the basis of these results, it was decided to collect the water at site F25 and pump it back to the tailings basin, to avoid the dispersion of the contaminant plume. The CN concentration was weekly monitored. Cyanide at F25 showed marked variations (0.1-2 mg/l) from February 1998 to March 1999, and then increased from 3 mg/l to 12 mg/l. At the other sites, CN was only occasionally detected (up to 23 μg/l) being mostly below the detection limit.

The climatic conditions, and the progressive filling of the tailings impoundment must be taken into consideration to understand the CN variations at F25. The geochemistry of CN depends on several factors, pH and temperature being among the most important. Free cyanide is mostly present as CN^- at pH>10 (90% CN^- at 20°C), while the HCN species increases with decreasing pH, and it is dominant at pH<8 (95% HCN at 20°C). Mixing of the seepage from the impoundment with shallow groundwater causes a decrease in pH at 7.7, at this condition the dominant HCN volatilizes (as shown by the strong cyanide smell). Fluctuations in the water table dilute the seepage more or less efficiently. Since long periods of dry weather as well as high temperatures during the summer season caused increases in CN concentrations at the impoundment due to evaporation, a more contaminated plume was observed throughout 1999.

5 CONCLUSIONS

The impact of a small-scale, open-pit Au mining on the aquatic environment was studied by monitoring the water quality prior to and during exploitation. The local groundwater quality prior to Au mining (1997) was poor (high TDS, elevate concentrations of SO_4, Al, Fe, Mn, Zn), due to acid drainage derived from the weathering of altered volcanic rocks hosting sulfide minerals. Oxidized-ore processing via cyanidation is ongoing, and the chemical characteristics of the waters downstream from the tailings basin have not changed over three-years of exploitation. Contaminant runoff after rainy periods was not observed, while storm events caused a significant dilution. When the hydrostatic pressure at the impoundment increased significantly, a contaminant plume extending about 200 m downstream was observed. The contaminated water was pumped back to the impoundment to contain the dispersion of the plume. Periodic monitoring allowed pinpointing of the contaminant plume and a prompt device to prevent the effects of pollution. However, despite solution recycling and cyanide recovery, extreme CN, Hg, As and Cu contents at the impoundment might represent a potential hazard in the future.

REFERENCES

Assorgia, A., Barca, S., Farris, M., Rizzo, R. & C. Spano (1994) Le successioni sedimentarie e vulcaniche Cenozoiche del distretto Monastir-Furtei. *L'Industria Mineraria* 15: 7-13.

Bowen G., Dussek C. and Hamilton R.M. (1994) Groundwater pollution resulting from the abandonment of Wheal Jane Mine in Cornwall. In Proc. 3rd Conf. Groundwater Pollution, Technical Services, London.

Cidu, R., Caboi, R., Fanfani, L. & F. Frau (1997) Acid drainage from sulfides hosting gold mineralization (Furtei, Sardinia). *Environm. Geol.* 30: 231-237.

Garbarino, C., Grillo, S.M., Melis, F., Pretti, S., Uras, I. & M. Fiori 1990. Paragenetic assemblages of ore minerals in an acid sulphate type gold occurrence at Serrenti-Furtei, Sardinia, Italy. In E.A. Ladeira (ed) *Brazil gold '91*: 143-150. Rotterdam: Balkema.

Madau, G., Pinna, G.P., Humphries, B. & A. Orrù (1996) Furtei gold project: environmental baseline study and management plan. In R. Ciccu (ed) *Proceedings SWEMP'96*: 189-196. Cagliari: DIGITA University of Cagliari.

Marcus, J.J. (ed) 1997. *Mining environmental handbook.* London: Imperial College Press.

Pecorini, G. 1966. Sull'età 'oligocenica' del vulcanismo al bordo orientale della fossa tettonica del Campidano (Sardegna). *Rend. Acc. Naz. Lincei* 40:1058-1065.

Ruggieri, G., Lattanzi, P., Luxoro, S.S., Dessì, R., Benvenuti, M. & G. Tanelli (1997) Geology, mineralogy, and fluid inclusion data of the Furtei high-sulfidation gold deposit, Sardinia, Italy. *Econ. Geol.* 92: 1-19.

Environmental Issues and Management of Waste in Energy and Mineral Production, Singhal & Mehrotra (eds)
© 2000 Balkema, Rotterdam, ISBN 90 5809 085 X

Development of environmental, social and economic indicators for a mining city in Bolivia

Anne-Marie Fleury & Richard Poulin
Laval University, Quebec City, Que., Canada

ABSTRACT: Although indicators have become widely used and talked about, not much work has been done on micro-economic (or local) indicators. The aim of this paper is to describe how local indicators were developed to measure the impacts of the mining industry in Potosi, a Bolivian city. The resulting set of indicators is specific to the project but could be used as an example for other micro-economic indicator projects. The objective of the article is to describe the approach used to develop indicators for a particular theme (in this case small-scale mining) at a local level. The article takes a brief look at what an indicator is followed by a description of the objective and particulars of the project. The creation of the database is then discussed as well as the construction of the indicators from this database, examples are given. How sustainable development was looked at for the project is also briefly discussed.

Key words: indicators, local, micro-economic, small-scale mining, Potosi

1 INTRODUCTION

1.1 *Indicators*

Much work has been done on indicators in the past decade, especially since the Rio Earth summit in 1992 that underlined the need to develop indicators as a tool for measuring sustainable development. A variety of definitions and approaches have been used by different organisations, but most of the work done is macro-economic and involves whole countries. Although some work exists on a local or micro-economic level, pre-determined sets of indicators that can be used globally are difficult to create as they are very specific to the region they are developed for.

Most definitions of indicators underline two characteristics;

1. the main use of an indicator is to represent and summarise a greater quantity of information on a particular theme.
2. indicators are a good tool for measuring tendencies in a system.

A simple definition that underlines the first characteristic is that «an indicator is a measure that summarises information pertinent to a particular phenomenon» (McQueen & Noak 1988, Gallopin 1997). Physically an indicator is a variable; it can be an index, a ratio or a direct measure. When a certain phenomenon is being studied, large quantities of information may exist on the theme, the advantage of working with indicators is that they are representative variables that are easier to manipulate than the whole information set or database. This makes them a good tool for measuring tendencies and changes in the studied phenomenon. The definition used by the Canadian International Development Agency (CIDA) emphasises this quality: an indicator «is an index [..] that describes a state or a situation and determines the changes brought to this state or situation during a given period. Indicators examine closely the results of development projects and initiatives.»

Bossel (1996), Moldan (1997) and Haberl & Schandl (1999) describe three functions for indicators; the simplification, the quantification and the communication of information. Simplifying and quantifying means that a part of the information is necessarily lost, this makes the selection of indicators very important. How effectively the chosen set of indicators represents (or communicates) the studied phenomenon varies according to the choice of which information is retained in the simplification process.

When choosing and constructing the indicators, it is important to keep in mind the objective behind the creation of the indicator set as well as their intended users. A clear definition of these factors as well as a constant awareness of them throughout the selection process is necessary so as to insure the usefulness and relevance of the chosen set of indicators.

The greatest limiting factor in developing indicators that communicate effectively the phenomenon in question is the availability and quality of the information from which they are created. Availability of data determines which indicators can be developed and the quality of the data determines the quality of the indicator. When working in local areas in developing countries, limited availability of data can be problematic. It is important to obtain as much information relevant to the topic as possible and to insure oneself of the integrity of the data.

1.2 *Indicators for a Bolivian mining city, Potosi.*

The objective of the project in this case was to develop indicators to measure the impact of the mining industry on the environment, population and economy in the city of Potosi, Bolivia. The city has a population of 120,000, it has been a mining one since its foundation 500 years ago. The Cerro Rico mountain in Potosi is a polymetallic mineral reserve that was initially mined in the colonial era for its rich deposits of silver. In the last few decades, tin exploitation on a fairly large scale by the national company Comibol was the main mining activity on the mountain. After the drop in tin pricés in the 1980's, large-scale mining activity ceased and was replaced by small-scale co-operative mining of zinc and silver deposits. The co-operatives are in actual fact loose affiliations of independent miners, they regroup about 4000 miners who exploit at 80 different mining sites on the mountain. The concentration of the mineral is done separately by one of the 40 small, privately run concentration plants in the city who buy the mineral from the co-operatives. As is usually the case when so many participants are involved, control and monitoring in the industry is difficult. The social and economic implications are important, the environmental implications are considerable.

The purpose behind developing indicators to measure the impact of mining in Potosi was mainly to communicate an image of the current situation in the city with respect to the environment, the urban population and the local economy. A second objective was to have a tool able to demonstrate the impacts that different projects and initiatives in the region will have. The intended users of these indicators included the Canada-Bolivia project for environmental control and prevention in mining (that sponsored this project), the Bolivian government, as well as any NGO's or groups conducting development projects (especially mining or environment projects) in the city.

2 THE DATABASE

2.1 *Creating the database*

The creation of the database is a key step when developing indicators. When working on a macro-economic level predetermined lists of example indicators exist, the World Bank, Unesco and CIDA are institutions that have constructed such lists. In this case, the selection of the indicators can be done from the lists before the compilation of the necessary information is done. The indicators are selected according to the needs of the project and the required information is then collected, this is a top to bottom approach. In the case of local indicators no predetermined lists exist, and a bottom to top approach must be taken. On a local scale, a comprehensive set of all the information relevant to the studied topic can be compiled without using extensive resources. Once a fairly complete database has been created, the most representative indicators are selected.

Planning which elements will be included in the database is important in order to gather a broad set of relevant data. When defining this plan, the input of as many interested parties as possible is beneficial. This insures that the database is complete and can comply with the different perspectives the interested parties may have. Once the plan is defined, direct compilation of the data on-site is imperative in order to develop a certain value judgement that can only be acquired with local experience. The initial plan must also be flexible enough to include changes that may be brought about once the work on-site has begun.

2.2 *The database created to measure mining impact in Potosi*

The information in the database for the Potosi project was initially divided into three categories, mining impact on the environment, on the population and on the economy. Many elements were included in the plan, for example the quantity and nature of solid mining wastes, the contamination of rivers, the quality of the dust near mining related sites, the number of people working in the mines and concentration plants, the average wages of the workers, the poverty index in the city, the health of the miners and the contribution of mining to the gross domestic product.

Over nine months were spent in Potosi compiling the information for the database and several adjustments to the initial plan were deemed necessary. The presentation of the information was re-structured to include five sections that incorporated information from the three original categories. These are co-operative exploitation, mineral concentration plants, solid mining wastes, water, and population.

The first two sections were mostly descriptive of the industry. Specific information on the co-operative miners and concentration plant workers was included as well as economic data such as taxes paid by the mining industry and profits and costs involved.

The next two categories involved environmental information. Solid mining wastes dealt with information on the quantities of tailings and waste rock as well as their potential to produce acid rock drainage (ARD). Information on dust levels near concentration plants and the tailings dam was included in this section too. Different sources of river contamination such as concentration plant effluents and ARD were considered in the water section as well as the quality of the river water and sediments.

The last section dealt mostly with statistical data on the population of Potosi. This includes information on poverty, urban infrastructures, literacy, as well economical information such as the active population, average wages, most important industries and the contributions of these industries to the gross internal product.

Throughout the data compilation, many information sources were used. In a city where the main industries are mostly informal, information is rare and can be unclear and conflicting. Using multiple sources of information is a good way to cross check and verify the information so as to insure data integrity.

Information from various projects underway or completed in Potosi were used as well as statistical information from the National Statistics Institute and other sources. With the available resources, as much information as possible was collected first hand. Data from the co-operatives and concentration plants was compiled and water and mining waste samples were taken. Certain things proved to be impossible to verify completely, and different information sources sometimes gave irreconcilably different data on certain points. In such cases a judgement was required to determine the more probable value or to conclude that no value could be fixed.

3 DEVELOPING THE INDICATORS

3.1 *Approach for selecting indicators*

Once the database has been created, the information used to develop the indicators must be selected from it. Different approaches exist to select this information.

Mitchell (1996) describes three approaches for the selection of indicators from the database, the figure 1 illustrates them. The first approach involves developing numerous specific indicators that are directly connected to the data. With this approach, a large amount of indicators represent the data in a very precise way. They do not summarise the data much, however very little information is lost. Another disadvantage is that such indicator sets are difficult to manipulate because of their size. It is a useful approach for users who are familiar with the studied theme and are interested in detailed information.

The second approach involves combining different data into information groups and creating composite indicators from them. These indicators are calculated using much data, they can be difficult to manipulate and recreate but they have the advantage of summarising concisely the information in the database.

The third approach is a combination of the first two approaches. This approach involves developing certain simple composite indicators and others directly from key information. The resulting indicators are relatively simple to manipulate, they are easy to calculate and they summarise the information in the database. One of the disadvantages of this approach is that it involves

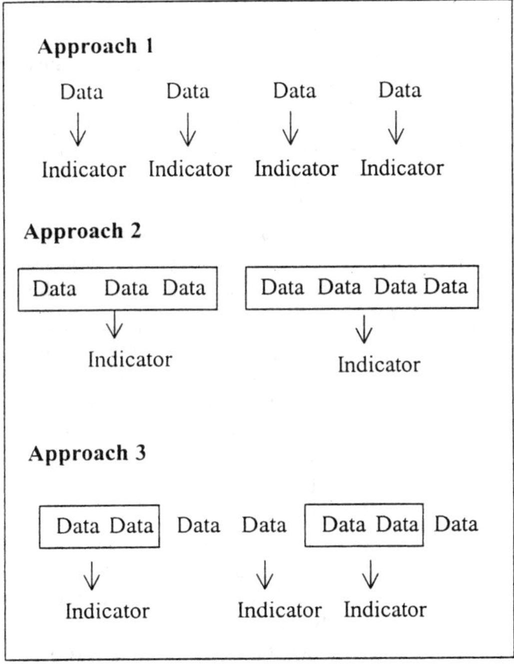

Figure 1 Different approaches for indicator selection, adapted from Mitchell (1996).

some subjectivity. The choice of which data constitutes key information during the indicator selection process involves a value judgement and therefore depends on the one(s) making the indicator selection.

3.2 *Indicators selected for measuring mining impact in Potosi.*

For measuring mining impact in Potosi, the third approach was considered to be the most suited. The indicators for the project were created to represent the impact of the small-scale mining activity in the city, therefore they had to be concise and summarising. They also are meant to eventually serve as a tool to measure changes in the city, and must be easy to calculate and update. Indicators created with the third approach satisfied these criteria.

Although the entire database was used, the indicators were developed mostly from the last three sections. The indicators were grouped under the categories environment, population and economy. The table 1 lists examples of indicators chosen in each of the categories.

An overall picture of the impacts of mining in Potosi was obtained through the indicator set. It was found that mining is the most important industry in the city. It employs directly about 17% of the active population but creates a multitude of jobs indirectly as it constitutes the most important independent industry in the city (i.e: that generates capital from outside the local population).

The industry was found to function with a high a degree of manual labour and safety standards were very low, especially in the mines. The ensuing mining pollution was found to be far-reaching. The indicators developed to measure the environmental impact of mining in Potosi compared air, and water quality to background levels in the region and measured the amount and nature of solid mine wastes. The indicators on water quality most dramatically represented the extent of mining pollution, certain heavy metal concentrations were found to be 100,000 times higher than in background levels.

Conditions for the local population were found to be fairly the same as in other Bolivian cities with the exception of climatic difficulties associated with the high altitude location of the city (4000m). However indicators on the health of the population demonstrated that the inhabitants of Potosi have a level of health inferior to that of average Bolivians, especially with respect to pulmonary diseases. Both the high altitude and the poor working conditions in the mines contribute to this circumstance.

Table 1 Examples of Indicators used to measure mining impact in Potosi

Category	Example of indicator
Environment	• (t of mining residue discharged daily) / (population in contact with residues) • (concentration of Zn in dust near a concentration plant) / (concentration of Zn in dust outside the city)
Population	• % of active population working in the mining industry • (poverty index in the city of Potosi) /(average poverty index of cities in Bolivia) • number of accidents in the mines per year • (frequency of pulmonary diseases among miners)/ (average frequency of pulmonary diseases in Bolivia)
Economy	• contribution of mining in Potosi to the GDP • salaries generated by the population through mining

3.3 *Evaluation of the selected indicators*

Each indicator in the set must be evaluated with respect to specific criteria. The criteria are considered throughout the selection process, indicators that don't satisfy the criteria are rejected. A separate evaluation should be done as well following the selection process in order to identify the strengths and weaknesses of each indicator.

Criteria are often similar for any indicator set, in the case of this particular project four criteria were used. The indicators must be:

- relevant to the studied phenomenon
- easy to calculate.
- expressed clearly
- sensitive to change

Not all the selected indicators satisfied fully the criteria. Some indicators were not directly relevant to the topic. For example, an indicator on the contribution of mining of the whole Potosi department to the Bolivian GDP is not directly relevant. This indicator brings interesting information on mining in the region and consequently was retained in the indicator set despite its failure to fully satisfy the first criterion.

Other indicators such as the frequency of accidents in the mines were very difficult to calculate and did not meet the second criterion.

The clearness of an indicator has to do with the form in which it is expressed. This criterion is important during the indicator selection, the choice of expressing the information as a percentages, a difference or a ratio was made according to which resulting variable was easiest to relate to. For instance, to compare life expectancy for the citizens of Potosi with the average life expectancy in Bolivia, it is much more meaningful to express this as a difference equal to 5.74 years than to express it as a ratio equal to 0.9.

The indicators expressed as ratios did not always satisfy the criterion on sensitivity to change. It was important to consider how a change in either of variables involved affected the value of the indicator. For example, when dealing with (t of mining residue discharged daily)/(population in contact with the residues), an increase in the population would decrease the indicator value without reflecting any true change in the amount of mining pollution. It was important to identify when this was the case in order to avoid misinterpretation of these indicators.

4 SUSTAINABLE DEVELOPMENT

4.1 *Definition*

The concept of sustainable development has been discussed extensively over the past few years, yet many different definitions of the term exist. It is necessary to clearly state which definition is used when discussing sustainable development with respect to a region.

For the Potosi project, the Brundtland definition was used. It states that sustainable development is the capacity to satisfy present needs without compromising the ability of future generations to satisfy their needs. Three principles are elaborated with respect to this definition:

- intra-generation equity or the sharing of benefits between different levels of society
- ecological integrity or the conservation and protection of the environment
- inter-generation equity or the sharing of benefits between the present and the future

The degree to which the mining industry respects each of these principles was considered in order to measure its contribution to the sustainable development of the city.

4.2 *Contribution of the mining industry to the sustainable development of Potosi*

The first principle was examined by estimating the distribution of material benefits generated from the mining industry. The different groups directly involved in the industry include the miners in the cooperatives, the workers in the concentration plants and the owners of these concentration plants. Among these groups it was found that the revenues generated are unevenly distributed, the owners of the concentration plants benefited far more than the other groups. Other benefits for parties that weren't directly involved such as non-mining jobs generated in the city by the industry can be considered as well. In this case however, the substantial inequity between the different concerned parties affirms the failure of the mining industry in Potosi to respect the intra-generation principle of sustainable development.

The indicators on the environment clearly demonstrated that ecological integrity is not maintained by the mining industry in Potosi. The externalities imposed on the local population, particularly by the concentration plants that emit residues directly into the river, are considerable. The second principle of sustainable development is not respected.

One way to discourage the imposition of externalities by an industry is through taxation. The taxation on the mining industry in Potosi is an ad valorem one where the tax is imposed on the selling value of the concentrated mineral (only the concentration plants pay tax). This tax is the same regardless of fluctuations in production costs or environmental investments that might be undertaken, it provides no incentive to implement ecological measures. An alternative tax such as a tax on profit or profitability (where only the return on the investment is taxed) would be a way to encourage environmental investments and contribute more to the ecological integrity of the industry.

Inter-generation equity is perhaps the key principle to sustainable development. Nappi & Poulin (1998) state that for sustainable development to exist, the stock of capital (manufactured and natural) must be maintained from one generation to the next. The substitutability of natural capital for manufactured capital is questionable, especially when dealing with a non-renewable resource, but can be considered as a means to maintain overall stock. The natural capital that a mineral resource represents can be maintained in two ways,

- by finding new reserves of equal worth
- by reinvesting the capital generated by the exploitation into another capital source.

In a limited area like Potosi there exists little chance of discovering new mineral reserves, so the first option is not a possibility.

The capital generated by the depletion of the mineral reserve could be partially reinvested into an exhaustion fund. This fund would be used to encourage alternative industries that could serve as a substitute source of capital once the non-renewable resource has been exhausted. This would be a way to maintain total stock (Daly 1990, Nappi & Poulin 1998) and encourage the mining industry to contribute more to intergeneration equity. At the present however, no reinvestment of the capital generated through mining is made into other industries such as tourism for example. It can therefore be concluded that the mining industry also fails to respect the third principle of Brundtland.

5 CONCLUSION

The aim of this paper was to describe the approach used to develop indicators to measure mining impact in the city of Potosi. Five basic steps were used to create the appropriate indicator set.

The first step consists in defining clearly the objective as well as the intended users of the indicators. This helps to insure that the resulting indicator set is both relevant and useful.

The second step involves planning out which elements to include in the database. It is advantageous at this point to incorporate the input of many interested parties, this provides different perspectives and broadens the scope of the database.

The actual compiling of the information for the database should include on-site work. This helps to develop a certain value judgement with respect to importance of different elements in the studied phenomenon. It also provides a certain perspective that might account for appropriate changes in the structure of the database. In addition, it is important to use various data sources to be able to cross-check and verify information and thus insure data integrity.

Once the database is created and all the available, relevant information is included, the indicators are selected from it. One of the described approaches to indicator selection can be used. When direct and simple composite indicators are created from key information, it is important to be aware of the subjectivity introduced by this method. Since a certain value judgement is used to identify key data, on-site experience becomes an additional advantage when using this approach.

Lastly, an evaluation of the selected indicators with respect to a specific criteria set is made. Criteria such as relevance to the studied theme, facility to calculate, clearness and sensitivity to change can be used. Although criteria should be kept in mind throughout the indicator selection process, the evaluation is important because it allows the users to be aware of the strengths and weaknesses of the indicators.

In the case of the project described in this paper, the indicators were developed to measure mining impact in the city of Potosi, as opposed to measuring sustainable development. The question of sustainable development was considered separately and with respect to the mining industry's contribution to it. This represents a difference with most projects that develop indicators specifically to measure sustainable development.

Not much work has been done on local indicators to date. This paper described the steps that were used to develop local indicators to measure mining impacts in a developing country, it does not provide a predetermined set of indicators to be used globally, the set is specific to its region. As concern for environmental and social impacts of mining activity increase and the acceptance of indicators as an effective tool to represent such impacts grows, other similar projects might arise. The indicator development process described here could serve as framework. It would also be very interesting to compare this approach with others that might be developed.

ACKNOWLEDGEMENTS

The Potosi project was an integral part of a larger project funded by CIDA to promote environmentally friendly practices in the Bolivian mining industry. The project is undertaken jointly by the Québec Ministry of natural resources (MRN), the Bolivian Vice Ministry of mining and metallurgy (VMMM) and Ministry of sustainable development as well as the national mining company, Comibol.

The support of the MRN, particularly that of George Cockburn, Marc Arpin and Marc Bélanger was considerable. On the Bolivian side, the vice-minister René Renjel, Carlos Feraudy and all the VMMM provided essential assistance in the realisation of the project. Last but not least, the help of Fernando Llanos at the Universidad Tomas Frias in Potosi was invaluable.

REFERENCES

Bromely, Daniel W. 1991. *Environment and Economy. Property Rights and Public Policy* Basil Blackwell Inc.

Chevalier, S., Choiniere, R., Bernier, L. & al. 1992. *User Guide to 40 Community Health Indicator.* Community Health Division, Health and Welfare Canada. Ottawa

CIDA – Canadian International Development Agency., 1997. *Guide des indicateurs tenant compte des écarts entre les hommes et les femmes*. Ottawa

Fleury, A.M. 2000. Thesis for a masters *Indicateurs pour mesurer l'impact minier sur l'environnement, la population et l'économie dans la ville de Potosi en Bolivie.* Laval University

Gallopin, Gilberto Carlos. 1997 Indicators and Their Use : Information for Decision-making. *Sustainability Indicators : Report of the Project on Indicators of Sustainable Development.* Pages 11-27 John Wiley & sons Ltd.

Haberl, Helmut et Schandl, Heinz 1999. Indicators of sustainable land use concepts for the analysis of society-nature interrelations and implications for sustainable development. *Environmental Management and Health* Volume 10 Issue 3

Holland. Leigh, Montfort University, Leicester, UK 1997. The Role of Expert Working Parties in the Successful Design and Implementation of Sustainability Indicators. *European Environment*, vol. 7, pages 39-45, John Wiley & Sons, Ltd.

McQueen, D., & Noak, H. 1988. Health Promotion Indicators : Current status, issues and problems. *Health Promotion 3*, pages 117-125

Mitchell, Gordon, The Environment Centre, University of Leeds, UK 1996. Problems and Fundamentals of Sustainable Development Indicators. *Sustainable Development* Vol.4, pages 1-11 John Wiley & Sons Ltd.

Nappi, C. & Poulin, R. 1998 Sustainable Development and Mine Management. *Non Renewable Resources Journal*, Vol. 7, Issue 4 pages 265-271

OECD Organization for Economic Cooperation and Development 1993. *Core Set of Indicators for Environmental Performance Reviews. A Synthesis Report by the Group on the State of the Environment*. Paris.

UDAPSO - Unidad de Analisis de Politicas Sociales & INE - Instituto Nacional de Estadistica. 1995 *Mapa de Pobreza, Una Guia Para La Accion Social* Bolivian Ministry of Human Development. La Paz

World Commission on Environment and Development (WCED) 1987. Our Common Future. Oxford University Press. Oxford.

Environmental Issues and Management of Waste in Energy and Mineral Production, Singhal & Mehrotra (eds)
© 2000 Balkema, Rotterdam, ISBN 90 5809 085 X

Regulations enforcing better environment protection in Polish coal mines

L.Gawlik
Mineral and Energy Economy Research Institute, Krakow & State Agency for Restructuring Hard Coal Industry, Katowice, Poland

U.Lorenz
Mineral and Energy Economy Research Institute, Krakow, Poland

ABSTRACT: Polish coal mining industry is unprofitable. In 1998 year the Council of Ministers launched the "Reform of coal mining industry in Poland in the years 1998 – 2002". This "Reform" assumes a deep restructuring of the industry with the aim to make it economically efficient and sounder to environment. After one year of the "Reform" realization it occurred that mainly because of the dramatic change in the coal market in Poland, some additional legal regulation are needed to achieve the aims of the reform. In the "Correction" of the reform, in spite of deeper technical restructuring of the industry, including additional mines closing, a special stress is put on regulations concerning investments for environmental protection. The paper describes general regulations concerning environment protection in mineral industry and the specific solutions proposed for coal mines in restructuring process in a wide view of the present economic situation of coal mines.

1 INTRODUCTION

Till the 1980 year the complex legal regulation concerning environment protection did not exist in Poland. Under the international pressure and in connection with worsening state of environment that started to threaten peoples living conditions, the protection and shaping of the environment act was passed. It was the first trial to regulate the fundamental problems of environmental protection and establish the unified state policy in this complicated interdisciplinary field.

The act regulates all the issues connected with each particular element of environment, including air protection, protection against noise and vibrations, protection against water pollution, protection of verdure in towns and villages, protection against electromagnetic radiation. The law introduces fees for use of environment, which are paid by all enterprises when they use water, pollute air, cut down tress, use of ploughland for other purposes or store or dump away wastes. The fees are one of the main sources of income of the National Environmental Protection Found. The Fund was established with the aim to support investments for environment protection and water management.

After collapse of communist regime, the main goal of the Ministry of Environment Protection, Natural Resources and Forestry became working up a long-term ecological policy of the country. Finally this document was passed in the Parliament in 1991 year as a demonstration both the society's and the authorities' will to improve the state of environment in Poland.

As a result of "Ecological Policy of Poland" realization and adjusting Polish law towards the rules that function in the European Union, many new bills were passed or amendments to a bill were introduced in last few years. Now there exist a set of bills that regulate conditions of environment and resources use. These are:

- Water law (1974),
- Atomic law (1986),
- Geological and mining law (1994),
- Land development act (1994),
- Building law (1994),
- Ploughland and forestry protection act (1995),
- Cleanness and tidiness supporting act (1996),
- Amended protection and shaping of the environment act (1997),
- Wastes law (1997).

All these acts regulate the economic activity conditions of enterprises in the aspect of their impact on environment.

To force the proper environment and resources management and influence the decision taken by independent enterprises towards such activities that decrease negative impact on environment, the system of fees paid for use of environment and the system of fines for disobeying limits of emission is established.

2 SYSTEM OF ENVIRONMENTAL FEES AND FINES IN COAL MINING INDUSTRY

Underground coal mining is unavoidable connected with the intervention to natural geological and hydrological structures. Mining activity additionally impacts the environment by water pollution, storage of mining wastes and tailings and by mining damages that arise on the surface of ground. The damages of minor negative impact on environment are caused by pollution of air by harmful substances, noise and cutting of trees and bushes.

The system of fees consists of:

1 fees paid for use of natural resources i.e. exploitation fee and fees for a special use of water and water installations (water consumption, dumping industrial wastes),

2 fees for economic use of environment i.e.

- fees for pollutants emission to the air, dependent on the kind and quantity of pollutants,
- fees for wastes storage, dependent on harmfulness and quantity of stored wastes,
- fees for cutting of trees,
- fees for using ploughland and forestry for other purposes.

Environmental fees are a significant part of coal production. In 1998 year 4,3% of total cost of production were cost of environmental protection. Over 95% of all ecological fees paid by coal mining industry are fees for dumping wastes to the rivers (especially dumping saline waters) and for wastes and tailings storage.

Fees are paid for using environment according to specified rules. The system of fines is foreseen for those who disobey the rules and exceed the limits of emission. Fines are paid in addition to fees. According to the law, ecological fines are paid for:

- dumping industrial wastes which do not comply with specified quality (fines for dumping to rivers underground water containing too high load of salt are the most significant for coal mining industry),
- exceeding quality or quantity of substances allowed to be add to the air,
- exceeding allowed level of noise,
- storing or dumping wastes in other places than specified or in a way different than allowed,
- devastation of verdure in consequence of improper technical or chemical intervention,
- cutting trees without allowance.

Expensiveness of fines is a strong incentive to minimize harmful impact on environment both by more careful management and by financing new investments for environment protection (Kudełko 1995).

Additionally it has to be said that all fines, including ecological fines are accounted to unjustified costs in Polish accountant system and therefore are taxable. For this reason fines effect the economic results of enterprises very strongly.

3 ENVIRONMENT PROTECTION IN RESTRUCTIRING OF COAL MINING INDUSTRY

In June 1998 year the Polish Council of Ministers approved the governmental program entitled "Reform of coal mining industry in Poland in the years 1998 – 2002". This middle-term economic program defines the state policy in coal mining industry. It creates the legal instruments for coal mining restructuring and indicates the necessity of financial support from state budget to accomplish the program. The program completion is needed (Karbownik & Pawełczyk 1998):

- to allow the industry to stop generate enormous financial losses,
- to help mining industry to stop maintain itself on cost of cooperating enterprises and to be a threaten of bankruptcy for other firms in the region,
- to stop the industry negatively influence the public finance,
- stop the difficult economic and financial situation that is a seed of social discontent.

Coal mining industry till the economic transition of the country was unprofitable. During the eight years of restructuring it has not become viable. The main reasons of the failure of all restructuring processes are the excessive capacities of the industry, higher than possibilities of coal sale both inland and for export, and high level of employment in coal mines. In consequence costs of coal production are higher than achievable coal prices. Such a situation incurs high debt of the industry. At the end of 1997 year loss after tax was PLN –3383,5 mln (~US$ 970 mln) and the liabilities reached the level of PLN 13.346,0 mln (~US$ 3800 mln). The liabilities for environmental fees and fines made over 32% of the total indebtedness.

The main goals of the "Reform" are:

- to adjust the mining enterprises to effective operation in condition of market economy,
- to retain competitiveness of Polish coal in the domestic market,
- to satisfy domestic demand and economically justified export for coal till at least 2010 year.

The additional goals are: meeting the environment needs and achieving competitiveness in the free market conditions for all energy sources according to the European Unions directives. Main actions foreseen are closing excessive mining capacities and decreasing employment in the industry. In the process of the reform a financial support of the state budget is declared. The sum of PLN 7180.5 mln (~US$ 2050 mln) subsides in the years 1998 –

2002 will be used mainly for supporting the process of miners' dismissal (63,3%) and for technical closing of mines (20,7%). 2,4% of the total sum of subsidies i.e. PLN 168 mln (~US$ 48 mln) have been assigned for removal of occurring surface damages as a result of past exploitation (Chaber et al. 1998).

The "Reform" assumes that mining companies will undertake realization of new projects aimed to decreasing on harmful impact of the industry on environment. These are:

- construction of underground waters desalination plants and installation of deep re-injection,
- increase of waste rock and tailings location in the underground openings,
- intensification of land reclamation and management,
- change of exploitation system to the one that minimizes occurring surface damages.

It is assumed that completion of ecological investments in mining industry will lead (Chaber & Krogulski 1997) to:

- additional utilization and management of 1240 tons of salt from underground waters per day,
- management of additional 2620 thousand tons of mining wastes and tailings per year,
- reclamation and management of 1,5 mln m^2 of land left after mining operations per year,
- gradual stopping accumulation of overdue works connected with mining damages removal.

To stop the increase of financial costs of the industry the reform assumes the financial restructuring of mines which consists in partial cancellation of liabilities and also partial postponement of liabilities payment till after reaching viability by mines.

It is evaluated that according to the special act, passed in November 1998, about PLN 9000 mln (~US$ 2200) of liabilities will be cancelled by the end of 2000 year. Out of this sum over 44% are the liabilities of the National Environmental Protection Fund and its provincial subsidiaries. These are ecological fees and fines. Out of total sum of over PLN 3200 mln (~US$ 800 mln) postponed payments of liabilities, 45% are prolonged payments of environmental fees.

Financial restructuring foreseen in the reform of the mining industry substantially burdens the National Environmental Protection Fund, causing shortness of capital. As a result its possibilities of supporting the investments for environment protection decreased. The Fund decreased also its financial support of ecological projects in mining industry. In 1998 year the National Environmental Protection Fund supported the mining industry with only 69% of planned sum and in 7 months of 1999 – only 0,2%.

4 ECOLOGY IN THE CORRECTION TO THE COAL MINING INDUSTRY REFORM

In 1999 year the market conditions in coal sale worsened. Demand for coal in Poland decreased, causing lower sale and lower income of coal mining industry comparing to the plans assumed in the coal mining industry reform. The terms of achieving viability of the industry occurred to be impended. After one year of the "Reform" the corrections were necessary.

The "Correction" of the reform was agreed by the Council of Ministers in December 1999 year. It assumes deeper technical restructuring of the industry, including additional mines closure and additional dismissal of miners. With the same financial support from the state budget as in the "Reform" the "Correction" assume higher costs of restructuring carried by mines. Those costs are connected mainly with additional reduction of employment in mines. Therefore the time of reaching profitability by the industry will be delayed. The "Correction" assumes it in 2002 year.

Within 7 months of 1999 year mining industry, because of difficulties in gaining funds, realized only 11% of planned ecological investments. It has to be said that "Reform" assumed that the investments would be financed from the own financial sources of mines. The priority of mine was to gather funds for miners salaries and the investment needs were neglected.

The "Correction" introduces new incentives for realization of those ecological projects that lead to obeying rules of environment protection in mines. It is decided that the "Sector analysis and assessment of the environment in coal mining industry" will be worked out. For each mine a detailed timetable of decreasing environmental pollution quotas will be established in such way that the mine should fulfill all the requirements for environment protection not later than in 2005 year.

Financial restructuring of mines consisting in cancellation and postponements of liabilities payments will be possible on condition of realization the scheduled investments with a fixed time limits. The authorities may decide to close a mine in case it is not able to fulfill the requirements of environment protection. This is a new regulation, as by now the closure of mine was decided only in case of persistent lack of profitability.

The "Correction" of the governmental program of coal mine industry restructuring ascertains that the improvement of the environment is an important goal of the reform.

5 CONCLUSIONS

Environment protection is an important element of the current state policy in Poland. Many years of neglecting this problem has to be compensated. The set of regulations concerning environment protection that have been passed within the last few years gives the good base for the modern environment and resources management.

Mining operation is connected with dumping saline waters, storage of mining wastes and occurring mining damages of the surface of ground. Decreasing of that impact on environment is one of the goals of governmental coal mining reform. Difficult financial situation of mines hinders realization of projects aimed to improve impact on environment.

The "Correction" of the reform gives strong incentives to promote those mines where environmental protection limits are obeyed. A full realization of the "Correction" gives a chance of conversion the industry into a modern and environmentally sound branch of national economy.

REFERENCES

Chaber M. & Krogulski K. 1997. Ecological fees as the fundamental legal instruments of environment protection. Widomości Górnicze No.5.

Chaber M. et al. 1998. Environmental aspects of hard coal mines closure in Poland. Proceedings of the Fifth International Symposium on Environmental Issues and Waste Management in Energy and Mineral Production (SWEPM'98): 441-447. Rotterdam: Balkema.

Karbownik A. & Pawełczyk E. 1998. Coal mining industry reform in Poland in the years 1998 – 2002. Synthesis of the governmental program. Biuletyn PARG No 7-8.

Kudełko M. 1995. Ecological costs in the cost structure of coal mining industry. Studies and monographs. No 36. Krakow: PAN.

Environmental Issues and Management of Waste in Energy and Mineral Production, Singhal & Mehrotra (eds)
 ISBN 90 5809 085 X

Study on the sustainable development for Shenfu-Dongsheng coal mining area

Zhenqi Hu
China University of Mining and Technology, Beijing, People's Republic of China

Jiang Jianwu & Lu Zhiren
Shenhua Group Company Limited, People's Republic of China

Gu Hehe
China University of Mining and Technology, Xuzhou, People's Republic of China

ABSTRACT Shenhua Group Co. Ltd is an important coal company in western of China. One of its coal mining areas is Shenfu-Dongsheng, which locates the continuous areas of Shanxi and Inner Mongolia. This is a region with sensitive ecological environment. Development of coal resources in this region may aggravate the environment. This paper introduced the ecological and geological conditions of Shenfu-Dongsheng coal mining area and the necessary of sustainable development for this area. The targets of sustainable development for Shenfu-Dongsheng coal mining area were presented based on the principle of sustainable development. Some actions for achieving the targets were explored. Ecological restoration of mining areas was the key action. The detail treatments of ecological restoration were also introduced in this paper.

1 INTRODUCTION

Some documents and books such as "Agenda 21", "Our Common Future", etc. have revealed the fact — sustainable development has become the common developing strategy for every countries in the world, which is the only choice for human beings' living and development. The China government has signed in the document "Agenda 21" at the 1992 Earth Summit and responded to this in 1994, by publishing the White Paper entitled "Agenda 21 in China" (The State Councile of China, 1994), which shows that the developing strategy in China is changing onto the sustainable development from traditional development.

Coal is the most attractive energy resource in China, accounting for approximately 75% of energy consumption. Coal output was 1.298 billion tons in 1995, with 1.4 billion tons for 2000. Coal industry in China plays an important role in China's economic and social development, thus, the China coal industry should follow the sustainable development strategy (Hu, 1997).

Shenfu-Dongsheng mining area is of rich coal resources with very good mining conditions, but it locates the region with fragile natural and ecological environment, poor transportation and local economics. The conflicts between coal resource development and ecological environment protection is very serious. Thus the study of sustainable development for Shenfu-Dongsheng mining area is very significant, which could play an important role in protection of the local environment while the coal resource is excavating.

2 NECESSITY OF SUSTAINABLE DEVELOPMENT OF SHENFU-DONGSHENG COAL MINING AREA (Lei, 1998)

Shenfu-Dongsheng coal field is located between the Maowusu Desert in the middle reach of the Yellow River and the Loess Plateau, which locates the continuous areas of Shanxi and Inner Mongolia. The average elevation is +1200m. About 85% of the area is desert or potential desert, with wind erosion standing at 2500t/km^2/a, water erosion at 5000t/km^2/a. Area of wind erosion is about 92% of total mining area while water erosion area accounts for 77%. Water and wind erosion occurs simultaneously in time and space. The climate is continental and semiarid. The average annual rainfall is 396.8 mm, mainly happens during June to September, accounting for 76% of total annual rainfall. 60~70% of the rainfall produces surface runoff. The annual average evaporation is 1319 mm, which is 3~4 times of annual rainfall. The average wind speed per year is 2.5~3.6m/s. The natural increase speed of desert land is 0.5% per year. The fragile natural and ecological environment has been responsible for persisting poverty plaguing the region, directly hindering the development of coal resources and local economic growth.

Authorized by the State Council, Shenhua Group in 1986 began to develop one of the top eight coal fields in the world, the Shenfu-Dongsheng Coal Field. Covering an area of 31200 km^2, the coal field has a known deposit of 223.6 billion tons. Furthermore, the coal seams are stable and shallow, with low content of ash, sulfur and phosphor but high calorie. More important, they are clean, ideal for power generation and gas production. Because of high quality, the coal field is listed as a major energy supplier in the early 21st century in the country's Ninth Five-Year Plan for National Economic and Social Development and Long-Term Targets Through the Year 2010.

Current developing mining area in Shenfu-Dongsheng is only a small part of the coal field, ranging of 3432 km^2. Three developing periods were divided. The designed mine scales are 10 Mt/a in the first period, 36 Mt/a in the second period and 66 Mt/a in the third period. The desert land is 1956 km^2, accounting for the 57% of the total mining area. Coal resource development could accelerate the desertification and soil and water loss in this area. It is studied that the first and second development periods of this mining area could produce 632 million tons of spoil and wastes. The increased erosion amount is 168~198 million tons and 129.64 km^2 of land will become desert if without any treatments. The increased desert land is 1.54 times of natural desertification. Considering the natural desertification, the increased desert land will reach 213.9 km^2 by 2000, which means that 85.5% of total developed land will become desert. The sand amount running off into rivers by wind erosion was 34.8 Mt/a before coal development. The increased sand amount is 4.79 Mt/a due to the first and second development periods of this mining area. Therefore, the land desertification is very serious in the mining area and the development of coal resources in this region may aggravate the environment. Necessary treatments are needed for sustainable development in this region.

3 TARGETS OF SUSTAINABLE DEVELOPMENT OF SHENFU-DONGSHENG COAL MINING AREA

3.1 Economic development targets

- By 2000: output from the mining area is 25 Mt/a and 6.4 Mt/a for overseas export
  By 2005: output from the mining area is 40 Mt/a and 7.5 Mt/a for overseas export
  By 2010: output from the mining area is 60 Mt/a and 10.0 Mt/a for overseas export
- 1996~2000: 12.83 billion yuan incomes and 1.201 billion yuan profit
  2001~2005: 57.49 billion yuan incomès and 1.836 billion yuan profit
  2006~2010: 112.64 billion yuan incomes and 15.08 billion yuan profit

3.2 Environmental protection

- By 2000: 132480 ha of land should be renovated, the renovating degree reaches 37.45%. Seven programs will be implemented as emphasis with an area of 1670 km^2, accounting for 43% of total mine area. The renovated degree of the area reaches the 51%.
- By 2010: 231250 ha of land should be renovated, the renovating degree reaches 65.37%. Vegetation coverage of mine area will reach 55.69%. The rate of waste water treatment reaches 95% and its reuse rate is more than 50%. Reclamation rate is more than 80%.

3.3 Social development target

All the staff of coal mine will be educated from or higher than middle school and the technical staff will be 30% of total number of staff. Living area will reach 16m^2 per person. Average salary of staff is more than \$1600 (U.S.) per person annually.

4 ACTION OF SUSTAINABLE DEVELOPMENT AND ITS EFFECTIVENESS

(1) Establish the policy of paying equal attention to both resource development and environmental protection.

Based on the conditions of Shenfu-Dongsheng mining area and the requirements of national sustainable development, Shenhua Group has been advocating the policy of paying equal attention to both resource development and environmental protection ever since the beginning. Proceeding from the long-term interest of the company and the need to preserve the potential for economic growth for the local people, the company has strictly implemented all the policies and regulations issued by the central and local government on environmental protection.

(2) Researching, planning and funding

The company has invested huge amounts of human, material and financial resources in the sustainable development program, establishing a series of research projects on possible damages to the ecological environment by mining operation. A three-step approach in cleaning up the environment at Shenfu-Dongsheng Coal Field was adopted. The area under planning comes to 3214 km^2, involving an estimated investment of 1.48 billion yuan. To ensure the consistency of work and effective implementation of planning, the company establishes, with approval from the Ministry of Finance, a special fund for environmental protection by drawing 0.45 yuan annually out of sales income per ton of coal. In 1994, the Environmental Clean-up Program for Shenfu-Dongsheng Coal Area was listed as one of the priority projects in "China's Agenda 21", a commitment made by the government to follow sustainable development.

(3) Fortifying sand and building shelter belts

Sand storm is very serious problem in this area. In winter and spring, the sand storm flows into the mining area mainly from the two wind-doorway (Wulanmulun and Huwushu) and three sandy sources (Hala, Qilianta and Majiata). Thus adopted principle of treatment is "give the prominence to the stress and set up the treatments based on the natural condition". The seven areas, ranging 190 km^2, were chosen as the stress of treatment, which were the two wind-doorway, three sandy lands (Hala, Qilianta and Majiata), and the areas along the annular road of mines and the railroad of mines.

At the two wind-doorway, three shelter belts of native trees were built for cutting down the wind power and stopping the sand storm. In the three sandy lands, there were many moving sand hills. The typical problem was wind erosion. A sand-stabilizing project using rattan nets was adapted and trees (such as poplar, willow), shrubs and grass were planted in the net. By the end of 1996, a total of 125 km^2 of area had been equipped with sheltering facilities against sand storms, with 35 million meters of sand barriers built, 16 million pits of arbor, 3 million trees and 726.7 ha of grass planted. The treated area is 125% of the first-step planned area and is increasing as the speed of 10 km^2 per year.

Along the 124 km mine road, two side of road ranging a width of 100~500m were built with the rattan net and tree shelter belts (about 50000 trees), which stops sand storm and protects the road. To increase the survival rate of planted vegetation, several water supply stations were built along the road, which made the survival rate of shrub and grass was 70% and planted trees were 90%.

Baoshen railway with 172 km passes the edges of two deserts (Kubuqi and Maowushu). The length of railway in the desert region is 81 km, accounting for 47% of total length of railway. A 120m of width along the railway were built with biological protection measures. In the years, 9 million meters of netted sand barriers, 4.6 million shrubs and 55200 trees along the railway were built. About 10 km of green barrier along both side of the railway was formed, which stops the sand storm and protects the railway.

(4) Protection of fountainhead and watershed management

Wulanmulun river is a branch of Yellow River. It runs across the mining area with a length of 75 km. It is a seasonal river with high content of sand and has a fast rise of water level during the rainstorm season of July and August, which may directly effect coal mines. Thus 24 km of the river course have been dredged up and a dam for a reservoir with a capacity for 236000 m^3 was built, thereby effectively forestalling soil erosion and damages that could be done to the area by floods.

There are five ditches around the Wulanmulun river. Three of the ditches are the fountainheads of coal mines and local people. Because of the high content of sand, more than one million yuan is needed for sand-clearing annually. A biological treatment for sand-solidifying is taken as a main measure and some necessary civil engineering as supplements. With the completion of the project to protect fountainheads, sand contents in water have fallen from 14-16 kg/m^3 before treatment to the present 0.15 kg/m^3.

(5) Land reclamation

According the requirement of reclamation, open pit mines usually adopt the method of mining by layers and damaged land is reclaimed while the coal is excavating. The excavated overburden should be backfilled from bottom to top and from large (pieces) to small by layers. Then the top is leveled and covered with a 1~1.5 m of soil. Finally, trees and grass are planted. The reclamation rate reaches 50%.

(6) Waste treatment and pollution control

Four waste water treatment plants with a daily capacity of 23000 tons were built and a central heating facility is installed. Industrial chimneys are equipped with water dust scrubbers or static dust scrubbers. Industrial wastes and rubbish are collected and treated as a special dump. Experiment for reclaiming open pits and waste dumps have also been successful.

After 10 years of strenuous efforts, Shenhua Group has built up a massive modern energy supply base in the depth of what used to be a desert. Amazingly, the quality of the environment around the area has not deteriorated; instead, an oasis has appeared miraculously. The number of days shrouded in sand storms during winter and spring has decreased by three fourths from a decade ago, and over 50% of areas targeted for environmental clean-up is covered with vegetation. Soil erosion along the Ulanmulun river (the main tributary within the coal mine), which used to be very serious, has been brought under effective control. Statistics show that over the years, the company has invested a total of 249 million yuan in environmental protection. A multi-shelter protection system has been built. The air and water quality has been improved and protected. About 5.18 million tons of sand running off into the Yellow rive is decreased annually and vegetation coverage in increased from 14% before treatment to 39%. In special treatment area, the coverage even reaches more than 65%. The damage of sand storm is obviously abated. Numbers of wildlife is increased. The accommodative capacity of the environment is effectively expanded. And the capacity of sustainable development is built up. Based on the developed evaluation model with 13 affecting factors, the degree of sustainable development of Shenfu-Dongsheng mining area is 0.72, which reaches the step of Prime Sustainable Development.

5 CLOSING REMARKS

The sustainable development program in Shenfu-Dongsheng mining area has not only improved the quality of life but has also brought in considerable economic benefits. Aside from generating income from the newly planted forest and grasslands, a cleaner environment also saves over 2 million yuan annually from clearing away sand around fountainheads and along roads. In addition, the quality of coal has been improved remarkably and economic benefits are increased due to the reduction of more than 92% of sand contents in open pits and coal dumps. The actions of sustainable development have also produced great social benefits. Mitigated soil erosion has reduced the volume of silt flowing into the Yellow River. Control of sandstorm has improved the natural conditions for agricultural and stockbreeding production. As a result, local people actively participate in environmental protection. Finally, the success of sustainable development at Shenfu-Dongsheng coal mine serves as an example for the entire Yellow River Valley region to clean up the environment and for the country's coal industry to seek sustainable development. Based on the plan of sustainable development, the vegetation coverage will expand to 55.7% and reduce sand flow to 58.9% of annual bed-load caliber in the mining area by the year 2006. The sustainable development degree will reach 0.86 by 2010, a step of sustainable development. By that time, the coal field will emerge as an environmentally-friendly energy enterprise with a pleasant and clean environment and with northwest Chinese characteristics.

ACKNOWLEDGEMENT

This paper is supported by Ministry of Coal Industry, National Natural Science Foundation of China, and Trans-century Outstanding Young Scientist Program by Ministry of Education.

REFERENCES

Hu, Zhenqi, Hehe Gu, etc. 1997. Major Environmental Problems and Control Techniques in China's Coal-mining Areas, published in the Proceedings of the Fourth International Conference on Technologies and Combustion for a Clean Environment, Lisbon, Portugal, 7-10 July 1997.

Lei, Jinglian (eds). 1998. Desert, Coal field and Green Land, Shenhua Group Co. Ltd. (in Chinese)

The State Council of China. 1994. Agenda 21 in China, China Environmental Science Publishing House (in Chinese)

Environmental Issues and Management of Waste in Energy and Mineral Production, Singhal & Mehrotra (eds)
 ISBN 90 5809 085 X

Present situation and development of land reclamation in China's metallurgical mines

Zhou Jian
National Bureau of Metallurgical Industry, People's Republic of China

Lei Pingxi
Maanshan Institute of Mining Research, People's Republic of China

ABSTRACT: This paper analyzes the destruction of land resources induced by the development of metallurgical mine, summarizes the advances in the technology, theory, application and concept of land reclamation in metallurgical mines, and forecasts the development trend of China's metallurgical mines.

Keywords Metallurgical mine, Land reclamation, Reclamation technique, Development trend

1. DEVELOPMENT OF METALLURGICAL MINES AND DESTRUCTION OF LAND RESOURCES

China is a large nation producing iron-steel, and is also a large nation producing and consuming metallurgical ore products. Except iron mines, China's metallurgical mines include the mines of eight auxiliary material (such as manganese, chromium, magnesite, fluorite, clay, limestone, quartz, dolomite). At present, the total mining-stripping amount have reached more than 7 hundred million tons per year and the largest productivity is near to 10 hundred million/year, occupying the first place in the world. The metallurgical mines are one of the largest land occupiers in the mining industrial system. To form an ore productivity of 10,000 tons, it is necessary to occupy land of 35 thousand m^2, of which the mining area approximately occupies 25%, dumping site 20%, tailing pond 10%, industrial district 10%, other and abandoned land 35%. A large scale development of metallurgical ore products will inevitably cause the occupation and destruction of large amount of land and seriously affect the environment in the mining area, as shown in Table 1.

The land destruction caused by the mining activity of metallurgical mines are mainly shown on occupation of a large amount of land for dumping site, waste rock pond and tailings pond and open-pit mining area and surface subsidence caused by underground mining which causes the destruction of surface soil, the failure of original draining-irrigating condition and the earth's surface breaking, etc. In addition, due to the pollution by dust, waste air and harmful elements and the destruction of underground water system caused by mining operation, the soil fertility and agricultural productivity are reduced and the calamity of landslide, collapse and mud-rock flow occur, occupying a large amount of land.

2. TECHNICAL STUDY OF LAND RECLAMATION IN METALLURGICAL MINES

Entering 90s, 20th century, the durable economic development in China restricted by the two important problems of the environment and land resources has been followed with wide interest, at the same time, the durable development of mining industry has also caused deep thinking in the mining industrial circles. A key factor of durable development in mining area is the durable utilization of land and its environment. The mining specialists have also recognized that the land reclamation is an important phase in the later stage of mining engineering, which is also the inevitable requirement of mining industry development.

Over the years, particularly in recent 10 years, the land reclamation techniques of metallurgical mines in China have developed from the simple engineering treatment to the reclamation technique system combing various forms with various ways and various methods of filling reclamation, ecological engineering reclamation and biological reclamation in which the majority of study is on reclamation program technique and biological reclamation technique.

Table 1. Effect of metallurgical mine mining on the environment

Production area	Characteristics			
	Land	Atmosphere	Water resources	Hydrogeology
Open-pit mining area	Making a requisition of agricultural and forest land, destructing the vegetation and fertile land, reducing productivity, destructing the area ecological landscape in the various degrees, occupying land caused by calamity of landslide, collapse and mud-rock flow, etc.	Dust and waste air pollution	Waste water pollution, excessive harmful elements and waste acid and alkali drained into water system	Destruction of regional hydrogeology and formation of extensive funnel area
Tailing Pond		Waste air pollution	Drained waste water containing suspension substance and chemical agents from the mill.	Variation caused by tailings of hydrogeological state
Slag dumping site		Strong dust induced by wind erosion	Drained waster water containing high-density suspension substances	Land swamping in slag dumping pile and its adjacent area, destruction of hydrogeological state
Industrial district		Waste air and dust of boiler room, thermoelectricity plant, and loading facilities.	Drained industrial waste water containing lubricating oil, oil products, toxic substances and others.	

Reclamation program technique. The reclamation program technique which belongs to the special program in total program of land utilization is key of success or failure of land reclamation. The particularity of land reclamation program in mining area lies in the particularity of reclamation targets, and the difficulty of reclamation program lies in the difference in nature of reclamation targets. Therefore, before the reclamation program is initiated, the area investigation and evaluation for reclamation condition are very important. As a result, the mathematical quantification analysis methods of fuzzy comprehensive judgment and grey gathering analysis. etc. have been introduced in practice progressively.

Based on the mining program, the prediction of land destruction form, the law of landforms variety and landscape pattern after the mining are one of the key techniques in overall program. The environmental system engineering, the theory of slope stability and sinking theory of land caused by mining have become the support of reclamation technique and the strategic decision of reclamation and utilization direction and the structure optimization of reclamation land utilization and the measures of reclamation engineering are core contents of reclamation program. The advanced program methods of graduation analysis, linear program and grey decision etc. have been put forward in practice progressively.

Biological reclamation technique. The biological reclamation technology includes the reclamation technique of ecological engineering, the biological technique of soil improvement and the rehabilitation technique of speeding vegetation. At present, these methods are mainly in-site testing, its specific directions are:

- The selection of reclamation material, the optimization of vegetation breed and proportion of nutrition;
- The racial structure and community characteristic, distribution of root system, and growing law of vegetation;
- The moving law of heavy metal and salt composition;
- The characteristics of land backfilled by waste material, different planting condition of arbor, bush and grassed.
- The plane, vertical and time structure problems in biological chain engineering and ecological reconstruction.

Reclamation engineering technique. The reclamation techniques of filling by stripped rock, tailings, industrial refuse and living refuse and covering by soil are also a focal point of study, in which the filling form, filling thickness, compacting technology, covering thickness of soil, planting breed and improving soil will constitute the technical system of this direction.

The lands of back-fill reclamation have been

mainly used for building engineering, and partial for planting grass and agricultural crops.

Theoretical study of land reclamation. The basic theory study of land reclamation is embodied in macroscopic and microscopic aspects. Macroscopically, the theory study of reclamation decision is more systematic and it possesses more definite recognition for characteristics, contents, targets and criteria. In addition, it possesses larger guiding sense for the set-up of quantification model of reclamation decision factors, the decision language model, the optimization structure model of land use and the measuring index system of comprehensive effect for reclamation scheme selection. At present, the basic problems of concept, intension, boundary and its course system on land reclamation are under discussion. Microscopically, the study in respect of mechanism of cultivated land destruction caused by mining and soil profile reconstruction theory is in progress today, and the soil profile reconstruction theory of "stripping by layers, crisscross backfilling" and the quantification model for evaluation of reclaimed soil productivity are put forward. Besides, on the basis of guiding the reclamation practice by using system engineering theory and ecological engineering theory, the viewpoint of landscape ecology is introduced, and the six theories of landscape ecological reconstruction in historic site are put forward: the theory of restoration of the former state; the theory of the dissipation structure; the optimization theory of landscape pattern; the theory of diversity and difference in nature, the theory of the combination of outside conditions in the light of local conditions with the theory of natural coordination and durable development. Thus, the land reclamation in mining area is combined with the ecological reconstruction, expanding the intension of land reclamation in mining area.

In addition, with the cooperation of the research organization of land reclamation of China's metallurgical mines with other nations (such as Australia. etc), the study work, the training of specialized subject, the academic exchange of land reclamation on land reclamation engineering projects have been carried out, making a contribution to the international cooperation of the study and practice land reclamation of China's metallurgical mines with other countries.

3. PRACTICE OF LAND RECLAMATION IN METALLURGICAL MINES

At the beginning of 1960s,the land reclamation of China's metallurgical mines was put forward, but the land reclamation engineering was backward in technique, simple in method and slow in development. Along with the development of national economy and gradual higher requirements on environmental protection, the Chinese government and mine enterprises began to lay stress on the land reclamation in mining area. Particularly after publishing the regulations of land reclamation in 1988,the land reclamation work began to move on a lawful road. With the cooperation of the scientific research institutions with the mines, the study and practice in situ of land reclamation of mine have been developed, and the new phase of land reclamation of China's metallurgical mines has been created under the principle of the comprehensive administration and overall consideration of economy, society and ecological benefits. According to an investigation report of the geographical research institute of Chinese Academy of Sciences, the rate of land reclamation has reached 23%, with some mines reaching more than 85%. The reclaimed land is mainly used for forestation and forming greenbelt, partially for foresting, agriculture and building, thus obtaining a better environmental and economical benefit, and in the management and administration, formation and method and the way of the reclamation, much valuable experience has been accumulated, such as:

The tailings discharged from the old tailing pond of Shuichang Iron Mine of Mining Co., Capital Iron and Steel Co. reached 67000 thousand m^3 and after it was closed, the environment pollution of dust flying caused serious harmful effect on the agricultural production and people life in locality. In order to improve the environment, to prevent and control the water and soil erosion and to restrain dust flying, through the cooperation of the special personnel with scientific research institutions, and on the basis of overall analysis of weather condition and soil condition of tailings pond in locality, the sea buckthorn which had strong resistance to wind-sand, salt and alkali, barren and arid was selected .In Less than two years, a greenbelt of 800 thousand m^2 with a survival rate of more than 95% has been formed, covering the overall tailing pond. By comparison with the traditional vegetation by covering soil, the cost of covering soil has been reduced by more than 10 million Yuan (R.M.B), thus opening a new quick, economical and feasible way for preventing-controlling sand-soil environment.

The waste rocks piled up in the dumping site of more than 1300 thousand m^2 in Daye Iron Mine, Wuhan Iron and Steel Co. are hard marble and dioirite. It is characterized by barren soil structure, serious deficiency in organic substance, poor water holding capacity and great reclamation difficulty. First, through the cooperation with scientific research institution, planting trees and grasses in an area of 33 thousand m^2 was tested.

The testing results have shown that the planting of arbor, bush and grass in hard rock dumping site is feasible, providing successful experience in the land reclamation in hard rock dumping site of metallurgical mines.

At present, the forest of more than 600 thousand m^2 is formed, effectively utilizing the dumping site, beautifying the environment and promoting the good recycle of ecological environment.

A part of site through engineering treatment has been used as the maintenance field of trucks, container monitoring station and chicken raising field, etc.

In Ganjingzi Limestone Mine of Anshan Iron and Steel Co., at the beginning of 1974,the topsoil began to be covered on the surface of dumping site in a planned way, the training base of football, chicken raising field and storehouse base were constructed and the arbor-bush economy forest, fruit trees, grape and vegetables were planted, the total reclaimed area utilized in mining area has been near to 1000 thousand m^2 with a reclamation rate of more than 70%.

In Mining Co. of Panzihua Iron and Steed Co., due to a heat and arid weather (dry season up to 185 days), large evaporation capacity, the reclamation and vegetation were very difficult. Through taking various measures, such as all-round planning, scientific testing, leveling and covering soil, storing water and preserving soil moisture, sowing seeds, vegetation, etc. over the last 10 years, the reclaimed area of dumping site and mining area has reached 800 thousand m^2.

In Nanshan Iron Mine of Maanshan Iron and Steel Co., the reclaimed land area of dumping site is 509 thousand m^2 planting 55 thousand trees. From 1993 to 1996,the orchard of 58.3 thousand m^2 constructed in which 7000 grape,500 persimmon and peach trees were planted. The cultural activity center and the learning field organized by old peoples have been constructed in low-lying tailings pond, and a sulphuric acid plant with an annual capacity of 40 thousand ton has been constructed on waste land of 31 thousand m^2. The availability of the reclaimed waste land of this mine has reached 70%.

At present, in the large-and medium-sized metallurgical mines of Baotou Iron and Steel Co., Anshan Iron and Steel Co., Tang shan Iron and Steel Co., Benxi Iron and Steel Co., Hanxing Iron and Steel Co., and Hainan province, the various reclamation experience and the techniques of various reclamation directions have been obtained, and the rate of land reclamation and the rate of forestation, as well as the benefits in respect of economy, society and ecology have also been improved remarkably.

Although the land reclamation of metallurgical mines has developed rapidly, compared with the developed countries, there are still remarkable difference in and the scientific research of reclamation, reclamation technique, and reclamation scale and the laws, regulations and policies of reclamation are not very perfect. Particularly, the shortage in reclamation fund seriously affects the development of land reclamation in metallurgical mines.

4. DEVELOPMENT TREND OF LAND RECLAMATION IN METALLURGICAL MINES

4.1 Uninterrupted Perfection of Policy-Law System and Technological Regulation of Land Reclamation

After the new land management rules issued by Chinese government, the corresponding laws and regulations of land reclamation will be also revised, and relevant means, detailed rules and regulations will also be issued. As a result, the problems in the organization, fund, management, check and accept, and evaluation of land reclamation will be solved substantively. A perfect mechanism of land reclamation will be formed in China inevitably.

4.2 Further Development of the Traditional Land Reclamation Field

With the aggravation of environmental destruction and affecting degree caused by the mining operation and the strengthening of the environmental protection consciousness and understanding for durable development strategy, the land reclamation of mine has been seen as a comprehensive problem of rehabilitation and reproduction of landscape productivity of the land destructed by mining, or the destructed land should be rehabilitated to a predetermined earth's surface form and productivity. Through creating conditions, the field is made to come into a technological process of durative, new and different utilization. Therefore, the study target of land reclamation in mining area is no longer only to consider the directly destructed land caused by mining operation (such as digging, occupying and subsiding land),and the ecological reconstruction and rehabilitation process appear to be even more important. In this process, it is necessary to possess a professional contingent of land reclamation and high starting-level study, because it is involved with a comprehensive system engineering which includes various courses of biology and ecology, hydrogeology, soil improvement, culture of crop, irrigation and water conservation, environmental protection, and landscape aesthetics, etc.

As the study of land reclamation is deepened, the monophylelic or pluralistic coupled relationship of personnel and land, personnel and nature, resources and environment, etc. will attract the attention of theory research circle inevitably and its purpose is to promote the further development of land reclamation work in mining area.

4.3 Agricultural Reclamation— Main Direction of Land Reclamation in Mining Area

The new land management rules issued by the Chinese government clearly provides that the reclaimed land should be given priority to agriculture, which is determined due to China's national conditions, Because metallurgical mines are located in the remote mountainous district, after the mining area is developed and with the input of goods and materials, electricity and information, and flowing-into of a large number of nonagricultural population, the original pattern of mining area will have undergo a radical change, aggravating the contradiction between person and land. With a large number of staff and workers dismissed for less people and high efficiency, the contradiction between person and land will be more intensive. This situation has arisen in many old mining areas.

Due to little high-new technological content of agricultural reclamation and small reclamation benefit, the agricultural utilization of reclaimed land in mining area is restricted. Therefore, in future, the study on the agricultural reclamation with high-efficiency and high-productivity will be more important and imperative. The key techniques to improve the benefit of agricultural reclamation are:

- The biological and engineering techniques to improve the reclaimed soil;
- The administration and management techniques of reclaimed land, and introduction of agricultural industrialization and formation of scale effect;
- The techniques to improve crop breed by using the genetic-gene engineering and its application and dissemination.

4.4 Speeding Industrialization Process of Land Reclamation

To make the land reclamation in mining area to move along the path of good recycle and self-development, the industrialization of reclamation will be inevitable selected.

The patent techniques and products formed through the development and research have formed a high-new technical industry which are not only able to increase output value of reclamation and rate of land reclamation by a wide margin, but are also able to provide a large number of fund for land reclamation, thus making the land reclamation in mining area to enter the path of self-development.

In future, the development of new-type reclamation products will be on the basis of the biological fertilizer speeding soil mellowing and introduction of improved varieties, the microorganism being able to rapidly weather waste material and the covering material being able to increase survival rate of vegetation, etc. Today, the high-new technical research of land reclamation based on the biological technique is an important research object overseas, and many products suitable for land reclamation in mining area have been developed. On the basis of using the advanced international techniques for reference, the land reclamation industry of China's metallurgical mines will also developed rapidly.

With the research personnel of other related subjects taking part in the field of land reclamation and the deepening of the theory of land reclamation and technical research, the course system and the theoretical framework of land reclamation will be set up gradually. Because China is a large country that possesses rich mineral resources and a great population, the potentiality in land reclamation is vast.

With the mankind entering 21st century, the land reclamation of mining area in China will have an all-round and rapid development, the study of land reclamation will be deepened and the rate of land reclamation will be raised remarkably.

In future, the land reclamation and the ecological reconstruction in mining area will become an important part of durable development in China's metallurgical mines and will obtain an energetic support and close attention of overall society.

Environmental Issues and Management of Waste in Energy and Mineral Production, Singhal & Mehrotra (eds)
© 2000 Balkema, Rotterdam, ISBN 90 5809 085 X

Environmental problems of mining in the Ural region of Russia

Alexander V. Khokhryakov & Alexander M. Olkhovsky
Urals State Academy of Mining and Geology, Russia

ABSTRACT: This paper presents the description of environmental situation in one of the oldest mining areas of Russia located at both slopes of the Ural mountains – the geographical border between Europe and Asia. The level of environmental disturbance and the contribution of mining industry into total environmental impact create new challenges for the development of mining in the region.

1 INTRODUCTION

Mineral raw material complex of the Urals and the branches connected with it compose the basis of industrial and economic potential of the region. Major mining enterprises of Russia are concentrated here Mining and processing of iron ore takes place in the cities of Kachkanar, Nizhny Tagil, Kushva. Copper mining and smelting takes place in Kirovgrad, bauxite mining is carried out in Severouralsk. Not far from Severouralsk there is the city of Krasnoturyinsk with a big aluminium plant. This is also the place where mining of iron ores copper ores and gold is carried out. The city of Karpinsk is famous for its coal production, the biggest in the Middle Urals and as for Asbest, it is the city where the largest in the world asbetos production is located. And this list is not complete.

At the same time more than two-centuries intensive exploitation of the Urals' interior became one of the main reasons of serious, often irreversible disturbances of the environment in the majority of mining regions of the Urals.

However, it is impossible to consider environmental consequences of the activity of mining industry separately from the existing environmental situation formed by all branches of the Ural industry.

According to official evaluations [1] the industrial zone of the Urals relates to the areas of «very sharp environmental situation» and the Urals region is characterized as a whole by «very high pollution of the air and water environment». Let's consider the summarized figures characterizing the industrial impact for the environment in the Ural region (tables 1, 2, 3).

Absolute values of the given figures are rather high and their relative decrease by 1999 is connected mainly with reducing volumes of production over the period (table 1).

This very bad effect on the environment inevitably tells on the health of the population, particularly taking into consideration economic decrease (tables 4, 5, 6).

Figure 1. The main mining centres of the Middle Urals

Table 1. Relative impact of different industries on the environment in the Ural region*

Branches of industry	Factors affecting the environment				
	water consumption	waste water discharge	wastes formation		discharges into the atmosphere
			volume	mass	
Ferrous metallurgy	13.3	23.5	5.8	46.1	30.7
fuel and power industry	34.6	3.1	3.0	8.7	23.6
non-ferrous metallurgy	6.4	5.7	3.1	16.2	17.3
chemical industry	2.3	7.8	0.6	0.2	0.3
machine building	6.4	6.7	11.9	0.3	2.5
timber industry	2.0	2.5	11.0	0.1	1.2
construction materials industry	2.0	2.0	1.2	26.4	3.3
transport and means of communication	1.6	0.8	3.3	0.1	6.8
automobiles	-	-	-	-	11.6
public utilities	22.9	44.1	49.1	0.1	1.6
microbiological and medical industry	2.9	2.5	0.3	0.2	-
agricultural industry	4.8	0.5	3.9	1.1	0.8
other branches	0.8	0.8	6.9	0.5	0.3
total	100%	100%	100%	100%	100%

* per cent to the total impact

Table 2. The industrial impact for the environment in the Ural region

Years	1985	1990	1995	1999
Discharges into the atmosphere, th/t a year	2770	2745	1565	1442
Discharges of industrial waste waters, mln. cub.m. a year	1900	1820	1846	1855
Mass of discharged substances into drainage, th.t/year	850	723	565	536
Industrial water consumption, mln.cub. m/year	no data	1207	1005	970
Disturbed lands at the beginning of a year, th.ha	68	65	59	72
Amount of industrial wastes, mln.t/year	520	410	235	152

Table 3. Main industrial enterprises of the Ural region and their contribution to the air pollution

Name of an enterprise	location	type of activities	discharges in per cent to regional amount
JSC «Nizhnetagilsky metallugical complex»	city of Nizhny Tagil	steel and cast iron production	10.8
JSC «Bogoslovsky aluminium plant»	city of Krasnoturyinsk	alluminium production	10.7
Reftinsky Hydro-Electric Power Station	city of Asbest	electric power production	10.7
JSC «Sredneuralsky copper-smelting plant»	city of Revda	copper smelting	10.2
JSC «Ural aluminium plant»	city of Kamensk-Uralsky	aluminium production	6.8
JSC «Svyatogor» - copper smelting plant	city of Krasnouralsk	copper smelting	4.0
JSC «Kachkanarsky mining dressing complex»	city of Kachkanar	iron ore mining and dressing	3.9
Ekaterinburg automobile transport	city of Ekaterinburg	automobile transport	3.1
JSC «Kirovogradsky copper smelting complex»	city of Kirovograd	copper smelting	2.8
JSC «Sukholozhsktsement» (cement production)	city of Sukhoy Log	cement production	2.7
JSC «Novolyalinskiy cellulose paper mill»	city of Novaya Lyalya	cellulose and paper production	2.6
JSC «Uralmash» (machine building)	city of Ekaterinburg	machine-building	2.6
JSC «Visokogorsky GOK»	city of Nizhny Tagil	iron ore mining and dressing	2.5
JSC «Kamensk-Uralsky alluminum plant» production	city of Kamensk-Uralsky	alluminum	2.1
JSC «Serovsky Plant of ferroalloys»	city of Serov	ferroalloys production	2.1
JSC «Sredneuralsky Hydro – Electric Power Station»	Verkhnyaya Pishma	electric power production	1.6
JSC «Rezhevskoy nickel plant»	city of Rezh	nickel smelting	1.2
Total:			80.5 %

Table 4. Death-rate and birth-rate indices dynamics in the Sverdlovsky region, 1991-1996 (per 1000 people a year)

index	birth-rate	death-rate
1991	10.8	11.5
1992	9.6	12.7
1993	8.5	15.2
1994	9.0	17.3
1995	8.6	15.5
1996	8.3	14.7

Table 5. Cases of malignant tumour and death-rate because of cancer in Sverdlovsk region in 1960-1996 (number of cases per 100 000 people a year)

index	death cases	sick cases
1960	123	174
1970	128	186
1980	164	247
1990	196	282
1996	209	300

What is the contribution of enterprises of mineral raw material complex into the general situation of ecological disturbances in the Urals? As the majority of mining enterprises are connected with the enterprises of processing branches in a real and organizational aspect let's try to answer this question taking mining-industrial centres of the Sverdlovsk region as an example .

Table 6. Structure of sick cases among the population able to work in the Ural region in 1999

Group of diseases	sick cases per 100 working population	sick- leave days per 100 working population	per cent to the total amount of sick cases
organs of breathing	23.7	217.9	39
osseous and muscular system	8.0	124.0	14
heart - vascular system	4.3	78.2	6
organs of digestion	2.6	40.4	5

2 DISCHARGES INTO THE ATMOSPHERE IN THE MAIN MINING CENTRES

NIZHNY TAGIL. Annually 160 thousand tons of polluting substances are discharged into the atmosphere. The proportion of mining enterprises (mainly Vysokogorsky Iron Ore Co) totals in average 48% of the general volume of discharges, the main discharged substances being carbon oxide and sulphuric anhydride.

KIROVGRAD. 86 th.t a year is discharged into the atmosphere including mining enterprises (Kirovograd metallurgical company) - 5,8 % of the total amount of discharges.

CITY OF ASBEST. 265 th.t is discharged annually. The share of asbestos mining industry - 0.2% of the total amount of discharges. The degree of catching is from 48 %. Main polluting substances - fluorine hydrogene, carbon oxide, asbestos dust.

KRASNOTURINSK.78 th.t is discharged annually. Discharges of mining enterprises (Bogoslovsky iron ore mine, Turinsky copper mine) total 2,.4% of the total amount of discharges in the town.

KACHKANAR. Discharges into the atmosphere - 238 th. t a year, of them Kachkanar Iron Ore Co. - 99,4% including sulphur anhydride - 99%, carbon oxide -100%, nitrogen oxide - 87%.

SEVEROURALSK. Total discharges in the town amount 1.3 th. t a year including by bauxite mining industry (SBM) - 72 %.

KUSHVA. Total discharge in the town - 77 th.t a year including mining enterprises (Goroblagodatsky Iron Ore Mine, Volkovsky copper mine) - 96%. Main discharged substances - sulphuric anhydride, nitrogen oxides, dust.

KARPINSK. 18.4 th.t a year is discharged into the atmosphere including mining industry - over 9%.

KRASNOURALSK Annual total discharge is 87 th.t. and of these 79% belongs to the copper production company «Svyatogor».

As a whole discharges of mining enterprises according to reports data amounted to 415 th.t a year or 26% of the total quantity of discharges into the atmosphere over the Sverdlovsk region, the main share is given by Kachkanar, Nizhniy Tagil, Kushva, Krasnouralsk, where main Ural iron and copper ores mining and processing enterprises are located.

3 DISCHARGE OF INDUSTRIAL WASTE WATERS

If we compare the available figures of discharges by the enterprises of mining industry with the total volume of polluted waste waters, discharged without clearing in the Ural region (117 mln.cub. m a year) we'll have the following:

Nizhny Tagil - 7,5%; Asbest -7,1 %; Severouralsk- 9 %; Krasnoturinsk - 1,5 %; Kachkanar - 18,9%; Kushva - 6%; Karpinsk - 3%.

In total the discharges by mining enterprises amount to 58% of the total volume of polluted waste waters, discharged without purifying by enterprises of the Ural region.

It should be noted that annually 220 mln.cub.m. of waters are discharged in drainage of mining workings that total 49% of the whole underground water catchment of the region. In this case 85% of all pumped out waters from mining workings is discharged without use.

4 LAND DISTURBANCE

By 1999 in the region 72 th.hectares. of land had been disturbed by all kinds of industrial activity. Of them approximately 57% were made by mining, processing and exploration of mineral deposits.

Of all the lands disturbed by enterprises of mineral raw material complex about 24% were dumps of overburdens, tailings and other wastes of mining production.

Table 7. Major mining industry waste dumps at the territory of Ural region (according to the State Environmental Protection Committee)

Name of a depository	City	Type of wastes	Amount of wastes mln.t.
Soryinsky slime depository JSC «Svyatogor»	Krasnouralsk	copper-bearing slime	16.6
Slime storage JSC «Khrompik»	Pervouralsk	slime with six-valent Cr	6.8
Slime depository JSC « Kirovogradsky metallurgical plant»	Kirovograd	copper-bearing slime	30.2
Proving grounds of arsenious cake JSC «Sredneuralsky copper smelting plant»	Revda	arsenious cake	0.2
Malishevsky mine office	Asbest	beryllium - bearing overburden and dressing wastes	15.7
Cheremshansky slime depository JSC «Visokogorsky GOK»	Nizhny Tagil	wet dressing wastes	42.1
Tailings depository JSC «Uralsbest»	Asbest	asbest-bearing dressing	519.5
Slag depository number JSC «Bogoslovsky aliminium plant»	Krasnoturyinsk	wastes dressing and aluminium production wastes, bauxite slime	56.9
Tailings storage of Kachkanarsky Iron Ore Co.	Kachkanar	dressing wastes	950.5

5 INDUSTRIAL WASTES ACCUMULATION

Industrial wastes are for the Urals one of the main ecological problems.

At present about 7,9 mlrd tons of different industrial wastes are located in the Sverdlovky region which are the reason of the negative effect on the environment and the greater part of the wastes are result of the mining enterprises activity (fig. 2,3; table 7). Though the absolute amount of the resulting wastes reduces every year (from 310 mln. t. in 1990 to 152 mln. t. in 1999), the problem doesn't become less serious . The same can be said about the mining industry, its wastes according to different estimations are about 88% from the total amount of industrial wastes, which are formed in the Sverdlovsky region every year.

In 1990 the amount of overburden and enclosing rocks in dumps and the amount of dressing wastes was 220 mln.t., in 1996 - 118 mln.t., in 1999 - 134 mln.t.

Considerable reduction during the last 8 years is the sign of production volumes reduction in mining industry , which is caused by both economic reasons and by exhaustion of many deposits resources.

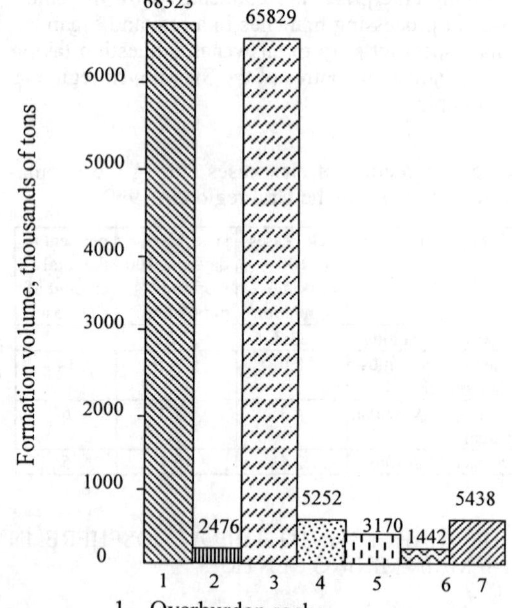

Figure 2. Industrial wastes formation in the Sverdlovsky region in 1999

CONCLUSIONS

Taking into consideration the above mentioned figures, you can say that, the contribution of enterprises of raw- mineral complex to the negative impact on the environment in the Sverdlovky region is rather big: discharges into the atmosphere reach 26% from the total volume of discharges in the region, wastes water discharge is 58% from the total volume, water consumption - 10% from the total industrial water consumption, underground water catching is 49% from the total amount, land disturbances are 57% from the total amount, wastes formation - more than 88% from the total amount of industrial wastes formed annually (fig.4).

The figures mentioned above show that the ecological safety of the Urals to a considerable degree depends on the level of ecological safety and the control over the ecological situation in the mining industrial complex.

Quickly changing economic situation makes it necessary to change the structural pattern of the Ural mining industry.

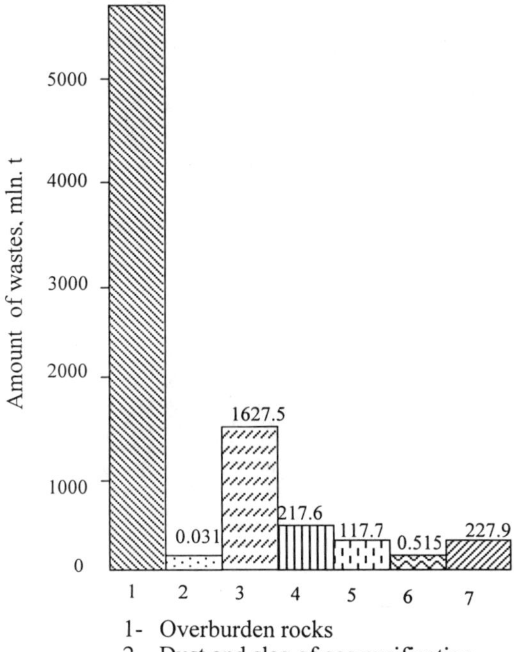

1- Overburden rocks
2- Dust and slag of gas purification
3- Dressing wastes
4- Ash slag of thermal electric stations
5- Metallurgical slag
6- Agricultural wastes
7- Other types of wastes

Figure 3. Amount of accumulated industrial wastes in the Ural region at the end of 1999, mln.t

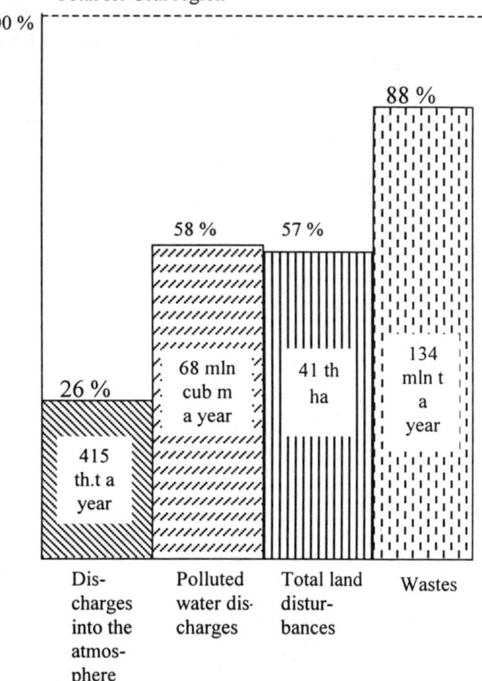

Dis-charges into the atmos-phere | Polluted water dis-charges | Total land distur-bances | Wastes

Figure 4. Contribution of mining industry to the environmental impact in the Ural region

Taking into account practical experience gained in the coal mining industry of Russia and also local conditions the following trends in changing the stuctural pattern in the Urals can be considered as the main ones:

- renovation of mining enterprises potential based on new deposits with favourable conditions for the development, including technogenious deposits;
- improving and regulating the existing mining complex ;
- sanation of the stable working enterprises, products of which are in great demand;
- consecutive elimination of unprofitable enterprises with no prospects;
- wide application of new, highly efficient technologies in mining and mineral processing.

REFERENCES

The Governmental Reports on the environmental condition and factors affecting the population's health in the Sverdlovsk region of Russia in 1995-1999.

Galchev F.I., Silyutin S. A. Changes in coal industry structural pattern and market of coking coals. Gorny Zhurnal (The Mining Magazine of Russia), № 11, 1997.

Khokhryakov A.V. About investments into gold mining in the Urals. Mineral Resources of Russia, № 3, 1997.

Environmental Issues and Management of Waste in Energy and Mineral Production, Singhal & Mehrotra (eds)
© 2000 Balkema, Rotterdam, ISBN 90 5809 085 X

The challenge of sustainable small (and medium) scale mining in Guyana

Karen Livan
Environmental Division, Guyana Geology and Mines Commission, Guyana

ABSTRACT: Gold, diamond and bauxite mining in Guyana are of great economic importance. Roughly 25% of the gold and all of the diamond currently produced are from medium and small scale operations. Small and medium scale operators, are hydraulicking larger volumes of material, with attendant challenges for water and tailings management; and developing steep, dangerous mining faces, where orebodies are found at greater depths. This is occurring in an environmental of competing land use from forestry and ecotourism and greater demands for biodiversity conservation.

Better defined geological resources and deeper deposits are dictating that the alternative use of heavy earth-moving equipment provides multiple benefits, and is more environmentally sound. However, capital and operating costs are higher.

In developing institutional capacity for environmental management, Guyana Geology and Mines Commission is collaborating with international agencies, and is the principal beneficiary of the four year CIDA project - Guyana Environmental Capacity Development Mining Project.

INTRODUCTION

At the turn of the century, and from the late 1930's to the present time, mineral production in Guyana, principally gold, diamond and bauxite production, has been important to the economy. About 20-25% of the gold production and all of the diamond production are from medium and small scale Guyanese operators, mainly using suction dredges and hydraulicking. The remaining gold production is from large scale open pit mining at Omai. Bauxite production is on a large scale, but a crucial part of the bauxite sector, refractory grade bauxite production, has declined sharply from the 1980's. Almost all of the gold, diamond and bauxite produced is exported: the total value of gold and bauxite exports in 1998 was US$203 million. Stone and sand production are increasing in importance. Stone is produced for local construction, road building and sea defences; silica sand for local construction and road building, and to a lesser extent, for export.

Sustained exploration, and vigorous mineral promotion by the Guyana Geology and Mines Commission (GGMC) from the 1980's, have resulted in large scale operations at Omai Gold Mines from 1992/93, as well as the significant development of geological resources of several hundred thousand ounces of gold at each of thirteen prospects at the end of 1997. There is active exploration for palaeoplacer and primary diamond deposits, and base metal deposits, as well as interest in kaolin and silica sand development for export. With these developments, it is expected that large and medium scale mineral production will diversify with the possible introduction of underground mining, and that medium scale mining will progress from the current principal hydraulicking/dredge mining to open pit style operations.

Environmental protection in mining has become increasingly important: its importance was forcibly highlighted in Guyana by the environmental accident at Omai Gold Mines in August 1995, when the tailings dam was breached, discharging 2.9 million cubic metres of cyanide bearing effluent into the Omai River. In response, the GGMC is seeking to develop institutional capacity in environmental management, including impact assessment, emergency response, and monitoring.

GGMC's MANDATE FOR ENVIRONMENTAL PROTECTION IN MINING

In addition to its responsibilities for undertaking geological surveys, GGMC's mandate includes promoting mineral resources development, monitoring and regulating the Mining Industry, offering Technical Assistance to the Mining Sector, and promoting and monitoring environmental protection and sustainablility during mining. Increasingly, from the 1980's GGMC has focused its effort on issues relating to Mining and the Environment.

HEIGHTENED ENVIRONMENTAL AWARENESS, FROM THE 1980'S

In the 1980's the Guyana Geology and Mines Commission (GGMC), in the absence of an Environmental Protection Act, began to take proactive interest in environmental protection, and this interest has continued, to date. The Environmental Protection Act was subsequently promulgated in 1996. In the 1990's, there was greater public awareness and concern over the environmental impact of mining, particularly river dredge mining, as suction dredge units had grown considerably in capacity from the mid-1980's, facilitated by the incorporation of a winch to operate the suction (gravel pump) end of the dredge. The slurry from the gravel pump was processed by a wide, riffled sluice mounted on a barge (*Watkin and Woolford, 1992*).

Within the last six years or so, from the early 1990's, as the rivers were perceived to be depleted, dredge mining has shifted from the river to the land. The ore is hydraulicked into a sump, and then delivered to a riffled sluice by gravel pumps.

Awareness of environmental protection in Guyana was fostered by international events such as the United Nations' landmark *'Earth Summit'* conference on the Environment and Development, held in Rio de Janeiro in 1992 with its resultant plan of action, *Agenda 21*, for *'sustainable development'*; as well as increasing multiple resource utilization in mining and forestry and more recently; ecotourism. To this milieu was added articulated concerns of indigenous people of the Amerindian communities.

GGMC'S APPROACH TO ENVIRONMENTAL MANAGEMENT

GGMC's approach has been multi-faceted, dealing with dredge mining and its effects; evaluation and enhancement of processing technologies employed by Small Scale Miners; and the drafting and adoption of what are essentially "*Codes of Best Practices*" embodied in an Environmental Management Agreement for small scale/artisanal miners. At the same time the Commission recognised that for effectively evaluating the effects of small scale dredge mining on the environment, environmental baseline data had to be collected, and accordingly, embarked on a field campaign for collecting water quality data on the major drainage basins where artisanal mining was traditionally carried out. Water samples, however, were not chemically analysed for mercury, heavy metals or other potentially polluting elements.

As environmental protection has been highlighted, structural changes have been introduced within GGMC, with the creation of a Unit, subsequently upgraded to a Division in 1998, with specific responsibilities for environmental management and protection in mining.

Demonstration of New Techniques

Aided by the Technical Assistance Group (TAG) of the Commonwealth Secretariat, two complementary studies were commissioned by the GGMC in 1991 and 1993 respectively, (1) to review the environmental aspects of dredge and small pit mining operations; and (2) to evaluate the technological and operational aspects of dredge and small scale open pit gold and diamond mining in Guyana.

In 1992, the GGMC contracted the Hydro-meteorological Service of the Ministry of Agriculture to undertake hydrological surveys in the major rivers in the principal mining districts, to determine hydro-chemical characteristics, sediment load and mineral transport of the rivers.

The GGMC also sought to introduce and encourage the local manufacture and use of simple mercury retorts for the recovery of mercury from amalgam. Gold produced by local miners is recovered with mercury.

GGMC, with assistance from the Canadian Executive Service Organisation (CESO) in 1993, finalized the preliminary design for a mobile demonstration gravity plant that incorporates gravity separation equipment other than the riffled sluice box, for enhanced recovery of gold, particularly fine gold (less than 250 microns). Alternatively, it has been proposed that the primary sluice box, including the Brazilian type 'carpeted sluice' can be used in conjunction with other secondary gravity processing equipment for enhanced recovery of fine gold (*Skeete and Seffon, 1993*). Enhanced recovery, it is reasoned, will increase profitability, and so pay for the added cost of environmental protection.

Test work will compare the recoveries from each process, while the use of other gravity separation equipment (jigs, spirals, shaking tables, centrifugal separators) is expected to yield a higher grade of concentrate, requiring less mercury for amalgamation, or altogether eliminating the need for the use of mercury.

Environmental Management Agreement and Plan

The studies and surveys undertaken were intended to provide background environmental data, and more specifically, to inform the Environmental Management Agreement (EMA) that was being drawn up by the GGMC, in close consultation with the Guyana Gold and Diamond Miners' Association (GGDMA), for *small and medium scale* operators in Guyana. The draft EMA was reviewed and vetted by UN Inter-regional Adviser on Mining and the Environment, Mr. Barry Middleton.

An Environmental Impact Assessment and Environment Management Plan were made part of the requirements for each application for *large scale* mining and quarry licences. These were submitted after review by GGMC, to the Guyana Agency for Health, Environment and Food Policy (GAHEF) and latterly to its successor, the Environmental Protection Agency (EPA), for approval.

With the enactment (in 1991) of the (new) Mining Act 1989, and the Environmental Protection Act in 1996, GGMC is updating the Mining regulations, and making these reflect greater responsibility for environmental protection. GGMC is receiving technical assistance from the Economic and Legal Advisory Service (ECLAS) of the

Commonwealth Secretariat for the finalisation of the new mining regulations. Regulations for the Environmental Act will soon be drafted by the Environmental Protection Agency, and GGMC will assist the EPA in drafting regulations that specifically address issues of Mining and the Environment.

IMPORTANCE OF SMALL AND MEDIUM SCALE ARTISANAL GOLD AND DIAMOND MINING IN GUYANA

Because of the importance - numerically (in 1998, there were approximately 11,000 small and 2,000 medium scale prospecting and mining properties), wide geographical distribution, and significant scale of cumulative production - of small (and medium) scale artisanal gold and diamond mining activities, over the years, GGMC had directed much of its resources to supporting the development of, and giving technical direction to artisanal small and medium scale mining. This effort targets the expansion and enhancement of their activities and resultant increases in mineral production, and it has been supported by the formal introduction of the medium Scale Mining Sector in 1992.

For these operators, who essentially mine alluvial and eluvial deposits, mercury amalgamation is universally used for gold recovery.

IN ENVIRONMENTAL MANAGEMENT

The "Omai spill" environmental accident that occurred in August 1995, accentuated the need for institutional capacity building in environmental management in Guyana, and at the GGMC.

The GGMC is working towards building institutional capacity for Environmental Management for small, medium and large scale mining, quarrying and sand pit operations, including Environmental Impact Assessment; Establishment of Guidelines and Standards; Monitoring; Enforcement; Risk Assessment; Emergency Response and Public Awareness. This capacity development is undergirded by two current projects of Technical and Financial Assistance - Guyana Environmental Capacity Development Mining Project, sponsored by the Canadian government, executed by the Canadian Agency for Development Assistance, CIDA and managed by the Canadian Centre for Mineral and Energy Technology (CANMET); and to a lesser extent the Environmental Management Program sponsored by the Inter-American Development Bank (IDB). Both projects emphasise training, and GGMC will be the principal beneficiary of the 4-year CIDA project, which will be financed to the tune of CAD $3.795 million. EPA, Guyana Gold and Diamond Miners Association (GGDMA), and to a lesser extent University of Guyana (UG) and the Institute of Applied Science and Technology (IAST), will also benefit. The IDB project, which focuses on the Mining and Forestry sectors, will primarily benefit the EPA, while also strengthening capacity at the GGMC and the Guyana Forestry Commission (GFC). Environmental regulations for mining will be developed under the IDB project, which also has public education as an important objective.

Another collaborative effort is co-operation with the international project that is being jointly undertaken by the British Geological Survey and Intermediate Technology of the UK, funded by the Development for International Development (DFID). This project is addressing the problem of low gold recoveries by small scale/artisanal miners, estimated at about 35%, by improving their mineral processing technology. An initial scoping survey reviewed gold recovery methods, their applicability to small scale mining and included a simple cost benefit analysis. The socio-economic aspects of small scale mining were also convened (*Styles, 1999*).

NEW CHALLENGES

As medium and large scale gold and diamond producers are expanding their operations, the need for environmental management is becoming more apparent. Removal of large volumes of material by hydraulicking gives rise to the development of steep, ofttimes dangerous, mine faces. Ore is diluted by slumped overburden, often with a high clay content. Discharge of large volumes of tailings and process water, as overburden is processed with the ore, results in high levels of suspended solids which can cause problems for communities downstream. The introduction of undesirable practices in the use of mercury to amalgamate the whole ore or the sluice box concentrate extends the exposure to mercury and the risk of contamination.

The Medium Scale Sector is poised to grow in sophistication to adopt open pit mining techniques using heavy earth-moving equipment, gold recovery using gravity separation systems other than the sluice box and reducing or eliminating the use of mercury. This represents a "great leap forward," occurring against a backdrop of prevailing low gold prices and a poor international investment climate.

Guyana Geology and Mines Commission, the Gold and Diamond Miners Association and the Multi-stakeholder Project, GENCAPD, are expanding the concept of field Demonstrations, from the initial applications for Small Scale miners using sluice box technology, to putting together a project proposal for technical assistance for a medium scale project, that will not only work towards improving mineral recoveries, but will also treat with safe and efficient mining methods, deposit evaluation, mine planning, environmental management and rehabilitation. As has been proven elsewhere, it is intended to show that more environmentally friendly mining techniques can also be bring benefits of efficiency and economy.

In terms of the behavioural aspect four new approaches to environmental management are desirable and altogether worthwhile - the goal of constantly aiming for self-improvement in all aspects of Environmental and Occupational Health and safety concerns; the goal of planning for and undertaking reclamation; the goal of self-regulation; and the goal of exercising care, in spite of harsh economic conditions, for the environmental legacy that the Mining Sector will leave to the next generation. In short, in the words of UN Resident Representative to Guyana, Mr. Richard Oliver, "*taking pride in one's environment, and working towards economic and environmental sustainability.*"

These goals apply equally to Small Scale dredge operators, and demonstrations geared for them through GENCAPD, will not only focus on the improvement of sluice box recovery, but also on water and tailings management. The first trial demonstration by GENCAPD was held at the Mahdia Mining area in September 1999. Consultant to GENCAPD Dr. Randy Clarkson reported a 20% increase in gold recovery after two sluices were re-fitted with angle iron and spread on the sluice box floor.

GGMC is working with miners to undertake field trials to improve and optimise sluice box recovery, and to find affordable ways of increasing gold recovery, especially of the fine size fractions (less than 250 micron size) by introducing new gravity processing equipment - Centrifugal bowl concentrators, jigs, and spiral concentrators. For GGMC, GENCAPD, and the British Department for International Development (DFID) supported project, field demonstrations and a training programme will be the strategies used to show the technical feasibility and economic viability of improving gold recovery through processing modifications, and to facilitate the adoption of the new techniques by miners.

CONCLUSION

The Mining Industry continues to play a vital role in the socio-economic development of Guyana. Small and Medium Scale Mining are important because of their significant cumulative production of gold and diamonds, and their potential for much larger production, as a more structured approach is employed, mining known reserves are worked by open pit methods, and the sluice box technology is augmented or updated, to obtain higher gold recovery.

Small and Medium Scale operations, because of the large number of individual operations, often itinerant, and prevalent hydraulicking techniques employed, with sluicing and amalgamation as the principal gold recovery methods, pose special challenges for Environmental protection. Increasing production capacity is increasing environmental disturbance, and other stakeholders - Ecotourism Industry, Foresters, Hinterland Communities, including indigenous efforts are demanding more and more environmental sustainability from mineral producers.

Always proactive in environmental protection in mining, GGMC is employing a multi-faceted approach that underscores Institutional Strengthening through Human Resource and Systems Development; improving mineral recoveries through technically and economically feasible process modifications; introducing better mining practices, tailings and water management systems, and fostering heightened awareness of the need and responsibility for environmental management in mining throughout the Mining Sector.

BIBLIOGRAPHY

Bank of Guyana, Half Yearly Report and Statistical Bulletin, 1999.

Skeete, D. & Sefton, Dr. V.B. : The Design and Development of a (1993) Demonstration Plant for enhanced gold recovery in small scale operations - Paper presented at the Guyana National Conference on Mining and Quarrying (Guyana Geology and Mines Commission Library).

Styles, M (1999): Recovering lost gold; DFID Earthworks Issue 9, November 1999, p6.

Watkin, E.M and Woolford, W.H (1992) : Alluvial Gold Mining in Guyana - Environmental Aspects (Guyana Geology and Mines Commission Library).

Environmental Issues and Management of Waste in Energy and Mineral Production, Singhal & Mehrotra (eds)
© 2000 Balkema, Rotterdam, ISBN 90 5809 085 X

Self-regulation and environmental management in the mining industry in Zimbabwe – A survey of issues

Oliver Maponga
Institute of Mining Research, University of Zimbabwe, Harare, Zimbabwe

ABSTRACT: This paper investigates the interaction of the command and control and self-regulation approaches in environmental management in the mining industry in Zimbabwe and explores the feasibility of using economic instruments to strengthen the current system. After a brief discussion of the command and control system through a review of their strengths and weaknesses in the first section, the paper analyses self-regulation in environmental management in the Zimbabwean mining industry and highlights constraints and opportunities of the approach. By drawing on lessons from other countries, the paper provides recommendations on how the current system could be improved. The paper concludes that collaboration between government and mining companies in improving environmental management systems. A proactive approach by industry is suggested as a way of improving environmental management in Zimbabwe.

1. INTRODUCTION:

The pressure on mining companies to comply with, or put in place good environmental programmes has increased exponentially since the Rio Summit in 1992. An environmental gold rush emerged in the 1990s as a dramatic change in attitude towards environmental issues has emerged. Communities have been at the forefront in pushing for more environmental accountability by the users of land resources including minerals. The globalization of minerals trade and pressure from shareholders have almost influenced change towards environmental responsibility by mining companies.

Mining companies, through their associations, have responded to these challenges by developing charters and codes of conduct to minimize environmental contamination from their activities. The founding of the International Council on Metals and the Environment (ICME) in 1991, the adoption of the Berlin Guidelines in 1991 and the evolution of internal environmental management systems are a few of the initiatives referred to here. The promotion of self-regulation and adherence to home environmental regulations where host regulations are inadequate is part of the overall commitment by mining companies to improve their image in delivering socially acceptable environmental products and initiatives. Yet the non-binding nature of most of these regulations exposes them to potential violation by some signatories.

The use of economic incentives to augment these other initiatives is now common in most developed countries. Yet adoption and application is limited in developing nations because of several factors including the lack of the information base, competitive markets for environmental instruments such as bonds, a strong legal structure and in certain instances political feasibility. The use of regulations is common in most nations including developing nations such as Zimbabwe. Yet some companies in these developing nations, usually subsidiaries of multi-nationals, have adopted self-regulation mechanisms usually involving policies of parent companies.

In the midst of this environmental gold rush during the 1990s a result of the weaknesses of the command and control system, mining companies have increasingly relied on self-regulation. As a proactive approach, self-regulation involves companies taking environmental responsibility as they usually factor in concerns of communities into operational planning. A supportive government system enhances the effectiveness of self-regulation initiatives.

2 THE SELF-REGULATION FRAMEWORK

Since the late 1980s companies have adopted environmental management systems (EMS) as a form of self-regulation to enhance their reputation as responsible environmental citizens. Environmental Management Systems are important as they provide a consistent framework across operations and ensure that resources are directed into areas that provide the most environmental benefit. As a management tool EMS allocates management and employee responsibilities relevant to the organisation's activities and legal requirements. Usually based on ISO standards, a typical EMS document describes the plan, do actions, check and review components - Table 1. The check and review components are the subject of an annual environmental report often made available to the public for comment and critique. This helps companies assess attainment of objectives and compliance of initiatives with community expectations. As a continuous process with an important feed back mechanism an EMS ensures that any new concerns are factored into plans for future actions and reviews.

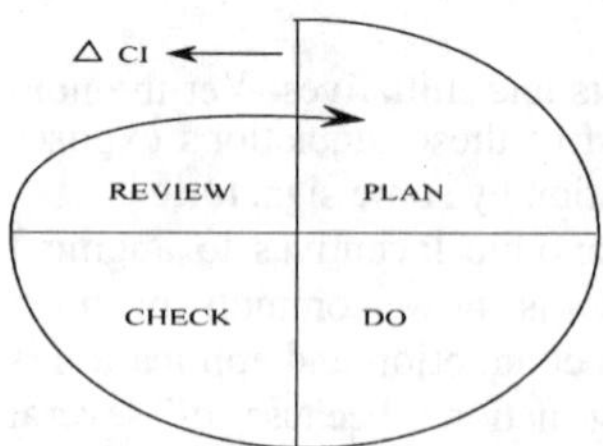

Figure 1: Simple Environmental Management System

The continuous improvement (CI) aspect of the EMS cycle is important as it ensures that annual reporting is adhered to and improvements are made as the annual report cites successes and failures. Issues and actions typically specified in the do component of a gold mining company include those shown in Appendix 1.

The priority areas in a typical EMS often include many of the following issues –

- An environmental strategic plan which integrates systems into the business unit and operations management to ensure adequate resources are allocated to achieve key objectives. The plan is linked to the annual business process.
- Environmental risk assessment and management procedures to identify significant risks to be addressed in the business plan process.

Table 1: Components of a Simplified EMS

Plan	i. The initial environmental review
	ii. Environmental policy
	iii. Record of environmental legislation
	iv. Assessment and record of significant environmental effects
	v. Organisation of responsibilities
	vi. Objectives and targets
	vii. Management program
Do	i. Management and manual procedures
	ii. Operational control
	iii. Records
Check	i. Monitoring against standards, targets and procedures
	ii. Record of results
	iii. Audits of processes and compliance
Review	i. Management system review procedures of the whole EMS
	ii. Technical certification, eg ISO 14001
	iii. Environmental Review Board which includes the CEO
	iv. Corporate sign-off of review (endorsement)
	v. Management commitment to EMS improvement

- Description of the environmental audit process and its timing. As a self-assessment tool, the audit evaluates environmental compliance, environmental performance, legal compliance and performance objectives.
- The frequency and structure of reporting procedures. The description includes all non-compliance incidents, rectification programs and progress on implementing EMS.
- A comparison with international standards for environmental management in important. Global standards provide a benchmark for companies. The comparison is designed to encourage commitment, implementation of robust planning, identification of environmental competency and training needs and incorporation of environmental risks and potential impacts into the whole planning process.

Setting up an elaborate EMS can be a lengthy process which depends on a multitude of factors including the nature of the company's operations, resources committed to the process and the organisation's commitment. For example, Westmin Talc (Netherlands), one of about 100 companies to have achieved ISO 14001 in the Netherlands in 1999 reports that it took them six years to set up an elaborate EMS.

Environmental management systems normally include the following components;

The use and strict adherence to recognised standard methods of monitoring, the use of approved measurement tools, standard reporting forms and the accreditation of on-site planning and analytical methods are all important components of a good EMS. Yet prime responsibility for the success of EMS is with the operator.

3 ENVIRONMENTAL MANAGEMENT - THE ZIMBABWEAN FRAMEWORK

Environmental regulations provide the backbone of good environmental performance in the mining industry around the world by providing a general framework for mining companies and individuals to follow in pursuit of environmental stewardship. Zimbabwe's situation is no exception. As a traditional or prescriptive approach, regulations set minimum discharge levels and general conditions to be satisfied in so far as pollution levels are concerned. In their many forms, regulations often involve standards to be met and bans on discharge of certain levels of contaminants. The effectiveness of the regulatory approach is enhanced by a strong administrative machinery as well as an effective judicial system to facilitate its operation especially the prosecution of violators. It is arguable that by providing conditioning policies and policing of the behaviour of companies, the regulatory approach levels the playing field for all operators.

Environmental issues in the mining Industry in Zimbabwe are governed by several pieces of legislation. The ones exerting the greatest influence on the management of the environment are:

- Mines and Minerals Act (Chapter 165),
- Natural Resources Act (Chapter 150),
- Atmospheric Pollution Prevention Act (Chapter 318)
- Water Act (14/1979),
- Parks and Wild Life Act (14/1975),
- Hazardous Substances Act (Chapter 322), and
- Factories Act (for smelters and refineries not located at mine sites),
- Mining of Alluvial Gold (Public Streams) Regulations of 1991

Each of these pieces of legislation has components affecting environmental management issues in the Zimbabwean minerals industry by providing prescriptions and minimum standards expected by government. These prescriptions often conform to international standards. Yet the Mines and Minerals Act (Chapter 165) (and relevant amendments) has overall authority on issues affecting the exploration for, extraction of and processing of minerals.

In addition to these legislative instruments, the Government of Zimbabwe introduced an elaborate EIA Policy in 1994. Table 3 of the policy contains a list of prescribed activities requiring Environmental Impact Assessment before commencement. The list includes mining and quarrying activities, petroleum mining and processing and industrial activities such as smelters and refineries. According to the policy, a prospectus has to be submitted whenever a prescribed activity is being considered. The prospectus should describe the project and the proposed environmental management and mitigation measures. Through its internal structures, the Ministry then determines the necessity of an EIA for the nominated project. The policy stipulates that a preliminary EIA (PEIA) report be submitted for projects assessed to require an environmental impact assessment. As a preliminary document, the report is based on existing information and is required to indicate main impacts, their severity, and identify opportunities to minimize, monitor and manage the impacts. The impacts identified as significant are analysed in more detail as part of a detailed EIA (DEIA). Public consultation is integral to the EIA process in Zimbabwe. Project proponents are required to conduct public consultation during the preparation of both PEIA and EIA reports. As a safe guard, the Minister and the Natural Resources Board are mandated to conduct public hearings to verify EIA reports. Further, all EIA reports for prescribed activities are available for public review and comment. The only exception is where market-sensitive information is part of the report and the proponent would like it to remain confidential.

The Environmental Management Bill (1998) currently under consideration is an initiative designed to bring the Zimbabwean environmental regulation system in line with that in existence in most countries. The intention is to streamline all environmental regulations through an Environment Act in-order to overcome the problems identified by Maponga and Mutemererwa (1995) as contributing to the weakness of the current regulatory system.

Sustainability is key to the proposed bill as it seeks to address both inter and intra generational equity while promoting resource extraction and utilisation. The major objectives of the Bill are,

- to provide for the integrated management of the environment and the conservation and sustainable utilization of natural resources,
- to provide for the prevention and control of pollution,
- to establish the Environment and Natural Resources Board,
- to repeal the Natural Resources Act and

- to provide for matters connected with managing the environment.

The administration of the bill is through the Minister with assistance from the Environment and Natural Resources Board. The Bill is organic as it allows for the periodic review of national environmental management and natural resources management policies. To enhance efficiency, the Board has powers to appoint specific committees to deal with particular environmental problems and recommend changes to regulations, approaches and actions.

Section 34 specifies projects which cannot proceed without an EIA. Through a monitoring system supervised by the Board, the Bill ensures that all activities in place prior to the Act comply with the Act's specifications. For projects requiring EIAs, the Bill requires that environmental audit and management plans be clearly spelt out as outlined in Appendix 3. On pollution, the bill specifies that a penalty of Z$5,000 be paid by the polluter and also the courts are mandated to impose a fine of Z$2,000 per day for the duration of the pollution. The introduction of pollution permits and licences to dump waste and the decentralisation of power to local Authorities is designed to improve the Bill's effectiveness. The decentralisation of authority empowers communities and should theoretically improve the environmental delivery system as communities are empowered to refuse projects they perceive to be detrimental to their environment. Overall authority of the Board ensures national uniformity in the delivery system. Further, the use of accredited laboratories in the analysis water, air and soil samples for levels of contamination safeguards against pollution.

When complete, the Environmental Law in Zimbabwe will modernise the country's environmental delivery system through harmonization of regulations that are, at times, contradictory. Yet implementation of the prescriptions of the new law may suffer from constraints in the administrative machinery.

In addition to the proposed law, the adoption of the International Organisation for Standardisation (ISO) guidelines by the Zimbabwean industry generally strengthens systems of environmental management. The standards provide guidelines on conducting environmental assessments of sites and organizations through systematically identifying and evaluating environmental aspects and issues and determining, where possible, their business consequences. Although the International Standard does not provide guidance on how to carryout initial environmental reviews, environmental audits, environmental impact assessments and environmental performance evaluations, the guidelines cover the roles and responsibilities of the parties to the assessment and the phases of the assessment process. Although subscription to these guidelines is voluntary, the guidelines provide a respectable starting point for companies keen on ISO certification.

On the strength of existing regulations, the Zimbabwean minerals industry has gone a long way in creating a reasonably good infrastructure. The combination of the Environmental Management Bill and the adoption of ISO certification should, on paper, ensure improved environmental management in the Zimbabwean minerals industry.

4 THE EXTENT OF SELF-REGULATION IN THE ZIMBABWEAN MINING INDUSTRY

A significant number of Zimbabwean mining companies has developed management systems to minimise and contain environmental contamination. A survey conducted on 10 of the largest mining companies in Zimbabwe showed that 80 per cent had, on paper, policies designed to tackle environmental problems at their operations. Yet the schemes varied from company to company in terms of their scope and operation. Generally, companies with environmental policies and codes of conduct are mostly subsidiaries of multi-nationals domiciled outside Zimbabwe. A few local companies have some public statement about their approach to environmental issues.

In generally the reviewed codes provide both general and specific guidelines to be followed. Companies such as Delta Gold, Zimplats, Anglo American Corporation Zimbabwe, Rio Tinto and a few others have group environmental policies which provide general guidelines on how environmental issues should be dealt with within the company. The adoption of British Standards BS 7750:1994 is common among the surveyed companies.

The codes of practice generally make a commitment by companies to adhere to environmental regulations in Zimbabwe. For example AAC (Zimbabwe)'s code of practice specifies that all new projects should meet legislated environmental quality standards, and where local standards are absent, appropriate international standards are set as targets.

Yet, apart from these pronouncements, self-regulation beyond statements is not very evident in the activities of the mining operations reviewed for this report. Analysts argue that at least two issues may account for the lack of self-regulation methods in Zimbabwe. The issues are,

- relatively limited knowledge of self-regulation, and
- lack of appropriate incentives

First, mining companies in Zimbabwe argue that a few fiscal incentives exist to encourage mining companies to expend additional resources to improve environmental programmes at their mineral extraction and processing operations. Companies argue that governments have to provide monetary incentives for companies to do more than just comply with regulations. Yet it could be argued compliance with current regulations is not uniform among companies due to a poor government monitoring system. Furthermore, although companies demonstrated some knowledge of international standards and conventions, their compliance with these could not be ascertained. To prove their commitment to self-regulation, companies have to be proactive and demonstrate to government their self-regulation policies and then request for concessions. Government needs to be educated on these initiatives as a first step. Through this approach, companies can influence policy to their advantage.

The survey showed that there is limited knowledge about the moral responsibility of companies to sustainable utilisation of natural resources. While sustainable is a concept accepted by Zimbabwean mining companies, commitment of resources for this cause is limited to only a few. To improve knowledge and appreciation, mining and mining-related professional associations should take the lead and encourage their members to invest in self-regulation mechanisms. The examples cited elsewhere in this paper illustrate the fruits of such actions. Industry-generated codes of conduct have motivated companies to adopt internal mechanisms to improve environmental compliance. The Chamber of Mines should spearhead the dissemination of information on international practices and codes to its members and other stakeholders. Industry initiatives are the cornerstone of self-regulation systems around the world.

One of the key ingredients to a successful self-regulation system is its acceptation by all employees and their commitment to uphold company policy. Employee education through multi-skilling and multitasking is integral to a successful self-regulation system.

5 SELF-REGULATION MODELS – LESSONS FOR ZIMBABWE

The experience of Australian, Canadian and American mining companies provide models for companies in other nations to follow. The models provide a pool of resources from which to learn from. The models applied by these companies are constantly evolving as companies attempt to keep pace with changing times. Though not fool proof, self-regulation models are an improvement over the traditional approaches of mining companies. In the late 1980s mining companies were vocal against environmental regulations. They perceived them as unnecessary costs designed to reduce profitability of their operations. The early view was that sustainable development and profitability were not compatible. In the latter part of the 1990s the majority of the minerals industry embraced sustainable development as part of the process of resource extraction and utilisation. The commitment towards environmental stewardship is now part of visions of western mining companies. The bottom-line to good business performance enshrines environmental responsibility, safety, teamwork and leadership. The bottom-line has factored environmental indicators as companies have accepted sustainable development as a guiding principle. Companies have developed sustainable development ladders with set targets from year to year and measure their performance on the basis of these ladders. As part of the self-regulation process companies have integrated environmental management with other management and accounting tools in a process involving all employees through education to enable continued compliance. Environmental management is no longer an add-on but an integral component to company operations as mining companies strive to achieve ISO 14001 certification. It's a race aimed at creating a good public image.

The western world has taken the lead in self-regulation. In Australia, the Australian Minerals Industry Code for Environmental Management encapsulating the industry's commitment to demonstrated environmental performance and community involvement originated from the initiatives of mining companies through their representative bodies. Introduced in 1996, the code is voluntary. Signatories to the code commit to public reporting of company environmental performance and implementing the code's principles after two years of membership. Internal audits against the Code and reporting of findings are mandatory. Companies have to outline environmental action plans, set targets and internal standards. The mechanisms are very specific and site specific targets on factors such as depth and grade of ore. Companies make a commitment to continual improvement through the International Standard of Environmental Management - the ISO 14001. As a system of continual improvement to enhance the environmental management system, the series is for the purpose of achieving improvements in overall environmental performance. The code requires public reporting and compliance after two years of subscription to the code. Public environmental reporting stimulates change and improves relationships with regulators and local communities. Transparency with communities through open communication enhances confidence between the industry and communities.

Some Australian companies have adopted the International Environmental Rating System (IERS) and commit to conforming to its requirements through internal audits based on the IERS specifications.

In addition to subscription to sustainable development and moral responsibility, companies adopt self-regulation for economic reasons. As an insurance against future litigation, self-regulation minimises or totally eliminates such costs. Litigation against companies on cases of environmental negligence is common as communities have become better informed and the courts have become sympathetic to views of such communities.

6 CONCLUSION

The paper has shown that self-regulation is not a very common approach in the Zimbabwean mining industry.

To improve their image and ensure sustainable use of resources, Zimbabwean mining companies have to be proactive and put in place policies to ameliorate the adverse environmental effects of minerals extraction. As a first step companies have to strive to comply with minimum standards as a process of best practice management. Compliance with national standards is an essential component of the application of best practice. In this approach, all relevant standards, codes and guidelines as stipulated in the relevant laws should adhered to as a minimum requirement in best practice management.

While the legislative framework is an important factor, imposing national standards might not yield the desired results under the present climate, quality management where industry takes the lead has been shown to be the best way to improve environmental management by mining companies in other countries. The application of best practice works extremely well where companies take the lead.

The experience of Western companies has shown that cooperative outcomes-based approach to environmental management delivers better long-term outcomes than command and control approach. Confidence between communities, stake-holders and regulators is paramount in improving overall environmental initiatives. The cases cited in this paper have shown that through networks industry can take the lead in sustainable use of mineral resources. Ecologically sustainable development - using, conserving and enhancing the community's natural resources in such a way that ecological processes, on which life depends, are maintained and quality of life for both present and future generations is increased - is unavoidable in the 1990s. Through ecologically sustainable development, problems related to the finite nature of resources are reduced drastically and inter and intra generational equity is promoted.

By setting internal standards companies carry out qualitative risk assessments of threats to the organisation, control and regulate them. The role of authority in this approach is to ensure that the responsibility is carried out.

Through government policies according favourable tax concessions to complying companies, mining companies and individuals could be encouraged to move beyond best available technology (BAT) to best possible technology (BPT). Tax rebates on innovation of technology and on importation of BPT would be a way forward. Encouragement through economic incentives such as differential taxes on environmentally sound products and technologies will achieve better results. Economic incentives are often regarded as a fool-proof fall back position in environmental issues. Operationally it may not be that easy to manage such a system in the developing world. Rather than opt for a fully market driven approach, the Zimbabwean government should experiment with a form of a 'carrot and stick' to entice companies to invest in BPTs. The use of bonds is common in many countries around the world.

Once governments encourage better performance that enables companies to go beyond standards, better results will be achieved. Industry has necessary technical and management capacity and business incentive to do so. The biggest challenge in Zimbabwe relates to policies to deal with contamination emanating from the activities of small scale mines.

7 REFERENCES

Sinding, K., (1999) Environmental Impact Assessment and Management in the Mining Industry, *Natural Resources Forum*, 23, 57-63.

Ostensson, O., (1998), UNCTAD: Commodity Resources – Sustainable Development. *Journal of Mineral Policy, Business and Environment, Raw Matetials Report,* 13(2), 34-40.

UNCTAD (1993), *Mineral Resources and Sustainable Development*. TD/B/CN.1/16

Auty, R.M., (1998), Mining as a Generator of Wealth: Potential Conflicts and Solutions. *Journal of Mineral Policy, Business and Environment, Raw Matetials Report*, 13(2), 4-12.

Hodge, I., (1995), *Environmental Economics*. London:Macmillan Press Ltd.

KCGM ,(1998), *Annual Environmental Report*.

WMC Resources ,(1997), *Annual Environmental Report*.

Pearce, D.; Barbier, E.; and Markandya, A. (1990), *Sustainable Development: Economics and Environment in the Third World*. London: Earthscan Publications

Ministry of Environment and Tourism (1994) *Environmental Impact Assessment Policy*, Harare:Zimbabwe.

Ministry of Mines, Environment and Tourism (1998) *Draft Environmental Management Bill – Working Document*, Harare:Zimbabwe.

APPENDIX 1: EMS ISSUES FOR A GOLD MINING COMPANY

i. Cyanide management in tailings storage areas
ii. Acid mine drainage
iii. Groundwater effect due to seepage from tailings
iv. Impact on biodiversity
v. Water and energy conservation
vi. Research on behaviour of cyanide tailings
vii. Monitor, remedy acid mine drainage
viii. Implement best practice design
ix. Minimise clearance and rehabilitate land
x. Pursue eco-efficiency
xi. Exploration
xii. Impact on bio-diversity
xiii. Water and energy conservation
xiv. Community concern about impacts
xv. Minimise land clearing through best-practice exploration and rehabilitate land
xvi. Pursue eco-efficiency targets
xvii. Develop resource conservation strategies at project planning stage: community and government consultation and social impact assessment and community development programs
xviii. Conduct EIAs for proposed operations

APPENDIX 2: ACTIVITIES REQURING AN ENVIRONMENTAL IMPACT ASSESSMENT

1. Agriculture
2. Dams and man made lakes
3. Drainage and irrigation
4. Forestry
5. Housing developments
6. Industry
7. Infrastructure
8. Mining and quarrying
9. Petroleum
10. Power generation and transmission
11. Tourist, resort and recreational development
12. Waste treatment and disposal
13. Water supply

APPENDIX 3: PROCEDURE FOR PROJECT REQUIRING EIA REPORTS

A developer shall, before implementing any project for which an environmental impact assessment is required under subsection (1), submit to the Minister, a prospectus stating –

(a) a description of the activity;
(b) the activities that shall be undertaken in the implementation of the activity;
(c) the likely impact of those activities on the environment;
(d) the number of people to be employed for purposes of implementing the activity;
(e) the segment or segments of the environment likely to be affected in the implementation of the activity;
(f) such other matters as the Minister may in writing require from the developer or any other person who the Director reasonably believes has information relating to the project.

Environmental Issues and Management of Waste in Energy and Mineral Production, Singhal & Mehrotra (eds)
 ISBN 90 5809 085 X

A new proposal for the Brazilian mining environmental regulation system

R.O. Neto
Minerar Consulting and Projects Ltd, Brazil

J.C. Koppe & J.F.C.L. Costa
Mining Engineering Department, Federal University of Rio Grande do Sul, Brazil

ABSTRACT: This paper introduces a new proposal for the environmental legislation, which regulates the Brazilian mining operations. The system in use is inefficient, requiring adjustments to ensure better results for both the community and industry. This study is structured as follows: i. evaluation on the bureaucratic structure, including documentation procedures as well as their analysis; ii. preparation of a reference term to be followed in the project; iii. definition of an evaluation criteria and proposition of taxation incentives for an efficient environmental management; iv. evaluation among governmental departments which one is the most adequate to deal with environmental licensing. Positive results are expected from these propositions, as the whole process will be more agile, creating incentives to the miner implement the mitigation measures.

1 INTRODUCTION

After more than fifteen years of existence, the practical results on the Brazilian environmental legislation are still very incipient, mainly in the mining sector. The bureaucracy in the system is the main reason as it is inoperative and inefficient, leading into a state of inertia (Antunes 1993). Paperwork has become an end in itself. The implementations of the mitigating and compensatory measures, essential to the maintenance of the environmental balance, have been left behind. The necessary solutions to this problem are suggested in what follows.

2 BUREAUCRATIC STRUCTURE

It is suggested a simplification in documentation and the analysis of the required legal documents in order to speed up the environmental license issuing. The change implies the centralization of analysis by only one governmental institution (ITGE 1988). Presently, an entrepreneur has to submit the forms to two different institutions: a mining one (DNPM – National Department of Mining Production) and to an environmental one (FEPAM – State Foundation of Environmental Protection), and both have no connection with each other.

We suggest that Mining Department centralize the procedures creating an internal environmental commission.

The following tables and figures display the differences between the present system and the new proposition.

Table 1 – Stages for Mining and Environmental Licensing.

STAGE	DESCRIPTION
1	Submit application to DNPM and Previous License application (LP) to FEPAM
2	FEPAM issues LP
3	Plant Licensing (LI) application
4	FEPAM issues LI
5	Turn LI over to DNPM
6	DNPM issues mining authorization
7	Turn DNPM authorization over to FEPAM
8	FEPAM issues LO

Note: LP is a previous environmental license; LI is an environmental license for the mining installation; LO is an environmental license for the mining operation

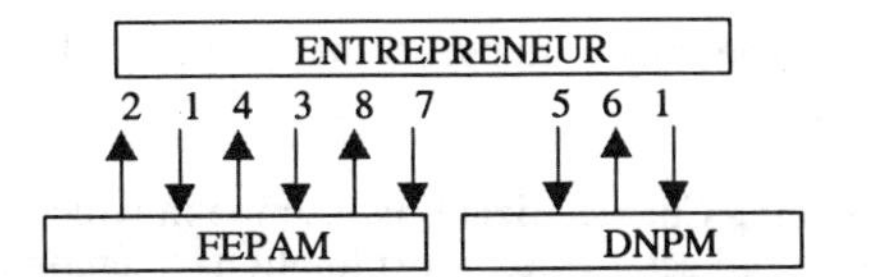

Figure 01- Chart with the stages for Mining and Environmental Licensing.

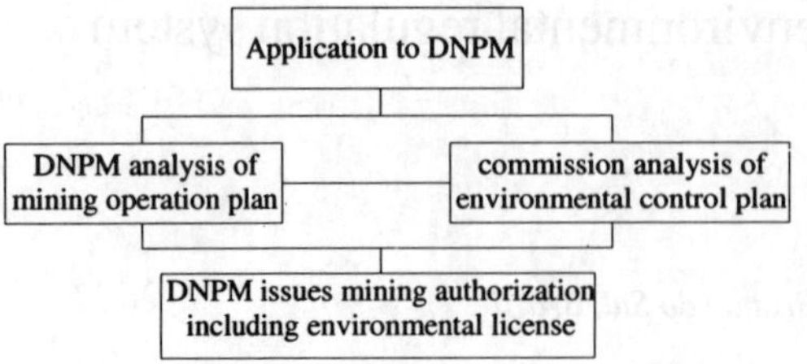

Figure 02 – The proposed proceedings for environmental mining legislation.

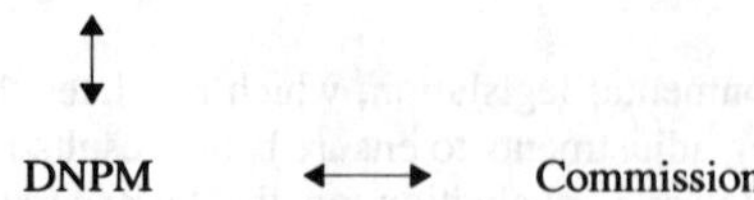

Figure 03 - The stages in the new proposition.

In this new way, the company receives a mining authorization including the environmental license issued by a single institution. The process is simplified and there is no need to apply to two independent institutions. Authorization issuing will be obtained faster.

The environment regulation institution must process the analysis faster to meet the company's legally established deadlines. It is suggested that the institution follow a 30 days limit to process the application.

3 REFERENCE TERM FOR PREPARING MINING ENVIRONMENTAL STUDIES

The proposition for a reference term involves a technical regulation to environmental work whose objective is to define the items to be presented and the detail level. The items in the report would be prepared according to the Reference Term. The form should address the particularities of each kind of deposit and mine size (Allister et al. 1994).

Table 2 displays the basic changes proposed for the Reference Term (Oliveira Neto 1999), describing the main advantages of their implementation.

The main items at the proposed Reference Term are:

i. Mining Characterization: describe in which step is the project, i.e. exploration, engineering, implementation, or operation. Also details about the ore, mining operation method, plant operation system, monthly production, and the company's main objective.

Table 2 -The basic proposition for the Reference Term and the related advantages.

PROPOSED CHANGE	ADVANTAGES
• Establishment of the so-called Environmental Plan;	• The analysis carried out by the institution consultants and technicians will be simplified;
• Regulation of the forms items;	• It will simplify the inspection of the implementation of measures;
• Mining plan with the associated environmental measures;	• Efficient planning;
• Cost estimates of the environmental measures.	• Cost control elements to the miner.

ii. Influence area: definition of area direct and indirectly influenced by environmental impacts.

iii. Environmental diagnosis: the environmental context where the area is inserted, i. e. physical environment, biotic environment and anthropic environment (Kopezinski 1997).

iv. Environmental impact assessment: identification and characterization of the negative impacts on the environment. The impacts may be classified into main, relevant, potential (high risk) and secondary (low risk). Positive impacts must also be reported. The description should contain physical, biotic and anthropic aspects.

v. Mitigation and compensation measures: each identified impact must contain its associated mitigation or compensation measure(s), described as follows:

- Suggested solution;
- Which objective or the impact it refers to;
- The location of compensatory or mitigation measures;
- Details of how the solution must be implemented;
- A time table to implement the solutions;
- Association of the solution adopted with the mine planning.

Tonso (1996) proposes a connection between mining operations and the environmental measures. The following items are recommended to be included:

vi. Reclamation of damaged areas: this item should describe the steps in implementation of environmental measures after mining ceases. The land should be used similarly as before or may be suitable for a different future utilization, making the land surface steady and preserving water quality.

vii. Cost estimation: to establish a monthly or yearly budget related to environmental measures to be added to the operational expenditures.

viii. Illustrations to be attached: to enhance the visualization and understanding of the measures.

The level of importance of each item were classified in three levels as follows:

- Level 0 (unnecessary), its presentation is not required;
- Level 1 (necessary), it must be presented briefly;
- Level 2 (indispensable), it must be detailed described. The agency should refuse a project where this item is omitted.

Tables 03 and 04 summarize the use of the reference term for various types of mines and materials.

Table 03 – Level of detail for the reference term item 1 and 2 suggested for surface mining.

SURFACE MINING					
Crushed rock	Coating materials	Sand	Limestone	Metals	Coal
level 2	level 2	level 2	Level 2	level 2	Level 2
level 1	level 0	level 0	Level 1	level 2	Level 2

Table 04 – Level of detail for the reference term item 1 and 2 suggested for underground mining.

UNDERGROUND MINING		
Metals	Coal	Gemstones
level 2	Level 2	Level 2
level 2	level 2	level 2

4 EVALUATION CRITERIA AND INCENTIVES

It is required a criteria to audit the environmental performance of a given company. Such proposition was described in what follows was adapted from the Australian system presented in the *Environmental Protection Regulation 1995 – Environmental Performance Criteria* (DME 1995).

This proposition is simple and based on the comparison between the programmed and the executed. Each stage has its performance level measured that would correspond to a certain tax discount offered to the company, named Financial Compensation for Mining Industry.

The environmental management levels can be established under the following criteria:

Stage 01: the stage of preparation including environmental control equipment acquisition, seedling purchase, land preparation, location of preservation areas, etc.

Stage 02: starting the implementation of main measures.

Stage 03: implementation of 50% of all previously established environmental measures.

Stage 04: actual implementation and operational stage of the measures in level 03.

Stage 05: implementation of the remaining measures (100%).

Stage 06: operational stage of all implemented and previously established measures.

As a final proposition, the following tax incentives can be established as a result of the environmental management performance evaluation. The proposition still requires further investigation in case it receives acceptance by the community.

Stage 01: 05% tax incentives;
Stage 02: 10% tax incentives;
Stage 03: 15% tax incentives;
Stage 04: 20% tax incentives;
Stage 05: 25% tax incentives;
Stage 06: 30% tax incentives;

These incentives represent the government counterpart in the environment reclamation process.

5 CONCLUSION

The change in the bureaucratic structure and in the governmental institution inter-relations (such as DNPM and FEPAM) unifies both environmental and mineral licensing. It simplifies the procedures as well as documentation establishing deadlines that the institutions should follow, making the necessary actions public and faster.

The environmental studies and reports must highlights essential points, such as the cost-benefit one, making them more objective and applicable by the mining industry. The basic rules can be obtained from a reference term, which includes the priorities to every kind of mining.

The procedures and criteria to be established for an effective environmental management in mining industry provide a permanent assessment along with incentives to the entrepreneur.

6 REFERENCES

Antunes, R., 1993. Between the inoperative and inefficient, *Ores Magazine*, n°182, p. 26-28. Brazil.

Allister, L., Beil S. & Cox A., 1994. An Economic and Public Policy Perspective on Mine Site Reclamation, *in Proceedings of the Third International Conference on Environmental Issues and Waste Management in Energy and Mineral Production,* p. 21-24,Perth, Western Australia.

DME-Department of Mines and Energy, 1995. *Environmental Management for Mining in Queensland*: p. 1-22, Austrália.

ITGE - Institut Technologic Geomining of the España, 1988. *Environment and Mining: GeoEnviromental Engineering*, Madrid: p. 1-15. Span.

Kopezinski, I., 1997. *Physical evaluation of the damage areas in mining*, Class Notes (SGS 803): p.7-96, Geotecnia Department-USP, São Carlos, SP, Brazil.

Oliveira Neto, R., 1999. *Avaliação do sistema de licenciamento ambiental vigente na mineração. Uma nova proposta de metodologia e procedimentos(Environmental mining legislation evaluation. A new proposal)*. Master dissertation, Federal University of Rio Grande do Sul, Porto Alegre, Brazil, 112 p (unpublished).

Tonso, S., 1996. *Reclamation start in the mining plan*, Ores Magazine: p. 31-32 São Paulo, Brazil, n. 209.

Environmental Issues and Management of Waste in Energy and Mineral Production, Singhal & Mehrotra (eds)

Evaluation of ecological consequences of coal mine closure in Kuzbass coal region

E.L.Schastlivcev, L.P.Barannic, B.I.Ovdenko & A.A.Bykov
Kemerovo Scientific Center of Siberian Branch of Russian Academy of Sciences, Russia

ABSTRACT: Kemerovo region (otherwise called Kuzbass) is the most industrially developed and urbanized region of Siberia, Russia. The main industrial branch of Kuzbass is coal output. Open pits and underground mines of Kuzbass produce about 40% of total amount of coal in Russia and more than 70% of coking coal. In the current process of the coal industry's restructuring, the closing of many unprofitable coal enterprises is associated with radical changes in their influence on the environment. The task to provide a probable forecast of ecological consequence of mine closure is both practically significant and complicated. In order to find some scientific approach to solve named problem the authors made in the paper the first attempts to analyze of accessible closed mines data in Kuzbass, to classify coal mines (working and closed) with respect to there negative influence on soil, water and atmosphere and to obtain some numerical estimates of possible bounds of this influence.

INTRODUCTION

In the State Report " The currant condition of the Environment of the Russian Federation" Kuzbass is characterized by such ecological problems as: soil destruction by mining development, pollution of soils, loss of fertile soils, exhaustion and pollution of surface and underground waters, atmospheric pollution and degradation of forestry. Coal output industry makes a significant contribution in this problem.

There are 71 coal enterprises in Kuzbass and this number is added form 47 underground mines and 24 open pits. The middle depth of underground mines workings is about 300 m and the maximum depth attains up to 600 m. The sum annual amount of produced coal along last five years is slightly eliminated and is approximately equal to 95 millions ton.

Wile working a coal mine has a rather large number of direct sources that cause local and regional industrial loading on soil, water and atmosphere. The permissible limits of these sources emissions and amount of hard wastes are established at design stage and recurrently verified by Territorial Environmental Authorities. An enterprise is responsible for validity the established limits and could be subjected to financial penalty for there excess.

At the present time 15 underground mines are at the different stages of closing and 10 are preparing to this process. With the closure of mines, of course, the most part of direct sources of soil, atmospheric and water pollution terminates their emission. In just the same way the State environmental control is canceled. But it does not mean, that the negative influence of closed or conserved coal enterprises on the environment come to an end completely. Moreover, it must be taken into account that the negative influence will intensify in the future at greater rates.

The authors are going to present below some information relating to coal mining's influence on different natural layers (soil, water, atmosphere) of Kemerovo region and interdependency of these influences. As far as possible we'll try to emphasize the happened and expected changes of influences after main closure.

SOIL

The first and main negative effect of coal output is a violation of the geological environment. From the start of the intensive exploitation of coal and other deposited fossil fuels about 100,000 hectares have been destroyed throughout Kemerovo region. These soils are situated in densely populated areas, and represent in a number of cases up to 15-20% of the urban territory. The South portion of Kuzbass that is a territory of the most intensive coal output is shown in Figure 1. The areas of destroyed soils surround

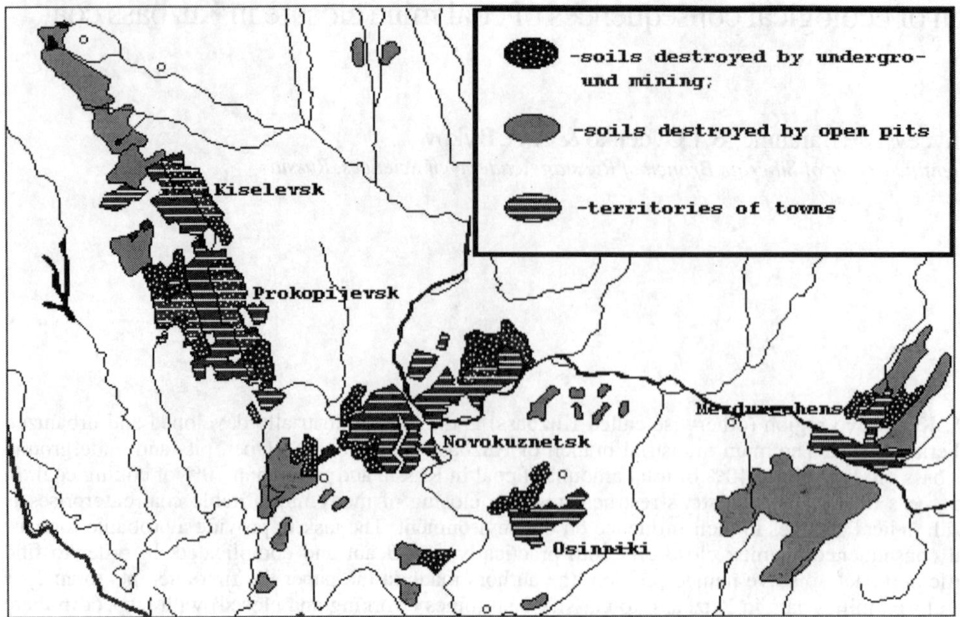

Figure 1. Fragment of the Kemerovo region's map of destroyed soils around south industrial towns. One can see that some areas of soils are destroyed by underground mines are intersected with urban territories.

areas of industrial towns. Moreover, in the cases of Kiselevsk and Prokopijevsk underground mines workings lay under the residential sites and restrict multistory buildings. Note, that for this paper the figures 1 and 4 were obtained by using GIS-system "Ecological atlas of Kemerovo region" (Bykov et al. 1996) and figures 5 and 6 with the help of atmospheric dispersion program ERA-1.0 (one can find more information at the WWW-page www.nsk.su/~lp).

The coal mine closure could be an impetus of the noticeable increase of the surface area of violated lands (approximately by 30 thousands hectares). Lifted to the surface of the land the huge masses of deeply mined rocks (about 8 cubic km) sooner or later will spread along the adjoining territory, polluting fertile lands with heavy metals, sulfurous and many other toxic oxidation products, that are dangerous for the biosphere and the population. Formed at other geological epochs, deeply mined rocks stand for a substratum alien to the existing biosphere, so they will inevitably change the nature of the established biogeochemical processes. Besides, migration of elements and compounds of these mining rocks, will, perhaps, transform the local geochemical changes to regional ones. In this connection it is necessary to effectuate geochemical monitoring of all violated territories, particularly of the closed or conserved coal enterprises.

Within the last years the work on the remediation of violated lands, which was conducted rather unsatisfactory before, now has practically been stopped. There are several reasons for that: a general deterioration of economic conditions of coal enterprises; imperfection of legislative and normative base in the field of land use and remediation. Yet the main reason is the ignorance of economic mechanism of nature use at the design of new mining enterprises and the reconstruction of the existing ones according to the currant coal industry's restructuring. Exactly this reason could increase afterwards an ecological risk of projects in coal mine closure, reduces a cost-performance of existing enterprises. Notwithstanding the imperfection of the normative and legislative basis, the incorporated and acting standards of payments for violated lands and storage of hard waste are capable to reduce a general cost-performance of coal mining, if the cardinal principles of nature use are ignored. On the contrary, by taking account of standard payments at the stage of design works and selecting technological schemes of stripping and coal mining, they are to ensure the reduction of the risk of the ecological investment, providing an increase in the economic effect on account of minimization of the damage to the environment. In this connection we think that the guideline of passage to balance and rational land use, remediation and recovering efficiency of lands, is an improvement of the economic mechanism of land-use and rational utilization of coal industry waste. For this not only the improvement of economic penalties is necessary, but also the improvement of the privilege and bonus systems of rational land-use; which is now lacking completely in the state legislative base

WATER

Significant drain's change of small rivers and depreciation of underground waters determine the impact of coal output on water regime of the geographical territory. The daily spillway of underground mines and open pits in Kuzbass makes up more 1 million m^3 of water that is pumping from workings. The drainage of deposits leads to the reduction of water stores, to drying up of wells with drinking water and so on. About 200 small rivers either disappeared completely or shortened their length throughout all the territory of mining works in Kuzbass. The total length of hydrographic net became shorter by 365 km. As a result the areas of approximately 3000 hectares is now transforming into a dried up zone with degraded plant community.

Coal mining industry on the whole in Kuzbass is a water pollution source of the second significance after municipal industry. Water escapes that are produced by coal production takes 28% of the total amount of crude wastewater in the region (SECKR 1999). In this connection water's volume emitted without cleaning from underground mines much more than one from open pits. Moreover, a qualitative analysis of wastewater (SECKR 1999) displays that at the underground mining the water's wastes are more polluted that at the open coal output. In particular, suspended solids in 2.5, oil productions in 3, phenol in 10, chlorides in 4 times respectively.

Figure 3. Outcrop of underground water onto land surface near flooding coal mine. 2 years after stop pumping.

Coal mines closure by using flood causes further negative influence on water regime of the nearby geographical territory. A rising of the underground water's level is detected in a number of cases when flooding workings is located in low place closure to small rivers. It leads to water filtration through surface level of soil into basements of industrial and dwelling houses that could be a cause of building

Figure 2. An example of damaged building near the closed underground mine. 3 years after mine closure.

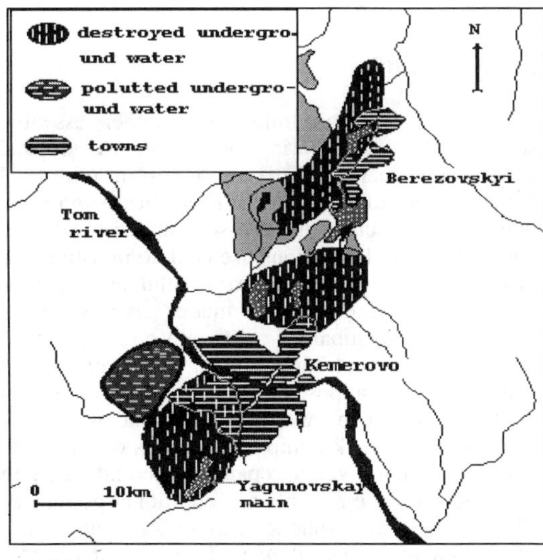

Figure 4. Destroyed and polluted underground water around Kemerovo

damage. Described effects take places for flooding coal mines around Novokuznetsk, Belovo and Kemerovo. An example for a mine of Belovo region one can see in Figure 2. As a usual in this cases boggy sites are formed nearby. In particular, in the effective area of the flooding coal mines near Kemerovo the total square of new boggy lands is equal to 12 hectares. A pattern is shown in Figure 3. In a year this area will be occupied with boggy plants and will become unfitness for farming or another purposes.

The next inauspicious item of mines closure is a high probability of further spreading of underground water pollution. Constant water pumping from workings in an operating mine is a cause of forming water table depression. The bounds of such areas around Kemerovo are shown in Figure 4. Once water pumping stops the mine will fill with surrounded water until the natural level is re-established. The first, water will be contaminated by mine underground wastes and by remains of mine equipment. The second, in the industrially developed region there is a high probability of flooding workings with water that is polluted by another enterprises. For example, Yagunovskaya main (closed 2 year ago) was situated to the South of Kemerovo (Figure 4). And to the East is a group of large chemical plants that are a source of essential pollution for not only atmosphere and Tom River, but also for underground water. And now some chemical pollutants, that can not be fully cleaned by filtration through land, are forthcoming into depressive zone and leads to deterioration of drinking water in boreholes of the some part of Kemerovo and the settlement near former Yagunovskaya main.

ATMOSPHERE

An operating coal mine is a rather essential source of atmospheric pollution. A typical underground main in Kuzbass has from 1 up to 3 boiler stations of divers capacity for heat supply not only coal production but also surrounded residential area. All the boiler stations use coal combustion and produce standard atmospheric pollutants such as NOx, SO_2, CO and particular matter (coal ash). Coal mining is accompanied with gaseous pollutant emission from ventilation shafts. The main part of emitted gas is composed with methane (Tailakov et al. 1999). When the weather is dry and windy the coal stores and rock dumps are sources of particular matter. Auxiliary works, repair service and transport machines produce a number of additional pollutants. The survey of atmospheric emissions inventory data of Kemerovo mines allow to find the intervals of possible annual emission rates for the prevalent pollutants. These intervals are given in Table 1.

Table 1. Possible bounds of annual emissions in atmosphere for underground mines in Kuzbass

Pollutant	Annual emission (ton)	
	min	max
NOx	14.5	372
SO_2	32	315
CO	138.2	1048
HF	0.0008	0.007
CH	0.005	1.472
Methane	20	28,000
MgO_2	0.004	0.049
Pb	0.00001	0.00037
Soot	0.00004	0.00052
Particular matter	74	1673

Atmospheric pollution with NOx, methane and particular matter could be more than Russian air quality standards (AQS). For example, graphical results of prediction short-term maximum ground-level concentration of methane emitted from ventilation shafts of Pervomaiskaya main (near Berezovskyi town) is shown in Figure 5. Inventory data are presented in Tailakov et al. 1999. Results of finding the maximum concentrations field are obtained by using program ERA-1.0 (mentioned above) on the base of Russian regulatory atmospheric dispersion model (Berlyand et al. 1974) with screening technique along wind speed (0.5-10 m/sec) and wind directions toward residential area.

Once a mine has closed a number of atmospheric pollution sources stop their emission. The first, these sources are boiler stations, except of a case when

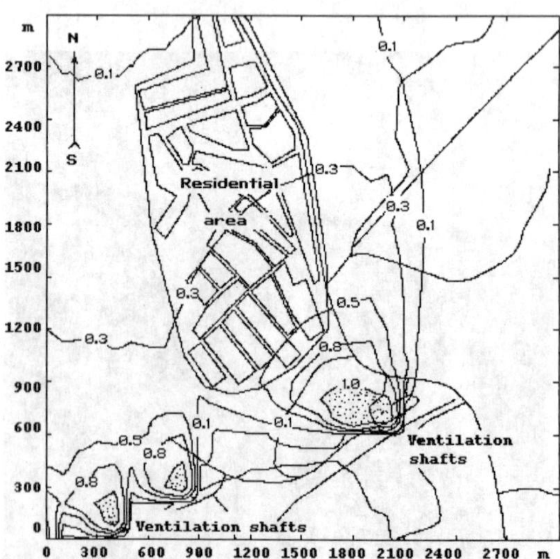

Figure 5. Isolines of sort-term ground-level concentration of methane emitted from ventilation shafts. Russian AQS for methane is 50.0 mg/m^3. Initial gas velocity equals to 5 m/sec.

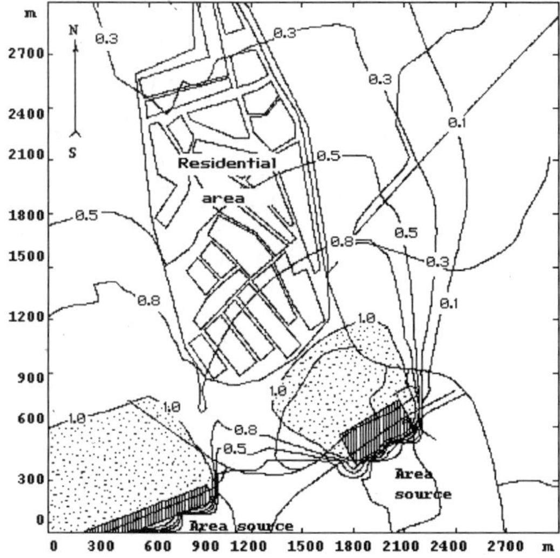

Figure 6. Possible atmospheric pollution with methane under the same emission rates that in Figure 5 but in the case of disorganized area sources with vertical gas velocity 1 m/sec. Areas with pollution more than AQS are dotted.

some of them remain for heat supply of residential settlement. The second are coal stores, sources of coal transport and reaper service. Furthermore the surface fans, which extract gas and air from the workings through ventilation shafts, are turned off and the methane emission terminates for a while. At the same time the water pumping stops and the mine workings will fill whit water. It leads to increase in pressure and in some interval mining gas could escape from old workings into atmosphere. Emission mechanisms and pathways of gas forthcoming are considered in details in Burrell 1999. It should be noted that ventilation shafts (when fans are working) could be considered as point sources with stack height about 5 m and rather high initial gas velocity (6-10 m/sec). Gas emission from closed mine originates in conditions that are rather more unfavorable from atmospheric dispersion point of view. Under the influence of water rising main gas can be forthcoming into atmosphere through several random sources throghout all the area over the workings with low vertical velocity. Figure 6 shows the difference (with figure 5) in the field of concentrations of methane emitted uniformly from ground-level area sources instead of elevated point sources with high vertical velocity. The sum emission rates in both cases are equal. It should be noted in addition that, according to Russian sanitary norm, methane is not a gas of high dangerous with respect to human health (AQS for methane is 50 mg/m^3 in comparison with AQS for SO_2 equals to 0.5 mg/m^3). But it could be revised under the last international requirements (Schults 1999).

CONCLUSION

The above preliminary results of investigations on coal mine closure, that have been conducted in the laboratory during last 2 years, reflects that the authors are only at the stage of data collection and there understanding. But the basic aim is to provide a methodology and some forecast system for getting a priory estimate of ecological consequences of coal mine closure. To research this problem it is necessary to set up the environmental monitoring of the geographic areas that are destroyed and polluted by coal enterprises (Gritsko et al. 1998).

REFERENCES

Burrell R. 1999. Recent experiences of surface gas emissions in UK coalfields. *Proc. of the Inl. Conf. Regional environmental problems of coal mining industry in the transition to sustainable development*: 174-188. Kemerovo: Kuzbassvuzizdat.

Bykov A.A, Pushkin S.G. & Schastlivtsev E.L. 1996. Using of GIS-technology for long-term environmental planning in Kuzbass coal region. *Proc. 26th Inl. Symp. on Computer Application in the Mineral Industries.* 497-501. The Pennsylvania State University. Littleton: Society for Mining, Metallurgy and Exploration, Inc.

Berlyand M.E. et al. 1974. Basic principles of determination of atmospheric pollution in the USSR. *Proc. Techn. Conf. on Observ. and Measurement of Atmos.Pollution. WMO No308. Helsinki 1974* : 125-133. Geneva: WMO.

Tailakov O.V., Polevshikov G.Y., Bykov A.A. et al. 1999. Economical and ecological assessment of coal mine methane recovery and utilization project in Kuznetsk coal basin of Russia. *Proc. of the Int. Coalbed Methane Symp. May 1999:* 439-447. Tiscalusa: The University of Alabama.

Schults K.H. 1999. Impact of coal mine methane on regional ecologies and economies: exploring opportunities in the Kuzbass. *Proc. of the Inl. Conf. Regional environmental problems of coal mining industry in the transition to sustainable development*: 188-201. Kemerovo: Kuzbassvuzizdat.

State Environmental Committee of Kemerovo Region (SECKR). 1999. *Yearly report on protection of the environment in Kemerovo Region.* Kemerovo: State Environmental Committee.

Gritsko G.I., Schastlivtsev E.L., Bykov A.A. et al. 1998. Basic principles of working out distributed cadastral survey of Kemerovo region natural resources on the basis of GIS technologies. *Computational Technologies J.* Vol.3. No5: 28-36. Novosibirsk: Siberian branch of Russian Academy of Sciences.

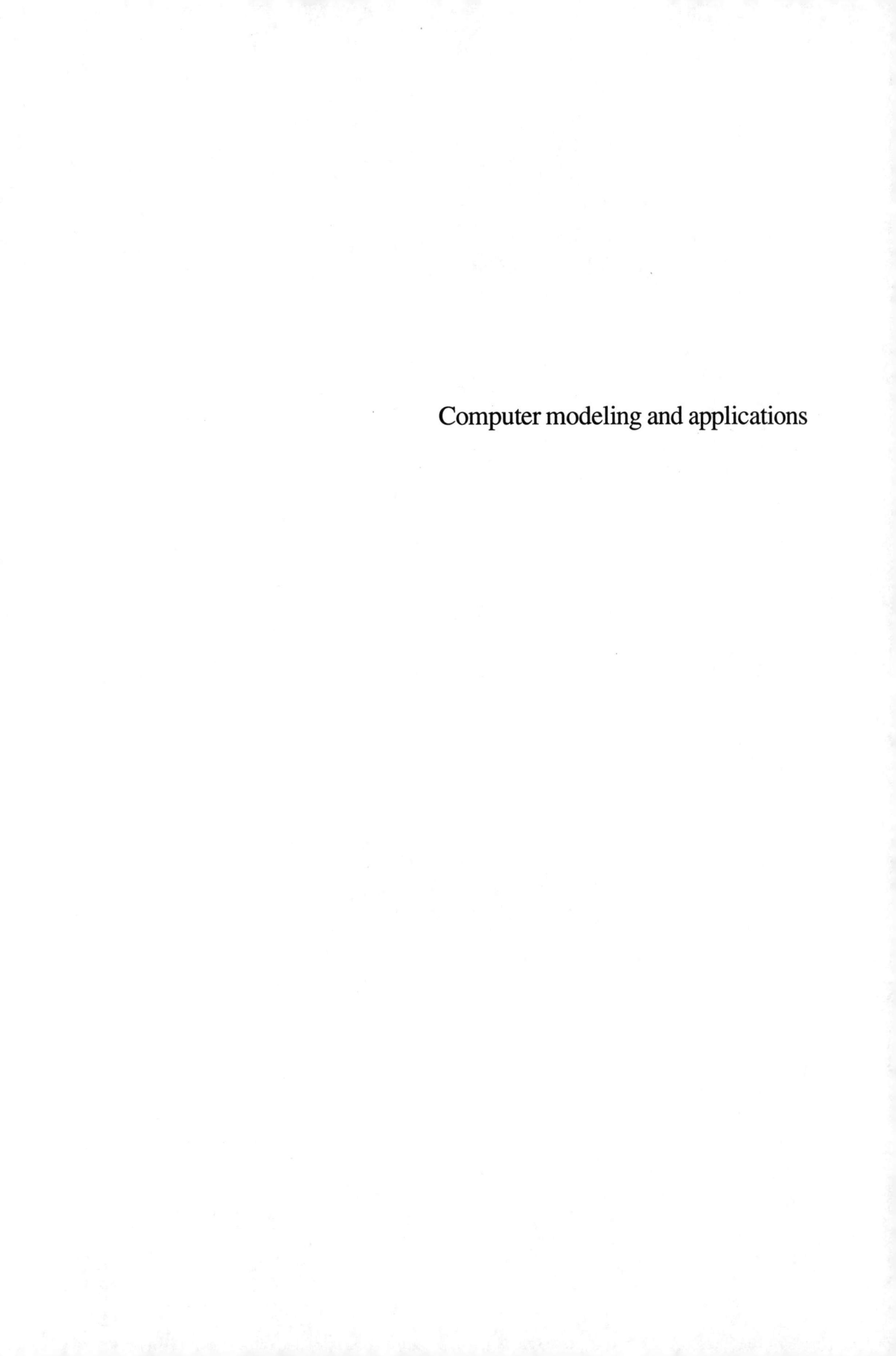

Computer modeling and applications

Environmental Issues and Management of Waste in Energy and Mineral Production, Singhal & Mehrotra (eds)
© 2000 Balkema, Rotterdam, ISBN 90 5809 085 X

Determination of preventive maintenance schedule using statistical process control

H. Ankara, H. Ipek & A. Konuk
Osmangazi University, Eskisehir, Turkey

ABSTRACT: The jump in levels presented in the control charts occurred due to the annual preventive maintenance schedules. However, after four months the plant went back to its level prior to preventive maintenance. This suggests that the use of preventive maintenance in four months period is appropriate. This was also supported by the stratification at the end of the first four months period. One week of four preventive maintenance periods per year can be better than four weeks of one preventive maintenance period

1 INTRODUCTION

By planing the maintenance schedule in the mining industry, the mathematical models developed for analyzing mining systems are divided into two techniques; deterministic and non-deterministic techniques. The accurate modeling of most mining cases in non-deterministic techniques, results in complex models with no direct analytic solution, such as, stochastic and non-stochastic models. Therefore, the deterministic techniques are generally employed as a practical choice of representing the true features, such as simulation and quality control (Yingling 1992, Ankara 1997). However, the weakness of the deterministic techniques as tool for system analysis is evident, because the deterministic techniques only give data from which one may infer some of the major properties of statistical distributions (Ankara 1997)

Maintenance actions are divided into two groups; preventive maintenance and corrective maintenance.

Preventive maintenance is used to denote all kinds of repair actions directed towards the improvement of performance characteristics of the systems.

Corrective maintenance is used to denote all total repair actions directed towards discovering the source of the failure and the subsequent repair.

If a production control of mineral processing is based on the current Quality Control data, the distribution of the grades of the produced concentrates is characterized by certain parameters and a relatively small rate of grades. When equipment in mineral processing plants performs abnormally, the rate of grade decreasing relatively. To determine the exact moment of malfunction, samples of concentrates are periodically taken to carry out the necessary measurements. If the problem is determined, the process must be stopped for inspection and possibly for repairing.

Statistical process control aims at examining whether a process proceed in a normal way or not. If there is a abnormality in a process, then it aims at determining the reasons for this abnormality and eliminating these sources.

Control charts are among the most effective means for process control system via statistical methods in an economical and secure way.

In this study, Etibank Emet colemanite plant data were used. In this plant, general maintenance is made generally during the fifth mount and it takes approximately one month.

2 THE QUALITY CONTROL FOR MESURABLE CHARACTERISTICS

For the quality control of the measurable properties of products the following control graphics were used:

- Range
- Standard deviation
- Mean

Each control graph contain a center line (CL) and upper (UCL) and lower (LCL) control limits which can occur randomly around the center line. In the drawing of control charts, the control limits are determined so that they are ± 3 standard deviations away from the central line.

2.1 *Decision on required number of subgroups before control limits are calculated*

The determination of the minimum number of subgroups required before control limits are calculated is a comprise between a desire to obtain the guidance given by average and control limits as soon as possible after the start of collecting data and a desire that the guidance be as reliable as possible. The fewer the subgroups used, the sooner the information thus obtained will provide a basis for action but the less the assurance that this basis for action is sound.

On statistical grounds it is desirable that control limits be based on at least 25 subgroups. Moreover, experience indicates that the first few subgroups obtained when a control charts is initiated may not be representative of what is measured later; the mere act of taking and recording measurements is a sometimes responsible for a change in the pattern of variation.

However, where subgroups are obtained slowly there is a natural desire on the part of those who initiated the control charts to draw some conclusion from them within a reasonable time. This impatience for an answer frequently leads to the policy of making preliminary calculation of control limits from the first 8 or 10 subgroups, with subsequent modification of limits as more subgroups are obtained. (Grant & Leavenworth 1985)

2.2 *Range control charts*

The difference between the largest (X_{max}) and smallest (X_{min}) values of samples obtained with the examplification of a production process is called Range (R).

$$R = X_{max} - X_{min} \tag{1}$$

Charts which are drawn to follow the range are called "Range Control Chart" or "R-Control Chart".

The mean of range is calculated with the following equation.

$$\overline{R} = \frac{\sum_{i=1}^{n} R_i}{n} \tag{2}$$

R-control charts are calculated with the following equation.

$$UCL_{\overline{R}} = \overline{R} + 3\sigma_{\overline{R}} \tag{3}$$

$$CL_{\overline{R}} = \overline{R} \tag{4}$$

$$LCL_{\overline{R}} = \overline{R} - 3\sigma_{\overline{R}} \tag{5}$$

Where $\sigma_R = R_i$ is the standard deviation of the range distribution and its value is unknown. The standard deviation of range is determined with the help of calculated coefficients (D_3 and D_4) depending on the number of sample. These coefficients;

$$D_3 = -3\sigma_R \tag{6}$$

$$D_4 = +3\sigma_R \tag{7}$$

The D_3 and D_4 coefficients were listed at the Table 1.

In this situation, to calculated R-control charts following equations can be used.

$$UCL_{\overline{R}} = D_4 \overline{R} \tag{8}$$

$$CL_{\overline{R}} = \overline{R} \tag{9}$$

$$LCL_{\overline{R}} = D_3 \overline{R} \tag{10}$$

In the use of range mean control charts, it is preferred that the number of observations in subgroup is $n \geq 4$. If the number of observations in subgroup is $n>15$, standard deviation mean charts are used.

2.3 *Standard deviation control charts*

If the number of observations in subgroup is $n>15$, it would be ineffective to study the change in quality with the range. In such cases, the change has to be examined and measured with standard deviation mean chart (S-Control Chart).

$$\overline{S} = \frac{\sum_{i=1}^{n} S_i}{n} \tag{11}$$

With the help of coefficients which were obtained depending on the number of sample, the standard deviation control limits can be determined.

$$UCL_{\overline{S}} = B_4 \overline{S} \tag{12}$$

$$CL_{\overline{S}} = \overline{S} \tag{13}$$

$$LCL_{\overline{S}} = B_3 \overline{S} \tag{14}$$

The D_3 and D_4 coefficients were listed at the Table 1.

2.4 *Mean control charts*

In quality control only the range or control of standard deviation is insufficient, in some cases the mean value should also be controlled. Charts designed for the control of the mean quality level are called "Mean Control Charts" or "X-control Charts"

When the sum of the mean X values divided by the number of subgroups, following equation can be obtained.

$$\overline{\overline{X}} = \frac{\sum_{i=1}^{n} \overline{X}_i}{n} \quad (15)$$

Standard deviation of distribution of means can be obtained following equation.

$$\sigma_{\overline{X}} = \frac{\sigma}{\sqrt{n}} \quad (16)$$

Where σ is standard deviation of population and when its value is not known and in cases where $n \geq 30$, standard deviation of sample can be taken as equal to that of standard deviation of population.

According to the average of means, control limits can be determined the following equations.

$$UCL_{\overline{X}} = \overline{\overline{X}} + 3\sigma_X \quad (17)$$

$$CL_{\overline{X}} = \overline{\overline{X}} \quad (18)$$

$$LCL_{\overline{X}} = \overline{\overline{X}} - 3\sigma_X \quad (19)$$

When the number of sample is less than 30 ($n<30$), standard deviation of sample can not be taken as equal to standard deviation of population so that range is accepted as measurement of variation. In this case control limits can be determined with the help of A_2 coefficient which is calculated in respect to number of sample before.

$$UCL_{\overline{X}} = \overline{\overline{X}} + A_2 \overline{R} \quad (20)$$

$$CL_{\overline{X}} = \overline{\overline{X}} \quad (21)$$

$$LCL_{\overline{X}} = \overline{\overline{X}} - A_2 \overline{R} \quad (22)$$

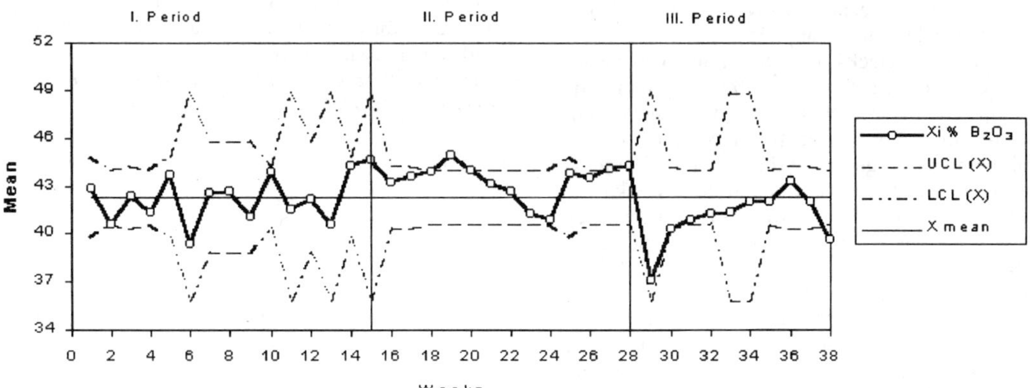

Figure 1. X control chart for the % B_2O_3 of –25+3 colemanite concentrate

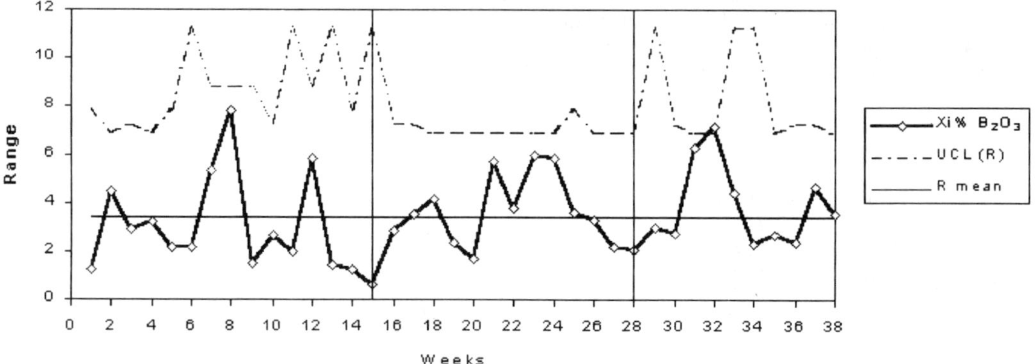

Figure 2. R control chart for the % B_2O_3 of –25+3 colemanite concentrate

Table 1. The coefficients for control charts

Number of Sample	D_3	D_4	B_3	B_4	A_2
2	0	3,267	0	3,267	1,881
3	0	2,575	0	2,586	1,023
4	0	2,282	0	2,266	0,729
5	0	2,115	0	2,089	0,577
6	0	2,004	0,030	1,970	0,483
7	0,076	1,924	0,118	1,882	0,419
8	0,136	1,864	0,185	1,815	0,373
9	0,184	1,816	0,239	1,761	0,337
10	0,223	1,777	0,284	1,716	0,308
11	0,256	1,744	0,321	1,679	0,285
12	0,284	1,716	0,354	1,646	0,266

3 THE ASSUMPTION OF PREVENTIVE MAINTENANCE BY CONTROL CHARTS

In this study, Etibank Emet colemanite plant data were used.. In this plant, -100 + 25 mm, -25 + 3 mm and -3 + 0.2 mm colemanite concentrates are produced by washing screening. Only at –100 + 25 mm additionally hand sorting is applied

-25 + 3 mm (38 weeks) production line was investigated because artifact at the -100 + 25 mm production line and neglectable of capacity at the –3 mm production line. This concentrator is shut off to general maintenance generally during May. The periods of maintenance is approximately four weeks. At the last ten weeks of year low grade concentrates were worked. This results were not taken into concentration.

In this study , upper control limits, lower control limits and center lines of range and mean control charts were calculated. As a subgroups were taken six work day in the week. Results were giving at the Figure1 and Figure 2 respectively .The mean of grades (% B_2O_3) were listed at the Table 2.

As seen following charts, charts can be tree periods. The first period determines control limits and mean of grades before the preventive maintenance, the second period determines control limits and mean of grades after the preventive maintenance and the third period determines back to the first period.

Table 2. The mean grade of % B_2O_3 of plant.

	Feed	-25+3 mm	-100+25 mm
I. Period	33,00	42,07	44,48
II. Period	33,05	43,34	45,20
III. Period	33,10	41,65	44,21

4 CONCLUSION

The plant equipment are always influenced by varying outside factors such as feed, the service conditions, partial repairs, etc. As a result of all these concentrate grades might be decreased

One week of four preventive maintenance periods per year can be better that four weeks of one preventive maintenance period.

REFERENCES

Ankara, H. 1997. Availability analysis for truck-shovel system in surface mining. Ph.D. Thesis. M.E.T.U. Ankara, Turkey.

Ankara, H. $ Bilir, K. 1995. Quality control in krible plant, *Madencilikte Bilgisayar Uygulamarı Senpozyumu.*, Izmir, Turkey (in Turkish)

Gertsbakh, I. B. 1977. *Models of preventive maintenance*. New York: North-Holland publishing Co.

Grant, E.L. $ Leavenworth, R. S. 1985. *Statistical quality control.* Singapore: McGraw- Hill.

Ipek, H., Ankara, H. & Ozdag, H. 1999. The application of statistical process control, *Minerals Eng.* 12: 827-835

Konuk, A. & Onder, S. 1999. *Mining statistics*, Eskisehir, Turkey: Osmangazi University

Yingling, J. C. 1992. Cycles and systems, *Mining engineering handbook.* 1:783-805

Environmental Issues and Management of Waste in Energy and Mineral Production, Singhal & Mehrotra (eds)

CO_2 emission forecast and economic modeling for Alberta under the Kyoto Agreement on global warming

Akihiro Hachiya & Samuel Frimpong
University of Alberta, Edmonton, Alb., Canada

ABSTRACT Green house gases have been a major concern within the international community over the last two decades due to their potential link to global warming. The Kyoto Agreement presents a serious economic challenge to Canada and Alberta. In this paper, the authors use the MRT5 methodology to formulate a comprehensive forecast model to predict CO_2 emissions for Alberta within 1999 and 2012. Economic modeling of CO_2 disposal is carried out based on capacity optimization, capital and operating costs requirements and the fundamental profiles of the Canadian and Alberta economies. The results show that CO_2 emissions in Alberta will increase from current levels of 122 to 215 million tonnes in 2012. The annualized equivalent cost for CO2 disposal ranges between $1.8 and $3.0 billion for the period from 1999 to 2012. The expected cost of disposal within this period is between $56.03 and $64.17 per tonne of CO_2, the energy price will increase from $0.043 per kWh to between $0.045 and $0.0464 per kWh and the disposal unit cost per tonne is between $56.03 .and $64.17.

1 INTRODUCTION

Global warming has been a major concern since the 1980's as a result of the continuous growth in emissions of green house gases (GHG). Canada has made a commitment to reduce GHG emissions by 6 percent below 1990 level by the year 2012 in the Kyoto Agreement [Gunter et al., 1998]. This means a net reduction of 25 % for Canada in 1997. Alberta emitted 28.2 percent of the total CO_2 emissions in Canada in 1990 and 30.5 percent in 1995 with an increasing potential in the future [Jacques et al., 1997]. Available data show that CO_2 emissions constitute about 80 percent of all GHG, followed by CH_4 (about 18 %) for Alberta [Jacques et al., 1997]. Thus, in order to control GHG emissions, appropriate technology must be designed to deal with CO_2 emissions. Sources of CO_2 emissions in Alberta include power generation from stationary fuel combustion (70.9%), mobile fuel combustion (14.8 %), industrial process (12.5 %), and agriculture (1.4 %). Stationary fuel combustion includes power generation, industry, commercial, residential, and agricultural uses. Mobile fuel combustion includes cars, air, rail and marine. Industrial process includes upstream oil and gas, cement and lime production.

Research initiatives are being taken to develop technology for capturing, utilizing and disposing CO_2 emissions [Government of Alberta, 1997]. These disposal technologies include enhanced oil recovery (EOR), coal-bed methane recovery (CMR) technology, CO_2/O_2 recycle combustion, and co-transport medium in pipelines and biofixation technology [Government of Alberta, 1997]. Research has also shown that reduction in CO_2 emissions could be achieved by using long-term CO_2 disposal and storage technology. These disposal and storage options include underground technology (such as, depleted natural gas and oil reservoirs and aquifers) and ocean bed technology [Bachu et al., 1996;Macdonald and Gunter, 1996].

This study focuses on especially the CO_2 emission forecast and economics for long-term disposal and storage of CO_2 emissions in deep aquifers in Alberta based on the Wabamum (TransAlta Utilities). The CO_2 is subsequently liquefied, transported and injected into the Glauconitic aquifer. It makes a significant contribution toward Alberta government's economic policy on global warming. Using the Wabamun results, a scaled model developed is to simulate the total CO_2 economics in Alberta.

2 ECONOMETRIC MODEL OF CO_2 EMISSIONS

2.1 *Mathematical Modeling*

The CO_2 emission model of Alberta is an econometric model designed to look at aggregate CO_2 emissions in Alberta by multiple regression function. While a number of factors contributed to Green-House Gas (GHG), emissions have increased largely due to an increase in economic activities [Jacques et al., 1997]. The rapid worldwide population growth and industrial growth rates have increased CO_2 emissions. Long-run potential economic growth is largely determined by growth in the fundamental determinants of the level of economic output including: labour force, the capital stock and productivity. Furthermore, about 31.7% of CO_2 was emitted from power generation sector in 1995 increasing energy consumption one of the biggest factors affecting CO_2 emissions [Jacques et al, 1997]. In this study the following factors have been chosen as the major determinants of CO_2 emissions:

- previous year CO_2 Emission Rate, x_1
- population Growth Rate, x_2
- industrial Growth Rate, x_3
- energy consumption, x_4
- technological progress, x_5.

The amount of CO_2 emissions increases gradually in the world, in Canada and in Alberta.
The previous year's amount of CO_2 emission will be the basis and CO_2 increases are determined by economic and technical factors.

The CO_2 emissions have increased as population increases [Jaques. el., 1997]. Industries have contributed to CO_2 emissions directly by burning fossil fuels and producing products such as cement, oil and coal. Eeconomy growth implies industrial growth and that results in CO_2 emission growth. Industries contribute about 24 percent of total CO_2 emissions in Alberta other than power generation sector [Jaques. el., 1997]. In Alberta, power generation is the biggest source of CO_2 emissions sharing 15 % of Alberta total CO_2 emission in 1995 [Jaques,et 1997]. Furthermore, 97 percent of electricity is generated by fossil fuel based power stations so that the energy consumption of electricity plays very significant role in the CO_2 emission forecast [Macdonald, Donner and Nikiforuk, 1996]. Technological progress contributes the CO_2 sequestration directly so that exact CO_2 emission forecast is obtained by offsetting this factor.

$$CO2 = \beta_0 + \beta_1 x_1 + \beta_2 x_2 + \beta_3 x_3 + \beta_4 x_4 + \beta_5 x_5 + \varepsilon_i \qquad (1)$$

The quantity of CO_2 emitted per period is *CO2* given as equation (1)

X_1, X_2, X_3, X_4, and X_5 are the respective previous year's emission rate, rates of population and industrial growth, energy consumption and technological progress. βs are the regression coefficients. The mean square error associated with equation (1) must be minimized, as in equation (2), to obtain appropriate CO_2 forecast.

By taking the first derivatives of equation (2) with respect to s , and setting them equal to zero. As a result, we obtain following equations in matrix form as shown as equation (3).

$$\textit{Minimize} \quad L_S = \sum_{i=1}^{n} \varepsilon_i^2 = \sum_{i=1}^{n}\left(CO2 - \hat{\beta}_0 - \sum_{j=1}^{5} \hat{\beta}_j x_{ij} \right)^2 \qquad (2)$$

$$CO2 = \mathbf{X}\beta + \varepsilon \qquad (3)$$

$$CO2 = \begin{bmatrix} CO2_1 \\ CO2_2 \\ \bullet \\ \bullet \\ \bullet \\ CO2_n \end{bmatrix} \quad \mathbf{X} = \begin{bmatrix} 1 & x_{11} & x_{12} & x_{13} & x_{14} & x_{15} \\ 1 & \bullet & x_{22} & \bullet & \bullet & \bullet \\ \bullet & \bullet & \bullet & \bullet & \bullet & \bullet \\ \bullet & \bullet & \bullet & \bullet & \bullet & \bullet \\ \bullet & \bullet & \bullet & \bullet & \bullet & \bullet \\ 1 & x_{n1} & \bullet & \bullet & \bullet & x_{n5} \end{bmatrix} \quad \beta = \begin{bmatrix} \beta_1 \\ \beta_2 \\ \bullet \\ \bullet \\ \bullet \\ \beta_n \end{bmatrix} \quad \varepsilon = \begin{bmatrix} \varepsilon_1 \\ \varepsilon_2 \\ \bullet \\ \bullet \\ \bullet \\ \varepsilon_n \end{bmatrix}$$

The input data for this experiment is shown as Table 1. Columns 2, 3, 4, 5 and 6 are respectively the previous year's CO_2 emission rate, population growth rate, industrial growth rate, energy consumption and technological progress. The data for population growth, industrial growth and energy consumption is provided by Statistics Canada, National Energy Board and Alberta Energy respectively. The population growth rate is the rate of increase of population in Alberta between the first and end of year in percent.

The industrial growth rate expresses the average annual economic growth rate of industry sector in percent. The energy consumption expresses the total electricity use in Alberta in a year in peta joule [PJ]. Technical progress is assumed to be the improvement of the rate of heat efficiency of fossil fuel-fired power plant because that is the biggest CO_2 source in Alberta. It is zero at the year 1988. The rate is estimated based on collected information from IEA (International Energy Agency), Tokyo Electric Power Co. and Trans Alta and discussion with local utility companies. Output data is the six regression coefficients, β_0, β_1, β_2, β_3, β_4 and β_5 as discussed above.

2.2 *Validation and Results from MRT5 Model*

The resulting MRT5 model for CO_2 emission forecast for Alberta is in equation (4). To measure the model adequacy the R-squared is estimated. The R-squared is a measurement of the amount of reduction in the variability of CO_2 emissions (CO2) obtained using regressor variables. As the value is close to 1, the model fits adequately.

$$CO2 = -58738+0.678x_1+4200x_2+1475x_3+44.4x_4+0.678x_5 \quad (4)$$

The R-squared for obtained model is 0.997 and this shows good accuracy of the model. Figure 1 shows actual and projected CO_2 emissions from 1988 to 1996 in Alberta. As shown, projected values almost trace actual data. The projected data is estimated by using equation (4). The input data is completely as the same as data used as input data for multiple regression shown as Table1. The maximum percentage difference between actual data and projected data is 2.25% of actual data for the year 1996.

Table 1 Input Data for CO_2 Forecast Model

Year	Previous Year [kt]	Pop.growth Rate[%]	Industrial Growth Rate [%]	Enrgy Consumption [PJs]	Technical Progress
1988	123500	1.56	0.8	2064	0
1989	124300	1.95	0.8	2109	-0.05
1990	125800	1.99	0.8	2158	-0.1
1991	125994	1.69	0.8	2199	-0.1
1992	126753	1.3	4.3	2258	-0.15
1993	130548	1.44	4.3	2323	-0.15
1994	135861	1.39	4.3	2401	-0.2
1995	141174	1.24	4.3	2474	-0.2
1996	147246	1.23	4.3	2557	-0.25
1997	151041	1.13	2.5	2592	-0.25
1998	2010	1.07	2.5	2704	-0.3
1999	2011	1	2.5	2784	-0.3
2000	2012	0.93	2.5	2861	-0.35
2001	164313	0.88	2.5	2904	-0.35
2002	168532	0.84	2.5	2953	-0.4
2003	170478	0.78	2.5	3021	-0.4
2004	174564	0.72	2.5	3089	-0.45
2005	177180	0.7	2.5	3157	-0.45
2006	181888	0.65	2.5	3224	-0.5
2007	184923	0.6	2.5	3292	-0.5
2008	189790	0.55	2.5	3360	-0.55
2009	192977	0.53	2.5	3427	-0.55
2010	198029	1	2.5	3495	-0.6
2011	203525	0.9	2.9	3563	-0.6
2012	210440	0.89	2.9	3630	-0.6

2.3 *Projection Results*

Projected CO_2 emission in Alberta is estimated by using the obtained model shown in equation (4).

The input data for the projection are given in Table 1. Figure 2 shows the projected CO_2 emissions in Alberta up to 2012. As shown, CO_2 emission gradually increases until 2012. The flat rate indicates Alberta's Kyoto target. At the end of target year 2012, emission rate will become 215Mt and this is 1.7times greater than the rate of 1990. As shown, if CO_2 emission keeps increasing the excess CO_2, will be 102.1 Mt, which is difference between CO_2 emission rate at the year 2012 and the Kyoto target. The capacity of CO_2 disposal requires that amount of storage space at the year 2012.

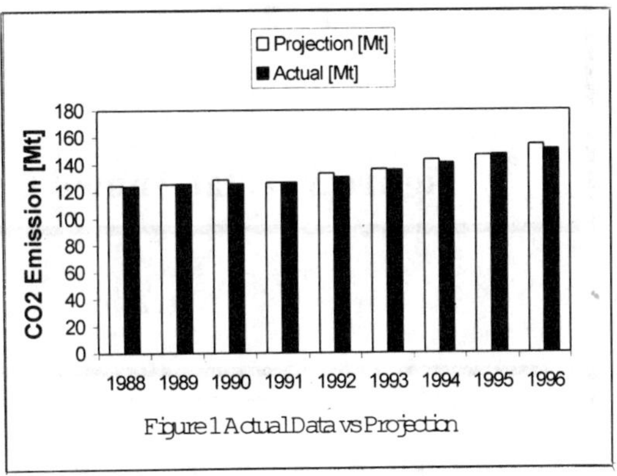

Figure 1 Actual Data vs Projection

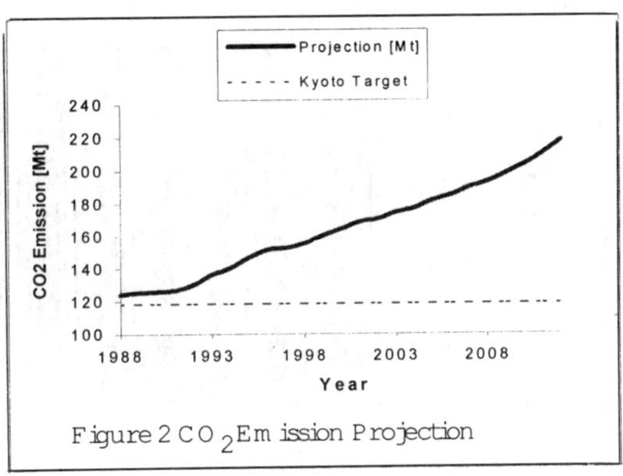

Figure 2 CO_2 Emission Projection

2.4 *CO_2 Sequestration Strategy Models*

In order to achieve the Kyoto target, CO_2 sequestration strategies are required to dispose excess CO_2 in land aquifers in Alberta. According to the Kyoto agreement, Canada must reduce CO_2 emissions up to 6% less than level of 1990, which is about 400 million tonnes for Canada and 118.4 million tonnes for

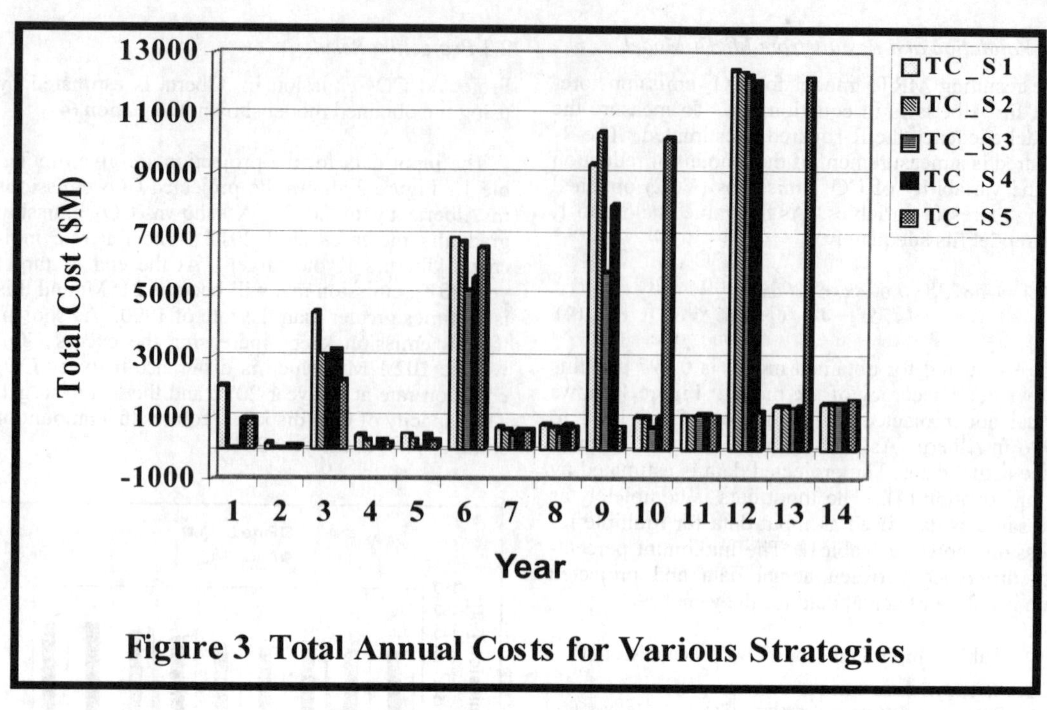

Figure 3 Total Annual Costs for Various Strategies

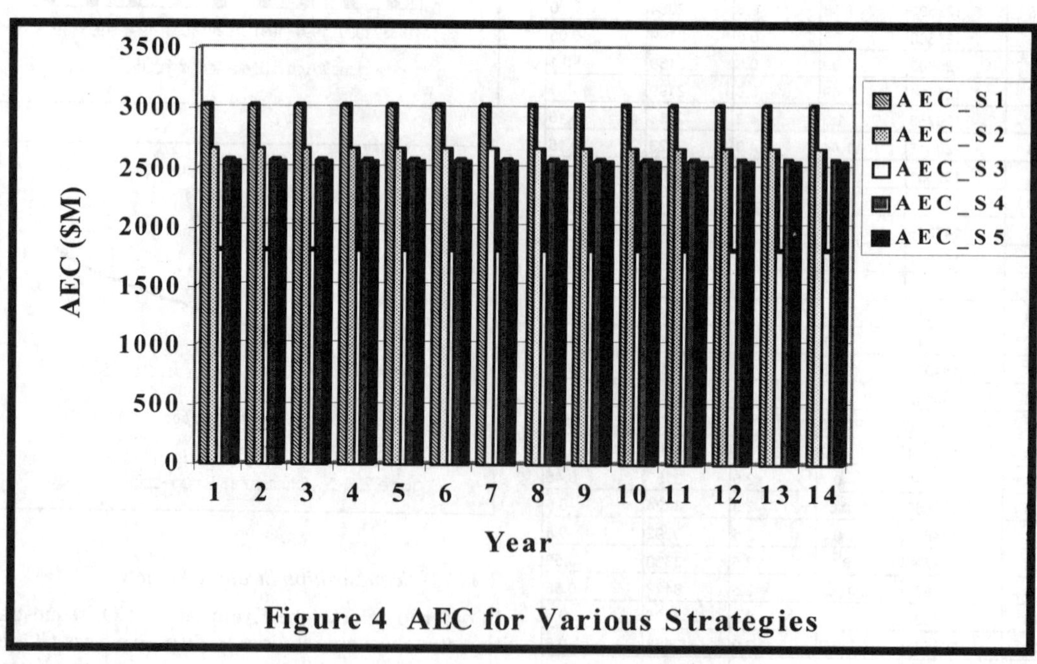

Figure 4 AEC for Various Strategies

Alberta. Here, five possible CO_2 sequestration strategy models are proposed based on the projection results shown as Figure 2.

2.5 *Sequestration Strategy 1*

The first CO_2 sequestration strategy is a gradual reduction strategy from the year 2000. In this model, to meet this Kyoto commitment, CO_2 will be reduced every year gradually from 2000 to 2012. CO_2 emission rate can be estimated by the following equation obtained from the projection model (5).

$$S_T 1 = 162578 - 3.43x_6 \qquad [x_6 = 1,2....,13] \qquad (5)$$

2.6 *Sequestration Strategy 2*

The strategy 1 starts CO_2 disposal from the year 2000 but this seems to be very hard to achieve because there is no time for preparation. Therefore, strategy 2 takes three years for preparation. The sequestration project starts from the year 2002 using a linear CO_2 sequestration strategy shown as equation (6). For year 2000 to 2001, the equation (4) is used because there is no reduction for these periods.

$$St2 = 171424 - 4.82x_7 \qquad [x_7 = 1,2....,11] \qquad (6)$$

2.7 *Sequestration Strategy 3*

This strategy gives the government and industries enough time to start the CO_2 sequestration project starting from the year 2005. Disposal technology can be well developed and enough feasibility study carried out before the project starts. To meet the Kyoto target, CO_2 must be reduced The model for 1999 to 2004, there is no reduction process so that the equation (4) is used. For between 2005 and 2012, the model is following:

$$St3 = 182347 - 7.99x_8 \qquad [x_8 = 1,2....,8] \qquad (7)$$

This strategy requires initial capacity of 50,000kt for the first four years to keep up with rapid increase of excess CO_2. At the year 2009, the capacity is expanded to 105,000 kt as well as other strategies.

2.8 *Sequestration Strategy*

This strategy presents the government and industries a tough challenge because the allowed CO_2 emission is fixed with 1999 level for first two years. This means CO2 disposal or storage project must be started from 2000. This strategy also allows the time for preparation for upcoming operation from the year 2002. From the year 2002, CO_2 must be reduced with the rate of 4020 kt per year. The model for this strategy is given by the CO_2 projection model equation (4) and the reduction rate. For 1999 to 2001, the amount of CO_2 emission is fixed as 162578 kt. After 2001, the model is given by the following equation (8):

$$St4 = 162578 - 4020x_9 \qquad [x_9 = 1,2....,13] \qquad (8)$$

X_9 is the period starting from 2001.

2.9 *Sequestration Strategy 5*

In this strategy, CO_2 reduction starts from the year 2000. The CO_2 emission rate is fixed the emission level of 1999 for the first four years. From the year 2006, CO_2 reduction rate is reduced. From the year 2005, CO_2 reduction must be increased at the rate of 5520 kt per year by this strategy. The model for this strategy is given by the CO_2 projection model equation (4) and the reduction rate. For 1999 to 2004, the amount of CO_2 emission is fixed as 162578kt. From 2005, the reduction rate is shown in the following equation (9):

$$St5 = 162578 - 5520x_{10} \qquad [x_{10} = 1,2....,8] \qquad (9)$$

3 ECONOMICS FOR ACDS

3.1 *Definition of the Economic Model of ACDS*

The economic model of the CO_2 disposal system comprises the capital and operating costs required to design, build, operate and maintain the system within a specified period. The economic model will also be affected by the periodic quantity of excess CO_2 and the fundamental economic parameters like rates of interest, inflation, escalation and taxation.

The total disposal cost function, TC_{CO2}, is given in equation (10) as

$$TC_{CO2} = \phi\left[\psi_1(\gamma_i), \psi_2(\tau_j), \psi_3(\rho_k), \psi_4(\lambda_l), \psi_5(\xi_m)\right] \quad [i=1,n_1; j=1,n_2; k=1,n_3; l=1,n_4; m=1,n_5] \qquad (10)$$

γ_i constitute the determinants of the capital investment in the project, which include the engineering design, construction, equipment and major parts, replacement, working capital, and a contingency factor. τ_j constitutes the determinants of the operating cost. The long-term periodic capital expenditures, CC, and operating costs, OC, for building, operating, maintaining and managing the disposal system are given in equations (11) and (12)

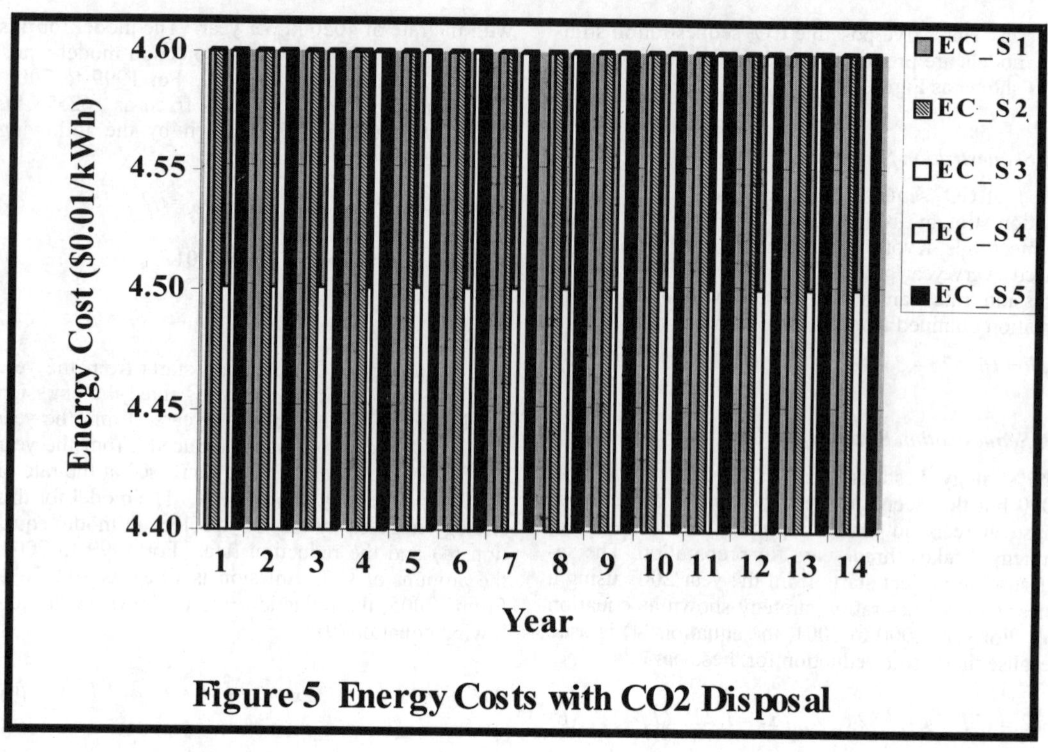

Figure 5 Energy Costs with CO2 Disposal

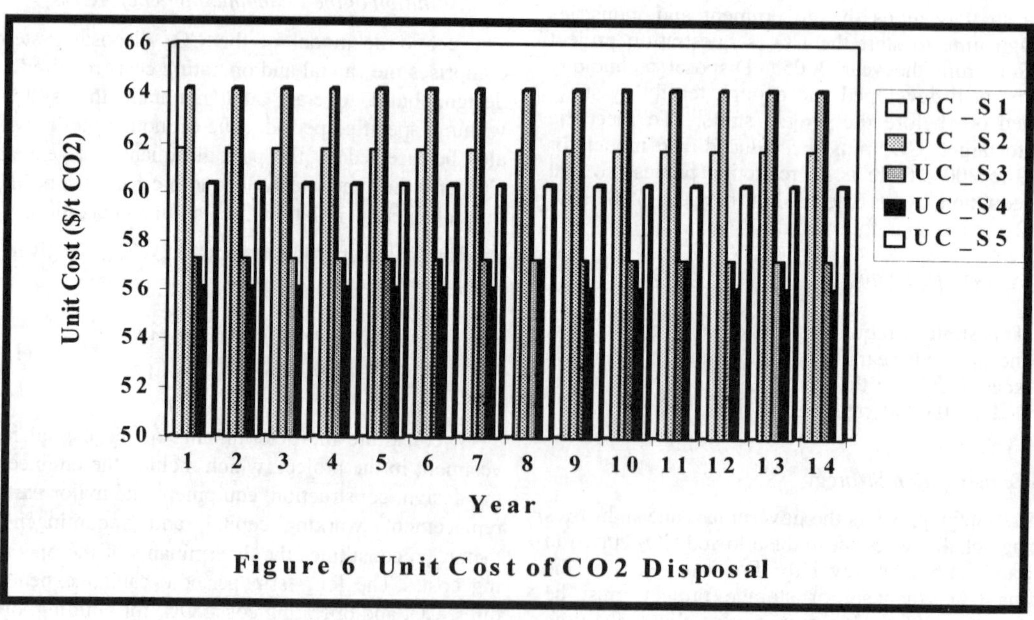

Figure 6 Unit Cost of CO2 Disposal

$$CC = C_0 + C_1X + C_2X^2 + C_3X^3 + \cdots\cdots + C_nX^n \quad (11)$$

$$OC = OC_1X + OC_2X^2 + OC_3X^3 + \cdots\cdots + OC_nX^n \quad (12)$$

$$X = (1+\rho_A)^{-1} \quad (13)$$

The sum of equations (11) and (12) is the value of the total costs of the CO_2 disposal and storage under aquifers. If all the periodic capital expenditures are the same and that of the operating costs are also the same, then these equations are geometric series, and the present worth can be written as

$$PV(TC) = (X^n\text{-}1/X\text{-}1)(C_0+OC_1/1+\rho_A) \quad (14)$$

It must be noted that any course of action must be carefully weighed to ensure a balance in energy cost and a sustainable eco-system. The life of the equipment used in the disposal system is expected to exceed the project duration. The above capital and operating costs are estimated based on the case study for Wabamun Thermal Power Plant. The costs are then converted from Wabamun scale to the Alberta scale.

3.2 *Quantitative AEC Model*

The Annual Equivalent Cost (AEC) criterion provides a basis for measuring investment value by determining equal payments on an annual basis. By this method, we can estimate the unit cost of CO_2 aquifer disposal for each year

The AEC function is defined by following:

$$AEC = A = \frac{iPV(TC)(1+i)^m}{\left[(1+i)^m - 1\right]} \quad (15)$$

PV is present value and i is interest rate.

In order to obtain a unit cost, in this case energy cost per kWh, we may proceed according to the following steps [Gentry and O'Nell 1984, Park, 1997]:

- Determine the number of units of CO_2 to be disposed each year over the project life.
- Identify the cash flow series associated with disposal over the project life.
- Calculate the PV(TC) of the project cash flow series at a given interest rate and then determine the AEC.
- Divide the AEC by the number of units of CO_2 to be disposed during each year.

3.3 *Validation of the Economic Model for the Wabamun Plant*

The Wabamun plant is the coal-fired power plant located 65km west of Edmonton generating 546MW and emits 4.6Mt of CO_2. The capital cost of the recovery plant at the Wabamun includes engineering cost, construction, equipment, major parts, contingency, and working. The conditions of the aquifer is assumed to be with permeability of 100md and pressure of 12.4Mpa. The number of sites for the Wabamun plant is four for service and one for stand-by. The injection sites are 5km from the Wabamun plant. Table 2 contains the determinans of capital cost for the Wabamun plant. Table 3 also shows the determinans of the operating cost for the Wabamun plant.

The total capital cost for the Wabamun plant is \$467.20M as in Table 2 and the total annual operating cost for the Wabamun plant is \$65.72M as illustrated in Table 4. The fixed and variable cost components are \$26.29M and \$8603 per kt, respectively.

Table 2 Capital Costs for the Wabamun Plant

CO2 Separation Plant Costs [\$]		Liquefaction Plant Costs	[\$M]	Pumps and Pipelines Costs [\$M]		Injection Plant Cost [\$M]	
CO2[US\$/t]	1.685	Compressors	77	Pumps	1.26	Pumps	3.5
Lifetime CO2 [t]	137520000	Eng.Design	7.7	Pipelines	6.25	Injector-wells	3.65
		Contingency	15.4	Eng.Design	0.75	Eng.Design	0.72
				Contingency	1.50	Construction	1.43
Total [US\$]	231721200						
Total [C\$]	348045242						
Total [\$M]	348.05		100.1		9.76		9.30
Total CC[\$M]	**467.20**						

Table 3 Operating Costs for the Wabamun Plant

	Capture	Liquefaction	Transportation	Injection	Other	Total
Base energy[kW]	65800	48772	251	1500		116323
Efficiency	0.8	0.8	0.8	0.8		
Total Energy[kW]	82250	60965	313.75	1875		
Unit cost [\$/kW]	0.043	0.043	0.043	0.043		
Total OC [\$M]	30.98	22.96	0.12	0.71	10.95	65.72
Fixed OC [\$M]	26.29	Variable OC[\$M]	39.43	Unit variable [\$/kt]	8603	

3.4 *Validation of the Scaled AEC Model for Alberta*

As discussed above, five CO_2 sequestration strategies are assumed and the capacity and capacity expansion of CO_2 disposal system are estimated based on these assumptions. In order to convert the capital cost from Wabamun scale to Alberta scale, these capacities are divided by Wabamun capacity. These values called capacity factor, are the number of Wabamun plant scale disposal system for Alberta. The amount of CO_2 disposed varies depending on year and strat

egy. To adjust the operating cost for the amount of CO_2 disposed in Alberta, the variable operating cost estimation is applied other than fixed operating cost. Based on the comparison of the Wabamun operating cost, it is assumed that 40% of the operating cost of the Wabamun plant is fixed and 60% is variable.

In this study, the fixed operating cost is based on the capacity of CO_2 disposal system. Similar to capital cost, this fixed cost is estimated by multiplying the capacity factor by the Wabamun scale. These cost are adjusted for escalation using 0.34% for capital cost and 0.16% for operating cost respectively. The escalation rate for the capital cost is the average escalation rate of pump, compressor, construction and petroleum engineering equipment from 1987 to 1998 provided by Statistics Canada [Statistics Canada, 1990, 1994, 1998]. The escalation rate for the operating cost is the average escalation rate of electricity because electricity cost contributes the greatest proportion of the total operation cost. This data is also from 1987 to 1998 provided by Statistics Canada [Statistics Canada, 1990, 1994, 1998]. The annual equivalent costs (AEC s) are estimated using the equation (15). The interest rate adjusted for inflation is 6.55%. Interest rate is provided by the Bank of Canada and inflation rate is provided by Statistics Canada [Bank of Canada, 1999 and Statistics Canada, 1990, 1994, 1998]. In this study, average rates of the year 1998 are used for both rates. The results for the economic analysis for the five strategies are indicated in Figures 3, 4, 5, and 6.

Figure 3 and 4 illustrates the respective total annual and the annual equivalent costs over the duration. The total annual costs comprise the annual operating and capital costs. The annual equivalent costs is the annualized equivalent of the total cost over the duration using a discount r to g percent. The annualized equivalent ranges between $2,500 M to about $3,000 M.

Figures 5 and 6 illustrate the respective impact g CO_2 disposal on energy price and the unit cost g CO_2 disposal. The energy cost increases from the current price g $0.043/kWh is between $0.045 and $0.046/kWh. The unit cost g CO_2 disposal also ranges from $56 and $64/ton g CO_2 (about US $38 and $44/ton g CO_2). By dividing AEC by the annualized excess CO_2, the unit disposal cost is estimated. Similarly, by dividing AEC by the total energy consumption of electricity, the unit CO_2 disposal cost per kWh of generated electricity. By adding this to present electricity price, the total electricity cost with CO_2 disposal cost can be estimated. Change in electricity prices with CO_2 shows the

price of $0.0464 per kWh. The electricity price increase about 7.9%

with CO_2 disposal. The unit CO_2 disposal cost is between $61.72 per tonne.

The Strategy 2 takes three years for preparation so that disposal is started from the year 2002. Similar to Strategy 1, the AEC and unit costs for The AEC value is $2,661M. The electricity price including CO_2 disposal cost is $0.0460 per kWh. The electricity price increases about 7.0% with CO_2 disposal. The unit CO_2 disposal cost is $64.17 per tonne.

Strategy 3 gives the government and industries enough time to start the CO_2 sequestration project starting from the year 2005. This strategy requires the smallest amount of CO_2 disposal among five strategies so that the base economics is very small compared to other strategies. The AEC for the Strategy 3 is $1804M. This strategy has much lower AEC because the CO_2 sequestration starts from the year 2005. As a result, the total amount of excess CO_2 in the project life is also small, which is 441.39 kt. That small amount of excess CO_2 reduces the variable operating cost, which depends on the amount of disposed CO_2. This result also shows that option 17 has the lowest price of $0.0450 per kWh. The electricity price increases just 4.7% with CO_2 disposal. The unit CO_2 disposal is $57.23 per tonne.. This strategy shows the lowest AEC, electricity price, and unit cost so that this option is the optimum option. The AEC for the Strategy 4 is $2561M and the electricity price for this strategy is $0.0459 per kWh. The electricity price increases 6.7% with CO_2 disposal. The unit CO_2 disposal cost $56.03 per tonne.

The amount of excess CO_2 in the sequestration strategy 5 is 573.32 kt which is similar to that of strategy 2. The AEC for the Strategy 5 is $2546M and $2894M. The electricity price is $0.0459 per kWh. The electricity price increases 6.77% with CO_2 disposal. The unit CO_2 disposal cost is $60.29.

4 CONCLUSIONS

CO_2 emission forecast model is developed using multiple regression technique. The model includes previous year's CO_2 emission rate, population growth rate, industrial growth rate, energy consumption and technological progress as repressor variables. The model shows high accuracy with the R-squared of 99.7%. By using the model and collected data, the future CO_2 emissions are projected. The CO_2 emission gradually increases until 2012 and at the end of target year 2012, emission rate will become 215Mt. The CO_2 sequestration strategy models are proposed based on the CO_2 emission projection results for later economic and risk analysis based on projection results. The economic modeling for the Wabamun Thermal Power Plant is carried out for five sequestration strategies. The capital cost of ACDS for the Wabamun plant is between 467 million dollars and the operating cost for the Wabamun plant is 65.72 million dollars to dispose 4584 kt of CO_2 in a year. Based on these Wabamun result, a scaled model for CO_2 disposal options and their respective economic models are thoroughly examined for the total CO_2 emissions in Alberta by using annual equivalent cost (AEC) model.. The sequestration strategy 3 is the optimum strategy with AEC of $1804M because of its lowest cost and the 4-year preparation period before, which is four years before operation.

5 REFERENCES

1. Bachu, S., Gunter, W.D., Law, D., and Perkins, E.H., 1996, *Aquifer Disposal of Carbon Dioxide*; © Hitchon Geochemical Service Ltd., Edmonton, Canada
2. Gentry, D.W. and O'Nell T.J., 1984, *"Mine Investment Analysis"* ©American Institute of Mining, Metallurgical and Petroleum Engineers, Inc.
3. Government of Alberta, 1997, *Alberta Progress Report on the National Action Program on Climate Change*; © Government of Alberta, Edmonton, Canada
4. Gunter, W.D., Wong, S., Cheel, D.B., and Sjostrom, G., 1998, Large CO_2 Sinks: Their Role in the Mitigation of Greenhouse Gases from an International, Canadian and Alberta Perspective; *Unpublished Report*, Alberta Research Council, Edmonton
5. Jacques, A., Neitzert, F. and Bolieau P., 1997, *Trends in Canada's Greenhouse Gas Emissions 1990-1995*; © Environment Canada, Canada
6. Macdonald, D. E., Donner, J. and Nikiforuk, A., 1996 *"Full Fuel Cycle Emission Analysis for Electric Power Generation Options and Its Application in a Market-Based Economy"* The Third International Conference on Carbon Dioxide Removal, September 9-11, 1996, Boston, MA, U.S.A.
7. Park C.S., 1997, *"Contemporary Engineering Economics Second Edition"* ©Addison-Wesley Publishing Company, Inc.
8. Statistics Canada, 1990, "Population Projections for Canada Provinces and Territories 1989-2011" ©Minister of Supply and Services Canada
9. Statistics Canada, 1994, "Population Projections for Canada Provinces and Territories 1993-2016" ©Minister of Supply and Services Canada
10. Statistics Canada, 1998, "Population Projections for Canada Provinces and Territories 1997-2020" ©Minister of Supply and Services Canada

Environmental Issues and Management of Waste in Energy and Mineral Production, Singhal & Mehrotra (eds)

Spatial indexing of environmental sensitivity for selection of industrial sites

Hilary I. Inyang & George Fisher
Center for Environmental Engineering, Science and Technology (CEEST), University of Massachusetts, Lowell, Mass., USA

ABSTRACT: Site selection for industrial activities has increasingly involved the use of rational and quantitative techniques for ranking of optional sites. In this paper, the use of a generic air pollution model with environmental attributes, to index a region is presented. The indexing methodology is formatted such that decision makers can change the magnitudes of weighting factors and derive spatially distributed sensitivity indices and pollution indices that correspond to air pollutant concentrations. Using data for the City of Lowell, Massachusetts, USA, a generic air pollutant dispersion model is applied. The results show the optimal location of an industrial facilty for the set of factors considered.

1. INTRODUCTION

The necessity for close proximity between the sources of raw materials or managed wastes and processing installations (to minimize material haulage costs) has resulted in an increase in the number of facilities such as incinerators, wastewater treatment plants, and solid waste transfer stations located in densely populated and other sensitive environments. Within broad regions that include such environments, it is necessary to select the best industrial site for a set of objectives and constraints. An objective could be the minimization of pollution impacts while constraints could be dense population, large number of surface water bodies, presence of endangered species habitat and unfavorable meteorological conditions in the region.

It is necessary to use rational and quantitative techniques in ranking potential sites for waste-generating facilities in order to improve objectivity through minimization of arbitrariness in the selection and weigghting of site ranking factors. As part of environmental impact assessment projects and research to support facility siting projects and permit applications, a variety of impact assessment methods and site selection methodologies have been developed by several workers. Perhaps the most common method is the Multi-attribute Utility Analysis Model which involves the development of an aggregate score for alternative sites on a set of factors. This method has been described by Call and Merkhofer (1988) for hazardous waste sites in the United States. Each of the ranking factors is usually a determinant of environmental pollution and/or human health risk. Determination of the score of each site on a particular factor requires data on the sensitivity of an ambient region around each site and level of risk-determining stress that will be imposed on that region by a facility at each alternative site.

Call and Miller (1990) have described several approaches for automating decisions using a variety of attributes developed for alternatives. Models that relate environmental and human health risk factors and facility operational conditions have been described by Schwartz et al. (1988). Essentially, the use of such models requires data that enable the appreciation of both the spatial and temporal variabilities in environmental conditions and hence, potential risks within the ambient region. A number of investigators (Hanson and Hargrave, 1996; Latour and Reiling, 1994; and Whitehead and Blomquist, 1991) have developed indexing information and methods that enable spatio-temporal analysis of the effects of locating facilities at a site. Although most of the ecological classification systems were not developed for specific facilities, they can provide useful information for facility site selection if their spatial scales are small enough for project impact evaluation.

Smith and Carpenter (1996) have adapted information from Avers et al. (1994) to classify spatial scales into four categories based on the size of ecological units and potential applications. The ecoregion is the largest spatial class. This class could be continental in size, and its application is in long term, broad-based environmental planning and assessment. The land unit is the smallest spatial scale and applies to specific project impact assessments such as the power plant siting that is the focus of this paper. The other two spatial scales: subregion and landscape are intermediate between the ecoregion and land unit scales. They are very commonly used in non-point source pollution studies such as those performed by Daniels and McTernan, (1995), Bui et al., (1996), and Lin et al., (1999). The division of target regions into spatial segments for environmental assessments is helpful in regulatory decision-making within the approach discussed by Wright et al. (1993).

2. THE LOWELL INCINERATOR SITING EXAMPLE

A hypothetical coal-burning electric power plant has been selected as the planned facility to illustrate this site ranking and selection approach. The plant is intended to furnish up to 750 megawatts of electricity. As shown in the map of the City of Lowell presented in Figure 1, three alternative sites in the southwest of the City have been identified for the power plant. Generally, it is known that air pollution which has secondary impacts on water quality, aesthetics, and human health of the City is potentially the greatest threat. The objective is to select one site from the three alternative sites that would have the least negative impact on the region if the power plant is sited there.

It is known that most of the time, the wind blows to the northeast of the city. The population of the region shown is about 100,000. A river, schools, residential homes and cultural facilities are also found in the City.

3. ENVIRONMENTAL SENSITIVITY ASSESSMENT

Environmental sensitivity assessment is conducted to determine the instrinsic sensitivity of various spatial units of the City before consideration of the impacts of the facility planned. For the purpose of this paper, the City area was divided into 35 grid

Figure 1. A map of the City of Lowell showing the three alternative sites for the hypothetical power plant

squares of equal dimensions. The evaluation factors considered are presented in Table 1. Each of the spatial units (squares) has attributes which can be represented as numerical scores on the chosen evaluation factors. However, it should be noted that not all the factors apply uniformly to all the squares because some sensitive features such as water bodies and schools are absent from some of the squares.

As an illustration using a grid square (Figure 2) that is larger than those shown in Figure 1, data for the relevant factors are two schools, one college, downtown shopping area and historic buildings. United States census data indicate that the population of people in this square is 5,000 per square mile.

If the entire area was to be populated, a score of 3.0 would be assigned. However, about 12% of the area is covered by surface water. So the population score is determined as 0.88 x 3.0 = 2.64. From census data, the split of the population was 25% children, and 20% elderly. Therefore for children, the score is computed as 0.25 x 2.64 = 0.66 and for the elderly, it is 0.20 x 2.64 = 0.53.

If the grid square were to be totally covered by water, it would be assigned a score of 3.0. However, only 12% of the square is water. Therefore, its score is 0.12 x 3.0 = 0.36. Each school and each college are valued at 4.0 and 2.0 respectively. The presence of a shopping area or historic buildings in the downtown area also received scores of 1.0 each. A summary of the scores for the grid square in Figure 2 is provided in Table 3. Similar determinations were made for the 35 squares resulting in the environmental sensitivity scores shown in Figure 1.

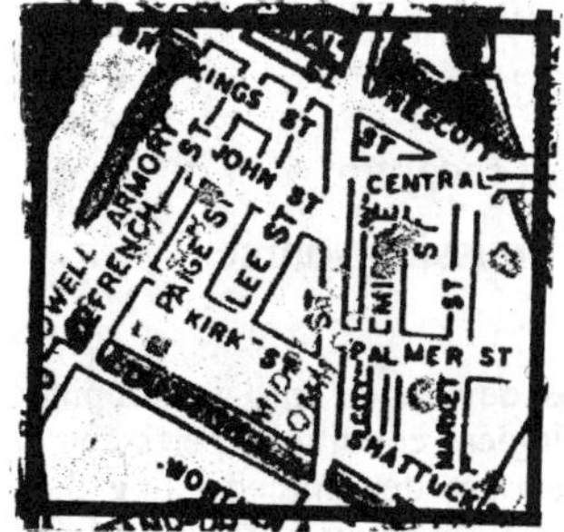

Figure 2. The grid square used for illustration

4. IMPACT ANALYSIS

4.1 *Air Pollution Model*

The generic Gaussian Plume Equation (GPE) which forms the basis for most of the air pollution dispersion models that are currently applied to practical problems, was used to evaluate each of the alternative sites. The GPE describes the diffusion of a gas in the atmosphere of known meteorological factors, notably wind speed. General formulations of this model are described in standard air pollution textbooks. The basic equation is shown below.

$$C = \frac{Q}{2\pi\,\sigma_y\sigma_z u}\exp\left[-\frac{y^2}{2\sigma_y^{\,2}}-\frac{(z-H)^2}{2\sigma_z^{\,2}}\right]$$

C = Concentration at a point (x, y, z)

Q = Emission source strength

Table 1. Selected environmental sensitivity factors and their justification

Environmental Sensitivity Factor	Justification
1. Population Density	Determines number of people that could be affected by population
2. Surface Water Area	For assessment of water supply pollution
3. Number of Schools	Indicates a high number of children in the location
4. Number of Hospitals	Indicates a high concentration of people who could be more sensitive to pollution
5. Public Areas, Parks, etc.	Large gathering of people who could be exposed; also aesthetic damages
6. Agricultural Land	Secondary damage to crops
7. Historic Buildings	Negative impacts on aesthetics

σ = Statistical dispersion coefficient

u = Wind speed

x, y, z = Location of receptor

H = Height of emission source

This basic model was developed into a computer program and meteorological conditions pertinent to Lowell, Massachusetts were determined using data from U.S. EPA, National Weather Service and the National Oceanic and Atmospheric Administration. The coverages of the plume over the City when the hypothetical power plant is located at each of the three alternative sites were determined. The coverages of the plumes are illustrated for the three alternative sites in Figure 3.

4.2 *Indexing of Potential Pollution*

Air pollution concentrations over the squares for which environmental sensitivity indices were estimated were used to develop potential pollution indices for each of the three alternative sites. For each plume, the plume area that exceeded the U.S. National Ambient Air Quality Standards in terms of pollutant concentrations were determined. Then outer points of the plume corresponding to 50%, 25% and 10% of the standards were also delineated. Thus each plume was divided into four areas, all of which were color-coded. The color codes are exceedance (red), 50% (orange), 25% (yellow) and 10% (green). In the black and white maps shown in Figure 3, these color codes appear as a dark grey plume core for exceedance, a grey zone representing both the 50% and 25% of the standard, and a dark rim representing 10% of standard.

For each square, the environmental sensitivity factors were then multiplied by the pollution factors corresponding to 2.0 for exceedance, 1.0 for 50%, 0.5 for 25%, and 0.2 for 10%. This was done only for the squares affected by the plumes for each alternative site. Upon summing up the City index, now modified due to pollution of sum spatial units (squares), the total score was determined for each of the considered sites. The best site is taken to be the one with the least difference between total score and the total sensitivity index. This difference is called the pollution index.

Table 3. Total grid score for the grid shown in Figure 2

FACTOR	SCORE
Population	2.64
Children	0.66
Elderly	0.53
Surface Water	0.36
Schools	8.00
Colleges	2.00
Shopping Area	1.00
Historic Buildings	1.00
TOTAL	16.19

5. SITE SELECTION

This analysis shows alternative sites 1, 2 and 3 have total scores of 309.9, 311.7 and 303.7 respectively. Considering that the total sensitivity index without the power plant is 269.2, the pollution indices for alternative sites 1, 2 and 3 are 40.7, 42.5, and 34.7, respectively. This implies that out of the three sites considered, the best site for the hypothetical power plant is site 3.

6. CONCLUSIONS, UTILITY AND LIMITATIONS OF THIS APPROACH

The facility site selection methodology described and illustrated herein is useful for comparison of alternative sites objectively. By moving the contaminant source (facility site) around, the potential impacts on a region of fixed dimensions can be evaluated. Opportunity is allowed for subjectivity in the selection and rating of factors by the operators of this methodology or citizen groups that could be potentially affected by the planned facility. This methodology can be operated through the use of Geographic Information Systems (GIS)

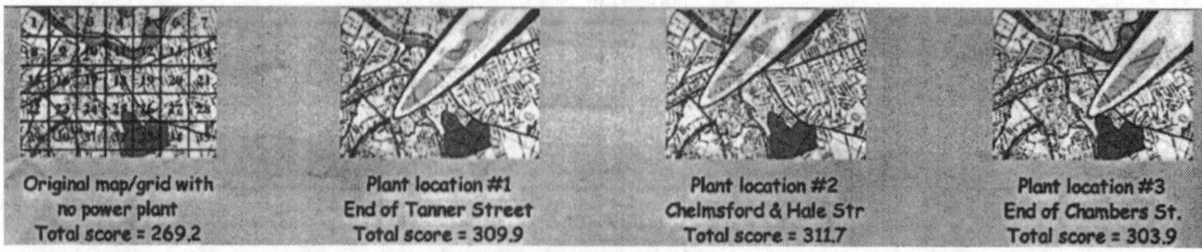

Figure 3. Plumes from the hypothetical power plant at each of the alternative sites and their respective total impact scores

which allows visualization of the results at various spatial and temporal scales.

In the approach described herein, both the rating of factors (independent of site) and the scores of each site on each factor are aggregated. This makes it somewhat difficult to apply this system to a case in which several alternative sites are considered. Work is in progress to eliminate this limitation by developing a set of statistical equations for use in automating the operation of the methodology. In the evolving approach, the numerical scale of each factor is more distinctly differentiated from the score of an option on each factor.

It should also be noted that the region for which the impacts are to be assessed must be defined. For example, in Figure 3, some portions of the contaminant plume lie outside the region of interest. This has a reducing effect on the pollution index for the target region.

7. REFERENCES

Avres, P.E., D.T. Cleland, W.H. McNab, M.E. Jensen, R.G. Bailey, T. King, C.B. Goudey and W.E. Russell. 1994. National hierachical framework of ecological units. U.S. Dept. Of Agric. Foreign Service, Washington DC.

Bui, E.N., K.R.J. Smettem, C.R. Moran and J. Williams. 1996. Use of soil survey information to assess regional salination risk using geographic information systems. J. Environ. Qual. 25: 433-439.

Call, H.J. and M.W. Merkhofer. 1988. A multi-attribute utility analysis model for ranking superfund sites. Proc. 9th National Conf. Superfund'88, Washington DC, Washington DC, Nov. 28-30, 1988: 44-54.

Daniels, B.T. and W.F. McTernan. 1995. A spatially varied index method to define relative risks associated with pesticide contaminatio of groundwaters. J. of Env. Systems. 24(2): 111-135.

Hanson, D.S. and B. Hargrave. 1996. Development of a multi-level ecological classification system for the state of Minnesota. Env. Monitoring and Assessment. 39: 75-84.

Latour, J.B. and R. Reiling. 1994. Comparative environmental threat analysis: three case studies. Env. Monitoring and Assessment. 29: 109-125.

Lin, H.S., H.D. Scott, K.F. Steele, and H.I. Inyang. 1999. Agricultural chemicals in the alluvial aquifer of a typical county of the Arkansas Delta. Env. Monitoring and Assessment. 58: 151-172.

Schwartz, S.I., R.A. McBride and R.L. Powell. 1988. Models for aiding hazardous waste facility siting decisions. J. of Env. Systems. 18(2): 97-122.

Smith, M. and C. Carpenter. 1996. Application of the USDA Forest Service national hierachical framework of ecological units at the sub-regional level: the New England-New York example. Env. Monitoring and Assessment. 39: 187-198.

Whitehead, J.C. and G.C. Blomquist. 1991. Measuring contingent values for wetlands: effects of information about related environmental goods. Water Resources Research. 27(10): 2523-2531.

Wright, F.G., H.I. Inyang and V.B. Myers. 1993. Risk reduction through regulatory control of waste disposal facility siting. J. of Env. Systems. 22(1): 27-35.

Environmental Issues and Management of Waste in Energy and Mineral Production, Singhal & Mehrotra (eds)

Hydrogeological modeling of mining operations at the Diavik Diamonds Project

Ken Kuchling
Diavik Diamond Mines Incorporated, Calgary, Alb., Canada

Don Chorley & Willy Zawadzki
Golder Associates Limited, Vancouver, B.C., Canada

ABSTRACT: Diavik Diamond Mines Inc. proposes to develop a diamond mining project at Lac de Gras in the Northwest Territories. As part of the Environmental Assessment, and mine design, estimates of mine water inflow quantity and quality were required. This paper describes the field investigations and numerical modelling studies that were completed in order to evaluate the groundwater conditions at site. The Diavik Diamonds Project is located in the Canadian Shield within the region of continuous permafrost, however mining operations will be located in unfrozen ground within the confines of a diked-off portion of Lac de Gras. Field investigations used to characterise the hydrogeological regime at site consist of extensive packer testing supplemented with a limited program of borehole flowmeter testing, borehole temperature logging and borehole camera imaging. This information was used to develop a conceptual hydrogeological model for the site, which in turn was modelled numerically using MODFLOW and MT3DMS to predict groundwater inflow volumes and water quality with time. Results indicate that the total mine inflows are expected to range up to 9,600 m^3/day with TDS concentrations gradually increasing in time to maximum levels of about 440 mg/L. The modelling also showed that lake water circulating through the rock mass will eventually comprise more than 80% of the mine water handled.

1 INTRODUCTION

The Diavik Diamonds Project is located at Lac de Gras, 300 kilometres northeast of Yellowknife, N.W.T, shown in Figure 1. The project is a joint venture between two Canadian companies, Aber Resources Ltd. (40%) of Toronto and Diavik Diamond Mines Inc. (60%) of Yellowknife, a subsidiary of Rio Tinto plc of London, England.

The project will entail the mining of four diamondiferous kimberlite pipes, termed the A154S, A154N, A418, and A21 pipes. The total kimberlite resource stands at about 37 million tonnes with an average grade of 3.6 carats per tonne, for a total resource of about 133 million carats. The proposed mine plan consists of the initial development of three open pit mines after which underground mining operations will continue beneath two of the pits (A418 and A154 pits). The estimated mineable ore reserve is approximately 26 million tonnes yielding an estimated 102 million carats over the 20 year mine life. Twenty million tonnes of the reserve will be mined with open pit methods while the remaining 6 million tonnes will be recovered with underground methods. Mining operations will extend to a depth of 400 metres below the lake bottom.

The Diavik project is located within the region of continuous permafrost, about 250 kilometres south of the Arctic Circle. The four kimberlite pipes are all located within the footprint of Lac de Gras, approximately 100 to 800 metres from the shoreline of east island. Therefore unfrozen ground conditions will be encountered in all the pits during mining.

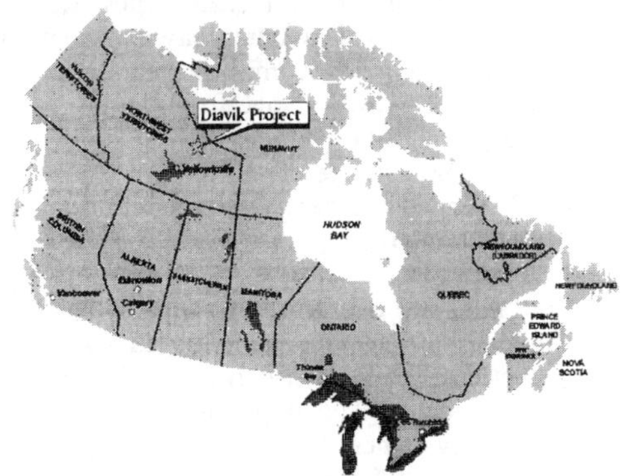

Figure 1: Project Location

Water retention dikes will be used to isolate the pit areas from the lake (Figure 2) with the consequence that groundwater will flow through the unfrozen rock. A key part of the project environmental assessment and mine design was prediction of the quantity and quality of mine water. The focus of this paper is on the hydrogeology of the A418 and A154 pit areas (Figure 2).

Figure2: Project layout

2 GEOLOGICAL SETTING AND PERMAFROST CONDITIONS

The regional geologic setting for the Lac de Gras area is within syntectonic and post-tectonic intrusive rocks of the Slave Province. The main Archean rock units consist of sedimentary greywacke metaturbidites, biotite tonalite, 2-mica granite, and granodiorite. Proterozoic diabase dike swarms also appear regionally. The open pits will be developed in competent granitic country rocks with Rock Quality Designation (RQD) in the range of 95%. Existing fractures and local joint systems will therefore govern the hydrogeology of the country rock mass.

The formation of the kimberlite pipes are relatively recent geological events, having been emplaced approximately 50-60 million years ago. For comparison, the host rocks vary from 2.5 to 2.7 billion years in age and the younger diabase dikes at 1.3 to 2.6 billion years in age. The kimberlite pipes are cylindrically shaped and near vertical, with diameters of about 80-120 metres.

The pit areas are overlain with approximately 10 metres of overburden, consisting of an average of eight (8) metres of bouldery till and two (2) metres of lake bed sediments. The upper few metres of the bedrock can be weathered with some open joints being infilled with silty fines.

Permafrost develops in areas where the heat loss from the ground during winter exceeds the combined energy gain during the summer and energy radiated by the local geothermal gradient (i.e. heat radiating upwards from depth). Permafrost generally develops under dry land masses where the ground surface is exposed to prolonged cold air temperatures. The average annual air temperature at the Diavik site is –12°C. Beneath bodies of water which do not entirely freeze to depth (>1.5 metres deep at the Diavik site), taliks, or thawed zones, will be present. In these locations, the lake water and lakebed will only cool to temperatures in the range of 0°C to +1°C in winter and this relatively "warm" temperature will prevent the development of permafrost in the ground. At the Diavik site the permafrost depth has been measured to a vertical depth of about 380 metres below the east island and the ground temperatures are typically in the range of -5°C. Permafrost depths measured under the various small islands in the lake range from 100 metres to being non-existent, depending on the size of the island. Along the lake shoreline, permafrost has been shown to extend sub-vertically downwards from the lake edge contact forming a "bulb" shaped zone under the islands. Hydrogeologically, permafrost is considered to be impervious.

3 HYDROGEOLOGICAL INVESTIGATIONS

In order to characterize the hydrogeology of the site, three successive geotechnical/hydrogeological field programs were conducted between the years 1996 and 1998. The locations of these boreholes are shown in Figure 3.

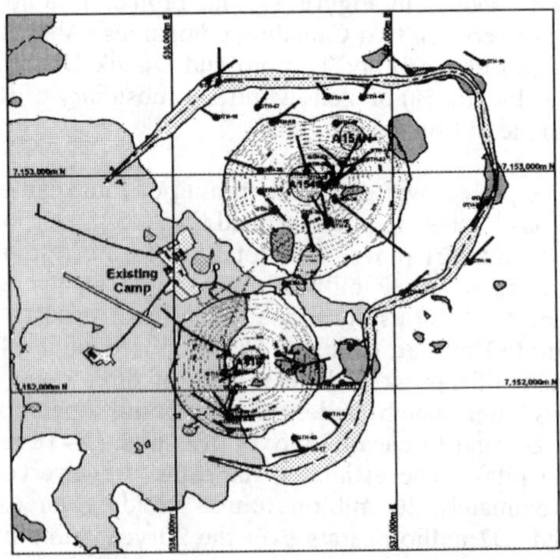

Figure 3: Geotechnical Borehole Locations

On average they were drilled at inclinations of –55° to vertical depths of 245 metres although two holes were extended down to 600 metres. Since the pit areas are under the lake, drilling was from the ice and restricted to the winter seasons only (February to mid-April). These core holes were logged geotechnically, including the collection of oriented core data. Packer testing was also undertaken in these holes. The goal of the hydrogeological investigations were to assess the hydraulic conductivity of the rock mass as a whole, and to determine the major factors controlling water flow, i.e. are the main flow paths along the natural jointing in the rock or mainly along major structures such as broken zones or faults. Knowledge of the groundwater flow system would help develop the conceptual hydrogeological model and aid in design of water collection or mitigation measures if needed. A 1000-metre long bulk sample decline was excavated to the A154S pipe (Figure 3) and this development also provided opportunity to assess hydrogeological conditions on a somewhat larger scale, although only 600 metres of the decline was actually in unfrozen rock beneath the lake. The length could be extended to 800 metres if boreholes drilled out from the face are included.

Four types of hydrogeological field investigation methods were used at site. They consisted of; (i) packer testing, (ii) borehole flowmeter testing, (iii) borehole temperature logging, and (iv) borehole video imaging.

Packer testing: Approximately two hundred shallow and deep HQ and NQ sized core holes were drilled as part of the geotechnical and hydrogeological investigations for the pit walls and the water retention dikes (pit wall holes only are shown in Figure 3). Most of the geotechnical holes were packer tested using constant head tests although some falling head tests were also conducted. In total approximately 600 packer tests were completed in the country rock over packer intervals ranging from 40 metres down to 3 metres. Packer testing intervals associated with the water retention dike design were generally short, in the range of 3 to 10 metres in both the bedrock and overburden. Packer tests associated with the pit investigations were done over 40 metre intervals in the bedrock only. The deepest packer test at site was done at a vertical depth of 570 metres below lake bottom.

Heat pulse flowmeter logging: This geophysical technique utilizes a down-the-hole probe that can measure water flow rates along a borehole. A small diameter submersible pump is used to lower the head in the borehole in order to induce upward water flow from depth. The flowmeter tool is lowered to various locations in the hole in order to take flow rate measurements. As cumulative flows are measured at various points in the hole, water-bearing zones can be identified by changes in borehole flow quantity. The water-bearing zone would be somewhere between a "high flow" reading and the previous "low flow" reading. Several boreholes were tested with this method however only one provided good results. In the unsuccessful holes, extremely permeable zones near the top of the hole yielded almost all of the water influx and consequently very little upward flow could be induced. Figure 4 shows a typical plot from the flowmeter log. The maximum and minimum values were determined by taking several readings at each depth increment.

Resistivity/temperature logging: This geophysical technique also utilizes a down-the-hole probe that measures both water temperature and fluid resistivity. Changes in water temperature along the borehole are signs of natural seepage or diffusion of groundwater into the hole. No flow is induced in the hole for this test in order to prevent mixing of water and the smearing of temperature signatures. The geophysical probe is lowered down the hole at a constant rate of about three metres per minute with a reading automatically taken every 3-cm. Due to the low Total Dissolved Solids (TDS) in the lake water and groundwater at the site, resistivity values were too low to be measured and therefore did not provide any useful information. Temperature readings generally measured borehole fluid temperatures of between

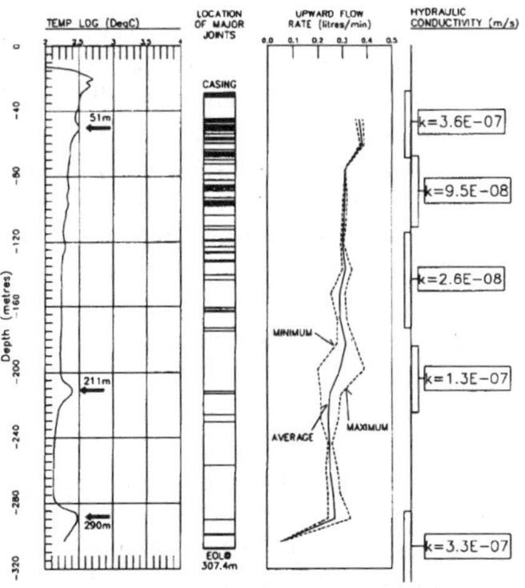

Figure 4: Example borehole geophysical log

2°C to 3°C, with the water bearing zones showing a slight temperature increase, in the range of 0.1°C to 0.5°C. Figure 4 shows a typical plot of a temperature log. Four boreholes were tested with this tool.

Borehole video imaging: This technique utilizes a SeeSnake down-the-hole camera. The self-illuminating camera provided black and white images along hole lengths in the range of 300 metres. After a 3 to 6 hour hole flushing period, the camera was lowered down the hole at a rate of about three metres per minute. Elapsed time and depth along hole are tracked with a pulley wheel equipped with an encoder and are displayed on the video image and recorded on videotape. At 300 metres the depth accuracy was usually to within 40 cm and was never off more than 70 cm. Five holes were examined with the camera. The goal of the video imaging was to examine water bearing zones identified by the packer tests and temperature logs. This was to determine whether the flow was from a highly broken zone or from an area with widely spaced single fractures. The images also helped to compare the in-situ rock mass with geotechnical core logs. It became apparent that in some instances, broken zones in the core appeared in-situ as tight fractures. Conversely several single fractures were encountered with centimeter wide joint apertures, something that would not be determined by examining core.

4 HYDROGEOLOGICAL CHARACTERIZATION

For hydrogeological characterization purposes, the country rock at Diavik was initially considered to consist of two domains, fractured rock zones and weakly fractured rocks, that can be differentiated based on the frequency, spacing and connectivity of the fractures. This differentiation is similar to the approach used by researchers at the Whiteshell Research Area (Stevenson et. al., 1996). A fractured rock zone was defined as a zone with enhanced fracture density and permeability relative to the background, that is spatially continuous over the scale of hundreds of metres or more. These zones are significant hydrogeological features that are important to groundwater flow because of their size, hydraulic conductivity, and connectivity. A zone of broken rock that is not significantly permeable due to sealing or a limited lateral extent would not be classified as a fractured rock zone. Highly permeable single fractures, on an individual basis, are not considered to be significant hydrogeological features, as they are not expected to extend over such large distances. Weakly fractured rock is defined as a volume of rock that contains relatively infrequent open fractured that are generally poorly connected. Weakly fractured rock is, on average, significantly less permeable than the fractured rock zones. By comparing the drill core logs and core photographs with packer tests and geophysical data, it was readily apparent the some of the highly permeable zones in the rock mass consisted of open, single joints. Broken rock zones identified in the core were not necessarily significantly water bearing. Possibly the broken zones consisted of random joints that happen to be closely spaced rather than being associated with a major fault structure. In fact very few fault zones were ever identified in the drill core or during construction of the bulk sample decline. Major water inflows encountered in the decline were mainly associated with single open joints. Figure 5 is a photo of a 1-cm wide vertical open joint that was encountered in cover drill hole in the decline. This joint was by far the most significant water bearing structure intercepted over the 600-metre decline length and resulted in temporary flooding at the decline face. Grouting eventually sealed off the uncontrolled water flow (grout can be seen in the lower portion of the fracture). Most of the rock domain at Diavik is therefore comprised of weakly fractured rock. Fractured rock zones appear to be rare and widely spaced.

From a hydrogeological modeling perspective, the conclusions from the field investigations confirmed that the most reasonable approach would be to model

Figure 5: In-situ open fracture

the rock mass as a single hydrostratigraphic unit using an average hydraulic conductivity rather than try to interpret distinct major water bearing horizons within the country rock. The next step in the hydrogeological characterization was to determine the average hydraulic conductivity and evaluate spatial trends in the data.

Figure 6 provides a histogram of the hydraulic conductivity (K) data derived from packer tests in the granitic country rock. The data appears to follow a lognormal distribution and range over several orders of magnitude, from 10^{-3} m/s to 10^{-9}m/s, with a mean of 2x10-7 m/s. Table 1 provides a summary of the measured hydrogeological parameters in the various stratigraphic units.

Researchers have shown that at various sites in the Canadian Shield and at other locations worldwide, rock mass permeability decreases with increasing depth. (Davison et al., 1994 a & b; Stevenson et al., 1996 a & b; Ophori et al., 1994 & 1996; Raven et al., 1987; and Burgess, 1979).

In order to assess whether there is a similar depth dependent trend at the Diavik site, average K values were calculated for successive 100 metre vertical intervals. The results are shown in Figure 7. The error bars show the maximum and minimum values over that interval. A general trend of decreasing K with depth is apparent in the data.

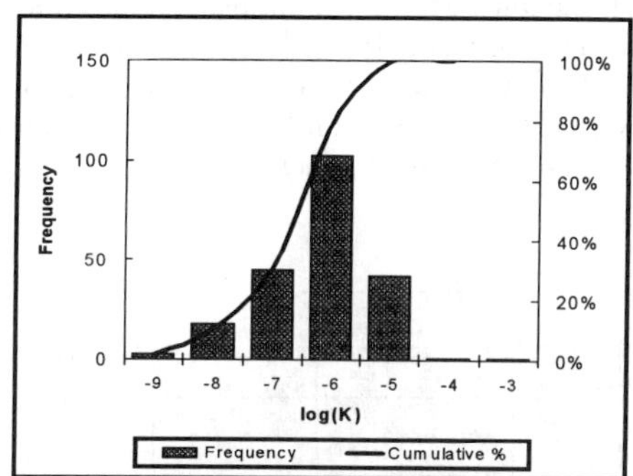

Figure 6: Packer test results

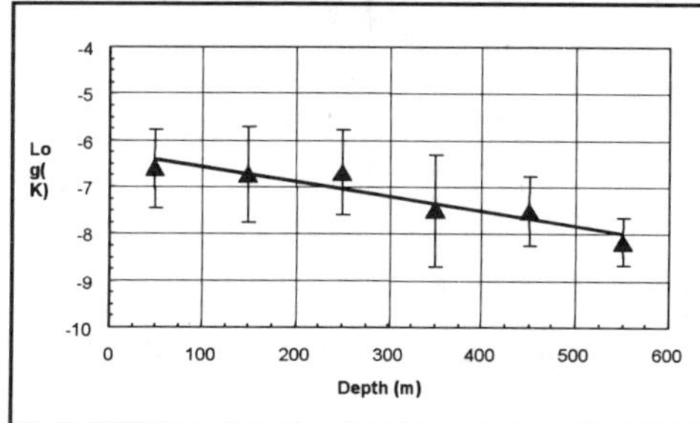

Figure 7: Hydraulic conductivity with depth

Table 1: Hydraulic Conductivity (K) Summary

Hydrogeological Unit	Average Measured K Values	Calibrated K
Lakebed sediments & till	4×10^{-5} m/s	*Not modeled*
Country Rock	2×10^{-7} m/s	5×10^{-7} m/s
Kimberlite	4×10^{-7} m/s	3×10^{-6} m/s
Diabase Dike	5×10^{-8} m/s	5×10^{-8} m/s

Hydraulic conductivity data in the kimberlite was collected mainly from cover holes in the bulk sample. Packer tests were not completed vertically down the kimberlite pipe due to difficulties with borehole collapse. The average K in the kimberlite was measured at $4x10^{-7}$ m/s.

In the lakebed overburden soils, the range of K values measured was between $7x10^{-3}$ and $2x10^{-8}$ m/s, with an average of $4x10^{-5}$ m/s. Since the K for the overburden soils was greater than that of the underlying bedrock, it would not be a controlling factor in groundwater recharge. Consequently, for simplification purposes, the overburden layer was excluded from the groundwater model.

Typically in the Canadian Shield, the concentration of Total Dissolved Solids (TDS) in the groundwater increases with depth. As part of the environmental baseline study, water quality sampling was conducted in the upper 350 metres of the rock mass at the Diavik site. Attempts to take deeper water samples were unsuccessful. Consequently the site data was supplemented with water quality samples collected at the Echo Bay Lupin Mine at depths in the range of 800-1300 metres. The Lupin Mine is about 100 kilometres north of the Diavik site. The combined water quality results (TDS) are plotted in Figure 8, which also presents water quality results collected by researchers elsewhere in the Canadian Shield. Two TDS versus depth profiles were developed; ① a profile based on the Canadian Shield data, and ② the Diavik profile based on the local and Lupin data. The Diavik profile estimates that at 500 metres depth the TDS concentration is 1000 mg/L and at 1000 metre depth the TDS concentration would be 6,400 mg/L. At 1600 metres the concentration would be 100,000 mg/L. Since mining operations are only planned to a depth of about 400 metres, a significant amount of groundwater upwell-

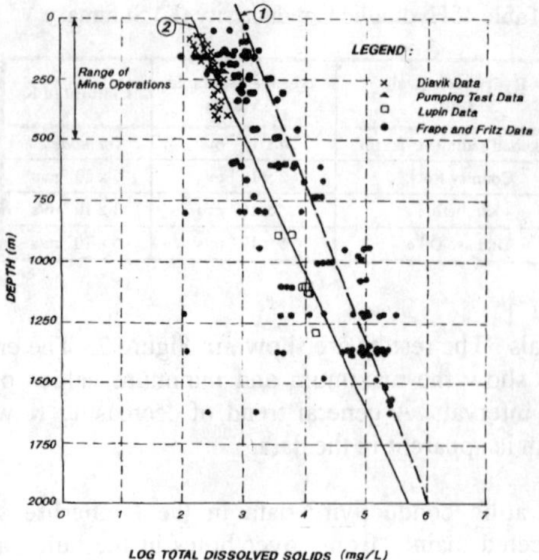

Figure 8: TDS vs. depth profile

ing could introduce quantities of high TDS water into the mine, which could affect the environmental impacts of the operation as well as capacity of the project water treatment plant.

5 GROUNDWATER MODELLING

Initially a preliminary modeling study was done to evaluate the type of model that should be used. Due to increasing TDS concentrations with depth it was possible that variable fluid densities could potentially influence the modeling results. A two-dimensional numerical model representing a vertical cross-section through A418 mine was used to compare flow and transport predictions that included variable density with predictions based on constant density. The model was constructed using FEFLOW (Diersch, 1998), which is a finite element code capable of simulating density and viscosity coupled groundwater flow, transport of solutes, and heat flow in three-dimensional porous media under a variety of boundary conditions. Due to the relatively low TDS concentrations and because of the relatively high hydraulic gradients that will be induced by mining, density effects were deemed to be negligible. Consequently the decision was made to use a single fluid density approach.

A detailed three-dimensional flow and transport model was constructed using Visual ModflowTM (Waterloo Hydrogeologic, 1999), which is a graphical pre- and post-processor for MODFLOW and MT3DMS model codes. MODFLOW is a finite difference code that was developed by the United States Geological Survey (McDonald and Harbaugh, 1988) to simulate transient groundwater in three-dimensional porous medium. MT3DMS is a MODFLOW companion code developed for the United States Environmental Protection Agency ((Zheng and Wang, 1998) which is capable of simulating three-dimensional, transient transport of dissolved chemicals in groundwater, using the grid and hydraulic heads calculated by MODFLOW. Both models assume that changes in the densities of a solute have insignificant effects on groundwater flow and solute transport.

The MODFLOW model grid for the A418/A154 pit area, shown in Figure 9, encompassed an area of 4.8 km by 5.3 km to a depth of about 1500 metres. It consisted of 108 columns, 125 rows, and 26 layers for a total of approximately 350,000 grid blocks. To provide sufficient resolution for model predictions and to ensure the stability of numerical solution, the grid block size is about 30 meters in the vicinity of the mines and increases towards the model boundaries. A separate, similar sized model was created for the A21 pit area.

Model calibration was completed by comparing actual inflows measured in cover holes drilled ahead of the bulk sample decline with inflows predicted by the model. The model could not be calibrated to the total water inflow recorded in the decline since grouting was regularly used to control inflows. The calibration process resulted in some minor changes to the

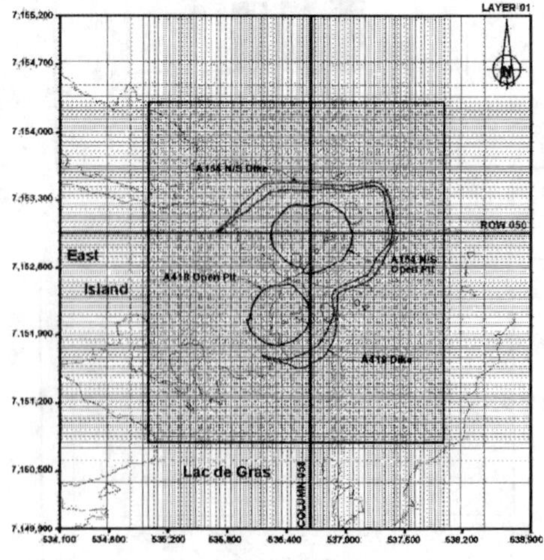

Figure 9: MODFLOW grid layout

initial hydrogeological parameters. Calibrated parameters are summarized in Table 1.

The groundwater model simulated the project development sequence, including construction of the water retention dikes and the staged excavation of the open pits and underground mines. Changes in the extent of mine components were incorporated in the model by automatically adjusting model boundaries every two years during the twenty-year mine life.

The predicted A418/A154 groundwater inflow quantities are shown in Figure 10, illustrating how inflow quantities increase with time as the pits and underground mines are deepened. The plot also shows corresponding TDS concentration in the mine water. Due to the lower permeability at depth, the amount of deep-seated brackish groundwater welling up is relatively minor. The TDS concentrations are averaged over the entire mine, combining lower TDS water from the upper portions of the pit with higher TDS water ingress near the pit bottom. The peak groundwater flows into the A154 and A418 open pit and underground workings is about 9,600 m^3/day with an average TDS concentration of 440 mg/L.

Figures 11 and 12 provide MODFLOW cross-sections through the A418/A154 pit areas showing hydraulic head and TDS concentration contours at Year 10 and Year 20 of the mine life.

In order to observe to what extent lake water migrates into the pit, an MT3D model run was done whereby the groundwater TDS concentrations were set to 0 mg/L and the lake water was arbitrarily set to 100 mg/L. By determining the mine inflow water quality with time, one is able to see how the percent lake water reporting to the pit increases with time. Figure 13 shows that the contribution of lake water to the total mine inflow increases rapidly in the first 5 years, and then levels off at about 85%

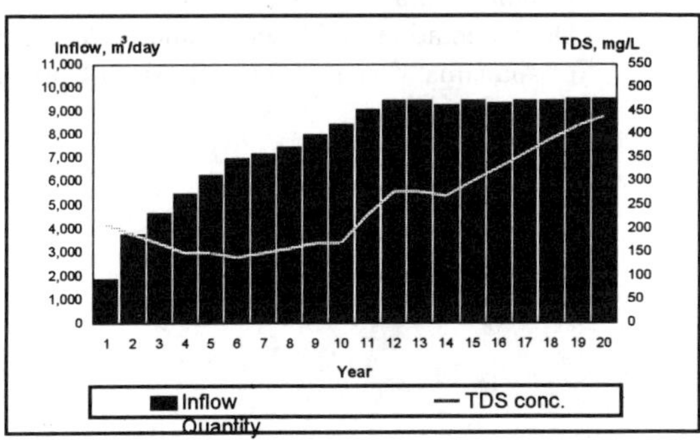

Figure 10: Mine inflow quantity and TDS concentration

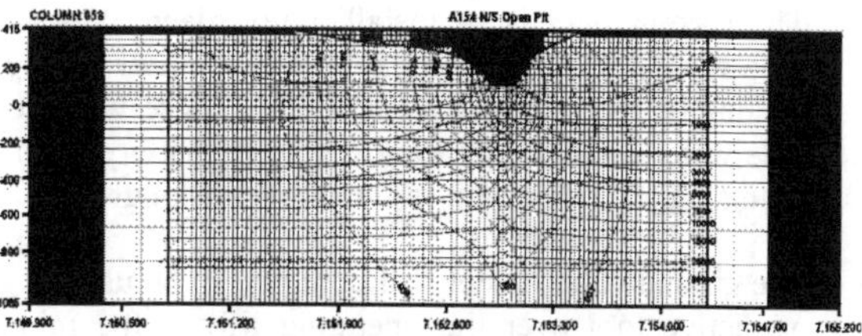

Figure 11: Hydraulic heads at Year 10

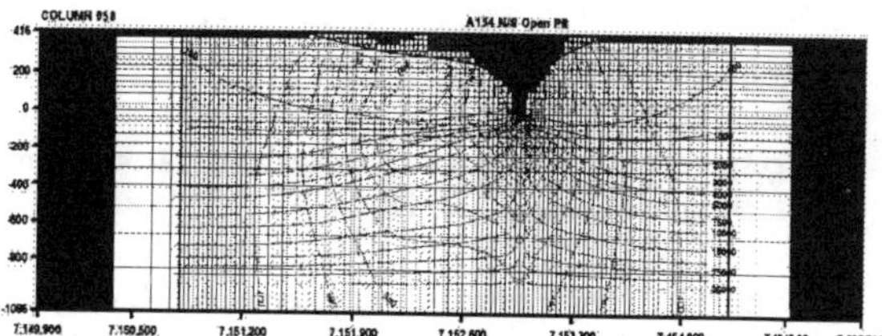

Figure 12: Hydraulic heads at Year 20

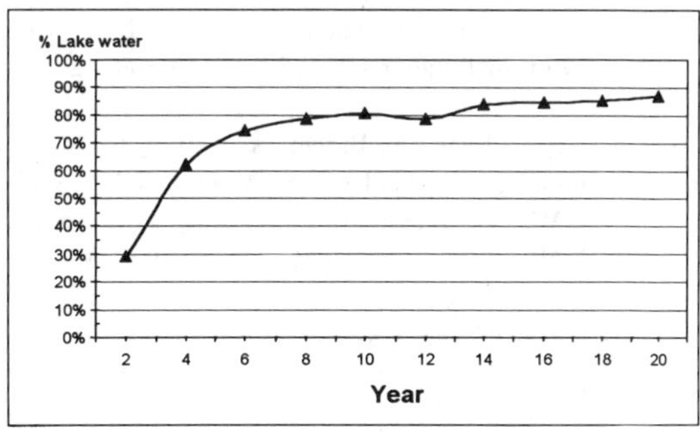

Figure 13: Percent lake water infiltration

6 CONCLUSIONS

Based on the hydrogeological investigations conducted at the Diavik site, it is reasonable to model the large scale rock mass as a single, hydrostratigraphic. A single fluid density approach is also valid even though the groundwater exhibits increasing TDS concentrations with depth.

Borehole temperature logging provides a relatively quick and simple way to assess the frequency and position of potential water bearing zones.

The current plan is to install a few deep Westbay type water sampling wells at site to monitor for upwelling of groundwater from depth. These will provide water quality data long before actually encountering higher TDS water in the mine workings.

The groundwater model will continue to be used as a predictive tool over the operating life of the mine. The hydraulic regime at site will be closely monitored and on-going model calibration will take place.

7 REFERENCES

Burgess, A.A. 1979. Pluton Hydrogeology. Atomic Energy of Canada, Ltd. TR-73.

Davison, C.C., A. Brown, R.A. Everitt, M. Gascoyne, E.T. Kozak, G.S. Lodha, C.D. Martin, N.M. Soonawala, D.R. Stevenson, G.A. Thorne, and S.H. Whitaker. 1994a. The Disposal of Canada's Nuclear Fuel Waste: Site Screening and Site Evaluation Technology. AECL-10713, Whiteshell Laboratories, Pinawa, Manitoba.

Davison, C.C., Chan, T., Brown, A., Gascoyne, M., Kamineni, D.C., Lodha, G.S., Melnyk, T.W., Nakka, B.W., O'Connor, P.A., Ophori, D.U., Sheier, N.W., Soonawalla, N.M., Stanchell, F.W., Stevenson, D.R., Thorne, G.A., Vandergraaf, T.T., Vilks, P., and Whitaker, S.H. 1994b. The Disposal of Canada's Nuclear Fuel Waste: The Geosphere Model for Postclosure Assessment. AECL-10719, Whiteshell Laboratories, Pinawa, Manitoba.

Diersch, H.G. 1998. FEFLOW Interactive, Graphics-Based Finite-Element Simulation System for Modeling Groundwater Flow, Contaminant Mass and Heat Transport Processes. WASY Institute for Water Resources Planning and System Research Ltd., Berlin, Germany.

Frape, S. K. and Fritz, P. 1997. Geochemical trends for groundwaters from the Canadian Shield; in Saline Water and Gases in Crystalline Rocks. Editors: Fritz, P. and Frape, S. K. Geological Association of Canada Special Paper 33, p. 19-38.

McDonald, M.G., and A.W. Harbaugh. 1988. A modular three-dimensional finite difference groundwater flow model. Techniques of Water-Resources Investigations, 06-A1, U.S. Geol. Surv., 528 p.

Stevenson, D.R., A. Brown, C.C. Davison, M. Gascoyne, R.G. McGregor, D.U. Ophori, N.W. Scheier, F.W. Stanchell, G.A. Thorne and D.K. Tomsons. 1996a. A Revised Conceptual Hydrogeologic Model of a Crystalline Rock Environment, Whiteshell Research Area, Southeastern Manitoba, Canada. AECL-11331, Whiteshell Laboratories, Pinawa, Manitoba.

Stevenson, D.R., E.T. Kozak, M. Gascoyne and R.A. Broadfoot. 1996b. Hydrogeologic Characteristics of Domains of Sparsely Fractured Rock in the Granitic Lac du Bonnet Batholith, Southeastern Manitoba, Canada. AECL-11558, Whiteshell Laboratories, Pinawa, Manitoba.

Ophori, D.U. and T.Chan. 1994. Regional Groundwater Flow in the Atikokan Research Area; Simulation of 18O and 3H Distributions. AECL-11083. Manitoba.

Ophori, D.U., Brown, A., Chan, T., Davison, C.C., Gascoyne, M., Scheier, N.W., Stanchell, F.W., and Stevenson, D.R., 1996. Revised Model of Regional Groundwater Flow in the Whiteshell Research Area. AECL Whiteshell Laboratories, Pinawa, Manitoba.

Waterloo Hydrogeologic Inc. 1999. Users Manual for Visual MODFLOW v. 2.8.1 – The Fully Integrated, Three-Dimensional, Graphical Modeling Environment for Professional Groundwater Flow and Contaminant Transport Modeling. Waterloo Hydrogeologic Inc., Waterloo, ON.

Zheng, C. and P.P. Wang. 1998. MT3DMS, a Modular Three-Dimensional Multispecies Transport Model for Simulation of Advection, Dispersion and Chemical Reactions of Contaminants in Groundwater Systems, Documentation and User's Guide. University of Alabama. prepared for U.S. Army Corps of Engineers.

Environmental Issues and Management of Waste in Energy and Mineral Production, Singhal & Mehrotra (eds)
© 2000 Balkema, Rotterdam, ISBN 90 5809 085 X

Elements of a resource-orientated analysis of the material flow of aluminium: A systems analysis approach

W. Kuckshinrichs, H.-G. Schwarz, P. Zapp & J.-F. Hake[1]
Programme Group Systems Analysis and Technology Evaluation (STE), Forschungszentrum Jülich, Germany

ABSTRACT: The production, processing and use of aluminium initiates direct and indirect material and energy flows. Resulting environmental damages justify efforts to reduce the flows. The Collaborative Research Centre (SFB) 525 „A Resource-Orientated Analysis of Metallic Raw Material Flows" aims to analyze existing options and to develop operative recommendations for a resource-sensitive utilization of metallic raw materials. The analysis is carried out using computer-based models. The models reduce the complexity of the material flows to its main elements, taking into account its technical, economical and ecological aspects on a different level. They are applied to set up balances of present material and energy flows. Using scenario techniques, trends and plausible developments are demonstrated. Its effects on material and energy flows, choice of production location, and resource productivity are to be analyzed. The paper concentrates first on methodological aspects of modelling the material flows of aluminium. It focuses on a process chain model and a model of the global aluminium industry. Moreover, model-based calculations of energy use and greenhouse gas emissions of present and future aluminium flows are presented.

1. INTRODUCTION

Worldwide, scientists and politicians are discussing the concept of sustainable development. At the moment, the discussion is shifting to resource and environmentally sensitive products and industries. Besides energy and its conversion, now nonferrous metals and the metals industry are not excluded (Stern, 1995; Shinya, 1998; Sánchez, 1998; Schwarz, 1999b). Experts agree, that for any development path to be sustainable economic, ecological and social aspects must be kept in mind, as well as aspects of intergenerative and intragenerative justice (fig. 1). It is agreed, too, that for more practical purposes there is a need to differentiate strategic rules, which are already formulated on a highly aggregated level, with respect to special aspects of products and production sectors, in order to formulate operative recommendations.

The long-term goal of the Collaborative Research Centre (SFB) 525 „Resource-orientated analysis of metallic raw material flows" is the identification of options for resource and environmental-sensitive supplying and processing of metallic raw materials considering technical developments and economic and ecological aims. The reproduction of interdisciplinary interactions is regarded as an important aspect of a resource-orientated analysis.

The scope of the analysis of SFB 525 ranges from deposit evaluation to extraction through the processing and smelting of mineral resources up to manufacturing and utilization. The recycling processes for supplying secondary resources after the use phase are also analysed and assessed as an integral part of the raw materials supply. Transportation, supply of final energy and utilization or disposal of the most important waste flows along the process chain are included in the investigation. This procedure makes it possible to analyse both environmental and economical aspects of the technical process chain, as well as selected social aspects.

As part of SFB 525, STE is developing and using computer-based tools within the framework of systems analysis. The combination and use of these tools is illustrated in fig. 2. The STE tools comprise an information system, a model to analyse material and energy flows in process chains, and a model to analyse the global metal industry. Additionally, experts from Aachen Technical University are developing tools for modelling single processes.

[1] The paper was produced within the framework of Collaborative Research Centre (SFB) 525 „Resource-orientated analysis of metallic raw material flows". Thanks are due to the German Research Council (DFG) for financial support.

Starting from the inventory of the status quo and the analysis of the corresponding resource utilization, scenarios will be designed to reflect a range of plausible developments. This includes basic trends of technical and organizational progress to be compared with goal-orientated scenarios considering interventions, e.g. by political decision makers. The potential effects on the material flows, the energy consumption, the choice of location for mining and processing plants and resource productivity will be examined for all scenarios.

This paper concentrates on methodological aspects of analysing the material flow of aluminium with the process chain model and the model of the global aluminium industry. Moreover, model-based calculations of energy demand and greenhouse gas emissions of present and future aluminium flows are presented.

2. MODELLING OF THE MATERIAL FLOW OF ALUMINIUM

2.1 *Global material flow*

The production of aluminium is divided into bauxite mining, production of alumina, and production of

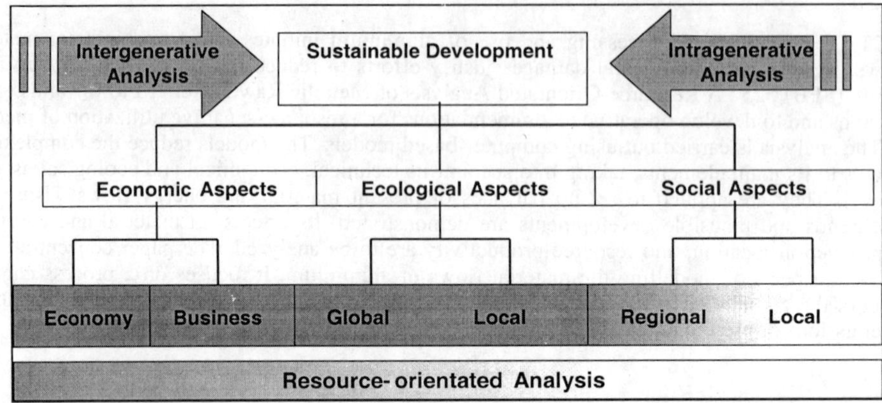

Fig. 1: Methodological aspects of sustainable development concepts

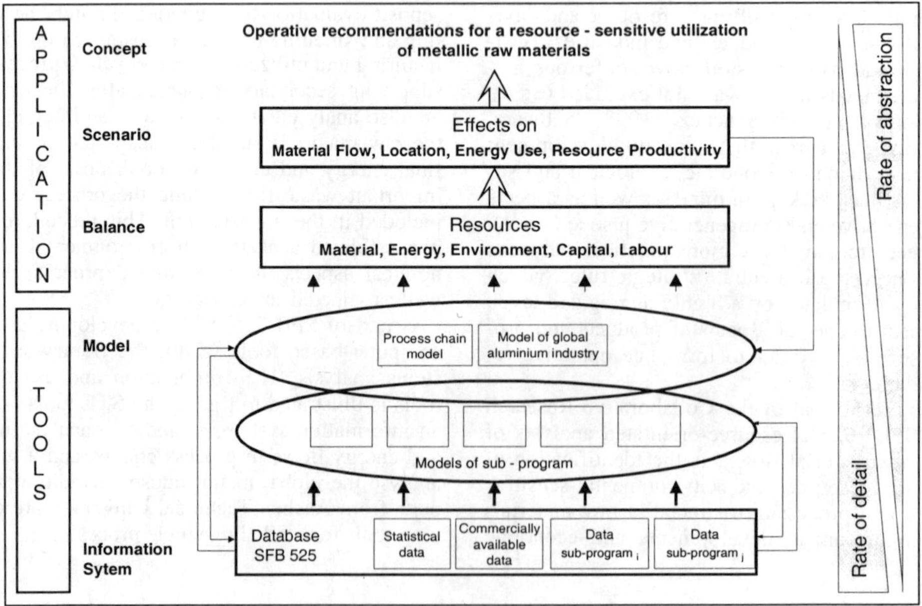

Fig 2: Tools and their application by SFB 525

primary and secondary aluminium. Mining of bauxite is tied to regions with natural deposits, whereas the production of primary aluminium is related to regions with sufficient and cheap electricity. The geographical distribution of production at each value-adding stage is the result of weighing up the competitive advantages and disadvantages of competing locations. Further basic aspects comprise geopolitical stability (especial for mining activities), environmental policy, infrastructure, and level of training of employees.

Production and demand figures are statistically fairly well covered. Fig. 3 gives an overview for 1995 and shows regional shares of bauxite reserves, production at different value-adding stages, and demand. Moreover, fig. 3 points to the aspects of global integration and international division of production of the aluminium industry. Different regional shares for production and consumption support this argument.

The production and processing of aluminium and use of aluminium-based products initiates direct and indirect material and energy flows, which in contrast to production figures are basically only available on a company level. These flows comprise red mud, industrial effluents, primary energy carriers, and greenhouse gases, e.g. CO_2 and perfluorinated hydrocarbons.

To quantify material and energy flows of the aluminium industry on a national or regional level model-based approaches may be used. It has to be taken into consideration that material and energy flows are determined technically, but motivated economically. Whereas the refining of bauxite to alumina reflects a technical production process, the decision to produce at a special location is taken considering economic aspects.

To analyse the material and energy flows of aluminium two computer-based models have been developed. The models are independent, but complement one another, as they look at material and energy flows from different viewpoints and focus on either technical or economic aspects.

2.2 *The process chain model*

Aims of process chain analysis

The process chain model serves for the presentation and evaluation of material and energy flows arising from the use of nonferrous metals and their environmental impact. Due to a very detailed presentation of technologies it offers the opportunity to investigate the effects of technology variations in the first place. In contrast to many other studies in the field of process chain analysis of raw materials (Shen, 1981; Tellus, 1992; Habersatter, 1995; EAA, 1996; BGR, 1997) the work of SFB 525 puts the main emphasis on the high disintegration for the presentation of sequences within the processes, rather than differences within geographical regions. The aggregation level of the various process modules is lower than in other studies. Possible questions may be:

- Which processes have the biggest share in the consumption of resources?

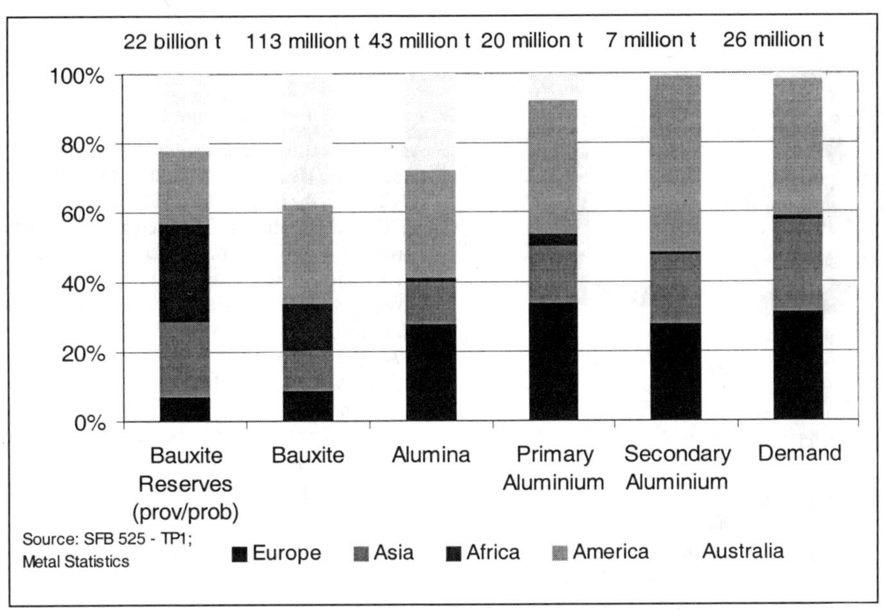

Fig. 3: Bauxite reserves, production, and consumption of aluminium and its regional shares

- With which process alternatives can the material and energy flow be reduced, while maintaining the same level of production?
- How must closed-loop material recycling be designed to reduce raw material input?
- What effects does an increase of the recycling quotas have on material and energy use?
- How does technical progress affect material and energy use and emissions?

Other questions can be answered, which refer to special regional features. Therefore, the location-independent modules have to be assigned to the different regions. The results show the influence of varied import and export structures.

Structure of a process chain for the German packaging sector

Within the work of SFB 525 further methodological aspects of the process chain analysis have been developed for the use of aluminium in the German packaging sector. The packaging sector is well documented and various aspects of process chain analysis, such as consideration of different alloying qualities, mixing of primary and secondary aluminium, or recycling concepts, can be modelled. At the same time the variety of products in this sector is smaller than for example in the transport sector. A mixing of the different methodological approaches is decreased by the limitated product complexity.

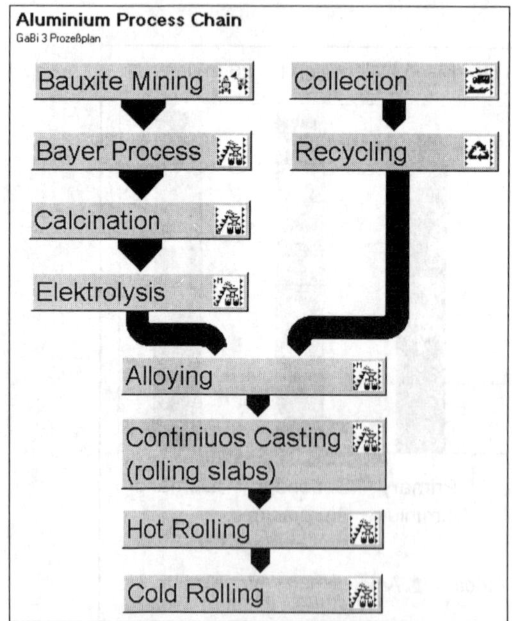

Fig. 4: Simplified structure of the aluminium process chain in the packaging sector

Inputs and outputs of the entire process chain are collected for the base year in the inventory analysis. The chosen process chain starts with the mining of the raw material bauxite according to the German imports, considers the production of alumina in the Bayer process with subsequent calcination, the production of aluminium from alumina in the electrolysis, the mixing of primary and secondary aluminium in the process step of alloying, up to the semifinished product foils, through the production of transportable formats, and finally hot and cold rolling. In the field of secondary production, the used packages are collected and mechanically or thermally recycled. The metal-containing prematerials are used to produce alloys. Figure 4 shows a simplified structure of the process chain.

In between the various production processes transport processes take place, which are also included in the analyses. Also taken into consideration are side chains for electricity production to obtain the necessary final energy for the processes. The production of intermediate materials for the Bayer process includes soda (NaOH) and lime (CaO) production. At the electrolysis level, anode, cathode, aluminium fluoride (AlF_3) and petroleum coke production is included in the balance. At the moment processes for waste treatment concentrate on red mud treatment, which is the biggest output along the process chain.

The second stage of manufacture and also the use phase are important elements in the process chain, but have not yet been included. First aspects in the analysis have already been made (Bauer et al., 2000).

To obtain a process chain, a process library has been set up, which includes all processes of the aluminium flow. Technology-specific and location-independent modules were determined. The technical status of the processes is considered in the classification of different technology levels. The processes are divided into old technologies (old), present technologies (PT), and the latest available technologies (NT). Furthermore, technical options for future use (FT) are presented. To keep the number of possible process modules small, those processes which differ just insignificantly are combined. The calculation of average modules of the site-independent processes is in a production-weighted manner.

At the moment about 170 processes directly connected with the production and recycling of aluminium, about 30 processes concerning intermediate inputs or outputs and about 120 modules describing transport and energy supplying processes are included in the process library. Additionally, mineralogical information on 170 deposits is included.

Contrast to other studies

The process chain, developed within the framework of SFB 525, differs greatly from other projects (Tellus, 1992; Hydro Aluminium a.s., 1994; EAA, 1996; BGR, 1997; Gielen, 1997; Reuter, 1998). Most of the studies do not directly include the mineralogical qualities of the deposits. It is therefore impossible to obtain detailed information. In the mining field many studies model regions. Different geological qualities in one region cannot be adequately considered. Local mining is mainly influenced by the geological structure rather than the region in which it is located.

If no mineralogical deposit information is known, the modelling of the Bayer process can only be done with average parameters, irrespective of the real geological structure. Also it is normally not considered whether the bauxite used is of gibbsitic quality or has a higher share of boehmite, which is important for the selection of the technology needed.

Quite a lot of studies about aluminium are comparisons of products which simulate specific locations in the process chain. Within the framework of SFB 525 production-weighted worldwide averages have been modelled by using site-independent processes, which can easily be modified for country-specific analysis.

An analysis of the primary process chain is performed by most of the studies. Often the scope ends here. The secondary production chain is not always integrated. One goal of SFB 525 is to give a complete overview of the production and recycling processes. In the module group 'alloying' a mixing of primary and secondary aluminium takes place. From this point on the process chain is not only modelled according to its condition but also according to its alloying contents. This approach is new. Now it is possible to deal with all aspects of the variety of aluminium alloys. Similarly, the production of semi-finished products is modelled showing the respective qualities.

Recycling of aluminium is not taken into consideration in many studies or only modelled in a highly aggregated way. For this production chain model different modules of recycling processes for the various sources of secondary raw material coming from the use phase are developed. Moreover, the recycling of new scrap is also taken into consideration.

2.3 *Model of the global aluminium industry*

The model of the global aluminium industry permits an integrated analysis of the technical, geographical, ecological and economic dimension of the material flow of primary aluminium (Schwarz, 1999a; Schwarz et al., 2000). Questions which can be answered by the model are:

- How will the production geography of the worldwide flow of primary aluminium change in the future?
- What about the trade flows?
- Which technologies will be successful in competition and will old plants, upgraded and expanded old plants, or new plants be used for future production?
- How will the absolute and specific final and primary energy consumption and emissions of greenhouse gas develop worldwide and in the regions?

Additionally, the effects of instruments of material flow management can be assessed, for example:

- What are the effects of national or global energy or carbon dioxide taxes on production geography and greenhouse gas emissions?
- What happens when import duties on bauxite, alumina or primary aluminium are perceptibly reduced or raised?

The model of the global aluminium industry answers these questions through simulation of a competitive market for primary aluminium and intermediate products. The flow of primary aluminium is seen as economically determined. Human needs and market processes are responsible for the extent, the technical conditions, and the ecological effects of the material flow. The demand for primary aluminium induces the material flow. Different technologies and regions are in competition to satisfy the demand for primary aluminium and intermediate products. The market process decides which technologies and regions will win the competition and therefore what quantity of emissions will be generated.

Model classification

The model of the global aluminium industry is a further development of existing approaches (Brown et al., 1983; Nichols et al., 1992; Manne et al., 1994). It is computer-based and uses a linear programming technique. The demand for primary aluminium is assumed to be given and its extent is independent of price from the model's perspective. The model is static. All the model parameters and all the variables (production and transport quantities, investment) refer to the year 2010. The reference year 1995 influences the model results only indirectly. The assessment of model parameters for the year 2010 is based on figures for 1995 or a period including 1995. The implemented mathematical algorithm calculates a long-term market equilibrium with perfect competition (Takayama et al., 1971; Landwehr, 1997). It is therefore assumed that the producers of bauxite, alumina and primary aluminium maximize their profits decentrally. They accept the market price as given and adjust their production quantities according to their marginal costs.

Material flow modelling

The model of the global aluminium industry considers the global flow of primary aluminium. The manufacture of primary aluminium in the remaining process chain, reusage through recycling and, if necessary, waste disposal is additionally part of the total flow of aluminium.

Technical transformations within the material flow are reflected in three process steps: mining, refining and smelting (Fig. 5). The chemical and mineralogical heterogeneity of bauxite is reflected in 18 bauxite types. The different technical options for the three process steps are integrated into a few technologies. Six technologies are distinguished for refining and four technologies for smelting. The model separates four kinds of investment strategies: old plants, upgrading, brownfield expansion and greenfield expansion. Upgrading of old plants means that the technical parameters of the plants are improved by changing components. Outmoded technologies are usually replaced by modern technology. Within the model, upgrading is not connected with capacity expansion. It is the brownfield expansion which leads to the capacity expansion of existing plants. Greenfield expansion means that new plants are built. The reason for considering the different investment strategies is, on the one hand, that old plants and new plants do not have the same technical parameters even if they use the same technology. Usually the new plant will be superior. On the other hand, new plants as well as brownfield expansion and upgrading cause investment costs, whereas the (former) investment costs of old plants are not relevant from the model's perspective ('sunk costs').

The model considers three transportation steps: mine to refinery, refinery to smelter and smelter to final demand. Means of transport are truck, rail, conveyor belt, inland and overseas ship. The model considers 15 world regions: USA, Canada, Brazil, Rest of Latin America, Germany, Rest of European Union, Rest of Western Europe, Former Soviet Union, Rest of Eastern Europe, Africa, Near and Middle East, India, China, Rest of Asia and Australia.

2.4 *Comparison of model characteristics*

The models considered in this paper are both process-based, complementing one another, but with different viewpoints concerning material and energy flows. Therefore, their contribution to the resource-orientated analysis differs in scale and direction.

Whereas the process chain model concentrates on the detailed reproduction of technical processes, the model of the global aluminium industry follows an economic approach. Both tools show material and energy flows along the production chain of aluminium, using the concept of process description via inputs and outputs. At the moment, the model of the global aluminium industry follows the production chain of aluminium from bauxite mining to production of primary aluminium on a global scale. As important side chains, the supply of final energy carriers, of anodes and of soda are included. Moreover, the process chain model follows the production chain up to the supply of secondary resources. Ad-

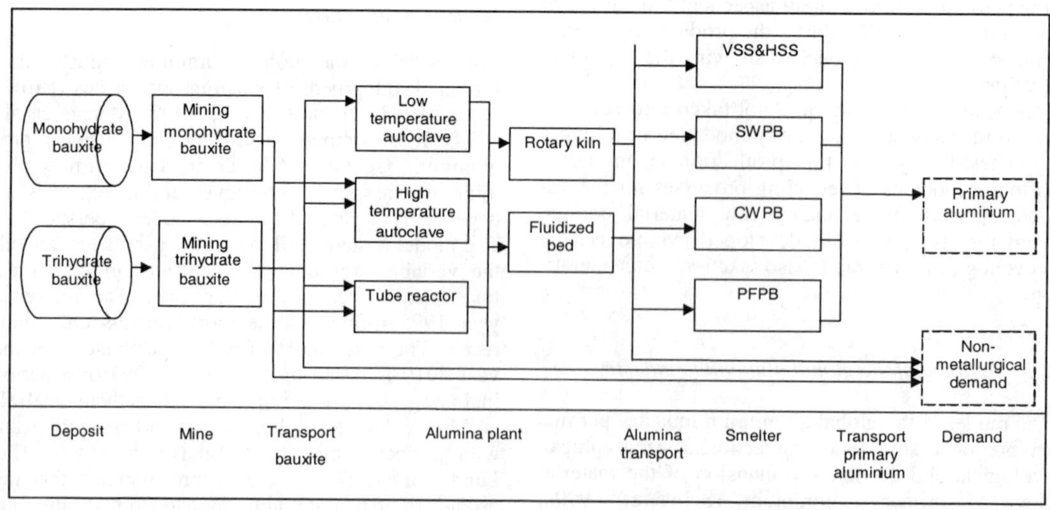

Fig. 5.: Simplified structure of the model of the global aluminium industry

Table 1: Characteristics of the model approaches

	Model of the global aluminium industry	Process chain model
Concept	Market simulation, cost minimization	Simulation of integrated technical processes
	Absolute aluminium trade balance	Specific input-output data
	Aggregated technologies	Disaggregated technologies
Systems boundary	Global	Regional / country-specific / site-independent
	Bauxite mining to primary aluminium	Bauxite mining up to processing and casting including recycling of secondary resources
Exoge-nous parame-ters	Demand	
	Technical parameters	Technical parameters
	Cost parameters Duties	
Model results	Regional and value-added stages: Production, investment	
	Demand for main input	Demand for main input
	Other inputs: e.g. energy, caustic soda, anodes	Other inputs: e.g. energy, caustic soda, petrol coke, pitch, lime, anodes, cathodes, AlF_3
	Emissions: CO_2, CF_4, C_2F_6	Emissions: e.g. CO_2, CF_4, C_2F_6, HF, red mud, industrial effluents

ditionally, it includes further important side chains such as handling of red mud and supply of aluminium fluoride (AlF_3). The process chain model follows the concept of a site-independent tool, which can easily be modified for country-specific or regional-specific studies based on a higher disintegration of technical processes.

The following table shows a comparison of important model characteristics. It focuses on the model's concept, the system's boundary, the exogenous parameters, and the model results.

- Whereas the process chain model simulates technical processes based on specific input-output data, the model of the global aluminium industry simulates a competitive aluminium market, using full aluminium trade balances.
- Whereas the process chain model exclusively uses technical parameters as exogenous parameters, demand for primary aluminium and cost parameters including duties and taxes constitute further important exogenous parameters of the model of the global aluminium industry.
- Whereas the process chain model assesses the specific demand for preproducts of the aluminium chain as well as the specific demand of a range of side chains and important emissions, the model of the global aluminium industry restricts itself to determining absolute balances of the production chain up to primary aluminium, and fewer, but important side chains. Moreover, as an economic model, it determines regional production and investment allocation.

3. RESOURCE-ORIENTATED ANALYSIS FROM A MODEL PERSPECTIVE

Offering the opportunity to analyze a system from a technical or an economic viewpoint, and to balance both energy use and emissions, the models are used first for inventory analysis. In the following, model-based scenario calculations for material and energy flows are presented.

3.1 *Inventory Analysis*

With respect to different questions for the models, to different model characteristics, and the resulting different system boundaries, the models are used to quantify different complementary balances. The process chain model focuses on the production of aluminium for packaging purposes in Germany whereas the model of the global aluminium industry concentrates on the worldwide production of primary aluminium. For the global model, 1995 serves as the base year, but 1997 for the process chain model.

To give a better understanding of energy use and CO_2 emissions calculated for the base year 1995, some reflections are useful. The specific final energy consumption within the material flow just depends on the energy efficiency of the technologies used for mining, refining and smelting. The primary energy consumption is also influenced by the efficiency of the primary energy conversion, especially the efficiency of the electricity supply for electrolysis. Furthermore, a differentiation between the contract mix and the regional mix for electricity supply results in different calculations.

The figures of overall efficiencies for regional electricity production and the associated CO_2-emission factors were assessed by project partners (Briem et al., 1999). Contract mix and regional mix are differentiated for smelting. The contract mix considers the energy carriers for electricity production used in accordance with existing contracts for smelting or for self-generated electricity. The regional mix on the other hand reflects the average shares of the energy carrier for the regional production of electricity. The contract mix is characterized by a high share of hydro-based electricity production on the world average. In 1995 this share was about 60%. This is not surprising. Cheap hydro-based electricity is the important locating factor for smelters. The share of hydro-based electricity for the regional mix was much lower, about 30%. Fossil-fuelled power plants had, in contrast, a much higher share with more than 56%. For the contract mix, the share was less than 37%.

Process chain model

In a weak point analysis those processes can be identified, which exert a major influence on the use of resources. One of the resources considered is the energy demand (see Fig 2). Geographical and time boundaries are given to calculate the starting condition. In the field of collecting and recycling secondary raw Material, data about the distribution of the technologies used cannot be gathered from the base year, but from 1997. Therefore this year is chosen as the starting period. A first analysis show the German situation, i.e. the mixing of primary and secondary aluminium, the production of the semifinished products and the scrap recycling only considers the German sites. The process chain sequence of primary aluminium production includes the imports and exports of bauxite, alumina and primary aluminium.

At the moment, the calculations are based on the assumption of regional mix for electricity supply. Electrolysis has the highest primary energy demand along the entire process chain during the production of aluminium foils as well as for another product. Per tonne of aluminium foil 175 GJ primary energy is needed, which corresponds to 143 GJ/t primary aluminium. Besides that the influence of the other processes will be analysed as well. In a first step, an order of the processes relative to the primary energy demand was collected. The process variations on one level (e.g. CWPB, PFPB) were combined. Side chains were analysed separately. The second most important process after electrolysis is the Bayer process, which with about 25 GJ/t aluminium foils needs far less primary energy. The subsequent calcination has the third highest primary energy demand with about 15 GJ/t foil. Hot and cold rolling,. anode production and the alloying process need 6.5 – 4 GJ/t foil. In the recycling sector the demand for the reuse of total packaging is identified (1 GJ/t foil) and not only the demand for that part which can be reused to produce foils. Mining, continuous casting, cathode production and red mud deposition need the lowest amount of primary energy and may be neglected.

Decisive for these results is the distribution structure of the exporting countries with the associated technologies and the chosen power mixes used to convert the final energy demand into primary energy. Beside the analysis of primary energy demand other questions such as the value of emissions occurring or the fluctuation rate of single inputs or outputs in one level of the process chain can be answered. To answer other questions a new process chain using the process library must be first built up.

Model of the global aluminium industry

The results are based on the assumption of contract mix for electricity supply. Hydro-based electricity has the highest overall efficiency and is free of carbon dioxide emissions. Therefore the global primary energy consumption and carbon dioxide emissions are lower for the contract mix. In 1995 the flow of primary aluminium consumed 2.4% of total world electricity production. Smelting itself was responsible for about 95% of the consumption of electricity in the material flow. In 1995 the world average specific primary energy consumption was about 160 GJ and the world average carbon dioxide emissions was 10.7 tonnes per tonne of primary aluminium for the contract mix (the figures for the regional mix were assessed to be 200 GJ and 13.2 tonnes). As the worldwide consumption of primary energy in 1995 was assessed at about 343,000 PJ, the share of the flow of primary aluminium was about 1%. Respectively, flow of primary aluminium initiates about 1% of energy-based global carbon dioxide emissions.

3.2 *Scenario*

As in the inventory analysis for the initial situation, the two approaches are used against the background of the different model characteristics in different ways for future projections of the aluminium supply and the resulting material and energy flows. The aim of work with the process chain model is the preparation and analysis of a scenario analysing the impacts of technical progress on material and energy flows.

This is implemented by altering the process chain of a base year by replacing individual technologies. This procedure enables, for example, old or currently used technologies to be replaced by the latest available technologies. A scenario assuming a range of primary aluminium demand is calculated using the model of the global aluminium industry. It is thus possible to analyze the impacts of different levels of demand on the production geography, the technology mix and the resulting material and energy flows.

The target year for both scenarios is 2010.

Process chain model

Using the process chain model and considering the process structure it is appropriate to answer questions concerning technical problems first. Therefore, in a first scenario only technical variations are investigated. Until 2010 neither a totally new structure of today's technical concepts nor the introduction of totally new technologies is likely. Instead, an assimilation of existing technologies is assumed. For the calculation the demand for packaging and the import structure according to 1997 has not (yet) been changed. Furthermore, some assumptions concerning technical change are presented.

The chosen time horizon of 15 years is a very short period for mining activities. Mining in one deposit normally takes much longer than that (30-40 years). Hence, for Germany no import change is assumed, the deposits will be the same for both years. Technical progress in mining processes is expressed in such a way, that mining activities approach those of the best deposits.

The mineralogical qualities will still determine the choice of Bayer technology. Changes in alumina production only result from the replacement of old rotary kilns by new fluidized bed furnaces.

The developments of the last few years have shown that most of the old Soederberg electrolysis plants have been replaced by modern prebaked technology (Pawlek, 1998). This trend will continue for the scenario calculations for 2010. It is assumed that all Soederberg plants will be replaced by PFPB plants (Quinkertz, 1999).

The next level is the alloying step. Right now, new furnaces types are being developed in this area. Nevertheless, it is assumed that existing concepts will be optimized by 2010 and their energy demand will decrease due to the use of synergetic effects.

Processes for semifinished products have been optimized over a long period. An improvement up to the year 2010 can be reached by replacing the technologies used today with alternative technologies, such as continuous casting processes.

In the process chain sequence, recycling of secondary raw materials is shifted towards modern technologies. Composite material recycling and pyrolysis are currently the latest technologies in use.

For the area of energy supply two effects have to be considered for the scenario calculations for 2010. One is the shift in the mix of the primary energy carriers, the other is the change in the energy de-

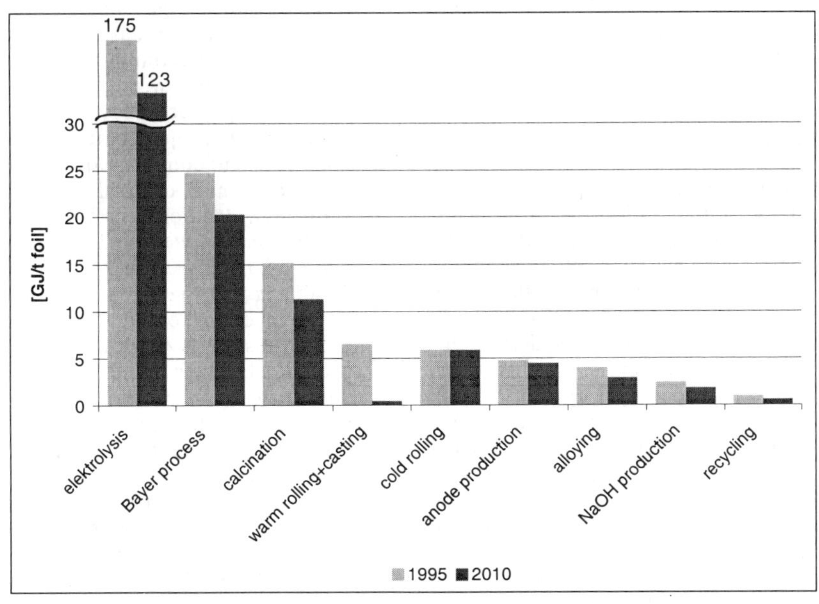

Figure 6: Primary energy demand per tonne aluminium foil

mand. The construction of new power plants means building plants with the latest technical standard (NT) (Briem, 1999).

A first look has been taken to the change of the energy demand following out of the suggested technical variations. To make the different final energy carriers comparable along the entire process chain they have been transformed into primary energy. Figure 6 shows the decrease of the primary energy demand for the most important processes per tonne aluminium foils produced.

The decrease of the primary energy demand is influenced by a combination of four single effects: improvement of existing technologies, replacement of technologies, technical related changes of material flows and an improvement in the energy supply. The influence of either of this four effects can differ between the various process steps. For example, the energy demand of the electrolysis decreases by 30 %. The mayor influence for this is the reduction of the material flow due to the exchange of the warm rolling process by an continous rolling process. The material demand of primary aluminium is lower. The second highest effect is combined with the improvement of the energy supplying technologies. The replacement of electrolysis technology has only a minor effect in this example.

Therefore every result has to be analysed carefully to yield to a better understanding of the complex system. Beside the energy demand other effects can be investigated such as emissions or material changes. In reality influences can occure which prevent the total scoop of the technical potential estimated in this example. Most of the time this are economic reasons, which are not considered in the process chain model. Economic models are used to investigate this effects.

Model of global aluminium industry

As a critical model parameter, assumptions on future demand of primary aluminium have great influence on model results because the implemented material balance equations lead to an equalization of supply and demand. Moreover, projections of the future global demand for primary aluminium are difficult. The reason is that the growth rates of the demand for primary aluminium are dependent on the future development of the world economy. Competition with other raw materials and the recycling quota additionally influence demand. Therefore, future demand for primary aluminium is varied to reflect a range of possible developments.

Three cases are distinguished. Firstly, the base case assumes a moderate growth rate of primary aluminium of 1.9% per year from 1995 to 2010. This is the growth rate which was valid for the period from 1980 to 1995. The low-demand case assumes a growth rate of 1.4% per year and the high-demand case a growth rate of 2.4%. The assumptions for future technological developments are similar to the assumptions for process chain modelling.

Though we use the model to analyse future energy use and greenhouse gas emissions of global primary aluminium industry, the scenario calculations allow to deduce several fundamental trends:

- Only gradual change in production geography: The reason for this is the good competitive situation of the old capacity. By means of upgrading, old capacity achieves almost the same technical parameters as new plants. At the same time upgrading is much cheaper. Therefore upgrading of old plants is often an economic alternative to building new plants even if the price situation of one region is disadvantageous, e. g. because of high energy prices or wages. The reason for changes in production geography is not so much the old capacity being superseded by new but rather additional demand which cannot be satisfied by existing capacity.
- Shifting of production of alumina to regions with significant bauxite production: The cost of bauxite is crucial for alumina production. On a worldwide average the bauxite costs in 1995 were about one third of total operating costs. The transport costs are essential for the regional differences in the costs of bauxite. The longer the transport distances and the lower the alumina content of the bauxite, the higher the specific transport costs are. This explains the trend for locating refineries close to mines and why only bauxite with a high alumina content is traded between regions. Because of the remarkably high capital costs of refining the shifting of production is relatively low.
- Shifting of production of primary aluminium to regions with cheap electricity for smelting: Alumina costs are dominant for smelting with a share of one third of total operating costs. Nevertheless, electricity costs which were responsible for one quarter of the costs are more important for the location of smelters. The reason for this is that the variation in prices of electricity is much higher than that of alumina. This is not surprising. Hydropower has a natural competitive advantage and electricity is only to a limited extent tradable. The existence of old plants in disadvantageous regions is usually not endangered. The total costs of new plants in privileged regions are mostly higher than the operating costs of existing plants in disadvantageous regions. But the building of new plants and capacity expansions of existing plants are usually realized in regions with low electricity costs so that the

production shares of these regions are gradually increasing.

- Declining interregional trade of bauxite and growing trade of alumina and primary aluminium: The changes in production geography have a strong influence on interregional trade. The share of total regional net exports in world production is decreasing for bauxite whereas it is stable or slightly increasing for alumina and primary aluminium. Instead of bauxite, the manufactured goods - alumina and primary aluminium - are traded.
- Changes in technology mix of primary aluminium and alumina production: In 1995 Soederberg plants produced more than 30% of world primary aluminium. Soederberg plants have high energy requirements. This leads to high operating costs. Therefore Soederberg plants will be upgraded with PFPB or shut down. CWPB and SWPB plants will also lose importance in comparison to PFPB plants. Autoclave technology will further dominate the digestion of bauxite. Tube reactors with a share of 2% of world alumina production in 1995 will enlarge their share only to a very limited extent. In 1995 the rotary kiln was used for calcination of 26% of world alumina. Because of higher energy requirements and therefore cost disadvantages the rotary kiln will be superseded by the fluidized bed in the near future.

The expansion of demand for primary aluminium will have great influence on the absolute figures of primary energy consumption and greenhouse gases as well. In the base case, assuming demand increase of 1.9% per year, absolute use of primary energy and CO_2 emissions increase only moderately. Perfluorinated hydrocarbons decrease considerably. In the low case, assuming demand increase of 1.4% per year, absolute primary energy use and CO_2 emissions decrease moderately, whereas perfluorinated hydrocarbons decrease by 44%. In the high case, assuming demand increase of 2.4% per year, absolute primary energy use and CO_2 emissions increase perceptibly. But even in the case of high demand, perfluorinated hydrocarbons decrease considerably.
In all cases, specific figures show a uniform decreasing trend. The specific consumption of electricity for smelting will decrease because of Soederberg technology being superseded by modern PFPB technology. There will also be a decline in specific consumption for other energy carriers.

The world average specific primary energy consumption in the material flow will decrease considerably (fig. 7). Reduced consumption for smelting accounts for more than 70% of this reduction. There are three reasons for this. In the first place, the higher efficiency of electricity generated from fossil fuels leads to a reduced consumption of primary energy. Secondly, modern PFPB plants use less electricity than the Soederberg plants they replace. And thirdly, the production of primary aluminium is being shifted to regions where the production of electricity is mainly based on hydropower. The shift itself is economically determined. There is a strongly marked (inverse) correlation between the share of hydropower for electricity production in one region and the regional tariffs for electricity for electrolysis.

The specific CO_2 emissions will decline significantly for the same reasons than the primary energy

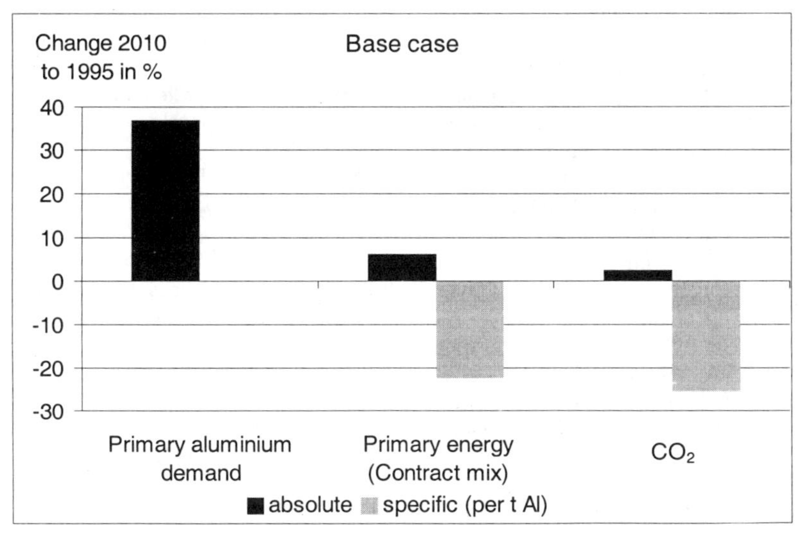

Fig. 7: Results of the base case

consumption. A decline in specific perfluorinated hydrocarbon emissions is very likely. The modern PFPB plants have much fewer anode effects. Therefore fewer perfluorinated hydrocarbons are generated.

4. CONCLUSIONS

The paper describes important elements in a resource-orientated overall consideration of the aluminium material flow, which is the object of Collaborative Research Centre 525. These elements comprise computer-based models for analysing the material and energy flows in process chains and for analysing the global aluminium industry, as well as their application in scenarios.

Methodological aspects of modelling the aluminium material flow are discussed in detail. The model approaches are characterized in a comparative manner and their potential elucidated with respect to application for an inventory analysis of a base year and for scenario calculations. It thus becomes apparent that the models complement each other in a resource-orientated overall consideration.

The basic results obtained by the inventory analyses performed with the two approaches for the base year 1995 are discussed.

Basic assumptions for the projection of future developments of the aluminium material flow are discussed. With respect to the scenario results and interpretations, major emphasis is placed in this paper on the model of the global aluminium industry.

5. REFERENCES

Bauer, C. et al. 1999. Einbindung von Nutzungsaspekten in die Stoffstromanalyse metallischer Rohstoffe. *To be published.*

Briem, S. et al. 1999. *Primärenergiebedarf und spezifische CO_2-Emission der globalen Stromerzeugung 1995 und 2010 - Methodik und Ergebnisse des Energiemodells.* Interner Bericht IB 99-12. Lehrstuhl für Reaktorsicherheit und -technik, RWTH Aachen.

Brown, M. et al. 1983. *Worldwide investment analysis – The case of aluminium.* World Bank Staff Working Papers 603.

BGR (Bundesanstalt für Geowissenschaften und Rohstoffe) 1997. *Stoffmengenflüsse und Energiebedarf bei der Gewinnung ausgewählter mineralischer Rohstoffe. Teil Aluminium.* Hannover.

EAA (European Aluminium Association) 1996.: *Ecological profile report for the european aluminium industry.* Bruxells.

Gielen, D.J. & A. van Dril 1997. *The basic metal industry and its energy use.* Netherlands Energy Research Foundation ECN, Petten/NL.

Habersatter, K. et al. 1995. Ökoinventare für Verpackungen. *BUWAL, Schriftenreihe Umwelt* No. 250/I+II.

Hausberg, J. et al. 1998. Global evaluation of geological constraints on the availability of mineral resources exemplary for the case of bauxite deposits. *Berichte der Deutschen Mineralogischen Gesellschaft, Beihefte zum European J. Mineralogy* 10(1): 120.

Hydro Aluminium a.s. 1994. *The Norwegian aluminium industry and the local environment. Summary report.* Elkem Aluminium ANS, Sor-Norge Aluminium A/S.

Labys, W. C. et al. (eds.) 1989. *Quantitative methods for market-oriented economic analysis over space and time.* Aldershot (UK).

Landwehr, U. 1997. *A model of the international aluminium flow.* Forschungszentrum Jülich, Interner Bericht STE – IB – 7/97.

Legarth, J. B. 1996. Sustainable metal resources management – the need for industrial development: efficiency improvement demands on metal resource management to enable a (sustainable) supply until 2050. *Journal of Cleaner Production* 4 (2): 97-104.

Manne, A. & L. Mathiesen 1994. The impact of unilateral OECD carbon taxes on the location of aluminium smelting. *International Journal of Global Energy Issues* 6(1/2): 52-61.

Nichols, R. K. et al. 1992. The corporate aluminium model: A GAMS-based model for optimal long range planning and investment analysis by aluminium companies. *Light Metals*: 43-58.

Pawlek, R.: *Primary aluminium smelters and producers of the world.* Aluminium-Verlag, Düsseldorf.

Quinkertz, R. et al. 1999. Primärenergieaufwand und kumulierte Emissionen verschiedener Elektrolysesysteme zur Aluminiumherstellung. *Erzmetall* 52(7/8): 393-402.

Reuter, M.A. 1998. The simulation of industrial ecosystems. *Minerals Engineering* 11(10): 891-918.

Sánchez, L. E. 1998. Industry response to the challenge of sustainability: The case of the Canadian nonferrous mining sector. *Environmental Management* 22(4): 521-531.

Schwarz, H.-G. 1999a. *Weltweiter Stoffstrom des Primäraluminiums: Grundlegende Entwicklungstendenzen.* Forschungszentrum Jülich,

Interner Bericht STE – IB - 2/99.

Schwarz, H.-G. & B. Krüger & W. Kuckshinrichs 2000. Studies on the global material flow of primary aluminium. *To be published in ALUMINIUM.*

Schwarz, L. H. 1999b. Sustainability: The materials role. *Metallurgical and Materials Transactions B* 30B(April): 157-170.

Shen, S. 1981. *Energy and material flows in the production of primary aluminium.* Prepared by Argonne National Laboratory for U.S. Department of Energy, ANL/CNSV-21.

Shinya, W. M. 1998. Canada's new minerals and metals policy. *Resources Policy* 24(2): 95-104.

Stern, D. I. 1995. The contribution of the mining sector to sustainability in developing countries. *Ecological Economics* 13: 53-63.

Takayama, T. & G. Judge 1971. *Spatial and temporal price and allocation models.* Amsterdam/London.

Tellus 1992. *Packaging study.* Vol. II. Prepared for U.S. Environmental Protection Agency.

Vita, D. Di 1997. Macroeconomic effects of the recycling of waste derived from imported non-renewable raw materials. *Resources Policy* 23(4): 179-186.

World Bank 1994. *Market outlook for major energy products, metals and minerals.* Washington.

Environmental Issues and Management of Waste in Energy and Mineral Production, Singhal & Mehrotra (eds)
© 2000 Balkema, Rotterdam, ISBN 90 5809 085 X

The Brazilian program of high pressure water jet to cut ornamental rocks

Carlos Tadeu Lauand, Guillermo Ruperto Martín Cortés, Lineu Azuaga Ayres da Silva & Wildor Theodoro Hennies
Mining Engineering Department, Polytechnic School of University of São Paulo, Brazil

Raimondo Ciccu
Dipartimento di Geoingegneria e Tecnologie Ambientali, Universitá degli Studi de Cagliari, Italy

ABSTRACT: After a period of some years of first studies of one of the Authors, a group was organized to acquire, install and operate, a modern High Pressure Water Jet Machining System as a New Module of the Rock Mechanics Laboratory at the Mining Engineering Department, on the Polytechnic School of the University of São Paulo. This Module is a XY table totally automated with a CAD/CAM System to realize with High Pressure Abrasive Water Jet nozzle cutting and create pieces from the most different kind of materials. The main objective, in the beginning of the researches, is the application of this advanced and environmental friendly technology cutting system for studies of the behaviour of Brazilian ornamental rocks mainly marbles and granites. The first phase of this Program, that was the installation of the system, is be realized, last year with the financial help of FAPESP the São Paulo State Agency for Research and will here be described. The future aims of the Project is to create Human Resources as Masters and Phi. D. familiarized and competent in the advanced productive processes, which to now are practically unknown in Brazil. This aims are investigation about the behaviour of rocks and other material to this kind of technology and its confrontation to the conventional ones.

1 INTRODUCTION

Since the middle of the 90's, one of the A. are searching in the area of rock excavation. In the last years a new searching field, that is the non conventional excavation techniques is the main aim Hennies, 1996). The non conventional techniques of rock excavation has a great future potential, than there can constitute alternative processes to the now conventional used, substituting there with advantages as more environmental friendly procedures.

In this researches many different aspects can be evaluate, and to now, papers were produced to relate noise of actual mining equipment, the electro-hydraulic or plasma blasting, water jet technology and application of jet piercing were some of the aspects treated in presented papers.

The noise problem was treated as human comfort need in the mining industry and some used equipment were analyzed in a cooperation research with the University of Cagliari (Massassi, Hennies and Eston, 1997) during the MPES'96 in Ostrava, Czech Republic.

In this same year, during the 17 World Mining Congress at Acapulco, Mexico was presented a paper about the electro-hydraulic principle or plasma blasting that use in near urban quarry areas is been tested as alternative to the conventional blasting with explosives (Stellin Jr., Hennies, Silva and Costa, 1997). This, study is the Phi. D. work that will be concluded in the next year.

In the middle of 1998, a first help was made for the studies at a Master level in the advanced Abrasive High Pressure Water Jet Technology for cutting ornamental rocks, as marbles and granites. In the same year, a second financial help to the studies at a Phi. D. level and additionally, a thematic project for the acquisition, installation and operation of a modern Modulus to the good equipped Rock Mechanics Laboratory, of the Mining Engineering Department of the Polytechnic School of the São Paulo University, was presented to FAPESP, Foundation of Research of São Paulo State. This Modulus is a Water Jet Machining System of High Pressure. The financial help of FAPESP, in the solicited projects, and the recognizance of the familiarization in the advanced techniques, with a promising future was decisive, for the success of this knowledge actualization and to a program, that if will realized, is now presented with what is do, and the aims to be attained that are presented in the article.

Finally, in 1999, a contribution was presented about the extraction of red granite blocks Capão Bonito (Hennies, Stellin Jr. and Cretelli, 1999) using

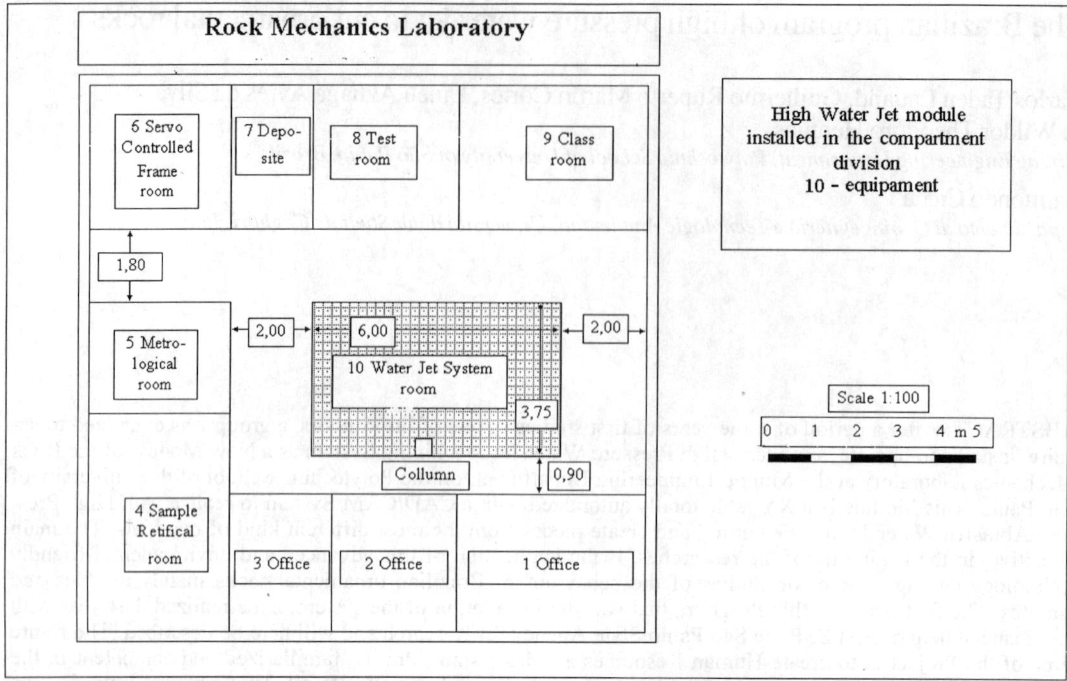

Figure 1 Plan of the Rock Mechanics Laboratory on Mining Engineering Department at EPUSP

the non conventional jet piercing technology now used to win the ornamental rock. This was presented at the 8th International Mine Planning and Equipment Selection Symposium, that was in Dnipropetrovsk, at Ukraine.

2 THE RESEARCH TEAM

To a success of the new research project of ornamental rock cutting aided by the high pressure abrasive water jet a team of researchers, evolved with this aim, was organized with professors and students of undergraduate and graduate levels.

This research team is now composed by three full professors, two graduating level students and one undergraduate level student.

Full Professor Lineu Azuaga Ayres da Silva, participate in this group as head of the Laboratory of Rock Mechanics and also as head of the Mining Engineering Department.

Full Professor Wildor Theodoro Hennies is the head of the new thematic project and is the adviser of the Master Program of the graduate students Carlos Tadeu Lauand and Magno Levi A. Mendes, and adviser of the Phi. D. Program of Guillermo Ruperto Martín Cortés, and also adviser of Scientific Initiation Program of the undergraduate mining engineering student Aldo Aparecido de Souza Junior. Additionally, a preliminary study as a end work undergraduate level was also advised by Prof. Hennies and presented in December of 1999 by the mining engineering student Oswaldo Nico Menta Simonsen.

The main objective of the work of graduate student Carlos Tadeu Lauand as electronic engineer is to study the problems with performance of the Jet Machining system and the associated software, for the optimization of the Brazilian dimension stones. Master Eng. Geologist Guillermo Ruperto Martín Cortés Phi. D. studies will be make special attention to the influence of the mineralogy and geological aspects of the Brazilian dimension stones qualitatively and quantitatively. Graduate student Magno Levi A. Mendes work is to use pre mixed water jets to cut steel to be applied for repair or maintenance of oil and gas ducts.

Finally, full Professor Raimondo Ciccu, recognized specialist in the field of water jet technology for ornamental rock cutting will be assort the team in the program to be developed in the next years. In the second week of December 1999, Prof. Ciccu visited our Mining Department and there were some discussion about the program for the next years.

As referred above, the program for the acquisition of a advanced equipment to cut ornamental rocks was proposed to the São Paulo research institution in December of 1998, and was approved in May 1999 when there was classified as a thematic Project. At this time the work begins for the acquisition, installation and final operation of the Water jet

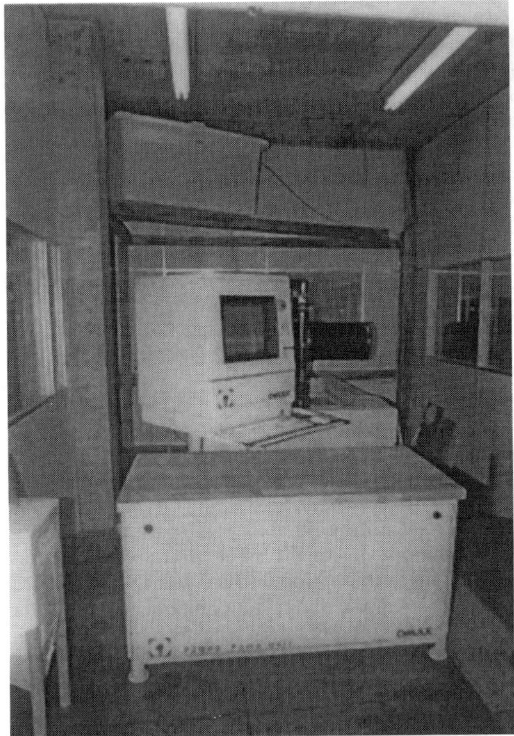

Figure 2 The high pressure abrasive water jet system

Machining System as the additional Module of the Rock Mechanics Laboratory of EPUSP.

3 THE ABRASIVE HIGH PRESSURE WATER JET MODULUS

The Figure 1, show a schematic view of the actual area on the Rock Mechanics Laboratory of EPUSP, in which a room of about 6 meters long by 3.75 meters wide and a high of 3.60 meters, has a anti acoustical division to receive the advanced abrasive high pressure water jet machining system.

The acquisition of this advanced machining system, its importation from USA and final installation in Brazil to initiate the operation was realized during the second semester of 1999.

The work procedure of the system consists in realize the design with a well known tool in engineering that is the computer aided design CAD, and next the equipment realize this design by the control also by computer of the manufacture of the part CAM. A controlled erosion on the target material, with the creation of very fine kerfs of the order of millimetres in wide and controlled deep with high dislocation velocities ends in the manufacture of high precision and reproducibility parts.

The target material to be cut with the erosion can be of the most different nature as metals, ceramics, composites, foods, glasses, rocks and others.

The bought Jet Machining Centre is from OMAX Corporation of the United States, Seattle, the Model is the 2652A, that means that the maximum dimension of table work area is 26" (660 mm) by 52" (1320 mm) the greatest area size of the slab that can be machined.

The system is constitute of four major components that are: the control system of the process, the high pressure water pump, the precision XY table, and the abrasive water jet nozzle system.

The control system of the process is realized by a standard PC computer that activated a software program (CAD/CAM) which also control the jet and other components. This system put on and out the water pump and move the abrasive water jet of the cutting tool in a precision route. The manufacture system is in this aspect totally automated as it has the CAD/CAM oriented program, to make first the design of the final piece (CAD), and than the manufacture of this piece in the equipment (CAM).

The water pump, elevate the water pressure to be utilized by the system. All pump operations are normally controlled by the control system in the PC.

The abrasive water jet nozzle consists of water at high pressure that is forced to pass in a sapphire orifice and flow at velocities over 750 meters per second. This high velocity beam make a suction of abrasive from a plastic feeding tube which is added to the current of high water jet.

Figure 2 shows a photography of the assemblage of equipment of this advanced abrasive high pressure water jet machining system installed at the area shown in the former figure.

The components of the modern high pressure abrasive water jet machining system acquired as Module for the Laboratory are; the control system by PC, a high pressure pump, the precision XY table and the abrasive water jet system.

Figure 3 PC equipment controller Cabin

4 THE PC CONTROL SYSTEM

The PC control system (Figure 3) is a software that is compatible with the personal computer of IBM.

The computer that came with the equipment is able to activate the furnished control software for the design and construction of pieces.

The parallel port of the computer controls movement of the abrasive jet and also turns the water pump off and on as needed.

The computer is mounted in a controller cabinet with a hinged arm, which allows the cabinet to be positioned as desired. The cabinet also shields the computer from any dust and water spray in the environment. Transparent membranes protect the keyboard and the mouse.

Normally the controller cabin has two plugs: one power plug and one emergency plug. It must be every time connect the power plug to linked the equipment. When the power plug is linked, the computer is initiated first and then supplied power to the linking of pump and piece manufacture table.

The emergency plug immediately disconnect the water pump and the dislocation of the abrasive water jet. The computer is not affected by de emergency plug and continues to send pulses for the position of the controller. In a emergency stop the system loss its original position. Because this, the emergency stop plug only must be used in real emergency situations.

If the space bar button is used during the operation the piece machining there give a stop in the movement of the abrasive jet and stop the water pump. The cut can in these mode easily restarted. This method is the recommended one to interrupt the operation by reasons others than that of security.

5 HIGH PRESSURE WATER PUMP

A complete documentation and maintenance instructions are furnished with a ultra high pressure water pump and its regulator.

The pump will furnish the water at high pressure necessary to the system. The pump that comes with the system has 20 HP.

The pump pressure is furnished by its regulator. A system installed permitted to work in full water pressure 275.6 MPa (40,000 psi) or lower if desired until 206.7 (30,000 psi). This pressure regulator is installed by the manufacturer, but can be changed by the operator of the machining system. The adjustments of the pressure regulator is a simple procedure and is explained in the pump documentation.

The pump is started on or out by the PC computer of the system controller as explained above. The pump can be stopped during the cut, and is restarted when the jet nozzle was moved to a new position.

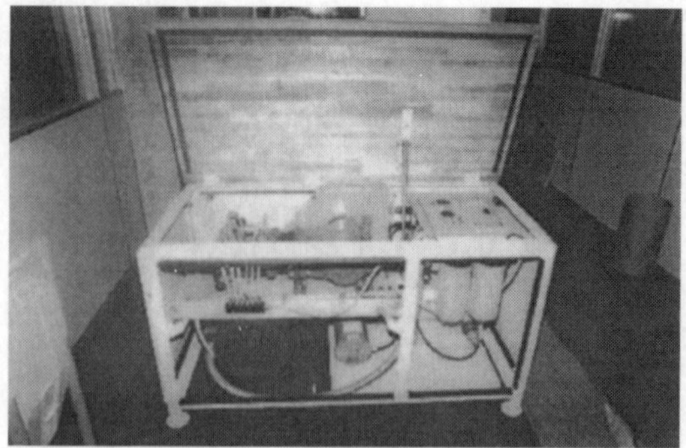

Figure 4 The system's high pressure water pump

Normally, the pump is acoustically sealed to noise reduction (at a level of 65 dB).

The water pump may not be retired from its cabin.

Important, to the durability of spare and components, is the water quality. For the maximum durability of there the option chosen was to use distiled water, practically without salt components and water box and tubes of polypropylene to avoid any contamination during its conduction to the pump.

At the pump there is a little reservoir were a little normal pump pressure the water through two filters before it goes to the high pressure pump.

Normal treated water is used to fill the catcher tank of the XY table, to work with the nozzle in the underwater cut method, for avoiding noise that is also of 65 dB in the environmental when the cut is done outside at the room atmosphere.

6 PRECISION X-Y TABLE

The high precision X-Y table installed in the rock mechanics laboratory of the Mining Engineering shown in figure 5.

Figure 5 The precision X-Y table

The Precision X-Y table is a rigid frame that rests on the water tank. The X-Y table is welded from large, heavy-wall structural aluminium tubes with castings that support either end of the X beam. A series of stainless-steel slats supports the part that is being machined.

The Y beam is connected to the carriage that moves along the X beam.

Both X and Y carriages ride on pre loaded linear ball bearings mounted on opposite sides of a round aluminium tube. The carriages are moved by pre loaded ball screws mounted along the bottom of the tubes.

The screws of the spheres are guided by steel beds of servo motors mounted in the inner part of the tubes. The dislocation set and direction mechanism is evolved and sealed by urethane to avoid the water and abrasive particles.

The servo motors are motors without brush with adjust position. Every command step to the servo motor give a dislocation of approximately 0.127 mm.

7 THE ABRASIVE WATER JET SYSTEM

The abrasive water jet system combines a high speed water jet flow with a flow of abrasive garnet to create the abrasive jet flow that cut through a wide range of materials (Figure 6).

The abrasive is stored in a hopper besides the moving cutting head in the X-Y table. A air compressor furnished the amount of controlled air regulated by a valve and feed a very well controlled beam of abrasive in the feeding line where it flows to the nozzle of the jewel.

In the abrasive jet nozzle, high pressure water is forced by the sapphire orifice to form a strait beam moving at 752 meters per second (2,743 Km/h). These beam cause the suction that conduct the abrasive and air through the plastic mixture tube.

The water, the abrasive and the air than enter in the tungsten carbide tube in where they combine to form a cutting beam with a speed of 305 m per second (1,100 Km/h). These beam is directed against the material to be cut.

Figure 6 The abrasive feeding system of the waterjet

In figure 7 is shown the jewel, mixing tube and the nozzle of the equipment.

Resumed, this is the equipment that was acquired as new Module to cut with high pressure abrasive water jet in the Rock Mechanics Laboratory of USP.

8 BRAZILIAN RESEARCH PROGRAM

The Brazilian research program has the aim to in the next years by the help of the international specialist domain the advanced technique of manufacture and the behaviour of the most different materials in where it can be applied.

The beginning studies is to research the behaviour of ornamental rocks hard such as granites and quartzite or more soft as marble and slates normally used as tiles for cover inside and outside walls and floors as civil construction materials. Also the study for metal cutting, more specifically steel with a pre

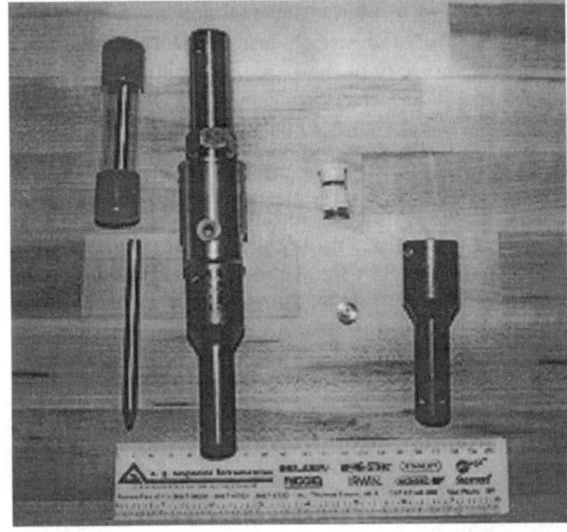

Figure 7 Jewel, mixing tube and nozzle of the system

Table 1 Projects to be developed in next years.

Target	Project	Year
Acquisition	Machinery	1999
Installation	Machinery	January 2000
1st Research	Undergraduate level	2000
2nd Research	Graduated level M. Sc.	2001
3rd Research	Other materials than Rocks	2001
4th Research	Graduated level Phi. D.	2003

mixed water jet is also one of the targets.

A complete Program of this aims to be attained in the near future is presented below in table 1 were the essential target are presented of any now programming by persons involved.

In November 1999, a visit make to a Brazilian industry called as ICAN, located at the city of São Carlos, permitted to Author to see in operation on of this system but there is a greater system, and worked with all kind of materials, not only rocks but also ceramics, metals and other products as plastics.

This equipment has a 100 HP pump, and its main application service is for the national automobilist industry in cutting steel alloy parts.

This aim is realized with first studies that are now beginning after the installation that was made in January of 2000 and the initial tests on Brazilian dimension rocks.

As mentioned above the research is be done in the most different levels, undergraduate preliminary investigations of knowledge of the technical, more advanced interpretation of first experimental work of graduate studies in the Master level, and also innovations that can be made to work better, with the Phd. D. level research.

As the new jet machining system works with a very large kind of materials, it is proposed to search also in metals, composites, and other materials that are used commonly in the Brazilian industry.

9 ACKNOWLEDGMENT

The AA. will take advantage of this opportunity to let here register their most sincere acknowledgment to FAPESP, the São Paulo State Research Foundation in the recognition of the financial help without which is were impossible to take the Brazilian industry in the knowledge of the behaviour of this important advanced equipment, for a more friendly environmental working establishment and also for the future sustainable industrial development.

REFERENCES

HENNIES; Wildor Theodoro. Rock Excavation: a critical review. In: INTERNATIONAL SYMPOSIUM ON MINE PLANNING AND EQUIPMENT SELECTION, 5., São Paulo, 1996. *Mine Planning and Equipment Selection 1996*: Proceedings. Rotterdam, Balkema, 1996. P.419-24.

MASSASSI, G.; HENNIES, W. T.; ESTON, S. M. de. Noise associated with mining excavation equipment. In: INTERNATIONAL SYMPOSIUM ON MINE PLANNING AND EQUIPMENT SELECTION, 6., Ostrava, 1997. *Mine planning and equipment selection 1997*: proceedings. Roterdam, A.A. Balkema, 1997. p.897-901

STELLIN JÚNIOR, A.; HENNIES, W. T.; SILVA, C. M. M.; COSTA, E. G.. The electro-hydraulic principle as blasting alternative in urban areas. In: WORLD MINING CONGRESS, 17./CONVENCION NACIONAL DE MINERIA. 22., Acapulco, 1997. *Technical papers*. Acapulco, Associacion de Ingenieros de Minas, Metalurgistas y Geologos do Mexico, 1997. p.49-56

HENNIES, W. T.; STELLIN JUNIOR, A.; CRETELLI, C. Jet piercing application for red granite block mining in São Paulo, Brazil. In: INTERNATIONAL SYMPOSIUM ON MINE PLANNING AND EQUIPMENT SELECTION, 8., Dnipropetrovsk, 1999. *Mine Planning and Equipment Selection 1999*: Proceedings. National Mining University of Ukraine NMUU Ukraine, 1999. P.21-26

Environmental Issues and Management of Waste in Energy and Mineral Production, Singhal & Mehrotra (eds)
© 2000 Balkema, Rotterdam, ISBN 90 5809 085 X

Development of an air attenuation model for noise prediction in surface mines and quarries

K. Pathak
Department of Mining Engineering, Indian School of Mines, Dhanbad, India

S. Durucan & A. Korre
T.H. Huxley School of Environment, Earth Sciences and Engineering, Royal School of Mines, Imperial College of Science, Technology and Medicine, London, UK

ABSTRACT: Surface mining and quarrying produces a complex noise field. The noise generated by mining operations changes with time and is affected by the changing meteorological conditions in situ. Theoretical research on the attenuation of noise in air has formed the basis for the development of an air attenuation model. The methodology developed by the authors accounts for the prevailing temperature and relative humidity conditions in the area. The model also includes the evaluation of ray path curvature directly from long-term experimental wind velocity and temperature data, or indirectly in terms of the Net Solar Index and the wind velocity. This paper describes the air attenuation model, which was developed as an integral part of an environmental noise prediction system.

1 INTRODUCTION

The prediction of environmental noise around an industrial complex involves the determination of sound power level of the noise sources, evaluation of attenuation factors and the combined analysis of these for a given noise propagation path. As illustrated in Figure 1, the environmental noise prediction system developed by the authors has four main components, namely the source power model, the ground attenuation model, the air attenuation model and the pit-wall attenuation model.

The source power model predicts the sound power level of the noise sources based on the theory of Equivalent Acoustic Centres where a number of distributed sources of noise are represented by an equivalent point source (Durucan et al, 1998) The attenuation models account for ground absorption, air absorption and the effects of pit-walls and provide the attenuation of sound energy under the prevailing mining and environmental conditions.

The noise prediction system developed integrates the source power model and the attenuation models such that:

$$L_p = L_w - A_s - A_g - A_a - A_b + D_c + R_c \quad (1)$$

Where L_p = sound pressure level at the receiver's location; L_w = sound power level of the equivalent acoustic centre of the work zones; A_s = attenuation due to spreading; A_g = attenuation due to ground absorption; A_a = attenuation due to air absorption; A_b = attenuation due to pit slopes; D_c = directivity corrections; R_c = reflection correction.

This paper reviews the theory on the effects of atmospheric conditions on environmental noise and describes the air attenuation model, which was developed as an integral part of an environmental noise prediction system.

2 METEOROLOGICAL EFFECTS ON SOUND PROPOGATION

Environmental noise prediction around a mining complex involves fixing the sound power of the machinery at a certain level and applying the theories of sound attenuation to evaluate noise. The environmental noise levels near a mining complex are the long-term averages of various types of noise and the actual propagation conditions prevailing during the period under consideration. Therefore, collection of statistical noise data from an existing mining site is important before proceeding with noise predictions.

The propagation characteristics of acoustic waves through the atmosphere modify the overall noise field at a specific location. The density, relative humidity and temperature of any specific volume of air in the atmosphere vary with the time of day and the day of the year. Sound pressure levels at a receiver fluctuate due to these effects. Thus, noise prediction and control systems need to model the complex behaviour of acoustic waves under realistic atmospheric conditions.

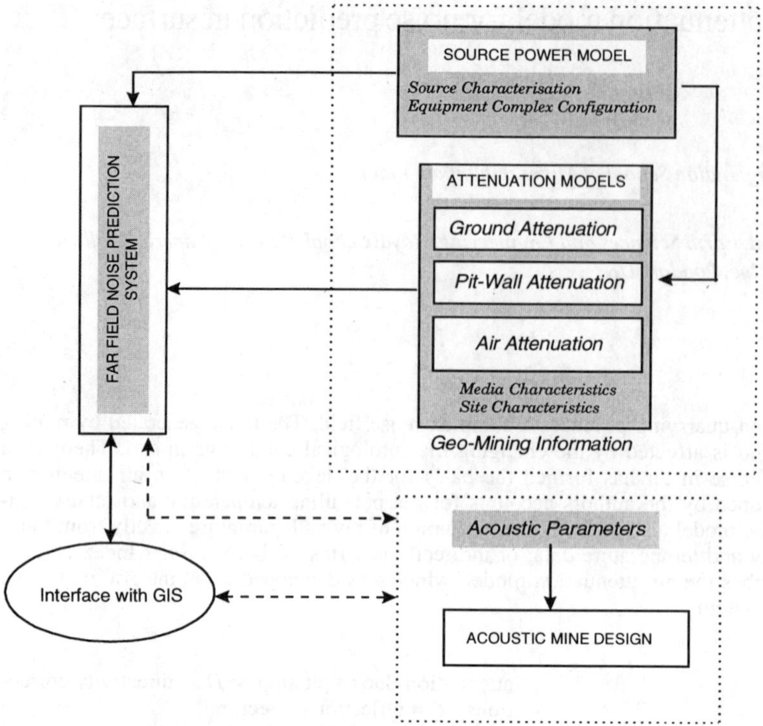

Figure 1. Main components of the environmental noise prediction system developed.

Atmospheric absorption, refraction and scattering by turbulence are the most significant meteorological effects on sound propagation. Change of refractive conditions of the media modifies the radius of curvature of the ray path between a source and a receiver. The weather's effect on sound pressure level is based on refraction. Upward bending of sound rays gives lower noise levels and downward bending offers higher noise levels. The parameter for characterising the curvature of near-horizontal sound rays due to refraction is given as (Gutenberg, 1942):

$$\frac{1}{R} = \frac{\frac{10}{\sqrt{T}}\frac{\delta T}{\delta z} + \frac{\delta u}{\delta z}}{c(1+\frac{u}{c})^2} \tag{2}$$

where R = radius of curvature; T = temperature (K); c = velocity of sound (m/s); u =component of the wind vector in the direction of the ray (m/s).

Increase in the values of $1/R$ shows an increase in the sound level. The normalised curvature is taken as:

$$k \cong {(0.6 + \Delta T + \Delta u)}/{3.2} \tag{3}$$

where ΔT = temperature difference in ºC or K; Δu = difference in the wind velocity component (m/s) at a meteorological standard height of 0.5 m and 10 m.

The values of k vary during the day and throughout the year due to changes in the meteorological conditions. Negative values of k mean upward bending of the sound ray. This is usually expected during the summer when the temperature decreases with height. Upwind conditions in winter are compensated with a temperature increase with height. This phenomenon is more pronounced in aircraft noise. In mining noise, the vertical gradient is less important as the level difference between the source and the receiver is relatively small.

In statistical noise estimation some of the data collected over a period of time is normally discarded before calculating k (Larsson, 1992). If the minimum value of the hourly k on some days is less than -3.0, measurements from those days are rejected. If one wishes to use the statistical method of analysing the meteorological data for noise prediction purposes, continuous monitoring for a long period is necessary as one requires at least seven days of significant data after rejecting the less significant data sets. In order to account for the meteorological effects on noise propagation at any specific site, one needs to determine the meteorological parameters of the site from the measured data over the area or collect it from the nearest meteorological station. Along with the meteorological data, source-receiver position is very important for noise prediction. From the meteorological and source-receiver location data the radius of curvature of the sound ray ($1/R$) can be calculated and the distribution of $1/R$ over a longer period can be evaluated.

The L_{eqA} and L_{max} are thus calculated for different days and processed to provide the best statistical information for the *base period* of noise prediction.

2.1 *Estimation of curvature without meteorological profile measurements*

In the method stated above for calculating k, wind velocity and temperature profiles both near the source and the receivers are necessary as the input data. In the absence of such measured and statistically processed data, the following methodology can be used:

$$v(z) = \frac{u_*}{\kappa} \ln \frac{z}{z_0} \tag{4}$$

where $v(z)$ = wind speed at height z; κ = von Karman constant ($\approx$0.4); z_0 = roughness length that can be read from Table 1. The friction velocity u_* is determined from the measured values of wind speeds at different heights (z_m) as:

$$u_* = \frac{\kappa v(z_m)}{\ln \frac{z_m}{z_0}} \tag{5}$$

For a wind speed difference between a receiver height of 10 m and 0.5 m, Δv is given by:

$$\Delta v = v(10) - v(0.5) = \frac{u_*}{\kappa}\left(\ln \frac{10}{z_0} - \ln \frac{0.5}{z_0} \right) \tag{6}$$

Δu for calculation of radius of curvature k in equation 3 can be evaluated from the measurement of the angle ϕ between the direction of wind and the line joining the source and the receiver using equation:

$$\Delta u = \Delta v \cos(\phi) \tag{7}$$

ΔT for 0.5-10 m height can be determined from the elevation of the sun and wind velocity, using the method developed by Turner (1964). Elevation of the sun is characterised by the net solar index (NSI). The values of NSI are defined based on rules listed in Table 2 (Turner, 1964).

Table 1. Aerodynamic roughness length for some typical terrain types (after Stull, 1988 and Oke, 1987).

Surface	Aerodynamic Roughness Length, z_0 (m)
Ice, mud flats	1×10^{-5}
Snow: flat ground to farmland	$0.1\text{-}11 \times 10^{-4}$
Water: Calm to open sea	$1\text{-}10 \times 10^{-4}$
Sand, desert	5×10^{-4}
Grass: cut or short ($\approx$0.02-0.1m)	0.003-0.02
Grass; long($\approx$0.1-1m)	0.02-0.1
Farmland: open to hedged	0.02-0.1
European average	0.09
Wooded country	0.1-0.5
Dense woodland	0.3-1
Urban areas	0.4-3
Mountains	2-80

Table 2. Values of Net Solar Index (after Turner, 1964).

1. If the cloud cover is 8/8 and the cloud base is less than 2500 m NSI is equal to 0 for day or night.
2. During the period between sunset and sunrise,
   a) if the total cloud cover is less than or equal to 3/8, NSI = -2.
   b) if the total cloud cover is greater than 3/8, NSI= -1.
3. During daytime for cloud cover less than or equal to 4/8:
   a) if the solar altitude is less than or equal to 150, NSI = 1.
   b) if the solar altitude is between 150 and 350, NSI = 2.
   c) if the solar altitude is between 350 and 600, NSI = 3.
   d) if the solar altitude is greater than 600, NSI = 4.
4. During daytime for cloud coverage greater than 4/8:
   a) if the cloud base is less than 2500 m subtract 2 from NSI.
   b) if the cloud base is higher than 2500 m subtract 1 from NSI.
   c) if NSI is found to be less than 1 it is considered as 1.

For known values of NSI, ΔT can be obtained from Table 3. In the past, this technique was applied to airport noise analysis and has yielded good results for positive values of k.

3 MODEL DEVELOPMENT

3.1 *The sound propagation parameter*

Larsson and Israelsson (1991) have defined a sound propagation parameter, which is used to characterise the noise level as:

$$W = \frac{k(z_s + z_r)^2}{d} \tag{8}$$

where k = normalised radius of curvature; z_s = source height; z_r = receiver height; d = distance between source and receiver.

If W is nearly equal to zero, the sound pressure level of that area will be very sensitive to the weather conditions. Once the cumulative distribution of k over a longer period is established it can be used for general planning purposes.

Table 3. Values of ΔT (°C) under different conditions.

Wind Speed m/s	NSI=4	NSI=3	NSI=2	NSI=1	NSI=0	NSI=-1	NSI=-2
<0.8	-2.5	-2.5	-1.3	-0.2	-0.1	1.0	4.1
0.8 - 1.8	-2.5	-1.3	-1.3	-0.2	-0.1	1.0	4.1
1.8 - 2.9	-2.5	-1.3	-1.3	-0.2	-0.1	1.0	4.1
2.9 - 3.3	-1.3	-1.3	-0.2	-0.1	-0.1	0.0	1.0
3.4 - 3.9	-1.3	-1.3	-0.2	-0.1	-0.1	-0.1	0.0
4.0 - 4.9	-1.3	-0.2	-0.2	-0.1	-0.1	-0.1	0.0
5.0 - 5.4	-0.2	-0.2	-0.1	-0.1	-0.1	-0.1	0.0
5.5 - 5.9	-0.2	-0.2	-0.1	-0.1	-0.1	-0.1	-0.1
>5.9	-0.2	-0.1	-0.1	-0.1	-0.1	-0.1	-0.1

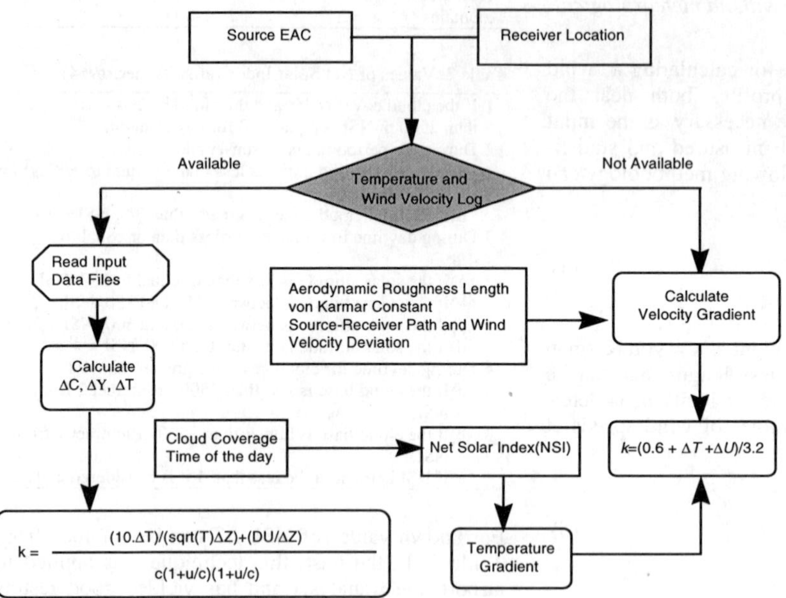

Figure 2. Methodology used to determine the radius of curvature of acoustic path from source to receiver (METEOK.CPP).

In a large and deep surface mine the meteorological parameters are the dominant factors in affecting the pit-to-surface propagation of noise. The ray path of noise propagation in such cases cannot be assumed to be a straight line and therefore the determination of the radius of curvature of the ray path is necessary. In the air attenuation model described in this paper, the effects of the meteorological parameters were incorporated using the theory discussed in Section 2. The computer code METEOK.CPP, which is outlined in Figure 2, was developed for this purpose. Figure 3 presents a sample output of this module.

As noted in the output file the module reads the input data from a number of files. These files store the processed meteorological data such as the temperatures and the wind velocity monitored continuously by the commercially available data acquisition systems. These data are processed to provide the best statistical information for the *base period* of noise prediction. The base period of noise prediction is the duration of measurement of equivalent continuous sound pressure level. The sample output shows the ray path curvature to determine four hourly L_{eq} Therefore, in this case the input data files provided the most significant and representative values of the temperatures and wind velocities of the surface mine at the pit top and pit bottom. In the absence of such statistically measured data, the module determines the values of k by using the theory discussed in Section 2.1 and Table 3.

For shallow depth quarries and surface mines the ray path curvature is not very important since the height differences of the source and the receiver are not very large. However, for long distance propagation of noise, attenuation due to air must take the ray path curvature into account. The values of k influence the sound pressure level in the free field.

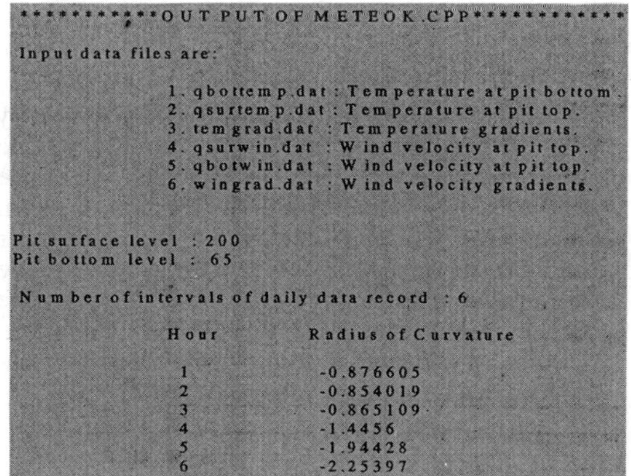

Figure 3. Sample output of module METEOK.CPP.

3.2 *Absorption of sound by air*

A propagating sound wave loses energy because of heat conduction, radiation, viscosity and diffusion. According to the classical absorption theory such loss is proportional to the square of frequency and does not depend on humidity. Molecular absorption in still homogeneous air has an exponential effect and is proportional to the distance from the source. The constant of proportionality is referred to as the attenuation coefficient, α. Thus,

$$A_a = \alpha \frac{r}{1000} \tag{9}$$

where A_a = excess attenuation due to air absorption at a distance r from the source.

Kneser (1933) theoretically evaluated the attenuation coefficient due to air absorption as a function of frequency, humidity and temperature. His findings were experimentally verified by Knudsen (1933). ISO 3891 provides corrections for air absorption for use in noise prediction models. German Standards (VDI 2714/2720) calculate air absorption as a function of the distance from the source to the receiver. ANSI 1978 provides the necessary tables for calculating air attenuation at different temperatures and humidity under octave bands.

Sutherland et al (1974) determined the coefficients of air attenuation for different meteorological conditions and tabulated these as shown in Table 4. The effects of meteorological conditions on attenuation due to air absorption are illustrated in Figure 4. The air attenuation model developed by the authors uses the values in Table 4 for prediction purposes as this enables the selection of most appropriate coefficients for the prevailing ambient temperature and relative humidity. Figure 5 illustrates attenuation due to air absorption at different source-receiver distances for 15 and 30°C.

4 MODEL APPLICATION

The principles and the theory discussed in this paper were used in developing an air attenuation model for use in the mining and quarrying environments. Figure 6 illustrates the structure of this model. Extracting the values of α from Table 4, air absorption (A_a) at a distance r can be determined using Equation 8.

In the model, the distance r to the receiver is measured from the co-ordinates of the Equivalent Acoustic Centre of a noise zone (Durucan et al, 1998). The air absorption rate for different relative humidities (*RH*) and temperatures (*T*) are retrieved from the input data files provided by the user.

The model developed has been validated through field experiments and the prediction of air attenuation over a limestone quarry in Ireland. The floor of the quarry was dipping towards the north at a gradient of 1 in 10. The face was similar to a box cut with an advancing face of 105 m. The resultant sound field is a semi-reverberant field formed by the direct, reflected and scattered sound energies. There were four main operating zones in the quarry. These are:

- Portable Crusher Zone (PC)
- Deep Face Zone (DF)
- Main Crusher Zone (MC)
- Fixed Plant Zone (FP)

The Portable Crusher Zone is located towards the

Table 4. Attenuation due to atmospheric absorption (after Sutherland et al, 1974).

Humidity	Temp.	α (dB per 1,000m)							
(%)	(°C)	63 Hz	125 Hz	250 Hz	500 Hz	1 kHz	2 kHz	4 kHz	8 kHz
25	15	0.2	0.6	1.3	2.4	5.9	19.3	66.9	198.0
	20	0.2	0.6	1.5	2.6	5.4	15.5	53.7	180.5
	25	0.2	0.6	1.6	3.1	5.6	13.5	43.6	153.4
	30	0.1	0.5	1.7	3.7	6.5	13.0	37.0	128.2
50	15	0.1	0.4	1.2	2.4	4.3	10.3	33.2	118.4
	20	0.1	0.4	1.2	2.8	5.0	10.0	28.1	97.4
	25	0.1	0.3	1.2	3.2	6.2	10.8	25.6	82.2
	30	0.1	0.3	1.1	3.4	7.4	12.8	25.4	72.4
75	15	0.1	0.3	1.0	2.4	4.5	8.7	23.7	81.6
	20	0.1	0.3	0.9	2.7	5.5	9.6	22.0	69.1
	25	0.1	0.2	0.9	2.8	6.5	11.5	22.4	61.5
	30	0.1	0.2	0.8	2.7	7.4	14.2	24.0	58.4

western side of the quarry, positioned on the pit floor at a level of 62 m. The crusher is comprised of a primary jaw crusher, secondary cone crusher and the vibrating screen. Materials are conveyed by 1.2 m wide conveyor belts. The main generators of noise are the crushers, screen and the conveyor drive. The crusher is fed by a front-end loader of 6 m^3 bucket capacity. On average, two trucks were moving in the work zone while another one is being loaded by a 4.6 m^3 hydraulic shovel. The zone is surrounded by benches with an overall slope angle of 60 to 70 degrees on three sides. The average height of the quarry walls was 23 m. On one side there were loose soil heaps separating the zone from the main crusher zone. Figure 7 illustrates the surface plan of the test quarry.

The sound field of this zone is characterised by the reflected sound from the side walls. In order to determine the sound power level of this work zone under steady state operating conditions seven monitoring stations were selected within the zone and the sound pressure levels were measured. The air attenuation of the noise generated in each zone was modeled separately for each zone. Figure 8 illustrates a GIS output of the predicted air attenuation values for the portable crusher zone.

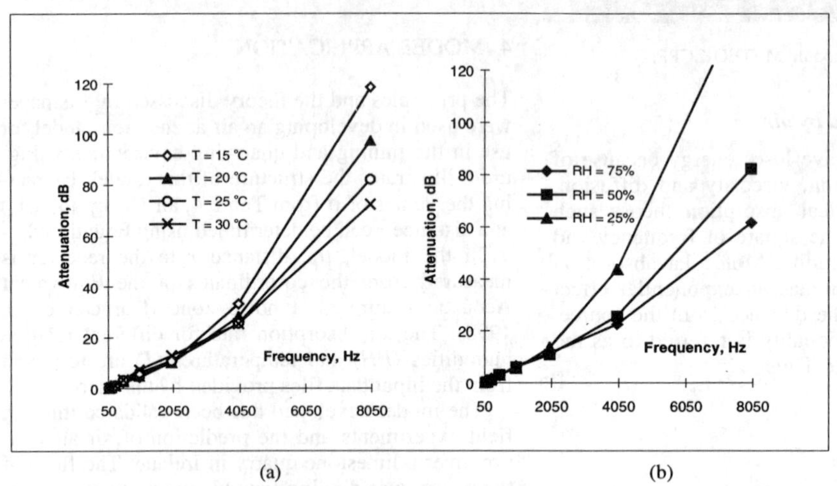

Figure 4. Attenuation due to air absorption over a distance of 1000 m, (a) at 50 % relative humidity, (b) at 25°C (drawn after Sutherland et al, 1974).

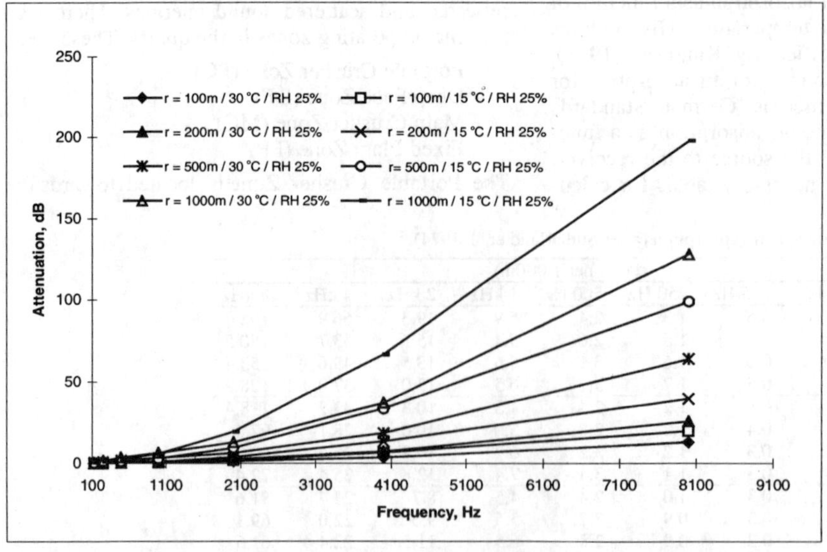

Figure 5. Air attenuation at different atmospheric conditions (drawn after Sutherland et al, 1974).

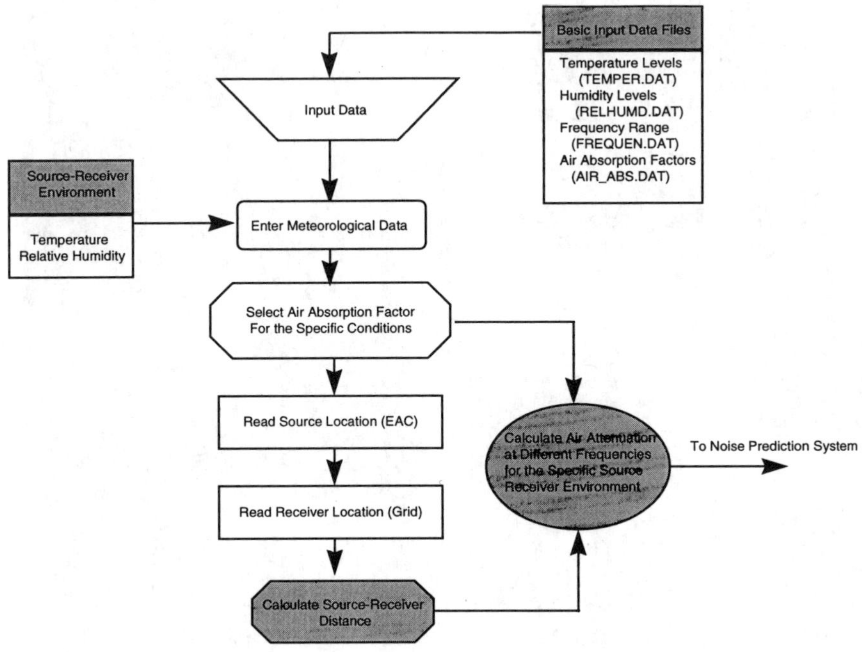

Figure 6. Air attenuation model (AIRABFIN.CPP)

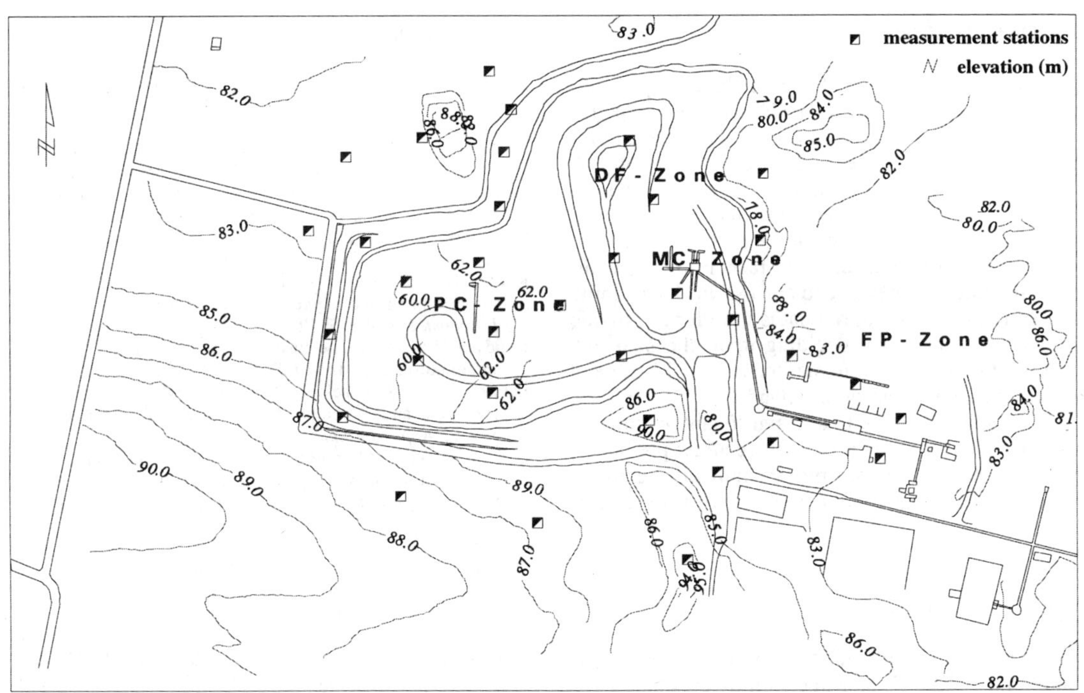

Figure 7. General layout of the limestone quarry in Ireland.

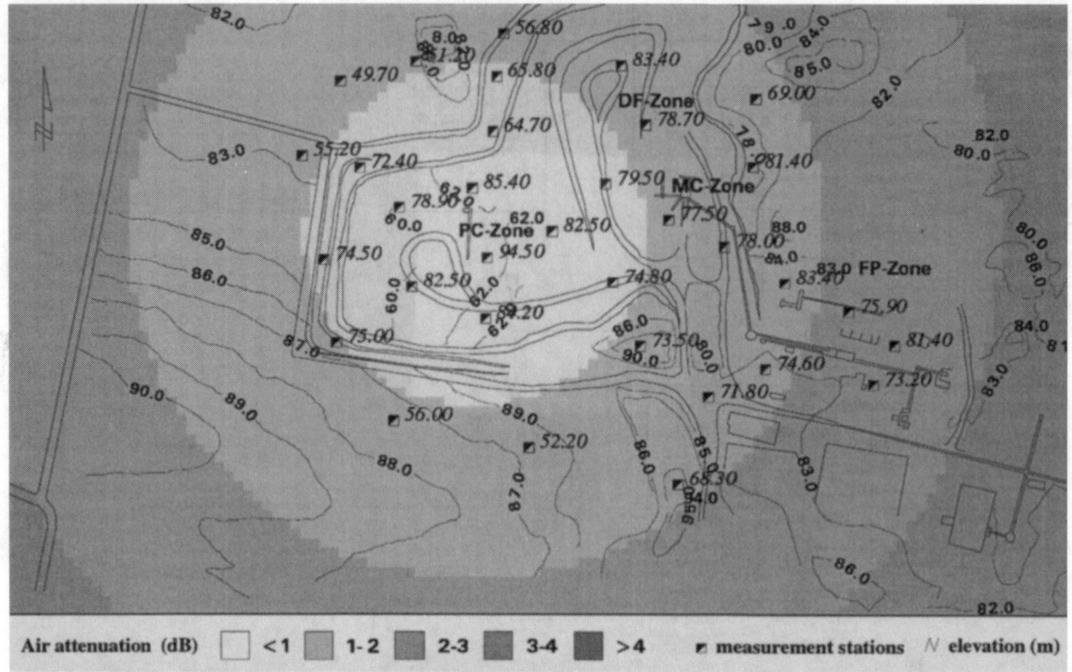

Figure 8. Predicted air attenuation values for the portable crusher zone.

5 CONCLUSIONS

The authors have adopted the methodology developed by Sutherland et al (1974) and developed a model to predict attenuation due to absorption in air considering temperature and relative humidity. A procedure has been developed to determine the ray path curvature due to temperature and wind velocity gradient between the source at the pit bottom and the receiver positions outside a surface mine. A computer module was developed for use in determining the radii of curvature of ray paths in order to provide the statistical basis for deciding upon the *base period* of noise prediction in surface mines. The model was validated in a quarry in Ireland.

The authors believe that a significant volume of statistical data is necessary for an accurate prediction of air attenuation taking into account all the meteorological and climatological changes in the atmosphere.

6 ACKNOWLEDGEMENTS

The authors wish to thank Mr. Brendan O'Reilly of Outokumpu Zinc Tara Mines Limited, Ireland for his invaluable assistance during the field validation tests of the model developed.

7 REFERENCES

ANSI S1.26 1978. *Method for the calculation of the absorption of sound by the atmosphere*. New York: American National Standard.

Durucan, S., K. Pathak, & S. Kunimatsu 1998. Development of a source characterisation model for the prediction of environmental noise near surface mines and quarries. *Proc 27th APCOM*, London, 18-23 April, IMM.

Gutenberg, B. 1942. Propagation of sound waves in the atmosphere. *J. Acoust. Soc. Am.* 13: 151-155.

Kneser, H.O. 1933. Interpretation of the anomalous sound absorption in air and oxygen in terms of molecular collisions. *J. Acoust. Soc. Am.* 5: 122.

Knudsen, V.O. 1933. The absorption of sound in air, in oxygen and in nitrogen. *J. Acoust. Soc. Am.* 5: 112.

Larsson, C. & S. Israelsson 1991. Effects of meteorological conditions and source height on sound propagation near the ground, *Applied Acoustics*, 33: 109-121.

Larsson, C. 1992. Meteorological effects on long range sound propagation: Some experimental results. *Proc. 5th Int. Symp. on Long Range Sound Propagation*, Milton Keynes, England.

Oke, T.R. 1987. *Boundary layer climates*. London: Methuen.

Stull, R.B. 1988. *An introduction to boundary layer meteorology*. Dordrecht: Kluwer Academic Publishers.

Sutherland, L.C., J.F. Piercy, H.E. Bass & L. B Evans 1974. Method for calculating the absorption of sound by the atmosphere. *J. of Acoust. Soc. Am.* 56, Supplement 1 (Abstr.)

Turner, D.B. 1964. A diffusion model for an urban area. *J. Appl. Meteorology*. 3: 83-91.

Environmental Issues and Management of Waste in Energy and Mineral Production, Singhal & Mehrotra (eds)
 ISBN 90 5809 085 X

Determination of ventilation system for mines by genetic algorithm*

Q.X.Yun & K.M.Huang
Xian University of Architecture and Technology, People's Republic of China

ABSTRACT: This paper presents a new approach to determine the ventilation system for mines by means of genetic algorithm. The new approach consists of three phases: establishment of ventilation networks by genetic algorithm, verification of ventilation network by graph theory, evaluation of network by ventilation calculation. The new approach has been successfully used in Xishiment underground mine.

1 INTRODUCTION

Ventilation is a key issue to control the environmental condition in underground mine. By means of ventilation the quality of air can be improved and the harmful contents can be eliminated in underground working faces. Particularly it is essential for the safety in underground coal mine because underground gas has to be attenuated and exhausted by ventilation. Therefore, ventilation has been a permanent subject for mining engineering.

In past years a lot of work has been done for the computation of ventilation parameters such as air pressure and air volume. However, it is still weak to determine the ventilation network with the combination of openings. In other words, it is easy to deal with the parameter optimization while it is difficult for the structural optimization.

In order to determine a ventilation network, traditional method is to compare a few feasible alternatives and select the best one as the final decision. However, because of the limitation of feasible alternatives due to their time-consuming and strenuous comparison, there may be missing for the optimal alternative. Therefore, this paper presents a new approach to determine ventilation system by means of genetic algorithm. The new approach consists of three phases:

1. Establish ventilation network by means of genetic algorithm. In this phase a variety of networks are generated by evolutionary computation. The networks represent different combination of openings such as shafts, drifts and adits.
2. Verify the ventilation network by graph theory.

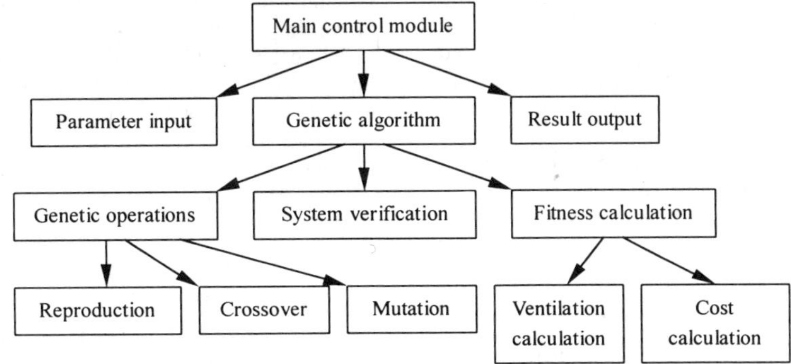

Figure 1. Flow chart of new approach

*This project is supported by National Natural Science Foundation of China (№59874019)

The purpose of this phase is to verify the network generated in the previous phase so that the network is feasible from the view of mining engineering.

3. Evaluate the network by ventilation calculation. This phase is used to calculate air pressure and air volume for the feasible ventilation network in order to evaluate the economics of system.

These phases are related and coordinated with each other as in Figure 1.

The new approach has been successfully used in Xishimen underground mine to improve its ventilation system. The application proves that new approach is reasonable and reliable.

2 GENETIC ALGORITHM

Genetic algorithm is one kind of evolutionary computation which imitate the procedure of evolution and heredity in biology. The fundamental principle of genetic algorithm is also Darwinian natural selection in nature – " the survival of the fitness". Based on a set of initial feasible alternatives, the final optimal result can be reached after a lot of iterations which include reproduction, crossover and mutation (Goldberg 1989).

The procedure of genetic algorithm is composed of coding, generating of initial population, calculation of fitness, genetic operation and termination (Yun et al 1997).

2.1 *Coding*

Coding is the representation of problem with the help of binary strings. To represent different ventilation systems with different combination of openings, a set of binary strings are used where each bit represents an opening. When the bit is equal to 1 it means that the corresponding opening is necessary for ventilation, otherwise 0 means that the opening should be deleted. For example, binary string 100101 illustrates that openings No.1, No.4 and No.6 will exist for ventilation while openings No.2, No.3 and No.5 are not necessary. Each bit in a string is similar to a gene in a chromosome which is the carrier of genetic information.

To simplify the algorithm, the length of string are fixed, where some 0 digit in front of the string means that no opening exists for the ventilation system.

2.2 *Generation of initial population*

According to the structure of binary string, a group of initial individuals are generated usually by random method. In order to improve the quality of initial population, some initial individuals can result from crossover and mutation based on a few initial random individuals.

The population size M is an important parameter in genetic algorithm. Usually the population size is a constant. Therefore, there will be M individuals in the initial population. All the individuals must be verified as described at section 3 later after its generating. Since it is time-consuming for verification and fitness calculation for each individual, M is usually controlled within 100.

2.3 *Calculation of fitness*

In genetic algorithm, fitness is the driving force for evolution. Fitness is also the objective function for ventilation system. Here the value of fitness includes two parts: one is the construction cost to set up the necessary openings, the other is the operation cost of ventilation.

The construction cost is calculated for each opening based on its specified cost. The operation cost results from the computation of ventilation networks described at section 4 later.

2.4 *Genetic operations*

1. Reproduction. Reproduction is used to copy some excellent individuals and transit them into next generation. Therefore, the quality of next generation will be better than previous one. To select the copied individuals, fitness-proportion selection is used. The probability of an individual i to be copied p(i) is as follows:

$$p(i) = f_i / \sum_{i=1}^{m} f_i \qquad (1)$$

where:

f_i—fitness of individual i.

With the help of fitness—proportion selection, the higher fitness the individual has, the greater probability it can be copied. Meanwhile, a few individuals with poor quality can still be copied to improve the variation of population.

2. Crossover. Crossover is essential to generate new individuals in genetic algorithm. During crossover new individual is generated by ways of exchanging their useful elements from parents. It is a random selection both for the parents and crossover points. An example of crossover is illustrated as follows:

Parent 1: 0101011 to Child 1: 0110000
Parent 2: 1010000 Child 2: 1001011

3. Mutation. This operator is also used to generate new individuals but more violently than crossover. By means of changing a bit from 0 to 1 or from 1 to 0, a new individual is emerged with some new genes. The selection of mutated bit is random.

2.5 *Termination*

After each genetic operation, the quality of population will be improved one generation after another generation. Eventually, it approximates to the optimum solution. As soon as the following criterion is satisfied, genetic algorithm will terminate:

$$\left|f_{opt} - f_{ave}\right| \leq \varepsilon \tag{2}$$

where:

f_{opt}— fitness of the optimal individual in a population;

f_{ave} — average of fitness in a population;

ε — a small specified value.

3 VERIFICATION OF VENTILATION NETWORK

A feasible ventilation network should meet the following requirements from the view of mining engineering:

1. It is a connected network without separate subnetwork. In other words, any node in network can be reached by another node through some arcs of network.
2. There is no end opening where the end-node of opening does not connect with any node. It means that there should be cyclic air current everywhere to improve the ventilation.
3. The network should connect with surface. In order to improve the ventilation there should be an entrance with an exit at lease in the network. Figure 2 illustrates a network which is not a feasible one.

Figure 2. Unfeasible ventilation network

To check the ventilation network if it meets the requirements, there are three methods in the new approach.

3.1 *Depth-first search*

This method is used to check the network whether it is connected one or not. In graph theory depth-first search is a general ergodic method to reach all nods of network (Stanat & McAllister 1977).

At the beginning, source node of network is the first one to be accessed. After source node is pushed into table L, all its conjoint nodes will be pushed into stock S. Then, a conjoint node V is pulled out from stock S and moved it to table L. All conjoint nodes with V will be found out in sequence and stock S only saves the new conjoint nodes which do not exist in table L and stock S. Repeat the procedure until stock S is empty. When stock S is empty, it means that the network is a connected one.

Figure 3 illustrates the depth-first search. Suppose that node A is the first one to push in table L, then table L and stock S become as follows:

L = [A]
S = [F, B, G]

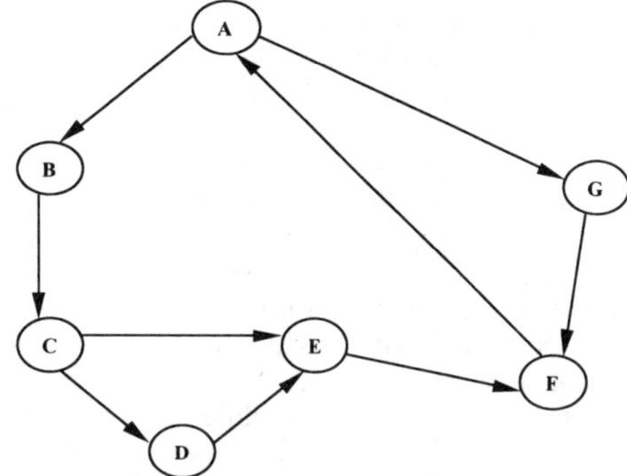

Figure 3. Depth-first search

After G is pulled out from stock S and pushed into table L, the list of L and S turn out to be :

L = [A, G]
S = [F, B]

Since the conjoint node F for G has already in stock S, B is the next nodes to be pulled out from stock S and moved into table L. The conjoint node of B is node C. Since node C is not in the list of L or S, it will be pushed into stock S. The new lists of L and S are as follows:

L = [A, G, B]
S = [F, C]

The 4-th iteration will change the lists of L and S as:

L = [A, G, B, C]
S = [F, D, E]

The 5-th iteration will result in:

L = [A, G, B, C, E]
S = [F, D]

The 6-th iteration will made the lists as:

L =[A, G, B, C,E, D]
S = [F]

Finally the list of stock S becomes empty after node F is moved to table L, i.e.:

L = [A, G, B, C, E, D, F]

Therefore, the network is proved to be a connected one.

3.2 *Degree of node*

Degree is used to check the node whether it is an end node. According to the definition in graph theory, degree of node is the number of arcs incident to or from the node. If the degree of node is equal to 1, it means that the node is an end node; otherwise the degree should be equal to or more than 2.

For example in Figure 3, degree of node A is equal to 3 while degree of node B is 2 and so on.

3.3 *Module*

In ventilation system, it is necessary for some openings to exist such as entrance and exit for air current as well as some special accesses for ventilation. A module is used to check these openings if they exist in network.

Module is a special string where some bits are always equal to 1 while others can be any digit. The fixed bits (equal to 1) represent the necessary openings for ventilation. As soon as a ventilation network is generated by genetic algorithm, the corresponding string is compared with module to check if the necessary openings exist.

For example, suppose module M=[*1**11**] where * implies the bit can be 1 or 0. Then a network N=[11011101] is satisfied the requirement.

3.4 *Matrix*

Usually matrix is used to represent network for computer to deal with in convenience. Incidence matrix is the common one in use as follows (Lawler 1976) :

$$A=\begin{bmatrix} a_{11} & \cdots & a_{1j} & \cdots & a_{1m} \\ \cdots & & \cdots & & \cdots \\ a_{i1} & \cdots & a_{ij} & \cdots & a_{im} \\ \cdots & & \cdots & & \cdots \\ a_{n1} & \cdots & a_{nj} & \cdots & a_{nm} \end{bmatrix} \tag{3}$$

where:

m—number of arcs;

n—number of nodes;

$$a_{ij}=\begin{cases} +1 & \text{if arc j is incident to node i} \\ -1 & \text{if arc j is incident from node j} \\ 0 & \text{otherwise} \end{cases}$$

In order to safe the memory of computer, incidence matrix A can be substituted by two index matrices D and E as follows:

$$D=\begin{bmatrix} d_{11} & \cdots & d_{1j} & \cdots & d_{1m} \\ d_{21} & \cdots & d_{2j} & \cdots & d_{2m} \end{bmatrix} \tag{4}$$

$$E=\begin{bmatrix} e_{11} & e_{12} & \cdots & e_{1P} \\ \cdots & \cdots & \cdots & \cdots \\ e_{i1} & e_{i2} & \cdots & e_{iP} \\ \cdots & \cdots & \cdots & \cdots \\ e_{n1} & e_{n2} & \cdots & e_{nP} \end{bmatrix} \tag{5}$$

where:

d_{1j}—source node of arc j;

d_{2j}—sink node of arc j;

e_{ij}—arc j which is incident with node i;

p—the maximum degree of nodes.

For example, if m=33 and n=22 the size of A will be m·n=721. On the other hand, the size of D plus E will be 2·m+n·p=154 for p=4. Obviously, the memory of computer can save 78.8% if matrix A is substituted by matrices D and E. The larger the m is, the more safe the memory will be.

4 VENTILATION CALCULATION

Air volume and air pressure are the main parameters to calculate the cost for ventilation as follows:

$$\text{Cost} = c\cdot H\cdot Q\cdot\eta \tag{6}$$

where:

c — specified cost;

H — air pressure;

Q — air volume;

η — power efficiency.

In order to calculate air volume and pressure, ventilation network can be divided into three sections: air entrance section, air exit section and air user section. The first two sections constitute natural ventilation network where air volume and pressure will be calculated in terms of natural resistance of opening, while in the third section air volume is assigned by working requirement such as blasting and safety. For example, in Figure 4 where air volume Q_8, Q_9, Q_{10}, Q_{14} and Q_{15} are assigned by working requirements, then arc 5 ~ 17 and 23 constitute air user section whereas the remaining part is natural ventilation section. To deal with ventilation calculation, the new approach uses an integrated method as follows.

4.1 *Establish natural ventilation network*

Step 1. Delete all arcs which have been assigned some fixed air volume. In Figure 4 they are arcs 8, 9, 10, 14, 15.

Step 2. Find the nodes which degree is equal to 1 and delete their incident arcs. In Figure 4 they are nodes (8), (9), (6) and arcs 7, 17, 16.

Step 3. Repeat step 2 again until there is no node with degree 1. In Figure 4 it is node (7) and arc 5.

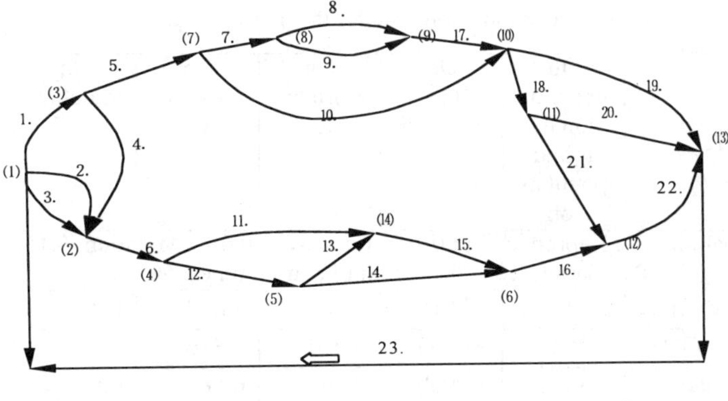

Figure 4. Ventilation network

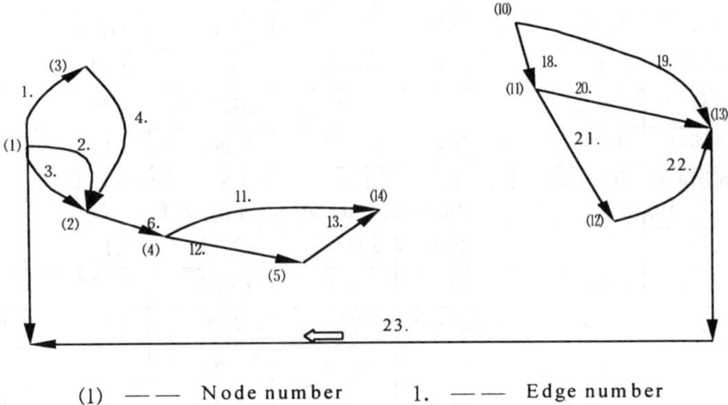

Figure 5. Natural ventilation network.

Finally, a natural ventilation network is established as in Figure 5.

4.2 *Solution of natural ventilation network*

There is conservation law for air volume as:

$$\sum_{j=1}^{m} a_{kj} Q_j = q_k \quad (k=1,2,\ldots,n-1) \qquad (7)$$

where:

m— number of arcs in section;
n — number of nodes in section;
a_{ij}— elements of incident matrix in section;
Q_j— air volume in arc j;
q_k— air volume for node k.

According to the law of ventilation resistance, there is:

$$h_j = R_j \,|\, Q_j \,|\, Q_j \quad (j=1,2,\ldots,m) \qquad (8)$$

where:

h_j— ventilation resistance of arc j;
R_j— resistance parameter of arc j.

There is also conservation law for air pressure as

$$\sum_{j=1}^{m} c_{ij} h_j = \sum_{j=1}^{m} c_{ij} (H_{fj} + H_{ej}) \qquad (9)$$

$$(i=1,2,\ldots,b)$$

where:

b — number of independent circle, b=m-n+1;
c_{ij}— elements of circle matrix where i represents number of circle and j is the arc constituting the circle;
h_j, H_{fj}, H_{ej}—ventilation resistance, mechanical air pressure and potential pressure respectively for arc j.

Equation (9) can be transformed as:

$$\sum_{i=1}^{n-1} a_{ij} p_i = h_j - H_{fj} - H_{ej} \quad (j=1,2,\ldots,m) \qquad (10)$$

where:

p_i—total air pressure of node i.

Table 1. Computation time for the new approach.

	Item	On generation of initial population	On fitness calculation	On reproduction	On crossover	On mutation	Total run time
Probability of : Reproduction : 0.10 Crossover: 0.65 Mutation: 0.15 Population: 100 Maximum iteration times: 50							
1	Time (second)	2.04	790.15	0.16	14.92	0.84	808.23
	Percentage (%)	0.252	97.763	0.020	1.846	0.104	
2	Time (second)	2.31	354.09	0.29	6.58	0.83	364.10
	Percentage (%)	0.634	97.251	0.080	1.807	0.228	
3	Time (second)	1.98	1151.10	0.39	17.13	2.17	1172.88
	Percentage (%)	0.169	98.143	0.033	1.461	0.185	
4	Time (second)	2.36	470.51	0.21	5.73	0.69	479.55
	Percentage (%)	0.492	98.115	0.044	1.195	0.144	
5	Time (second)	2.64	508.22	0.22	10.94	0.76	522.84
	Percentage (%)	0.505	97.204	0.042	2.092	0.145	
Probability of : Reproduction : 0.20 Crossover: 0.65 Mutation: 0.15 Population: 100 Maximum iteration times: 200							
1	Time (second)	2.14	1300.02	2.09	12.93	3.12	1320.52
	Percentage (%)	0.162	98.448	0.158	0.979	0.236	
2	Time (second)	2.36	1782.93	1.73	39.20	3.46	1829.74
	Percentage (%)	0.129	97.442	0.095	2.142	0.189	
3	Time (second)	1.76	1496.42	1.56	18.87	3.27	1521.88
	Percentage (%)	0.116	98.327	0.103	1.240	0.215	
4	Time (second)	2.04	1814.19	1.15	101.14	2.17	1920.69
	Percentage (%)	0.106	94.455	0.060	5.266	0.113	
5	Time (second)	2.20	1950.82	1.76	15.55	3.31	1973.75
	Percentage (%)	0.111	98.838	0.089	0.788	0.168	
Probability of : Reproduction : 0.10 Crossover: 0.75 Mutation: 0.15 Population:100 Maximum iteration times: 50							
1	Time (second)	2.47	753.11	0.33	32.45	1.14	789.50
	Percentage (%)	0.313	95.391	0.042	4.110	0.144	
2	Time (second)	2.20	1234.80	0.24	22.75	0.99	1261.09
	Percentage (%)	0.174	97.915	0.019	1.804	0.079	
3	Time (second)	2.14	513.73	0.17	21.19	0.94	538.22
	Percentage (%)	0.398	95.450	0.032	3.937	0.175	
4	Time (second)	1.82	457.95	0.32	18.09	0.72	478.95
	Percentage (%)	0.380	95.615	0.067	3.777	0.150	
5	Time (second)	2.25	384.95	0.17	6.92	0.79	395.08
	Percentage (%)	0.570	97.436	0.043	1.752	0.200	

Substituted by equation (8), it turns out to be:

$$\sum_{i=1}^{n-1} a_{ij} p_i = R_j | Q_j | Q_j - H_{fj} - H_{ej} \quad (11)$$

$$(j=1,2,\ldots,m)$$

Combined equation (7) with (11), one can obtain the solution of ventilation by similar linear method after setting:

$$c_j = 1/(R_j | Q_j |) \quad (j=1,2,\ldots,m) \quad (12)$$

The solution of ventilation becomes:

$$Q_j = (\sum_{i=1}^{n-1} a_{ij} p_i + H_{fj} + H_{ej}) c_j \quad (13)$$

$$(j=1,2,\ldots,m)$$

Substituting equation (12) into equation (7), there will be:

$$\sum_{i=1}^{n-1} \sum_{j=1}^{m} a_{kj} a_{ij} c_j p_i = q_k - \sum_{j=1}^{m} a_{kj} c_j (H_{fj} + H_{ej}) \quad (14)$$

$$(k=1,2,\ldots,n-1)$$

By means of equation (14), total air pressure of node p_i can be obtained approximately. After a set of iterations to correct c_j and H_{fj}, the true p_i will be reached.

4.3 *Adjust the air pressure*

The air pressure do not meet the conservation law in air user section. Therefore, critical path method is used to adjust their pressure. The basic formula to determine the pressure is:

$$p_{ik} = \sum_{j=0}^{i} (h_j - H_j) \quad (15)$$

where:

p_{ik} — air pressure of node i according to path k;

h_j, H_j,— ventilation resistance and mechanical air pressure respectively for arc j according to path k.

After the critical path is found from air entrance to air exit, the total air volume and total air pressure for the ventilation system can be obtained.

5 APPLICATION

The new approach has been successfully used for the improvement of ventilation system in Xishimen underground iron ore mine of china. The mine has a history of more than 25 years, and there are many leakage from working face to surface because of the irregular mining. Therefore it is necessary to modify the ventilation system.

By means of genetic algorithm, a new ventilation system is proposed and it is under construction. The parameters of genetic algorithm are as follows:

Length of binary string L = 33 bits
Number of individuals in population M = 50
Reproduction probability $P_r = 0.3$
Crossover probability $P_c = 0.4$
Mutation probability $P_m = 0.01$
Maximum iterations of a run N=100.

Table 1 illustrates the computation time of the new approach for 15 runs. It is shown that about 97~98% of the total computation time is used for fitness calculation, mostly for ventilation calculation. The second item for time consumption is crossover, which takes about 1~2% of total time. In general, the time consumption for the new approach is acceptable, about 7~20 minutes for each run.

Since crossover takes more time than reproduction and mutation in genetic algorithm, the probability of crossover emerge as the most important parameter to be studied. From table 1 it is shown that probability of crossover will be better if it takes 0.65.

Because of the characteristics of ventilation system, most of individuals in population are not feasible alternatives from the view of mining engineering. Figure 6 represents the percentage of feasible alternatives from the initial population in 100 separate runs respectively. It is shown that only 1% individuals in the initial population will be feasible alternatives as potential ventilation network.

In order to study the impact of genetic operators on the success of the evolution for ventilation network, a pair of individuals are picked up for crossover, its crossover point is randomly selected until a feasible individual emerges. Then, the ratio between the number of random selection and the length of binary string represents the success ratio of crossover. Similar method is also used to study the success ratio of mutation. Figure 7 and Figure 8 are the result of success ratio for crossover and mutation. It is shown that the success ratio is about 50%.

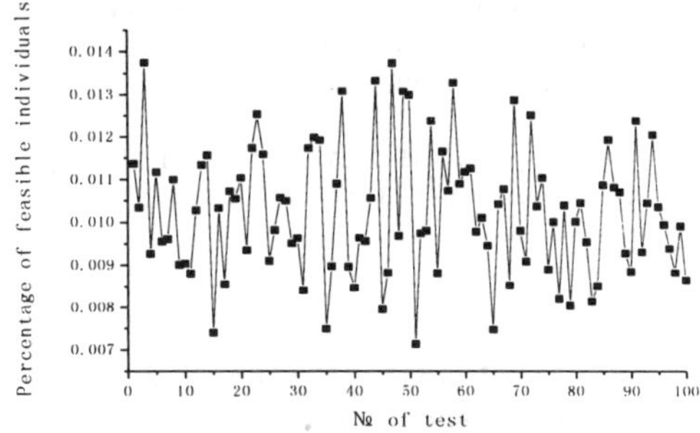

Figure 6. Percentage of feasible alternatives in the initial population

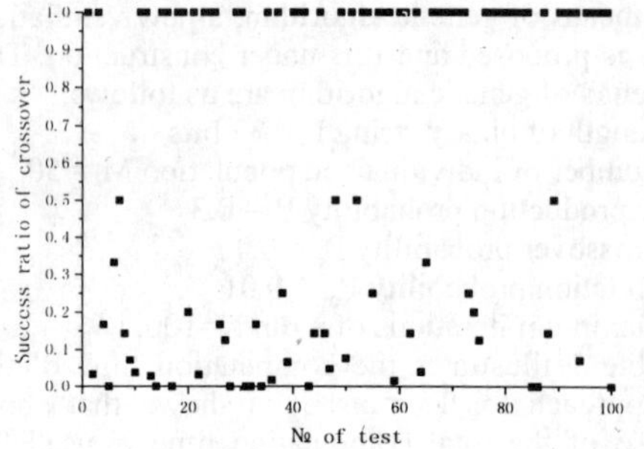

Figure 7. The result of success ratio for crossover

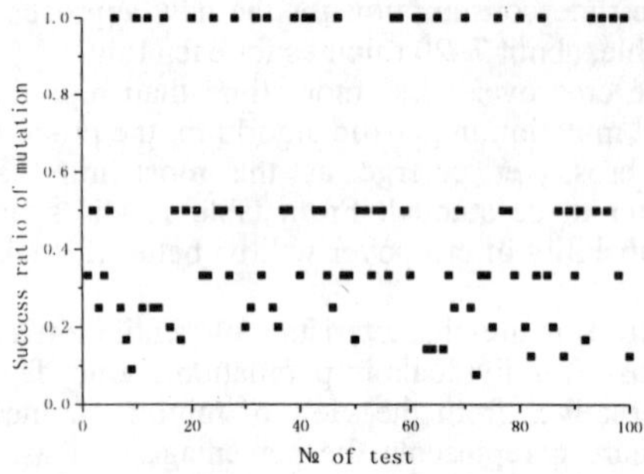

Figure 8. The result of success ratio for mutation

6 CONCLUSIONS

Genetic algorithm is a useful method to deal with structural optimization as well as parameter optimization by means of adaptive search. It is proved to be a potential prospective approach for the complex and complicate mining problems.

The new approach presented in this paper is composed of three phases: generating network by genetic algorithm, verifying network by graph theory and evaluating network by ventilation calculation. This method can reach the optimum solution by means of heuristic search in a broad space.

In this paper some parameters are also studied to show their impact on genetic algorithm, such as the generation of the initial population, success ratio of crossover and mutation as well as the time consumption.

The new approach is proved by application to be reliable and can be used for other mines as well.

REFERENCES

Goldberg, D. E. 1989. Genetic algorithms—in search, optimization and machine learning. New York: Addison-Wesley Publishing Company.

Lawler, E. L. 1976. Combinatorial optimization: networks and matroids. New York: Holt, Rinehart and Winston.

Stanat, D. F. & McAllister, D. F. 1977. Discrete mathematics in computer science. New York: Prentice-Hall Inc.

Yun, Q. X. et al 1997. Genetic algorithms and genetic programming—an optimization technique by search. Beijing: Metallurgical Industry Publishing Company. (Chinese)

Environmental Issues and Management of Waste in Energy and Mineral Production, Singhal & Mehrotra (eds)
© 2000 Balkema, Rotterdam, ISBN 90 5809 085 X

Pollutants migration in groundwater below a phosphogypsum storage site

M. Zairi & M.J. Rouis
Département de Géologie, Ecole Nationale d'Ingénieurs de Sfax, Tunisia

ABSTRACT: In the most of the developing countries, solid wastes storage continue to be uncontrolled. It's operated consists, in the most cases, by filling natural or artificial excavations or simply by land raising without any protection measurement against ground water and soil contamination.The north coasts development project of Sfax city contains a new projected beach. The south limit of this beach is a few meters from the NPK phosphogypsum storage site. In fact, the phosphogypsum which is a solid by product of phosphoric acid industry, is stored since 30 years on natural soil without any protective organ. This study was conducted in order to assess the effects of this waste disposal on ground water quality. This paper presents the results of an investigation carried out in the site. Geologic and hydrogeological properties of the site are defined. Phosphate, fluoride and cadmium concentrations were measured in ground water samples collected from boreholes on the disposal site. It was found that most of samples were characterised with a high concentrations and a very low pH values.
Simulations with a two dimensional finite element model for pollutants migration in saturated porous media of the dispersion of phosphates, fluoride and cadmium in the site are particularly analysed.

1 INTRODUCTION

Regardless to their origin, solid wastes contain always directly dangerous substances. The contamination is essentially due to the leaching process which causes pollutants to move and to be dispersed in the natural medium. The waste storage water and the rain waters have a crucial role in the leaching process. These waters infiltrate through wastes and will be directly in contact with their different chemical compounds. Many dissolution reactions take place and cause the increase of the amount of dissolved cations and anions. These lasts migrate with the water or leachate through the waste pile to finally reach the water table.

Among the dissolved substances, some dangerous compounds are transported. When arrived to the water table, pollutants begin a migration which causes the contaminated zone to progressively extents. Area of polluted zone may come very large and will be of many times that of the storage site (Freeze & Cherry, 1979). The pollutants migration produces a contamination panache that elongates in the ground water flow direction.

Quantitative prediction of pollutants migration is now of great interest in the field of waste management and storage. Numerical modelling of contaminant transport through porous media is an efficient tool to make this prediction possible. Site selection and design of protective requirements for landfills and storage facilities are highly related to the behaviour of contaminants in such structures. This behaviour may be well defined by numerical simulation.

At Sfax city, the phosphogypsum produced by the NPK processing plant was stored, during an approximate period of 30 years, directly on the soil surface.

A local water table is formed in the phosphogypsum embankment by storage and rain waters infiltration. In this study, the ground water contamination related to the phosphogypsum storage is considered. The pollutants dispersion in the studied area is analysed using a two dimensional finite element model.

2 NPK PHOSPHOGYPSUM STORAGE SITE STUDY

2.1 *Waste characterisation*

Fertilisers industry development in Tunisia causes a phosphoric acid production growth. This production is based on the treatment of the natural phosphates by the sulphuric acid giving a solid by product: the phosphogypsum. this last is issued from the filtration of the solution of phosphate attack by the sulphuric acid. It's mainly constituted by calcium sulphate, free phosphoric acid, a lot of heavy metals salts such as cadmium, zinc and lead. It's also highly concentrated with phosphates and fluorides. In fact, the two main components of Tunisian natural phosphate are the fluor-apatite $(Ca_3(PO_4)_2)_3CaF_2$ and the calcium carbonate $CaCO_3$. The phosphoric acid production is conducted at 80°C by a dihydrate processing with the sulphuric acid (Manguin, 1978).

The chemical reaction gives a solid phase product ($CaSO_4$, $2H_2O$), suspended in a solution of phosphoric acid (H_3PO_4), which contains about 28% by weight P_2O_5 (Chakchouk & Trabelsi, 1989). Secondary reaction by-products are iron, aluminium, calcium and magnesium sulphates, together with significant quantities of hydro-fluoridric acid. The mixture is vacuum-filtered; the solid phase is washed, the rinse water is added to phosphogypsum and pumped to the storage basins. This process produces 5 tons of phosphogypsum for each ton of phosphoric acid.

Many chemical analysis were conducted on several phosphogypsum samples (Rouis et al., 1990), they confirmed the existence of many toxic substances in great quantities in this refuse. Table 1 presents concentrations of the main contaminants in the phosphogypsum. If compared to Quebec standards for contaminated soils and materials (MENVIQ, 1988), the phosphogypsum is highly contaminated by zinc and molybdenum and slightly contaminated by cadmium and mercury. Its also highly polluted with fluoride and phosphates.

The radioactive elements presence in natural phosphate rocks causes the phosphogypsum to be a low level radioactive waste. This radioactivity is related to uranium 238 and thorium 232 decomposition. Radioactive elements presence in phosphogypsum depends on the type of treatment of natural phosphate rocks. In fact, according to this treatment and oxidation and reduction conditions, radioactive elements are divided between the phosphogypsum and the phosphoric acid (Rutherford et al, 1994).

2.2 *NPK phosphogypsum storage site characterization*

The NPK phosphogypsum was stockpiled during a period of 30 years directly on the soil surface. The phosphogypsum pile have the shape of a terril surrounded by a peripheral plate particularly extended in the north and east directions. The phosphogypsum covered area is approximately 1,551,900 m^2. The phosphogypsum thickness ranges from 5 to 8.5 m in the central terril and from 0 to 3.5 m in the peripheral plate. The total volume of phosphogypsum

Table 1. Average contaminants concentrations in Sfax phosphogypsum (Rouis et al., 1990).

Compound	Concentration (mg/kg)	Quebec standards (mg/kg)
P_2O_5	31000	-
F	40000	-
Zn	315	100
Cd	39.8	1.5
Hg	14.5	0.2
Ni	15.4	50
Fe	58.4	-
Cu	5.9	50
Mo	5	2
Co	7.6	15

stored during the site operating is about 5,372,200 m^3.

The phosphogypsum is directly bonded on two types of soils, the first is a conchiferous sand, the second is a conchiferous algal mud. Below these two layers, the lithological sequence is essentially constituted by a brownish silty clay and a yellow fine sand. The inner part of the lithologic sequence is a 12 m deep compact red clay. A discontinuous limestone crust is locally encountered in the investigated zone.

The piezometric map (Fig. 1) shows a water table with a divergent flow regime; the central terril constitute a piezometric dome from which a multidirectional flow occurs. In fact, two preferential flow directions to the sea are noted. The presence of a water pond at the west of the central terril is related to a locally water table outcrop.

The flow patterns analysis indicates the presence of two water mixture zones. The first between the phosphogypsum water table and the industrial residual waters which are rejected in the zone of the water table outcrop. The second is situated at the south of the central terril and constitute the zone of water convergence from the phosphogypsum water table and the phreatic one.

2.3 *Site water quality*

In order to characterize the pollution level in the site, an investigation were conducted, it consists essentially in the execution of 31 boreholes (Fig. 1) having a depth ranging from 1 to 20m. These boreholes are then used for water table sampling. Water samples are collected using a hand auger. An approximate volume of 1 litre was taken from each hole and sealed in plastic bottles. The chemical analytical techniques used to charecterize the water table quality are:

- electro-chemical analyse using a specific fluoride electrode:
- visible spectrophotometer for phosphate;
- atomic absorption spectrometer for heavy metals;

- mercury using a mercury analyser with a wave length of 253.7 nm.

The results of chemical analyses show that ground water at the NPK phosphogypsum storage site are highly polluted by fluoride, phosphate and heavy metals. In table 2 are presented minimal, maximal and average water table concentrations of these constituents. Water samples revealed a concentrations superior to those permitted by Tunisian standard for residual waters discharge (MET[7]). Measured pH values are strongly acid, allowing a high mobility for heavy metals which are in their ionic free state within a such acidic system.

The three compounds: phosphate, fluoride and cadmium are considered to be exclusively generated by phosphogypsum in the investigated area. Their distribution in the investigated zone is particularly examined. These compounds concentrations distribution obtained from chemical analysis are presented on figure 2. Maximum concentrations are recorded in the central terril with a values of about 20 g/L for phosphates, 200 mg/l for fluoride and ranging from 5 to 10 mg/l for cadmium. This is due to the high phosphogypsum thickness and consequently the much available matter for dissolution and the longer flow time through the waste stockpile. At the peripheral plate, high concentrations zones are coinciding to those of slight hydraulic gradient. The pollutants concentration distribution in the site is directly related to water table flow path. At the two water mixture zones, the lesser concentrations plume is recorded with values of 5 to 7 g/L for Po4, about 50 mg/l for fluoride and from 0.02 to 1 mg/l for cadmium. In fact, in these zones, dilution by industrial residual waters and those of the phreatic one causes the concentration decrease. If compared to Tunisian standards for residual water discharge (MET, 1988), the analysed waters are highly contaminated by phosphates, fluoride and cadmium.

Table 2 Ground water contaminants concentrations at the NPK phosphogypsum site.

Compound concentration (mg/l)	Minimal	Maximal	Average	Tunisian standard
pH	1.7	8.4	3.4	6.5 to 9
F	0.43	336.3	87.25	5
Po4	10	23250	10100	0.1
Cd	0.01	10.25	1.2	0.005
Hg	0.0016	0.0031	0.0022	0.001
Fe	0.5	50.0	18.22	1
Zn	0.41	41.0	15.71	10

3 POLLUTANT MIGRATION MODELLING

One form of the governing equation of two dimensional pollutant migration in saturated, homogenous and isotropic porous media is given by:

$$\frac{\partial C}{\partial t} = D^{*}_{x}\frac{\partial^2 C}{\partial x^2} + D^{*}_{y}\frac{\partial^2 C}{\partial y^2} - V^{*}_{x}\frac{\partial C}{\partial x} - V^{*}_{y}\frac{\partial C}{\partial y} \quad (1)$$

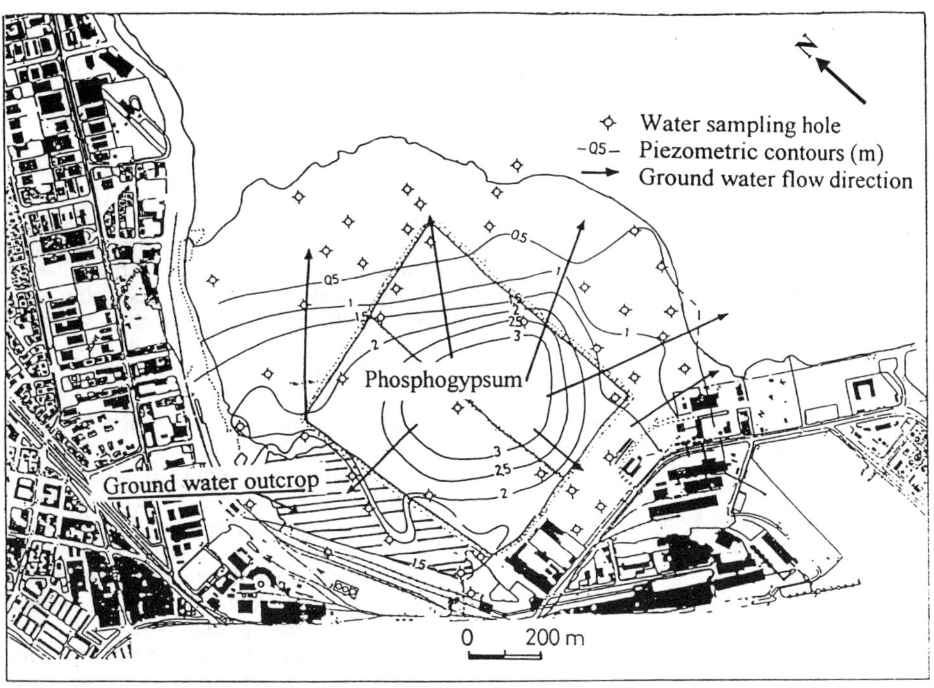

Figure 1 NPK Phosphogypsum storage site piezometric map.

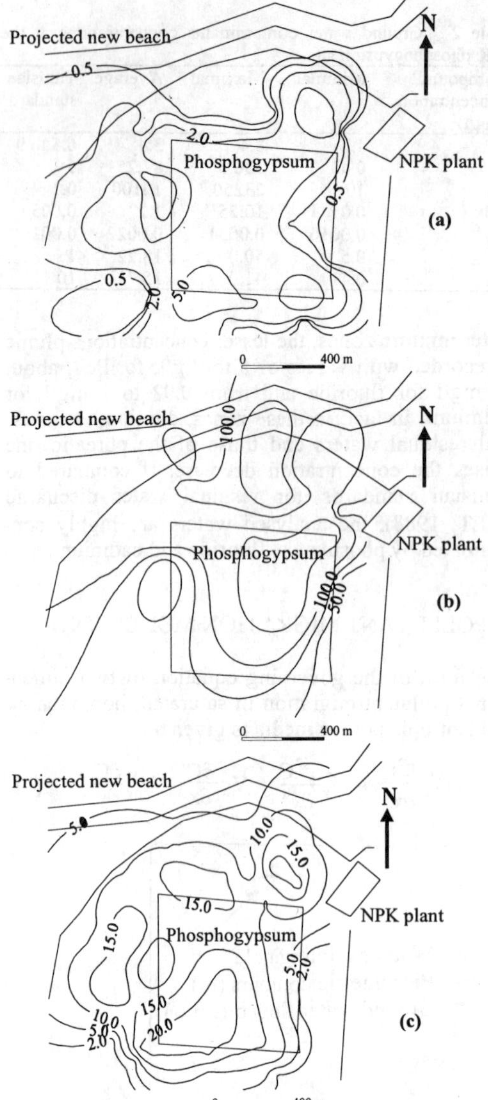

Figure 2 Measured isoconcentrations maps for a: cadmium (mg/l), b: fluoride (mg/l) and c: phosphates (g/l).

C = concentration of the considered contaminant in a point of co-ordinates x and y at time t; Vx,Vy = water flow velocities in X and Y directions; Dx,Dy = hydrodynamic dispersion coefficients of contaminant in X and Y directions; $D^*=D/R_d$, $V^*=V/R_d$; R_d is the retardation factor, defined by Bear (1972, in Auriault and Lewandowska, 1993) and which is given by:

$$R_d = 1 + \rho K_d / n \qquad (2)$$

n: medium porosity; ρ: soil dry density; K_d: distribution coefficient.

In the case of conservative species, which have no interaction with porous medium particles, the retardation factor is equal to 1. For species adsorbed by soil particles the value of this factor depends on the type of the considered element and the soil adsorption capacity, expressed by the distribution or partitioning factor (K_d).

The two-dimensional numerical model for the migration of contaminants in saturated porous media, developed by Zairi (1996) and named MIGPO, is used to simulate the behaviour of pollutants in the NPK phosphogypsum storage site. In this model the finite element method based on the Galerkin formulation (Zienkiewicz, 1977) and using eight nodes quadrilateral isoparametric elements, was used to formulate the numerical description of the dispersive-diffusive mass transport equation. The finite element spatial discretization of the governing equation of pollutants migration in porous media have given a first order differential equation in time:

$$[K]\{C\} + [M]\left\{\frac{\partial C}{\partial t}\right\} = \{F\} \quad \text{for} \quad t > t_0 \qquad (3)$$

In the last equation, the boundary conditions are presented by the vector F and typical matrixes elements are given by:

$$K_{ij} = \sum_e \int_{A^e} \left\{ D^*_x \frac{\partial N_i}{\partial x}\frac{\partial N_j}{\partial x} + D^*_y \frac{\partial N_i}{\partial y}\frac{\partial N_j}{\partial y} + V^*_x N_i \frac{\partial N_j}{\partial x} + V^*_y N_i \frac{\partial N_j}{\partial y} \right\} dA^e \qquad (4)$$

$$M_{ij} = \sum_e \int_{A^e} N_i N_j dA^e \qquad (5)$$

Where: Ni, Nj: elementary approximation functions; A^e: element cross sectional area.

The resolution of equation (3) consists to find a functions set that satisfy (3) at each time t and for the initial conditions at the time t_0. This is done by replacing the time derivatives by a finite difference approximation.

The validation of MIGPO results was done by comparisons with the results of Ogata (1970) analytic solution and Rowe and Booker (1985) semi-analytic solution together with experimental results. These comparisons has demonstrated the reliability of the MIGPO model by giving a good similarities between it's results and those of the others solutions.

3.1 *Model predictions*

In the NPK phosphogypsum storage site discretization, only the domain between the central terril and the sea is considered. This domain, having an approximate area of 550,000 m^2, is divided into 56 elements having 219 nodes. Four zones with different hydrodynamic properties are considered. Table 3 presents the dispersion coefficients (Dx,Dy), velocities (Vx,Vy) and longitudinal (αl) and transversal (αt) dispersivities for each zone.

Many values of the retardation factor and the longitudinal and transversal dispersivities were examined during the simulation process. The retained

Table 3. Hydrodynamic parameters for simulation of pollutants dispersion at the NPK phosphogypsum site.

Zone	1	2	3	4
D_x(m^2/year)	80	145	80	145
D_y(m^2/year)	32.5	45	32.5	45
V_x(m/year)	3.1	3.65	3.1	3.65
V_y(m/year)	1.65	0.55	0	0
α_l (m)	28.13	39.91	25.81	39.73
α_t (m)	3.9	11.56	10.48	12.33

ones (Table 3) are those giving a good similarity between the measured concentrations on the site and those calculated for a simulation period of 30 years, which is the time passed from the starting of phosphogypsum storage.

The water flow velocities in the site (Vx and Vy) are calculated from field measured hydraulic gradient and results of laboratory permeability tests on intact soil samples. The flow velocities are used together with longitudinal and transversal dispersivities to calculate dispersion coefficients (Dx and Dy). To model the ground water flow at the NPK phosphogypsum storage site, a steady state analysis was used. This allowed a calculation of the water flux at the model limits and thus the mass flux along these limits. The limit of the central terril is considered as an imposed entering mass flux, the value of this flux is calculated from the water flux multiplied by an average concentration depending on the simulated element and assuming that the ground water at this location is at a chemical equilibrium with solid phase (phosphogypsum). The mass flux is also considered to vary with the portion of the phosphogypsum limit part. In fact, the terril water is more concentrated in it's central part than in the north and south ones, so the flux inequalities.

3.2 *Simulation results*

3.2.1 *Phosphates*

For phosphates dispersion simulations, the west domain limit is considered of an imposed concentration of 30 g/L along a portion of 250 m at the north of the terril. The south domain limit and the west one at the south of the terril are maintained at a fixed concentration of 1 g/L. The total phosphate entering flux is of 155000 g/year. All interactions between phosphates and soil solid particles are neglected and a retardation factor of 1 is considered.

The simulated phosphates concentration distribution for a time period of 30 years (Fig. 3-a) shows a well likeness to that established from analytic data. The concentration differences between the two maps is negligible. The concentration zoning observed on the field is recognised on the simulated concentrations map. On this last, the amount of phosphates in the water table ranges from 1 to 10 g/L at the mixture zone in the south of the terril. Waters with the highest phosphate concentrations are located at the central terril and the area at it's east. The observed and calculated concentrations at the north of the terril are comparable and range from 5 to 30 g/L with a decreasing values in the sea direction

The provisional phosphates distribution map for a time period of 60 years (Fig. 3-b), shows a concentration range from 5 to 20 g/L at the south of the terril. On the domain limit, this concentration is greater than 10 g/L (less than 5 g/L for a time period of 30 years). High concentrations, more than 70 g/L, are recorded at the north part of the east limit of the terril. At the north of the terril, phosphate concentration ranges from 50 g/L to 10 g/L at the farther points to the phosphogypsum stockpile. The concentration front of 20 g/L is found to be displaced 200 m, measured perpendicularly to isoconcentration curves, during 30 years. This gives an average velocity of 6.7 m/a for phosphate concentration front migration in the site.

3.2.2 *Fluoride*

For the fluoride behaviour simulations, the west domain limit is considered of an imposed concentration of 70 mg/l along a portion of 250 m at the north of the terril and null for the remaining part of the limit. The fluoride is assumed to be a conservative species, all it's interactions with the soil solid phase are neglected. The total entering fluoride flux in the do-

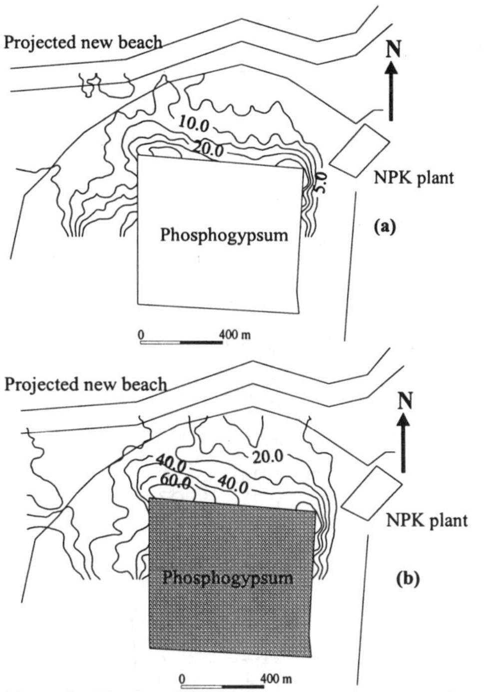

Figure 3 Simulated Po_4 (g/l) concentrations maps for a: 30 years and b: 60 years.

main is fixed to 1297000 mg/year.

The fluoride simulated isoconcentration map for a time period of 30 years (Fig. 4-a), shows a good correlation with that established from water table analysis data. The area located at the east of the terril have the more fluoride concentrated water with more than 100 mg/l. The mixture zone at the south of the terril presents a concentration of about 10 mg/l. At the north of the terril, simulated concentrations are slightly different from that observed on the field. In fact, the measured anomaly of 100 mg/l is not recognised by simulations. Concentrations evolution in time is examined for a period of 60 years, assuming that the initial and boundary conditions are invariable. The fluoride isoconcentration map for this period (Fig. 4-b) shows a similar zoning to that of 30 years. However, concentrations are higher. they are greater than 300 mg/l in the zone situated at the east of the phosphogypsum terril. In the mixture zone, concentration is greater than 10 mg/l. At the north of the terril concentrations range from 50 to 200 mg/l.

3.2.3 *Cadmium*

In the cadmium dispersion simulation, the south and east domain limits are maintained at an imposed null concentrations. In fact, the measured values on these limits are very slight (less than 0,02 mg/l). The west limit at the north of the terril is considered of an imposed 1 mg/l cadmium concentration along the first 250 m and null for the remaining part. The total cadmium entering flux equals 40880 mg/year. This flux is unequally distributed on the terril limits, with the highly values at the east one. The cadmium is allowed to be partially retained by soil particles, so a retardation factor of 1.5 is considered in this element dispersion simulation. Although, this value is slight, it has permitted a well identity between the measured and calculated concentrations for a period of 30 years.

The simulated cadmium distribution map for 30 years (Fig. 5-a) shows a similar zoning to that obtained from water table samples analysis. An area of slight concentrations located at the south of the terril (mixture zone), an area of high concentrations in the east of the terril and a zone of intermediary concentrations at it's north are clearly identified on the two maps.

The provisional cadmium distribution map for a period of 60 years (Fig. 5-b) indicates an increasing concentrations in the whole domain. The highest concentration is about 10 mg/l and it's located at the east of the phosphogypsum terril.

The concentration front of 5 mg/l at the east of the terril have moved 185 m in the sea direction during 30 years, this gives an average velocity of 6.2 m/a for cadmium concentration front migration in the domain.

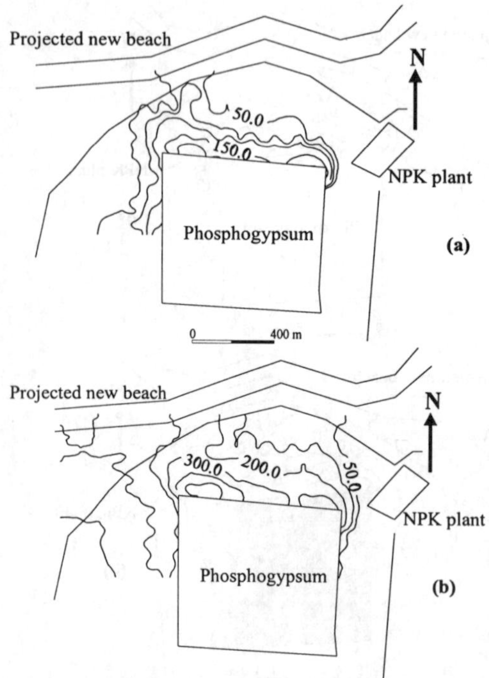

Figure 4 Simulated fluoride (mg/l) isoconcentrations maps for a: 30 years and b: 60 years.

4 CONCLUSION

The uncontrolled phosphogypsum storage at the NPK site has led to an alarming level of ground water contamination. This contamination is essentially due to the leaching process by the phosphogypsum storage water and the rain water. Heavy metals concentration and acidity of ground water are very high. The ionic form of heavy metals favoured by the ground water acidic pH values, is a factor for their rapid migration.

The water table below and around the NPK phosphogypsum terril is particularly high contaminated by phosphate, fluoride and cadmium. The spatial distribution of this pollution is related to the hydrodynamic regime of the site water table. This fact is demonstrated by the good correspondence between the zones of high concentrations and those of a slight hydraulic gradient. However, the phosphogypsum thickness have a significant role so that highly concentrations are encountered in the central terril, which have the maximum phosphogypsum thickness. This pollution is actually at the location of the projected new beach at the east of the terril, and has enhanced an extension in the direction of this area at the north of the terril.

The numeric model for the NPK phosphogypsum storage site have permitted to recognise the actual distribution of the phreatic water table contamination

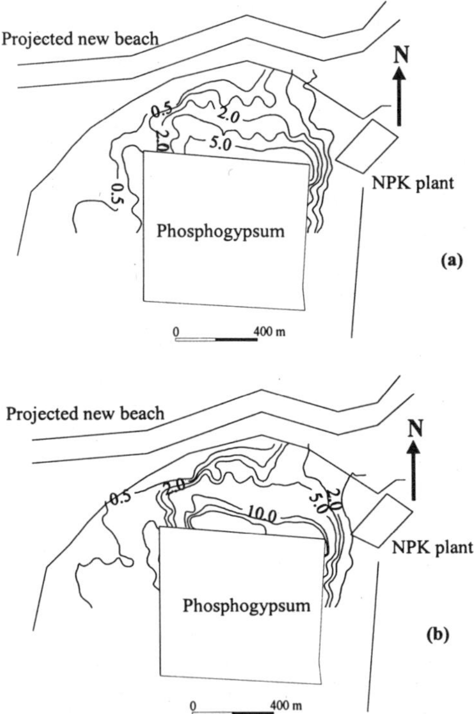

Figure 5 Simulated cadmium (mg/l) isoconcentrations maps for a: 30 years and b: 60 years.

by phosphates, fluoride and cadmium. The calculated average velocities for these contaminants fronts are high (about 6 m/year) and they indicate a great danger coming from the fast pollutants dispersion in the site.

The provisional simulations has also demonstrated a general increasing, in the whole domain, of concentrations of water table with the simulated compounds; phosphates, fluoride and cadmium. This contamination continue to propagate and occupy more and more extended areas. If no corrective action to stop contaminants migration is executed, The water table concentration with these three compounds at the new beach location will be, in 30 years, of 1 mg/l for cadmium, greater than 50 mg/l for fluoride and about 20 g/L for phosphates. These waters will, by underground flow reach the sea and constitute a potential risk for fauna and flora.

The execution of a protection system, for the projected new beach, in order to stop pollutants migration from phosphogypsum through the water table such as a cut off wall or a vertical liner, the covering of this last and the decontamination of the actual polluted waters and soils, are crucial to make the integration of the NPK phosphogypsum storage site to the Sfax north coasts development project successful and of no health risks.

REFERENCES

Auriault, J.L., and Lewandowska, 1993. Homogenisation analysis of diffusion and adsorption macrotransport in porous media: macrotransport in the absence of advection. *Géotechnique*, Vol. 43, N°3, 457-469.

Bear, J., 1972. *Dynamics of fluids in porous media*. Elsevier, New York, 1972.

Chakchouk, M., Trabelsi, F., 1989. Contribution à l'étude de la récupération des eaux de rejet de l'usine SIAPE A., *Rapport interne*, ENIS, Tunisia.

Freeze, R.A., Cherry, J.A., 1979. *Ground water*, New Jersey, Prentice-Hall.

Mangin, S.,1978. Utilisation du phosphogypse en technique routière, *Bulletin de liaison des laboratoires des ponts et chaussées*, Spécial VII, le phosphogypse, 7-13.

Ministère de l'Economie Tunisienne (MET), 1988. Norme N.T. 106.002.

Ministry of Environment of Québec (MENVIQ),, 1988. Grille des critères indicatifs de la contamination des sols et de l'eau souterraine. Québec, Canada.

Ogata, A., 1970. Theory of dispersion in a granular medium, Fluid movement in earth materials, *USA Geological survey,, Professional paper 411-I.*

Rouis, M.J., Ben Salah, A., Ballivy, G.,1990. Phosphogypsum management in Tunisia: environmental problem and required solutions, *Proceedings of the Third international symposium on phosphogypsum, Orlando, vol. 1, 1990*, 87-105.

Rowe, R.K. and Booker, J.R., 1985 1-D pollutant migration in soils of finite depth, *Journal of geotechnical engineering*, 1985, Vol. 111, N°4, April 1985.

Rutherford, P.M., Dudas, M.J., Samek, R.A., 1994 Environmental impact of phosphogypsum, *The Science of the total environment*, , Vol. 149, 1-38.

Zairi, M., 1996 Modélisation mathématique de la migration des polluants dans les milieux poreux: application aux cas des sites de stockage de phosphogypse de Sfax. *Thèse de doctorat de spécialité, Ecole Nationale d'Ingénieurs Sfax*, Tunisie.

Zienkiewicz, O.C., 1977. *The finite element method in engineering science*. Mc Graw-Hill, New York.

Environmental Issues and Management of Waste in Energy and Mineral Production, Singhal & Mehrotra (eds)
© 2000 Balkema, Rotterdam, ISBN 90 5809 085 X

Methods of the connected geomonitoring for modeling a condition of natural systems in regions with intensive technogeny influence

M.A.Zhuravkov & A.N.Zemskov
Department of Mechanics-Mathematics, Belarus State University, Belarus

K.R.Al-Momani
Department of Operation and Maintenance, Jordan University of Science and Technology, Jordan

ABSTRACT: In paper the results of performance of research works on building up of system of the connected geomonitoring for modeling and study of a state of natural systems as a whole and their separate elements are presented. The necessity of the complex approach to study of a state a component of natural environment (ecosystem) is proved. The structure of system is described. The methods of mathematical modeling and forecast of development of ecosystems and state of their dynamic balance are submitted. The results of performance of some researches for regions of the regions of excavation of deposits of potash salts in Russia and Belarus and area included two major sites: Jordan University of science and technology (JUST) and Al-Hussan Industrial city (HIC) are presented.

1 GENERAL CONDITIONS

In process of development of a human society active intervention of the people in various components of an environment more and more grows, more and more intensively natural resources are involved in economic activity. However natural resources are not boundless. With each year power, water, wood, soil etc. resources become more and more scarce. The increase of volume of production of raw and fuel- energy resources manages all more expensive. There is a problem of rational use of natural resources.

The quantity of various hard, liquid and gaseous waste of ability to live of the people steadily grows. In result environment of man live (ecosystem) in the increasing scales «becomes soiled» (pollution of water sources, exhaustion of soil, destruction of woods, radioactive pollution, electromagnetic fields etc.).

The remark. Under ecosystem we shall understand geophysical environment in aggregate with objects of other nature and problems of stability of an environment. Geophysical environment we shall understand any composition of hard, liquid and gaseous bodies cooperating among themselves at mechanical movement in force fields of the Earth.

Today technogeny influence on a nature exceeds its recovering potential, that entails irreversible changes of an environment not only local, but also regional scale. Last circumstances result in infringement of dynamic balance between the various parties of ecosystem of region, that is expressed in inability of natural environment to carry out functions, peculiar to it of an exchange of substances and energy, to support conditions necessary for existence and development for life (ecological crisis).

The scales of modern activity of mankind have not similarity in a history of the Earth. For only one year the production of various minerals achieves 100 milliard tons. Flinging open ground, the man annually moves weight of ground approximately three times superior all volcanic products, rising from bowels of the Earth for the same term, and in 200 times more, than communicates in the sea and oceans by the current waters.

It is obvious that to supervise and to operate an ecological state of region now probably only at purposeful study of a state of not separate elements of ecosystem of region, but their set at the obligatory account of their interaction.

The detached analysis of separate elements of ecosystem of region can result in return results. Technogeny influence on ecosystem of region it is impossible to consider as safe until all its separate consequences are appreciated. The intervention in life of natural environment is especially felt by consideration of the large-scale projects (large-scale mining works, construction of dams, meliorative work etc.). In this case absence of system of forecasting of the remote consequences of influence on natural object is especially visible.

For achievement of purposes of accurate study, modeling and prognoses of state of ecosystem of region is necessary building up regional complex system of the connected geoecological monitoring. The basic purpose of performance of works should be, on the basis of correct of the all-round analysis of a modern geoecological state of ecosystem of considered region, authentic forecast of probable evolutionary development of a state of ecosystem of region depending on technogeny influence on it.

Very interesting the results of question given to the companies of chemical products sector and manufacture of rubber and plastic products sector explains the company strategy in reducing the materials used in production and the procedures followed to protect the environment in Jordan during last several years (table 1*).

Table 1.

Question	Yes %		No %		unspecified %	
	995	I1996	995	I1996	995	I199
Whether raw materials are used to reduce pollutants in the process of production without affecting the product itself	46.1	37.0	52.9	63.0	1.0	0
Whether the final product is designed for reducing risks of pollution	51.0	44.0	48.0	56.0	1.0	0
Whether change in the production process is carried out through improvement of efficiency and reducing pollution resulting from the production process	67.3	61.0	30.8	38.0	1.9	1.0
Whether the equipment and instruments are used in the optimum manner	98.0	94.0	1.0	6.0	1.0	0
Whether maintenance and modification on equipment and instruments are carried out for improving the production process	99.0	98.0	0	1.0	1.0	1.0
Whether rationalizing use of water through use of advanced technology in the production process	69.2	74	28.8	24.0	2.0	2.0
Whether specific characteristics of liquid pollutants are defined	55.8	40.0	37.5	53.0	6.7	7.0
Whether values of organic and chemical pollutants loads resulting from industrial wastes are evaluated	46.2	32.0	46.2	61	7.6	7.0
If the enterprise does not have a defining and treatment unit system - is the reason attributed to the fact that the system constitute a financial burden	19.2	17.0	64.4	68.0	16.4	15.0
If the enterprise willing to participate in establishing a central dumping area for dangerous and poisonous wastes	45.2	43.0	51.0	53.0	3.8	4.0

(*) - materials of Department of Statistics (Jordan)

The analysis of the table shows, that last years the attention to consequences of influence of industrial activity on natural environment begins to grow and affects planning of manufacture. Though, as it is visible from the table, the economic and technical factors nevertheless for the present are prevailing.

2 CONNECTED GEOECOLOGICAL MONITORING

The structure of system of the connected geoecological monitoring (SCGM) can be presented as set of large blocks (Zhuravkov et al. 1997b, 1998a.), among which:

- System-airspace;
- Surveying;
- Meteorological;
- Hydrological and hydro-geological;
- Structural-geomorphological;
- Engineering- and structural-geological;
- Geodynamic and geomechanical;
- Physics-geological;
- Seismological;
- Radiometric;
- Engineering-ecological etc.

The connected geoecological monitoring requires radical change of methodology of carried out works. First of all complex of works should be systematic organized. The system organization means representative classification of works, revealing hierarchical subordination between them. The analysis of last allows most representatively to decide questions of a choice minimally of necessary complex of methods, ways, means and technologies. For example, the ecological maintenance can be realized only on the basis of the connected processing of results of works on geochemical, soil&geobotany, physics-geological, geodynamic, geomechanical, structural-geological, hydro-geological, engineering-geological, hydrological, meteorological, surveying and system-space directions.

On a degree of modern use and industrial realization recommended directions of works and the separate groups of methods can be incorporated in three groups:

The first group is characterized by rather complete application of new methods and technologies - further perfection and to rationalization is subject 10 - 30 % of used methods and technologies (for example, engineering-geodetic researches).

In the second group the directions and sections are incorporated, in which structure the new methods are used unsufficiently and from 30 up to 60 % of used methods is subject to replacement or radical reorganization. To this group it is possible to relate preparation of materials of remote sounding on hard copies, hydrological and meteorological researches, structural-geomorphological researches, reinterpretation of the geology-geophysical information, study of tecnogeny, geodynamic and geomechanical processes, ecological monitoring in a real time scale etc.

The third group includes directions and sections, which in research-and-production activity are used extremely unsufficiently or practically are not applied - there is a necessity for introduction from 60% up to 100% of new methods and technologies. In this group are included preparation of materials of remote sounding on magnetic carriers, engineering-geological interpretation (decipherment), connected processing of results of engineering-geological, hydro-geological, design, geophysical, system-space and other works, structural decipherment, study natural and natural-tecnogeny of geodynamic processes etc.

3 GEOINFORMATION TECHNOLOGIES

To carry out modeling and researches it is impossible without construction of the integrated connected information model of region. Only basing on such model the performance of modeling and forecasting of a state of ecosystem as a whole is possible, taking into account interinfluence of its separate components.

Basis of construction of information systems and databases for all sections of Earth Sciences is the set of the approaches, among which, for example, methods of remote sounding, GIS- and GPS-technology etc.

Set of methods of remote sounding, GIS- and GPS-technologies we shall further name as geoinformation technologies.

The geoinformation technologies underlie transformations during reception and data processing, performance of various model researches and calculations. On the basis of geoinformation technologies the opportunity of creation of the integrated digital model of the Earth is created.

On the basis of geoinformation technologies the performance of complex researches on complete study of a state of the certain regions as a whole and their components (wood, water sources, underground space and so on) is possible.

Owing to development of computer and web technologies the procedure of reception, storage, transfer and uses of the geodata undergo essential changes.

The remark. The geodata - data set about components of ecosystem.

Some experts name geoinformation technologies as technologies of «5S» (GPS, DPS, RS, GIS, ES), including ground geodetic measurement, surveying, photogrametry, remote sounding, mathematical and computer modeling, satellite technologies, expert systems, modern communication technologies etc.

The geoinformation technologies should satisfy to the following requirements:

- Should realize procedures from work with the initial received data, processing up to their complete analysis, ordering and formalization;
- To allow to work with various types of the data (data of satellite technologies, ground technologies, physical, chemical analysis etc.);
- To form the integrated and object-guided bases including graphic (topological), symbolical and attributive data;
- To have the functions allowing automatically to take semantic and the unsemantic information from the digital connected projects;
- To have functions for processing, performance of researches and realization of modeling with use of the given information bases;
- To have functions for convenient visualization and display of the data (interface of display of the data).

The remark. As the digital connected project (further simply « the project ») we shall name the information bases, logically interconnected by determined rules describing various objects of investigated region, and models determining their interrelation.

4 MODELLING OF DEVELOPMENT OF ECOSYSTEM OF REGIONS

There are changes of various character owing to natural and technogeny factors in functioning natural components of ecosystem of region. Not always these changes carry positive character. Frequently (and last years in most cases) technogeny influence on components of ecosystem carries sharply negative character. Equilibrium state of ecosystem is broken, that can result in approach sharply of negative (pathological) situations of various spatial and temporary extents. So, for example, in practical activity of the mining enterprises a number of cases is known, when increase of the quantitative characteristics of volume conducts to disproportionate growth of economic parameters. So, the increase of productivity of work of fans of the main airing at 10 % by escalating blades of fans conducts to increase of the charge of the electric power on 20-30%.

Obviously, what to execute an estimation of stability and ecological balance of ecosystem of region, of conditions of destruction of ecological balance it is necessary to have the "instrument" for modeling both general state of ecosystem of region, and state of its separate elements. It is natural, that as the basic "model instrument" effective use of the approaches of mathematical and computer modeling is. Having in stock the advanced system such of "solvers" for modeling a state of various ecosystem's elements in aggregate with the integrated digital model of region are possible are to carried out by the forecast and solution of a wide range of tasks.

The approaches of the theory of stability and catastrophes, method of neurons networks are perspective for development of methods of modeling.

Let's describe one of the possible approaches to development such of «solver» (Zhuravkov et al. 1997a.).

Let's consider of ecosystem with a set of some factors "X".

For example, "X" - population in the given region; air environment; soil resources etc.

For the given factors we shall enter a parameter "A", determining system of life-support and restoration of those factors of ecosystem.

For the mentioned factors accordingly "A" - system of life-support of the population; resources necessary for restoration of air environment (green plantings); measures for restoration "of vital force" of soil (recultivation, melioration, mineral fertilizers etc).

Let ecosystem under the factor "X" at presence of system of life-support "A" is reproduced with speed proportional to product $A \cdot X$, so

$$dX/dt = k_1 A \cdot X. \qquad (1)$$

Here k_1 is constant.

The remark. The symbols "X" and "A" are used for a designation as actually of factors, and their measures.

Similarly let "B" is complex provoking destruction and damage in ecosystem on the factor "X".

For the designated factors "B" accordingly are factors determining mortality the population; measures resulting in pollution of air; measures resulting in an exhaustion, deterioration of a condition of oil.

Let's assume, that the speed of destruction and damage of ecosystem under the factor "X" is proportional to a measure of system X and two constant k_3 and B, so

$$dX/dt = -k_3 B \cdot X. \qquad (2)$$

Then, the evolution of ecosystem under the factor "X" in this case can be described by the equation:

$$dX/dt = (k_1 A - k_3 B) X. \qquad (3)$$

At $k_1 A > k_3 B$ the equation (3) predicts exponential steady development of the factor "X", and at $k_1 A < k_3 B$ is exponential recession, destruction of ecosystem under the factor "X".

Thus, for stabilization of ecological system it is necessary to observe the certain conformity between system of life-support "A" and complex "B" responsible for destruction of ecosystem under the factor "X".

The described model though is rather primitive, but worker.

Let's complicate model. Let's enter into consideration the factor "Y", interconnected with the factor "X". Let's assume, that the speed of destruction and damage of ecosystem under the factor "Y" is proportional to a measure of system Y and two constant k_3 and B.

For example, for the factor "X" is "population" then "Y" is "illnesses caused by ecologically adverse conditions"; if "X" is "air" then "Y" is "harmful emissions in an atmosphere"; if "X" is "soil" then "Y" is "production of minerals, cultivation of products etc".

Let restoration of ecosystem under the factor "X" submits to the law (1) and we shall assume that potential opportunities of "A" for performance of a complex of restoration measures "X" are not limited (that is measures "A" it is enough for complete restoration of ecosystem under the factor "X").

At the entered assumptions the destruction of ecosystem under the factor "X" comes as a result of interaction of the factors "X" and "Y". Potential opportunities of the factor "Y" there is enough that if not to accept of any measures then ecosystem under the factor "X" collapses («perishes»).

Concerning the factor "Y" is prospective, that ecosystem degradation under the factor "Y" according to the law of a type (2):

$$\frac{dY}{dt} = -k_3 BY, \qquad (4)$$

where k_3 and B are constants.

For example, the restoration of an environment submits to the law (1) (factor "X"). Concerning minerals is prospective, that their stocks are deflected only as a result of production according to the law (4) (factor "Y"). The interaction between set "X" of nature-guarded measures and restoration of environment on the one hand and set "Y" by production of minerals with another occurs at the expense of extraction of minerals. Therefore the stocks of minerals are exhausted and volume of performance of a complex of nature-guarded measures grows owing to its proportionality to volumes of mining works.

As the factors "X" and "Y" cooperate and interdependence the general speed of results of their interaction is proportional to product of their measures:

$$dX/dt = -k_2 X \cdot Y, \qquad dY/dt = k_2 X \cdot Y, \qquad (5)$$

where k_2 is constant.

Uniting the equations (1) - (5) we shall receive the following nonlinear system of the equations describing evolution of considered ecosystem under the factors "X" and "Y":

$$\begin{cases} \dot{X} = k_1 AX - k_2 XY \\ \dot{Y} = k_2 XY - k_3 BY \end{cases}, \qquad (6)$$

The equations (6) represent the classical Lotki-Volterra's equations.

The untrivial stationary (steady) solution of the equation (6) has a kind

$$X_s = k_3 B/k_2, \qquad Y_s = k_1 A/k_2. \tag{7}$$

The solution (7) is the more optimal solution for the steady balanced state of ecosystem in time (that is "of peace coexistence of the factors *"X"* and *"Y"*). If for quantitative parameters "X" and *"Y"* from the beginning to adhere to meanings (7), then stationary balance in time in this case will be observed according to the considered determined model.

For modeling of ecosystem more really to consider stochastic models with introduction of casual fluctuations. According to it in the determined system (6) are entered casual fluctuations and for research of stability of a stationary state we shall present "X" and "Y" as the following sums:

$$X = X_s + x, \qquad Y = Y_s + y. \tag{8}$$

Substituting (8) in (6) and accepting, that increments x and y are small values so their products can be neglected, we shall receive of linearization equations

$$\dot{x} = -k_3 B \cdot y, \qquad \dot{y} = k_1 A \cdot x, \tag{9}$$

which describing behavior of small deviations from a stationary state.

Let's transform system (9) by exception y:

$$\ddot{x} + k_1 k_3 ABx = 0. \tag{10}$$

The equation (10) represents the equation of the classical mechanics describing harmonic fluctuations of system.

The phase trajectories of the equation (6) in space X,Y are the closed curves with the continuously varied period located in a vicinity of the stationary solution - of a point $S = (X_s, Y_s)$. The space $X;Y$ is filled with infinite set of such phase trajectories with the own period. The trajectory which is taking place through any initial point represents complete dynamic evolution of ecosystem. Thus, if the system is rejected from a stationary condition S there are not fading fluctuations about a point (X_s, Y_s) which amplitude depends on an initial deviation from S.

The state of ecosystem under the factors *"X"* and *"Y"* on the basis of the given scheme can be described as follows:

If to begin from small meanings of measures of the factors *"X"* and *"Y"* (for example, production of minerals and complex of natural-guarded measures) then efficiency of a small complex of restoration measures *"X"* at insignificant destructive influence of the factor *"Y"* can create visibility of efficiency of such volume of measures and further at increase of influence of the factor *"Y"* (the small volumes of production of minerals do not require expenses on nature-guarded measure or those are very insignificant). However significant and unguided increase of influence of the factor *"Y"* requires of exponential increase of influence of the factor *"X"*. But that at achievement of the certain borders becomes impossible and in a result we have sharp deterioration of a state of ecosystem. That in turn requires reduction of influence of the factor *"Y"* (unguided increase of mineral production requires insupportable large increase of volumes of nature-guarded measures and results in sharp deterioration of a state of an environment, that requires strict restriction and reduction of volumes of production). Thus, has a place return to an initial point.

The probable management of a state and balance in ecosystem is shown on figure 1.

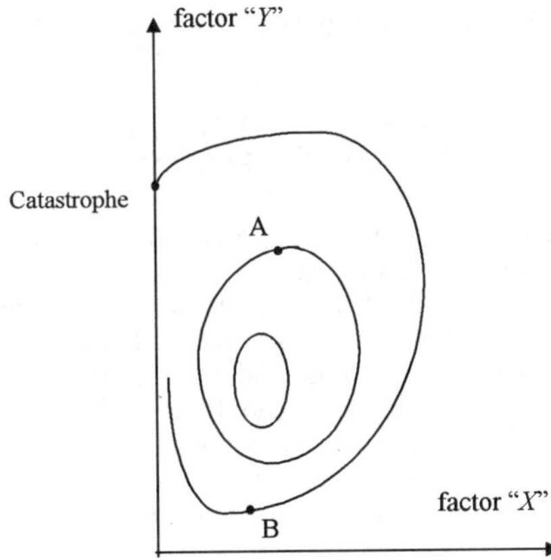

Figure 1. Probable schemes of development of state of region's ecosystem on base on various strategies of management of factors *"X"* и *"Y"*.

Let task is the stabilization of an ecological situation in region with intensive technogeny influence.

If an initial condition is the situation *"A"* the small cycle ensuring a steady state of ecosystem of region on the factors *"X"* and *"Y"* is possible. Last means that for ecological stability of environment it is necessary being set an interval of change of parameter *"X"* or *"Y"* to define an interval of change of second parameter and strictly to watch that both parameters did not fall outside the limits these intervals.

The remark. It is possible to determine such initial condition, by analogy to representations of stability in the mechanics, as "points of unconditional stability"

For example, if to proceed from "of an interval of volumes of emissions in an atmosphere" $[Y - \Delta Y; Y + \Delta Y]$ that, according to the certain techniques, it is necessary to count up necessary "volume of green plantings" $\left[X-\Delta X; X+\Delta X\right]$. The precise tracking observance of a presence of the factors in the given intervals results that the phase trajectory corresponds to a curve which is taking place through a point "*A*". The initial point "*A*" is defined by initial quantitative parameters of the factors "*X*" and "*Y*".

Completely other picture is observed if an initial state is the situation "*B*"- state "of critical stability". The given situation is characterized by that the sharp increase of parameters of the factor "*Y*" requires fast and significant changes of parameters of the factor "*X*" that not always it is possible to execute and can result system in a state "of ecological accident".

For example the sharp substantial growth of amount of the consumers of water in region requires radical reconsideration of politics in the field of maintenance of region by water (sources of water, system of clearing etc.), that frequently it is impossible to execute.

5 SOME EXAMPLES OF PRACTICAL IMPLEMENTATION OF APPROACH

On the base on complex approach the researches were executed for some regions.

One of the area of studies included two major sites: Jordan University of science and technology (JUST) and Al-Hussan Industrial city (HIC).

JUST and HIC lay between Irbid and Ramtha. JUST has 8500 day time students and area about 1093 hectars. HIC has area near 427,000 m^2. There are 56 factory with 4000 employee distributed as follows:

№	Type of industry	Number of factories	Type of destructive influence
1	Food	2	Air pollution
2	Chemical & Fertilizers	13	Air & Soil pollution
3	Engineering	16	
4	Plastic & Rubber	10	Air pollution
5	Wearing & Spinning	12	
6	Constructional	1	
7	Furniture	3	
8	Pharmaceutics	1	
9	Lather	1	Air pollution

The most pollution due to industry is related directly to the following industries: chemical and fertilizer factory

As objects of researches the regions of excavation of deposits of potash salts in Russia and Belarus were chosen also. The given regions include directly areas of excavation of deposits (underground mines) and localities which are taking place in the given regions.

Primary for performance of applied researches the construction of mathematical computer model of considered regions was made on the basis of geoinformation technologies (Zhuravkov et al. 1997 b.).

1. One of parameters of pollution of territories is the areas of haloes of chronic pollution of a snow cover around of them (determined, for example, by results of interpretation of the satellite television images). Proceeding from a population and industrial potential of Soligorsk industrial region and of territorial - industrial "Berezniki-Solikamsk" was designed that the areas of haloes of chronic pollution make accordingly about 1000 and 2000 square kilometers. Established as a result of processing the satellite images the area of such halo for a complex " Berezniki-Solikamsk" makes 2250 square kilometers.

2. By results of processing the executed researches is established that the basic sources of pollution of an environment in industrial regions and cities are:

- automotive transport;
- the large industrial enterprises;
- the small enterprises (usually settling down in urban feature).

The researches have shown that the emissions of large industrial enterprises make about 75 % in common balance of atmospheric pollution of region.

The share of adverse influence of a vehicle is continuously increased by an environment in total volume of all kinds of antropogeneous pollution of the large cities and can make 40 and more than percents.

Comparison of the given medical researches on morbidity and demographic parameters in advanced industrial regions on the one hand both quantitative and qualitative parameters of emissions of motor transport have shown their close interrelation on the certain kinds of morbidity.

3. In conditions when the antropogeneous enforcement on the limited natural resources amplifies, it is necessary to operate only with closed systems. Therefore the clearing industrial and waste water should be considered as a valuable resource having in rich potential of application for replacement of water in underground layers, for the industrial purposes, and, at the appropriate safety measures, for household needs.

As is known, up to 80 % of illnesses are by and large connected to quality of water. The interesting researches are executed under a management prof. A. Kulberg (Fedin 1996). Water reacts to any adverse external influences by means of inclusion of the mechanism of genetic updating of seaweed - algy constantly which is taking place in water. Algoy are let out "the molecules of an alarm" which quantity identifies in the generalized kind a degree of danger of pollution of water for the man. Thereof weakens and that and collapses of immune system of protection of the man.

4. One of the factors that needs to be considered during the layout of urban topology and high intensity industrial region topology (for example, street directions, spacing and width, building height restrictions etc.) is the effect of the air pollution associated with the automotive transport that use routes in this area and emissions of industrial enterprises that are taking place in this region. Although the pollution is generated at road and street level, its effect can be widespread due to interaction of the pollutant dispersion and diffusion with the wind speed and direction. The pollutant level is explicitly dependent on both area topology and wind speed and direction. In order to study the effect of an urban dispersion computer modeling would have to be performed on based on techniques of Computational Fluid Mechanics and using of computer model of area. The meteorological information as represented by a wind rose (wind speed and direction) is used to calculate pollutant levels as a function of area topology variables (street depth and width, etc.). As variables of function's pollutant are used the pollutant source in conjuction with a traffic scenario.

5. The important consequence of underground production of minerals is the influence on a earth surface and undersurface areas. On the basis of the approaches of the connected geomonitoring the complex of modeling of a deforming state of a surface and undersurface areas, engineering structures located in the given zones is developed (Zhuravkov et al. 1999).

6 CONCLUSIONS

1. To supervise and to operate an ecological state of region now probably only at purposeful study of a state of set of elements of ecosystem of region at the obligatory account of their interaction. Technogeny influence on ecosystem of region it is impossible to consider as safe until all its separate consequences are appreciated.

2. For achievement of purposes of accurate study, modeling and prognoses of state of ecosystem of region is necessary building up regional complex system of the connected geoecological monitoring. The structure of system of the connected geoecological monitoring can be presented as set of large blocks.

3. For carry out modeling and researches it is necessary construction of the integrated connected information model of region. Basis of construction of information systems and databases for all sections of Earth Sciences is the set of the approaches, among which, for example, methods of remote sounding, GIS- and GPS-technology etc.

4. Geoinformation technologies are technologies of «5S» (GPS, DPS, RS, GIS, ES), including ground geodetic measurement, surveying, photogrametry, remote sounding, mathematical and computer modeling, satellite technologies, expert systems, modern communication technologies etc.

5. Perspective for development of methods of modeling a state of ecosystems are approaches of the theory of stability and catastrophes.

REFERENCES

Zhuravkov, M., Andreyko, S. & Zemskov, A. 1997. Modeling stability ecosystem on the basis of the approaches of the theory of stability and catastrophes. *Computer applications and operations research in the mineral industries. Proceedings APCOM'97.Moskow:* 502 – 506. Moskow: Moskow State Mining University.

Zhuravkov, M.A. & Smuchnik, A.D. 1997. *The design of geomonitoring systems for regions of large-scale familiarization of underground space.* Minsk: Belarus Academy of Science of Safety of life-activity.

Zhuravkov, M., Tyashkevich, I.& Belyaev, B. 1998. Estimation of condition of geospheres. *Environmental Issues and Waste Management in Energy and Mineral Production. Proceedings SWEMP'98. Ankara:* 221 – 225. Rotterdam: Balkema.

Zhuravkov, M. & Nevelson, I. 1999. Methods of computer modeling of processes of strata movement. *Reports of Int. Cong. "Mine surveying under the conditions of market economy". Nessebar*: 90 - 98.

Fedin E. 1996. I have given birth you, I you and ... *The Moscow news. January 4.*

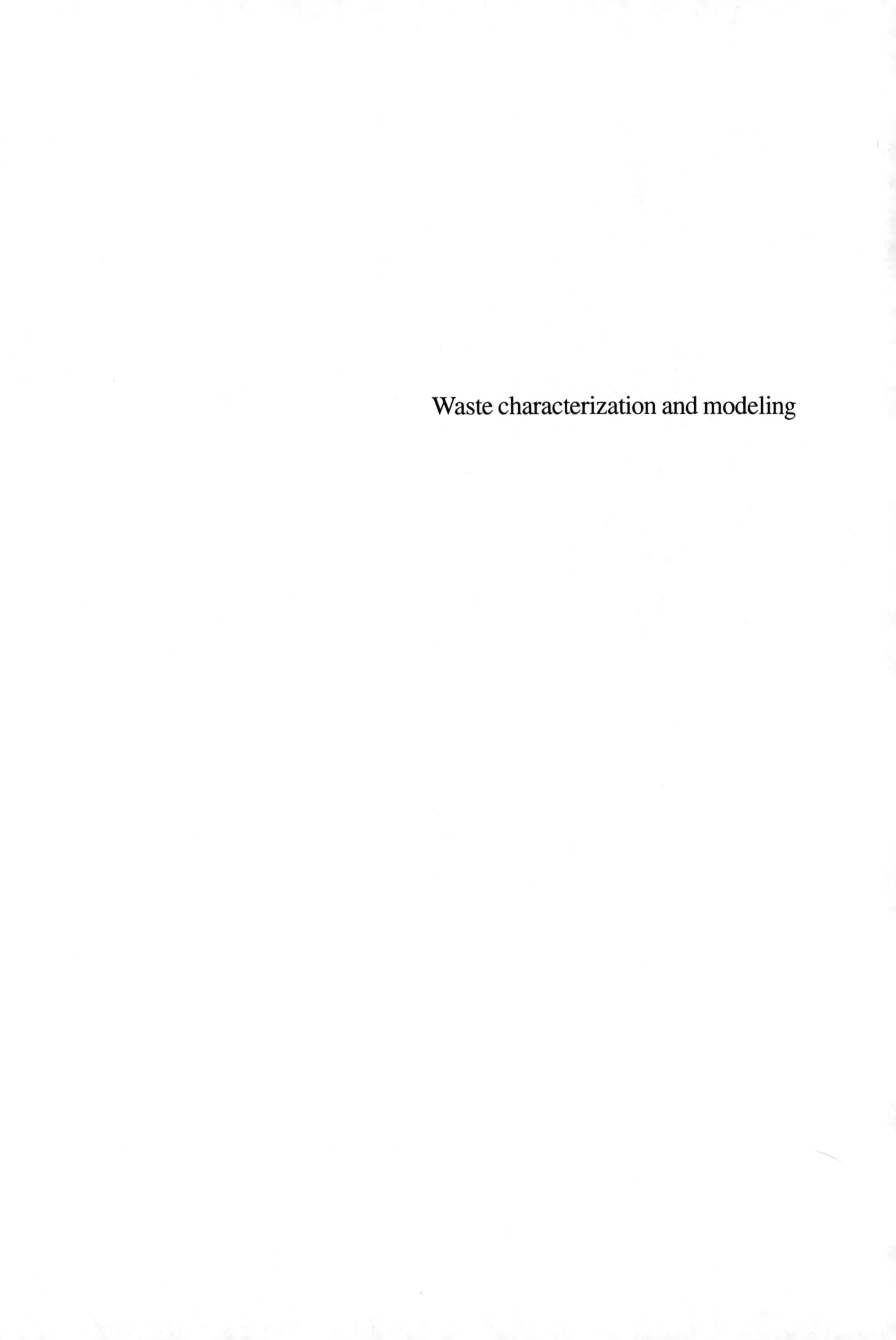

Waste characterization and modeling

Environmental Issues and Management of Waste in Energy and Mineral Production, Singhal & Mehrotra (eds)
© 2000 Balkema, Rotterdam, ISBN 90 5809 085 X

Recovery of chromium from industrial wastes

M.P.Antony, V.D.Tathavadkar, C.C.Clavert & A.Jha
Department of Materials, University of Leeds, UK

M.Wilkinson
Elementis Chromium, Eaglescliff, Stockston-on-Tees, UK

ABSTRACT: Sodium chromate is produced via the soda ash roasting reaction of chromite ore with sodium carbonate. After the reaction, nearly 15 % of chromium oxide, remains unreacted and ends up in the waste stream for landfill. High purity chromium metal is produced via the aluminothermic reduction of Cr_2O_3. The slag generated contains around 8% of chromium oxide. The Cr_2O_3 exists in the residue either as a solid solution with α-Al_2O_3 or with γ-Fe_2O_3. The residue from the sodium chromate process and the slag from chromium metal production were analyzed both physically and chemically. The analytical scanning electron microscopic technique was used extensively for determining the mineral phase compositions. The kinetics of chromium extraction from the residue by using the soda ash roasting process under oxidizing atmosphere was investigated. The kinetics of leaching of Cr^{3+} ions via aqueous phase from spinel phase was also studied by treating the waste into acid solutions with different concentration.

1 INTRODUCTION

Chromium chemicals are produced by the alkali roasting of chromite ores with sodium carbonate. Sodium carbonate is mixed with the chromite ore and heated at temperatures around 1100 °C in air. Sodium oxide produced as a result of the decomposition reaction of Na_2CO_3 in the presence of chromite ore combines with Cr_2O_3 to form sodium chromate in an oxidising atmosphere (1). The over all reaction of iron chromite can be written as shown in equation 1.

$$2Na_2CO_3 + FeO.Cr_2O_3 + 1.75\ O_2 \rightarrow 2Na_2CrO_4 + 0.5\ Fe_2O_3 + 2CO_2 \qquad (1)$$

Water soluble sodium chromate is extractedt. In the extraction process, alumina and magnesia are two critical components which are invariably abundant with the chromite ores. Alumina and magnesia in the presence of silica stabilise the spinel phase $Mg(Cr,Al)_2O_4$ in preference to chromate. As a result, the efficiency of the extraction process drops considerably. The maximum extraction efficiency of Cr_2O_3 achieved by using the above mentioned process is approximately 80%. Unreacted chromium oxide remains with the refractory oxides and ends up in the waste stream. High purity chromium metal is produced by the aluminothermic reduction of chromium oxide. The slag contains approximately 8% Cr_2O_3.

Experiments were carried out to study the recovery of Cr from waste residue under oxidising, inert, and partially reducing atmospheres. The kinetics of chromium extraction from the waste by using the soda -ash roasting process under oxidizing condition was investigated. The dissolution of Cr from spinel via aqueous phase medium was also studied by treating the waste into acid solutions with different concentration (2).

2 EXPERIMENTAL PROCEDURE

2.1 *Physical and Chemical Characterization*

The residue and the slag used in the present study were obtained from two chromium producing companies in the United Kingdom, Elementis Chromium and London Scandinavian Metallurgical Company, respectively. The waste residue containing chromium was characterized by using physical methods. Physical characterization included measurement of particle size by laser scattering technique, surface area by nitrogen adsorption method, density by pycnometry using helium gas and identification of various phases by X-ray powder diffraction method. The distribution of chromium present in different phases was determined by using scanning electron microscopic (SEM) techniques and Electron Probe Microanalyzer (EPMA).

The chemical assay of the two compounds was carried out by quantitative chemical analysis. Residue was fused with sodium peroxide in a nickel crucible. Sodium peroxide was used for the destruction of the matrix completely. The fused mass was digested in the boiling water. Small quantity of ammonium carbonate was added to the boiling solution for removing iron (Fe) and aluminum (Al) as their hydroxide precipitates. The solution was then filtered. The filtrate containing sodium chromate was converted to sodium dichromate by adding sulphuric acid. The sodium dichromate thus formed was analyzed for the estimation of chromium. The precipitate obtained was dissolved in the concentrated hydrochloric acid and iron and aluminum were determined by using the atomic absorption spectrophotometer (AAS) having hollow cathode lamps with resonance lines at wavelengths 386.0 and 309.3 nm respectively. Three methods were employed for the chemical analysis of chromium namely, volumetric titration, UV-visible spectrophotometry and atomic absorption spectrophotometry. These three methods were standardized by preparing a large number of standards, which were also analyzed to obtain a calibration curve. In the volumetric analysis, a known amount of the standard ferrous ammonium sulphate was added to the filtrate containing sodium dichromate. The excess ferrous ammonium sulphate was back-titrated with the standard sodium dichromate solution. Chromium was determined colourimetrically as chromate. The transmittance of the solution was measured at 365 – 370 nm. Amount of chromium present in a dilute solution was also determined by using AAS at a wavelength 428.9 nm.

2.2 *Thermogravimetric Study*

Thermogravimetric experiments were carried out with the residue in order to understand the various reaction steps taking place during the soda ash roasting reaction. Known amounts of the waste residue were mixed with the stoichiometric amounts of sodium carbonate and then heated to 1100°C at a heating rate of 5°C per minute in a thermobalance in air, argon, and argon – 5% hydrogen atmospheres. The weight changes were recorded as a function of time and temperature. A gas flow of 500 ml per minute was maintained during the experiment.

2.3 *Kinetics of formation of sodium chromate*

Experiments were carried out to find out the loading capacity of sodium carbonate required for achieving a maximum conversion of Cr_2O_3 to sodium chromate. The residue was thoroughly mixed with different quantities of sodium carbonate starting from 20 % to 200 % of the amount required to react stoichiometrically with the Cr_2O_3 present in the residue. The samples were heated in a temperature-controlled furnace at 1150°C for two hours. The reaction product containing sodium chromate was extracted using water and the aqueous solution was analyzed for chromium quantitatively.

A series of roasting experiments were conducted to examine the effects of time and temperature of roasting on the formation of sodium chromate. Residue was mixed with Na_2CO_3 and heated at 1000°C, 1100°C and 1200°C from 30 minutes to 150 minutes at intervals of 30 minutes. Sodium chromate formed was leached in water and analyzed volumetrically each time.

The chemical composition of waste residue is 14 % Cr_2O_3, 42 % Fe_2O_3, 22 % Al_2O_3, 15 % MgO and 1.5 % each of SiO_2 and Na_2O. The slag contains 8.6 % Cr_2O_3, 88.2 % Al_2O_3, 2.3 % K_2O, 0.6 % CaO, 0.1 % TiO_2. The major difference between the slag and the residue is that the slag is rich in alumna, whereas the residue from soda ash roasting process contains more Fe_2O_3. It would be economical if it is possible to recover chromium from the combined mixture of wastes rather than the individual sources. For co-extraction of chromium equal amounts of residue and slag were mixed and heated to 1200C for six hours. Various phases formed in the residue before and after heat treatment and the distribution of chromium in different phases have been examined using the scanning electron microscopic technique. Soda ash roasting of the combined residue was carried out at 1100°C and 1200°C.

2.4 *Acid leaching study*

Preliminary experiments were carried out to extract chromium or iron from the residue by digesting it with sulphuric acid (H_2SO_4). The objective was to establish optimum conditions for the removal of iron or chromium preferentially. An Eh-pH diagram as shown in figure 1 was constructed for Fe-Cr-H_2O system at room temperature to find out the E_h-$_pH$ region wherein Fe or Cr could be selectively dissolved in H_2SO_4.

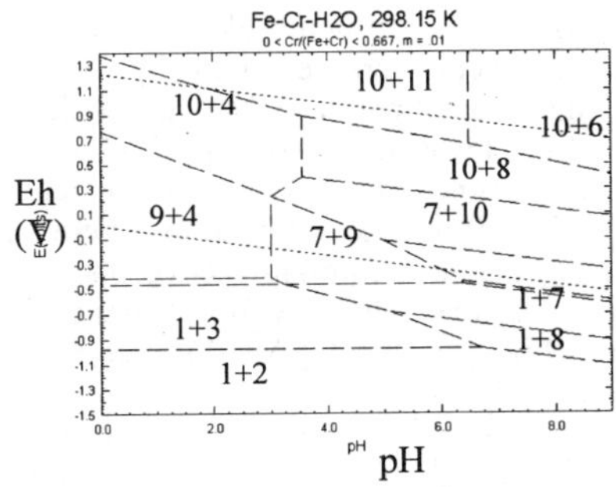

1-Fe(s) 2-Cr(s) 3-Cr^{2+} 4-Cr^{3+} 5-Cr_2O_3(s) 6-CrO_4^{2-} 7-$FeCr_2O_4$(s) 8-Cr_2O_3(s) 9-Fe^{2+} 10-Fe_2O_3 11-$HCrO^{4-}$

Figure 1: Eh-pH diagram of Fe-Cr-O-H system

A region of pH from 3 to 5 and Eh – 0.4 to 0.1 volts could be identified wherein Fe^{2+} in solution and $FeCr_2O_4$ solid are thermodynamically stable. However, the kinetics of dissolution is very slow in this region. Known amounts of residue were dissolved in concentrated H_2SO_4, 1.5 M H_2SO_4, 1.0 M H_2SO_4 and 0.1 M H_2SO_4 in glass beakers from 30 minutes to 180 minutes. Dissolution experiments were carried out at room temperature and boiling temperature of H_2SO_4. The amount of iron and chromium present in the solution was determined by using AAS.

3 RESULTS AND DISCUSSIONS

The particle size of the residue was measured to be between 15 and 18 µm. Surface area and density were determined to be 3.657 m^2 / g and 3.81 g / cm^3 respectively. X-ray diffraction pattern of the residue is shown in figure 2. The residue consists of (Mg.Fe) $(Cr.Al)_2$ O_4 spinel phase. γ- Fe_2O_3 which has got a defective spinel structure co-exist in the matrix as a single phase. This fact was confirmed by the microstructure analysis using SEM and EPMA. It was observed that the gray chromite grains consist of a skeleton of iron and alumina-rich spinel phase. Some grains have an outer iron rich spinel rim formed by the roasting reaction and a core having unreacted chromite phase. The remarkable observation was that the residue consists of iron-rich spinel phase. There are no separate phases of iron oxides present in the residue. Major phase present in the slag was identified to be that of alumina. Small particles of chromium metal is also entrapped in the matrix. It was confirmed by energy dispersive X-ray analysis that the Al_2O_3 grains contain small amounts of Cr ions. Examination of the heat treated waste mixture revealed that Al^{3+} ions present in the slag had diffused into the chromite grains of the residue.

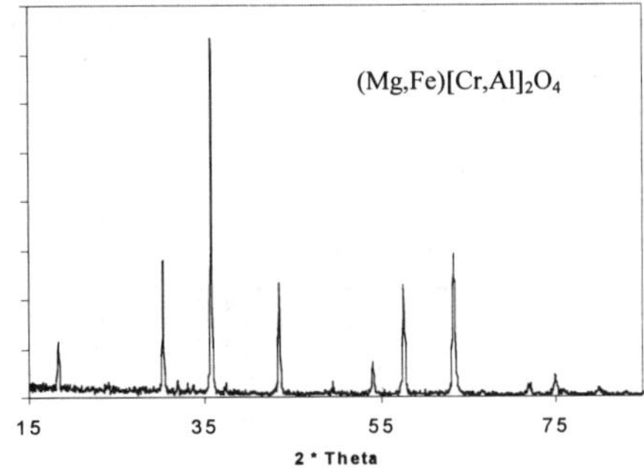

Figure 2: X-Ray diffraction pattern of residue

The changes in weight of the residue as a function of temperature when heated in a thermobalance under air, argon and argon-hydrogen atmospheres are shown in figure 3. A weight loss of 0.27 % during heating up to 145°C (≈ 25 minutes) due to the removal of moisture was recorded. The slope of the curve decreased after 145 °C. From 145°C to 600 °C a weight loss of 0.221% (≈ 91 minutes) was recorded. The weight loss was due to the removal of water of crystallization. Sodium carbonate started decomposing around 600°C to form sodium oxide and carbon dioxide. The sodium oxide formed combined with the Cr_2O_3 to form sodium chromite and sodium chromate. A weight loss of 0.57 % (≈ 47 minutes) was recorded due to the removal of carbon dioxide. The rate of weight loss was less when the sample was heated in air because of the parallel oxidation reaction of chromite to chromate. A weight gain of 0.1% (≈ 45 minutes) was recorded after 880°C due to the oxidation of sodium chromite to sodium chromate when heated in air. There was no oxidation reaction when the sample was heated in argon atmosphere. Iron oxides started getting reduced when the sample was heated in argon – hydrogen atmospheres contributing to a further increase in weight loss.

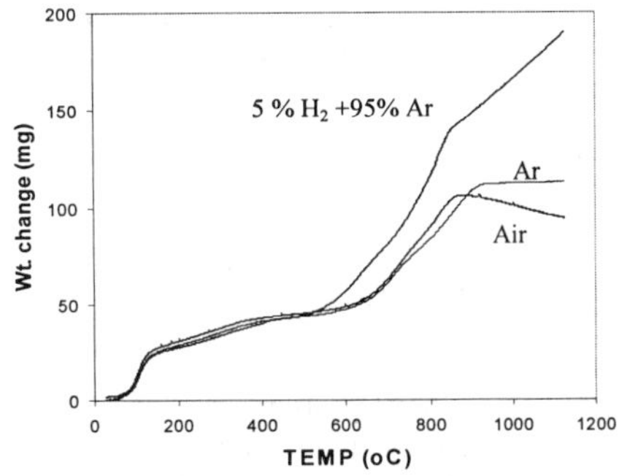

Figure 3 : Observed weight loss during isochronal heating of the residue in different atmospheres

Table 1. Estimation of sodium carbonate required for the formation of sodium chromate

Fraction of Na_2CO_3 required to react stoichiometrically with the Cr_2O_3	Percentage of chromium extracted
0.2	15.9
0.4	38.8
0.6	57.2
0.8	78.9
1.0	79.1
1.2	78.6
1.4	79.6
1.6	78.4
1.8	78.3
2.0	78.6

The percentage of chromium extracted by heating the residue with different quantities of sodium carbonate is given in table 1.

It is evident from the results that the percentage of chromium extraction increased with the increase in amounts of sodium carbonate up to 80 % of the stoichiometric amount required for the reaction. There was no further increase in the yield of water-soluble chromate by the addition of more sodium carbonate. Excess sodium carbonate might be consumed by the side reactions to form sodium aluminate and sodium ferrite. Also higher sodium carbonate additions lead the formation of melts which fill the micropores with a liquid sodium chromate –sodium carbonate which also envelops the chromite grains and prevents the diffusion of oxygen through liquid film to reach the reaction sites (3).

The effects of roasting time and temperature on the extent of chromium recovery is represented in figure 4.

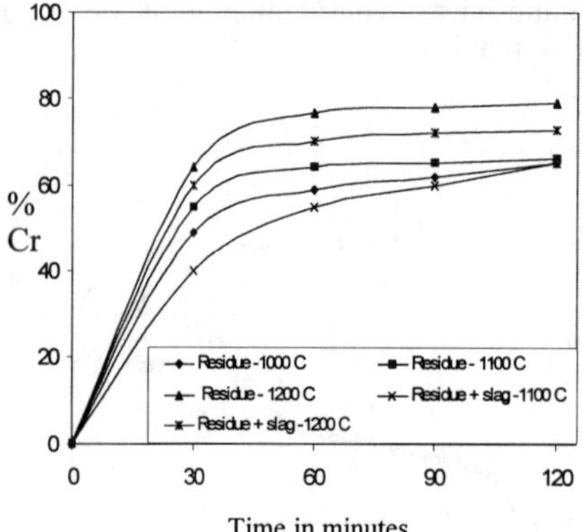

Figure 4: Soda ash roasting reaction of residue at different time and temperatures

Results of the experiments clearly revealed that chromium recovery increased with the rise in both roasting time and temperature. Maximum chromium recovery of 78% was achieved at 1200°C in 90 minutes of heating time. The reaction rate decreased remarkably after 90 minutes of heating. There was no improvement in the chromium recovery afterwards. The lower reaction rate may be attributed to the presence of highly stable $MgAlCrO_4$ chromate compound. The yield of extractable chromium was less when the mixture of residue and slag heated with sodium carbonate. Some of the sodium carbonate reacted with slag containing more alumina to form sodium aluminate and thereby decrease the efficiency of chromium recovery. The overall reactions taking place during sodium chromate formation can be summarized below. The reaction takes place through many intermediate steps.

$$3Na_2CO_3 \longrightarrow 3Na_2O + CO_2 \quad (2)$$

$$2(Fe,Mg)O.(Cr,Al)_2O_3 + 3Na_2O + 6O_2 \longrightarrow$$

$$2Na_2CrO_4 + 2MgAlCrO_4 + Fe_2O_3 + Na_2Al_2O_4 \quad (3)$$

Heat treatment of the residue with the slag and subsequent soda ash roasting of the mixture did not improve the efficiency of chromium extraction.

Results of the acid leaching experiments revealed that the attack of residue by H_2SO_4 is a very complex process. Approximately 49% of iron and 4% of chromium could be digested in concentrated H_2SO_4 at room temperature after six hours of stirring of the residue in the acid. In boiling acid condition, 70% of iron and 90% of chromium had leached out to the acid in six hours. Neither iron nor chromium could be completely removed by leaching the residue with concentrated H_2SO_4. When the experiments were conducted in 1.0 M and 1.5 M H_2SO_4, the chromium and iron concentration in solution were 1.7% and 0.6% respectively. The concentration of chromium remained same irrespective of the time of dissolution. This may be due to the uniform protonic attack of the chromite grain by dilute acids. The percentage Fe dissolved in 0.1 M H_2SO_4 as a function of time is shown in figure 5.

A steady increase in concentration of iron from 0.27 % in 30 minutes to 0.73 % in 180 minutes was observed in the solution where as the concentration of chromium remained unaltered at 0.5%. It can be inferred from the results of the experiments that major portion of iron present in the residue can be removed by way of washing the residue with dilute sulphuric acid in successive steps. This will lead to an increase in chromium content in the residue. Subsequent soda ash roasting of the residue would improve the extraction efficiency of chromium.

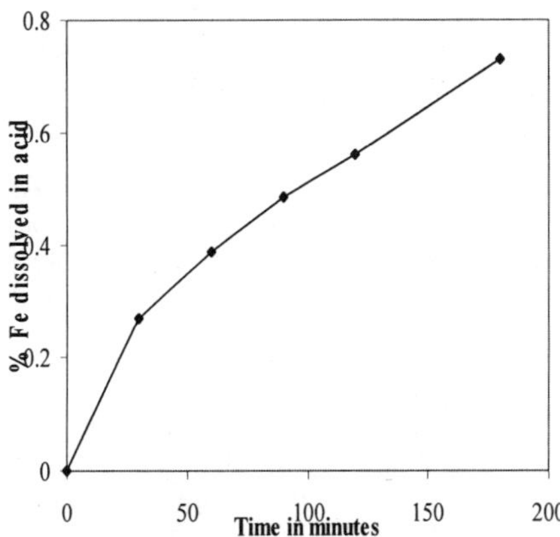

Figure 5: Percentage Fe dissolved in acid vs time

4 CONCLUSIONS

The following conclusions can be drawn from the results of the experiments.

1. The residue consists of an iron-rich chromite spinel. Separate phases of iron oxides could not be identified. This observation indicates that the residue retains the parent crystal structure of chromite ore which in turn suggests that chromium diffuses out of the spinel matrix and react with sodium oxide to form sodium chromate or sodium chromite.
2. Thermogravimetric study reveals that the presence of oxygen is essential for the formation of sodium chromate.
3. Addition of sodium carbonate in excess of the amount required for stoichiometric reaction does not increase the conversion of chromium into a water soluble sodium chromate compound.
4. Soda ash reaction is rapid at the beginning and stops practically after 90 minutes of heating. The complete recovery of chromium could not be achieved. This may be due to the stabilisation of more stable $Mg(Al.Cr)_2O_4$ phase.

5 REFERENCES

1. Tel'pish, V.V. and Vil'nyanskii, Ya. E. 1969: *J. Appl. Chem. USSR.* 42: 1118 - 1120.
2. Clay, J. C, Pearce J.F & Tretheway, B.H. 1950. *J. Soc. Chem. Ind.* 69: 275 - 281.
3. Vil'nyanskii , Ya. E. & Pudovkina, O.N. 1947. *J. Appl. Chem. USSR*. 20: 794 -799.

Environmental Issues and Management of Waste in Energy and Mineral Production, Singhal & Mehrotra (eds)
© 2000 Balkema, Rotterdam, ISBN 90 5809 085 X

Characterization of free silica in underground gold mines in Minas Gerais, Brazil

R.Z.L.Cançado & A.E.C.Peres
Federal University of Minas Gerais, UFMG, Belo Horizonte, Brazil

ABSTRACT: Silicosis has been described for centuries. Regarding clinical aspects, the relevance and risk of this pneumoconiosis arise from the fact of being a chronical disease, irreversible in evolution and not presenting any specific treatment. Beyond these characteristics, other parameters may be taken into consideration to express its importance concerning public health to a point that, even where occupational health has reached desirable levels, silicosis still constitutes a serious challenge. The geological conditions of Minas Gerais state, Brazil, mainly the so-called *Iron Quadrangle*, favored the commissioning of several mining companies, underground gold mines representing an important segment. The intrinsic features of these activities render them potential generators of environmental dust containing free silica. The present work describes the analytical methodology utilized and presents results of field monitoring in one representative company.

1 INTRODUCTION

The geological characteristics of the Iron Quadrangle region, in the state of Minas Gerais, Brazil, have attracted for many years a significant number of mining and metallurgical companies, most of them dealing with iron and, in a second position, gold. In the year of 1994, the Brazilian government, concerned with activities prone to generating environmental pollution, created programs of environmental risks prevention and respiratory system protection. The importance of these programs arises from the fact that workers that perform their activities under unhealthy conditions, inhaling dusts containing free silica, may acquire, during their working period of life, a pneumoconiosis known as silicosis, a chronical and irreversible disease, that still lacks of specific treatment. Strict care is required for the correct assessing of free silica in environmental dusts, in order that the analytical results represent a true picture of the environment under investigation. The following facts are worth mentioning, according to Cançado (1996) and FUNDACENTRO (1989):

- knowledge of the raw materials utilized in the processes, the unit operations, the final products, the controlling procedures adopted, etc.;
- the correct preparation of the cassettes, mainly regarding its sealing and the initial weighing of the collecting filter;
- the field operation concerning the calibration of the pump flow rate, the definition of using or not using the cyclone (total dust and/or breathable dust), the sampling time, the collected mass, the filters shaft transport, among others;
- the definition of the analytical method to be utilized for quantifying the mass content of free silica in the sample (x-ray diffratometry and infrared spectroscopy).

The present work aimed at studying techniques for quantifying free silica in environmental dusts, by means of instrumental analytical methods, emphasizing the standard sample and unknown sample preparations, the mass of dust collected in environments under risk, the origin of the dusts and the maximum acceptable percentual analytical errors.

2 SAMPLING INSTRUMENTS

The sampling instruments are the tools for collecting the dusts for quantitative assessing, according to existing standards. Sampling is performed with the help of a sampling apparatus consisting of:

- sampling pump;
- filters holder, containing a filter compatible with the kind of dust to be collected;
- a particles splitter, necessary for collecting dusts in a specific size range.

For evaluating the degree of exposure of the worker, the sampling apparatus must simulate human breathing, permitting the passage of the air loaded with particles (with the help of the sampling pump) and retaining on the collecting filter (with the use of the particles splitter) the particles that would probably deposit on the breathing system, including lungs

3 DETERMINATION OF CRYSTALLINE SILICA BY X-RAY DIFFRATOMETRY

3.1 *Methodology*

Quantitative and qualitative determinations of crystalline silica were performed according to the methodology proposed by Cançado (1996), FUNDACENTRO (1989) and Padilha et al. (1984):

- the dust sample is collected on the PVC membrane filter;
- the mass of the sample is determined as the difference between the mass of the filter after collection and that of the virgin filter (gravimetric analysis);
- the filter is placed in a porcelain crucible and then calcined in a muffle at 800°C, for two hours;
- an X-ray diffratometry scanning is performed just to verify the presence of crystalline free silica phases (quartz, tridimite and cristobalite), as well as interfering phases (micas, feldspar, graphite, etc.);
- after the confirmation of the presence of silica in the sample and the definition of the peak to be utilized for the analysis, the quantitative assessing of the sample is performed by means of comparison with the standard calibration curve adequate for each case;
- the calibration curve is established from standards containing fluorite (CaF_2), as internal standard, and the free silica polymorphic species under analysis; it is prepared taking into consideration the relation between the intensity of the fluorite peak, as a function of the mass of the corresponding silica standard, yielding a straight line crossing, as closely as possible, the origin of the axes (errors larger than 10% for standards with mass of the polymorphic free silica species over 0.04 mg indicate that another set of standards must be prepared).

3.2 *Results evaluation*

- Sampling volume

$$Qm = (Qi + Qf)/2 \quad (3.1)$$

Qm = average sampling flow rate [L/min]
Qi = initial sampling flow rate [L/min]
Qf = final sampling flow rate [L/min]

The final sampled volume is given by:

$$Va = Qm \times Ta \quad (3.2)$$

Va = sampling volume [L]
Ta = sampling time [min]

Considering that the tolerance limits are normally expressed in [mg/m³]:

$$Va = (Qm \times ta)/1000 \ [m^3] \quad (3.3)$$

- Dust concentration
-

$$Cs = m/Va \quad (3.4)$$

Cs = dust concentration [mg/m³]
m = mass of the collected sample [mg]
Va = sampling volume [m³]

- Tolerance limit calculation (TL)

Total dust:

$$TL = 24/(\%SiO_2 + 3) \ [mg/m^3] \quad (3.5)$$

Breathable dust:

$$TL = 8/(\%SiO_2 + 2) \ [mg/m^3] \quad (3.6)$$

- Risk evaluation (Cs)

$Cs \geq TL$: risk situation
$Cs < TL$: no risk

Results of the evaluation of crystalline free silica in environmental dusts collected in an underground gold mine in full operation are presented in table 1.

4 SIZE ANALYSIS

4.1 *Samples preparation*

The samples are collected in the field, according to current standards of environmental assessment and occupational health recommended by FUNDACENTRO (1989). The collected samples preparation methodology is:

- the filter is placed in a porcelain crucible covered with a lid and taken to muffle at room temperature (or temperature below 300°C if it had been previously utilized);
- the temperature is adjusted to 800°C, the filter being burned during the heating period. After reaching this temperature, a period of two hours is recommended for totally calcining the filter;
- the calcined residue is transfered to a 50 mL erlenmeyer flask by means of the addition of a few milimeters of water, the crucible being scrapped with a glass rod;
- the flask is hand shaken and placed in an ultrasonic bath for 30 minutes for dispersing the agglomerant;

Table 1. Evaluation of crystalline free silica in environmental dusts collected in an underground gold mine in full operation.

Sample	Average flowrate (L/min)	Sampling time (min)	Sampled volume (m^3)	Mass of sample (mg)	Mass of silica (mg)	SiO_2 %	Concentration (mg/m^3)	Tolerance limit (mg/m^3)
1	1.75	220	0.385	0.38	0.02	5.3	0.99	1.10
2	1.57	263	0.413	0.21	0.02	9.5	0.51	1.92
3	1.67	139	0.232	1.00	0.06	6.0	4.31	1.00
4	1.63	154	0.251	6.00	0.48	8.0	23.90	2.18
5	1.42	127	0.180	1.00	0.10	10.0	5.56	1.85
6	1.69	253	0.428	1.21	0.12	9.9	2.83	0.67
7	1.69	360	0.608	1.74	0.17	9.5	2.86	0.70
8	1.67	215	0.358	1.54	0.11	7.1	4.30	2.38
9	1.63	360	0.587	1.21	0.12	9.9	2.06	0.68
10	1.67	1890	0.301	1.81	0.16	8.8	6.01	2.03
11	1.65	360	0.594	1.60	0.08	5.0	2.69	3.00
12	1.71	132	0.226	2.10	0.12	5.7	9.29	1.05
13	1.61	360	0.580	6.00	0.48	8.0	10.34	2.18
14	1.50	186	0.279	4.10	0.34	8.3	14.70	2.76
15	1.50	221	0.332	2.07	0.04	1.9	6.43	4.90

- the material is collected by means of a dropper, placed onto a cylindrical piece of polished metal or acrylic and left in a dryer for 24 hours;
- the samples are coated with a carbon conducting film via a vacuum evaporator (metallizer);
- the sample shaft of the scanning electron microscope is connected to the support by means of a conductive carbon paint.

4.2 *Image analysis and interpretation*

Images processing is a technique that consists of obtaining computer aided information on a sample. The computer converts the image into numerical shape, a process known as image digitalization, that divides the image in sevaral tiny regions known as pixels (Picture Elements), that occupy a position inside a matrix. Each pixel is identified by a line and column numerical reference corresponding to its position in the matrix.

The software "Image-Pro Plus", a program in Windows environment that performs advanced images processing and characterization, was utilized. The main features of the sofware are:

- plotting and counting objects automatically;
- measuring properties such as: projection area, perimeter, diameters (maximum, minimum, average) and radius;
- classifying and presenting objects by range of values of the measured property, using colors contrast to enhance visualization;
- eliminating measurement imperfections, grouping, separating, inserting and hiding particles;
- presenting the results numerically as tables and/or graphs.

Results of size analysis performed in a sample collected in an underground gold mine are presented in table 2.

5 CONCLUSIONS

- Environmental dusts generated by gold mining companies, located in the Iron Quadrangle of the state of Minas Gerais, Brazil, present as major phases quartz, iron oxides and muscovite and as less abundant phases, if present, feldspars and calcite.
- The linearity of the calibration curves by the standard samples utilized in the X-ray diffratometry analyses is highly enhanced if the curves are split into two, one for lower grades and the other for higher grades, as observed from the calculation of the analytical method maximum percent errors.
- The maximum percent errors presented by the fitting equation are:

  i. 10% for quartz-fluorite curves, with low quartz concentrations.

  ii. 5% for the quartz-fluorite curve, with high quartz concentrations.
- The sample preparation methodology was adequate to provide particles dispersion.
- The images analysis software "Image-Pro Plus" utilized for treating data from the scanning electron microscopy, aiming at the determination of average particle diameters, proved to be a reliable tool, faster than the manual analysis method.

Table 2. Results of size analysis performed in a sample collected in an underground gold mine.

Size range (µm)	Number of particles	Average diameter (µm)	Mass mg	Mass %	Cum. % passing
<0.20	746	0.20	8.28E-09	0.22	0.22
0.20 - 0.50	1045	0.37	7.27E-08	1.95	2.18
0.50 - 1.00	724	0.68	3.21E-07	8.63	10.80
1.00 - 2.00	297	1.33	9.64E-07	25.90	36.70
2.00 - 5.00	67	2.60	1.63E-06	43.85	80.55
5.00 - 8.00	2	6.39	7.24E-07	19.45	100.00
total	2881	-	3.72E-06	100.00	-

REFERENCES

Cançado, R.Z.L. 1996. Characterization of silica in environmental dust. Ph.D. thesis. Belo Horizonte: UFMG. (in Portuguese)

FUNDACENTRO/MTb 1989. Standards for dusts evaluation and laboratory analysis methodology. São Paulo. (in Portuguese)

Padilha, A.F. 1984. Instrumental analytical Techniques applied to geology.: 2-44. São Paulo: Edgard Blücher. (in Portuguese)

Environmental Issues and Management of Waste in Energy and Mineral Production, Singhal & Mehrotra (eds)
© 2000 Balkema, Rotterdam, ISBN 90 5809 085 X

Deformation behaviors of oil sands mine waste soils: Compaction effects

Hua Li & David C. Sego
Department of Civil and Environmental Engineering, University of Alberta, Edmonton, Alb., Canada

ABSTRACT: Overburden excavation for oil sands mining results in a large volume of soil for disposal. The soil undergoes little compaction thus it has a large potential for volume change after deposition when subjected to loading, or when inundated by re-establishment of the ground water system. The large volume change can be substantial decreased through compaction during placement. The deformation behavior is studied using compression and wetting tests. It was found that the applied compaction pressure has an important influence on the deformation behavior. The compaction effect could be neglected when the future overburden stress is greater than compaction pressure.

1 INTRODUCTION

Overburden excavation for oil sands mining results in a large volume of soil for disposal. The soils having undergone little compaction during disposal may have poor and highly variable geotechnical characteristics. These dump materials are usually called non-engineered fills.

The non-engineered fills are placed in a loose unsaturated condition due to lack of systematic compaction during their placement. They have a large potential to undergo volume change after deposition under load, or when submerged by water. Where buildings or highway are founded on such non-engineered fills, large ground movements can be expected to occur and it may be necessary to improve the fill prior to subsequent construction by the use of an appropriate ground improvement technique thereby limiting and controlling subsequent ground movements.

The deformation behavior of compacted soils is also important in embankment design and has been studied by a number of researchers (Lee and Haley 1968, Nwaboukei and Lovell 1986, Hausmann 1990). Most studies only focused on the behavior in the neighborhood of optimum moisture content. But the deformations of non-compacted soils are generally not studied. In addition, dumped waste soils may vary over a large range of moisture content being placed from almost dry to very wet. The change in their deformation properties with this water content range is of interest.

Laboratory tests were carried out on a fine-grained overburden soil. The tests include one dimensional compression, pneumatic compaction, and wetting tests. All the tests were performed using the same test cylinder to avoid sample disturbance. The effects of compaction on the soil deformation behavior are studied by comparing the difference of the compression curves and the void ratio at various applied compressive stress levels.

2 SAMPLE PREPARATION

Pleistocene lacustrine clay (P_l) was used in this study. The soil was acquired from the overburden soils on the lease of Syncrude Canada Ltd., Fort McMurray, Alberta. Its specific gravity is 2.69. The plastic limit of the soil is 16.9%, and the liquid limit is 37%. The soil consists of 34% clay, 44% silt and 22% sand. The particle size distribution can be found in a paper by Li and Sego (1999a).

Air-dried soil was mixed to a certain water content and stored in a moisture room for at least 48 hours prior to use in this study. The water content range was from 1% at the air-dried condition to 46%, which is about 1.25 times the liquid limit (w_L). According to Burland (1990), the properties of such samples at a water content of w_L to $1.5w_L$ are independent of the natural and in-situ state of a soil and therefore are representative of the inherent nature of the soil.

Clods were formed during the soil mixing and preparation. The clod size has a dominant influence on the initial void ratio of each sample. The clod size increases as the prepared water content increases from the dry condition. The observed maximum size is over 10 mm. The size would decrease as the water content approaches the liquid limit, where the soil became visually saturated. The clod size effect has also been observed by many researchers during preparation of soil samples. It is believed that the surface tension of water holds the soil particles together thus forming the clods.

During the loading process some samples were flooded via the porous top and bottom platens after different vertical stresses had been applied. The wetting collapse was measured for these samples and its relationship to the initial water content discussed.

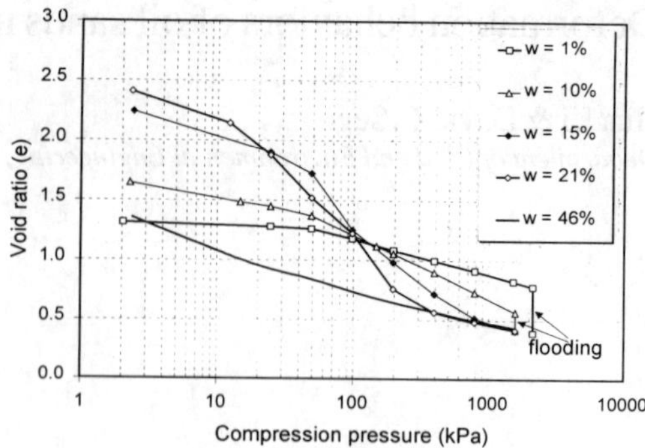

Figure 1. Void ratio versus compression pressure of as-dumped soil prepared at different water content

3 COMPRESSION BEHAVIOR OF AS-DUMPED SOIL

Soil samples were mixed at water contents of 1%, 10%, 15%, 21% and 46%. These were allowed to moisture condition for 48 hours prior to being placed into the oedometer. The loose soils mixed at different water content were placed in a rigid confining ring with a diameter at 63.4 mm. Compression tests were conducted by applying pneumatic pressure to the samples following application of a small initial seating stress of 2.4 kPa. The applied stresses ranged from 2.4 kPa to 1569 kPa with the increment doubling its previous applied pressure. The next load was only applied after the soil deformed by less than 0.005mm in last two hours of the previous load increment. Free drainage was allowed during the compression test. The maximum applied test stress represents about 80 meters of deposited soil, that is expected to cover the practical range of overburden stress found within typical waste dumps.

The compression test was carried out without any compaction effort being applied to the sample to simulate end dumping over the edge of a dump face. The initial void ratio of each sample varied with its different initial water content. As shown in Figure 1, the initial void ratio of air-dried soil ($w = 1\%$, $e_0 = 1.31$) and wet soil ($w = 46\%$, $e_0 = 1.36$) both are smaller compared with those at intermediate moisture contents. The initial void ratio was the greatest at 2.42 when the initial water content was 21%. The large void ratio contributed by the larger clods, which form when the soil is mixed at moisture contents near the plastic limit of the soil. The air-dried soil only has soil particles and no clods, while the very wet soil at 46% has no clods as the water content is greater than the liquid limit of the soil.

The one-dimensional compression curves for the different samples are shown in Fig. 1. When the compression stress increased, the deformation of air-dried sample ($w = 1\%$) occurs slowly. The reason is the higher friction between particles due to lack lubrication by water as well as the resistance of the particles to compression due to higher internal matric suction in the soil. Although the void ratio is small and close for both the air-dried soil and very wet soil, their particle arrangement are different. The air-dried soil has the voids filled with air and the wet soil has mostly water filled voids. Therefore their deformation response is different.

The sample with initial water content of 46% is more compressible at the lower vertical stresses than the samples with less water content (Fig. 1). Large amounts of water lubricate the soil particles, and contribute to the ease of squeezing out this water at lower stresses. Therefore, the sample with initially higher water contents generally will have a smaller void ratio, or the sample achieves greater density.

The void ratio of the soil decreases as the compression stress increases (Fig.2). The sample with greater water content resulted in smaller void ratio (or high dry density) once the applied stress increased above 98 kPa. The greater amount of water helps to decrease the friction between particles, which allows for closer packing of the particles. Extra water in the sample may be squeezed out at high-applied stresses. The saturation line (S=100%) represents where the sample would become saturated under the given vertical stress. When the applied stress increases to 1569 kPa, the sample with water content of 15% was saturated (Fig.2). In other words, 15% water content is enough to allow the soil particles to move to the saturate condition under a vertical stress at 1569 kPa.

One may observe higher matric suction in an air-dried fine-grained soil. According to Delage and Graham (1996), constant suctions are observed at low water contents, showing that the density has little influence on the suction at these lower moisture contents. This is probably because the compression is concentrated in the air-filled voids of the soil,

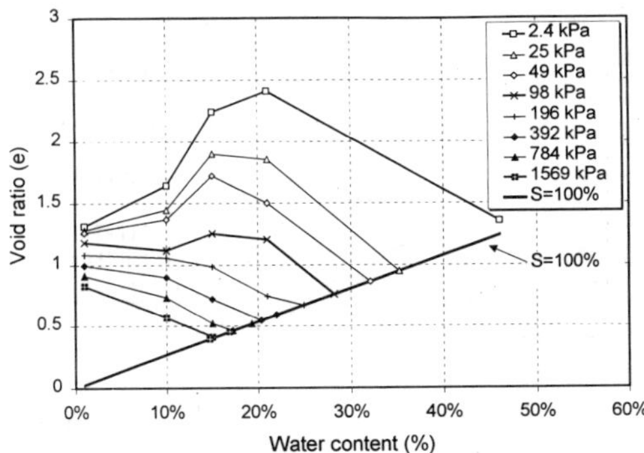

Figure 2. Void ratio progress of as-dumped soil prepared at different water contents

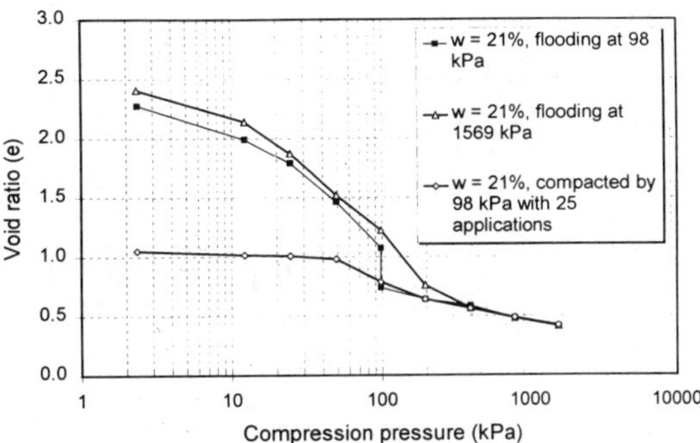

Figure 3. Influence of compaction pressure on initial void ratio for the 21% initial water content sample

while smaller water-filled pores remain unchanged. The large matric suction within the air-dried soil will allow for larger collapse deformation once the sample is flooded.

With the increase of water content from 1% to 10%, 15% and 21%, the compression curve moves closer to the compression curve for the sample initially prepared at 46% moisture content once the applied stress is above 100 kPa. The samples all approach the same curve as the 46% moisture content sample but at different applied stresses. The wetter the soil, the smaller the pressure required to approach the unique compression curve (Fig. 1). For example the curve at 15% reaches the same void ratio as the curve at 46% with a stress of 900 kPa while the curve at 21% only required 400 kPa. For samples at water content of 10% and 1%, the maximum stress (1569 kPa) was not enough to eliminate the remaining air-filled voids within these samples.

4 COMPRESSION BEHAVIOR OF DUMPED SOIL AFTER COMPACTION

The deformation properties of the sample after being subjected to different levels of compaction effort are of interest. To eliminate the effect of clods, the dumped soil was initially compacted under a stress of 98 kPa and its post-compaction compression behavior was studied. The compaction tests were performed by applying a pneumatic pressure of 98 kPa directly to the sample in the test cylinder. The pressure was manually increased and then decreased 25 times with each pressure application lasting about 2 seconds. The rate was designed to model the slow speed of haul truck travelling at 4 to 6 km/hr across the surface of a overburden waste dump.

Compression tests were carried out immediately after the compaction test using the equipment and procedure as previous outlined. The results for the 21% initial moisture content sample are shown in Figure 3. The initial void ratio and the void ratio below the applied compaction pressure of 98 kPa is substantially reduced compared to the as dumped samples which had a large potential to undergo volume change. This confirms the importance of nominal compaction for engineered fill as well as waste dumps.

Figure 4 shows the compression results of samples prepared with different initial water contents and subjected to initial compaction of 98 kPa pressure. When the vertical stress is higher than the as-compacted pre-stress (98 kPa), the specimens exhibit a more compressible behavior. The soil with higher water content results in higher compressibility above the as-compacted pre-stress because of the lubrication effect of water. This agrees with the previous finding from the as-dump soil.

The initial void ratio of soil after compaction at different water content and its change under compression is shown in Fig. 5. The compaction curves in Fig. 6 present the same data in a more conventional plot. An optimum moisture content around 26% was found after compaction where its initial void ratio was a minimum (or the dry density is maximum). Compression tests show the void ratio decrease faster on the wet side after 98 kPa stress was applied for all the initial void ratio. This agrees with the results in Fig. 2, i.e. the static load (98 kPa) is more effective on wet samples because it is easier to move the soil particles closer together with the lubrication of water. It is also interesting that the compaction effect disappears if the subsequent applied stress is higher than the applied compaction pressure.

Figure 5 also shows that the compacted samples have the same post compaction void ratio for all water content less than 17%. This also means there is a constant dry density at water contents lower than 17% (Fig. 6). Faure and De Mata (1994) and Li and

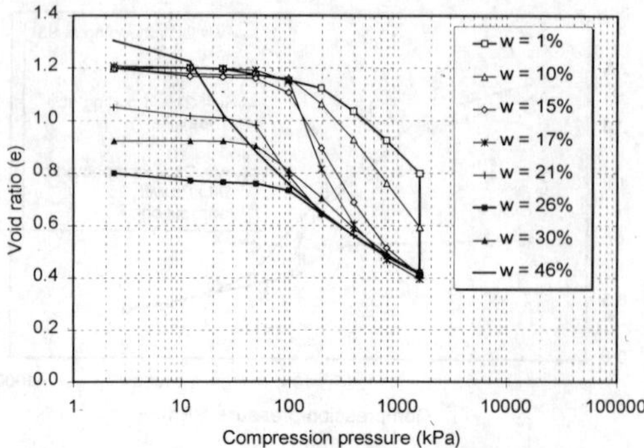

Figure 4. Compression curves of compacted soil prepared at different initial water content (compaction pressure at 98 kPa)

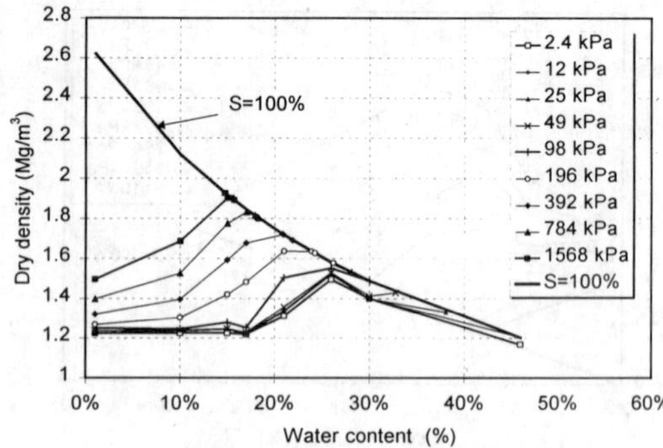

Figure 6. Dry density progress of compacted soil prepared at different initial water contents (compaction pressure at 98 kPa)

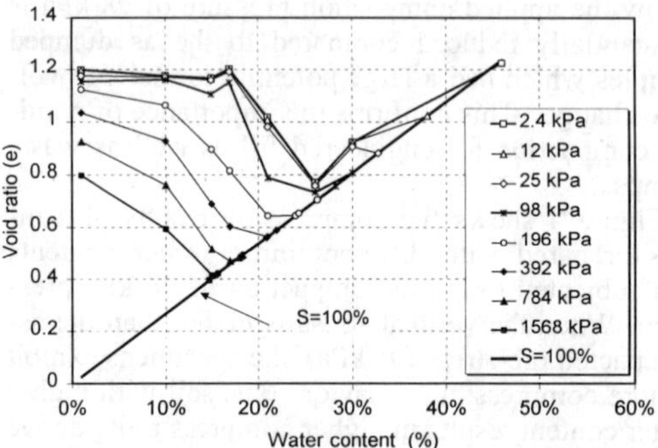

Figure 5. Void ratio progress of compacted soil prepared at different initial water contents (compaction pressure at 98 kPa)

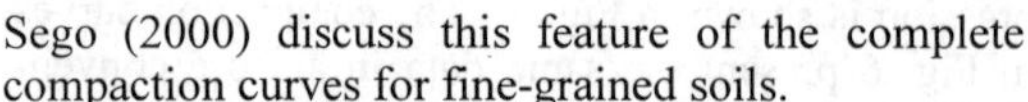

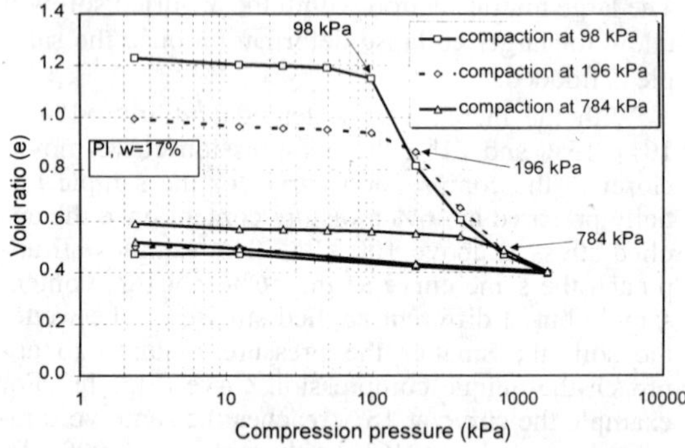

Figure 7. Deformation behavior of w=17% soil with different compaction effects

Sego (2000) discuss this feature of the complete compaction curves for fine-grained soils.

Different initial compaction pressures were used to compact samples at 17% water content. Compression tests then were carried out and the data is presented in Figure 7. The applied compaction pressure is again located at the maximum curvature of the compression curve and controls the deformation behavior of the compacted fill. Consequently, for embankment design in which as-compacted compressibility is important, the as-compacted pre-stress should be determined or can be used to effectively control the deformation behavior under future increasing applied stresses.

5 COMPACTION EFFECTS DURING LOADING COMPRESSION TEST

The initial void ratios of soil with and without pre-compaction are compared in Figure 8. The figure shows the initial void ratio has been substantially decreased by the application of a small compaction effort at 98 kPa. The largest decrease occurs in the midrange of the water content where the void ratio is most affected by the clod effect as previous described.

It is also interesting that the compaction effect disappears if the subsequent static compression stress is higher than the applied compaction pressure. Once the applied stress increases to the previous compaction pressure (98 kPa), the difference in void ratio between samples with and without compaction became insignificant. The curves for 1569 kPa showed no difference that means there is no compaction effect at all at this stress level.

It is observed that the curve with 98 kPa applied compaction pressure is different from the compression curve with 98 kPa applied compression stress (Fig. 8). The compaction curve shows a minimum void ratio (or the maximum dry density) at a water

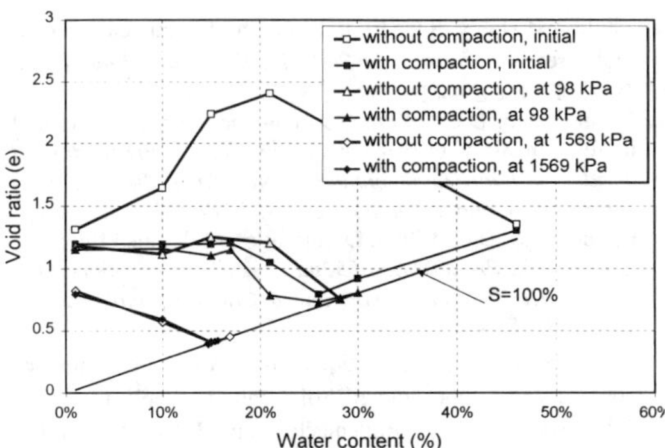

Figure 8. The void ratio comparison with and without pre-compaction at different applied vertical stress

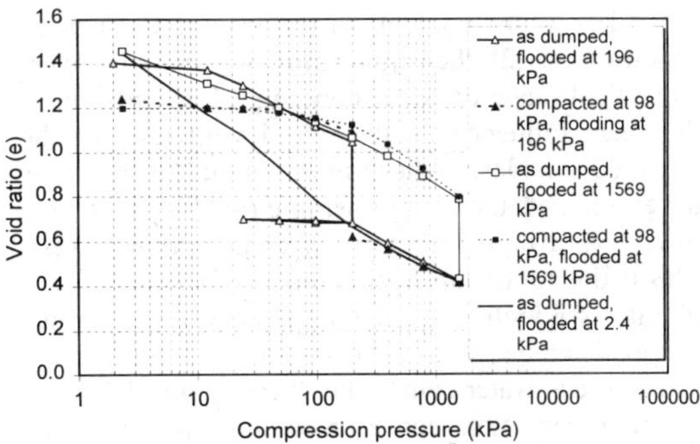

Figure 9. Wetting tests for air-dried soil with or without compaction effort

content, while in the compression test the sample with greater water content always results in smaller void ratio (or high dry density) once the pressure increased above 98 kPa. It is because the compression test could squeeze out extra water under the sustained stress compared to the 2 seconds of the applied compaction pressure. The increased water content decreases the friction between particles, which allows for closer packing of the individual particles.

6 COMPACTION EFFECT ON WETTING COLLAPSE

Non-engineered fills such as dumped overburden soils are likely to be susceptible to collapse associated with future inundation by water. This could occur due to the rise of a ground water table that previously had been drawn down or due to water infiltrating from the surface via cracks. Vulnerability to collapse depends on overburden type, placement method, moisture content, stress level and stress history. Under moderate stresses a poorly compacted mudstone and sandstone backfill underwent collapse compressions of up to 2% when the fill was submerged by a rising water table (Charles et al, 1984). A poorly compacted stiff clay fill suffered up to 6% collapse compression on inundation from surface trenches (Charles 1992). Lawton et al (1989) studied the degree of relative density as a guide for clayey sand or silt, and concluded that a relative density of between 85 and 90% was needed to prevent structural damage of buildings from collapse settlement due to inundation. There is no doubt the dumped soil with little or no compaction would have potential for substantial settlement on inundation.

Air-dried samples were used to determine if the deformation of the soil is sensitive to the loading-wetting sequence. Figure 9 shows the collapse behavior of the air-dried soil when it was flooded after application of different applied vertical stresses of 2.4, 196 and 1586 kPa. The void ratios after loading-wetting or wetting-loading sequences are generally the same. This establishes that the lacustrine clay used in this study is independent of the loading-wetting sequence. Therefore, the double oedometer procedure proposed by Jennings and Knight (1957) is appropriate.

Figure 9 shows there is a difference in the deformation behavior when the soil is compacted or not. The wetting collapse of the compacted sample is smaller than the as-dumped sample within the range of applied compaction pressure. Figure 9 also show two compacted samples (with 98 kPa, 25 applications) had the same wetting collapse as the soils without compaction when the flooding took place at 196 kPa and 1569 kPa. This indicates that the magnitude of wetting collapse is not affected by pre-compaction once the overburden stress increases beyond the pre-compaction pressure. Therefore, choosing an appropriate initial compaction pressure that will cover the subsequent overburden stress is very important to control the future wetting collapse of dumped waste soil.

Related tests by Li and Sego (1999b) has shown that once the compression pressure exceeds the compaction pressure, the collapse potential decreases with either initial water content increases or compression pressure increases.

7 CONCLUSIONS

The compression behavior of a fine-grained soil prepared over a wide range of water content was studied. The water content varies from the air-dried condition to 1.25 the liquid limit (w_L), which covers a practical range of water contents. Clod size has a dominant influence on the initial void ratio of the prepared samples. After the voids due to the clods are eliminated, the wettest sample (46%) reached the lowest void ratio for each different applied stress. This is due to the lubrication effect of the water. An ultimate void ratio of a given soil under the overburden load can be estimated from the void ratio - pressure curve at water content between w_L and $1.5w_L$.

The compaction can substantially decrease the initial void ratio in non-engineered fill. But the compaction effects disappear if the subsequent static compression pressure is higher than the applied compaction pressure. Consequently, for embankment design in which as-compacted compressibility is important, the as-compacted pre-stress should be determined.

The collapse of the tested soil is independent of loading – wetting sequence. It is also shown that the wetting collapse is not affected by compaction if the wetting take place at a higher vertical stress than the applied compaction pressure. Therefore, choosing an appropriate compaction pressure that covers the subsequent loading level is very important in controlling the future wetting collapse of dumped waste soil.

REFERENCES

Burland, J.B. 1990. On the compressibility and shear strength of natural clays. *Geotechnique* 40(3): 329-378.

Charles, J.A., Hughes, D.B. and Burford, D. 1984. The effect of a rise of water table on the settlement of backfill at Horsley restored opencast coal mining site, 1973-1983. *Proceedings of 3rd International Conference on Ground Movements and Structures, Cardiss*: 423-442.

Charles, J.A. 1992. The causes, magnitudes and control of ground movements in fills. In J.D. Geddes (Ed.), *Ground Movements and Structures*: 3-28. Vol. 4. Pentech Press.

Delage, P. and Graham J. 1996. Mechanical behavior of unsaturated soils: Understanding the behavior of unsaturated soils requires reliable conceptual models. In E.E. Alonso and P. Delage (Eds), *Unsaturated Soils*: 1223-1256. Balkema: Rotterdam.

Faure, A.G. and Da Mata, J.D.V. 1994. Penetration resistance value along compaction curves. *Journal of Geotechnical Engineering, ASCE* 120(1): 46-59.

Hausmann, M.R. 1990. *Engineering Principles of Ground Modification*, McGraw-Hill Inc.: New York.

Jennings, J.E. and Knight, K. 1957. The prediction of total heave from the double oedometer test. *Transactions, Symposium on expansive clays*: 13-19. South African Institute of Civil Engineering.

Lawton, E.C., Fragaszy, R.J. and Hardcastle, J.H. 1989. Collapse of compacted clayey sand. *Journal of Geotechnical Engineering, ASCE* 115(9): 1252-1267.

Lee, K.L. and Haley, S.C. 1968. Strength of compacted clay at high pressure. *Journal of Soil Mechanics and Foundation Division, ASCE* 94(1): 1303-1332.

Li, H. and Sego, D.C. 1999a. Soil compaction parameters and its relationship with soil physical properties. *Proceeding of 52nd Canadian Geotechnical Conference, Regina, Saskatchewan, Canada, October 25-27, 1999.* 517-524.

Li, H. and Sego, D.C. 1999b. Deformation of dumped fined-grained soil. *Proceeding of 52nd Canadian Geotechnical Conference, Regina, Saskatchewan, Canada, October 25-27, 1999.* 453-460.

Li, H. and Sego, D. C. 2000. Equation for Complete Compaction Curve of Fine-grained Soils and Its Applications. In D.W. Shanklin, K.R. Rademacher, and J.R. Talbot (Eds), *Constructing and Controlling Compaction of Earth Fills, ASTM Special Technical Publication 1384*. American Society for Testing and Materials, West Conshohocken, PA.

Nwaboukei, S.O. and Lovell, C.W. 1986. Compressibility and settlement of compacted fills. In R.N. Yong and F.C. Townsend (Eds.), *Consolidation of Soils: Testing and Evaluation, ASTM STP 892*: 184-202. American Society for Testing and Materials, West Conshohocken, PA.

Environmental Issues and Management of Waste in Energy and Mineral Production, Singhal & Mehrotra (eds)
 ISBN 90 5809 085 X

Development and field implementation of a new generation penetrometric sampling system

D.A.O'Neill
New England Foundation Company Incorporated, Quincy, Mass., USA

S.Durucan, A.Korre & S.Ahmed
T.H.Huxley School of Environment, Earth Sciences and Engineering, Royal School of Mines, Imperial College of Science, Technology and Medicine, London, UK

ABSTRACT: This paper describes the development of a second generation penetrometric sampling system as an environmental assessment tool for the characterisation of groundwater quality. The system improvements included an increase in the maximum rate of groundwater sampling capacity and the development of a new multiparametric flow cell using mini- and microsensors to obtain more accurate values of a greater number of physico-chemical parameters in real-time. The system was tested extensively, both in the laboratory and in the field around an active municipal solid waste landfill. The field data collected was used to characterise the spatial distribution of the groundwater quality parameters around the landfill site.

1 INTRODUCTION

Environmental legislation worldwide calls for increased monitoring of landfill, mining and other industrial sites in order to protect water resources. Such monitoring is essential for all types of landfills and mine sites at all stages, commencing from the measurement of "background" levels of water quality parameters of both surface and ground waters prior to the construction of landfills, tailings dams, and waste dumps to until well after the final closure of the facility in order to prevent and/or mitigate pollution. Groundwater resources characterisation of a potential industrial site and appropriate positioning and spatial referencing of the sensors in a continuous monitoring scheme at the baseline studies stage of the EIA necessitates a more direct and cost effective approach wherein accurately referenced geotechnical and hydrological data can be obtained.

Comprehensive characterisation of sites relative to groundwater quality requires a vast amount of data that in reality can vary both spatially and with time. Spatial characterisation must be three dimensional since contaminant concentrations can vary significantly in the horizontal (x-y) plane, given pollutant sources of limited dimensions and the directional nature of the groundwater movement, and in the vertical (z) plane, as a function of the stratigraphy and groundwater conditions at the site and the nature of the contamination. Variations with time are also important, particularly for assessing worsening contaminant conditions or the effectiveness of remediation measures.

Penetrometric sampling systems are ideal for addressing the 3D spatial characterisation needs since the truck-mounted equipment can be readily mobilised at different locations, at each of which they can be used to obtain data relatively rapidly with respect to vertical variations. In their current configurations, however, they should be considered for characterising the sites at one given point in time; evaluation of conditions at any point in time thereafter requires a new set of measurements.

Among the other advantages of using penetrometric systems for sampling groundwater are: the capacity to identify appropriate sampling strata during the sampling process; elimination of the time necessary to drill, develop and purge a borehole; elimination of the cost associated with disposal of drilling fluid; and the possibility of obtaining a groundwater sample of the highest possible integrity (O'Neill et al. 1995; Robertson et al. 1995). Several versions of these so-called "direct push" instruments exist, these are the Envirocone®; the Geoprobe (Geoprobe, 1995); the ARA cone sipper (Bratton et al., 1995) and the BAT Enviroprobe (Rad et al., 1988), each of which functions according to a different concept. The Multifunctional Envirocone® Test System (METS) is one of the only devices providing measurement of tip resistance and pore pressure during penetration between sampling depths. In addition, this device, equipped with a flushing capability, does not need to be removed from the ground prior to advancing the instrument to the next sampling depth. Data obtained with systems such as METS are also very useful in the design of subsequent monitoring well installation.

One of the principal problems with such devices is associated with the very small volumes obtainable, especially in medium-low to low permeability soils. Evaluating a profile of the variation in contaminant concentration at five different depths in, for example, silty fine sands could take up to four or five days which is simply too long. Thus, it was necessary to develop a second generation system (METS II) with improvements including an increase in the maximum rate of groundwater sampling possible and the development of a new multiparametric flow cell based on mini- and microsensors to obtain more accurate values of a greater number of physico-chemical parameters of the groundwater in real-time.

Considering that the greater the number of relevant parameters measured, the greater the validity of the subsequent analyses, and also bearing in mind the known interference difficulties and inaccuracies associated with using ion selective electrodes, efforts were aimed at incorporating a rapid, relatively inexpensive method for determining multiple inorganic ions and cations using a very small sample (less than 20 to 40 ml).

2 DEVELOPMENT OF THE NEW GENERATION PENETROMETRIC SAMPLING SYSTEM (METS II)

An earlier version of the MET system revolved around the use of the first generation Envirocone® developed by Ismes S.p.A., Seriate (BG), Italy, and is fully described by O'Neill et al. (1995). Recent work at Ismes focused on the further development of the various components of this penetrometric testing equipment to render its deployment in the field more productive. The schematic diagram of the METS II provided in Figure 1 illustrates the equipment and identifies the geotechnical parameters and chemical components that can be measured.

The modifications discussed herein were designed principally to address the very low rate of groundwater sampling of the first generation system (<1 to 3ml/min in silty clays to approximately 20 ml/min in coarse sand). The approach was two-pronged: on the one hand the Envirocone® was modified to maximise the flow of the water sample; and on the other hand, a completely new multiparametric flow cell was constructed based on mini and micro-sensors with sensitive tip dimensions more compatible with the volume of sample with which they were in contact.

2.1 *Development of the penetrometric tip for improved sample volume*

Improvement of the sampling flow volume required significant modification of the Envirocone® sampling tip, while the pump module remained essentially the same. The new Envirocone® II essentially permits sampling from a sinterised steel filtered zone (approximately 44mm in diameter) that is 40 cm long, four times longer than that of the older version. In order to effect this increase in the filter length without increasing the overall length of the penetrometric tip and pump module, the diameter of the module that houses the inclinometer had to be reduced. Thus the inclinometer had to be miniaturised. A new electrolevel inclinometer was designed and installed. The sensor, which requires a significant amount of signal conditioning, was equipped with surface mounted electronics. The sensor was rigidly fixed to the penetrometer tip.

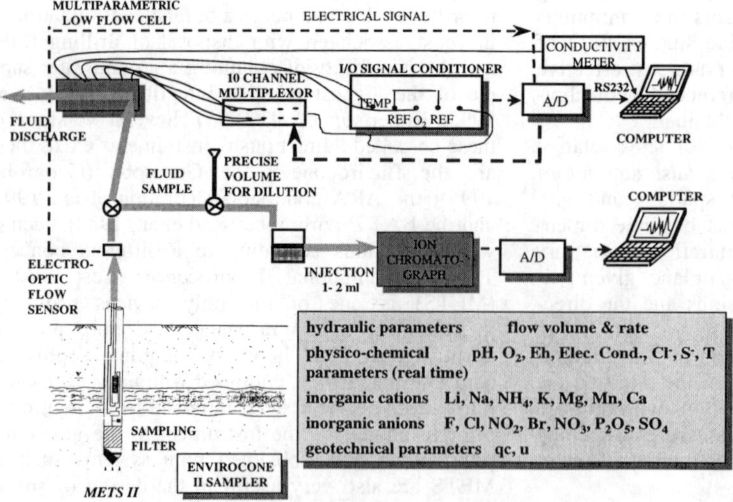

Figure 1. Schematic diagram of the METS II.

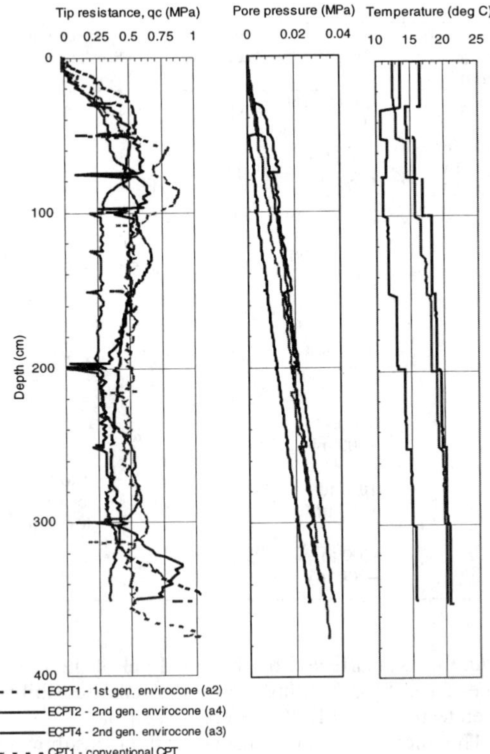

Figure 2. Penetrometric data from tests in medium sand.

The penetrometric sampler was tested for leakage in the laboratory in a specially prepared cell at pressures equal to up to 5 bar. The prototype was also tested in the field conditions to a depth of 3.5 m, acquiring penetrometric data in order to ensure the functioning of the complete system. The way in which the test site was constructed resulted in a very heterogeneous lateral and vertical distribution of sand density. The penetration tip resistance (q_c), penetration pore pressure (u) and temperature (T) obtained during these initial tests are presented in Figure 2. A conventional cone penetration test (CPT1) was also performed to validate the Envirocone® data (ECPT1-4). The test program revealed the necessity of incorporating in the acquisition program the temperature sensor offsets which are unique to each sensor.

The sampling rate of the new generation penetrometric tip was evaluated in the laboratory (with 7 ft of tubing) and in the penetrometer truck (with 75 ft of tubing). Both series of tests were performed with the sampler tip in a 10 cm diameter cylinder of water. The laboratory tests demonstrated that both the old and the new tip are capable of sampling on the order of 300 ml/min – a value indicative of the internal Envirocone® pump itself. During the tests in the truck, once again performed in a cylinder of water, thus presenting none of the hydraulic resistance that soil provides, the Envirocone II pumped slightly faster than the 1st generation instrument (40 ml/min as opposed to 35 ml/min).

Sampling rates were measured under actual field testing conditions in a medium sand with little fine sand, the penetration parameters of which are shown in Figure 2. The advantage of the new design is evident in Figure 3. Even in this relatively permeable soil ($k \approx 3x10^{-2}$ cm/sec), the pump rate of the 2nd generation penetrometric tip (42 ml/min) is equal to almost twice that of the 1st generation (23 ml/min). In soils of lower permeability the pump rate of the new design is expected to even more notably outperform the 1st generation sampler.

2.2 *Development of the mini and micro-sensor-based multiparametric flow cell*

Micro- and mini-sensors with the required specifications were available commercially for the measurement of pH, dissolved oxygen (DO), temperature (T), electrical conductivity (EC), and chlorides, as was a reliable reference electrode. For determining levels of suphides and redox, on the other hand, micro-sensors were designed and constructed using 75 micron-diameter Ag and Pt wires, high electrical resistance glass, stainless steel hypodermic needles and electrically-insulating resins. The general characteristics of the sensors that are presented in Table 1 demonstrate significant improvement relative to the previous sensors used in the METS multiparametric flow cell in that their dimensions are significantly smaller. Also, the ranges, accuracies and response times reported are believed to be much more relevant to the use in micro-flow cell conditions than those typically reported for the larger sensors, which are about 7mm to 15 mm in diameter, since immersion in 20ml to 30 ml of sample fluid is generally assumed.

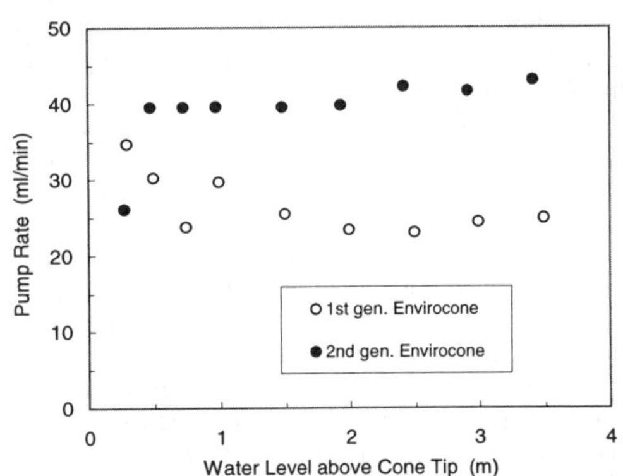

Figure 3. Sampling rate test in medium sand with little fine sand.

Table 1. General characteristics of the new micro and mini-sensors.

Sensor	Type	Unit measured	External Diameter (mm)	Immersed Depth (mm)	Range	Accuracy	Stirring effect	Response Time (sec)
Temperature	heat conductive resin (commercial)	pA	3	1 to 3	-40 to +150 °C	± 0.2 °C		< 5s
Electrical conductivity	2 plate platinum electrode (commercial)	S/cm		flow through - dead volume 2,1 ml	0 to 200 mS/cm	0.1μS/cm	-	5
Reference	Ag/AgCl (commercial)		4.5	<1	–	–	–	–
Dissolved oxygen	amperometric Pt wire, Au tip (commercial, also custom version in future)	pA	3	0.1	10 to 1500 pA	0.1 mg/l	2 to 5 %	< 5
pH	glass membrane, Ag wire in NaCl gel potentiometric (commercial)	mV	0.9	3	2 to 12	± 0.1	< 1%	< 10
Chlorides	Solid polymeric membrane, potentiometric (commercial)	mV	1.35	3	0.1 mM to 1000 mM	–	< 1%	< 5
Total sulphur (S^{II}) (combined)	potentiometric, Ag/Ag_2S (Custom)	mV	1	3	1μM to 1000 μM	–	< 1%	1 to 5 (> 20 at low conc.)
Reduction potential (combined)	potentiometric, Pt wire (Custom)	mV	1	3	-200 to +200 mV	± 20 mV	< 1%	10

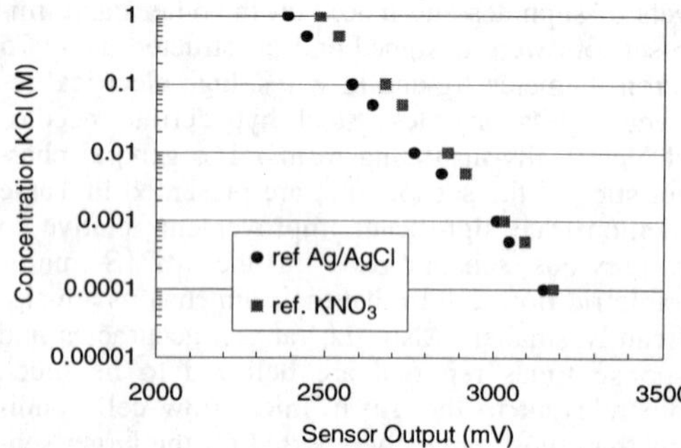

Figure 4. Effect of Ag/AgCl reference on chloride sensor output.

All the micro- and mini-sensors were fully calibrated in the laboratory and their accuracy, precision and potential interferences were verified through the same A/D converter that was used in the first phase of the field work. In anticipation of possible on site testing conditions, the sensors were characterised with respect to effect of temperature (5 C to 40 C) and drift (periods ranging from five hours to a couple of days). All the sensors demonstrated excellent accuracy and stability, including the dissolved oxygen, the ion selective chloride and the sulphide sensors. It was found that the pH sensor is not consistently affected by temperature, possibly due to compensating effects given its very small size. Dissolved oxygen, Cl^-, and conductivity vary linearly with temperature. It has been demonstrated that the Cl^- sensor, as evidenced by the data presented in Figure 4, is unaffected by the AgCl gel of the reference sensor by exhibiting almost the same sensibility when tested with a KNO_3 reference sensor.

The micro- and mini-sensors for the measurement of pH, DO, T, EC, chlorides, sulphides, nitrates and redox, along with the reference electrode, are housed in a plexiglass flow-through cell designed to expose the sensitive portions of each of the sensors to a flow of sampled groundwater in a channel which is 4 mm in diameter. A separate teflon sleeve was designed and machined for each fragile sensor in order to protect the glass and epoxy-filled hypodermic needles which are only 0.9mm to 4mm in diameter.

The approximate disposition of the sensors in the plexiglass flow cell is indicated schematically in Figure 5, except for the nitrate ion selective sensor which was installed between the dissolved oxygen and chloride sensors. The plexiglass cell is mounted at an angle on a HDPE board along with two 3-way valves used to direct the groundwater sample flow either towards the sensors or towards a sample container for further laboratory analyses. The stainless steel ball valves also permit easy access to the flow cell for calibration and decontamination of the sensors. The upwards angle of the plexiglass cell aids in the removal of eventual air bubbles.

The electronic conditioning and data acquisition system, illustrated schematically in Figure 1, includes a high-precision signal conditioner that directly accepts the signals from the reference sensor, temperature and amperometric (DO) sensors, a multiplexor for up to ten potentiometric sensors, a conductivity meter to provide the input signal and conversion for the electrical conductivity and, finally,

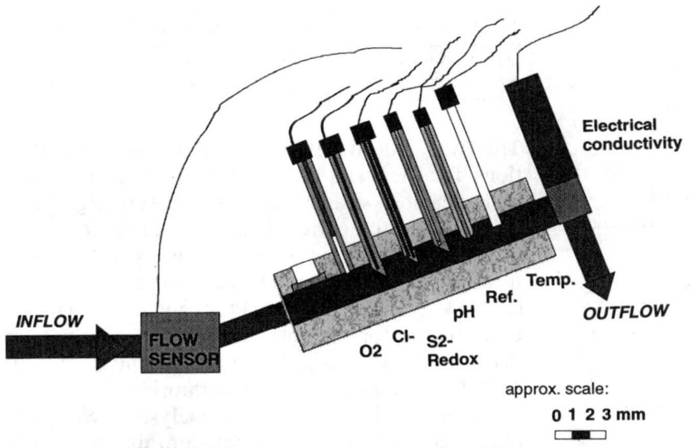

Figure 5. Schematic diagram of the low volume multiparametric micro-sensor flow cell.

a high precision analog/digital board (A/D) to convert all the signals. The analog signal from the signal conditioning unit is connected directly to the A/D converter. The A/D converter, controlled by a computer, also emits the digital signal for automatic channel selection from the multiplexor. These electronics permit a complete scan of all sensors approximately every 3 sec.

"MicroCella", the custom software developed using LabView for monitoring and recording data from the multiparametric low flow cell, permits real-time graphic representations of the output from all the sensors in appropriate units, either as a function of elapsed time and flow volume. The scale of the individual graphs can be readily adjusted while data is being collected in order to better identify sample arrival. The software also permits calibration of each of the sensors before and at any time during sampling along with the storing of the calibration history for any sensor. The menu-driven package, organised in modules relative to sensor calibration, multiplexor control, data acquisition, data elaboration and data presentation, presents the user with a series of panels relative to the function of interest. The program is designed to run on a Pentium-based portable computer, connected by serial cable to the A/D converter.

2.3 *Implementation of ulterior chemical component analyses*

In assessing the practicability of utilising sensors for the measurement of additional ions and cations useful in the detection of typical pollutants, it was concluded that, while it was worthwhile to attempt to measure such ions as Cl^- and S^{2-} via sensors, it was better to employ a dual column ion chromatograph for other ions such as ammonium, nitrates and nitrites for the following reasons:

- sensor measurement of several ions and cations requires pre-treatment of the sample.
- sensor measurements tend to be prone to interference from other anions and cations;
- with a very small sample (less than 3 ml) it is possible to obtain upwards of seven important anions and seven cations;
- ion chromatography permits a wide dynamic range (on the order of a few ppb upwards of several hundreds of ppm) while ensuring a high quality, precise and repeatable measurement.

the relatively greater time for chromatographic analysis can be resolved considering the actual sampling rates of METS II. The Dionex 120 dual column chromatograph was selected for its durability and reliability. A hydraulic circuit was designed in order to obtain precise volumes in situ in order to permit accurate dilutions.

3 CHARACTERISATION OF A MUNICIPAL SOLID WASTE LANDFILL SITE

3.1 *Field measurements*

Systematic field validation and site characterisation work with the truck-mounted METS II unit was effected at a municipal solid waste (MSW) landfill located in a flatlands area approximately 50 km west of the seacoast in Italy. The landfill dimensions were approximately 400 m in the north-south (y) direction and by 650 m in the east-west (x) direction. The groundwater flow was northwest to southeast while the groundwater level varied between 1.8m to 3m below ground surface. In general, the soil profile was as shown in Table 2.

The solid waste was placed at a depth below the original ground surface equal to about 7 m. The field work was performed between February and April

Table 2. Soil profile around the landfill site.

Stratum thickness	Soil Type
1-3m	Fill and topsoil
0-1m	silty Clay
1m	silty fine Sand
2-7m	silty Clay
1-2m	silty fine Sand interbedded with 1-2m thick strata of silty Clay to a depth of about 20m to 22m.

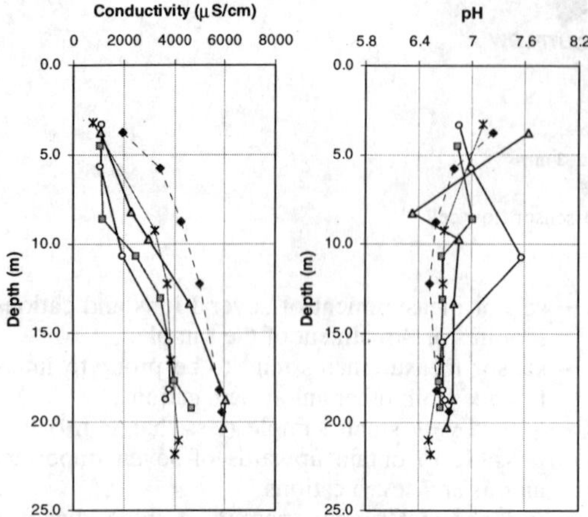

Figure 6. Penetrometric data from the field validation of the Envirocone® II at the Italian landfill.

1999 and seven vertical locations were characterised in order to tests the system and to provide data for the spatial assessment of the groundwater quality in the area. Using the tip resistance and pore pressure measurements during penetration, the silty fine sand layers were chosen as the strata from which groundwater was sampled. Figure 6 presents some of the EC profiles obtained at depth in the vicinity of the MSW landfill.

3.2 *Statistical and spatial data analysis*

In addition to simple statistical analysis, multivariate statistical analysis techniques were utilised aiming to develop an understanding of the interrelationships amongst the groundwater quality parameters measured. The results of the field tests of the Envirocone II system were analysed by means of principal component and factor analysis tools. All variables which were measured above the detection limits of the sensors in the field and those that were provided from later analysis in the laboratory for each sampling depth and for penetration were used (Table 3). PCA analysis was used to indicate the number of important factors that lie within the data set. From this analysis it was possible to recognise that there is one predominant process that controls the distribution of the values for the 14 variables.

Table 3. Parameters used for the PCA and factor analysis.

pH	Cond	Li	Na	NH_4	K	Mg
Ca	Fl	Cl	Br	NO_2	S	HCO_3

This was indeed confirmed by the simple correlations for the pairs of variables were a number of high Pearson-product correlations were significant, while the partial correlations adjusted for the effects of all other variables were relatively small. Kaiser's measure of sampling adequacy (over-all MSA = 0.8438) revealed that the PCA and factor analysis was reliable for the data and that the only variable that was not sufficiently represented by the PCA/factor analysis solution is fluoride.

From the simple statistical analysis it was possible to recognise the simple relationships such as the positive correlation between calcium, chloride and the absence of correlation between pH and NO_2 with other variables. What was more interesting to examine are the multivariate dimensions, the factors in other words, that represent the variance hidden in the data set and try to relate them with processes that occur in nature and to decide whether these are specific to the data set examined or it is possible to generalise to a wider range of data sources. Factor analysis was used as the next step after the PCA. The results of the factor analysis are summarised in Tables 4 and 5 and Figure 7. It was shown that after the eighth factor the eigenvalues became negative. The reason was that a number of variables were well correlated and therefore retaining additional factors would not be sensible. Three factors were chosen for further calculations and it was decided to explore two factor rotation techniques: the orthogonal varimax rotation and the oblique promax rotation. By looking at the root mean square off-diagonal residuals, it was confirmed that the factor solution generalises for all variables.

As shown from the scatterplots of the orthogonal factor loadings (Figure 8) and, as further confirmed from the correlations between the three factors after oblique rotation, there was a negative correlation between FACTORs 1 and 2 (r=-0.57) and negligible correlation's between the other two pairs of factors.

Table 4. Eigenvalues and proportions of variance for the factor analysis of the data from the Italian Landfill.

Eigenvalues of the Reduced Correlation Matrix: Total = 11.23887 Average = 0.8028

	1	2	3	4	5	6	7
Eigenvalue	8.830	0.978	0.710	0.382	0.339	0.174	0.097
Difference	7.852	0.268	0.328	0.044	0.165	0.077	0.065
Proportion	0.786	0.087	0.063	0.034	0.030	0.016	0.009
Cumulative	0.786	0.873	0.936	0.970	1.000	1.016	1.024
	8	9	10	11	12	13	14
Eigenvalue	0.032	-0.002	-0.018	-0.045	-0.061	-0.076	-0.101
Difference	0.034	0.016	0.027	0.016	0.015	0.024	
Proportion	0.003	0.000	-0.002	-0.004	-0.006	-0.007	-0.009
Cumulative	1.027	1.027	1.025	1.021	1.016	1.009	1

It is a fact that oblique rotation is adding complexities in reporting results, but since three factors were enough to describe the data, promax rotation with varimax pre-rotation method was employed. The results, as shown in the factor structure matrix, strengthen the association of the factors with those variables that load significantly on them and reduce even more the values for the variables that were weakly related with the unrotated factors. The scatterplots of the semipartial correlations of the variables with the factors are shown in Figure 7 along with the reference axis correlations and the corresponding angles.

Another change that became apparent with the different rotation methods was in the relative amount of variance explained by each factor, which was more uniformly distributed among the factors after oblique rotation. The simplicity of the structure was assessed from factor loadings, where for each factor there were few excellent and very good, and many low correlations between variables and factors. Table 7 summarises the factor analysis solution attributing to each factor those variables that correlate best with them. The variables are arranged in descending loading order and a loading value of 0.35 was selected as the interpretation criteria.

The data collected by the penetrometric sampling system has spatial reference since the geographical xy co-ordinates of the penetration point are known as well as the depth of penetration and the tagged samples. Each penetration sample provides a number of profiles similar to well-logs and this is repeated at several locations in a site to be characterised (Figure 6). For each variable measured the data has 3D reference which can be utilised in geostatistical analysis for the estimation of the measured variables in areas between different penetration points. At each sampling point, water quality parameters were collected from 6 different positions in the depth providing a total of 30 observations.

A three dimensional view of water quality around the landfill site was established utilising the real world co-ordinates of the penetration points and the depth of the spot samples collected for each. Three variables: conductivity, pH and dissolved oxygen were selected for interpretation on the basis that all sampling points had measurable values for these parameters. The average shape of the landfill was introduced in the plot with a rectangular shape as illustrated in Figure 8 and the total area that was selected for the initial study was 1400x1400m to 24 m depth.

The results of the preliminary analysis are illustrated in Figure 9 for conductivity. The variograms were calculated in two dimensions, the horizontal plane and the x-z plane. There are still difficulties with the modelling aspects of the variograms due to the inherent characteristics of the sampling, i.e. short distances between samples in depth and longer distances on the x-y plane. A number of options are currently being studied to overcome these problems utilising surrogate information about the site including geology - stratigraphy of the sediments around the landfill and additional information provided on a continuous profile during the penetration of the Envirocone II system such as the pore pressure and tip resistance.

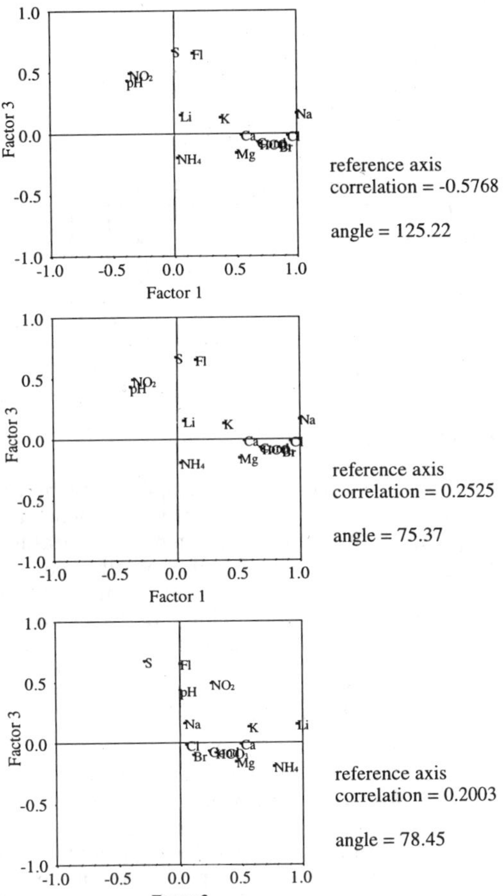

Figure 7. Factor pattern plots after promax rotation for the Northern Italian Landfill penetrometric data.

Table 7. Variables attributed to each factor

FACTOR 1	FACTOR 2	FACTOR 3
Na	Li	S
Cl	NH4	Fl
Br	K	NO_2
HCO_3		pH
Cond		
Ca		
Mg		

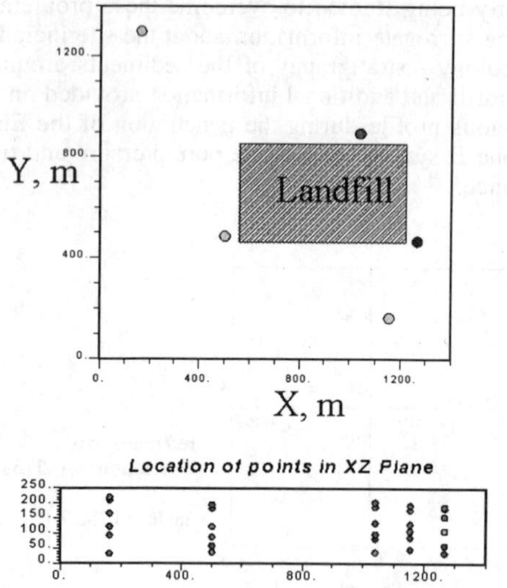

Figure 8. Layout of the penetration samples in 3D around the Italian Landfill.

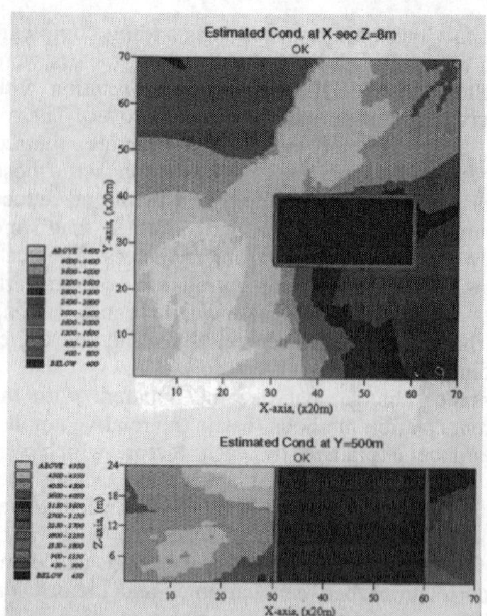

Figure 9. Kriging estimates of conductivity values around the Italian Landfill.

4 CONCLUSIONS

The METS II penetrometric sampling system was developed to enhance the speed of groundwater sampling and the number of components measured during and immediately after the sampling process. The four-fold increase in the length of the filter zone more than doubled the sampling rate in medium to low permeability soils. The integrity of the groundwater sample is very high since the sample, obtained with a positive displacement pump, comes into contact only with stainless steel and PTFE and is exposed to the atmosphere for only the moments of introduction into the sampling bottle. Sampling bottles could be developed which would limit even this exposure.

The multiparametric flow cell equipped with mini and micro-sensors was developed to permit the real time measurement of pH, EC, DO, T, redox, Cl^-, S^{2+} and NO_3. The sensor system proved reliable only for the physico-chemical parameters. The ion selective sensors for Cl^- and NO_3^- were far to sensitive to the presence of other ions as well as temperature. The flowcell derived parameters were then accompanied by a large data set of ion and cation concentrations obtained with a dual column ion chromatograph. The advantages of using the chromatograph included the very small sample necessary -less than 3 ml for a complete set of data- and the speed with which these data could be obtained –between 45 min and one hour.

The system was successfully utilised to characterise a landfill. The soil profile at the site consisted of numerous strata of silty clay interbedded with layers of silty sand which were frequently less than 1m thick. Thus, each location required approximately two days in order to sample between 100 to 200 ml at five different depths up to 22m below ground surface. The thin strata of silty fine sand were successfully located with the Envirocone II. The arrival of the high quality sample was successfully announced by the plateaus achieved by the curves of pH, EC, DO and temperature.

5 ACKNOWLEDGEMENTS

The research reported in this paper was carried out as part of a European Commission funded project, Contract No: BRPR-CT97-0377. The authors wish to thank Ismes S.p.A of Italy and the European Commission for their support of the project.

6 REFERENCES

Bratton, J. & J. Shinn 1995. CPT probes under current research at Applied Research Associates, in-house publication.

Geoprobe 1995. *Newsletter*, Salina, KS, USA.

O'Neill, D.A., G. Baldi & A. Della Torre 1995. The multifunctional Envirocone test system, *Advances in Site Investigation Practices*, Institution of Civil Engineers, London.

Rad, N.S., S. Sollie, T. Lunne & B.A.Tortensson 1988. A new offshore soil investigation tool for measuring the in situ coefficient of permeability and sampling pore water and gas, *Proc. BOSS Conference*, 409-417.

Robertson, P., T. Lunne & J. Powell 1995. Applications of penetration testing for geo-environmental purposes, *Advances in Site Investigation Practice*, Institution of Civil Engineers, London.

Environmental Issues and Management of Waste in Energy and Mineral Production, Singhal & Mehrotra (eds)
© 2000 Balkema, Rotterdam, ISBN 90 5809 085 X

Study on anisotropic rock mass unloading mechanics in nuclear waste repository in salt mine

Ha Qiu-Ling
China Three Gorges Project Corporation Technical Committee, Beijing, People's Republic of China

ABSTRACT: There exist both stress concentration(loading area) and unloading in the secondary stress field in the process of rock mass excavation. There exist substantial difference of mechanics property under unloading condition with that for the joints of rock mass under loading. The current rock mass mechanics is based on loading, which cannot comply with engineering physical occurrence and leads to the substantial difference between the theoretical analysis and the actual monitoring results. Therefore, the author has put forward the new concept of anisotropic rock mass unloading mechanics which has proved to keep the theoretical analysis results in compliance with that of actual monitoring results.

The above study is based on the geologic condition of rock mass with steep angle joints, Which is different from that of the repository in salt for nuclear waste treatment. But the unloading process is similar with slight different unloading direction. It will be helpful to the further understanding and evaluation of the respository safety to apply the anisotropic rock mass unloading mechanics to the study on the stability of rock mass in the repository in salt for nuclear waste treatment.

1 GENERAL DESCRIPTION

There exist both stress concentration (loading area) and unloading in the secondary stress field in the process of rock mass excavation. There exists substantial difference of rock mechanic property. The current rock mechanics is based on loading. It can be concluded the loading rock mechanics would not comply with the actual unloading process.The analytical result is differ-ent from actual monitoring result.

Based on 10-year engineering experience and anisotropic rock mass unloading mechanics study, the author has put forward the new concept of "anisotropic rock mass unloading mechanics", which has proved to be correct by the monitoring data of so many projects as the Phyllis tunnel rock mass stability analysis, the underground powerhouse deformation anaysis in the huge Ertan Hydro-power station, Jingchuan open Nikel mine high slope and the permanent shiplock engineering of Three Gorges Project. Many papers and books have been published in recent years.[3]、[4]、[5]、[6]、[7]

In November/1998, USA Scholar Hamid Maleki has published the paper "An Analysis of Deformation and Failure Mechanisms in a Repository in Salt" in Brasil-South America rock mechanics workshop. We try to apply our new theory of unloading rock mechanics to the rock mass stability analysis of this repository in salt for nuclear waste treatment.

2 THE INTRODUCTION TO THE PROBLEMS

Generally speaking, the underground rock excavation cause unloading. With the changing secondary stress filed, there will be stress concentration in tangential direction, while there happens unloading in radial direction. The unloading Scope will vary with different underground project scale, dimensions, ground stress and construction procedure.

There exist many joints and strata in rock mass. The mechanic property of rock mass joints are substantially different for loading and unloading. With loading case, the rock mass structure face and strata will maintain good stress-bearing capacity and higher mechanic parameters, but with large unloading condition, there will be rapid decreasing mechanic parameters especially when tensile stress appears with the occurrence of bad rock mass joints. The mechanic property is shown in Fig 1.

Loading status

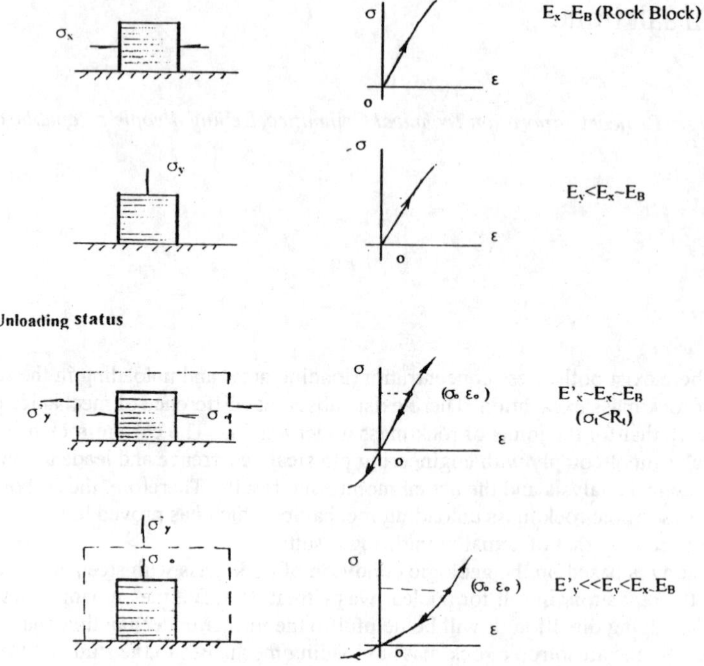

Fig 1 The mechanics property under loading and unloading condition of horizontal strata

As is shown in the above diagram, the mechanics property is similar to each other with both loading and unloading, i.e., $Emx \approx Emx' \approx E_B$. when rock mass tensile stress is less than rock mass limited tension($\sigma_t < R_t$)with the case that the horizontal direction. Rock mass characteristics is approximately the same as rock mass mechanics property, i.e., $Emx \approx Eb$. However, there will be substantially different mechanics property in vertical direction under loading and Unloading condition, i.e., $Emy' << Emy < Eb$. The current rock mass mechanics is established on loading condition which includes rock mass mechanics tests and parameter values, the constitutive stress—deformation relationship study, computational mechanics and computer software as well. Therefore, the present mechanics may be designated as "loading rock mass mechanics".

With the unloading process of rock mass excavation, unloading Mechanics tests should be done on rock mass, the corresponding Unloading computation mechanics and software shall be developed. Therefore, the corresponding mechanics may be named as "Unloading Rock Mass Mechanics."

In different rock Engineering sections, the mechanics analysis and computations should be carried out respectively for loading and unloading cases. The theoretical analysis should comply with the actual engineering mechanic conditions. giving a reasonable explaination.

3 THE KEY POINT OF ANISOTROPIC ROCK MASS NON-LINEAR MECHANICS.

3.1 Rock mass scale effects and its anisotropics.

In addition to rock mass strength, the key factor to decide strata rock mass strength is strata structure, the thickness of Rock mass strata varies greatly. The scale effect shall be much sensitive when rock engineering scale is very large compared with geological strata thickness. Such is the case for thin Phyllis. There would not be obvious scale effectand the rock Mechanics parameters approach test results when rock engineering scale is small compared with the rock strata thickness.

The horizontal mechanics parameter Emx approaches that of rock mass E_B, which will lead to large mechanic parameters. There will exist obvious anisotropic characteristics with the lower vertical rock mass parameter Emg.

3.2 There exist apparent non—linear constitutive stress deformation relationship with rock mass unloading in vertical direction.

The strata mechanics condition is poor and shows apparent Anisotropics because of horizontal strata structure. The anisotropics would be even more obvious with unloading case.

3.3 The analytical non-linear unloading mechanics models and computer software need to be established.

The above-mentioned rock mass models and software are based on loading condition. Some rock mass will be in a state of unloading and there shall even happen tensile stress areas. The analytical non-linear unloading mechanics models and computer software shall be developed considering anisotropic property of rock mass and non-linear unloading stress-deformation relationship. The loading and unloading mechanics shall be applied respectively to different engineering sections with different mechanic conditions. The theoretical analysis results should comply with the actual engineering geologic states.

3.4 Proceed with the sensitivity analysis of rock mass deformation to different rock tensile strength

The rock mass tensile strength has little impact on rock mass deformation for traditional loading rock mechanics. On the contrary, the rock mass tensile strength is much sensitive to the rock deformation under unloading condition. Therefore, it is imperative to do sensitive analysis of the rock deformation to different tensile strength at initial stage of project construction since it is very difficult to decide the orresponding value.

3.5 Unloading damage deformation analysis

The initial deformation will proceed soon after rock excavation after unloading is completed. Thereafter, the rock deformation will develop from external rock mass to internal rock mass as time goes on. The rock deformation will further damage the structure face, which will lead to a worse rock mass, a lower rock mass rigidity and further deformation as well. Such deformation may be named as unloading damage deformation, which deformation will develop with time going. The above deformation concept is somewhat different from the current creep deformation. The current study on creep deformation is based on soft rock mass with loading condition. To be exact, the

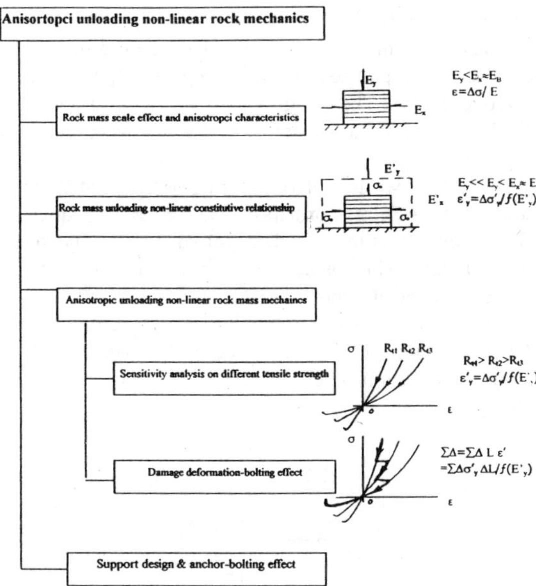

Fig 2 The anisotropic rock mass unloading non-linear mechanics

time-going deformation includes rock damage deformation and rock creep deformation.

3.6 support design and anchor bolting effect analysis

The support design and anchor bolting effect will be more advanced than traditional rule-of-thumb method, which are based on new rock mechanics and stress-deformation analysis results with more clear support design concept and the capability of doing comparisons on different schemes before making final support design.

4 THE ROCK MASS DEFORMATION AND STABILITY ANALYSIS IN THE REPOSITORY IN SALT FOR NUCLEAR WASTE TREATMENT

4.1 general description

In 1998,Hamid Maleki and Lokash Chaturvedi from USA have published the paper "an analysis of deformation and failure mechanisms in a repository in salt" on the fifth rock mechanics conference in Brazil and south America in 1998[8].

The above repository is situated about 45km east of Carlsbad, New Mexico. The underground depth 655m,the evaporation strata 610m,the mine column dimension 30×90m,the chamber size 10×4m (B×H). The E140 drift was mined in 1983,following by the decision to proceed with mining pane[ ] between 1986 and 1988. The design verification was done on the basis of monitoring deformation and creep after 5 years.

The above waste isolation pilot plant repository is located in the Salado formation, a horizontally bedded formation consisting of halite, anhydrite, polyhydrite and clay layer.The immediate roof is a 2-m-thick layer and the immediate floor is a 1-to 1.5-m-thick salt layer. the pure halite is overlain by clay G and anhydrite layer B, which is 10cm thick .there is a prominent 1-m-thick non-salt bed below the storage level, as shown in Fig.3

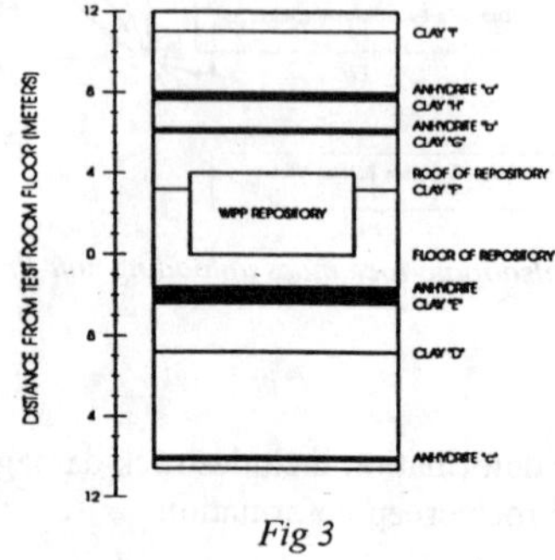

Fig 3

Mechanic properties of salt and anhydrite

Rock	compressive strength, Mpa	Young's Modulus, Gpa	Poisson's ratio
Salt(hydrite)	14-30	31	0.25
Hard anhydrite	100	75	0.35

Note: Poisson's ratio may be wrong

Clay test data aren't too many, the anhydrite is very hard, salt(halite) is hard too. The moisture and water leakage is found by drilling clay rock mass.

The rock support are undertaken year after year,.

No support for repository roof was undertaken before the year 1986;

9000 rock bolts have been installed for the E140 drift between 1986 and 1988;

the class-75 anchor bolts long in 2-3m are mined in the year 1990;

the systematic supports, including wire mesh and anchor are carried out in panel one and House one in the year 1991 for the second time.

The cables and resin bolts are installed comprehensively in E140 drift and panel one between 1992 and 1996.

The third-time support was reinforced in House 7,area S1950.the monitoring company has implemented large-scale monitoring, which includes deformation and anchor bolt loading.

The monitoring company has done detailed observations including deformationg and anchor bolt loads.The monitoring data indicates the quantity and quality of support S is not efficient enough on controlling the strata deformation.

The main monitoring data are as shown in Fig 3/4/5/6/7 in this paper. The roof /floor convergence deformation value in 1993/1994/1995/1996/1998 is 5.73cm, 6.23cm, 6.88cm, 6.62cm and 6.01cm respectively, with the average value being 6.27cm, the total convergence value in 5 years is 31.47cm.

The maximum horizontal deformation of 13.5cm according to literature [8].

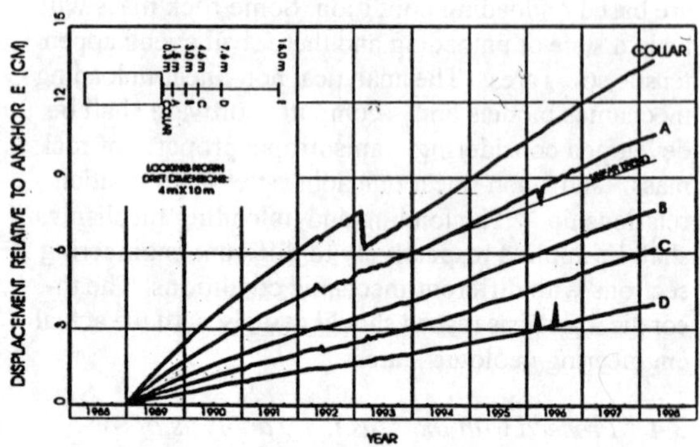

Fig 4

It is concluded that the stability can be enhanced only by modifying the repository chamber type. it is proposed that 10×4m square chamber should be changed into 5×4m square chamber.

4.2 stress and deformation analysis

4.2.1 the value of parameters

1. engineering underground depth 655m,ground stress 16Mpa approximately.
2. Rock mass deformation modulus:

Horizontal direction: E_X = 30Gpa, μ =0.25

Vertical loading direction: E_Y = 18Gpa, μ =0.35

Vertical unloading direction: unloading Non-linear

3. Rock mass single axis stress:

Horizontal direction: R_{cx}=30Mpa

Vertical direction: R_{cy}=20Mpa

4. Rock mass single axis tensile stress:
Horizontal direction: R_{tx}=3Mpa
Vertical direction: R_{ty}=0.4Mpa

4.2.2 preliminary analysis results on secondary unloading stress field after excavation

1) The side-wall stress concentration is as high as 60Mpa, the main compression strength overpass rock mass compression strength. A large scale destructive plastic area by compression & shear, the exists also tensile strength(0.19Mpa).

2) The roof unloading scope is over 16m(σ_c=14Mpa), accompanied by large tensile strength reaches even 1.5Mpa. The stress district can't be enhanced even after anchor bolting.

By analysis the stress and strength, it is found the side wall compression and roof tensile value is too high, which leads to the obvious insatiable cavern.

4.2.3 unloading deformation analysis

The convergence value on the roof and sidewall would be very small by the theoretical analysis, i.e, 1.7cm and 0.22cm respectively. There would not be apparent improvement in controlling the deformation even after anchor-bolting.

The convergence value of the roof and sidewall would be 9.45cm(R_t=0.4Mpa) and 1.65cm respectively to theoretical analysis, which is 6~8 times bigger than the value by applying load mechanics. There is no apparent enhancement in the roof, the floor and both sides after anchor-bolting, The unloading deformation analysis results comply with the actual monitoring data in the WIPP repository (the annual convergence value is about 6.72cm).

Such a big unloading deformation shall lead to a even bigger damage deformation by destructuring the rock mass gradually into the inner rock mass. the development speed ratio would be over 1m per year on the basis of monitoring data about the internal rock mass damage.

The salt molecular structure creeping with long-term need to be taken into consideration because of its soft property.

The actual monitoring in the WIPP repository indicates that the mechanic property and stability must be re-evaluated by apply in the loading rock mass mechanics theory.

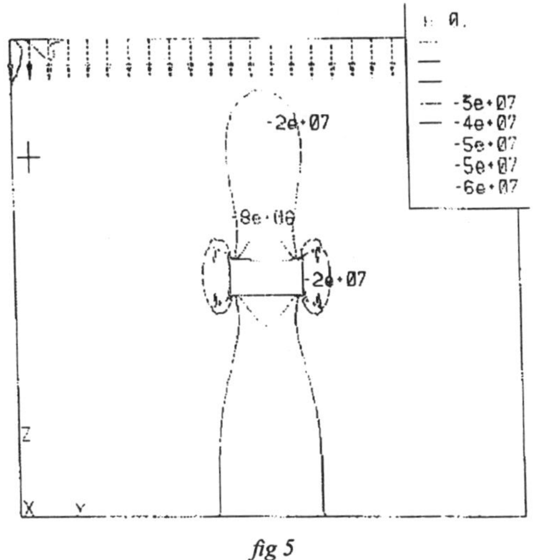

fig 5

5 CONCLUSION

5.1 The anisotropic charateristic is obvious since the nuclear repository in salt is having a geologic condition with horizontal strata structure.

5.2 In rock mass excavation stress field, the tangential direction is loading, the axial direction is unloading. The arch and floor is accompanied with large unloading area and tensile strength. The side wall stress concentration is much bigger than salt strength-bearing capacity which is much unfavorable to the chamber stress.

5.3 There exists a substantial difference between the two results figured out respectively by applying the traditional loading rock mass mechanics and the recommended new unloading rock mass mechanics, while the latter theoretical analysis result is almost equal to the actual monitoring result. it is concluded the unloading rock mass mechanics complies with the actual engineering unloading mechanics. It is suggested to re-evaluate WIPP repository with the unloading rock mechanics theory. The author believes that the WIPP support design is not efficient in controlling the deformation. The chamber dimension(5×4m) is recommended, which is the result of our multi-scheme study and rock support design by applying the unloading rock mechanics to difference chamber types.

5.4 the prospect analysis

the current WIPP facilities have been put into use for several decades. In the near future, the nuclear facilities shall be out of operation. It is imperative to construct WIPP repository based on even more reliable safety. If the deteriorating global environment.

It is improper to evaluate WIPP and decide on rock support by applying the current loading rock mechanics. The Anisotropic Unloading rock mechanics is applicable to various rock mass engineering, especially for WIPP safety assessment.

REFERENCE

[1] Ha Qiuling, The Application Of ansiotropic Rock Mass Unloading Mechanics In Underground Engineering, Norwegian Underground Engineering Engnieering Association, 1997.
[2] Ha Qiuling, The Anisotropic Rock Mass Mechanics In Mining Slope Treatment, The 15th Mining Conference In Turkey, 1997.
[3] Ha Qiuling, Liu Guolin, A Study On Engineering Geology Of Rock Mass Slope Unloading Mechanics, China Construction Industry Publishing House, Beijing, 1997.
[4] Ha Qiuling, Li Jianlin, A Study On The Macro-Mechanics Parameters Of Rock Slope Unloading Mass, China Construction Industry Publishing House, Beijing,1997.
[5] Ha Qiuling Zhang yongxin, A study on the Anisotropic Unloading Non-linear Rock Mechanic in Rock Mass Slope, China Construction.
[6] Ha Qiuling, Liu Guoling, Li jianling , Zhang Yongxin, The Joint Unloading non-linear Rock Mechanics, China Construction Industry Publishing House Beijing, 1998.
[7] Ha Qiuling, The Anisotropic Unloading non-linear Rock Mechanics, The Brazil-South America Rock Mechanics Association 1998.
[8] Hamid Maleki, Lokesh Chaturvedi, An Analysis of Deformation and Failure Mechanisms in a Repository in Salt, The Brazil-South America Rock Mechanics Association 1998.

Environmental Issues and Management of Waste in Energy and Mineral Production, Singhal & Mehrotra (eds)
© 2000 Balkema, Rotterdam, ISBN 90 5809 085 X

Properties of fly ash stabilized haul road construction materials

Dwayne D.Tannant & Vivek Kumar
University of Alberta, Edmonton, Alb., Canada

ABSTRACT: Larger haul trucks are being used at surface mines in Canada thus requiring better haul roads to carry heavier loads. The availability of good quality aggregate to build haul roads is limited for prairie coal mines. However, most of these mines are located adjacent to coal-fired electrical power plants, which produce by-product fly ash as a waste. Fly ash can be used to increase strength and stiffness of soil and road bases.

Unconfined compressive strength tests conducted on various mixtures of fly ash, kiln dust, mine spoil, and coal seam partings showed that the cementing characteristics of unclassified fly ash from central Alberta coals was low. However, the addition of cement kiln dust, which is high in CaO, enabled the fly ash to exhibit significant cementing action.

Mixtures of fly ash, kiln dust, and mine spoil or coal seam partings had unconfined compressive strengths of about 1MPa and elastic moduli of about 350MPa after 14 to 28 days. This compares favourably with compacted mine spoil or coal seam partings which have estimated unconfined compressive strengths of less than 0.4MPa. Thus fly ash stabilized mine spoil or coal seam partings were found to have potential for use in constructing haul road base and sub-base layers since maximum tire pressures on the running surface are less than 0.7MPa.

1 INTRODUCTION

Alberta obtains more than 90% of its electrical power from coal burning power plants situated next to coal strip mines. Approximately 26Mt of coal (Coal Association of Canada 1999) are burned each year in Alberta, generating about 2.4Mt of fly ash and 0.6Mt of bottom ash (Joshi 1999). Most of the ash is hauled from the power plants back into the mined-out areas for disposal in landfills. While efforts are underway to establish markets and other end uses for ash, the disposal of millions of tonnes of ash each year remains a problem.

Heavy, large capacity trucks are used to haul coal from the mine to the power plants. These trucks can achieve gross vehicle weights of 4000kN. The tire pressures used to support the weight of the truck and the coal are typically in the range of 600kPa to 690kPa. The haul roads must be constructed from materials with sufficient bearing capacity and stiffness to maintain road serviceability. Haul roads that are in poor condition can detrimentally influence mining by (1) reducing productivity by increasing rolling resistance, (2) increasing costs due to more road and truck maintenance.

A typical Alberta coal mine has about 10km of permanent haul road and various temporary roads. These roads are typically constructed from locally available materials, usually found on the mine property itself. These materials tend to consist of glacial till and sedimentary rock such as siltstone, mudstone, and sandstone found associated with the coal. Pit-run or crushed gravel is often placed on the top surface of the roads.

Choice of construction material for any layer of a haul road depends on the maximum stress level in that layer in two ways. First, the strength of the material should be greater than the maximum stress. Secondly, but more importantly, the material's Young's modulus should be high enough that the strain induced in the material by a haul truck is less than a critical strain limit. According to various authors (Thompson and Visser 1997; Morgan et al. 1994) the critical strain limit for a layer to continue to act as a beam, thus continue to give necessary support, ranges between 1500 to 2000 micro-strains.

The focus of this paper is to evaluate whether the addition of fly ash to mine spoil or coal seam partings can improve the physical properties of these materials and thus improve their performance as construction materials in a haul road. The successful use of fly ash to stabilize soils elsewhere (Hobeda 1984) suggests that their use in haul roads should have multiple benefits, including diverting some ash from landfills to road construction and improving mining productivity.

2 FLY ASH STABILIZED HAUL ROAD CONSTRUCTION MATERIAL

2.1 *Fly ash properties*

Properties of fly ash depend on many factors. Type of coal burned is the single most important factor influencing the type of fly ash generated. Type of collector used also affects the fly ash quality.

The lime (CaO) content of fly ash dictates the cementing properties of fly ash as it is essential for a pozzolanic reaction to form cementitious compounds with the soil particles.

Type of coal – Cementing capability of fly ash decreases with increase in the rank of coal from which ash is produced. So, fly ash produced from bituminous coals generally has lower cementing capability (due to insufficient free lime) than ash from sub-bituminous coal (DiGioia et al. 1979). Fly ash from lignite shows good cementing properties without addition of lime, whereas fly ash from bituminous and sub-bituminous coals (types of coal present in Alberta) may require additional lime.

Collection system – Torrey (1978) reports that fly ash collected by electrostatic precipitators (ESP) has 38% more CaO and 58% less carbon than ash collected by mechanical collectors. Moreover the former is finer than the later. Thus, fly ash collected by ESP is more reactive and consequentially, more suitable as haul road construction material than fly ash collected by mechanical collectors.

2.2 *Lime-fly ash-aggregate mix (LFA) for road construction*

Lime-Fly ash-Aggregate mix (LFA) has been widely used for road construction. Torrey (1978) reports use of fly ash, lime, cement, cement and aggregate concrete mixes in ratios of 30:9:1:57:40 to surface a road with heavy traffic. Torrey (1978) also reports use of LFA for base and sub-base layers of roads. Vaeg (1984) discusses successful use of LFA as a road construction material in Sweden. Fly ash and LFA have various desirable properties as a haul road construction material.

Low cost and high availability - Fly ash, a major component in a LFA mix is a low cost material that is abundant near prairie coal mines. In fact, it has become a liability for the utility companies after the enforcement of strict environmental regulations for disposal of fly ash (Kennedy et al. 1981). Fly ash disposable costs offset part of the costs incurred in transporting fly ash from a power plant to a road construction site. Moreover, rail wagon or haul trucks transporting coal from a mine to a power plant can carry ash on their return trip.

Cementing properties – Siliceous or alumino siliceous materials, present in fly ash, when in finely divided form, react with alkali or alkaline earth products such as CaO, to produce cementitious products, in presence of water. The alkali or alkaline earth products, if not present in the fly ash, can be added. The cementing reaction, which is time dependent, increases strength and decreases compressibility, permeability and frost susceptibility. Thus suitability of fly ash or LFA for road construction improves with the age of curing.

Low density – Compacted LFA has a low density (1800-2000kg/m^3) and can be used as a low density fill.

Compressive strength – LFA can have fairly high compressive strength compared to most soils used alone (Tables 1 & 2). The strength of the LFA depends on the amount of fly ash that is used. Thus, LFA may be used to build roads designed for large haul trucks.

Low compressibility - One desirable property of a haul road construction material is that it should not settle much on repeated loading. The compressibility of fly ash is very low compared to other construction materials. The compression index and recompression index for fly ash varies from 0.10 to 0.25 and from 0.02 to 0.04, respectively (DiGioia et al. 1979).

Environmentally safe - Various studies done by the DiGioia et al. (1979) and the Kennedy et al. (1981) on the leaching properties of fly ash, demonstrate that there is little danger of leaching toxins from fly ash fill into water bodies in particular and the environment on the whole. Bituminous coal has an average of 0.3% free lime and 0.5% soluble sulphate. Rohrman (1971) reports that fly ash leachate is alkaline (pH 6.2 to 11.5), whereas Theis and Marley (1976) reports acidic nature of fly ash

Table 1. Compressive strength of fly ash-lime-aggregate mixes (Meyers, et al. 1976).

Curing period	Compressive strength (MPa)
7 Days	2.76 – 3.10
28 Days	3.79 – 4.13
2 Years	≈ 10

Table 2. Compressive strength (MPa) of hydrated fly ash (Meyers, et al. 1976).

	Moisture content		
Curing period	20%	40%	60%
3 Days	9.49	3.46	1.36
7 Days	11.24	4.18	1.71
14 Days	14.72	4.94	2.06

leachate. An acidic leachate can be expected to have higher concentration of trace elements. Fortunately, Pluth et. al. (1981), after testing fly ash from five different Alberta coal mines, report that the leachate of the fly ash is alkaline. Nevertheless, accurate prediction of leachate quality needs assessment of site-specific soil attenuation parameters.

3 UNCONFINED STRENGTH AND MODULUS OF FLY ASH-SOIL MIXTURES

Unconfined compression tests were performed on various combinations of fly ash, kiln dust and road building aggregates at different ages of curing to measure compressive strength and modulus.

Previous studies (DiGioia et. al. 1979 and Hobeda 1984) have evaluated fly ash and fly ash mixed with cement and aggregates for use in commercial roads. The fly ash used in these tests was usually refined and the test objectives were to prove applicability in commercial roads. The objective of the testing described here is to find suitable combinations of fly ash, cementing material and aggregates for use in mine haul roads. Thus unclassified (unrefined) fly ash was used and the aggregates selected for the tests were the typical road building materials used for mine haul roads. The choice of materials was such that the test results have direct application for mine haul roads.

Careful mixed and proportioning of ingredients and water is a precondition to achieve desirable mix properties. Deviation of mix ratios from optimum may result in a large drop in strength and other properties.

3.1 *Component properties*

The first batch of fly ash, mine spoil and coal seam partings were obtained from the Sundance power plant and adjoining Highvale Mine on 20th of May 1999. Sundance delivered a second batch of fly ash on 20th of August 1999. The first batch of kiln dust was picked up from the Lafarge facility in Edmonton on 4th of June 1999 and Lafarge delivered a second batch on 29th of July 1999.

The unclassified fly ash from the Sundance power plant contains about 15% CaO. The unclassified fly ash differs from the fly ash that is sold by Lafarge. The commercial fly ash is cleaned (via a cyclone) to remove ash particles and larger particles, creating a Class Ci fly ash.

Cement kiln dust was obtained from the Lafarge Exshaw plant near Canmore. This material is currently a waste byproduct from cement production. The two batches of kiln dust differed in their ability to activate cementing capabilities of fly ash as discussed later.

Partings between seam 3 and seam 4 of Highvale Mine was used as an aggregate. It is a dark grey siltstone. Particle size analysis was done on dried partings that passed through 1" (25.4mm) sieve according to ASTM D422–63 (Fig. 1). More than 90% of the particles passed through ¾" (19mm) sieve and more than 50% passed through 3/8" (9.5mm) sieve. Less than 5% of the material had particle diameter less than 0.85mm.

Mine spoil is the material found between seam 1 and the topsoil. It is a yellow to light brown silt. Particle size analysis was done on the dried portion of the material that passed through a 1" (25.4mm) sieve according to ASTM D422–63 (Fig. 1). More than 95% of the particles passed through a ¾" (19mm) sieve and slightly less than 50% passed through a 4.25mm sieve but approximately 25% of the particles had diameter smaller than 0.85mm. The mine spoil had a specific gravity of 2.62 (ASTM D854).

3.2 *Moisture-density relationships*

Optimum moisture content is required for maximum compaction, which is essential for proper strength generation in a LFA mix. Moreover, moisture

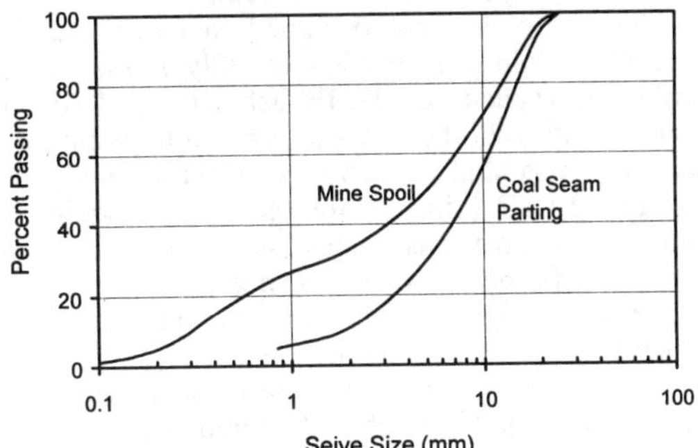

Figure 1 Particle size distributions.

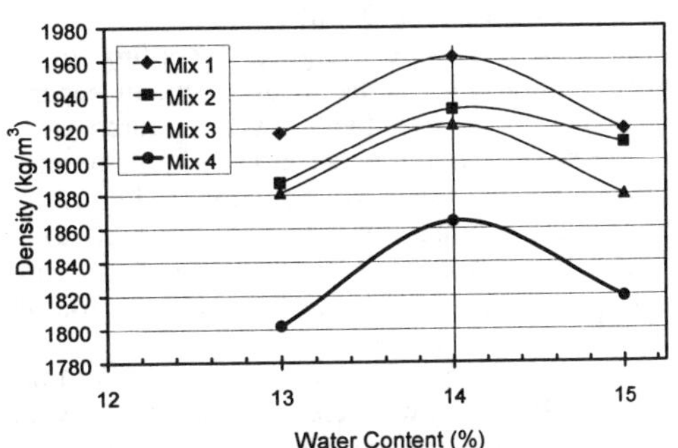

Figure 2 Moisture-density relationship for LFA mixes with mine spoil.

present in the mixture is essential for the pozzolanic reaction, which creates cementing action. Water contents were determined by conducting moisture-density tests following ASTM D558. A series of standard proctor compaction tests were performed and the measured moisture content was plotted against density. For subsequent strength testing, the water content corresponding to maximum density was chosen for the sample preparation. Figure 2 shows moisture-density relationship for LFA mixes with mine spoil.

The optimum moisture contents for maximum compaction density of the coal partings and of the mine spoil were also measured and were found to be 0.5% to 1% lower than that for LFA. This may be attributed to presence of fly ash in LFA mixes.

Haul road construction materials are typically mixed at moisture contents 2 to 3% lower than optimum because it is easier to place and spread drier material (Lay, 1990).

3.3 *Sample preparation*

The relative proportions of the different ingredients used for the tests were based on a literature review, suggestions by Lafarge, and results from previous tests. Fly ash and kiln dust were dry and fine grained, so no pretreatment was required.

This study was not designed to determine the amount of kiln dust needed to fully mobilize the cementing capacity of the fly ash. The goal was to focus on utilization of fly ash while minimizing the use of potentially more expensive additives containing CaO (kiln dust or lime). The choice of percent kiln dust was somewhat arbitrary in these tests and further research is needed in order to determine the optimal amount of kiln dust needed to achieve the desired strengths.

The parting material was air-dried for 5 days, then sieved and crushed. Material was passing through a ¾" (19mm) sieve was used for testing. The mine spoil was cohesive and moist. It was air dried for 15 days. Dried lumps were broken to bring the grading below ¾" (19mm).

The materials were weighed according to the desired proportions (Table 3) and kept in separate containers. The ingredients were poured in a concrete mixer (adding water last) and mixed until a homogenous mixture was obtained (Figure 3). Approximate mixing times were around 20 to 25 minutes. The mixer was stopped frequently to carve sticking material behind blades and at the bottom the mixer.

Cylindrical forms of diameter 101.6mm (4") and height 203.2mm (8") were cleaned and interior walls were oiled. Then the mixture was placed in the forms and compacted in three layers by rodding following the procedures in given ASTM D698. The

Table 3. Weights ratio of dry ingredients.

Mix	Mix No.	Ingredients (%)				
		Fly* Ash	Kiln* Dust	Seam* Parting	Mine* Spoil	Water
Fly Ash Only		100	-	-	-	17
Fly Ash + Kiln Dust	1	90	10	-	-	18
Fly Ash + Kiln Dust	2	80	20	-	-	17
FA + KD + Parting	1	12.5	3.5	84	-	11
FA + KD + Parting	2	16	4	80	-	13
FA + KD + Parting	3	20	4	76	-	13
FA + KD + Parting	4	25	5	70	-	14
FA + KD + Spoil	1	12.5	3.5	-	84	14
FA + KD + Spoil	2	16	4	-	80	14
FA + KD + Spoil	3	20	4	-	76	14
FA + KD + Spoil	4	25	5	-	70	14

* dry percentage.

Figure 3. Concrete mixer and compacted samples.

Table 4. Density achieved during sample preparation.

Mix	Density (kg/m^3)	No. of Samples
Fly Ash Only	1530	12
Fly Ash + Kiln Dust (mix 1)	1650	9
Fly Ash + Kiln Dust (mix 2)	1635	14
Fly Ash + Kiln Dust + Parting (mix 1)	1825	12
Fly Ash + Kiln Dust + Parting (mix 2)	1850	14
Fly Ash + Kiln Dust + Parting (mix 3)	1850	14
Fly Ash + Kiln Dust + Parting (mix 4)	1760	14
Fly Ash + Kiln Dust + Spoil (mix 1)	1960	14
Fly Ash + Kiln Dust + Spoil (mix 2)	1930	14
Fly Ash + Kiln Dust + Spoil (mix 3)	1920	14
Fly Ash + Kiln Dust + Spoil (mix 4)	1870	14

densities achieved by this compaction are summarized in the Table 4.

The samples were kept in a moist room with

100% humidity and allowed to cure. The samples were taken out of the forms after 2 days by forcing compressed air through a small hole in the bottom of the forms. The samples were then returned to the moist room until the day of testing.

The samples were either capped with sulfur or ends were dressed to make the ends flat (the choice was dictated by the strength of the sample; the weak samples could not be capped because the samples broke while being taken out of the capping mould).

3.4 *Test procedure*

Uniaxial compression tests were performed following ASTM D5102. The tests were initially planned to be taken after 3, 7, 14, and 28 (if required) days of curing. But the sample with 100% fly ash was found to be very weak so the 14-day test was cancelled and a test after 56 days of curing was done instead. The compression test were done with an automatic data acquisition system (data logger) except one set of tests (7-day test of fly ash and kiln dust (mix 1)) in which readings were observed manually. Compression rates ranging from 0.1 to 0.5mm/min were used but changing compression rates within this range did not seem to have any effect on the results. Changes in compression rates were made to keep the total loading time below 15 minutes so as to minimize creep effects.

Higher sampling rates were used for higher compression rates to get a relatively consistent rate of sampling with respect to strain. Sampling rates ranged between 2 and 15 samples per minute, with 10 samples per minute being the most common value. Lower limit on sample rate was guided by the fact that there should be enough data points to capture the complete nature of stress-strain curve and at the same time record the maximum stress.

3.5 *Results*

A series of tests were conducted to evaluate, in a general manner, the cementing action of the fly ash. Test specimens were made from fly ash with and without addition of kiln dust. The fly ash compacts well and can achieve minor strength simply by mechanical interlock between the compacted particles.

Compressive strengths and Young's modulus obtained from the tests are shown in Figures 4 & 5. These figures show that the addition of kiln dust is needed in order to achieve significant strengths (>1MPa) from the fly ash. With kiln dust, the fly ash shows time-depend strength and stiffness improvements. These results confirmed expectations that fly ash alone from central Alberta coals does not possess sufficient CaO to be self-cementing.

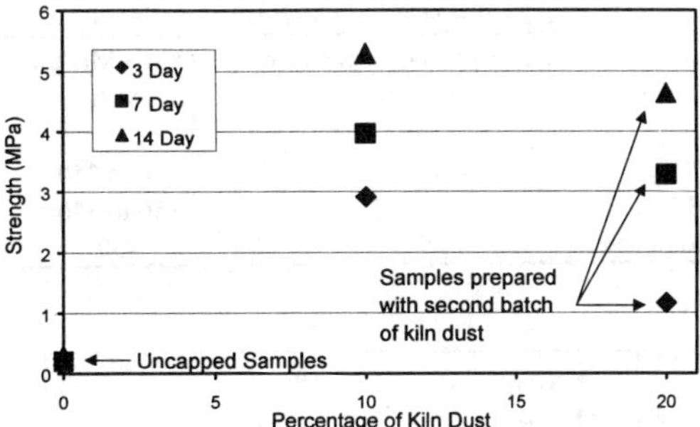

Figure 4 Unconfined compressive strength of fly ash - kiln dust samples.

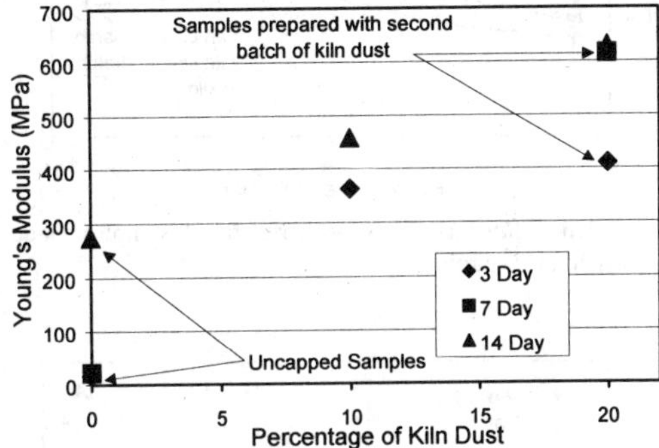

Figure 5 Young's modulus of fly ash - kiln dust samples.

Because only a few tests were performed, it is not possible to state the optimal kiln dust/fly ash ratio that creates the best "binder" for soil improvement. For subsequent tests where the binder was mixed with the soils, the kiln dust/fly ash ratio was selected between 0.2:1 and 0.28:1.

Figures 6 to 9 show the compressive strengths and Young's modulus obtained from the tests involving mixtures of mine spoil or partings and binder (fly ash plus kiln dust). Most points shown on these figures are averages of three separate tests.

Given the variability in the test results, it appears the specimens made with crushed coal seam partings (siltstone) had properties that were similar to those made from the mine spoil (silt).

The strengths achieved in the stabilized soils are less than those for the fly ash-kiln dust samples, but are sufficiently high after a period of curing to be useful for road construction. While not a comprehensive suite of test data, the results show (Table 5) that strength and stiffness increase over a 28-day period and that mixes with high binder content perform best. Strengths in the order of 0.6MPa are obtained after 7 days while 28-day strengths are about 0.8 to 1MPa depending on the

Table 5. Typical stabilized silt or siltstone properties.

Age	Unconfined Compressive Strength (MPa)	Young's Modulus (MPa)
7 Day	0.4 to 0.6	150 to 250
28 Day	0.8 to 1	150 to 350
No binder	0.2	<50

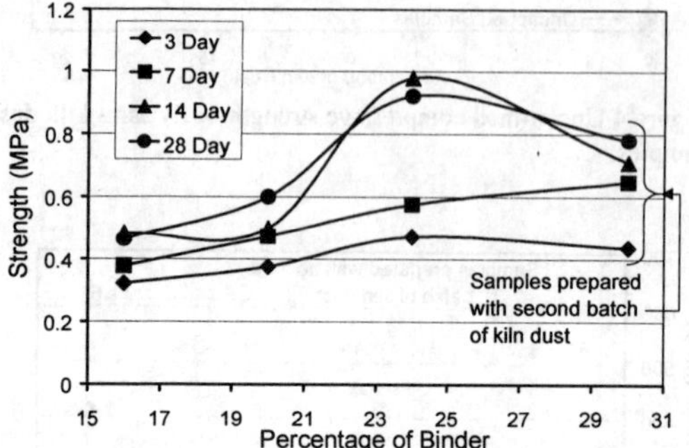

Figure 6 Unconfined compressive strength of samples made with crushed mine parting.

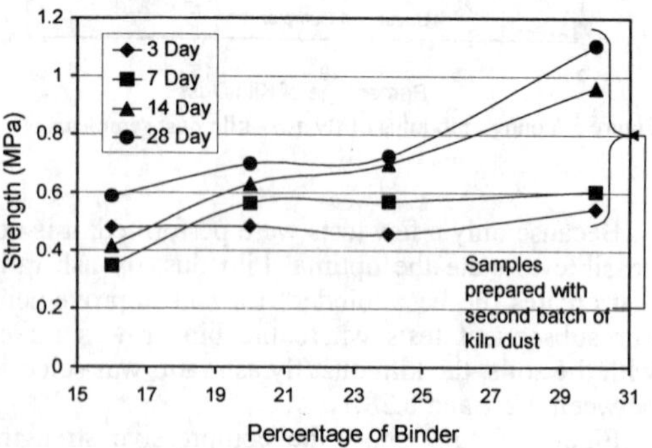

Figure 7 Unconfined compressive strength of samples made with crushed mine spoil.

binder content. Unconfined compression tests conducted on cylinders of compacted silt or crushed siltstone gave compressive strengths of about 0.2 MPa. Therefore, the addition of the fly ash-kiln dust binder can significantly improve the strength (and stiffness) of these materials. The Young's modulus generally increased with longer curing period and higher percentage of binder.

The fly ash and the kiln dust were delivered in two batches. Both batches of fly ash seemed to have similar properties but the two batches of kiln dust appeared to differ. Samples with mine parting gave better results with the first batch of kiln dust, whereas mine spoil gave better results with second

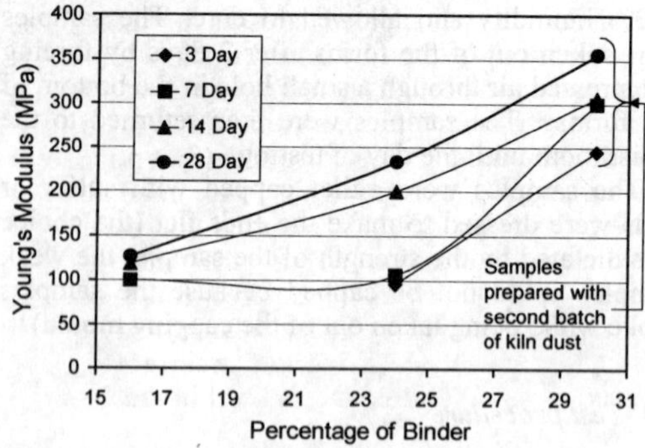

Figure 8 Young's modulus of samples made with crushed mine parting.

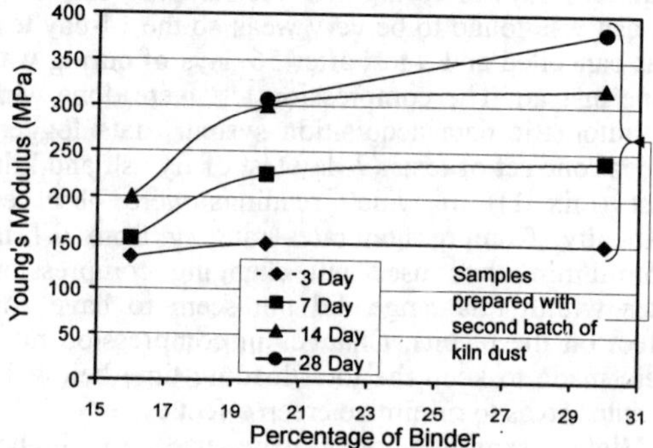

Figure 9 Young's modulus of samples made with crushed mine spoil.

batch of kiln dust. The compressive strength should increase with percentage of fly ash Figures 4 & 6 show a decrease in compressive strength for the highest percentage of binder.

It should not be concluded that compressive strength peaks at 20% fly ash because the samples with parting (mix 4), spoil (mix 4) and kiln dust (mix 2), were prepared with the second batch of kiln dust. For specimens made with coal partings, mix 4 had compressive strengths less than that of mix 3, which is reverse to the general trend (Fig. 6). This may be attributed to a different type of kiln dust used to prepare mix 4 with parting. This may also be reason for lower strength of samples with kiln dust (20%) with fly ash than that of samples with kiln dust (10%) with fly ash. But this anomaly was not observed in samples with spoil.

4 DESIGN CRITERIA FOR HAUL ROADS

The bearing capacity or the compressive strength of a material that can be used in a layer of a road depends on the anticipated stresses in that layer. Moreover, strain caused by a haul truck at any point in the road cross-section should not be greater than a critical strain limit. Stresses at any layer of a road beneath a tire depend on the tire size (foot print area) and inflation pressure. Thus the modulus of elasticity and the strength required for a suitable and inflation pressure. Thus the modulus of elasticity and the strength required for a suitable construction material for a particular layer of a haul road can be calculated using the strain criteria and the maximum anticipated stress level in that layer of the road.

Inflation pressures for large haul truck tires vary between 600 to 690kPa. Application of these pressure over the foot print of the tire creates stresses bulbs within the road. The stresses cause by the tire decrease with depth into the road. Ignoring dynamic effects, the inflation pressure gives the maximum stress that the surface course of a road must carry and can be used to select appropriate materials for surface course construction.

Figure 10 shows a typical cross-section for a haul road. Roads are constructed on a pre-existing or constructed sub-grade. The various layers above the sub-grade can be of various thicknesses although typical values are indicated in Figure 10. The quality (and the cost) of materials used to construct roads tends to increase towards the surface of the road.

The bearing capacity of dense gravel or very dense sand and gravel is about 800 to 1100kPa (Wade 1989) and hence is used as a surface course construction material.

The stresses in layers below the surface course can be calculated using stress models or application of elastic theory. For example, a simple assumption is that a tire creates a uniform circular load over an isotropic, homogeneous elastic half space. While different layers in actual roads have different stiffness and strength properties, this assumption was used to examine the stresses beneath a typical tire (Fig. 11) with an inflation pressure, *p*. A typical haul truck tire has a foot print area of about 1.13 m^2 giving an equivalent diameter, *w* of 1.2 m.

Figure 11 shows that the stresses in the base layer, which typically starts at 0.3 to 0.6m below the road surface will be about 0.65 to 0.9 times the tire pressure or about 0.3 to 0.65MPa. Similarly, a typical sub-base begins at a depth of roughly 1.5m. Therefore, the sub-base experiences about 0.2 times the tire pressure or about 0.1 to 0.2MPa. Based on a strength criterion, appropriate construction materials for the base and sub-base need to have bearing capacities that exceed the expected stresses as shown in Table 6.

Another important criterion for haul road design is a strain limit for each layer. A road can not adequately support haul trucks when strains

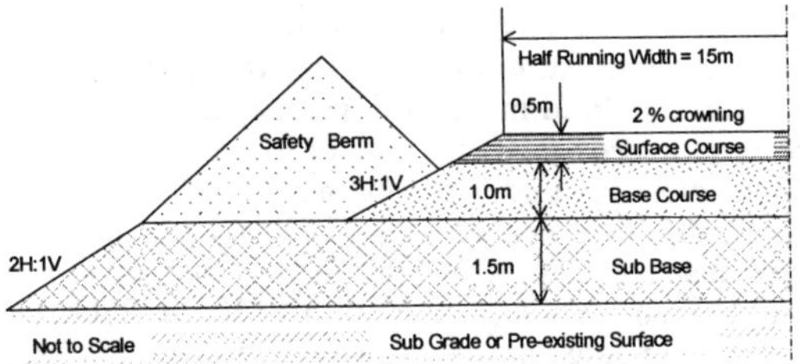

Figure 10 Typical cross-section of a haul road.

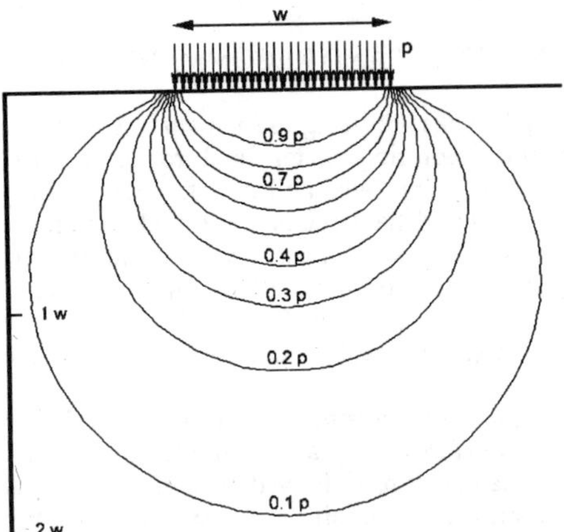

Figure 11 Stress bulbs below a circular pressure distribution.

Table 6 Minimum bearing capacity and Young's modulus of haul road construction materials.

	Thickness (m)	Bearing Capacity (MPa)	Young's Modulus (MPa)
Surface Course	0.3 to 0.6	0.7 to 0.9	-
Base Course	1.0	0.3 to 0.65	150 to 350
Sub-base	1.5	0.1 to 0.2	100 to 150

exceeded a critical strain limit. Morgan et al. (1994) found that the critical strain limit was about 1500 micro-strain while Thompson and Visser (1997) noted that the limit was around 2000 micro-strain. The estimated stresses in each layer (noted above) can be used to estimate the minimum Young's modulus that will ensure strains are less than the critical strain limit. For example, a 1m thick base course subjected to stresses of 0.3 to 0.65MPa requires a Young's modulus greater than 150 to 350MPa.

Construction materials such as crushed gravel and pit run gravel satisfy the above criteria for surface and bases courses, respectively and are being currently used at various surface mines. However, these materials may not be located near mines and have to be hauled to the mine for road construction. Fly ash stabilized mine partings or mine spoil could be used to replace pit run gravel for base layers and could also be used for sub-base layers. Table 6 shows that addition of about 20% fly ash and 5% kiln dust yield strengths after 7 days that are 0.4 to 0.6MPa and Young's moduli of 150 to 250MPa. These properties are tentatively sufficient for base of fly ash, kiln dust, and soil to generate even better material properties.

5 CONCLUSION AND RECOMMENDATIONS

Larger haul trucks are being used at surface mines in Canada thus requiring better haul roads to carry heavier loads. The availability of good quality aggregate to build haul roads is limited for prairie coal mines. However, most of these mines are located adjacent to coal-fired electrical power plants and their waste by-product, fly ash. Fly ash can be used to increase strength and stiffness of soil and road bases.

Unconfined compressive strength tests conducted on various mixtures of fly ash, kiln dust, mine spoil, and coal seam partings showed that the cementing characteristics of unclassified fly ash from central Alberta coals was low. However, the addition of cement kiln dust, which is high in CaO, enabled the fly ash to exhibit significant cementing action.

Mixtures of fly ash, kiln dust, and mine spoil or coal seam partings had unconfined compressive strengths of about 0.4 to 0.6MPa after 7 days and 0.8 to 1MPa after 28 days. The elastic moduli of these materials were 150 to 350MPa after 14 to 28 days. This compares favourably with compacted mine spoil or coal seam partings which have estimated unconfined compressive strengths of less than 0.4MPa. Thus fly ash stabilized mine spoil or coal seam partings were found to have potential for use in constructing haul road base and sub-base layers since stresses induced in the base layer will be less than about 0.3 to 0.65MPa. Furthermore, the compacted fly ash stabilized soils had Young's moduli that were high enough to meet strain criteria for haul road construction.

More work is needed to better define the amount of CaO needed to fully activate the cementing characteristics of fly ash from Alberta's prairie coals. Alternatives to kiln dust as a source for CaO can also be tested. Once a cheap, yet effective binder is developed, it has good potential for use as an additive in road construction near coal-fired power plants in Alberta.

One problem with binder use in road construction is that the strength and stiffness of the stabilized soils are time-dependent. Road construction procedures and the length of time that the road must wait before being placed into service are an important consideration. However, where opportunities occur, construction of a haul road test section using fly ash is recommended. This would permit monitoring over a period of time and would significantly increase the level of confidence in the applicability of fly ash as a haul road construction material.

ACKNOWLEDGEMENTS

The authors wish to thank Tony Smith and Ron Lyle from TransAlta Utilities Corp. and Rick Ketcheson and Al Slessor from Lafarge Canada Inc. for their support. SMART (Surface Mining Association for Research & Technology) has encouraged work on haul road design.

REFERENCES

Coal Association of Canada, 1999. Web Page: http://www.coal.ca/stats.htm#domesticconsumption.

DiGioia, A.M., McLaren, R.J. & Taylor, L.R. 1979. *Fly Ash Structural Fill Handbook.* EA-1281. Research Project 1156-1. Monroeville, GAI Consultants.

Hobeda, P. 1984. *Use of Waste Material from Coal Combustion in Road Construction.* NTIS.

Joshi R.C. 1999. Personal communication.

Lay, M.G. 1990. *Handbook of Road Technology.* Vol. 1. New York: Gordon and Breach Science Publishers.

Kennedy, F.M., Schroeder, A.C. & Veitch, J.D. 1981. *Economics of Ash Disposal at Coal-fired Power Plants.* PB82-192535. NTIS.

Meyers, J.F., Pichumani, R. & Kapples, B.S. 1976. *Fly Ash as a Highway Construction Material – A Manual.* 76-16 Implementation Package. U.S. Department of Transportation, Federal Highway Administration.

Morgan, J.R., Tucker, J.S. & McInnes, D.B. 1994. A mechanistic design approach for unsealed mine haul roads. *Pavement Design and Performance in Road Construction* 1412: 69-81.

Pluth, D.J., Gwyer, B.D. & Robertson, J.A. 1981. Physical characteristics and chemistry of fly ash. *Proceedings: Coal Ash and Reclamation. Alberta Land Conservation and Reclamation Council* RRTAC 81-3: 68-80.

Rohrman, F.A. 1971 (August). Analyzing the effect of fly Ash on water pollution. *Power*: 76-77

Theis, T.L. & Marley, J.J. 1976. *The Contamination of Groundwater by Heavy Metals from Land Disposal of Fly Ash.* US Energy Research and Development Administration. Contract No. E (11-1) – 2727. NTIS

Thompson, R.J. and Visser, A.T. 1997. A mechanistic structural design procedure for surface mine haul roads. *International Journal of Surface Mining, Reclamation and Environment* 11: 121-128.

Torrey, S. 1978. *Coal Ash Utilization – Fly Ash, Bottom Ash and Slag.* Pollution Technology Review No. 48. New Jersey: NDC.

Wade, N.H., *Design Manual for Surface Mine Haul Roads.* Monenco Consultants Limited, Calgary, Alberta. 1989.

Author index